A Matrix Algebra Approach to Artificial Intelligence

Xian-Da Zhang

A Matrix Algebra Approach to Artificial Intelligence

 Springer

Xian-Da Zhang (Deceased)
Department of Automation
Tsinghua University
Beijing, Beijing, China

ISBN 978-981-15-2769-2   ISBN 978-981-15-2770-8 (eBook)
https://doi.org/10.1007/978-981-15-2770-8

© Springer Nature Singapore Pte Ltd. 2020
This work is subject to copyright. All rights are reserved by the Publisher, whether the whole or part of the material is concerned, specifically the rights of translation, reprinting, reuse of illustrations, recitation, broadcasting, reproduction on microfilms or in any other physical way, and transmission or information storage and retrieval, electronic adaptation, computer software, or by similar or dissimilar methodology now known or hereafter developed.
The use of general descriptive names, registered names, trademarks, service marks, etc. in this publication does not imply, even in the absence of a specific statement, that such names are exempt from the relevant protective laws and regulations and therefore free for general use.
The publisher, the authors, and the editors are safe to assume that the advice and information in this book are believed to be true and accurate at the date of publication. Neither the publisher nor the authors or the editors give a warranty, expressed or implied, with respect to the material contained herein or for any errors or omissions that may have been made. The publisher remains neutral with regard to jurisdictional claims in published maps and institutional affiliations.

This Springer imprint is published by the registered company Springer Nature Singapore Pte Ltd.
The registered company address is: 152 Beach Road, #21-01/04 Gateway East, Singapore 189721, Singapore

Preface

Human intelligence is the intellectual prowess of humans, which is marked by four basic and important abilities: learning ability, cognition (acquiring and storing knowledge) ability, generalization ability, and computation ability. Correspondingly, artificial intelligence (AI) also consists of four basic and important methods: machine learning (learning intelligence), neural network (cognitive intelligence), support vector machines (generalization intelligence), and evolutionary computation (computational intelligence).

The development of AI is built on mathematics. For example, multivariant calculus deals with the aspect of numerical optimization, which is the driving force behind most machine learning algorithms. The main math applications in AI are matrix algebra, optimization, and mathematical statistics, but the latter two are usually described and applied in the form of matrix. Therefore, matrix algebra is a vast mathematical tool of fundamental importance in most AI subjects.

The aim of this book is to provide the solid matrix algebra theory and methods for four basic and important AI fields, including machine learning, neural networks, support vector machines, and evolutionary computation.

Structure and Contents

The book consists of two parts.

Part I (Introduction to Matrix Algebra) provides fundamentals of matrix algebra and contains Chaps. 1 through 5. Chapter 1 presents the basic operations and performances of matrices, followed by a description of vectorization of matrix and matricization of vector. Chapter 2 is devoted to matrix differential as an important and effective tool in gradient computation and optimization. Chapter 3 is concerned with convex optimization theory and methods by focusing on gradient/subgradient methods in smooth and nonsmooth convex optimizations, and constrained convex optimization. Chapter 4 describes singular value decomposition (SVD) together with Tikhonov regularization and total least squares for solving over-determined matrix equations, followed by the Lasso and LARS methods for solving underdetermined matrix equations. Chapter 5 is devoted to the eigenvalue decomposition (EVD), the generalized eigenvalue decomposition, the Rayleigh quotient, and the generalized Rayleigh quotient.

Part II (Artificial Intelligence) focuses on machine learning, neural networks, support vector machines (SVMs), and evolutionary computation from the perspective of matrix algebra. This part is the main body of the book and consists of the following four chapters.

Chapter 6 (Machine Learning) presents first the basic theory and methods in machine learning including single-objective optimization, feature selection, principal component analysis and canonical correlation analysis together with supervised, unsupervised, and semi-supervised learning and active learning. Then, this chapter highlights topics and advances in machine learning: graph machine learning, reinforcement learning, Q-learning, and transfer learning.

Chapter 7 (Neural Networks) describes optimization problem, activation functions, and basic neural networks. The core part of this chapter are topics and advances in neural networks: convolutional neural networks (CNNs), dropout learning, autoencoders, extreme learning machine (ELM), graph embedding, network embedding, graph neural networks (GNNs), batch normalization networks, and generative adversarial networks (GANs).

Chapter 8 (Support Vector Machines) discusses the support vector machine regression and classification, and the relevance vector machine.

Chapter 9 (Evolutionary Computation) is concerned primarily with multiobjective optimization, multiobjective simulated annealing, multiobjective genetic algorithms, multiobjective evolutionary algorithms, evolutionary programming, differential evolution together with ant colony optimization, artificial bee colony algorithms, and particle swarm optimization. In particular, this chapter highlights also topics and advances in evolutionary computation: Pareto optimization theory, noisy multiobjective optimization, and opposition-based evolutionary computation.

Part I uses some revised content and materials from my book *Matrix Analysis and Applications* (Cambridge University Press, 2017), but there are considerable differences in content and book objective. This book is concentrated on the applications of the matrix algebra approaches in AI in Part II (561 pages of text), compared to Part I which is only 205 pages of text. This book is also related to my previous book *Linear Algebra in Signal Processing* (in Chinese, Science Press, Beijing, 1997; Japanese translation, Morikita Press, Tokyo, 2008) in some ideas.

Features and Contributions

- The first book on matrix algebra methods and applications in artificial intelligence.
- Introduces the machine learning tree, the neural network tree, and the evolutionary tree.

- Presents the solid matrix algebra theory and methods for four core AI areas: Machine Learning, Neural Networks, Support Vector Machines, and Evolutionary Computation.
- Highlights selected topics and advances of machine learning, neural networks, and evolutionary computation.
- Summarizes about 80 AI algorithms so that readers can further understand and implement AI methods.

Audience

This book is widely suitable for scientists, engineers, and graduate students in many disciplines, including but not limited to artificial intelligence, computer science, mathematics, engineering, etc.

Acknowledgments

I want to thank more than 40 of my former doctoral students and more than 20 graduate students for their cooperative research in intelligent signal and information processing and pattern recognition.

I am grateful to the several anonymous internal and external experts invited by Springer for their comments and suggestions. I am very grateful to my editor, Dr. Celine Lanlan Chang, Executive Editor, Computer Science, for her patience, understanding, and help in the course of my writing this book.

Finally, I would also like to thank my wife Xiao-Ying Tang for her consistent understanding and support for my work, teaching, and research in the past near 50 years.

Beijing, China Xian-Da Zhang
November, 2019

A Note from the Family of Dr. Zhang

It is with a heavy heart we share with you that the author of this book, our beloved father, passed away before this book was published. We want to share with you some inspirational thoughts about his life's journey and how passionate he was about learning and teaching.

Our father endured many challenges in his life. During his high school years, the Great Chinese Famine occurred, but hunger could not deter him from studying. During his college years, our father's family was facing poverty, so he sought to help alleviate this by undergoing his studies at a military institution, where free meals were provided. Not long after, his tenacity to learn would be tested again, as the Cultural Revolution started, closing all the universities in China. He was assigned to work in a remote factory, but his perseverance to learn endured as he continued his education by studying hard secretly during off-hours.

After all the universities were reopened, our father left us to continue his study in a different city. He obtained his PhD degree at the age of 41 and became a professor at Tsinghua University in 1992.

Our father has taught and mentored many students both professionally and personally throughout his life. He is even more passionate in sharing his ideas and knowledge through writing as he believes that books, with its greater reach, will benefit many more people. He had been working for more than 12 h a day before he was admitted into the hospital. He planned to take a break after he finishes three books this year. Unfortunately, he could not handle such a heavy workload that he asked our help in editing this book in our last conversation.

Our father has lived a life with purpose. He discovered his great passion when he was young: learning and teaching. For him, it was even something to die for. He told our mom once that he would rather live fewer years to produce a high-quality book. Despite the numerous challenges and hardships he faced throughout his life, he never stopped learning and teaching. He self-studied Artificial Intelligence in the past few years and completed this book before his final days.

We sincerely hope you will enjoy reading his final work, as we believe that he would undoubtedly have been very happy to inform and teach and share with you his latest learning.

Fremont, CA, USA Yewei Zhang and Wei Wei (Daughter and Son In Law)
Philadelphia, PA, USA Zhang and John Zhang (Son and Grand Son)
April, 2020

Contents

Part I   Introduction to Matrix Algebra

1 Basic Matrix Computation 3
   1.1   Basic Concepts of Vectors and Matrices 3
         1.1.1 Vectors and Matrices 3
         1.1.2 Basic Vector Calculus 6
         1.1.3 Basic Matrix Calculus 7
   1.2   Sets and Linear Mapping 10
         1.2.1 Sets 10
         1.2.2 Linear Mapping 12
   1.3   Norms 14
         1.3.1 Vector Norms 14
         1.3.2 Matrix Norms 18
   1.4   Random Vectors 20
         1.4.1 Statistical Interpretation of Random Vectors 20
         1.4.2 Gaussian Random Vectors 24
   1.5   Basic Performance of Matrices 27
         1.5.1 Quadratic Forms 27
         1.5.2 Determinants 27
         1.5.3 Matrix Eigenvalues 30
         1.5.4 Matrix Trace 32
         1.5.5 Matrix Rank 33
   1.6   Inverse Matrices and Moore–Penrose Inverse Matrices 35
         1.6.1 Inverse Matrices 35
         1.6.2 Left and Right Pseudo-Inverse Matrices 39
         1.6.3 Moore–Penrose Inverse Matrices 40
   1.7   Direct Sum and Hadamard Product 42
         1.7.1 Direct Sum of Matrices 43
         1.7.2 Hadamard Product 43

	1.8	Kronecker Products	46
		1.8.1 Definitions of Kronecker Products	46
		1.8.2 Performance of Kronecker Products	47
	1.9	Vectorization and Matricization	48
		1.9.1 Vectorization and Commutation Matrix	48
		1.9.2 Matricization of Vectors	50
	References		53
2	**Matrix Differential**		**55**
	2.1	Jacobian Matrix and Gradient Matrix	55
		2.1.1 Jacobian Matrix	55
		2.1.2 Gradient Matrix	57
		2.1.3 Calculation of Partial Derivative and Gradient	59
	2.2	Real Matrix Differential	61
		2.2.1 Calculation of Real Matrix Differential	62
		2.2.2 Jacobian Matrix Identification	64
		2.2.3 Jacobian Matrix of Real Matrix Functions	68
	2.3	Complex Gradient Matrices	71
		2.3.1 Holomorphic Function and Complex Partial Derivative	72
		2.3.2 Complex Matrix Differential	75
		2.3.3 Complex Gradient Matrix Identification	82
	References		87
3	**Gradient and Optimization**		**89**
	3.1	Real Gradient	89
		3.1.1 Stationary Points and Extreme Points	90
		3.1.2 Real Gradient of $f(\mathbf{X})$	93
	3.2	Gradient of Complex Variable Function	95
		3.2.1 Extreme Point of Complex Variable Function	95
		3.2.2 Complex Gradient	98
	3.3	Convex Sets and Convex Function Identification	100
		3.3.1 Standard Constrained Optimization Problems	100
		3.3.2 Convex Sets and Convex Functions	102
		3.3.3 Convex Function Identification	104
	3.4	Gradient Methods for Smooth Convex Optimization	106
		3.4.1 Gradient Method	106
		3.4.2 Projected Gradient Method	108
		3.4.3 Convergence Rates	110
	3.5	Nesterov Optimal Gradient Method	112
		3.5.1 Lipschitz Continuous Function	113
		3.5.2 Nesterov Optimal Gradient Algorithms	115
	3.6	Nonsmooth Convex Optimization	118
		3.6.1 Subgradient and Subdifferential	119
		3.6.2 Proximal Operator	123
		3.6.3 Proximal Gradient Method	128

3.7	Constrained Convex Optimization		132
	3.7.1	Penalty Function Method	133
	3.7.2	Augmented Lagrange Multiplier Method	135
	3.7.3	Lagrange Dual Method	137
	3.7.4	Karush–Kuhn–Tucker Conditions	139
	3.7.5	Alternating Direction Method of Multipliers	144
3.8	Newton Methods		147
	3.8.1	Newton Method for Unconstrained Optimization	147
	3.8.2	Newton Method for Constrained Optimization	150
References			153

4 Solution of Linear Systems ... 157

4.1	Gauss Elimination		157
	4.1.1	Elementary Row Operations	158
	4.1.2	Gauss Elimination for Solving Matrix Equations	159
	4.1.3	Gauss Elimination for Matrix Inversion	160
4.2	Conjugate Gradient Methods		162
	4.2.1	Conjugate Gradient Algorithm	163
	4.2.2	Biconjugate Gradient Algorithm	164
	4.2.3	Preconditioned Conjugate Gradient Algorithm	165
4.3	Condition Number of Matrices		167
4.4	Singular Value Decomposition (SVD)		171
	4.4.1	Singular Value Decomposition	171
	4.4.2	Properties of Singular Values	173
	4.4.3	Singular Value Thresholding	174
4.5	Least Squares Method		175
	4.5.1	Least Squares Solution	175
	4.5.2	Rank-Deficient Least Squares Solutions	178
4.6	Tikhonov Regularization and Gauss–Seidel Method		179
	4.6.1	Tikhonov Regularization	179
	4.6.2	Gauss–Seidel Method	181
4.7	Total Least Squares Method		184
	4.7.1	Total Least Squares Solution	184
	4.7.2	Performances of TLS Solution	189
	4.7.3	Generalized Total Least Square	191
4.8	Solution of Under-Determined Systems		193
	4.8.1	ℓ_1-Norm Minimization	193
	4.8.2	Lasso	196
	4.8.3	LARS	197
References			199

5 Eigenvalue Decomposition ... 203

5.1	Eigenvalue Problem and Characteristic Equation		203
	5.1.1	Eigenvalue Problem	203
	5.1.2	Characteristic Polynomial	205

5.2	Eigenvalues and Eigenvectors	206
	5.2.1	Eigenvalues	206
	5.2.2	Eigenvectors	208
5.3	Generalized Eigenvalue Decomposition (GEVD)	210
	5.3.1	Generalized Eigenvalue Decomposition	210
	5.3.2	Total Least Squares Method for GEVD	214
5.4	Rayleigh Quotient and Generalized Rayleigh Quotient	214
	5.4.1	Rayleigh Quotient	215
	5.4.2	Generalized Rayleigh Quotient	216
	5.4.3	Effectiveness of Class Discrimination	217
References	220

Part II Artificial Intelligence

6 Machine Learning ... 223

6.1	Machine Learning Tree	223
6.2	Optimization in Machine Learning	226
	6.2.1	Single-Objective Composite Optimization	226
	6.2.2	Gradient Aggregation Methods	230
	6.2.3	Coordinate Descent Methods	232
	6.2.4	Benchmark Functions for Single-Objective Optimization	236
6.3	Majorization-Minimization Algorithms	241
	6.3.1	MM Algorithm Framework	242
	6.3.2	Examples of Majorization-Minimization Algorithms	244
6.4	Boosting and Probably Approximately Correct Learning	247
	6.4.1	Boosting for Weak Learners	248
	6.4.2	Probably Approximately Correct Learning	250
6.5	Basic Theory of Machine Learning	253
	6.5.1	Learning Machine	253
	6.5.2	Machine Learning Methods	254
	6.5.3	Expected Performance of Machine Learning Algorithms	256
6.6	Classification and Regression	256
	6.6.1	Pattern Recognition and Classification	257
	6.6.2	Regression	259
6.7	Feature Selection	260
	6.7.1	Supervised Feature Selection	261
	6.7.2	Unsupervised Feature Selection	264
	6.7.3	Nonlinear Joint Unsupervised Feature Selection	266
6.8	Principal Component Analysis	269
	6.8.1	Principal Component Analysis Basis	269
	6.8.2	Minor Component Analysis	270
	6.8.3	Principal Subspace Analysis	271

	6.8.4	Robust Principal Component Analysis	276
	6.8.5	Sparse Principal Component Analysis	278
6.9	Supervised Learning Regression		281
	6.9.1	Principle Component Regression	282
	6.9.2	Partial Least Squares Regression	285
	6.9.3	Penalized Regression	289
	6.9.4	Gradient Projection for Sparse Reconstruction	293
6.10	Supervised Learning Classification		296
	6.10.1	Binary Linear Classifiers	296
	6.10.2	Multiclass Linear Classifiers	298
6.11	Supervised Tensor Learning (STL)		301
	6.11.1	Tensor Algebra Basics	301
	6.11.2	Supervised Tensor Learning Problems	307
	6.11.3	Tensor Fisher Discriminant analysis	309
	6.11.4	Tensor Learning for Regression	311
	6.11.5	Tensor K-Means Clustering	314
6.12	Unsupervised Clustering		314
	6.12.1	Similarity Measures	316
	6.12.2	Hierarchical Clustering	320
	6.12.3	Fisher Discriminant Analysis (FDA)	324
	6.12.4	K-Means Clustering	327
6.13	Spectral Clustering		330
	6.13.1	Spectral Clustering Algorithms	330
	6.13.2	Constrained Spectral Clustering	333
	6.13.3	Fast Spectral Clustering	335
6.14	Semi-Supervised Learning Algorithms		337
	6.14.1	Semi-Supervised Inductive/Transductive Learning	338
	6.14.2	Self-Training	340
	6.14.3	Co-training	341
6.15	Canonical Correlation Analysis		343
	6.15.1	Canonical Correlation Analysis Algorithm	344
	6.15.2	Kernel Canonical Correlation Analysis	347
	6.15.3	Penalized Canonical Correlation Analysis	352
6.16	Graph Machine Learning		354
	6.16.1	Graphs	354
	6.16.2	Graph Laplacian Matrices	358
	6.16.3	Graph Spectrum	360
	6.16.4	Graph Signal Processing	363
	6.16.5	Semi-Supervised Graph Learning: Harmonic Function Method	368
	6.16.6	Semi-Supervised Graph Learning: Min-Cut Method	371
	6.16.7	Unsupervised Graph Learning: Sparse Coding Method	377
6.17	Active Learning		378
	6.17.1	Active Learning: Background	379
	6.17.2	Statistical Active Learning	380

	6.17.3	Active Learning Algorithms	382
	6.17.4	Active Learning Based Binary Linear Classifiers	383
	6.17.5	Active Learning Using Extreme Learning Machine	384
6.18	Reinforcement Learning		387
	6.18.1	Basic Concepts and Theory	387
	6.18.2	Markov Decision Process (MDP)	391
6.19	Q-Learning		394
	6.19.1	Basic Q-Learning	394
	6.19.2	Double Q-Learning and Weighted Double Q-Learning	396
	6.19.3	Online Connectionist Q-Learning Algorithm	399
	6.19.4	Q-Learning with Experience Replay	400
6.20	Transfer Learning		402
	6.20.1	Notations and Definitions	403
	6.20.2	Categorization of Transfer Learning	406
	6.20.3	Boosting for Transfer Learning	410
	6.20.4	Multitask Learning	411
	6.20.5	EigenTransfer	414
6.21	Domain Adaptation		417
	6.21.1	Feature Augmentation Method	418
	6.21.2	Cross-Domain Transform Method	421
	6.21.3	Transfer Component Analysis Method	423
References			427

7 Neural Networks ... 441
7.1 Neural Network Tree ... 441
7.2 From Modern Neural Networks to Deep Learning ... 444
7.3 Optimization of Neural Networks ... 445
7.3.1 Online Optimization Problems ... 446
7.3.2 Adaptive Gradient Algorithm ... 447
7.3.3 Adaptive Moment Estimation ... 449
7.4 Activation Functions ... 452
7.4.1 Logistic Regression and Sigmoid Function ... 452
7.4.2 Softmax Regression and Softmax Function ... 454
7.4.3 Other Activation Functions ... 456
7.5 Recurrent Neural Networks ... 460
7.5.1 Conventional Recurrent Neural Networks ... 460
7.5.2 Backpropagation Through Time (BPTT) ... 463
7.5.3 Jordan Network and Elman Network ... 467
7.5.4 Bidirectional Recurrent Neural Networks ... 468
7.5.5 Long Short-Term Memory (LSTM) ... 471
7.5.6 Improvement of Long Short-Term Memory ... 474
7.6 Boltzmann Machines ... 477
7.6.1 Hopfield Network and Boltzmann Machines ... 477
7.6.2 Restricted Boltzmann Machine ... 481

		7.6.3	Contrastive Divergence Learning	484

- 7.6.3 Contrastive Divergence Learning 484
- 7.6.4 Multiple Restricted Boltzmann Machines 487
- 7.7 Bayesian Neural Networks 490
  - 7.7.1 Naive Bayesian Classification 490
  - 7.7.2 Bayesian Classification Theory 491
  - 7.7.3 Sparse Bayesian Learning 493
- 7.8 Convolutional Neural Networks 496
  - 7.8.1 Hankel Matrix and Convolution 498
  - 7.8.2 Pooling Layer 504
  - 7.8.3 Activation Functions in CNNs 507
  - 7.8.4 Loss Function 510
- 7.9 Dropout Learning 514
  - 7.9.1 Dropout for Shallow and Deep Learning 515
  - 7.9.2 Dropout Spherical K-Means 518
  - 7.9.3 DropConnect 520
- 7.10 Autoencoders 524
  - 7.10.1 Basic Autoencoder 525
  - 7.10.2 Stacked Sparse Autoencoder 532
  - 7.10.3 Stacked Denoising Autoencoders 535
  - 7.10.4 Convolutional Autoencoders (CAE) 538
  - 7.10.5 Stacked Convolutional Denoising Autoencoder . 539
  - 7.10.6 Nonnegative Sparse Autoencoder 540
- 7.11 Extreme Learning Machine 542
  - 7.11.1 Single-Hidden Layer Feedforward Networks with Random Hidden Nodes 543
  - 7.11.2 Extreme Learning Machine Algorithm for Regression and Binary Classification 545
  - 7.11.3 Extreme Learning Machine Algorithm for Multiclass Classification 549
- 7.12 Graph Embedding 551
  - 7.12.1 Proximity Measures and Graph Embedding 552
  - 7.12.2 Multidimensional Scaling 557
  - 7.12.3 Manifold Learning: Isometric Map 559
  - 7.12.4 Manifold Learning: Locally Linear Embedding 560
  - 7.12.5 Manifold Learning: Laplacian Eigenmap 563
- 7.13 Network Embedding 567
  - 7.13.1 Structure and Property Preserving Network Embedding ... 568
  - 7.13.2 Community Preserving Network Embedding 569
  - 7.13.3 Higher-Order Proximity Preserved Network Embedding 573
- 7.14 Neural Networks on Graphs 576
  - 7.14.1 Graph Neural Networks (GNNs) 577
  - 7.14.2 DeepWalk and GraphSAGE 579
  - 7.14.3 Graph Convolutional Networks (GCNs) 583

	7.15	Batch Normalization Networks	588
		7.15.1 Batch Normalization	588
		7.15.2 Variants and Extensions of Batch Normalization	593
	7.16	Generative Adversarial Networks (GANs)	598
		7.16.1 Generative Adversarial Network Framework	598
		7.16.2 Bidirectional Generative Adversarial Networks	602
		7.16.3 Variational Autoencoders	604
	References		607
8	**Support Vector Machines**		**617**
	8.1	Support Vector Machines: Basic Theory	617
		8.1.1 Statistical Learning Theory	618
		8.1.2 Linear Support Vector Machines	621
	8.2	Kernel Regression Methods	624
		8.2.1 Reproducing Kernel and Mercer Kernel	624
		8.2.2 Representer Theorem and Kernel Regression	628
		8.2.3 Semi-Supervised and Graph Regression	630
		8.2.4 Kernel Partial Least Squares Regression	632
		8.2.5 Laplacian Support Vector Machines	633
	8.3	Support Vector Machine Regression	635
		8.3.1 Support Vector Machine Regressor	635
		8.3.2 ϵ-Support Vector Regression	636
		8.3.3 ν-Support Vector Machine Regression	639
	8.4	Support Vector Machine Binary Classification	641
		8.4.1 Support Vector Machine Binary Classifier	642
		8.4.2 ν-Support Vector Machine Binary Classifier	645
		8.4.3 Least Squares SVM Binary Classifier	647
		8.4.4 Proximal Support Vector Machine Binary Classifier	649
		8.4.5 SVM-Recursive Feature Elimination	651
	8.5	Support Vector Machine Multiclass Classification	653
		8.5.1 Decomposition Methods for Multiclass Classification	653
		8.5.2 Least Squares SVM Multiclass Classifier	656
		8.5.3 Proximal Support Vector Machine Multiclass Classifier	659
	8.6	Gaussian Process for Regression and Classification	660
		8.6.1 Joint, Marginal, and Conditional Probabilities	661
		8.6.2 Gaussian Process	661
		8.6.3 Gaussian Process Regression	663
		8.6.4 Gaussian Process Classification	666
	8.7	Relevance Vector Machine	667
		8.7.1 Sparse Bayesian Regression	668
		8.7.2 Sparse Bayesian Classification	672
		8.7.3 Fast Marginal Likelihood Maximization	673
	References		677

9 Evolutionary Computation ... 681
- 9.1 Evolutionary Computation Tree ... 681
- 9.2 Multiobjective Optimization ... 683
  - 9.2.1 Multiobjective Combinatorial Optimization ... 684
  - 9.2.2 Multiobjective Optimization Problems ... 686
- 9.3 Pareto Optimization Theory ... 691
  - 9.3.1 Pareto Concepts ... 691
  - 9.3.2 Fitness Selection Approach ... 698
  - 9.3.3 Nondominated Sorting Approach ... 700
  - 9.3.4 Crowding Distance Assignment Approach ... 702
  - 9.3.5 Hierarchical Clustering Approach ... 703
  - 9.3.6 Benchmark Functions for Multiobjective Optimization ... 704
- 9.4 Noisy Multiobjective Optimization ... 707
  - 9.4.1 Pareto Concepts for Noisy Multiobjective Optimization ... 708
  - 9.4.2 Performance Metrics for Approximation Sets ... 712
- 9.5 Multiobjective Simulated Annealing ... 714
  - 9.5.1 Principle of Simulated Annealing ... 714
  - 9.5.2 Multiobjective Simulated Annealing Algorithm ... 718
  - 9.5.3 Archived Multiobjective Simulated Annealing ... 721
- 9.6 Genetic Algorithm ... 723
  - 9.6.1 Basic Genetic Algorithm Operations ... 724
  - 9.6.2 Genetic Algorithm with Gene Rearrangement Clustering ... 729
- 9.7 Nondominated Multiobjective Genetic Algorithms ... 733
  - 9.7.1 Fitness Functions ... 733
  - 9.7.2 Fitness Selection ... 734
  - 9.7.3 Nondominated Sorting Genetic Algorithms ... 736
  - 9.7.4 Elitist Nondominated Sorting Genetic Algorithm ... 739
- 9.8 Evolutionary Algorithms (EAs) ... 740
  - 9.8.1 $(1+1)$ Evolutionary Algorithm ... 741
  - 9.8.2 Theoretical Analysis on Evolutionary Algorithms ... 742
- 9.9 Multiobjective Evolutionary Algorithms ... 743
  - 9.9.1 Classical Methods for Solving Multiobjective Optimization Problems ... 743
  - 9.9.2 MOEA Based on Decomposition (MOEA/D) ... 745
  - 9.9.3 Strength Pareto Evolutionary Algorithm ... 749
  - 9.9.4 Achievement Scalarizing Functions ... 754
- 9.10 Evolutionary Programming ... 758
  - 9.10.1 Classical Evolutionary Programming ... 758
  - 9.10.2 Fast Evolutionary Programming ... 759
  - 9.10.3 Hybrid Evolutionary Programming ... 761
- 9.11 Differential Evolution ... 763
  - 9.11.1 Classical Differential Evolution ... 763
  - 9.11.2 Differential Evolution Variants ... 765

9.12	Ant Colony Optimization		767
	9.12.1	Real Ants and Artificial Ants	768
	9.12.2	Typical Ant Colony Optimization Problems	771
	9.12.3	Ant System and Ant Colony System	773
9.13	Multiobjective Artificial Bee Colony Algorithms		776
	9.13.1	Artificial Bee Colony Algorithms	776
	9.13.2	Variants of ABC Algorithms	778
9.14	Particle Swarm Optimization		780
	9.14.1	Basic Concepts	780
	9.14.2	The Canonical Particle Swarm	781
	9.14.3	Genetic Learning Particle Swarm Optimization	783
	9.14.4	Particle Swarm Optimization for Feature Selection	785
9.15	Opposition-Based Evolutionary Computation		788
	9.15.1	Opposition-Based Learning	788
	9.15.2	Opposition-Based Differential Evolution	790
	9.15.3	Two Variants of Opposition-Based Learning	791
References			795
Index			805

List of Notations

$\forall$	For all
$\|$	Such that
$\ni$	Such that
$\exists$	There exists
$\nexists$	There does not exist
$\wedge$	Logical AND
$\vee$	Logical OR
$\|A\|$	Cardinality of a set A
$A \Rightarrow B$	"condition A results in B" or "A implies B"
$A \subseteq B$	A is a subset of B
$A \subset B$	A is a proper subset of B
$A = B$	Sets $A = B$
$A \cup B$	Union of sets A and B
$A \cap B$	Intersection of sets A and B
$A \cap B = \emptyset$	Sets A and B are disjoint
$A + B$	Sum set of sets A and B
$A - B$	The set of elements of A that are not in B
$X \setminus A$	Complement of the set A in the set X
$A \succ B$	Set A dominates set B: every $\mathbf{f}(\mathbf{x}_2) \in B$ is dominated by at least one $\mathbf{f}(\mathbf{x}_1) \in A$ such that $\mathbf{f}(\mathbf{x}_1) <_{IN} \mathbf{f}(\mathbf{x}_2)$ (for minimization) or $\mathbf{f}(\mathbf{x}_1) >_{IN} \mathbf{f}(\mathbf{x}_2)$ (for maximization)
$A \succeq B$	A weakly dominates B: for every $\mathbf{f}(\mathbf{x}_2) \in B$ and at least one $\mathbf{f}(\mathbf{x}_1) \in A$ $\mathbf{f}(\mathbf{x}_1) \leq_{IN} \mathbf{f}(\mathbf{x}_2)$ (for minimization) or $\mathbf{f}(\mathbf{x}_1) \geq_{IN} \mathbf{f}(\mathbf{x}_2)$ (for maximization)
$A \succ\succ B$	A strictly dominates B: every $\mathbf{f}(\mathbf{x}_2) \in B$ is strictly dominated by at least one $\mathbf{f}(\mathbf{x}_1) \in A$ such that $f_i(\mathbf{x}_1) <_{IN} f_i(\mathbf{x}_2), \forall i = \{1, \ldots, m\}$ (for minimization) or $f_i(\mathbf{x}_1) >_{IN} f_i(\mathbf{x}_2), \forall i = \{1, \ldots, m\}$ (for maximization)
$A \parallel B$	Sets A and B are incomparable: neither $A \succeq B$ nor $B \succeq A$
$A \triangleright B$	A is better than B: every $\mathbf{f}(\mathbf{x}_2) \in B$ is weakly dominated by at least one $\mathbf{f}(\mathbf{x}_1) \in A$ and $A \neq B$

$\mathrm{AGG}_{\mathrm{mean}}(z)$	Mean aggregate function of z
$\mathrm{AGG}_{\mathrm{LSTM}}(z)$	LSTM aggregate function of z
$\mathrm{AGG}_{\mathrm{pool}}(z)$	Pooling aggregate function of z
$\mathbb{C}$	Complex numbers
$\mathbb{C}^n$	Complex n-vector
$\mathbb{C}^{m \times n}$	Complex $m \times n$ matrix
$\mathbb{C}[x]$	Complex polynomial
$\mathbb{C}[x]^{m \times n}$	Complex $m \times n$ polynomial matrix
$\mathbb{C}^{I \times J \times K}$	Complex third-order tensors
$\mathbb{C}^{I_1 \times \cdots \times I_N}$	Complex N-order tensor
$\mathbb{K}$	Real or complex number
$\mathbb{K}^n$	Real or complex n-vector
$\mathbb{K}^{m \times n}$	Real or complex $m \times n$ matrix
$\mathbb{K}^{I \times J \times K}$	Real or complex third-order tensor
$\mathbb{K}^{I_1 \times \cdots \times I_N}$	Real or complex N-order tensor
$G(V, E, \mathbf{W})$	Graph with vertex set V, edge set E and adjacency matrix $\mathbf{W}$
$\mathcal{N}(v)$	Neighbors of a vertex (node) v
$\mathrm{PReLU}(z)$	Parametric rectified linear unit activation function of z
$\mathrm{ReLU}(z)$	Rectified linear unit activation function of z
$\mathbb{R}$	Real number
$\mathbb{R}^n$	Real n-vectors ($n \times 1$ real matrix)
$\mathbb{R}^{m \times n}$	Real $m \times n$ matrix
$\mathbb{R}[x]$	Real polynomial
$\mathbb{R}[x]^{m \times n}$	Real $m \times n$ polynomial matrix
$\mathbb{R}^{I \times J \times K}$	Real third-order tensors
$\mathbb{R}^{I_1 \times \cdots \times I_N}$	Real N-order tensor
$\mathbb{R}_+$	Nonnegative real numbers, nonnegative orthant
$\mathbb{R}_{++}$	Positive real number
$\sigma(z)$	Sigmoid activation function of z
$\mathrm{softmax}(z)$	Softmax activation function of z
$\mathrm{softplus}(z)$	Softplus activation function of z
$\mathrm{softsign}(z)$	Softsign activation function of z
$\tanh(z)$	Tangent (tanh) hyperbolic activation function of z
$T : V \to W$	Mapping the vectors in V to corresponding vectors in W
$T^{-1} : W \to V$	Inverse mapping of the one-to-one mapping $T : V \to W$
$X_1 \times \cdots \times X_n$	Cartesian product of n sets $X_1, \ldots, X_n$
$\{(\mathbf{x}_i, y_i = +1)\}$	Set of training data vectors $\mathbf{x}_i$ belonging to the classes $(+)$
$\{(\mathbf{x}_i, y_i = -1)\}$	Set of training data vectors $\mathbf{x}_i$ belonging to the classes $(-)$
$\mathbf{1}_n$	n-dimensional summing vector with all entries 1
$\mathbf{0}_n$	n-dimensional zero vector with all zero entries
$\mathbf{e}_i$	Base vector whose ith entry equal to 1 and others being zero
$\mathbf{x} \sim N(\bar{\mathbf{x}}, \mathbf{\Gamma}_x)$	Gaussian vector with mean vector $\bar{\mathbf{x}}$ and covariance matrix $\mathbf{\Gamma}_x$
$\|\mathbf{x}\|_0$	ℓ_0-norm: number of nonzero entries of vector $\mathbf{x}$
$\|\mathbf{x}\|_1$	ℓ_1-norm of vector $\mathbf{x}$
$\|\mathbf{x}\|_2$	Euclidean form of vector $\mathbf{x}$

List of Notations

$\|\mathbf{x}\|_p$	ℓ_p-norm or Hölder norm of vector $\mathbf{x}$
$\|\mathbf{x}\|_*$	Nuclear norm of vector $\mathbf{x}$
$\|\mathbf{x}\|_\infty$	ℓ_∞-norm of vector $\mathbf{x}$
$\langle \mathbf{x}, \mathbf{y} \rangle$	Inner product of vectors $\mathbf{x}$ and $\mathbf{y}$
$d(\mathbf{x}, \mathbf{y})$	Distance or dissimilarity between vectors $\mathbf{x}$ and $\mathbf{y}$
$N_\epsilon(\mathbf{x})$	ϵ-neighborhood of vector $\mathbf{x}$
$\rho(\mathbf{x}, \mathbf{y})$	Correlation coefficient between two random vectors $\mathbf{x}$ and $\mathbf{y}$
$\mathbf{x} \in A$	$\mathbf{x}$ belongs to the set A, i.e., $\mathbf{x}$ is an element or member of A
$\mathbf{x} \notin A$	$\mathbf{x}$ is not an element of the set A
$\mathbf{x} \circ \mathbf{y} = \mathbf{x}\mathbf{y}^H$	Outer product of vectors $\mathbf{x}$ and $\mathbf{y}$
$\mathbf{x} \perp \mathbf{y}$	Vector orthogonal
$\mathbf{x} > 0$	Positive vector with components $x_i > 0, \forall i$
$\mathbf{x} \geq 0$	Nonnegative vector with components $x_i \geq 0, \forall i$
$\mathbf{x} \geq \mathbf{y}$	Vector elementwise inequality $x_i \geq y_i, \forall i$
$\mathbf{x} \succ \mathbf{x}'$	$\mathbf{x}$ domains (or outperforms) $\mathbf{x}'$: $\mathbf{f}(\mathbf{x}) < \mathbf{f}(\mathbf{x}')$ for minimization
$\mathbf{x} \succ \mathbf{x}'$	$\mathbf{x}$ domains (or outperforms) $\mathbf{x}'$: $\mathbf{f}(\mathbf{x}) > \mathbf{f}(\mathbf{x}')$ for maximization
$\mathbf{x} \succeq \mathbf{x}'$	$\mathbf{x}$ weakly dominates $\mathbf{x}'$: $\mathbf{f}(\mathbf{x}) \leq \mathbf{f}(\mathbf{x}')$ for minimization
$\mathbf{x} \succeq \mathbf{x}'$	$\mathbf{x}$ weakly dominates $\mathbf{x}'$: $\mathbf{f}(\mathbf{x}) \geq \mathbf{f}(\mathbf{x}')$ for maximization
$\mathbf{x} \succ\succ \mathbf{x}'$	$\mathbf{x}$ strictly dominates $\mathbf{x}'$: $f_i(\mathbf{x}) < f_i(\mathbf{x}'), \forall i$ for minimization
$\mathbf{x} \succ\succ \mathbf{x}'$	$\mathbf{x}$ strictly dominates $\mathbf{x}'$: $f_i(\mathbf{x}) > f_i(\mathbf{x}'), \forall i$ for maximization
$\mathbf{x} \parallel \mathbf{x}'$	$\mathbf{x}$ and $\mathbf{x}'$ are incomparable, i.e., $\mathbf{x} \not\succeq \mathbf{x}' \wedge \mathbf{x}' \not\succeq \mathbf{x}$
$\mathbf{f}(\mathbf{x}) = \mathbf{f}(\mathbf{x}')$	$f_i(\mathbf{x}) = f_i(\mathbf{x}'), \forall i = 1, \ldots, m$
$\mathbf{f}(\mathbf{x}) \neq \mathbf{f}(\mathbf{x}')$	$f_i(\mathbf{x}) \neq f_i(\mathbf{x}')$, for at least one $i \in \{1, \ldots, m\}$
$\mathbf{f}(\mathbf{x}) \leq \mathbf{f}(\mathbf{x}')$	$f_i(\mathbf{x}) \leq f_i(\mathbf{x}'), \forall i = 1, \ldots, m$
$\mathbf{f}(\mathbf{x}) < \mathbf{f}(\mathbf{x}')$	$\forall i = 1, \ldots, m : f_i(\mathbf{x}) \leq f_i(\mathbf{x}') \wedge \exists j \in \{1, \ldots, m\} : f_j(\mathbf{x}) < f_j(\mathbf{x}')$
$\mathbf{f}(\mathbf{x}) \geq \mathbf{f}(\mathbf{x}')$	$f_i(\mathbf{x}) \geq f_i(\mathbf{x}'), \forall i = 1, \ldots, m$
$\mathbf{f}(\mathbf{x}) > \mathbf{f}(\mathbf{x}')$	$\forall i = 1, \ldots, m : f_i(\mathbf{x}) \geq f_i(\mathbf{x}') \wedge \exists j \in \{1, \ldots, m\} : f_j(\mathbf{x}) > f_j(\mathbf{x}')$
$\mathbf{f}(\mathbf{x}_1) <_{IN} \mathbf{f}(\mathbf{x}_2)$	Interval order relation: $\forall i = 1, \ldots, m : \underline{f}_i(\mathbf{x}_1) \leq \underline{f}_i(\mathbf{x}_2) \wedge \overline{f}_i(\mathbf{x}_1) \leq \overline{f}_i(\mathbf{x}_2) \wedge \exists j \in \{1, \ldots, m\} : \underline{f}_j(\mathbf{x}_1) \neq \underline{f}_j(\mathbf{x}_2) \vee \overline{f}_j(\mathbf{x}_1) \neq \overline{f}_j(\mathbf{x}_2)$
$\mathbf{f}(\mathbf{x}_1) >_{IN} \mathbf{f}(\mathbf{x}_2)$	Interval order relation: $\forall i = 1, \ldots, m : \underline{f}_i(\mathbf{x}_1) \geq \underline{f}_i(\mathbf{x}_2) \wedge \overline{f}_i(\mathbf{x}_1) \geq \overline{f}_i(\mathbf{x}_2) \wedge \exists j \in \{1, \ldots, m\} : \underline{f}_j(\mathbf{x}_1) \neq \underline{f}_j(\mathbf{x}_2) \vee \overline{f}_j(\mathbf{x}_1) \neq \overline{f}_j(\mathbf{x}_2)$
$\mathbf{f}(\mathbf{x}_1) \leq_{IN} \mathbf{f}(\mathbf{x}_2)$	Weak interval order relation: $\forall i \in \{1, \ldots, m\} : \underline{f}_i(\mathbf{x}_1) \leq \underline{f}_i(\mathbf{x}_2) \wedge \overline{f}_i(\mathbf{x}_1) \leq \overline{f}_i(\mathbf{x}_2)$
$\mathbf{f}(\mathbf{x}_1) \geq_{IN} \mathbf{f}(\mathbf{x}_2)$	Weak interval order relation: $\forall i \in \{1, \ldots, m\} : \underline{f}_i(\mathbf{x}_1) \geq \underline{f}_i(\mathbf{x}_2) \wedge \overline{f}_i(\mathbf{x}_1) \geq \overline{f}_i(\mathbf{x}_2)$
$\mathbf{A}^T$	Transpose of matrix $\mathbf{A}$
$\mathbf{A}^H$	Complex conjugate transpose of matrix $\mathbf{A}$
$\mathbf{A}^{-1}$	Inverse of nonsingular matrix $\mathbf{A}$
$\mathbf{A}^\dagger$	Moore–Penrose inverse of matrix $\mathbf{A}$

$\mathbf{A}^*$	Conjugate of matrix $\mathbf{A}$
$\mathbf{A} \succ 0$	Positive definite matrix
$\mathbf{A} \succeq 0$	Positive semi-definite matrix
$\mathbf{A} \prec 0$	Negative definite matrix
$\mathbf{A} \preceq 0$	Negative semi-definite matrix
$\mathbf{A} > 0$	Positive (or elementwise positive) matrix
$\mathbf{A} \geq 0$	Nonnegative (or elementwise nonnegative) matrix
$\mathbf{A} \geq \mathbf{B}$	Matrix elementwise inequality $a_{ij} \geq b_{ij}, \forall i, j$
$\mathbf{I}_n$	$n \times n$ Identity matrix
$\mathbf{O}_n$	$n \times n$ Null matrix
$|\mathbf{A}|$	Determinant of matrix $\mathbf{A}$
$\|\mathbf{A}\|_1$	Maximum absolute column-sum norm of matrix $\mathbf{A}$
$\|\mathbf{A}\|_2 = \|\mathbf{A}\|_{\text{spec}}$	Spectrum norm of matrix $\mathbf{A}$
$\|\mathbf{A}\|_F$	Frobenius norm of matrix $\mathbf{A}$
$\|\mathbf{A}\|_\infty$	Max norm of $\mathbf{A}$: absolute maximum of all entries of $\mathbf{A}$
$\|\mathbf{A}\|_{\mathbf{G}}$	Mahalanobis norm of matrix $\mathbf{A}$
$\|\mathbf{A}\|_*$	Nuclear norm, called also the trace norm, of matrix $\mathbf{A}$
$\mathbf{A} \oplus \mathbf{B}$	Direct sum of an $m \times m$ matrix $\mathbf{A}$ and an $n \times n$ matrix $\mathbf{B}$
$\mathbf{A} \odot \mathbf{B}$	Hadamard product (or elementwise product) of $\mathbf{A}$ and $\mathbf{B}$
$\mathbf{A} \oslash \mathbf{B}$	Elementwise division of matrices $\mathbf{A}$ and $\mathbf{B}$
$\mathbf{A} \otimes \mathbf{B}$	Kronecker product of matrices $\mathbf{A}$ and $\mathbf{B}$
$\langle \mathbf{A}, \mathbf{B} \rangle$	Inner (or dot) product of $\mathbf{A}$ and $\mathbf{B}$: $\langle \mathbf{A}, \mathbf{B} \rangle = \langle \text{vec}(\mathbf{A}), \text{vec}(\mathbf{B}) \rangle$
$\rho(\mathbf{A})$	Spectral radius of matrix $\mathbf{A}$
$\text{cond}(\mathbf{A})$	Condition number of matrix $\mathbf{A}$
$\text{diag}(\mathbf{A})$	Diagonal function of $\mathbf{A} = [a_{ij}]$: $\sum_{i=1}^n |a_{ii}|^2$
$\mathbf{Diag}(\mathbf{A})$	Diagonal matrix consisting of diagonal entries of $\mathbf{A}$
$\text{eig}(\mathbf{A})$	Eigenvalues of the Hermitian matrix $\mathbf{A}$
$\text{Gr}(n, r)$	Grassmann manifold
$\text{rvec}(\mathbf{A})$	Row vectorization of matrix $\mathbf{A}$
$\text{off}(\mathbf{A})$	Off function of $\mathbf{A} = [a_{ij}]$: $\sum_{i=1, i \neq j}^m \sum_{j=1}^n |a_{ij}|^2$
$\text{tr}(\mathbf{A})$	Trace of matrix $\mathbf{A}$
$\text{vec}(\mathbf{A})$	Vectorization of matrix $\mathbf{A}$

List of Figures

Fig. 1.1	Classification of vectors	5
Fig. 6.1	Machine learning tree	225
Fig. 6.2	Three-way array (left) and a tensor modeling a face (right)	301
Fig. 6.3	An MDP models the synchronous interaction between agent and environment	391
Fig. 6.4	One neural network as a collection of interconnected units, where i, j, and k are neurons in output layer, hidden layer, and input layer, respectively	399
Fig. 6.5	Transfer learning setup: storing knowledge gained from solving one problem and applying it to a different but related problem	405
Fig. 6.6	The transformation mapping. (**a**) The symmetric transformation (T_S and T_T) of the source-domain feature set $X_S = \{\mathbf{x}_{S_i}\}$ and target-domain feature set $X_T = \{\mathbf{x}_{T_i}\}$ into a common latent feature set $X_C = \{\mathbf{x}_{C_i}\}$. (**b**) The asymmetric transformation T_T of the source-domain feature set X_S to the target-domain feature set X_T	407
Fig. 7.1	Neural network tree	444
Fig. 7.2	General structure of a regular unidirectional three-layer RNN with hidden state recurrence, where z^{-1} denotes a delay line. Left is RNN with a delay line, and right is RNN unfolded in time for two time steps	462
Fig. 7.3	General structure of a regular unidirectional three-layer RNN with output recurrence, where z^{-1} denotes a delay line. Left is RNN with a delay line, and right is RNN unfolded in time for two time steps	463

Fig. 7.4	Illustration of Jordan network	469
Fig. 7.5	Illustration of Elman network	469
Fig. 7.6	Illustration of the Bidirectional Recurrent Neural Network (BRNN) unfolded in time for three time steps. **Upper:** output units, **Middle:** forward and backward hidden units, **Lower:** input units	470
Fig. 7.7	A comparison between Hopfield network and Boltzmann machine. **Left**: is Hopfield network in which a fully connected network of six binary thresholding neural units. Every unit is connected with data. **Right**: is Boltzmann machine whose model splits into two parts: visible units and hidden units (shaded nodes). The dashed line is used to highlight the model separation	478
Fig. 7.8	A comparison between **Left** Boltzmann machine and **Right** restricted Boltzmann machine. In the restricted Boltzmann machine there are no connections between hidden units (shaded nodes) and no connections between visible units (unshaded nodes)	481
Fig. 7.9	A toy example using four different pooling techniques. **Left:** resulting activations within a given pooling region. **Right:** pooling results given by four different pooling techniques. If $\lambda = 0.4$ is taken then the mixed pooling result is 1.46. The mixed pooling and the stochastic pooling can represent multi-modal distributions of activations within a region	508
Fig. 7.10	An example of a thinned network produced by applying dropout to a full-connection neural network. Crossed units have been dropped. The dropout probabilities of the first, second, and third layers are 0.4, 0.6, and 0.4, respectively,	516
Fig. 7.11	An example of the structure of DropConnect. The dotted lines show dropped connections	521
Fig. 7.12	MLP with one hidden layer for auto association. The shape of the whole network is similar to an hourglass	525
Fig. 7.13	Illustration of the architecture of the basic autoencoder with "encoder" and "decoder" networks for high-level feature learning. The leftmost layer of the whole network is called the input layer, and the rightmost layer the output layer. The middle layer of nodes is called the hidden layer, because its values are not observed in the training set. The circles labeled "+1" are called bias units, and correspond to the intercept term **b**. I_i denotes the index i, $i = 1,\ldots,n$	526

List of Figures

Fig. 7.14 Autoencoder training flowchart. Encoder f encodes the input $\mathbf{x}$ to $\mathbf{h} = f(\mathbf{x})$, and the decoder g decodes $\mathbf{h} = f(\mathbf{x})$ to the output $\mathbf{y} = \hat{\mathbf{x}} = g(f(\mathbf{x}))$ so that the reconstruction error $L(\mathbf{x}, \mathbf{y})$ is minimized 527

Fig. 7.15 Stacked autoencoder scheme with 2 hidden layers, where $\mathbf{h}^{(d)} = f(\mathbf{W}_1^{(d)} \mathbf{x}^{(d)} + \mathbf{b}_1^{(d)})$ and $\mathbf{y}^{(d)} = f(\mathbf{W}_2^{(d)} \mathbf{h}^{(d)} + \mathbf{b}_2^{(d)})$, $d = 1, 2$ with $\mathbf{x}^{(1)} = \mathbf{x}$, $\mathbf{x}^{(2)} = \mathbf{y}^{(1)}$ and $\hat{\mathbf{x}} = \mathbf{y}^{(2)}$ 533

Fig. 7.16 Comparison between the two network architectures for a batch data: (**a**) the network with no BatchNorm layer; (**b**) the same network as in (**a**) with a BatchNorm layer inserted after the fully connected layer $\mathbf{W}$. All the layer parameters have exactly the same value in both networks, and the two networks have the same loss function $\mathcal{L}$, i.e., $\hat{\mathcal{L}} = \mathcal{L}$ 590

Fig. 7.17 The basic structure of generative adversarial networks (GANs). The generator G uses a random noise $\mathbf{z}$ to generate a synthetic data $\hat{\mathbf{x}} = G(\mathbf{z})$, and the discriminator D tries to identify whether the synthesized data $\hat{\mathbf{x}}$ is a real data $\mathbf{x}$, i.e., making a real or fake inference 599

Fig. 7.18 Bidirectional generative adversarial network (BiGAN) is a combination of a standard GAN and an Encoder. The generator G from the standard GAN framework maps latent samples $\mathbf{z}$ to generated data $G(\mathbf{z})$. An encoder E maps data $\mathbf{x}$ to the output $E(\mathbf{x})$. Both the generator tuple $(G(\mathbf{z}), \mathbf{z})$ and the encoder tuple $(\mathbf{x}, E(\mathbf{x}))$ act, in a bidirectional way, as inputs to the discriminator D 603

Fig. 7.19 A variational autoencoder (VAEs) connected with a standard generative adversarial network (GAN). Unlike the BiGAN in which the encoder's output $E(\mathbf{x})$ is one of two bidirectional inputs, the autoencoder E here executes variation $\mathbf{z} \sim E(\mathbf{x})$ that is used as the random noise in GAN 604

Fig. 8.1 Three points in $\mathbb{R}^2$, shattered by oriented lines. The arrow points to the side with the points labeled black 620

Fig. 8.2 The decision directed acyclic graphs (DDAG) for finding the best class out of four classes 656

Fig. 9.1 Evolutionary computation tree 683

Fig. 9.2 Pareto efficiency and Pareto improvement. Point A is an inefficient allocation between preference criterions f_1 and f_2 because it does not satisfy the constraint curve of f_1 and f_2. Two decisions to move from Point A to Points C and D would be a Pareto improvement, respectively. They improve both f_1 and f_2, without making anyone else worse off. Hence, these two moves would be a Pareto improvement and be Pareto optimal, respectively. A move from Point A to Point B would not be a Pareto improvement because it decreases the cost f_1 by increasing another cost f_2, thus making one side better off by making another worse off. The move from any point that lies under the curve to any point on the curve cannot be a Pareto improvement due to making one of two criterions f_1 and f_2 worse 695

Fig. 9.3 Illustration of degeneracy due to crossover (from [16]). Here, asterisk denotes three cluster centers (1.1, 1.0), (2.2, 2.0), (3.4, 1.2) for chromosome $\mathbf{m}_1$, + denotes three cluster centers (3.2, 1.4), (1.8, 2.2), (0.5, 0.7) for the chromosome $\mathbf{m}_2$, open triangle denotes the chromosome obtained by crossing $\mathbf{m}_1$ and $\mathbf{m}_2$, and open circle denotes the chromosome obtained by crossing $\mathbf{m}_1$ and $\mathbf{m}'_2$ with three cluster centers (0.5, 0.7), (1.8, 2.2), (3.2, 1.4) 732

Fig. 9.4 Geometric representation of opposite number of a real number x 788

Fig. 9.5 The opposite point and the quasi-opposite point. Given a solution $\mathbf{x}_i = [x_{i1}, \ldots, x_{iD}]^T$ with $x_{ij} \in [a_j, b_j]$ and $M_{ij} = (a_j + b_j)/2$, then x^o_{ij} and x^q_{ij} are the jth elements of the opposite point $\mathbf{x}^o_i$ and the quasi-opposite point $\mathbf{x}^q_i$ of $\mathbf{x}_i$, respectively. (a) When $x_{ij} > M_{ij}$. (b) When $x_{ij} < M_{ij}$ 792

Fig. 9.6 The opposite, quasi-opposite, and quasi-reflected points of a solution (point) $\mathbf{x}_i$. Given a solution $\mathbf{x}_i = [x_{i1}, \ldots, x_{iD}]^T$ with $x_{ij} \in [a_j, b_j]$ and $M_{ij} = (a_j + b_j)/2$, then x^o_{ij}, x^q_{ij}, and x^{qr}_{ij} are the jth elements of the opposite point $\mathbf{x}^o_i$, the quasi-opposite point $\mathbf{x}^q_i$, and the quasi-reflected opposite point $\mathbf{x}^{qr}_i$ of $\mathbf{x}_i$, respectively. (a) When $x_{ij} > M_{ij}$. (b) When $x_{ij} < M_{ij}$ 794

List of Tables

Table 1.1	Quadratic forms and positive definiteness of a Hermitian matrix **A**	28
Table 2.1	Symbols of real functions	56
Table 2.2	Differential matrices and Jacobian matrices of trace functions	67
Table 2.3	Differentials and Jacobian matrices of determinant functions	69
Table 2.4	Matrix differentials and Jacobian matrices of real functions	70
Table 2.5	Differentials and Jacobian matrices of matrix functions	71
Table 2.6	Forms of complex-valued functions	73
Table 2.7	Nonholomorphic and holomorphic functions	74
Table 2.8	Complex gradient matrices of trace functions	82
Table 2.9	Complex gradient matrices of determinant functions	83
Table 2.10	Complex matrix differential and complex Jacobian matrix	86
Table 3.1	Extreme-point conditions for the complex variable functions	98
Table 3.2	Proximal operators of several typical functions	127
Table 6.1	Similarity values and clustering results	322
Table 6.2	Distance matrix in Example 6.4	323
Table 6.3	Distance matrix after first clustering in Example 6.4	323
Table 6.4	Distance matrix after second clustering in Example 6.4	323
Table 9.1	Relation comparison between objective vectors and approximation sets	711

List of Algorithms

Algorithm 3.1	Gradient descent algorithm and its variants	107
Algorithm 3.2	Nesterov (first) optimal gradient algorithm	116
Algorithm 3.3	Nesterov algorithm with adaptive convexity parameter	117
Algorithm 3.4	Nesterov (third) optimal gradient algorithm	118
Algorithm 3.5	FISTA algorithm with fixed step	132
Algorithm 3.6	Davidon–Fletcher–Powell (DFP) quasi-Newton method	149
Algorithm 3.7	Broyden–Fletcher–Goldfarb–Shanno (BFGS) quasi-Newton method	149
Algorithm 3.8	Newton algorithm via backtracking line search	150
Algorithm 3.9	Feasible-start Newton algorithm	151
Algorithm 3.10	Infeasible-start Newton algorithm	153
Algorithm 4.1	Conjugate gradient algorithm	164
Algorithm 4.2	Biconjugate gradient algorithm	165
Algorithm 4.3	PCG algorithm with preprocessor	166
Algorithm 4.4	PCG algorithm without preprocessor	167
Algorithm 4.5	TLS algorithm for minimum norm solution	187
Algorithm 4.6	Least angle regressions (LARS) algorithm with Lasso modification	199
Algorithm 5.1	Lanczos algorithm for GEVD	212
Algorithm 5.2	Tangent algorithm for computing the GEVD	213
Algorithm 5.3	GEVD algorithm for singular matrix **B**	213
Algorithm 6.1	Stochastic gradient (SG) method	229
Algorithm 6.2	SVRG method for minimizing an empirical risk R_n	232
Algorithm 6.3	SAGA method for minimizing an empirical risk R_n	232
Algorithm 6.4	APPROX coordinate descent method	236
Algorithm 6.5	LogitBoost (two classes)	250
Algorithm 6.6	Adaptive boosting (AdaBoost) algorithm	252
Algorithm 6.7	Gentle AdaBoost	252

Algorithm 6.8	Feature clustering algorithm	266
Algorithm 6.9	Spectral projected gradient (SPG) algorithm for nonlinear joint unsupervised feature selection	268
Algorithm 6.10	Robust PCA via accelerated proximal gradient	278
Algorithm 6.11	Sparse principal component analysis (SPCA) algorithm	281
Algorithm 6.12	Principal component regression algorithm	284
Algorithm 6.13	Nonlinear iterative partial least squares (NIPALS) algorithm	286
Algorithm 6.14	Simple nonlinear iterative partial least squares regression algorithm	289
Algorithm 6.15	GPSR-BB algorithm	296
Algorithm 6.16	k-nearest neighbor (kNN)	299
Algorithm 6.17	Alternating projection algorithm for the supervised tensor learning	309
Algorithm 6.18	Tensor learning algorithm for regression	313
Algorithm 6.19	Random K-means algorithm	328
Algorithm 6.20	Normalized spectral clustering algorithm of Ng et al.	331
Algorithm 6.21	Normalized spectral clustering algorithm of Shi and Malik	332
Algorithm 6.22	Constrained spectral clustering for two-way partition	335
Algorithm 6.23	Constrained spectral clustering for K-way partition	336
Algorithm 6.24	Nyström method for matrix approximation	336
Algorithm 6.25	Fast spectral clustering algorithm	337
Algorithm 6.26	Self-training algorithm	340
Algorithm 6.27	Propagating 1-nearest neighbor clustering algorithm	341
Algorithm 6.28	Co-training algorithm	343
Algorithm 6.29	Canonical correlation analysis (CCA) algorithm	348
Algorithm 6.30	Penalized (sparse) canonical component analysis (CCA) algorithm	353
Algorithm 6.31	Manifold regularization algorithms	368
Algorithm 6.32	Harmonic function algorithm for semi-supervised graph learning	371
Algorithm 6.33	ℓ_1 directed graph construction algorithm	378
Algorithm 6.34	Active learning	383
Algorithm 6.35	Co-active learning	383
Algorithm 6.36	Active learning for finding opposite pair close to separating hyperplane	384
Algorithm 6.37	Active learning-extreme learning machine (AL-ELM) algorithm	387
Algorithm 6.38	Q-learning	396
Algorithm 6.39	Double Q-learning	397
Algorithm 6.40	Weighted double Q-learning	398
Algorithm 6.41	Modified connectionist Q learning algorithm	400
Algorithm 6.42	Deep Q-learning with experience replay	401
Algorithm 6.43	TrAdaBoost algorithm	412

List of Algorithms

Algorithm 6.44	SVD-based alternating structure optimization algorithm....	413
Algorithm 6.45	Structural correspondence learning algorithm	414
Algorithm 6.46	EigenCluster: a unified framework for transfer learning	417
Algorithm 6.47	ℓ_p-norm multiple kernel learning algorithm	422
Algorithm 6.48	Heterogeneous feature augmentation.........................	422
Algorithm 6.49	Information-theoretic metric learning	424
Algorithm 7.1	ADAGRAD algorithm for minimizing $\{\langle \mathbf{u}, \mathbf{x}\rangle + \frac{1}{2}\langle \mathbf{x}, \mathbf{H}\mathbf{x}\rangle + \lambda\|\mathbf{x}\|_2\}$	450
Algorithm 7.2	ADAM algorithm for stochastic optimization................	451
Algorithm 7.3	CD_1 fast RBM learning algorithm............................	487
Algorithm 7.4	Backpropagation algorithm for the basic autoencoder........	531
Algorithm 7.5	Extreme learning machine (ELM) algorithm	546
Algorithm 7.6	ELM algorithm for regression and binary classification.....	549
Algorithm 7.7	ELM multiclass classification algorithm	551
Algorithm 7.8	DeepWalk(G, w, d, γ, t)..	580
Algorithm 7.9	SkipGram($\Phi, \mathcal{W}_{v_i}, w$)	581
Algorithm 7.10	GraphSAGE embedding generation algorithm...............	583
Algorithm 7.11	Batch renormalization, applied to activation $\mathbf{y}$ over a mini-batch..	595
Algorithm 7.12	Minibatch stochastic gradient descent training of GANs....	601
Algorithm 8.1	PSVM binary classification algorithm.......................	650
Algorithm 8.2	SVM-recursive feature elimination (SVM-RFE) algorithm..	653
Algorithm 8.3	LS-SVM multiclass classification algorithm	658
Algorithm 8.4	PSVM multiclass classification algorithm	660
Algorithm 8.5	Gaussian process regression algorithm	666
Algorithm 9.1	Roulette wheel fitness selection algorithm	700
Algorithm 9.2	Fast-nondominated-sort(P)	701
Algorithm 9.3	Crowding-distance-assignment (I)	703
Algorithm 9.4	Single-objective simulated annealing.........................	717
Algorithm 9.5	Multiobjective simulated annealing	721
Algorithm 9.6	Archived multiobjective simulated annealing (AMOSA) algorithm	722
Algorithm 9.7	Genetic algorithm with gene rearrangement (GAGR) clustering algorithm	730
Algorithm 9.8	Nondominated sorting genetic algorithm (NSGA)	738
Algorithm 9.9	$(1+1)$ evolutionary algorithm	741
Algorithm 9.10	Multiobjective evolutionary algorithm based on decomposition (MOEA/D).....................................	748
Algorithm 9.11	MOEA/D-DE algorithm for Solving (9.9.5)	750
Algorithm 9.12	SPEA algorithm	752
Algorithm 9.13	Improved strength Pareto evolutionary algorithm (SPEA2)..	755

Algorithm 9.14	Differential evolution (DE)	764
Algorithm 9.15	Framework of artificial bee colony (ABC) algorithm	779
Algorithm 9.16	Particle swarm optimization (PSO) algorithm	783
Algorithm 9.17	Genetic learning PSO (GL-PSO) algorithm	785
Algorithm 9.18	Hybrid particle swarm optimization with local search (HPSO-LS)	787
Algorithm 9.19	Opposition-based differential evolution (ODE)	791

Part I
Introduction to Matrix Algebra

Chapter 1
Basic Matrix Computation

In science and engineering, we often encounter the problem of solving a system of linear equations. Matrices provide the most basic and useful mathematical tool for describing and solving such systems. As the introduction to matrix algebra, this chapter presents the basic operations and performance of matrices, followed by a description of vectorization of matrix and matricization of vector.

1.1 Basic Concepts of Vectors and Matrices

First we introduce the basic concepts and notation for vectors and matrices.

1.1.1 Vectors and Matrices

A real *system of linear equations*

$$\left.\begin{array}{c} a_{11}x_1 + \cdots + a_{1n}x_n = b_1 \\ \vdots \\ a_{m1}x_1 + \cdots + a_{mn}x_n = b_m \end{array}\right\} \qquad (1.1.1)$$

can be simply rewritten using real *vector* and *matrix* symbols as a real *matrix equation*

$$\mathbf{Ax} = \mathbf{b}, \qquad (1.1.2)$$

where

$$\mathbf{A} = \begin{bmatrix} a_{11} & \cdots & a_{1n} \\ \vdots & \ddots & \vdots \\ a_{m1} & \cdots & a_{mn} \end{bmatrix} \in \mathbb{R}^{m \times n}, \quad \mathbf{x} = \begin{bmatrix} x_1 \\ \vdots \\ x_n \end{bmatrix} \in \mathbb{R}^n, \quad \mathbf{b} = \begin{bmatrix} b_1 \\ \vdots \\ b_m \end{bmatrix} \in \mathbb{R}^m.$$
(1.1.3)

If a system of linear equations is complex valued, then $\mathbf{A} \in \mathbb{C}^{m \times n}$, $\mathbf{x} \in \mathbb{C}^n$, and $\mathbf{b} \in \mathbb{C}^m$. Here, $\mathbb{R}$ (or $\mathbb{C}$) denotes the set of real (or complex) numbers, and $\mathbb{R}^m$ (or $\mathbb{C}^m$) represents the set of all real (or complex) m-dimensional column vectors, while $\mathbb{R}^{m \times n}$ (or $\mathbb{C}^{m \times n}$) is the set of all $m \times n$ real (or complex) matrices.

An m-dimensional *row vector* $\mathbf{x} = [x_1, \ldots, x_m]$ is represented as $\mathbf{x} \in \mathbb{R}^{1 \times m}$ or $\mathbf{x} \in \mathbb{C}^{1 \times m}$. To save writing space, an m-dimensional column vector is usually written as the transposed form of a row vector, denoted $\mathbf{x} = [x_1, \ldots, x_m]^T$, where T denotes "transpose."

There are three different types of vector in science and engineering [18]:

- *Physical vector:* Its elements are physical quantities with magnitude and direction, such as a displacement vector, a velocity vector, an acceleration vector, and so forth.
- *Geometric vector:* A directed line segment or arrow can visualize a physical vector. Such a representation is known as a geometric vector. For example, $\mathbf{v} = \overrightarrow{AB}$ represents the directed line segment with initial point A and terminal point B.
- *Algebraic vector:* A geometric vector needs to be represented in algebraic form in order to operate. For a geometric vector $\mathbf{v} = \overrightarrow{AB}$ on a plane, if its initial point is $A = (a_1, a_2)$ and its terminal point is $B = (b_1, b_2)$, then the geometric vector $\mathbf{v} = \overrightarrow{AB}$ can be represented in an algebraic form $\mathbf{v} = \begin{bmatrix} b_1 - a_1 \\ b_2 - a_2 \end{bmatrix}$. Such a geometric vector described in algebraic form is known as an algebraic vector.

The vectors encountered often in practical applications are physical vectors, while geometric vectors and algebraic vectors are, respectively, the visual representation and the algebraic form of physical vectors. Algebraic vectors provide a computational tool for physical vectors.

Depending on different types of element values, algebraic vectors can be divided into the following three types:

- *Constant vector:* All entries take real constant numbers or complex constant numbers, e.g., $\mathbf{a} = [1, 5, 2]^T$.
- *Function vector:* Its entries take function values, e.g., $\mathbf{x} = [x^1, \ldots, x^n]^T$.
- *Random vector:* It uses random variables or signals as entries, e.g., $\mathbf{x}(n) = [x_1(n), \ldots, x_m(n)]^T$ where $x_1(n), \ldots, x_m(n)$ are m random variables or signals.

1.1 Basic Concepts of Vectors and Matrices

Fig. 1.1 Classification of vectors

Vectors
- Physical vectors
- Geometric vectors
- Algebraic vectors
  - Constant vectors
  - Function vectors
  - Random vectors

Figure 1.1 summarizes the classification of vectors.

A vector all of whose components are equal to zero is called a *zero vector* and is denoted as $\mathbf{0} = [0, \ldots, 0]^T$.

An $n \times 1$ vector $\mathbf{x} = [x_1, \ldots, x_n]^T$ with only one nonzero entry $x_i = 1$ constitutes a *basis vector*, denoted $\mathbf{e}_i$; e.g.,

$$\mathbf{e}_1 = \begin{bmatrix} 1 \\ 0 \\ 0 \\ \vdots \\ 0 \end{bmatrix}, \quad \mathbf{e}_2 = \begin{bmatrix} 0 \\ 1 \\ 0 \\ \vdots \\ 0 \end{bmatrix}, \quad \ldots, \quad \mathbf{e}_n = \begin{bmatrix} 0 \\ 0 \\ 0 \\ \vdots \\ 1 \end{bmatrix}. \tag{1.1.4}$$

In modeling physical problems, the matrix $\mathbf{A}$ is usually the symbolic representation of a physical system (e.g., a linear system, a filter, or a learning machine).

An $m \times n$ matrix $\mathbf{A}$ is called a *square matrix* if $m = n$, a *broad matrix* for $m < n$, and a *tall matrix* for $m > n$.

The *main diagonal* of an $n \times n$ matrix $\mathbf{A} = [a_{ij}]$ is the segment connecting the top left to the bottom right corner. The entries located on the main diagonal, $a_{11}, a_{22}, \ldots, a_{nn}$, are known as the (main) *diagonal elements*.

An $n \times n$ matrix $\mathbf{A} = [a_{ij}]$ is called a *diagonal matrix, identity matrix*, and *zero matrix* if all entries off the main diagonal are zero, and some of all diagonal entries take nonzero values, all diagonal entries take 1 or 0, respectively; i.e.,

$$\mathbf{D} = \mathbf{Diag}(d_{11}, \ldots, d_{nn}), \tag{1.1.5}$$

$$\mathbf{I} = \mathbf{Diag}(1, \ldots, 1), \tag{1.1.6}$$

$$\mathbf{O} = \mathbf{Diag}(0, \ldots, 0). \tag{1.1.7}$$

Clearly, an $n \times n$ identity matrix $\mathbf{I}$ can be represented as $\mathbf{I} = [\mathbf{e}_1, \ldots, \mathbf{e}_n]$ using basis vectors.

In this book, we use often the following matrix symbols.

$\mathbf{A}(i, :)$ means the ith row of $\mathbf{A}$.

$\mathbf{A}(:, j)$ means the jth column of $\mathbf{A}$.

$\mathbf{A}(p:q, r:s)$ means the $(q-p+1) \times (s-r+1)$ *submatrix* consisting of the pth row to the qth row and the rth column to the sth column of $\mathbf{A}$. For example,

$$\mathbf{A}(2:5, 1:3) = \begin{bmatrix} a_{21} & a_{22} & a_{23} \\ a_{31} & a_{32} & a_{33} \\ a_{41} & a_{42} & a_{43} \\ a_{51} & a_{52} & a_{53} \end{bmatrix}.$$

A matrix $\mathbf{A}$ is an $m \times n$ *block matrix* if it can be represented in the form

$$\mathbf{A} = [\mathbf{A}_{ij}] = \begin{bmatrix} \mathbf{A}_{11} & \mathbf{A}_{12} & \cdots & \mathbf{A}_{1n} \\ \mathbf{A}_{21} & \mathbf{A}_{22} & \cdots & \mathbf{A}_{2n} \\ \vdots & \vdots & \ddots & \vdots \\ \mathbf{A}_{m1} & \mathbf{A}_{m2} & \cdots & \mathbf{A}_{mn} \end{bmatrix}.$$

1.1.2 Basic Vector Calculus

The *vector addition* of $\mathbf{u} = [u_1, \ldots, u_n]^T$ and $\mathbf{v} = [v_1, \ldots, v_n]^T$ is defined as

$$\mathbf{u} + \mathbf{v} = [u_1 + v_1, \ldots, u_n + v_n]^T. \tag{1.1.8}$$

Vector addition obeys the commutation law and the associative law:

- *Commutative law:* $\mathbf{u} + \mathbf{v} = \mathbf{v} + \mathbf{u}$.
- *Associative law:* $(\mathbf{u} + \mathbf{v}) \pm \mathbf{w} = \mathbf{u} + (\mathbf{v} \pm \mathbf{w}) = (\mathbf{u} \pm \mathbf{w}) + \mathbf{v}$.

The *vector multiplication* of an $n \times 1$ vector $\mathbf{u}$ by a scalar α is given by

$$\alpha \mathbf{u} = [\alpha u_1, \ldots, \alpha u_n]^T. \tag{1.1.9}$$

The basic property of vector multiplication by a scalar is that it obeys the distributive law:

$$\alpha(\mathbf{u} + \mathbf{v}) = \alpha \mathbf{u} + \alpha \mathbf{v}. \tag{1.1.10}$$

Given two real or complex $n \times 1$ vectors $\mathbf{u} = [u_1, \ldots, u_n]^T$ and $\mathbf{v} = [v_1, \ldots, v_n]^T$, their *inner product* (also called *dot product* or *scalar product*), denoted $\mathbf{u} \cdot \mathbf{v}$ or $\langle \mathbf{u}, \mathbf{v} \rangle$, is defined as the real number

$$\mathbf{u} \cdot \mathbf{v} = \langle \mathbf{u}, \mathbf{v} \rangle = \mathbf{u}^T \mathbf{v} = u_1 v_1 + \cdots + u_n v_n = \sum_{i=1}^{n} u_i v_i, \tag{1.1.11}$$

1.1 Basic Concepts of Vectors and Matrices

or the complex number

$$\mathbf{u} \cdot \mathbf{v} = \langle \mathbf{u}, \mathbf{v} \rangle = \mathbf{u}^H \mathbf{v} = u_1^* v_1 + \cdots + u_n^* v_n = \sum_{i=1}^{n} u_i^* v_i, \qquad (1.1.12)$$

where $\mathbf{u}^H$ is the *complex conjugate transpose* (or *Hermitian conjugate*) of $\mathbf{u}$.

Note that if $\mathbf{u}$ and $\mathbf{v}$ are two row vectors, then $\mathbf{u} \cdot \mathbf{v} = \mathbf{u}\mathbf{v}^T$ for real vectors and $\mathbf{u} \cdot \mathbf{v} = \mathbf{u}\mathbf{v}^H$ for complex vectors.

The *outer product* (or *cross product*) of an $m \times 1$ real vector and an $n \times 1$ real vector, denoted $\mathbf{u} \circ \mathbf{v}$, is defined as the $m \times n$ real matrix

$$\mathbf{u} \circ \mathbf{v} = \mathbf{u}\mathbf{v}^T = \begin{bmatrix} u_1 v_1 & \cdots & u_1 v_n \\ \vdots & \vdots & \vdots \\ u_m v_1 & \cdots & u_m v_n \end{bmatrix}; \qquad (1.1.13)$$

if $\mathbf{u}$ and $\mathbf{v}$ are complex, then their outer product is the $m \times n$ complex matrix given by

$$\mathbf{u} \circ \mathbf{v} = \mathbf{u}\mathbf{v}^H = \begin{bmatrix} u_1 v_1^* & \cdots & u_1 v_n^* \\ \vdots & \vdots & \vdots \\ u_m v_1^* & \cdots & u_m v_n^* \end{bmatrix}. \qquad (1.1.14)$$

1.1.3 Basic Matrix Calculus

Definition 1.1 (Matrix Transpose) If $\mathbf{A} = [a_{ij}]$ is an $m \times n$ matrix, then the *matrix transpose* $\mathbf{A}^T$ is an $n \times m$ matrix with the (i, j)th entry $[\mathbf{A}^T]_{ij} = a_{ji}$. The *conjugate* of $\mathbf{A}$ is represented as $\mathbf{A}^*$ and is an $m \times n$ matrix with (i, j)th entry $[\mathbf{A}^*]_{ij} = a_{ij}^*$, while the *conjugate* or *Hermitian transpose* of $\mathbf{A}$, denoted $\mathbf{A}^H \in \mathbb{C}^{n \times m}$, is defined as

$$\mathbf{A}^H = (\mathbf{A}^*)^T = \begin{bmatrix} a_{11}^* & \cdots & a_{m1}^* \\ \vdots & \ddots & \vdots \\ a_{1n}^* & \cdots & a_{mn}^* \end{bmatrix}. \qquad (1.1.15)$$

Definition 1.2 (Hermitian Matrix) An $n \times n$ real (or complex) matrix satisfying $\mathbf{A}^T = \mathbf{A}$ (or $\mathbf{A}^H = \mathbf{A}$) is called a *symmetric matrix* (or *Hermitian matrix*).

There are the following relationships between the transpose and conjugate transpose of a matrix:

$$\mathbf{A}^H = (\mathbf{A}^*)^T = (\mathbf{A}^T)^*. \qquad (1.1.16)$$

For an $m \times n$ block matrix $\mathbf{A} = [\mathbf{A}_{ij}]$, its conjugate transpose $\mathbf{A}^H = [\mathbf{A}_{ji}^H]$ is an $n \times m$ block matrix:

$$\mathbf{A}^H = \begin{bmatrix} \mathbf{A}_{11}^H & \cdots & \mathbf{A}_{m1}^H \\ \vdots & \ddots & \vdots \\ \mathbf{A}_{1n}^H & \cdots & \mathbf{A}_{mn}^H \end{bmatrix}.$$

The simplest algebraic operations with matrices are the addition of two matrices and the multiplication of a matrix by a scalar.

Definition 1.3 (Matrix Addition) Given two $m \times n$ matrices $\mathbf{A} = [a_{ij}]$ and $\mathbf{B} = [b_{ij}]$, *matrix addition* $\mathbf{A} + \mathbf{B}$ is defined by $[\mathbf{A} + \mathbf{B}]_{ij} = a_{ij} + b_{ij}$. Similarly, *matrix subtraction* $\mathbf{A} - \mathbf{B}$ is defined as $[\mathbf{A} - \mathbf{B}]_{ij} = a_{ij} - b_{ij}$.

By using this definition, it is easy to verify that the addition and subtraction of two matrices obey the following rules:

- *Commutative law:* $\mathbf{A} + \mathbf{B} = \mathbf{B} + \mathbf{A}$.
- *Associative law:* $(\mathbf{A} + \mathbf{B}) \pm \mathbf{C} = \mathbf{A} + (\mathbf{B} \pm \mathbf{C}) = (\mathbf{A} \pm \mathbf{C}) + \mathbf{B}$.

Definition 1.4 (Matrix Product) The *matrix product* of an $m \times n$ matrix $\mathbf{A} = [a_{ij}]$ and an $r \times s$ matrix $\mathbf{B} = [b_{ij}]$, denoted $\mathbf{AB}$, exists only when $n = r$ and is an $m \times s$ matrix with entries

$$[\mathbf{AB}]_{ij} = \sum_{k=1}^{n} a_{ik}b_{kj}, \quad i = 1, \ldots, m; \; j = 1, \ldots, s.$$

In particular, if $\mathbf{B} = \alpha \mathbf{I}$, then $[\alpha \mathbf{A}]_{ij} = \alpha a_{ij}$. If $\mathbf{B} = \mathbf{x} = [x_1, \ldots, x_n]^T$, then $[\mathbf{Ax}]_i = \sum_{j=1}^{n} a_{ij}x_j$ for $i = 1, \ldots, m$.

The matrix product obeys the following rules of operation:

1. *Associative law of multiplication:* If $\mathbf{A} \in \mathbb{C}^{m \times n}$, $\mathbf{B} \in \mathbb{C}^{n \times p}$, and $\mathbf{C} \in \mathbb{C}^{p \times q}$, then $\mathbf{A}(\mathbf{BC}) = (\mathbf{AB})\mathbf{C}$.
2. *Left distributive law of multiplication:* For two $m \times n$ matrices $\mathbf{A}$ and $\mathbf{B}$, if $\mathbf{C}$ is an $n \times p$ matrix, then $(\mathbf{A} \pm \mathbf{B})\mathbf{C} = \mathbf{AC} \pm \mathbf{BC}$.
3. *Right distributive law of multiplication:* If $\mathbf{A}$ is an $m \times n$ matrix, while $\mathbf{B}$ and $\mathbf{C}$ are two $n \times p$ matrices, then $\mathbf{A}(\mathbf{B} \pm \mathbf{C}) = \mathbf{AB} \pm \mathbf{AC}$.
4. If α is a scalar and $\mathbf{A}$ and $\mathbf{B}$ are two $m \times n$ matrices, then $\alpha(\mathbf{A} + \mathbf{B}) = \alpha\mathbf{A} + \alpha\mathbf{B}$.

Note that the product of two matrices generally does not satisfy the commutative law, namely $\mathbf{AB} \neq \mathbf{BA}$.

Another important operation on a square matrix is its inversion.

Put $\mathbf{x} = [x_1, \ldots, x_n]^T$ and $\mathbf{y} = [y_1, \ldots, y_n]^T$. The matrix-vector product $\mathbf{Ax} = \mathbf{y}$ can be regarded as a *linear transform* of the vector $\mathbf{x}$, where the $n \times n$ matrix $\mathbf{A}$ is called the *linear transform matrix*. Let $\mathbf{A}^{-1}$ denote the *linear inverse transform* of

1.1 Basic Concepts of Vectors and Matrices

the vector $\mathbf{y}$ onto $\mathbf{x}$. If $\mathbf{A}^{-1}$ exists, then one has

$$\mathbf{x} = \mathbf{A}^{-1}\mathbf{y}. \tag{1.1.17}$$

This can be viewed as the result of using $\mathbf{A}^{-1}$ to premultiply the original linear transform $\mathbf{A}\mathbf{x} = \mathbf{y}$, giving $\mathbf{A}^{-1}\mathbf{A}\mathbf{x} = \mathbf{A}^{-1}\mathbf{y} = \mathbf{x}$, which means that the *linear inverse transform matrix* $\mathbf{A}^{-1}$ must satisfy $\mathbf{A}^{-1}\mathbf{A} = \mathbf{I}$. Similarly, we have

$$\mathbf{x} = \mathbf{A}^{-1}\mathbf{y} \Rightarrow \mathbf{A}\mathbf{x} = \mathbf{A}\mathbf{A}^{-1}\mathbf{y} \equiv \mathbf{y}, \quad \forall \mathbf{y} \neq \mathbf{0}.$$

This implies that $\mathbf{A}^{-1}$ must also satisfy $\mathbf{A}\mathbf{A}^{-1} = \mathbf{I}$. Then, the inverse matrix can be defined below.

Definition 1.5 (Inverse Matrix) Let $\mathbf{A}$ be an $n \times n$ matrix. The matrix $\mathbf{A}$ is said to be invertible if there is an $n \times n$ matrix $\mathbf{A}^{-1}$ such that $\mathbf{A}\mathbf{A}^{-1} = \mathbf{A}^{-1}\mathbf{A} = \mathbf{I}$, and $\mathbf{A}^{-1}$ is referred to as the *inverse matrix* of $\mathbf{A}$.

The following are properties of the conjugate, transpose, conjugate transpose, and inverse matrices.

1. The matrix conjugate, transpose, and conjugate transpose satisfy the distributive law:

$$(\mathbf{A} + \mathbf{B})^* = \mathbf{A}^* + \mathbf{B}^*, \quad (\mathbf{A} + \mathbf{B})^T = \mathbf{A}^T + \mathbf{B}^T, \quad (\mathbf{A} + \mathbf{B})^H = \mathbf{A}^H + \mathbf{B}^H.$$

2. The transpose, conjugate transpose, and inverse matrix of product of two matrices satisfy the following relationship:

$$(\mathbf{A}\mathbf{B})^T = \mathbf{B}^T\mathbf{A}^T, \quad (\mathbf{A}\mathbf{B})^H = \mathbf{B}^H\mathbf{A}^H, \quad (\mathbf{A}\mathbf{B})^{-1} = \mathbf{B}^{-1}\mathbf{A}^{-1}$$

in which both $\mathbf{A}$ and $\mathbf{B}$ are assumed to be invertible.

3. Each of the symbols for the conjugate, transpose, and conjugate transpose can be exchanged with the symbol for the inverse:

$$(\mathbf{A}^*)^{-1} = (\mathbf{A}^{-1})^* = \mathbf{A}^{-*}, \quad (\mathbf{A}^T)^{-1} = (\mathbf{A}^{-1})^T = \mathbf{A}^{-T}, \quad (\mathbf{A}^H)^{-1} = (\mathbf{A}^{-1})^H = \mathbf{A}^{-H}.$$

4. For any $m \times n$ matrix $\mathbf{A}$, both the $n \times n$ matrix $\mathbf{B} = \mathbf{A}^H\mathbf{A}$ and the $m \times m$ matrix $\mathbf{C} = \mathbf{A}\mathbf{A}^H$ are Hermitian matrices.

An $n \times n$ matrix $\mathbf{A}$ is *nonsingular* if and only if the matrix equation $\mathbf{A}\mathbf{x} = \mathbf{0}$ has only the zero solution $\mathbf{x} = \mathbf{0}$. If $\mathbf{A}\mathbf{x} = \mathbf{0}$ exists for any nonzero solution $\mathbf{x} \neq \mathbf{0}$, then the matrix $\mathbf{A}$ is *singular*.

For an $n \times n$ matrix $\mathbf{A} = [\mathbf{a}_1, \ldots, \mathbf{a}_n]$, the matrix equation $\mathbf{A}\mathbf{x} = \mathbf{0}$ is equivalent to

$$\mathbf{a}_1 x_1 + \cdots + \mathbf{a}_n x_n = \mathbf{0}. \tag{1.1.18}$$

If the matrix equation $\mathbf{Ax} = \mathbf{0}$ has a zero solution vector only, then the column vectors of $\mathbf{A}$ are linearly independent, and thus the matrix $\mathbf{A}$ is said to be nonsingular.

To summarize the above discussions, the nonsingularity of an $n \times n$ matrix $\mathbf{A}$ can be determined in any of the following three ways:

- its column vectors are linearly independent;
- the matrix equation $\mathbf{Ax} = \mathbf{b}$ exists a unique nonzero solution;
- the matrix equation $\mathbf{Ax} = \mathbf{0}$ has only a zero solution.

1.2 Sets and Linear Mapping

The set of all n-dimensional vectors with real (or complex) components is called a real (or complex) n-dimensional vector space, denoted $\mathbb{R}^n$ (or $\mathbb{C}^n$). In real-world artificial intelligence problems, we are usually given N n-dimensional real data vectors $\{\mathbf{x}_1, \ldots, \mathbf{x}_N\}$ that belong to a subset X other than the whole set $\mathbb{R}^n$. Such a subset is known as a vector subspace in n-dimensional vector space $\mathbb{R}^n$, denoted as $\{\mathbf{x}_1, \ldots, \mathbf{x}_N\} \in X \subset \mathbb{R}^n$. In this section, we present the sets, the vector subspaces, and the linear mapping of one vector subspace onto another.

1.2.1 Sets

As the name implies, a *set* is a collection of elements.

A set is usually denoted by $S = \{\cdot\}$; inside the braces are the elements of the set S. If there are only a few elements in the set S, these elements are written out within the braces, e.g., $S = \{a, b, c\}$.

To describe the composition of a more complex set mathematically, the symbol "|" is used to mean "such that." For example, $S = \{\mathbf{x} | P(\mathbf{x}) = 0\}$ reads "the element $\mathbf{x}$ in set S such that $P(\mathbf{x}) = 0$." A set with only one element α is called a *singleton*, denoted $\{\alpha\}$.

The following are several common notations for set operations:

- $\forall$ denotes "for all $\cdots$";
- $\mathbf{x} \in A$ reads "$\mathbf{x}$ belongs to the set A", i.e., $\mathbf{x}$ is an element or member of A;
- $\mathbf{x} \notin A$ means that $\mathbf{x}$ is not an element of the set A;
- $\ni$ denotes "such that";
- $\exists$ denotes "there exists";
- $\nexists$ denotes "there does not exist";
- $A \Rightarrow B$ reads "condition A results in B" or "A implies B."

For example, the expression

$$\exists \theta \in V \ni \mathbf{x} + \theta = \mathbf{x} = \theta + \mathbf{x}, \ \forall \mathbf{x} \in V$$

1.2 Sets and Linear Mapping

denotes the trivial statement "there is a zero element θ in the set V such that the addition $\mathbf{x} + \theta = \mathbf{x} = \theta + \mathbf{x}$ holds for all elements $\mathbf{x}$ in V."

Let A and B be two sets; then the sets have the following basic relations.

The notation $A \subseteq B$ reads "the set A is contained in the set B" or "A is a *subset* of B," which implies that each element in A is an element in B, namely $x \in A \Rightarrow x \in B$.

If $A \subset B$, then A is called a *proper subset* of B. The notation $B \supset A$ reads "B contains A" or "B is a *superset* of A." The set with no elements is denoted by $\emptyset$ and is called the *null set*.

The notation $A = B$ reads "the set A equals the set B," which means that $A \subseteq B$ and $B \subseteq A$, or $\mathbf{x} \in A \Leftrightarrow \mathbf{x} \in B$ (any element in A is an element in B, and vice versa). The negation of $A = B$ is written as $A \neq B$, implying that A does not belong to B, neither does B belong to A.

The union of A and B is denoted as $A \cup B$. If $X = A \cup B$, then X is called the *union set* of A and B. The union set is defined as follows:

$$X = A \cup B = \{\mathbf{x} \in X | \mathbf{x} \in A \text{ or } \mathbf{x} \in B\}. \quad (1.2.1)$$

In other words, the elements of the union set $A \cup B$ consist of the elements of A and the elements of B.

The intersection of both sets A and B is represented by the notation $A \cap B$ and is defined as follows:

$$X = A \cap B = \{\mathbf{x} \in X | \mathbf{x} \in A \text{ and } \mathbf{x} \in B\}. \quad (1.2.2)$$

The set $X = A \cap B$ is called the *intersection set* of A and B. Each element of the intersection set $A \cap B$ consists of elements common to both A and B. Especially, if $A \cap B = \emptyset$ (null set), then the sets A and B are known as the *disjoint sets*.

The notation $Z = A + B$ means the sum set of sets A and B and is defined as follows:

$$Z = A + B = \{\mathbf{z} = \mathbf{x} + \mathbf{y} \in Z | \mathbf{x} \in A, \mathbf{y} \in B\}, \quad (1.2.3)$$

namely, an element $\mathbf{z}$ of the *sum set* $Z = A + B$ consists of the sum of the element $\mathbf{x}$ in A and the element $\mathbf{y}$ in B.

The set-theoretic difference of the sets A and B, denoted "$A - B$," is also termed the *difference set* and is defined as follows:

$$X = A - B = \{\mathbf{x} \in X | \mathbf{x} \in A, \text{ but } \mathbf{x} \notin B\}. \quad (1.2.4)$$

That is to say, the difference set $A - B$ is the set of elements of A that are not in B. The difference set $A - B$ is also sometimes denoted by the notation $X = A \setminus B$. For example, $\{\mathbb{C}^n \setminus \mathbf{0}\}$ denotes the set of nonzero vectors in complex n-dimensional vector space.

The *relative complement* of A in X is defined as

$$A^c = X - A = X \setminus A = \{\mathbf{x} \in X \,|\, \mathbf{x} \notin A\}. \tag{1.2.5}$$

Example 1.1 For the sets

$$A = \{1, 5, 3\}, \quad B = \{3, 4, 5\},$$

we have

$$A \cup B = \{1, 5, 3, 4\}, \quad A \cap B = \{5, 3\}, \quad A + B = \{4, 9, 8\},$$

$$A - B = A \setminus B = \{1\}, \quad B - A = B \setminus A = \{4\}.$$

If $\mathbf{x} \in X$ and $\mathbf{y} \in Y$, then the set of all ordered pairs $(\mathbf{x}, \mathbf{y})$ is denoted by $X \times Y$ and is termed the *Cartesian product* of the sets X and Y, and is defined as

$$X \times Y = \{(\mathbf{x}, \mathbf{y}) \,|\, \mathbf{x} \in X, \, \mathbf{y} \in Y\}. \tag{1.2.6}$$

Similarly, $X_1 \times \cdots \times X_n$ denotes the Cartesian product of n sets $X_1, \ldots, X_n$, and its elements are the *ordered n-ples* $(\mathbf{x}_1, \ldots, \mathbf{x}_n)$:

$$X_1 \times \cdots \times X_n = \{(\mathbf{x}_1, \ldots, \mathbf{x}_n) \,|\, \mathbf{x}_1 \in X_1, \ldots, \mathbf{x}_n \in X_n\}. \tag{1.2.7}$$

1.2.2 Linear Mapping

Consider the transformation between vectors in two vector spaces. In mathematics, mapping is a synonym for mathematical function or for morphism.

A *mapping* $T : V \to W$ represents a rule for transforming the vectors in V to corresponding vectors in W. The subspace V is said to be the *initial set* or *domain* of the mapping T and W its *final set* or *codomain*.

When $\mathbf{v}$ is some vector in the vector space V, $T(\mathbf{v})$ is referred to as the *image* of the vector $\mathbf{v}$ under the mapping T, or the value of the mapping T at the point $\mathbf{v}$, whereas $\mathbf{v}$ is known as the *original image* of $T(\mathbf{v})$.

If $T(\mathbf{v})$ represents a collection of transformed outputs of all vectors $\mathbf{v}$ in V, i.e.,

$$T(V) = \{T(\mathbf{v}) \,|\, \mathbf{v} \in V\}, \tag{1.2.8}$$

then $T(V)$ is said to be the *range* of the mapping T, denoted by

$$\text{Im}(T) = T(V) = \{T(\mathbf{v}) \,|\, \mathbf{v} \in V\}. \tag{1.2.9}$$

1.2 Sets and Linear Mapping

Definition 1.6 (Linear Mapping) A mapping T is called a *linear mapping* or *linear transformation* in the vector space V if it satisfies the linear relationship

$$T(c_1\mathbf{v}_1 + c_2\mathbf{v}_2) = c_1 T(\mathbf{v}_1) + c_2 T(\mathbf{v}_2) \tag{1.2.10}$$

for all $\mathbf{v}_1, \mathbf{v}_2 \in V$ and all scalars c_1, c_2.

A linear mapping has the following basic properties: if $T : V \to W$ is a linear mapping, then

$$T(\mathbf{0}) = \mathbf{0} \quad \text{and} \quad T(-\mathbf{x}) = -T(\mathbf{x}). \tag{1.2.11}$$

If $f(\mathbf{X}, \mathbf{Y})$ is a scalar function with real matrices $\mathbf{X} \in \mathbb{R}^{n \times n}$ and $\mathbf{Y} \in \mathbb{R}^{n \times n}$ as variables, then in linear mapping notation, the function can be denoted by the Cartesian product form $f : \mathbb{R}^{n \times n} \times \mathbb{R}^{n \times n} \to \mathbb{R}$.

An interesting special application of linear mappings is the electronic amplifier $\mathbf{A} \in \mathbb{C}^{n \times n}$ with high fidelity (Hi-Fi). By Hi-Fi, it means that there is the following linear relationship between any input signal vector $\mathbf{u}$ and the corresponding output signal vector $\mathbf{Au}$ of the amplifier:

$$\mathbf{Au} = \lambda \mathbf{u}, \tag{1.2.12}$$

where $\lambda > 1$ is the amplification factor or gain. The equation above is a typical characteristic equation of a matrix.

Definition 1.7 (One-to-One Mapping) $T : V \to W$ is a *one-to-one mapping* if it is either *injective mapping* or *surjective mapping*, i.e., $T(\mathbf{x}) = T(\mathbf{y})$ implies $\mathbf{x} = \mathbf{y}$ or distinct elements have distinct images.

A one-to-one mapping $T : V \to W$ has an *inverse mapping* $T^{-1} : W \to V$. The inverse mapping T^{-1} restores what the mapping T has done. Hence, if $T(\mathbf{v}) = \mathbf{w}$, then $T^{-1}(\mathbf{w}) = \mathbf{v}$, resulting in $T^{-1}(T(\mathbf{v})) = \mathbf{v}, \forall \mathbf{v} \in V$ and $T(T^{-1}(\mathbf{w})) = \mathbf{w}, \forall \mathbf{w} \in W$.

If $\mathbf{u}_1, \ldots, \mathbf{u}_p$ are the input vectors of a system T in engineering, then $T(\mathbf{u}_1), \ldots, T(\mathbf{u}_p)$ can be viewed as the output vectors of the system. The criterion for identifying whether a system is linear is: if the system input is the linear expression $\mathbf{y} = c_1\mathbf{u}_1 + \cdots + c_p\mathbf{u}_p$, then the system is said to be linear only if its output satisfies the linear expression $T(\mathbf{y}) = T(c_1\mathbf{u}_1 + \cdots + c_p\mathbf{u}_p) = c_1 T(\mathbf{u}_1) + \cdots + c_p T(\mathbf{u}_p)$. Otherwise, the system is nonlinear.

The following are intrinsic relationships between a linear space and a linear mapping.

Theorem 1.1 ([4, p. 29]) *Let V and W be two vector spaces, and let $T : V \to W$ be a linear mapping. Then the following relationships are true:*

- *if M is a linear subspace in V, then $T(M)$ is a linear subspace in W;*
- *if N is a linear subspace in W, then the linear inverse transform $T^{-1}(N)$ is a linear subspace in V.*

For a given linear transformation $\mathbf{y} = \mathbf{Ax}$, if our task is to obtain the output vector $\mathbf{y}$ from the input vector $\mathbf{x}$ given a transformation matrix $\mathbf{A}$, then $\mathbf{Ax} = \mathbf{y}$ is said to be a *forward problem*. Conversely, the problem of finding the input vector $\mathbf{x}$ from the output vector $\mathbf{y}$ is known as an *inverse problem*. Clearly, the essence of the forward problem is a matrix-vector calculation, while the essence of the inverse problem is to solve a matrix equation.

1.3 Norms

Many problems in artificial intelligence need to solve optimization problems in which vector and/or matrix norms are the cost function terms to be optimized.

1.3.1 Vector Norms

Definition 1.8 (Vector Norm) Let V be a real or complex vector space. Given a vector $\mathbf{x} \in V$, the mapping function $p(\mathbf{x}) : V \to \mathbb{R}$ is called the *vector norm* of $\mathbf{x} \in V$, if for all vectors $\mathbf{x}, \mathbf{y} \in V$ and any scalar $c \in \mathbb{K}$ (here $\mathbb{K}$ denotes $\mathbb{R}$ or $\mathbb{C}$), the following three *norm axioms* hold:

- *Nonnegativity:* $p(\mathbf{x}) \geq 0$ and $p(\mathbf{x}) = 0 \Leftrightarrow \mathbf{x} = \mathbf{0}$.
- *Homogeneity:* $p(c\mathbf{x}) = |c| \cdot p(\mathbf{x})$ is true for all complex constant c.
- *Triangle inequality:* $p(\mathbf{x} + \mathbf{y}) \leq p(\mathbf{x}) + p(\mathbf{y})$.

In a real or complex inner product space V, the vector norms have the following properties [4].

1. $\|\mathbf{0}\| = 0$ and $\|\mathbf{x}\| > 0$, $\forall \mathbf{x} \neq \mathbf{0}$.
2. $\|c\mathbf{x}\| = |c| \|\mathbf{x}\|$ holds for all vector $\mathbf{x} \in V$ and any scalar $c \in \mathbb{K}$.
3. *Polarization identity:* For real inner product spaces we have

$$\langle \mathbf{x}, \mathbf{y} \rangle = \frac{1}{4}\big(\|\mathbf{x} + \mathbf{y}\|^2 - \|\mathbf{x} - \mathbf{y}\|^2\big), \ \forall \, \mathbf{x}, \mathbf{y}, \tag{1.3.1}$$

and for complex inner product spaces we have

$$\langle \mathbf{x}, \mathbf{y} \rangle = \frac{1}{4}\big(\|\mathbf{x} + \mathbf{y}\|^2 - \|\mathbf{x} - \mathbf{y}\|^2 - j\|\mathbf{x} + j\mathbf{y}\|^2 + j\|\mathbf{x} - j\mathbf{y}\|^2\big), \ \forall \, \mathbf{x}, \mathbf{y}. \tag{1.3.2}$$

4. *Parallelogram law*

$$\|\mathbf{x} + \mathbf{y}\|^2 + \|\mathbf{x} - \mathbf{y}\|^2 = 2\big(\|\mathbf{x}\|^2 + \|\mathbf{y}\|^2\big), \ \forall \, \mathbf{x}, \mathbf{y}. \tag{1.3.3}$$

1.3 Norms

5. *Triangle inequality*

$$\|\mathbf{x}+\mathbf{y}\| \leq \|\mathbf{x}\| + \|\mathbf{y}\|, \quad \forall\, \mathbf{x}, \mathbf{y} \in V. \tag{1.3.4}$$

6. *Cauchy–Schwartz inequality*

$$|\langle \mathbf{x}, \mathbf{y} \rangle| \leq \|\mathbf{x}\| \cdot \|\mathbf{y}\|. \tag{1.3.5}$$

The equality $|\langle \mathbf{x}, \mathbf{y} \rangle| = \|\mathbf{x}\| \cdot \|\mathbf{y}\|$ holds if and only if $\mathbf{y} = c\mathbf{x}$, where c is some nonzero complex constant.

The following are several common norms of constant vectors.

- ℓ_0-*norm*

$$\|\mathbf{x}\|_0 \stackrel{\text{def}}{=} \sum_{i=1}^{m} x_i^0, \quad \text{where } x_i^0 = \begin{cases} 1, & \text{if } x_i \neq 0; \\ 0, & \text{if } x_i = 0. \end{cases}$$

That is,

$$\|\mathbf{x}\|_0 = \text{number of nonzero entries of } \mathbf{x}. \tag{1.3.6}$$

- ℓ_1-*norm*

$$\|\mathbf{x}\|_1 \stackrel{\text{def}}{=} \sum_{i=1}^{m} |x_i| = |x_1| + \cdots + |x_m|, \tag{1.3.7}$$

i.e., $\|\mathbf{x}\|_1$ is the sum of absolute (or modulus) values of nonzero entries of $\mathbf{x}$.

- ℓ_2-*norm* or the *Euclidean norm*

$$\|\mathbf{x}\|_2 \stackrel{\text{def}}{=} \|\mathbf{x}\|_E = \sqrt{|x_1|^2 + \cdots + |x_m|^2}. \tag{1.3.8}$$

- ℓ_∞-*norm*

$$\|\mathbf{x}\|_\infty \stackrel{\text{def}}{=} \max\{|x_1|, \ldots, |x_m|\}. \tag{1.3.9}$$

- ℓ_p-*norm* or *Hölder norm* [19]

$$\|\mathbf{x}\|_p \stackrel{\text{def}}{=} \left(\sum_{i=1}^{m} |x_i|^p\right)^{1/p}, \quad p \geq 1. \tag{1.3.10}$$

The ℓ_0-norm does not satisfy the homogeneity $\|c\mathbf{x}\|_0 = |c| \cdot \|\mathbf{x}\|_0$, and thus is a *quasi-norm*, while the ℓ_p-norm is also a quasi-norm if $0 < p < 1$ but a norm if $p \geq 1$.

Clearly, when $p = 1$ or $p = 2$, the ℓ_p-norm reduces to the ℓ_1-norm or the ℓ_2-norm, respectively.

In artificial intelligence, ℓ_0-norm is usually used as a measure of sparse vectors, i.e.,

$$\text{sparse vector } \mathbf{x} = \arg\min_{\mathbf{x}} \|\mathbf{x}\|_0. \tag{1.3.11}$$

However, the ℓ_0-norm is difficult to be optimized. Importantly, the ℓ_1-norm can be regarded as a relaxed form of the ℓ_0-norm, and is easy to be optimized:

$$\text{sparse vector } \mathbf{x} = \arg\min_{\mathbf{x}} \|\mathbf{x}\|_1. \tag{1.3.12}$$

Here are several important applications of the Euclidean norm.

- Measuring the size or length of a vector:

$$\text{size}(\mathbf{x}) = \|\mathbf{x}\|_2 = \sqrt{x_1^2 + \cdots + x_m^2}, \tag{1.3.13}$$

which is called the *Euclidean length*.
- Defining the ϵ-*neighborhood* of a vector $\mathbf{x}$:

$$N_\epsilon(\mathbf{x}) = \{\mathbf{y} \mid \|\mathbf{y} - \mathbf{x}\|_2 \leq \epsilon\}, \quad \epsilon > 0. \tag{1.3.14}$$

- Measuring the distance between vectors $\mathbf{x}$ and $\mathbf{y}$:

$$d(\mathbf{x}, \mathbf{y}) = \|\mathbf{x} - \mathbf{y}\|_2 = \sqrt{(x_1 - y_1)^2 + \cdots + (x_m - y_m)^2}. \tag{1.3.15}$$

This is known as the *Euclidean distance*.
- Defining the *angle* θ ($0 \leq \theta \leq 2\pi$) between vectors $\mathbf{x}$ and $\mathbf{y}$:

$$\theta \stackrel{\text{def}}{=} \arccos\left(\frac{\langle \mathbf{x}, \mathbf{y} \rangle}{\sqrt{\langle \mathbf{x}, \mathbf{x} \rangle}\sqrt{\langle \mathbf{y}, \mathbf{y} \rangle}}\right) = \arccos\left(\frac{\mathbf{x}^H \mathbf{y}}{\|\mathbf{x}\|\|\mathbf{y}\|}\right). \tag{1.3.16}$$

A vector with unit Euclidean length is known as a *normalized* (or *standardized*) *vector*. For any nonzero vector $\mathbf{x} \in \mathbb{C}^m$, $\mathbf{x}/\langle \mathbf{x}, \mathbf{x} \rangle^{1/2}$ is the normalized version of the vector and has the same direction as $\mathbf{x}$.

The norm $\|\mathbf{x}\|$ is said to be a *unitary invariant norm* if $\|\mathbf{U}\mathbf{x}\| = \|\mathbf{x}\|$ holds for all vectors $\mathbf{x} \in \mathbb{C}^m$ and all unitary matrices $\mathbf{U} \in \mathbb{C}^{m \times m}$ such that $\mathbf{U}^H \mathbf{U} = \mathbf{I}$.

Proposition 1.1 ([15]) *The Euclidean norm $\|\cdot\|_2$ is unitary invariant.*

1.3 Norms

If the inner product $\langle \mathbf{x}, \mathbf{y} \rangle = \mathbf{x}^H \mathbf{y} = 0$, then the angle between the vectors $\theta = \pi/2$, from which we have the following definition on orthogonality of two vectors.

Definition 1.9 (Orthogonal) Two constant vectors $\mathbf{x}$ and $\mathbf{y}$ are said to be orthogonal, denoted by $\mathbf{x} \perp \mathbf{y}$, if their inner product $\langle \mathbf{x}, \mathbf{y} \rangle = \mathbf{x}^H \mathbf{y} = 0$.

Definition 1.10 (Inner Product of Function Vectors) Let $\mathbf{x}(t), \mathbf{y}(t)$ be two function vectors in the complex vector space $\mathbb{C}^n$, and let the definition field of the function variable t be $[a, b]$ with $a < b$. Then the *inner product of the function vectors* $\mathbf{x}(t)$ and $\mathbf{y}(t)$ is defined as

$$\langle \mathbf{x}(t), \mathbf{y}(t) \rangle \stackrel{\text{def}}{=} \int_a^b \mathbf{x}^H(t)\mathbf{y}(t) dt. \tag{1.3.17}$$

The angle between function vectors is defined as follows:

$$\cos \theta \stackrel{\text{def}}{=} \frac{\langle \mathbf{x}(t), \mathbf{y}(t) \rangle}{\sqrt{\langle \mathbf{x}(t), \mathbf{x}(t) \rangle} \sqrt{\langle \mathbf{y}(t), \mathbf{y}(t) \rangle}} = \frac{\int_a^b \mathbf{x}^H(t) \mathbf{y}(t) dt}{\|\mathbf{x}(t)\| \cdot \|\mathbf{y}(t)\|}, \tag{1.3.18}$$

where $\|\mathbf{x}(t)\|$ is the *norm of the function vector* $\mathbf{x}(t)$ and is defined as follows:

$$\|\mathbf{x}(t)\| \stackrel{\text{def}}{=} \left(\int_a^b \mathbf{x}^H(t) \mathbf{x}(t) dt \right)^{1/2}. \tag{1.3.19}$$

Clearly, if the inner product of two function vectors is equal to zero, i.e.,

$$\int_a^b \mathbf{x}^H(t) \mathbf{y}(t) dt = 0,$$

then the angle $\theta = \pi/2$. Hence, two function vectors are said to be orthogonal in $[a, b]$, denoted $\mathbf{x}(t) \perp \mathbf{y}(t)$, if their inner product is equal to zero.

The following proposition shows that, for any two orthogonal vectors, the square of the norm of their sum is equal to the sum of the squares of the respective vector norms.

Proposition 1.2 *If $\mathbf{x} \perp \mathbf{y}$, then $\|\mathbf{x} + \mathbf{y}\|^2 = \|\mathbf{x}\|^2 + \|\mathbf{y}\|^2$.*

Proof From the axioms of vector norms it is known that

$$\|\mathbf{x} + \mathbf{y}\|^2 = \langle \mathbf{x} + \mathbf{y}, \mathbf{x} + \mathbf{y} \rangle = \langle \mathbf{x}, \mathbf{x} \rangle + \langle \mathbf{x}, \mathbf{y} \rangle + \langle \mathbf{y}, \mathbf{x} \rangle + \langle \mathbf{y}, \mathbf{y} \rangle. \tag{1.3.20}$$

Since $\mathbf{x}$ and $\mathbf{y}$ are orthogonal, we have $\langle \mathbf{x}, \mathbf{y} \rangle = E\{\mathbf{x}^T \mathbf{y}\} = 0$. Moreover, from the axioms of inner products it is known that $\langle \mathbf{y}, \mathbf{x} \rangle = \langle \mathbf{x}, \mathbf{y} \rangle = 0$. Substituting this result into Eq. (1.3.20), we immediately get

$$\|\mathbf{x}+\mathbf{y}\|^2 = \langle \mathbf{x}, \mathbf{x} \rangle + \langle \mathbf{y}, \mathbf{y} \rangle = \|\mathbf{x}\|^2 + \|\mathbf{y}\|^2.$$

This completes the proof of the proposition. ∎

This proposition is also referred to as the *Pythagorean theorem*.

On the *orthogonality* of two vectors, we have the following conclusive remarks [40].

- *Mathematical definitions:* Two vectors **x** and **y** are orthogonal if their inner product is equal to zero, i.e., $\langle \mathbf{x}, \mathbf{y} \rangle = 0$.
- *Geometric interpretation:* If two vectors are orthogonal, then their angle is $\pi/2$, and the projection of one vector onto the other vector is equal to zero.
- *Physical significance:* When two vectors are orthogonal, each vector contains no components of the other, that is, there exist no interactions or interference between these vectors.

1.3.2 Matrix Norms

The inner product and norms of vectors can be easily extended to the inner product and norms of matrices.

For two $m \times n$ complex matrices $\mathbf{A} = [\mathbf{a}_1, \ldots, \mathbf{a}_n]$ and $\mathbf{B} = [\mathbf{b}_1, \ldots, \mathbf{b}_n]$, stack them, respectively, into the following $mn \times 1$ vectors according to their columns:

$$\mathbf{a} = \text{vec}(\mathbf{A}) = \begin{bmatrix} \mathbf{a}_1 \\ \vdots \\ \mathbf{a}_n \end{bmatrix}, \quad \mathbf{b} = \text{vec}(\mathbf{B}) = \begin{bmatrix} \mathbf{b}_1 \\ \vdots \\ \mathbf{b}_n \end{bmatrix},$$

where the elongated vector $\text{vec}(\mathbf{A})$ is the vectorization of the matrix **A**. We will discuss the vectorization of matrices in detail in Sect. 1.9.

The inner product of two $m \times n$ matrices **A** and **B**, denoted $\langle \mathbf{A}, \mathbf{B} \rangle | \mathbb{C}^{m \times n} \times \mathbb{C}^{m \times n} \to \mathbb{C}$, is defined as the inner product of two elongated vectors:

$$\langle \mathbf{A}, \mathbf{B} \rangle = \langle \text{vec}(\mathbf{A}), \text{vec}(\mathbf{B}) \rangle = \sum_{i=1}^{n} \mathbf{a}_i^H \mathbf{b}_i = \sum_{i=1}^{n} \langle \mathbf{a}_i, \mathbf{b}_i \rangle, \tag{1.3.21}$$

or equivalently written as

$$\langle \mathbf{A}, \mathbf{B} \rangle = (\text{vec}\, \mathbf{A})^H \text{vec}(\mathbf{B}) = \text{tr}(\mathbf{A}^H \mathbf{B}), \tag{1.3.22}$$

where $\text{tr}(\mathbf{C})$ represents the trace function of a square matrix **C**, defined as the sum of its diagonal entries.

1.3 Norms

Let $\mathbf{a} = [a_{11}, \ldots, a_{m1}, a_{12}, \ldots, a_{m2}, \ldots, a_{1n}, \ldots, a_{mn}]^T = \text{vec}(\mathbf{A})$ be an $mn \times 1$ elongated vector of the $m \times n$ matrix $\mathbf{A}$. The ℓ_p-norm of the matrix $\mathbf{A}$ uses the ℓ_p-norm definition of the elongated vector $\mathbf{a}$ as follows:

$$\|\mathbf{A}\|_p \stackrel{\text{def}}{=} \|\mathbf{a}\|_p = \|\text{vec}(\mathbf{A})\|_p = \left(\sum_{i=1}^{m} \sum_{j=1}^{n} |a_{ij}|^p \right)^{1/p}. \tag{1.3.23}$$

Since this kind of matrix norm is represented by the matrix entries, it is named the entrywise norm. The following are three typical entrywise matrix norms:

1. ℓ_1-norm ($p = 1$)

$$\|\mathbf{A}\|_1 \stackrel{\text{def}}{=} \sum_{i=1}^{m} \sum_{j=1}^{n} |a_{ij}|. \tag{1.3.24}$$

2. *Frobenius norm* ($p = 2$)

$$\|\mathbf{A}\|_F \stackrel{\text{def}}{=} \left(\sum_{i=1}^{m} \sum_{j=1}^{n} |a_{ij}|^2 \right)^{1/2} \tag{1.3.25}$$

is the most common matrix norm. Clearly, the Frobenius norm is an extension of the Euclidean norm of the vector to the elongated vector $\mathbf{a} = [a_{11}, \ldots, a_{m1}, \ldots, a_{1n}, \ldots, a_{mn}]^T$.

3. *Max norm* or ℓ_∞-norm ($p = \infty$)

$$\|\mathbf{A}\|_\infty = \max_{i=1,\ldots,m;\, j=1,\ldots,n} \{|a_{ij}|\}. \tag{1.3.26}$$

Another commonly used matrix norm is the induced matrix norm, $\|\mathbf{A}\|_2$, defined as

$$\|\mathbf{A}\|_2 = \sqrt{\lambda_{\max}(\mathbf{A}^H \mathbf{A})} = \sigma_{\max}(\mathbf{A}), \tag{1.3.27}$$

where $\lambda_{\max}(\mathbf{A}^H \mathbf{A})$ and $\sigma_{\max}(\mathbf{A})$ are the maximum eigenvalue of $\mathbf{A}^H \mathbf{A}$ and the maximum singular value of $\mathbf{A}$, respectively.

Because $\sum_{i=1}^{m} \sum_{j=1}^{n} |a_{ij}|^2 = \text{tr}(\mathbf{A}^H \mathbf{A})$, the Frobenius norm can be also written in the form of the trace function as follows:

$$\|\mathbf{A}\|_F \stackrel{\text{def}}{=} \langle \mathbf{A}, \mathbf{A} \rangle^{1/2} = \sqrt{\text{tr}(\mathbf{A}^H \mathbf{A})} = \sqrt{\sigma_1^2 + \ldots + \sigma_k^2}, \tag{1.3.28}$$

where $k = \text{rank}(\mathbf{A}) \leq \min\{m, n\}$ is the rank of $\mathbf{A}$. Clearly, we have

$$\|\mathbf{A}\|_2 \leq \|\mathbf{A}\|_F. \tag{1.3.29}$$

Given an $m \times n$ matrix $\mathbf{A}$, its Frobenius norm weighted by a positive definite matrix $\mathbf{\Omega}$, denoted $\|\mathbf{A}\|_\Omega$, is defined by

$$\|\mathbf{A}\|_\Omega = \sqrt{\operatorname{tr}(\mathbf{A}^H \mathbf{\Omega} \mathbf{A})}. \tag{1.3.30}$$

This norm is usually called the *Mahalanobis norm*.

The following are the relationships between the inner products and norms [15].

- *Cauchy–Schwartz inequlity*

$$|\langle \mathbf{A}, \mathbf{B} \rangle|^2 \leq \|\mathbf{A}\|^2 \|\mathbf{B}\|^2. \tag{1.3.31}$$

The equals sign holds if and only if $\mathbf{A} = c\mathbf{B}$, where c is a complex constant.

- *Pythagoras' theorem*

$$\langle \mathbf{A}, \mathbf{B} \rangle = 0 \quad \Rightarrow \quad \|\mathbf{A} + \mathbf{B}\|^2 = \|\mathbf{A}\|^2 + \|\mathbf{B}\|^2. \tag{1.3.32}$$

- *Polarization identity*

$$\operatorname{Re}(\langle \mathbf{A}, \mathbf{B} \rangle) = \frac{1}{4} \left(\|\mathbf{A} + \mathbf{B}\|^2 - \|\mathbf{A} - \mathbf{B}\|^2 \right), \tag{1.3.33}$$

$$\operatorname{Re}(\langle \mathbf{A}, \mathbf{B} \rangle) = \frac{1}{2} \left(\|\mathbf{A} + \mathbf{B}\|^2 - \|\mathbf{A}\|^2 - \|\mathbf{B}\|^2 \right), \tag{1.3.34}$$

where $\operatorname{Re}(\langle \mathbf{A}, \mathbf{B} \rangle)$ represents the real part of the inner product $\langle \mathbf{A}, \mathbf{B} \rangle$.

1.4 Random Vectors

In science and engineering applications, the measured data are usually random variables. A vector with random variables as its entries is called a random vector.

In this section, we discuss the statistics and properties of random vectors by focusing on Gaussian random vectors.

1.4.1 Statistical Interpretation of Random Vectors

In the statistical interpretation of random vectors, the first-order and second-order statistics of random vectors are the most important.

Definition 1.11 (Inner Product of Random Vectors) Let both $\mathbf{x}(\xi)$ and $\mathbf{y}(\xi)$ be $n \times 1$ random vectors of variable ξ. Then the *inner product of random vectors* $\mathbf{x}(\xi)$ and $\mathbf{y}(\xi)$ is defined as follows:

$$\langle \mathbf{x}(\xi), \mathbf{y}(\xi) \rangle \stackrel{\text{def}}{=} E\{\mathbf{x}^H(\xi)\mathbf{y}(\xi)\}, \tag{1.4.1}$$

1.4 Random Vectors

where E is the expectation operator $E\{\mathbf{x}(\xi)\} = [E\{x_1(\xi)\}, \ldots, E\{x_n(\xi)\}]^T$ and the function variable ξ may be time t, circular frequency f, angular frequency ω or space parameter s, and so on.

The square of the *norm of a random vector* $\mathbf{x}(\xi)$ is defined as

$$\|\mathbf{x}(\xi)\|^2 \stackrel{\text{def}}{=} E\{\mathbf{x}^H(\xi)\mathbf{x}(\xi)\}. \tag{1.4.2}$$

Given a random vector $\mathbf{x}(\xi) = [x_1(\xi), \ldots, x_m(\xi)]^T$, its *mean vector* $\boldsymbol{\mu}_x$ is defined as

$$\boldsymbol{\mu}_x = E\{\mathbf{x}(\xi)\} \stackrel{\text{def}}{=} \begin{bmatrix} E\{x_1(\xi)\} \\ \vdots \\ E\{x_m(\xi)\} \end{bmatrix} = \begin{bmatrix} \mu_1 \\ \vdots \\ \mu_m \end{bmatrix}, \tag{1.4.3}$$

where $E\{x_i(\xi)\} = \mu_i$ represents the mean of the ith random variable $x_i(\xi)$.

The *autocorrelation matrix* of the random vector $\mathbf{x}(\xi)$ is defined by

$$\mathbf{R}_x \stackrel{\text{def}}{=} E\{\mathbf{x}(\xi)\mathbf{x}^H(\xi)\} = \begin{bmatrix} r_{11} & \cdots & r_{1m} \\ \vdots & \ddots & \vdots \\ r_{m1} & \cdots & r_{mm} \end{bmatrix}, \tag{1.4.4}$$

where r_{ii} ($i = 1, \ldots, m$) denotes the autocorrelation function of the random variable $x_i(\xi)$:

$$r_{ii} \stackrel{\text{def}}{=} E\{x_i(\xi)x_i^*(\xi)\} = E\{|x_i(\xi)|^2\}, \quad i = 1, \ldots, m, \tag{1.4.5}$$

whereas r_{ij} represents the cross-correlation function of $x_i(\xi)$ and $x_j(\xi)$:

$$r_{ij} \stackrel{\text{def}}{=} E\{x_i(\xi)x_j^*(\xi)\}, \quad i, j = 1, \ldots, m, \ i \neq j. \tag{1.4.6}$$

Clearly, the autocorrelation matrix is a complex conjugate symmetric (i.e., Hermitian).

The *autocovariance matrix* of the random vector $\mathbf{x}(\xi)$, denoted $\mathbf{C}_x$, is defined as

$$\mathbf{C}_x = \text{cov}(\mathbf{x}, \mathbf{x}) \stackrel{\text{def}}{=} E\{(\mathbf{x}(\xi) - \boldsymbol{\mu}_x)(\mathbf{x}(\xi) - \boldsymbol{\mu}_x)^H\} \tag{1.4.7}$$

$$= \begin{bmatrix} c_{11} & \cdots & c_{1m} \\ \vdots & \ddots & \vdots \\ c_{m1} & \cdots & c_{mm} \end{bmatrix}, \tag{1.4.8}$$

where the diagonal entries

$$c_{ii} \stackrel{\text{def}}{=} E\{|x_i(\xi) - \mu_i|^2\} = \sigma_i^2, \quad i = 1, \ldots, m \tag{1.4.9}$$

denote the variance σ_i^2 of the random variable $x_i(\xi)$, whereas the other entries

$$c_{ij} \stackrel{\text{def}}{=} E\{[x_i(\xi) - \mu_i][x_j(\xi) - \mu_j]^*\} = E\{x_i(\xi)x_j^*(\xi)\} - \mu_i\mu_j^* = c_{ji}^* \tag{1.4.10}$$

express the covariance of the random variables $x_i(\xi)$ and $x_j(\xi)$. The autocovariance matrix is also Hermitian.

There is the following relationship between the autocorrelation and autocovariance matrices:

$$\mathbf{C}_x = \mathbf{R}_x - \boldsymbol{\mu}_x \boldsymbol{\mu}_x^H. \tag{1.4.11}$$

By generalizing the autocorrelation and autocovariance matrices, one obtains the *cross-correlation matrix* of the random vectors $\mathbf{x}(\xi)$ and $\mathbf{y}(\xi)$,

$$\mathbf{R}_{xy} \stackrel{\text{def}}{=} E\{\mathbf{x}(\xi)\mathbf{y}^H(\xi)\} = \begin{bmatrix} r_{x_1,y_1} & \cdots & r_{x_1,y_m} \\ \vdots & \ddots & \vdots \\ r_{x_m,y_1} & \cdots & r_{x_m,y_m} \end{bmatrix}, \tag{1.4.12}$$

and the *cross-covariance matrix*

$$\mathbf{C}_{xy} \stackrel{\text{def}}{=} E\{[\mathbf{x}(\xi) - \boldsymbol{\mu}_x][\mathbf{y}(\xi) - \boldsymbol{\mu}_y]^H\} = \begin{bmatrix} c_{x_1,y_1} & \cdots & c_{x_1,y_m} \\ \vdots & \ddots & \vdots \\ c_{x_m,y_1} & \cdots & c_{x_m,y_m} \end{bmatrix}, \tag{1.4.13}$$

where $r_{x_i,y_j} \stackrel{\text{def}}{=} E\{x_i(\xi)y_j^*(\xi)\}$ is the cross-correlation of the random vectors $x_i(\xi)$ and $y_j(\xi)$ and $c_{x_i,y_j} \stackrel{\text{def}}{=} E\{[x_i(\xi) - \mu_{x_i}][y_j(\xi) - \mu_{y_j}]^*\}$ is the cross-covariance of $x_i(\xi)$ and $y_j(\xi)$.

From Eqs. (1.4.12) and (1.4.13) it is easily known:

$$\mathbf{C}_{xy} = \mathbf{R}_{xy} - \boldsymbol{\mu}_x \boldsymbol{\mu}_y^H. \tag{1.4.14}$$

In real-world applications, a data vector $\mathbf{x} = [x(0), x(1), \ldots, x(N-1)]^T$ with nonzero mean μ_x usually needs to undergo a zero-mean normalization:

$$\mathbf{x} \leftarrow \mathbf{x} = [x(0) - \mu_x, x(1) - \mu_x, \ldots, x(N-1) - \mu_x]^T,$$

where $\mu_x = \frac{1}{N} \sum_{n=0}^{N-1} x(n)$.

1.4 Random Vectors

After zero-mean normalization, the correlation matrices and covariance matrices are equal, i.e., $\mathbf{R}_x = \mathbf{C}_x$ and $\mathbf{R}_{xy} = \mathbf{C}_{xy}$.

Some properties of these matrices are as follows.

1. The autocorrelation matrix is Hermitian, i.e., $\mathbf{R}_x^H = \mathbf{R}_x$.
2. The autocorrelation matrix of the linear combination vector $\mathbf{y} = \mathbf{Ax} + \mathbf{b}$ satisfies $\mathbf{R}_y = \mathbf{A}\mathbf{R}_x\mathbf{A}^H$.
3. The cross-correlation matrix is not Hermitian but satisfies $\mathbf{R}_{xy}^H = \mathbf{R}_{yx}$.
4. $\mathbf{R}_{(x_1+x_2)y} = \mathbf{R}_{x_1 y} + \mathbf{R}_{x_2 y}$.
5. If $\mathbf{x}$ and $\mathbf{y}$ have the same dimension, then $\mathbf{R}_{x+y} = \mathbf{R}_x + \mathbf{R}_{xy} + \mathbf{R}_{yx} + \mathbf{R}_y$.
6. $\mathbf{R}_{Ax, By} = \mathbf{A}\mathbf{R}_{xy}\mathbf{B}^H$.

The degree of correlation of two random variables $x(\xi)$ and $y(\xi)$ can be measured by their *correlation coefficient*:

$$\rho_{xy} \stackrel{\text{def}}{=} \frac{E\{(x(\xi) - \bar{x})(y(\xi) - \bar{y})^*\}}{\sqrt{E\{|x(\xi) - \bar{x}|^2\} E\{|y(\xi) - \bar{y}|^2\}}} = \frac{c_{xy}}{\sigma_x \sigma_y}. \qquad (1.4.15)$$

Here $c_{xy} = E\{(x(\xi) - \bar{x})(y(\xi) - \bar{y})^*\}$ is the cross-covariance of the random variables $x(\xi)$ and $y(\xi)$, while σ_x^2 and σ_y^2 are, respectively, the variances of $x(\xi)$ and $y(\xi)$. Applying the Cauchy–Schwartz inequality to Eq. (1.4.15), we have

$$0 \leq |\rho_{xy}| \leq 1. \qquad (1.4.16)$$

The correlation coefficient ρ_{xy} measures the degree of similarity of two random variables $x(\xi)$ and $y(\xi)$.

- The closer ρ_{xy} is to zero, the weaker the similarity of the random variables $x(\xi)$ and $y(\xi)$ is.
- The closer ρ_{xy} is to 1, the more similar $x(\xi)$ and $y(\xi)$ are.
- The case $\rho_{xy} = 0$ means that there are no correlated components between the random variables $x(\xi)$ and $y(\xi)$. Thus, if $\rho_{xy} = 0$, the random variables $x(\xi)$ and $y(\xi)$ are said to be uncorrelated. Since this uncorrelation is defined in a statistical sense, it is usually said to be a *statistical uncorrelation*.
- It is easy to verify that if $x(\xi) = c\, y(\xi)$, where c is a complex number, then $|\rho_{xy}| = 1$. Up to a fixed amplitude scaling factor $|c|$ and a phase $\phi(c)$, the random variables $x(\xi)$ and $y(\xi)$ are the same, so that $x(\xi) = c \cdot y(\xi) = |c|e^{j\phi(c)} y(\xi)$. Hence, if $|\rho_{xy}| = 1$, then random variables $x(\xi)$ and $y(\xi)$ are said to be completely correlated or *coherent*.

Definition 1.12 (Uncorrelated) Two random vectors $\mathbf{x}(\xi) = [x_1(\xi), \ldots, x_m(\xi)]^T$ and $\mathbf{y}(\xi) = [y_1(\xi), \ldots, y_n(\xi)]^T$ are said to be *statistically uncorrelated* if their cross-covariance matrix $\mathbf{C}_{xy} = \mathbf{O}_{m \times n}$ or, equivalently, $\rho_{x_i, y_j} = 0, \forall i, j$.

In contrast with the case for a constant vector or function vector, an $m \times 1$ random vector $\mathbf{x}(\xi)$ and an $n \times 1$ random vector $\mathbf{y}(\xi)$ are said to be orthogonal if any entry of $\mathbf{x}(\xi)$ is orthogonal to each entry of $\mathbf{y}(\xi)$, namely

$$r_{x_i,y_j} = E\{x_i(\xi)y_j(\xi)\} = 0, \ \forall i = 1,\ldots,m, j = 1,\ldots,n; \quad (1.4.17)$$

or $\mathbf{R}_{xy} = E\{\mathbf{x}(\xi)\mathbf{y}^H(\xi)\} = \mathbf{O}_{m\times n}$.

Definition 1.13 (Orthogonality of Random Vectors) An $m \times 1$ random vector $\mathbf{x}(\xi)$ and an $n \times 1$ random vector $\mathbf{y}(\xi)$ are orthogonal each other, denoted by $\mathbf{x}(\xi) \perp \mathbf{y}(\xi)$, if their cross-correlation matrix is equal to the $m \times n$ null matrix $\mathbf{O}$, i.e.,

$$\mathbf{R}_{xy} = E\{\mathbf{x}(\xi)\mathbf{y}^H(\xi)\} = \mathbf{O}. \quad (1.4.18)$$

Note that for the zero-mean normalized $m \times 1$ random vector $\mathbf{x}(\xi)$ and $n \times 1$ normalized random vector $\mathbf{y}(\xi)$, their statistical uncorrelation and orthogonality are equivalent, as their cross-covariance and cross-correlation matrices are equal, i.e., $\mathbf{C}_{xy} = \mathbf{R}_{xy}$.

1.4.2 Gaussian Random Vectors

Definition 1.14 (Gaussian Random Vector) If each entry $x_i(\xi), i = 1,\ldots,m$, is Gaussian random variable, then the random vector $\mathbf{x} = [x_1(\xi),\ldots,x_m(\xi)]^T$ is called a *Gaussian random vector*.

Let $\mathbf{x} \sim N(\bar{\mathbf{x}}, \mathbf{\Gamma}_x)$ denote a real Gaussian or normal random vector with the mean vector $\bar{\mathbf{x}} = [\bar{x}_1,\ldots,\bar{x}_m]^T$ and covariance matrix $\mathbf{\Gamma}_x = E\{(\mathbf{x}-\bar{\mathbf{x}})(\mathbf{x}-\bar{\mathbf{x}})^T\}$. If each entry of the Gaussian random vector is *independent identically distributed* (iid), then its covariance matrix $\mathbf{\Gamma}_x = E\{(\mathbf{x}-\bar{\mathbf{x}})(\mathbf{x}-\bar{\mathbf{x}})^T\} = \mathbf{Diag}(\sigma_1^2,\ldots,\sigma_m^2)$, where $\sigma_i^2 = E\{(x_i - \bar{x}_i)^2\}$ is the variance of the Gaussian random variable x_i.

Under the condition that all entries are statistically independent of each other, the probability density function of a Gaussian random vector $\mathbf{x} \sim N(\bar{\mathbf{x}}, \mathbf{\Gamma}_x)$ is the *joint probability density function* of its m random variables, i.e.,

$$\begin{aligned}
f(\mathbf{x}) &= f(x_1,\ldots,x_m) \\
&= f(x_1)\cdots f(x_m) \\
&= \frac{1}{\sqrt{2\pi\sigma_1^2}}\exp\left(-\frac{(x_1-\bar{x}_1)^2}{2\sigma_1^2}\right)\cdots\frac{1}{\sqrt{2\pi\sigma_m^2}}\exp\left(-\frac{(x_m-\bar{x}_m)^2}{2\sigma_m^2}\right) \\
&= \frac{1}{(2\pi)^{m/2}\sigma_1\cdots\sigma_m}\exp\left(-\frac{(x_1-\bar{x}_1)^2}{2\sigma_1^2}-\cdots-\frac{(x_m-\bar{x}_m)^2}{2\sigma_m^2}\right)
\end{aligned}$$

1.4 Random Vectors

or

$$f(\mathbf{x}) = \frac{1}{(2\pi)^{m/2}|\mathbf{\Gamma}_x|^{1/2}} \exp\left(-\frac{1}{2}(\mathbf{x}-\bar{\mathbf{x}})^T \mathbf{\Gamma}_x^{-1}(\mathbf{x}-\bar{\mathbf{x}})\right). \tag{1.4.19}$$

If the entries are not statistically independent of each other, then the probability density function of the Gaussian random vector $\mathbf{x} \sim N(\bar{\mathbf{x}}, \mathbf{\Gamma}_x)$ is also given by Eq. (1.4.19), but the exponential term becomes [25, 29]

$$(\mathbf{x}-\bar{\mathbf{x}})^T \mathbf{\Gamma}_x^{-1}(\mathbf{x}-\bar{\mathbf{x}}) = \sum_{i=1}^{m}\sum_{j=1}^{m}[\mathbf{\Gamma}_x^{-1}]_{i,j}(x_i - \mu_i)(x_j - \mu_j), \tag{1.4.20}$$

where $[\mathbf{\Gamma}_x^{-1}]_{ij}$ represents the (i, j)th entry of the inverse matrix $\mathbf{\Gamma}_x^{-1}$ and $\mu_i = E\{x_i\}$ is the mean of the random variable x_i.

The *characteristic function* of a real Gaussian random vector is given by

$$\Phi_{\mathbf{x}}(\omega_1, \ldots, \omega_m) = \exp\left(j\boldsymbol{\omega}^T \boldsymbol{\mu}_x - \frac{1}{2}\boldsymbol{\omega}^T \mathbf{\Gamma}_x \boldsymbol{\omega}\right), \tag{1.4.21}$$

where $\boldsymbol{\omega} = [\omega_1, \ldots, \omega_m]^T$.

If $x_i \sim CN(\mu_i, \sigma_i^2)$, then $\mathbf{x} = [x_1, \ldots, x_m]^T$ is called a *complex Gaussian random vector*, denoted $\mathbf{x} \sim CN(\boldsymbol{\mu}_x, \mathbf{\Gamma}_x)$, where $\boldsymbol{\mu}_x = [\mu_1, \ldots, \mu_m]^T$ and $\mathbf{\Gamma}$ are, respectively, the mean vector and the covariance matrix of the random vector $\mathbf{x}$. If $x_i = u_i + jv_i$ and the random vectors $[u_1, v_1]^T, \ldots, [u_m, v_m]^T$ are statistically independent of each other, then the probability density function of a complex Gaussian random vector $\mathbf{x}$ is given by Poularikas [29, p. 35-5]

$$f(\mathbf{x}) = f(x_1)\cdots f(x_m)$$

$$= \left(\pi^m \prod_{i=1}^{m}\sigma_i^2\right)^{-1} \exp\left(-\sum_{i=1}^{m}\frac{1}{\sigma_i^2}|x_i - \mu_i|^2\right)$$

$$= \frac{1}{\pi^m |\mathbf{\Gamma}_x|} \exp\left(-(\mathbf{x}-\boldsymbol{\mu}_x)^H \mathbf{\Gamma}_x^{-1}(\mathbf{x}-\boldsymbol{\mu}_x)\right), \tag{1.4.22}$$

where $\mathbf{\Gamma}_x = \mathbf{Diag}(\sigma_1^2, \ldots, \sigma_m^2)$. The characteristic function of the complex Gaussian random vector $\mathbf{x}$ is determined by

$$\Phi_{\mathbf{x}}(\boldsymbol{\omega}) = \exp\left(j\operatorname{Re}(\boldsymbol{\omega}^H \boldsymbol{\mu}_x) - \frac{1}{4}\boldsymbol{\omega}^H \mathbf{\Gamma}_x \boldsymbol{\omega}\right). \tag{1.4.23}$$

A Gaussian random vector $\mathbf{x}$ has the following important properties.

- The probability density function of $\mathbf{x}$ is completely described by its mean vector and covariance matrix.

- If two Gaussian random vectors **x** and **y** are statistically uncorrelated, then they are also statistically independent.
- Given a Gaussian random vector **x** with mean vector $\boldsymbol{\mu}_x$ and covariance matrix $\boldsymbol{\Gamma}_x$, the random vector **y** obtained by the linear transformation $\mathbf{y}(\xi) = \mathbf{A}\mathbf{x}(\xi)$ is also a Gaussian random vector, and its probability density function is given by

$$f(\mathbf{y}) = \frac{1}{(2\pi)^{m/2}|\boldsymbol{\Gamma}_y|^{1/2}} \exp\left(-\frac{1}{2}(\mathbf{y}-\boldsymbol{\mu}_y)^T \boldsymbol{\Gamma}_y^{-1}(\mathbf{y}-\boldsymbol{\mu}_y)\right) \tag{1.4.24}$$

for real Gaussian random vectors and

$$f(\mathbf{y}) = \frac{1}{\pi^m |\boldsymbol{\Gamma}_y|} \exp\left(-(\mathbf{y}-\boldsymbol{\mu}_y)^H \boldsymbol{\Gamma}_y^{-1}(\mathbf{y}-\boldsymbol{\mu}_y)\right) \tag{1.4.25}$$

for complex Gaussian random vectors.

The statistical expression of a real white Gaussian noise vector is given by

$$E\{\mathbf{x}(t)\} = \mathbf{0} \quad \text{and} \quad E\{\mathbf{x}(t)\mathbf{x}^T(t)\} = \sigma^2 \mathbf{I}. \tag{1.4.26}$$

But, this expression is not available for a complex white Gaussian noise vector.

Consider a complex Gaussian random vector $\mathbf{x}(t) = [x_1(t), \ldots, x_m(t)]^T$. If $x_i(t)$, $i = 1, \ldots, m$, have zero mean and the same variance σ^2, then the real part $x_{R,i}(t)$ and the imaginary part $x_{I,i}(t)$ are two real white Gaussian noises that are statistically independent and have the same variance. This implies that

$$E\{x_{R,i}(t)\} = 0, \quad E\{x_{I,i}(t)\} = 0,$$

$$E\{x_{R,i}^2(t)\} = E\{x_{I,i}^2(t)\} = \frac{1}{2}\sigma^2,$$

$$E\{x_{R,i}(t)x_{I,i}(t)\} = 0,$$

$$E\{x_i(t)x_i^*(t)\} = E\{x_{R,i}^2(t)\} + E\{x_{I,i}^2(t)\} = \sigma^2.$$

From the above conditions we know that

$$E\{x_i^2(t)\} = E\{(x_{R,i}(t) + j x_{I,i}(t))^2\}$$
$$= E\{x_{R,i}^2(t)\} - E\{x_{I,i}^2(t)\} + j 2 E\{x_{R,i}(t)x_{I,i}(t)\}$$
$$= \frac{1}{2}\sigma^2 - \frac{1}{2}\sigma^2 + 0 = 0.$$

Since $x_1(t), \ldots, x_m(t)$ are statistically uncorrelated, we have

$$E\{x_i(t)x_k(t)\} = 0, \quad E\{x_i(t)x_k^*(t)\} = 0, \quad i \neq k.$$

Therefore, the statistical expression of the complex white Gaussian noise vector $\mathbf{x}(t)$ is given by

$$E\{\mathbf{x}(t)\} = \mathbf{0}, \quad E\{\mathbf{x}(t)\mathbf{x}^H(t)\} = \sigma^2 \mathbf{I}, \quad E\{\mathbf{x}(t)\mathbf{x}^T(t)\} = \mathbf{O}. \qquad (1.4.27)$$

1.5 Basic Performance of Matrices

For a multivariate representation with mn components, we need some scalars to describe the basic performance of an $m \times n$ matrix.

1.5.1 Quadratic Forms

The *quadratic form* of an $n \times n$ matrix $\mathbf{A}$ is defined as $\mathbf{x}^H \mathbf{A} \mathbf{x}$, where $\mathbf{x}$ may be any $n \times 1$ nonzero vector. In order to ensure the uniqueness of the definition of quadratic form $(\mathbf{x}^H \mathbf{A} \mathbf{x})^H = \mathbf{x}^H \mathbf{A} \mathbf{x}$, the matrix $\mathbf{A}$ is required to be Hermitian or complex conjugate symmetric, i.e., $\mathbf{A}^H = \mathbf{A}$. This assumption ensures also that any quadratic form function is real-valued. One of the basic advantages of a real-valued function is its suitability for comparison with a zero value.

If $\mathbf{x}^H \mathbf{A} \mathbf{x} > 0$, then the quadratic form of $\mathbf{A}$ is said to be a positive definite function, and the Hermitian matrix $\mathbf{A}$ is known as a *positive definite matrix*. Similarly, one can define the positive semi-definiteness, negative definiteness, and negative semi-definiteness of a Hermitian matrix, as shown in Table 1.1.

A Hermitian matrix $\mathbf{A}$ is said to be *indefinite matrix* if $\mathbf{x}^H \mathbf{A} \mathbf{x} > 0$ for some nonzero vectors $\mathbf{x}$ and $\mathbf{x}^H \mathbf{A} \mathbf{x} < 0$ for other nonzero vectors $\mathbf{x}$.

1.5.2 Determinants

The *determinant* of an $n \times n$ matrix $\mathbf{A}$, denoted $\det(\mathbf{A})$ or $|\mathbf{A}|$, is defined as

$$\det(\mathbf{A}) = |\mathbf{A}| = \begin{vmatrix} a_{11} & a_{12} & \cdots & a_{1n} \\ a_{21} & a_{22} & \cdots & a_{2n} \\ \vdots & \vdots & \ddots & \vdots \\ a_{n1} & a_{n2} & \cdots & a_{nn} \end{vmatrix}. \qquad (1.5.1)$$

Let $\mathbf{A}_{ij}$ be the $(n-1) \times (n-1)$ submatrix obtained by removing the ith row and the jth column from the $n \times n$ matrix $\mathbf{A} = [a_{ij}]$. The determinant $A_{ij} = \det(\mathbf{A}_{ij})$ of the remaining matrix $\mathbf{A}_{ij}$ is known as the *cofactor* of the entry a_{ij}. In particular, when $j = i$, A_{ii} is known as the *principal minor* of $\mathbf{A}$. The cofactor A_{ij} is related

Table 1.1 Quadratic forms and positive definiteness of a Hermitian matrix $\mathbf{A}$

Quadratic forms	Symbols	Positive definiteness
$\mathbf{x}^H \mathbf{A} \mathbf{x} > 0$	$\mathbf{A} \succ 0$	$\mathbf{A}$ is a positive definite matrix
$\mathbf{x}^H \mathbf{A} \mathbf{x} \geq 0$	$\mathbf{A} \succeq 0$	$\mathbf{A}$ is a positive semi-definite matrix
$\mathbf{x}^H \mathbf{A} \mathbf{x} < 0$	$\mathbf{A} \prec 0$	$\mathbf{A}$ is a negative definite matrix
$\mathbf{x}^H \mathbf{A} \mathbf{x} \leq 0$	$\mathbf{A} \preceq 0$	$\mathbf{A}$ is a negative semi-definite matrix

to the determinant of the submatrix $\mathbf{A}_{ij}$ as follows:

$$A_{ij} = (-1)^{i+j} \det(\mathbf{A}_{ij}). \tag{1.5.2}$$

The determinant of an $n \times n$ matrix $\mathbf{A}$ can be computed by

$$\det(\mathbf{A}) = a_{i1}A_{i1} + \cdots + a_{in}A_{in} = \sum_{j=1}^{n} a_{ij}(-1)^{i+j} \det(\mathbf{A}_{ij}) \tag{1.5.3}$$

or

$$\det(\mathbf{A}) = a_{1j}A_{1j} + \cdots + a_{nj}A_{nj} = \sum_{i=1}^{n} a_{ij}(-1)^{i+j} \det(\mathbf{A}_{ij}). \tag{1.5.4}$$

Hence the determinant of $\mathbf{A}$ can be recursively calculated: an nth-order determinant can be computed from the $(n-1)$th-order determinants, while each $(n-1)$th-order determinant can be calculated from the $(n-2)$th-order determinants, and so forth.

For a 3×3 matrix $\mathbf{A}$, its determinant is recursively given by

$$\det(\mathbf{A}) = \det \begin{bmatrix} a_{11} & a_{12} & a_{13} \\ a_{21} & a_{22} & a_{23} \\ a_{31} & a_{32} & a_{33} \end{bmatrix} = a_{11}A_{11} + a_{12}A_{12} + a_{13}A_{13}$$

$$= a_{11}(-1)^{1+1} \begin{vmatrix} a_{22} & a_{23} \\ a_{32} & a_{33} \end{vmatrix} + a_{12}(-1)^{1+2} \begin{vmatrix} a_{21} & a_{23} \\ a_{31} & a_{33} \end{vmatrix} + a_{13}(-1)^{1+3} \begin{vmatrix} a_{21} & a_{22} \\ a_{31} & a_{33} \end{vmatrix}$$

$$= a_{11}(a_{22}a_{33} - a_{23}a_{32}) - a_{12}(a_{21}a_{33} - a_{23}a_{31}) + a_{13}(a_{21}a_{33} - a_{22}a_{31}).$$

This is the *diagonal method* for a third-order determinant.

A matrix with nonzero determinant is known as a *nonsingular matrix*.

Determinant Equalities [20]
1. If two rows (or columns) of a matrix $\mathbf{A}$ are exchanged, then the value of $\det(\mathbf{A})$ remains unchanged, but the sign is changed.
2. If some row (or column) of a matrix $\mathbf{A}$ is a linear combination of other rows (or columns), then $\det(\mathbf{A}) = 0$. In particular, if some row (or column) is

proportional or equal to another row (or column), or there is a zero row (or column), then $\det(\mathbf{A}) = 0$.
3. The determinant of an identity matrix is equal to 1, i.e., $\det(\mathbf{I}) = 1$.
4. Any square matrix $\mathbf{A}$ and its transposed matrix $\mathbf{A}^T$ have the same determinant, i.e., $\det(\mathbf{A}) = \det(\mathbf{A}^T)$; however, $\det(\mathbf{A}^H) = (\det(\mathbf{A}^T))^*$.
5. The determinant of a Hermitian matrix is real-valued, since

$$\det(\mathbf{A}) = \det(\mathbf{A}^H) = (\det(\mathbf{A}))^*. \tag{1.5.5}$$

6. The determinant of the product of two square matrices is equal to the product of their determinants, i.e.,

$$\det(\mathbf{AB}) = \det(\mathbf{A})\det(\mathbf{B}), \quad \mathbf{A}, \mathbf{B} \in \mathbb{C}^{n \times n}. \tag{1.5.6}$$

7. For any constant c and any $n \times n$ matrix $\mathbf{A}$, $\det(c \cdot \mathbf{A}) = c^n \cdot \det(\mathbf{A})$.
8. If $\mathbf{A}$ is nonsingular, then $\det(\mathbf{A}^{-1}) = 1/\det(\mathbf{A})$.
9. For matrices $\mathbf{A}_{m \times m}, \mathbf{B}_{m \times n}, \mathbf{C}_{n \times m}, \mathbf{D}_{n \times n}$, the determinant of the block matrix for nonsingular $\mathbf{A}$ is given by

$$\det\begin{bmatrix} \mathbf{A} & \mathbf{B} \\ \mathbf{C} & \mathbf{D} \end{bmatrix} = \det(\mathbf{A})\det(\mathbf{D} - \mathbf{C}\mathbf{A}^{-1}\mathbf{B}) \tag{1.5.7}$$

and for nonsingular $\mathbf{D}$ we have

$$\det\begin{bmatrix} \mathbf{A} & \mathbf{B} \\ \mathbf{C} & \mathbf{D} \end{bmatrix} = \det(\mathbf{D})\det(\mathbf{A} - \mathbf{B}\mathbf{D}^{-1}\mathbf{C}). \tag{1.5.8}$$

10. The determinant of a triangular (upper or lower triangular) matrix $\mathbf{A}$ is equal to the product of its main diagonal entries:

$$\det(\mathbf{A}) = \prod_{i=1}^{n} a_{ii}.$$

The determinant of a diagonal matrix $\mathbf{A} = \mathbf{Diag}(a_{11}, \ldots, a_{nn})$ is also equal to the product of its diagonal entries.

Here we give a proof of Eq. (1.5.7):

$$\det\begin{bmatrix} \mathbf{A} & \mathbf{B} \\ \mathbf{C} & \mathbf{D} \end{bmatrix} = \det\left(\begin{bmatrix} \mathbf{A} & \mathbf{O} \\ \mathbf{C} & \mathbf{D} - \mathbf{C}\mathbf{A}^{-1}\mathbf{B} \end{bmatrix} \begin{bmatrix} \mathbf{I} & \mathbf{A}^{-1}\mathbf{B} \\ \mathbf{O} & \mathbf{I} \end{bmatrix}\right)$$

$$= \det(\mathbf{A})\det(\mathbf{D} - \mathbf{C}\mathbf{A}^{-1}\mathbf{B}).$$

We can prove Eq. (1.5.8) in a similar way.

The following are two determinant inequalities [20].

- The determinant of a positive definite matrix **A** is larger than 0, i.e., $\det(\mathbf{A}) > 0$.
- The determinant of a positive semi-definite matrix **A** is larger than or equal to 0, i.e., $\det(\mathbf{A}) \geq 0$.

1.5.3 Matrix Eigenvalues

For an $n \times n$ matrix **A**, if the linear algebraic equation

$$\mathbf{Au} = \lambda \mathbf{u} \tag{1.5.9}$$

has a nonzero $n \times 1$ solution vector **u**, then the scalar λ is called an eigenvalue of the matrix **A**, and **u** is its eigenvector corresponding to λ.

An equivalent form of the matrix equation (1.5.9) is

$$(\mathbf{A} - \lambda \mathbf{I})\mathbf{u} = \mathbf{0}. \tag{1.5.10}$$

Only condition for this equation having nonzero solution vector **u** is that the determinant of the matrix $\mathbf{A} - \lambda \mathbf{I}$ is equal to zero, i.e.,

$$\det(\mathbf{A} - \lambda \mathbf{I}) = 0. \tag{1.5.11}$$

This equation is known as the *characteristic equation* of the matrix **A**.

The characteristic equation (1.5.11) reflects the following two facts:

- If (1.5.11) holds for $\lambda = 0$, then $\det(\mathbf{A}) = 0$. This implies that as long as a matrix **A** has a zero eigenvalue, this matrix must be singular.
- All the eigenvalues of a zero matrix are zero, and for any singular matrix there exists at least one zero eigenvalue. Clearly, if all n diagonal entries of an $n \times n$ singular matrix **A** contain a subtraction of the same scalar $x \neq 0$ that is not an eigenvalue of **A** then the matrix $\mathbf{A} - x\mathbf{I}$ must be nonsingular, since $|\mathbf{A} - x\mathbf{I}| \neq 0$.

Let $\text{eig}(\mathbf{A})$ represent the eigenvalues of the matrix **A**. The basic properties of eigenvalues are listed below:

1. For $\mathbf{A}_{m \times m}$ and $\mathbf{B}_{m \times m}$, $\text{eig}(\mathbf{AB}) = \text{eig}(\mathbf{BA})$ because

$$\mathbf{ABu} = \lambda \mathbf{u} \Rightarrow (\mathbf{BA})\mathbf{Bu} = \lambda \mathbf{Bu} \Rightarrow \mathbf{BAu}' = \lambda \mathbf{u}',$$

where $\mathbf{u}' = \mathbf{Bu}$. However, if the eigenvector of **AB** corresponding to λ is **u**, then the eigenvector of **BA** corresponding to the same λ is $\mathbf{u}' = \mathbf{Bu}$.

2. If $\text{rank}(\mathbf{A}) = r$, then the matrix **A** has at most r different eigenvalues.
3. The eigenvalues of the inverse matrix satisfy $\text{eig}(\mathbf{A}^{-1}) = 1/\text{eig}(\mathbf{A})$.

1.5 Basic Performance of Matrices

4. Let **I** be the identity matrix; then

$$\text{eig}(\mathbf{I} + c\mathbf{A}) = 1 + c \cdot \text{eig}(\mathbf{A}), \qquad (1.5.12)$$

$$\text{eig}(\mathbf{A} - c\mathbf{I}) = \text{eig}(\mathbf{A}) - c. \qquad (1.5.13)$$

Since $\mathbf{u}^H \mathbf{u} > 0$ for any nonzero vector $\mathbf{u}$, from $\lambda = \mathbf{u}^H \mathbf{A}\mathbf{u}/(\mathbf{u}^H \mathbf{u})$ it is directly known that positive definite and nonpositive definite matrices can be described by their eigenvalues as follows.

- *Positive definite matrix:* Its all eigenvalues are positive real numbers.
- *Positive semi-definite matrix:* Its all eigenvalues are nonnegative.
- *Negative definite matrix:* Its all eigenvalues are negative.
- *Negative semi-definite matrix:* Its all eigenvalues are nonpositive.
- *Indefinite matrix:* It has both positive and negative eigenvalues.

If **A** is a positive definite or positive semi-definite matrix, then

$$\det(\mathbf{A}) \leq \prod_i a_{ii}. \qquad (1.5.14)$$

This inequality is called the *Hadamard inequality*.

The characteristic equation (1.5.11) suggests two methods for improving the numerical stability and the accuracy of solutions of the matrix equation $\mathbf{Ax} = \mathbf{b}$, as described below [40].

- *Method for improving the numerical stability:* Consider the matrix equation $\mathbf{Ax} = \mathbf{b}$, where **A** is usually positive definite or nonsingular. However, owing to noise or errors, **A** may sometimes be close to singular. We can alleviate this difficulty as follows. If λ is a small positive number, then $-\lambda$ cannot be an eigenvalue of **A**. This implies that the characteristic equation $|\mathbf{A} - x\mathbf{I}| = |\mathbf{A} - (-\lambda)\mathbf{I}| = |\mathbf{A} + \lambda\mathbf{I}| = 0$ cannot hold for any $\lambda > 0$, and thus the matrix $\mathbf{A} + \lambda\mathbf{I}$ must be nonsingular. Therefore, if we solve $(\mathbf{A} + \lambda\mathbf{I})\mathbf{x} = \mathbf{b}$ instead of the original matrix equation $\mathbf{Ax} = \mathbf{b}$, and λ takes a very small positive value, then we can overcome the singularity of **A** to improve greatly the numerical stability of solving $\mathbf{Ax} = \mathbf{b}$. This method for solving $(\mathbf{A} + \lambda\mathbf{I})\mathbf{x} = \mathbf{b}$, with $\lambda > 0$, instead of $\mathbf{Ax} = \mathbf{b}$ is the well-known Tikhonov regularization method for solving nearly singular matrix equations.
- *Method for improving the accuracy:* For a matrix equation $\mathbf{Ax} = \mathbf{b}$, with the data matrix **A** nonsingular but containing additive interference or observation noise, if we choose a very small positive scalar λ to solve $(\mathbf{A} - \lambda\mathbf{I})\mathbf{x} = \mathbf{b}$ instead of $\mathbf{Ax} = \mathbf{b}$, then the influence of the noise of the data matrix **A** on the solution vector **x** will be greatly decreased. This is the basis of the well-known total least squares (TLS) method.

1.5.4 Matrix Trace

Definition 1.15 (Trace) The sum of the diagonal entries of an $n \times n$ matrix $\mathbf{A}$ is known as its *trace*, denoted tr($\mathbf{A}$):

$$\mathrm{tr}(\mathbf{A}) = a_{11} + \cdots + a_{nn} = \sum_{i=1}^{n} a_{ii}. \tag{1.5.15}$$

Clearly, for a random signal $\mathbf{x} = [x_1, \ldots, x_n]^T$, the trace of its autocorrelation matrix $\mathbf{R}_x$, $\mathrm{trace}(\mathbf{R}_x) = \sum_{i=1}^{n} E\{|x_i|^2\}$, denotes the energy of the random signal.

The following are some properties of the matrix trace.

Trace Equality [20]
1. If both $\mathbf{A}$ and $\mathbf{B}$ are $n \times n$ matrices, then $\mathrm{tr}(\mathbf{A} \pm \mathbf{B}) = \mathrm{tr}(\mathbf{A}) \pm \mathrm{tr}(\mathbf{B})$.
2. If $\mathbf{A}$ and $\mathbf{B}$ are $n \times n$ matrices and c_1 and c_2 are constants, then $\mathrm{tr}(c_1 \cdot \mathbf{A} \pm c_2 \cdot \mathbf{B}) = c_1 \cdot \mathrm{tr}(\mathbf{A}) \pm c_2 \cdot \mathrm{tr}(\mathbf{B})$. In particular, $\mathrm{tr}(c\mathbf{A}) = c \cdot \mathrm{tr}(\mathbf{A})$.
3. $\mathrm{tr}(\mathbf{A}^T) = \mathrm{tr}(\mathbf{A})$, $\mathrm{tr}(\mathbf{A}^*) = (\mathrm{tr}(\mathbf{A}))^*$ and $\mathrm{tr}(\mathbf{A}^H) = (\mathrm{tr}(\mathbf{A}))^*$.
4. If $\mathbf{A} \in \mathbb{C}^{m \times n}$, $\mathbf{B} \in \mathbb{C}^{n \times m}$, then $\mathrm{tr}(\mathbf{AB}) = \mathrm{tr}(\mathbf{BA})$.
5. If $\mathbf{A}$ is an $m \times n$ matrix, then $\mathrm{tr}(\mathbf{A}^H \mathbf{A}) = 0$ implies that $\mathbf{A}$ is an $m \times n$ zero matrix.
6. $\mathbf{x}^H \mathbf{A} \mathbf{x} = \mathrm{tr}(\mathbf{A}\mathbf{x}\mathbf{x}^H)$ and $\mathbf{y}^H \mathbf{x} = \mathrm{tr}(\mathbf{x}\mathbf{y}^H)$.
7. The trace of an $n \times n$ matrix is equal to the sum of its eigenvalues, namely $\mathrm{tr}(\mathbf{A}) = \lambda_1 + \cdots + \lambda_n$.
8. The trace of a block matrix satisfies

$$\mathrm{tr}\begin{bmatrix} \mathbf{A} & \mathbf{B} \\ \mathbf{C} & \mathbf{D} \end{bmatrix} = \mathrm{tr}(\mathbf{A}) + \mathrm{tr}(\mathbf{D}),$$

where $\mathbf{A} \in \mathbb{C}^{m \times m}$, $\mathbf{B} \in \mathbb{C}^{m \times n}$, $\mathbf{C} \in \mathbb{C}^{n \times m}$ and $\mathbf{D} \in \mathbb{C}^{n \times n}$.
9. For any positive integer k, we have

$$\mathrm{tr}(\mathbf{A}^k) = \sum_{i=1}^{n} \lambda_i^k. \tag{1.5.16}$$

By the trace equality $\mathrm{tr}(\mathbf{UV}) = \mathrm{tr}(\mathbf{VU})$, it is easy to see that

$$\mathrm{tr}(\mathbf{A}^H \mathbf{A}) = \mathrm{tr}(\mathbf{A}\mathbf{A}^H) = \sum_{i=1}^{n} \sum_{j=1}^{n} a_{ij} a_{ij}^* = \sum_{i=1}^{n} \sum_{j=1}^{n} |a_{ij}|^2. \tag{1.5.17}$$

Interestingly, if we substitute $\mathbf{U} = \mathbf{A}, \mathbf{V} = \mathbf{BC}$ and $\mathbf{U} = \mathbf{AB}, \mathbf{V} = \mathbf{C}$ into the trace equality $\mathrm{tr}(\mathbf{UV}) = \mathrm{tr}(\mathbf{VU})$, we obtain

$$\mathrm{tr}(\mathbf{ABC}) = \mathrm{tr}(\mathbf{BCA}) = \mathrm{tr}(\mathbf{CAB}). \tag{1.5.18}$$

1.5 Basic Performance of Matrices

Similarly, if we let $\mathbf{U} = \mathbf{A}, \mathbf{V} = \mathbf{BCD}$ or $\mathbf{U} = \mathbf{AB}, \mathbf{V} = \mathbf{CD}$ or $\mathbf{U} = \mathbf{ABC}, \mathbf{V} = \mathbf{D}$, respectively, we obtain

$$\mathrm{tr}(\mathbf{ABCD}) = \mathrm{tr}(\mathbf{BCDA}) = \mathrm{tr}(\mathbf{CDAB}) = \mathrm{tr}(\mathbf{DABC}). \tag{1.5.19}$$

Moreover, if $\mathbf{A}$ and $\mathbf{B}$ are $m \times m$ matrices, and $\mathbf{B}$ is nonsingular, then

$$\mathrm{tr}(\mathbf{BAB}^{-1}) = \mathrm{tr}(\mathbf{B}^{-1}\mathbf{AB}) = \mathrm{tr}(\mathbf{ABB}^{-1}) = \mathrm{tr}(\mathbf{A}). \tag{1.5.20}$$

The Frobenius norm of an $m \times n$ matrix $\mathbf{A}$ can be also defined using the traces of the $m \times m$ matrix $\mathbf{A}^H \mathbf{A}$ or that of the $n \times n$ matrix $\mathbf{AA}^H$, as follows [22, p. 10]:

$$\|\mathbf{A}\|_F = \sqrt{\mathrm{tr}(\mathbf{A}^H \mathbf{A})} = \sqrt{\mathrm{tr}(\mathbf{AA}^H)}. \tag{1.5.21}$$

1.5.5 Matrix Rank

Theorem 1.2 ([36]) *Among a set of p-dimensional (row or column) vectors, there are at most p linearly independent (row or column) vectors.*

Theorem 1.3 ([36]) *For an $m \times n$ matrix $\mathbf{A}$, the number of linearly independent rows and the number of linearly independent columns are the same.*

From this theorem we have the following definition of the rank of a matrix.

Definition 1.16 (Rank) The *rank* of an $m \times n$ matrix $\mathbf{A}$ is defined as the number of its linearly independent rows or columns.

It needs to be pointed out that the matrix rank gives only the number of linearly independent rows or columns, but it gives no information on the locations of these independent rows or columns.

According to the rank, we can classify the matrix as follows.

- If $\mathrm{rank}(\mathbf{A}_{m \times n}) < \min\{m, n\}$, then $\mathbf{A}$ is known as a *rank-deficient matrix*.
- If $\mathrm{rank}(\mathbf{A}_{m \times n}) = m\ (< n)$, then the matrix $\mathbf{A}$ is a *full row rank matrix*.
- If $\mathrm{rank}(\mathbf{A}_{m \times n}) = n\ (< m)$, then the matrix $\mathbf{A}$ is called a *full column rank matrix*.
- If $\mathrm{rank}(\mathbf{A}_{n \times n}) = n$, then $\mathbf{A}$ is said to be a *full-rank matrix* (or nonsingular matrix).

The matrix equation $\mathbf{A}_{m \times n} \mathbf{x}_{n \times 1} = \mathbf{b}_{m \times 1}$ is said to be a *consistent equation*, if it has at least one exact solution. A matrix equation with no exact solution is said to be an *inconsistent equation*.

A matrix $\mathbf{A}$ with $\mathrm{rank}(\mathbf{A}) = r_A$ has r_A linearly independent column vectors. All linear combinations of the r_A linearly independent column vectors constitute a vector space, called the *column space* or the *range* or the *manifold* of $\mathbf{A}$.

The column space $\text{Col}(\mathbf{A}) = \text{Col}(\mathbf{a}_1, \ldots, \mathbf{a}_n)$ or the range $\text{Range}(\mathbf{A}) = \text{Range}(\mathbf{a}_1, \ldots, \mathbf{a}_n)$ is r_A-dimensional. Hence the rank of a matrix can be defined by using the dimension of its column space or range, as described below.

Definition 1.17 (Dimension of Column Space) The dimension of the column space $\text{Col}(\mathbf{A})$ or the range $\text{Range}(\mathbf{A})$ of an $m \times n$ matrix $\mathbf{A}$ is defined by the rank of the matrix, namely

$$r_A = \dim(\text{Col}(\mathbf{A})) = \dim(\text{Range}(\mathbf{A})). \tag{1.5.22}$$

The following statements about the rank of the matrix $\mathbf{A}$ are equivalent:

- $\text{rank}(\mathbf{A}) = k$;
- there are k and not more than k columns (or rows) of $\mathbf{A}$ that combine a linearly independent set;
- there is a $k \times k$ submatrix of $\mathbf{A}$ with nonzero determinant, but all the $(k+1) \times (k+1)$ submatrices of $\mathbf{A}$ have zero determinant;
- the dimension of the column space $\text{Col}(\mathbf{A})$ or the range $\text{Range}(\mathbf{A})$ equals k;
- $k = n - \dim[\text{Null}(\mathbf{A})]$, where $\text{Null}(\mathbf{A})$ denotes the null space of the matrix $\mathbf{A}$.

Theorem 1.4 ([36]) *The rank of the product matrix* $\mathbf{AB}$ *satisfies the inequality*

$$\text{rank}(\mathbf{AB}) \leq \min\{\text{rank}(\mathbf{A}), \text{rank}(\mathbf{B})\}. \tag{1.5.23}$$

Lemma 1.1 *If premultiplying an $m \times n$ matrix $\mathbf{A}$ by an $m \times m$ nonsingular matrix $\mathbf{P}$, or postmultiplying it by an $n \times n$ nonsingular matrix $\mathbf{Q}$, then the rank of $\mathbf{A}$ is not changed, namely* $\text{rank}(\mathbf{PAQ}) = \text{rank}(\mathbf{A})$.

Here are properties of the rank of a matrix.

- The rank is a positive integer.
- The rank is equal to or less than the number of columns or rows of the matrix.
- Premultiplying any matrix $\mathbf{A}$ by a full column rank matrix or postmultiplying it by a full row rank matrix, then the rank of the matrix $\mathbf{A}$ remains unchanged.

Rank Equalities
1. If $\mathbf{A} \in \mathbb{C}^{m \times n}$, then $\text{rank}(\mathbf{A}^H) = \text{rank}(\mathbf{A}^T) = \text{rank}(\mathbf{A}^*) = \text{rank}(\mathbf{A})$.
2. If $\mathbf{A} \in \mathbb{C}^{m \times n}$ and $c \neq 0$, then $\text{rank}(c\mathbf{A}) = \text{rank}(\mathbf{A})$.
3. If $\mathbf{A} \in \mathbb{C}^{m \times m}$, $\mathbf{B} \in \mathbb{C}^{m \times n}$, and $\mathbf{C} \in \mathbb{C}^{n \times n}$ are nonsingular, then

$$\text{rank}(\mathbf{AB}) = \text{rank}(\mathbf{B}) = \text{rank}(\mathbf{BC}) = \text{rank}(\mathbf{ABC}).$$

That is, after premultiplying and/or postmultiplying by a nonsingular matrix, the rank of $\mathbf{B}$ remains unchanged.
4. For $\mathbf{A}, \mathbf{B} \in \mathbb{C}^{m \times n}$, $\text{rank}(\mathbf{A}) = \text{rank}(\mathbf{B})$ if and only if there exist nonsingular matrices $\mathbf{X} \in \mathbb{C}^{m \times m}$ and $\mathbf{Y} \in \mathbb{C}^{n \times n}$ such that $\mathbf{B} = \mathbf{XAY}$.
5. $\text{rank}(\mathbf{AA}^H) = \text{rank}(\mathbf{A}^H\mathbf{A}) = \text{rank}(\mathbf{A})$.
6. If $\mathbf{A} \in \mathbb{C}^{m \times m}$, then $\text{rank}(\mathbf{A}) = m \Leftrightarrow \det(\mathbf{A}) \neq 0 \Leftrightarrow \mathbf{A}$ is nonsingular.

1.6 Inverse Matrices and Moore–Penrose Inverse Matrices

Matrix inversion is an important aspect of matrix calculus. In particular, the matrix inversion lemma is often used in science and engineering. In this section we discuss the inverse of a full-rank square matrix, the pseudo-inverse of a non-square matrix with full row (or full column) rank, and the inversion of a rank-deficient matrix.

1.6.1 Inverse Matrices

A nonsingular matrix $\mathbf{A}$ is said to be invertible, if its inverse $\mathbf{A}^{-1}$ exists so that $\mathbf{A}^{-1}\mathbf{A} = \mathbf{A}\mathbf{A}^{-1} = \mathbf{I}$.

On the nonsingularity or invertibility of an $n \times n$ matrix $\mathbf{A}$, the following statements are equivalent [15].

- $\mathbf{A}$ is nonsingular.
- $\mathbf{A}^{-1}$ exists.
- $\text{rank}(\mathbf{A}) = n$.
- All rows of $\mathbf{A}$ are linearly independent.
- All columns of $\mathbf{A}$ are linearly independent.
- $\det(\mathbf{A}) \neq 0$.
- The dimension of the range of $\mathbf{A}$ is n.
- The dimension of the null space of $\mathbf{A}$ is equal to zero.
- $\mathbf{A}\mathbf{x} = \mathbf{b}$ is a consistent equation for every $\mathbf{b} \in \mathbb{C}^n$.
- $\mathbf{A}\mathbf{x} = \mathbf{b}$ has a *unique solution* for every $\mathbf{b}$.
- $\mathbf{A}\mathbf{x} = \mathbf{0}$ has only the *trivial solution* $\mathbf{x} = \mathbf{0}$.

The inverse matrix $\mathbf{A}^{-1}$ has the following properties [2, 15].

1. $\mathbf{A}^{-1}\mathbf{A} = \mathbf{A}\mathbf{A}^{-1} = \mathbf{I}$.
2. $\mathbf{A}^{-1}$ is unique.
3. The determinant of the inverse matrix is equal to the reciprocal of the determinant of the original matrix, i.e., $|\mathbf{A}^{-1}| = 1/|\mathbf{A}|$.
4. The inverse matrix $\mathbf{A}^{-1}$ is nonsingular.
5. The inverse matrix of an inverse matrix is the original matrix, i.e., $(\mathbf{A}^{-1})^{-1} = \mathbf{A}$.
6. The inverse matrix of a Hermitian matrix $\mathbf{A} = \mathbf{A}^H$ satisfies $(\mathbf{A}^H)^{-1} = (\mathbf{A}^{-1})^H = \mathbf{A}^{-1}$. That is to say, the inverse matrix of any Hermitian matrix is a Hermitian matrix as well.
7. $(\mathbf{A}^*)^{-1} = (\mathbf{A}^{-1})^*$.

8. If **A** and **B** are invertible, then $(\mathbf{AB})^{-1} = \mathbf{B}^{-1}\mathbf{A}^{-1}$.
9. If $\mathbf{A} = \mathbf{Diag}(a_1, \ldots, a_m)$ is a diagonal matrix, then its inverse matrix

$$\mathbf{A}^{-1} = \mathbf{Diag}(a_1^{-1}, \ldots, a_m^{-1}).$$

10. Let **A** be nonsingular. If **A** is an orthogonal matrix, then $\mathbf{A}^{-1} = \mathbf{A}^T$, and if **A** is a unitary matrix, then $\mathbf{A}^{-1} = \mathbf{A}^H$.

Lemma 1.2 *Let* **A** *be an* $n \times n$ *invertible matrix, and* **x** *and* **y** *be two* $n \times 1$ *vectors such that* $(\mathbf{A} + \mathbf{xy}^H)$ *is invertible; then*

$$(\mathbf{A} + \mathbf{xy}^H)^{-1} = \mathbf{A}^{-1} - \frac{\mathbf{A}^{-1}\mathbf{xy}^H\mathbf{A}^{-1}}{1 + \mathbf{y}^H\mathbf{A}^{-1}\mathbf{x}}. \tag{1.6.1}$$

Lemma 1.2 is called the *matrix inversion lemma*, and was presented by Sherman and Morrison [37] in 1950.

The matrix inversion lemma can be extended to an inversion formula for a sum of matrices:

$$\begin{aligned}(\mathbf{A} + \mathbf{UBV})^{-1} &= \mathbf{A}^{-1} - \mathbf{A}^{-1}\mathbf{UB}(\mathbf{B} + \mathbf{BVA}^{-1}\mathbf{UB})^{-1}\mathbf{BVA}^{-1} \\ &= \mathbf{A}^{-1} - \mathbf{A}^{-1}\mathbf{U}(\mathbf{I} + \mathbf{BVA}^{-1}\mathbf{U})^{-1}\mathbf{BVA}^{-1}\end{aligned} \tag{1.6.2}$$

or

$$(\mathbf{A} - \mathbf{UV})^{-1} = \mathbf{A}^{-1} + \mathbf{A}^{-1}\mathbf{U}(\mathbf{I} - \mathbf{VA}^{-1}\mathbf{U})^{-1}\mathbf{VA}^{-1}. \tag{1.6.3}$$

The above formula is called *Woodbury formula* due to the work of Woodbury in 1950 [38].

Taking $\mathbf{U} = \mathbf{u}$, $\mathbf{B} = \beta$, and $\mathbf{V} = \mathbf{v}^H$, the Woodbury formula gives the result

$$(\mathbf{A} + \beta\mathbf{uv}^H)^{-1} = \mathbf{A}^{-1} - \frac{\beta}{1 + \beta\mathbf{v}^H\mathbf{A}^{-1}\mathbf{u}}\mathbf{A}^{-1}\mathbf{uv}^H\mathbf{A}^{-1}. \tag{1.6.4}$$

In particular, if letting $\beta = 1$, then Eq. (1.6.4) reduces to formula (1.6.1), the matrix inversion lemma of Sherman and Morrison.

As a matter of fact, before Woodbury obtained the inversion formula (1.6.2), Duncan [8] in 1944 and Guttman [12] in 1946 had obtained the following inversion formula:

$$(\mathbf{A} - \mathbf{UD}^{-1}\mathbf{V})^{-1} = \mathbf{A}^{-1} + \mathbf{A}^{-1}\mathbf{U}(\mathbf{D} - \mathbf{VA}^{-1}\mathbf{U})^{-1}\mathbf{VA}^{-1}. \tag{1.6.5}$$

This formula is usually called the *Duncan–Guttman inversion formula* [28].

1.6 Inverse Matrices and Moore–Penrose Inverse Matrices

In addition to the Woodbury formula, the inverse matrix of a sum of matrices also has the following forms [14]:

$$(\mathbf{A} + \mathbf{UBV})^{-1} = \mathbf{A}^{-1} - \mathbf{A}^{-1}(\mathbf{I} + \mathbf{UBVA}^{-1})^{-1}\mathbf{UBVA}^{-1} \quad (1.6.6)$$

$$= \mathbf{A}^{-1} - \mathbf{A}^{-1}\mathbf{UB}(\mathbf{I} + \mathbf{VA}^{-1}\mathbf{UB})^{-1}\mathbf{VA}^{-1} \quad (1.6.7)$$

$$= \mathbf{A}^{-1} - \mathbf{A}^{-1}\mathbf{UBV}(\mathbf{I} + \mathbf{A}^{-1}\mathbf{UBV})^{-1}\mathbf{A}^{-1} \quad (1.6.8)$$

$$= \mathbf{A}^{-1} - \mathbf{A}^{-1}\mathbf{UBVA}^{-1}(\mathbf{I} + \mathbf{UBVA}^{-1})^{-1}. \quad (1.6.9)$$

The following are inversion formulas for block matrices.
When the matrix $\mathbf{A}$ is invertible, one has [1]:

$$\begin{bmatrix} \mathbf{A} & \mathbf{U} \\ \mathbf{V} & \mathbf{D} \end{bmatrix}^{-1} = \begin{bmatrix} \mathbf{A}^{-1} + \mathbf{A}^{-1}\mathbf{U}(\mathbf{D} - \mathbf{VA}^{-1}\mathbf{U})^{-1}\mathbf{VA}^{-1} & -\mathbf{A}^{-1}\mathbf{U}(\mathbf{D} - \mathbf{VA}^{-1}\mathbf{U})^{-1} \\ -(\mathbf{D} - \mathbf{VA}^{-1}\mathbf{U})^{-1}\mathbf{VA}^{-1} & (\mathbf{D} - \mathbf{VA}^{-1}\mathbf{U})^{-1} \end{bmatrix}. \quad (1.6.10)$$

If the matrices $\mathbf{A}$ and $\mathbf{D}$ are invertible, then [16, 17]

$$\begin{bmatrix} \mathbf{A} & \mathbf{U} \\ \mathbf{V} & \mathbf{D} \end{bmatrix}^{-1} = \begin{bmatrix} (\mathbf{A} - \mathbf{UD}^{-1}\mathbf{V})^{-1} & -\mathbf{A}^{-1}\mathbf{U}(\mathbf{D} - \mathbf{VA}^{-1}\mathbf{U})^{-1} \\ -\mathbf{D}^{-1}\mathbf{V}(\mathbf{A} - \mathbf{UD}^{-1}\mathbf{V})^{-1} & (\mathbf{D} - \mathbf{VA}^{-1}\mathbf{U})^{-1} \end{bmatrix} \quad (1.6.11)$$

or [8]

$$\begin{bmatrix} \mathbf{A} & \mathbf{U} \\ \mathbf{V} & \mathbf{D} \end{bmatrix}^{-1} = \begin{bmatrix} (\mathbf{A} - \mathbf{UD}^{-1}\mathbf{V})^{-1} & -(\mathbf{A} - \mathbf{UD}^{-1}\mathbf{V})^{-1}\mathbf{UD}^{-1} \\ -(\mathbf{D} - \mathbf{VA}^{-1}\mathbf{U})^{-1}\mathbf{VA}^{-1} & (\mathbf{D} - \mathbf{VA}^{-1}\mathbf{U})^{-1} \end{bmatrix}. \quad (1.6.12)$$

For a $(m + 1) \times (m + 1)$ nonsingular Hermitian matrix $\mathbf{R}_{m+1}$, it can always be written in a block form:

$$\mathbf{R}_{m+1} = \begin{bmatrix} \mathbf{R}_m & \mathbf{r}_m \\ \mathbf{r}_m^H & \rho_m \end{bmatrix}, \quad (1.6.13)$$

where ρ_m is the $(m + 1, m + 1)$th entry of $\mathbf{R}_{m+1}$, and $\mathbf{R}_m$ is an $m \times m$ Hermitian matrix. In order to compute $\mathbf{R}_{m+1}^{-1}$ using $\mathbf{R}_m^{-1}$, let

$$\mathbf{Q}_{m+1} = \begin{bmatrix} \mathbf{Q}_m & \mathbf{q}_m \\ \mathbf{q}_m^H & \alpha_m \end{bmatrix} \quad (1.6.14)$$

be the inverse matrix of $\mathbf{R}_{m+1}$. Then we have

$$\mathbf{R}_{m+1}\mathbf{Q}_{m+1} = \begin{bmatrix} \mathbf{R}_m & \mathbf{r}_m \\ \mathbf{r}_m^H & \rho_m \end{bmatrix} \begin{bmatrix} \mathbf{Q}_m & \mathbf{q}_m \\ \mathbf{q}_m^H & \alpha_m \end{bmatrix} = \begin{bmatrix} \mathbf{I}_m & \mathbf{0}_m \\ \mathbf{0}_m^H & 1 \end{bmatrix} \quad (1.6.15)$$

which gives the following four equations:

$$\mathbf{R}_m \mathbf{Q}_m + \mathbf{r}_m \mathbf{q}_m^H = \mathbf{I}_m, \tag{1.6.16}$$

$$\mathbf{r}_m^H \mathbf{Q}_m + \rho_m \mathbf{q}_m^H = \mathbf{0}_m^H, \tag{1.6.17}$$

$$\mathbf{R}_m \mathbf{q}_m + \mathbf{r}_m \alpha_m = \mathbf{0}_m, \tag{1.6.18}$$

$$\mathbf{r}_m^H \mathbf{q}_m + \rho_m \alpha_m = 1. \tag{1.6.19}$$

If $\mathbf{R}_m$ is invertible, then from Eq. (1.6.18), it follows that

$$\mathbf{q}_m = -\alpha_m \mathbf{R}_m^{-1} \mathbf{r}_m. \tag{1.6.20}$$

Substitute this result into Eq. (1.6.19) to yield

$$\alpha_m = \frac{1}{\rho_m - \mathbf{r}_m^H \mathbf{R}_m^{-1} \mathbf{r}_m}. \tag{1.6.21}$$

After substituting Eq. (1.6.21) into Eq. (1.6.20), we can obtain

$$\mathbf{q}_m = \frac{-\mathbf{R}_m^{-1} \mathbf{r}_m}{\rho_m - \mathbf{r}_b^H \mathbf{R}_m^{-1} \mathbf{r}_m}. \tag{1.6.22}$$

Then, substitute Eq. (1.6.22) into Eq. (1.6.16):

$$\mathbf{Q}_m = \mathbf{R}_m^{-1} - \mathbf{R}_m^{-1} \mathbf{r}_m \mathbf{q}_m^H = \mathbf{R}_m^{-1} + \frac{\mathbf{R}_m^{-1} \mathbf{r}_m (\mathbf{R}_m^{-1} \mathbf{r}_m)^H}{\rho_m - \mathbf{r}_m^H \mathbf{R}_m^{-1} \mathbf{r}_m}. \tag{1.6.23}$$

In order to simplify (1.6.21)–(1.6.23), put

$$\mathbf{b}_m \stackrel{\text{def}}{=} [b_0^{(m)}, b_1^{(m)}, \ldots, b_{m-1}^{(m)}]^T = -\mathbf{R}_m^{-1} \mathbf{r}_m, \tag{1.6.24}$$

$$\beta_m \stackrel{\text{def}}{=} \rho_m - \mathbf{r}_m^H \mathbf{R}_m^{-1} \mathbf{r}_m = \rho_m + \mathbf{r}_m^H \mathbf{b}_m. \tag{1.6.25}$$

Then (1.6.21)–(1.6.23) can be, respectively, simplified to

$$\alpha_m = \frac{1}{\beta_m}, \quad \mathbf{q}_m = \frac{1}{\beta_m} \mathbf{b}_m, \quad \mathbf{Q}_m = \mathbf{R}_m^{-1} + \frac{1}{\beta_m} \mathbf{b}_m \mathbf{b}_m^H.$$

Substituting the these results into (1.6.15), it is immediately seen that

$$\mathbf{R}_{m+1}^{-1} = \mathbf{Q}_{m+1} = \begin{bmatrix} \mathbf{R}_m^{-1} & \mathbf{0}_m \\ \mathbf{0}_m^H & 0 \end{bmatrix} + \frac{1}{\beta_m} \begin{bmatrix} \mathbf{b}_m \mathbf{b}_m^H & \mathbf{b}_m \\ \mathbf{b}_m^H & 1 \end{bmatrix}. \tag{1.6.26}$$

This formula for calculating the $(m + 1) \times (m + 1)$ inverse $\mathbf{R}_{m+1}$ from the $m \times m$ inverse $\mathbf{R}_m$ is called the *block inversion lemma* for Hermitian matrices [24].

1.6.2 Left and Right Pseudo-Inverse Matrices

From a broader perspective, any $n \times m$ matrix $\mathbf{G}$ may be called the inverse of a given $m \times n$ matrix $\mathbf{A}$, if the product of $\mathbf{G}$ and $\mathbf{A}$ is equal to the identity matrix $\mathbf{I}$.

Definition 1.18 (Left Inverse, Right Inverse [36]) The matrix $\mathbf{L}$ satisfying $\mathbf{LA} = \mathbf{I}$ but not $\mathbf{AL} = \mathbf{I}$ is called the *left inverse* of the matrix $\mathbf{A}$. Similarly, the matrix $\mathbf{R}$ satisfying $\mathbf{AR} = \mathbf{I}$ but not $\mathbf{RA} = \mathbf{I}$ is said to be the *right inverse* of $\mathbf{A}$.

A matrix $\mathbf{A} \in \mathbb{C}^{m \times n}$ has a left inverse only when $m \geq n$ and a right inverse only when $m \leq n$, respectively. It should be noted that the left or right inverse of a given matrix $\mathbf{A}$ is usually not unique. Let us consider the conditions for a unique solution of the left and right inverse matrices.

Let $m > n$ and let the $m \times n$ matrix $\mathbf{A}$ have full column rank, i.e., $\text{rank}(\mathbf{A}) = n$. In this case, the $n \times n$ matrix $\mathbf{A}^H \mathbf{A}$ is invertible. It is easy to verify that

$$\mathbf{L} = (\mathbf{A}^H \mathbf{A})^{-1} \mathbf{A}^H \qquad (1.6.27)$$

satisfies $\mathbf{LA} = \mathbf{I}$ but not $\mathbf{AL} = \mathbf{I}$. This type of left inverse matrix is uniquely determined, and is usually called the *left pseudo-inverse matrix* of $\mathbf{A}$.

On the other hand, if $m < n$ and $\mathbf{A}$ has the full row rank, i.e., $\text{rank}(\mathbf{A}) = m$, then the $m \times m$ matrix $\mathbf{AA}^H$ is invertible. Define

$$\mathbf{R} = \mathbf{A}^H (\mathbf{AA}^H)^{-1}. \qquad (1.6.28)$$

It is easy to see that $\mathbf{R}$ satisfies $\mathbf{AR} = \mathbf{I}$ but $\mathbf{RA} \neq \mathbf{I}$. The matrix $\mathbf{R}$ is also uniquely determined and is usually called the *right pseudo-inverse matrix* of $\mathbf{A}$.

The left pseudo-inverse matrix is closely related to the least squares solution of over-determined equations, while the right pseudo-inverse matrix is closely related to the least squares minimum norm solution of under-determined equations.

Consider the recursion $\mathbf{F}_m = [\mathbf{F}_{m-1}, \mathbf{f}_m]$ of an $n \times m$ matrix $\mathbf{F}_m$. Then, the left pseudo-inverse $\mathbf{F}_m^{\text{left}} = (\mathbf{F}_m^H \mathbf{F}_m)^{-1} \mathbf{F}_m^H$ can be recursively computed by Zhang [39]

$$\mathbf{F}_m^{\text{left}} = \begin{bmatrix} \mathbf{F}_{m-1}^{\text{left}} - \mathbf{F}_{m-1}^{\text{left}} \mathbf{f}_m \mathbf{e}_m^H \Delta_m^{-1} \\ \mathbf{e}_m^H \Delta_m^{-1} \end{bmatrix}, \qquad (1.6.29)$$

where $\mathbf{e}_m = (\mathbf{I}_n - \mathbf{F}_{m-1} \mathbf{F}_{m-1}^{\text{left}}) \mathbf{f}_m$ and $\Delta_m^{-1} = (\mathbf{f}_m^H \mathbf{e}_m)^{-1}$; the initial recursion value is $\mathbf{F}_1^{\text{left}} = \mathbf{f}_1^H / (\mathbf{f}_1^H \mathbf{f}_1)$. Similarly, the right pseudo-inverse matrix $\mathbf{F}_m^{\text{right}} =$

$\mathbf{F}_m^H(\mathbf{F}_m\mathbf{F}_m^H)^{-1}$ has the following recursive formula [39]:

$$\mathbf{F}_m^{\text{right}} = \begin{bmatrix} \mathbf{F}_{m-1}^{\text{right}} - \Delta_m \mathbf{F}_{m-1}^{\text{right}} \mathbf{f}_m \mathbf{c}_m \\ \Delta_m \mathbf{c}_m^H \end{bmatrix}. \tag{1.6.30}$$

Here $\mathbf{c}_m^H = \mathbf{f}_m^H(\mathbf{I}_n - \mathbf{F}_{m-1}\mathbf{F}_{m-1}^\dagger)$ and $\Delta_m = \mathbf{c}_m^H \mathbf{f}_m$. The initial recursion value is $\mathbf{F}_1^{\text{right}} = \mathbf{f}_1^H/(\mathbf{f}_1^H\mathbf{f}_1)$.

1.6.3 Moore–Penrose Inverse Matrices

Given an $m \times n$ rank-deficient matrix $\mathbf{A}$, regardless of the size of m and n but with rank$(\mathbf{A}) = k < \min\{m, n\}$. The inverse of an $m \times n$ rank-deficient matrix is said to be its generalized inverse matrix, denoted as $\mathbf{A}^\dagger$, that is an $n \times m$ matrix.

The generalized inverse of a rank-deficient matrix $\mathbf{A}$ should satisfy the following four conditions.

1. If $\mathbf{A}^\dagger$ is the generalized inverse of $\mathbf{A}$, then $\mathbf{A}\mathbf{x} = \mathbf{y} \Rightarrow \mathbf{x} = \mathbf{A}^\dagger \mathbf{y}$. Substituting $\mathbf{x} = \mathbf{A}^\dagger \mathbf{y}$ into $\mathbf{A}\mathbf{x} = \mathbf{y}$, we have $\mathbf{A}\mathbf{A}^\dagger \mathbf{y} = \mathbf{y}$, and thus $\mathbf{A}\mathbf{A}^\dagger \mathbf{A}\mathbf{x} = \mathbf{A}\mathbf{x}$. Since this equation should hold for any given nonzero vector $\mathbf{x}$, $\mathbf{A}^\dagger$ must satisfy the condition:

$$\mathbf{A}\mathbf{A}^\dagger\mathbf{A} = \mathbf{A}. \tag{1.6.31}$$

2. Given any $\mathbf{y} \neq \mathbf{0}$, the solution equation of the original matrix equation, $\mathbf{x} = \mathbf{A}^\dagger \mathbf{y}$, can be written as $\mathbf{x} = \mathbf{A}^\dagger \mathbf{A}\mathbf{x}$, yielding $\mathbf{A}^\dagger \mathbf{y} = \mathbf{A}^\dagger \mathbf{A}\mathbf{A}^\dagger \mathbf{y}$. Since $\mathbf{A}^\dagger \mathbf{y} = \mathbf{A}^\dagger \mathbf{A}\mathbf{A}^\dagger \mathbf{y}$ should hold for any given nonzero vector $\mathbf{y}$, the second condition

$$\mathbf{A}^\dagger\mathbf{A}\mathbf{A}^\dagger = \mathbf{A}^\dagger \tag{1.6.32}$$

must be satisfied as well.

3. If an $m \times n$ matrix $\mathbf{A}$ is of full column rank or full row rank, we certainly hope that the generalized inverse matrix $\mathbf{A}^\dagger$ will include the left and right pseudo-inverse matrices as two special cases. Because the left and right pseudo-inverse matrix $\mathbf{L} = (\mathbf{A}^H\mathbf{A})^{-1}\mathbf{A}^H$ and $\mathbf{R} = \mathbf{A}^H(\mathbf{A}\mathbf{A}^H)^{-1}$ of the $m \times n$ full column rank matrix $\mathbf{A}$ satisfy, respectively, $\mathbf{A}\mathbf{L} = \mathbf{A}(\mathbf{A}^H\mathbf{A})^{-1}\mathbf{A}^H = (\mathbf{A}\mathbf{L})^H$ and $\mathbf{R}\mathbf{A} = \mathbf{A}^H(\mathbf{A}\mathbf{A}^H)^{-1}\mathbf{A} = (\mathbf{R}\mathbf{A})^H$, in order to guarantee that $\mathbf{A}^\dagger$ exists uniquely for any $m \times n$ matrix $\mathbf{A}$, the following two conditions must be added:

$$\mathbf{A}\mathbf{A}^\dagger = (\mathbf{A}\mathbf{A}^\dagger)^H, \quad \mathbf{A}^\dagger\mathbf{A} = (\mathbf{A}^\dagger\mathbf{A})^H. \tag{1.6.33}$$

1.6 Inverse Matrices and Moore–Penrose Inverse Matrices

Definition 1.19 (Moore–Penrose Inverse [26]) Let $\mathbf{A}$ be any $m \times n$ matrix. An $n \times m$ matrix $\mathbf{A}^\dagger$ is said to be the *Moore–Penrose inverse* of $\mathbf{A}$ if $\mathbf{A}^\dagger$ meets the following four conditions (usually called the *Moore–Penrose conditions*):

(a) $\mathbf{A}\mathbf{A}^\dagger\mathbf{A} = \mathbf{A}$;
(b) $\mathbf{A}^\dagger\mathbf{A}\mathbf{A}^\dagger = \mathbf{A}^\dagger$;
(c) $\mathbf{A}\mathbf{A}^\dagger$ is an $m \times m$ Hermitian matrix, i.e., $\mathbf{A}\mathbf{A}^\dagger = (\mathbf{A}\mathbf{A}^\dagger)^H$;
(d) $\mathbf{A}^\dagger\mathbf{A}$ is an $n \times n$ Hermitian matrix, i.e., $\mathbf{A}^\dagger\mathbf{A} = (\mathbf{A}^\dagger\mathbf{A})^H$.

Remarks From the projection viewpoint, Moore [23] showed in 1935 that the generalized inverse matrix $\mathbf{A}^\dagger$ of an $m \times n$ matrix $\mathbf{A}$ must meet two conditions, but these conditions are not convenient for practical use. After two decades, Penrose [26] in 1955 presented the four conditions (a)–(d) stated above. In 1956, Rado [32] showed that the four conditions of Penrose are equivalent to the two conditions of Moore. Therefore the conditions (a)–(d) are called the Moore–Penrose conditions, and the generalized inverse matrix satisfying the Moore–Penrose conditions is referred to as the Moore–Penrose inverse of $\mathbf{A}$.

It is easy to know that the inverse $\mathbf{A}^{-1}$, the left pseudo-inverse matrix $(\mathbf{A}^H\mathbf{A})^{-1}\mathbf{A}^H$, and the right pseudo-inverse matrix $\mathbf{A}^H(\mathbf{A}\mathbf{A}^H)^{-1}$ are special examples of the Moore–Penrose inverse $\mathbf{A}^\dagger$.

The Moore–Penrose inverse of any $m \times n$ matrix $\mathbf{A}$ can be uniquely determined by Boot [5]

$$\mathbf{A}^\dagger = (\mathbf{A}^H\mathbf{A})^\dagger\mathbf{A}^H \qquad (\text{if } m \geq n) \qquad (1.6.34)$$

or [10]

$$\mathbf{A}^\dagger = \mathbf{A}^H(\mathbf{A}\mathbf{A}^H)^\dagger \qquad (\text{if } m \leq n). \qquad (1.6.35)$$

From Definition 1.19 it is easily seen that $\mathbf{A}^\dagger = (\mathbf{A}^H\mathbf{A})^\dagger\mathbf{A}^H$ and $\mathbf{A}^\dagger = \mathbf{A}^H(\mathbf{A}\mathbf{A}^H)^\dagger$ meet the four Moore–Penrose conditions.

For an $m \times n$ matrix $\mathbf{A}$ with rank r, where $r \leq \min(m, n)$, there are the following three methods for computing the Moore–Penrose inverse matrix $\mathbf{A}^\dagger$.

1. *Equation-solving method* [26]
   - Solve the matrix equations $\mathbf{A}\mathbf{A}^H\mathbf{X}^H = \mathbf{A}$ and $\mathbf{A}^H\mathbf{A}\mathbf{Y} = \mathbf{A}^H$ to yield the solutions $\mathbf{X}^H$ and $\mathbf{Y}$, respectively.
   - Compute the generalized inverse matrix $\mathbf{A}^\dagger = \mathbf{X}\mathbf{A}\mathbf{Y}$.

2. *Full-rank decomposition method:* If $\mathbf{A} = \mathbf{FG}$, where $\mathbf{F}_{m \times r}$ is of full column rank and $\mathbf{G}_{r \times n}$ is of full row rank, then $\mathbf{A} = \mathbf{FG}$ is called the *full-rank decomposition* of the matrix $\mathbf{A}$. By Searle [36], a matrix $\mathbf{A} \in \mathbb{C}^{m \times n}$ with rank($\mathbf{A}$) $= r$ has the full-rank decomposition $\mathbf{A} = \mathbf{FG}$. Therefore, if $\mathbf{A} = \mathbf{FG}$ is a full-rank decomposition of the $m \times n$ matrix $\mathbf{A}$, then $\mathbf{A}^\dagger = \mathbf{G}^\dagger\mathbf{F}^\dagger = \mathbf{G}^H(\mathbf{G}\mathbf{G}^H)^{-1}(\mathbf{F}^H\mathbf{F})^{-1}\mathbf{F}^H$.

3. *Recursive methods:* Block the matrix $\mathbf{A}_{m\times n}$ into $\mathbf{A}_k = [\mathbf{A}_{k-1}, \mathbf{a}_k]$, where $\mathbf{A}_{k-1}$ consists of the first $k-1$ columns of $\mathbf{A}$ and $\mathbf{a}_k$ is the kth column of $\mathbf{A}$. Then, the Moore–Penrose inverse $\mathbf{A}_k^\dagger$ of the block matrix $\mathbf{A}_k$ can be recursively calculated from $\mathbf{A}_{k-1}^\dagger$. When $k = n$, we get the Moore–Penrose inverse matrix $\mathbf{A}^\dagger$. Such a recursive algorithm was presented by Greville in 1960 [11].

From [20, 22, 30, 31] and [33], the Moore–Penrose inverses $\mathbf{A}^\dagger$ have the following properties.

1. For an $m \times n$ matrix $\mathbf{A}$, its Moore–Penrose inverse $\mathbf{A}^\dagger$ is uniquely determined.
2. The Moore–Penrose inverse of the complex conjugate transpose matrix $\mathbf{A}^H$ is given by $(\mathbf{A}^H)^\dagger = (\mathbf{A}^\dagger)^H = \mathbf{A}^{\dagger H} = \mathbf{A}^{H\dagger}$.
3. The generalized inverse of a Moore–Penrose inverse matrix is equal to the original matrix, namely $(\mathbf{A}^\dagger)^\dagger = \mathbf{A}$.
4. If $c \neq 0$, then $(c\mathbf{A})^\dagger = c^{-1}\mathbf{A}^\dagger$.
5. If $\mathbf{D} = \mathbf{Diag}(d_{11}, \ldots, d_{nn})$, then $\mathbf{D}^\dagger = \mathbf{Diag}(d_{11}^\dagger, \ldots, d_{nn}^\dagger)$, where $d_{ii}^\dagger = d_{ii}^{-1}$ (if $d_{ii} \neq 0$) or $d_{ii}^\dagger = 0$ (if $d_{ii} = 0$).
6. The Moore–Penrose inverse of an $m \times n$ zero matrix $\mathbf{O}_{m\times n}$ is an $n \times m$ zero matrix, i.e., $\mathbf{O}_{m\times n}^\dagger = \mathbf{O}_{n\times m}$.
7. If $\mathbf{A}^H = \mathbf{A}$ and $\mathbf{A}^2 = \mathbf{A}$, then $\mathbf{A}^\dagger = \mathbf{A}$.
8. If $\mathbf{A} = \mathbf{BC}$, $\mathbf{B}$ is of full column rank, and $\mathbf{C}$ is of full row rank, then

$$\mathbf{A}^\dagger = \mathbf{C}^\dagger \mathbf{B}^\dagger = \mathbf{C}^H(\mathbf{CC}^H)^{-1}(\mathbf{B}^H\mathbf{B})^{-1}\mathbf{B}^H.$$

9. $(\mathbf{AA}^H)^\dagger = (\mathbf{A}^\dagger)^H \mathbf{A}^\dagger$ and $(\mathbf{AA}^H)^\dagger(\mathbf{AA}^H) = \mathbf{AA}^\dagger$.
10. If the matrices $\mathbf{A}_i$ are mutually orthogonal, i.e., $\mathbf{A}_i^H \mathbf{A}_j = \mathbf{O}$, $i \neq j$, then $(\mathbf{A}_1 + \cdots + \mathbf{A}_m)^\dagger = \mathbf{A}_1^\dagger + \cdots + \mathbf{A}_m^\dagger$.
11. Regarding the ranks of generalized inverse matrices, one has $\mathrm{rank}(\mathbf{A}^\dagger) = \mathrm{rank}(\mathbf{A}) = \mathrm{rank}(\mathbf{A}^H) = \mathrm{rank}(\mathbf{A}^\dagger \mathbf{A}) = \mathrm{rank}(\mathbf{AA}^\dagger) = \mathrm{rank}(\mathbf{AA}^\dagger\mathbf{A}) = \mathrm{rank}(\mathbf{A}^\dagger\mathbf{AA}^\dagger)$.
12. The Moore–Penrose inverse of any matrix $\mathbf{A}_{m\times n}$ can be determined by $\mathbf{A}^\dagger = (\mathbf{A}^H\mathbf{A})^\dagger \mathbf{A}^H$ or $\mathbf{A}^\dagger = \mathbf{A}^H(\mathbf{AA}^H)^\dagger$.

1.7 Direct Sum and Hadamard Product

This section discusses two special operations of matrices: the direct sum of two or more matrices and the Hadamard product of two matrices.

1.7.1 Direct Sum of Matrices

Definition 1.20 (Direct Sum [9]) The *direct sum* of an $m \times m$ matrix $\mathbf{A}$ and an $n \times n$ matrix $\mathbf{B}$, denoted $\mathbf{A} \oplus \mathbf{B}$, is an $(m+n) \times (m+n)$ matrix and is defined as follows:

$$\mathbf{A} \oplus \mathbf{B} = \begin{bmatrix} \mathbf{A} & \mathbf{O}_{m \times n} \\ \mathbf{O}_{n \times m} & \mathbf{B} \end{bmatrix}. \tag{1.7.1}$$

From the above definition, it is easily shown that the direct sum of matrices has the following properties [15, 27]:

1. If c is a constant, then $c\,(\mathbf{A} \oplus \mathbf{B}) = c\mathbf{A} \oplus c\mathbf{B}$.
2. The direct sum does not satisfy exchangeability: $\mathbf{A} \oplus \mathbf{B} \neq \mathbf{B} \oplus \mathbf{A}$ unless $\mathbf{A} = \mathbf{B}$.
3. If $\mathbf{A}, \mathbf{B}$ are two $m \times m$ matrices and $\mathbf{C}$ and $\mathbf{D}$ are two $n \times n$ matrices, then

$$(\mathbf{A} \pm \mathbf{B}) \oplus (\mathbf{C} \pm \mathbf{D}) = (\mathbf{A} \oplus \mathbf{C}) \pm (\mathbf{B} \oplus \mathbf{D}),$$

$$(\mathbf{A} \oplus \mathbf{C})(\mathbf{B} \oplus \mathbf{D}) = \mathbf{AB} \oplus \mathbf{CD}.$$

4. If $\mathbf{A}, \mathbf{B}, \mathbf{C}$ are $m \times m$, $n \times n$, $p \times p$ matrices, respectively, then

$$\mathbf{A} \oplus (\mathbf{B} \oplus \mathbf{C}) = (\mathbf{A} \oplus \mathbf{B}) \oplus \mathbf{C} = \mathbf{A} \oplus \mathbf{B} \oplus \mathbf{C}.$$

5. If $\mathbf{A}_{m \times m}$ and $\mathbf{B}_{n \times n}$ are, respectively, orthogonal matrices, then $\mathbf{A} \oplus \mathbf{B}$ is an $(m+n) \times (m+n)$ orthogonal matrix.
6. The complex conjugate, transpose, complex conjugate transpose, and inverse matrices of the direct sum of two matrices are given by

$$(\mathbf{A} \oplus \mathbf{B})^* = \mathbf{A}^* \oplus \mathbf{B}^*, \quad (\mathbf{A} \oplus \mathbf{B})^T = \mathbf{A}^T \oplus \mathbf{B}^T,$$

$$(\mathbf{A} \oplus \mathbf{B})^H = \mathbf{A}^H \oplus \mathbf{B}^H, \quad (\mathbf{A} \oplus \mathbf{B})^{-1} = \mathbf{A}^{-1} \oplus \mathbf{B}^{-1} \text{ (if } \mathbf{A}^{-1} \text{ and } \mathbf{B}^{-1} \text{ exist).}$$

7. The trace, rank, and determinant of the direct sum of N matrices are as follows:

$$\mathrm{tr}\left(\bigoplus_{i=0}^{N-1} \mathbf{A}_i\right) = \sum_{i=0}^{N-1} \mathrm{tr}(\mathbf{A}_i), \quad \mathrm{rank}\left(\bigoplus_{i=0}^{N-1} \mathbf{A}_i\right) = \sum_{i=0}^{N-1} \mathrm{rank}(\mathbf{A}_i), \quad \left|\bigoplus_{i=0}^{N-1} \mathbf{A}_i\right| = \prod_{i=0}^{N-1} |\mathbf{A}_i|.$$

1.7.2 Hadamard Product

Definition 1.21 (Hadamard Product) The *Hadamard product* of two $m \times n$ matrices $\mathbf{A} = [a_{ij}]$ and $\mathbf{B} = [b_{ij}]$ is denoted $\mathbf{A} \odot \mathbf{B}$ and is also an $m \times n$ matrix, each entry of which is defined as the product of the corresponding entries of the two matrices:

$$[\mathbf{A} \odot \mathbf{B}]_{ij} = a_{ij}b_{ij}. \tag{1.7.2}$$

That is, the Hadamard product is a mapping $\mathbb{R}^{m \times n} \times \mathbb{R}^{m \times n} \to \mathbb{R}^{m \times n}$.

The Hadamard product is also known as the *Schur product* or the *elementwise product*. Similarly, the *elementwise division* of two $m \times n$ matrices $\mathbf{A} = [a_{ij}]$ and $\mathbf{B} = [b_{ij}]$, denoted $\mathbf{A} \oslash \mathbf{B}$, is defined as

$$[\mathbf{A} \oslash \mathbf{B}]_{ij} = a_{ij}/b_{ij}. \tag{1.7.3}$$

The following theorem describes the positive definiteness of the Hadamard product and is usually known as the *Hadamard product theorem* [15].

Theorem 1.5 *If two $m \times m$ matrices $\mathbf{A}$ and $\mathbf{B}$ are positive definite (positive semi-definite), then their Hadamard product $\mathbf{A} \odot \mathbf{B}$ is positive definite (positive semi-definite) as well.*

Corollary 1.1 (Fejer Theorem [15]) *An $m \times m$ matrix $\mathbf{A} = [a_{ij}]$ is positive semi-definite if and only if*

$$\sum_{i=1}^{m}\sum_{j=1}^{m} a_{ij}b_{ij} \geq 0$$

holds for all $m \times m$ positive semi-definite matrices $\mathbf{B} = [b_{ij}]$.

The following theorems describe the relationship between the Hadamard product and the matrix trace.

Theorem 1.6 ([22, p. 46]) *Let $\mathbf{A}, \mathbf{B}, \mathbf{C}$ be $m \times n$ matrices, $\mathbf{1} = [1, \ldots, 1]^T$ be an $n \times 1$ summing vector and $\mathbf{D} = \mathbf{Diag}(d_1, \ldots, d_m)$, where $d_i = \sum_{j=1}^{n} a_{ij}$; then*

$$\operatorname{tr}\left(\mathbf{A}^T(\mathbf{B} \odot \mathbf{C})\right) = \operatorname{tr}\left((\mathbf{A}^T \odot \mathbf{B}^T)\mathbf{C}\right), \tag{1.7.4}$$

$$\mathbf{1}^T \mathbf{A}^T (\mathbf{B} \odot \mathbf{C})\mathbf{1} = \operatorname{tr}(\mathbf{B}^T \mathbf{D}\mathbf{C}). \tag{1.7.5}$$

Theorem 1.7 ([22, p. 46]) *Let $\mathbf{A}, \mathbf{B}$ be two $n \times n$ positive definite square matrices, and $\mathbf{1} = [1, \ldots, 1]^T$ be an $n \times 1$ summing vector. Suppose that $\mathbf{M}$ is an $n \times n$ diagonal matrix, i.e., $\mathbf{M} = \mathbf{Diag}(\mu_1, \ldots, \mu_n)$, while $\mathbf{m} = \mathbf{M1}$ is an $n \times 1$ vector. Then one has*

$$\operatorname{tr}(\mathbf{A}\mathbf{M}\mathbf{B}^T\mathbf{M}) = \mathbf{m}^T(\mathbf{A} \odot \mathbf{B})\mathbf{m}, \tag{1.7.6}$$

$$\operatorname{tr}(\mathbf{A}\mathbf{B}^T) = \mathbf{1}^T(\mathbf{A} \odot \mathbf{B})\mathbf{1}, \tag{1.7.7}$$

$$\mathbf{M}\mathbf{A} \odot \mathbf{B}^T\mathbf{M} = \mathbf{M}(\mathbf{A} \odot \mathbf{B}^T)\mathbf{M}. \tag{1.7.8}$$

1.7 Direct Sum and Hadamard Product

From the above definition, it is known that Hadamard products obey the exchange law, the associative law, and the distributive law of the addition:

$$\mathbf{A} \odot \mathbf{B} = \mathbf{B} \odot \mathbf{A}, \tag{1.7.9}$$

$$\mathbf{A} \odot (\mathbf{B} \odot \mathbf{C}) = (\mathbf{A} \odot \mathbf{B}) \odot \mathbf{C}, \tag{1.7.10}$$

$$\mathbf{A} \odot (\mathbf{B} \pm \mathbf{C}) = \mathbf{A} \odot \mathbf{B} \pm \mathbf{A} \odot \mathbf{C}. \tag{1.7.11}$$

The properties of Hadamard products are summarized below [22].

1. If $\mathbf{A}, \mathbf{B}$ are $m \times n$ matrices, then

$$(\mathbf{A} \odot \mathbf{B})^T = \mathbf{A}^T \odot \mathbf{B}^T, \quad (\mathbf{A} \odot \mathbf{B})^H = \mathbf{A}^H \odot \mathbf{B}^H, \quad (\mathbf{A} \odot \mathbf{B})^* = \mathbf{A}^* \odot \mathbf{B}^*.$$

2. The Hadamard product of a matrix $\mathbf{A}_{m \times n}$ and a zero matrix $\mathbf{O}_{m \times n}$ is given by
$\mathbf{A} \odot \mathbf{O}_{m \times n} = \mathbf{O}_{m \times n} \odot \mathbf{A} = \mathbf{O}_{m \times n}$.
3. If c is a constant, then $c (\mathbf{A} \odot \mathbf{B}) = (c\mathbf{A}) \odot \mathbf{B} = \mathbf{A} \odot (c \mathbf{B})$.
4. The Hadamard product of two positive definite (positive semi-definite) matrices $\mathbf{A}, \mathbf{B}$ is positive definite (positive semi-definite) as well.
5. The Hadamard product of the matrix $\mathbf{A}_{m \times m} = [a_{ij}]$ and the identity matrix $\mathbf{I}_m$ is an $m \times m$ diagonal matrix, i.e.,

$$\mathbf{A} \odot \mathbf{I}_m = \mathbf{I}_m \odot \mathbf{A} = \mathbf{Diag}(\mathbf{A}) = \mathbf{Diag}(a_{11}, \ldots, a_{mm}).$$

6. If $\mathbf{A}, \mathbf{B}, \mathbf{D}$ are three $m \times m$ matrices and $\mathbf{D}$ is a diagonal matrix, then

$$(\mathbf{DA}) \odot (\mathbf{BD}) = \mathbf{D}(\mathbf{A} \odot \mathbf{B})\mathbf{D}.$$

7. If $\mathbf{A}, \mathbf{C}$ are two $m \times m$ matrices and $\mathbf{B}, \mathbf{D}$ are two $n \times n$ Matrices, then

$$(\mathbf{A} \oplus \mathbf{B}) \odot (\mathbf{C} \oplus \mathbf{D}) = (\mathbf{A} \odot \mathbf{C}) \oplus (\mathbf{B} \odot \mathbf{D}).$$

8. If $\mathbf{A}, \mathbf{B}, \mathbf{C}, \mathbf{D}$ are all $m \times n$ matrices, then

$$(\mathbf{A} + \mathbf{B}) \odot (\mathbf{C} + \mathbf{D}) = \mathbf{A} \odot \mathbf{C} + \mathbf{A} \odot \mathbf{D} + \mathbf{B} \odot \mathbf{C} + \mathbf{B} \odot \mathbf{D}.$$

9. If $\mathbf{A}, \mathbf{B}, \mathbf{C}$ are $m \times n$ matrices, then $\mathrm{tr}\left(\mathbf{A}^T (\mathbf{B} \odot \mathbf{C})\right) = \mathrm{tr}\left((\mathbf{A}^T \odot \mathbf{B}^T)\mathbf{C}\right)$.

The Hadamard (i.e., elementwise) products of two matrices are widely used in machine learning, neural networks, and evolutionary computations.

1.8 Kronecker Products

The Hadamard product described in the previous section is a special product of two matrices. In this section, we discuss another special product of two matrices: the Kronecker products. The Kronecker product is also known as the *direct product* or *tensor product* [20].

1.8.1 Definitions of Kronecker Products

Kronecker products are divided into right and left Kronecker products.

Definition 1.22 (Right Kronecker Product [3]) Given an $m \times n$ matrix $\mathbf{A} = [\mathbf{a}_1, \ldots, \mathbf{a}_n]$ and another $p \times q$ matrix $\mathbf{B}$, their *right Kronecker product* $\mathbf{A} \otimes \mathbf{B}$ is an $mp \times nq$ matrix defined by

$$\mathbf{A} \otimes \mathbf{B} = [a_{ij}\mathbf{B}]_{i=1, j=1}^{m,n} = \begin{bmatrix} a_{11}\mathbf{B} & a_{12}\mathbf{B} & \cdots & a_{1n}\mathbf{B} \\ a_{21}\mathbf{B} & a_{22}\mathbf{B} & \cdots & a_{2n}\mathbf{B} \\ \vdots & \vdots & \ddots & \vdots \\ a_{m1}\mathbf{B} & a_{m2}\mathbf{B} & \cdots & a_{mn}\mathbf{B} \end{bmatrix}. \quad (1.8.1)$$

Definition 1.23 (Left Kronecker Product [9, 34]) For an $m \times n$ matrix $\mathbf{A}$ and a $p \times q$ matrix $\mathbf{B} = [\mathbf{b}_1, \ldots, \mathbf{b}_q]$, their *left Kronecker product* $\mathbf{A} \otimes \mathbf{B}$ is an $mp \times nq$ matrix defined by

$$[\mathbf{A} \otimes \mathbf{B}]_{\text{left}} = [\mathbf{A}b_{ij}]_{i=1, j=1}^{p,q} = \begin{bmatrix} \mathbf{A}b_{11} & \mathbf{A}b_{12} & \cdots & \mathbf{A}b_{1q} \\ \mathbf{A}b_{21} & \mathbf{A}b_{22} & \cdots & \mathbf{A}b_{2q} \\ \vdots & \vdots & \ddots & \vdots \\ \mathbf{A}b_{p1} & \mathbf{A}b_{p2} & \cdots & \mathbf{A}b_{pq} \end{bmatrix}. \quad (1.8.2)$$

Clearly, the left or right Kronecker product is a mapping $\mathbb{R}^{m \times n} \times \mathbb{R}^{p \times q} \to \mathbb{R}^{mp \times nq}$. It is easily seen that the left and right Kronecker products have the following relationship: $[\mathbf{A} \otimes \mathbf{B}]_{\text{left}} = \mathbf{B} \otimes \mathbf{A}$. Since the right Kronecker product form is the one generally adopted, this book uses the right Kronecker product hereafter unless otherwise stated.

In particular, when $n = 1$ and $q = 1$, the Kronecker product of two matrices reduces to the Kronecker product of two column vectors $\mathbf{a} \in \mathbb{R}^m$ and $\mathbf{b} \in \mathbb{R}^p$:

$$\mathbf{a} \otimes \mathbf{b} = [a_i \mathbf{b}]_{i=1}^m = \begin{bmatrix} a_1 \mathbf{b} \\ \vdots \\ a_m \mathbf{b} \end{bmatrix}. \quad (1.8.3)$$

The result is an $mp \times 1$ vector. Evidently, the outer product of two vectors $\mathbf{x} \circ \mathbf{y} = \mathbf{xy}^T$ can be also represented using the Kronecker product as $\mathbf{x} \circ \mathbf{y} = \mathbf{x} \otimes \mathbf{y}^T$.

1.8.2 Performance of Kronecker Products

Summarizing results from [3, 6] and other literature, Kronecker products have the following properties.

1. The Kronecker product of any matrix and a zero matrix is equal to the zero matrix, i.e., $\mathbf{A} \otimes \mathbf{O} = \mathbf{O} \otimes \mathbf{A} = \mathbf{O}$.
2. If α and β are constants, then $\alpha \mathbf{A} \otimes \beta \mathbf{B} = \alpha\beta(\mathbf{A} \otimes \mathbf{B})$.
3. The Kronecker product of an $m \times m$ identity matrix and an $n \times n$ identity matrix is equal to an $mn \times mn$ identity matrix, i.e., $\mathbf{I}_m \otimes \mathbf{I}_n = \mathbf{I}_{mn}$.
4. For matrices $\mathbf{A}_{m \times n}, \mathbf{B}_{n \times k}, \mathbf{C}_{l \times p}, \mathbf{D}_{p \times q}$, we have

$$(\mathbf{AB}) \otimes (\mathbf{CD}) = (\mathbf{A} \otimes \mathbf{C})(\mathbf{B} \otimes \mathbf{D}). \tag{1.8.4}$$

5. For matrices $\mathbf{A}_{m \times n}, \mathbf{B}_{p \times q}, \mathbf{C}_{p \times q}$, we have

$$\mathbf{A} \otimes (\mathbf{B} \pm \mathbf{C}) = \mathbf{A} \otimes \mathbf{B} \pm \mathbf{A} \otimes \mathbf{C}, \tag{1.8.5}$$

$$(\mathbf{B} \pm \mathbf{C}) \otimes \mathbf{A} = \mathbf{B} \otimes \mathbf{A} \pm \mathbf{C} \otimes \mathbf{A}. \tag{1.8.6}$$

6. The inverse and generalized inverse matrix of Kronecker products satisfy

$$(\mathbf{A} \otimes \mathbf{B})^{-1} = \mathbf{A}^{-1} \otimes \mathbf{B}^{-1}, \quad (\mathbf{A} \otimes \mathbf{B})^\dagger = \mathbf{A}^\dagger \otimes \mathbf{B}^\dagger. \tag{1.8.7}$$

7. The transpose and the complex conjugate transpose of Kronecker products are given by

$$(\mathbf{A} \otimes \mathbf{B})^T = \mathbf{A}^T \otimes \mathbf{B}^T, \quad (\mathbf{A} \otimes \mathbf{B})^H = \mathbf{A}^H \otimes \mathbf{B}^H. \tag{1.8.8}$$

8. The rank of the Kronecker product is

$$\text{rank}(\mathbf{A} \otimes \mathbf{B}) = \text{rank}(\mathbf{A})\text{rank}(\mathbf{B}). \tag{1.8.9}$$

9. The determinant of the Kronecker product

$$\det(\mathbf{A}_{n \times n} \otimes \mathbf{B}_{m \times m}) = (\det \mathbf{A})^m (\det \mathbf{B})^n. \tag{1.8.10}$$

10. The trace of the Kronecker product is given by

$$\text{tr}(\mathbf{A} \otimes \mathbf{B}) = \text{tr}(\mathbf{A})\text{tr}(\mathbf{B}). \tag{1.8.11}$$

11. For matrices $\mathbf{A}_{m\times n}, \mathbf{B}_{m\times n}, \mathbf{C}_{p\times q}, \mathbf{D}_{p\times q}$, we have

$$(\mathbf{A}+\mathbf{B})\otimes(\mathbf{C}+\mathbf{D}) = \mathbf{A}\otimes\mathbf{C} + \mathbf{A}\otimes\mathbf{D} + \mathbf{B}\otimes\mathbf{C} + \mathbf{B}\otimes\mathbf{D}. \tag{1.8.12}$$

12. For matrices $\mathbf{A}_{m\times n}, \mathbf{B}_{p\times q}, \mathbf{C}_{k\times l}$, it is true that

$$(\mathbf{A}\otimes\mathbf{B})\otimes\mathbf{C} = \mathbf{A}\otimes(\mathbf{B}\otimes\mathbf{C}). \tag{1.8.13}$$

13. For matrices $\mathbf{A}_{m\times n}, \mathbf{B}_{p\times q}$, we have $\exp(\mathbf{A}\otimes\mathbf{B}) = \exp(\mathbf{A})\otimes\exp(\mathbf{B})$.
14. Let $\mathbf{A}\in\mathbb{C}^{m\times n}$ and $\mathbf{B}\in\mathbb{C}^{p\times q}$ then [22, p. 47]

$$\mathbf{K}_{pm}(\mathbf{A}\otimes\mathbf{B}) = (\mathbf{B}\otimes\mathbf{A})\mathbf{K}_{qn}, \tag{1.8.14}$$

$$\mathbf{K}_{pm}(\mathbf{A}\otimes\mathbf{B})\mathbf{K}_{nq} = \mathbf{B}\otimes\mathbf{A}, \tag{1.8.15}$$

$$\mathbf{K}_{pm}(\mathbf{A}\otimes\mathbf{B}) = \mathbf{B}\otimes\mathbf{A}, \tag{1.8.16}$$

$$\mathbf{K}_{mp}(\mathbf{B}\otimes\mathbf{A}) = \mathbf{A}\otimes\mathbf{B}, \tag{1.8.17}$$

where $\mathbf{K}$ is a commutation matrix (see Sect. 1.9.1).

1.9 Vectorization and Matricization

Consider the operators that transform a matrix into a vector or vice versa.

1.9.1 Vectorization and Commutation Matrix

The function or operator that transforms a matrix into a vector is known as the *vectorization* of the matrix.

The *column vectorization* of a matrix $\mathbf{A}\in\mathbb{R}^{m\times n}$, denoted $\mathrm{vec}(\mathbf{A})$, is a linear transformation that arranges the entries of $\mathbf{A} = [a_{ij}]$ as an $mn\times 1$ vector via column stacking:

$$\mathrm{vec}(\mathbf{A}) = [a_{11},\ldots,a_{m1},\ldots,a_{1n},\ldots,a_{mn}]^T. \tag{1.9.1}$$

A matrix $\mathbf{A}$ can be also arranged as a row vector by row stacking:

$$\mathrm{rvec}(\mathbf{A}) = [a_{11},\ldots,a_{1n},\ldots,a_{m1},\ldots,a_{mn}]. \tag{1.9.2}$$

This is known as the *row vectorization* of the matrix.

1.9 Vectorization and Matricization

For instance, given a matrix $\mathbf{A} = \begin{bmatrix} a_{11} & a_{12} \\ a_{21} & a_{22} \end{bmatrix}$, then $\text{vec}(\mathbf{A}) = [a_{11}, a_{21}, a_{12}, a_{22}]^T$ and $\text{rvec}(\mathbf{A}) = [a_{11}, a_{12}, a_{21}, a_{22}]$.

The column vectorization is usually called the vectorization simply.

Clearly, there exist the following relationships between the vectorization and the row vectorization of a matrix:

$$\text{rvec}(\mathbf{A}) = (\text{vec}(\mathbf{A}^T))^T, \quad \text{vec}(\mathbf{A}^T) = (\text{rvec } \mathbf{A})^T. \tag{1.9.3}$$

One obvious fact is that, for a given $m \times n$ matrix $\mathbf{A}$, the two vectors $\text{vec}(\mathbf{A})$ and $\text{vec}(\mathbf{A}^T)$ contain the same entries but the orders of their entries are different. Interestingly, there is a unique $mn \times mn$ permutation matrix that can transform $\text{vec}(\mathbf{A})$ into $\text{vec}(\mathbf{A}^T)$. This permutation matrix is called the *commutation matrix*, denoted as $\mathbf{K}_{mn}$, and is defined by

$$\mathbf{K}_{mn} \text{vec}(\mathbf{A}) = \text{vec}(\mathbf{A}^T). \tag{1.9.4}$$

Similarly, there is an $nm \times nm$ permutation matrix transforming $\text{vec}(\mathbf{A}^T)$ into $\text{vec}(\mathbf{A})$. Such a commutation matrix, denoted $\mathbf{K}_{nm}$, is defined by

$$\mathbf{K}_{nm} \text{vec}(\mathbf{A}^T) = \text{vec}(\mathbf{A}). \tag{1.9.5}$$

From (1.9.4) and (1.9.5) it can be seen that $\mathbf{K}_{nm} \mathbf{K}_{mn} \text{vec}(\mathbf{A}) = \mathbf{K}_{nm} \text{vec}(\mathbf{A}^T) = \text{vec}(\mathbf{A})$. Since this formula holds for any $m \times n$ matrix $\mathbf{A}$, we have $\mathbf{K}_{nm} \mathbf{K}_{mn} = \mathbf{I}_{mn}$ or $\mathbf{K}_{mn}^{-1} = \mathbf{K}_{nm}$.

The $mn \times mn$ commutation matrix $\mathbf{K}_{mn}$ has the following properties [21].

1. $\mathbf{K}_{mn} \text{vec}(\mathbf{A}) = \text{vec}(\mathbf{A}^T)$ and $\mathbf{K}_{nm} \text{vec}(\mathbf{A}^T) = \text{vec}(\mathbf{A})$, where $\mathbf{A}$ is an $m \times n$ matrix.
2. $\mathbf{K}_{mn}^T \mathbf{K}_{mn} = \mathbf{K}_{mn} \mathbf{K}_{mn}^T = \mathbf{I}_{mn}$, or $\mathbf{K}_{mn}^{-1} = \mathbf{K}_{nm}$.
3. $\mathbf{K}_{mn}^T = \mathbf{K}_{nm}$.
4. $\mathbf{K}_{mn}$ can be represented as a Kronecker product of the basic vectors:

$$\mathbf{K}_{mn} = \sum_{j=1}^{n} (\mathbf{e}_j^T \otimes \mathbf{I}_m \otimes \mathbf{e}_j).$$

5. $\mathbf{K}_{1n} = \mathbf{K}_{n1} = \mathbf{I}_n$.
6. $\mathbf{K}_{nm} \mathbf{K}_{mn} \text{vec}(\mathbf{A}) = \mathbf{K}_{nm} \text{vec}(\mathbf{A}^T) = \text{vec}(\mathbf{A})$.
7. Eigenvalues of the commutation matrix $\mathbf{K}_{nn}$ are 1 and -1 and their multiplicities are, respectively, $\frac{1}{2}n(n+1)$ and $\frac{1}{2}n(n-1)$.
8. The rank of the commutation matrix is given by $\text{rank}(\mathbf{K}_{mn}) = 1 + d(m-1, n-1)$, where $d(m, n)$ is the greatest common divisor of m and n, $d(n, 0) = d(0, n) = n$.

9. $\mathbf{K}_{mn}(\mathbf{A} \otimes \mathbf{B})\mathbf{K}_{pq} = \mathbf{B} \otimes \mathbf{A}$, and thus can be equivalently written as $\mathbf{K}_{mn}(\mathbf{A} \otimes \mathbf{B}) = (\mathbf{B} \otimes \mathbf{A})\mathbf{K}_{qp}$, where $\mathbf{A}$ is an $n \times p$ matrix and $\mathbf{B}$ is an $m \times q$ matrix. In particular, $\mathbf{K}_{mn}(\mathbf{A}_{n \times n} \otimes \mathbf{B}_{m \times m}) = (\mathbf{B} \otimes \mathbf{A})\mathbf{K}_{mn}$.
10. $\mathrm{tr}(\mathbf{K}_{mn}(\mathbf{A}_{m \times n} \otimes \mathbf{B}_{m \times n})) = \mathrm{tr}(\mathbf{A}^T \mathbf{B}) = \left(\mathrm{vec}(\mathbf{B}^T)\right)^T \mathbf{K}_{mn} \mathrm{vec}(\mathbf{A})$.

An $mn \times mn$ commutation matrix can be constructed as follows. First let

$$\mathbf{K}_{mn} = \begin{bmatrix} \mathbf{K}_1 \\ \vdots \\ \mathbf{K}_m \end{bmatrix}, \quad \mathbf{K}_i \in \mathbb{R}^{n \times mn}, i = 1, \ldots, m; \tag{1.9.6}$$

then the (i, j)th entry of the first submatrix $\mathbf{K}_1$ is given by

$$K_1(i, j) = \begin{cases} 1, & j = (i-1)m + 1, \ i = 1, \ldots, n, \\ 0, & \text{otherwise.} \end{cases} \tag{1.9.7}$$

Next, the ith submatrix $\mathbf{K}_i$ ($i = 2, \ldots, m$) is constructed from the $(i-1)$th submatrix $\mathbf{K}_{i-1}$ as follows:

$$\mathbf{K}_i = [\mathbf{0}, \mathbf{K}_{i-1}(1 : mn - 1)], \quad i = 2, \ldots, m, \tag{1.9.8}$$

where $\mathbf{K}_{i-1}(1 : mn-1)$ denotes a submatrix consisting of the first $(mn-1)$ columns of the $n \times m$ submatrix $\mathbf{K}_{i-1}$.

1.9.2 Matricization of Vectors

Matricization refers to a mapping that transform a vector into a matrix. The operation for transforming an mn-dimensional column vector $\mathbf{a} = [a_1, \ldots, a_{mn}]^T$ into an $m \times n$ matrix $\mathbf{A}$ is known as the *matricization of column vector* or *unfolding of column vector*, denoted $\mathrm{unvec}_{m,n}(\mathbf{a})$, and is defined as

$$\mathbf{A}_{m \times n} = \mathrm{unvec}_{m,n}(\mathbf{a}) = \begin{bmatrix} a_1 & a_{m+1} & \cdots & a_{m(n-1)+1} \\ a_2 & a_{m+2} & \cdots & a_{m(n-1)+2} \\ \vdots & \vdots & \ddots & \vdots \\ a_m & a_{2m} & \cdots & a_{mn} \end{bmatrix}, \tag{1.9.9}$$

where

$$A_{ij} = a_{i+(j-1)m}, \quad i = 1, \ldots, m, \ j = 1, \ldots, n. \tag{1.9.10}$$

Similarly, the operation for transforming an $1 \times mn$ row vector $\mathbf{b} = [b_1, \ldots, b_{mn}]$ into an $m \times n$ matrix $\mathbf{B}$ is called the *matricization of row vector* or *unfolding of row*

1.9 Vectorization and Matricization

vector, denoted unrvec$_{m,n}$(**b**), and is defined as

$$\mathbf{B}_{m \times n} = \text{unrvec}_{m,n}(\mathbf{b}) = \begin{bmatrix} b_1 & b_2 & \cdots & b_n \\ b_{n+1} & b_{n+2} & \cdots & b_{2n} \\ \vdots & \vdots & \ddots & \vdots \\ b_{(m-1)n+1} & b_{(m-1)n+2} & \cdots & b_{mn} \end{bmatrix}. \quad (1.9.11)$$

This is equivalently represented in element form as

$$B_{ij} = b_{j+(i-1)n}, \quad i = 1, \ldots, m, \ j = 1, \ldots, n. \quad (1.9.12)$$

It can be seen from the above definitions that there are the following relationships between matricization (unvec) and column vectorization (vec) or row vectorization (rvec):

$$\begin{bmatrix} A_{11} & \cdots & A_{1n} \\ \vdots & \ddots & \vdots \\ A_{m1} & \cdots & A_{mn} \end{bmatrix} \underset{\text{unvec}}{\overset{\text{vec}}{\rightleftarrows}} [A_{11}, \ldots, A_{m1}, \ldots, A_{1n}, \ldots, A_{mn}]^T,$$

$$\begin{bmatrix} A_{11} & \cdots & A_{1n} \\ \vdots & \ddots & \vdots \\ A_{m1} & \cdots & A_{mn} \end{bmatrix} \underset{\text{unvec}}{\overset{\text{rvec}}{\rightleftarrows}} [A_{11}, \ldots, A_{1n}, \ldots, A_{m1}, \ldots, A_{mn}],$$

which can be written as

$$\text{unvec}_{m,n}(\mathbf{a}) = \mathbf{A}_{m \times n} \quad \Leftrightarrow \quad \text{vec}(\mathbf{A}_{m \times n}) = \mathbf{a}_{mn \times 1}, \quad (1.9.13)$$

$$\text{unrvec}_{m,n}(\mathbf{b}) = \mathbf{B}_{m \times n} \quad \Leftrightarrow \quad \text{rvec}(\mathbf{B}_{m \times n}) = \mathbf{b}_{1 \times mn}. \quad (1.9.14)$$

The vectorization operator has the following properties [7, 13, 22].

1. The vectorization of a transposed matrix is given by $\text{vec}(\mathbf{A}^T) = \mathbf{K}_{mn} \text{vec}(\mathbf{A})$ for $\mathbf{A} \in \mathbb{C}^{m \times n}$.
2. The vectorization of a matrix sum is given by $\text{vec}(\mathbf{A} + \mathbf{B}) = \text{vec}(\mathbf{A}) + \text{vec}(\mathbf{B})$.
3. The vectorization of a Kronecker product is given by Magnus and Neudecker [22, p. 184]:

$$\text{vec}(\mathbf{X} \otimes \mathbf{Y}) = (\mathbf{I}_m \otimes \mathbf{K}_{qp} \otimes \mathbf{I}_n)(\text{vec}(\mathbf{X}) \otimes \text{vec}(\mathbf{Y})). \quad (1.9.15)$$

4. The trace of a matrix product is given by

$$\text{tr}(\mathbf{A}^T \mathbf{B}) = (\text{vec}(\mathbf{A}))^T \text{vec}(\mathbf{B}), \quad (1.9.16)$$

$$\text{tr}(\mathbf{A}^H \mathbf{B}) = (\text{vec}(\mathbf{A}))^H \text{vec}(\mathbf{B}), \quad (1.9.17)$$

$$\text{tr}(\mathbf{ABC}) = (\text{vec}(\mathbf{A}))^T (\mathbf{I}_p \otimes \mathbf{B}) \text{vec}(\mathbf{C}), \quad (1.9.18)$$

while the trace of the product of four matrices is determined by Magnus and Neudecker [22, p. 31]:

$$\text{tr}(\mathbf{ABCD}) = (\text{vec}(\mathbf{D}^T))^T (\mathbf{C}^T \otimes \mathbf{A}) \text{vec}(\mathbf{B}) = (\text{vec}(\mathbf{D}))^T (\mathbf{A} \otimes \mathbf{C}^T) \text{vec}(\mathbf{B}^T).$$

5. The Kronecker product of two vectors $\mathbf{a}$ and $\mathbf{b}$ can be represented as the vectorization of their outer product $\mathbf{ba}^T$ as follows:

$$\mathbf{a} \otimes \mathbf{b} = \text{vec}(\mathbf{ba}^T) = \text{vec}(\mathbf{b} \circ \mathbf{a}). \tag{1.9.19}$$

6. The vectorization of the Hadamard product is given by

$$\text{vec}(\mathbf{A} \odot \mathbf{B}) = \text{vec}(\mathbf{A}) \odot \text{vec}(\mathbf{B}) = \mathbf{Diag}(\text{vec}(\mathbf{A})) \text{vec}(\mathbf{B}), \tag{1.9.20}$$

where $\mathbf{Diag}(\text{vec}(\mathbf{A}))$ is a diagonal matrix whose entries are the vectorization function $\text{vec}(\mathbf{A})$.

7. The relation of the vectorization of the matrix product $\mathbf{A}_{m \times p} \mathbf{B}_{p \times q} \mathbf{C}_{q \times n}$ to the Kronecker product is given by Schott [35, p. 263]:

$$\text{vec}(\mathbf{ABC}) = (\mathbf{C}^T \otimes \mathbf{A}) \text{vec}(\mathbf{B}), \tag{1.9.21}$$

$$\text{vec}(\mathbf{ABC}) = (\mathbf{I}_q \otimes \mathbf{AB}) \text{vec}(\mathbf{C}) = (\mathbf{C}^T \mathbf{B}^T \otimes \mathbf{I}_m) \text{vec}(\mathbf{A}), \tag{1.9.22}$$

$$\text{vec}(\mathbf{AC}) = (\mathbf{I}_p \otimes \mathbf{A}) \text{vec}(\mathbf{C}) = (\mathbf{C}^T \otimes \mathbf{I}_m) \text{vec}(\mathbf{A}). \tag{1.9.23}$$

Example 1.2 Consider the solution of the matrix equation $\mathbf{AXB} = \mathbf{C}$, where $\mathbf{A} \in \mathbb{R}^{m \times n}$, $\mathbf{X} \in \mathbb{R}^{n \times p}$, $\mathbf{B} \in \mathbb{R}^{p \times q}$, and $\mathbf{C} \in \mathbb{R}^{m \times q}$. By using the vectorization function property $\text{vec}(\mathbf{AXB}) = (\mathbf{B}^T \otimes \mathbf{A}) \text{vec}(\mathbf{X})$, the vectorization $\text{vec}(\mathbf{AXB}) = \text{vec}(\mathbf{C})$ of the original matrix equation can be, in Kronecker product form, rewritten as [33]:

$$(\mathbf{B}^T \otimes \mathbf{A}) \text{vec}(\mathbf{X}) = \text{vec}(\mathbf{C}),$$

and thus $\text{vec}(\mathbf{X}) = (\mathbf{B}^T \otimes \mathbf{A})^\dagger \text{vec}(\mathbf{C})$. Then by matricizing $\text{vec}(\mathbf{X})$, we get the solution matrix $\mathbf{X}$ of the original matrix equation $\mathbf{AXB} = \mathbf{C}$.

Brief Summary of This Chapter

- This chapter focuses on basic operations and performance of matrices, which constitute the foundation of matrix algebra.
- Trace, rank, positive definiteness, Moore–Penrose inverse matrices, Kronecker product, Hadamard product (elementwise product), and vectorization of matrices are frequently used in artificial intelligence.

References

1. Banachiewicz, T.: Zur Berechungung der Determinanten, wie auch der Inverse, und zur darauf basierten Auflösung der Systeme linearer Gleichungen. Acta Astron. Sér. C **3**, 41–67 (1937)
2. Barnett, S.: Matrices: Methods and Applications. Clarendon Press, Oxford (1990)
3. Bellman, R.: Introduction to Matrix Analysis, 2nd edn. McGraw-Hill, New York (1970)
4. Berberian, S.K.: Linear Algebra. Oxford University Press, New York (1992)
5. Boot, J.: Computation of the generalized inverse of singular or rectangular matrices. Am. Math Mon. **70**, 302–303 (1963)
6. Brewer, J.W.: Kronecker products and matrix calculus in system theory. IEEE Trans. Circuits Syst. **25**, 772–781 (1978)
7. Brookes, M.: Matrix Reference Manual (2011). https://www.ee.ic.ac.uk/hp/staff/dmb/matrix/intro.html
8. Duncan, W.J.: Some devices for the solution of large sets of simultaneous linear equations. Lond. Edinb. Dublin Philos. Mag. J. Sci. Ser. 7 **35**, 660–670 (1944)
9. Graybill, F.A.: Matrices with Applications in Statistics. Wadsworth International Group, Balmont (1983)
10. Graybill, F.A., Meyer, C.D., Painter, R.J.: Note on the computation of the generalized inverse of a matrix. SIAM Rev. **8**(4), 522–524 (1966)
11. Greville, T.N.E.: Some applications of the pseudoinverse of a matrix. SIAM Rev. **2**, 15–22 (1960)
12. Guttman, L.: Enlargement methods for computing the inverse matrix. Ann. Math. Stat. **17**, 336–343 (1946)
13. Henderson, H.V., Searle, S.R.: The vec-permutation matrix, the vec operator and Kronecker products: a review. Linear Multilinear Alg. **9**, 271–288 (1981)
14. Hendeson, H.V., Searle, S.R.: On deriving the inverse of a sum of matrices. SIAM Rev. **23**, 53–60 (1981)
15. Horn, R.A., Johnson, C.R.: Matrix Analysis. Cambridge University Press, Cambridge (1985)
16. Hotelling, H.: Some new methods in matrix calculation. Ann. Math. Stat. **14**, 1–34 (1943)
17. Hotelling, H.: Further points on matrix calculation and simultaneous equations. Ann. Math. Stat. **14**, 440–441 (1943)
18. Johnson, L.W., Riess, R.D., Arnold, J.T.: Introduction to Linear Algebra, 5th edn. Prentice-Hall, New York (2000)
19. Lancaster, P., Tismenetsky, M.: The Theory of Matrices with Applications, 2nd edn. Academic, New York (1985)
20. Lütkepohl, H.: Handbook of Matrices. Wiley, New York (1996)
21. Magnus, J.R., Neudecker, H.: The commutation matrix: some properties and applications. Ann. Stat. **7**, 381–394 (1979)
22. Magnus, J.R., Neudecker, H.: Matrix Differential Calculus with Applications in Statistics and Econometrics, rev. edn. Wiley, Chichester (1999)
23. Moore, E.H.: General analysis, part 1. Mem. Am. Philos. Soc. **1**, 1 (1935)
24. Noble, B., Danniel, J.W.: Applied Linear Algebra, 3rd edn. Prentice-Hall, Englewood Cliffs (1988)
25. Papoulis, A.: Probability, Random Variables and Stochastic Processes. McGraw-Hill, New York (1991)
26. Penrose, R.A.: A generalized inverse for matrices. Proc. Cambridge Philos. Soc. **51**, 406–413 (1955)
27. Phillip, A., Regalia, P.A., Mitra, S.: Kronecker products, unitary matrices and signal processing applications. SIAM Rev. **31**(4), 586–613 (1989)
28. Piegorsch, W.W., Casella, G.: The early use of matrix diagonal increments in statistical problems. SIAM Rev. **31**, 428–434 (1989)
29. Poularikas, A.D.: The Handbook of Formulas and Tables for Signal Processing. CRC Press, New York (1999)

30. Price, C.: The matrix pseudoinverse and minimal variance estimates. SIAM Rev. **6**, 115–120 (1964)
31. Pringle, R.M., Rayner, A.A.: Generalized Inverse of Matrices with Applications to Statistics. Griffin, London (1971)
32. Rado, R.: Note on generalized inverse of matrices. Proc. Cambridge Philos. Soc. **52**, 600–601 (1956)
33. Rao, C.R., Mitra, S.K.: Generalized Inverse of Matrices and its Applications. Wiley, New York (1971)
34. Regalia, P.A., Mitra, S.K.: Kronecker products, unitary matrices and signal processing applications. SIAM Rev. **31**(4), 586–613 (1989)
35. Schott, J.R.: Matrix Analysis for Statistics. Wiley, New York (1997)
36. Searle, S.R.: Matrix Algebra Useful for Statistics. Wiley, New York (1982)
37. Sherman, J., Morrison, W.J.: Adjustment of an inverse matrix corresponding to a change in one element of a given matrix. Ann. Math. Stat. **21**, 124–127 (1950)
38. Woodbury, M.A.: Inverting modified matrices. Memorandum Report 42, Statistical Research Group, Princeton (1950)
39. Zhang, X.D.: Numerical computations of left and right pseudo inverse matrices (in Chinese). Kexue Tongbao **7**(2), 126 (1982)
40. Zhang, X.D.: Matrix Analysis and Applications. Cambridge University Press, Cambridge (2017)

Chapter 2
Matrix Differential

The matrix differential is a generalization of the multivariate function differential. The matrix differential (including the matrix partial derivative and gradient) is an important operation tool in matrix algebra and optimization in machine learning, neural networks, support vector machine and evolutionary computation. This chapter is concerned with the theory and methods of matrix differential.

2.1 Jacobian Matrix and Gradient Matrix

In this section we discuss the partial derivatives of real functions. Table 2.1 summarizes the symbols of the real functions.

2.1.1 Jacobian Matrix

First, we introduce definitions on partial derivations and Jacobian matrix.

1. *Row partial derivative operator* with respect to an $m \times 1$ vector is defined as

$$\nabla_{\mathbf{x}^T} \stackrel{\text{def}}{=} \frac{\partial}{\partial \mathbf{x}^T} = \left[\frac{\partial}{\partial x_1}, \ldots, \frac{\partial}{\partial x_m} \right], \tag{2.1.1}$$

and *row partial derivative vector* of real scalar function $f(\mathbf{x})$ with respect to its $m \times 1$ vector variable $\mathbf{x}$ is given by

$$\nabla_{\mathbf{x}^T} f(\mathbf{x}) = \frac{\partial f(\mathbf{x})}{\partial \mathbf{x}^T} = \left[\frac{\partial f(\mathbf{x})}{\partial x_1}, \ldots, \frac{\partial f(\mathbf{x})}{\partial x_m} \right]. \tag{2.1.2}$$

Table 2.1 Symbols of real functions

Function type	Variable $\mathbf{x} \in \mathbb{R}^m$	Variable $\mathbf{X} \in \mathbb{R}^{m \times n}$
Scalar function $f \in \mathbb{R}$	$f(\mathbf{x}) : \mathbb{R}^m \to \mathbb{R}$	$f(\mathbf{X}) : \mathbb{R}^{m \times n} \to \mathbb{R}$
Vector function $\mathbf{f} \in \mathbb{R}^p$	$\mathbf{f}(\mathbf{x}) : \mathbb{R}^m \to \mathbb{R}^p$	$\mathbf{f}(\mathbf{X}) : \mathbb{R}^{m \times n} \to \mathbb{R}^p$
Matrix function $\mathbf{F} \in \mathbb{R}^{p \times q}$	$\mathbf{F}(\mathbf{x}) : \mathbb{R}^m \to \mathbb{R}^{p \times q}$	$\mathbf{F}(\mathbf{X}) : \mathbb{R}^{m \times n} \to \mathbb{R}^{p \times q}$

2. *Row partial derivative operator* with respect to an $m \times n$ matrix $\mathbf{X}$ is defined as

$$\nabla_{(\text{vec}\,\mathbf{X})^T} \stackrel{\text{def}}{=} \frac{\partial}{\partial(\text{vec}\,\mathbf{X})^T} = \left[\frac{\partial}{\partial x_{11}}, \ldots, \frac{\partial}{\partial x_{m1}}, \ldots, \frac{\partial}{\partial x_{1n}}, \ldots, \frac{\partial}{\partial x_{mn}}\right], \tag{2.1.3}$$

and row partial derivative vector of real scalar function $f(\mathbf{X})$ with respect to its matrix variable $\mathbf{X} \in \mathbb{R}^{m \times n}$ is given by

$$\nabla_{(\text{vec}\,\mathbf{X})^T} f(\mathbf{X}) = \frac{\partial f(\mathbf{X})}{\partial(\text{vec}\,\mathbf{X})^T} = \left[\frac{\partial f(\mathbf{X})}{\partial x_{11}}, \ldots, \frac{\partial f(\mathbf{X})}{\partial x_{m1}}, \ldots, \frac{\partial f(\mathbf{X})}{\partial x_{1n}}, \ldots, \frac{\partial f(\mathbf{X})}{\partial x_{mn}}\right]. \tag{2.1.4}$$

3. *Jacobian operator* with respect to an $m \times n$ matrix $\mathbf{X}$ is defined as

$$\nabla_{\mathbf{X}^T} \stackrel{\text{def}}{=} \frac{\partial}{\partial \mathbf{X}^T} = \begin{bmatrix} \frac{\partial}{\partial x_{11}} & \cdots & \frac{\partial}{\partial x_{m1}} \\ \vdots & \ddots & \vdots \\ \frac{\partial}{\partial x_{1n}} & \cdots & \frac{\partial}{\partial x_{mn}} \end{bmatrix}, \tag{2.1.5}$$

and *Jacobian matrix* of the real scalar function $f(\mathbf{X})$ with respect to its matrix variable $\mathbf{X} \in \mathbb{R}^{m \times n}$ is given by

$$\mathbf{J} = \nabla_{\mathbf{X}^T} f(\mathbf{X}) = \frac{\partial f(\mathbf{X})}{\partial \mathbf{X}^T} = \begin{bmatrix} \frac{\partial f(\mathbf{X})}{\partial x_{11}} & \cdots & \frac{\partial f(\mathbf{X})}{\partial x_{m1}} \\ \vdots & \ddots & \vdots \\ \frac{\partial f(\mathbf{X})}{\partial x_{1n}} & \cdots & \frac{\partial f(\mathbf{X})}{\partial x_{mn}} \end{bmatrix} \in \mathbb{R}^{n \times m}. \tag{2.1.6}$$

There is the following relationship between the Jacobian matrix and the row partial derivative vector:

$$\nabla_{(\text{vec}\,\mathbf{X})^T} f(\mathbf{X}) = \text{rvec}(\mathbf{J}) = \left(\text{vec}(\mathbf{J}^T)\right)^T. \tag{2.1.7}$$

This important relation is the basis of the Jacobian matrix identification.

2.1 Jacobian Matrix and Gradient Matrix

As a matter of fact, the Jacobian matrix is more useful than the row partial derivative vector.

The following theorem provides a specific expression for the Jacobian matrix of a $p \times q$ real-valued matrix function $\mathbf{F}(\mathbf{X})$ with $m \times n$ matrix variable $\mathbf{X}$.

Definition 2.1 (Jacobian Matrix [6]) Let the vectorization of a $p \times q$ matrix function $\mathbf{F}(\mathbf{X})$ be given by

$$\text{vec } \mathbf{F}(\mathbf{X}) \stackrel{\text{def}}{=} [f_{11}(\mathbf{X}), \ldots, f_{p1}(\mathbf{X}), \ldots, f_{1q}(\mathbf{X}), \ldots, f_{pq}(\mathbf{X})]^T \in \mathbb{R}^{pq}. \quad (2.1.8)$$

Then the $pq \times mn$ Jacobian matrix of $\mathbf{F}(\mathbf{X})$ is defined as

$$\mathbf{J} = \nabla_{(\text{vec}\mathbf{X})^T} \mathbf{F}(\mathbf{X}) \stackrel{\text{def}}{=} \frac{\partial \text{ vec}\mathbf{F}(\mathbf{X})}{\partial (\text{vec}\mathbf{X})^T} \in \mathbb{R}^{pq \times mn} \quad (2.1.9)$$

whose specific expression $\mathbf{J}$ is given by

$$\mathbf{J} = \begin{bmatrix} \frac{\partial f_{11}}{\partial (\text{vec }\mathbf{X})^T} \\ \vdots \\ \frac{\partial f_{p1}}{\partial (\text{vec }\mathbf{X})^T} \\ \vdots \\ \frac{\partial f_{1q}}{\partial (\text{vec }\mathbf{X})^T} \\ \vdots \\ \frac{\partial f_{pq}}{\partial (\text{vec }\mathbf{X})^T} \end{bmatrix} = \begin{bmatrix} \frac{\partial f_{11}}{\partial x_{11}} & \cdots & \frac{\partial f_{11}}{\partial x_{m1}} & \cdots & \frac{\partial f_{11}}{\partial x_{1n}} & \cdots & \frac{\partial f_{11}}{\partial x_{mn}} \\ \vdots & \vdots & \vdots & \vdots & \vdots & \vdots & \vdots \\ \frac{\partial f_{p1}}{\partial x_{11}} & \cdots & \frac{\partial f_{p1}}{\partial x_{m1}} & \cdots & \frac{\partial f_{p1}}{\partial x_{1n}} & \cdots & \frac{\partial f_{p1}}{\partial x_{mn}} \\ \vdots & \vdots & \vdots & \vdots & \vdots & \vdots & \vdots \\ \frac{\partial f_{1q}}{\partial x_{11}} & \cdots & \frac{\partial f_{1q}}{\partial x_{m1}} & \cdots & \frac{\partial f_{1q}}{\partial x_{1n}} & \cdots & \frac{\partial f_{1q}}{\partial x_{mn}} \\ \vdots & \vdots & \vdots & \vdots & \vdots & \vdots & \vdots \\ \frac{\partial f_{pq}}{\partial x_{11}} & \cdots & \frac{\partial f_{pq}}{\partial x_{m1}} & \cdots & \frac{\partial f_{pq}}{\partial x_{1n}} & \cdots & \frac{\partial f_{pq}}{\partial x_{mn}} \end{bmatrix}. \quad (2.1.10)$$

2.1.2 Gradient Matrix

The partial derivative operator in column form is referred to as the gradient vector operator.

Definition 2.2 (Gradient Vector Operators) The *gradient vector operators* with respect to an $m \times 1$ vector $\mathbf{x}$ and to an $m \times n$ matrix $\mathbf{X}$ are, respectively, defined as

$$\nabla_{\mathbf{x}} \stackrel{\text{def}}{=} \frac{\partial}{\partial \mathbf{x}} = \left[\frac{\partial}{\partial x_1}, \ldots, \frac{\partial}{\partial x_m} \right]^T \quad (2.1.11)$$

and

$$\nabla_{\text{vec }\mathbf{X}} \stackrel{\text{def}}{=} \frac{\partial}{\partial \text{ vec }\mathbf{X}} = \left[\frac{\partial}{\partial x_{11}}, \ldots, \frac{\partial}{\partial x_{1n}}, \ldots, \frac{\partial}{\partial x_{m1}}, \ldots, \frac{\partial}{\partial x_{mn}}\right]^T. \quad (2.1.12)$$

Definition 2.3 (Gradient Matrix Operator) The *gradient matrix operator* with respect to an $m \times n$ matrix $\mathbf{X}$, denoted as $\nabla_{\mathbf{X}} = \frac{\partial}{\partial \mathbf{X}}$, is defined as

$$\nabla_{\mathbf{X}} \stackrel{\text{def}}{=} \frac{\partial}{\partial \mathbf{X}} = \begin{bmatrix} \frac{\partial}{\partial x_{11}} & \cdots & \frac{\partial}{\partial x_{1n}} \\ \vdots & \ddots & \vdots \\ \frac{\partial}{\partial x_{m1}} & \cdots & \frac{\partial}{\partial x_{mn}} \end{bmatrix}. \quad (2.1.13)$$

Definition 2.4 (Gradient Vectors) The *gradient vectors* of functions $f(\mathbf{x})$ and $f(\mathbf{X})$ are, respectively, defined as

$$\nabla_{\mathbf{x}} f(\mathbf{x}) \stackrel{\text{def}}{=} \left[\frac{\partial f(\mathbf{x})}{\partial x_1}, \ldots, \frac{\partial f(\mathbf{x})}{\partial x_m}\right]^T = \frac{\partial f(\mathbf{x})}{\partial \mathbf{x}}, \quad (2.1.14)$$

$$\nabla_{\text{vec }\mathbf{X}} f(\mathbf{X}) \stackrel{\text{def}}{=} \left[\frac{\partial f(\mathbf{X})}{\partial x_{11}}, \ldots, \frac{\partial f(\mathbf{X})}{\partial x_{m1}}, \ldots, \frac{\partial f(\mathbf{X})}{\partial x_{1n}}, \ldots, \frac{\partial f(\mathbf{X})}{\partial x_{mn}}\right]^T. \quad (2.1.15)$$

Definition 2.5 (Gradient Matrix) The *gradient matrix* of the function $f(\mathbf{X})$ is defined as

$$\nabla_{\mathbf{X}} f(\mathbf{X}) = \begin{bmatrix} \frac{\partial f(\mathbf{X})}{\partial x_{11}} & \cdots & \frac{\partial f(\mathbf{X})}{\partial x_{1n}} \\ \vdots & \ddots & \vdots \\ \frac{\partial f(\mathbf{X})}{\partial x_{m1}} & \cdots & \frac{\partial f(\mathbf{X})}{\partial x_{mn}} \end{bmatrix} = \frac{\partial f(\mathbf{X})}{\partial \mathbf{X}}. \quad (2.1.16)$$

For a real matrix function $\mathbf{F}(\mathbf{X}) \in \mathbb{R}^{p \times q}$ with matrix variable $\mathbf{X} \in \mathbb{R}^{m \times n}$, its gradient matrix is defined as

$$\nabla_{\mathbf{X}} \mathbf{F}(\mathbf{X}) = \frac{\partial (\text{vec }\mathbf{F}(\mathbf{X}))^T}{\partial \text{ vec}\mathbf{X}} = \left(\frac{\partial \text{vec}\mathbf{F}(\mathbf{X})}{\partial (\text{vec }\mathbf{X})^T}\right)^T. \quad (2.1.17)$$

Comparing Eq. (2.1.16) with Eq. (2.1.6), we have

$$\nabla_{\mathbf{X}} f(\mathbf{X}) = \mathbf{J}^T. \quad (2.1.18)$$

Similarly, comparing Eq. (2.1.17) with Eq. (2.1.9) gives

$$\nabla_{\mathbf{X}} \mathbf{F}(\mathbf{X}) = \mathbf{J}^T. \quad (2.1.19)$$

2.1 Jacobian Matrix and Gradient Matrix

That is to say, the gradient matrix of a real scalar function $f(\mathbf{X})$ or a real matrix function $\mathbf{F}(\mathbf{X})$ is equal to the transpose of respective Jacobian matrix.

An obvious fact is that, given a real scalar function $f(\mathbf{x})$, its gradient vector is directly equal to the transpose of the partial derivative vector. In this sense, the partial derivative in row vector form is a covariant form of the gradient vector, so the row partial derivative vector is also known as the *cogradient vector*. Similarly, the Jacobian matrix is sometimes called the *cogradient matrix*. The cogradient is a *covariant operator* [3] that itself is not the gradient, but is related to the gradient.

For this reason, the partial derivative operator $\partial/\partial \mathbf{x}^T$ and the Jacobian operator $\partial/\partial \mathbf{X}^T$ are known as the (row) partial derivative operator, the covariant form of the gradient operator or the cogradient operator.

The direction of the negative gradient $-\nabla_{\mathbf{x}} f(\mathbf{x})$ is known as the *gradient flow direction* of the function $f(\mathbf{x})$ at the point $\mathbf{x}$, and is expressed as

$$\dot{\mathbf{x}} = -\nabla_{\mathbf{x}} f(\mathbf{x}) \quad \text{or} \quad \dot{\mathbf{X}} = -\nabla_{\text{vec}\mathbf{X}} f(\mathbf{X}). \tag{2.1.20}$$

From the definition formula of the gradient vector, we have the following conclusive remarks:

- In the gradient flow direction, the function $f(\mathbf{x})$ decreases at the *maximum descent rate*. On the contrary, in the opposite direction (i.e., the positive gradient direction), the function increases at the *maximum ascent rate*.
- Each component of the gradient vector gives the rate of change of the scalar function $f(\mathbf{x})$ in the component direction.

2.1.3 Calculation of Partial Derivative and Gradient

The *gradient computation* of a real function with respect to its matrix variable has the following properties and rules [5]:

1. If $f(\mathbf{X}) = c$, where c is a real constant and $\mathbf{X}$ is an $m \times n$ real matrix, then the gradient $\partial c/\partial \mathbf{X} = \mathbf{O}_{m \times n}$.
2. *Linear rule:* If $f(\mathbf{X})$ and $g(\mathbf{X})$ are two real-valued functions of the matrix variable $\mathbf{X}$, and c_1 and c_2 are two real constants, then

$$\frac{\partial [c_1 f(\mathbf{X}) + c_2 g(\mathbf{X})]}{\partial \mathbf{X}} = c_1 \frac{\partial f(\mathbf{X})}{\partial \mathbf{X}} + c_2 \frac{\partial g(\mathbf{X})}{\partial \mathbf{X}}. \tag{2.1.21}$$

3. *Product rule:* If $f(\mathbf{X}), g(\mathbf{X})$, and $h(\mathbf{X})$ are real-valued functions of the matrix variable $\mathbf{X}$, then

$$\frac{\partial (f(\mathbf{X})g(\mathbf{X}))}{\partial \mathbf{X}} = g(\mathbf{X}) \frac{\partial f(\mathbf{X})}{\partial \mathbf{X}} + f(\mathbf{X}) \frac{\partial g(\mathbf{X})}{\partial \mathbf{X}} \tag{2.1.22}$$

and

$$\frac{\partial(f(\mathbf{X})g(\mathbf{X})h(\mathbf{X}))}{\partial \mathbf{X}} = g(\mathbf{X})h(\mathbf{X})\frac{\partial f(\mathbf{X})}{\partial \mathbf{X}} + f(\mathbf{X})h(\mathbf{X})\frac{\partial g(\mathbf{X})}{\partial \mathbf{X}}$$
$$+ f(\mathbf{X})g(\mathbf{X})\frac{\partial h(\mathbf{X})}{\partial \mathbf{X}}. \quad (2.1.23)$$

4. *Quotient rule:* If $g(\mathbf{X}) \neq 0$, then

$$\frac{\partial(f(\mathbf{X})/g(\mathbf{X}))}{\partial \mathbf{X}} = \frac{1}{g^2(\mathbf{X})}\left(g(\mathbf{X})\frac{\partial f(\mathbf{X})}{\partial \mathbf{X}} - f(\mathbf{X})\frac{\partial g(\mathbf{X})}{\partial \mathbf{X}}\right). \quad (2.1.24)$$

5. *Chain rule:* If $\mathbf{X}$ is an $m \times n$ matrix and $y = f(\mathbf{X})$ and $g(y)$ are, respectively, the real-valued functions of the matrix variable $\mathbf{X}$ and of the scalar variable y, then

$$\frac{\partial g(f(\mathbf{X}))}{\partial \mathbf{X}} = \frac{\mathrm{d}g(y)}{\mathrm{d}y}\frac{\partial f(\mathbf{X})}{\partial \mathbf{X}}. \quad (2.1.25)$$

As an extension, if $g(\mathbf{F}(\mathbf{X})) = g(\mathbf{F})$, where $\mathbf{F} = [f_{kl}] \in \mathbb{R}^{p \times q}$ and $\mathbf{X} = [x_{ij}] \in \mathbb{R}^{m \times n}$, then the chain rule is given by Petersen and Petersen [7]

$$\left[\frac{\partial g(\mathbf{F})}{\partial \mathbf{X}}\right]_{ij} = \frac{\partial g(\mathbf{F})}{\partial x_{ij}} = \sum_{k=1}^{p}\sum_{l=1}^{q}\frac{\partial g(\mathbf{F})}{\partial f_{kl}}\frac{\partial f_{kl}}{\partial x_{ij}}. \quad (2.1.26)$$

When computing the partial derivative of the functions $f(\mathbf{x})$ and $f(\mathbf{X})$, it is necessary to make the following basic assumption.

Independence Assumption Given a real-valued function f, we assume that the vector variable $\mathbf{x} = [x_i]_{i=1}^{m} \in \mathbb{R}^m$ and the matrix variable $\mathbf{X} = [x_{ij}]_{i=1,j=1}^{m,n} \in \mathbb{R}^{m \times n}$ do not themselves have any special structure; namely, the entries of $\mathbf{x}$ (and $\mathbf{X}$) are independent.

The independence assumption can be expressed in mathematical form as follows:

$$\frac{\partial x_i}{\partial x_j} = \delta_{ij} = \begin{cases} 1, & i = j; \\ 0, & \text{otherwise.} \end{cases} \quad (2.1.27)$$

$$\frac{\partial x_{kl}}{\partial x_{ij}} = \delta_{ki}\delta_{lj} = \begin{cases} 1, & k = i \text{ and } l = j; \\ 0, & \text{otherwise.} \end{cases} \quad (2.1.28)$$

These expressions on independence are the basic formulas for partial derivative computation.

2.2 Real Matrix Differential

Example 2.1 Consider the real function

$$f(\mathbf{X}) = \mathbf{a}^T \mathbf{X}\mathbf{X}^T \mathbf{b} = \sum_{k=1}^{m}\sum_{l=1}^{m} a_k \left(\sum_{p=1}^{n} x_{kp} x_{lp}\right) b_l, \quad \mathbf{X} \in \mathbb{R}^{m\times n},\ \mathbf{a},\mathbf{b} \in \mathbb{R}^{n\times 1}.$$

Using Eq. (2.1.28), we can see easily that

$$\left[\frac{\partial f(\mathbf{X})}{\partial \mathbf{X}^T}\right]_{ij} = \frac{\partial f(\mathbf{X})}{\partial x_{ji}} = \sum_{k=1}^{m}\sum_{l=1}^{m}\sum_{p=1}^{n} \frac{\partial a_k x_{kp} x_{lp} b_l}{\partial x_{ji}}$$

$$= \sum_{k=1}^{m}\sum_{l=1}^{m}\sum_{p=1}^{n} \left[a_k x_{lp} b_l \frac{\partial x_{kp}}{\partial x_{ji}} + a_k x_{kp} b_l \frac{\partial x_{lp}}{\partial x_{ji}}\right]$$

$$= \sum_{i=1}^{m}\sum_{l=1}^{m}\sum_{j=1}^{n} a_j x_{li} b_l + \sum_{k=1}^{m}\sum_{i=1}^{m}\sum_{j=1}^{n} a_k x_{ki} b_j$$

$$= \sum_{i=1}^{m}\sum_{j=1}^{n} \left[\mathbf{X}^T \mathbf{b}\right]_i a_j + \left[\mathbf{X}^T \mathbf{a}\right]_i b_j,$$

which yields, respectively, the Jacobian matrix and the gradient matrix as follows:

$$\mathbf{J} = \mathbf{X}^T(\mathbf{b}\mathbf{a}^T + \mathbf{a}\mathbf{b}^T) \quad \text{and} \quad \nabla_\mathbf{X} f(\mathbf{X}) = \mathbf{J}^T = (\mathbf{b}\mathbf{a}^T + \mathbf{a}\mathbf{b}^T)\mathbf{X}.$$

Example 2.2 Let $\mathbf{F}(\mathbf{X}) = \mathbf{AXB}$, where $\mathbf{A} \in \mathbb{R}^{p\times m}, \mathbf{X} \in \mathbb{R}^{m\times n}, \mathbf{B} \in \mathbb{R}^{n\times q}$. We have

$$\frac{\partial f_{kl}}{\partial x_{ij}} = \frac{\partial [\mathbf{AXB}]_{kl}}{\partial x_{ij}} = \frac{\partial \left(\sum_{u=1}^{m}\sum_{v=1}^{n} a_{ku} x_{uv} b_{vl}\right)}{\partial x_{ij}} = b_{jl} a_{ki}$$

$$\Rightarrow \nabla_\mathbf{X}(\mathbf{AXB}) = \mathbf{B} \otimes \mathbf{A}^T \quad \Rightarrow \quad \mathbf{J} = (\nabla_\mathbf{X}(\mathbf{AXB}))^T = \mathbf{B}^T \otimes \mathbf{A}.$$

That is, the $pq \times mn$ Jacobian matrix is $\mathbf{J} = \mathbf{B}^T \otimes \mathbf{A}$, and the $mn \times pq$ gradient matrix is given by $\nabla_\mathbf{X}(\mathbf{AXB}) = \mathbf{B} \otimes \mathbf{A}^T$.

2.2 Real Matrix Differential

Although direct computation of partial derivatives $\partial f_{kl}/\partial x_{ji}$ or $\partial f_{kl}/\partial x_{ij}$ can be used to find the Jacobian matrices or the gradient matrices of many matrix functions, for more complex functions (such as the inverse matrix, the Moore–Penrose inverse matrix and the exponential functions of a matrix), direct computation of their partial derivatives is more complicated and difficult. Hence, naturally we want to have an easily remembered and effective mathematical tool for computing the Jacobian matrices and the gradient matrices of real scalar functions and real matrix functions. Such a mathematical tool is the matrix differential.

2.2.1 Calculation of Real Matrix Differential

The differential of an $m \times n$ matrix $\mathbf{X} = [x_{ij}]$ is known as the *matrix differential*, denoted $d\mathbf{X}$, which is still $m \times n$ matrix and is defined as $d\mathbf{X} = [dx_{ij}]_{i=1,j=1}^{m,n}$.

Consider the differential of a trace function $\mathrm{tr}(\mathbf{U})$. We have

$$d(\mathrm{tr}\,\mathbf{U}) = d\left(\sum_{i=1}^{n} u_{ii}\right) = \sum_{i=1}^{n} du_{ii} = \mathrm{tr}(d\mathbf{U}),$$

namely $d(\mathrm{tr}\,\mathbf{U}) = \mathrm{tr}(d\mathbf{U})$.

The matrix differential of the matrix product $\mathbf{UV}$ is given in element form:

$$[d(\mathbf{UV})]_{ij} = d\left([\mathbf{UV}]_{ij}\right) = d\left(\sum_{k} u_{ik} v_{kj}\right) = \sum_{k} d(u_{ik} v_{kj})$$

$$= \sum_{k} \left((du_{ik})v_{kj} + u_{ik}dv_{kj}\right) = \sum_{k} (du_{ik})v_{kj} + \sum_{k} u_{ik}dv_{kj}$$

$$= [(d\mathbf{U})\mathbf{V}]_{ij} + [\mathbf{U}(d\mathbf{V})]_{ij}.$$

Then, we have the matrix differential $d(\mathbf{UV}) = (d\mathbf{U})\mathbf{V} + \mathbf{U}(d\mathbf{V})$.

Common computation formulas for matrix differential are given as follows [6, pp. 148–154]:

1. The differential of a constant matrix is a zero matrix, namely $d\mathbf{A} = \mathbf{O}$.
2. The matrix differential of the product $\alpha \mathbf{X}$ is given by $d(\alpha \mathbf{X}) = \alpha\, d\mathbf{X}$.
3. The matrix differential of a transposed matrix is equal to the transpose of the original matrix differential, namely $d(\mathbf{X}^T) = (d\mathbf{X})^T$.
4. The matrix differential of the sum (or difference) of two matrices is given by $d(\mathbf{U} \pm \mathbf{V}) = d\mathbf{U} \pm d\mathbf{V}$. More generally, we have $d(a\mathbf{U} \pm b\mathbf{V}) = a \cdot d\mathbf{U} \pm b \cdot d\mathbf{V}$.
5. The matrix differentials of the functions $\mathbf{UV}$ and $\mathbf{UVW}$, where $\mathbf{U} = \mathbf{F}(\mathbf{X})$, $\mathbf{V} = \mathbf{G}(\mathbf{X})$, $\mathbf{W} = \mathbf{H}(\mathbf{X})$, are, respectively, given by

$$d(\mathbf{UV}) = (d\mathbf{U})\mathbf{V} + \mathbf{U}(d\mathbf{V}) \tag{2.2.1}$$

$$d(\mathbf{UVW}) = (d\mathbf{U})\mathbf{VW} + \mathbf{U}(d\mathbf{V})\mathbf{W} + \mathbf{UV}(d\mathbf{W}). \tag{2.2.2}$$

If $\mathbf{A}$ and $\mathbf{B}$ are constant matrices, then $d(\mathbf{AXB}) = \mathbf{A}(d\mathbf{X})\mathbf{B}$.

2.2 Real Matrix Differential

6. The differential of the matrix trace $d(\operatorname{tr} \mathbf{X})$ is equal to the trace of the matrix differential $d\mathbf{X}$, namely

$$d(\operatorname{tr} \mathbf{X}) = \operatorname{tr}(d\mathbf{X}). \tag{2.2.3}$$

In particular, the differential of the trace of the matrix function $\mathbf{F}(\mathbf{X})$ is given by $d(\operatorname{tr} \mathbf{F}(\mathbf{X})) = \operatorname{tr}(d\mathbf{F}(\mathbf{X}))$.

7. The differential of the determinant of $\mathbf{X}$ is given by

$$d|\mathbf{X}| = |\mathbf{X}|\operatorname{tr}(\mathbf{X}^{-1}d\mathbf{X}). \tag{2.2.4}$$

In particular, the differential of the determinant of the matrix function $\mathbf{F}(\mathbf{X})$ is computed by $d|\mathbf{F}(\mathbf{X})| = |\mathbf{F}(\mathbf{X})|\operatorname{tr}(\mathbf{F}^{-1}(\mathbf{X})d\mathbf{F}(\mathbf{X}))$.

8. The matrix differential of the Kronecker product is given by

$$d(\mathbf{U} \otimes \mathbf{V}) = (d\mathbf{U}) \otimes \mathbf{V} + \mathbf{U} \otimes d\mathbf{V}. \tag{2.2.5}$$

9. The matrix differential of the Hadamard product is computed by

$$d(\mathbf{U} \odot \mathbf{V}) = (d\mathbf{U}) \odot \mathbf{V} + \mathbf{U} \odot d\mathbf{V}. \tag{2.2.6}$$

10. The matrix differential of the inverse matrix is given by

$$d(\mathbf{X}^{-1}) = -\mathbf{X}^{-1}(d\mathbf{X})\mathbf{X}^{-1}. \tag{2.2.7}$$

11. The differential of the vectorization function $\operatorname{vec} \mathbf{X}$ is equal to the vectorization of the matrix differential, i.e.,

$$d \operatorname{vec} \mathbf{X} = \operatorname{vec}(d\mathbf{X}). \tag{2.2.8}$$

12. The differential of the matrix logarithm is given by

$$d \log \mathbf{X} = \mathbf{X}^{-1}d\mathbf{X}. \tag{2.2.9}$$

In particular, $d \log \mathbf{F}(\mathbf{X}) = \mathbf{F}^{-1}(\mathbf{X})\, d\mathbf{F}(\mathbf{X})$.

13. The matrix differentials of $\mathbf{X}^\dagger$, $\mathbf{X}^\dagger \mathbf{X}$, and $\mathbf{X}\mathbf{X}^\dagger$ are given by

$$\begin{aligned} d(\mathbf{X}^\dagger) = &\; -\mathbf{X}^\dagger(d\mathbf{X})\mathbf{X}^\dagger + \mathbf{X}^\dagger(\mathbf{X}^\dagger)^T(d\mathbf{X}^T)(\mathbf{I} - \mathbf{X}\mathbf{X}^\dagger) \\ &\; + (\mathbf{I} - \mathbf{X}^\dagger\mathbf{X})(d\mathbf{X}^T)(\mathbf{X}^\dagger)^T\mathbf{X}^\dagger, \end{aligned} \tag{2.2.10}$$

$$d(\mathbf{X}^\dagger \mathbf{X}) = \mathbf{X}^\dagger(d\mathbf{X})(\mathbf{I} - \mathbf{X}^\dagger\mathbf{X}) + \left(\mathbf{X}^\dagger(d\mathbf{X})(\mathbf{I} - \mathbf{X}^\dagger\mathbf{X})\right)^T, \tag{2.2.11}$$

$$d(\mathbf{X}\mathbf{X}^\dagger) = (\mathbf{I} - \mathbf{X}\mathbf{X}^\dagger)(d\mathbf{X})\mathbf{X}^\dagger + \left((\mathbf{I} - \mathbf{X}\mathbf{X}^\dagger)(d\mathbf{X})\mathbf{X}^\dagger\right)^T. \tag{2.2.12}$$

2.2.2 Jacobian Matrix Identification

In multivariate calculus, the multivariate function $f(x_1, \ldots, x_m)$ is said to be differentiable at the point $(x_1, \ldots, x_m)$, if a change in $f(x_1, \ldots, x_m)$ can be expressed as

$$\Delta f(x_1, \ldots, x_m) = f(x_1 + \Delta x_1, \ldots, x_m + \Delta x_m) - f(x_1, \ldots, x_m)$$
$$= A_1 \Delta x_1 + \cdots + A_m \Delta x_m + O(\Delta x_1, \ldots, \Delta x_m),$$

here $A_1, \ldots, A_m$ are independent of $\Delta x_1, \ldots, \Delta x_m$, respectively, and $O(\Delta x_1, \ldots, \Delta x_m)$ denotes the second-order and the higher-order terms in $\Delta x_1, \ldots, \Delta x_m$. In this case, the partial derivative $\partial f/\partial x_1, \ldots, \partial f/\partial x_m$ must exist, and

$$\frac{\partial f}{\partial x_1} = A_1, \quad \ldots \quad , \quad \frac{\partial f}{\partial x_m} = A_m.$$

The linear part of the change $\Delta f(x_1, \ldots, x_m)$,

$$A_1 \Delta x_1 + \cdots + A_m \Delta x_m = \frac{\partial f}{\partial x_1} dx_1 + \cdots + \frac{\partial f}{\partial x_m} dx_m,$$

is said to be the *differential* or *first-order differential* of the multivariate function $f(x_1, \ldots, x_m)$ and is denoted by

$$df(x_1, \ldots, x_m) = \frac{\partial f}{\partial x_1} dx_1 + \cdots + \frac{\partial f}{\partial x_m} dx_m. \qquad (2.2.13)$$

The sufficient condition for a multivariate function $f(x_1, \ldots, x_m)$ to be differentiable at the point $(x_1, \ldots, x_m)$ is that the partial derivatives $\partial f/\partial x_1, \ldots, \partial f/\partial x_m$ exist and are continuous.

Equation (2.2.13) provides two Jacobian matrix identification methods.

- For a scalar function $f(\mathbf{x})$ with variable $\mathbf{x} = [x_1, \ldots, x_m]^T \in \mathbb{R}^m$, if regarding the elements $x_1, \ldots, x_m$ as m variables, and using Eq. (2.2.13), then we can directly obtain the differential of the scalar function $f(\mathbf{x})$ as follows:

$$df(\mathbf{x}) = \frac{\partial f(\mathbf{x})}{\partial x_1} dx_1 + \cdots + \frac{\partial f(\mathbf{x})}{\partial x_m} dx_m = \begin{bmatrix} \frac{\partial f(\mathbf{x})}{\partial x_1} & \cdots & \frac{\partial f(\mathbf{x})}{\partial x_m} \end{bmatrix} \begin{bmatrix} dx_1 \\ \vdots \\ dx_m \end{bmatrix}$$

or

$$df(\mathbf{x}) = \frac{\partial f(\mathbf{x})}{\partial \mathbf{x}^T} d\mathbf{x}, \qquad (2.2.14)$$

2.2 Real Matrix Differential

where $\frac{\partial f(\mathbf{x})}{\partial \mathbf{x}^T} = \left[\frac{\partial f(\mathbf{x})}{\partial x_1}, \ldots, \frac{\partial f(\mathbf{x})}{\partial x_m}\right]$ and $d\mathbf{x} = [dx_1, \ldots, dx_m]^T$. If denoting the row vector $\mathbf{A} = \frac{\partial f(\mathbf{x})}{\partial \mathbf{x}^T}$, then the first-order differential in (2.2.14) can be represented as a trace:

$$df(\mathbf{x}) = \frac{\partial f(\mathbf{x})}{\partial \mathbf{x}^T}d\mathbf{x} = \mathbf{A}d\mathbf{x} = \operatorname{tr}(\mathbf{A}\,d\mathbf{x})$$

because $\mathbf{A}d\mathbf{x}$ is a scalar, and for any scalar α we have $\alpha = \operatorname{tr}(\alpha)$. This shows that there is an equivalence relationship between the Jacobian matrix of the scalar function $f(\mathbf{x})$ and its matrix differential as follows:

$$df(\mathbf{x}) = \operatorname{tr}(\mathbf{A}\,d\mathbf{x}) \quad \Leftrightarrow \quad \mathbf{J} = \frac{\partial f(\mathbf{x})}{\partial \mathbf{x}^T} = \mathbf{A}. \qquad (2.2.15)$$

In other words, if the differential of the function $f(\mathbf{x})$ is denoted as $df(\mathbf{x}) = \operatorname{tr}(\mathbf{A}\,d\mathbf{x})$, then the matrix $\mathbf{A}$ is just the Jacobian matrix of the function $f(\mathbf{x})$.

- For a scalar function $f(\mathbf{X})$ with variable $\mathbf{X} = [\mathbf{x}_1, \ldots, \mathbf{x}_n] \in \mathbb{R}^{m\times n}$, if denoting $\mathbf{x}_j = [x_{1j}, \ldots, x_{mj}]^T, j = 1, \ldots, n$, then Eq. (2.2.13) becomes

$$df(\mathbf{X}) = \left[\frac{\partial f(\mathbf{X})}{\partial x_{11}} \cdots \frac{\partial f(\mathbf{X})}{\partial x_{m1}} \cdots \frac{\partial f(\mathbf{X})}{\partial x_{1n}} \cdots \frac{\partial f(\mathbf{X})}{\partial x_{mn}}\right]\begin{bmatrix}dx_{11}\\ \vdots\\ dx_{m1}\\ \vdots\\ dx_{1n}\\ \vdots\\ dx_{mn}\end{bmatrix},$$

$$= \frac{\partial f(\mathbf{X})}{\partial (\operatorname{vec}\mathbf{X})^T}\,d\operatorname{vec}\mathbf{X} = \nabla_{(\operatorname{vec}\mathbf{X})^T} f(\mathbf{X})\,d\operatorname{vec}\mathbf{X}. \qquad (2.2.16)$$

By the relationship between the row partial derivative vector and the Jacobian matrix in Eq. (2.1.7), $\nabla_{(\operatorname{vec}\mathbf{X})^T} f(\mathbf{X}) = \left(\operatorname{vec}(\mathbf{J}^T)\right)^T$, Eq. (2.2.16) can be written as

$$df(\mathbf{X}) = (\operatorname{vec}\mathbf{A}^T)^T d(\operatorname{vec}\mathbf{X}), \qquad (2.2.17)$$

where

$$\mathbf{A} = \mathbf{J} = \frac{\partial f(\mathbf{X})}{\partial \mathbf{X}^T} = \begin{bmatrix}\frac{\partial f(\mathbf{X})}{\partial x_{11}} & \cdots & \frac{\partial f(\mathbf{X})}{\partial x_{m1}}\\ \vdots & \ddots & \vdots\\ \frac{\partial f(\mathbf{X})}{\partial x_{1n}} & \cdots & \frac{\partial f(\mathbf{X})}{\partial x_{mn}}\end{bmatrix} \qquad (2.2.18)$$

is the Jacobian matrix of the scalar function $f(\mathbf{X})$. Using the relationship between the vectorization operator vec and the trace function $\mathrm{tr}(\mathbf{B}^T\mathbf{C}) = (\mathrm{vec}\,\mathbf{B})^T\mathrm{vec}\,\mathbf{C}$, and letting $\mathbf{B} = \mathbf{A}^T$ and $\mathbf{C} = \mathrm{d}\mathbf{X}$, then Eq. (2.2.17) can be expressed in the trace form as

$$\mathrm{d}f(\mathbf{X}) = \mathrm{tr}(\mathbf{A}\mathrm{d}\mathbf{X}). \tag{2.2.19}$$

This can be regarded as the canonical form of the differential of a scalar function $f(\mathbf{X})$.

The above discussion shows that once the matrix differential of a scalar function $\mathrm{d}f(\mathbf{X})$ is expressed in its canonical form, we can identify the Jacobian matrix and/or the gradient matrix of the scalar function $f(\mathbf{X})$, as stated below.

Proposition 2.1 *If a scalar function $f(\mathbf{X})$ is differentiable at the point $\mathbf{X}$, then the Jacobian matrix $\mathbf{A}$ can be directly identified as follows [6]:*

$$\mathrm{d}f(\mathbf{x}) = \mathrm{tr}(\mathbf{A}\mathrm{d}\mathbf{x}) \quad \Leftrightarrow \quad \mathbf{J} = \mathbf{A}, \tag{2.2.20}$$

$$\mathrm{d}f(\mathbf{X}) = \mathrm{tr}(\mathbf{A}\mathrm{d}\mathbf{X}) \quad \Leftrightarrow \quad \mathbf{J} = \mathbf{A}. \tag{2.2.21}$$

Proposition 2.1 motivates the following effective approach to directly identifying the Jacobian matrix $\mathbf{J} = \mathrm{D}_{\mathbf{X}}f(\mathbf{X})$ of the scalar function $f(\mathbf{X})$:

1. Find the differential $\mathrm{d}f(\mathbf{X})$ of the real function $f(\mathbf{X})$, and denote it in the canonical form as $\mathrm{d}f(\mathbf{X}) = \mathrm{tr}(\mathbf{A}\,\mathrm{d}\mathbf{X})$.
2. The Jacobian matrix is directly given by $\mathbf{A}$.

The following are two main points in applying Proposition 2.1:

- Any scalar function $f(\mathbf{X})$ can always be written in the form of a trace function, because $f(\mathbf{X}) = \mathrm{tr}(f(\mathbf{X}))$.
- No matter where $\mathrm{d}\mathbf{X}$ appears initially in the trace function, we can place it in the rightmost position via the trace property $\mathrm{tr}(\mathbf{C}(\mathrm{d}\mathbf{X})\mathbf{B}) = \mathrm{tr}(\mathbf{B}\mathbf{C}\,\mathrm{d}\mathbf{X})$, giving the canonical form $\mathrm{d}f(\mathbf{X}) = \mathrm{tr}(\mathbf{A}\,\mathrm{d}\mathbf{X})$.

It has been shown [6] that the Jacobian matrix $\mathbf{A}$ is uniquely determined: if there are $\mathbf{A}_1$ and $\mathbf{A}_2$ such that $\mathrm{d}f(\mathbf{X}) = \mathbf{A}_i \mathrm{d}\mathbf{X}$, $i = 1, 2$, then $\mathbf{A}_1 = \mathbf{A}_2$.

Since the gradient matrix is the transpose of the Jacobian matrix for a given real function $f(\mathbf{X})$, Proposition 2.1 implies in addition that

$$\mathrm{d}f(\mathbf{X}) = \mathrm{tr}(\mathbf{A}\mathrm{d}\mathbf{X}) \quad \Leftrightarrow \quad \nabla_{\mathbf{X}}f(\mathbf{X}) = \mathbf{A}^T. \tag{2.2.22}$$

Because the Jacobian matrix $\mathbf{A}$ is uniquely determined, the gradient matrix is uniquely determined as well.

2.2 Real Matrix Differential

Jacobian Matrices of Trace Functions

Example 2.3 The differential of the trace $\text{tr}(\mathbf{X}^T\mathbf{AX})$ is given by

$$\begin{aligned}
\text{d}\,\text{tr}(\mathbf{X}^T\mathbf{AX}) &= \text{tr}\left(\text{d}(\mathbf{X}^T\mathbf{AX})\right) = \text{tr}\left((\text{d}\mathbf{X})^T\mathbf{AX} + \mathbf{X}^T\mathbf{A}\text{d}\mathbf{X}\right) \\
&= \text{tr}\left((\text{d}\mathbf{X})^T\mathbf{AX}\right) + \text{tr}(\mathbf{X}^T\mathbf{A}\text{d}\mathbf{X}) = \text{tr}\left((\mathbf{AX})^T\text{d}\mathbf{X}\right) + \text{tr}(\mathbf{X}^T\mathbf{A}\text{d}\mathbf{X}) \\
&= \text{tr}\left(\mathbf{X}^T(\mathbf{A}^T + \mathbf{A})\text{d}\mathbf{X}\right),
\end{aligned}$$

which yields the gradient matrix

$$\frac{\partial\,\text{tr}(\mathbf{X}^T\mathbf{AX})}{\partial\,\mathbf{X}} = \left(\mathbf{X}^T(\mathbf{A}^T + \mathbf{A})\right)^T = (\mathbf{A} + \mathbf{A}^T)\mathbf{X}. \tag{2.2.23}$$

Similarly, we can compute the differential matrices and Jacobian matrices of other typical trace functions.

Table 2.2 summarizes the differential matrices and Jacobian matrices of several typical trace functions [6].

Table 2.2 Differential matrices and Jacobian matrices of trace functions

$f(\mathbf{X})$	Differential $\text{d}f(\mathbf{X})$	Jacobian matrix $\mathbf{J} = \partial f(\mathbf{X})/\partial\mathbf{X}$
$\text{tr}(\mathbf{X})$	$\text{tr}(\mathbf{I}\text{d}\mathbf{X})$	$\mathbf{I}$
$\text{tr}(\mathbf{X}^{-1})$	$-\text{tr}(\mathbf{X}^{-2}\text{d}\mathbf{X})$	$-\mathbf{X}^{-2}$
$\text{tr}(\mathbf{AX})$	$\text{tr}(\mathbf{A}\text{d}\mathbf{X})$	$\mathbf{A}$
$\text{tr}(\mathbf{X}^2)$	$2\text{tr}(\mathbf{X}\text{d}\mathbf{X})$	$2\mathbf{X}$
$\text{tr}(\mathbf{X}^T\mathbf{X})$	$2\text{tr}(\mathbf{X}^T\text{d}\mathbf{X})$	$\mathbf{X}^T$
$\text{tr}(\mathbf{X}^T\mathbf{AX})$	$\text{tr}\left(\mathbf{X}^T(\mathbf{A}+\mathbf{A}^T)\text{d}\mathbf{X}\right)$	$\mathbf{X}^T(\mathbf{A}+\mathbf{A}^T)$
$\text{tr}(\mathbf{XAX}^T)$	$\text{tr}\left((\mathbf{A}+\mathbf{A}^T)\mathbf{X}^T\text{d}\mathbf{X}\right)$	$(\mathbf{A}+\mathbf{A}^T)\mathbf{X}^T$
$\text{tr}(\mathbf{XAX})$	$\text{tr}((\mathbf{AX}+\mathbf{XA})\text{d}\mathbf{X})$	$\mathbf{AX}+\mathbf{XA}$
$\text{tr}(\mathbf{AX}^{-1})$	$-\text{tr}\left(\mathbf{X}^{-1}\mathbf{AX}^{-1}\text{d}\mathbf{X}\right)$	$-\mathbf{X}^{-1}\mathbf{AX}^{-1}$
$\text{tr}(\mathbf{AX}^{-1}\mathbf{B})$	$-\text{tr}\left(\mathbf{X}^{-1}\mathbf{BAX}^{-1}\text{d}\mathbf{X}\right)$	$-\mathbf{X}^{-1}\mathbf{BAX}^{-1}$
$\text{tr}\left((\mathbf{X}+\mathbf{A})^{-1}\right)$	$-\text{tr}\left((\mathbf{X}+\mathbf{A})^{-2}\text{d}\mathbf{X}\right)$	$-(\mathbf{X}+\mathbf{A})^{-2}$
$\text{tr}(\mathbf{XAXB})$	$\text{tr}((\mathbf{AXB}+\mathbf{BXA})\text{d}\mathbf{X})$	$\mathbf{AXB}+\mathbf{BXA}$
$\text{tr}(\mathbf{XAX}^T\mathbf{B})$	$\text{tr}\left((\mathbf{AX}^T\mathbf{B}+\mathbf{A}^T\mathbf{X}^T\mathbf{B}^T)\text{d}\mathbf{X}\right)$	$\mathbf{AX}^T\mathbf{B}+\mathbf{A}^T\mathbf{X}^T\mathbf{B}^T$
$\text{tr}(\mathbf{AXX}^T\mathbf{B})$	$\text{tr}\left(\mathbf{X}^T(\mathbf{BA}+\mathbf{A}^T\mathbf{B}^T)\text{d}\mathbf{X}\right)$	$\mathbf{X}^T(\mathbf{BA}+\mathbf{A}^T\mathbf{B}^T)$
$\text{tr}(\mathbf{AX}^T\mathbf{XB})$	$\text{tr}\left((\mathbf{BA}+\mathbf{A}^T\mathbf{B}^T)\mathbf{X}^T\text{d}\mathbf{X}\right)$	$(\mathbf{BA}+\mathbf{A}^T\mathbf{B}^T)\mathbf{X}^T$

Here $\mathbf{A}^{-2} = \mathbf{A}^{-1}\mathbf{A}^{-1}$

Jacobian Matrices of Determinant Functions

Consider the Jacobian matrix identification of typical determinant functions.

Example 2.4 For the nonsingular matrix $\mathbf{XX}^T$, we have

$$\begin{aligned}
\mathrm{d}|\mathbf{XX}^T| &= |\mathbf{XX}^T|\operatorname{tr}\left((\mathbf{XX}^T)^{-1}\mathrm{d}(\mathbf{XX}^T)\right) \\
&= |\mathbf{XX}^T|\left(\operatorname{tr}\left((\mathbf{XX}^T)^{-1}(\mathrm{d}\mathbf{X})\mathbf{X}^T\right) + \operatorname{tr}\left((\mathbf{XX}^T)^{-1}\mathbf{X}(\mathrm{d}\mathbf{X})^T\right)\right) \\
&= |\mathbf{XX}^T|\left(\operatorname{tr}\left(\mathbf{X}^T(\mathbf{XX}^T)^{-1}\mathrm{d}\mathbf{X}\right) + \operatorname{tr}\left(\mathbf{X}^T(\mathbf{XX}^T)^{-1}\mathrm{d}\mathbf{X}\right)\right) \\
&= \operatorname{tr}\left(2|\mathbf{XX}^T|\mathbf{X}^T(\mathbf{XX}^T)^{-1}\mathrm{d}\mathbf{X}\right).
\end{aligned}$$

By Proposition 2.1, we get the gradient matrix

$$\frac{\partial |\mathbf{XX}^T|}{\partial \mathbf{X}} = 2|\mathbf{XX}^T|(\mathbf{XX}^T)^{-1}\mathbf{X}. \tag{2.2.24}$$

Similarly, set $\mathbf{X} \in \mathbb{R}^{m \times n}$. If $\operatorname{rank}(\mathbf{X}) = n$, i.e., $\mathbf{X}^T\mathbf{X}$ is invertible, then

$$\mathrm{d}|\mathbf{X}^T\mathbf{X}| = \operatorname{tr}\left(2|\mathbf{X}^T\mathbf{X}|(\mathbf{X}^T\mathbf{X})^{-1}\mathbf{X}^T\mathrm{d}\mathbf{X}\right), \tag{2.2.25}$$

and hence $\partial |\mathbf{X}^T\mathbf{X}|/\partial \mathbf{X} = 2|\mathbf{X}^T\mathbf{X}|\mathbf{X}(\mathbf{X}^T\mathbf{X})^{-1}$.

Similarly, we can compute the differential matrices and Jacobian matrices of other typical determinant functions.

Table 2.3 summarizes the real matrix differentials and the Jacobian matrices of several typical determinant functions.

2.2.3 Jacobian Matrix of Real Matrix Functions

Let $f_{kl} = f_{kl}(\mathbf{X})$ be the entry of the kth row and lth column of the real matrix function $\mathbf{F}(\mathbf{X})$; then $\mathrm{d}f_{kl}(\mathbf{X}) = [\mathrm{d}\mathbf{F}(\mathbf{X})]_{kl}$ represents the differential of the scalar function $f_{kl}(\mathbf{X})$ with respect to the variable matrix $\mathbf{X}$. From Eq. (2.2.16) we have

$$\mathrm{d}f_{kl}(\mathbf{X}) = \begin{bmatrix} \frac{\partial f_{kl}(\mathbf{X})}{\partial x_{11}} & \cdots & \frac{\partial f_{kl}(\mathbf{X})}{\partial x_{m1}} & \cdots & \frac{\partial f_{kl}(\mathbf{X})}{\partial x_{1n}} & \cdots & \frac{\partial f_{kl}(\mathbf{X})}{\partial x_{mn}} \end{bmatrix} \begin{bmatrix} \mathrm{d}x_{11} \\ \vdots \\ \mathrm{d}x_{m1} \\ \vdots \\ \mathrm{d}x_{1n} \\ \vdots \\ \mathrm{d}x_{mn} \end{bmatrix}$$

2.2 Real Matrix Differential

Table 2.3 Differentials and Jacobian matrices of determinant functions

$f(\mathbf{X})$	Differential $df(\mathbf{X})$	Jacobian matrix $\mathbf{J} = \partial f(\mathbf{X})/\partial \mathbf{X}$
$|\mathbf{X}|$	$|\mathbf{X}|\operatorname{tr}(\mathbf{X}^{-1}d\mathbf{X})$	$|\mathbf{X}|\mathbf{X}^{-1}$
$\log|\mathbf{X}|$	$\operatorname{tr}(\mathbf{X}^{-1}d\mathbf{X})$	$\mathbf{X}^{-1}$
$|\mathbf{X}^{-1}|$	$-|\mathbf{X}^{-1}|\operatorname{tr}(\mathbf{X}^{-1}d\mathbf{X})$	$-|\mathbf{X}^{-1}|\mathbf{X}^{-1}$
$|\mathbf{X}^2|$	$2|\mathbf{X}|^2\operatorname{tr}(\mathbf{X}^{-1}d\mathbf{X})$	$2|\mathbf{X}|^2\mathbf{X}^{-1}$
$|\mathbf{X}^k|$	$k|\mathbf{X}|^k\operatorname{tr}(\mathbf{X}^{-1}d\mathbf{X})$	$k|\mathbf{X}|^k\mathbf{X}^{-1}$
$|\mathbf{X}\mathbf{X}^T|$	$2|\mathbf{X}\mathbf{X}^T|\operatorname{tr}\left(\mathbf{X}^T(\mathbf{X}\mathbf{X}^T)^{-1}d\mathbf{X}\right)$	$2|\mathbf{X}\mathbf{X}^T|\mathbf{X}^T(\mathbf{X}\mathbf{X}^T)^{-1}$
$|\mathbf{X}^T\mathbf{X}|$	$2|\mathbf{X}^T\mathbf{X}|\operatorname{tr}\left((\mathbf{X}^T\mathbf{X})^{-1}\mathbf{X}^Td\mathbf{X}\right)$	$2|\mathbf{X}^T\mathbf{X}|(\mathbf{X}^T\mathbf{X})^{-1}\mathbf{X}^T$
$\log|\mathbf{X}^T\mathbf{X}|$	$2\operatorname{tr}\left((\mathbf{X}^T\mathbf{X})^{-1}\mathbf{X}^Td\mathbf{X}\right)$	$2(\mathbf{X}^T\mathbf{X})^{-1}\mathbf{X}^T$
$|\mathbf{A}\mathbf{X}\mathbf{B}|$	$|\mathbf{A}\mathbf{X}\mathbf{B}|\operatorname{tr}\left(\mathbf{B}(\mathbf{A}\mathbf{X}\mathbf{B})^{-1}\mathbf{A}d\mathbf{X}\right)$	$|\mathbf{A}\mathbf{X}\mathbf{B}|\mathbf{B}(\mathbf{A}\mathbf{X}\mathbf{B})^{-1}\mathbf{A}$
$|\mathbf{X}\mathbf{A}\mathbf{X}^T|$	$|\mathbf{X}\mathbf{A}\mathbf{X}^T|\operatorname{tr}\left((\mathbf{A}\mathbf{X}^T(\mathbf{X}\mathbf{A}\mathbf{X}^T)^{-1}\right.$ $\left.+(\mathbf{X}\mathbf{A})^T(\mathbf{X}\mathbf{A}^T\mathbf{X}^T)^{-1})d\mathbf{X}\right)$	$|\mathbf{X}\mathbf{A}\mathbf{X}^T|\left(\mathbf{A}\mathbf{X}^T(\mathbf{X}\mathbf{A}\mathbf{X}^T)^{-1}\right.$ $\left.+(\mathbf{X}\mathbf{A})^T(\mathbf{X}\mathbf{A}^T\mathbf{X}^T)^{-1}\right)$
$|\mathbf{X}^T\mathbf{A}\mathbf{X}|$	$|\mathbf{X}^T\mathbf{A}\mathbf{X}|\operatorname{tr}\left(((\mathbf{X}^T\mathbf{A}\mathbf{X})^{-T}(\mathbf{A}\mathbf{X})^T\right.$ $\left.+(\mathbf{X}^T\mathbf{A}\mathbf{X})^{-1}\mathbf{X}^T\mathbf{A})d\mathbf{X}\right)$	$|\mathbf{X}^T\mathbf{A}\mathbf{X}|\left((\mathbf{X}^T\mathbf{A}\mathbf{X})^{-T}(\mathbf{A}\mathbf{X})^T\right.$ $\left.+(\mathbf{X}^T\mathbf{A}\mathbf{X})^{-1}\mathbf{X}^T\mathbf{A}\right)$

or

$$d\operatorname{vec}\mathbf{F}(\mathbf{X}) = \mathbf{A}d\operatorname{vec}\mathbf{X}, \qquad (2.2.26)$$

where

$$d\operatorname{vec}\mathbf{F}(\mathbf{X}) = [df_{11}(\mathbf{X}), \ldots, df_{p1}(\mathbf{X}), \ldots, df_{1q}(\mathbf{X}), \ldots, df_{pq}(\mathbf{X})]^T, \qquad (2.2.27)$$

$$d\operatorname{vec}\mathbf{X} = [dx_{11}, \ldots, dx_{m1}, \ldots, dx_{1n}, \ldots, dx_{mn}]^T, \qquad (2.2.28)$$

$$\mathbf{A} = \begin{bmatrix} \frac{\partial f_{11}(\mathbf{X})}{\partial x_{11}} & \cdots & \frac{\partial f_{11}(\mathbf{X})}{\partial x_{m1}} & \cdots & \frac{\partial f_{11}(\mathbf{X})}{\partial x_{1n}} & \cdots & \frac{\partial f_{11}(\mathbf{X})}{\partial x_{mn}} \\ \vdots & \vdots & \vdots & \vdots & \vdots & \vdots & \vdots \\ \frac{\partial f_{p1}(\mathbf{X})}{\partial x_{11}} & \cdots & \frac{\partial f_{p1}(\mathbf{X})}{\partial x_{m1}} & \cdots & \frac{\partial f_{p1}(\mathbf{X})}{\partial x_{1n}} & \cdots & \frac{\partial f_{p1}(\mathbf{X})}{\partial x_{mn}} \\ \vdots & \vdots & \vdots & \vdots & \vdots & \vdots & \vdots \\ \frac{\partial f_{1q}(\mathbf{X})}{\partial x_{11}} & \cdots & \frac{\partial f_{1q}(\mathbf{X})}{\partial x_{m1}} & \cdots & \frac{\partial f_{1q}(\mathbf{X})}{\partial x_{1n}} & \cdots & \frac{\partial f_{1q}(\mathbf{X})}{\partial x_{mn}} \\ \vdots & \vdots & \vdots & \vdots & \vdots & \vdots & \vdots \\ \frac{\partial f_{pq}(\mathbf{X})}{\partial x_{11}} & \cdots & \frac{\partial f_{pq}(\mathbf{X})}{\partial x_{m1}} & \cdots & \frac{\partial f_{pq}(\mathbf{X})}{\partial x_{1n}} & \cdots & \frac{\partial f_{pq}(\mathbf{X})}{\partial x_{mn}} \end{bmatrix}$$

$$= \frac{\partial \operatorname{vec}\mathbf{F}(\mathbf{X})}{\partial(\operatorname{vec}\mathbf{X})^T}. \qquad (2.2.29)$$

In other words, the matrix $\mathbf{A}$ is the Jacobian matrix $\mathbf{J} = \frac{\partial \text{vec}\mathbf{F}(\mathbf{X})}{\partial (\text{vec}\,\mathbf{X})^T}$ of the matrix function $\mathbf{F}(\mathbf{X})$.

Let $\mathbf{F}(\mathbf{X}) \in \mathbb{R}^{p \times q}$ be a matrix function including $\mathbf{X}$ and $\mathbf{X}^T$ as variables, where $\mathbf{X} \in \mathbb{R}^{m \times n}$.

Theorem 2.1 ([6]) *Given a matrix function* $\mathbf{F}(\mathbf{X}) : \mathbb{R}^{m \times n} \to \mathbb{R}^{p \times q}$, *then its* $pq \times mn$ *Jacobian matrix can be identified as follows:*

$$\mathrm{d}\mathbf{F}(\mathbf{X}) = \mathbf{A}(\mathrm{d}\mathbf{X})\mathbf{B} + \mathbf{C}(\mathrm{d}\mathbf{X}^T)\mathbf{D},$$

$$\Leftrightarrow \mathbf{J} = \frac{\partial \,\text{vec}\,\mathbf{F}(\mathbf{X})}{\partial (\text{vec}\,\mathbf{X})^T} = (\mathbf{B}^T \otimes \mathbf{A}) + (\mathbf{D}^T \otimes \mathbf{C})\mathbf{K}_{mn}, \qquad (2.2.30)$$

the $mn \times pq$ *gradient matrix can be determined from*

$$\nabla_{\mathbf{X}}\mathbf{F}(\mathbf{X}) = \frac{\partial (\text{vec}\,\mathbf{F}(\mathbf{X}))^T}{\partial \,\text{vec}\,\mathbf{X}} = (\mathbf{B} \otimes \mathbf{A}^T) + \mathbf{K}_{nm}(\mathbf{D} \otimes \mathbf{C}^T). \qquad (2.2.31)$$

Table 2.4 summarizes the matrix differentials and Jacobian matrices of some real functions.

Table 2.5 lists some matrix functions and their Jacobian matrices.

Example 2.5 Let $\mathbf{F}(\mathbf{X}, \mathbf{Y}) = \mathbf{X} \otimes \mathbf{Y}$ be the Kronecker product of two matrices $\mathbf{X} \in \mathbb{R}^{p \times m}$ and $\mathbf{Y} \in \mathbb{R}^{n \times q}$. Consider the matrix differential $\mathrm{d}\mathbf{F}(\mathbf{X}, \mathbf{Y}) = (\mathrm{d}\mathbf{X}) \otimes \mathbf{Y} + \mathbf{X} \otimes (\mathrm{d}\mathbf{Y})$. By the vectorization formula $\text{vec}(\mathbf{X} \otimes \mathbf{Y}) = (\mathbf{I}_m \otimes \mathbf{K}_{qp} \otimes \mathbf{I}_n)(\text{vec}\,\mathbf{X} \otimes \text{vec}\,\mathbf{Y})$, we have

$$\text{vec}(\mathrm{d}\mathbf{X} \otimes \mathbf{Y}) = (\mathbf{I}_m \otimes \mathbf{K}_{qp} \otimes \mathbf{I}_n)(\mathrm{d}\,\text{vec}\mathbf{X} \otimes \text{vec}\,\mathbf{Y})$$

$$= (\mathbf{I}_m \otimes \mathbf{K}_{qp} \otimes \mathbf{I}_n)(\mathbf{I}_{pm} \otimes \text{vec}\,\mathbf{Y})\mathrm{d}\,\text{vec}\mathbf{X}, \qquad (2.2.32)$$

$$\text{vec}(\mathbf{X} \otimes \mathrm{d}\mathbf{Y}) = (\mathbf{I}_m \otimes \mathbf{K}_{qp} \otimes \mathbf{I}_n)(\text{vec}\,\mathbf{X} \otimes \mathrm{d}\,\text{vec}\mathbf{Y})$$

$$= (\mathbf{I}_m \otimes \mathbf{K}_{qp} \otimes \mathbf{I}_n)(\text{vec}\,\mathbf{X} \otimes \mathbf{I}_{nq})\mathrm{d}\,\text{vec}\mathbf{Y}. \qquad (2.2.33)$$

Table 2.4 Matrix differentials and Jacobian matrices of real functions

Functions	Matrix differential	Jacobian matrix
$f(x) : \mathbb{R} \to \mathbb{R}$	$\mathrm{d}f(x) = A\mathrm{d}x$	$A \in \mathbb{R}$
$f(\mathbf{x}) : \mathbb{R}^m \to \mathbb{R}$	$\mathrm{d}f(\mathbf{x}) = \mathbf{A}\mathrm{d}\mathbf{x}$	$\mathbf{A} \in \mathbb{R}^{1 \times m}$
$f(\mathbf{X}) : \mathbb{R}^{m \times n} \to \mathbb{R}$	$\mathrm{d}f(\mathbf{X}) = \text{tr}(\mathbf{A}\mathrm{d}\mathbf{X})$	$\mathbf{A} \in \mathbb{R}^{n \times m}$
$\mathbf{f}(\mathbf{x}) : \mathbb{R}^m \to \mathbb{R}^p$	$\mathrm{d}\mathbf{f}(\mathbf{x}) = \mathbf{A}\mathrm{d}\mathbf{x}$	$\mathbf{A} \in \mathbb{R}^{p \times m}$
$\mathbf{f}(\mathbf{X}) : \mathbb{R}^{m \times n} \to \mathbb{R}^p$	$\mathrm{d}\mathbf{f}(\mathbf{X}) = \mathbf{A}\mathrm{d}(\text{vec}\mathbf{X})$	$\mathbf{A} \in \mathbb{R}^{p \times mn}$
$\mathbf{F}(\mathbf{x}) : \mathbb{R}^m \to \mathbb{R}^{p \times q}$	$\mathrm{d}\,\text{vec}\mathbf{F}(\mathbf{x}) = \mathbf{A}\mathrm{d}\mathbf{x}$	$\mathbf{A} \in \mathbb{R}^{pq \times m}$
$\mathbf{F}(\mathbf{X}) : \mathbb{R}^{m \times n} \to \mathbb{R}^{p \times q}$	$\mathrm{d}\mathbf{F}(\mathbf{X}) = \mathbf{A}(\mathrm{d}\mathbf{X})\mathbf{B} + \mathbf{C}(\mathrm{d}\mathbf{X}^T)\mathbf{D}$	$(\mathbf{B}^T \otimes \mathbf{A}) + (\mathbf{D}^T \otimes \mathbf{C})\mathbf{K}_{mn} \in \mathbb{R}^{pq \times mn}$

2.3 Complex Gradient Matrices

Table 2.5 Differentials and Jacobian matrices of matrix functions

F(X)	dF(X)	Jacobian matrix
$\mathbf{X}^T\mathbf{X}$	$\mathbf{X}^T d\mathbf{X} + (d\mathbf{X}^T)\mathbf{X}$	$(\mathbf{I}_n \otimes \mathbf{X}^T) + (\mathbf{X}^T \otimes \mathbf{I}_n)\mathbf{K}_{mn}$
$\mathbf{X}\mathbf{X}^T$	$\mathbf{X}(d\mathbf{X}^T) + (d\mathbf{X})\mathbf{X}^T$	$(\mathbf{I}_m \otimes \mathbf{X})\mathbf{K}_{mn} + (\mathbf{X} \otimes \mathbf{I}_m)$
$\mathbf{A}\mathbf{X}^T\mathbf{B}$	$\mathbf{A}(d\mathbf{X}^T)\mathbf{B}$	$(\mathbf{B}^T \otimes \mathbf{A})\mathbf{K}_{mn}$
$\mathbf{X}^T\mathbf{B}\mathbf{X}$	$\mathbf{X}^T\mathbf{B}\,d\mathbf{X} + (d\mathbf{X}^T)\mathbf{B}\mathbf{X}$	$\mathbf{I} \otimes (\mathbf{X}^T\mathbf{B}) + ((\mathbf{B}\mathbf{X})^T \otimes \mathbf{I})\mathbf{K}_{mn}$
$\mathbf{A}\mathbf{X}^T\mathbf{B}\mathbf{X}\mathbf{C}$	$\mathbf{A}(d\mathbf{X}^T)\mathbf{B}\mathbf{X}\mathbf{C} + \mathbf{A}\mathbf{X}^T\mathbf{B}(d\mathbf{X})\mathbf{C}$	$((\mathbf{B}\mathbf{X}\mathbf{C})^T \otimes \mathbf{A})\mathbf{K}_{mn} + \mathbf{C}^T \otimes (\mathbf{A}\mathbf{X}^T\mathbf{B})$
$\mathbf{A}\mathbf{X}\mathbf{B}\mathbf{X}^T\mathbf{C}$	$\mathbf{A}(d\mathbf{X})\mathbf{B}\mathbf{X}^T\mathbf{C} + \mathbf{A}\mathbf{X}\mathbf{B}(d\mathbf{X}^T)\mathbf{C}$	$(\mathbf{B}\mathbf{X}^T\mathbf{C})^T \otimes \mathbf{A} + (\mathbf{C}^T \otimes (\mathbf{A}\mathbf{X}\mathbf{B}))\mathbf{K}_{mn}$
$\mathbf{X}^{-1}$	$-\mathbf{X}^{-1}(d\mathbf{X})\mathbf{X}^{-1}$	$-(\mathbf{X}^{-T} \otimes \mathbf{X}^{-1})$
$\mathbf{X}^k$	$\sum_{j=1}^{k} \mathbf{X}^{j-1}(d\mathbf{X})\mathbf{X}^{k-j}$	$\sum_{j=1}^{k} (\mathbf{X}^T)^{k-j} \otimes \mathbf{X}^{j-1}$
$\log \mathbf{X}$	$\mathbf{X}^{-1} d\mathbf{X}$	$\mathbf{I} \otimes \mathbf{X}^{-1}$
$\exp(\mathbf{X})$	$\sum_{k=0}^{\infty} \frac{1}{(k+1)!} \sum_{j=0}^{k} \mathbf{X}^j (d\mathbf{X}) \mathbf{X}^{k-j}$	$\sum_{k=0}^{\infty} \frac{1}{(k+1)!} \sum_{j=0}^{k} (\mathbf{X}^T)^{k-j} \otimes \mathbf{X}^j$

Hence, the Jacobian matrices with respect to the variable matrices $\mathbf{X}$ and $\mathbf{Y}$ are, respectively, given by

$$\mathbf{J}_\mathbf{X}(\mathbf{X} \otimes \mathbf{Y}) = (\mathbf{I}_m \otimes \mathbf{K}_{qp} \otimes \mathbf{I}_n)(\mathbf{I}_{pm} \otimes \text{vec}\,\mathbf{Y}), \quad (2.2.34)$$

$$\mathbf{J}_\mathbf{Y}(\mathbf{X} \otimes \mathbf{Y}) = (\mathbf{I}_m \otimes \mathbf{K}_{qp} \otimes \mathbf{I}_n)(\text{vec}\,\mathbf{X} \otimes \mathbf{I}_{nq}). \quad (2.2.35)$$

The analysis and examples in this section show that the first-order real matrix differential is indeed an effective mathematical tool for identifying the Jacobian matrix and the gradient matrix of a real function. And this tool is simple and easy to master.

2.3 Complex Gradient Matrices

In many engineering applications, observed data are usually complex. In these cases, the objective function of an optimization problem is a real-valued function of a complex vector or matrix. Hence, the gradient of the real objective function with respect to the complex vector or matrix variable is a complex vector or complex matrix. This complex gradient has the following two forms:

- *Complex gradient:* the gradient of the objective function with respect to the complex vector or matrix variable itself;
- *Conjugate gradient:* the gradient of the objective function with respect to the complex conjugate vector or matrix variable.

2.3.1 Holomorphic Function and Complex Partial Derivative

Before discussing the complex gradient and conjugate gradient, it is necessary to recall the relevant facts about complex functions.

Definition 2.6 (Complex Analytic Function [4]) Let $D \subseteq \mathbb{C}$ be the definition domain of the function $f : D \to \mathbb{C}$. The function $f(z)$ with complex variable z is said to be a *complex analytic function* in the domain D if $f(z)$ is *complex differentiable*, namely $\lim_{\Delta z \to 0} \frac{f(z + \Delta z) - f(z)}{\Delta z}$ exists for all $z \in D$.

In the standard framework of complex functions, a complex function $f(z)$ (where $z = x + jy$) is written in the real polar coordinates $r \stackrel{\text{def}}{=} (x, y)$ as $f(r) = f(x, y)$.

The terminology "complex analytic function" is commonly replaced by the completely synonymous terminology "*holomorphic function*." It is noted that a (real) analytic complex function is (real) in the real-variable x-domain and y-domain, but is not necessarily holomorphic in the complex variable domain $z = x + jy$, i.e., it may be complex nonanalytic.

A complex function $f(z)$ can always be expressed in terms of its real part $u(x, y)$ and imaginary part $v(x, y)$ as

$$f(z) = u(x, y) + jv(x, y),$$

where $z = x + jy$ and both $u(x, y)$ and $v(x, y)$ are real functions.

For a holomorphic scalar function, the following four statements are equivalent [2]:

1. The complex function $f(z)$ is a holomorphic (i.e., complex analytic) function.
2. The derivative $f'(z)$ of the complex function exists and is continuous.
3. The complex function $f(z)$ satisfies the *Cauchy–Riemann condition*

$$\frac{\partial u}{\partial x} = \frac{\partial v}{\partial y} \quad \text{and} \quad \frac{\partial v}{\partial x} = -\frac{\partial u}{\partial y}. \tag{2.3.1}$$

4. All derivatives of the complex function $f(z)$ exist, and $f(z)$ has a convergent power series.

The Cauchy–Riemann condition is also called the *Cauchy–Riemann equations*. The function $f(z) = u(x, y) + jv(x, y)$ is a holomorphic function only when both the real functions $u(x, y)$ and $v(x, y)$ satisfy the *Laplace equations* at the same time:

$$\frac{\partial^2 u(x, y)}{\partial x^2} + \frac{\partial^2 u(x, y)}{\partial y^2} = 0 \quad \text{and} \quad \frac{\partial^2 v(x, y)}{\partial x^2} + \frac{\partial^2 v(x, y)}{\partial y^2} = 0. \tag{2.3.2}$$

2.3 Complex Gradient Matrices

A real function $g(x, y)$ is called a *harmonic function*, if it satisfies the Laplace equation

$$\frac{\partial^2 g(x, y)}{\partial x^2} + \frac{\partial^2 g(x, y)}{\partial y^2} = 0. \tag{2.3.3}$$

A complex function $f(z) = u(x, y) + jv(x, y)$ is not a holomorphic function if any of two real functions $u(x, y)$ and $v(x, y)$ does not meet the Cauchy–Riemann condition or the Laplace equations.

Although the power function z^n, the exponential function e^z, the logarithmic function $\ln z$, the sine function $\sin z$, and the cosine function $\cos z$ are holomorphic functions, i.e., analytic functions in the complex plane, many commonly used functions are not holomorphic. A natural question to ask is whether there is a general representation form $f(z, \cdot)$ instead of $f(z)$ such that $f(z, \cdot)$ is always holomorphic. The key to solving this problem is to adopt $f(z, z^*)$ instead of $f(z)$, as shown in Table 2.6.

In the framework of complex derivatives, the *formal partial derivatives* of complex numbers are defined as

$$\frac{\partial}{\partial z} = \frac{1}{2}\left(\frac{\partial}{\partial x} - j\frac{\partial}{\partial y}\right), \quad \frac{\partial}{\partial z^*} = \frac{1}{2}\left(\frac{\partial}{\partial x} + j\frac{\partial}{\partial y}\right). \tag{2.3.4}$$

The formal partial derivatives above were presented by Wirtinger [8] in 1927, so they are sometimes called *Wirtinger partial derivatives*.

On the partial derivatives of the complex variable $z = x + jy$, a basic assumption is the independence of its real and imaginary parts:

$$\frac{\partial x}{\partial y} = 0 \quad \text{and} \quad \frac{\partial y}{\partial x} = 0. \tag{2.3.5}$$

Table 2.6 Forms of complex-valued functions

Function	Variables $z, z^* \in \mathbb{C}$	Variables $\mathbf{z}, \mathbf{z}^* \in \mathbb{C}^m$	Variables $\mathbf{Z}, \mathbf{Z}^* \in \mathbb{C}^{m \times n}$
$f \in \mathbb{C}$	$f(z, z^*)$	$f(\mathbf{z}, \mathbf{z}^*)$	$f(\mathbf{Z}, \mathbf{Z}^*)$
	$f : \mathbb{C} \times \mathbb{C} \to \mathbb{C}$	$f : \mathbb{C}^m \times \mathbb{C}^m \to \mathbb{C}$	$f : \mathbb{C}^{m \times n} \times \mathbb{C}^{m \times n} \to \mathbb{C}$
$\mathbf{f} \in \mathbb{C}^p$	$\mathbf{f}(z, z^*)$	$\mathbf{f}(\mathbf{z}, \mathbf{z}^*)$	$\mathbf{f}(\mathbf{Z}, \mathbf{Z}^*)$
	$\mathbf{f} : \mathbb{C} \times \mathbb{C} \to \mathbb{C}^p$	$\mathbf{f} : \mathbb{C}^m \times \mathbb{C}^m \to \mathbb{C}^p$	$\mathbf{f} : \mathbb{C}^{m \times n} \times \mathbb{C}^{m \times n} \to \mathbb{C}^p$
$\mathbf{F} \in \mathbb{C}^{p \times q}$	$\mathbf{F}(z, z^*)$	$\mathbf{F}(\mathbf{z}, \mathbf{z}^*)$	$\mathbf{F}(\mathbf{Z}, \mathbf{Z}^*)$
	$\mathbf{F} : \mathbb{C} \times \mathbb{C} \to \mathbb{C}^{p \times q}$	$\mathbf{F} : \mathbb{C}^m \times \mathbb{C}^m \to \mathbb{C}^{p \times q}$	$\mathbf{F} : \mathbb{C}^{m \times n} \times \mathbb{C}^{m \times n} \to \mathbb{C}^{p \times q}$

From both the definition of the partial derivative and the above independence assumption, it is easy to conclude that

$$\frac{\partial z}{\partial z^*} = \frac{\partial x}{\partial z^*} + j\frac{\partial y}{\partial z^*} = \frac{1}{2}\left(\frac{\partial x}{\partial x} + j\frac{\partial x}{\partial y}\right) + j\frac{1}{2}\left(\frac{\partial y}{\partial x} + j\frac{\partial y}{\partial y}\right)$$
$$= \frac{1}{2}(1+0) + j\frac{1}{2}(0+j),$$
$$\frac{\partial z^*}{\partial z} = \frac{\partial x}{\partial z} - j\frac{\partial y}{\partial z} = \frac{1}{2}\left(\frac{\partial x}{\partial x} - j\frac{\partial x}{\partial y}\right) - j\frac{1}{2}\left(\frac{\partial y}{\partial x} - j\frac{\partial y}{\partial y}\right) = \frac{1}{2}(1-0) - j\frac{1}{2}(0-j).$$

That is to say,

$$\frac{\partial z}{\partial z^*} = 0 \quad \text{and} \quad \frac{\partial z^*}{\partial z} = 0. \tag{2.3.6}$$

Equation (2.3.6) reveals a basic result in the theory of complex variables: the complex variable z and the complex conjugate variable z^* are independent variables.

Therefore, when finding the *complex partial derivative* $\nabla_z f(z, z^*)$ and the *complex conjugate partial derivative* $\nabla_{z^*} f(z, z^*)$, the complex variable z and the complex conjugate variable z^* can be regarded as two independent variables:

$$\nabla_z f(z, z^*) = \left.\frac{\partial f(z, z^*)}{\partial z}\right|_{z^*=\text{const}}, \quad \nabla_{z^*} f(z, z^*) = \left.\frac{\partial f(z, z^*)}{\partial z^*}\right|_{z=\text{const}}. \tag{2.3.7}$$

This implies that when any nonholomorphic function $f(z)$ is written as $f(z, z^*)$, it becomes holomorphic, because, for a fixed z^* value, the function $f(z, z^*)$ is analytic on the whole complex plane $z = x + jy$, and for a fixed z value, the function $f(z, z^*)$ is analytic on the whole complex plane $z^* = x - jy$, see, e.g., [2].

Table 2.7 is a comparison between the nonholomorphic and holomorphic representation forms of complex functions.

The function $f(z) = |z|^2$ itself is not a holomorphic function with respect to z, but $f(z, z^*) = |z|^2 = zz^*$ is holomorphic, because its first-order partial derivatives $\partial |z|^2/\partial z = z^*$ and $\partial |z|^2/\partial z^* = z$ exist and are continuous.

Table 2.7 Nonholomorphic and holomorphic functions

Functions	Nonholomorphic	Holomorphic
Coordinates	$\begin{cases} r \stackrel{\text{def}}{=} (x, y) \in \mathbb{R} \times \mathbb{R} \\ z = x + jy \end{cases}$	$\begin{cases} r \stackrel{\text{def}}{=} (z, z^*) \in \mathbb{C} \times \mathbb{C} \\ z = x + jy, \ z^* = x - jy \end{cases}$
Representation	$f(r) = f(x, y)$	$f(c) = f(z, z^*)$

2.3 Complex Gradient Matrices

The following are common formulas and rules for the complex partial derivatives:

1. The conjugate partial derivative of the complex conjugate function $\frac{\partial f^*(z,z^*)}{\partial z^*}$:

$$\frac{\partial f^*(z, z^*)}{\partial z^*} = \left(\frac{\partial f(z, z^*)}{\partial z}\right)^*. \quad (2.3.8)$$

2. The partial derivative of the conjugate of the complex function $\frac{\partial f^*(z,z^*)}{\partial z}$:

$$\frac{\partial f^*(z, z^*)}{\partial z} = \left(\frac{\partial f(z, z^*)}{\partial z^*}\right)^*. \quad (2.3.9)$$

3. *Complex differential rule*

$$df(z, z^*) = \frac{\partial f(z, z^*)}{\partial z} dz + \frac{\partial f(z, z^*)}{\partial z^*} dz^*. \quad (2.3.10)$$

4. *Complex chain rule*

$$\frac{\partial h(g(z, z^*))}{\partial z} = \frac{\partial h(g(z, z^*))}{\partial g(z, z^*)} \frac{\partial g(z, z^*)}{\partial z} + \frac{\partial h(g(z, z^*))}{\partial g^*(z, z^*)} \frac{\partial g^*(z, z^*)}{\partial z}, \quad (2.3.11)$$

$$\frac{\partial h(g(z, z^*))}{\partial z^*} = \frac{\partial h(g(z, z^*))}{\partial g(z, z^*)} \frac{\partial g(z, z^*)}{\partial z^*} + \frac{\partial h(g(z, z^*))}{\partial g^*(z, z^*)} \frac{\partial g^*(z, z^*)}{\partial z^*}. \quad (2.3.12)$$

2.3.2 Complex Matrix Differential

Consider the complex matrix function $\mathbf{F}(\mathbf{Z})$ and the holomorphic complex matrix function $\mathbf{F}(\mathbf{Z}, \mathbf{Z}^*)$.

On *holomorphic complex matrix functions*, the following statements are equivalent [1]:

- The matrix function $\mathbf{F}(\mathbf{Z})$ is a holomorphic function of the complex matrix variable $\mathbf{Z}$.
- The complex matrix differential $d\,\text{vec}\mathbf{F}(\mathbf{Z}) = \frac{\partial\,\text{vec}\mathbf{F}(\mathbf{Z})}{\partial(\text{vec}\mathbf{Z})^T} d\,\text{vec}\mathbf{Z}$.
- For all $\mathbf{Z}$, $\frac{\partial\,\text{vec}\mathbf{F}(\mathbf{Z})}{\partial(\text{vec}\,\mathbf{Z}^*)^T} = \mathbf{O}$ (zero matrix) holds.
- For all $\mathbf{Z}$, $\frac{\partial\,\text{vec}\mathbf{F}(\mathbf{Z})}{\partial(\text{vecRe}\mathbf{Z})^T} + j\frac{\partial\,\text{vec}\mathbf{F}(\mathbf{Z})}{\partial(\text{vecIm}\mathbf{Z})^T} = \mathbf{O}$ holds.

The complex matrix function $\mathbf{F}(\mathbf{Z}, \mathbf{Z}^*)$ is obviously a holomorphic function, and its matrix differential is

$$d\,\text{vec}\mathbf{F}(\mathbf{Z}, \mathbf{Z}^*) = \frac{\partial\,\text{vec}\mathbf{F}(\mathbf{Z}, \mathbf{Z}^*)}{\partial(\text{vec}\mathbf{Z})^T} d\,\text{vec}\mathbf{Z} + \frac{\partial\,\text{vec}\mathbf{F}(\mathbf{Z}, \mathbf{Z}^*)}{\partial(\text{vec}\mathbf{Z}^*)^T} d\,\text{vec}\mathbf{Z}^*. \quad (2.3.13)$$

The complex matrix differential $d\mathbf{Z} = [dZ_{ij}]_{i=1,j=1}^{m,n}$ has the following properties [1]:

1. *Transpose:* $d\mathbf{Z}^T = d(\mathbf{Z}^T) = (d\mathbf{Z})^T$.
2. *Hermitian transpose:* $d\mathbf{Z}^H = d(\mathbf{Z}^H) = (d\mathbf{Z})^H$.
3. *Conjugate:* $d\mathbf{Z}^* = d(\mathbf{Z}^*) = (d\mathbf{Z})^*$.
4. *Linearity (additive rule):* $d(\mathbf{Y} + \mathbf{Z}) = d\mathbf{Y} + d\mathbf{Z}$.
5. *Chain rule:* If $\mathbf{F}$ is a function of $\mathbf{Y}$, while $\mathbf{Y}$ is a function of $\mathbf{Z}$, then

$$d\,\mathrm{vec}\mathbf{F} = \frac{\partial\,\mathrm{vec}\mathbf{F}}{\partial(\mathrm{vec}\mathbf{Y})^T} d\,\mathrm{vec}\mathbf{Y} = \frac{\partial\,\mathrm{vec}\mathbf{F}}{\partial(\mathrm{vec}\mathbf{Y})^T} \frac{\partial\,\mathrm{vec}\mathbf{Y}}{\partial(\mathrm{vec}\mathbf{Z})^T} d\,\mathrm{vec}\mathbf{Z},$$

where $\frac{\partial\,\mathrm{vec}\mathbf{F}}{\partial(\mathrm{vec}\mathbf{Y})^T}$ and $\frac{\partial\,\mathrm{vec}\mathbf{Y}}{\partial(\mathrm{vec}\mathbf{Z})^T}$ are the *normal complex partial derivative* and the *generalized complex partial derivative*, respectively.

6. *Multiplication rule:*

$$d(\mathbf{UV}) = (d\mathbf{U})\mathbf{V} + \mathbf{U}(d\mathbf{V})$$

$$d\,\mathrm{vec}(\mathbf{UV}) = (\mathbf{V}^T \otimes \mathbf{I}) d\,\mathrm{vec}\mathbf{U} + (\mathbf{I} \otimes \mathbf{U}) d\,\mathrm{vec}\mathbf{V}.$$

7. *Kronecker product:* $d(\mathbf{Y} \otimes \mathbf{Z}) = d\mathbf{Y} \otimes \mathbf{Z} + \mathbf{Y} \otimes d\mathbf{Z}$.
8. *Hadamard product:* $d(\mathbf{Y} \odot \mathbf{Z}) = d\mathbf{Y} \odot \mathbf{Z} + \mathbf{Y} \odot d\mathbf{Z}$.

Let us consider the relationship between the complex matrix differential and the complex partial derivative.

First, the complex differential rule for scalar variables,

$$df(z, z^*) = \frac{\partial f(z, z^*)}{\partial z} dz + \frac{\partial f(z, z^*)}{\partial z^*} dz^* \qquad (2.3.14)$$

is easily extended to a complex differential rule for the multivariate real scalar function $f(\cdot) = f((z_1, z_1^*), \ldots, (z_m, z_m^*))$:

$$df(\cdot) = \frac{\partial f(\cdot)}{\partial z_1} dz_1 + \cdots + \frac{\partial f(\cdot)}{\partial z_m} dz_m + \frac{\partial f(\cdot)}{\partial z_1^*} dz_1^* + \cdots + \frac{\partial f(\cdot)}{\partial z_m^*} dz_m^*$$

$$= \frac{\partial f(\cdot)}{\partial \mathbf{z}^T} d\mathbf{z} + \frac{\partial f(\cdot)}{\partial \mathbf{z}^H} d\mathbf{z}^*. \qquad (2.3.15)$$

Here, $d\mathbf{z} = [dz_1, \ldots, dz_m]^T$, $d\mathbf{z}^* = [dz_1^*, \ldots, dz_m^*]^T$. This complex differential rule is the basis of the complex matrix differential.

In particular, if $f(\cdot) = f(\mathbf{z}, \mathbf{z}^*)$, then

$$df(\mathbf{z}, \mathbf{z}^*) = \frac{\partial f(\mathbf{z}, \mathbf{z}^*)}{\partial \mathbf{z}^T} d\mathbf{z} + \frac{\partial f(\mathbf{z}, \mathbf{z}^*)}{\partial \mathbf{z}^H} d\mathbf{z}^*$$

2.3 Complex Gradient Matrices

or

$$df(\mathbf{z}, \mathbf{z}^*) = D_{\mathbf{z}} f(\mathbf{z}, \mathbf{z}^*) \, d\mathbf{z} + D_{\mathbf{z}^*} f(\mathbf{z}, \mathbf{z}^*) \, d\mathbf{z}^*, \qquad (2.3.16)$$

where

$$D_{\mathbf{z}} f(\mathbf{z}, \mathbf{z}^*) = \left. \frac{\partial f(\mathbf{z}, \mathbf{z}^*)}{\partial \mathbf{z}^T} \right|_{\mathbf{z}^* = \text{const}} = \left[\frac{\partial f(\mathbf{z}, \mathbf{z}^*)}{\partial z_1}, \ldots, \frac{\partial f(\mathbf{z}, \mathbf{z}^*)}{\partial z_m} \right], \qquad (2.3.17)$$

$$D_{\mathbf{z}^*} f(\mathbf{z}, \mathbf{z}^*) = \left. \frac{\partial f(\mathbf{z}, \mathbf{z}^*)}{\partial \mathbf{z}^H} \right|_{\mathbf{z} = \text{const}} = \left[\frac{\partial f(\mathbf{z}, \mathbf{z}^*)}{\partial z_1^*}, \ldots, \frac{\partial f(\mathbf{z}, \mathbf{z}^*)}{\partial z_m^*} \right] \qquad (2.3.18)$$

are, respectively, the *cogradient vector* and the *conjugate cogradient vector* of the real scalar function $f(\mathbf{z}, \mathbf{z}^*)$, while

$$D_{\mathbf{z}} = \frac{\partial}{\partial \mathbf{z}^T} \stackrel{\text{def}}{=} \left[\frac{\partial}{\partial z_1}, \ldots, \frac{\partial}{\partial z_m} \right], \quad D_{\mathbf{z}^*} = \frac{\partial}{\partial \mathbf{z}^H} \stackrel{\text{def}}{=} \left[\frac{\partial}{\partial z_1^*}, \ldots, \frac{\partial}{\partial z_m^*} \right] \qquad (2.3.19)$$

are termed the *cogradient operator* and the *conjugate cogradient operator* of complex vector variable $\mathbf{z} \in \mathbb{C}^m$, respectively.

For $\mathbf{z} = \mathbf{x} + j\mathbf{y} = [z_1, \ldots, z_m]^T \in \mathbb{C}^m$ with $\mathbf{x} = [x_1, \ldots, x_m]^T \in \mathbb{R}^m$, $\mathbf{y} = [y_1, \ldots, y_m]^T \in \mathbb{R}^m$, due to the independence between the real part x_i and the imaginary part y_i, if applying the complex partial derivative operators

$$D_{z_i} = \frac{\partial}{\partial z_i} = \frac{1}{2} \left(\frac{\partial}{\partial x_i} - j \frac{\partial}{\partial y_i} \right), \quad D_{z_i^*} = \frac{\partial}{\partial z_i^*} = \frac{1}{2} \left(\frac{\partial}{\partial x_i} + j \frac{\partial}{\partial y_i} \right), \qquad (2.3.20)$$

to each element of the row vector $\mathbf{z}^T = [z_1, \ldots, z_m]$, then we obtain the following *complex cogradient operator*:

$$D_{\mathbf{z}} = \frac{\partial}{\partial \mathbf{z}^T} = \frac{1}{2} \left(\frac{\partial}{\partial \mathbf{x}^T} - j \frac{\partial}{\partial \mathbf{y}^T} \right) \qquad (2.3.21)$$

and the *complex conjugate cogradient operator*

$$D_{\mathbf{z}^*} = \frac{\partial}{\partial \mathbf{z}^H} = \frac{1}{2} \left(\frac{\partial}{\partial \mathbf{x}^T} + j \frac{\partial}{\partial \mathbf{y}^T} \right). \qquad (2.3.22)$$

Similarly, the *complex gradient operator* and the *complex conjugate gradient operator* are, respectively, defined as

$$\nabla_{\mathbf{z}} = \frac{\partial}{\partial \mathbf{z}} \stackrel{\text{def}}{=} \left[\frac{\partial}{\partial z_1}, \ldots, \frac{\partial}{\partial z_m} \right]^T, \quad \nabla_{\mathbf{z}^*} = \frac{\partial}{\partial \mathbf{z}^*} \stackrel{\text{def}}{=} \left[\frac{\partial}{\partial z_1^*}, \ldots, \frac{\partial}{\partial z_m^*} \right]^T. \qquad (2.3.23)$$

Hence, the *complex gradient vector* and the *complex conjugate gradient vector* of the real scalar function $f(\mathbf{z}, \mathbf{z}^*)$ are, respectively, defined as

$$\nabla_\mathbf{z} f(\mathbf{z}, \mathbf{z}^*) = \left.\frac{\partial f(\mathbf{z}, \mathbf{z}^*)}{\partial \mathbf{z}}\right|_{\mathbf{z}^* = \text{const vector}} = (\mathrm{D}_\mathbf{z} f(\mathbf{z}, \mathbf{z}^*))^T, \qquad (2.3.24)$$

$$\nabla_{\mathbf{z}^*} f(\mathbf{z}, \mathbf{z}^*) = \left.\frac{\partial f(\mathbf{z}, \mathbf{z}^*)}{\partial \mathbf{z}^*}\right|_{\mathbf{z} = \text{const vector}} = (\mathrm{D}_{\mathbf{z}^*} f(\mathbf{z}, \mathbf{z}^*))^T. \qquad (2.3.25)$$

On the other hand, by mimicking the complex partial derivative operator to each element of the complex vector $\mathbf{z} = [z_1, \ldots, z_m]^T$, the complex gradient operator and the complex conjugate gradient operator are obtained as follows:

$$\nabla_\mathbf{z} = \frac{\partial}{\partial \mathbf{z}} = \frac{1}{2}\left(\frac{\partial}{\partial \mathbf{x}} - \mathrm{j}\frac{\partial}{\partial \mathbf{y}}\right), \quad \nabla_{\mathbf{z}^*} = \frac{\partial}{\partial \mathbf{z}^*} = \frac{1}{2}\left(\frac{\partial}{\partial \mathbf{x}} + \mathrm{j}\frac{\partial}{\partial \mathbf{y}}\right). \qquad (2.3.26)$$

By the above definitions, it is easy to obtain

$$\frac{\partial \mathbf{z}^T}{\partial \mathbf{z}} = \frac{\partial \mathbf{x}^T}{\partial \mathbf{z}} + \mathrm{j}\frac{\partial \mathbf{y}^T}{\partial \mathbf{z}} = \frac{1}{2}\left(\frac{\partial \mathbf{x}^T}{\partial \mathbf{x}} - \mathrm{j}\frac{\partial \mathbf{x}^T}{\partial \mathbf{y}}\right) + \mathrm{j}\frac{1}{2}\left(\frac{\partial \mathbf{y}^T}{\partial \mathbf{x}} - \mathrm{j}\frac{\partial \mathbf{y}^T}{\partial \mathbf{y}}\right) = \mathbf{I}_{m \times m},$$

and

$$\frac{\partial \mathbf{z}^T}{\partial \mathbf{z}^*} = \frac{\partial \mathbf{x}^T}{\partial \mathbf{z}^*} + \mathrm{j}\frac{\partial \mathbf{y}^T}{\partial \mathbf{z}^*} = \frac{1}{2}\left(\frac{\partial \mathbf{x}^T}{\partial \mathbf{x}} + \mathrm{j}\frac{\partial \mathbf{x}^T}{\partial \mathbf{y}}\right) + \mathrm{j}\frac{1}{2}\left(\frac{\partial \mathbf{y}^T}{\partial \mathbf{x}} + \mathrm{j}\frac{\partial \mathbf{y}^T}{\partial \mathbf{y}}\right) = \mathbf{O}_{m \times m}.$$

Summarizing the above results and their conjugate, transpose and complex conjugate transpose, we have the following important results:

$$\frac{\partial \mathbf{z}^T}{\partial \mathbf{z}} = \mathbf{I}, \quad \frac{\partial \mathbf{z}^H}{\partial \mathbf{z}^*} = \mathbf{I}, \quad \frac{\partial \mathbf{z}}{\partial \mathbf{z}^T} = \mathbf{I}, \quad \frac{\partial \mathbf{z}^*}{\partial \mathbf{z}^H} = \mathbf{I}, \qquad (2.3.27)$$

$$\frac{\partial \mathbf{z}^T}{\partial \mathbf{z}^*} = \mathbf{O}, \quad \frac{\partial \mathbf{z}^H}{\partial \mathbf{z}} = \mathbf{O}, \quad \frac{\partial \mathbf{z}}{\partial \mathbf{z}^H} = \mathbf{O}, \quad \frac{\partial \mathbf{z}^*}{\partial \mathbf{z}^T} = \mathbf{O}. \qquad (2.3.28)$$

The above results show that the complex vector variable $\mathbf{z}$ and its complex conjugate vector variable $\mathbf{z}^*$ can be viewed as two independent variables. This important fact is not surprising because the angle between $\mathbf{z}$ and $\mathbf{z}^*$ is $\pi/2$, i.e., they are orthogonal to each other. Hence, we can summarize the rules for using the cogradient operator and gradient operator below.

- When using the complex cogradient operator $\partial/\partial \mathbf{z}^T$ or the complex gradient operator $\partial/\partial \mathbf{z}$, the complex conjugate vector variable $\mathbf{z}^*$ can be handled as a constant vector.
- When using the complex conjugate cogradient operator $\partial/\partial \mathbf{z}^H$ or the complex conjugate gradient operator $\partial/\partial \mathbf{z}^*$, the vector variable $\mathbf{z}$ can be handled as a constant vector.

2.3 Complex Gradient Matrices

Now, we consider the real scalar function $f(\mathbf{Z}, \mathbf{Z}^*)$ with variables $\mathbf{Z}, \mathbf{Z}^* \in \mathbb{C}^{m \times n}$. Performing the vectorization of $\mathbf{Z}$ and $\mathbf{Z}^*$, respectively, from Eq. (2.3.16) we get the first-order complex differential rule for the real scalar function $f(\mathbf{Z}, \mathbf{Z}^*)$:

$$\begin{aligned} \mathrm{d}f(\mathbf{Z}, \mathbf{Z}^*) &= \frac{\partial f(\mathbf{Z}, \mathbf{Z}^*)}{\partial (\mathrm{vec}\mathbf{Z})^T} \mathrm{dvec}\mathbf{Z} + \frac{\partial f(\mathbf{Z}, \mathbf{Z}^*)}{\partial (\mathrm{vec}\mathbf{Z}^*)^T} \mathrm{dvec}\mathbf{Z}^* \\ &= \frac{\partial f(\mathbf{Z}, \mathbf{Z}^*)}{\partial (\mathrm{vec}\mathbf{Z})^T} \mathrm{dvec}\mathbf{Z} + \frac{\partial f(\mathbf{Z}, \mathbf{Z}^*)}{\partial (\mathrm{vec}\mathbf{Z}^*)^T} \mathrm{dvec}\mathbf{Z}^*, \end{aligned} \qquad (2.3.29)$$

where

$$\frac{\partial f(\mathbf{Z}, \mathbf{Z}^*)}{\partial (\mathrm{vec}\,\mathbf{Z})^T} = \left[\frac{\partial f(\mathbf{Z}, \mathbf{Z}^*)}{\partial Z_{11}}, \ldots, \frac{\partial f(\mathbf{Z}, \mathbf{Z}^*)}{\partial Z_{m1}}, \ldots, \frac{\partial f(\mathbf{Z}, \mathbf{Z}^*)}{\partial Z_{1n}}, \ldots, \frac{\partial f(\mathbf{Z}, \mathbf{Z}^*)}{\partial Z_{mn}} \right],$$

$$\frac{\partial f(\mathbf{Z}, \mathbf{Z}^*)}{\partial (\mathrm{vec}\,\mathbf{Z}^*)^T} = \left[\frac{\partial f(\mathbf{Z}, \mathbf{Z}^*)}{\partial Z_{11}^*}, \ldots, \frac{\partial f(\mathbf{Z}, \mathbf{Z}^*)}{\partial Z_{m1}^*}, \ldots, \frac{\partial f(\mathbf{Z}, \mathbf{Z}^*)}{\partial Z_{1n}^*}, \ldots, \frac{\partial f(\mathbf{Z}, \mathbf{Z}^*)}{\partial Z_{mn}^*} \right].$$

We define the complex cogradient vector and the complex conjugate cogradient vector as

$$\mathrm{D}_{\mathrm{vec}\,\mathbf{Z}} f(\mathbf{Z}, \mathbf{Z}^*) = \frac{\partial f(\mathbf{Z}, \mathbf{Z}^*)}{\partial (\mathrm{vec}\,\mathbf{Z})^T}, \quad \mathrm{D}_{\mathrm{vec}\,\mathbf{Z}^*} f(\mathbf{Z}, \mathbf{Z}^*) = \frac{\partial f(\mathbf{Z}, \mathbf{Z}^*)}{\partial (\mathrm{vec}\,\mathbf{Z}^*)^T}, \qquad (2.3.30)$$

and the complex gradient vector and the complex conjugate gradient vector of the function $f(\mathbf{Z}, \mathbf{Z}^*)$ as

$$\nabla_{\mathrm{vec}\,\mathbf{Z}} f(\mathbf{Z}, \mathbf{Z}^*) = \frac{\partial f(\mathbf{Z}, \mathbf{Z}^*)}{\partial \,\mathrm{vec}\,\mathbf{Z}}, \quad \nabla_{\mathrm{vec}\,\mathbf{Z}^*} f(\mathbf{Z}, \mathbf{Z}^*) = \frac{\partial f(\mathbf{Z}, \mathbf{Z}^*)}{\partial \,\mathrm{vec}\,\mathbf{Z}^*}. \qquad (2.3.31)$$

The conjugate gradient vector $\nabla_{\mathrm{vec}\,\mathbf{Z}^*} f(\mathbf{Z}, \mathbf{Z}^*)$ has the following properties [1]:

1. The conjugate gradient vector of the function $f(\mathbf{Z}, \mathbf{Z}^*)$ at an extreme point is equal to the zero vector, i.e., $\nabla_{\mathrm{vec}\,\mathbf{Z}^*} f(\mathbf{Z}, \mathbf{Z}^*) = \mathbf{0}$.
2. The conjugate gradient vector $\nabla_{\mathrm{vec}\,\mathbf{Z}^*} f(\mathbf{Z}, \mathbf{Z}^*)$ and the negative conjugate gradient vector $-\nabla_{\mathrm{vec}\,\mathbf{Z}^*} f(\mathbf{Z}, \mathbf{Z}^*)$ point in the direction of the steepest ascent and steepest descent of the function $f(\mathbf{Z}, \mathbf{Z}^*)$, respectively.
3. The step length of the steepest increase slope is $\|\nabla_{\mathrm{vec}\,\mathbf{Z}^*} f(\mathbf{Z}, \mathbf{Z}^*)\|_2$.
4. The conjugate gradient vector $\nabla_{\mathrm{vec}\,\mathbf{Z}^*} f(\mathbf{Z}, \mathbf{Z}^*)$ and the negative conjugate gradient vector $-\nabla_{\mathrm{vec}\,\mathbf{Z}^*} f(\mathbf{Z}, \mathbf{Z}^*)$ can be used separately as update in gradient ascent algorithms and gradient descent algorithms.

Furthermore, the *complex Jacobian matrix* and the *complex conjugate Jacobian matrix* of the real scalar function $f(\mathbf{Z}, \mathbf{Z}^*)$ are, respectively, as follows:

$$\mathbf{J}_{\mathbf{Z}} \stackrel{\text{def}}{=} \left.\frac{\partial f(\mathbf{Z}, \mathbf{Z}^*)}{\partial \mathbf{Z}^T}\right|_{\mathbf{Z}^* = \text{const matrix}} = \begin{bmatrix} \frac{\partial f(\mathbf{Z},\mathbf{Z}^*)}{\partial Z_{11}} & \cdots & \frac{\partial f(\mathbf{Z},\mathbf{Z}^*)}{\partial Z_{m1}} \\ \vdots & \ddots & \vdots \\ \frac{\partial f(\mathbf{Z},\mathbf{Z}^*)}{\partial Z_{1n}} & \cdots & \frac{\partial f(\mathbf{Z},\mathbf{Z}^*)}{\partial Z_{mn}} \end{bmatrix}, \quad (2.3.32)$$

$$\mathbf{J}_{\mathbf{Z}^*} \stackrel{\text{def}}{=} \left.\frac{\partial f(\mathbf{Z}, \mathbf{Z}^*)}{\partial \mathbf{Z}^H}\right|_{\mathbf{Z} = \text{const matrix}} = \begin{bmatrix} \frac{\partial f(\mathbf{Z},\mathbf{Z}^*)}{\partial Z_{11}^*} & \cdots & \frac{\partial f(\mathbf{Z},\mathbf{Z}^*)}{\partial Z_{m1}^*} \\ \vdots & \ddots & \vdots \\ \frac{\partial f(\mathbf{Z},\mathbf{Z}^*)}{\partial Z_{1n}^*} & \cdots & \frac{\partial f(\mathbf{Z},\mathbf{Z}^*)}{\partial Z_{mn}^*} \end{bmatrix}. \quad (2.3.33)$$

Similarly, the complex gradient matrix and the complex conjugate gradient matrix of the real scalar function $f(\mathbf{Z}, \mathbf{Z}^*)$ are, respectively, given by

$$\nabla_{\mathbf{Z}} f(\mathbf{Z}, \mathbf{Z}^*) \stackrel{\text{def}}{=} \left.\frac{\partial f(\mathbf{Z}, \mathbf{Z}^*)}{\partial \mathbf{Z}}\right|_{\mathbf{Z}^* = \text{const matrix}} = \begin{bmatrix} \frac{\partial f(\mathbf{Z},\mathbf{Z}^*)}{\partial Z_{11}} & \cdots & \frac{\partial f(\mathbf{Z},\mathbf{Z}^*)}{\partial Z_{1n}} \\ \vdots & \ddots & \vdots \\ \frac{\partial f(\mathbf{Z},\mathbf{Z}^*)}{\partial Z_{m1}} & \cdots & \frac{\partial f(\mathbf{Z},\mathbf{Z}^*)}{\partial Z_{mn}} \end{bmatrix},$$
(2.3.34)

and

$$\nabla_{\mathbf{Z}^*} f(\mathbf{Z}, \mathbf{Z}^*) \stackrel{\text{def}}{=} \left.\frac{\partial f(\mathbf{Z}, \mathbf{Z}^*)}{\partial \mathbf{Z}^*}\right|_{\mathbf{Z} = \text{const matrix}} = \begin{bmatrix} \frac{\partial f(\mathbf{Z},\mathbf{Z}^*)}{\partial Z_{11}^*} & \cdots & \frac{\partial f(\mathbf{Z},\mathbf{Z}^*)}{\partial Z_{1n}^*} \\ \vdots & \ddots & \vdots \\ \frac{\partial f(\mathbf{Z},\mathbf{Z}^*)}{\partial Z_{m1}^*} & \cdots & \frac{\partial f(\mathbf{Z},\mathbf{Z}^*)}{\partial Z_{mn}^*} \end{bmatrix}.$$
(2.3.35)

Summarizing the definitions above, there are the following relations among the various complex partial derivatives of the real scalar function $f(\mathbf{Z}, \mathbf{Z})$:

- The conjugate gradient (or cogradient) vector is equal to the complex conjugate of the gradient (or cogradient) vector; and the conjugate Jacobian (or gradient) matrix is equal to the complex conjugate of the Jacobian (or gradient) matrix.
- The gradient (or conjugate gradient) vector is equal to the transpose of the cogradient (or conjugate cogradient) vector, namely

$$\nabla_{\text{vec}\,\mathbf{Z}} f(\mathbf{Z}, \mathbf{Z}^*) = \mathrm{D}_{\text{vec}\,\mathbf{Z}}^T f(\mathbf{Z}, \mathbf{Z}^*), \quad \nabla_{\text{vec}\,\mathbf{Z}^*} f(\mathbf{Z}, \mathbf{Z}^*) = \mathrm{D}_{\text{vec}\,\mathbf{Z}^*}^T f(\mathbf{Z}, \mathbf{Z}^*).$$
(2.3.36)

2.3 Complex Gradient Matrices

- The cogradient (or conjugate cogradient) vector is equal to the transpose of the vectorization of Jacobian (or conjugate Jacobian) matrix:

$$D_{\text{vec}\,\mathbf{Z}} f(\mathbf{Z}, \mathbf{Z}) = \left(\text{vec}\,\mathbf{J}_{\mathbf{Z}}\right)^T, \quad D_{\text{vec}\,\mathbf{Z}^*} f(\mathbf{Z}, \mathbf{Z}) = \left(\text{vec}\,\mathbf{J}_{\mathbf{Z}^*}\right)^T. \quad (2.3.37)$$

- The gradient (or conjugate gradient) matrix is equal to the transpose of the Jacobian (or conjugate Jacobian) matrix:

$$\nabla_{\mathbf{Z}} f(\mathbf{Z}, \mathbf{Z}^*) = \mathbf{J}_{\mathbf{Z}}^T \quad \text{and} \quad \nabla_{\mathbf{Z}^*} f(\mathbf{Z}, \mathbf{Z}^*) = \mathbf{J}_{\mathbf{Z}^*}^T. \quad (2.3.38)$$

Here are the rules of operation for the complex gradient.

1. If $f(\mathbf{Z}, \mathbf{Z}^*) = c$ (a constant), then its gradient matrix and conjugate gradient matrix are equal to the zero matrix, namely $\partial c / \partial \mathbf{Z} = \mathbf{O}$ and $\partial c / \partial \mathbf{Z}^* = \mathbf{O}$.
2. *Linear rule:* If $f(\mathbf{Z}, \mathbf{Z}^*)$ and $g(\mathbf{Z}, \mathbf{Z}^*)$ are scalar functions, and c_1 and c_2 are complex numbers, then

$$\frac{\partial (c_1 f(\mathbf{Z}, \mathbf{Z}^*) + c_2 g(\mathbf{Z}, \mathbf{Z}^*))}{\partial \mathbf{Z}^*} = c_1 \frac{\partial f(\mathbf{Z}, \mathbf{Z}^*)}{\partial \mathbf{Z}^*} + c_2 \frac{\partial g(\mathbf{Z}, \mathbf{Z}^*)}{\partial \mathbf{Z}^*}.$$

3. *Multiplication rule:*

$$\frac{\partial f(\mathbf{Z}, \mathbf{Z}^*) g(\mathbf{Z}, \mathbf{Z}^*)}{\partial \mathbf{Z}^*} = g(\mathbf{Z}, \mathbf{Z}^*) \frac{\partial f(\mathbf{Z}, \mathbf{Z}^*)}{\partial \mathbf{Z}^*} + f(\mathbf{Z}, \mathbf{Z}^*) \frac{\partial g(\mathbf{Z}, \mathbf{Z}^*)}{\partial \mathbf{Z}^*}.$$

4. *Quotient rule:* If $g(\mathbf{Z}, \mathbf{Z}^*) \neq 0$, then

$$\frac{\partial f/g}{\partial \mathbf{Z}^*} = \frac{1}{g^2(\mathbf{Z}, \mathbf{Z}^*)} \left[g(\mathbf{Z}, \mathbf{Z}^*) \frac{\partial f(\mathbf{Z}, \mathbf{Z}^*)}{\partial \mathbf{Z}^*} - f(\mathbf{Z}, \mathbf{Z}^*) \frac{\partial g(\mathbf{Z}, \mathbf{Z}^*)}{\partial \mathbf{Z}^*} \right].$$

If $h(\mathbf{Z}, \mathbf{Z}^*) = g(\mathbf{F}(\mathbf{Z}, \mathbf{Z}^*), \mathbf{F}^*(\mathbf{Z}, \mathbf{Z}^*))$, then the quotient rule becomes

$$\frac{\partial h(\mathbf{Z}, \mathbf{Z}^*)}{\partial \text{vec}\,\mathbf{Z}} = \frac{\partial g(\mathbf{F}(\mathbf{Z}, \mathbf{Z}^*), \mathbf{F}^*(\mathbf{Z}, \mathbf{Z}^*))}{\partial (\text{vec}\,\mathbf{F}(\mathbf{Z}, \mathbf{Z}^*))^T} \frac{\partial (\text{vec}\,\mathbf{F}(\mathbf{Z}, \mathbf{Z}^*))^T}{\partial \text{vec}\,\mathbf{Z}}$$
$$+ \frac{\partial g(\mathbf{F}(\mathbf{Z}, \mathbf{Z}^*), \mathbf{F}^*(\mathbf{Z}, \mathbf{Z}^*))}{\partial (\text{vec}\,\mathbf{F}^*(\mathbf{Z}, \mathbf{Z}^*))^T} \frac{\partial (\text{vec}\,\mathbf{F}^*(\mathbf{Z}, \mathbf{Z}^*))^T}{\partial \text{vec}\,\mathbf{Z}}, \quad (2.3.39)$$

and

$$\frac{\partial h(\mathbf{Z}, \mathbf{Z}^*)}{\partial \text{vec}\,\mathbf{Z}^*} = \frac{\partial g(\mathbf{F}(\mathbf{Z}, \mathbf{Z}^*), \mathbf{F}^*(\mathbf{Z}, \mathbf{Z}^*))}{\partial (\text{vec}\,\mathbf{F}(\mathbf{Z}, \mathbf{Z}^*))^T} \frac{\partial (\text{vec}\mathbf{F}(\mathbf{Z}, \mathbf{Z}^*))^T}{\partial \text{vec}\,\mathbf{Z}^*}$$
$$+ \frac{\partial g(\mathbf{F}(\mathbf{Z}, \mathbf{Z}^*), \mathbf{F}^*(\mathbf{Z}, \mathbf{Z}^*))}{\partial (\text{vec}\mathbf{F}^*(\mathbf{Z}, \mathbf{Z}^*))^T} \frac{\partial (\text{vec}\,\mathbf{F}^*(\mathbf{Z}, \mathbf{Z}^*))^T}{\partial \text{vec}\,\mathbf{Z}^*}. \quad (2.3.40)$$

2.3.3 Complex Gradient Matrix Identification

If letting $\mathbf{A} = \mathrm{D}_{\mathbf{Z}} f(\mathbf{Z}, \mathbf{Z}^*)$ and $\mathbf{B} = \mathrm{D}_{\mathbf{Z}^*} f(\mathbf{Z}, \mathbf{Z}^*)$, then

$$\frac{\partial f(\mathbf{Z}, \mathbf{Z}^*)}{\partial (\mathrm{vec}\,\mathbf{Z})^T} = \mathrm{rvec}\mathrm{D}_{\mathbf{Z}} f(\mathbf{Z}, \mathbf{Z}^*) = \mathrm{rvec}\mathbf{A} = (\mathrm{vec}(\mathbf{A}^T))^T, \tag{2.3.41}$$

$$\frac{\partial f(\mathbf{Z}, \mathbf{Z}^*)}{\partial (\mathrm{vec}\,\mathbf{Z})^H} = \mathrm{rvec}\mathrm{D}_{\mathbf{Z}^*} f(\mathbf{Z}, \mathbf{Z}^*) = \mathrm{rvec}\mathbf{B} = (\mathrm{vec}(\mathbf{B}^T))^T. \tag{2.3.42}$$

Hence, Eq. (2.3.29) can be rewritten as

$$\mathrm{d}f(\mathbf{Z}, \mathbf{Z}^*) = (\mathrm{vec}(\mathbf{A}^T))^T \mathrm{dvec}\,\mathbf{Z} + (\mathrm{vec}(\mathbf{B}^T))^T \mathrm{dvec}\mathbf{Z}^*. \tag{2.3.43}$$

Using $\mathrm{tr}(\mathbf{C}^T\mathbf{D}) = (\mathrm{vec}\,\mathbf{C})^T \mathrm{vec}\mathbf{D}$, Eq. (2.3.43) can be written as

$$\mathrm{d}f(\mathbf{Z}, \mathbf{Z}^*) = \mathrm{tr}(\mathbf{A}\mathrm{d}\mathbf{Z} + \mathbf{B}\mathrm{d}\mathbf{Z}^*). \tag{2.3.44}$$

Proposition 2.2 *Given a scalar function* $f(\mathbf{Z}, \mathbf{Z}^*) : \mathbb{C}^{m \times n} \times \mathbb{C}^{m \times n} \to \mathbb{C}$, *its complex Jacobian and gradient matrices can be, respectively, identified by*

$$\mathrm{d}f(\mathbf{Z}, \mathbf{Z}^*) = \mathrm{tr}(\mathbf{A}\mathrm{d}\mathbf{Z} + \mathbf{B}\mathrm{d}\mathbf{Z}^*) \Leftrightarrow \begin{cases} \mathbf{J}_{\mathbf{Z}} = \mathbf{A}, \\ \mathbf{J}_{\mathbf{Z}^*} = \mathbf{B}; \end{cases} \tag{2.3.45}$$

$$\mathrm{d}f(\mathbf{Z}, \mathbf{Z}^*) = \mathrm{tr}(\mathbf{A}\mathrm{d}\mathbf{Z} + \mathbf{B}\mathrm{d}\mathbf{Z}^*) \Leftrightarrow \begin{cases} \nabla_{\mathbf{Z}} f(\mathbf{Z}, \mathbf{Z}^*) = \mathbf{A}^T, \\ \nabla_{\mathbf{Z}^*} f(\mathbf{Z}, \mathbf{Z}^*) = \mathbf{B}^T. \end{cases} \tag{2.3.46}$$

That is to say, the complex gradient matrix and the complex conjugate gradient matrix are identified as the transposes of the matrices $\mathbf{A}$ *and* $\mathbf{B}$, *respectively.*

Table 2.8 lists the complex gradient matrices of several trace functions.
Table 2.9 lists the complex gradient matrices of several determinant functions.

Table 2.8 Complex gradient matrices of trace functions

$f(\mathbf{Z}, \mathbf{Z}^*)$	$\mathrm{d}f$	$\partial f/\partial \mathbf{Z}$	$\partial f/\partial \mathbf{Z}^*$
$\mathrm{tr}(\mathbf{AZ})$	$\mathrm{tr}(\mathbf{A}\mathrm{d}\mathbf{Z})$	$\mathbf{A}^T$	$\mathbf{O}$
$\mathrm{tr}(\mathbf{AZ}^H)$	$\mathrm{tr}(\mathbf{A}^T \mathrm{d}\mathbf{Z}^*)$	$\mathbf{O}$	$\mathbf{A}$
$\mathrm{tr}(\mathbf{ZAZ}^T\mathbf{B})$	$\mathrm{tr}((\mathbf{AZ}^T\mathbf{B} + \mathbf{A}^T\mathbf{Z}^T\mathbf{B}^T)\mathrm{d}\mathbf{Z})$	$\mathbf{B}^T\mathbf{ZA}^T + \mathbf{BZA}$	$\mathbf{O}$
$\mathrm{tr}(\mathbf{ZAZB})$	$\mathrm{tr}((\mathbf{AZB} + \mathbf{BZA})\mathrm{d}\mathbf{Z})$	$(\mathbf{AZB} + \mathbf{BZA})^T$	$\mathbf{O}$
$\mathrm{tr}(\mathbf{ZAZ}^*\mathbf{B})$	$\mathrm{tr}(\mathbf{AZ}^*\mathbf{B}\mathrm{dd}\mathbf{Z} + \mathbf{BZA}\mathrm{dd}\mathbf{Z}^*)$	$\mathbf{B}^T\mathbf{Z}^H\mathbf{A}^T$	$\mathbf{A}^T\mathbf{Z}^T\mathbf{B}^T$
$\mathrm{tr}(\mathbf{ZAZ}^H\mathbf{B})$	$\mathrm{tr}(\mathbf{AZ}^H\mathbf{B}\mathrm{d}\mathbf{Z} + \mathbf{A}^T\mathbf{Z}^T\mathbf{B}^T\mathrm{d}\mathbf{Z}^*)$	$\mathbf{B}^T\mathbf{Z}^*\mathbf{A}^T$	$\mathbf{BZA}$
$\mathrm{tr}(\mathbf{AZ}^{-1})$	$-\mathrm{tr}(\mathbf{Z}^{-1}\mathbf{AZ}^{-1}\mathrm{d}\mathbf{Z})$	$-\mathbf{Z}^{-T}\mathbf{A}^T\mathbf{Z}^{-T}$	$\mathbf{O}$
$\mathrm{tr}(\mathbf{Z}^k)$	$k\,\mathrm{tr}(\mathbf{Z}^{k-1}\mathrm{d}\mathbf{Z})$	$k\,(\mathbf{Z}^T)^{k-1}$	$\mathbf{O}$

2.3 Complex Gradient Matrices

Table 2.9 Complex gradient matrices of determinant functions

$f(\mathbf{Z}, \mathbf{Z}^*)$	$\mathrm{d}f$	$\partial f/\partial \mathbf{Z}$	$\partial f/\partial \mathbf{Z}^*$
$|\mathbf{Z}|$	$|\mathbf{Z}|\mathrm{tr}(\mathbf{Z}^{-1}\mathrm{d}\mathbf{Z})$	$|\mathbf{Z}|\mathbf{Z}^{-T}$	$\mathbf{O}$
$|\mathbf{Z}\mathbf{Z}^T|$	$2|\mathbf{Z}\mathbf{Z}^T|\mathrm{tr}(\mathbf{Z}^T(\mathbf{Z}\mathbf{Z}^T)^{-1}\mathrm{d}\mathbf{Z})$	$2|\mathbf{Z}\mathbf{Z}^T|(\mathbf{Z}\mathbf{Z}^T)^{-1}\mathbf{Z}$	$\mathbf{O}$
$|\mathbf{Z}^T\mathbf{Z}|$	$2|\mathbf{Z}^T\mathbf{Z}|\mathrm{tr}((\mathbf{Z}^T\mathbf{Z})^{-1}\mathbf{Z}^T\mathrm{d}\mathbf{Z})$	$2|\mathbf{Z}^T\mathbf{Z}|\mathbf{Z}(\mathbf{Z}^T\mathbf{Z})^{-1}$	$\mathbf{O}$
$|\mathbf{Z}\mathbf{Z}^*|$	$|\mathbf{Z}\mathbf{Z}^*|\mathrm{tr}(\mathbf{Z}^*(\mathbf{Z}\mathbf{Z}^*)^{-1}\mathrm{d}\mathbf{Z}$ $+ (\mathbf{Z}\mathbf{Z}^*)^{-1}\mathbf{Z}\mathrm{d}\mathbf{Z}^*)$	$|\mathbf{Z}\mathbf{Z}^*|(\mathbf{Z}^H\mathbf{Z}^T)^{-1}\mathbf{Z}^H$	$|\mathbf{Z}\mathbf{Z}^*|\mathbf{Z}^T(\mathbf{Z}^H\mathbf{Z}^T)^{-1}$
$|\mathbf{Z}^*\mathbf{Z}|$	$|\mathbf{Z}^*\mathbf{Z}|\mathrm{tr}((\mathbf{Z}^*\mathbf{Z})^{-1}\mathbf{Z}^*\mathrm{d}\mathbf{Z}$ $+ \mathbf{Z}(\mathbf{Z}^*\mathbf{Z})^{-1}\mathrm{d}\mathbf{Z}^*)$	$|\mathbf{Z}^*\mathbf{Z}|\mathbf{Z}^H(\mathbf{Z}^T\mathbf{Z}^H)^{-1}$	$|\mathbf{Z}^*\mathbf{Z}|(\mathbf{Z}^T\mathbf{Z}^H)^{-1}\mathbf{Z}^T$
$|\mathbf{Z}\mathbf{Z}^H|$	$|\mathbf{Z}\mathbf{Z}^H|\mathrm{tr}(\mathbf{Z}^H(\mathbf{Z}\mathbf{Z}^H)^{-1}\mathrm{d}\mathbf{Z}$ $+ \mathbf{Z}^T(\mathbf{Z}^*\mathbf{Z}^T)^{-1}\mathrm{d}\mathbf{Z}^*)$	$|\mathbf{Z}\mathbf{Z}^H|(\mathbf{Z}^*\mathbf{Z}^T)^{-1}\mathbf{Z}^*$	$|\mathbf{Z}\mathbf{Z}^H|(\mathbf{Z}\mathbf{Z}^H)^{-1}\mathbf{Z}$
$|\mathbf{Z}^H\mathbf{Z}|$	$|\mathbf{Z}^H\mathbf{Z}|\mathrm{tr}((\mathbf{Z}^H\mathbf{Z})^{-1}\mathbf{Z}^H\mathrm{d}\mathbf{Z}$ $+ (\mathbf{Z}^T\mathbf{Z}^*)^{-1}\mathbf{Z}^T\mathrm{d}\mathbf{Z}^*)$	$|\mathbf{Z}^H\mathbf{Z}|\mathbf{Z}^*(\mathbf{Z}^T\mathbf{Z}^*)^{-1}$	$|\mathbf{Z}^H\mathbf{Z}|\mathbf{Z}(\mathbf{Z}^H\mathbf{Z})^{-1}$
$|\mathbf{Z}^k|$	$k|\mathbf{Z}|^k\mathrm{tr}(\mathbf{Z}^{-1}\mathrm{d}\mathbf{Z})$	$k|\mathbf{Z}|^k\mathbf{Z}^{-T}$	$\mathbf{O}$

If $\mathbf{f}(\mathbf{z}, \mathbf{z}^*) = [f_1(\mathbf{z}, \mathbf{z}^*), \ldots, f_n(\mathbf{z}, \mathbf{z}^*)]^T$ is an $n \times 1$ complex vector function with $m \times 1$ complex vector variable, then

$$\begin{bmatrix} \mathrm{d}f_1(\mathbf{z}, \mathbf{z}^*) \\ \vdots \\ \mathrm{d}f_n(\mathbf{z}, \mathbf{z}^*) \end{bmatrix} = \begin{bmatrix} \mathrm{D}_\mathbf{z} f_1(\mathbf{z}, \mathbf{z}^*) \\ \vdots \\ \mathrm{D}_\mathbf{z} f_n(\mathbf{z}, \mathbf{z}^*) \end{bmatrix} \mathrm{d}\mathbf{z} + \begin{bmatrix} \mathrm{D}_{\mathbf{z}^*} f_1(\mathbf{z}, \mathbf{z}^*) \\ \vdots \\ \mathrm{D}_{\mathbf{z}^*} f_n(\mathbf{z}, \mathbf{z}^*) \end{bmatrix} \mathrm{d}\mathbf{z}^*,$$

which can simply be written as

$$\mathrm{d}\mathbf{f}(\mathbf{z}, \mathbf{z}^*) = \mathbf{J}_\mathbf{z}\mathrm{d}\mathbf{z} + \mathbf{J}_{\mathbf{z}^*}\mathrm{d}\mathbf{z}^*, \tag{2.3.47}$$

where $\mathrm{d}\mathbf{f}(\mathbf{z}, \mathbf{z}^*) = [\mathrm{d}f_1(\mathbf{z}, \mathbf{z}^*), \ldots, \mathrm{d}f_n(\mathbf{z}, \mathbf{z}^*)]^T$, while

$$\mathbf{J}_\mathbf{z} = \frac{\partial \mathbf{f}(\mathbf{z}, \mathbf{z}^*)}{\partial \mathbf{z}^T} = \begin{bmatrix} \frac{\partial f_1(\mathbf{z},\mathbf{z}^*)}{\partial z_1} & \cdots & \frac{\partial f_1(\mathbf{z},\mathbf{z}^*)}{\partial z_m} \\ \vdots & \ddots & \vdots \\ \frac{\partial f_n(\mathbf{z},\mathbf{z}^*)}{\partial z_1} & \cdots & \frac{\partial f_n(\mathbf{z},\mathbf{z}^*)}{\partial z_m} \end{bmatrix}, \tag{2.3.48}$$

and

$$\mathbf{J}_{\mathbf{z}^*} = \frac{\partial \mathbf{f}(\mathbf{z}, \mathbf{z}^*)}{\partial \mathbf{z}^H} = \begin{bmatrix} \frac{\partial f_1(\mathbf{z},\mathbf{z}^*)}{\partial z_1^*} & \cdots & \frac{\partial f_1(\mathbf{z},\mathbf{z}^*)}{\partial z_m^*} \\ \vdots & \ddots & \vdots \\ \frac{\partial f_n(\mathbf{z},\mathbf{z}^*)}{\partial z_1^*} & \cdots & \frac{\partial f_n(\mathbf{z},\mathbf{z}^*)}{\partial z_m^*} \end{bmatrix} \tag{2.3.49}$$

are, respectively, the complex Jacobian matrix and the complex conjugate Jacobian matrix of the vector function $\mathbf{f}(\mathbf{z}, \mathbf{z}^*)$.

For a $p \times q$ matrix function $\mathbf{F}(\mathbf{Z}, \mathbf{Z}^*)$ with $m \times n$ complex matrix variable $\mathbf{Z}$, if $\mathbf{F}(\mathbf{Z}, \mathbf{Z}^*) = [\mathbf{f}_1(\mathbf{Z}, \mathbf{Z}^*), \ldots, \mathbf{f}_q(\mathbf{Z}, \mathbf{Z}^*)]$, then $\mathrm{d}\mathbf{F}(\mathbf{Z}, \mathbf{Z}^*) = [\mathrm{d}\mathbf{f}_1(\mathbf{Z}, \mathbf{Z}^*), \ldots, \mathrm{d}\mathbf{f}_q(\mathbf{Z}, \mathbf{Z}^*)]$, and (2.3.47) holds for the vector functions $\mathbf{f}_i(\mathbf{Z}, \mathbf{Z}^*), i = 1, \ldots, q$. This implies that

$$\begin{bmatrix} \mathrm{d}\mathbf{f}_1(\mathbf{Z}, \mathbf{Z}^*) \\ \vdots \\ \mathrm{d}\mathbf{f}_q(\mathbf{Z}, \mathbf{Z}^*) \end{bmatrix} = \begin{bmatrix} \mathrm{D}_{\mathrm{vec}\,\mathbf{Z}}\mathbf{f}_1(\mathbf{Z}, \mathbf{Z}^*) \\ \vdots \\ \mathrm{D}_{\mathrm{vec}\,\mathbf{Z}}\mathbf{f}_q(\mathbf{Z}, \mathbf{Z}^*) \end{bmatrix} \mathrm{dvec}\mathbf{Z} + \begin{bmatrix} \mathrm{D}_{\mathrm{vec}\,\mathbf{Z}^*}\mathbf{f}_1(\mathbf{Z}, \mathbf{Z}^*) \\ \vdots \\ \mathrm{D}_{\mathrm{vec}\,\mathbf{Z}^*}\mathbf{f}_q(\mathbf{Z}, \mathbf{Z}^*) \end{bmatrix} \mathrm{dvec}\mathbf{Z}^*, \tag{2.3.50}$$

where $\mathrm{D}_{\mathrm{vec}\,\mathbf{Z}}\mathbf{f}_i(\mathbf{Z}, \mathbf{Z}^*) = \frac{\partial \mathbf{f}_i(\mathbf{Z},\mathbf{Z}^*)}{\partial (\mathrm{vec}\mathbf{Z})^T} \in \mathbb{C}^{p \times mn}$ and $\mathrm{D}_{\mathrm{vec}\,\mathbf{Z}^*}\mathbf{f}_i(\mathbf{Z}, \mathbf{Z}^*) = \frac{\partial \mathbf{f}_i(\mathbf{Z},\mathbf{Z}^*)}{\partial (\mathrm{vec}\mathbf{Z}^*)^T} \in \mathbb{C}^{p \times mn}$.

Equation (2.3.50) can be simply rewritten as

$$\mathrm{dvec}\mathbf{F}(\mathbf{Z}, \mathbf{Z}^*) = \mathbf{A}\mathrm{dvec}\mathbf{Z} + \mathbf{B}\mathrm{dvec}\mathbf{Z}^* \in \mathbb{C}^{pq}, \tag{2.3.51}$$

where

$$\mathrm{dvec}\mathbf{Z} = [\mathrm{d}Z_{11}, \ldots, \mathrm{d}Z_{m1}, \ldots, \mathrm{d}Z_{1n}, \ldots, \mathrm{d}Z_{mn}]^T,$$
$$\mathrm{dvec}\mathbf{Z}^* = [\mathrm{d}Z_{11}^*, \ldots, \mathrm{d}Z_{m1}^*, \ldots, \mathrm{d}Z_{1n}^*, \ldots, \mathrm{d}Z_{mn}^*]^T,$$
$$\mathrm{dvec}\mathbf{F}(\mathbf{Z}, \mathbf{Z}^*)) = [\mathrm{d}f_{11}(\mathbf{Z}, \mathbf{Z}^*), \ldots, \mathrm{d}f_{p1}(\mathbf{Z}, \mathbf{Z}^*), \ldots, \mathrm{d}f_{1q}(\mathbf{Z}, \mathbf{Z}^*),$$
$$\ldots, \mathrm{d}f_{pq}(\mathbf{Z}, \mathbf{Z}^*)]^T,$$

$$\mathbf{A} = \begin{bmatrix} \frac{\partial f_{11}(\mathbf{Z},\mathbf{Z}^*)}{\partial Z_{11}} & \cdots & \frac{\partial f_{11}(\mathbf{Z},\mathbf{Z}^*)}{\partial Z_{m1}} & \cdots & \frac{\partial f_{11}(\mathbf{Z},\mathbf{Z}^*)}{\partial Z_{1n}} & \cdots & \frac{\partial f_{11}(\mathbf{Z},\mathbf{Z}^*)}{\partial Z_{mn}} \\ \vdots & \vdots & \vdots & \vdots & \vdots & \vdots & \vdots \\ \frac{\partial f_{p1}(\mathbf{Z},\mathbf{Z}^*)}{\partial Z_{11}} & \cdots & \frac{\partial f_{p1}(\mathbf{Z},\mathbf{Z}^*)}{\partial Z_{m1}} & \cdots & \frac{\partial f_{p1}(\mathbf{Z},\mathbf{Z}^*)}{\partial Z_{1n}} & \cdots & \frac{\partial f_{p1}(\mathbf{Z},\mathbf{Z}^*)}{\partial Z_{mn}} \\ \vdots & \vdots & \vdots & \vdots & \vdots & \vdots & \vdots \\ \frac{\partial f_{1q}(\mathbf{Z},\mathbf{Z}^*)}{\partial Z_{11}} & \cdots & \frac{\partial f_{1q}(\mathbf{Z},\mathbf{Z}^*)}{\partial Z_{m1}} & \cdots & \frac{\partial f_{1q}(\mathbf{Z},\mathbf{Z}^*)}{\partial Z_{1n}} & \cdots & \frac{\partial f_{1q}(\mathbf{Z},\mathbf{Z}^*)}{\partial Z_{mn}} \\ \vdots & \vdots & \vdots & \vdots & \vdots & \vdots & \vdots \\ \frac{\partial f_{pq}(\mathbf{Z},\mathbf{Z}^*)}{\partial Z_{11}} & \cdots & \frac{\partial f_{pq}(\mathbf{Z},\mathbf{Z}^*)}{\partial Z_{m1}} & \cdots & \frac{\partial f_{pq}(\mathbf{Z},\mathbf{Z}^*)}{\partial Z_{1n}} & \cdots & \frac{\partial f_{pq}(\mathbf{Z},\mathbf{Z}^*)}{\partial Z_{mn}} \end{bmatrix}$$

$$= \frac{\partial \,\mathrm{vec}\,\mathbf{F}(\mathbf{Z}, \mathbf{Z}^*)}{\partial (\mathrm{vec}\,\mathbf{Z})^T} = \mathbf{J}_{\mathrm{vec}\,\mathbf{Z}},$$

2.3 Complex Gradient Matrices

and

$$\mathbf{B} = \begin{bmatrix} \frac{\partial f_{11}(\mathbf{Z},\mathbf{Z}^*)}{\partial Z_{11}^*} & \cdots & \frac{\partial f_{11}(\mathbf{Z},\mathbf{Z}^*)}{\partial Z_{m1}^*} & \cdots & \frac{\partial f_{11}(\mathbf{Z},\mathbf{Z}^*)}{\partial Z_{1n}^*} & \cdots & \frac{\partial f_{11}(\mathbf{Z},\mathbf{Z}^*)}{\partial Z_{mn}^*} \\ \vdots & \vdots & \vdots & \vdots & \vdots & \vdots & \vdots \\ \frac{\partial f_{p1}(\mathbf{Z},\mathbf{Z}^*)}{\partial Z_{11}^*} & \cdots & \frac{\partial f_{p1}(\mathbf{Z},\mathbf{Z}^*)}{\partial Z_{m1}^*} & \cdots & \frac{\partial f_{p1}(\mathbf{Z},\mathbf{Z}^*)}{\partial Z_{1n}^*} & \cdots & \frac{\partial f_{p1}(\mathbf{Z},\mathbf{Z}^*)}{\partial Z_{mn}^*} \\ \vdots & \vdots & \vdots & \vdots & \vdots & \vdots & \vdots \\ \frac{\partial f_{1q}(\mathbf{Z},\mathbf{Z}^*)}{\partial Z_{11}^*} & \cdots & \frac{\partial f_{1q}(\mathbf{Z},\mathbf{Z}^*)}{\partial Z_{m1}^*} & \cdots & \frac{\partial f_{1q}(\mathbf{Z},\mathbf{Z}^*)}{\partial Z_{1n}^*} & \cdots & \frac{\partial f_{1q}(\mathbf{Z},\mathbf{Z}^*)}{\partial Z_{mn}^*} \\ \vdots & \vdots & \vdots & \vdots & \vdots & \vdots & \vdots \\ \frac{\partial f_{pq}(\mathbf{Z},\mathbf{Z}^*)}{\partial Z_{11}^*} & \cdots & \frac{\partial f_{pq}(\mathbf{Z},\mathbf{Z}^*)}{\partial Z_{m1}^*} & \cdots & \frac{\partial f_{pq}(\mathbf{Z},\mathbf{Z}^*)}{\partial Z_{1n}^*} & \cdots & \frac{\partial f_{pq}(\mathbf{Z},\mathbf{Z}^*)}{\partial Z_{mn}^*} \end{bmatrix}$$

$$= \frac{\partial \, \text{vec}\, \mathbf{F}(\mathbf{Z}, \mathbf{Z}^*)}{\partial (\text{vec}\, \mathbf{Z}^*)^T} = \mathbf{J}_{\text{vec}\, \mathbf{Z}^*}.$$

The complex gradient matrix and the complex conjugate gradient matrix of the matrix function $\mathbf{F}(\mathbf{Z}, \mathbf{Z}^*)$ are, respectively, defined as

$$\nabla_{\text{vec}\, \mathbf{Z}} \mathbf{F}(\mathbf{Z}, \mathbf{Z}^*) = \frac{\partial (\text{vec}\, \mathbf{F}(\mathbf{Z}, \mathbf{Z}^*))^T}{\partial \, \text{vec}\, \mathbf{Z}} = (\mathbf{J}_{\text{vec}\, \mathbf{Z}})^T, \qquad (2.3.52)$$

$$\nabla_{\text{vec}\, \mathbf{Z}^*} \mathbf{F}(\mathbf{Z}, \mathbf{Z}^*) = \frac{\partial (\text{vec}\, \mathbf{F}(\mathbf{Z}, \mathbf{Z}^*))^T}{\partial \, \text{vec}\, \mathbf{Z}^*} = (\mathbf{J}_{\text{vec}\, \mathbf{Z}^*})^T. \qquad (2.3.53)$$

Proposition 2.3 *For a complex matrix function $\mathbf{F}(\mathbf{Z}, \mathbf{Z}^*) \in \mathbb{C}^{p \times q}$ with $\mathbf{Z}, \mathbf{Z}^* \in \mathbb{C}^{m \times n}$, we have its complex Jacobian matrix and the complex gradient matrix:*

$$\text{dvec}\mathbf{F}(\mathbf{Z}, \mathbf{Z}^*) = \mathbf{A}\text{dvec}\, \mathbf{Z} + \mathbf{B}\text{dvec}\mathbf{Z}^* \Leftrightarrow \begin{cases} \mathbf{J}_{\text{vec}\mathbf{Z}} = \mathbf{A}, \\ \mathbf{J}_{\text{vec}\mathbf{Z}^*} = \mathbf{B}, \end{cases} \qquad (2.3.54)$$

$$\text{dvec}\mathbf{F}(\mathbf{Z}, \mathbf{Z}^*) = \mathbf{A}\text{dvec}\mathbf{Z} + \mathbf{B}\text{dvec}\mathbf{Z}^* \Leftrightarrow \begin{cases} \nabla_{\text{vec}\mathbf{Z}}\mathbf{F}(\mathbf{Z}, \mathbf{Z}^*) = \mathbf{A}^T, \\ \nabla_{\text{vec}\mathbf{Z}^*}\mathbf{F}(\mathbf{Z}, \mathbf{Z}^*) = \mathbf{B}^T. \end{cases} \qquad (2.3.55)$$

If $d(\mathbf{F}(\mathbf{Z}, \mathbf{Z}^*)) = \mathbf{A}(d\mathbf{Z})\mathbf{B} + \mathbf{C}(d\mathbf{Z}^*)\mathbf{D}$, then the vectorization result is given by

$$\text{dvec}\, \mathbf{F}(\mathbf{Z}, \mathbf{Z}^*) = (\mathbf{B}^T \otimes \mathbf{A})\text{dvec}\, \mathbf{Z} + (\mathbf{D}^T \otimes \mathbf{C})\text{dvec}\mathbf{Z}^*.$$

By Proposition 2.3 we have the following identification formula:

$$d\mathbf{F}(\mathbf{Z}, \mathbf{Z}^*) = \mathbf{A}(d\mathbf{Z})\mathbf{B} + \mathbf{C}(d\mathbf{Z}^*)\mathbf{D} \Leftrightarrow \begin{cases} \mathbf{J}_{\text{vec}\, \mathbf{Z}} = \mathbf{B}^T \otimes \mathbf{A}, \\ \mathbf{J}_{\text{vec}\, \mathbf{Z}^*} = \mathbf{D}^T \otimes \mathbf{C}. \end{cases} \qquad (2.3.56)$$

Table 2.10 Complex matrix differential and complex Jacobian matrix

Function	Matrix differential	Jacobian matrix
$f(z, z^*)$	$df(z, z^*) = a dz + b dz^*$	$\frac{\partial f}{\partial z} = a$, $\frac{\partial f}{\partial z^*} = b$
$f(\mathbf{z}, \mathbf{z}^*)$	$df(\mathbf{z}, \mathbf{z}^*) = \mathbf{a}^T d\mathbf{z} + \mathbf{b}^T d\mathbf{z}^*$	$\frac{\partial f}{\partial \mathbf{z}^T} = \mathbf{a}^T$, $\frac{\partial f}{\partial \mathbf{z}^H} = \mathbf{b}^T$
$f(\mathbf{Z}, \mathbf{Z}^*)$	$df(\mathbf{Z}, \mathbf{Z}^*) = \mathrm{tr}(\mathbf{A} d\mathbf{Z} + \mathbf{B} d\mathbf{Z}^*)$	$\frac{\partial f}{\partial \mathbf{Z}^T} = \mathbf{A}$, $\frac{\partial f}{\partial \mathbf{Z}^H} = \mathbf{B}$
$\mathbf{F}(\mathbf{Z}, \mathbf{Z}^*)$	$d\,\mathrm{vec}\mathbf{F} = \mathbf{A}\,d\,\mathrm{vec}\mathbf{Z} + \mathbf{B}\,d\,\mathrm{vec}\mathbf{Z}^*$	$\frac{\partial \mathrm{vec}\mathbf{F}}{\partial (\mathrm{vec}\mathbf{Z})^T} = \mathbf{A}$, $\frac{\partial \mathrm{vec}\mathbf{F}}{\partial (\mathrm{vec}\mathbf{Z}^*)^T} = \mathbf{B}$
	$d\mathbf{F} = \mathbf{A}(d\mathbf{Z})\mathbf{B} + \mathbf{C}(d\mathbf{Z}^*)\mathbf{D}$	$\frac{\partial \mathrm{vec}\mathbf{F}}{\partial (\mathrm{vec}\mathbf{Z})^T} = \mathbf{B}^T \otimes \mathbf{A}$, $\frac{\partial \mathrm{vec}\mathbf{F}}{\partial (\mathrm{vec}\mathbf{Z}^*)^T} = \mathbf{D}^T \otimes \mathbf{C}$
	$d\mathbf{F} = \mathbf{A}(d\mathbf{Z})^T\mathbf{B} + \mathbf{C}(d\mathbf{Z}^*)^T\mathbf{D}$	$\frac{\partial \mathrm{vec}\mathbf{F}}{\partial (\mathrm{vec}\mathbf{Z})^T} = (\mathbf{B}^T \otimes \mathbf{A})\mathbf{K}_{mn}$, $\frac{\partial \mathrm{vec}\mathbf{F}}{\partial (\mathrm{vec}\mathbf{Z}^*)^T} = (\mathbf{D}^T \otimes \mathbf{C})\mathbf{K}_{mn}$

Similarly, if $d\mathbf{F}(\mathbf{Z}, \mathbf{Z}^*) = \mathbf{A}(d\mathbf{Z})^T\mathbf{B} + \mathbf{C}(d\mathbf{Z}^*)^T\mathbf{D}$, then we have the result

$$d\,\mathrm{vec}\mathbf{F}(\mathbf{Z}, \mathbf{Z}^*) = (\mathbf{B}^T \otimes \mathbf{A}) d\,\mathrm{vec}\mathbf{Z}^T + (\mathbf{D}^T \otimes \mathbf{C}) d\,\mathrm{vec}\mathbf{Z}^H$$

$$= (\mathbf{B}^T \otimes \mathbf{A})\mathbf{K}_{mn} d\,\mathrm{vec}\mathbf{Z} + (\mathbf{D}^T \otimes \mathbf{C})\mathbf{K}_{mn} d\,\mathrm{vec}\mathbf{Z}^*,$$

where we have used the vectorization property $\mathrm{vec}\mathbf{X}^T_{m \times n} = \mathbf{K}_{mn}\mathrm{vec}\mathbf{X}$. By Proposition 2.3, the following identification formula is obtained:

$$d\mathbf{F}(\mathbf{Z}, \mathbf{Z}^*) = \mathbf{A}(d\mathbf{Z})^T\mathbf{B} + \mathbf{C}(d\mathbf{Z}^*)^T\mathbf{D} \Leftrightarrow \begin{cases} \mathbf{J}_{\mathrm{vec}\mathbf{Z}} = (\mathbf{B}^T \otimes \mathbf{A})\mathbf{K}_{mn}, \\ \mathbf{J}_{\mathrm{vec}\mathbf{Z}^*} = (\mathbf{D}^T \otimes \mathbf{C})\mathbf{K}_{mn}. \end{cases}$$
(2.3.57)

The above equation shows that, as in the vector case, the key to identifying the gradient matrix and conjugate gradient matrix of a matrix function $\mathbf{F}(\mathbf{Z}, \mathbf{Z}^*)$ is to write its matrix differential into the canonical form $d\mathbf{F}(\mathbf{Z}, \mathbf{Z}^*) = \mathbf{A}(d\mathbf{Z})^T\mathbf{B} + \mathbf{C}(d\mathbf{Z}^*)^T\mathbf{D}$.

Table 2.10 lists the corresponding relationships between the first-order complex matrix differential and the complex Jacobian matrix, where $\mathbf{z} \in \mathbb{C}^m$, $\mathbf{Z} \in \mathbb{C}^{m \times n}$, $\mathbf{F} \in \mathbb{C}^{p \times q}$.

Brief Summary of This Chapter

This chapter presents the matrix differential for real and complex matrix functions. The matrix differential is a powerful tool for finding the gradient vectors/matrices that are key for update in optimization, as will see in the next chapter.

References

1. Brookes, M.: Matrix Reference Manual (2011). https://www.ee.ic.ac.uk/hp/staff/dmb/matrix/intro.html
2. Flanigan, F.: Complex Variables: Harmonic and Analytic Functions, 2nd edn. Dover, New York (1983)
3. Frankel, T.: The Geometry of Physics: An Introduction (with Corrections and Additions). Cambridge University Press, Cambridge (2001)
4. Kreyszig, E.: Advanced Engineering Mathematics, 7th edn. Wiley, New York (1993)
5. Lütkepohl, H.: Handbook of Matrices. Wiley, New York (1996)
6. Magnus, J.R., Neudecker, H.: Matrix Differential Calculus with Applications in Statistics and Econometrics, rev. edn. Wiley, Chichester (1999)
7. Petersen, K.B., Petersen, M.S.: The Matrix Cookbook (2008). https://matrixcookbook.com
8. Wirtinger, W.: Zur formalen theorie der funktionen von mehr komplexen veränderlichen. Math. Ann. **97**, 357–375 (1927)

Chapter 3
Gradient and Optimization

Many machine learning methods involve solving complex optimization problems, e.g., neural networks, support vector machines, and evolutionary computation. Optimization theory mainly considers (1) the existence conditions for an extremum value (gradient analysis); (2) the design of optimization algorithms and convergence analysis. This chapter focuses on convex optimization theory and methods by focusing on gradient/subgradient methods in smooth and nonsmooth convex optimizations and constrained convex optimization.

3.1 Real Gradient

Consider an unconstrained minimization problem for $f(\mathbf{x}) : \mathbb{R}^n \to \mathbb{R}$ described by

$$\min_{\mathbf{x} \in S} f(\mathbf{x}), \tag{3.1.1}$$

where $S \in \mathbb{R}^n$ is a subset of the n-dimensional vector space $\mathbb{R}^n$; the vector variable $\mathbf{x} \in S$ is called the *optimization vector* and is chosen so as to fulfill Eq. (3.1.1). The function $f : \mathbb{R}^n \to \mathbb{R}$ is known as the *objective function*; it represents the cost or price paid by selecting the optimization vector $\mathbf{x}$, and thus is also called the *cost function*. Conversely, the negative cost function $-f(\mathbf{x})$ can be understood as the *value function* or *utility function* of $\mathbf{x}$. Hence, solving the optimization problem (3.1.1) corresponds to minimizing the cost function or maximizing the value function. That is to say, the minimization problem $\min_{\mathbf{x} \in S} f(\mathbf{x})$ and the maximization problem $\max_{\mathbf{x} \in S} \{-f(\mathbf{x})\}$ of the negative objective function are equivalent.

The above optimization problem is called an *unconstrained optimization problem* due to having no constraint conditions.

Most nonlinear programming methods for solving unconstrained optimization problems are based on the idea of relaxation and approximation [25].

- *Relaxation:* The sequence $\{a_k\}_{k=0}^{\infty}$ is known as a *relaxation sequence*, if $a_{k+1} \leq a_k$, $\forall k \geq 0$. Hence, in the process of solving iteratively the optimization problem (3.1.1) it is necessary to generate a relaxation sequence of the cost function $f(\mathbf{x}_{k+1}) \leq f(\mathbf{x}_k)$, $k = 0, 1, \ldots$
- *Approximation:* Approximating an objective function means using a simpler objective function instead of the original objective function.

Therefore, by relaxation and approximation, we can achieve the following.

1. If the objective function $f(\mathbf{x})$ is lower bounded in the definition domain $S \in \mathbb{R}^n$, then the sequence $\{f(\mathbf{x}_k)\}_{k=0}^{\infty}$ must converge.
2. In any case, we can improve upon the initial value of the objective function $f(\mathbf{x})$.
3. The minimization of a nonlinear objective function $f(\mathbf{x})$ can be implemented by numerical methods to a sufficiently high approximation accuracy.

3.1.1 Stationary Points and Extreme Points

Definition 3.1 (Stationary Point) In optimization, a *stationary point* or *critical point* of a differentiable function with one variable is a point in the domain of function where its derivative is zero. In other words, at a stationary point the function stops increasing or decreasing (hence the name).

A stationary point could be a saddle point, however, is usually expected to be a global minimum or maximum point of the objective function $f(\mathbf{x})$.

Definition 3.2 (Global Minimum Point) A point $\mathbf{x}^*$ in the vector subspace $S \in \mathbb{R}^n$ is known as a *global minimum point* of the function $f(\mathbf{x})$ if

$$f(\mathbf{x}^*) \leq f(\mathbf{x}), \quad \forall \mathbf{x} \in S, \mathbf{x} \neq \mathbf{x}^*. \tag{3.1.2}$$

A global minimum point is also called an absolute minimum point. The value of the function at this point, $f(\mathbf{x}^*)$, is called the *global minimum* or absolute minimum of the function $f(\mathbf{x})$ in the vector subspace S.

Let $\mathcal{D}$ denote the definition domain of a function $f(\mathbf{x})$, where $\mathcal{D}$ may be the whole vector space $\mathbb{R}^n$ or its some subset. If

$$f(\mathbf{x}^*) < f(\mathbf{x}), \quad \forall \mathbf{x} \in \mathcal{D}, \tag{3.1.3}$$

then $\mathbf{x}^*$ is said to be a *strict global minimum point* or a strict absolute minimum point of the function $f(\mathbf{x})$.

3.1 Real Gradient

The ideal goal of minimization is to find the global minimum of a given objective function. However, this desired object is often difficult to achieve due to the following two main reasons.

- It is usually difficult to know the global or complete information about a function $f(\mathbf{x})$ in the definition domain $\mathcal{D}$.
- It is usually impractical to design an algorithm for identifying a global extreme point, because it is almost impossible to compare the value $f(\mathbf{x}^*)$ with all values of the function $f(\mathbf{x})$ in the definition domain $\mathcal{D}$.

In contrast, it is much easier to obtain local information about an objective function $f(\mathbf{x})$ in the vicinity of some point $\mathbf{c}$. At the same time, it is much simpler to design an algorithm for comparing the function value at some point $\mathbf{c}$ with other function values at the points near $\mathbf{c}$. Hence, most minimization algorithms can find only a local minimum point $\mathbf{c}$; the value of the objective function at $\mathbf{c}$ is the minimum of its values in the neighborhood of $\mathbf{c}$.

Definition 3.3 (Open Neighborhood, Closed Neighborhood) Given a point $\mathbf{c} \in \mathcal{D}$ and a positive number r, the set of all points $\mathbf{x}$ satisfying $\|\mathbf{x} - \mathbf{c}\|_2 < r$ is said to be an *open neighborhood* with radius r of the point $\mathbf{c}$, denoted by

$$B_o(\mathbf{c}; r) = \{\mathbf{x} | \mathbf{x} \in \mathcal{D}, \ \|\mathbf{x} - \mathbf{c}\|_2 < r\}. \tag{3.1.4}$$

If

$$B_c(\mathbf{c}; r) = \{\mathbf{x} | \mathbf{x} \in \mathcal{D}, \ \|\mathbf{x} - \mathbf{c}\|_2 \leq r\}, \tag{3.1.5}$$

then $B_c(\mathbf{c}; r)$ is known as a *closed neighborhood* of $\mathbf{c}$.

Consider a scalar $r > 0$, and let $\mathbf{x} = \mathbf{c} + \Delta \mathbf{x}$ be a point in the definition domain $\mathcal{D}$.

- If

$$f(\mathbf{c}) \leq f(\mathbf{c} + \Delta \mathbf{x}), \quad \forall\, 0 < \|\Delta \mathbf{x}\|_2 \leq r, \tag{3.1.6}$$

then the point $\mathbf{c}$ and the function value $f(\mathbf{c})$ are called a *local minimum point* and a *local minimum* (value) of the function $f(\mathbf{x})$, respectively.
- If

$$f(\mathbf{c}) < f(\mathbf{c} + \Delta \mathbf{x}), \quad \forall\, 0 < \|\Delta \mathbf{x}\|_2 \leq r, \tag{3.1.7}$$

then the point $\mathbf{c}$ and the function value $f(\mathbf{c})$ are known as a *strict local minimum point* and a *strict local minimum* of the function $f(\mathbf{x})$, respectively.
- If

$$f(\mathbf{c}) \geq f(\mathbf{c} + \Delta \mathbf{x}), \quad \forall\, 0 < \|\Delta \mathbf{x}\|_2 \leq r, \tag{3.1.8}$$

then the point **c** and the function value $f(\mathbf{c})$ are called a *local maximum point* and a *local maximum* (value) of the function $f(\mathbf{x})$, respectively.
- If

$$f(\mathbf{c}) > f(\mathbf{c} + \Delta\mathbf{x}), \quad \forall 0 < \|\Delta\mathbf{x}\|_2 \leq r, \tag{3.1.9}$$

then the point **c** and the function value $f(\mathbf{c})$ are known as a *strict local maximum point* and a *strict local maximum* of the function $f(\mathbf{x})$, respectively.
- If

$$f(\mathbf{c}) \leq f(\mathbf{x}), \quad \forall \mathbf{x} \in \mathcal{D}, \tag{3.1.10}$$

then the point **c** and the function value $f(\mathbf{c})$ are said to be a *global minimum point* and a *global minimum* of the function $f(\mathbf{x})$ in the definition domain $\mathcal{D}$, respectively.
- If

$$f(\mathbf{c}) < f(\mathbf{x}), \quad \forall \mathbf{x} \in \mathcal{D}, \mathbf{x} \neq \mathbf{c}, \tag{3.1.11}$$

then the point **c** and the function value $f(\mathbf{c})$ are referred to as a strict global minimum point and a *strict global minimum* of the function $f(\mathbf{x})$ in the definition domain $\mathcal{D}$, respectively.
- If

$$f(\mathbf{c}) \geq f(\mathbf{x}), \quad \forall \mathbf{x} \in \mathcal{D}, \tag{3.1.12}$$

then the point **c** and the function value $f(\mathbf{c})$ are said to be a *global maximum point* and a *global maximum* of the function $f(\mathbf{x})$ in the definition domain $\mathcal{D}$, respectively.
- If

$$f(\mathbf{c}) > f(\mathbf{x}), \quad \forall \mathbf{x} \in \mathcal{D}, \mathbf{x} \neq \mathbf{c}, \tag{3.1.13}$$

then the point **c** and the function value $f(\mathbf{c})$ are referred to as a *strict global maximum point* and a *strict global maximum* of the function $f(\mathbf{x})$ in the definition domain $\mathcal{D}$, respectively.

The minimum points and the maximum points are collectively called the *extreme points* of the function $f(\mathbf{x})$, and the minimal value and the maximal value are known as the *extrema* or extreme values of $f(\mathbf{x})$.

In particular, if some point $\mathbf{x}_0$ is a unique local extreme point of a function $f(\mathbf{x})$ in the neighborhood $B(\mathbf{c}; r)$, then it is said to be an *isolated local extreme point*.

3.1.2 Real Gradient of $f(\mathbf{X})$

In practical applications, it is impossible to compare directly the value of an objective function $f(\mathbf{x})$ at some point with all possible values in its neighborhood. Fortunately, the Taylor series expansion provides a simple method for coping with this difficulty.

The second-order Taylor series approximation of $f(\mathbf{x})$ at the point $\mathbf{c}$ is given by

$$f(\mathbf{c} + \Delta\mathbf{x}) = f(\mathbf{c}) + (\nabla f(\mathbf{c}))^T \Delta\mathbf{x} + \frac{1}{2}(\Delta\mathbf{x})^T \mathbf{H}[f(\mathbf{c})]\Delta\mathbf{x}, \quad (3.1.14)$$

where

$$\nabla f(\mathbf{c}) = \frac{\partial f(\mathbf{c})}{\partial \mathbf{c}} = \left.\frac{\partial f(\mathbf{x})}{\partial \mathbf{x}}\right|_{\mathbf{x}=\mathbf{c}}, \quad (3.1.15)$$

$$\mathbf{H}[f(\mathbf{c})] = \frac{\partial^2 f(\mathbf{c})}{\partial \mathbf{c} \partial \mathbf{c}^T} = \left.\frac{\partial^2 f(\mathbf{x})}{\partial \mathbf{x} \partial \mathbf{x}^T}\right|_{\mathbf{x}=\mathbf{c}}, \quad (3.1.16)$$

are, respectively, the gradient vector and the Hessian matrix of $f(\mathbf{x})$ at the point $\mathbf{c}$.

Definition 3.4 (Neighborhood) A *neighborhood* $B(\mathbf{X}_*; r)$ of a real function $f(\mathbf{X}) : \mathbb{R}^{m \times n} \to \mathbb{R}$ with center point $\text{vec}\mathbf{X}_*$ and the radius r is defined as

$$B(\mathbf{X}_*; r) = \{\mathbf{X} | \mathbf{X} \in \mathbb{R}^{m \times n}, \|\text{vec}\mathbf{X} - \text{vec}\mathbf{X}_*\|_2 < r\}. \quad (3.1.17)$$

The second-order Taylor series approximation formula of the function $f(\mathbf{X})$ at the point $\mathbf{X}_*$ is given by

$$\begin{aligned}
f(\mathbf{X}_* + \Delta\mathbf{X}) &= f(\mathbf{X}_*) + \left(\frac{\partial f(\mathbf{X}_*)}{\partial \text{vec}\mathbf{X}_*}\right)^T \text{vec}(\Delta\mathbf{X}) \\
&\quad + \frac{1}{2}(\text{vec}(\Delta\mathbf{X}))^T \frac{\partial^2 f(\mathbf{X}_*)}{\partial \text{vec}\mathbf{X}_* \partial (\text{vec}\mathbf{X}_*)^T} \text{vec}(\Delta\mathbf{X}) \\
&= f(\mathbf{X}_*) + \left(\nabla_{\text{vec}\mathbf{X}_*} f(\mathbf{X}_*)\right)^T \text{vec}(\Delta\mathbf{X}) \\
&\quad + \frac{1}{2}(\text{vec}(\Delta\mathbf{X}))^T \mathbf{H}[f(\mathbf{X}_*)]\text{vec}(\Delta\mathbf{X}), \quad (3.1.18)
\end{aligned}$$

where

$$\nabla_{\text{vec}\mathbf{X}_*} f(\mathbf{X}_*) = \left.\frac{\partial f(\mathbf{X})}{\partial \text{vec}\mathbf{X}}\right|_{\mathbf{X}=\mathbf{X}_*} \in \mathbb{R}^{mn},$$

$$\mathbf{H}[f(\mathbf{X}_*)] = \left.\frac{\partial^2 f(\mathbf{X})}{\partial \text{vec}\mathbf{X} \partial (\text{vec }\mathbf{X})^T}\right|_{\mathbf{X}=\mathbf{X}_*} \in \mathbb{R}^{mn \times mn},$$

are the gradient vector and the Hessian matrix of $f(\mathbf{X})$ at the point $\mathbf{X}_*$, respectively.

Theorem 3.1 (First-Order Necessary Condition [28]) *If $\mathbf{x}_*$ is a local extreme point of a function $f(\mathbf{x})$ and $f(\mathbf{x})$ is continuously differentiable in the neighborhood $B(\mathbf{x}_*; r)$ of the point $\mathbf{x}_*$, then*

$$\nabla_{\mathbf{x}_*} f(\mathbf{x}_*) = \left. \frac{\partial f(\mathbf{x})}{\partial \mathbf{x}} \right|_{\mathbf{x}=\mathbf{x}_*} = \mathbf{0}. \tag{3.1.19}$$

Theorem 3.2 (Second-Order Necessary Conditions [21, 28]) *If $\mathbf{x}_*$ is a local minimum point of $f(\mathbf{x})$ and the second-order gradient $\nabla_{\mathbf{x}}^2 f(\mathbf{x})$ is continuous in an open neighborhood $B_o(\mathbf{x}_*; r)$ of $\mathbf{x}_*$, then*

$$\nabla_{\mathbf{x}_*} f(\mathbf{x}_*) = \left. \frac{\partial f(\mathbf{x})}{\partial \mathbf{x}} \right|_{\mathbf{x}=\mathbf{x}_*} = \mathbf{0} \quad \text{and} \quad \nabla_{\mathbf{x}_*}^2 f(\mathbf{x}_*) = \left. \frac{\partial^2 f(\mathbf{x})}{\partial \mathbf{x} \partial \mathbf{x}^T} \right|_{\mathbf{x}=\mathbf{x}_*} \succeq 0, \tag{3.1.20}$$

i.e., the gradient vector of $f(\mathbf{x})$ at the point $\mathbf{x}_*$ is a zero vector and the Hessian matrix $\nabla_{\mathbf{x}}^2 f(\mathbf{x})$ at the point $\mathbf{x}_*$ is a positive semi-definite matrix.

Theorem 3.3 (Second-Order Sufficient Condition [21, 28]) *Let $\nabla_{\mathbf{x}}^2 f(\mathbf{x})$ be continuous in an open neighborhood of $\mathbf{x}_*$. If the conditions*

$$\nabla_{\mathbf{x}_*} f(\mathbf{x}_*) = \left. \frac{\partial f(\mathbf{x})}{\partial \mathbf{x}} \right|_{\mathbf{x}=\mathbf{x}_*} = \mathbf{0} \quad \text{and} \quad \nabla_{\mathbf{x}_*}^2 f(\mathbf{x}_*) = \left. \frac{\partial^2 f(\mathbf{x})}{\partial \mathbf{x} \partial \mathbf{x}^T} \right|_{\mathbf{x}=\mathbf{x}_*} \succ 0 \tag{3.1.21}$$

are satisfied, then $\mathbf{x}_*$ is a strict local minimum point of the objection function $f(\mathbf{x})$. Here $\nabla_{\mathbf{x}}^2 f(\mathbf{x}_*) \succ 0$ means that the Hessian matrix $\nabla_{\mathbf{x}_*}^2 f(\mathbf{x}_*)$ at the point $\mathbf{x}_*$ is a positive definite matrix.

Comparing Eq. (3.1.18) with Eq. (3.1.14), we have the following conditions similar to Theorems 3.1–3.3.

1. *First-order necessary condition for a minimizer:* If $\mathbf{X}_*$ is a local extreme point of a function $f(\mathbf{X})$, and $f(\mathbf{X})$ is continuously differentiable in the neighborhood $B(\mathbf{X}_*; r)$ of the point $\mathbf{X}_*$, then

$$\nabla_{\text{vec}\mathbf{X}_*} f(\mathbf{X}_*) = \left. \frac{\partial f(\mathbf{X})}{\partial \text{vec}\mathbf{X}} \right|_{\mathbf{X}=\mathbf{X}_*} = \mathbf{0}. \tag{3.1.22}$$

2. *Second-order necessary conditions for a local minimum point:* If $\mathbf{X}_*$ is a local minimum point of $f(\mathbf{X})$, and the Hessian matrix $\mathbf{H}[f(\mathbf{X})]$ is continuous in an open neighborhood $B_o(\mathbf{X}_*; r)$, then

$$\nabla_{\text{vec}\mathbf{X}_*} f(\mathbf{X}_*) = \mathbf{0} \quad \text{and} \quad \left. \frac{\partial^2 f(\mathbf{X})}{\partial \text{vec}\mathbf{X} \partial (\text{vec}\mathbf{X})^T} \right|_{\mathbf{X}=\mathbf{X}_*} \succeq 0. \tag{3.1.23}$$

3. *Second-order sufficient conditions for a strict local minimum point:* It is assumed that $\mathbf{H}[f(\mathbf{x})]$ is continuous in an open neighborhood of $\mathbf{X}_*$. If the conditions

$$\nabla_{\text{vec}\mathbf{X}_*} f(\mathbf{X}_*) = \mathbf{0} \quad \text{and} \quad \frac{\partial^2 f(\mathbf{X})}{\partial \text{vec}\mathbf{X}\partial(\text{vec}\mathbf{X})^T}\bigg|_{\mathbf{X}=\mathbf{X}_*} \succ 0 \qquad (3.1.24)$$

are satisfied, then $\mathbf{X}_*$ is a strict local minimum point of $f(\mathbf{X})$.

4. *Second-order necessary conditions for a local maximum point:* If $\mathbf{X}_*$ is a local maximum point of $f(\mathbf{X})$, and the Hessian matrix $\mathbf{H}[f(\mathbf{X})]$ is continuous in an open neighborhood $B_o(\mathbf{X}_*; r)$, then

$$\nabla_{\text{vec}\mathbf{X}_*} f(\mathbf{X}_*) = \mathbf{0} \quad \text{and} \quad \frac{\partial^2 f(\mathbf{X})}{\partial \text{vec}\mathbf{X}\partial(\text{vec}\mathbf{X})^T}\bigg|_{\mathbf{X}=\mathbf{X}_*} \preceq 0. \qquad (3.1.25)$$

5. *Second-order sufficient conditions for a strict local maximum point:* Assume that $\mathbf{H}[f(\mathbf{X})]$ is continuous in an open neighborhood of $\mathbf{X}_*$. If the conditions

$$\nabla_{\text{vec}\mathbf{X}_*} f(\mathbf{X}_*) = \mathbf{0} \quad \text{and} \quad \frac{\partial^2 f(\mathbf{X})}{\partial \text{vec}\mathbf{X}\partial(\text{vec}\mathbf{X})^T}\bigg|_{\mathbf{X}=\mathbf{X}_*} \prec 0 \qquad (3.1.26)$$

are satisfied, then $\mathbf{X}_*$ is a strict local maximum point of $f(\mathbf{X})$.

The point $\mathbf{X}_*$ is only a saddle point of $f(\mathbf{X})$, if

$$\nabla_{\text{vec}\mathbf{X}_*} f(\mathbf{X}_*) = \mathbf{0} \quad \text{and} \quad \frac{\partial^2 f(\mathbf{X})}{\partial \text{vec}\mathbf{X}\partial(\text{vec}\mathbf{X})^T}\bigg|_{\mathbf{X}=\mathbf{X}_*} \text{ is indefinite.}$$

3.2 Gradient of Complex Variable Function

Consider gradient analysis of a real-valued function with complex vector variable.

3.2.1 Extreme Point of Complex Variable Function

Consider the unconstrained optimization of a real-valued function $f(\mathbf{Z}, \mathbf{Z}^*) : \mathbb{C}^{m \times n} \times \mathbb{C}^{m \times n} \to \mathbb{R}$.

From the first-order differential

$$\begin{aligned} \mathrm{d}f(\mathbf{Z}, \mathbf{Z}^*) &= \left(\frac{\partial f(\mathbf{Z}, \mathbf{Z}^*)}{\partial \text{vec }\mathbf{Z}}\right)^T \text{vec}(\mathrm{d}\mathbf{Z}) + \left(\frac{\partial f(\mathbf{Z}, \mathbf{Z}^*)}{\partial \text{vec }\mathbf{Z}^*}\right)^T \text{vec}(\mathrm{d}\mathbf{Z}^*) \\ &= \left[\frac{\partial f(\mathbf{Z}, \mathbf{Z}^*)}{\partial (\text{vec}\mathbf{Z})^T}, \frac{\partial f(\mathbf{Z}, \mathbf{Z}^*)}{\partial (\text{vec }\mathbf{Z}^*)^T}\right]\begin{bmatrix}\text{vec}(\mathrm{d}\mathbf{Z}) \\ \text{vec}(\mathrm{d}\mathbf{Z})^*\end{bmatrix} \end{aligned} \qquad (3.2.1)$$

and the second-order differential

$$d^2 f(\mathbf{Z}, \mathbf{Z}^*) = \begin{bmatrix} \text{vec}(d\mathbf{Z}) \\ \text{vec}(d\mathbf{Z}^*) \end{bmatrix}^H \begin{bmatrix} \frac{\partial^2 f(\mathbf{Z},\mathbf{Z}^*)}{\partial (\text{vec}\,\mathbf{Z}^*)\partial (\text{vec}\,\mathbf{Z})^T} & \frac{\partial^2 f(\mathbf{Z},\mathbf{Z}^*)}{\partial (\text{vec}\,\mathbf{Z}^*)\partial (\text{vec}\,\mathbf{Z}^*)^T} \\ \frac{\partial^2 f(\mathbf{Z},\mathbf{Z}^*)}{\partial (\text{vec}\,\mathbf{Z})\partial (\text{vec}\,\mathbf{Z})^T} & \frac{\partial^2 f(\mathbf{Z},\mathbf{Z}^*)}{\partial (\text{vec}\,\mathbf{Z})\partial (\text{vec}\,\mathbf{Z}^*)^T)} \end{bmatrix} \begin{bmatrix} \text{vec}(d\mathbf{Z}) \\ \text{vec}(d\mathbf{Z}^*) \end{bmatrix}$$
(3.2.2)

we get the second-order series approximation of $f(\mathbf{Z}, \mathbf{Z}^*)$ at the point $\mathbf{C}$ given by

$$f(\mathbf{Z}, \mathbf{Z}^*) = f(\mathbf{C}, \mathbf{C}^*) + (\nabla f(\mathbf{C}, \mathbf{C}^*))^T \text{vec}(\Delta \tilde{\mathbf{C}})$$
$$+ \frac{1}{2}(\text{vec}(\Delta \tilde{\mathbf{C}}))^H \mathbf{H}[f(\mathbf{C}, \mathbf{C}^*)]\text{vec}(\Delta \tilde{\mathbf{C}}),$$
(3.2.3)

where

$$\Delta \tilde{\mathbf{C}} = \begin{bmatrix} \Delta \mathbf{C} \\ \Delta \mathbf{C}^* \end{bmatrix} = \begin{bmatrix} \mathbf{Z} - \mathbf{C} \\ \mathbf{Z}^* - \mathbf{C}^* \end{bmatrix} \in \mathbb{C}^{2m \times n},$$
(3.2.4)

$$\nabla f(\mathbf{C}, \mathbf{C}^*) = \begin{bmatrix} \frac{\partial f(\mathbf{Z},\mathbf{Z}^*)}{\partial (\text{vec}\,\mathbf{Z})} \\ \frac{\partial f(\mathbf{Z},\mathbf{Z}^*)}{\partial (\text{vec}\,\mathbf{Z}^*)} \end{bmatrix}_{\mathbf{Z}=\mathbf{C}} \in \mathbb{C}^{2mn \times 1},$$
(3.2.5)

$$\mathbf{H}[f(\mathbf{C}, \mathbf{C}^*)] = \begin{bmatrix} \frac{\partial^2 f(\mathbf{Z},\mathbf{Z}^*)}{\partial (\text{vec}\,\mathbf{Z}^*)\partial (\text{vec}\,\mathbf{Z})^T} & \frac{\partial^2 f(\mathbf{Z},\mathbf{Z}^*)}{\partial (\text{vec}\,\mathbf{Z}^*)\partial (\text{vec}\,\mathbf{Z}^*)^T} \\ \frac{\partial^2 f(\mathbf{Z},\mathbf{Z}^*)}{\partial (\text{vec}\,\mathbf{Z})\partial (\text{vec}\,\mathbf{Z})^T} & \frac{\partial^2 f(\mathbf{Z},\mathbf{Z}^*)}{\partial (\text{vec}\,\mathbf{Z})\partial (\text{vec}\,\mathbf{Z}^*)^T} \end{bmatrix}_{\mathbf{Z}=\mathbf{C}}$$

$$= \begin{bmatrix} \mathbf{H}_{\mathbf{Z}^*\mathbf{Z}} & \mathbf{H}_{\mathbf{Z}^*\mathbf{Z}^*} \\ \mathbf{H}_{\mathbf{ZZ}} & \mathbf{H}_{\mathbf{ZZ}^*} \end{bmatrix}_{\mathbf{Z}=\mathbf{C}} \in \mathbb{C}^{2mn \times 2mn}.$$
(3.2.6)

Notice that

$$\nabla f(\mathbf{C}, \mathbf{C}^*) = \mathbf{0}_{2mn \times 1} \Leftrightarrow \left.\frac{\partial f(\mathbf{Z}, \mathbf{Z}^*)}{\partial \text{vec}\,\mathbf{Z}}\right|_{\mathbf{Z}=\mathbf{C}} = \mathbf{0}_{mn \times 1},$$

$$\left.\frac{\partial f(\mathbf{Z}, \mathbf{Z}^*)}{\partial \text{vec}\,\mathbf{Z}^*}\right|_{\mathbf{Z}^*=\mathbf{C}^*} = \mathbf{0}_{mn \times 1} \Leftrightarrow \left.\frac{\partial f(\mathbf{Z}, \mathbf{Z}^*)}{\partial \mathbf{Z}^*}\right|_{\mathbf{Z}^*=\mathbf{C}^*} = \mathbf{O}_{m \times n}.$$

By Eq. (3.2.3) we have the following conditions on extreme points, similar to Theorems 3.1–3.3.

1. *First-order necessary condition for a local extreme point:* If $\mathbf{C}$ is a local extreme point of a function $f(\mathbf{Z}, \mathbf{Z}^*)$, and $f(\mathbf{Z}, \mathbf{Z}^*)$ is continuously differentiable in the neighborhood $B(\mathbf{C}; r)$ of the point $\mathbf{C}$, then

$$\nabla f(\mathbf{C}, \mathbf{C}^*) = \mathbf{0} \quad \text{or} \quad \left.\frac{\partial f(\mathbf{Z}, \mathbf{Z}^*)}{\partial \mathbf{Z}^*}\right|_{\mathbf{Z}=\mathbf{C}} = \mathbf{O}.$$
(3.2.7)

3.2 Gradient of Complex Variable Function

2. *Second-order necessary conditions for a local minimum point:* If $\mathbf{C}$ is a local minimum point of $f(\mathbf{Z}, \mathbf{Z}^*)$, and the Hessian matrix $\mathbf{H}[f(\mathbf{Z}, \mathbf{Z}^*)]$ is continuous in an open neighborhood $B_o(\mathbf{C}; r)$ of the point $\mathbf{C}$, then

$$\left.\frac{\partial f(\mathbf{Z}, \mathbf{Z}^*)}{\partial \mathbf{Z}^*}\right|_{\mathbf{Z}=\mathbf{C}} = \mathbf{O} \quad \text{and} \quad \mathbf{H}[f(\mathbf{C}, \mathbf{C}^*)] \succeq 0. \tag{3.2.8}$$

3. *Second-order sufficient conditions for a strict local minimum point:* Suppose that $\mathbf{H}[f(\mathbf{Z}, \mathbf{Z}^*)]$ is continuous in an open neighborhood of $\mathbf{C}$. If the conditions

$$\left.\frac{\partial f(\mathbf{Z}, \mathbf{Z}^*)}{\partial \mathbf{Z}^*}\right|_{\mathbf{Z}=\mathbf{C}} = \mathbf{O} \quad \text{and} \quad \mathbf{H}[f(\mathbf{C}, \mathbf{C}^*)] \succ 0 \tag{3.2.9}$$

are satisfied, then $\mathbf{C}$ is a strict local minimum point of $f(\mathbf{Z}, \mathbf{Z}^*)$.

4. *Second-order necessary conditions for a local maximum point:* If $\mathbf{C}$ is a local maximum point of $f(\mathbf{Z}, \mathbf{Z}^*)$, and the Hessian matrix $\mathbf{H}[f(\mathbf{Z}, \mathbf{Z}^*)]$ is continuous in an open neighborhood $B_o(\mathbf{C}; r)$ of the point $\mathbf{C}$, then

$$\left.\frac{\partial f(\mathbf{Z}, \mathbf{Z}^*)}{\partial \mathbf{Z}^*}\right|_{\mathbf{Z}=\mathbf{C}} = \mathbf{O} \quad \text{and} \quad \mathbf{H}[f(\mathbf{C}, \mathbf{C}^*)] \preceq 0. \tag{3.2.10}$$

5. *Second-order sufficient conditions for a strict local maximum point:* Suppose that $\mathbf{H}[f(\mathbf{Z}, \mathbf{Z}^*)]$ is continuous in an open neighborhood of $\mathbf{C}$, if the conditions

$$\left.\frac{\partial f(\mathbf{Z}, \mathbf{Z}^*)}{\partial \mathbf{Z}^*}\right|_{\mathbf{Z}=\mathbf{C}} = \mathbf{O} \quad \text{and} \quad \mathbf{H}[f(\mathbf{C}, \mathbf{C}^*)] \prec 0 \tag{3.2.11}$$

are satisfied, then $\mathbf{C}$ is a strict local maximum point of $f(\mathbf{Z}, \mathbf{Z}^*)$.

If $\left.\frac{\partial f(\mathbf{Z}, \mathbf{Z}^*)}{\partial \mathbf{Z}^*}\right|_{\mathbf{Z}=\mathbf{C}} = \mathbf{O}$, but the Hessian matrix $\mathbf{H}[f(\mathbf{C}, \mathbf{C}^*)]$ is indefinite, then $\mathbf{C}$ is only a saddle point of $f(\mathbf{Z}, \mathbf{Z}^*)$.

Table 3.1 summarizes the extreme-point conditions for three complex variable functions.

The Hessian matrices in Table 3.1 are defined as follows:

$$\mathbf{H}[f(c, c^*)] = \left.\begin{bmatrix} \frac{\partial^2 f(z, z^*)}{\partial z^* \partial z} & \frac{\partial^2 f(z, z^*)}{\partial z^* \partial z^*} \\ \frac{\partial^2 f(z, z^*)}{\partial z \partial z} & \frac{\partial^2 f(z, z^*)}{\partial z \partial z^*} \end{bmatrix}\right|_{z=c} \in \mathbb{C}^{2 \times 2},$$

$$\mathbf{H}[f(\mathbf{c}, \mathbf{c}^*)] = \left.\begin{bmatrix} \frac{\partial^2 f(\mathbf{z}, \mathbf{z}^*)}{\partial \mathbf{z}^* \partial \mathbf{z}^T} & \frac{\partial^2 f(\mathbf{z}, \mathbf{z}^*)}{\partial \mathbf{z}^* \partial \mathbf{z}^H} \\ \frac{\partial^2 f(\mathbf{z}, \mathbf{z}^*)}{\partial \mathbf{z} \partial \mathbf{z}^T} & \frac{\partial^2 f(\mathbf{z}, \mathbf{z}^*)}{\partial \mathbf{z} \partial \mathbf{z}^H} \end{bmatrix}\right|_{\mathbf{z}=\mathbf{c}} \in \mathbb{C}^{2n \times 2n},$$

Table 3.1 Extreme-point conditions for the complex variable functions

Functions	$f(z, z^*): \mathbb{C} \times \mathbb{C} \to \mathbb{R}$	$f(\mathbf{z}, \mathbf{z}^*): \mathbb{C}^n \times \mathbb{C}^n \to \mathbb{R}$	$f(\mathbf{Z}, \mathbf{Z}^*): \mathbb{C}^{m \times n} \times \mathbb{C}^{m \times n} \to \mathbb{R}$
Stationary point	$\left.\frac{\partial f(z, z^*)}{\partial z^*}\right\|_{z=c} = 0$	$\left.\frac{\partial f(\mathbf{z}, \mathbf{z}^*)}{\partial \mathbf{z}^*}\right\|_{\mathbf{z}=\mathbf{c}} = \mathbf{0}$	$\left.\frac{\partial f(\mathbf{Z}, \mathbf{Z}^*)}{\partial \mathbf{Z}^*}\right\|_{\mathbf{Z}=\mathbf{C}} = \mathbf{O}$
Local minimum	$\mathbf{H}[f(c, c^*)] \succeq 0$	$\mathbf{H}[f(\mathbf{c}, \mathbf{c}^*)] \succeq 0$	$\mathbf{H}[f(\mathbf{C}, \mathbf{C}^*)] \succeq 0$
Strict local minimum	$\mathbf{H}[f(c, c^*)] \succ 0$	$\mathbf{H}[f(\mathbf{c}, \mathbf{c}^*)] \succ 0$	$\mathbf{H}[f(\mathbf{C}, \mathbf{C}^*)] \succ 0$
Local maximum	$\mathbf{H}[f(c, c^*)] \preceq 0$	$\mathbf{H}[f(\mathbf{c}, \mathbf{c}^*)] \preceq 0$	$\mathbf{H}[f(\mathbf{C}, \mathbf{C}^*)] \preceq 0$
Strict local maximum	$\mathbf{H}[f(c, c^*)] \prec 0$	$\mathbf{H}[f(\mathbf{c}, \mathbf{c}^*)] \prec 0$	$\mathbf{H}[f(\mathbf{C}, \mathbf{C}^*)] \prec 0$
Saddle point	$\mathbf{H}[f(c, c^*)]$ indef.	$\mathbf{H}[f(\mathbf{c}, \mathbf{c}^*)]$ indef.	$\mathbf{H}[f(\mathbf{C}, \mathbf{C}^*)]$ indef.

$$\mathbf{H}[f(\mathbf{C}, \mathbf{C}^*)] = \left.\begin{bmatrix} \frac{\partial^2 f(\mathbf{Z}, \mathbf{Z}^*)}{\partial(\text{vec}\mathbf{Z}^*)\partial(\text{vec}\mathbf{Z})^T} & \frac{\partial^2 f(\mathbf{Z}, \mathbf{Z}^*)}{\partial(\text{vec}\mathbf{Z}^*)\partial(\text{vec}\mathbf{Z}^*)^T} \\ \frac{\partial^2 f(\mathbf{Z}, \mathbf{Z}^*)}{\partial(\text{vec}\mathbf{Z})\partial(\text{vec}\mathbf{Z})^T} & \frac{\partial^2 f(\mathbf{Z}, \mathbf{Z}^*)}{\partial(\text{vec}\mathbf{Z})\partial(\text{vec}\mathbf{Z}^*)^T} \end{bmatrix}\right|_{\mathbf{Z}=\mathbf{C}} \in \mathbb{C}^{2mn \times 2mn}.$$

3.2.2 Complex Gradient

Given a real-valued objective function $f(\mathbf{w}, \mathbf{w}^*)$ or $f(\mathbf{W}, \mathbf{W}^*)$, the gradient analysis of its unconstrained minimization problem can be summarized as follows.

1. The conjugate gradient matrix determines a closed solution of the minimization problem.
2. The sufficient conditions of local minimum points are determined by the conjugate gradient vector and the Hessian matrix of the objective function.
3. The negative direction of the conjugate gradient vector determines the steepest-descent iterative algorithm for solving the minimization problem.
4. The Hessian matrix gives the Newton algorithm for solving the minimization problem.

In the following, we discuss the above gradient analysis.

Closed Solution of the Unconstrained Minimization Problem

By letting the conjugate gradient vector (or matrix) be equal to a zero vector (or matrix), we can find a closed solution of the given unconstrained minimization problem.

Example 3.1 For an over-determined matrix equation $\mathbf{A}\mathbf{z} = \mathbf{b}$, define the log-likelihood function

$$l(\hat{\mathbf{z}}) = C - \frac{1}{\sigma^2}\mathbf{e}^H\mathbf{e} = C - \frac{1}{\sigma^2}(\mathbf{b} - \mathbf{A}\hat{\mathbf{z}})^H(\mathbf{b} - \mathbf{A}\hat{\mathbf{z}}), \tag{3.2.12}$$

3.2 Gradient of Complex Variable Function

where C is a real constant. Then, the conjugate gradient of the log-likelihood function is given by

$$\nabla_{\hat{\mathbf{z}}^*} l(\hat{\mathbf{z}}) = \frac{\partial l(\hat{\mathbf{z}})}{\partial \mathbf{z}^*} = \frac{1}{\sigma^2} \mathbf{A}^H \mathbf{b} - \frac{1}{\sigma^2} \mathbf{A}^H \mathbf{A} \hat{\mathbf{z}}.$$

By setting $\nabla_{\hat{\mathbf{z}}^*} l(\hat{\mathbf{z}}) = \mathbf{0}$, we obtain $\mathbf{A}^H \mathbf{A} \mathbf{z}_{\text{opt}} = \mathbf{A}^H \mathbf{b}$, where $\mathbf{z}_{\text{opt}}$ is the solution for maximizing the log-likelihood function $l(\hat{\mathbf{z}})$. Hence, if $\mathbf{A}^H \mathbf{A}$ is nonsingular, then

$$\mathbf{z}_{\text{opt}} = (\mathbf{A}^H \mathbf{A})^{-1} \mathbf{A}^H \mathbf{b}. \tag{3.2.13}$$

This is the ML solution of the matrix equation $\mathbf{A}\mathbf{z} = \mathbf{b}$. Clearly, the ML solution and the LS solution are the same for the matrix equation $\mathbf{A}\mathbf{z} = \mathbf{b}$.

Steepest Descent Direction of Real Objective Function

There are two choices in determining a stationary point $\mathbf{C}$ of an objective function $f(\mathbf{Z}, \mathbf{Z}^*)$ with a complex matrix as variable:

$$\left. \frac{\partial f(\mathbf{Z}, \mathbf{Z}^*)}{\partial \mathbf{Z}} \right|_{\mathbf{Z}=\mathbf{C}} = \mathbf{O}_{m \times n} \quad \text{or} \quad \left. \frac{\partial f(\mathbf{Z}, \mathbf{Z}^*)}{\partial \mathbf{Z}^*} \right|_{\mathbf{Z}^*=\mathbf{C}^*} = \mathbf{O}_{m \times n}. \tag{3.2.14}$$

The question is which gradient to select when designing a learning algorithm for an optimization problem. To answer this question, it is necessary to introduce the definition of the curvature direction.

Definition 3.5 (Direction of Curvature [14]) If $\mathbf{H}$ is the Hessian operator acting on a nonlinear function $f(\mathbf{x})$, a vector $\mathbf{p}$ is said to be

1. the *direction of positive curvature* of the function f, if $\mathbf{p}^H \mathbf{H} \mathbf{p} > 0$;
2. the *direction of zero curvature* of the function f, if $\mathbf{p}^H \mathbf{H} \mathbf{p} = 0$;
3. the *direction of negative curvature* of the function f, if $\mathbf{p}^H \mathbf{H} \mathbf{p} < 0$.

The scalar $\mathbf{p}^H \mathbf{H} \mathbf{p}$ is called the *curvature* of the function f along the direction $\mathbf{p}$.

The *curvature direction* is the direction of the maximum rate of change of the objective function.

Theorem 3.4 ([6]) *Let $f(\mathbf{z})$ be a real-valued function of the complex vector $\mathbf{z}$. By regarding $\mathbf{z}$ and $\mathbf{z}^*$ as independent variables, the curvature direction of the real objective function $f(\mathbf{z}, \mathbf{z}^*)$ is given by the conjugate gradient vector $\nabla_{\mathbf{z}^*} f(\mathbf{z}, \mathbf{z}^*)$.*

Theorem 3.4 shows that each component of $\nabla_{\mathbf{z}^*} f(\mathbf{z}, \mathbf{z}^*)$, or $\nabla_{\text{vec}\mathbf{Z}^*} f(\mathbf{Z}, \mathbf{Z}^*)$, gives the rate of change of the objective function $f(\mathbf{z}, \mathbf{z}^*)$, or $f(\mathbf{Z}, \mathbf{Z}^*)$, in this direction:

- The conjugate gradient vector $\nabla_{\mathbf{z}^*} f(\mathbf{z}, \mathbf{z}^*)$, or $\nabla_{\text{vec}\mathbf{Z}^*} f(\mathbf{Z}, \mathbf{Z}^*)$, gives the fastest growing direction of the objective function;
- The negative gradient vector $-\nabla_{\mathbf{z}^*} f(\mathbf{z}, \mathbf{z}^*)$, or $-\nabla_{\text{vec}\mathbf{Z}^*} f(\mathbf{Z}, \mathbf{Z}^*)$, provides the steepest decreasing direction of the objective function.

Hence, the negative direction of the conjugate gradient is used as the updating direction in minimization algorithm:

$$\mathbf{z}_k = \mathbf{z}_{k-1} - \mu \nabla_{\mathbf{z}^*} f(\mathbf{z}), \quad \mu > 0. \quad (3.2.15)$$

That is, the correction amount $-\mu \nabla_{\mathbf{z}^*} f(\mathbf{z})$ of alternate solutions in the iterative process is proportional to the negative conjugate gradient of the objective function.

Because the direction of the negative conjugate gradient vector always points in the decreasing direction of an objection function, this type of learning algorithm is called the *gradient descent algorithm* or *steepest descent algorithm*.

The constant μ in a steepest descent algorithm is referred to as the *learning step* and determines the rate at which the alternate solution converges to the optimization solution.

3.3 Convex Sets and Convex Function Identification

In the above section we have discussed unconstrained optimization problems. This section presents constrained optimization. The basic idea of solving a constrained optimization problem is to transform it into a unconstrained optimization problem.

3.3.1 Standard Constrained Optimization Problems

Consider the standard form of constrained optimization problems

$$\min_{\mathbf{x}} f_0(\mathbf{x}) \quad \text{subject to} \quad f_i(\mathbf{x}) \leq 0, \ i = 1, \ldots, m, \ \mathbf{A}\mathbf{x} = \mathbf{b} \quad (3.3.1)$$

which can be written as

$$\min_{\mathbf{x}} f_0(\mathbf{x}) \quad \text{subject to} \quad f_i(\mathbf{x}) \leq 0, i = 1, \ldots, m, \ h_i(\mathbf{x}) = 0, i = 1, \ldots, q.$$
$$(3.3.2)$$

The variable $\mathbf{x}$ in constrained optimization problems is called the *optimization variable* or *decision variable* and the function $f_0(\mathbf{x})$ is the *objective function* or cost function, while

$$f_i(\mathbf{x}) \leq 0, \ \mathbf{x} \in \mathcal{I}, \quad \text{and} \quad h_i(\mathbf{x}) = 0, \ \mathbf{x} \in \mathcal{E}, \quad (3.3.3)$$

are known as the *inequality constraints* and the *equality constraints*, respectively; $\mathcal{I}$ and $\mathcal{E}$ are, respectively, the definition domains of the inequality constraint function

3.3 Convex Sets and Convex Function Identification

and the equality constraint function, i.e.,

$$\mathcal{I} = \bigcap_{i=1}^{m} \text{dom } f_i \quad \text{and} \quad \mathcal{E} = \bigcap_{i=1}^{q} \text{dom } h_i. \tag{3.3.4}$$

The inequality constraints and the equality constraints are collectively called *explicit constraints*. An optimization problem without explicit constraints (i.e., $m = q = 0$) reduces to an unconstrained optimization problem.

The m inequality constraints $f_i(\mathbf{x}) \leq 0$, $i = 1, \ldots, m$ and the q equality constraints $h_j(\mathbf{x}) = 0$, $j = 1, \ldots, q$, denote $m + q$ strict requirements or stipulations restricting the possible choices of $\mathbf{x}$. An objective function $f_0(\mathbf{x})$ represents the costs paid by selecting $\mathbf{x}$. Conversely, the negative objective function $-f_0(\mathbf{x})$ can be understood as the values or benefits got by selecting $\mathbf{x}$. Hence, solving the constrained optimization problem (3.3.2) amounts to choosing $\mathbf{x}$ given $m + q$ strict requirements such that the cost is minimized or the value is maximized.

The *optimal solution* of the constrained optimization problem (3.3.2), denoted p^*, is defined as the *infimum* of the objective function $f_0(\mathbf{x})$:

$$p^* = \inf \{ f_0(\mathbf{x}) | f_i(\mathbf{x}) \leq 0, \ i = 1, \ldots, m; \ h_j(\mathbf{x}) = 0, \ j = 1, \ldots, q \}. \tag{3.3.5}$$

If $p^* = \infty$, then the constrained optimization problem (3.3.2) is said to be *infeasible*, i.e., no point $\mathbf{x}$ can meet the $m + q$ constraint conditions. If $p^* = -\infty$, then the constrained optimization problem (3.3.2) is *lower unbounded*. The following are the key steps in solving the constrained optimization (3.3.2).

1. Search for the feasible points of a given constrained optimization problem.
2. Search for the point such that the constrained optimization problem reaches the optimal value.
3. Avoid turning the original constrained optimization into a lower unbounded problem.

The point $\mathbf{x}$ meeting all the inequality constraints and equality constraints is called a *feasible point*. The set of all feasible points is called the *feasible domain* or *feasible set*, denoted by $\mathcal{F}$, and is defined as

$$\mathcal{F} \stackrel{\text{def}}{=} \mathcal{I} \cap \mathcal{E} = \{ \mathbf{x} | f_i(\mathbf{x}) \leq 0, \ i = 1, \ldots, m; \ h_j(\mathbf{x}) = 0, \ j = 1, \ldots, q \}. \tag{3.3.6}$$

Any point not in the feasible set is referred to as an *infeasible point*.

The intersection set of the definition domain $\text{dom } f_0$ of the objective function f_0 and its feasible domain $\mathcal{F}$ is defined as

$$\mathcal{D} = \text{dom } f_0 \cap \bigcap_{i=1}^{m} \text{dom } f_i \cap \bigcap_{j=1}^{q} \text{dom } h_j = \text{dom } f_0 \cap \mathcal{F} \tag{3.3.7}$$

that is known as the *definition domain* of a constrained optimization problem.

A feasible point **x** is optimal if $f_0(\mathbf{x}) = p^*$. It is usually difficult to solve a constrained optimization problem. When the number of decision variables in **x** is large, it is particularly difficult. This is due to the following reasons [19].

1. A constrained optimization problem may be occupied by a local optimal solution in its definition domain.
2. It is very difficult to find a feasible point.
3. The stopping criteria in general unconstrained optimization algorithms usually fail in a constrained optimization problem.
4. The convergence rate of a constrained optimization algorithm is usually poor.
5. The size of numerical problem can make a constrained minimization algorithm either completely stop or wander, so that these algorithms cannot converge in the normal way.

The above difficulties in solving constrained optimization problems can be overcome by using the convex optimization technique. In essence, *convex optimization* is the minimization (or maximization) of a convex (or concave) function under the constraint of a convex set. Convex optimization is a fusion of optimization, convex analysis, and numerical computation.

3.3.2 Convex Sets and Convex Functions

Definition 3.6 (Convex Set) A set $S \in \mathbb{R}^n$ is *convex*, if the line connecting any two points $\mathbf{x}, \mathbf{y} \in S$ is also in the set S, namely

$$\mathbf{x}, \mathbf{y} \in S, \ \theta \in [0, 1] \quad \Rightarrow \quad \theta \mathbf{x} + (1 - \theta)\mathbf{y} \in S. \tag{3.3.8}$$

Many familiar sets are convex, e.g., the *unit ball* $S = \{\mathbf{x} \,|\, \|\mathbf{x}\|_2 \leq 1\}$. However, the unit sphere $S = \{\mathbf{x} \,|\, \|\mathbf{x}\|_2 = 1\}$ is not a convex set because the line connecting two points on the sphere is obviously not itself on the sphere.

Letting $S_1 \subseteq \mathbb{R}^n$ and $S_2 \subseteq \mathbb{R}^m$ be two convex sets, and $\mathcal{A}(\mathbf{x}) : \mathbb{R}^n \to \mathbb{R}^m$ be the linear operator such that $\mathcal{A}(\mathbf{x}) = \mathbf{A}\mathbf{x} + \mathbf{b}$. A convex set has the following important properties [25, Theorem 2.2.4].

1. The *intersection* $S_1 \cap S_2 = \{\mathbf{x} \in \mathbb{R}^n \,|\, \mathbf{x} \in S_1, \mathbf{x} \in S_2\}$ $(m = n)$ is a convex set.
2. The *sum of sets* $S_1 + S_2 = \{\mathbf{z} = \mathbf{x} + \mathbf{y} \,|\, \mathbf{x} \in S_1, \mathbf{y} \in S_2\}$ $(m = n)$ is a convex set.
3. The *direct sum* $S_1 \oplus S_2 = \{(\mathbf{x}, \mathbf{y}) \in \mathbb{R}^{n+m} \,|\, \mathbf{x} \in S_1, \mathbf{y} \in S_2\}$ is a convex set.
4. The *conic hull* $\mathcal{K}(S_1) = \{\mathbf{z} \in \mathbb{R}^n \,|\, \mathbf{z} = \beta \mathbf{x}, \mathbf{x} \in S_1, \beta \geq 0\}$ is a convex set.
5. The *affine image* $\mathcal{A}(S_1) = \{\mathbf{y} \in \mathbb{R}^m \,|\, \mathbf{y} = \mathcal{A}(\mathbf{x}), \mathbf{x} \in S_1\}$ is a convex set.
6. The *inverse affine image* $\mathcal{A}^{-1}(S_2) = \{\mathbf{x} \in \mathbb{R}^n \,|\, \mathbf{x} = \mathcal{A}^{-1}(\mathbf{y}), \mathbf{y} \in S_2\}$ is a convex set.
7. The following *convex hull* is a convex set:

$$\text{conv}(S_1, S_2) = \{\mathbf{z} \in \mathbb{R}^n \,|\, \mathbf{z} = \alpha \mathbf{x} + (1 - \alpha)\mathbf{y}, \mathbf{x} \in S_1, \mathbf{y} \in S_2; \alpha \in [0, 1]\}.$$

3.3 Convex Sets and Convex Function Identification

Given a vector $\mathbf{x} \in \mathbb{R}^n$ and a constant $\rho > 0$, then

$$B_o(\mathbf{x}, \rho) = \{\mathbf{y} \in \mathbb{R}^n \,|\, \|\mathbf{y} - \mathbf{x}\|_2 < \rho\}, \tag{3.3.9}$$

$$B_c(\mathbf{x}, \rho) = \{\mathbf{y} \in \mathbb{R}^n \,|\, \|\mathbf{y} - \mathbf{x}\|_2 \leq \rho\}, \tag{3.3.10}$$

are called the *open ball* and the *closed ball* with center $\mathbf{x}$ and radius ρ.

A convex set $S \subseteq \mathbb{R}^n$ is known as a *convex cone* if all rays starting at the origin and all segments connecting any two points of these rays are still within the convex set, i.e.,

$$\mathbf{x}, \mathbf{y} \in S, \ \lambda, \mu \geq 0 \ \Rightarrow \ \lambda\mathbf{x} + \mu\mathbf{y} \in S. \tag{3.3.11}$$

The *nonnegative orthant* $\mathbb{R}^n_+ = \{\mathbf{x} \in \mathbb{R}^n \,|\, \mathbf{x} \geq \mathbf{0}\}$ is a convex cone. The set of positive semi-definite matrices $\mathbf{X} \succeq \mathbf{0}$, $S^n_+ = \{\mathbf{X} \in \mathbb{R}^{n \times n} \,|\, \mathbf{X} \succeq \mathbf{0}\}$, is a convex cone as well, since the positive combination of any number of positive semi-definite matrices is still positive semi-definite. Hence, S^n_+ is called the *positive semi-definite cone*.

Definition 3.7 (Affine Function) The vector function $\mathbf{f}(\mathbf{x}) : \mathbb{R}^n \to \mathbb{R}^m$ is known as an *affine function* if it has the form

$$\mathbf{f}(\mathbf{x}) = \mathbf{A}\mathbf{x} + \mathbf{b}. \tag{3.3.12}$$

Similarly, the matrix function $\mathbf{F}(\mathbf{x}) : \mathbb{R}^n \to \mathbb{R}^{p \times q}$ is called an affine (matrix) function, if it has the form

$$\mathbf{F}(\mathbf{x}) = \mathbf{A}_0 + x_1\mathbf{A}_1 + \cdots + x_n\mathbf{A}_n, \tag{3.3.13}$$

where $\mathbf{A}_i \in \mathbb{R}^{p \times q}$. An affine function sometimes is roughly referred to as a linear function.

Definition 3.8 (Strictly Convex Function [25]) Given a convex set $S \in \mathbb{R}^n$ and a function $f : S \to \mathbb{R}$, then we have the following.

- $f : \mathbb{R}^n \to \mathbb{R}$ is a *convex function* if and only if $S = \mathrm{dom} f$ is a convex set, and for all vectors $\mathbf{x}, \mathbf{y} \in S$ and each scalar $\alpha \in (0, 1)$, the function satisfies the *Jensen inequality*

$$f(\alpha\mathbf{x} + (1 - \alpha)\mathbf{y}) \leq \alpha f(\mathbf{x}) + (1 - \alpha)f(\mathbf{y}). \tag{3.3.14}$$

- The function $f(\mathbf{x})$ is known as a *strictly convex function* if and only if $S = \mathrm{dom} f$ is a convex set and, for all vectors $\mathbf{x}, \mathbf{y} \in S$ and each scalar $\alpha \in (0, 1)$, the function satisfies the inequality

$$f(\alpha\mathbf{x} + (1 - \alpha)\mathbf{y}) < \alpha f(\mathbf{x}) + (1 - \alpha)f(\mathbf{y}). \tag{3.3.15}$$

A strongly convex function has the following three equivalent definitions.

1. The function $f(\mathbf{x})$ is strongly convex if [25]

$$f(\alpha\mathbf{x} + (1-\alpha)\mathbf{y}) \leq \alpha f(\mathbf{x}) + (1-\alpha)f(\mathbf{y}) - \frac{\mu}{2}\alpha(1-\alpha)\|\mathbf{x} - \mathbf{y}\|_2^2 \quad (3.3.16)$$

holds for all vectors $\mathbf{x}, \mathbf{y} \in S$ and scalars $\alpha \in [0, 1]$ and $\mu > 0$.

2. The function $f(\mathbf{x})$ is strongly convex if [2]

$$(\nabla f(\mathbf{x}) - \nabla f(\mathbf{y}))^T(\mathbf{x} - \mathbf{y}) \geq \mu\|\mathbf{x} - \mathbf{y}\|_2^2 \quad (3.3.17)$$

holds for all vectors $\mathbf{x}, \mathbf{y} \in S$ and some scalar $\mu > 0$.

3. The function $f(\mathbf{x})$ is strongly convex if [25]

$$f(\mathbf{y}) \geq f(\mathbf{x}) + [\nabla f(\mathbf{x})]^T(\mathbf{y} - \mathbf{x}) + \frac{\mu}{2}\|\mathbf{y} - \mathbf{x}\|_2^2 \quad (3.3.18)$$

for some scalar $\mu > 0$.

In the above three definitions, the constant $\mu \, (> 0)$ is called the *convexity parameter* of the strongly convex function $f(\mathbf{x})$.

The following is the relationship between convex functions, strictly convex functions, and strongly convex functions:

$$\text{strongly convex function} \Rightarrow \text{strictly convex function} \Rightarrow \text{convex function.} \quad (3.3.19)$$

Definition 3.9 (Quasi-Convex [33]) A function $f(\mathbf{x})$ is *quasi-convex* if, for all vectors $\mathbf{x}, \mathbf{y} \in S$ and a scalar $\alpha \in [0, 1]$, the inequality

$$f(\alpha\mathbf{x} + (1-\alpha)\mathbf{y}) \leq \max\{f(\mathbf{x}), f(\mathbf{y})\} \quad (3.3.20)$$

holds. A function $f(\mathbf{x})$ is *strongly quasi-convex* if, for all vectors $\mathbf{x}, \mathbf{y} \in S$, $\mathbf{x} \neq \mathbf{y}$ and the scalar $\alpha \in (0, 1)$, the strict inequality

$$f(\alpha\mathbf{x} + (1-\alpha)\mathbf{y}) < \max\{f(\mathbf{x}), f(\mathbf{y})\} \quad (3.3.21)$$

holds. A function $f(\mathbf{x})$ is referred to as *strictly quasi-convex* if the strict inequality (3.3.21) holds for all vectors $\mathbf{x}, \mathbf{y} \in S$, $f(\mathbf{x}) \neq f(\mathbf{y})$ and a scalar $\alpha \in (0, 1)$.

3.3.3 Convex Function Identification

Consider an objective function defined in the convex set S, $f(\mathbf{x}) : S \to \mathbb{R}$, a natural question to ask is how to determine whether the function is convex.

3.3 Convex Sets and Convex Function Identification

Convex-function identification methods are divided into first-order gradient and second-order gradient identification methods.

The following are the first-order necessary and sufficient conditions for identifying the convex function.

First-Order Necessary and Sufficient Conditions

Theorem 3.5 ([32]) Let $f : S \to \mathbb{R}$ be a first differentiable function defined in the convex set S in the n-dimensional vector space $\mathbb{R}^n$, then for all vectors $\mathbf{x}, \mathbf{y} \in S$, we have

$$f(\mathbf{x}) \text{ convex} \quad \Leftrightarrow \quad \langle \nabla_{\mathbf{x}} f(\mathbf{x}) - \nabla_{\mathbf{x}} f(\mathbf{y}), \mathbf{x} - \mathbf{y} \rangle \geq 0,$$

$$f(\mathbf{x}) \text{ strictly convex} \quad \Leftrightarrow \quad \langle \nabla_{\mathbf{x}} f(\mathbf{x}) - \nabla_{\mathbf{x}} f(\mathbf{y}), \mathbf{x} - \mathbf{y} \rangle > 0, \ \mathbf{x} \neq \mathbf{y},$$

$$f(\mathbf{x}) \text{ strongly convex} \quad \Leftrightarrow \quad \langle \nabla_{\mathbf{x}} f(\mathbf{x}) - \nabla_{\mathbf{x}} f(\mathbf{y}), \mathbf{x} - \mathbf{y} \rangle \geq \mu \|\mathbf{x} - \mathbf{y}\|_2^2.$$

Theorem 3.6 ([3]) If the function $f : S \to \mathbb{R}$ is differentiable in a convex definition domain, then f is convex if and only if

$$f(\mathbf{y}) \geq f(\mathbf{x}) + \langle \nabla_{\mathbf{x}} f(\mathbf{x}), \mathbf{y} - \mathbf{x} \rangle. \tag{3.3.22}$$

Second-Order Necessary and Sufficient Conditions

Theorem 3.7 ([21]) Let $f : S \to \mathbb{R}$ be a function defined in the convex set $S \in \mathbb{R}^n$, and second differentiable, then $f(\mathbf{x})$ is a convex function if and only if the Hessian matrix is positive semi-definite, namely

$$\mathbf{H}_{\mathbf{x}}[f(\mathbf{x})] = \frac{\partial^2 f(\mathbf{x})}{\partial \mathbf{x} \partial \mathbf{x}^T} \succeq 0, \quad \forall \mathbf{x} \in S. \tag{3.3.23}$$

Remark Let $f : S \to \mathbb{R}$ be a function defined in the convex set $S \in \mathbb{R}^n$, and second differentiable; then $f(\mathbf{x})$ is strictly convex if and only if the Hessian matrix is positive definite, namely

$$\mathbf{H}_{\mathbf{x}}[f(\mathbf{x})] = \frac{\partial^2 f(\mathbf{x})}{\partial \mathbf{x} \partial \mathbf{x}^T} \succ 0, \quad \forall \mathbf{x} \in S, \tag{3.3.24}$$

whereas the sufficient condition for a strict minimum point requires that the Hessian matrix is positive definite at one point $\mathbf{c}$, Eq. (3.3.24) requires that the Hessian matrix is positive definite at all points in the convex set S.

It should be pointed out that, apart from the ℓ_0-norm, the vector norms for all other ℓ_p $(p \geq 1)$,

$$\|\mathbf{x}\|_p = \left(\sum_{i=1}^{n} |x_i|^p \right)^{1/p}, \ p \geq 1; \quad \|\mathbf{x}\|_\infty = \max_i |x_i| \tag{3.3.25}$$

are convex functions.

3.4 Gradient Methods for Smooth Convex Optimization

Convex optimization is divided into first-order algorithms and second-order algorithms. Focusing upon smoothing functions, this section presents first-order algorithms for smooth convex optimization: the gradient method, the gradient projection method, and the convergence rates.

3.4.1 Gradient Method

Optimization method generates a relaxation sequence

$$\mathbf{x}_{k+1} = \mathbf{x}_k + \mu_k \Delta \mathbf{x}_k, \quad k = 1, 2, \ldots \tag{3.4.1}$$

to search for the optimal point $\mathbf{x}_{\text{opt}}$. In the above equation, $k = 1, 2, \ldots$ represents the iterative number and $\mu_k \geq 0$ is the step size or step length of the kth iteration; the concatenated symbol of Δ and $\mathbf{x}$, $\Delta \mathbf{x}$, denotes a vector in $\mathbb{R}^n$ known as the *step direction* or *search direction* of the objective function $f(\mathbf{x})$.

From the first-order approximation expression to the objective function at the point $\mathbf{x}_k$

$$f(\mathbf{x}_{k+1}) \approx f(\mathbf{x}_k) + (\nabla f(\mathbf{x}_k))^T \Delta \mathbf{x}_k, \tag{3.4.2}$$

it follows that if

$$(\nabla f(\mathbf{x}_k))^T \Delta \mathbf{x}_k < 0, \tag{3.4.3}$$

then $f(\mathbf{x}_{k+1}) < f(\mathbf{x}_k)$ for minimization problems. Hence the search direction $\Delta \mathbf{x}_k$ such that $(\nabla f(\mathbf{x}_k))^T \Delta \mathbf{x}_k < 0$ is called the *descent step* or *descent direction* of the objective function $f(\mathbf{x})$ at the kth iteration.

Obviously, in order to make $(\nabla f(\mathbf{x}_k))^T \Delta \mathbf{x}_k < 0$, we should take

$$\Delta \mathbf{x}_k = -\nabla f(\mathbf{x}_k) \cos \theta, \tag{3.4.4}$$

where $0 \leq \theta < \pi/2$ is the acute angle between the descent direction and the negative gradient direction $-\nabla f(\mathbf{x}_k)$.

If $\theta = 0$, then $\Delta \mathbf{x}_k = -\nabla f(\mathbf{x}_k)$. That is to say, the negative gradient direction of the objective function f at the point $\mathbf{x}$ is directly taken as the search direction. In this case, the length of the descent step $\|\Delta \mathbf{x}_k\|_2 = \|\nabla f(\mathbf{x}_k)\|_2 \geq \|\nabla f(\mathbf{x}_k)\|_2 \cos \theta$ takes the maximum value, and thus the descent direction $\Delta \mathbf{x}_k$ has the maximal descent step or rate, and the corresponding descent algorithm is called the *steepest descent*

3.4 Gradient Methods for Smooth Convex Optimization

method:

$$\mathbf{x}_{k+1} = \mathbf{x}_k - \mu_k \nabla f(\mathbf{x}_k), \quad k = 1, 2, \ldots \tag{3.4.5}$$

The *steepest descent direction* $\Delta \mathbf{x} = -\nabla f(\mathbf{x})$ uses only first-order gradient information about the objective function $f(\mathbf{x})$. If second-order information, i.e., the Hessian matrix $\nabla^2 f(\mathbf{x}_k)$ of the objective function is used, then we may search for a better search direction. In this case, the optimal descent direction $\Delta \mathbf{x}$ is the solution minimizing the second-order Taylor approximation function of $f(\mathbf{x})$:

$$\min_{\Delta \mathbf{x}} \left\{ f(\mathbf{x} + \Delta \mathbf{x}) = f(\mathbf{x}) + (\nabla f(\mathbf{x}))^T \Delta \mathbf{x} + \frac{1}{2}(\Delta \mathbf{x})^T \nabla^2 f(\mathbf{x}) \Delta \mathbf{x} \right\}. \tag{3.4.6}$$

At the optimal point, the gradient vector of the objective function with respect to the parameter vector $\Delta \mathbf{x}$ has to equal zero, i.e.,

$$\frac{\partial f(\mathbf{x} + \Delta \mathbf{x})}{\partial \Delta \mathbf{x}} = \nabla f(\mathbf{x}) + \nabla^2 f(\mathbf{x}) \Delta \mathbf{x} = \mathbf{0}$$

$$\Leftrightarrow \quad \Delta \mathbf{x}_{\text{nt}} = -\left(\nabla^2 f(\mathbf{x})\right)^{-1} \nabla f(\mathbf{x}), \tag{3.4.7}$$

where $\Delta \mathbf{x}_{\text{nt}}$ is called the *Newton step* or the *Newton descent direction* and the corresponding optimization method is the *Newton* or *Newton–Raphson method*.

Algorithm 3.1 summarizes the gradient descent algorithm and its variants.

Algorithm 3.1 Gradient descent algorithm and its variants

1. **input:** An initial point $\mathbf{x}_1 \in \text{dom } f$ and the allowed error $\epsilon > 0$. Put $k = 1$.
2. **repeat**
3. Compute the gradient $\nabla f(\mathbf{x}_k)$ (and the Hessian matrix $\nabla^2 f(\mathbf{x}_k)$).
4. Choose the descent direction

$$\Delta \mathbf{x}_k = \begin{cases} -\nabla f(\mathbf{x}_k) & \text{(steepest descent method)}, \\ -\left(\nabla^2 f(\mathbf{x}_k)\right)^{-1} \nabla f(\mathbf{x}_k) & \text{(Newton method)}. \end{cases}$$

5. Choose a step $\mu_k > 0$, and update $\mathbf{x}_{k+1} = \mathbf{x}_k + \mu_k \Delta \mathbf{x}_k$.
6. **exit if** $|f(\mathbf{x}_{k+1}) - f(\mathbf{x}_k)| \leq \epsilon$.
7. **return** $k \leftarrow k + 1$.
8. **output:** $\mathbf{x} \leftarrow \mathbf{x}_k$.

3.4.2 Projected Gradient Method

The variable vector **x** is unconstrained in the gradient algorithms, i.e., $\mathbf{x} \in \mathbb{R}^n$. When choosing a constrained variable vector $\mathbf{x} \in C$, where $C \subset \mathbb{R}^n$, then the update formula in the gradient algorithm should be replaced by a formula containing the projected gradient:

$$\mathbf{x}_{k+1} = \mathcal{P}_C(\mathbf{x}_k - \mu_k \nabla f(\mathbf{x}_k)). \tag{3.4.8}$$

This algorithm is called the *projected gradient method* or the *gradient projection method*. In (3.4.8), $\mathcal{P}_C(\mathbf{y})$ is known as the projection of point **y** onto the set C, and is defined as

$$\mathcal{P}_C(\mathbf{y}) = \arg\min_{\mathbf{x} \in C} \frac{1}{2} \|\mathbf{x} - \mathbf{y}\|_2^2. \tag{3.4.9}$$

The projection operator can be equivalently expressed as

$$\mathcal{P}_C(\mathbf{y}) = \mathbf{P}_C \mathbf{y}, \tag{3.4.10}$$

where $\mathbf{P}_C$ is the projection matrix onto the subspace C. If C is the column space of the matrix **A**, then

$$\mathbf{P}_A = \mathbf{A}(\mathbf{A}^T \mathbf{A})^{-1} \mathbf{A}^T. \tag{3.4.11}$$

Theorem 3.8 ([25]) *If C is a convex set, then there exists a unique projection $\mathbf{P}_C(\mathbf{y})$.*

In particular, if $C = \mathbb{R}^n$, i.e., the variable vector **x** is unconstrained, then the projection operator is equal to the identity matrix, namely $\mathcal{P}_C = \mathbf{I}$, and thus

$$\mathcal{P}_{\mathbb{R}^n}(\mathbf{y}) = \mathbf{P}\mathbf{y} = \mathbf{y}, \quad \forall \mathbf{y} \in \mathbb{R}^n.$$

In this case, the projected gradient algorithm simplifies to the gradient algorithm.

The following are projections of the vector **x** onto several typical sets [35].

1. The projection onto the affine set $C = \{\mathbf{x} | \mathbf{A}\mathbf{x} = \mathbf{b}\}$ with $\mathbf{A} \in \mathbb{R}^{p \times n}$, rank$(\mathbf{A}) = p$ is

$$\mathcal{P}_C(\mathbf{x}) = \mathbf{x} + \mathbf{A}^T(\mathbf{A}\mathbf{A}^T)^{-1}(\mathbf{b} - \mathbf{A}\mathbf{x}). \tag{3.4.12}$$

If $p \ll n$ or $\mathbf{A}\mathbf{A}^T = \mathbf{I}$, then the projection $\mathcal{P}_C(\mathbf{x})$ gives a low cost result.

3.4 Gradient Methods for Smooth Convex Optimization

2. The projection onto the *hyperplane* $C = \{\mathbf{x} | \mathbf{a}^T \mathbf{x} = b\}$ (where $\mathbf{a} \neq \mathbf{0}$) is

$$\mathcal{P}_C(\mathbf{x}) = \mathbf{x} + \frac{b - \mathbf{a}^T \mathbf{x}}{\|\mathbf{a}\|_2^2} \mathbf{a}. \quad (3.4.13)$$

3. The projection onto the *nonnegative orthant* $C = \mathbb{R}_+^n$ is

$$\mathcal{P}_C(\mathbf{x}) = (\mathbf{x})^+ \iff [(\mathbf{x})^+]_i = \max\{x_i, 0\}. \quad (3.4.14)$$

4. The projection onto the *half-space* $C = \{\mathbf{x} | \mathbf{a}^T \mathbf{x} \leq b\}$ (where $\mathbf{a} \neq \mathbf{0}$):

$$\mathcal{P}_C(\mathbf{x}) = \begin{cases} \mathbf{x} + \dfrac{b - \mathbf{a}^T \mathbf{x}}{\|\mathbf{a}\|_2^2} \mathbf{a}, & \text{if } \mathbf{a}^T \mathbf{x} > b; \\ \mathbf{x}, & \text{if } \mathbf{a}^T \mathbf{x} \leq b. \end{cases} \quad (3.4.15)$$

5. The projection onto the *rectangular set* $C = [\mathbf{a}, \mathbf{b}]$ (where $a_i \leq x_i \leq b_i$) is

$$\mathcal{P}_C(\mathbf{x}) = \begin{cases} a_i, & \text{if } x_i \leq a_i; \\ x_i, & \text{if } a_i \leq x_i \leq b_i; \\ b_i, & \text{if } x_i \geq b_i. \end{cases} \quad (3.4.16)$$

6. The projection onto the *second-order cones* $C = \{(\mathbf{x}, t) | \|\mathbf{x}\|_2 \leq t, \mathbf{x} \in \mathbb{R}^n\}$ is

$$\mathcal{P}_C(\mathbf{x}) = \begin{cases} (\mathbf{x}, t), & \text{if } \|\mathbf{x}\|_2 \leq t; \\ \dfrac{t + \|\mathbf{x}\|_2}{2\|\mathbf{x}\|_2} \begin{bmatrix} \mathbf{x} \\ t \end{bmatrix}, & \text{if } -t < \|\mathbf{x}\|_2 < t; \\ (\mathbf{0}, 0), & \text{if } \|\mathbf{x}\|_2 \leq -t, \mathbf{x} \neq \mathbf{0}. \end{cases} \quad (3.4.17)$$

7. The projection onto the *Euclidean ball* $C = \{\mathbf{x} | \|\mathbf{x}\|_2 \leq 1\}$ is

$$\mathcal{P}_C(\mathbf{x}) = \begin{cases} \dfrac{1}{\|\mathbf{x}\|_2} \mathbf{x}, & \text{if } \|\mathbf{x}\|_2 > 1; \\ \mathbf{x}, & \text{if } \|\mathbf{x}\|_2 \leq 1. \end{cases} \quad (3.4.18)$$

8. The projection onto the ℓ_1-*norm ball* $C = \{\mathbf{x} | \|\mathbf{x}\|_1 \leq 1\}$ is

$$\mathcal{P}_C(\mathbf{x})_i = \begin{cases} x_i - \lambda, & \text{if } x_i > \lambda; \\ 0, & \text{if } -\lambda \leq x_i \leq \lambda; \\ x_i + \lambda, & \text{if } x_i < -\lambda. \end{cases} \quad (3.4.19)$$

Here $\lambda = 0$ if $\|\mathbf{x}\|_1 \leq 1$; otherwise, λ is the solution of the equation

$$\sum_{i=1}^n \max\{|x_i| - \lambda, 0\} = 1.$$

9. The projection onto the *positive semi-definite cones* $C = \mathbb{S}_+^n$ is

$$\mathcal{P}_C(\mathbf{X}) = \sum_{i=1}^n \max\{0, \lambda_i\}\mathbf{q}_i\mathbf{q}_i^T, \qquad (3.4.20)$$

where $\mathbf{X} = \sum_{i=1}^n \lambda_i \mathbf{q}_i \mathbf{q}_i^T$ is the eigenvalue decomposition of $\mathbf{X}$.

3.4.3 Convergence Rates

By the *convergence rate* it refers to the number of iterations needed for an optimization algorithm to make the estimated error of the objective function achieve the required accuracy, or given an iteration number K, the accuracy that an optimization algorithm reaches. The inverse of the convergence rate is known as the *complexity* of an optimization algorithm.

Let $\mathbf{x}^*$ be a local or global minimum point. The estimation error of an optimization algorithm is defined as the difference between the value of an objective function at iteration point $\mathbf{x}_k$ and its value at the global minimum point, i.e.,

$$\delta_k = f(\mathbf{x}_k) - f(\mathbf{x}^*).$$

We are naturally interested in the convergence problems of an optimization algorithm:

- Given an iteration number K, what is the designed accuracy $\lim_{1 \leq k \leq K} \delta_k$?
- Given an allowed accuracy ϵ, how many iterations does the algorithm need to achieve the designed accuracy $\min_k \delta_k \leq \epsilon$?

When analyzing the convergence problems of optimization algorithms, we often focus on the speed of the updating sequence $\{\mathbf{x}_k\}$ at which the objective function argument converges to its ideal minimum point $\mathbf{x}^*$. In numerical analysis, the speed at which a sequence reaches its limit is called the convergence rate.

Q-Convergence Rate

Assume that a sequence $\{\mathbf{x}_k\}$ converges to $\mathbf{x}^*$. If there are a real number $\alpha \geq 1$ and a positive constant μ independent of the iterations k such that

$$\mu = \lim_{k \to \infty} \frac{\|\mathbf{x}_{k+1} - \mathbf{x}^*\|_2}{\|\mathbf{x}_k - \mathbf{x}^*\|_2^\alpha}, \qquad (3.4.21)$$

3.4 Gradient Methods for Smooth Convex Optimization

then $\{\mathbf{x}_k\}$ is said to have an *α-order Q-convergence rate*. The Q-convergence rate means the *Quotient convergence rate*. It has the following typical values [29]:

1. When $\alpha = 1$, the Q-convergence rate is called the limit-convergence rate of $\{\mathbf{x}_k\}$:

$$\mu = \lim_{k \to \infty} \frac{\|\mathbf{x}_{k+1} - \mathbf{x}^*\|_2}{\|\mathbf{x}_k - \mathbf{x}^*\|_2}. \tag{3.4.22}$$

According to the value of μ, the limit-convergence rate of the sequence $\{\mathbf{x}_k\}$ can be divided into three types:

- *sublinear convergence rate*, $\alpha = 1$, $\mu = 1$;
- *linear convergence rate*, $\alpha = 1$, $\mu \in (0, 1)$;
- *superlinear convergence rate*, $\alpha = 1$, $\mu = 0$ or $1 < \alpha < 2$, $\mu = 0$.

2. When $\alpha = 2$, we say that the sequence $\{\mathbf{x}_k\}$ has a *quadratic convergence rate*.
3. When $\alpha = 3$, the sequence $\{\mathbf{x}_k\}$ is said to have a *cubic convergence rate*.

If $\{\mathbf{x}_k\}$ is sublinearly convergent, and

$$\lim_{k \to \infty} \frac{\|\mathbf{x}_{k+2} - \mathbf{x}_{k+1}\|_2}{\|\mathbf{x}_{k+1} - \mathbf{x}_k\|_2} = 1,$$

then the sequence $\{\mathbf{x}_k\}$ is said to have a *logarithmic convergence rate*.

Sublinear convergence comprises a class of slow convergence rates; the linear convergence comprises a class of fast convergence; and the superlinear convergence and quadratic convergence comprise, respectively, classes of very fast and extremely fast convergence. When designing an optimization algorithm, we often require it at least to be linearly convergent, preferably to be quadratically convergent. The ultra-fast cubic convergence rate is in general difficult to achieve.

Local Convergence Rate

The Q-convergence rate is a limited convergence rate. When we evaluate an optimization algorithm, a practical question is: how many iterations does it need to achieve the desired accuracy? The answer depends on the local convergence rate of the output objective function sequence of a given optimization algorithm.

The *local convergence rate* of the sequence $\{\mathbf{x}_k\}$ is denoted r_k and is defined by

$$r_k = \left\| \frac{\mathbf{x}_{k+1} - \mathbf{x}^*}{\mathbf{x}_k - \mathbf{x}^*} \right\|_2. \tag{3.4.23}$$

The complexity of an optimization algorithm is defined as the inverse of the local convergence rate of the updating variable.

The following is a classification of local convergence rates [25].

- *Sublinear rate:* This convergence rate is given by a power function of the iteration number k and is usually denoted as

$$f(\mathbf{x}_k) - f(\mathbf{x}^*) \leq \epsilon = O\left(\frac{1}{\sqrt{k}}\right). \qquad (3.4.24)$$

- *Linear rate:* This convergence rate is expressed as an exponential function of the iteration number k, and is usually defined as

$$f(\mathbf{x}_k) - f(\mathbf{x}^*) \leq \epsilon = O\left(\frac{1}{k}\right). \qquad (3.4.25)$$

- *Quadratic rate:* This convergence rate is a bi-exponential function of the iteration number k and is usually measured by

$$f(\mathbf{x}_k) - f(\mathbf{x}^*) \leq \epsilon = O\left(\frac{1}{k^2}\right). \qquad (3.4.26)$$

For example, in order to achieve the approximation accuracy $f(\mathbf{x}_k) - f(\mathbf{x}^*) \leq \epsilon = 0.0001$, optimization algorithms with sublinear, linear, and quadratic rates need to run about 10^8, 10^4, and 100 iterations, respectively.

Theorem 3.9 ([25]) *Let $\epsilon = f(\mathbf{x}_k) - f(\mathbf{x}^*)$ denote the estimation error of the objective function given by the updating sequence of the gradient method $\mathbf{x}_{k+1} = \mathbf{x}_k - \alpha \nabla f(\mathbf{x}_k)$. Given a convex function $f(\mathbf{x})$, the upper bound of the estimation error $\epsilon = f(\mathbf{x}_k) - f(\mathbf{x}^*)$ is given by*

$$f(\mathbf{x}_k) - f(\mathbf{x}^*) \leq \frac{2L\|\mathbf{x}_0 - \mathbf{x}^*\|_2^2}{k+4}. \qquad (3.4.27)$$

This theorem shows that the local convergence rate of the gradient method $\mathbf{x}_{k+1} = \mathbf{x}_k - \alpha \nabla f(\mathbf{x}_k)$ is the linear rate $O(1/k)$. Although this is a fast convergence rate, it is not by far optimal, as will be seen in the next section.

3.5 Nesterov Optimal Gradient Method

Let $Q \subset \mathbb{R}^n$ be a convex set in the vector space $\mathbb{R}^n$. Consider the unconstrained optimization problem $\min_{\mathbf{x} \in Q} f(\mathbf{x})$.

3.5.1 Lipschitz Continuous Function

Definition 3.10 (Lipschitz Continuous [25]) An objective function $f(\mathbf{x})$ is said to be *Lipschitz continuous* in the definition domain Q if

$$|f(\mathbf{x}) - f(\mathbf{y})| \leq L\|\mathbf{x} - \mathbf{y}\|_2, \quad \forall \mathbf{x}, \mathbf{y} \in Q \tag{3.5.1}$$

holds for some *Lipschitz constant* $L > 0$. Similarly, we say that the gradient vector $\nabla f(\mathbf{x})$ of a differentiable function $f(\mathbf{x})$ is Lipschitz continuous in the definition domain Q if

$$\|\nabla f(\mathbf{x}) - \nabla f(\mathbf{y})\|_2 \leq L\|\mathbf{x} - \mathbf{y}\|_2, \quad \forall \mathbf{x}, \mathbf{y} \in Q \tag{3.5.2}$$

holds for some Lipschitz constant $L > 0$.

Hereafter a *Lipschitz continuous function* with the Lipschitz constant L will be denoted as an *L-Lipschitz continuous function*.

The function $f(\mathbf{x})$ is said to be continuous at the point $\mathbf{x}_0$, if $\lim_{\mathbf{x} \to \mathbf{x}_0} f(\mathbf{x}) = f(\mathbf{x}_0)$. When we say that $f(\mathbf{x})$ is a continuous function, this means the $f(\mathbf{x})$ is continuous at every point $\mathbf{x} \in Q$.

A Lipschitz continuous function $f(\mathbf{x})$ must be a continuous function, but a continuous function is not necessarily a Lipschitz continuous function.

An everywhere differentiable function is called a *smooth function*. A smooth function must be continuous, but a continuous function is not necessarily smooth. A typical example occurs when the continuous function has a "sharp" point at which it is non-differentiable. Hence, a Lipschitz continuous function is not necessarily differentiable. However, a function $f(\mathbf{x})$ with Lipschitz continuous gradient in the definition domain Q must be smooth in Q, because the definition of a Lipschitz continuous gradient stipulates that $f(\mathbf{x})$ is differentiable in the definition domain.

In convex optimization, the notation $\mathcal{C}_L^{k,p}(Q)$ (where $Q \subseteq \mathbb{R}^n$) denotes the Lipschitz continuous function class with the following properties [25]:

- The function $f \in \mathcal{C}_L^{k,p}(Q)$ is k times continuously differentiable in Q.
- The pth-order derivative of the function $f \in \mathcal{C}_L^{k,p}(Q)$ is L-Lipschitz continuous:

$$\|f^{(p)}(\mathbf{x}) - f^{(p)}(\mathbf{y})\| \leq L\|\mathbf{x} - \mathbf{y}\|_2, \quad \forall \mathbf{x}, \mathbf{y} \in Q.$$

If $k \neq 0$, then $f \in \mathcal{C}_L^{k,p}(Q)$ is said to be a differentiable function. Obviously, it always has $p \leq k$. If $q > k$ then $\mathcal{C}_L^{q,p}(Q) \subseteq \mathcal{C}_L^{k,p}(Q)$. For example, $\mathcal{C}_L^{2,1}(Q) \subseteq \mathcal{C}_L^{1,1}(Q)$.

The following are three common function classes $\mathcal{C}_L^{k,p}(Q)$:

1. $f(\mathbf{x}) \in \mathcal{C}_L^{0,0}(Q)$ is L-Lipschitz continuous but non-differentiable in Q;
2. $f(\mathbf{x}) \in \mathcal{C}_L^{1,0}(Q)$ is *L-Lipschitz continuously differentiable* in Q, but its gradient is not;
3. $f(\mathbf{x}) \in \mathcal{C}_L^{1,1}(Q)$ is L-Lipschitz continuously differentiable in Q, and its gradient $\nabla f(\mathbf{x})$ is L-Lipschitz continuous in Q.

The basic property of the $\mathcal{C}_L^{k,p}(Q)$ function class is that if $f_1 \in \mathcal{C}_{L_1}^{k,p}(Q)$, $f_2 \in \mathcal{C}_{L_2}^{k,p}(Q)$ and $\alpha, \beta \in \mathbb{R}$, then

$$\alpha f_1 + \beta f_2 \in \mathcal{C}_{L_3}^{k,p}(Q),$$

where $L_3 = |\alpha| L_1 + |\beta| L_2$.

Among all Lipschitz continuous functions, $\mathcal{C}_L^{1,1}(Q)$, with the Lipschitz continuous gradient, is the most important function class, and is widely applied in convex optimization.

On the $\mathcal{C}_L^{1,1}(Q)$ function class, one has the following two lemmas [25].

Lemma 3.1 *The function $f(\mathbf{x})$ belongs to $\mathcal{C}_L^{2,1}(\mathbb{R}^n)$ if and only if*

$$\|f''(\mathbf{x})\|_F \leq L, \quad \forall \mathbf{x} \in \mathbb{R}^n. \tag{3.5.3}$$

Lemma 3.2 *If $f(\mathbf{x}) \in \mathcal{C}_L^{1,1}(Q)$, then*

$$|f(\mathbf{y}) - f(\mathbf{x}) - \langle \nabla f(\mathbf{x}), \mathbf{y} - \mathbf{x} \rangle| \leq \frac{L}{2}\|\mathbf{y} - \mathbf{x}\|_2^2, \quad \forall \mathbf{x}, \mathbf{y} \in Q. \tag{3.5.4}$$

Lemma 3.2 is a key inequality for analyzing the convergence rate of a $\mathcal{C}_L^{1,1}$ function $f(\mathbf{x})$ in gradient algorithms. By Lemma 3.2, it follows [37] that

$$f(\mathbf{x}_{k+1}) \leq f(\mathbf{x}_k) + \langle \nabla f(\mathbf{x}_k), \mathbf{x}_{k+1} - \mathbf{x}_k \rangle + \frac{L}{2}\|\mathbf{x}_{k+1} - \mathbf{x}_k\|_2^2$$

$$\leq f(\mathbf{x}_k) + \langle \nabla f(\mathbf{x}_k), \mathbf{x} - \mathbf{x}_k \rangle + \frac{L}{2}\|\mathbf{x} - \mathbf{x}_k\|_2^2 - \frac{L}{2}\|\mathbf{x} - \mathbf{x}_{k+1}\|_2^2$$

$$\leq f(\mathbf{x}) + \frac{L}{2}\|\mathbf{x} - \mathbf{x}_k\|_2^2 - \frac{L}{2}\|\mathbf{x} - \mathbf{x}_{k+1}\|_2^2.$$

Put $\mathbf{x} = \mathbf{x}^*$ and $\delta_k = f(\mathbf{x}_k) - f(\mathbf{x}^*)$; then

$$0 \leq \frac{L}{2}\|\mathbf{x}^* - \mathbf{x}_{k+1}\|_2^2 \leq -\delta_{k+1} + \frac{L}{2}\|\mathbf{x}^* - \mathbf{x}_k\|_2^2$$

$$\leq \cdots \leq -\sum_{i=1}^{k+1} \delta_i + \frac{L}{2}\|\mathbf{x}^* - \mathbf{x}_0\|_2^2.$$

3.5 Nesterov Optimal Gradient Method

From the estimation errors of the projected gradient algorithm $\delta_1 \geq \delta_2 \geq \cdots \geq \delta_{k+1}$, it follows that $-(\delta_1 + \cdots + \delta_{k+1}) \leq -(k+1)\delta_{k+1}$, and thus the above inequality can be simplified to

$$0 \leq \frac{L}{2}\|\mathbf{x}^* - \mathbf{x}_{k+1}\|_2^2 \leq -(k+1)\delta_k + \frac{L}{2}\|\mathbf{x}^* - \mathbf{x}_0\|_2^2,$$

from which the upper bound of the convergence rate of the projected gradient algorithm is given by Yu et al. [37]

$$\delta_k = f(\mathbf{x}_k) - f(\mathbf{x}^*) \leq \frac{L\|\mathbf{x}^* - \mathbf{x}_0\|_2^2}{2(k+1)}. \tag{3.5.5}$$

This shows that, as for the (basic) gradient method, the local convergence rate of the projected gradient method is $O(1/k)$.

3.5.2 Nesterov Optimal Gradient Algorithms

Theorem 3.10 ([24]) *Let $f(\mathbf{x})$ be a convex function with L-Lipschitz gradient. If the updating sequence $\{\mathbf{x}_k\}$ meets the condition*

$$\mathbf{x}_k \in \mathbf{x}_0 + \mathrm{Span}\{\mathbf{x}_0, \ldots, \mathbf{x}_{k-1}\},$$

then the lower bound of the estimation error $\epsilon = f(\mathbf{x}_k) - f(\mathbf{x}^*)$ achieved by any first-order optimization method is given by

$$f(\mathbf{x}_k) - f(\mathbf{x}^*) \geq \frac{3L\|\mathbf{x}_0 - \mathbf{x}^*\|_2^2}{32(k+1)^2}, \tag{3.5.6}$$

where $\mathrm{Span}\{\mathbf{u}_0, \ldots, \mathbf{u}_{k-1}\}$ denotes the linear subspace spanned by $\mathbf{u}_0, \ldots, \mathbf{u}_{k-1}$, $\mathbf{x}_0$ is the initial value of the gradient method, and $f(\mathbf{x}^*)$ denotes the minimal value of the function f.

Theorem 3.10 shows that the optimal convergence rate of any first-order optimization method is the quadratic rate $O(1/k^2)$.

Since the convergence rate of gradient methods is the linear rate $O(1/k)$, and the convergence rate of the optimal first-order optimization methods is the quadratic rate $O(1/k^2)$, the gradient methods are far from optimal.

The *heavy ball method* (HBM) can efficiently improve the convergence rate of the gradient methods.

The HBM is a two-step method: let $\mathbf{y}_0$ and $\mathbf{x}_0$ be two initial vectors and let α_k and β_k be two positive valued sequences; then the first-order method for solving an unconstrained minimization problem $\min_{\mathbf{x} \in \mathbb{R}^n} f(\mathbf{x})$ can use two-step updates [31]:

$$\left.\begin{aligned} \mathbf{y}_k &= \beta_k \mathbf{y}_{k-1} - \nabla f(\mathbf{x}_k), \\ \mathbf{x}_{k+1} &= \mathbf{x}_k + \alpha_k \mathbf{y}_k. \end{aligned}\right\} \quad (3.5.7)$$

In particular, letting $\mathbf{y}_0 = \mathbf{0}$, then the above two-step updates can be rewritten as the one-step update

$$\mathbf{x}_{k+1} = \mathbf{x}_k - \alpha_k \nabla f(\mathbf{x}_k) + \beta_k (\mathbf{x}_k - \mathbf{x}_{k-1}), \quad (3.5.8)$$

where $\mathbf{x}_k - \mathbf{x}_{k-1}$ is called the *momentum*. As seen in (3.5.7), the HBM treats the iterations as a point mass with momentum $\mathbf{x}_k - \mathbf{x}_{k-1}$, which thus tends to continue moving in the direction $\mathbf{x}_k - \mathbf{x}_{k-1}$.

Let $\mathbf{y}_k = \mathbf{x}_k + \beta_k (\mathbf{x}_k - \mathbf{x}_{k-1})$, and use $\nabla f(\mathbf{y}_k)$ instead of $\nabla f(\mathbf{x}_k)$. Then the update Eq. (3.5.7) becomes

$$\left.\begin{aligned} \mathbf{y}_k &= \mathbf{x}_k + \beta_k (\mathbf{x}_k - \mathbf{x}_{k-1}), \\ \mathbf{x}_{k+1} &= \mathbf{y}_k - \alpha_k \nabla f(\mathbf{y}_k). \end{aligned}\right\} \quad (3.5.9)$$

This is the basic form of the optimal gradient method proposed by Nesterov in 1983 [24], usually called the *Nesterov (first) optimal gradient algorithm* and shown in Algorithm 3.2.

Algorithm 3.2 Nesterov (first) optimal gradient algorithm [24]

1. **input:** Lipschitz constant L and convexity parameter guess μ.
2. **initialization:** Choose $\mathbf{x}_{-1} = \mathbf{0}$, $\mathbf{x}_0 \in \mathbb{R}^n$ and $\alpha_0 \in (0, 1)$. Set $k = 0$, $q = \mu/L$.
3. **repeat**
4.   Compute $\alpha_{k+1} \in (0, 1)$ from the equation $\alpha_{k+1}^2 = (1 - \alpha_{k+1})\alpha_k^2 + q\alpha_{k+1}$.
5.   Set $\beta_k = \frac{\alpha_k(1-\alpha_k)}{\alpha_k^2 + \alpha_{k+1}}$.
6.   Compute $\mathbf{y}_k = \mathbf{x}_k + \beta_k(\mathbf{x}_k - \mathbf{x}_{k-1})$.
7.   Compute $\mathbf{x}_{k+1} = \mathbf{y}_k - \alpha_{k+1} \nabla f(\mathbf{y}_k)$.
8.   **exit if** $\mathbf{x}_k$ converges.
9.   **return** $k \leftarrow k + 1$.
10. **output:** $\mathbf{x} \leftarrow \mathbf{x}_k$.

It is easily seen that the Nesterov optimal gradient algorithm is intuitively like the heavy ball formula (3.5.7) but it is not identical; i.e., it uses $\nabla f(\mathbf{y})$ instead of $\nabla f(\mathbf{x})$.

In the Nesterov optimal gradient algorithm, the estimation sequence $\{\mathbf{x}_k\}$ is an *approximation solution sequence*, and $\{\mathbf{y}_k\}$ is a searching point sequence.

3.5 Nesterov Optimal Gradient Method

Theorem 3.11 ([24]) *Let f be a convex function with L-Lipschitz gradient. The Nesterov optimal gradient algorithm achieves*

$$f(\mathbf{x}_k) - f(\mathbf{x}^*) \leq \frac{CL\|\mathbf{x}_k - \mathbf{x}^*\|_2^2}{(k+1)^2}. \qquad (3.5.10)$$

Clearly, the convergence rate of the Nesterov optimal gradient method is the optimal rate for the first-order gradient methods up to constants. Hence, the Nesterov gradient method is indeed an optimal first-order minimization method.

It has been shown [18] that the use of a strong convexity constant μ^* is very effective in reducing the number of iterations of the Nesterov algorithm. A Nesterov algorithm with *adaptive convexity parameter*, using a decreasing sequence $\mu_k \to \mu^*$, was proposed in [18]. It is shown in Algorithm 3.3.

Algorithm 3.3 Nesterov algorithm with adaptive convexity parameter [18]
---
1. **input:** Lipschitz constant L, and the convexity parameter guess $\mu^* \leq L$.
2. **initialization:** $\mathbf{x}_0, \mathbf{v}_0 = \mathbf{x}_0, \gamma_0 > 0, \beta > 1, \mu_0 \in [\mu^*, \gamma_0)$. Set $k = 0, \theta \in [0, 1], \mu_+ = \mu_0$.
3. **repeat**
4. $\quad \mathbf{d}_k = \mathbf{v}_k - \mathbf{x}_k$.
5. $\quad \mathbf{y}_k = \mathbf{x}_k + \theta_k \mathbf{d}_k$.
6. $\quad$ **exit if** $\nabla f(\mathbf{y}_k) = \mathbf{0}$.
7. $\quad$ Steepest descent step: $\mathbf{x}_{k+1} = \mathbf{y}_k - \nu \nabla f(\mathbf{y}_k)$. If L is known, then $\nu \geq 1/L$.
8. $\quad$ If $(\gamma_k - \mu^*) < \beta(\mu_+ - \mu^*)$, then choose $\mu_+ \in [\mu^*, \gamma/\beta]$.
9. $\quad$ Compute $\tilde{\mu} = \frac{\|\nabla f(\mathbf{y}_k)\|^2}{2[f(\mathbf{y}_k) - f(\mathbf{x}_{k+1})]}$.
10. $\quad$ If $\mu_+ \geq \tilde{\mu}$, then $\mu_+ = \max\{\mu^*, \tilde{\mu}/10\}$.
11. $\quad$ Compute α_k as the largest root of $A\alpha^2 + B\alpha + C = 0$.
12. $\quad \gamma_{k+1} = (1 - \alpha_k)\gamma_k + \alpha_k \mu_+$.
13. $\quad \mathbf{v}_{k+1} = \frac{1}{\gamma_{k+1}}\left((1 - \alpha_k)\gamma_k \mathbf{v}_k + \alpha_k(\mu_+ \mathbf{y}_k - \nabla f(\mathbf{y}_k))\right)$.
14. **return** $k \leftarrow k + 1$.
15. **output:** $\mathbf{y}_k$ as an optimal solution.

The coefficients A, B, C in Step 11 are given by

$$G = \gamma_k \left(\frac{\mu}{2}\|\mathbf{v}_k - \mathbf{y}_k\|^2 + (\nabla f(\mathbf{y}_k))^T(\mathbf{v}_k - \mathbf{y}_k)\right),$$

$$A = G + \frac{1}{2}\|\nabla f(\mathbf{y}_k)\|^2 + (\mu - \gamma_k)(f(\mathbf{x}_k) - f(\mathbf{y}_k)),$$

$$B = (\mu - \gamma_k)\left(f(\mathbf{x}_{k+1}) - f(\mathbf{x}_k)\right) - \gamma_k\left(f(\mathbf{y}_k) - f(\mathbf{x}_k)\right) - G,$$

$$C = \gamma_k\left(f(\mathbf{x}_{k+1}) - f(\mathbf{x}_k)\right), \quad \text{with } \mu = \mu_+.$$

It is suggested that $\gamma_0 = L$ if L is known, $\mu_0 = \max\{\mu^*, \gamma_0/100\}$ and $\beta = 1.02$. The Nesterov (first) optimal gradient algorithm has two limitations:

- $\mathbf{y}_k$ may be out of the definition domain Q, and thus the objective function $f(\mathbf{x})$ is required to be well defined at every point in Q;
- this algorithm is available only for the minimization of the Euclidean norm $\|\mathbf{x}\|_2$.

To cope with these limitations, Nesterov developed two other optimal gradient algorithms [26, 27].

Definite the ith element of the $n \times 1$ mapping vector $V_Q(\mathbf{u}, \mathbf{g})$ as [26]

$$V_Q^{(i)}(\mathbf{u}, \mathbf{g}) = u_i e^{-g_i} \left(\sum_{j=1}^{n} u_j e^{-g_j} \right)^{-1}, \quad i = 1, \cdots, n. \tag{3.5.11}$$

Algorithm 3.4 shows the *Nesterov third optimal gradient algorithm*.

Algorithm 3.4 Nesterov (third) optimal gradient algorithm [26]

1. **input:** Choose $\mathbf{x}_0 \in \mathbb{R}^n$ and $\alpha \in (0, 1)$.
2. **initialization:** $\mathbf{y}_0 = \operatorname{argmin}_{\mathbf{x}} \left\{ \frac{L}{\alpha} d(\mathbf{x}) + \frac{1}{2} \left(f(\mathbf{x}_0) + \langle f'(\mathbf{x}_0), \mathbf{x} - \mathbf{x}_0 \rangle \right) : \mathbf{x} \in Q \right\}$. Set $k = 0$.
3. **repeat**
4. Find $\mathbf{z}_k = \operatorname{argmin}_{\mathbf{x}} \left\{ \frac{L}{\alpha} d(\mathbf{x}) + \sum_{i=0}^{k} \frac{i+1}{2} \left(f(\mathbf{x}_i) + \langle \nabla f(\mathbf{x}_i), \mathbf{x} - \mathbf{x}_i \rangle \right) : \mathbf{x} \in Q \right\}$.
5. Set $\tau_k = \frac{2}{k+3}$ and $\mathbf{x}_{k+1} = \tau_k \mathbf{z}_k + (1 - \tau_k) \mathbf{y}_k$.
6. Use (3.5.11) to compute $\hat{\mathbf{x}}_{k+1}(i) = V_Q^{(i)} \left(\mathbf{z}_k, \frac{\alpha}{L} \tau_k \nabla f(\mathbf{x}_{k+1}) \right), i = 1, \ldots, n$.
7. Set $\mathbf{y}_{k+1} = \tau_k \hat{\mathbf{x}}_{k+1} + (1 - \tau_k) \mathbf{y}_k$.
8. **exit if** $\mathbf{x}_k$ converges.
9. **return** $k \leftarrow k + 1$.
10. **output:** $\mathbf{x} \leftarrow \mathbf{x}_k$.

In Algorithm 3.4, $d(\mathbf{x})$ is a *prox-function* of the domain Q. This function has two different choices [26]:

$$d(\mathbf{x}) = \ln n + \sum_{i=1}^{n} x_i \ln |x_i|, \quad \text{for } \|\mathbf{x}\|_1 = \sum_{i=1}^{n} |x_i|, \tag{3.5.12}$$

$$d(\mathbf{x}) = \frac{1}{2} \sum_{i=1}^{n} \left(x_i - \frac{1}{n} \right)^2, \quad \text{for } \|\mathbf{x}\|_2 = \left(\sum_{i=1}^{n} x_i^2 \right)^{1/2}. \tag{3.5.13}$$

3.6 Nonsmooth Convex Optimization

Gradient methods require that the gradient $\nabla f(\mathbf{x})$ exists at the point $\mathbf{x}$, while the Nesterov optimal gradient method requires further that the objective function has L-Lipschitz continuous gradient. Therefore, gradient methods in general and

3.6 Nonsmooth Convex Optimization

the Nesterov optimal gradient method in particular are available only for smooth convex optimization. As an important extension of gradient methods, this section focuses upon the proximal gradient method which is available for nonsmooth convex optimization.

3.6.1 Subgradient and Subdifferential

The following are two typical examples of nonsmooth convex minimization.

1. *Basis pursuit denoising* seeks a sparse near-solution to an under-determined matrix equation [8]

$$\min_{\mathbf{x}\in\mathbb{R}^n}\left\{\|\mathbf{x}\|_1 + \frac{\lambda}{2}\|\mathbf{A}\mathbf{x} - \mathbf{b}\|_2^2\right\}. \tag{3.6.1}$$

This problem is essentially equivalent to the Lasso (sparse least squares) problem [34]

$$\min_{\mathbf{x}\in\mathbb{R}^n} \frac{\lambda}{2}\|\mathbf{A}\mathbf{x} - \mathbf{b}\|_2^2 \quad \text{subject to} \quad \|\mathbf{x}\|_1 \leq K. \tag{3.6.2}$$

2. *Robust principal component analysis* approximates an input data matrix $\mathbf{D}$ to a sum of a low-rank matrix $\mathbf{L}$ and a sparse matrix $\mathbf{S}$:

$$\min\left\{\|\mathbf{L}\|_* + \lambda\|\mathbf{S}\|_1 + \frac{\gamma}{2}\|\mathbf{L} + \mathbf{S} - \mathbf{D}\|_F^2\right\}. \tag{3.6.3}$$

The key challenge of the above minimization problems is the non-smoothness induced by the non-differentiable ℓ_1 norm $\|\cdot\|_1$ and nuclear norm $\|\cdot\|_*$.

Consider the following *combinatorial optimization problem*:

$$\min_{\mathbf{x}\in E}\{F(\mathbf{x}) = f(\mathbf{x}) + h(\mathbf{x})\}, \tag{3.6.4}$$

where

- $E \subset \mathbb{R}^n$ is a finite-dimensional real vector space;
- $h : E \to \mathbb{R}$ is a convex function, but is non-differentiable or nonsmooth in E;
- $f : \mathbb{R}^n \to \mathbb{R}$ is a continuous smooth convex function, and its gradient is L-Lipschitz continuous:

$$\|\nabla f(\mathbf{x}) - \nabla f(\mathbf{y})\|_2 \leq L\|\mathbf{x} - \mathbf{y}\|_2, \quad \forall \mathbf{x}, \mathbf{y} \in \mathbb{R}^n.$$

To address the challenge of non-smoothness, there are two problems to be solved:

1. how to cope with the non-smoothness;
2. how to design a Nesterov-like optimal gradient method for nonsmooth convex optimization.

Because the gradient vector of a nonsmooth function $h(\mathbf{x})$ does not exist everywhere, neither a gradient method nor the Nesterov optimal gradient method is available. A natural question to ask is whether a nonsmooth function has some class of "generalized gradient" similar to gradient.

For a twice continuous differentiable function $f(\mathbf{x})$, its second-order approximation is given by

$$f(\mathbf{x} + \Delta\mathbf{x}) \approx f(\mathbf{x}) + (\nabla f(\mathbf{x}))^T \Delta\mathbf{x} + (\Delta\mathbf{x})^T \mathbf{H} \Delta\mathbf{x}.$$

If the Hessian matrix $\mathbf{H}$ is positive semi-definite or positive definite, then we have the inequality

$$f(\mathbf{x} + \Delta\mathbf{x}) \geq f(\mathbf{x}) + (\nabla f(\mathbf{x}))^T \Delta\mathbf{x},$$

or

$$f(\mathbf{y}) \geq f(\mathbf{x}) + (\nabla f(\mathbf{x}))^T (\mathbf{y} - \mathbf{x}), \quad \forall \mathbf{x}, \mathbf{y} \in \text{dom } f(\mathbf{x}). \quad (3.6.5)$$

Although a nonsmooth function $h(\mathbf{x})$ does not have a gradient vector $\nabla h(\mathbf{x})$, it is possible to find another vector $\mathbf{g}$ instead of the gradient vector $\nabla f(\mathbf{x})$ such that the inequality (3.6.5) still holds.

Definition 3.11 (Subgradient, Subdifferential [25]) A vector $\mathbf{g} \in \mathbb{R}^n$ is said to be a *subgradient vector* of the function $h : \mathbb{R}^n \to \mathbb{R}$ at the point $\mathbf{x} \in \mathbb{R}^n$ if

$$h(\mathbf{y}) \geq h(\mathbf{x}) + \mathbf{g}^T (\mathbf{y} - \mathbf{x}), \quad \forall \mathbf{x}, \mathbf{y} \in \text{dom } h. \quad (3.6.6)$$

The set of all subgradient vectors of the function h at the point $\mathbf{x}$ is known as the *subdifferential* of the function h at the point $\mathbf{x}$, denoted $\partial h(\mathbf{x})$, and is defined as

$$\partial h(\mathbf{x}) \stackrel{\text{def}}{=} \left\{ \mathbf{g} \big| h(\mathbf{y}) \geq h(\mathbf{x}) + \mathbf{g}^T (\mathbf{y} - \mathbf{x}), \forall \mathbf{y} \in \text{dom } h \right\}. \quad (3.6.7)$$

Theorem 3.12 ([25]) *The optimal solution of a nonsmooth function $h(\mathbf{x})$ given by $h(\mathbf{x}^*) = \min_{\mathbf{x} \in \text{dom } h} h(\mathbf{x})$ if and only if $\partial h(\mathbf{x}^*)$ is subdifferential of $h(\mathbf{x})$, i.e., $\mathbf{0} \in \partial h(\mathbf{x}^*)$.*

When $h(\mathbf{x})$ is differentiable, we have $\partial h(\mathbf{x}) = \{\nabla h(\mathbf{x})\}$, as the gradient vector of a smooth function is unique, so we can view the gradient operator ∇h of a convex and differentiable function as a point-to-point mapping, i.e., ∇h maps each point

3.6 Nonsmooth Convex Optimization

$\mathbf{x} \in \text{dom } h$ to the point $\nabla h(\mathbf{x})$. In contrast, the subdifferential operator ∂h, defined in Eq. (3.6.7), of a closed proper convex function h can be viewed as a point-to-set mapping, i.e., ∂h maps each point $\mathbf{x} \in \text{dom } h$ to the set $\partial h(\mathbf{x})$.[1] Any point $\mathbf{g} \in \partial h(\mathbf{x})$ is called a subgradient of h at $\mathbf{x}$. Generally speaking, a function $h(\mathbf{x})$ may have one or several subgradient vectors at some point $\mathbf{x}$.

The function $h(\mathbf{x})$ is said to be *subdifferentiable* at the point $\mathbf{x}$ if it at least has one subgradient vector. More generally, the function $h(\mathbf{x})$ is said to be subdifferentiable in the definition domain $\text{dom } h$, if it is subdifferentiable at all points $\mathbf{x} \in \text{dom } h$.

The following are several examples of subdifferentials [25].

1. Let $f(x) = |x|, x \in \mathbb{R}$. Then $\partial f(0) = [-1, 1]$ since $f(x) = \max_{-1 \leq g \leq 1} g \cdot x$.
2. For function $f(\mathbf{x}) = \sum_{i=1}^{n} |\mathbf{a}_i^T \mathbf{x} - b_i|$, if denoting

$$I_-(\mathbf{x}) = \{i | \langle \mathbf{a}_i, \mathbf{x} \rangle - b_i < 0\},$$
$$I_+(\mathbf{x}) = \{i | \langle \mathbf{a}_i, \mathbf{x} \rangle - b_i > 0\},$$
$$I_0(\mathbf{x}) = \{i | \langle \mathbf{a}_i, \mathbf{x} \rangle - b_i = 0\},$$

then

$$\partial h(\mathbf{x}) = \sum_{i \in I_+(\mathbf{x})} \mathbf{a}_i - \sum_{i \in I_-(\mathbf{x})} \mathbf{a}_i + \sum_{i \in I_0(\mathbf{x})} [-\mathbf{a}_i, \mathbf{a}_i]. \quad (3.6.8)$$

3. For function $f(x) = \max_{1 \leq i \leq n} x_i$, denoting $I(x) = \{i : x_i = f(x)\}$, then

$$\partial f(x) = \begin{cases} \text{Conv}\{e_i | i \in I(x)\}, & x \neq 0; \\ \text{Conv}\{e_1, \ldots, e_n\}, & x = 0. \end{cases} \quad (3.6.9)$$

Here $e_i = 1$ for $i \in I(x)$ and $e_i = 0$ for $i \notin I(x)$, and

$$\text{Conv}\{e_1, \ldots, e_n\} = \left\{ x = \sum_{i=1}^{n} \alpha_i e_i \,\bigg|\, \alpha_i \geq 0, \sum_{i=1}^{n} \alpha_i = 1 \right\} \quad (3.6.10)$$

is a convex set.

4. The subdifferentials of Euclidean norm $f(\mathbf{x}) = \|\mathbf{x}\|_2$ are given by

$$\partial \|\mathbf{x}\|_2 = \begin{cases} \{\mathbf{x}/\|\mathbf{x}\|_2\}, & \mathbf{x} \neq \mathbf{0}; \\ \{\mathbf{y} \in \mathbb{R}^n | \|\mathbf{y}\|_2 \leq 1\}, & \mathbf{x} = \mathbf{0}, \mathbf{y} \neq \mathbf{0}. \end{cases} \quad (3.6.11)$$

---

[1] A proper convex function $f(x)$ takes values in the extended real number line such that $f(x) < +\infty$ for at least one x and $f(x) > -\infty$ for every x.

5. The subdifferentials of ℓ_1-norm $f(\mathbf{x}) = \|\mathbf{x}\|_1 = \sum_{i=1}^{n} |x_i|$ are given by

$$\partial \|\mathbf{x}\|_1 = \begin{cases} \{\mathbf{x} \in \mathbb{R}^n | \max_{1 \le i \le n} |x_i| < 1\}, & \mathbf{x} = \mathbf{0}; \\ \sum_{i \in I_+(\mathbf{x})} e_i - \sum_{i \in I_-(\mathbf{x})} e_i + \sum_{i \in I_0(\mathbf{x})} [-e_i, e_i], & \mathbf{x} \ne \mathbf{0}. \end{cases} \quad (3.6.12)$$

Here $I_+(\mathbf{x}) = \{i | x_i > 0\}$, $I_-(\mathbf{x}) = \{i | x_i < 0\}$ and $I_0(\mathbf{x}) = \{i | x_i = 0\}$.

The basic properties of the subdifferential are as follows [4, 25].

- *Convexity:* $\partial h(\mathbf{x})$ is always a closed convex set, even if $h(\mathbf{x})$ is not convex.
- *Nonempty and boundedness:* If $\mathbf{x} \in \text{int}(\text{dom } h)$, i.e., $\mathbf{x}$ is an interior point in the domain of h, then the subdifferential $\partial h(\mathbf{x})$ is nonempty and bounded.
- *Nonnegative factor:* If $\alpha > 0$, then $\partial(\alpha h(\mathbf{x})) = \alpha \partial h(\mathbf{x})$.
- *Subdifferential:* If h is convex and differentiable at the point $\mathbf{x}$, then the subdifferential is a singleton $\partial h(\mathbf{x}) = \{\nabla h(\mathbf{x})\}$, namely its gradient is its unique subgradient. Conversely, if h is a convex function and $\partial h(\mathbf{x}) = \{\mathbf{g}\}$, then h is differentiable at the point $\mathbf{x}$ and $\mathbf{g} = \nabla h(\mathbf{x})$.
- *Minimum point of non-differentiable function:* The point $\mathbf{x}^*$ is a minimum point of the convex function h if and only if h is subdifferentiable at $\mathbf{x}^*$ and

$$\mathbf{0} \in \partial h(\mathbf{x}^*). \quad (3.6.13)$$

This condition is known as the first-order optimality condition of the nonsmooth convex function $h(\mathbf{x})$. If h is differentiable, then the first-order optimality condition $\mathbf{0} \in \partial h(\mathbf{x})$ simplifies to $\nabla h(\mathbf{x}) = \mathbf{0}$.
- *Subdifferential of a sum of functions:* If $h_1, \ldots, h_m$ are convex functions, then the subdifferential of $h(\mathbf{x}) = h_1(\mathbf{x}) + \cdots + h_m(\mathbf{x})$ is given by

$$\partial h(\mathbf{x}) = \partial h_1(\mathbf{x}) + \cdots + \partial h_m(\mathbf{x}).$$

- *Subdifferential of affine transform:* If $\phi(\mathbf{x}) = h(\mathbf{A}\mathbf{x} + \mathbf{b})$, then the subdifferential $\partial \phi(\mathbf{x}) = \mathbf{A}^T \partial h(\mathbf{A}\mathbf{x} + \mathbf{b})$.
- *Subdifferential of pointwise maximal function:* Let h be a pointwise maximal function of the convex functions $h_1, \ldots, h_m$, i.e., $h(\mathbf{x}) = \max_{i=1,\ldots,m} h_i(\mathbf{x})$; then

$$\partial h(\mathbf{x}) = \text{conv} \left(\cup \{\partial h_i(\mathbf{x}) | h_i(\mathbf{x}) = h(\mathbf{x})\} \right).$$

That is to say, the subdifferential of a pointwise maximal function $h(\mathbf{x})$ is the convex hull of the union set of subdifferentials of the "active function" $h_i(\mathbf{x})$ at the point $\mathbf{x}$.

The subgradient vector $\mathbf{g}$ of the function $h(\mathbf{x})$ is denoted as $\mathbf{g} = \tilde{\nabla} h(\mathbf{x})$.

3.6.2 Proximal Operator

Consider the combinatorial optimization problem

$$\min_{\mathbf{x} \in C} \sum_{i=1}^{P} f_i(\mathbf{x}), \tag{3.6.14}$$

where $C_i = \text{dom } f_i(\mathbf{x})$ for $i = 1, \ldots, P$ is a closed convex set of the m-dimensional Euclidean space $\mathbb{R}^m$ and $C = \bigcap_{i=1}^{P} C_i$ is the intersection of these closed convex sets.

The types of intersection C can be divided into the following three cases [7]:

- The intersection C is nonempty and "small" (all members of C are quite similar).
- The intersection C is nonempty and "large" (the differences between the members of C are large).
- The intersection C is empty, which means that the imposed constraints of intersecting sets are mutually contradictory.

It is difficult to solve the combinatorial optimization problem (3.6.14) directly. But, if

$$f_1(\mathbf{x}) = \|\mathbf{x} - \mathbf{x}_0\|, \quad f_i(\mathbf{x}) = I_{C_i}(\mathbf{x}) = \begin{cases} 0, & \mathbf{x} \in C_i; \\ +\infty, & \mathbf{x} \notin C_i, \end{cases}$$

then the original combinatorial optimization problem can be divided into separate problems

$$\min_{\mathbf{x} \in \bigcap_{i=2}^{P} C_i} \|\mathbf{x} - \mathbf{x}_0\|. \tag{3.6.15}$$

Different from the combinatorial optimization problem (3.6.14), the separated optimization problems (3.6.15) can be solved using the projection method. In particular, when the C_i are convex sets, the projection of a convex objective function onto the intersection of these convex sets is closely related to the proximal operator of the objective function.

Definition 3.12 (Proximal Operator) The *proximal operator* of a convex function $h(\mathbf{x})$ is defined as [23]

$$\mathbf{prox}_h(\mathbf{u}) = \arg\min_{\mathbf{x}} \left\{ h(\mathbf{x}) + \frac{1}{2} \|\mathbf{x} - \mathbf{u}\|_2^2 \right\} \tag{3.6.16}$$

or

$$\mathbf{prox}_{\mu h}(\mathbf{u}) = \arg\min_{\mathbf{x}} \left\{ h(\mathbf{x}) + \frac{1}{2\mu} \|\mathbf{x} - \mathbf{u}\|_2^2 \right\} \tag{3.6.17}$$

with scale parameter $\mu > 0$.

In particular, for a convex function $h(\mathbf{X})$, its proximal operator is defined as

$$\mathbf{prox}_{\mu h}(\mathbf{U}) = \arg\min_{\mathbf{X}} \left\{ h(\mathbf{X}) + \frac{1}{2\mu} \|\mathbf{X} - \mathbf{U}\|_F^2 \right\}. \qquad (3.6.18)$$

The proximal operator has the following important properties [35].

1. *Existence and uniqueness:* The proximal operator $\mathbf{prox}_h(\mathbf{u})$ always exists, and is unique for all $\mathbf{x}$.
2. *Subgradient characterization:* There is the following correspondence between the proximal mapping $\mathbf{prox}_h(\mathbf{u})$ and the subgradient $\partial h(\mathbf{x})$:

$$\mathbf{x} = \mathbf{prox}_h(\mathbf{u}) \quad \Leftrightarrow \quad \mathbf{x} - \mathbf{u} \in \partial h(\mathbf{x}). \qquad (3.6.19)$$

3. *Nonexpansive mapping:* The proximal operator $\mathbf{prox}_h(\mathbf{u})$ is a nonexpansive mapping with constant 1: if $\mathbf{x} = \mathbf{prox}_h(\mathbf{u})$ and $\hat{\mathbf{x}} = \mathbf{prox}_h(\hat{\mathbf{u}})$, then

$$(\mathbf{x} - \hat{\mathbf{x}})^T (\mathbf{u} - \hat{\mathbf{u}}) \geq \|\mathbf{x} - \hat{\mathbf{x}}\|_2^2.$$

4. *Proximal operator of a separable sum function:* If $h : \mathbb{R}^{n_1} \times \mathbb{R}^{n_2} \to \mathbb{R}$ is a separable sum function, i.e., $h(\mathbf{x}_1, \mathbf{x}_2) = h_1(\mathbf{x}_1) + h_2(\mathbf{x}_2)$, then

$$\mathbf{prox}_h(\mathbf{x}_1, \mathbf{x}_2) = (\mathbf{prox}_{h_1}(\mathbf{x}_1), \mathbf{prox}_{h_2}(\mathbf{x}_2)).$$

5. *Scaling and translation of argument:* If $h(\mathbf{x}) = f(\alpha \mathbf{x} + \mathbf{b})$, where $\alpha \neq 0$, then

$$\mathbf{prox}_h(\mathbf{x}) = \frac{1}{\alpha} \left(\mathbf{prox}_{\alpha^2 f}(\alpha \mathbf{x} + \mathbf{b}) - \mathbf{b} \right).$$

6. *Proximal operator of the conjugate function:* If $h^*(\mathbf{x})$ is the conjugate function of the function $h(\mathbf{x})$, then, for all $\mu > 0$, the proximal operator of the conjugate function is given by

$$\mathbf{prox}_{\mu h^*}(\mathbf{x}) = \mathbf{x} - \mu \mathbf{prox}_{h/\mu}(\mathbf{x}/\mu).$$

If $\mu = 1$, then the above equation simplifies to

$$\mathbf{x} = \mathbf{prox}_h(\mathbf{x}) + \mathbf{prox}_{h^*}(\mathbf{x}). \qquad (3.6.20)$$

This decomposition is called the *Moreau decomposition*.

An operator closely related to the proximal operator is the soft thresholding operator of one real variable.

3.6 Nonsmooth Convex Optimization

Definition 3.13 (Soft Thresholding Operator) The action of the *soft thresholding operator* on a real variable $x \in \mathbb{R}$, denoted $\mathcal{S}_\tau[x]$ or soft(x, τ), is defined as

$$\text{soft}(x, \tau) = \mathcal{S}_\tau[x] = \begin{cases} x - \tau, & x > \tau; \\ 0, & |x| \leq \tau; \\ x + \tau, & x < -\tau. \end{cases} \tag{3.6.21}$$

Here $\tau > 0$ is called the *soft threshold value* of the real variable x. The action of the soft thresholding operator can be equivalently written as

$$\text{soft}(x, \tau) = (x - \tau)_+ - (-x - \tau)_+ = \max\{x - \tau, 0\} - \max\{-x - \tau, 0\}$$
$$= (x - \tau)_+ + (x + \tau)_- = \max\{x - \tau, 0\} + \min\{x + \tau, 0\}. \tag{3.6.22}$$

This operator is also known as the *shrinkage operator*, because it can reduce the variable x, the elements of the vector $\mathbf{x}$, and the matrix $\mathbf{X}$ to zero, thereby shrinking the range of elements. Hence, the action of the soft thresholding operator is sometimes written as [1, 5]:

$$\text{soft}(x, \tau) = (|x| - \tau)_+ \text{sign}(x) = (1 - \tau/|x|)_+ x. \tag{3.6.23}$$

The soft thresholding operation on a real vector $\mathbf{x} \in \mathbb{R}^n$, denoted soft$(\mathbf{x}, \tau)$, is defined as a vector with entries

$$\text{soft}(\mathbf{x}, \tau)_i = \max\{x_i - \tau, 0\} + \min\{x_i + \tau, 0\}$$
$$= \begin{cases} x_i - \tau, & x_i > \tau; \\ 0, & |x_i| \leq \tau; \\ x_i + \tau, & x_i < -\tau. \end{cases} \tag{3.6.24}$$

The soft thresholding operation on a real matrix $\mathbf{X} \in \mathbb{R}^{m \times n}$, denoted soft$(\mathbf{X})$, is defined as an $m \times n$ real matrix with entries

$$\text{soft}(\mathbf{X}, \tau)_{ij} = \max\{x_{ij} - \tau, 0\} + \min\{x_{ij} + \tau, 0\}$$
$$= \begin{cases} x_{ij} - \tau, & x_{ij} > \tau; \\ 0, & |x_{ij}| \leq \tau; \\ x_{ij} + \tau, & x_{ij} < -\tau. \end{cases} \tag{3.6.25}$$

Given a function $h(\mathbf{x})$, our goal is to find an explicit expression for

$$\mathbf{prox}_{\mu h}(\mathbf{u}) = \arg\min_{\mathbf{x}} \left\{ h(\mathbf{x}) + \frac{1}{2\mu} \|\mathbf{x} - \mathbf{u}\|_2^2 \right\}.$$

Theorem 3.13 ([25, p. 129]) *Let $\mathbf{x}^*$ be an optimal solution of the minimization problem $\min_{\mathbf{x}\in\text{dom}\,\phi}\phi(\mathbf{x})$. If the function $\phi(\mathbf{x})$ is subdifferentiable, then $\mathbf{x}^* = \arg\min_{\mathbf{x}\in\text{dom}\,\phi}\phi(\mathbf{x})$ or $\phi(\mathbf{x}^*) = \min_{\mathbf{x}\in\text{dom}\,\phi}\phi(\mathbf{x})$ if and only if $\mathbf{0}\in\partial\phi(\mathbf{x}^*)$.*

By the above theorem, the first-order optimality condition of the function

$$\phi(\mathbf{x}) = h(\mathbf{x}) + \frac{1}{2\mu}\|\mathbf{x}-\mathbf{u}\|_2^2 \tag{3.6.26}$$

is given by

$$\mathbf{0}\in\partial h(\mathbf{x}^*) + \frac{1}{\mu}(\mathbf{x}^*-\mathbf{u}), \tag{3.6.27}$$

because $\partial\frac{1}{2\mu}\|\mathbf{x}-\mathbf{u}\|_2^2 = \frac{1}{\mu}(\mathbf{x}-\mathbf{u})$. Hence, if and only if $\mathbf{0}\in\partial h(\mathbf{x}^*) + \mu^{-1}(\mathbf{x}^*-\mathbf{u})$, we have

$$\mathbf{x}^* = \mathbf{prox}_{\mu h}(\mathbf{u}) = \arg\min_{\mathbf{x}}\left\{h(\mathbf{x}) + \frac{1}{2\mu}\|\mathbf{x}-\mathbf{u}\|_2^2\right\}. \tag{3.6.28}$$

From Eq. (3.6.27) we have

$$\mathbf{0}\in\mu\partial h(\mathbf{x}^*) + (\mathbf{x}^*-\mathbf{u}) \Leftrightarrow \mathbf{u}\in(I+\mu\partial h)\mathbf{x}^* \Leftrightarrow (I+\mu\partial h)^{-1}\mathbf{u}\in\mathbf{x}^*,$$

where I is the identical operator such that $I\mathbf{x}=\mathbf{x}$.

Since $\mathbf{x}^*$ is only one point, $(I+\mu\partial h)^{-1}\mathbf{u}\in\mathbf{x}^*$ should read as $(I+\mu\partial h)^{-1}\mathbf{u}=\mathbf{x}^*$ and thus

$$\mathbf{0}\in\mu\partial h(\mathbf{x}^*) + (\mathbf{x}^*-\mathbf{u}) \Leftrightarrow \mathbf{x}^* = (I+\mu\partial h)^{-1}\mathbf{u}$$
$$\Leftrightarrow \mathbf{prox}_{\mu h}(\mathbf{u}) = (I+\mu\partial h)^{-1}\mathbf{u}.$$

This shows that the proximal operator $\mathbf{prox}_{\mu h}$ and the subdifferential operator ∂h are related as follows:

$$\mathbf{prox}_{\mu h} = (I+\mu\partial h)^{-1}. \tag{3.6.29}$$

The (point-to-point) mapping $(I+\mu\partial h)^{-1}$ is known as the *resolvent* of the subdifferential operator ∂h with parameter $\mu>0$, so the proximal operator $\mathbf{prox}_{\mu h}$ is the resolvent of the subdifferential operator ∂h.

Notice that the subdifferential $\partial h(\mathbf{x})$ is a point-to-set mapping, for which neither direction is unique, whereas the proximal operation $\mathbf{prox}_{\mu h}(\mathbf{u})$ is a point-to-point mapping: $\mathbf{prox}_{\mu h}(\mathbf{u})$ maps any point $\mathbf{u}$ to a unique point $\mathbf{x}$.

3.6 Nonsmooth Convex Optimization

Table 3.2 Proximal operators of several typical functions

Functions	Proximal operators
$h(\mathbf{x}) = \|\mathbf{x}\|_1$	$(\mathbf{prox}_{\mu h}(\mathbf{u}))_i = \begin{cases} u_i - \mu, & u_i > \mu; \\ 0, & \|u_i\| \le \mu; \\ u_i + \mu, & u_i < -\mu. \end{cases}$
$h(\mathbf{x}) = \|\mathbf{x}\|_2$	$\mathbf{prox}_{\mu h}(\mathbf{u}) = \begin{cases} (1 - \mu/\|\mathbf{u}\|_2)\mathbf{u}, & \|\mathbf{u}\|_2 \ge \mu; \\ 0, & \|\mathbf{u}\|_2 < \mu. \end{cases}$
$h(\mathbf{x}) = \mathbf{a}^T\mathbf{x} + b$	$\mathbf{prox}_{\mu h}(\mathbf{u}) = \mathbf{u} - \mu \mathbf{a}$
$h(\mathbf{x}) = \frac{1}{2}\mathbf{x}^T\mathbf{A}\mathbf{x} + \mathbf{b}^T\mathbf{x}$	$\mathbf{prox}_{\mu h}(\mathbf{x}) = \mathbf{u} + (\mathbf{A} + \mu^{-1}\mathbf{I})^{-1}(\mathbf{b} - \mathbf{A}\mathbf{u})$
$h(\mathbf{x}) = -\sum_{i=1}^{n} \log x_i$	$(\mathbf{prox}_{\mu h}(\mathbf{u}))_i = \frac{1}{2}\left(u_i + \sqrt{u_i^2 + 4\mu}\right)$
$h(\mathbf{x}) = I_C(\mathbf{x})$	$\mathbf{prox}_{\mu h}(\mathbf{u}) = \mathcal{P}_C(\mathbf{u}) = \arg\min_{\mathbf{x} \in C} \|\mathbf{x} - \mathbf{u}\|_2$
$h(\mathbf{x}) = \sup_{\mathbf{y} \in C} \mathbf{y}^T\mathbf{x}$	$\mathbf{prox}_{\mu h}(\mathbf{u}) = \mathbf{u} - \mu \mathcal{P}_C(\mathbf{u}/\mu)$
$h(\mathbf{x}) = \phi(\mathbf{x} - \mathbf{z})$	$\mathbf{prox}_{\mu h}(\mathbf{u}) = \mathbf{z} + \mathbf{prox}_{\mu \phi}(\mathbf{u} - \mathbf{z})$
$h(\mathbf{x}) = \phi(\mathbf{x}/\rho)$	$\mathbf{prox}_h(\mathbf{u}) = \rho \mathbf{prox}_{\phi/\rho^2}(\mathbf{u}/\rho)$
$h(\mathbf{x}) = \phi(-\mathbf{x})$	$\mathbf{prox}_{\mu h}(\mathbf{u}) = -\mathbf{prox}_{\mu \phi}(-\mathbf{u})$
$h(\mathbf{x}) = \phi^*(\mathbf{x})$	$\mathbf{prox}_{\mu h}(\mathbf{u}) = \mathbf{u} - \mathbf{prox}_{\mu \phi}(\mathbf{u})$

Example 3.2 For the quadratic function $h(\mathbf{x}) = \frac{1}{2}\mathbf{x}^T\mathbf{A}\mathbf{x} - \mathbf{b}^T\mathbf{x} + \mathbf{c}$, where $\mathbf{A} \in \mathbb{R}^{n \times n}$ is positive definite. The first-order optimality condition of $\mathbf{prox}_{\mu h}(\mathbf{u})$ is

$$\frac{\partial}{\partial \mathbf{x}}\left(\frac{1}{2}\mathbf{x}^T\mathbf{A}\mathbf{x} - \mathbf{b}^T\mathbf{x} + \mathbf{c} + \frac{1}{2\mu}\|\mathbf{x} - \mathbf{u}\|_2^2\right) = \mathbf{A}\mathbf{x} - \mathbf{b} + \frac{1}{\mu}(\mathbf{x} - \mathbf{u}) = \mathbf{0},$$

with $\mathbf{x} = \mathbf{x}^* = \mathbf{prox}_{\mu h}(\mathbf{u})$. Hence, we have

$$\mathbf{A}\mathbf{x}^* - \mathbf{b} + \mu^{-1}(\mathbf{x}^* - \mathbf{u}) = \mathbf{0} \Leftrightarrow (\mathbf{A} + \mu^{-1}\mathbf{I})\mathbf{x}^* = \mu^{-1}\mathbf{u} + \mathbf{b}$$
$$\Leftrightarrow \mathbf{x}^* = (\mathbf{A} + \mu^{-1}\mathbf{I})^{-1}[(\mathbf{A} + \mu^{-1}\mathbf{I})\mathbf{u} + \mathbf{b} - \mathbf{A}\mathbf{u}]$$

which gives directly the result

$$\mathbf{x}^* = \mathbf{prox}_{\mu h}(\mathbf{u}) = \mathbf{u} + (\mathbf{A} + \mu^{-1}\mathbf{I})^{-1}(\mathbf{b} - \mathbf{A}\mathbf{u}).$$

Table 3.2 lists the proximal operators of several typical functions [10, 30].

3.6.3 Proximal Gradient Method

Consider a typical form of *nonsmooth convex optimization*,

$$\min_{\mathbf{x}} J(\mathbf{x}) = f(\mathbf{x}) + h(\mathbf{x}), \tag{3.6.30}$$

where $f(\mathbf{x})$ is a convex, smooth (i.e., differentiable), and L-Lipschitz function, and $h(\mathbf{x})$ is a convex but nonsmooth function (such as $\|\mathbf{x}\|_1$, $\|\mathbf{X}\|_*$, and so on).

Quadratic Approximation

Consider the quadratic approximation of an L-Lipschitz smooth function $f(\mathbf{x})$ around the point $\mathbf{x}_k$:

$$f(\mathbf{x}) = f(\mathbf{x}_k) + (\mathbf{x} - \mathbf{x}_k)^T \nabla f(\mathbf{x}_k) + \frac{1}{2}(\mathbf{x} - \mathbf{x}_k)^T \nabla^2 f(\mathbf{x}_k)(\mathbf{x} - \mathbf{x}_k)$$

$$\approx f(\mathbf{x}_k) + (\mathbf{x} - \mathbf{x}_k)^T \nabla f(\mathbf{x}_k) + \frac{L}{2}\|\mathbf{x} - \mathbf{x}_k\|_2^2,$$

where $\nabla^2 f(\mathbf{x}_k)$ is approximated by a diagonal matrix $L\mathbf{I}$.

Minimize $J(\mathbf{x}) = f(\mathbf{x}) + h(\mathbf{x})$ via iteration to yield

$$\mathbf{x}_{k+1} = \arg\min_{\mathbf{x}} \{f(\mathbf{x}) + h(\mathbf{x})\}$$

$$\approx \arg\min_{\mathbf{x}} \left\{ f(\mathbf{x}_k) + (\mathbf{x} - \mathbf{x}_k)^T \nabla f(\mathbf{x}_k) + \frac{L}{2}\|\mathbf{x} - \mathbf{x}_k\|_2^2 + h(\mathbf{x}) \right\}$$

$$= \arg\min_{\mathbf{x}} \left\{ h(\mathbf{x}) + \frac{L}{2}\left\| \mathbf{x} - \left(\mathbf{x}_k - \frac{1}{L}\nabla f(\mathbf{x}_k)\right)\right\|_2^2 \right\}$$

$$= \mathbf{prox}_{L^{-1}h}\left(\mathbf{x}_k - \frac{1}{L}\nabla f(\mathbf{x}_k)\right).$$

In practical applications, the Lipschitz constant L of $f(\mathbf{x})$ is usually unknown. Hence, a question is how to choose L in order to accelerate the convergence of $\mathbf{x}_k$ arises. To this end, let $\mu = 1/L$ and consider the fixed point iteration

$$\mathbf{x}_{k+1} = \mathbf{prox}_{\mu h}\left(\mathbf{x}_k - \mu \nabla f(\mathbf{x}_k)\right). \tag{3.6.31}$$

This iteration is called the *proximal gradient method* for solving the nonsmooth convex optimization problem, where the step μ is chosen to equal some constant or is determined by linear searching.

Forward-Backward Splitting

To derive a proximal gradient algorithm for nonsmooth convex optimization, let $A = \partial h$ and $B = \nabla f$ denote the subdifferential operator and the gradient operator,

3.6 Nonsmooth Convex Optimization

respectively. Then, the first-order optimality condition for the objective function $h(\mathbf{x}) + f(\mathbf{x})$, $\mathbf{0} \in (\partial h + \nabla f)\mathbf{x}^*$, can be written in operator form as $\mathbf{0} \in (A + B)\mathbf{x}^*$, and thus we have

$$\mathbf{0} \in (A + B)\mathbf{x}^* \Leftrightarrow (I - \mu B)\mathbf{x}^* \in (I + \mu A)\mathbf{x}^*$$
$$\Leftrightarrow (I + \mu A)^{-1}(I - \mu B)\mathbf{x}^* = \mathbf{x}^*. \quad (3.6.32)$$

Here $(I - \mu B)$ is a forward operator and $(I + \mu A)^{-1}$ is a backward operator.

The backward operator $(I + \mu \partial h)^{-1}$ is sometimes known as the resolvent of ∂h with parameter μ.

By $\mathbf{prox}_{\mu h} = (I + \mu A)^{-1}$ and $(I - \mu B)\mathbf{x}_k = \mathbf{x}_k - \mu \nabla f(\mathbf{x}_k)$, Eq. (3.6.32) can be written as the fixed point iteration

$$\mathbf{x}_{k+1} = (I + \mu A)^{-1}(I - \mu B)\mathbf{x}_k = \mathbf{prox}_{\mu h}\left(\mathbf{x}_k - \mu \nabla f(\mathbf{x}_k)\right). \quad (3.6.33)$$

If we let

$$\mathbf{y}_k = (I - \mu B)\mathbf{x}_k = \mathbf{x}_k - \mu \nabla f(\mathbf{x}_k), \quad (3.6.34)$$

then Eq. (3.6.33) reduces to

$$\mathbf{x}_{k+1} = (I + \mu A)^{-1}\mathbf{y}_k = \mathbf{prox}_{\mu h}(\mathbf{y}_k). \quad (3.6.35)$$

In other words, the proximal gradient algorithm can be split two iterations:

- forward iteration: $\mathbf{y}_k = (I - \mu B)\mathbf{x}_k = \mathbf{x}_k - \mu \nabla f(\mathbf{x}_k)$;
- backward iteration: $\mathbf{x}_{k+1} = (I + \mu A)^{-1}\mathbf{y}_k = \mathbf{prox}_{\mu h}(\mathbf{y}_k)$.

The forward iteration is an explicit iteration which is easily computed, whereas the backward iteration is an implicit iteration.

Example 3.3 For the ℓ_1-norm function $h(\mathbf{x}) = \|\mathbf{x}\|_1$, the backward iteration is

$$\mathbf{x}_{k+1} = \mathbf{prox}_{\mu \|\mathbf{x}\|_1}(\mathbf{y}_k) = \mathrm{soft}_{\|\cdot\|_1}(\mathbf{y}_k, \mu)$$
$$= [\mathrm{soft}_{\|\cdot\|_1}(y_k(1), \mu), \ldots, \mathrm{soft}_{\|\cdot\|_1}(y_k(n), \mu)]^T \quad (3.6.36)$$

can be obtained by the *soft thresholding operation* [15]:

$$x_{k+1}(i) = (\mathrm{soft}_{\|\cdot\|_1}(\mathbf{y}_k, \mu))_i = \mathrm{sign}(y_k(i)) \max\{|y_k(i)| - \mu, 0\}, \quad (3.6.37)$$

for $i = 1, \ldots, n$. Here $x_{k+1}(i)$ and $y_k(i)$ are the ith entries of the vectors $\mathbf{x}_{k+1}$ and $\mathbf{y}_k$, respectively.

Example 3.4 For the ℓ_2-norm function $h(\mathbf{x}) = \|\mathbf{x}\|_2$, the backward iteration is

$$\mathbf{x}_{k+1} = \mathbf{prox}_{\mu\|\mathbf{x}\|_2}(\mathbf{y}_k) = \text{soft}_{\|\cdot\|_2}(\mathbf{y}_k, \mu)$$
$$= [\text{soft}_{\|\cdot\|_2}(y_k(1), \mu), \ldots, \text{soft}_{\|\cdot\|_2}(y_k(n), \mu)]^T, \quad (3.6.38)$$

where

$$\text{soft}_{\|\cdot\|_2}(y_k(i), \mu) = \max\{|y_k(i)| - \mu, 0\} \frac{y_k(i)}{\|\mathbf{y}_k\|_2}, \quad i = 1, \ldots, n. \quad (3.6.39)$$

Example 3.5 For the nuclear norm of matrix $\|\mathbf{X}\|_* = \sum_{i=1}^{\min\{m,n\}} \sigma_i(\mathbf{X})$, the corresponding proximal gradient method becomes

$$\mathbf{X}_k = \mathbf{prox}_{\mu\|\cdot\|_*}\left(\mathbf{X}_{k-1} - \mu\nabla f(\mathbf{X}_{k-1})\right). \quad (3.6.40)$$

If $\mathbf{W} = \mathbf{U}\mathbf{\Sigma}\mathbf{V}^T$ is the SVD of the matrix $\mathbf{W} = \mathbf{X}_{k-1} - \mu\nabla f(\mathbf{X}_{k-1})$, then

$$\mathbf{prox}_{\mu\|\cdot\|_*}(\mathbf{W}) = \mathbf{U}\mathcal{D}_\mu(\mathbf{\Sigma})\mathbf{V}^T, \quad (3.6.41)$$

where the *singular value thresholding* (SVT) operation $\mathcal{D}_\mu(\mathbf{\Sigma})$ is defined as

$$[\mathcal{D}_\mu(\mathbf{\Sigma})]_i = \begin{cases} \sigma_i(\mathbf{X}) - \mu, & \text{if } \sigma_i(\mathbf{X}) > \mu, \\ 0, & \text{otherwise.} \end{cases} \quad (3.6.42)$$

Example 3.6 If the nonsmooth function $h(\mathbf{x}) = I_C(\mathbf{x})$ is an indicator function, then the proximal gradient iteration $\mathbf{x}_{k+1} = \mathbf{prox}_{\mu h}(\mathbf{x}_k - \mu\nabla f(\mathbf{x}_k))$ reduces to the gradient projection iteration

$$\mathbf{x}_{k+1} = \mathcal{P}_C\left(\mathbf{x}_k - \mu\nabla f(\mathbf{x}_k)\right). \quad (3.6.43)$$

A comparison between other gradient methods and the proximal gradient method is summarized below [38].

- The update formulas are as follows.

$$\text{general gradient method: } \mathbf{x}_{k+1} = \mathbf{x}_k - \mu\nabla f(\mathbf{x}_k);$$

$$\text{Newton method: } \mathbf{x}_{k+1} = \mathbf{x}_k - \mu\mathbf{H}^{-1}(\mathbf{x}_k)\nabla f(\mathbf{x}_k);$$

$$\text{proximal gradient method: } \mathbf{x}_{k+1} = \mathbf{prox}_{\mu h}\left(\mathbf{x}_k - \mu\nabla f(\mathbf{x}_k)\right).$$

General gradient methods and the Newton method use a low-level (explicit) update, whereas the proximal gradient method uses a high-level (implicit) operation $\mathbf{prox}_{\mu h}$.

3.6 Nonsmooth Convex Optimization

- General gradient methods and the Newton method are available only for smooth and unconstrained optimization problems, while the proximal gradient method is available for smooth or nonsmooth and/or constrained or unconstrained optimization problems.
- The Newton method allows modest-sized problems to be addressed; the general gradient methods are applicable for large-size problems, and, sometimes, distributed implementations, and the proximal gradient method is available for all large-size problems and distributed implementations.

In particular, the proximal gradient method includes the general gradient method, the projected gradient method, and the iterative soft thresholding method as special cases.

1. *Gradient method:* If $h(\mathbf{x}) = 0$, due to $\mathbf{prox}_h(\mathbf{x}) = \mathbf{x}$, the proximal gradient algorithm (3.6.31) reduces to the general gradient algorithm $\mathbf{x}_k = \mathbf{x}_{k-1} - \mu_k \nabla f(\mathbf{x}_{k-1})$. That is to say, the gradient algorithm is a special case of the proximal gradient algorithm when the nonsmooth convex function $h(\mathbf{x}) = 0$.
2. *Projected gradient method:* For $h(\mathbf{x}) = I_C(\mathbf{x})$, an indicator function, the minimization problem (3.6.4) becomes the unconstrained minimization $\min_{\mathbf{x} \in C} f(\mathbf{x})$. Because $\mathbf{prox}_h(\mathbf{x}) = \mathcal{P}_C(\mathbf{x})$, the proximal gradient algorithm reduces to

$$\mathbf{x}_k = \mathcal{P}_C \left(\mathbf{x}_{k-1} - \mu_k \nabla f(\mathbf{x}_{k-1}) \right) \tag{3.6.44}$$

$$= \arg\min_{\mathbf{u} \in C} \left\| \mathbf{u} - \mathbf{x}_{k-1} + \mu_k \nabla f(\mathbf{x}_{k-1}) \right\|_2^2. \tag{3.6.45}$$

This is just the projected gradient method.

3. *Iterative soft thresholding method:* When $h(\mathbf{x}) = \|\mathbf{x}\|_1$, the minimization problem (3.6.4) becomes the unconstrained minimization problem $\min \{ f(\mathbf{x}) + \|\mathbf{x}\|_1 \}$. In this case, the proximal gradient algorithm becomes

$$\mathbf{x}_k = \mathbf{prox}_{\mu_k h} \left(\mathbf{x}_{k-1} - \mu_k \nabla f(\mathbf{x}_{k-1}) \right). \tag{3.6.46}$$

This is called the *iterative soft thresholding method*, for which

$$\mathbf{prox}_{\mu h}(\mathbf{u})_i = \begin{cases} u_i - \mu, & u_i > \mu, \\ 0, & -\mu \leq u_i \leq \mu, \\ u_i + \mu, & u_i < -\mu. \end{cases}$$

From the viewpoint of convergence, the proximal gradient method is suboptimal.

A Nesterov-like proximal gradient algorithm developed by Beck and Teboulle [1] is called the *fast iterative soft thresholding algorithm* (FISTA), as shown in Algorithm 3.5.

Algorithm 3.5 FISTA algorithm with fixed step [1]
1. **input:** The Lipschitz constant L of $\nabla f(\mathbf{x})$.
2. **initialization:** $\mathbf{y}_1 = \mathbf{x}_0 \in \mathbb{R}^n$, $t_1 = 1$.
3. **repeat**
4.   Compute $\mathbf{x}_k = \mathbf{prox}_{L^{-1}h}\left(\mathbf{y}_k - \frac{1}{L}\nabla f(\mathbf{y}_k)\right)$.
5.   Compute $t_{k+1} = \frac{1}{2}(1 + \sqrt{1 + 4t_k^2})$.
6.   Compute $\mathbf{y}_{k+1} = \mathbf{x}_k + \left(\frac{t_k - 1}{t_{k+1}}\right)(\mathbf{x}_k - \mathbf{x}_{k-1})$.
7.   **exit if** $\mathbf{x}_k$ is converged.
8.   **return** $k \leftarrow k + 1$.
9. **output:** $\mathbf{x} \leftarrow \mathbf{x}_k$.

Theorem 3.14 ([1]) *Let $\{\mathbf{x}_k\}$ and $\{\mathbf{y}_k\}$ be two sequences generated by the FISTA algorithm; then, for any iteration $k \geq 1$,*

$$F(\mathbf{x}_k) - F(\mathbf{x}^*) \leq \frac{2L\|\mathbf{x}_k - \mathbf{x}^*\|_2^2}{(k+1)^2}, \quad \forall \mathbf{x}^* \in X_*,$$

where $\mathbf{x}^*$ and X_* represent, respectively, the optimal solution and the optimal solution set of $\min (F(\mathbf{x}) = f(\mathbf{x}) + h(\mathbf{x}))$.

Theorem 3.14 shows that to achieve the ϵ-optimal solution $F(\bar{\mathbf{x}}) - F(\mathbf{x}^*) \leq \epsilon$, the FISTA algorithm needs at most $\lceil C/\sqrt{\epsilon} - 1 \rceil$ iterations, where $C = \sqrt{2L\|\mathbf{x}_0 - \mathbf{x}^*\|_2^2}$. Since it has the same fast convergence rate as the optimal first-order algorithm, the FISTA is indeed an optimal algorithm.

3.7 Constrained Convex Optimization

Consider the *constrained minimization problem*

$$\min_{\mathbf{x}} f_0(\mathbf{x}) \quad \text{subject to} \quad f_i(\mathbf{x}) \leq 0, i = 1, \ldots, m; \; h_j(\mathbf{x}) = 0, j = 1, \ldots, q. \tag{3.7.1}$$

If the objective function $f_0(\mathbf{x})$ and the inequality constraint functions $f_i(\mathbf{x})$, $i = 1, \ldots, m$ are convex, and the equality constraint functions $h_j(\mathbf{x})$ have the affine form $\mathbf{h}(\mathbf{x}) = \mathbf{Ax} - \mathbf{b}$, then Eq. (3.7.1) is called a *constrained convex optimization problem*.

The basic idea in solving a constrained optimization problem is to transform it into an unconstrained optimization problem. The transformation methods have the following three types:

- the Lagrange multiplier method for equality or inequality constrains;
- the penalty function method for equality constraints; and
- the augmented Lagrange multiplier method for both equality and inequality constrains.

3.7 Constrained Convex Optimization

Because the Lagrange multiplier method is just a simpler example of the augmented Lagrange multiplier method, we mainly focus on the penalty function method and the augmented Lagrange multiplier method in this section.

3.7.1 Penalty Function Method

The *penalty function method* is a widely used constrained optimization method, and its basic idea is simple: by using the penalty function and/or the barrier function, a constrained optimization becomes an unconstrained optimization of the composite function consisting of the original objective function and the constraint conditions.

The penalty function method transforms the original constrained optimization problem into the following unconstrained optimization problem:

$$\min_{\mathbf{x}\in\mathcal{S}} \{L_\rho(\mathbf{x}) = f_0(\mathbf{x}) + \rho p(\mathbf{x})\}, \tag{3.7.2}$$

where the coefficient ρ is a penalty parameter that reflects the intensity of "punishment" via the weighting of the *penalty function* $p(\mathbf{x})$; the greater ρ is, the greater the value of the penalty term.

The main property of the penalty function is as follows: if $p_1(\mathbf{x})$ is the penalty for the closed set $\mathcal{F}_1$, and $p_2(\mathbf{x})$ is the penalty for the closed set $\mathcal{F}_2$, then $p_1(\mathbf{x}) + p_2(\mathbf{x})$ is the penalty for the intersection $\mathcal{F}_1 \cap \mathcal{F}_2$.

The following are two common penalty functions.

- *Exterior penalty function:*

$$p(\mathbf{x}) = \rho_1 \sum_{i=1}^{m} \left(\max\{0, f_i(\mathbf{x})\} \right)^r + \rho_2 \sum_{j=1}^{q} |h_j(\mathbf{x})|^2, \tag{3.7.3}$$

where r is usually 1 or 2.
- *Interior penalty function:*

$$p(\mathbf{x}) = \rho_1 \sum_{i=1}^{m} -\frac{1}{f_i(\mathbf{x})} \log(-f_i(\mathbf{x})) + \rho_2 \sum_{j=1}^{q} |h_j(\mathbf{x})|^2. \tag{3.7.4}$$

If $f_i(\mathbf{x}) \leq 0, \forall i = 1, \ldots, m$ and $h_j(\mathbf{x}) = 0, \forall j = 1, \ldots, q$, then $p(\mathbf{x}) = 0$, i.e., the exterior penalty function has no effect on the interior points in the feasible set $\text{int}(\mathcal{F})$. On the contrary, if, for some iteration point $\mathbf{x}_k$, the inequality constraint $f_i(\mathbf{x}_k) > 0, i \in \{1, \ldots, m\}$, and/or $h_j(\mathbf{x}_k) \neq 0, j \in \{1, \ldots, q\}$, then the penalty term $p(\mathbf{x}_k) \neq 0$, that is, any point outside the feasible set $\mathcal{F}$ is "punished" in exterior penalty function defined in (3.7.3).

The role of interior penalty function in (3.7.4) is equivalent to building a fence on the feasible set boundary $\text{bnd}(\mathcal{F})$, and thus blocks any point on $\text{bnd}(\mathcal{F})$, whereas the points in the relative feasible interior set relint($\mathcal{F}$) are slightly punished. The interior penalty function is also known as the *barrier function*.

The following are three typical barrier functions for a closed set $\mathcal{F}$ [25]:

- *Power-function barrier function:* $\phi(\mathbf{x}) = \sum_{i=1}^{m} -\frac{1}{(f_i(\mathbf{x}))^p}$, $p \geq 1$;
- *Logarithmic barrier function:* $\phi(\mathbf{x}) = -\frac{1}{f_i(\mathbf{x})} \sum_{i=1}^{m} \log(-f_i(\mathbf{x}))$;
- *Exponential barrier function:* $\phi(\mathbf{x}) = \sum_{i=1}^{m} \exp\left(-\frac{1}{f_i(\mathbf{x})}\right)$.

In other words, the logarithmic barrier function in (3.7.4) can be replaced by the power-function barrier function or the exponential barrier function.

When $p = 1$, the power-function barrier function $\phi(\mathbf{x}) = \sum_{i=1}^{m} -\frac{1}{f_i(\mathbf{x})}$ is called the *inverse barrier function* and was presented by Carroll in 1961 [9]; while

$$\phi(\mathbf{x}) = \mu \sum_{i=1}^{m} \frac{1}{\log(-f_i(\mathbf{x}))} \tag{3.7.5}$$

is known as the classical *Fiacco–McCormick logarithmic barrier function* [12], where μ is the *barrier parameter*.

A comparison of the features of the external and interior penalty functions are as follows.

- The exterior penalty function method is usually known as the *penalty function method*, and the interior penalty function method is customarily called the *barrier method*.
- The exterior penalty function method punishes all points outside of the feasible set, and its solution satisfies all the inequality constraints $f_i(\mathbf{x}) \leq 0, i = 1, \ldots, m$, and all the equality constraints $h_j(\mathbf{x}) = 0, j = 1, \ldots, q$, and is thus an exact solution to the constrained optimization problem. In other words, the exterior penalty function method is an optimal design scheme. In contrast, the interior penalty function (or barrier) method blocks all points on the boundary of the feasible set, and the found solution satisfies only the strict inequality $f_i(\mathbf{x}) < 0, i = 1, \ldots, m$, and the equality constraints $h_j(\mathbf{x}) = 0, j = 1, \ldots, q$, and hence is only an approximate solution. That is to say, the interior penalty function method is a suboptimal design scheme.
- The exterior penalty function method can be started using an unfeasible point, and its convergence is slow, whereas the interior penalty function method requires the initial point to be a feasible interior point, so its selection is difficult; but it has a good convergence and approximation performance.

In evolutionary computations the exterior penalty function method is normally used, while the interior function method is NP-hard owing to the feasible initial point search.

Engineering designers, especially process controllers, prefer to use the interior penalty function method, because this method allows the designers to observe the changes in the objective function value corresponding to the design points in the feasible set in the optimization process. However, this facility cannot be provided by any exterior penalty function method.

In the strict sense of the penalty function classification, all the above kinds of penalty function belong to the "death penalty": the infeasible solution points $\mathbf{x} \in S \setminus F$ (the difference set of the search space S and the feasible set F) are completely ruled out by the penalty function $p(\mathbf{x}) = +\infty$. If the feasible search space is convex, or it is the rational part of the whole search space, then this death penalty works very well [22]. However, for genetic algorithms and evolutionary computations, the boundary between the feasible set and the infeasible set is unknown and thus it is difficult to determine the precise position of the feasible set. In these cases, other penalty functions should be used [36]: static, dynamic, annealing, adaptive, or co-evolutionary penalties.

3.7.2 Augmented Lagrange Multiplier Method

The *augmented Lagrange multiplier method* transforms the constrained optimization problem (3.7.1) into an unconstrained optimization problem with Lagrange function

$$\mathcal{L}(\mathbf{x}, \boldsymbol{\lambda}, \boldsymbol{\nu}) = f_0(\mathbf{x}) + \sum_{i=1}^{m} \lambda_i f_i(\mathbf{x}) + \sum_{j=1}^{q} \nu_j h_j(\mathbf{x})$$
$$= f_0(\mathbf{x}) + \boldsymbol{\lambda}^T \mathbf{f}(\mathbf{x}) + \boldsymbol{\nu}^T \mathbf{h}(\mathbf{x}), \qquad (3.7.6)$$

where $\mathcal{L}(\mathbf{x}, \boldsymbol{\lambda}, \boldsymbol{\nu})$ is known as the augmented Lagrange function, $\boldsymbol{\lambda} = [\lambda_1, \ldots, \lambda_m]^T$ and $\boldsymbol{\nu} = [\nu_1, \ldots, \nu_q]^T$ are, respectively, called the *Lagrange multiplier vector* and the *penalty parameter vector*, whereas $\mathbf{f}(\mathbf{x}) = [f_1(\mathbf{x}), \ldots, f_m(\mathbf{x})]^T$ and $\mathbf{h}(\mathbf{x}) = [h_1(\mathbf{x}), \ldots, h_q(\mathbf{x})]^T$ are the inequality constraint vector and the equality constraint vector, respectively.

The unconstrained optimization problem

$$\min_{\mathbf{x}} \left\{ \mathcal{L}(\mathbf{x}, \boldsymbol{\lambda}, \boldsymbol{\nu}) = f_0(\mathbf{x}) + \sum_{i=1}^{m} \lambda_i f_i(\mathbf{x}) + \sum_{j=1}^{q} \nu_j h_j(\mathbf{x}) \right\} \qquad (3.7.7)$$

is known as the *primal problem* for constrained optimization in (3.7.1). The primal problem is difficult to solve, because the general solution vector $\mathbf{x}$ may violate one of the inequality constraints $\mathbf{f}(\mathbf{x}) < \mathbf{0}$ and/or the equality constraints $\mathbf{h}(\mathbf{x}) = \mathbf{A}\mathbf{x} = \mathbf{0}$.

The augmented Lagrange function includes the Lagrange function and the penalty function as two special examples.

- If $\boldsymbol{v} = \mathbf{0}$, then the augmented Lagrange function simplifies to the Lagrange function:

$$\mathcal{L}(\mathbf{x}, \boldsymbol{\lambda}, \mathbf{0}) = \mathcal{L}(\mathbf{x}, \boldsymbol{\lambda}) = f_0(\mathbf{x}) + \sum_{i=1}^{m} \lambda_i f_i(\mathbf{x}).$$

- If $\boldsymbol{\lambda} = \mathbf{0}$ and $\boldsymbol{v} = \rho \mathbf{h}(\mathbf{x})$ with $\rho > 0$, then the augmented Lagrange function reduces to the penalty function:

$$\mathcal{L}(\mathbf{x}, \mathbf{0}, \boldsymbol{v}) = \mathcal{L}(\mathbf{x}, \boldsymbol{v}) = f_0(\mathbf{x}) + \rho \sum_{i=1}^{q} |h_i(\mathbf{x})|^2.$$

The above two facts show that the augmented Lagrange multiplier method combines both the Lagrange multiplier method and the penalty function method.

For the standard constrained optimization problem (3.7.1), the *feasible set* is defined as the set of points satisfying all the inequality and equality constraints, namely

$$\mathcal{F} = \left\{ \mathbf{x} \big| f_i(\mathbf{x}) \leq 0, i = 1, \ldots, m,\ h_j(\mathbf{x}) = 0, j = 1, \ldots, q \right\}. \tag{3.7.8}$$

Therefore, the augmented Lagrange multiplier method becomes to solve the dual (optimization) problem

$$\min_{\mathbf{x} \in \mathcal{F}} \left\{ L_D(\boldsymbol{\lambda}, \boldsymbol{v}) = f_0(\mathbf{x}) + \sum_{i=1}^{m} \lambda_i f_i(\mathbf{x}) + \sum_{j=1}^{q} v_j h_j(\mathbf{x}) \right\}, \tag{3.7.9}$$

where $L_D(\boldsymbol{\lambda}, \boldsymbol{v})$ is called the dual function.

Proposition 3.1 *Let* $p^* = f_0(\mathbf{x}^*) = \min_{\mathbf{x}} L(\mathbf{x}, \boldsymbol{\lambda}, \boldsymbol{v})$ *be the optimal solution of the primal problem and* $d^* = L_D(\boldsymbol{\lambda}^*, \boldsymbol{v}^*) = \min_{\boldsymbol{\lambda} \leq \mathbf{0}, \boldsymbol{v}} L(\mathbf{x}, \boldsymbol{\lambda}, \boldsymbol{v})$ *be the optimal solution of the dual problem. Then* $d^* = L_D(\boldsymbol{\lambda}^*, \boldsymbol{v}^*) \leq p^* = f_0(\mathbf{x}^*) = \min_{\mathbf{x}} L(\mathbf{x}, \boldsymbol{\lambda}, \boldsymbol{v})$.

Proof In Lagrange function $L(\mathbf{x}, \boldsymbol{\lambda}, \boldsymbol{v})$, for any feasible point $\mathbf{x} \in \mathcal{F}$, due to $\boldsymbol{\lambda} \geq \mathbf{0}$, $f_i(\mathbf{x}) < 0$ and $h_j(\mathbf{x}) = 0$ for all i, j, we have

$$\sum_{i=1}^{m} \lambda_i f_i(\mathbf{x}) + \sum_{j=1}^{q} v_j h_j(\mathbf{x}) \leq 0.$$

3.7 Constrained Convex Optimization

From this inequality it follows that

$$L(\mathbf{x}, \boldsymbol{\lambda}, \boldsymbol{v}) = f_0(\mathbf{x}) + \sum_{i=1}^{m} \lambda_i f_i(\mathbf{x}) + \sum_{j=1}^{q} v_j h_j(\mathbf{x}) \leq f_0(\mathbf{x}^*) = \min_{\mathbf{x}} L(\mathbf{x}, \boldsymbol{\lambda}, \boldsymbol{v}).$$

Hence, for any feasible point $\mathbf{x} \in \mathcal{F}$, we have the following result:

$$L_D(\boldsymbol{\lambda}^*, \boldsymbol{v}^*) = \min_{\boldsymbol{\lambda} \leq 0, \boldsymbol{v}} L(\mathbf{x}, \boldsymbol{\lambda}, \boldsymbol{v}) \leq \min_{\mathbf{x}} L(\mathbf{x}, \boldsymbol{\lambda}, \boldsymbol{v}) = f_0(\mathbf{x}^*).$$

That is to say, $d^* = L_D(\boldsymbol{\lambda}^*, \boldsymbol{v}^*) \leq p^* = f_0(\mathbf{x}^*)$. ∎

3.7.3 Lagrange Dual Method

Consider how to solve the dual minimization problem

$$\min \mathcal{L}_D(\boldsymbol{\lambda}, \boldsymbol{v}) = \min_{\boldsymbol{\lambda} \geq 0, \boldsymbol{v}} \left\{ f_0(\mathbf{x}) + \sum_{i=1}^{m} \lambda_i f_i(\mathbf{x}) + \sum_{j=1}^{q} v_j h_j(\mathbf{x}) \right\}. \quad (3.7.10)$$

Owing to the nonnegativity of the Lagrange multiplier vector $\boldsymbol{\lambda}$, the augmented Lagrange function $\mathcal{L}(\mathbf{x}, \boldsymbol{\lambda}, \boldsymbol{v})$ may tend to negative infinity when some λ_i equals a very large positive number. Hence, we first need to maximize the original augmented Lagrange function to get

$$J_1(\mathbf{x}) = \max_{\boldsymbol{\lambda} \geq 0, \boldsymbol{v}} \left\{ f_0(\mathbf{x}) + \sum_{i=1}^{m} \lambda_i f_i(\mathbf{x}) + \sum_{j=1}^{q} v_j h_j(\mathbf{x}) \right\}. \quad (3.7.11)$$

The problem with using the unconstrained maximization (3.7.11) is that it is impossible to avoid *violation of the constraint* $f_i(\mathbf{x}) > 0$. This may result in $J_1(\mathbf{x})$ equalling positive infinity; namely,

$$J_1(\mathbf{x}) = \begin{cases} f_0(\mathbf{x}), & \text{if } \mathbf{x} \text{ meets all original constraints;} \\ (f_0(\mathbf{x}), +\infty), & \text{otherwise.} \end{cases} \quad (3.7.12)$$

From the above equation it follows that in order to minimize $f_0(\mathbf{x})$ subject to all inequality and equality constraints, we should minimize $J_1(\mathbf{x})$ to get the *primal cost function*

$$J_P(\mathbf{x}) = \min_{\mathbf{x}} J_1(\mathbf{x}) = \min_{\mathbf{x}} \max_{\boldsymbol{\lambda} \geq 0, \boldsymbol{v}} \mathcal{L}(\mathbf{x}, \boldsymbol{\lambda}, \boldsymbol{v}). \quad (3.7.13)$$

This is a *minimax problem*, whose solution is the *supremum* of the Lagrange function $\mathcal{L}(\mathbf{x}, \boldsymbol{\lambda}, \boldsymbol{v})$, namely

$$J_P(\mathbf{x}) = \sup\left(f_0(\mathbf{x}) + \sum_{i=1}^{m} \lambda_i f_i(\mathbf{x}) + \sum_{i=1}^{q} v_i h_i(\mathbf{x}) \right). \tag{3.7.14}$$

From (3.7.14) and (3.7.12) it follows that the optimal value of the original constrained minimization problem is given by

$$p^* = J_P(\mathbf{x}^*) = \min_{\mathbf{x}} f_0(\mathbf{x}) = f_0(\mathbf{x}^*), \tag{3.7.15}$$

which is simply known as the *optimal primal value*.

But, the minimization of a nonconvex objective function cannot be converted into the minimization of another convex function. Hence, if $f_0(\mathbf{x})$ is a convex function, then even if we designed an optimization algorithm that can find a local extremum $\tilde{\mathbf{x}}$ of the original cost function, there is no guarantee that $\tilde{\mathbf{x}}$ is a global extremum point.

Fortunately, the minimization of a convex function $f(\mathbf{x})$ and the maximization of the concave function $-f(\mathbf{x})$ are equivalent. On the basis of this dual relation, it is easy to obtain a dual method for solving the optimization problem of a nonconvex objective function: convert the minimization of a nonconvex objective function into the maximization of a concave objective function.

For this purpose, construct another objective function from the Lagrange function $\mathcal{L}(\mathbf{x}, \boldsymbol{\lambda}, \boldsymbol{v})$:

$$J_2(\boldsymbol{\lambda}, \boldsymbol{v}) = \min_{\mathbf{x}} \mathcal{L}(\mathbf{x}, \boldsymbol{\lambda}, \boldsymbol{v})$$

$$= \min_{\mathbf{x}} \left\{ f_0(\mathbf{x}) + \sum_{i=1}^{m} \lambda_i f_i(\mathbf{x}) + \sum_{i=1}^{q} v_i h_i(\mathbf{x}) \right\}. \tag{3.7.16}$$

From the above equation it is known that

$$\min_{\mathbf{x}} \mathcal{L}(\mathbf{x}, \boldsymbol{\lambda}, \boldsymbol{v}) = \begin{cases} \min_{\mathbf{x}} f_0(\mathbf{x}), & \text{if } \mathbf{x} \text{ meets all the original constraints,} \\ (-\infty, \min_{\mathbf{x}} f_0(\mathbf{x})), & \text{otherwise.} \end{cases}$$

Its maximization function

$$J_D(\boldsymbol{\lambda}, \boldsymbol{v}) = \max_{\boldsymbol{\lambda} \geq 0, \boldsymbol{v}} J_2(\boldsymbol{\lambda}, \boldsymbol{v}) = \max_{\boldsymbol{\lambda} \geq 0, \boldsymbol{v}} \min_{\mathbf{x}} \mathcal{L}(\mathbf{x}, \boldsymbol{\lambda}, \boldsymbol{v}) \tag{3.7.17}$$

is called the *dual objective function* for the original problem. This is a *maximin problem* for the Lagrange function $\mathcal{L}(\mathbf{x}, \boldsymbol{\lambda}, \boldsymbol{v})$.

3.7 Constrained Convex Optimization

Since the *maximin* of the Lagrange function $\mathcal{L}(\mathbf{x}, \boldsymbol{\lambda}, \boldsymbol{v})$ is its *infimum*, we have

$$J_{\mathrm{D}}(\boldsymbol{\lambda}, \boldsymbol{v}) = \inf \left(f_0(\mathbf{x}) + \sum_{i=1}^{m} \lambda_i f_i(\mathbf{x}) + \sum_{i=1}^{q} v_i h_i(\mathbf{x}) \right). \quad (3.7.18)$$

The dual objective function defined by Eq. (3.7.18) has the following characteristics.

- The dual function $J_{\mathrm{D}}(\boldsymbol{\lambda}, \boldsymbol{v})$ is an infimum of the augmented Lagrange function $\mathcal{L}(\mathbf{x}, \boldsymbol{\lambda}, \boldsymbol{v})$.
- The dual function $J_{\mathrm{D}}(\boldsymbol{\lambda}, \boldsymbol{v})$ is a maximizing objective function, and thus it is a value or utility function rather than a cost function.
- The dual function $J_{\mathrm{D}}(\boldsymbol{\lambda}, \boldsymbol{v})$ is *lower unbounded*: its lower bound is $-\infty$. Hence, $J_{\mathrm{D}}(\boldsymbol{\lambda}, \boldsymbol{v})$ is a concave function of the variable $\mathbf{x}$ even if $f_0(\mathbf{x})$ is not a convex function.

Theorem 3.15 ([28, p. 16]) *Any local minimum point $\mathbf{x}^*$ of an unconstrained convex optimization function $f(\mathbf{x})$ is a global minimum point. If the convex function $f(\mathbf{x})$ is differentiable, then the stationary point $\mathbf{x}^*$ such that $\partial f(\mathbf{x})/\partial \mathbf{x} = \mathbf{0}$ is a global minimum point of $f(\mathbf{x})$.*

Theorem 3.15 shows that any extreme point of the concave function is a global extreme point. Therefore, the algorithm design of the standard constrained minimization problem (3.7.1) becomes the design of an unconstrained maximization algorithm for the dual objective function $J_{\mathrm{D}}(\boldsymbol{\lambda}, \boldsymbol{v})$. Such a method is known as the *Lagrange dual method*.

3.7.4 Karush–Kuhn–Tucker Conditions

Proposition 3.1 shows that

$$d^* \leq \min_{\mathbf{x}} f_0(\mathbf{x}) = p^*. \quad (3.7.19)$$

The difference between the optimal primal value p^* and the optimal dual value d^*, denoted $p^* - d^*$, is known as the *duality gap* between the original minimization problem and the dual maximization problem.

Equation (3.7.19) gives the relationship between the maximin and the minimax of the augmented Lagrange function $\mathcal{L}(\mathbf{x}, \boldsymbol{\lambda}, \boldsymbol{v})$. In fact, for any nonnegative real-valued function $f(\mathbf{x}, \mathbf{y})$, there is the following inequality relation between the maximin and minimax:

$$\max_{\mathbf{x}} \min_{\mathbf{y}} f(\mathbf{x}, \mathbf{y}) \leq \min_{\mathbf{y}} \max_{\mathbf{x}} f(\mathbf{x}, \mathbf{y}). \quad (3.7.20)$$

If $d^* \leq p^*$, then the Lagrange dual method is said to have *weak duality*, while when $d^* = p^*$, we say that the Lagrange dual method satisfies *strong duality*.

Let $\mathbf{x}^*$ and $(\boldsymbol{\lambda}^*, \boldsymbol{v}^*)$ represent any original optimal point and dual optimal points with zero dual gap $\epsilon = 0$. Since $\mathbf{x}^*$ among all original feasible points $\mathbf{x}$ minimizes the augmented Lagrange objective function $\mathcal{L}(\mathbf{x}, \boldsymbol{\lambda}^*, \boldsymbol{v}^*)$, the gradient vector of $\mathcal{L}(\mathbf{x}, \boldsymbol{\lambda}^*, \boldsymbol{v}^*)$ at the point $\mathbf{x}^*$ must be equal to the zero vector, namely

$$\nabla f_0(\mathbf{x}^*) + \sum_{i=1}^{m} \lambda_i^* \nabla f_i(\mathbf{x}^*) + \sum_{j=1}^{q} v_j^* \nabla h_j(\mathbf{x}^*) = \mathbf{0}.$$

Therefore, the *Karush–Kuhn–Tucker (KKT) conditions* (i.e., the first-order necessary conditions) of the Lagrange dual unconstrained optimization problem are given by Nocedal and Wright [28] below:

$$\left.\begin{aligned}
f_i(\mathbf{x}^*) \leq 0, \quad & i = 1, \ldots, m \quad \text{(original inequality constraints)}, \\
h_j(\mathbf{x}^*) = 0, \quad & j = 1, \ldots, q \quad \text{(original equality constraints)}, \\
\lambda_i^* \geq 0, \quad & i = 1, \ldots, m \quad \text{(nonnegativity)}, \\
\lambda_i^* f_i(\mathbf{x}^*) = 0, \quad & i = 1, \ldots, m \quad \text{(complementary slackness)}, \\
\nabla f_0(\mathbf{x}^*) + \sum_{i=1}^{m} \lambda_i^* \nabla f_i(\mathbf{x}^*) + \sum_{j=1}^{q} v_j^* \nabla h_j(\mathbf{x}^*) = \mathbf{0} \quad & \text{(stationary point)}.
\end{aligned}\right\}$$

(3.7.21)

A point $\mathbf{x}$ satisfying the KKT conditions is called a KKT point.

Remark 1 The first KKT condition and the second KKT condition are the original inequality and equality constraint conditions, respectively.

Remark 2 The third KKT condition is the nonnegative condition of the Lagrange multiplier λ_i, which is a key constraint of the Lagrange dual method.

Remark 3 The fourth KKT condition (complementary slackness) is also called dual complementary and is another key constraint of the Lagrange dual method. This condition implies that, for a violated constraint $f_i(\mathbf{x}) > 0$, the corresponding Lagrange multiplier λ_i must equal zero, hence we can avoid completely any violated constraint. Thus the role of this condition is to establish a barrier $f_i(\mathbf{x}) = 0, i = 1, \ldots, m$ on the boundary of inequality constraints that prevents the occurrence of constraint violation $f_i(\mathbf{x}) > 0$.

Remark 4 The fifth KKT condition is the condition for there to be a stationary point at $\min_{\mathbf{x}} \mathcal{L}(\mathbf{x}, \boldsymbol{\lambda}, \boldsymbol{v})$.

Remark 5 If the inequality constraint $f_i(\mathbf{x}) \leq 0, i = 1, \ldots, m$, in the constrained optimization (3.7.1) becomes $c_i(\mathbf{x}) \geq 0, i = 1, \ldots, m$, then the Lagrange function

3.7 Constrained Convex Optimization

should be modified to

$$\mathcal{L}(\mathbf{x}, \boldsymbol{\lambda}, \boldsymbol{v}) = f_0(\mathbf{x}) - \sum_{i=1}^{m} \lambda_i c_i(\mathbf{x}) + \sum_{j=1}^{q} v_j h_j(\mathbf{x}),$$

and all inequality constrained functions $f_i(\mathbf{x})$ in the KKT condition formula (3.7.21) should be replaced by $-c_i(\mathbf{x})$.

In the following, we discuss the necessary modifications of the KKT conditions under some assumptions.

Definition 3.14 (Violated Constraint) For inequality constraints $f_i(\mathbf{x}) \le 0, i = 1, \ldots, m$, if $f_i(\bar{\mathbf{x}}) = 0$ at the point $\bar{\mathbf{x}}$, then the ith constraint is said to be an *active constraint* at the point $\bar{\mathbf{x}}$. If $f_i(\bar{\mathbf{x}}) < 0$, then the ith constraint is called an *inactive constraint* at the point $\bar{\mathbf{x}}$. If $f_i(\bar{\mathbf{x}}) > 0$, then the ith constraint is known as a *violated constraint* at the point $\bar{\mathbf{x}}$. The index set of all active constraints at the point $\bar{\mathbf{x}}$ is denoted $\mathcal{A}(\bar{\mathbf{x}}) = \{i \mid f_i(\bar{\mathbf{x}}) = 0\}$ and is referred to as the *active set* of the point $\bar{\mathbf{x}}$.

Let m inequality constraints $f_i(\mathbf{x}), i = 1, \ldots, m$, have k active constraints $f_{A_1}(\mathbf{x}^*), \ldots, f_{A_k}(\mathbf{x}^*)$ and $m - k$ inactive constraints at some KKT point $\mathbf{x}^*$.

In order to satisfy the complementarity in the KKT conditions $\lambda_i f_i(\mathbf{x}^*) = 0$, the Lagrange multipliers λ_i^* corresponding to the inactive constraints $f_i(\mathbf{x}^*) < 0$ must be equal to zero. This implies that the last KKT condition in (3.7.21) becomes

$$\nabla f_0(\mathbf{x}^*) + \sum_{i \in \mathcal{A}} \lambda_i^* \nabla f_i(\mathbf{x}^*) + \sum_{j=1}^{q} v_j^* \nabla h_j(\mathbf{x}^*) = \mathbf{0}$$

or

$$\begin{bmatrix} \frac{\partial f_0(\mathbf{x}^*)}{\partial x_1^*} \\ \vdots \\ \frac{\partial f_0(\mathbf{x}^*)}{\partial x_n^*} \end{bmatrix} + \begin{bmatrix} \frac{\partial h_1(\mathbf{x}^*)}{\partial x_1^*} & \cdots & \frac{\partial h_q(\mathbf{x}^*)}{\partial x_1^*} \\ \vdots & \ddots & \vdots \\ \frac{\partial h_1(\mathbf{x}^*)}{\partial x_n^*} & \cdots & \frac{\partial h_q(\mathbf{x}^*)}{\partial x_n^*} \end{bmatrix} \begin{bmatrix} v_1^* \\ \vdots \\ v_q^* \end{bmatrix} = - \begin{bmatrix} \frac{\partial f_{A1}(\mathbf{x}^*)}{\partial x_1^*} & \cdots & \frac{\partial f_{Ak}(\mathbf{x}^*)}{\partial x_1^*} \\ \vdots & \ddots & \vdots \\ \frac{\partial f_{A1}(\mathbf{x}^*)}{\partial x_n^*} & \cdots & \frac{\partial f_{Ak}(\mathbf{x}^*)}{\partial x_n^*} \end{bmatrix} \begin{bmatrix} \lambda_{A1}^* \\ \vdots \\ \lambda_{Ak}^* \end{bmatrix}$$

namely

$$\nabla f_0(\mathbf{x}^*) + (\mathbf{J}_h(\mathbf{x}^*))^T \boldsymbol{v}^* = -(\mathbf{J}_\mathcal{A}(\mathbf{x}^*))^T \boldsymbol{\lambda}_\mathcal{A}^*, \qquad (3.7.22)$$

where $\mathbf{J}_h(\mathbf{x}^*)$ is the Jacobian matrix of the equality constraint function $h_j(\mathbf{x}) = 0, j = 1, \ldots, q$ at the point $\mathbf{x}^*$ and

$$\boldsymbol{\lambda}_\mathcal{A}^* = [\lambda_{A1}^*, \ldots, \lambda_{Ak}^*] \in \mathbb{R}^k, \qquad (3.7.23)$$

and

$$\mathbf{J}_{\mathcal{A}}(\mathbf{x}^*) = \begin{bmatrix} \frac{\partial f_{\mathcal{A}1}(\mathbf{x}^*)}{\partial x_1^*} & \cdots & \frac{\partial f_{\mathcal{A}1}(\mathbf{x}^*)}{\partial x_n^*} \\ \vdots & \ddots & \vdots \\ \frac{\partial f_{\mathcal{A}k}(\mathbf{x}^*)}{\partial x_1^*} & \cdots & \frac{\partial f_{\mathcal{A}k}(\mathbf{x}^*)}{\partial x_n^*} \end{bmatrix} \in \mathbb{R}^{k \times n}, \quad (3.7.24)$$

are the Lagrange multiplier vector of the active constraint function and the Jacobian matrix, respectively.

Equation (3.7.22) shows that if the Jacobian matrix $\mathbf{J}_{\mathcal{A}}(\bar{\mathbf{x}})$ of the active constraint function at the feasible point $\bar{\mathbf{x}}$ is of full row rank, then the actively constrained Lagrange multiplier vector can be uniquely determined:

$$\boldsymbol{\lambda}_{\mathcal{A}}^* = -(\mathbf{J}_{\mathcal{A}}(\bar{\mathbf{x}})\mathbf{J}_{\mathcal{A}}(\bar{\mathbf{x}})^T)^{-1}\mathbf{J}_{\mathcal{A}}(\bar{\mathbf{x}}) \left(\nabla f_0(\bar{\mathbf{x}}) + (\mathbf{J}_h(\bar{\mathbf{x}}))^T \boldsymbol{v}^* \right). \quad (3.7.25)$$

In optimization algorithm design we always want strong duality to be set up. A simple method for determining whether strong duality holds is Slater's theorem.

The set of points satisfying only the strict inequality constraints $f_i(\mathbf{x}) < 0$ and the equality constraints $h_i(\mathbf{x}) = 0$ is denoted by

$$\text{relint}(\mathcal{F}) = \left\{ \mathbf{x} \middle| f_i(\mathbf{x}) < 0, i = 1, \ldots, m, \; h_j(\mathbf{x}) = 0, j = 1, \ldots, q \right\} \quad (3.7.26)$$

and is known as the *relative feasible interior set* or the *relative strictly feasible set*. The points in the feasible interior set are called *relative interior points*.

In an optimization process, the constraint restriction that iterative points should be in the interior of the feasible domain is known as the *Slater condition*. Slater's theorem says that if the Slater condition is satisfied, and the original inequality optimization problem (3.7.1) is a convex optimization problem, then the optimal value d^* of the dual unconstrained optimization problem (3.7.17) is equal to the optimal value p^* of the original optimization problem, i.e., strong duality holds.

The following summarizes the relationships between the original constrained optimization problem and the Lagrange dual unconstrained convex optimization problem.

- Only when the inequality constraint functions $f_i(\mathbf{x}), i = 1, \ldots, m$, are convex, and the equality constraint functions $h_j(\mathbf{x}), j = 1, \ldots, q$, are affine, can an original constrained optimization problem be converted into a dual unconstrained maximization problem by the Lagrange dual method.
- The maximization of a concave function is equivalent to the minimization of the corresponding convex function.
- If the original constrained optimization is a convex problem, then the points of the Lagrange objective function $\tilde{\mathbf{x}}$ and $(\tilde{\boldsymbol{\lambda}}, \tilde{\boldsymbol{v}})$ satisfying the KKT conditions are the original and dual optimal points, respectively. In other words, the optimal

3.7 Constrained Convex Optimization

solution $\mathbf{d}^*$ of the Lagrange dual unconstrained optimization is the optimal solution $\mathbf{p}^*$ of the original constrained convex optimization problem.
- In general, the optimal solution of the Lagrange dual unconstrained optimization problem is not the optimal solution of the original constrained optimization problem but only a ϵ-suboptimal solution, where $\epsilon = f_0(\mathbf{x}^*) - J_D(\boldsymbol{\lambda}^*, \boldsymbol{\nu}^*)$.

Consider the constrained optimization problem

$$\min_{\mathbf{x}} f(\mathbf{x}) \quad \text{subject to} \quad \mathbf{A}\mathbf{x} \leq \mathbf{h}, \mathbf{B}\mathbf{x} = \mathbf{b}. \tag{3.7.27}$$

Let the nonnegative vector $\mathbf{s} \geq \mathbf{0}$ be a *slack variable vector* such that $\mathbf{A}\mathbf{x} + \mathbf{s} = \mathbf{h}$. Hence, the inequality constraint $\mathbf{A}\mathbf{x} \leq \mathbf{h}$ becomes the equality constraint $\mathbf{A}\mathbf{x} + \mathbf{s} - \mathbf{h} = \mathbf{0}$. Taking the penalty function $\phi(\mathbf{g}(\mathbf{x})) = \frac{1}{2}\|\mathbf{g}(\mathbf{x})\|_2^2$, the augmented Lagrange objective function is given by

$$\mathcal{L}_\rho(\mathbf{x}, \mathbf{s}, \boldsymbol{\lambda}, \boldsymbol{\nu}) = f(\mathbf{x}) + \boldsymbol{\lambda}^T(\mathbf{A}\mathbf{x} + \mathbf{s} - \mathbf{h}) + \boldsymbol{\nu}^T(\mathbf{B}\mathbf{x} - \mathbf{b})$$
$$+ \frac{\rho}{2}\left(\|\mathbf{B}\mathbf{x} - \mathbf{b}\|_2^2 + \|\mathbf{A}\mathbf{x} + \mathbf{s} - \mathbf{h}\|_2^2\right), \tag{3.7.28}$$

where the two Lagrange multiplier vectors $\boldsymbol{\lambda} \geq \mathbf{0}$ and $\boldsymbol{\nu} \geq \mathbf{0}$ are nonnegative vectors and the penalty parameter $\rho > 0$.

The dual gradient ascent method for solving the original optimization problem (3.7.27) is given by

$$\mathbf{x}_{k+1} = \arg\min_{\mathbf{x}} \mathcal{L}_\rho(\mathbf{x}, \mathbf{s}_k, \boldsymbol{\lambda}_k, \boldsymbol{\nu}_k), \tag{3.7.29}$$

$$\mathbf{s}_{k+1} = \arg\min_{\mathbf{s} \geq \mathbf{0}} \mathcal{L}_\rho(\mathbf{x}_{k+1}, \mathbf{s}, \boldsymbol{\lambda}_k, \boldsymbol{\nu}_k), \tag{3.7.30}$$

$$\boldsymbol{\lambda}_{k+1} = \boldsymbol{\lambda}_k + \rho_k(\mathbf{A}\mathbf{x}_{k+1} + \mathbf{s}_{k+1} - \mathbf{h}), \tag{3.7.31}$$

$$\boldsymbol{\nu}_{k+1} = \boldsymbol{\nu}_k + \rho_k(\mathbf{B}\mathbf{x}_{k+1} - \mathbf{b}), \tag{3.7.32}$$

where the gradient vectors are

$$\frac{\partial \mathcal{L}_\rho(\mathbf{x}_{k+1}, \mathbf{s}_{k+1}, \boldsymbol{\lambda}_k, \boldsymbol{\nu}_k)}{\partial \boldsymbol{\lambda}_k} = \mathbf{A}\mathbf{x}_{k+1} + \mathbf{s}_{k+1} - \mathbf{h},$$

$$\frac{\partial \mathcal{L}_\rho(\mathbf{x}_{k+1}, \mathbf{s}_{k+1}, \boldsymbol{\lambda}_k, \boldsymbol{\nu}_k)}{\partial \boldsymbol{\nu}_k} = \mathbf{B}\mathbf{x}_{k+1} - \mathbf{b}.$$

Equations (3.7.29) and (3.7.30) are, respectively, the updates of the original variable $\mathbf{x}$ and the intermediate variable $\mathbf{s}$, whereas Eqs. (3.7.31) and (3.7.32) are the dual gradient ascent iterations of the Lagrange multiplier vectors $\boldsymbol{\lambda}$ and $\boldsymbol{\nu}$, taking into account the inequality constraint $\mathbf{A}\mathbf{x} \geq \mathbf{h}$ and the equality constraint $\mathbf{B}\mathbf{x} = \mathbf{b}$, respectively.

3.7.5 Alternating Direction Method of Multipliers

In applied statistics and machine learning we often encounter large-scale equality constrained optimization problems, where the dimension of $\mathbf{x} \in \mathbb{R}^n$ is very large. If the vector $\mathbf{x}$ may be decomposed into several subvectors, i.e., $\mathbf{x} = (\mathbf{x}_1, \ldots, \mathbf{x}_r)$, and the objective function may also be decomposed into

$$f(\mathbf{x}) = \sum_{i=1}^{r} f_i(\mathbf{x}_i),$$

where $\mathbf{x}_i \in \mathbb{R}^{n_i}$ and $\sum_{i=1}^{r} n_i = n$, then a large-scale optimization problem can be transformed into a few *distributed optimization problems*.

The *alternating direction multiplier method* (ADMM) is a simple and effective method for solving distributed optimization problems.

The ADMM method decomposes an optimization problem into smaller subproblems; then, their local solutions are restored or reconstructed into a large-scale optimization solution of the original problem.

The ADMM was proposed by Gabay and Mercier [16] and Glowinski and Marrocco [17] independently in the mid-1970s.

Corresponding to the composition of the objective function $f(\mathbf{x})$, the equality constraint matrix is blocked as follows:

$$\mathbf{A} = [\mathbf{A}_1, \ldots, \mathbf{A}_r], \quad \mathbf{A}\mathbf{x} = \sum_{i=1}^{r} \mathbf{A}_i \mathbf{x}_i.$$

Hence the augmented Lagrange objective function can be written as [5]

$$\mathcal{L}_\rho(\mathbf{x}, \boldsymbol{\lambda}) = \sum_{i=1}^{r} \mathcal{L}_i(\mathbf{x}_i, \boldsymbol{\lambda})$$

$$= \sum_{i=1}^{r} \left(f_i(\mathbf{x}_i) + \boldsymbol{\lambda}^T \mathbf{A}_i \mathbf{x}_i \right) - \boldsymbol{\lambda}^T \mathbf{b} + \frac{\rho}{2} \left\| \sum_{i=1}^{r} (\mathbf{A}_i \mathbf{x}_i) - \mathbf{b} \right\|_2^2.$$

Applying the dual ascent method to the augmented Lagrange objective function yields a decentralized algorithm for parallel computing [5]:

$$\mathbf{x}_i^{k+1} = \arg\min_{\mathbf{x}_i \in \mathbb{R}^{n_i}} \mathcal{L}_i(\mathbf{x}_i, \boldsymbol{\lambda}_k), \quad i = 1, \ldots, r, \qquad (3.7.33)$$

$$\boldsymbol{\lambda}_{k+1} = \boldsymbol{\lambda}_k + \rho_k \left(\sum_{i=1}^{r} \mathbf{A}_i \mathbf{x}_i^{k+1} - \mathbf{b} \right). \qquad (3.7.34)$$

3.7 Constrained Convex Optimization

Here the updates $\mathbf{x}_i$ ($i = 1, \ldots, r$) can be run independently in parallel. Because $\mathbf{x}_i, i = 1, \ldots, r$ are updated in an alternating or sequential manner, this augmented multiplier method is known as the "alternating direction" method of multipliers.

In practical applications, the simplest decomposition is the objective function decomposition of $r = 2$:

$$\min \{f(\mathbf{x}) + h(\mathbf{z})\} \quad \text{subject to} \quad \mathbf{Ax} + \mathbf{Bz} = \mathbf{b}, \tag{3.7.35}$$

where $\mathbf{x} \in \mathbb{R}^n, \mathbf{z} \in \mathbb{R}^m, \mathbf{A} \in \mathbb{R}^{p \times n}, \mathbf{B} \in \mathbb{R}^{p \times m}, \mathbf{b} \in \mathbb{R}^p$.

The augmented Lagrange cost function of the optimization problem (3.7.35) is given by

$$\mathcal{L}_\rho(\mathbf{x}, \mathbf{z}, \boldsymbol{\lambda}) = f(\mathbf{x}) + h(\mathbf{z}) + \boldsymbol{\lambda}^T (\mathbf{Ax} + \mathbf{Bz} - \mathbf{b})$$
$$+ \frac{\rho}{2} \|\mathbf{Ax} + \mathbf{Bz} - \mathbf{b}\|_2^2. \tag{3.7.36}$$

It is easily seen that the optimality conditions are divided into the original feasibility condition

$$\mathbf{Ax} + \mathbf{Bz} - \mathbf{b} = \mathbf{0} \tag{3.7.37}$$

and two dual feasibility conditions,

$$\mathbf{0} \in \partial f(\mathbf{x}) + \mathbf{A}^T \boldsymbol{\lambda} + \rho \mathbf{A}^T (\mathbf{Ax} + \mathbf{Bz} - \mathbf{b}) = \partial f(\mathbf{x}) + \mathbf{A}^T \boldsymbol{\lambda}, \tag{3.7.38}$$

$$\mathbf{0} \in \partial h(\mathbf{z}) + \mathbf{B}^T \boldsymbol{\lambda} + \rho \mathbf{B}^T (\mathbf{Ax} + \mathbf{Bz} - \mathbf{b}) = \partial h(\mathbf{z}) + \mathbf{B}^T \boldsymbol{\lambda}, \tag{3.7.39}$$

where $\partial f(\mathbf{x})$ and $\partial h(\mathbf{z})$ are the subdifferentials of the subobjective functions $f(\mathbf{x})$ and $h(\mathbf{z})$, respectively.

The updates of the ADMM for the optimization problem $\min \mathcal{L}_\rho(\mathbf{x}, \mathbf{z}, \boldsymbol{\lambda})$ are as follows:

$$\mathbf{x}_{k+1} = \arg\min_{\mathbf{x} \in \mathbb{R}^n} \mathcal{L}_\rho(\mathbf{x}, \mathbf{z}_k, \boldsymbol{\lambda}_k), \tag{3.7.40}$$

$$\mathbf{z}_{k+1} = \arg\min_{\mathbf{z} \in \mathbb{R}^m} \mathcal{L}_\rho(\mathbf{x}_{k+1}, \mathbf{z}, \boldsymbol{\lambda}_k), \tag{3.7.41}$$

$$\boldsymbol{\lambda}_{k+1} = \boldsymbol{\lambda}_k + \rho_k (\mathbf{A}\mathbf{x}_{k+1} + \mathbf{B}\mathbf{z}_{k+1} - \mathbf{b}). \tag{3.7.42}$$

The original feasibility cannot be strictly satisfied; its error

$$\mathbf{r}_k = \mathbf{A}\mathbf{x}_k + \mathbf{B}\mathbf{z}_k - \mathbf{b} \tag{3.7.43}$$

is known as the *original residual* (vector) in the kth iteration. Hence the update of the Lagrange multiplier vector can be simply written as

$$\lambda_{k+1} = \lambda_k + \rho_k \mathbf{r}_{k+1}. \tag{3.7.44}$$

Likewise, neither can dual feasibility be strictly satisfied. Because $\mathbf{x}_{k+1}$ is the minimization variable of $\mathcal{L}_\rho(\mathbf{x}, \mathbf{z}_k, \lambda_k)$, we have

$$\begin{aligned}\mathbf{0} &\in \partial f(\mathbf{x}_{k+1}) + \mathbf{A}^T \lambda_k + \rho \mathbf{A}^T (\mathbf{A}\mathbf{x}_{k+1} + \mathbf{B}\mathbf{z}_k - \mathbf{b}) \\ &= \partial f(\mathbf{x}_{k+1}) + \mathbf{A}^T (\lambda_k + \rho \mathbf{r}_{k+1} + \rho \mathbf{B}(\mathbf{z}_k - \mathbf{z}_{k+1})) \\ &= \partial f(\mathbf{x}_{k+1}) + \mathbf{A}^T \lambda_{k+1} + \rho \mathbf{A}^T \mathbf{B}(\mathbf{z}_k - \mathbf{z}_{k+1}).\end{aligned}$$

Comparing this result with the dual feasibility formula (3.7.38), it is easy to see that

$$\mathbf{s}_{k+1} = \rho \mathbf{A}^T \mathbf{B}(\mathbf{z}_k - \mathbf{z}_{k+1}) \tag{3.7.45}$$

is the error vector of the dual feasibility, and hence is known as the *dual residual* vector in the $(k+1)$th iteration.

The stopping criterion of the ADMM is as follows: the original residual and the dual residual in the $(k+1)$th iteration should be very small, namely [5]

$$\|\mathbf{r}_{k+1}\|_2 \leq \epsilon_{\text{pri}}, \quad \|\mathbf{s}_{k+1}\|_2 \leq \epsilon_{\text{dual}}, \tag{3.7.46}$$

where ϵ_{pri} and ϵ_{dual} are, respectively, the allowed disturbances of the primal feasibility and the dual feasibility.

If letting $\mathbf{v} = (1/\rho)\lambda$ be the Lagrange multiplier vector scaled by the ratio $1/\rho$, called the *scaled dual vector*, then Eqs. (3.7.40)–(3.7.42) become [5]

$$\mathbf{x}_{k+1} = \arg\min_{\mathbf{x} \in \mathbb{R}^n} \left\{ f(\mathbf{x}) + (\rho/2)\|\mathbf{A}\mathbf{x} + \mathbf{B}\mathbf{z}_k - \mathbf{b} + \mathbf{v}_k\|_2^2 \right\}, \tag{3.7.47}$$

$$\mathbf{z}_{k+1} = \arg\min_{\mathbf{z} \in \mathbb{R}^m} \left\{ h(\mathbf{z}) + (\rho/2)\|\mathbf{A}\mathbf{x}_{k+1} + \mathbf{B}\mathbf{z} - \mathbf{b} + \mathbf{v}_k\|_2^2 \right\}, \tag{3.7.48}$$

$$\mathbf{v}_{k+1} = \mathbf{v}_k + \mathbf{A}\mathbf{x}_{k+1} + \mathbf{B}\mathbf{z}_{k+1} - \mathbf{b} = \mathbf{v}_k + \mathbf{r}_{k+1}. \tag{3.7.49}$$

The scaled dual vector has an interesting interpretation [5]: from the residual at the kth iteration, $\mathbf{r}_k = \mathbf{A}\mathbf{x}_k + \mathbf{B}\mathbf{z}_k - \mathbf{b}$, it is easily seen that

$$\mathbf{v}_k = \mathbf{v}_0 + \sum_{i=1}^{k} \mathbf{r}^i. \tag{3.7.50}$$

That is, the scaled dual vector is the running sum of the original residuals of all k iterations.

Equations (3.7.47)–(3.7.49) are referred to as the scaled ADMM, while Eqs. (3.7.40)–(3.7.42) are the ADMM with no scaling.

3.8 Newton Methods

The first-order optimization algorithms use only the zero-order information $f(\mathbf{x})$ and the first-order information $\nabla f(\mathbf{x})$ about an objective function. It is known [20] that if the objective function is twice differentiable then the Newton method based on the Hessian matrix is of quadratic or more rapid convergence.

3.8.1 Newton Method for Unconstrained Optimization

For an unconstrained optimization $\min_{\mathbf{x} \in \mathbb{R}^n} f(\mathbf{x})$, if the Hessian matrix $\mathbf{H} = \nabla^2 f(\mathbf{x})$ is positive definite, then from the Newton equation

$$\nabla^2 f(\mathbf{x}) \Delta \mathbf{x} = -\nabla f(\mathbf{x}), \tag{3.8.1}$$

we get the Newton step $\Delta \mathbf{x} = -(\nabla^2 f(\mathbf{x}))^{-1} \nabla f(\mathbf{x})$ which results in the gradient descent algorithm

$$\mathbf{x}_{k+1} = \mathbf{x}_k - \mu_k (\nabla^2 f(\mathbf{x}_k))^{-1} \nabla f(\mathbf{x}_k) = \mathbf{x}_k - \mu_k \mathbf{H}^{-1} \nabla f(\mathbf{x}_k). \tag{3.8.2}$$

This is the well-known Newton method.

The Newton method may encounter the following two thorny issues.

- The Hessian matrix $\mathbf{H} = \nabla^2 f(\mathbf{x})$ is difficult to find.
- Even if $\mathbf{H} = \nabla^2 f(\mathbf{x})$ can be found, its inverse $\mathbf{H}^{-1} = (\nabla^2 f(\mathbf{x}))^{-1}$ may be numerically unstable.

There are the following three methods for resolving the above two thorny issues.

1. *Truncated Newton method:* Instead of using directly the inverse of the Hessian matrix, an iteration method for solving the Newton matrix equation $\nabla^2 f(\mathbf{x}) \Delta \mathbf{x}_{\text{nt}} = -\nabla f(\mathbf{x})$ is to find an approximate solution for the Newton step $\Delta \mathbf{x}_{\text{nt}}$. The iteration method for solving Eq. (3.8.1) approximately is known as the truncated Newton method [11], where the conjugate gradient algorithm and the preconditioned conjugate gradient algorithm are two popular algorithms for such a case. The truncated Newton method is especially useful for large-scale unconstrained and constrained optimization problems and interior-point methods.

2. *Modified Newton method:* When the Hessian matrix is not positive definite, the Newton matrix equation (3.8.1) can be modified to [14]

$$(\nabla^2 f(\mathbf{x}) + \mathbf{E})\Delta \mathbf{x}_{\text{nt}} = -\nabla f(\mathbf{x}), \quad (3.8.3)$$

where $\mathbf{E}$ is a positive semi-definite matrix that is usually taken as a diagonal matrix such that $\nabla^2 f(\mathbf{x}) + \mathbf{E}$ is symmetric positive definite. Such a method is called the modified Newton method. The typical modified Newton method takes $\mathbf{E} = \delta \mathbf{I}$, where $\delta > 0$ is small.

3. *Quasi-Newton method:* Using a symmetric positive definite matrix to approximate the Hessian matrix or its inverse, the Newton method is known as the quasi-Newton method. Consider the second-order Taylor series expansion of objective function $f(\mathbf{x})$ at point $\mathbf{x}_{k+1}$:

$$f(\mathbf{x}) \approx f(\mathbf{x}_{k+1}) + \nabla f(\mathbf{x}_{k+1}) \cdot (\mathbf{x} - \mathbf{x}_{k+1}) + \frac{1}{2}(\mathbf{x} - \mathbf{x}_{k+1})^T \nabla^2 f(\mathbf{x}_{k+1})(\mathbf{x} - \mathbf{x}_{k+1}).$$

Hence, we have $\nabla f(\mathbf{x}) \approx \nabla f(\mathbf{x}_{k+1}) + \mathbf{H}_{k+1} \cdot (\mathbf{x} - \mathbf{x}_{k+1})$. If taking $\mathbf{x} = \mathbf{x}_k$, then the quasi-Newton condition is given by

$$\mathbf{g}_{k+1} - \mathbf{g}_k \approx \mathbf{H}_{k+1} \cdot (\mathbf{x}_{k+1} - \mathbf{x}_k). \quad (3.8.4)$$

Letting $\mathbf{s}_k = \mathbf{x}_{k+1} - \mathbf{x}_k$ denote the Newton step and $\mathbf{y}_k = \mathbf{g}_{k+1} - \mathbf{g}_k = \nabla f(\mathbf{x}_{k+1}) - \nabla f(\mathbf{x}_k)$, then the quasi-Newton condition can be rewritten as

$$\mathbf{y}_k \approx \mathbf{H}_{k+1} \cdot \mathbf{s}_k \quad \text{or} \quad \mathbf{s}_k = \mathbf{H}_{k+1}^{-1} \cdot \mathbf{y}_k. \quad (3.8.5)$$

- *DFP (Davidon–Fletcher–Powell) method:* Use an $n \times n$ matrix $\mathbf{D}_k = \mathbf{D}(\mathbf{x}_k)$ to approximate the inverse of Hessian matrix, i.e., $\mathbf{D}_k \approx \mathbf{H}^{-1}$. In this setting, the update of $\mathbf{D}_k$ is given by Nesterov [25]:

$$\mathbf{D}_{k+1} = \mathbf{D}_k + \frac{\mathbf{s}_k \mathbf{s}_k^T}{\mathbf{y}_k^T \mathbf{s}_k} - \frac{\mathbf{D}_k \mathbf{y}_k \mathbf{y}_k^T \mathbf{D}_k}{\mathbf{y}_k^T \mathbf{D}_k \mathbf{y}_k}. \quad (3.8.6)$$

- *BFGS (Broyden–Fletcher–Goldfarb–Shanno) method:* Use a symmetric positive definite matrix to approximate the Hessian matrix, i.e., $\mathbf{B} = \mathbf{H}$. Then, the update of $\mathbf{B}_k$ is given by Nesterov [25]:

$$\mathbf{B}_{k+1} = \mathbf{B}_k + \frac{\mathbf{y}_k \mathbf{y}_k^T}{\mathbf{y}_k^T \mathbf{s}_k} - \frac{\mathbf{B}_k \mathbf{s}_k \mathbf{s}_k^T \mathbf{B}_k}{\mathbf{s}_k^T \mathbf{B}_k \mathbf{s}_k}. \quad (3.8.7)$$

An important property of the DFP and BFGS methods is that they preserve the positive definiteness of the matrices.

In the iterative process of the various Newton methods, it is usually required that a search is made of the optimal point along the direction of the straight line $\{\mathbf{x} +$

3.8 Newton Methods

Algorithm 3.6 Davidon–Fletcher–Powell (DFP) quasi-Newton method

1. **input:** Objective function $f(\mathbf{x})$, gradient vector $\mathbf{g}(\mathbf{x}) = \nabla f(\mathbf{x})$, accuracy threshold ϵ.
2. **initialization:** Take the initial value $\mathbf{x}_0$ and $\mathbf{D}_0 = \mathbf{I}$. Take $k = 0$.
3. Determine the search direction $\mathbf{p}_k = -\mathbf{D}_k \mathbf{g}_k$.
4. Determine the step length $\mu_k = \arg\min_{\mu \geq 0} f(\mathbf{x}_k + \mu \mathbf{p}_k)$.
5. Update the solution vector $\mathbf{x}_{k+1} = \mathbf{x}_k + \mu_k \mathbf{p}_k$.
6. Calculate the Newton step $\mathbf{s}_k = \mathbf{x}_{k+1} - \mathbf{x}_k$.
7. Calculate $\mathbf{g}_{k+1} = \mathbf{g}(\mathbf{x}_{k+1})$. If $\|\mathbf{g}_{k+1}\|_2 < \epsilon$ then stop iterations and output $\mathbf{x}^* = \mathbf{x}_k$. Otherwise, continue the next step.
8. Let $\mathbf{y}_k = \mathbf{g}_{k+1} - \mathbf{g}_k$.
9. Calculate $\mathbf{D}_{k+1} = \mathbf{D}_k + \frac{\mathbf{s}_k \mathbf{s}_k^T}{\mathbf{y}_k^T \mathbf{s}_k} - \frac{\mathbf{D}_k \mathbf{y}_k \mathbf{y}_k^T \mathbf{D}_k}{\mathbf{y}_k^T \mathbf{D}_k \mathbf{y}_k}$.
10. Update $k \leftarrow k + 1$ and return to Step 3.
11. **output:** Minimum point $\mathbf{x}^* = \mathbf{x}_k$ of $f(\mathbf{x})$.

Algorithm 3.7 Broyden–Fletcher–Goldfarb–Shanno (BFGS) quasi-Newton method

1. **input:** Objective function $f(\mathbf{x})$, gradient vector $\mathbf{g}(\mathbf{x}) = \nabla f(\mathbf{x})$, accuracy threshold ϵ.
2. **initialization:** Take the initial value $\mathbf{x}_0$ and $\mathbf{B}_0 = \mathbf{I}$. Take $k = 0$.
3. Solve $\mathbf{B}_k \mathbf{p}_k = -\mathbf{g}_k$ for determining the search direction $\mathbf{p}_k$.
4. Determine the step length $\mu_k = \arg\min_{\mu \geq 0} f(\mathbf{x}_k + \mu \mathbf{p}_k)$.
5. Update the solution vector $\mathbf{x}_{k+1} = \mathbf{x}_k + \mu_k \mathbf{p}_k$.
6. Calculate the Newton step $\mathbf{s}_k = \mathbf{x}_{k+1} - \mathbf{x}_k$.
7. Calculate $\mathbf{g}_{k+1} = \mathbf{g}(\mathbf{x}_{k+1})$. If $\|\mathbf{g}_{k+1}\|_2 < \epsilon$ then stop iterations and output $\mathbf{x}^* = \mathbf{x}_k$. Otherwise, continue the next step.
8. Let $\mathbf{y}_k = \mathbf{g}_{k+1} - \mathbf{g}_k$.
9. Calculate $\mathbf{B}_{k+1} = \mathbf{B}_k + \frac{\mathbf{y}_k \mathbf{y}_k^T}{\mathbf{y}_k^T \mathbf{s}_k} - \frac{\mathbf{B}_k \mathbf{s}_k \mathbf{s}_k^T \mathbf{B}_k}{\mathbf{s}_k^T \mathbf{B}_k \mathbf{s}_k}$.
10. Update $k \leftarrow k + 1$ and return to Step 3.
11. **output:** Minimum point $\mathbf{x}^* = \mathbf{x}_k$ of $f(\mathbf{x})$.

$\mu \Delta \mathbf{x} \,|\, \mu \geq 0\}$. This step is called *linear search*. However, this choice of procedure can only minimize the objective function approximately, i.e., make it sufficiently small. Such a search for approximate minimization is called an *inexact line search*. In this search the step size μ_k on the search line $\{\mathbf{x} + \mu \Delta \mathbf{x} \,|\, \mu \geq 0\}$ must decrease the objective function $f(\mathbf{x}_k)$ sufficiently to ensure that

$$f(\mathbf{x}_k + \mu \Delta \mathbf{x}_k) < f(\mathbf{x}_k) + \alpha \mu (\nabla f(\mathbf{x}_k))^T \Delta \mathbf{x}_k, \quad \alpha \in (0, 1). \tag{3.8.8}$$

The inequality condition (3.8.8) is sometimes called the *Armijo condition* [13, 20, 28]. In general, for a larger step size μ, the Armijo condition is usually not met. Hence, Newton algorithms can start the search from $\mu = 1$; if the Armijo condition is not met, then the step size μ needs to be reduced to $\beta \mu$, with the correction factor $\beta \in (0, 1)$. If for step size $\beta \mu$ the Armijo condition is still not met, then the step size needs to be reduced again until the Aemijo condition is met. Such a search method is known as *backtracking line search*.

Algorithm 3.8 shows the Newton algorithm via backtracking line search.

Algorithm 3.8 Newton algorithm via backtracking line search
1. **input:**
2. **initialization:** $x_1 \in \text{dom } f(x)$, and parameters $\alpha \in (0, 0.5), \beta \in (0, 1)$. Put $k = 1$.
3. **repeat**
4.   Compute $b_k = \nabla f(x_k)$ and $H_k = \nabla^2 f(x_k)$.
5.   Solve the Newton equation $H_k \Delta x_k = -b_k$.
6.   Update $x_{k+1} = x_k + \mu \Delta x_k$.
7.   **exit if** $|f(x_{k+1}) - f(x_k)| < \epsilon$.
8. **return** $k \leftarrow k + 1$.
9. **output:** $x \leftarrow x_k$.

The backtracking line search method can ensure that the objective function satisfies $f(x_{k+1}) < f(x_k)$ and the step size μ may not need to be too small.

If the Newton equation is solved by other methods in Step 2, then Algorithm 3.8 becomes the truncated Newton method, the modified Newton method, or the quasi-Newton method, respectively.

3.8.2 Newton Method for Constrained Optimization

First, consider the equality constrained optimization problem with a real variable:

$$\min_{x} f(x) \quad \text{subject to} \quad Ax = b, \tag{3.8.9}$$

where $f : \mathbb{R}^n \to \mathbb{R}$ is a convex function and is twice differentiable, while $A \in \mathbb{R}^{p \times n}$ and $\text{rank}(A) = p$ with $p < n$.

Let Δx_{nt} denote the *Newton search direction*. Then the second-order Taylor approximation of the objective function $f(x)$ is given by

$$f(x + \Delta x_{nt}) = f(x) + (\nabla f(x))^T \Delta x_{nt} + \frac{1}{2} (\Delta x_{nt})^T \nabla^2 f(x) \Delta x_{nt},$$

subject to $A(x + \Delta x_{nt}) = b$ and $A \Delta x_{nt} = 0$. In other words, the Newton search direction can be determined by the equality constrained optimization problem:

$$\min_{\Delta x_{nt}} \left\{ f(x) + (\nabla f(x))^T \Delta x_{nt} + \frac{1}{2} (\Delta x_{nt})^T \nabla^2 f(x) \Delta x_{nt} \right\}$$

subject to $A \Delta x_{nt} = 0$.

3.8 Newton Methods

Letting λ be the Lagrange multiplier vector multiplying the equality constraint $\mathbf{A}\Delta\mathbf{x}_{nt} = \mathbf{0}$, we have the Lagrange objective function

$$\mathcal{L}(\Delta\mathbf{x}_{nt}, \lambda) = f(\mathbf{x}) + (\nabla f(\mathbf{x}))^T \Delta\mathbf{x}_{nt} + \frac{1}{2}(\Delta\mathbf{x}_{nt})^T \nabla^2 f(\mathbf{x})\Delta\mathbf{x}_{nt} + \lambda^T \mathbf{A}\Delta\mathbf{x}_{nt}. \tag{3.8.10}$$

By the first-order optimal condition $\frac{\partial \mathcal{L}(\Delta\mathbf{x}_{nt}, \lambda)}{\partial \Delta\mathbf{x}_{nt}} = \mathbf{0}$ and the constraint condition $\mathbf{A}\Delta\mathbf{x}_{nt} = \mathbf{0}$, it is easily obtained that $\nabla f(\mathbf{x}) + \nabla^2 f(\mathbf{x})\Delta\mathbf{x}_{nt} + \mathbf{A}^T\lambda = \mathbf{0}$ and $\mathbf{A}\Delta\mathbf{x}_{nt} = \mathbf{0}$. These two equations can be merged into

$$\begin{bmatrix} \nabla^2 f(\mathbf{x}) & \mathbf{A}^T \\ \mathbf{A} & \mathbf{O} \end{bmatrix} \begin{bmatrix} \Delta\mathbf{x}_{nt} \\ \lambda \end{bmatrix} = \begin{bmatrix} -\nabla f(\mathbf{x}) \\ \mathbf{O} \end{bmatrix}. \tag{3.8.11}$$

As a stopping criterion for the Newton algorithm, one can use [3]

$$\lambda^2(\mathbf{x}) = (\Delta\mathbf{x}_{nt})^T \nabla^2 f(\mathbf{x}) \Delta\mathbf{x}_{nt}. \tag{3.8.12}$$

Algorithm 3.9 shows *feasible-start Newton algorithm*.

Algorithm 3.9 Feasible-start Newton algorithm [3]

1. **input:** $\mathbf{A} \in \mathbb{R}^{p \times n}$, $\mathbf{b} \in \mathbb{R}^p$, tolerance $\epsilon > 0$, parameters $\alpha \in (0, 0.5)$, $\beta \in (0, 1)$.
2. **initialization:** A feasible initial point $\mathbf{x}_1 \in \text{dom } f$ with $\mathbf{A}\mathbf{x}_1 = \mathbf{b}$. Put $k = 1$.
3. **repeat**
4.   Compute the gradient vector $\nabla f(\mathbf{x}_k)$ and the Hessian matrix $\nabla^2 f(\mathbf{x}_k)$.
4.   Use the preconditioned conjugate gradient algorithm to solve
$$\begin{bmatrix} \nabla^2 f(\mathbf{x}_k) & \mathbf{A}^T \\ \mathbf{A} & \mathbf{O} \end{bmatrix} \begin{bmatrix} \Delta\mathbf{x}_{nt,k} \\ \lambda_{nt} \end{bmatrix} = \begin{bmatrix} -\nabla f(\mathbf{x}_k) \\ \mathbf{O} \end{bmatrix}.$$
5.   Compute $\lambda^2(\mathbf{x}_k) = (\Delta\mathbf{x}_{nt,k})^T \nabla^2 f(\mathbf{x}_k) \Delta\mathbf{x}_{nt,k}$.
6.   **exit if** $\lambda^2(\mathbf{x}_k) < \epsilon$.
7.   Let $\mu = 1$.
8.   **while** not converged **do**
9.   Update $\mu \leftarrow \beta\mu$.
10.   **break if** $f(\mathbf{x}_k + \mu\Delta\mathbf{x}_{nt,k}) < f(\mathbf{x}_k) + \alpha\mu(\nabla f(\mathbf{x}_k))^T \Delta\mathbf{x}_{nt,k}$.
11.   **end while**
12. Update $\mathbf{x}_{k+1} \leftarrow \mathbf{x}_k + \mu\Delta\mathbf{x}_{nt,k}$.
13. **return** $k \leftarrow k + 1$.
14. **output:** $\mathbf{x} \leftarrow \mathbf{x}_k$.

However, in many cases it is not easy to find a feasible point as an initial point. If $\mathbf{x}_k$ is an infeasible point, consider the equality constrained optimization problem

$$\min_{\Delta \mathbf{x}_k} \left\{ f(\mathbf{x}_k + \Delta \mathbf{x}_k) = f(\mathbf{x}_k) + (\nabla f(\mathbf{x}_k))^T \Delta \mathbf{x}_k + \frac{1}{2} (\Delta \mathbf{x}_k)^T \nabla^2 f(\mathbf{x}_k) \Delta \mathbf{x}_k \right\}$$

subject to $\mathbf{A}(\mathbf{x}_k + \Delta \mathbf{x}_k) = \mathbf{b}$.

Let $\boldsymbol{\lambda}_{k+1} (= \boldsymbol{\lambda}_k + \Delta \boldsymbol{\lambda}_k)$ be the Lagrange multiplier vector corresponding to the equality constraint $\mathbf{A}(\mathbf{x}_k + \Delta \mathbf{x}_k) = \mathbf{b}$. Then the Lagrange objective function is

$$\mathcal{L}(\Delta \mathbf{x}_k, \boldsymbol{\lambda}_{k+1}) = f(\mathbf{x}_k) + (\nabla f(\mathbf{x}_k))^T \Delta \mathbf{x}_k + \frac{1}{2} (\Delta \mathbf{x}_k)^T \nabla^2 f(\mathbf{x}_k) \Delta \mathbf{x}_k$$
$$+ \boldsymbol{\lambda}_{k+1}^T (\mathbf{A}(\mathbf{x}_k + \Delta \mathbf{x}_k) - \mathbf{b}).$$

From the optimal conditions

$$\frac{\partial \mathcal{L}(\Delta \mathbf{x}_k, \boldsymbol{\lambda}_{k+1})}{\partial \Delta \mathbf{x}_k} = \mathbf{0} \quad \text{and} \quad \frac{\partial \mathcal{L}(\Delta \mathbf{x}_k, \boldsymbol{\lambda}_{k+1})}{\partial \boldsymbol{\lambda}_{k+1}} = \mathbf{0},$$

we have

$$\nabla f(\mathbf{x}_k) + \nabla^2 f(\mathbf{x}_k) \Delta \mathbf{x}_k + \mathbf{A}^T \boldsymbol{\lambda}_{k+1} = \mathbf{0}, \quad \mathbf{A} \Delta \mathbf{x}_k = -(\mathbf{A} \mathbf{x}_k - \mathbf{b}),$$

which can be merged into

$$\begin{bmatrix} \nabla^2 f(\mathbf{x}_k) & \mathbf{A}^T \\ \mathbf{A} & \mathbf{O} \end{bmatrix} \begin{bmatrix} \Delta \mathbf{x}_k \\ \boldsymbol{\lambda}_{k+1} \end{bmatrix} = - \begin{bmatrix} \nabla f(\mathbf{x}_k) \\ \mathbf{A} \mathbf{x}_k - \mathbf{b} \end{bmatrix}. \quad (3.8.13)$$

Substituting $\boldsymbol{\lambda}_{k+1} = \boldsymbol{\lambda}_k + \Delta \boldsymbol{\lambda}_k$ into the above equation, we have

$$\begin{bmatrix} \nabla^2 f(\mathbf{x}_k) & \mathbf{A}^T \\ \mathbf{A} & \mathbf{O} \end{bmatrix} \begin{bmatrix} \Delta \mathbf{x}_k \\ \Delta \boldsymbol{\lambda}_k \end{bmatrix} = - \begin{bmatrix} \nabla f(\mathbf{x}_k) + \mathbf{A}^T \boldsymbol{\lambda}_k \\ \mathbf{A} \mathbf{x}_k - \mathbf{b} \end{bmatrix}. \quad (3.8.14)$$

Algorithm 3.10 shows an infeasible-start Newton algorithm based on (3.8.14).

Algorithm 3.10 Infeasible-start Newton algorithm [3]

1. **input:** Any allowed tolerance $\epsilon > 0$, $\alpha \in (0, 0.5)$, and $\beta \in (0, 1)$.
2. **initialization:** An initial point $\mathbf{x}_1 \in \mathbb{R}^n$, any initial Lagrange multiplier $\boldsymbol{\lambda}_1 \in \mathbb{R}^p$. Put $k = 1$.
3. **repeat**
4.   Compute the gradient vector $\nabla f(\mathbf{x}_k)$ and the Hessian matrix $\nabla^2 f(\mathbf{x}_k)$.
5.   Compute $\mathbf{r}(\mathbf{x}_k, \boldsymbol{\lambda}_k) = \begin{bmatrix} \nabla f(\mathbf{x}_k) + \mathbf{A}^T \boldsymbol{\lambda}_k \\ \mathbf{A}\mathbf{x}_k - \mathbf{b} \end{bmatrix}$.
6.   **exit if** $\mathbf{A}\mathbf{x}_k = \mathbf{b}$ and $\|\mathbf{r}(\mathbf{x}_k, \boldsymbol{\lambda}_k)\|_2 < \epsilon$.
7.   Adopt the preconditioned conjugate gradient algorithm to solve the KKT equation (3.8.14), yielding the Newton step $(\Delta\mathbf{x}_k, \Delta\boldsymbol{\lambda}_k)$.
8.   Let $\mu = 1$.
9.   **while** not converged **do**
10.   Update $\mu \leftarrow \beta\mu$.
11.   **break if** $f(\mathbf{x}_k + \mu\Delta\mathbf{x}_{\text{nt},k}) < f(\mathbf{x}_k) + \alpha\mu(\nabla f(\mathbf{x}_k))^T \Delta\mathbf{x}_{\text{nt},k}$.
12.   **end while**
13.   Newton update
$$\begin{bmatrix} \mathbf{x}_{k+1} \\ \boldsymbol{\lambda}_{k+1} \end{bmatrix} = \begin{bmatrix} \mathbf{x}_k \\ \boldsymbol{\lambda}_k \end{bmatrix} + \mu \begin{bmatrix} \Delta\mathbf{x}_k \\ \Delta\boldsymbol{\lambda}_k \end{bmatrix}.$$
14. **return** $k \leftarrow k + 1$.
15. **output:** $\mathbf{x} \leftarrow \mathbf{x}_k$, $\boldsymbol{\lambda} \leftarrow \boldsymbol{\lambda}_k$.

Brief Summary of This Chapter

- This chapter mainly introduces the convex optimization theory and methods for single-objective minimization/maximization.
- Subgradient optimization algorithms have important applications in artificial intelligence.
- Multiobjective optimization involves with Pareto optimization theory that is the topic in Chap. 9 (Evolutionary Computation).

References

1. Beck, A., Teboulle, M.: A fast iterative shrinkage-thresholding algorithm for linear inverse problems. SIAM J. Imag. Sci. **2**(1), 183–202 (2009)
2. Bertsekas, D.P., Nedich, A., Ozdaglar, A.: Convex Analysis and Optimization. Athena Scientific, Nashua (2003)
3. Boyd, S., Vandenberghe, L.: Convex Optimization. Cambridge University Press, Cambridge (2004)
4. Boyd, S., Vandenberghe, L.: Subgradients. Notes for EE364b, Stanford University, Winter 2006–2007 (2008)
5. Boyd, S., Parikh, N., Chu, E., Peleato, B., Eckstein, J.: Distributed optimization and statistical learning via the alternating direction method of multipliers. Found. Trends Mach. Learn. **3**(1), 1–122 (2010)
6. Brandwood, D.H.: A complex gradient operator and its application in adaptive array theory. IEE Proc. F Commun. Radar Signal Process. **130**, 11–16 (1983)

7. Byrne, C., Censor, Y.: Proximity function minimization using multiple Bregman projections, with applications to split feasibility and Kullback–Leibler distance minimization. Ann. Oper. Res. **105**, 77–98 (2001)
8. Chen, S.S., Donoho, D.L., Saunders, M.A.: Atomic decomposition by basis pursuit. SIAM J. Sci. Comput. **20**(1), 33–61 (1998)
9. Carroll, C.W.: The created response surface technique for optimizing nonlinear restrained systems. Oper. Res. **9**, 169–184 (1961)
10. Combettes, P.L., Pesquet, J.C.: Proximal splitting methods in signal processing. In: Fixed-Point Algorithms for Inverse Problems in Science and Engineering, pp. 185–212. Springer, New York (2011)
11. Dembo, R.S., Steihaug, T.: Truncated-Newton algorithms for large-scale unconstrained optimization. Math Program. **26**, 190–212 (1983)
12. Fiacco, A.V., McCormick, G.P.: Nonlinear Programming: Sequential Unconstrained minimization Techniques. Wiley, New York (1968); or Classics Appl. Math. 4, SIAM, PA: Philadelphia, Reprint of the 1968 original, (1990)
13. Fletcher, R.: Practical Methods of Optimization, 2nd edn. Wiley, New York (1987)
14. Forsgren, A., Gill, P.E., Wright, M.H.: Interior methods for nonlinear optimization. SIAM Rev. **44**, 525–597 (2002)
15. Friedman, J., Hastie, T., Höeling, H., Tibshirani, R.: Pathwise coordinate optimization. Ann. Appl. Stat. **1**(2), 302–332 (2007)
16. Gabay, D., Mercier, B.: A dual algorithm for the solution of nonlinear variational problems via finite element approximations. Comput. Math. Appl. **2**, 17–40 (1976)
17. Glowinski, R., Marrocco, A.: Sur l'approximation, par elements finis d'ordre un, et la resolution, par penalisation-dualité, d'une classe de problems de Dirichlet non lineares. Revue Française d'Automatique, Informatique, et Recherche Opérationelle **9**, 41–76 (1975)
18. Gonzaga, C.C., Karas, E.W.: Fine tuning Nesterov's steepest descent algorithm for differentiable convex programming. Math. Program. Ser. A **138**, 141–166 (2013)
19. Hindi, H.: A tutorial on convex optimization. In: Proceedings of the 2004 American Control Conference, Boston, June 30-July 2, pp. 3252–3265 (2004)
20. Luenberger, D.: An Introduction to Linear and Nonlinear Programming, 2nd edn. Addison-Wesley, Boston (1989)
21. Magnus, J.R., Neudecker, H.: Matrix Differential Calculus with Applications in Statistics and Econometrics, rev. edn. Wiley, Chichester (1999)
22. Michalewicz, Z., Dasgupta, D., Le Riche R., Schoenauer M.: Evolutionary algorithms for constrained engineering problems. Comput. Ind. Eng. J. **30**, 851–870 (1996)
23. Moreau, J.J.: Fonctions convexes duales et points proximaux dans un espace Hilbertien. Rep. Paris Acad. Sci. Ser. A **255**, 2897–2899 (1962)
24. Nesterov, Y.: A method for solving a convex programming problem with rate of convergence $O(\frac{1}{k^2})$. Soviet Math. Dokl. **269**(3), 543–547 (1983)
25. Nesterov, Y.: Introductory Lectures on Convex Optimization: A Basic Course. Kluwer Academic, Boston (2004)
26. Nesterov, Y.: Smooth minimization of nonsmooth functions (CORE Discussion Paper 2003/12, CORE 2003). Math. Program. **103**(1), 127–152 (2005)
27. Nesterov, Y., Nemirovsky, A.: A general approach to polynomial-time algorithms design for convex programming. Report, Central Economical and Mathematical Institute, USSR Academy of Sciences, Moscow (1988)
28. Nocedal, J., Wright, S.J.: Numerical Optimization. Springer, New York (1999)
29. Ortega, J.M., Rheinboldt, W.C.: Iterative Solutions of Nonlinear Equations in Several Variables, pp. 253–255. Academic, New York (1970)
30. Parikh, N., Bord, S.: Proximal algorithms. Found. Trends Optim. **1**(3), 123–231 (2013)
31. Polyak, B.T.: Introduction to Optimization. Optimization Software, New York (1987)
32. Scutari, G., Palomar, D.P., Facchinei, F., Pang, J.S.: Convex optimization, game theory, and variational inequality theory. IEEE Signal Process. Mag. **27**(3), 35–49 (2010)
33. Syau, Y.R.: A note on convex functions. Int. J. Math. Math. Sci. **22**(3), 525–534 (1999)

References

34. Tibshirani, R.: Regression shrinkage and selection via the lasso. J. R. Stat. Soc. B **58**, 267–288 (1996)
35. Vandenberghe, L.: Lecture Notes for EE236C (Spring 2019), Section 6 The Proximal Mapping, UCLA (2019)
36. Yeniay, O., Ankara, B.: Penalty function methods for constrained optimization with genetic algorithms. Math. Comput. Appl. **10**(1), 45–56 (2005)
37. Yu, K., Zhang, T., Gong, Y.: Nonlinear learning using local coordinate coding. In: Advances in Neural Information Processing Systems, **22**, 2223–2231 (2009)
38. Zhang, X.D.: Matrix Analysis and Applications. Cambridge University Press, Cambridge (2017)

Chapter 4
Solution of Linear Systems

One of the most common problems in science and engineering is to solve the matrix equation $\mathbf{Ax} = \mathbf{b}$ for finding $\mathbf{x}$ when the data matrix $\mathbf{A}$ and the data vector $\mathbf{b}$ are given.

Matrix equations can be divided into three types.

1. *Well-determined equation:* If $m = n$ and $\text{rank}(\mathbf{A}) = n$, i.e., the matrix $\mathbf{A}$ is nonsingular, then the matrix equation $\mathbf{Ax} = \mathbf{b}$ is said to be well-determined.
2. *Under-determined equation:* The matrix equation $\mathbf{Ax} = \mathbf{b}$ is under-determined if the number of linearly independent equations is less than the number of independent unknowns.
3. *Over-determined equation:* The matrix equation $\mathbf{Ax} = \mathbf{b}$ is known as over-determined if the number of linearly independent equations is larger than the number of independent unknowns.

This chapter is devoted to describe singular value decomposition (SVD) together with Gauss elimination and conjugate gradient methods for solving well-determined matrix equations followed by Tikhonov regularization and total least squares for solving over-determined matrix equations, and the Lasso and LARS methods for solving under-determined matrix equations.

4.1 Gauss Elimination

First consider solution of well-determined equations. In well-determined equation, the number of independent equations and the number of independent unknowns are the same so that the solution of this system of equations is uniquely determined. The exact solution of a well-determined matrix equation $\mathbf{Ax} = \mathbf{b}$ is given by $\mathbf{x} = \mathbf{A}^{-1}\mathbf{b}$. A well-determined equation is a consistent equation.

In this section, we deal with Gauss elimination method and conjugate gradient methods for solving an over-determined matrix equation $\mathbf{Ax} = \mathbf{b}$ and finding the inverse of a singular matrix. This method performs only elementary row operations to the augmented matrix $\mathbf{B} = [\mathbf{A}, \mathbf{b}]$.

4.1.1 Elementary Row Operations

When solving an $m \times n$ system of linear equations, it is useful to reduce equations. The principle of reduction processes is to keep the solutions of the linear systems unchanged.

Definition 4.1 (Equivalent Systems) Two linear systems of equations in n unknowns are said to be *equivalent systems* if they have the same sets of solutions.

To transform a given $m \times n$ matrix equation $\mathbf{A}_{m \times n}\mathbf{x}_n = \mathbf{b}_m$ into an equivalent system, a simple and efficient way is to apply a sequence of elementary operations on the given matrix equation.

Definition 4.2 (Elementary Row Operations) The following three types of operation on the rows of a system of linear equations are called *elementary row operations*.

- *Type I:* Interchange any two equations, say the pth and qth equations; this is denoted by $R_p \leftrightarrow R_q$.
- *Type II:* Multiply the pth equation by a nonzero number α; this is denoted by $\alpha R_p \to R_p$.
- *Type III:* Add β times the pth equation to the qth equation; this is denoted by $\beta R_p + R_q \to R_q$.

Clearly, any above type of elementary row operation does not change the solution of a system of linear equations, so after elementary row operations, the reduced system of linear equations and the original system of linear equations are equivalent.

Definition 4.3 (Leading Entry) The leftmost nonzero entry of a nonzero row is called the *leading entry* of the row. If the leading entry is equal to 1 then it is said to be the *leading-1 entry*.

As a matter of fact, any elementary operation on an $m \times n$ system of equations $\mathbf{Ax} = \mathbf{b}$ is equivalent to the same type of elementary operation on the *augmented matrix* $\mathbf{B} = [\mathbf{A}, \mathbf{b}]$, where the column vector $\mathbf{b}$ is written alongside $\mathbf{A}$. Therefore, performing elementary row operations on a system of linear equations $\mathbf{Ax} = \mathbf{b}$ can be in practice implemented by using the same elementary row operations on the augmented matrix $\mathbf{B} = [\mathbf{A}, \mathbf{b}]$.

The discussions above show that if, after a sequence of elementary row operations, the augmented matrix $\mathbf{B}_{m \times (n+1)}$ becomes another simpler matrix $\mathbf{C}_{m \times (n+1)}$ then two matrices are *row equivalent* for the same solution of linear system.

For the convenience of solving a system of linear equations, the final row equivalent matrix should be of echelon form.

Definition 4.4 (Echelon Matrix) A matrix is said to be an *echelon matrix* if it has the following forms:

- all rows with all entries zero are located at the bottom of the matrix;
- the leading entry of each nonzero row appears always to the right of the leading entry of the nonzero row above;
- all entries below the leading entry of the same column are equal to zero.

Definition 4.5 (Reduced Row-Echelon Form Matrix [22]) An echelon matrix $\mathbf{B}$ is said to be a *reduced row-echelon form (RREF) matrix* if the leading entry of each nonzero row is a leading-1 entry, and each leading-1 entry is the only nonzero entry in the column in which it is located.

Theorem 4.1 *Any $m \times n$ matrix $\mathbf{A}$ is row equivalent to one and only one matrix in reduced row-echelon form.*

Proof See [27, Appendix A].

Definition 4.6 (Pivot Column [27, p. 15]) A *pivot position* of an $m \times n$ matrix $\mathbf{A}$ is the position of some leading entry of its echelon form. Each column containing a pivot position is called a *pivot column* of the matrix $\mathbf{A}$.

4.1.2 Gauss Elimination for Solving Matrix Equations

Elementary row operations can be used to solve matrix equations and perform matrix inversion.

Consider how to solve an $n \times n$ matrix equation $\mathbf{Ax} = \mathbf{b}$, where the inverse matrix $\mathbf{A}^{-1}$ of the matrix $\mathbf{A}$ exists. We hope that the solution vector $\mathbf{x} = \mathbf{A}^{-1}\mathbf{b}$ can be obtained by using elementary row operations.

First use the matrix $\mathbf{A}$ and the vector $\mathbf{b}$ to construct the $n \times (n+1)$ augmented matrix $\mathbf{B} = [\mathbf{A}, \mathbf{b}]$. Since the solution $\mathbf{x} = \mathbf{A}^{-1}\mathbf{b}$ can be written as a new matrix equation $\mathbf{Ix} = \mathbf{A}^{-1}\mathbf{b}$, we get a new augmented matrix $\mathbf{C} = [\mathbf{I}, \mathbf{A}^{-1}\mathbf{b}]$ associated with the solution $\mathbf{x} = \mathbf{A}^{-1}\mathbf{b}$. Hence, we can write the solution process for the two matrix equations $\mathbf{Ax} = \mathbf{b}$ and $\mathbf{x} = \mathbf{A}^{-1}\mathbf{b}$, respectively, as follows:

$$\text{matrix equations} \quad \mathbf{Ax} = \mathbf{b} \xrightarrow{\text{Elementary row operations}} \mathbf{x} = \mathbf{A}^{-1}\mathbf{b},$$

$$\text{augmented matrices} \quad [\mathbf{A}, \mathbf{b}] \xrightarrow{\text{Elementary row operations}} [\mathbf{I}, \mathbf{A}^{-1}\mathbf{b}].$$

This implies that, after suitable elementary row operations on the augmented matrix $[\mathbf{A}, \mathbf{b}]$, if the left-hand part of the new augmented matrix is an $n \times n$ identity matrix $\mathbf{I}$, then the $(n+1)$th column gives the solution $\mathbf{x} = \mathbf{A}^{-1}\mathbf{b}$ of the original equation $\mathbf{A}\mathbf{x} = \mathbf{b}$ directly. This method is called the *Gauss* or *Gauss–Jordan elimination method*.

Any $m \times n$ complex matrix equation $\mathbf{A}\mathbf{x} = \mathbf{b}$ can be written as

$$(\mathbf{A}_r + j\mathbf{A}_i)(\mathbf{x}_r + j\mathbf{x}_i) = \mathbf{b}_r + j\mathbf{b}_i, \tag{4.1.1}$$

where $\mathbf{A}_r, \mathbf{x}_r, \mathbf{b}_r$ and $\mathbf{A}_i, \mathbf{x}_i, \mathbf{b}_i$ are the real and imaginary parts of $\mathbf{A}, \mathbf{x}, \mathbf{b}$, respectively. Expanding the above equation to yield

$$\mathbf{A}_r\mathbf{x}_r - \mathbf{A}_i\mathbf{x}_i = \mathbf{b}_r, \tag{4.1.2}$$

$$\mathbf{A}_i\mathbf{x}_r + \mathbf{A}_r\mathbf{x}_i = \mathbf{b}_i. \tag{4.1.3}$$

The above equations can be combined into

$$\begin{bmatrix} \mathbf{A}_r & -\mathbf{A}_i \\ \mathbf{A}_i & \mathbf{A}_r \end{bmatrix} \begin{bmatrix} \mathbf{x}_r \\ \mathbf{x}_i \end{bmatrix} = \begin{bmatrix} \mathbf{b}_r \\ \mathbf{b}_i \end{bmatrix}. \tag{4.1.4}$$

Thus, m complex-valued equations with n complex unknowns become $2m$ real-valued equations with $2n$ real unknowns.

In particular, if $m = n$, then we have

$$\text{complex matrix equation } \mathbf{A}\mathbf{x} = \mathbf{b} \xrightarrow[\text{operations}]{\text{Elementary row}} \mathbf{x} = \mathbf{A}^{-1}\mathbf{b},$$

$$\text{augmented matrix } \begin{bmatrix} \mathbf{A}_r & -\mathbf{A}_i & \mathbf{b}_r \\ \mathbf{A}_i & \mathbf{A}_r & \mathbf{b}_i \end{bmatrix} \xrightarrow[\text{operations}]{\text{Elementary row}} \begin{bmatrix} \mathbf{I}_n & \mathbf{O}_n & \mathbf{x}_r \\ \mathbf{O}_n & \mathbf{I}_n & \mathbf{x}_i \end{bmatrix}.$$

This shows that if we write the $n \times (n+1)$ complex augmented matrix $[\mathbf{A}, \mathbf{b}]$ as a $2n \times (2n+1)$ real augmented matrix and perform elementary row operations to make its left-hand side become a $2n \times 2n$ identity matrix, then the upper and lower halves of the $(2n+1)$th column give, respectively, the real and imaginary parts of the complex solution vector $\mathbf{x}$ of the original complex matrix equation $\mathbf{A}\mathbf{x} = \mathbf{b}$.

4.1.3 Gauss Elimination for Matrix Inversion

Consider the inversion operation on an $n \times n$ nonsingular matrix $\mathbf{A}$. This problem can be modeled as an $n \times n$ matrix equation $\mathbf{A}\mathbf{X} = \mathbf{I}$ whose solution $\mathbf{X}$ is the inverse matrix of $\mathbf{A}$. It is easily seen that the augmented matrix of the matrix equation $\mathbf{A}\mathbf{X} = \mathbf{I}$ is $[\mathbf{A}, \mathbf{I}]$, whereas the augmented matrix of the solution equation $\mathbf{I}\mathbf{X} = \mathbf{A}^{-1}$ is

4.1 Gauss Elimination

[$\mathbf{I}, \mathbf{A}^{-1}$]. Hence, we have the following relations:

$$\text{matrix equation} \quad \mathbf{AX} = \mathbf{I} \quad \xrightarrow{\text{Elementary row operations}} \quad \mathbf{X} = \mathbf{A}^{-1},$$

$$\text{augmented matrix} \quad [\mathbf{A}, \mathbf{I}] \quad \xrightarrow{\text{Elementary row operations}} \quad [\mathbf{I}, \mathbf{A}^{-1}].$$

This result tells us that if we use elementary row operations on the $n \times 2n$ augmented matrix $[\mathbf{A}, \mathbf{I}]$ so that its left-hand part becomes an $n \times n$ identity matrix then the right-hand part yields the inverse $\mathbf{A}^{-1}$ of the given $n \times n$ matrix $\mathbf{A}$ directly. This operation is called the *Gauss elimination method for matrix inversion*.

Example 4.1 Use the Gauss elimination method to find the inverse of the matrix

$$\mathbf{A} = \begin{bmatrix} 1 & 1 & 2 \\ 3 & 4 & -1 \\ -1 & 1 & 1 \end{bmatrix}.$$

Perform elementary row operations on the augmented matrix of $[\mathbf{A}, \mathbf{I}]$ to yield the following results:

$$\begin{bmatrix} 1 & 1 & 2 & 1 & 0 & 0 \\ 3 & 4 & -1 & 0 & 1 & 0 \\ -1 & 1 & 1 & 0 & 0 & 1 \end{bmatrix} \xrightarrow{(-3)R_1+R_2 \to R_2} \begin{bmatrix} 1 & 1 & 2 & 1 & 0 & 0 \\ 0 & 1 & -7 & -3 & 1 & 0 \\ -1 & 1 & 1 & 0 & 0 & 1 \end{bmatrix} \xrightarrow{R_1+R_3 \to R_3}$$

$$\begin{bmatrix} 1 & 1 & 2 & 1 & 0 & 0 \\ 0 & 1 & -7 & -3 & 1 & 0 \\ 0 & 2 & 3 & 1 & 0 & 1 \end{bmatrix} \xrightarrow{(-1)R_2+R_1 \to R_1} \begin{bmatrix} 1 & 0 & 9 & 4 & -1 & 0 \\ 0 & 1 & -7 & -3 & 1 & 0 \\ 0 & 2 & 3 & 1 & 0 & 1 \end{bmatrix} \xrightarrow{(-2)R_2+R_3 \to R_3}$$

$$\begin{bmatrix} 1 & 0 & 9 & 4 & -1 & 0 \\ 0 & 1 & -7 & -3 & 1 & 0 \\ 0 & 0 & 17 & 7 & -2 & 1 \end{bmatrix} \xrightarrow{\frac{1}{17}R_3 \to R_3} \begin{bmatrix} 1 & 0 & 9 & 4 & -1 & 0 \\ 0 & 1 & -7 & -3 & 1 & 0 \\ 0 & 0 & 1 & \frac{7}{17} & \frac{-2}{17} & \frac{1}{17} \end{bmatrix} \xrightarrow{(-9)R_3+R_1 \to R_1}$$

$$\begin{bmatrix} 1 & 0 & 0 & \frac{5}{17} & \frac{1}{17} & \frac{-9}{17} \\ 0 & 1 & -7 & -3 & 1 & 0 \\ 0 & 0 & 1 & \frac{7}{17} & \frac{-2}{17} & \frac{1}{17} \end{bmatrix} \xrightarrow{7R_3+R_2 \to R_2} \begin{bmatrix} 1 & 0 & 0 & \frac{5}{17} & \frac{1}{17} & \frac{-9}{17} \\ 0 & 1 & 0 & \frac{-2}{17} & \frac{3}{17} & \frac{7}{17} \\ 0 & 0 & 1 & \frac{7}{17} & \frac{-2}{17} & \frac{1}{17} \end{bmatrix}.$$

That is to say,

$$\begin{bmatrix} 1 & 1 & 2 \\ 3 & 4 & -1 \\ -1 & 1 & 1 \end{bmatrix}^{-1} = \frac{1}{17} \begin{bmatrix} 5 & 1 & -9 \\ -2 & 3 & 7 \\ 7 & -2 & 1 \end{bmatrix}.$$

For the matrix equation

$$x_1 + x_2 + 2x_3 = 6,$$
$$3x_1 + 4x_2 - x_3 = 5,$$
$$-x_1 + x_2 + x_3 = 2,$$

its solution

$$\mathbf{x} = \mathbf{A}^{-1}\mathbf{b} = \begin{bmatrix} 1 & 1 & 2 \\ 3 & 4 & -1 \\ -1 & 1 & 1 \end{bmatrix}^{-1} \begin{bmatrix} 6 \\ 5 \\ 2 \end{bmatrix} = \frac{1}{17} \begin{bmatrix} 5 & 1 & -9 \\ -2 & 3 & 7 \\ 7 & -2 & 1 \end{bmatrix} \begin{bmatrix} 6 \\ 5 \\ 2 \end{bmatrix} = \begin{bmatrix} 1 \\ 1 \\ 2 \end{bmatrix}.$$

Now, we consider the inversion of an $n \times n$ nonsingular complex matrix $\mathbf{A}$. Then, its inversion can be modeled as the complex matrix equation $(\mathbf{A}_r + j\mathbf{A}_i)(\mathbf{X}_r + j\mathbf{X}_i) = \mathbf{I}$. This complex equation can be rewritten as

$$\begin{bmatrix} \mathbf{A}_r & -\mathbf{A}_i \\ \mathbf{A}_i & \mathbf{A}_r \end{bmatrix} \begin{bmatrix} \mathbf{X}_r \\ \mathbf{X}_i \end{bmatrix} = \begin{bmatrix} \mathbf{I}_n \\ \mathbf{O}_n \end{bmatrix}, \tag{4.1.5}$$

from which we get the following relation:

$$\text{complex matrix equation } \mathbf{A}\mathbf{X} = \mathbf{I} \xrightarrow{\text{Elementary row operations}} \mathbf{X} = \mathbf{A}^{-1},$$

$$\text{augmented matrix } \begin{bmatrix} \mathbf{A}_r & -\mathbf{A}_i & \mathbf{I}_n \\ \mathbf{A}_i & \mathbf{A}_r & \mathbf{O}_n \end{bmatrix} \xrightarrow{\text{Elementary row operations}} \begin{bmatrix} \mathbf{I}_n & \mathbf{O}_n & \mathbf{X}_r \\ \mathbf{O}_n & \mathbf{I}_n & \mathbf{X}_i \end{bmatrix}.$$

That is to say, after making elementary row operations on the $2n \times 3n$ augmented matrix to transform its left-hand side to a $2n \times 2n$ identity matrix, then the upper and lower halves of the $2n \times n$ matrix on the right give, respectively, the real and imaginary parts of the inverse matrix $\mathbf{A}^{-1}$ of the complex matrix $\mathbf{A}$.

4.2 Conjugate Gradient Methods

Given a matrix equation $\mathbf{Ax} = \mathbf{b}$, where $\mathbf{A} \in \mathbb{R}^{n \times n}$ is a nonsingular matrix. This matrix equation can be equivalently written as

$$\mathbf{x} = (\mathbf{I} - \mathbf{A})\mathbf{x} + \mathbf{b}, \tag{4.2.1}$$

which inspires the following iteration algorithm:

$$\mathbf{x}_{k+1} = (\mathbf{I} - \mathbf{A})\mathbf{x}_k + \mathbf{b}. \tag{4.2.2}$$

4.2 Conjugate Gradient Methods

This algorithm is known as the *Richardson iteration* and can be rewritten in the more general form

$$\mathbf{x}_{k+1} = \mathbf{M}\mathbf{x}_k + \mathbf{c}, \qquad (4.2.3)$$

where $\mathbf{M}$ is an $n \times n$ matrix, called the *iteration matrix*.

An iteration in the form of Eq. (4.2.3) is termed as a *stationary iterative method* and is not as effective as a *nonstationary iterative method*.

Nonstationary iteration methods comprise a class of iteration methods in which $\mathbf{x}_{k+1}$ is related to the former iteration solutions $\mathbf{x}_k, \mathbf{x}_{k-1}, \ldots, \mathbf{x}_0$. The most typical nonstationary iteration method is the *Krylov subspace method*, described by

$$\mathbf{x}_{k+1} = \mathbf{x}_0 + \mathcal{K}_k, \qquad (4.2.4)$$

where

$$\mathcal{K}_k = \mathrm{Span}(\mathbf{r}_0, \mathbf{A}\mathbf{r}_0, \ldots, \mathbf{A}^{k-1}\mathbf{r}_0) \qquad (4.2.5)$$

is called the kth iteration *Krylov subspace*; $\mathbf{x}_0$ is the initial iteration value and $\mathbf{r}_0$ denotes the initial residual vector.

Krylov subspace methods have various forms, of which the three most common are the conjugate gradient method, the biconjugate gradient method, and the preconditioned conjugate gradient method.

4.2.1 Conjugate Gradient Algorithm

The conjugate gradient method uses $\mathbf{r}_0 = \mathbf{A}\mathbf{x}_0 - \mathbf{b}$ as the initial residual vector. The applicable object of the conjugate gradient method is limited to the symmetric positive definite equation, $\mathbf{A}\mathbf{x} = \mathbf{b}$, where $\mathbf{A}$ is an $n \times n$ symmetric positive definite matrix.

The nonzero vector combination $\{\mathbf{p}_0, \mathbf{p}_1, \ldots, \mathbf{p}_k\}$ is said to be $\mathbf{A}$-orthogonal or $\mathbf{A}$-conjugate if

$$\mathbf{p}_i^T \mathbf{A} \mathbf{p}_j = 0, \quad \forall\, i \neq j. \qquad (4.2.6)$$

This property is known as $\mathbf{A}$-*orthogonality* or $\mathbf{A}$-*conjugacy*. Obviously, if $\mathbf{A} = \mathbf{I}$, then the $\mathbf{A}$-conjugacy reduces to general orthogonality.

All algorithms adopting the conjugate vector as the update direction are known as *conjugate direction algorithms*. If the conjugate vectors $\mathbf{p}_0, \mathbf{p}_1, \ldots, \mathbf{p}_{n-1}$ are not predetermined, but are updated by the gradient descent method in the updating process, then we say that the minimization algorithm for the objective function $f(\mathbf{x})$ is a *conjugate gradient algorithm*.

Algorithm 4.1 gives a conjugate gradient algorithm.

Algorithm 4.1 Conjugate gradient algorithm [16]
1. **input:** $\mathbf{A} = \mathbf{A}^T \in \mathbb{R}^{n \times n}$, $\mathbf{b} \in \mathbb{R}^n$, the largest iteration step number $k_{\max}$, the allowed error ϵ.
2. **initialization:** Choose $\mathbf{x}_0 \in \mathbb{R}^n$, and let $\mathbf{r} = \mathbf{A}\mathbf{x}_0 - \mathbf{b}$ and $\rho_0 = \|\mathbf{r}\|_2^2$.
3. **repeat**
4. If $k = 1$ then $\mathbf{p} = \mathbf{r}$. Otherwise, let $\beta = \rho_{k-1}/\rho_{k-2}$ and $\mathbf{p} = \mathbf{r} + \beta \mathbf{p}$.
5. $\mathbf{w} = \mathbf{A}\mathbf{p}$.
6. $\alpha = \rho_{k-1}/(\mathbf{p}^T \mathbf{w})$.
7. $\mathbf{x} = \mathbf{x} + \alpha \mathbf{p}$.
8. $\mathbf{r} = \mathbf{r} - \alpha \mathbf{w}$.
9. $\rho_k = \|\mathbf{r}\|_2^2$.
10. **exit if** $\sqrt{\rho_k} < \epsilon \|\mathbf{b}\|_2$ or $k = k_{\max}$.
11. **return** $k \leftarrow k + 1$.
12. **output:** $\mathbf{x} \leftarrow \mathbf{x}_k$.

From Algorithm 4.1 it can be seen that in the iteration process of the conjugate gradient algorithm, the solution of the matrix equation $\mathbf{A}\mathbf{x} = \mathbf{b}$ is given by

$$\mathbf{x}_k = \sum_{i=1}^{k} \alpha_i \mathbf{p}_i = \sum_{i=1}^{k} \frac{\langle \mathbf{r}_{i-1}, \mathbf{r}_{i-1} \rangle}{\langle \mathbf{p}_i, \mathbf{A}\mathbf{p}_i \rangle} \mathbf{p}_i, \qquad (4.2.7)$$

that is, $\mathbf{x}_k$ belongs to the kth Krylov subspace

$$\mathbf{x}_k \in \mathrm{Span}\{\mathbf{p}_1, \mathbf{p}_2, \ldots, \mathbf{p}_k\} = \mathrm{Span}\{\mathbf{r}_0, \mathbf{A}\mathbf{r}_0, \ldots, \mathbf{A}^{k-1}\mathbf{r}_0\}.$$

The fixed-iteration method requires the updating of the iteration matrix $\mathbf{M}$, but no matrix needs to be updated in Algorithm 4.1. Hence, the Krylov subspace method is also called the matrix-free method [24].

4.2.2 Biconjugate Gradient Algorithm

If the matrix $\mathbf{A}$ is not a real symmetric matrix, then we can use the *biconjugate gradient method* of Fletcher [13] for solving the matrix equation $\mathbf{A}\mathbf{x} = \mathbf{b}$. As the name implies, there are two search directions, $\mathbf{p}_i$ and $\bar{\mathbf{p}}_j$, that are $\mathbf{A}$-conjugate in this method:

$$\begin{aligned}
\bar{\mathbf{p}}_i^T \mathbf{A} \mathbf{p}_j &= \mathbf{p}_i^T \mathbf{A} \bar{\mathbf{p}}_j = 0, \quad i \neq j; \\
\bar{\mathbf{r}}_i^T \mathbf{r}_j &= \mathbf{r}_i^T \bar{\mathbf{r}}_j = 0, \quad i \neq j; \qquad (4.2.8) \\
\bar{\mathbf{r}}_i^T \mathbf{p}_j &= \mathbf{r}_i^T \bar{\mathbf{p}}_j = 0, \quad j < i.
\end{aligned}$$

Here, $\mathbf{r}_i$ and $\bar{\mathbf{r}}_j$ are the residual vectors associated with search directions $\mathbf{p}_i$ and $\bar{\mathbf{p}}_j$, respectively. Algorithm 4.2 shows a biconjugate gradient algorithm.

Algorithm 4.2 Biconjugate gradient algorithm [13]
1. **input:** The matrix $\mathbf{A}$.
2. **initialization:** $\mathbf{p}_1 = \mathbf{r}_1, \bar{\mathbf{p}}_1 = \bar{\mathbf{r}}_1$.
3. **repeat**
4. $\alpha_k = \bar{\mathbf{r}}_k^T \mathbf{r}_k / (\bar{\mathbf{p}}_k^T \mathbf{A} \mathbf{p}_k)$.
5. $\mathbf{r}_{k+1} = \mathbf{r}_k - \alpha_k \mathbf{A} \mathbf{p}_k$.
6. $\bar{\mathbf{r}}_{k+1} = \bar{\mathbf{r}}_k - \alpha_k \mathbf{A}^T \bar{\mathbf{p}}_k$.
7. $\beta_k = \bar{\mathbf{r}}_{k+1}^T \mathbf{r}_{k+1} / (\bar{\mathbf{r}}_k^T \mathbf{r}_k)$.
8. $\mathbf{p}_{k+1} = \mathbf{r}_{k+1} + \beta_k \mathbf{p}_k$.
9. $\bar{\mathbf{p}}_{k+1} = \bar{\mathbf{r}}_{k+1} + \beta_k \bar{\mathbf{p}}_k$.
10.   **exit if** $k = k_{\max}$.
11.   **return** $k \leftarrow k + 1$.
12. **output:** $\mathbf{x} \leftarrow \mathbf{x}_k + \alpha_k \mathbf{p}_k$, $\bar{\mathbf{x}}_{k+1} \leftarrow \bar{\mathbf{x}}_k + \alpha_k \bar{\mathbf{p}}_k$.

4.2.3 Preconditioned Conjugate Gradient Algorithm

Consider the symmetric indefinite saddle point problem

$$\begin{bmatrix} \mathbf{A} & \mathbf{B}^T \\ \mathbf{B} & \mathbf{O} \end{bmatrix} \begin{bmatrix} \mathbf{x} \\ \mathbf{q} \end{bmatrix} = \begin{bmatrix} \mathbf{f} \\ \mathbf{g} \end{bmatrix},$$

where $\mathbf{A}$ is an $n \times n$ real symmetric positive definite matrix, $\mathbf{B}$ is an $m \times n$ real matrix with full row rank m ($\leq n$), and $\mathbf{O}$ is an $m \times m$ zero matrix. Bramble and Pasciak [4] developed a preconditioned conjugate gradient iteration method for solving the above problem.

The basic idea of the *preconditioned conjugate gradient iteration* is as follows: through a clever choice of the scalar product form, the preconditioned saddle matrix becomes a symmetric positive definite matrix.

To simplify the discussion, we assume that the matrix equation with large condition number $\mathbf{A}\mathbf{x} = \mathbf{b}$ needs to be converted into a new symmetric positive definite equation. To this end, let $\mathbf{M}$ be a symmetric positive definite matrix that can approximate the matrix $\mathbf{A}$ and for which it is easier to find the inverse matrix. Hence, the original matrix equation $\mathbf{A}\mathbf{x} = \mathbf{b}$ is converted into $\mathbf{M}^{-1}\mathbf{A}\mathbf{x} = \mathbf{M}^{-1}\mathbf{b}$ such that the new and original matrix equations have the same solution. However, in the new matrix equation $\mathbf{M}^{-1}\mathbf{A}\mathbf{x} = \mathbf{M}^{-1}\mathbf{b}$ there is a hidden danger: $\mathbf{M}^{-1}\mathbf{A}$ is generally not either symmetric or positive definite even if both $\mathbf{M}$ and $\mathbf{A}$ are symmetric positive definite. Therefore, it is unreliable to use the matrix $\mathbf{M}^{-1}$ directly as the preprocessor of the matrix equation $\mathbf{A}\mathbf{x} = \mathbf{b}$.

Let $\mathbf{S}$ be the square root of a symmetric matrix $\mathbf{M}$, i.e., $\mathbf{M} = \mathbf{S}\mathbf{S}^T$, where $\mathbf{S}$ is symmetric positive definite. Now, use $\mathbf{S}^{-1}$ instead of $\mathbf{M}^{-1}$ as the *preprocessor* to convert the original matrix equation $\mathbf{A}\mathbf{x} = \mathbf{b}$ to $\mathbf{S}^{-1}\mathbf{A}\mathbf{x} = \mathbf{S}^{-1}\mathbf{b}$. If $\mathbf{x} = \mathbf{S}^{-T}\hat{\mathbf{x}}$, then the preconditioned matrix equation is given by

$$\mathbf{S}^{-1}\mathbf{A}\mathbf{S}^{-T}\hat{\mathbf{x}} = \mathbf{S}^{-1}\mathbf{b}. \tag{4.2.9}$$

Compared with the matrix $\mathbf{M}^{-1}\mathbf{A}$, which is not symmetric positive definite, $\mathbf{S}^{-1}\mathbf{A}\mathbf{S}^{-T}$ must be symmetric positive definite if $\mathbf{A}$ is symmetric positive definite. The symmetry of $\mathbf{S}^{-1}\mathbf{A}\mathbf{S}^{-T}$ is easily seen, its positive definiteness can be verified as follows: by checking the quadratic function it easily follows that $\mathbf{y}^T(\mathbf{S}^{-1}\mathbf{A}\mathbf{S}^{-T})\mathbf{y} = \mathbf{z}^T\mathbf{A}\mathbf{z}$, where $\mathbf{z} = \mathbf{S}^{-T}\mathbf{y}$. Because $\mathbf{A}$ is symmetric positive definite, we have $\mathbf{z}^T\mathbf{A}\mathbf{z} > 0, \forall \mathbf{z} \neq \mathbf{0}$, and thus $\mathbf{y}^T(\mathbf{S}^{-1}\mathbf{A}\mathbf{S}^{-T})\mathbf{y} > 0, \forall \mathbf{y} \neq \mathbf{0}$. That is to say, $\mathbf{S}^{-1}\mathbf{A}\mathbf{S}^{-T}$ must be positive definite.

At this point, the conjugate gradient method can be applied to solve the matrix equation (4.2.9) in order to get $\hat{\mathbf{x}}$, and then we can recover $\mathbf{x}$ via $\mathbf{x} = \mathbf{S}^{-T}\hat{\mathbf{x}}$.

Algorithm 4.3 shows a *preconditioned conjugate gradient* (PCG) algorithm with a preprocessor, developed in [37].

Algorithm 4.3 PCG algorithm with preprocessor [37]

1. **input:** $\mathbf{A}$, $\mathbf{b}$, preprocessor $\mathbf{S}^{-1}$, maximal number of iterations $k_{\max}$, and allowed error $\epsilon < 1$.
2. **initialization:** $k = 0$, $\mathbf{r}_0 = \mathbf{A}\mathbf{x} - \mathbf{b}$, $\mathbf{d}_0 = \mathbf{S}^{-1}\mathbf{r}_0$, $\delta_{\text{new}} = \mathbf{r}_0^T\mathbf{d}_0$, $\delta_0 = \delta_{\text{new}}$.
3. $\mathbf{q}_{k+1} = \mathbf{A}\mathbf{d}_k$.
4. $\alpha = \delta_{\text{new}}/(\mathbf{d}_k^T\mathbf{q}_{k+1})$.
5. $\mathbf{x}_{k+1} = \mathbf{x}_k + \alpha \mathbf{d}_k$.
6. If k can be divided exactly by 50, then $\mathbf{r}_{k+1} = \mathbf{b} - \mathbf{A}\mathbf{x}_k$. Otherwise, update $\mathbf{r}_{k+1} = \mathbf{r}_k - \alpha \mathbf{q}_{k+1}$.
7. $\mathbf{s}_{k+1} = \mathbf{S}^{-1}\mathbf{r}_{k+1}$.
8. $\delta_{\text{old}} = \delta_{\text{new}}$.
9. $\delta_{\text{new}} = \mathbf{r}_{k+1}^T\mathbf{s}_{k+1}$.
10. **exit if** $k = k_{\max}$ or $\delta_{\text{new}} < \epsilon^2 \delta_0$.
11. $\beta = \delta_{\text{new}}/\delta_{\text{old}}$.
12. $\mathbf{d}_{k+1} = \mathbf{s}_{k+1} + \beta \mathbf{d}_k$.
13. **return** $k \leftarrow k + 1$.
14. **output:** $\mathbf{x} \leftarrow \mathbf{x}_k$.

The use of a preprocessor can be avoided, because there are correspondences between the variables of the matrix equations $\mathbf{A}\mathbf{x} = \mathbf{b}$ and $\mathbf{S}^{-1}\mathbf{A}\mathbf{S}^{-T}\hat{\mathbf{x}} = \mathbf{S}^{-1}\mathbf{b}$, as follows [24]:

$$\mathbf{x}_k = \mathbf{S}^{-1}\hat{\mathbf{x}}_k, \quad \mathbf{r}_k = \mathbf{S}\hat{\mathbf{r}}_k, \quad \mathbf{p}_k = \mathbf{S}^{-1}\hat{\mathbf{p}}_k, \quad \mathbf{z}_k = \mathbf{S}^{-1}\hat{\mathbf{r}}_k.$$

On the basis of these correspondences, it is easy to develop a PCG algorithm without preprocessor [24], as shown in Algorithm 4.4.

A wonderful introduction to the conjugate gradient method is presented in [37].

Regarding the complex matrix equation $\mathbf{A}\mathbf{x} = \mathbf{b}$, where $\mathbf{A} \in \mathbb{C}^{n \times n}$, $\mathbf{x} \in \mathbb{C}^n$, $\mathbf{b} \in \mathbb{C}^n$, we can write it as the following real matrix equation:

$$\begin{bmatrix} \mathbf{A}_R & -\mathbf{A}_I \\ \mathbf{A}_I & \mathbf{A}_R \end{bmatrix} \begin{bmatrix} \mathbf{x}_R \\ \mathbf{x}_I \end{bmatrix} = \begin{bmatrix} \mathbf{b}_R \\ \mathbf{b}_I \end{bmatrix}. \tag{4.2.10}$$

Clearly, if $\mathbf{A} = \mathbf{A}_R + j\mathbf{A}_I$ is a Hermitian positive definite matrix, then Eq. (4.2.10) is a symmetric positive definite matrix equation. Hence, $(\mathbf{x}_R, \mathbf{x}_I)$ can be solved by

Algorithm 4.4 PCG algorithm without preprocessor [24]

1. **input:** $\mathbf{A} = \mathbf{A}^T \in \mathbb{R}^{n \times n}$, $\mathbf{b} \in \mathbb{R}^n$, maximal iteration number $k_{\max}$, and allowed error ϵ.
2. **initialization:** $\mathbf{x}_0 \in \mathbb{R}^n$, $\mathbf{r}_0 = \mathbf{A}\mathbf{x}_0 - \mathbf{b}$, $\rho_0 = \|\mathbf{r}\|_2^2$ and $\mathbf{M} = \mathbf{S}\mathbf{S}^T$. Put $k = 1$.
3. **repeat**
4. $\quad \mathbf{z}_k = \mathbf{M}\mathbf{r}_{k-1}$.
5. $\quad \tau_{k-1} = \mathbf{z}_k^T \mathbf{r}_{k-1}$.
6. $\quad$ If $k = 1$, then $\beta = 0$, $\mathbf{p}_1 = \mathbf{z}_1$. Otherwise, $\beta = \tau_{k-1}/\tau_{k-2}$, $\mathbf{p}_k \leftarrow \mathbf{z}_k + \beta \mathbf{p}_k$.
7. $\quad \mathbf{w}_k = \mathbf{A}\mathbf{p}_k$.
8. $\quad \alpha = \tau_{k-1}/(\mathbf{p}_k^T \mathbf{w}_k)$.
9. $\quad \mathbf{x}_{k+1} = \mathbf{x}_k + \alpha \mathbf{p}_k$.
10. $\quad \mathbf{r}_k = \mathbf{r}_{k-1} - \alpha \mathbf{w}_k$.
11. $\quad \rho_k = \mathbf{r}_k^T \mathbf{r}_k$.
12. $\quad$ **exit if** $\sqrt{\rho_k} < \epsilon \|\mathbf{b}\|_2$ or $k = k_{\max}$.
13. **return** $k \leftarrow k + 1$.
14. **output:** $\mathbf{x} \leftarrow \mathbf{x}_k$.

adopting the conjugate gradient algorithm or the preconditioned conjugate gradient algorithm.

The main advantage of gradient methods is that the computation of each iteration is very simple; however, their convergence is generally slow.

4.3 Condition Number of Matrices

In many applications in science and engineering, it is often necessary to consider an important problem: there exist some uncertainties or errors in the actual observation data, and, furthermore, numerical calculation of the data is always accompanied by error. What is the impact of these errors? Is a particular algorithm numerically stable for data processing?

In order to answer these questions, the following two concepts are extremely important:

- the numerical stability of various kinds of algorithms;
- the condition number or perturbation analysis of the problem of interest.

Given some application problem f where $d^* \in D$ is the data without noise or disturbance in a data group D. Let $f(d^*) \in F$ denote the solution of f, where F is a solution set. For observed data $d \in D$, we want to evaluate $f(d)$. Owing to background noise and/or observation error, $f(d)$ is usually different from $f(d^*)$. If $f(d)$ is "close" to $f(d^*)$, then the problem f is said to be a "*well-conditioned problem.*" On the contrary, if $f(d)$ is obviously different from $f(d^*)$ even when d is very close to d^*, then we say that the problem f is an "*ill-conditioned problem.*" If there is no further information about the problem f, the term "approximation" cannot describe the situation accurately.

In perturbation theory, a method or algorithm for solving a problem f is said to be *numerically stable* if its sensitivity to a perturbation is not larger than the sensitivity inherent in the original problem. More precisely, f is stable if the approximate solution $f(d)$ is close to the solution $f(d^*)$ without perturbation for all $d \in D$ close to d^*.

To mathematically describe numerical stability of a method or algorithm, we first consider the well-determined linear equation $\mathbf{Ax} = \mathbf{b}$, where the $n \times n$ coefficient matrix $\mathbf{A}$ has known entries and the $n \times 1$ data vector $\mathbf{b}$ is also known, while the $n \times 1$ vector $\mathbf{x}$ is an unknown parameter vector to be solved. Naturally, we are interested in the *numerical stability* of this solution: if the coefficient matrix $\mathbf{A}$ and/or the data vector $\mathbf{b}$ are perturbed, how will the solution vector $\mathbf{x}$ be changed? Does it maintain a certain stability? By studying the influence of perturbations of the coefficient matrix $\mathbf{A}$ and/or the data vector $\mathbf{b}$, we can obtain a numerical value, called the condition number, describing an important characteristic of the coefficient matrix $\mathbf{A}$.

For the convenience of analysis, we first assume that there is only a perturbation $\delta \mathbf{b}$ of the data vector $\mathbf{b}$, while the matrix $\mathbf{A}$ is stable. In this case, the exact solution vector $\mathbf{x}$ will be perturbed to $\mathbf{x} + \delta \mathbf{x}$, namely

$$\mathbf{A}(\mathbf{x} + \delta \mathbf{x}) = \mathbf{b} + \delta \mathbf{b}. \qquad (4.3.1)$$

This implies that

$$\delta \mathbf{x} = \mathbf{A}^{-1} \delta \mathbf{b}, \qquad (4.3.2)$$

since $\mathbf{Ax} = \mathbf{b}$.

By the property of the matrix norm, Eq. (4.3.2) gives

$$\|\delta \mathbf{x}\| \leq \|\mathbf{A}^{-1}\| \cdot \|\delta \mathbf{b}\|. \qquad (4.3.3)$$

Similarly, from the linear equation $\mathbf{Ax} = \mathbf{b}$ we get

$$\|\mathbf{b}\| \leq \|\mathbf{A}\| \cdot \|\mathbf{x}\|. \qquad (4.3.4)$$

From Eqs. (4.3.3) and (4.3.4) it is immediately clear that

$$\frac{\|\delta \mathbf{x}\|}{\|\mathbf{x}\|} \leq \left(\|\mathbf{A}\| \cdot \|\mathbf{A}^{-1}\| \right) \frac{\|\delta \mathbf{b}\|}{\|\mathbf{b}\|}. \qquad (4.3.5)$$

Next, consider the influence of simultaneous perturbations $\delta \mathbf{x}$ and $\delta \mathbf{A}$. In this case, the linear equation becomes

$$(\mathbf{A} + \delta \mathbf{A})(\mathbf{x} + \delta \mathbf{x}) = \mathbf{b}.$$

4.3 Condition Number of Matrices

From the above equation it can be derived that

$$\begin{aligned}
\delta \mathbf{x} &= \left((\mathbf{A} + \delta\mathbf{A})^{-1} - \mathbf{A}^{-1}\right)\mathbf{b} \\
&= \left[\mathbf{A}^{-1}\left(\mathbf{A} - (\mathbf{A} + \delta\mathbf{A})\right)(\mathbf{A} + \delta\mathbf{A})^{-1}\right]\mathbf{b} \\
&= -\mathbf{A}^{-1}\delta\mathbf{A}(\mathbf{A} + \delta\mathbf{A})^{-1}\mathbf{b} \\
&= -\mathbf{A}^{-1}\delta\mathbf{A}(\mathbf{x} + \delta\mathbf{x}).
\end{aligned} \quad (4.3.6)$$

Then we have

$$\|\delta\mathbf{x}\| \leq \|\mathbf{A}^{-1}\| \cdot \|\delta\mathbf{A}\| \cdot \|\mathbf{x} + \delta\mathbf{x}\|,$$

namely

$$\frac{\|\delta\mathbf{x}\|}{\|\mathbf{x} + \delta\mathbf{x}\|} \leq \left(\|\mathbf{A}\| \cdot \|\mathbf{A}^{-1}\|\right)\frac{\|\delta\mathbf{A}\|}{\|\mathbf{A}\|}. \quad (4.3.7)$$

Equations (4.3.5) and (4.3.7) show that the relative error of the solution vector $\mathbf{x}$ is proportional to the numerical value

$$\text{cond}(\mathbf{A}) = \|\mathbf{A}\| \cdot \|\mathbf{A}^{-1}\|, \quad (4.3.8)$$

where cond($\mathbf{A}$) is called the *condition number* of the matrix $\mathbf{A}$ with respect to its inverse and is sometimes denoted $\kappa(\mathbf{A})$.

Condition numbers have the following properties.

- cond($\mathbf{A}$) = cond($\mathbf{A}^{-1}$).
- cond($c\mathbf{A}$) = cond($\mathbf{A}$).
- cond($\mathbf{A}$) $\geq$ 1.
- cond($\mathbf{AB}$) $\leq$ cond($\mathbf{A}$)cond($\mathbf{B}$).

Here, we give the proofs of cond($\mathbf{A}$) $\geq$ 1 and cond($\mathbf{AB}$) $\leq$ cond($\mathbf{A}$)cond($\mathbf{B}$):

$$\text{cond}(\mathbf{A}) = \|\mathbf{A}\| \cdot \|\mathbf{A}^{-1}\| \geq \|\mathbf{A}\mathbf{A}^{-1}\| = \|\mathbf{I}\| = 1;$$

$$\text{cond}(\mathbf{AB}) = \|\mathbf{AB}\| \cdot \|(\mathbf{AB})^{-1}\| \leq \|\mathbf{A}\| \cdot \|\mathbf{B}\| \cdot (\|\mathbf{B}^{-1}\| \cdot \|\mathbf{A}^{-1}\|) = \text{cond}(\mathbf{A})\text{cond}(\mathbf{B}).$$

An orthogonal matrix $\mathbf{A}$ is perfectly conditioned in the sense that cond($\mathbf{A}$) = 1. The following are four common condition numbers.

- ℓ_1 *condition number*, denoted as $\text{cond}_1(\mathbf{A})$, is defined as

$$\text{cond}_1(\mathbf{A}) = \|\mathbf{A}\|_1 \|\mathbf{A}^{-1}\|_1. \quad (4.3.9)$$

- ℓ_2 *condition number*, denoted as $\operatorname{cond}_2(\mathbf{A})$, is defined as

$$\operatorname{cond}_2(\mathbf{A}) = \|\mathbf{A}\|_2 \|\mathbf{A}^{-1}\|_2 = \sqrt{\frac{\lambda_{\max}(\mathbf{A}^H\mathbf{A})}{\lambda_{\min}(\mathbf{A}^H\mathbf{A})}} = \frac{\sigma_{\max}(\mathbf{A})}{\sigma_{\min}(\mathbf{A})}, \qquad (4.3.10)$$

where $\lambda_{\max}(\mathbf{A}^H\mathbf{A})$ and $\lambda_{\min}(\mathbf{A}^H\mathbf{A})$ are the maximum and minimum eigenvalues of $\mathbf{A}^H\mathbf{A}$, respectively; and $\sigma_{\max}(\mathbf{A})$ and $\sigma_{\min}(\mathbf{A})$ are the maximum and minimum singular values of $\mathbf{A}$, respectively.

- ℓ_∞ *condition number*, denoted as $\operatorname{cond}_\infty(\mathbf{A})$, is defined as

$$\operatorname{cond}_\infty(\mathbf{A}) = \|\mathbf{A}\|_\infty \cdot \|\mathbf{A}^{-1}\|_\infty. \qquad (4.3.11)$$

- *Frobenius norm condition number*, denoted as $\operatorname{cond}_F(\mathbf{A})$, is defined as

$$\operatorname{cond}_F(\mathbf{A}) = \|\mathbf{A}\|_F \cdot \|\mathbf{A}^{-1}\|_F = \sqrt{\sum_i \sigma_i^2 \sum_j \sigma_j^{-2}}. \qquad (4.3.12)$$

Consider an over-determined linear equation $\mathbf{A}\mathbf{x} = \mathbf{b}$, where $\mathbf{A}$ is an $m \times n$ matrix with $m > n$. An over-determined equation has a unique linear least squares (LS) solution given by

$$\mathbf{A}^H \mathbf{A} \mathbf{x} = \mathbf{A}^H \mathbf{b}, \qquad (4.3.13)$$

i.e., $\mathbf{x}_{\mathrm{LS}} = (\mathbf{A}^H \mathbf{A})^{-1} \mathbf{A}^H \mathbf{b}$.

It should be noticed that when using the LS method to solve the matrix equation $\mathbf{A}\mathbf{x} = \mathbf{b}$, a matrix $\mathbf{A}$ with a larger condition number may result into a worse solution due to $\operatorname{cond}_2(\mathbf{A}^H \mathbf{A}) = (\operatorname{cond}_2(\mathbf{A}))^2$.

As a comparison, we consider using the *QR factorization* $\mathbf{A} = \mathbf{Q}\mathbf{R}$ (where $\mathbf{Q}$ is orthogonal and $\mathbf{R}$ is upper triangular) to solve an over-determined equation $\mathbf{A}\mathbf{x} = \mathbf{b}$ then

$$\operatorname{cond}(\mathbf{Q}) = 1, \quad \operatorname{cond}(\mathbf{A}) = \operatorname{cond}(\mathbf{Q}^H \mathbf{A}) = \operatorname{cond}(\mathbf{R}), \qquad (4.3.14)$$

since $\mathbf{Q}^H \mathbf{Q} = \mathbf{I}$. In this case we have $\mathbf{Q}^H \mathbf{A} \mathbf{x} = \mathbf{Q}^H \mathbf{b} \Rightarrow \mathbf{R}\mathbf{x} = \mathbf{Q}^H \mathbf{b}$. Since $\operatorname{cond}(\mathbf{A}) = \operatorname{cond}(\mathbf{R})$, the QR factorization method $\mathbf{R}\mathbf{x} = \mathbf{Q}^H \mathbf{b}$ has better numerical stability (i.e., a smaller condition number) than the least squares method $\mathbf{x}_{\mathrm{LS}} = (\mathbf{A}^H \mathbf{A})^{-1} \mathbf{A}^H \mathbf{b}$.

On the more effective methods than QR factorization for solving over-determined matrix equations, we will discuss them after introducing the singular value decomposition.

4.4 Singular Value Decomposition (SVD)

Beltrami (1835–1899) and Jordan (1838–1921) are recognized as the founders of singular value decomposition (SVD): Beltrami in 1873 published the first paper on SVD [2]. One year later, Jordan published his own independent derivation of SVD [23].

4.4.1 Singular Value Decomposition

Theorem 4.2 (SVD) *Let* $\mathbf{A} \in \mathbb{R}^{m \times n}$ *(or* $\mathbb{C}^{m \times n}$*), then there exist orthogonal (or unitary) matrices* $\mathbf{U} \in \mathbb{R}^{m \times m}$ *(or* $\mathbf{U} \in \mathbb{C}^{m \times m}$*) and* $\mathbf{V} \in \mathbb{R}^{n \times n}$ *(or* $\mathbf{V} \in \mathbb{C}^{n \times n}$*) such that*

$$\mathbf{A} = \mathbf{U}\boldsymbol{\Sigma}\mathbf{V}^T \ (or \ \mathbf{U}\boldsymbol{\Sigma}\mathbf{V}^H), \quad \boldsymbol{\Sigma} = \begin{bmatrix} \boldsymbol{\Sigma}_1 & \mathbf{O} \\ \mathbf{O} & \mathbf{O} \end{bmatrix}, \quad (4.4.1)$$

where $\mathbf{U} = [\mathbf{u}_1, \ldots, \mathbf{u}_m] \in \mathbb{C}^{m \times m}$, $\mathbf{V} = [\mathbf{v}_1, \ldots, \mathbf{v}_n] \in \mathbb{C}^{n \times n}$ *and* $\boldsymbol{\Sigma}_1 = \mathbf{Diag}(\sigma_1, \ldots, \sigma_l)$ *with diagonal entries*

$$\sigma_1 \geq \cdots \geq \sigma_r > \sigma_{r+1} = \ldots = \sigma_l = 0, \quad (4.4.2)$$

in which $l = \min\{m, n\}$ *and* $r = $ *rank of* $\mathbf{A}$.

This theorem was first shown by Eckart and Young [11] in 1939, but the proof by Klema and Laub [26] is simpler.

The nonzero values $\sigma_1, \ldots, \sigma_r$ together with the zero values $\sigma_{r+1} = \cdots = \sigma_l = 0$ are called the *singular values* of the matrix $\mathbf{A}$, and $\mathbf{u}_1, \ldots, \mathbf{u}_m$ and $\mathbf{v}_1, \ldots, \mathbf{v}_n$ are known as the *left-* and *right singular vectors*, respectively, while $\mathbf{U} \in \mathbb{C}^{m \times m}$ and $\mathbf{V} \in \mathbb{C}^{n \times n}$ are called the *left-* and *right singular vector matrices*, respectively.

Here are several useful explanations and remarks on singular values and SVD [46].

1. The SVD of a matrix $\mathbf{A}$ can be rewritten as the vector form

$$\mathbf{A} = \sum_{i=1}^{r} \sigma_i \mathbf{u}_i \mathbf{v}_i^H. \quad (4.4.3)$$

This expression is sometimes said to be the *dyadic decomposition* of $\mathbf{A}$ [16].

2. Suppose that the $n \times n$ matrix $\mathbf{V}$ is unitary. Postmultiply (4.4.1) by $\mathbf{V}$ to get $\mathbf{AV} = \mathbf{U}\boldsymbol{\Sigma}$, whose column vectors are given by

$$\mathbf{Av}_i = \begin{cases} \sigma_i \mathbf{u}_i, & i = 1, 2, \ldots, r; \\ 0, & i = r+1, r+2, \ldots, n. \end{cases} \quad (4.4.4)$$

3. Suppose that the $m \times m$ matrix $\mathbf{U}$ is unitary. Premultiply (4.4.1) by $\mathbf{U}^H$ to yield $\mathbf{U}^H \mathbf{A} = \mathbf{\Sigma} \mathbf{V}$, whose column vectors are given by

$$\mathbf{u}_i^H \mathbf{A} = \begin{cases} \sigma_i \mathbf{v}_i^T, & i = 1, 2, \ldots, r; \\ 0, & i = r+1, r+2, \ldots, n. \end{cases} \quad (4.4.5)$$

4. When the matrix rank $r = \text{rank}(\mathbf{A}) < \min\{m, n\}$, because $\sigma_{r+1} = \cdots = \sigma_h = 0$ with $h = \min\{m, n\}$, the SVD formula (4.4.1) can be simplified to

$$\mathbf{A} = \mathbf{U}_r \mathbf{\Sigma}_r \mathbf{V}_r^H, \quad (4.4.6)$$

where $\mathbf{U}_r = [\mathbf{u}_1, \ldots, \mathbf{u}_r]$, $\mathbf{V}_r = [\mathbf{v}_1, \ldots, \mathbf{v}_r]$, and $\mathbf{\Sigma}_r = \textbf{Diag}(\sigma_1, \ldots, \sigma_r)$. Equation (4.4.6) is called the *truncated singular value decomposition* or the *thin singular value decomposition* of the matrix $\mathbf{A}$. In contrast, Eq. (4.4.1) is known as the *full singular value decomposition*.

5. Premultiplying (4.4.4) by $\mathbf{u}_i^H$, and noting that $\mathbf{u}_i^H \mathbf{u}_i = 1$, it is easy to obtain

$$\mathbf{u}_i^H \mathbf{A} \mathbf{v}_i = \sigma_i, \quad i = 1, 2, \ldots, \min\{m, n\}, \quad (4.4.7)$$

which can be written in matrix form as

$$\mathbf{U}^H \mathbf{A} \mathbf{V} = \mathbf{\Sigma} = \begin{bmatrix} \mathbf{\Sigma}_1 & \mathbf{O} \\ \mathbf{O} & \mathbf{O} \end{bmatrix}, \quad \mathbf{\Sigma}_1 = \textbf{Diag}(\sigma_1, \ldots, \sigma_r). \quad (4.4.8)$$

Equations (4.4.1) and (4.4.8) are two definitive forms of SVD.

6. From (4.4.1) it follows that $\mathbf{AA}^H = \mathbf{U} \mathbf{\Sigma}^2 \mathbf{U}^H$. This shows that the singular value σ_i of an $m \times n$ matrix $\mathbf{A}$ is the positive square root of the corresponding nonnegative eigenvalue of the matrix product $\mathbf{AA}^H$.

7. If the matrix $\mathbf{A}_{m \times n}$ has rank r, then

   - the leftmost r columns of the $m \times m$ unitary matrix $\mathbf{U}$ constitute an orthonormal basis of the column space of the matrix $\mathbf{A}$, i.e., $\text{Col}(\mathbf{A}) = \text{Span}\{\mathbf{u}_1, \ldots, \mathbf{u}_r\}$;
   - the leftmost r columns of the $n \times n$ unitary matrix $\mathbf{V}$ constitute an orthonormal basis of the row space of $\mathbf{A}$ or the column space of $\mathbf{A}^H$, i.e., $\text{Row}(\mathbf{A}) = \text{Span}\{\mathbf{v}_1, \ldots, \mathbf{v}_r\}$;
   - the rightmost $n - r$ columns of $\mathbf{V}$ constitute an orthonormal basis of the null space of the matrix $\mathbf{A}$, i.e., $\text{Null}(\mathbf{A}) = \text{Span}\{\mathbf{v}_{r+1}, \ldots, \mathbf{v}_n\}$;
   - the rightmost $m - r$ columns of $\mathbf{U}$ constitute an orthonormal basis of the null space of the matrix $\mathbf{A}^H$, i.e., $\text{Null}(\mathbf{A}^H) = \text{Span}\{\mathbf{u}_{r+1}, \ldots, \mathbf{u}_m\}$.

4.4.2 Properties of Singular Values

Theorem 4.3 (Eckart–Young Theorem [11]) *If the singular values of* $\mathbf{A} \in \mathbb{C}^{m \times n}$ *are given by*

$$\sigma_1 \geq \sigma_2 \geq \cdots \geq \sigma_r \geq 0, \quad r = \text{rank}(\mathbf{A}),$$

then

$$\sigma_k = \min_{\mathbf{E} \in \mathbb{C}^{m \times n}} \left(\|\mathbf{E}\|_{\text{spec}} | \text{rank}(\mathbf{A} + \mathbf{E}) \leq k - 1 \right), \quad k = 1, \ldots, r, \quad (4.4.9)$$

and there is an error matrix $\mathbf{E}_k$ *with* $\|\mathbf{E}_k\|_{\text{spec}} = \sigma_k$ *such that*

$$\text{rank}(\mathbf{A} + \mathbf{E}_k) = k - 1, \quad k = 1, \ldots, r. \quad (4.4.10)$$

The Eckart–Young theorem shows that the singular value σ_k is equal to the minimum spectral norm of the error matrix $\mathbf{E}_k$ such that the rank of $\mathbf{A} + \mathbf{E}_k$ is $k - 1$.

An important application of the Eckart–Young theorem is to the best rank-k approximation of the matrix $\mathbf{A}$, where $k < r = \text{rank}(\mathbf{A})$.

Define

$$\mathbf{A}_k = \sum_{i=1}^{k} \sigma_i \mathbf{u}_i \mathbf{v}_i^H, \quad k < r; \quad (4.4.11)$$

then $\mathbf{A}_k$ is the solution of the optimization problem:

$$\mathbf{A}_k = \arg\min_{\text{rank}(\mathbf{X}) = k} \|\mathbf{A} - \mathbf{X}\|_F^2, \quad k < r, \quad (4.4.12)$$

and the squared approximation error is given by

$$\|\mathbf{A} - \mathbf{A}_k\|_F^2 = \sigma_{k+1}^2 + \cdots + \sigma_r^2. \quad (4.4.13)$$

Theorem 4.4 ([20, 21]) *Let* $\mathbf{A}$ *be an* $m \times n$ *matrix with singular values* $\sigma_1 \geq \cdots \geq \sigma_r$, *where* $r = \min\{m, n\}$. *If the* $p \times q$ *matrix* $\mathbf{B}$ *is a submatrix of* $\mathbf{A}$ *with singular values* $\gamma_1 \geq \cdots \geq \gamma_{\min\{p,q\}}$, *then*

$$\sigma_i \geq \gamma_i, \quad i = 1, \ldots, \min\{p, q\} \quad (4.4.14)$$

and

$$\gamma_i \geq \sigma_{i+(m-p)+(n-q)}, \quad i \leq \min\{p + q - m, p + q - n\}. \quad (4.4.15)$$

This is the *interlacing theorem for singular values*.

The singular values of a matrix are closely related to its norm, determinant, and condition number.

1. *Relationship between Singular Values and Norms:*

$$\|\mathbf{A}\|_2 = \sigma_1 = \sigma_{\max},$$

$$\|\mathbf{A}\|_F = \left(\sum_{i=1}^{m} \sum_{j=1}^{n} |a_{ij}|^2 \right)^{1/2} = \|\mathbf{U}^H \mathbf{A} \mathbf{V}\|_F = \|\mathbf{\Sigma}\|_F = \sqrt{\sigma_1^2 + \cdots + \sigma_r^2}.$$

2. *Relationship between Singular Values and Determinant:*

$$|\det(\mathbf{A})| = |\det(\mathbf{U}\mathbf{\Sigma}\mathbf{V}^H)| = |\det \mathbf{\Sigma}| = \sigma_1 \cdots \sigma_n. \tag{4.4.16}$$

If all the σ_i are nonzero, then $|\det(\mathbf{A})| \neq 0$, which shows that $\mathbf{A}$ is nonsingular. If there is at least one singular value $\sigma_i = 0 \, (i > r)$, then $\det(\mathbf{A}) = 0$, namely $\mathbf{A}$ is singular. This is the reason why the $\sigma_i, i = 1, \ldots, \min\{m, n\}$ are known as the singular values.

3. *Relationship between Singular Values and Condition Number:*

$$\text{cond}_2(\mathbf{A}) = \sigma_1 / \sigma_p, \quad p = \min\{m, n\}. \tag{4.4.17}$$

4.4.3 Singular Value Thresholding

Consider the truncated SVD of a low-rank matrix $\mathbf{W} \in \mathbb{R}^{m \times n}$:

$$\mathbf{W} = \mathbf{U}\mathbf{\Sigma}\mathbf{V}^T, \quad \mathbf{\Sigma} = \mathbf{Diag}(\sigma_1, \cdots, \sigma_r) \tag{4.4.18}$$

where $r = \text{rank}(\mathbf{W}) \ll \min\{m, n\}, \mathbf{U} \in \mathbb{R}^{m \times r}, \mathbf{V} \in \mathbb{R}^{n \times r}$.

Let the threshold value $\tau \geq 0$, then

$$\mathcal{D}_\tau(\mathbf{W}) = \mathbf{U}\mathcal{D}_\tau(\mathbf{\Sigma})\mathbf{V}^T \tag{4.4.19}$$

is called the *singular value thresholding* (SVT) of the matrix $\mathbf{W}$, where

$$\mathcal{D}_\tau(\mathbf{\Sigma}) = \text{soft}(\mathbf{\Sigma}, \tau) = \mathbf{Diag}\left((\sigma_1 - \tau)_+, \cdots, (\sigma_r - \tau)_+\right) \tag{4.4.20}$$

is known as the *soft thresholding*, and

$$(\sigma_i - \tau)_+ = \begin{cases} \sigma_i - \tau, & \text{if } \sigma_i > \tau; \\ 0, & \text{otherwise} \end{cases}$$

is the *soft thresholding operation*.

The relationships between the SVT and the SVD are as follows.

- If the soft threshold value $\tau = 0$, then the SVT is reduced to the truncated SVD (4.4.18).
- All of singular values are soft thresholded by the soft threshold value $\tau > 0$, which just changes the magnitudes of singular values, does not change the left and right singular vector matrices $\mathbf{U}$ and $\mathbf{V}$.

Theorem 4.5 ([6]) *For each soft threshold value $\mu > 0$ and the matrix $\mathbf{W} \in \mathbb{R}^{m \times n}$, the SVT operator obeys*

$$\mathbf{U}\text{soft}(\mathbf{\Sigma}, \mu)\mathbf{V}^T = \arg\min_{\mathbf{X}} \left(\mu \|\mathbf{X}\|_* + \frac{1}{2} \|\mathbf{X} - \mathbf{W}\|_F^2 \right), \tag{4.4.21}$$

$$\text{soft}(\mathbf{W}, \mu) = \arg\min_{\mathbf{X}} \left(\mu \|\mathbf{X}\|_1 + \frac{1}{2} \|\mathbf{X} - \mathbf{W}\|_F^2 \right), \tag{4.4.22}$$

where $\mathbf{U}\mathbf{\Sigma}\mathbf{V}^T$ is the SVD of $\mathbf{W}$, and the soft thresholding (shrinkage) operator

$$[\text{soft}(\mathbf{W}, \mu)]_{ij} = \begin{cases} w_{ij} - \mu, & w_{ij} > \mu; \\ w_{ij} + \mu, & w_{ij} < -\mu; \\ 0, & \text{otherwise.} \end{cases} \tag{4.4.23}$$

Here $w_{ij} \in \mathbb{R}$ is the (i, j)th entry of $\mathbf{W} \in \mathbb{R}^{m \times n}$.

4.5 Least Squares Method

In over-determined equation, the number of independent equations is larger than the number of independent unknowns, the number of independent equations appears surplus for determining the unique solution. An over-determined matrix equation $\mathbf{A}\mathbf{x} = \mathbf{b}$ has no exact solution and thus is an inconsistent equation that may in some cases have an approximate solution. There are four commonly used methods for solutions of matrix equations: Least squares method, Tikhonov regularization method, Gauss–Seidel method, and total least squares method. From this section we will introduce these four methods in turn.

4.5.1 Least Squares Solution

Consider an over-determined matrix equation $\mathbf{A}\mathbf{x} = \mathbf{b}$, where $\mathbf{b}$ is an $m \times 1$ data vector, $\mathbf{A}$ is an $m \times n$ data matrix, and $m > n$.

Suppose the data vector and the additive observation error or noise exist, i.e., $\mathbf{b} = \mathbf{b}_0 + \mathbf{e}$, where $\mathbf{b}_0$ and $\mathbf{e}$ are the errorless data vector and the additive error vector, respectively.

In order to resist the influence of the error on matrix equation solution, we use a correction vector $\Delta \mathbf{b}$ to "disturb" the data vector $\mathbf{b}$ for forcing $\mathbf{A}\mathbf{x} = \mathbf{b} + \Delta \mathbf{b}$ to compensate the uncertainty existing in the data vector $\mathbf{b}$ (noise or error). The idea for solving the matrix equation can be described by the optimization problem

$$\min_{\mathbf{x}} \left\{ \|\Delta \mathbf{b}\|^2 = \|\mathbf{A}\mathbf{x} - \mathbf{b}\|_2^2 = (\mathbf{A}\mathbf{x} - \mathbf{b})^T (\mathbf{A}\mathbf{x} - \mathbf{b}) \right\}. \tag{4.5.1}$$

Such a method is known as *ordinary least squares (OLS) method*, simply called the *least squares (LS) method*.

As a matter of fact, the correction vector $\Delta \mathbf{b} = \mathbf{A}\mathbf{x} - \mathbf{b}$ is just the error vector of both sides of the matrix equation $\mathbf{A}\mathbf{x} = \mathbf{b}$. Hence the central idea of the LS method is to find the solution $\mathbf{x}$ such that the sum of the squared error $\|\mathbf{A}\mathbf{x} - \mathbf{b}\|_2^2$ is minimized, namely

$$\hat{\mathbf{x}}_{\mathrm{LS}} = \arg\min_{\mathbf{x}} \ \|\mathbf{A}\mathbf{x} - \mathbf{b}\|_2^2. \tag{4.5.2}$$

In order to derive the analytic solution of $\mathbf{x}$, expand Eq. (4.5.1) to get

$$\phi = \mathbf{x}^T \mathbf{A}^T \mathbf{A} \mathbf{x} - \mathbf{x}^T \mathbf{A}^T \mathbf{b} - \mathbf{b}^T \mathbf{A} \mathbf{x} + \mathbf{b}^T \mathbf{b}.$$

Find the derivative of ϕ with respect to $\mathbf{x}$, and let the result equal zero, then

$$\frac{d\phi}{d\mathbf{x}} = 2\mathbf{A}^T \mathbf{A} \mathbf{x} - 2\mathbf{A}^T \mathbf{b} = 0.$$

That is to say, the solution $\mathbf{x}$ must meet

$$\mathbf{A}^T \mathbf{A} \mathbf{x} = \mathbf{A}^T \mathbf{b}. \tag{4.5.3}$$

The above equation is identifiable or unidentifiable depending on the rank of the $m \times n$ matrix $\mathbf{A}$.

- *Identifiable:* When $\mathbf{A}\mathbf{x} = \mathbf{b}$ is the over-determined equation, it has the unique solution

$$\mathbf{x}_{\mathrm{LS}} = (\mathbf{A}^T \mathbf{A})^{-1} \mathbf{A}^T \mathbf{b} \quad \text{if } \mathrm{rank}(\mathbf{A}) = n, \tag{4.5.4}$$

or

$$\mathbf{x}_{\mathrm{LS}} = (\mathbf{A}^T \mathbf{A})^{\dagger} \mathbf{A}^T \mathbf{b} \quad \text{if } \mathrm{rank}(\mathbf{A}) < n, \tag{4.5.5}$$

4.5 Least Squares Method

where $\mathbf{B}^\dagger$ denotes the Moore–Penrose inverse matrix of $\mathbf{B}$. In parameter estimation theory, the unknown parameter vector $\mathbf{x}$ is said to be uniquely identifiable, if it is uniquely determined.

- *Unidentifiable:* For an under-determined equation $\mathbf{Ax} = \mathbf{b}$, if $\text{rank}(\mathbf{A}) = m < n$, then different solutions of $\mathbf{x}$ give the same value of $\mathbf{Ax}$. Clearly, although the data vector $\mathbf{b}$ can provide some information about $\mathbf{Ax}$, we cannot distinguish different parameter vectors $\mathbf{x}$ corresponding to the same $\mathbf{Ax}$. Such an unknown vector $\mathbf{x}$ is said to be unidentifiable.

In parameter estimation, an estimate $\hat{\boldsymbol{\theta}}$ of the parameter vector $\boldsymbol{\theta}$ is known as an *unbiased estimator*, if its mathematical expectation is equal to the true unknown parameter vector, i.e., $E\{\hat{\boldsymbol{\theta}}\} = \boldsymbol{\theta}$. Further, an unbiased estimator is called the optimal unbiased estimator if it has the minimum variance. Similarly, for an over-determined matrix equation $\mathbf{A}\boldsymbol{\theta} = \mathbf{b}+\mathbf{e}$ with noisy data vector $\mathbf{b}$, if the mathematical expectation of the LS solution $\hat{\boldsymbol{\theta}}_{\text{LS}}$ is equal to the true parameter vector $\boldsymbol{\theta}$, i.e., $E\{\hat{\boldsymbol{\theta}}_{\text{LS}}\} = \boldsymbol{\theta}$, then $\hat{\boldsymbol{\theta}}_{\text{LS}}$ is called the *optimal unbiased estimator*.

Theorem 4.6 (Gauss–Markov Theorem) *Consider the set of linear equations*

$$\mathbf{Ax} = \mathbf{b} + \mathbf{e}, \qquad (4.5.6)$$

where the $m \times n$ matrix $\mathbf{A}$ and the $n \times 1$ vector $\mathbf{x}$ are, respectively, the constant matrix and the parameter vector; $\mathbf{b}$ is the $m \times 1$ vector with random error vector $\mathbf{e} = [e_1, \cdots, e_m]^T$, and its mean vector and covariance matrix are, respectively,

$$E\{\mathbf{e}\} = \mathbf{0}, \qquad \text{cov}(\mathbf{e}) = E\{\mathbf{e}\mathbf{e}^H\} = \sigma^2 \mathbf{I}.$$

Then the $n \times 1$ parameter vector $\mathbf{x}$ exists the optimal unbiased solution $\hat{\mathbf{x}}$, if and only if $\text{rank}(\mathbf{A}) = n$. In this case, the optimal unbiased solution is given by the LS solution

$$\hat{\mathbf{x}}_{\text{LS}} = (\mathbf{A}^H \mathbf{A})^{-1} \mathbf{A}^H \mathbf{b}, \qquad (4.5.7)$$

and its covariance

$$\text{var}(\hat{\mathbf{x}}_{\text{LS}}) \leq \text{var}(\tilde{\mathbf{x}}), \qquad (4.5.8)$$

where $\tilde{\mathbf{x}}$ is any other solution of the matrix equation $\mathbf{Ax} = \mathbf{b} + \mathbf{e}$.

Proof See e.g., [46].

In the Gauss–Markov theorem $\text{cov}(\mathbf{e}) = \sigma^2 \mathbf{I}$ implies that all components of the additive noise vector $\mathbf{e}$ are mutually uncorrelated, and have the same variance σ^2. Only in this case, the LS solution is unbiased and optimal.

4.5.2 Rank-Deficient Least Squares Solutions

In many science and engineering applications, it is usually necessary to use a low-rank matrix to approximate a noisy or disturbed matrix. The following theorem gives an evaluation of the quality of approximation.

Theorem 4.7 (Low-Rank Approximation) Let $\mathbf{A} = \sum_{i=1}^{p} \sigma_i \mathbf{u}_i \mathbf{v}_i^T$ be the SVD of $\mathbf{A} \in \mathbb{R}^{m \times n}$, where $p = \text{rank}(\mathbf{A})$. If $k < p$ and $\mathbf{A}_k = \sum_{i=1}^{k} \sigma_i \mathbf{u}_i \mathbf{v}_i^T$ is the rank-k approximation of $\mathbf{A}$, then the approximation quality can be measured by the Frobenius norm

$$\min_{\text{rank}(\mathbf{B})=k} \|\mathbf{A}-\mathbf{B}\|_F = \|\mathbf{A}-\mathbf{A}_k\|_F = \left(\sum_{i=k+1}^{q} \sigma_i^2 \right)^{1/2}, \quad q = \min\{m, n\}. \quad (4.5.9)$$

Proof See, e.g., [10, 21, 30].

For an over-determined and rank-deficient system of linear equations $\mathbf{Ax} = \mathbf{b}$, let the SVD of $\mathbf{A}$ be given by $\mathbf{A} = \mathbf{U}\mathbf{\Sigma}\mathbf{V}^H$, where $\mathbf{\Sigma} = \mathbf{Diag}(\sigma_1, \ldots, \sigma_r, 0, \ldots, 0)$. The LS solution

$$\hat{\mathbf{x}} = \mathbf{A}^\dagger \mathbf{b} = \mathbf{V}\mathbf{\Sigma}^\dagger \mathbf{U}^H \mathbf{b}, \quad (4.5.10)$$

where $\mathbf{\Sigma}^\dagger = \mathbf{Diag}(1/\sigma_1, \ldots, 1/\sigma_r, 0, \ldots, 0)$.

Equation (4.5.10) can be represented as

$$\mathbf{x}_{\text{LS}} = \sum_{i=1}^{r} (\mathbf{u}_i^H \mathbf{b}/\sigma_i) \mathbf{v}_i. \quad (4.5.11)$$

The corresponding minimum residual is given by

$$r_{\text{LS}} = \|\mathbf{Ax}_{\text{LS}} - \mathbf{b}\|_2 = \|[\mathbf{u}_{r+1}, \ldots, \mathbf{u}_m]^H \mathbf{b}\|_2. \quad (4.5.12)$$

Although, in theory, when $i > r$ the true singular values $\sigma_i = 0$, the computed singular values $\hat{\sigma}_i$, $i > r$, are not usually equal to zero and sometimes even have quite a large perturbation. In these cases, an estimate of the matrix rank r is required. In signal processing and system theory, the rank estimate $\hat{r}$ is usually called the *effective rank*.

Effective rank can be estimated by one of the following two common methods.

1. *Normalized Singular Value Method.* Compute the *normalized singular values* $\bar{\sigma}_i = \hat{\sigma}_i/\hat{\sigma}_1$, and select the largest integer i satisfying the criterion $\bar{\sigma}_i \geq \epsilon$ as an estimate of the effective rank $\hat{r}$. Obviously, this criterion is equivalent to choosing the maximum integer i satisfying

$$\hat{\sigma}_i \geq \epsilon \hat{\sigma}_1 \quad (4.5.13)$$

as $\hat{r}$; here ϵ is a very small positive number, e.g., $\epsilon = 0.1$ or $\epsilon = 0.05$.

2. *Norm Ratio Method.* Let an $m \times n$ matrix $\mathbf{A}_k$ be the rank-k approximation to the original $m \times n$ matrix $\mathbf{A}$. Define the *Frobenius norm ratio* as

$$\nu(k) = \frac{\|\mathbf{A}_k\|_F}{\|\mathbf{A}\|_F} = \frac{\sqrt{\sigma_1^2 + \cdots + \sigma_k^2}}{\sqrt{\sigma_1^2 + \cdots + \sigma_h^2}}, \quad h = \min\{m, n\}, \tag{4.5.14}$$

and choose the minimum integer k satisfying

$$\nu(k) \geq \alpha \tag{4.5.15}$$

as the effective rank estimate $\hat{r}$, where α is close to 1, e.g., $\alpha = 0.997$ or 0.998.

After the effective rank $\hat{r}$ is determined via the above two criteria,

$$\hat{\mathbf{x}}_{\mathrm{LS}} = \sum_{i=1}^{\hat{r}} (\hat{\mathbf{u}}_i^H \mathbf{b} / \hat{\sigma}_i) \hat{\mathbf{v}}_i \tag{4.5.16}$$

can be regarded as a reasonable approximation to the LS solution $\mathbf{x}_{\mathrm{LS}}$.

4.6 Tikhonov Regularization and Gauss–Seidel Method

The LS method is widely used for solving matrix equations and is applicable for many real-world applications in machine learning, neural networks, support vector machines, and evolutionary computation. But, due to its sensitive to the perturbation of the data matrix, the LS methods must be improved in these applications. A simple and efficient way is the well-known Tikhonov regularization.

4.6.1 Tikhonov Regularization

When $m = n$, and $\mathbf{A}$ is nonsingular, the solution of matrix equation $\mathbf{A}\mathbf{x} = \mathbf{b}$ is given by $\hat{\mathbf{x}} = \mathbf{A}^{-1}\mathbf{b}$; and when $m > n$ and $\mathbf{A}_{m \times n}$ is of full column rank, the solution of the matrix equation is $\hat{\mathbf{x}}_{\mathrm{LS}} = \mathbf{A}^{\dagger}\mathbf{b} = (\mathbf{A}^H\mathbf{A})^{-1}\mathbf{A}^H\mathbf{b}$.

The problem is: the data matrix $\mathbf{A}$ is often rank deficient in engineering applications. In these cases, the solution $\hat{\mathbf{x}} = \mathbf{A}^{-1}\mathbf{b}$ or $\hat{\mathbf{x}}_{\mathrm{LS}} = (\mathbf{A}^H\mathbf{A})^{-1}\mathbf{A}^H\mathbf{b}$ either diverges, or even exists, it is the bad approximation to the unknown vector $\mathbf{x}$. Even if we happen to find a reasonable approximation of $\mathbf{x}$, the error estimate $\|\mathbf{x} - \hat{\mathbf{x}}\| \leq \|\mathbf{A}^{-1}\| \|\mathbf{A}\hat{\mathbf{x}} - \mathbf{b}\|$ or $\|\mathbf{x} - \hat{\mathbf{x}}\| \leq \|\mathbf{A}^{\dagger}\| \|\mathbf{A}\hat{\mathbf{x}} - \mathbf{b}\|$ is very disappointing [32]. By observation, it is easily known that the problem lies in the inversion of the covariance matrix $\mathbf{A}^H\mathbf{A}$ of the rank-deficient data matrix $\mathbf{A}$.

As an improvement of the LS cost function $\frac{1}{2}\|\mathbf{Ax}-\mathbf{b}\|_2^2$, Tikhonov [39] in 1963 proposed the *regularized least squares cost function*

$$J(\mathbf{x}) = \frac{1}{2}\left(\|\mathbf{Ax}-\mathbf{b}\|_2^2 + \lambda\|\mathbf{x}\|_2^2\right), \tag{4.6.1}$$

where $\lambda \geq 0$ is called the *regularization parameter*.

The conjugate gradient of the cost function $J(\mathbf{x})$ with respect to the argument $\mathbf{x}$ is given by

$$\frac{\partial J(\mathbf{x})}{\partial \mathbf{x}^H} = \frac{\partial}{\partial \mathbf{x}^H}\left((\mathbf{Ax}-\mathbf{b})^H(\mathbf{Ax}-\mathbf{b}) + \lambda \mathbf{x}^H\mathbf{x}\right) = \mathbf{A}^H\mathbf{Ax} - \mathbf{A}^H\mathbf{b} + \lambda\mathbf{x}.$$

Let $\frac{\partial J(\mathbf{x})}{\partial \mathbf{x}^H} = \mathbf{0}$, then the *Tikhonov regularization solution*

$$\hat{\mathbf{x}}_{\text{Tik}} = (\mathbf{A}^H\mathbf{A} + \lambda\mathbf{I})^{-1}\mathbf{A}^H\mathbf{b}. \tag{4.6.2}$$

This method using $(\mathbf{A}^H\mathbf{A} + \lambda\mathbf{I})^{-1}$ instead of the direct inversion of the covariance matrix $(\mathbf{A}^H\mathbf{A})^{-1}$ is called the *Tikhonov regularization method* (or simply *regularized method*). In signal processing and image processing, the regularization method is sometimes known as the *relaxation method*.

The nature of Tikhonov regularization method is: by adding a very small disturbance λ to each diagonal entry of the covariance matrix $\mathbf{A}^H\mathbf{A}$ of rank-deficient matrix $\mathbf{A}$, the inversion of the singular covariance matrix $\mathbf{A}^H\mathbf{A}$ becomes the inversion of a nonsingular matrix $\mathbf{A}^H\mathbf{A} + \lambda\mathbf{I}$, thereby greatly improving the numerical stability of solving the rank-deficient matrix equation $\mathbf{Ax} = \mathbf{b}$.

Obviously, if the data matrix $\mathbf{A}$ is of full column rank, but exists the error or noise, we must adopt the opposite method to the Tikhonov regularization: adding a very small negative disturbance $-\lambda$ to each diagonal entry of the covariance matrix $\mathbf{A}^H\mathbf{A}$ for suppressing the influence of noise. Such a method using a very small negative disturbance matrix $-\lambda\mathbf{I}$ is called the *anti-Tikhonov regularization method* or *anti-regularized method*, and its solution is given by

$$\hat{\mathbf{x}} = (\mathbf{A}^H\mathbf{A} - \lambda\mathbf{I})^{-1}\mathbf{A}^H\mathbf{b}. \tag{4.6.3}$$

The total least square method is a typical anti-regularized method and will be discussed later.

When the regularization parameter λ varies in the definition interval $[0, \infty)$, the family of solutions for a regularized LS problem is known as its *regularization path*.

Tikhonov regularization solution has the following important properties [25]:

1. *Linearity:* The Tikhonov regularization LS solution $\hat{\mathbf{x}}_{\text{Tik}} = (\mathbf{A}^H\mathbf{A} + \lambda\mathbf{I})^{-1}\mathbf{A}^H\mathbf{b}$ is the linear function of the observed data vector $\mathbf{b}$.
2. *Limit characteristic when $\lambda \to 0$:* When the regularization parameter $\lambda \to 0$, the Tikhonov regularization LS solution converges to the ordinary LS solution

4.6 Tikhonov Regularization and Gauss–Seidel Method

or the Moore–Penrose solution $\lim_{\lambda \to 0} \hat{\mathbf{x}}_{\text{Tik}} = \hat{\mathbf{x}}_{\text{LS}} = \mathbf{A}^{\dagger}\mathbf{b} = (\mathbf{A}^H\mathbf{A})^{-1}\mathbf{A}^H\mathbf{b}$. The solution point $\hat{\mathbf{x}}_{\text{Tik}}$ has the minimum ℓ_2-norm among all the feasible points meeting $\mathbf{A}^H(\mathbf{A}\mathbf{x} - \mathbf{b}) = \mathbf{0}$:

$$\hat{\mathbf{x}}_{\text{Tik}} = \underset{\mathbf{A}^T(\mathbf{b}-\mathbf{A}\mathbf{x})=\mathbf{0}}{\arg \min} \ \|\mathbf{x}\|_2. \tag{4.6.4}$$

3. *Limit characteristic when* $\lambda \to \infty$: When $\lambda \to \infty$, the optimal Tikhonov regularization LS solution converges to a zero vector, i.e., $\lim_{\lambda \to \infty} \hat{\mathbf{x}}_{\text{Tik}} = \mathbf{0}$.
4. *Regularization path:* When the regularization parameter λ varies in $[0, \infty)$, the optimal solution of the Tikhonov regularization LS problem is the smooth function of the regularization parameter, i.e., when λ decreases to zero, the optimal solution converges to the Moore–Penrose solution; and when λ increases, the optimal solution converges to zero vector.

The Tikhonov regularization can effectively prevent the divergence of the LS solution $\hat{\mathbf{x}}_{\text{LS}} = (\mathbf{A}^T\mathbf{A})^{-1}\mathbf{A}^T\mathbf{b}$ when $\mathbf{A}$ is rank deficient, thereby improves obviously the convergence property of the LS algorithm and the alternative LS algorithm, and is widely applied.

4.6.2 Gauss–Seidel Method

A matrix equation $\mathbf{A}\mathbf{x} = \mathbf{b}$ can be rearranged as

$$
\begin{aligned}
a_{11}x_1 &= b_1 - a_{12}x_2 - \cdots - a_{1n}x_n \\
a_{21}x_1 + a_{22}x_2 &= b_2 \quad\ \, - \cdots - a_{2n}x_n \\
&\ \vdots \\
a_{(n-1)1}x_1 + a_{(n-2)}x_2 + \cdots + a_{(n-1)n}x_n &= b_{n-1} - a_{nn}x_n \\
a_{n1}x_1 + a_{n2}x_2 + \cdots + a_{nn}x_n &= b_n.
\end{aligned}
$$

Based on the above rearrangement, the Gauss–Seidel method begins with an initial approximation to the solution, $\mathbf{x}^{(0)}$, and then compute an update for the first element of $\mathbf{x}$:

$$x_1^{(1)} = \frac{1}{a_{11}} \left(b_1 - \sum_{j=2}^{n} a_{1j} x_j^{(0)} \right). \tag{4.6.5}$$

Continuing in this way for the other elements of $\mathbf{x}$, the Gauss–Seidel method gives

$$x_i^{(1)} = \frac{1}{a_{ii}} \left(b_i - \sum_{j=1}^{i-1} a_{ij} x_j^{(1)} - \sum_{j=i+1}^{n} a_{ij} x_j^{(0)} \right), \quad \text{for } i = 1, \ldots, n. \tag{4.6.6}$$

After getting the approximation $\mathbf{x}^{(1)} = [x_1^{(1)}, \ldots, x_n^{(1)}]^T$, we then continue this same kind of iteration for $\mathbf{x}^{(2)}, \mathbf{x}^{(3)}, \ldots$ until $\mathbf{x}$ converges.

The Gauss–Seidel method not only can find the solution of the matrix equation, but also are applicable for solving a nonlinear optimization problem.

Let $X_i \subseteq \mathbb{R}^{n_i}$ be the feasible set of $n_i \times 1$ vector $\mathbf{x}_i$. Consider the nonlinear minimization problem

$$\min_{\mathbf{x} \in X} \{ f(\mathbf{x}) = f(\mathbf{x}_1, \cdots, \mathbf{x}_m) \}, \qquad (4.6.7)$$

where $\mathbf{x} \in X = X_1 \times \cdots \times X_m \subseteq \mathbb{R}^n$ is the Cartesian product of a closed nonempty convex set $X_i \subseteq \mathbb{R}^{n_i}$, $i = 1, \cdots, m$, and $\sum_{i=1}^m n_i = n$.

Equation (4.6.7) is an unconstrained optimization problem with m coupled variable vectors. An efficient approach for solving this class of the *coupled optimization problems* is the *block nonlinear Gauss–Seidel method*, called simply the *GS method* [3, 18].

In every iteration of the GS method, $m - 1$ variable vectors are fixed, one remaining variable vector is minimized. This idea constitutes the basic framework of the GS method for solving nonlinear unconstrained optimization problem (4.6.7):

1. Initial $m - 1$ variable vectors $\mathbf{x}_i$, $i = 2, \cdots, m$, and let $k = 0$.
2. Find the solution of separated sub-optimization problem

$$\mathbf{x}_i^{k+1} = \arg \min_{\mathbf{y} \in X_i} f\left(\mathbf{x}_1^{k+1}, \cdots, \mathbf{x}_{i-1}^{k+1}, \mathbf{y}, \mathbf{x}_{i+1}^k, \cdots, \mathbf{x}_m^k\right), \quad i = 1, \cdots, m.$$
(4.6.8)

At the $(k + 1)$th iteration of updating $\mathbf{x}_i$, all $\mathbf{x}_1, \cdots, \mathbf{x}_{i-1}$ have been updated as $\mathbf{x}_1^{k+1}, \cdots, \mathbf{x}_{i-1}^{k+1}$, so these sub-vectors and $\mathbf{x}_{i+1}^k, \cdots, \mathbf{x}_m^k$ to be updated are fixed as the known vectors.

3. To test whether m variable vectors are all convergent. If convergent, then output the optimization results $(\mathbf{x}_1^{k+1}, \cdots, \mathbf{x}_m^{k+1})$; otherwise, let $k \leftarrow k + 1$, return to Eq. (4.6.8), and continue to iteration until the convergence criterion meets.

If the objective function $f(\mathbf{x})$ of the optimization (4.6.7) is the LS error function (for example, $\|\mathbf{A}\mathbf{x} - \mathbf{b}\|_2^2$), then the GS method is customarily called the *alternating least squares (ALS) method*.

Example 4.2 Consider the full-rank decomposition of an $m \times n$ known data matrix $\mathbf{X} = \mathbf{AB}$, where the $m \times r$ matrix $\mathbf{A}$ is of full column rank, and the $r \times n$ matrix $\mathbf{B}$ is of full row rank. Let the cost function of the matrix full-rank decomposition be

$$f(\mathbf{A}, \mathbf{B}) = \frac{1}{2} \|\mathbf{X} - \mathbf{AB}\|_F^2. \qquad (4.6.9)$$

4.6 Tikhonov Regularization and Gauss–Seidel Method

Then the ALS algorithm first initializes the matrix $\mathbf{A}$. At the $(k+1)$th iteration, from the fixed matrix $\mathbf{A}_k$ we update the LS solution of $\mathbf{B}$ as follows:

$$\mathbf{B}_{k+1} = (\mathbf{A}_k^T \mathbf{A}_k)^{-1} \mathbf{A}_k^T \mathbf{X}. \tag{4.6.10}$$

Next, from the transpose of matrix decomposition $\mathbf{X}^T = \mathbf{B}^T \mathbf{A}^T$, we can immediately update the LS solution of $\mathbf{A}^T$ as

$$\mathbf{A}_{k+1}^T = (\mathbf{B}_{k+1} \mathbf{B}_{k+1}^T)^{-1} \mathbf{B}_{k+1} \mathbf{X}^T. \tag{4.6.11}$$

The above two kinds of LS methods are performed alternately. Once the whole ALS algorithm is converged, the optimization results of matrix decomposition can be obtained.

The fact that the GS algorithm may not converge was observed by Powell in 1973 [34] who called it the *"circle phenomenon"* of the GS method. Recently, a lot of simulation experiences have shown [28, 31] that even converged, the iterative process of the ALS method is also very easy to fall into the *"swamp"*: an unusually large number of iterations leads to a very slow convergence rate. [28, 31].

A kind of simple and effective way for avoiding the circle and swamp phenomenon of the GS method is to make the Tikhonov regularization of the objective function of the optimization problem (4.6.7), namely the separated sub-optimization algorithm (4.6.8) is regularized as

$$\mathbf{x}_i^{k+1} = \arg\min_{\mathbf{y} \in X_i} \left\{ f(\mathbf{x}_1^{k+1}, \cdots, \mathbf{x}_{i-1}^{k+1}, \mathbf{y}, \mathbf{x}_{i+1}^k, \cdots, \mathbf{x}_m^k) + \frac{1}{2} \tau_i \|\mathbf{y} - \mathbf{x}_i^k\|_2^2 \right\}, \tag{4.6.12}$$

where $i = 1, \cdots, m$.

The above algorithm is called the *proximal point versions* of the GS methods [1, 3], abbreviated as PGS method.

The role of the regularization term $\|\mathbf{y} - \mathbf{x}_i^k\|_2^2$ is to force the updated vector $\mathbf{x}_i^{k+1} = \mathbf{y}$ close to $\mathbf{x}_i^k$, not to deviate too much, and thus avoid the violent shock of the iterative process in order to prevent the divergence of the algorithm.

The GS or PGS method is said to be well defined, if each sub-optimization problem has an optimal solution [18].

Theorem 4.8 ([18]) *If the PGS method is well defined, and the sequence $\{\mathbf{x}^k\}$ exists limit points, then every limit point $\bar{\mathbf{x}}$ of $\{\mathbf{x}^k\}$ is a critical point of the optimization problem (4.6.7).*

This theorem shows that the convergence performance of the PGS method is better than that of the GS method.

Many simulation experiments show [28] that under the condition achieving the same error, the iterative number of the GS method in swamp iteration is unusually large, whereas the PGS method tends to converge quickly. The PGS method is also called the *regularized Gauss–Seidel method* in some literature.

The ALS method and the regularized ALS method have important applications in nonnegative matrix decomposition and the tensor analysis.

4.7 Total Least Squares Method

For a matrix equation $\mathbf{A}_{m \times n} \mathbf{x}_n = \mathbf{b}_m$, all the LS method, the Tikhonov regularization method, and the Gauss–Seidel method give the solution of n parameters. However, by the rank-deficiency of the matrix $\mathbf{A}$ we know that the unknown parameter vector $\mathbf{x}$ contains only r independent parameters; the other parameters are linearly dependent on the r independent parameters. In many engineering applications, we naturally want to find the r independent parameters other than the n parameters containing redundancy components. In other words, we want to estimate only the principal parameters and to eliminate minor components. This problem can be solved via the low-rank total least squares (TLS) method.

The earliest ideas about TLS can be traced back to the paper of Pearson in 1901 [33] who considered the approximate method for solving the matrix equation $\mathbf{Ax} = \mathbf{b}$ when both $\mathbf{A}$ and $\mathbf{b}$ exist the errors. But, only in 1980, Golub and Van Loan [15] have first time made the overall analysis from the point of view of numerical analysis, and have formally known this method as the total least squares. In mathematical statistics, this method is called the orthogonal regression or errors-in-variables regression [14]. In system identification, the TLS method is called the characteristic vector method or the Koopmans–Levin method [42]. Now, the TLS method has been widely used in statistics, physics, economics, biology and medicine, signal processing, automatic control, system science, artificial intelligence, and many other disciplines and fields.

4.7.1 Total Least Squares Solution

Let $\mathbf{A}_0$ and $\mathbf{b}_0$ express unobservable error-free data matrix and error-free data vector, respectively. The actual observed data matrix and data vector are, respectively, given by

$$\mathbf{A} = \mathbf{A}_0 + \mathbf{E}, \quad \mathbf{b} = \mathbf{b}_0 + \mathbf{e}, \tag{4.7.1}$$

where $\mathbf{E}$ and $\mathbf{e}$ express the error data matrix and the error data vector, respectively.

The basic idea of the TLS is: not only use the correction vector $\Delta \mathbf{b}$ to perturb the data vector $\mathbf{b}$, but also use the correction matrix $\Delta \mathbf{A}$ to perturb the data matrix $\mathbf{A}$,

4.7 Total Least Squares Method

making the joint compensation for errors or noise in both $\mathbf{A}$ and $\mathbf{b}$:

$$\mathbf{b} + \Delta\mathbf{b} = \mathbf{b}_0 + \mathbf{e} + \Delta\mathbf{b} \to \mathbf{b}_0,$$
$$\mathbf{A} + \Delta\mathbf{A} = \mathbf{A}_0 + \mathbf{E} + \Delta\mathbf{A} \to \mathbf{A}_0,$$

so that the solution of the noisy matrix equation is transformed into the solution of the error-free matrix equation:

$$(\mathbf{A} + \Delta\mathbf{A})\mathbf{x} = \mathbf{b} + \Delta\mathbf{b} \ \Rightarrow \ \mathbf{A}_0\mathbf{x} = \mathbf{b}_0. \qquad (4.7.2)$$

Naturally, we want the correction data matrix and the correction data vectors are as small as possible. Hence the TLS problem can be expressed as the constrained optimization problem

$$\text{TLS:} \quad \min_{\Delta\mathbf{A},\Delta\mathbf{b},\mathbf{x}} \|[\Delta\mathbf{A},\Delta\mathbf{b}]\|_F^2 \quad \text{subject to} \quad (\mathbf{A}+\Delta\mathbf{A})\mathbf{x} = \mathbf{b}+\Delta\mathbf{b} \qquad (4.7.3)$$

or

$$\text{TLS:} \quad \min_{\mathbf{z}} \|\mathbf{D}\|_F^2 \quad \text{subject to} \quad \mathbf{D}\mathbf{z} = -\mathbf{B}\mathbf{z}, \qquad (4.7.4)$$

where $\mathbf{D} = [\Delta\mathbf{A}, \Delta\mathbf{b}]$, $\mathbf{B} = [\mathbf{A}, \mathbf{b}]$, and $\mathbf{z} = \begin{bmatrix} \mathbf{x} \\ -1 \end{bmatrix}$ is the $(n+1) \times 1$ vector.

Under the assumption $\|\mathbf{z}\|_2 = 1$, from $\|\mathbf{D}\|_2 \leq \|\mathbf{D}\|_F$ and $\|\mathbf{D}\|_2 = \sup_{\|\mathbf{z}\|_2=1} \|\mathbf{D}\mathbf{z}\|_2$, we have $\min_{\mathbf{z}} \|\mathbf{D}\|_F^2 = \min_{\mathbf{z}} \|\mathbf{D}\mathbf{z}\|_2^2$ and thus can rewrite (4.7.4) as

$$\text{TLS:} \quad \min_{\mathbf{z}} \|\mathbf{B}\mathbf{z}\|_2^2 \quad \text{subject to} \quad \|\mathbf{z}\| = 1 \qquad (4.7.5)$$

due to $\mathbf{D}\mathbf{z} = -\mathbf{B}\mathbf{z}$.

Case 1: Single Smallest Singular Value

The singular value σ_n of $\mathbf{B}$ is significantly larger than σ_{n+1}, i.e., the smallest singular value has only one. In this case, the TLS problem (4.7.5) is easily solved via the Lagrange multiplier method. To this end, define the objective function

$$J(\mathbf{z}) = \|\mathbf{B}\mathbf{z}\|_2^2 + \lambda(1 - \mathbf{z}^H\mathbf{z}), \qquad (4.7.6)$$

where λ is the Lagrange multiplier. Notice that $\|\mathbf{B}\mathbf{z}\|_2^2 = \mathbf{z}^H \mathbf{B}^H \mathbf{B} \mathbf{z}$, from $\frac{\partial J(\mathbf{z})}{\partial \mathbf{z}^*} = 0$ it follows that

$$\mathbf{B}^H \mathbf{B} \mathbf{z} = \lambda \mathbf{z}. \qquad (4.7.7)$$

This shows that we should select as the Lagrange multiplier the smallest eigenvalue $\lambda_{\min}$ of the matrix $\mathbf{B}^H\mathbf{B} = [\mathbf{A},\mathbf{b}]^H[\mathbf{A},\mathbf{b}]$ (i.e., the squares of the smallest singular value of **B**), while the TLS solution vector **z** is the eigenvector corresponding to the smallest eigenvalue λ_{min} of $[\mathbf{A}, \mathbf{b}]^H [\mathbf{A}, \mathbf{b}]$. In other words, the TLS solution vector $\begin{bmatrix} \mathbf{x} \\ -1 \end{bmatrix}$ is the solution of the Rayleigh quotient minimization problem

$$\min_{\mathbf{x}} \ J(\mathbf{x}) = \frac{\begin{bmatrix} \mathbf{x} \\ -1 \end{bmatrix}^H [\mathbf{A}, \mathbf{b}]^H [\mathbf{A}, \mathbf{b}] \begin{bmatrix} \mathbf{x} \\ -1 \end{bmatrix}}{\begin{bmatrix} \mathbf{x} \\ -1 \end{bmatrix}^H \begin{bmatrix} \mathbf{x} \\ -1 \end{bmatrix}} = \frac{\|\mathbf{A}\mathbf{x} - \mathbf{b}\|_2^2}{\|\mathbf{x}\|_2^2 + 1}. \tag{4.7.8}$$

Let the SVD of the $m \times (n+1)$ augmented matrix **B** be $\mathbf{B} = \mathbf{U\Sigma V}^H$, where the singular values are arranged in the order of $\sigma_1 \geq \cdots \geq \sigma_{n+1}$, and their corresponding right singular vectors are $\mathbf{v}_1, \cdots, \mathbf{v}_{n+1}$. According to the above analysis, the TLS solution is $\mathbf{z} = \mathbf{v}_{n+1}$, namely

$$\mathbf{x}_{TLS} = -\frac{1}{v(n+1, n+1)} \begin{bmatrix} v(1, n+1) \\ \vdots \\ v(n, n+1) \end{bmatrix}, \tag{4.7.9}$$

where $v(i, n+1)$ is the ith entry of $(n+1)$th column of **V**.

Remark If the augmented data matrix is given by $\mathbf{B} = [-\mathbf{b}, \mathbf{A}]$, then the TLS solution is provided by

$$\mathbf{x}_{TLS} = \frac{1}{v(1, n+1)} \begin{bmatrix} v(2, n+1) \\ \vdots \\ v(n+1, n+1) \end{bmatrix}. \tag{4.7.10}$$

Case 2: Multiple Smallest Singular Values
In this case, there are multiple smallest singular value of **B**, i.e., multiple small singular values are repeated or very close. Letting

$$\sigma_1 \geq \sigma_2 \geq \cdots \geq \sigma_p > \sigma_{p+1} \approx \cdots \approx \sigma_{n+1}, \tag{4.7.11}$$

then the right singular vector matrix $\mathbf{V}_1 = [\mathbf{v}_{p+1}, \mathbf{v}_{p+2}, \cdots, \mathbf{v}_{n+1}]$ will give $n - p + 1$ possible TLS solutions $\mathbf{x}_i = -\mathbf{y}_{p+i}/\alpha_{p+i}$, $i = 1, \cdots, n+1-p$. To find the unique TLS solution, one can make the Householder transformation **Q** to $\mathbf{V}_1$ such that

$$\mathbf{V}_1 \mathbf{Q} = \begin{bmatrix} \mathbf{y} & \vdots & \mathbf{x} \\ \cdots & \vdots & \cdots \\ \alpha & \vdots & 0 \cdots 0 \end{bmatrix}. \tag{4.7.12}$$

4.7 Total Least Squares Method

The unique TLS solution is the minimum norm solution given by $\hat{\mathbf{x}}_{\text{TLS}} = \mathbf{y}/\alpha$. This algorithm was proposed by Golub and Van Loan [15], as shown in Algorithm 4.5.

Algorithm 4.5 TLS algorithm for minimum norm solution
---
1. **input:** $\mathbf{A} \in \mathbb{C}^{m \times n}, \mathbf{b} \in \mathbb{C}^m, \alpha > 0$.
2. **repeat**
3.   Compute $\mathbf{B} = [\mathbf{A}, \mathbf{b}] = \mathbf{U}\mathbf{\Sigma}\mathbf{V}^H$, and save $\mathbf{V}$ and all singular values.
4.   Determine the number p of principal singular values.
5.   Put $\mathbf{V}_1 = [\mathbf{v}_{p+1}, \cdots, \mathbf{v}_{n+1}]$, and compute the Householder transformation

$$\mathbf{V}_1 \mathbf{Q} = \begin{bmatrix} \mathbf{y} & \vdots & \times \\ \cdots & \vdots & \cdots \\ \alpha & \vdots & 0 \cdots 0 \end{bmatrix}.$$

   where α is a scalar, and $\times$ denotes the uninterested block.

6.   **exit if** $\alpha \neq 0$.
7. **return** $p \leftarrow p - 1$
8. **output:** $\mathbf{x}_{\text{TLS}} = -\mathbf{y}/\alpha$.

The above minimum norm solution $\mathbf{x}_{\text{TLS}} = \mathbf{y}/\alpha$ contains n parameters other than p independent principal parameters because of rank$(\mathbf{A}) = p < n$.

In signal processing, system theory and artificial intelligence, the unique TLS solution with no redundant parameters is more interesting, which is the optimal LS approximate solution. To derive the optimal LS approximate solution, let the $m \times (n+1)$ matrix $\hat{\mathbf{B}} = \mathbf{U}\mathbf{\Sigma}_p \mathbf{V}^H$ be an optimal approximation with rank p of the augmented matrix $\mathbf{B}$, where $\mathbf{\Sigma}_p = \text{Diag}(\sigma_1, \cdots, \sigma_p, 0, \cdots, 0)$. Denote the $m \times (p+1)$ matrix $\hat{\mathbf{B}}_j^{(p)}$ as a submatrix among the $m \times (n+1)$ optimal approximate matrix $\hat{\mathbf{B}}$:

$$\hat{\mathbf{B}}_j^{(p)} : \text{sub-matrix of the } j\text{th to the } (j+p)\text{th columns of } \hat{\mathbf{B}}. \tag{4.7.13}$$

Clearly, there are $(n+1-p)$ submatrices $\hat{\mathbf{B}}_1^{(p)}, \hat{\mathbf{B}}_2^{(p)}, \cdots, \hat{\mathbf{B}}_{n+1-p}^{(p)}$.

As stated before, the fact that the efficient rank of $\mathbf{B}$ is equal to p means that p components are linearly independent in the parameter vector $\mathbf{x}$. Let the $(p+1) \times 1$ vector $\mathbf{a} = \begin{bmatrix} \mathbf{x}^{(p)} \\ -1 \end{bmatrix}$, where $\mathbf{x}^{(p)}$ is the column vector consisting of p linearly independent unknown parameters of the vector $\mathbf{x}$. Then, the original TLS problem becomes the solution of the following $(n+1-p)$ TLS problems:

$$\hat{\mathbf{B}}_j^{(p)} \mathbf{a} = 0, \quad j = 1, 2, \cdots, n+1-p \tag{4.7.14}$$

or equivalent to the solution of one synthetic TLS problem

$$\begin{bmatrix} \hat{\mathbf{B}}(1:p+1) \\ \vdots \\ \hat{\mathbf{B}}(n+1-p:n+1) \end{bmatrix} \mathbf{a} = \mathbf{0}, \qquad (4.7.15)$$

where $\hat{\mathbf{B}}(i:p+i) = \hat{\mathbf{B}}_i^{(p)}$ is defined in (4.7.13). It is not difficult to show that

$$\hat{\mathbf{B}}(i:p+i) = \sum_{k=1}^{p} \sigma_k \mathbf{u}_k (\mathbf{v}_k^i)^H, \qquad (4.7.16)$$

where $\mathbf{v}_k^i$ is a windowed segment of the kth column vector of $\mathbf{V}$, defined as

$$\mathbf{v}_k^i = [v(i,k), v(i+1,k), \cdots, v(i+p,k)]^T. \qquad (4.7.17)$$

Here $v(i,k)$ is the (i,k)th entry of $\mathbf{V}$.

According to the least square principle, finding the LS solution of Eq. (4.7.15) is equivalent to minimize the measure (or cost) function

$$\begin{aligned} f(\mathbf{a}) &= [\hat{\mathbf{B}}(1:p+1)\mathbf{a}]^H \hat{\mathbf{B}}(1:p+1)\mathbf{a} + [\hat{\mathbf{B}}(2:p+2)\mathbf{a}]^H \hat{\mathbf{B}}(2:p+2)\mathbf{a} \\ &\quad + \cdots + [\hat{\mathbf{B}}(n+1-p:n+1)\mathbf{a}]^H \hat{\mathbf{B}}(n+1-p:n+1)\mathbf{a} \\ &= \mathbf{a}^H \left[\sum_{i=1}^{n+1-p} [\hat{\mathbf{B}}(i:p+i)]^H \hat{\mathbf{B}}(i:p+i) \right] \mathbf{a}. \end{aligned} \qquad (4.7.18)$$

Define the $(p+1) \times (p+1)$ matrix

$$\mathbf{S}^{(p)} = \sum_{i=1}^{n+1-p} [\hat{\mathbf{B}}(i:p+i)]^H \hat{\mathbf{B}}(i:p+i), \qquad (4.7.19)$$

then the measure function can be simply written as

$$f(\mathbf{a}) = \mathbf{a}^H \mathbf{S}^{(p)} \mathbf{a}. \qquad (4.7.20)$$

The minimal variable $\mathbf{a}$ of the measure function $f(\mathbf{a})$ is determined by $\frac{\partial f(\mathbf{a})}{\partial \mathbf{a}^*} = 0$ as follows:

$$\mathbf{S}^{(p)} \mathbf{a} = \alpha \mathbf{e}_1 \qquad (4.7.21)$$

where $\mathbf{e}_1 = [1, 0, \cdots, 0]^T$, and the constant $\alpha > 0$ represents the error energy. From (4.7.19) and (4.7.16) we have

$$\mathbf{S}^{(p)} = \sum_{j=1}^{p} \sum_{i=1}^{n+1-p} \sigma_j^2 \mathbf{v}_j^i (\mathbf{v}_j^i)^H. \qquad (4.7.22)$$

Solving the matrix equation (4.7.21) is simple. Let $\mathbf{S}^{-(p)}$ be the inverse matrix $\mathbf{S}^{(p)}$. Then the solution vector $\mathbf{a}$ is only dependent of the first column of the inverse matrix $\mathbf{S}^{-(p)}$. It is easily known that the ith entry of $\mathbf{x}^{(p)} = [x_{\text{TLS}}(1), \cdots, x_{\text{TLS}}(p)]^T$ in the TLS solution vector $\mathbf{a} = \begin{bmatrix} \mathbf{x}^{(p)} \\ -1 \end{bmatrix}$ is given by

$$x_{\text{TLS}}(i) = -\mathbf{S}^{-(p)}(i, 1)/\mathbf{S}^{-(p)}(p+1, 1), \qquad i = 1, \cdots, p. \qquad (4.7.23)$$

This solution is known as the *optimal least square approximate solution*. Because the number of parameters of this solution and the efficient rank are the same, it is also called the *low-rank TLS solution* [5].

Notice that if the augmented matrix $\mathbf{B} = [-\mathbf{b}, \mathbf{A}]$, then

$$x_{\text{TLS}}(i) = \mathbf{S}^{-(p)}(i+1, 1)/\mathbf{S}^{-(p)}(1, 1), \qquad i = 1, 2, \cdots, p. \qquad (4.7.24)$$

In summary, given $\mathbf{A} \in \mathbb{C}^{m \times n}, \mathbf{b} \in \mathbb{C}^n$, the low-rank SVD-TLS algorithm consists of the following steps [5].

1. Compute the SVD $\mathbf{B} = [\mathbf{A}, \mathbf{b}] = \mathbf{U}\mathbf{\Sigma}\mathbf{V}^H$, and save $\mathbf{V}$.
2. Determine the efficient rank p of $\mathbf{B}$.
3. Use (4.7.22) and (4.7.17) to calculate $(p+1) \times (p+1)$ matrix $\mathbf{S}^{(p)}$.
4. Compute the inverse $\mathbf{S}^{-(p)}$ and the total least square solution

$$x_{\text{TLS}}(i) = -\mathbf{S}^{-(p)}(i, 1)/\mathbf{S}^{-(p)}(p+1, 1), \ i = 1, \cdots, p.$$

4.7.2 Performances of TLS Solution

The TLS has two interesting explanations: one is its geometric interpretation [15], and another is its closed solution [43].

Geometric Interpretation of TLS Solution

Let $\mathbf{a}_i^T$ be the ith row of the matrix $\mathbf{A}$, and b_i be the ith entry of the vector $\mathbf{b}$. Then the TLS solution $\mathbf{x}_{\text{TLS}}$ is the minimal vector such that

$$\min_{\mathbf{x}} \frac{\|\mathbf{A}\mathbf{x} - \mathbf{b}\|_2^2}{\|\mathbf{x}\|_2^2 + 1} = \sum_{i=1}^{n} \frac{|\mathbf{a}_i^T \mathbf{x} - b_i|^2}{\mathbf{x}^T \mathbf{x} + 1}, \qquad (4.7.25)$$

where $|\mathbf{a}_i^T\mathbf{x}-b_i|/(\mathbf{x}^T\mathbf{x}+1)$ is the distance from the point $\begin{pmatrix}\mathbf{a}_i\\b_i\end{pmatrix}\in\mathbb{C}^{n+1}$ to the nearest point in the subspace P_x, and the subspace P_x is defined as

$$P_x = \left\{\begin{pmatrix}\mathbf{a}\\b\end{pmatrix} : \mathbf{a}\in\mathbb{C}^{n\times 1}, b\in\mathbb{C}, b=\mathbf{x}^T\mathbf{a}\right\}. \tag{4.7.26}$$

Hence the TLS solution can be expressed by using the subspace P_x [15]: the square sum of the distance from the TLS solution point $\begin{pmatrix}\mathbf{a}_i\\b_i\end{pmatrix}$ to the subspace P_x is minimized.

Closed Solution of TLS Problems

If the singular values of the augmented matrix $\mathbf{B}$ are $\sigma_1 \geq \cdots \geq \sigma_{n+1}$, then the TLS solution can be expressed as [43]

$$\mathbf{x}_{\text{TLS}} = (\mathbf{A}^H\mathbf{A} - \sigma_{n+1}^2\mathbf{I})^{-1}\mathbf{A}^H\mathbf{b}. \tag{4.7.27}$$

Compared with the Tikhonov regularization method, the TLS is a kind of anti-regularization method and can be interpreted as a kind of least square method with noise removal: it first removes the noise affection term $\sigma_{n+1}^2\mathbf{I}$ from the covariance matrix $\mathbf{A}^T\mathbf{A}$, and then finds the inverse matrix of $\mathbf{A}^T\mathbf{A}-\sigma_{n+1}^2\mathbf{I}$ to get the LS solution.

Letting the noisy data matrix be $\mathbf{A} = \mathbf{A}_0 + \mathbf{E}$, then its covariance matrix $\mathbf{A}^H\mathbf{A} = \mathbf{A}_0^H\mathbf{A}_0 + \mathbf{E}^H\mathbf{A}_0 + \mathbf{A}_0^H\mathbf{E} + \mathbf{E}^H\mathbf{E}$. Obviously, when the error matrix $\mathbf{E}$ has zero mean, the mathematical expectation of the covariance matrix is given by

$$E\{\mathbf{A}^H\mathbf{A}\} = E\{\mathbf{A}_0^H\mathbf{A}_0\} + E\{\mathbf{E}^H\mathbf{E}\} = \mathbf{A}_0^H\mathbf{A}_0 + E\{\mathbf{E}^H\mathbf{E}\}.$$

If the column vectors of the error matrix are statistically independent, and have the same variance, i.e., $E\{\mathbf{E}^T\mathbf{E}\} = \sigma^2\mathbf{I}$, then the smallest eigenvalue $\lambda_{n+1} = \sigma_{n+1}^2$ of the $(n+1)\times(n+1)$ covariance matrix $\mathbf{A}^H\mathbf{A}$ is the square of the singular value of the error matrix $\mathbf{E}$. Because the square of singular value σ_{n+1}^2 happens to reflect the common variance σ^2 of each column vector of the error matrix, the covariance matrix $\mathbf{A}_0^H\mathbf{A}_0$ of the error-free data matrix can be retrieved from $\mathbf{A}^H\mathbf{A} - \sigma_{n+1}^2\mathbf{I}$, namely $\mathbf{A}^T\mathbf{A} - \sigma_{n+1}^2\mathbf{I} = \mathbf{A}_0^H\mathbf{A}_0$. In other words, the TLS method can effectively restrain the influence of the unknown error matrix.

It should be pointed out that the main difference between the TLS method and the Tikhonov regularization method for solving the matrix equation $\mathbf{A}_{m\times n}\mathbf{x}_n = \mathbf{b}_m$ is: the TLS solution can contain only $p = \text{rank}([\mathbf{A},\mathbf{b}])$ principal parameters, and exclude the redundant parameters; whereas the Tikhonov regularization method can only provide all n parameters including redundant parameters.

4.7.3 Generalized Total Least Square

The ordinary LS, the Tikhonov regularization, and the TLS method can be derived and explained by a unified theoretical framework [46].

Consider the minimization problem

$$\min_{\Delta \mathbf{A}, \Delta \mathbf{b}, \mathbf{x}} \left(\|[\Delta \mathbf{A}, \Delta \mathbf{b}]\|_F^2 + \lambda \|\mathbf{x}\|_2^2 \right) \quad (4.7.28)$$

subject to

$$(\mathbf{A} + \alpha \Delta \mathbf{A})\mathbf{x} = \mathbf{b} + \beta \Delta \mathbf{b},$$

where α and β are, respectively, the weighting coefficients of the perturbation $\Delta \mathbf{A}$ of the data matrix $\mathbf{A}$ and the perturbation $\Delta \mathbf{b}$ of the data vector $\mathbf{b}$, and λ is the Tikhonov regularization parameter. The above minimization problem is called the *generalized total least squares* (GTLS) problem [46].

The constraint condition $(\mathbf{A} + \alpha \Delta \mathbf{A})\mathbf{x} = (\mathbf{b} + \beta \Delta \mathbf{b})$ can be represented as

$$\left([\alpha^{-1} \mathbf{A}, \beta^{-1} \mathbf{b}] + [\Delta \mathbf{A}, \Delta \mathbf{b}] \right) \begin{bmatrix} \alpha \mathbf{x} \\ -\beta \end{bmatrix} = 0.$$

If letting $\mathbf{D} = [\Delta \mathbf{A}, \Delta \mathbf{b}]$ and $\mathbf{z} = \begin{bmatrix} \alpha \mathbf{x} \\ -\beta \end{bmatrix}$, then the above equation becomes

$$\mathbf{D}\mathbf{z} = -\left[\alpha^{-1} \mathbf{A}, \beta^{-1} \mathbf{b} \right] \mathbf{z}. \quad (4.7.29)$$

Under the assumption of $\mathbf{z}^H \mathbf{z} = 1$, then

$$\min \|\mathbf{D}\|_F^2 = \min \|\mathbf{D}\mathbf{z}\|_2^2 = \min \left\| [\alpha^{-1} \mathbf{A}, \beta^{-1} \mathbf{b}] \mathbf{z} \right\|_2^2,$$

and thus the solution to the GTLS problem (4.7.28) can be rewritten as

$$\hat{\mathbf{x}}_{\text{GTLS}} = \arg \min_{\mathbf{z}} \left(\frac{\|[\alpha^{-1} \mathbf{A}, \beta^{-1} \mathbf{b}]\mathbf{z}\|_2^2}{\mathbf{z}^H \mathbf{z}} + \lambda \|\mathbf{x}\|_2^2 \right). \quad (4.7.30)$$

Noticing that $\mathbf{z}^H \mathbf{z} = \alpha^2 \mathbf{x}^H \mathbf{x} + \beta^2$ and

$$[\alpha^{-1} \mathbf{A}, \beta^{-1} \mathbf{b}]\mathbf{z} = [\alpha^{-1} \mathbf{A}, \beta^{-1} \mathbf{b}] \begin{bmatrix} \alpha \mathbf{x} \\ -\beta \end{bmatrix} = \mathbf{A}\mathbf{x} - \mathbf{b},$$

the GTLS solution in (4.7.30) can be expressed as

$$\hat{\mathbf{x}}_{\text{GTLS}} = \arg\min_{\mathbf{x}} \left(\frac{\|\mathbf{A}\mathbf{x} - \mathbf{b}\|_2^2}{\alpha^2 \|\mathbf{x}\|_2^2 + \beta^2} + \lambda \|\mathbf{x}\|_2^2 \right). \tag{4.7.31}$$

The following are the comparison of the ordinary LS method, the Tikhonov regularization, and the TLS method.

1. *Comparison of optimization problems*
   - The ordinary least squares: $\alpha = 0, \beta = 1$ and $\lambda = 0$, which gives

   $$\min_{\Delta\mathbf{A}, \Delta\mathbf{b}, \mathbf{x}} \|\Delta\mathbf{b}\|_2^2 \quad \text{subject to} \quad \mathbf{A}\mathbf{x} = \mathbf{b} + \Delta\mathbf{b}. \tag{4.7.32}$$

   - The Tikhonov regularization: $\alpha = 0, \beta = 1$ and $\lambda > 0$, which gives

   $$\min_{\Delta\mathbf{A}, \Delta\mathbf{b}, \mathbf{x}} \left(\|\Delta\mathbf{b}\|_2^2 + \lambda \|\mathbf{x}\|_2^2 \right) \quad \text{subject to} \quad \mathbf{A}\mathbf{x} = \mathbf{b} + \Delta\mathbf{b}. \tag{4.7.33}$$

   - The total least squares: $\alpha = \beta = 1$ and $\lambda = 0$, which yields

   $$\min_{\Delta\mathbf{A}, \Delta\mathbf{b}, \mathbf{x}} \left(\|\Delta\mathbf{A}, \Delta\mathbf{b}\|_2^2 \right) \quad \text{subject to} \quad (\mathbf{A} + \Delta\mathbf{A})\mathbf{x} = \mathbf{b} + \Delta\mathbf{b}. \tag{4.7.34}$$

2. *Comparison of solution vectors:* When the weighting coefficients α, β and the Tikhonov regularization parameter λ take the suitable values, the GTLS solution (4.7.31) gives the following results:
   - $\hat{\mathbf{x}}_{\text{LS}} = (\mathbf{A}^H \mathbf{A})^{-1} \mathbf{A}^H \quad (\alpha = 0, \beta = 1, \lambda = 0)$,
   - $\hat{\mathbf{x}}_{\text{Tik}} = (\mathbf{A}^H \mathbf{A} + \lambda \mathbf{I})^{-1} \mathbf{A}^H \mathbf{b} \quad (\alpha = 0, \beta = 1, \lambda > 0)$,
   - $\hat{\mathbf{x}}_{\text{TLS}} = (\mathbf{A}^H \mathbf{A} - \lambda \mathbf{I})^{-1} \mathbf{A}^H \mathbf{b} \quad (\alpha = 1, \beta = 1, \lambda = 0)$.

3. *Comparison of perturbation methods*
   - Ordinary LS method: it uses the possible small correction term $\Delta\mathbf{b} = \mathbf{A}\mathbf{x} - \mathbf{b}$ to disturb the data vector $\mathbf{b}$ such that $\mathbf{b} - \Delta\mathbf{b} \approx \mathbf{b}_0$.
   - Tikhonov regularization method: it adds the same perturbation term $\lambda > 0$ to every diagonal entry of the matrix $\mathbf{A}^H \mathbf{A}$ for avoiding the numerical instability of the LS solution $(\mathbf{A}^H \mathbf{A})^{-1} \mathbf{A}^H \mathbf{b}$.
   - TLS method: by subtracting the perturbation matrix $\lambda \mathbf{I}$, the noise or disturbance in the covariance matrix of the original data matrix is restrained.

4. *Comparison of application ranges*
   - The LS method is applicable for the data matrix $\mathbf{A}$ with full column rank and the data vector $\mathbf{b}$ existing the iid Gaussian errors.

- The Tikhonov regularization method is applicable for the data matrix **A** with deficient column rank.
- The TLS method is applicable for the data matrix **A** with full column rank and both **A** and the data vector **b** existing the iid Gaussian error.

4.8 Solution of Under-Determined Systems

A full row rank under-determined matrix equation $\mathbf{A}_{m \times n} \mathbf{x}_{n \times 1} = \mathbf{b}_{m \times 1}$ with $m < n$ has infinitely many solutions. In these cases, letting $\mathbf{x} = \mathbf{A}^H \mathbf{y}$ then

$$\mathbf{A}\mathbf{A}^H \mathbf{y} = \mathbf{b} \Rightarrow \mathbf{y} = (\mathbf{A}\mathbf{A}^H)^{-1}\mathbf{b} \Rightarrow \mathbf{x} = \mathbf{A}^H (\mathbf{A}\mathbf{A}^H)^{-1}\mathbf{b}.$$

This unique solution is known as the minimum norm solution. However, such a solution has little application in engineering. We are interested in the sparse solution **x** with a small number of nonzero entries.

4.8.1 ℓ_1-Norm Minimization

To seek the sparsest solution with the fewest nonzero elements, we would ask the following two questions.

- Can it ever be unique?
- How to find the sparsest solution?

For any positive number $p > 0$, the ℓ_p-norm of the vector **x** is defined as

$$\|\mathbf{x}\|_p = \left(\sum_{i \in \text{support}(\mathbf{x})} |x_i|^p \right)^{1/p}. \tag{4.8.1}$$

Then the ℓ_0-norm of an $n \times 1$ vector **x** can be defined as

$$\|\mathbf{x}\|_0 = \lim_{p \to 0} \|\mathbf{x}\|_p^p = \lim_{p \to 0} \sum_{i=1}^n |x_i|^p = \sum_{i=1}^n 1(x_i \neq 0) = \#\{i | x_i \neq 0\}, \tag{4.8.2}$$

where $\#\{i | x_i \neq 0\}$ denotes the number of all nonzero elements of **x**. Thus if $\|\mathbf{x}\|_0 \ll n$, then **x** is sparse.

The core problem of sparse representation is ℓ_0-*norm minimization*

$$(P_0) \qquad \min_{\mathbf{x}} \|\mathbf{x}\|_0 \quad \text{subject to } \mathbf{b} = \mathbf{A}\mathbf{x}, \tag{4.8.3}$$

where $\mathbf{A} \in \mathbb{R}^{m \times n}, \mathbf{x} \in \mathbb{R}^n, \mathbf{b} \in \mathbb{R}^m$.

As the observation signal is usually contaminated by noise, the equality constraint in the above optimization problem should be relaxed to the ℓ_0-norm minimization with the inequality constraint $\epsilon \geq 0$ which allows a certain error perturbation

$$\min_{\mathbf{x}} \|\mathbf{x}\|_0 \quad \text{subject to} \quad \|\mathbf{A}\mathbf{x} - \mathbf{b}\|_2 \leq \epsilon. \tag{4.8.4}$$

A key term coined and defined in [9] that is crucial for the study of uniqueness is the sparsity of the matrix $\mathbf{A}$.

Definition 4.7 (Sparsity [9]) Given a matrix $\mathbf{A}$, its *sparsity*, denoted as $\sigma = \text{spark}(\mathbf{A})$, is defined as the smallest possible number such that there exists a subgroup of σ columns from $\mathbf{A}$ that are linearly dependent.

The sparsity gives a simple criterion for uniqueness of sparse solution of an under-determined system of linear equations $\mathbf{A}\mathbf{x} = \mathbf{b}$, as stated below.

Theorem 4.9 ([9, 17]) *If a system of linear equations $\mathbf{A}\mathbf{x} = \mathbf{b}$ has a solution $\mathbf{x}$ obeying $\|\mathbf{x}\|_0 < \text{spark}(\mathbf{A})/2$, this solution is necessarily the sparsest solution.*

The index set of nonzero elements of the vector $\mathbf{x} = [x_i, \cdots, x_n]^T$ is called its support, denoted by $\text{support}(\mathbf{x}) = \{i \,|\, x_i \neq 0\}$. The length of the support (i.e., the number of nonzero elements) is measured by ℓ_0-norm

$$\|\mathbf{x}\|_0 = |\text{support}(\mathbf{x})|. \tag{4.8.5}$$

A vector $\mathbf{x} \in \mathbb{R}^n$ is said to be K-*sparse*, if $\|\mathbf{x}\|_0 \leq K$, where $K \in \{1, \cdots, n\}$.

The set of the K-sparse vectors is denoted by

$$\Sigma_K = \left\{ \mathbf{x} \in \mathbb{R}^{n \times 1} \,\middle|\, \|\mathbf{x}\|_0 \leq K \right\}. \tag{4.8.6}$$

If $\hat{\mathbf{x}} \in \Sigma_K$, then the vector $\hat{\mathbf{x}} \in \mathbb{R}^n$ is known as the K-*sparse approximation*.

Clearly, there is a close relationship between ℓ_0-norm definition formula (4.8.5) and ℓ_p-norm definition formula (4.8.1): when $p \to 0$, $\|\mathbf{x}\|_0 = \lim_{p \to 0} \|\mathbf{x}\|_p^p$. Because $\|\mathbf{x}\|_p$ is the convex function if and only if $p \geq 1$, the ℓ_1-norm is the objective function closest to ℓ_0-norm. Then, from the viewpoint of optimization, the ℓ_1-norm is said to be the *convex relaxation* of the ℓ_0-norm. Therefore, the ℓ_0-norm minimization problem (P_0) in (4.8.3) can transformed into the following convexly relaxed ℓ_1-*norm minimization* problem:

$$(P_1) \qquad \min_{\mathbf{x}} \|\mathbf{x}\|_1 \quad \text{subject to} \quad \mathbf{b} = \mathbf{A}\mathbf{x}. \tag{4.8.7}$$

This is a convex optimization problem, because ℓ_1-norm $\|\mathbf{x}\|_1$, as the objective function, itself is the convex function, and the equality constraint $\mathbf{b} = \mathbf{A}\mathbf{x}$ is the affine function.

4.8 Solution of Under-Determined Systems

Due to observation noise, the equality constrained optimization problem (P_1) should be relaxed as the inequality constrained optimization problem below:

$$(P_{10}) \qquad \min_{\mathbf{x}} \|\mathbf{x}\|_1 \quad \text{subject to} \quad \|\mathbf{b} - \mathbf{A}\mathbf{x}\|_2 \leq \epsilon. \qquad (4.8.8)$$

The ℓ_1-norm minimization above is also called the *basis pursuit (BP)*. This is a *quadratically constrained linear programming (QCLP) problem*.

If $\mathbf{x}_1$ is the solution to (P_1), and $\mathbf{x}_0$ is the solution to (P_0), then [8]

$$\|\mathbf{x}_1\|_1 \leq \|\mathbf{x}_0\|_1, \qquad (4.8.9)$$

because $\mathbf{x}_0$ is only the feasible solution to (P_1), while $\mathbf{x}_1$ is the optimal solution to (P_1). The direct relationship between $\mathbf{x}_0$ and $\mathbf{x}_1$ is given by $\mathbf{A}\mathbf{x}_1 = \mathbf{A}\mathbf{x}_0$.

Similar to the inequality ℓ_0-norm minimization expression (4.8.4), the inequality ℓ_1-norm minimization expression (4.8.8) has also two variants:

- Since $\mathbf{x}$ is constrained as K-sparse vector, the inequality ℓ_1-norm minimization becomes the inequality ℓ_2-norm minimization

$$(P_{11}) \qquad \min_{\mathbf{x}} \frac{1}{2}\|\mathbf{b} - \mathbf{A}\mathbf{x}\|_2^2 \quad \text{subject to} \quad \|\mathbf{x}\|_1 \leq q. \qquad (4.8.10)$$

This is a *quadratic programming (QP) problem*.

- Using the Lagrange multiplier method, the inequality constrained ℓ_1-norm minimization problem (P_{11}) becomes

$$(P_{12}) \qquad \min_{\lambda,\mathbf{x}} \frac{1}{2}\|\mathbf{b} - \mathbf{A}\mathbf{x}\|_2^2 + \lambda\|\mathbf{x}\|_1. \qquad (4.8.11)$$

This optimization is called the *basis pursuit denoising* (BPDN) [7].

The optimization problems (P_{10}) and (P_{11}) are, respectively, called the *error constrained ℓ_1-norm minimization* and *ℓ_1-penalty minimization* [41]. The ℓ_1-penalty minimization is also known as the regularized ℓ_1 linear programming or ℓ_1-norm regularization least squares.

The Lagrange multiplier is known as the regularization parameter that controls the sparseness of the sparse solution; and the greater λ is, the more sparse the solution $\mathbf{x}$ is. When the regularization parameter λ is large enough, the solution $\mathbf{x}$ is a zero vector. With the gradual decrease of λ, the sparsity of the solution vector $\mathbf{x}$ gradually decreases. As λ is gradually reduced to 0, the solution vector $\mathbf{x}$ becomes the vector such that $\|\mathbf{b} - \mathbf{A}\mathbf{x}\|_2^2$ is minimized. That is to say, $\lambda > 0$ can balance the error squares sum cost function $\frac{1}{2}\|\mathbf{b} - \mathbf{A}\mathbf{x}\|_2^2$ and the ℓ_1-norm cost function $\|\mathbf{x}\|_1$ of the twin objectives

$$J(\lambda, \mathbf{x}) = \frac{1}{2}\|\mathbf{b} - \mathbf{A}\mathbf{x}\|_2^2 + \lambda\|\mathbf{x}\|_1. \qquad (4.8.12)$$

4.8.2 Lasso

The solution of a system of sparse equations is closely related to regression analysis. Minimizing the least squared error may usually lead to sensitive solutions. Many regularization methods were proposed to decrease this sensitivity. Among them, Tikhonov regularization [40] and Lasso [12, 38] are two widely known and cited algorithms, as pointed out in [44].

The problem of regression analysis is one of the fundamental problems within the many fields such as statistics, supervised machine learning, optimization, and so on. In order to reduce the computational complexity of solving directly the optimization problem (P_1), consider a *linear regression* problem: given an observed data vector $\mathbf{b} \in \mathbb{R}^m$ and an observed data matrix $\mathbf{A} \in \mathbb{R}^{m \times n}$, find a fitting coefficient vector $\mathbf{x} \in \mathbb{R}^n$ such that

$$\hat{b}_i = x_1 a_{i1} + x_2 a_{i2} + \cdots + x_n a_{in}, \quad i = 1, \ldots, m, \tag{4.8.13}$$

or

$$\hat{\mathbf{b}} = \sum_{i=1}^{n} x_i \mathbf{a}_i = \mathbf{A}\mathbf{x}, \tag{4.8.14}$$

where

$$\mathbf{x} = [x_1, \ldots, x_n]^T, \quad \mathbf{b} = [b_1, \ldots, b_m]^T, \quad \mathbf{A} = [\mathbf{a}_1, \ldots, \mathbf{a}_n] = \begin{bmatrix} a_{11} & \cdots & a_{1n} \\ \vdots & \ddots & \vdots \\ a_{m1} & \cdots & a_{mn} \end{bmatrix}.$$

As a preprocessing of the linear regression, it is assumed that

$$\sum_{i=1}^{m} b_i = 0, \quad \sum_{i=1}^{m} a_{ij} = 0, \quad \sum_{i=1}^{m} a_{ij}^2 = 1, \quad j = 1, \ldots, n, \tag{4.8.15}$$

and let the column vectors of the input matrix $\mathbf{A}$ be linearly independent. The preprocessed input matrix $\mathbf{A}$ is called an *orthonormal input matrix*, its column vector $\mathbf{a}_i$ is known as a *predictor*; and the vector $\mathbf{x}$ is simply called a coefficient vector.

Tibshirani [38] in 1996 has proposed the *least absolute shrinkage and selection operator (Lasso) algorithm* for solving the linear regression. The basic idea of the Lasso is: under the constraint that the ℓ_1-norm of the prediction vector does not

4.8 Solution of Under-Determined Systems

exceed an upper bound q, the squares sum of prediction errors is minimized, namely

$$\text{Lasso}: \quad \min_{\mathbf{x}} \|\mathbf{b} - \mathbf{A}\mathbf{x}\|_2^2$$

$$\text{subject to} \quad \|\mathbf{x}\|_1 = \sum_{i=1}^{n} |x_i| \le q. \tag{4.8.16}$$

Obviously, the Lasso model and the QP problem (4.8.10) have the exactly same form.

The bound q is a tuning parameter. When q is large enough, the constraint has no effect on $\mathbf{x}$ and the solution is just the usual multiple linear least squares regression of $\mathbf{x}$ on $a_{i1}, \ldots, a_{in}$ and $b_i, i = 1, \ldots, m$. However, for the smaller values of q ($q \ge 0$), some of the coefficients x_j will take zero. Choosing a suitable q will lead to a sparse coefficient vector $\mathbf{x}$.

The regularized Lasso problem is given by

$$\text{Regularized Lasso:} \quad \min_{\mathbf{x}} \|\mathbf{b} - \mathbf{A}\mathbf{x}\|_2^2 + \lambda \|\mathbf{x}\|_1. \tag{4.8.17}$$

The Lasso problem involves both the ℓ_1-norm constrained fitting for statistics and the data mining.

With the aid of two basic functions, the Lasso makes shrinkage and selection for linear regression.

- *Contraction function:* the Lasso shrinks the range of parameters to be estimated, only a small number of selected parameters are estimated at each step.
- *Selection function:* the Lasso would automatically select a very small part of variables for linear regression, yielding a spare solution.

The Lasso method achieves the better prediction accuracy by shrinkage and variable selection.

The Lasso has many generalizations, here are a few examples. In machine learning, sparse kernel regression is referred to as the *generalized Lasso* [36]. Multidimensional shrinkage-thresholding method is known as the *group Lasso* [35]. A sparse multiview feature selection method via low-rank analysis is called the *MRM-Lasso* (multiview rank minimization-based Lasso) [45]. *Distributed Lasso* solves the distributed sparse linear regression problems [29].

4.8.3 LARS

The LARS is the abbreviation of *least angle regressions* and is a stepwise regression method. *Stepwise regression* is also known as "forward stepwise regression." Given a collection of possible predictors, $X = \{\mathbf{x}_1, \ldots, \mathbf{x}_m\}$, the aim of LARS is to identify the fitting vector most correlated with the response vector, i.e., the angle between the two vectors is minimized.

The LARS algorithm contains the two basic steps.

1. The first step selects the fitting vector having largest absolute correlation with the response vector $\mathbf{y}$, say $\mathbf{x}_{j_1} = \max_i \{|\text{corr}(\mathbf{y}, \mathbf{x}_i)|\}$, and performs simple linear regression of $\mathbf{y}$ on $\mathbf{x}_{j_1}$: $\mathbf{y} = \beta_1 \mathbf{x}_{j_1} + \mathbf{r}$, where the residual vector $\mathbf{r}$ denotes the residuals of the remaining variables removing $\mathbf{x}_{j_1}$ from X. The residual vector $\mathbf{r}$ is orthogonal to $\mathbf{x}_{j_1}$. When some predictors happen to be highly correlated with $\mathbf{x}_j$ and thus are approximately orthogonal to $\mathbf{r}$, these predictors perhaps are eliminated, i.e., the residual vector $\mathbf{r}$ will not have information on these eliminating predictors.
2. In the second step, $\mathbf{r}$ is regarded as the new response. The LARS selects the fitting vector having largest absolute correlation with the new response $\mathbf{r}$, say $\mathbf{x}_{j_2}$, and performs simple linear regression $\mathbf{r} \leftarrow \mathbf{r} - \beta_2 \mathbf{x}_{j_2}$. Repeat this selection process until no predictor has any correlation with $\mathbf{r}$.

After k selection steps, one gets a set of predictors $\mathbf{x}_{j_1}, \ldots, \mathbf{x}_{j_k}$ that are then used in the usual way to construct a k-parameter linear model. This selection regression is an aggressive fitting technique that can be overly greedy [12]. The stepwise regression may be said to be of *under-regression* due to perhaps eliminating useful predictors that happen to be correlated with current predictor.

To avoid the under-regression in forward stepwise regression, the LARS algorithm allows "stagewise regression" to be implemented using fairly large steps.

Algorithm 4.6 shows the LARS algorithm developed in [12].

Stagewise regression, called also "forward stagewise regression," is a much more cautious version of stepwise regression, and creates a coefficient profile as follows: at each step it increments the coefficient of that variable most correlated with the current residuals by an amount $\pm\epsilon$, with the sign determined by the sign of the correlation, as shown below [19].

1. Start with $\mathbf{r} = \mathbf{y}$ and $\beta_1 = \beta_2 = \ldots = \beta_p = 0$.
2. Find the predictor $\mathbf{x}_j$ most correlated with $\mathbf{r}$.
3. Update $\beta_j \leftarrow \beta_j + \delta_j$, where $\delta_j = \epsilon \cdot \text{sign}[\text{corr}(\mathbf{r}, \mathbf{x}_j)]$;
4. Update $\mathbf{r} \leftarrow \mathbf{r} - \delta_j \mathbf{x}_j$, and repeat Steps 2 and 3 until no predictor has any correlation with $\mathbf{r}$.

Brief Summary of This Chapter

- This chapter mainly introduces singular value decomposition and several typical methods for solving the well-determined, over-determined, and under-determined matrix equations.
- Tikhonov regularization method and ℓ_1-norm minimization have wide applications in artificial intelligence.

- Generalized total least squares method includes the LS method, Tikhonov regularization method, and total least squares method as special examples.

Algorithm 4.6 Least angle regressions (LARS) algorithm with Lasso modification [12]

---

1. **input:** The data vector $\mathbf{b} \in \mathbb{R}^m$ and the input matrix $\mathbf{A} \in \mathbb{R}^{m \times n}$.
2. **initialization:** $\Omega_0 = \emptyset$, $\hat{\mathbf{b}} = \mathbf{0}$ and $\mathbf{A}_{\Omega_0} = \mathbf{A}$. Put $k = 1$.
3. **repeat**
4. Compute the correlation vector $\hat{\mathbf{c}}_k = \mathbf{A}_{\Omega_{k-1}}^T (\mathbf{b} - \hat{\mathbf{b}}_{k-1})$.
5. Update the active set $\Omega_k = \Omega_{k-1} \cup \{j^{(k)} \mid |\hat{c}_k(j)| = C\}$ with $C = \max\{|\hat{c}_k(1)|, \ldots, |\hat{c}_k(n)|\}$.
6. Update the input matrix $\mathbf{A}_{\Omega_k} = [s_j \mathbf{a}_j, \, j \in \Omega_k]$, where $s_j = \text{sign}(\hat{c}_k(j))$.
7. Find the direction of the current minimum angle

$$\mathbf{G}_{\Omega_k} = \mathbf{A}_{\Omega_k}^T \mathbf{A}_{\Omega_k} \in \mathbb{R}^{k \times k},$$

$$\alpha_{\Omega_k} = (\mathbf{1}_k^T \mathbf{G}_{\Omega_k} \mathbf{1}_k)^{-1/2},$$

$$\mathbf{w}_{\Omega_k} = \alpha_{\Omega_k} \mathbf{G}_{\Omega_k}^{-1} \mathbf{1}_k \in \mathbb{R}^k,$$

$$\boldsymbol{\mu}_k = \mathbf{A}_{\Omega_k} \mathbf{w}_{\Omega_k} \in \mathbb{R}^k.$$

8. Compute $\mathbf{b} = \mathbf{A}_{\Omega_k}^T \boldsymbol{\mu}_k = [b_1, \ldots, b_m]^T$ and estimate the coefficient vector

$$\hat{\mathbf{x}}_k = (\mathbf{A}_{\Omega_k}^T \mathbf{A}_{\Omega_k})^{-1} \mathbf{A}_{\Omega_k}^T = \mathbf{G}_{\Omega_k}^{-1} \mathbf{A}_{\Omega_k}^T.$$

9. Compute $\hat{\gamma} = \min_{j \in \Omega_k^c} \left\{ \dfrac{C - \hat{c}_k(j)}{\alpha_{\Omega_k} - b_j}, \dfrac{C + \hat{c}_k(j)}{\alpha_{\Omega_k} + b_j} \right\}^+$, $\tilde{\gamma} = \min_{j \in \Omega_k} \left\{ -\dfrac{x_j}{w_j} \right\}^+$, where w_j is the jth of entry

$\mathbf{w}_{\Omega_k} = [w_1, \ldots, w_n]^T$ and $\min\{\cdot\}^+$ denotes the positive minimum term. If there is not positive term then $\min\{\cdot\}^+ = \infty$.

10. If $\tilde{\gamma} < \hat{\gamma}$ then the fitted vector and the active set are modified as follows:

$$\hat{\mathbf{b}}_k = \hat{\mathbf{b}}_{k-1} + \tilde{\gamma} \boldsymbol{\mu}_k, \quad \Omega_k = \Omega_k - \{\tilde{j}\},$$

where the removed index $\tilde{j}$ is the index $j \in \Omega_k$ such that $\tilde{\gamma}$ is a minimum. Conversely, if $\hat{\gamma} < \tilde{\gamma}$

then $\hat{\mathbf{b}}_k$ and Ω_k are modified as follows:

$$\hat{\mathbf{b}}_k = \hat{\mathbf{b}}_{k-1} + \hat{\gamma} \boldsymbol{\mu}_k, \quad \Omega_k = \Omega_k \cup \{\hat{j}\},$$

where the added index $\hat{j}$ is the index $j \in \Omega_k$ such that $\hat{\gamma}$ is a minimum.

11. **exit** If some stopping criterion is satisfied.
12. **return** $k \leftarrow k + 1$.
13. **output:** coefficient vector $\mathbf{x} = \hat{\mathbf{x}}_k$.

References

1. Auslender, A.: Asymptotic properties of the Fenchel dual functional and applications to decomposition problems. J. Optim. Theory Appl. **73**(3), 427–449 (1992)
2. Beltrami, E.: Sulle funzioni bilineari, Giomale di Mathematiche ad Uso Studenti Delle Universita. **11**, 98–106 (1873). An English translation by D Boley is available as Technical Report 90-37, University of Minnesota, Department of Computer Science (1990)

3. Bertsekas, D.P.: Nonlinear Programming, 2nd edn. Athena Scientific, Nashua (1999)
4. Bramble, J, Pasciak, J.: A preconditioning technique for indefinite systems resulting from mixed approximations of elliptic problems. Math. Comput. **50**(181), 1–17 (1988)
5. Cadzow, J.A.: Spectral estimation: an overdetermined rational model equation approach. Proc. IEEE **70**, 907–938 (1982)
6. Cai, D., Zhang, C., He, S.: Unsupervised feature selection for multi-cluster data. In: Proceedings of the 16th ACM SIGKDD, July 25–28, Washington, pp. 333–342 (2010)
7. Chen, S.S., Donoho, D.L., Saunders, M.A.: Atomic decomposition by basis pursuit. SIAM Rev. **43**(1), 129–159 (2001)
8. Donoho, D.L.: For most large underdetermined systems of linear equations, the minimal ℓ^1 solution is also the sparsest solution. Commun. Pure Appl. Math. **59**, 797–829 (2006)
9. Donoho, D.L., Grimes, C.: Hessian eigenmaps: locally linear embedding techniques for high-dimensional data. Proc. Natl. Acad. Sci. **100**(10), 5591–5596 (2003)
10. Eckart, C., Young, G.: The approximation of one matrix by another of lower rank. Psychometrica **1**, 211–218 (1936)
11. Eckart, C., Young, G.: A Principal axis transformation for non-Hermitian matrices. Null Am. Math. Soc. **45**, 118–121 (1939)
12. Efron, B., Hastie, T., Johnstone, I., Tibshirani, R.: Least angle regression. Ann. Stat. **32**, 407–499 (2004)
13. Fletcher, R.: Conjugate gradient methods for indefinite systems. In: Watson, G.A. (ed.) Proceedings of Dundee Conference on Numerical Analysis, pp. 73–89. Springer, New York (1975)
14. Gleser, L.J.: Estimation in a multivariate "errors in variables" regression model: large sample results. Ann. Stat. **9**, 24–44 (1981)
15. Golub, G.H., Van Loan, C.F.: An analysis of the total least squares problem. SIAM J. Numer. Anal. **17**, 883–893 (1980)
16. Golub, G.H., Van Loan, C.F.: Matrix Computation, 2nd edn. The John Hopkins University Press, Baltimore (1989)
17. Gorodnitsky, I.F., Rao, B.D.: Sparse signal reconstruction from limited data using FOCUSS: a re-weighted norm minimization algorithm. IEEE Trans. Signal Process. **45**, 600–616 (1997)
18. Grippo, L., Sciandrone, M.: On the convergence of the block nonlinear Gauss-Seidel method under convex constraints. Oper. Res. Lett. **26**, 127–136 (1999)
19. Hastie, T., Taylor, J., Tibshiran, R., Walther, G.: Forward stagewise regression and the monotone lasso. Electron. J. Stat. **1**, 1–29 (2007)
20. Horn, R.A., Johnson, C.R.: Topics in Matrix Analysis. Cambridge University Press, Cambridge (1991)
21. Huffel, S.V., Vandewalle, J.: The Total Least Squares Problems: Computational Aspects and Analysis. Frontiers in Applied Mathematics, vol. 9. SIAM, Philadelphia (1991)
22. Johnson, L.W., Riess, R.D., Arnold, J.T.: Introduction to Linear Algebra, 5th edn. Prentice-Hall, New York (2000)
23. Jordan, C.: Memoire sur les formes bilineaires. J. Math. Pures Appl. Deuxieme Ser. **19**, 35–54 (1874)
24. Kelley, C.T.: Iterative Methods for Linear and Nonlinear Equations. Frontiers in Applied Mathematics, vol. 16. SIAM, Philadelphia (1995)
25. Kim, S.J., Koh, K., Lustig, M., Boyd, S., Gorinevsky, D.: An interior-point method for large-scale ℓ_1-regularized least squares. IEEE J. Sel. Topics Signal Process. **1**(4), 606–617 (2007)
26. Klema, V.C., Laub, A.J.: The singular value decomposition: its computation and some applications. IEEE Trans. Autom. Control **25**, 164–176 (1980)
27. Lay, D.C.: Linear Algebra and Its Applications, 2nd edn. Addison-Wesley, New York (2000)
28. Li, N., Kindermannb, S., Navasca, C.: Some convergence results on the regularized alternating least-squares method for tensor decomposition. Linear Algebra Appl. **438**(2), 796–812 (2013)
29. Mateos, G., Bazerque, J.A., Giannakis, G.B.: Distributed sparse linear regression. IEEE Trans. Signal Process. **58**(10), 5262–5276 (2010)

30. Mirsky, L.: Symmetric gauge functions and unitarily invariant norms. Quart. J. Math. Oxford **11**, 50–59 (1960)
31. Navasca, C., Lathauwer, L.D., Kindermann, S.: Swamp reducing technique for tensor decomposition. In: The 16th Proceedings of the European Signal Processing Conference, Lausanne, August 25–29 (2008)
32. Neumaier, A.: Solving ill-conditioned and singular linear systems: a tutorial on regularization. SIAM Rev. **40**(3), 636–666 (1998)
33. Pearson, K.: On lines and planes of closest fit to points in space. Philos. Mag. **2**, 559–572 (1901)
34. Powell, M.J.D.: On search directions for minimization algorithms. Math. Program. **4**, 193–201 (1973)
35. Puig, A.T., Wiesel, A., Fleury, G., Hero, A.O.: Multidimensional shrinkage-thresholding operator and group LASSO penalties. IEEE Signal Process. Lett. **18**(6), 343–346 (2011)
36. Roth, V.: The generalized LASSO. IEEE Trans. Neural Netw. **15**(1), 16–28 (2004)
37. Shewchuk, J.R.: An introduction to the conjugate gradient method without the agonizing pain. http://quake-papers/painless-conjugate-gradient-pics.ps
38. Tibshirani, R.: Regression shrinkage and selection via the lasso. J. R. Stat. Soc. B **58**, 267–288 (1996)
39. Tikhonov, A.: Solution of incorrectly formulated problems and the regularization method. Soviet Math. Dokl. **4**, 1035–1038 (1963)
40. Tikhonov, A.N., Arsenin, V.Y.: Solutions of Ill-Posed Problems. Wiley, New York (1977)
41. Tropp, J.A.: Just relax: convex programming methods for identifying sparse signals in noise. IEEE Trans. Inform. Theory **52**(3), 1030–1051 (2006)
42. Van Huffel, S., Vandewalle, J.: Analysis and properties of the generalized total least squares problem $Ax = b$ when some or all columns in A are subject to error. SIAM J. Matrix Anal. Appl. **10**, 294–315 (1989)
43. Wilkinson, J.H.: The Algebraic Eigenvalue Problem. Clarendon Press, Oxford (1965)
44. Xu, H., Caramanis, C., Mannor, S.: Robust regression and Lasso. IEEE Trans. Inform. Theory **56**(7), 3561–3574 (2010)
45. Yang, W., Gao, Y., Shi, Y., Cao, L.: MRM-Lasso: a sparse multiview feature selection method via low-rank analysis. IEEE Trans. Neural Netw. Learn. Syst. **26**(11), 2801–2815 (2015)
46. Zhang, X.D.: Matrix Analysis and Applications. Cambridge University Press, Cambridge (2017)

Chapter 5
Eigenvalue Decomposition

This chapter is devoted to another core subject of matrix algebra: the eigenvalue decomposition (EVD) of matrices, including various generalizations of EVD such as the generalized eigenvalue decomposition, the Rayleigh quotient, and the generalized Rayleigh quotient.

5.1 Eigenvalue Problem and Characteristic Equation

The eigenvalue problem not only is a very interesting theoretical problem but also has a wide range of applications in artificial intelligence.

5.1.1 Eigenvalue Problem

If $\mathcal{L}[\mathbf{w}] = \mathbf{w}$ holds for any nonzero vector $\mathbf{w}$, then $\mathcal{L}$ is called an *identity transformation*. More generally, if, when a linear operator acts on a vector, the output is a multiple of this vector, then the linear operator is said to have an *input reproducing* characteristic.

Definition 5.1 (Eigenvalue, Eigenvector) Suppose that a nonzero vector $\mathbf{u}$ is the input of the linear operator $\mathcal{L}$. If the output vector is the same as the input vector $\mathbf{u}$ up to a constant factor λ, i.e.,

$$\mathcal{L}[\mathbf{u}] = \lambda \mathbf{u}, \quad \mathbf{u} \neq \mathbf{0}, \tag{5.1.1}$$

then the vector $\mathbf{u}$ is known as an *eigenvector* of the linear operator $\mathcal{L}$ and the scalar λ is the corresponding *eigenvalue* of $\mathcal{L}$.

From the above definition, it is known that if each eigenvector **u** is regarded as an input of a linear time-invariant system $\mathcal{L}$, then the eigenvalue λ associated with eigenvector **u** is equivalent to the gain of the linear system $\mathcal{L}$ with **u** as the input. Only when the input of the system $\mathcal{L}$ is an eigenvector **u**, its output is the same as the input up to a constant factor. Thus an eigenvector can be viewed as description of the characteristics of the system, and hence is also called a characteristic vector. This is a physical explanation of eigenvectors in relation to linear systems.

If a linear transformation $\mathcal{L}[\mathbf{x}] = \mathbf{A}\mathbf{x}$, then its eigenvalue problem (5.1.1) can be written as

$$\mathbf{A}\mathbf{u} = \lambda \mathbf{u}, \quad \mathbf{u} \neq \mathbf{0}. \tag{5.1.2}$$

The scalar λ is known as an eigenvalue of the matrix **A**, and the vector **u** as an eigenvector associated with the eigenvalue λ. Equation (5.1.2) is sometimes called the *eigenvalue–eigenvector equation*.

As an example, we consider a linear time-invariant system $h(k)$ with transfer function $H(e^{j\omega}) = \sum_{k=-\infty}^{\infty} h(k) e^{-j\omega k}$. When inputting a complex exponential or harmonic signal $e^{j\omega n}$, the system output is given by

$$\mathcal{L}[e^{j\omega n}] = \sum_{k=-\infty}^{\infty} h(n-k) e^{j\omega k} = \sum_{k=-\infty}^{\infty} h(k) e^{j\omega(n-k)} = H(e^{j\omega}) e^{j\omega n}.$$

If the harmonic signal vector $\mathbf{u}(\omega) = [1, e^{j\omega}, \ldots, e^{j\omega(N-1)}]^T$ is the system input, then its output is given by

$$\mathcal{L} \begin{bmatrix} 1 \\ e^{j\omega} \\ \vdots \\ e^{j\omega(N-1)} \end{bmatrix} = H(e^{j\omega}) \begin{bmatrix} 1 \\ e^{j\omega} \\ \vdots \\ e^{j\omega(N-1)} \end{bmatrix} \Rightarrow \mathcal{L}[\mathbf{u}(\omega)] = H(e^{j\omega}) \mathbf{u}(\omega).$$

That is to say, the harmonic signal vector $\mathbf{u}(\omega) = [1, e^{j\omega}, \ldots, e^{j\omega(N-1)}]^T$ is an eigenvector of the linear time-invariant system, and the system transfer function $H(e^{j\omega})$ is the eigenvalue associated with $\mathbf{u}(\omega)$.

From Eq. (5.1.2) it is easily known that if $\mathbf{A} \in \mathbb{C}^{n \times n}$ is a Hermitian matrix then its eigenvalue λ must be a real number, and

$$\mathbf{A} = \mathbf{U}\mathbf{\Sigma}\mathbf{U}^H, \tag{5.1.3}$$

where $\mathbf{U} = [\mathbf{u}_1, \ldots, \mathbf{u}_n]^T$ is a unitary matrix, and $\mathbf{\Sigma} = \mathbf{Diag}(\lambda_1, \ldots, \lambda_n)$. Equation (5.1.3) is called the *eigenvalue decomposition* (EVD) of **A**.

Since an eigenvalue λ and its associated eigenvector **u** often appear in pairs, the two-tuple $(\lambda, \mathbf{u})$ is called an *eigenpair* of the matrix **A**. An eigenvalue may take a zero value, but an eigenvector cannot be a zero vector.

5.1 Eigenvalue Problem and Characteristic Equation

Equation (5.1.2) means that the linear transformation **Au** does not "change the direction" of the input vector **u**. Hence the linear transformation **Au** is a mapping "keeping the direction unchanged." In order to determine the vector **u**, we can rewrite (5.1.2) as

$$(\mathbf{A} - \lambda \mathbf{I})\mathbf{u} = \mathbf{0}. \tag{5.1.4}$$

If the above equation is assumed to hold for certain nonzero vectors **u**, then the only condition is that, for those vectors **u**, the determinant of the matrix $\mathbf{A} - \lambda \mathbf{I}$ is equal to zero:

$$|\mathbf{A} - \lambda \mathbf{I}| = 0. \tag{5.1.5}$$

Hence the eigenvalue problem solving consists of the following two steps:

- Find all scalars λ (eigenvalues) such that the matrix $|\mathbf{A} - \lambda \mathbf{I}| = 0$;
- Given an eigenvalue λ, find all nonzero vectors **u** satisfying $(\mathbf{A} - \lambda \mathbf{I})\mathbf{u} = \mathbf{0}$; they are the eigenvector(s) corresponding to the eigenvalue λ.

5.1.2 Characteristic Polynomial

As discussed above, the matrix $(\mathbf{A} - \lambda \mathbf{I})$ is singular if and only if its determinant $\det(\mathbf{A} - \lambda \mathbf{I}) = 0$, i.e.,

$$(\mathbf{A} - \lambda \mathbf{I}) \text{ singular} \quad \Leftrightarrow \quad \det(\mathbf{A} - \lambda \mathbf{I}) = 0, \tag{5.1.6}$$

where the matrix $\mathbf{A} - \lambda \mathbf{I}$ is called the *characteristic matrix* of **A**.

The determinant

$$p(x) = \det(\mathbf{A} - x\mathbf{I}) = \begin{vmatrix} a_{11} - x & a_{12} & \cdots & a_{1n} \\ a_{21} & a_{22} - x & \cdots & a_{2n} \\ \vdots & \vdots & \ddots & \vdots \\ a_{n1} & a_{n2} & \cdots & a_{nn} - x \end{vmatrix}$$
$$= p_n x^n + p_{n-1} x^{n-1} + \cdots + p_1 x + p_0 \tag{5.1.7}$$

is known as the *characteristic polynomial* of **A**, and

$$p(x) = \det(\mathbf{A} - x\mathbf{I}) = 0 \tag{5.1.8}$$

is said to be the *characteristic equation* of **A**.

The roots of the characteristic equation $\det(\mathbf{A} - x\mathbf{I}) = 0$ are known as the eigenvalues, characteristic values, latent values, the characteristic roots, or latent roots.

Obviously, computing the n eigenvalues λ_i of an $n \times n$ matrix $\mathbf{A}$ and finding the n roots of the nth-order characteristic polynomial $p(x) = \det(\mathbf{A} - x\mathbf{I}) = 0$ are two equivalent problems. An $n \times n$ matrix $\mathbf{A}$ generates an nth-order characteristic polynomial. Likewise, each nth-order polynomial can also be written as the characteristic polynomial of an $n \times n$ matrix.

Theorem 5.1 ([1]) *Any polynomial*

$$p(\lambda) = \lambda^n + a_1 \lambda^{n-1} + \cdots + a_{n-1}\lambda + a_n$$

can be written as the characteristic polynomial of the $n \times n$ matrix

$$\mathbf{A} = \begin{bmatrix} -a_1 & -a_2 & \cdots & -a_{n-1} & -a_n \\ -1 & 0 & \cdots & 0 & 0 \\ 0 & -1 & \cdots & 0 & 0 \\ \vdots & \vdots & \ddots & \vdots & \vdots \\ 0 & 0 & \cdots & -1 & 0 \end{bmatrix},$$

namely $p(\lambda) = \det(\lambda \mathbf{I} - \mathbf{A})$.

5.2 Eigenvalues and Eigenvectors

Given an $n \times n$ matrix $\mathbf{A}$, we consider the computation and properties of the eigenvalues and eigenvectors of $\mathbf{A}$.

5.2.1 Eigenvalues

Even if an $n \times n$ matrix $\mathbf{A}$ is real, some roots of its characteristic equation may be complex, and the root multiplicity can be arbitrary or even equal to n. These roots are collectively referred to as the eigenvalues of the matrix $\mathbf{A}$.

An eigenvalue λ of a matrix $\mathbf{A}$ is said to have *algebraic multiplicity* μ, if λ is a μ-multiple root of the characteristic polynomial $\det(\mathbf{A} - x\mathbf{I}) = 0$.

If the algebraic multiplicity of an eigenvalue λ is 1, then it is called a *single eigenvalue*. A nonsingle eigenvalue is said to be a *multiple eigenvalue*.

It is well known that any nth-order polynomial $p(x)$ can be written in the factorized form

$$p(x) = a(x - x_1)(x - x_2) \cdots (x - x_n). \tag{5.2.1}$$

5.2 Eigenvalues and Eigenvectors

The n roots of the characteristic polynomial $p(x)$, denoted $x_1, x_2, \ldots, x_n$, are not necessarily different from each other, and also are not necessarily real for a real matrix. For example, for the Givens rotation matrix

$$\mathbf{A} = \begin{bmatrix} \cos\theta & -\sin\theta \\ \sin\theta & \cos\theta \end{bmatrix},$$

its characteristic equation is

$$\det(\mathbf{A} - \lambda\mathbf{I}) = \begin{vmatrix} \cos\theta - \lambda & -\sin\theta \\ \sin\theta & \cos\theta - \lambda \end{vmatrix} = (\cos\theta - \lambda)^2 + \sin^2\theta = 0.$$

However, if θ is not an integer multiple of π, then $\sin^2\theta > 0$. In this case, the characteristic equation cannot give a real value for λ, i.e., the two eigenvalues of the rotation matrix are complex, and the two corresponding eigenvectors are complex vectors.

The eigenvalues of an $n \times n$ matrix (not necessarily Hermitian) $\mathbf{A}$ have the following properties [8].

1. An $n \times n$ matrix $\mathbf{A}$ has a total of n eigenvalues, where multiple eigenvalues count according to their multiplicity.
2. If a nonsymmetric real matrix $\mathbf{A}$ has complex eigenvalues and/or complex eigenvectors, then they must appear in the form of a complex conjugate pair.
3. If $\mathbf{A}$ is a real symmetric matrix or a Hermitian matrix, then its all eigenvalues are real numbers.
4. The eigenvalues of a diagonal matrix and a triangular matrix satisfy the following:
   - if $\mathbf{A} = \mathbf{Diag}(a_{11}, \ldots, a_{nn})$, then its eigenvalues are given by $a_{11}, \ldots, a_{nn}$;
   - if $\mathbf{A}$ is a triangular matrix, then all its diagonal entries are eigenvalues.
5. Given an $n \times n$ matrix $\mathbf{A}$:
   - if λ is an eigenvalue of $\mathbf{A}$, then λ is also an eigenvalue of $\mathbf{A}^T$;
   - if λ is an eigenvalue of $\mathbf{A}$, then λ^* is an eigenvalue of $\mathbf{A}^H$;
   - if λ is an eigenvalue of $\mathbf{A}$, then $\lambda + \sigma^2$ is an eigenvalue of $\mathbf{A} + \sigma^2\mathbf{I}$;
   - if λ is an eigenvalue of $\mathbf{A}$, then $1/\lambda$ is an eigenvalue of its inverse matrix $\mathbf{A}^{-1}$.
6. All eigenvalues of an *idempotent matrix* $\mathbf{A}^2 = \mathbf{A}$ take 0 or 1.
7. If $\mathbf{A}$ is an $n \times n$ real orthogonal matrix, then all its eigenvalues are located on the unit circle, i.e., $|\lambda_i(\mathbf{A})| = 1$ for all $i = 1, \ldots, n$.
8. The relationship between eigenvalues and the matrix singularity is as follows:
   - if $\mathbf{A}$ is singular, then it has at least one zero eigenvalue;
   - if $\mathbf{A}$ is nonsingular, then its all eigenvalues are nonzero.

9. The relationship between eigenvalues and trace: the sum of all eigenvalues of **A** is equal to its trace, namely $\sum_{i=1}^{n} \lambda_i = \text{tr}(\mathbf{A})$.
10. A Hermitian matrix **A** is positive definite (semidefinite) if and only if its all eigenvalues are positive (nonnegative).
11. The relationship between eigenvalues and determinant: the product of all eigenvalues of **A** is equal to its determinant, namely $\prod_{i=1}^{n} \lambda_i = \det(\mathbf{A}) = |\mathbf{A}|$.
12. The Cayley–Hamilton theorem: if $\lambda_1, \lambda_2, \ldots, \lambda_n$ are the eigenvalues of an $n \times n$ matrix **A**, then

$$\prod_{i=1}^{n} (\mathbf{A} - \lambda_i \mathbf{I}) = 0.$$

13. If the $n \times n$ matrix **B** is nonsingular, then $\lambda(\mathbf{B}^{-1}\mathbf{A}\mathbf{B}) = \lambda(\mathbf{A})$. If **B** is unitary, then $\lambda(\mathbf{B}^H \mathbf{A}\mathbf{B}) = \lambda(\mathbf{A})$.
14. The matrix products $\mathbf{A}_{m \times n} \mathbf{B}_{n \times m}$ and $\mathbf{B}_{n \times m} \mathbf{A}_{m \times n}$ have the same nonzero eigenvalues.
15. If an eigenvalue of a matrix **A** is λ, then the corresponding eigenvalue of the matrix polynomial $f(\mathbf{A}) = \mathbf{A}^n + c_1 \mathbf{A}^{n-1} + \cdots + c_{n-1} \mathbf{A} + c_n \mathbf{I}$ is given by

$$f(\lambda) = \lambda^n + c_1 \lambda^{n-1} + \cdots + c_{n-1} \lambda + c_n. \tag{5.2.2}$$

16. If λ is an eigenvalue of a matrix **A**, then e^λ is an eigenvalue of the matrix exponential function $e^\mathbf{A}$.

5.2.2 Eigenvectors

If a matrix $\mathbf{A}_{n \times n}$ is a complex matrix, and λ is one of its eigenvalues, then the vector **v** satisfying

$$(\mathbf{A} - \lambda \mathbf{I})\mathbf{v} = \mathbf{0} \quad \text{or} \quad \mathbf{A}\mathbf{v} = \lambda \mathbf{v} \tag{5.2.3}$$

is called the *right eigenvector* of the matrix **A** associated with the eigenvalue λ, while the vector **u** satisfying

$$\mathbf{u}^H(\mathbf{A} - \lambda \mathbf{I}) = \mathbf{0}^T \quad \text{or} \quad \mathbf{u}^H \mathbf{A} = \lambda \mathbf{u}^H \tag{5.2.4}$$

is known as the *left eigenvector* of **A** associated with the eigenvalue λ.

If a matrix **A** is Hermitian, then its all eigenvalues are real, and hence from Eq. (5.2.3) it is immediately known that $((\mathbf{A} - \lambda \mathbf{I})\mathbf{v})^T = \mathbf{v}^T (\mathbf{A} - \lambda \mathbf{I}) = \mathbf{0}^T$, yielding $\mathbf{v} = \mathbf{u}$; namely, the right and left eigenvectors of any Hermitian matrix are the same.

5.2 Eigenvalues and Eigenvectors

It is useful to compare the similarities and differences between the SVD and the EVD of a matrix.

- The SVD is available for any $m \times n$ (where $m \geq n$ or $m < n$) matrix, while the EVD is available only for square matrices.
- For an $n \times n$ non-Hermitian matrix $\mathbf{A}$, its kth singular value is defined as the spectral norm of the error matrix $\mathbf{E}_k$ making the rank of the original matrix $\mathbf{A}$ decreased by 1:

$$\sigma_k = \min_{\mathbf{E} \in \mathbb{C}^{m \times n}} \left\{ \|\mathbf{E}\|_{\text{spec}} : \text{rank}(\mathbf{A} + \mathbf{E}) \leq k - 1 \right\}, \quad k = 1, \ldots, \min\{m, n\}, \tag{5.2.5}$$

while its eigenvalues are defined as the roots of the characteristic polynomial $\det(\mathbf{A} - \lambda \mathbf{I}) = 0$. There is no inherent relationship between the singular values and the eigenvalues of the same square matrix, but each nonzero singular value of an $m \times n$ matrix $\mathbf{A}$ is the positive square root of some nonzero eigenvalue of the $n \times n$ Hermitian matrix $\mathbf{A}^H \mathbf{A}$ or the $m \times m$ Hermitian matrix $\mathbf{A} \mathbf{A}^H$.

- The left singular vector $\mathbf{u}_i$ and the right singular vector $\mathbf{v}_i$ of an $m \times n$ matrix $\mathbf{A}$ associated with the singular value σ_i are defined as the two vectors satisfying $\mathbf{u}_i^H \mathbf{A} \mathbf{v}_i = \sigma_i$, while the left and right eigenvectors are defined by $\mathbf{u}^H \mathbf{A} = \lambda_i \mathbf{u}^H$ and $\mathbf{A} \mathbf{v}_i = \lambda_i \mathbf{v}_i$, respectively. Hence, for the same $n \times n$ non-Hermitian matrix $\mathbf{A}$, there is no inherent relationship between its (left and right) singular vectors and its (left and right) eigenvectors. However, the left singular vector $\mathbf{u}_i$ and right singular vector $\mathbf{v}_i$ of a matrix $\mathbf{A} \in \mathbb{C}^{m \times n}$ are, respectively, the eigenvectors of the $m \times m$ Hermitian matrix $\mathbf{A} \mathbf{A}^H$ and of the $n \times n$ matrix $\mathbf{A}^H \mathbf{A}$.

From Eq. (5.1.2) it is easily seen that after multiplying an eigenvector $\mathbf{u}$ of a matrix $\mathbf{A}$ by any nonzero scalar μ, then $\mu \mathbf{u}$ is still an eigenvector of $\mathbf{A}$. In general it is assumed that eigenvectors have unit norm, i.e., $\|\mathbf{u}\|_2 = 1$.

Using eigenvectors we can introduce the condition number of any single eigenvalue.

Definition 5.2 (Condition Number of Eigenvalue [17, p. 93]) The condition number of a single eigenvalue λ of any matrix $\mathbf{A}$ is defined as

$$\text{cond}(\lambda) = \frac{1}{\cos \theta(\mathbf{u}, \mathbf{v})}, \tag{5.2.6}$$

where $\theta(\mathbf{u}, \mathbf{v})$ represents the acute angle between the left and right eigenvectors associated with the eigenvalue λ.

Definition 5.3 (Matrix Spectrum) The set of all eigenvalues $\lambda \in \mathbb{C}$ of a matrix $\mathbf{A} \in \mathbb{C}^{n \times n}$ is called the *spectrum* of the matrix $\mathbf{A}$, denoted $\lambda(\mathbf{A})$. The *spectral radius* of a matrix $\mathbf{A}$, denoted $\rho(\mathbf{A})$, is a nonnegative real number and is defined as

$$\rho(\mathbf{A}) = \max |\lambda| : \lambda \in \lambda(\mathbf{A}). \tag{5.2.7}$$

Definition 5.4 (Inertia) The *inertia* of a symmetric matrix $\mathbf{A} \in \mathbb{R}^{n \times n}$, denoted In($\mathbf{A}$), is defined as the triplet

$$\text{In}(\mathbf{A}) = (i_+(\mathbf{A}), i_-(\mathbf{A}), i_0(\mathbf{A})),$$

where $i_+(\mathbf{A}), i_-(\mathbf{A})$ and $i_0(\mathbf{A})$ are, respectively, the numbers of the positive, negative, and zero eigenvalues of $\mathbf{A}$ (each multiple eigenvalue is counted according to its multiplicity). Moreover, the quality $i_+(\mathbf{A}) - i_-(\mathbf{A})$ is known as the *signature* of $\mathbf{A}$.

5.3 Generalized Eigenvalue Decomposition (GEVD)

The EVD of an $n \times n$ single matrix can be extended to the EVD of a matrix pair or pencil consisting of two matrices. The EVD of a matrix pencil is called the generalized eigenvalue decomposition (GEVD). As a matter of fact, the EVD is a special example of the GEVD.

5.3.1 Generalized Eigenvalue Decomposition

The basis of the EVD is the eigensystem expressed by the linear transformation $\mathcal{L}[\mathbf{u}] = \lambda \mathbf{u}$: taking the linear transformation $\mathcal{L}[\mathbf{u}] = \mathbf{A}\mathbf{u}$, one gets the EVD $\mathbf{A}\mathbf{u} = \lambda \mathbf{u}$.

Now consider the generalization of the eigensystem composed of two linear systems $\mathcal{L}_a$ and $\mathcal{L}_b$ both of whose inputs are the vector $\mathbf{u}$, but the output $\mathcal{L}_a[\mathbf{u}]$ of the first system $\mathcal{L}_a$ is some constant times λ the output $\mathcal{L}_b[\mathbf{u}]$ of the second system $\mathcal{L}_b$. Hence the eigensystem is generalized as [11]

$$\mathcal{L}_a[\mathbf{u}] = \lambda \mathcal{L}_b[\mathbf{u}] \quad (\mathbf{u} \neq \mathbf{0}); \tag{5.3.1}$$

this is called a *generalized eigensystem*, denoted $(\mathcal{L}_a, \mathcal{L}_b)$. The constant λ and the nonzero vector $\mathbf{u}$ are known as the *generalized eigenvalue* and the *generalized eigenvector* of generalized eigensystem $(\mathcal{L}_a, \mathcal{L}_b)$, respectively.

In particular, if two linear transformations are, respectively, $\mathcal{L}_a[\mathbf{u}] = \mathbf{A}\mathbf{u}$ and $\mathcal{L}_b[\mathbf{u}] = \mathbf{B}\mathbf{u}$, then the generalized eigensystem becomes the *generalized eigenvalue decomposition* (GEVD):

$$\mathbf{A}\mathbf{u} = \lambda \mathbf{B}\mathbf{u}. \tag{5.3.2}$$

The two $n \times n$ matrices $\mathbf{A}$ and $\mathbf{B}$ compose a *matrix pencil* (or *matrix pair*), denoted $(\mathbf{A}, \mathbf{B})$. The constant λ and the nonzero vector $\mathbf{u}$ are, respectively, referred to as the generalized eigenvalue and the generalized eigenvector of the matrix pencil $(\mathbf{A}, \mathbf{B})$.

5.3 Generalized Eigenvalue Decomposition (GEVD)

A generalized eigenvalue and its corresponding generalized eigenvector are collectively called a *generalized eigenpair*, denoted $(\lambda, \mathbf{u})$. Equation (5.3.2) shows that the ordinary eigenvalue problem is a special example of the generalized eigenvalue problem when the matrix pencil is $(\mathbf{A}, \mathbf{I})$.

Although the generalized eigenvalue and the generalized eigenvector always appear in pairs, the generalized eigenvalue can be found independently. To do this, we rewrite the generalized characteristic equation (5.3.2) as $(\mathbf{A} - \lambda\mathbf{B})\mathbf{u} = \mathbf{0}$. In order to ensure the existence of a nonzero solution $\mathbf{u}$, the matrix $\mathbf{A} - \lambda\mathbf{B}$ cannot be nonsingular, i.e., the determinant must be equal to zero:

$$(\mathbf{A} - \lambda\mathbf{B}) \text{ singular} \quad \Leftrightarrow \quad \det(\mathbf{A} - \lambda\mathbf{B}) = 0. \tag{5.3.3}$$

The polynomial $\det(\mathbf{A} - \lambda\mathbf{B})$ is called the *generalized characteristic polynomial* of the matrix pencil $(\mathbf{A}, \mathbf{B})$, and $\det(\mathbf{A} - \lambda\mathbf{B}) = 0$ is known as the *generalized characteristic equation* of the matrix pencil $(\mathbf{A}, \mathbf{B})$. Hence, the matrix pencil $(\mathbf{A}, \mathbf{B})$ is often expressed as $\mathbf{A} - \lambda\mathbf{B}$.

The generalized eigenvalues of the matrix pencil $(\mathbf{A}, \mathbf{B})$ are all solutions z of the generalized characteristic equation $\det(\mathbf{A} - z\mathbf{B}) = 0$, including the generalized eigenvalues equal to zero. If the matrix $\mathbf{B} = \mathbf{I}$, then the generalized characteristic polynomial reduces to the characteristic polynomial $\det(\mathbf{A} - \lambda\mathbf{I}) = 0$ of the matrix $\mathbf{A}$. In other words, the generalized characteristic polynomial is a generalization of the characteristic polynomial, whereas the characteristic polynomial is a special example of the generalized characteristic polynomial of the matrix pencil $(\mathbf{A}, \mathbf{B})$ when $\mathbf{B} = \mathbf{I}$.

The generalized eigenvalues $\lambda(\mathbf{A}, \mathbf{B})$ are defined as

$$\lambda(\mathbf{A}, \mathbf{B}) = \{z \in \mathbb{C} | \det(\mathbf{A} - z\mathbf{B}) = 0\}. \tag{5.3.4}$$

Theorem 5.2 ([9]) *The pair $\lambda \in \mathbb{C}$ and $\mathbf{u} \in \mathbb{C}^n$ are, respectively, the generalized eigenvalue and the generalized eigenvector of the $n \times n$ matrix pencil $(\mathbf{A}, \mathbf{B})$ if and only if $|\mathbf{A} - \lambda\mathbf{B}| = 0$ and $\mathbf{u} \in \text{Null}(\mathbf{A} - \lambda\mathbf{B})$ with $\mathbf{u} \neq \mathbf{0}$.*

The following are the properties of the GEVD $\mathbf{Au} = \lambda\mathbf{Bu}$ [10, pp. 176–177].

1. If $\mathbf{A}$ and $\mathbf{B}$ are exchanged, then every generalized eigenvalue will become its reciprocal, but the generalized eigenvector is unchanged, namely

$$\mathbf{Au} = \lambda\mathbf{Bu} \quad \Rightarrow \quad \mathbf{Bu} = \frac{1}{\lambda}\mathbf{Au}.$$

2. If $\mathbf{B}$ is nonsingular, then the GEVD becomes the standard EVD of $\mathbf{B}^{-1}\mathbf{A}$:

$$\mathbf{Au} = \lambda\mathbf{Bu} \quad \Rightarrow \quad (\mathbf{B}^{-1}\mathbf{A})\mathbf{u} = \lambda\mathbf{u}.$$

3. If $\mathbf{A}$ and $\mathbf{B}$ are real positive definite matrices, then the generalized eigenvalues must be positive.

4. If **A** is singular then $\lambda = 0$ must be a generalized eigenvalue.
5. If both **A** and **B** are positive definite Hermitian matrices then the generalized eigenvalues must be real, and

$$\mathbf{u}_i^H \mathbf{A} \mathbf{u}_j = \mathbf{u}_i^H \mathbf{B} \mathbf{u}_j = 0, \quad i \neq j.$$

Strictly speaking, the generalized eigenvector **u** described above is called the *right generalized eigenvector* of the matrix pencil $(\mathbf{A}, \mathbf{B})$. The *left generalized eigenvector* corresponding to the generalized eigenvalue λ is defined as the column vector **v** such that

$$\mathbf{v}^H \mathbf{A} = \lambda \mathbf{v}^H \mathbf{B}. \tag{5.3.5}$$

Let both **X** and **Y** be nonsingular matrices; then from (5.3.2) and (5.3.5) one has

$$\mathbf{X}\mathbf{A}\mathbf{u} = \lambda \mathbf{X}\mathbf{B}\mathbf{u}, \quad \mathbf{v}^H \mathbf{A}\mathbf{Y} = \lambda \mathbf{v}^H \mathbf{B}\mathbf{Y}. \tag{5.3.6}$$

This shows that premultiplying the matrix pencil $(\mathbf{A}, \mathbf{B})$ by a nonsingular matrix **X** does not change the right generalized eigenvector **u** of the matrix pencil $(\mathbf{A}, \mathbf{B})$, while postmultiplying $(\mathbf{A}, \mathbf{B})$ by any nonsingular matrix **Y** does not change the left generalized eigenvector **v**.

Algorithm 5.1 uses a contraction mapping to compute the generalized eigenpairs $(\lambda, \mathbf{u})$ of a given $n \times n$ real symmetric matrix pencil $(\mathbf{A}, \mathbf{B})$.

Algorithm 5.1 Lanczos algorithm for GEVD [17, p. 298]

1. **input:** $n \times n$ real symmetric matrices **A**, **B**.
2. **initialization:** Choose $\mathbf{u}_1$ such that $\|\mathbf{u}_1\|_2 = 1$, and set $\alpha_1 = 0$, $\mathbf{z}_0 = \mathbf{u}_0 = \mathbf{0}$, $\mathbf{z}_1 = \mathbf{B}\mathbf{u}_1$, $i = 1$.
3. **repeat**
4. Compute
$$\mathbf{u} = \mathbf{A}\mathbf{u}_i - \alpha_i \mathbf{z}_{i-1}.$$
$$\beta_i = \langle \mathbf{u}, \mathbf{u}_i \rangle.$$
$$\mathbf{u} = \mathbf{u} - \beta_i \mathbf{z}_i.$$
$$\mathbf{w} = \mathbf{B}^{-1}\mathbf{u}.$$
$$\alpha_{i+1} = \sqrt{\langle \mathbf{w}, \mathbf{u} \rangle}.$$
$$\mathbf{u}_{i+1} = \mathbf{w}/\alpha_{i+1}.$$
$$\mathbf{z}_{i+1} = \mathbf{u}/\alpha_{i+1}.$$
$$\lambda_i = \beta_{i+1}/\alpha_{i+1}.$$
5. **exit if** $i = n$.
6. **return** $i \leftarrow i + 1$.
7. **output:** $(\lambda_i, \mathbf{u}_i), i = 1, \ldots, n.$

Algorithm 5.2 is a tangent algorithm for computing the GEVD of a symmetric positive definite matrix pencil.

When the matrix **B** is singular, Algorithms 5.1 and 5.2 will be unstable. The GEVD algorithm for the matrix pencil $(\mathbf{A}, \mathbf{B})$ with singular matrix **B** was proposed

5.3 Generalized Eigenvalue Decomposition (GEVD)

Algorithm 5.2 Tangent algorithm for computing the GEVD [4]

1. **input:** $n \times n$ real symmetric matrices $\mathbf{A}, \mathbf{B}$.
2. **repeat**
3. Compute $\mathbf{\Delta}_A = \text{Diag}(A_{11}, \ldots, A_{nn})^{-1/2}$, $\mathbf{A}_s = \mathbf{\Delta}_A \mathbf{A} \mathbf{\Delta}_A$, $\mathbf{B}_1 = \mathbf{\Delta}_A \mathbf{B} \mathbf{\Delta}_A$.
4. Calculate the Cholesky decomposition $\mathbf{R}_A^T \mathbf{R}_A = \mathbf{A}_s$, $\mathbf{R}_B^T \mathbf{R}_B = \mathbf{\Pi}^T \mathbf{B}_1 \mathbf{\Pi}$.
5. Solve the matrix equation $\mathbf{F}\mathbf{R}_B = \mathbf{A}\mathbf{\Pi}$ for $\mathbf{F} = \mathbf{A}\mathbf{\Pi}\mathbf{R}_B^{-1}$.
6. Compute the SVD $\mathbf{\Sigma} = \mathbf{V}\mathbf{F}\mathbf{U}^T$ of $\mathbf{F}$.
7. Compute $\mathbf{X} = \mathbf{\Delta}_A \mathbf{\Pi} \mathbf{R}_B^{-1} \mathbf{U}$.
8. exit if $\mathbf{AX} = \mathbf{BX}\mathbf{\Sigma}^2$.
9. **repeat**
10. **output:** $\mathbf{X}, \mathbf{\Sigma}$.

by Nour-Omid et al. [12], this is shown in Algorithm 5.3. The main idea of this algorithm is to introduce a shift factor σ such that $\mathbf{A} - \sigma \mathbf{B}$ is nonsingular.

Algorithm 5.3 GEVD algorithm for singular matrix $\mathbf{B}$ [12, 17]

1. **input:**
2. **initialization:** Choose a basis $\mathbf{w}$, $\mathbf{z}_1 = \mathbf{B}\mathbf{w}$, $\alpha_1 = \sqrt{\langle \mathbf{w}, \mathbf{z}_1 \rangle}$. Set $\mathbf{u}_0 = \mathbf{0}$, $i = 1$.
3. **repeat**
4. Compute
$$\mathbf{u}_i = \mathbf{w}/\alpha_i.$$
$$\mathbf{z}_i = (\mathbf{A} - \sigma\mathbf{B})^{-1}\mathbf{w}.$$
$$\mathbf{w} = \mathbf{w} - \alpha_i \mathbf{u}_{i-1}.$$
$$\beta_i = \langle \mathbf{w}, \mathbf{z}_i \rangle.$$
$$\mathbf{z}_{i+1} = \mathbf{B}\mathbf{w}.$$
$$\alpha_{i+1} = \sqrt{\langle \mathbf{z}_{i+1}, \mathbf{w} \rangle}.$$
$$\lambda_i = \beta_i / \alpha_{i+1}.$$
5. exit if $i = n$.
6. return $i \leftarrow i + 1$.
7. **output:** $(\lambda_i, \mathbf{u}_i)$, $i = 1, \ldots, n$.

Definition 5.5 (Equivalent Matrix Pencils) Two matrix pencils with the same generalized eigenvalues are known as *equivalent matrix pencils*.

From the definition of the generalized eigenvalue $\det(\mathbf{A} - \lambda\mathbf{B}) = 0$ and the properties of a determinant it is easy to see that

$$\det(\mathbf{XAY} - \lambda\mathbf{XBY}) = 0 \quad \Leftrightarrow \quad \det(\mathbf{A} - \lambda\mathbf{B}) = 0.$$

Therefore, premultiplying and/or postmultiplying one or more nonsingular matrices, the generalized eigenvalues of the matrix pencil are unchanged. This result can be described as the following proposition.

Proposition 5.1 *If $\mathbf{X}$ and $\mathbf{Y}$ are two nonsingular matrices, then the matrix pencils $(\mathbf{XAY}, \mathbf{XBY})$ and $(\mathbf{A}, \mathbf{B})$ are equivalent.*

The GEVD can be equivalently written as $\alpha \mathbf{Au} = \beta \mathbf{Bu}$ as well. In this case, the generalized eigenvalue is defined as $\lambda = \beta/\alpha$.

5.3.2 Total Least Squares Method for GEVD

As shown by Roy and Kailath [16], use of a least squares estimator can lead to some potential numerical difficulties in solving the generalized eigenvalue problem. To overcome this difficulty, a higher-dimensional ill-conditioned LS problem needs to be transformed into a lower-dimensional well-conditioned LS problem.

Without changing its nonzero generalized eigenvalues, a higher-dimensional matrix pencil can be transformed into a lower-dimensional matrix pencil by using a truncated SVD.

Consider the GEVD of the matrix pencil $(\mathbf{A}, \mathbf{B})$. Let the SVD of $\mathbf{A}$ be given by

$$\mathbf{A} = \mathbf{U\Sigma V}^H = [\mathbf{U}_1, \mathbf{U}_2] \begin{bmatrix} \boldsymbol{\Sigma}_1 & \mathbf{O} \\ \mathbf{O} & \boldsymbol{\Sigma}_2 \end{bmatrix} \begin{bmatrix} \mathbf{V}_1^H \\ \mathbf{V}_2^H \end{bmatrix}, \qquad (5.3.7)$$

where $\boldsymbol{\Sigma}_1$ is composed of p principal singular values. Under the condition that the generalized eigenvalues remain unchanged, we can premultiply the matrix pencil $\mathbf{A} - \gamma \mathbf{B}$ by $\mathbf{U}_1^H$ and postmultiply it by $\mathbf{V}_1$ to get [18]:

$$\mathbf{U}_1^H (\mathbf{A} - \gamma \mathbf{B}) \mathbf{V}_1 = \boldsymbol{\Sigma}_1 - \gamma \mathbf{U}_1^H \mathbf{B} \mathbf{V}_1. \qquad (5.3.8)$$

Hence the original $n \times n$ matrix pencil $(\mathbf{A}, \mathbf{B})$ becomes a $p \times p$ matrix pencil $(\boldsymbol{\Sigma}_1, \mathbf{U}_1^H \mathbf{B} \mathbf{V}_1)$. The GEVD of the lower-dimensional matrix pencil $(\boldsymbol{\Sigma}_1, \mathbf{U}_1^H \mathbf{B} \mathbf{V}_1)$ is called the total least squares (TLS) method for the GEVD of the higher-dimensional matrix pencil $(\mathbf{A}, \mathbf{B})$, which was developed in [18].

5.4 Rayleigh Quotient and Generalized Rayleigh Quotient

In physics and artificial intelligence, one often meets the maximum or minimum of the quotient of the quadratic function of a Hermitian matrix. This quotient has two forms: the Rayleigh quotient (also called the Rayleigh–Ritz ratio) of a Hermitian matrix and the generalized Rayleigh quotient (or generalized Rayleigh–Ritz ratio) of two Hermitian matrices.

5.4.1 Rayleigh Quotient

In his study of the small oscillations of a vibrational system, in order to find the appropriate generalized coordinates, Rayleigh [15] in the 1930s proposed a special form of quotient, later called the Rayleigh quotient.

Definition 5.6 (Rayleigh Quotient) The *Rayleigh quotient* or *Rayleigh–Ritz ratio* $R(\mathbf{x})$ of a Hermitian matrix $\mathbf{A} \in \mathbb{C}^{n \times n}$ is a scalar and is defined as

$$R(\mathbf{x}) = R(\mathbf{x}, \mathbf{A}) = \frac{\mathbf{x}^H \mathbf{A} \mathbf{x}}{\mathbf{x}^H \mathbf{x}}, \tag{5.4.1}$$

where $\mathbf{x}$ is a vector to be selected such that the Rayleigh quotient is maximized or minimized.

The Rayleigh quotient has the following important properties [2, 13, 14].

1. *Homogeneity:* If α and β are scalars, then $R(\alpha \mathbf{x}, \beta \mathbf{A}) = \beta R(\mathbf{x}, \mathbf{A})$.
2. *Shift invariance:* $R(\mathbf{x}, \mathbf{A} - \alpha \mathbf{I}) = R(\mathbf{x}, \mathbf{A}) - \alpha$.
3. *Orthogonality:* $\mathbf{x} \perp (\mathbf{A} - R(\mathbf{x})\mathbf{I})\mathbf{x}$.
4. *Boundedness:* When the vector $\mathbf{x}$ lies in the range of all nonzero vectors, the Rayleigh quotient $R(\mathbf{x})$ falls in a complex plane region (called the range of $\mathbf{A}$). This region is closed, bounded, and convex. If $\mathbf{A}$ is a Hermitian matrix such that $\mathbf{A} = \mathbf{A}^H$, then this region is a closed interval $[\lambda_1, \lambda_n]$.
5. *Minimum residual:* For all vectors $\mathbf{x} \neq \mathbf{0}$ and all scalars μ, we have $\|(\mathbf{A} - R(\mathbf{x})\mathbf{I})\mathbf{x}\| \leq \|(\mathbf{A} - \mu \mathbf{I})\mathbf{x}\|$.

Theorem 5.3 (Rayleigh–Ritz Theorem) *Let $\mathbf{A} \in \mathbb{C}^{n \times n}$ be a Hermitian matrix whose eigenvalues are arranged in increasing order:*

$$\lambda_{\min} = \lambda_1 \leq \lambda_2 \leq \cdots \leq \lambda_{n-1} \leq \lambda_n = \lambda_{\max}; \tag{5.4.2}$$

then

$$\max_{\mathbf{x} \neq \mathbf{0}} \frac{\mathbf{x}^H \mathbf{A} \mathbf{x}}{\mathbf{x}^H \mathbf{x}} = \max_{\mathbf{x}^H \mathbf{x} = 1} \frac{\mathbf{x}^H \mathbf{A} \mathbf{x}}{\mathbf{x}^H \mathbf{x}} = \lambda_{\max}, \quad \text{if } \mathbf{A}\mathbf{x} = \lambda_{\max} \mathbf{x}, \tag{5.4.3}$$

$$\min_{\mathbf{x} \neq \mathbf{0}} \frac{\mathbf{x}^H \mathbf{A} \mathbf{x}}{\mathbf{x}^H \mathbf{x}} = \min_{\mathbf{x}^H \mathbf{x} = 1} \frac{\mathbf{x}^H \mathbf{A} \mathbf{x}}{\mathbf{x}^H \mathbf{x}} = \lambda_{\min}, \quad \text{if } \mathbf{A}\mathbf{x} = \lambda_{\min} \mathbf{x}. \tag{5.4.4}$$

More generally, the eigenvectors and eigenvalues of $\mathbf{A}$ correspond, respectively, to the critical points and the critical values of the Rayleigh quotient $R(\mathbf{x})$.

This theorem has various proofs, see, e.g., [6, 7] or [3].

As a matter of fact, from the eigenvalue decomposition $\mathbf{A}\mathbf{x} = \lambda \mathbf{x}$ we have

$$\mathbf{A}\mathbf{x} = \lambda \mathbf{x} \quad \Leftrightarrow \quad \mathbf{x}^H \mathbf{A} \mathbf{x} = \lambda \mathbf{x}^H \mathbf{x} \quad \Leftrightarrow \quad \frac{\mathbf{x}^H \mathbf{A} \mathbf{x}}{\mathbf{x}^H \mathbf{x}} = \lambda. \tag{5.4.5}$$

That is to say, a Hermitian matrix $\mathbf{A} \in \mathbb{C}^{n \times n}$ has n Rayleigh quotients that are equal to n eigenvalues of $\mathbf{A}$, namely $R_i(\mathbf{x}, \mathbf{A}) = \lambda_i(\mathbf{A})$, $i = 1, \ldots, n$.

5.4.2 Generalized Rayleigh Quotient

Definition 5.7 (Generalized Rayleigh Quotient) Let $\mathbf{A}$ and $\mathbf{B}$ be two $n \times n$ Hermitian matrices, and let $\mathbf{B}$ be positive definite. The *generalized Rayleigh quotient* or *generalized Rayleigh–Ritz ratio* $R(\mathbf{x})$ of the matrix pencil $(\mathbf{A}, \mathbf{B})$ is a scalar, and is defined as

$$R(\mathbf{x}) = \frac{\mathbf{x}^H \mathbf{A} \mathbf{x}}{\mathbf{x}^H \mathbf{B} \mathbf{x}}, \qquad (5.4.6)$$

where $\mathbf{x}$ is a vector to be selected such that the generalized Rayleigh quotient is maximized or minimized.

In order to find the generalized Rayleigh quotient, we define a new vector $\tilde{\mathbf{x}} = \mathbf{B}^{1/2} \mathbf{x}$, where $\mathbf{B}^{1/2}$ denotes the square root of the positive definite matrix $\mathbf{B}$. Substitute $\mathbf{x} = \mathbf{B}^{-1/2} \tilde{\mathbf{x}}$ into the definition formula of the generalized Rayleigh quotient (5.4.6); then we have

$$R(\tilde{\mathbf{x}}) = \frac{\tilde{\mathbf{x}}^H \left(\mathbf{B}^{-1/2}\right)^H \mathbf{A} \left(\mathbf{B}^{-1/2}\right) \tilde{\mathbf{x}}}{\tilde{\mathbf{x}}^H \tilde{\mathbf{x}}}. \qquad (5.4.7)$$

This implies that the generalized Rayleigh quotient of the matrix pencil $(\mathbf{A}, \mathbf{B})$ is equivalent to the Rayleigh quotient of the matrix product $\mathbf{C} = (\mathbf{B}^{-1/2})^H \mathbf{A} (\mathbf{B}^{-1/2})$. By the Rayleigh–Ritz theorem, it is known that, when the vector $\tilde{\mathbf{x}}$ is selected as the eigenvector corresponding to the minimum eigenvalue $\lambda_{\min}$ of $\mathbf{C}$, the generalized Rayleigh quotient takes a minimum value $\lambda_{\min}$, while when the vector $\tilde{\mathbf{x}}$ is selected as the eigenvector corresponding to the maximum eigenvalue $\lambda_{\max}$ of $\mathbf{C}$, the generalized Rayleigh quotient takes the maximum value $\lambda_{\max}$.

Consider the EVD of the matrix product $(\mathbf{B}^{-1/2})^H \mathbf{A} (\mathbf{B}^{-1/2}) \tilde{\mathbf{x}} = \lambda \tilde{\mathbf{x}}$. If $\mathbf{B} = \sum_{i=1}^{n} \beta_i \mathbf{v}_i \mathbf{v}_i^H$ is the EVD of the matrix $\mathbf{B}$, then

$$\mathbf{B}^{1/2} = \sum_{i=1}^{n} \sqrt{\beta_i} \mathbf{v}_i \mathbf{v}_i^H \quad \text{and} \quad \mathbf{B}^{-1/2} = \sum_{i=1}^{n} \frac{1}{\sqrt{\beta_i}} \mathbf{v}_i \mathbf{v}_i^H..$$

This shows that $\mathbf{B}^{-1/2}$ is also a Hermitian matrix, so that $(\mathbf{B}^{-1/2})^H = \mathbf{B}^{-1/2}$.

Premultiplying $(\mathbf{B}^{-1/2})^H \mathbf{A} (\mathbf{B}^{-1/2}) \tilde{\mathbf{x}} = \lambda \tilde{\mathbf{x}}$ by $\mathbf{B}^{-1/2}$ and substituting $(\mathbf{B}^{-1/2})^H = \mathbf{B}^{-1/2}$, it follows that $\mathbf{B}^{-1} \mathbf{A} \mathbf{B}^{-1/2} \tilde{\mathbf{x}} = \lambda \mathbf{B}^{-1/2} \tilde{\mathbf{x}}$ or $\mathbf{B}^{-1} \mathbf{A} \mathbf{x} = \lambda \mathbf{x}$. Hence, the EVD of the matrix product $(\mathbf{B}^{-1/2})^H \mathbf{A} (\mathbf{B}^{-1/2})$ is equivalent to the EVD of $\mathbf{B}^{-1} \mathbf{A}$.

5.4 Rayleigh Quotient and Generalized Rayleigh Quotient

Because the EVD of $\mathbf{B}^{-1}\mathbf{A}$ is the GEVD of the matrix pencil $(\mathbf{A}, \mathbf{B})$, the above analysis can be summarized as follows: the conditions for the maximum and the minimum of the generalized Rayleigh quotient are:

$$R(\mathbf{x}) = \frac{\mathbf{x}^H \mathbf{A} \mathbf{x}}{\mathbf{x}^H \mathbf{B} \mathbf{x}} = \lambda_{\max}, \quad \text{if we choose } \mathbf{A}\mathbf{x} = \lambda_{\max} \mathbf{B}\mathbf{x}, \tag{5.4.8}$$

$$R(\mathbf{x}) = \frac{\mathbf{x}^H \mathbf{A} \mathbf{x}}{\mathbf{x}^H \mathbf{B} \mathbf{x}} = \lambda_{\min}, \quad \text{if we choose } \mathbf{A}\mathbf{x} = \lambda_{\min} \mathbf{B}\mathbf{x}. \tag{5.4.9}$$

That is to say, in order to maximize the generalized Rayleigh quotient, the vector $\mathbf{x}$ must be taken as the generalized eigenvector corresponding to the maximum generalized eigenvalue of the matrix pencil $(\mathbf{A}, \mathbf{B})$. In contrast, for the generalized Rayleigh quotient to be minimized, the vector $\mathbf{x}$ should be taken as the generalized eigenvector corresponding to the minimum generalized eigenvalue of the matrix pencil $(\mathbf{A}, \mathbf{B})$.

5.4.3 Effectiveness of Class Discrimination

Pattern recognition is widely applied in recognitions of human characteristics (such as a person's face, fingerprint, iris) and various radar targets (such as aircraft, ships). In these applications, the extraction of signal features is crucial. For example, when a target is considered as a linear system, the target parameter is a feature of the target signal.

The divergence is a measure of the distance or dissimilarity between two signals and is often used in feature discrimination and evaluation of the effectiveness of class discrimination.

Let Q denote the common dimension of the signal-feature vectors extracted by various methods. Assume that there are c classes of signals; the Fisher class separability measure, called simply the *Fisher measure*, between the ith and the jth classes is used to determine the ranking of feature vectors. Consider the class discrimination of c classes of signals. Let $\mathbf{v}_k^{(l)}$ denote the kth sample feature vector of the lth class of signals, where $l = 1, \ldots, c$ and $k = 1, \cdots, l_K$, with l_K the number of sample feature vectors in the lth signal class. Under the assumption that the prior probabilities of the random vectors $\mathbf{v}^{(l)} = \mathbf{v}_k^{(l)}$ are the same for $k = 1, \ldots, l_K$ (i.e., equal probabilities), the Fisher measure is defined as

$$m^{(i,j)} = \frac{\sum_{l=i,j} \left(\text{mean}_k(\mathbf{v}_k^{(l)}) - \text{mean}_l(\text{mean}(\mathbf{v}_k^{(l)})) \right)^2}{\sum_{l=i,j} \text{var}_k(\mathbf{v}_k^{(l)})}, \tag{5.4.10}$$

where:

- $\text{mean}_k(\mathbf{v}_k^{(l)})$ is the mean (centroid) of all signal-feature vectors of the lth class;
- $\text{var}(\mathbf{v}_k^{(l)})$ is the variance of all signal-feature vectors of the lth class;
- $\text{mean}_l(\text{mean}_k(\mathbf{v}_k^{(l)}))$ is the total sample centroid of all classes.

As an extension of the Fisher measure, consider the projection of all $Q \times 1$ feature vectors onto the $(c-1)$th dimension of class discrimination space.

Put $N = N_1 + \cdots + N_c$, where N_i represents the number of feature vectors of the ith class signal extracted in the training phase. Suppose that $\mathbf{s}_{i,k} = [s_{i,k}(1), \ldots, s_{i,k}(Q)]^T$ denotes the $Q \times 1$ feature vector obtained by the kth group of observation data of the ith signal in the training phase, while $\mathbf{m}_i = [m_i(1), \ldots, m_i(Q)]^T$ is the sample mean of the ith signal-feature vector, where

$$m_i(q) = \frac{1}{N_i} \sum_{k=1}^{N_i} s_{i,k}(q), \quad i = 1, \ldots, c, \ q = 1, \ldots, Q.$$

Similarly, let $\mathbf{m} = [m(1), \ldots, m(Q)]^T$ denote the ensemble mean of all the signal-feature vectors obtained from all the observation data, where

$$m(q) = \frac{1}{c} \sum_{i=1}^{c} m_i(q), \quad q = 1, \ldots, Q.$$

Then, we have *within-class scatter matrix* and *between-class scatter matrix* [5]:

$$\mathbf{S}_w \stackrel{\text{def}}{=} \frac{1}{c} \sum_{i=1}^{c} \left(\frac{1}{N_i} \sum_{k=1}^{N_i} (\mathbf{s}_{i,k} - \mathbf{m}_i)(\mathbf{s}_{i,k} - \mathbf{m}_i)^T \right), \tag{5.4.11}$$

$$\mathbf{S}_b \stackrel{\text{def}}{=} \frac{1}{c} \sum_{i=1}^{c} (\mathbf{m}_i - \mathbf{m})(\mathbf{m}_i - \mathbf{m})^T. \tag{5.4.12}$$

Define a criterion function

$$J(\mathbf{U}) \stackrel{\text{def}}{=} \frac{\prod_{\text{diag}} \mathbf{U}^T \mathbf{S}_b \mathbf{U}}{\prod_{\text{diag}} \mathbf{U}^T \mathbf{S}_w \mathbf{U}}, \tag{5.4.13}$$

where $\prod_{\text{diag}} \mathbf{A}$ denotes the product of diagonal entries of the matrix $\mathbf{A}$. As it is a measure of class discrimination ability, J should be maximized. We refer to Span($\mathbf{U}$) as the

5.4 Rayleigh Quotient and Generalized Rayleigh Quotient

class discrimination space, if

$$\mathbf{U} = \arg\max_{\mathbf{U}\in\mathbb{R}^{Q\times Q}} J(\mathbf{U}) = \frac{\prod_{\text{diag}} \mathbf{U}^T \mathbf{S}_b \mathbf{U}}{\prod_{\text{diag}} \mathbf{U}^T \mathbf{S}_w \mathbf{U}}. \qquad (5.4.14)$$

This optimization problem can be equivalently written as

$$[\mathbf{u}_1,\ldots,\mathbf{u}_Q] = \arg\max_{\mathbf{u}_i \in \mathbb{R}^Q} \left\{ \frac{\prod_{i=1}^{Q} \mathbf{u}_i^T \mathbf{S}_b \mathbf{u}_i}{\prod_{i=1}^{Q} \mathbf{u}_i^T \mathbf{S}_w \mathbf{u}_i} = \prod_{i=1}^{Q} \frac{\mathbf{u}_i^T \mathbf{S}_b \mathbf{u}_i}{\mathbf{u}_i^T \mathbf{S}_w \mathbf{u}_i} \right\} \qquad (5.4.15)$$

whose solution is given by

$$\mathbf{u}_i = \arg\max_{\mathbf{u}_i \in \mathbb{R}^Q} \frac{\mathbf{u}_i^T \mathbf{S}_b \mathbf{u}_i}{\mathbf{u}_i^T \mathbf{S}_w \mathbf{u}_i}, \quad i = 1,\ldots,Q. \qquad (5.4.16)$$

This is just the maximization of the generalized Rayleigh quotient. The above equation has a clear physical meaning: the column vectors of the matrix $\mathbf{U}$ constituting the optimal class discrimination subspace which should maximize the between-class scatter and minimize the within-class scatter, namely which should maximize the generalized Rayleigh quotient.

For c classes of signals, the optimal class discrimination subspace is $(c-1)$-dimensional. Therefore Eq. (5.4.16) is maximized only for $c-1$ generalized Rayleigh quotients. In other words, it is necessary to solve the following generalized eigenvalue problem:

$$\mathbf{S}_b \mathbf{u}_i = \lambda_i \mathbf{S}_w \mathbf{u}_i, \quad i = 1, 2, \ldots, c-1, \qquad (5.4.17)$$

for $c-1$ generalized eigenvectors $\mathbf{u}_1, \ldots, \mathbf{u}_{c-1}$. These generalized eigenvectors constitute the $Q \times (c-1)$ matrix

$$\mathbf{U}_{c-1} = [\mathbf{u}_1, \ldots, \mathbf{u}_{c-1}] \qquad (5.4.18)$$

whose columns span the optimal class discrimination subspace.

After obtaining the $Q \times (c-1)$ matrix $\mathbf{U}_{c-1}$, we can find its projection onto the optimal class discrimination subspace,

$$\mathbf{y}_{i,k} = \mathbf{U}_{c-1}^T \mathbf{s}_{i,k}, \quad i = 1, \ldots, c, \ k = 1, \ldots, N_i, \qquad (5.4.19)$$

for every signal-feature vector obtained in the training phase.

When there are only three classes of signals ($c = 3$), the optimal class discrimination subspace is a plane, and the projection of every feature vector onto the optimal class discrimination subspace is a point. These projections directly reflect the discrimination ability of different feature vectors in signal classification.

Brief Summary of This Chapter

This chapter focuses on the matrix algebra used widely in artificial intelligence: eigenvalue decomposition, general eigenvalue decomposition, Rayleigh quotients, generalized Rayleigh quotients, and Fisher measure.

References

1. Bellman, R.: Introduction to Matrix Analysis, 2nd edn. McGraw-Hill, New York (1970)
2. Chatelin, F.: Eigenvalues of Matrices. Wiley, New York (1993)
3. Cirrincione, G., Cirrincione, M., Herault J., et al.: The MCA EXIN neuron for the minor component analysis. IEEE Trans. Neural Netw. **13**(1), 160–187 (2002)
4. Drmac, Z.: A tangent algorithm for computing the generalized singular value decomposition. SIAM J. Numer. Anal. **35**(5), 1804–1832 (1998)
5. Duda, R.O., Hart, P.E., Stork, D.G.: Pattern Classification and Scene Analysis. Wiley, New York (1973)
6. Golub, G.H., Van Loan, C.F.: Matrix Computation, 2nd edn. The John Hopkins University Press, Baltimore (1989)
7. Helmke, U., Moore, J.B.: Optimization and Dynamical Systems. Springer, London (1994)
8. Horn, R.A., Johnson, C.R.: Matrix Analysis. Cambridge University Press, Cambridge (1985)
9. Huang, L.: Linear Algebra in System and Control Theory (in Chinese). Science Press, Beijing (1984)
10. Jennings, A., McKeown, J.J.: Matrix Computations. Wiley, New York (1992)
11. Johnson, D.H., Dudgeon, D.E.: Array Signal Processing: Concepts and Techniques. Prentice Hall, Englewood Cliffs (1993)
12. Nour-Omid, B., Parlett, B.N., Ericsson, T., Jensen, P.S.: How to implement the spectral transformation. Math. Comput. **48**, 663–673 (1987)
13. Parlett, B.N.: The Rayleigh quotient iteration and some generalizations for nonnormal matrices. Math. Comput. **28**(127), 679–693 (1974)
14. Parlett, B.N.: The Symmetric Eigenvalue Problem. Prentice-Hall, Englewood Cliffs (1980)
15. Rayleigh, L.: The Theory of Sound, 2nd edn. Macmillan, New York (1937)
16. Roy, R., Kailath, T.: ESPRIT - estimation of signal parameters via rotational invariance techniques. IEEE Trans. Acoust. Speech Signal Process. **37**, 297–301 (1989)
17. Saad, Y.: Numerical Methods for Large Eigenvalue Problems. Manchester University Press, New York (1992)
18. Zhang, X.D., Liang, Y.C.: Prefiltering-based ESPRIT for estimating parameters of sinusoids in non-Gaussian ARMA noise. IEEE Trans. Signal Process. **43**, 349–353 (1995)

Part II
Artificial Intelligence

Chapter 6
Machine Learning

Machine learning is a subset of artificial intelligence. This chapter presents first a machine learning tree, and then focuses on the matrix algebra methods in machine learning including single-objective optimization, feature selection, principal component analysis, and canonical correlation analysis together with supervised, unsupervised, and semi-supervised learning and active learning. More importantly, this chapter highlights selected topics and advances in machine learning: graph machine learning, reinforcement learning, Q-learning, and transfer learning.

6.1 Machine Learning Tree

Machine learning (ML) is a subset of artificial intelligence, which build a mathematical model based on sample data, known as "training data," in order to make predictions or decisions without being explicitly programmed to perform the task. Regarding learning, a good definition given by Mitchell [181] is

> A computer program is said to learn from experience E with respect to some class of tasks T and performance measure P if its performance at tasks in T, as measured by P, improves with experience E.

In machine learning, neural networks, support vector machines, and evolutionary computation, we are usually given a training set and a test set. By the *training set*, it will mean the union of the *labeled set* and the *unlabeled set* of examples available to machine learners. In comparison, *test set* consists of examples never seen before. Let $(X_l, Y_l) = \{(\mathbf{x}_1, y_1), \ldots, (\mathbf{x}_l, y_l)\}$ denote the labeled set, where $\mathbf{x}_i \in \mathbb{R}^D$ is the ith D-dimensional data vector and $y_i \in \mathbb{R}$ or $y_i \in \{1, \ldots, M\}$ is the corresponding label of the data vector $\mathbf{x}_i$. In regression problems, y_i is the regression or fitting of $\mathbf{x}_i$. In classification problems, y_i is the corresponding class label of $\mathbf{x}_i$ among the M classes of targets. The labeled data $\mathbf{x}_i, i = 1, \ldots, l$ are observed by the user, while $y_i, i = 1, \ldots, l$ are labeled by data labeling experts or supervisors. The unlabeled set is comprised of the data vectors and is denoted by $X_u = \{\mathbf{x}_1, \ldots, \mathbf{x}_u\}$.

Machine learning aims to establish a regressor or classifier through learning the training set, and then to evaluate the performance of the regressor or classifier through the test set.

According to the nature of training data, we can classify machine learning as follows.

1. *Regular or Euclidean structured data learning*

    - *Supervised learning:* Given a training set consisting of labeled data (i.e., example inputs and their desired outputs) $(X_{\text{train}}, Y_{\text{train}}) = \{(\mathbf{x}_1, y_1), \ldots, (\mathbf{x}_l, y_l)\}$, supervised learning learns a general rule that maps inputs to outputs. This is like a "teacher" or a supervisor (data labeling expert) giving a student a problem (finding the mapping relationship between inputs and outputs) and its solutions (labeled output data) and telling that student to figure out how to solve other, similar problems: finding the mapping from the features of unseen samples to their correct labels or target values in the future.
    - *Unsupervised learning:* In unsupervised learning, the training set consists of the unlabeled set only $X_{\text{train}} = X_u = \{\mathbf{x}_1, \ldots, \mathbf{x}_u\}$. The main task of the machine learner is to find the solutions on its own (i.e., patterns, structures, or knowledge in unlabeled data). This is like giving a student a set of patterns and asking him or her to figure out the underlying motifs that generated the patterns.
    - *Semi-supervised learning:* Given a training set $(X_{\text{train}}, Y_{\text{train}}) = \{(\mathbf{x}_1, y_1), \ldots, (\mathbf{x}_l y_l)\} \cup \{\mathbf{x}_{l+1}, \ldots, \mathbf{x}_{l+u}\}$ with $l \ll u$, i.e., we are given a small amount of labeled data together with a large amount of unlabeled data. Semi-supervised learning falls between unsupervised learning (without any labeled training data) and supervised learning (with completely labeled training data). Depending on how the data is labeled, semi-supervised learning can be divided into the following categories:

        - *Self-training* is a semi-supervised learning using its own predictions to teach itself.
        - *Co-training* is a weakly semi-supervised learning for multi-view data using the co-training setting, and use their own predictions to teach themselves.
        - *Active learning* is a semi-supervised learning where the learner has some active or participatory role in determining which data points it will ask to be labeled by an expert or teacher.

    - *Reinforcement learning:* Training data (in the form of rewards and punishments) is given only as feedback to an artificial intelligence agent in a dynamic environment. This feedback between the learning system and the interaction experience is useful to improve performance in the task being learned. The machine learning based on data feedback is called reinforcement learning. Q-learning is a popular model-free reinforcement learning and learns a reward or punishment function (action-value functions, simply called Q-function).

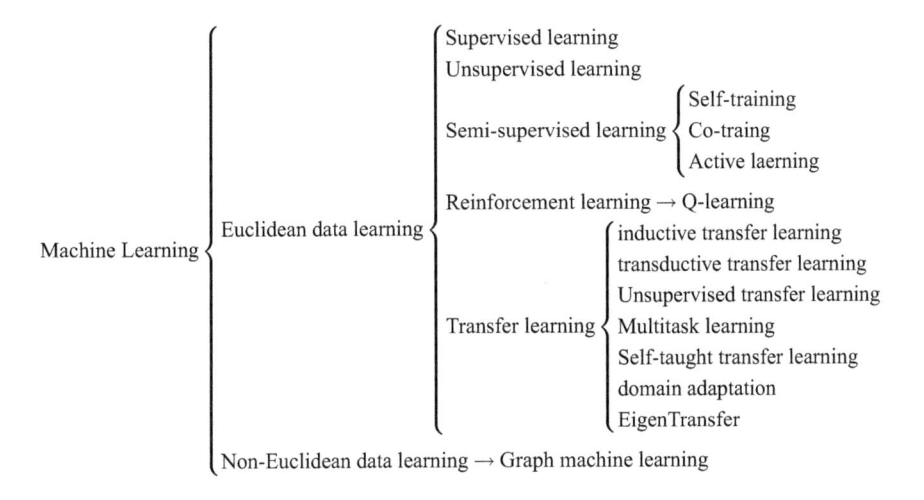

6.1 Machine Learning Tree

- *Transfer learning:* In many real-world applications, the data distribution changes or data are outdated, and thus it is necessary to apply transfer learning for considering transfer of knowledge from the source domain to the target domain. Transfer learning includes but not limited to inductive transfer learning, transductive transfer learning, unsupervised transfer learning, multitask learning, self-taught transfer learning, domain adaptation, and EigenTransfer.

2. *Graph machine learning* is an irregular or non-Euclidean structured data learning, and learns the structure of a graph, called also graph construction, from training samples in semi-supervised and unsupervised learning cases.

The above classification of machine learning can be vividly represented by a machine learning tree, as shown in Fig. 6.1.

The machine learning tree is so-called because Fig. 6.1 looks like a tree after rotating it 90° to the left.

There are two basic tasks of machine learning: classification (for discrete data) and prediction (for continuous data).

Deep learning is to learn the internal law and representation level of sample data. The information at different hierarchy levels obtained in the learning process is very helpful to the interpretation of data such as text, image, and voice. Deep learning is a complex machine learning algorithm, which has achieved much better results in speech and image recognition than previous related technologies.

Limited to space, this chapter will not discuss deep learning, but focuses only on supervised learning, unsupervised learning, reinforcement learning, and transfer learning.

Before dealing with machine learning in detail, it is necessary to start with preparation knowledge of machine learning: its optimization problems, majorization-minimization algorithms, and how to boost a weak learning algorithm to a strong learning algorithm.

$$\text{Machine Learning} \begin{cases} \text{Euclidean data learning} \begin{cases} \text{Supervised learning} \\ \text{Unsupervised learning} \\ \text{Semi-supervised learning} \begin{cases} \text{Self-training} \\ \text{Co-traing} \\ \text{Active laerning} \end{cases} \\ \text{Reinforcement learning} \to \text{Q-learning} \\ \text{Transfer learning} \begin{cases} \text{inductive transfer learning} \\ \text{transductive transfer learning} \\ \text{Unsupervised transfer learning} \\ \text{Multitask learning} \\ \text{Self-taught transfer learning} \\ \text{domain adaptation} \\ \text{EigenTransfer} \end{cases} \end{cases} \\ \text{Non-Euclidean data learning} \to \text{Graph machine learning} \end{cases}$$

Fig. 6.1 Machine learning tree

6.2 Optimization in Machine Learning

Two of the pillars of machine learning are matrix algebra and optimization. Matrix algebra involves the matrix-vector representation and modeling of currently available data, and optimization involves the numerical computation of parameters for a system designed to make decisions on yet unseen data.

Optimization problems arise throughout machine learning. On optimization in machine learning, the following questions are naturally asked and important [31]:

1. How do optimization problems arise in machine learning applications, and what makes them challenging?
2. What have been the most successful optimization methods for large-scale machine learning, and why?
3. What recent advances have been made in the design of algorithms, and what are open questions in this research area?

This section focuses upon the most successful optimization methods for large-scale machine learning that involves very large data sets and for which the number of model parameters to be optimized is also large.

6.2.1 Single-Objective Composite Optimization

In supervised machine learning, we are given the collection of a data set $\{(\mathbf{x}_1, y_1), \ldots, (\mathbf{x}_N, y_N)\}$, where $\mathbf{x}_i \in \mathbb{R}^d$ for each $i \in \{1, \ldots, N\}$ represents the feature vector of a target, and the scalar y_i is a label indicating whether a corresponding feature vector belongs ($y_i = 1$) or not ($y_i = -1$) to a particular class (i.e., topic of interest).

With a set of examples $\{(\mathbf{x}_i, y_i)\}_{i=1}^{N}$, define the *prediction function* with respect to the ith sample as

$$h(\mathbf{w}; \mathbf{x}_i) = \mathbf{w} \cdot \mathbf{x}_i = \langle \mathbf{w}, \mathbf{x}_i \rangle = \mathbf{w}^T \mathbf{x}_i, \qquad (6.2.1)$$

where the predictor $\mathbf{w} = [w_1, \ldots, w_d]^T \in \mathbb{R}^d$ is a d-dimension parameter vector of machine learning model that should have the good "generalization ability."

Construct a machine learning program in which the performance of a prediction function $\hat{y}_i = h(\mathbf{w}; \mathbf{x}_i)$ is measured by counting how often the program prediction $f(\mathbf{x}_i; \mathbf{w})$ differs from the correct prediction y_i, i.e., $\hat{y}_i \neq y_i$. Therefore, we need to search for a predictor function that minimizes the frequency of observed mispredictions, called the *empirical risk*:

$$R_n(h) = \frac{1}{N} \sum_{i=1}^{N} \mathbb{1}[h_i(\mathbf{x}_i) \neq y_i], \qquad (6.2.2)$$

6.2 Optimization in Machine Learning

where

$$\mathbb{1}[A] = \begin{cases} 1, & \text{if } A \text{ is true;} \\ 0, & \text{otherwise.} \end{cases} \quad (6.2.3)$$

The optimization in machine learning seeks to design a classifier/predictor $\mathbf{w}$ so that $\hat{y} = \langle \mathbf{w}, \mathbf{x} \rangle$ can provide the correct classification/prediction for any unknown input vector $\mathbf{x}$. The objective $f(\mathbf{w}; \mathbf{x})$ in machine learning can be represented as the following common form:

$$f(\mathbf{w}; \mathbf{x}) = l(\mathbf{w}; \mathbf{x}) + h(\mathbf{w}; \mathbf{x}), \quad (6.2.4)$$

where $l(\mathbf{w}; \mathbf{x})$ is the *loss function* and $h(\mathbf{w}; \mathbf{x})$ denotes the *regularization function*.

In this manner, the expected risk for a given $\mathbf{w}$ is the expected value of the loss function $l(\mathbf{w}; \mathbf{x})$ taken with respect to the distribution of $\mathbf{x}$ as follows [31]:

$$\text{(Expected Risk)} \qquad R(\mathbf{w}) = E\{l(\mathbf{w}; \mathbf{x})\} \quad (6.2.5)$$

since the true distribution of $\mathbf{x}$ is often unknown, the expected risk is commonly estimated by the empirical risk

$$\text{(Empirical Risk)} \qquad R_n(\mathbf{w}) = \frac{1}{N} \sum_{i=1}^{N} l_i(\mathbf{w}), \quad (6.2.6)$$

where $l_i(\mathbf{w}) = l(\mathbf{w}; \mathbf{x}_i)$.

If we were to try to minimize the empirical risk only, the method would be referred to as ERM (Empricial Risk Minimization). However, the empirical risk usually underestimates the true/expected risk by the *estimation error*, and thus we add a regularization term $h(w; x)$ to the composite function.

The corresponding *prediction error* is given by

$$e_i(\mathbf{w}) = e(\mathbf{w}; \mathbf{x}_i, y_i) = \mathbf{w}^T \mathbf{x}_i - y_i. \quad (6.2.7)$$

Then one can write the *empirical error* as the average of the prediction errors for all N samples $\{(\mathbf{x}_i, y_i)\}_{i=1}^{N}$:

$$\text{(Empirical Error)} \qquad R_{\text{emp}}(\mathbf{w}) = \frac{1}{N} \sum_{i=1}^{N} e_i(\mathbf{w}). \quad (6.2.8)$$

The loss function incurred by the parameter vector **w** can be represented as

$$l(\mathbf{w}) = \frac{1}{2N}\sum_{i=1}^{N}|e_i(\mathbf{w})|^2 = \frac{1}{2N}\sum_{i=1}^{N}(\mathbf{w}^T\mathbf{x}_i - y_i)^2. \quad (6.2.9)$$

The above machine learning tasks such as classification and regression can be represented as solving a "*composite*" (additive) *optimization* problems

$$\min_{\mathbf{w}} \left\{ f(\mathbf{w}) = l(\mathbf{w}) + h(\mathbf{w}) = \frac{1}{2N}\sum_{i=1}^{N}\left(\mathbf{w}^T\mathbf{x}_i - y_i\right)^2 + h(\mathbf{w}) \right\}, \quad (6.2.10)$$

where the objective function $f(\mathbf{x})$ is given by the sum of N component loss functions $f_i(\mathbf{w}) = |e_i(\mathbf{w})|^2 = (\mathbf{w}^T\mathbf{x}_i - y_i)^2$ and a possibly nonsmooth regularization function $h(\mathbf{w})$. It is assumed that each component loss function $f_i(\mathbf{w}; \mathbf{x}_i) : \mathbb{R}^d \times \mathbb{R}^d \to \mathbb{R}$ is continuously differentiable and the sum of N component loss functions $f(\mathbf{x})$ is strongly convex, while the regularization function $h : \mathbb{R}^d \to \mathbb{R}$ is proper, closed, and convex but not necessarily differentiable, and is used for preventing overfitting.

The composite optimization tasks exist widely in various artificial intelligence applications, e.g., image recognition, speech recognition, task allocation, text classification/processing, and so on.

To simplify notation, denote $\mathbf{g}_j(\mathbf{w}_k) = \mathbf{g}(\mathbf{w}_k; \mathbf{x}_j, y_j)$ and $f_j(\mathbf{w}_k) = f(\mathbf{w}_k; \mathbf{x}_j, y_j)$ hereafter. Optimization methods for machine learning fall into two broad categories [31]:

- *Stochastic method:* The prototypical stochastic method is the *stochastic gradient (SG) method* [214] defined by

$$\mathbf{w}_{k+1} \leftarrow \mathbf{w}_k - \alpha_k \nabla f_{i_k}(\mathbf{w}_k) \quad (6.2.11)$$

for all $k \in \{1, 2, \ldots\}$, the index i_k, corresponding to the sample pair $(\mathbf{x}_{i_k}, y_{i_k})$, is chosen randomly from $\{1, \ldots, N\}$ and α_k is a positive stepsize. Each iteration of this method involves only the computation of the gradient $\nabla f_{i_k}(\mathbf{w}_k)$ corresponding to one sample.

- *Batch method:* The simplest form of batch method is the steepest descent algorithm:

$$\mathbf{w}_{k+1} \leftarrow \mathbf{w}_k - \alpha_k \nabla R_n(\mathbf{w}_k) = \mathbf{w}_k - \frac{\alpha_k}{N}\sum_{i=1}^{N}\nabla f_i(\mathbf{w}_k). \quad (6.2.12)$$

The steepest descent algorithm is also referred to as the gradient, batch gradient, or full gradient method.

6.2 Optimization in Machine Learning

The gradient algorithms are usually represented as

$$\mathbf{w}_{k+1} = \mathbf{w}_k - \alpha_k \mathbf{g}(\mathbf{w}_k). \tag{6.2.13}$$

Traditional gradient-based methods are effective for solving small-scale learning problems, but in the context of large-scale machine learning one of the core strategies of interest is the SG method proposed by Robbins and Monro [214]:

$$\mathbf{g}(\mathbf{w}_k) \leftarrow \frac{1}{|S_k|} \sum_{i \in S_k} \nabla f_i(\mathbf{w}_k), \tag{6.2.14}$$

$$\mathbf{w}_{k+1} \leftarrow \mathbf{w}_k - \alpha_k \mathbf{g}(\mathbf{w}_k), \tag{6.2.15}$$

for all $k \in \{1, 2, \ldots\}$ and $S_k \subseteq \{1, \ldots, N\}$ is a small subset of sample pairs $(\mathbf{x}_i, y_i)$ chosen randomly from $\{1, \ldots, N\}$ and α_k is a positive stepsize.

Algorithm 6.1 shows a generalized SG (mini-batch) method.

Algorithm 6.1 Stochastic gradient (SG) method [31]
1. **input:** Sample pairs $(\mathbf{x}_j, y_j)$, $j = 1, \ldots, N$.
2. **initialization:** Choose an initial iterate $\mathbf{w}_1$.
3. **for** $k = 1, 2, \ldots$ **do**
4.   Generate a realization of the random variables $(\mathbf{x}_i^{(k)}, y_i^{(k)})$, where $i \in S_k$ and $S_k \subseteq \{1, \ldots, N\}$ is a small subset of samples.
5.   Compute a stochastic vector $\mathbf{g}(\mathbf{w}_k) \leftarrow \frac{1}{|S_k|} \sum_{i \in S_k} \nabla f(\mathbf{w}_k; \mathbf{x}_i^{(k)}, y_i^{(k)})$.
6.   Choose a stepsize $\alpha_k > 0$.
7.   Update the new iterate as $\mathbf{w}_{k+1} \leftarrow \mathbf{w}_k - \alpha_k \mathbf{g}(\mathbf{w}_k)$.
8.   **exit:** if $\mathbf{w}_{k+1}$ is converged.
9. **end for**
10. **output:** $\mathbf{w} = \frac{1}{k+1} \sum_{i=1}^{k+1} \mathbf{w}_i$.

The SG iteration is very inexpensive, requiring only the evaluation of $\mathbf{g}(\mathbf{w}_k, \xi_k)$. However, since SG generates noisy iterate sequences that tend to oscillate around minimizers during the optimization process, a natural idea is to compute a corresponding sequence of *iterate averages*

$$\tilde{\mathbf{w}}_{k+1} \leftarrow \frac{1}{k+1} \sum_{i=1}^{k+1} \mathbf{w}_i, \tag{6.2.16}$$

where the averaged sequence $\{\tilde{\mathbf{w}}_k\}$ has no effect on the computation of the SG iterate sequence $\{\mathbf{w}_k\}$. Iterate averages would automatically possess less noisy behavior.

6.2.2 Gradient Aggregation Methods

Consider a common single-objective convex optimization problem

$$\min_{\mathbf{x} \in \mathbb{R}^d} \{F(\mathbf{x}) = f(\mathbf{x}) + h(\mathbf{x})\}, \qquad (6.2.17)$$

where the first term $f(\mathbf{x})$ is smooth and the second term $h(\mathbf{x})$ is possibly nonsmooth, which allows for the modeling of constraints.

In many applications in optimization, pattern recognition, signal processing, and machine learning, $f(\mathbf{x})$ has an additional structure:

$$f(\mathbf{x}) = \frac{1}{N} \sum_{i=1}^{N} f_i(\mathbf{x}) \qquad (6.2.18)$$

which is the average of a number of convex functions f_i.

When N is large and when a solution with low to medium accuracy is sufficient, it is a common choice to perform classical stochastic methods, especially stochastic gradient descent (SGD) method for solving the optimization problem (6.2.17). The SGD method can be dated back to the 1951 seminal work of Robbins and Monro [214]. SGD selects an index $i \in \{1, \ldots, N\}$ uniformly at random, and then updates $\mathbf{x}$ using a stochastic estimate $\nabla f_i(\mathbf{x})$ of $\nabla f(\mathbf{x})$. Thanks to the computation of $\nabla f_i(\mathbf{x})$ that is N times cheaper than the computation of the full gradient $\nabla f(\mathbf{x})$, for optimization problems where N is very large, the per-iteration savings can be extremely large, spanning several orders of magnitude [151].

SG algorithms only use the current gradient information, while *gradient aggregation* is a method to reuse and/or revise previously computed gradient information. The goal of gradient aggregation is to achieve an improved convergence rate and a lower variance. In addition, if one maintains indexed gradient estimates in storage, then one can revise specific estimates as new information to be collected.

Three famous gradient aggregation methods are SAG, SVRG, and SAGA.

- *Stochastic average gradient* (SAG) method: If stochastic average gradient is used to replace iterate averages:

$$\text{SAG:} \quad \mathbf{g}(\mathbf{w}_k) = \frac{1}{N} \left(\nabla f_j(\mathbf{w}_k) - \nabla f_j(\mathbf{w}_{k-1}) + \sum_{i=1}^{N} \nabla f_i(\mathbf{w}_{[i]}) \right), \qquad (6.2.19)$$

one gets the well-known stochastic average gradient (SAG) method [156, 222]. In iteration k of SAG, $\nabla f_j(\mathbf{w}_k) = \nabla f(\mathbf{w}_k; \mathbf{x}_j, y_j), j \in \{1, \ldots, N\}$ is chosen from the current gradient at random, and $\nabla f_j(\mathbf{w}_{k-1}) = \nabla f(\mathbf{w}_{k-1}; \mathbf{x}_{[j]}, y_{[j]}), j \in \{1, \ldots, N\}$ is chosen from the past gradient at iteration $k-1$ randomly, while $\nabla f_i(\mathbf{w}_{[i]}) = \nabla f(\mathbf{w}_{[i]}; \mathbf{x}_{[i]}, y_{[i]})$ for all $i \in \{1, \ldots, N\}$, where $\mathbf{w}_{[i]}$ represents the latest iterate at which ∇f_i was evaluated.

6.2 Optimization in Machine Learning

- *Stochastic variance reduced gradient* (SVRG) method: Stochastic gradient descent (SGD) can be improved with the *variance reduction* (VR) technique. The SVRG method [133] adopts a constant learning rate to train parameters:

$$\text{SVRG:} \quad \tilde{\mathbf{g}}_j \leftarrow \nabla f_{i_j}(\tilde{\mathbf{w}}_j) - \left(\nabla f_{i_j}(\mathbf{w}_k) - \frac{1}{N} \sum_{i=1}^{N} \nabla f_i(\mathbf{w}_k) \right). \quad (6.2.20)$$

- *Stochastic average gradient aggregation* (SAGA) method: This method [72] is inspired both from SAG and SVRG. The stochastic gradient vector in SAGA is given by

$$\text{SAGA:} \quad \mathbf{g}_k \leftarrow \nabla f_j(\mathbf{w}_k) - \nabla f_j(\mathbf{w}_{[j]}) + \frac{1}{N} \sum_{i=1}^{N} \nabla f_i(\mathbf{w}_{[i]}), \quad (6.2.21)$$

where $j \in \{1, \ldots, N\}$ is chosen at random and the stochastic vectors are set by

$$\nabla f_j(\mathbf{w}_k) = \frac{\partial f(\mathbf{w}_k; \mathbf{x}_j, y_j)}{\partial \mathbf{w}_k}, \quad (6.2.22)$$

and

$$\nabla f_j(\mathbf{w}_{[j]}) = \left. \frac{\partial f(\mathbf{w}; \mathbf{x}_j, y_j)}{\partial \mathbf{w}} \right|_{\mathbf{w}=\mathbf{w}_{[j]}}, \quad (6.2.23)$$

$$\nabla f_i(\mathbf{w}_{[i]}) = \left. \frac{\partial f(\mathbf{w}; \mathbf{x}_i, y_i)}{\partial \mathbf{w}} \right|_{\mathbf{w}=\mathbf{w}_{[i]}}, \quad (6.2.24)$$

for all $i \in \{1, \ldots, N\}$, and an integer $j \in \{1, \ldots, N\}$ is chosen at random. $\mathbf{w}_{[i]}$ and $\mathbf{w}_{[j]}$ represent the latest iterates at which ∇f_i and ∇f_j were, respectively, evaluated.

A formal description of a few variants of SVRG is presented as Algorithm 6.2. SAGA method for minimizing an empirical risk E_n is shown in Algorithm 6.3. It has been shown [72] that

- Theoretical convergence rates for SAGA in the strongly convex case are better than those for SAG and SVRG,
- SAGA is applicable to non-strongly convex problems without modification.

Algorithm 6.2 SVRG method for minimizing an empirical risk R_n [31, 133]

1. **input:** Sample pairs $(\mathbf{x}_j, y_j)$, $j = 1, \ldots, N$, update frequency m and learning rate α.
2. **initialization:** Choose an initial iterate $\mathbf{w}_1 \in \mathbb{R}^d$.
3. **for** $k = 1, 2, \ldots$ **do**
4.   Compute $\nabla_i(\mathbf{w}_k) = \frac{\partial f(\mathbf{w}_k; \mathbf{x}_i, y_i)}{\partial \mathbf{w}_k}$.
5.   Compute the batch gradient $R_n(\mathbf{w}_k) = \frac{1}{N} \sum_{i=1}^{N} \nabla f_i(\mathbf{w}_k)$.
6.   Initialize $\tilde{\mathbf{w}}_1 \leftarrow \mathbf{w}_k$.
7.   **for** $j = 1, \ldots, m$ **do**
8.   Chose i_j uniformly from $\{1, \ldots, N\}$.
9.   Set $\tilde{\mathbf{g}}_j \leftarrow \nabla f_{i_j}(\tilde{\mathbf{w}}_j) - (\nabla f_{i_j}(\mathbf{w}_k) - R_n(\mathbf{w}_k))$.
10.   Set $\tilde{\mathbf{w}}_{j+1} \leftarrow \tilde{\mathbf{w}}_j - \alpha \tilde{\mathbf{g}}_j$.
11.   **end for**
12.   Option (a): Set $\mathbf{w}_{k+1} = \tilde{\mathbf{w}}_{m+1}$.
13.   Option (b): Set $\mathbf{w}_{k+1} = \frac{1}{m} \sum_{j=1}^{m} \tilde{\mathbf{w}}_{j+1}$.
14.   Option (c): Choose j uniformly from $\{1, \ldots, m\}$ and set $\mathbf{w}_{k+1} = \tilde{\mathbf{w}}_{j+1}$.
15.   **exit:** If $\mathbf{w}_{k+1}$ is converged.
16. **end for**
17. **output:** $\mathbf{w} = \mathbf{w}_{k+1}$.

Algorithm 6.3 SAGA method for minimizing an empirical risk R_n [31, 72]

1. **input:** Sample pairs $(\mathbf{x}_j, y_j)$, $j = 1, \ldots, N$, stepsize $\alpha > 0$.
2. **initialization:** Choose an initial iterate $\mathbf{w}_1 \in \mathbb{R}^d$.
3. **for** $i = 1, \ldots, N$ **do**
4.   Compute $\nabla f_i(\mathbf{w}_1) = \frac{\partial f(\mathbf{w}_1; \mathbf{x}_i, y_i)}{\partial \mathbf{w}_1}$.
5.   Store $\nabla f_i(\mathbf{w}_{[i]}) \leftarrow \nabla f_i(\mathbf{w}_1)$.
6. **end for**
7. **for** $k = 1, 2, \ldots$ **do**
8.   Choose j uniformly in $\{1, \ldots, N\}$.
9.   Compute $\nabla f_j(\mathbf{w}_k) = \frac{\partial f(\mathbf{w}_k; \mathbf{x}_j, y_j)}{\partial \mathbf{w}_k}$.
10.   Set $\mathbf{g}_k \leftarrow \nabla f_j(\mathbf{w}_k) - \nabla f_j(\mathbf{w}_{[j]}) + \frac{1}{n} \sum_{i=1}^{n} \nabla f_i(\mathbf{w}_{[i]})$.
11.   Store $\nabla f_j(\mathbf{w}_{[i]}) \leftarrow \nabla f_j(\mathbf{w}_k)$.
12.   Set $\mathbf{w}_{k+1} \leftarrow \mathbf{w}_k - \alpha \mathbf{g}_k$.
13.   **exit:** If $\mathbf{w}_{k+1}$ is converged.
14. **end for**
15. **output:** $\mathbf{w} = \mathbf{w}_{k+1}$.

6.2.3 Coordinate Descent Methods

Coordinate descent methods (CDMs) are among the first optimization schemes suggested for solving smooth unconstrained minimization problems and large-scale regression problems (see, e.g., [12, 25, 188]).

As the name suggests, the basic operation of coordinate descent methods takes steps along coordinate directions: the objective is minimized with respect to a single variable while all others are kept fixed, then other variables are updated similarly in an iterative manner.

6.2 Optimization in Machine Learning

The coordinate descent method for minimizing $f(\mathbf{w}) : \mathbb{R}^d \to \mathbb{R}$ is given by the iteration

$$\mathbf{w}_{k+1} \leftarrow \mathbf{w}_k - \alpha_k \nabla_{i_k} f(\mathbf{w}_k) \mathbf{e}_{i_k} \tag{6.2.25}$$

with

$$\nabla f_{i_k} f(\mathbf{w}_k) = \frac{\partial f(\mathbf{w}_k)}{\partial w_{i_k}}, \tag{6.2.26}$$

where w_{i_k} denotes the i_kth element of the parameter vector $\mathbf{w}_k = [w_1, \ldots, w_d]^T \in \mathbb{R}^d$ and $\mathbf{e}_{i_k}$ represents the i_kth *unit coordinate vector* (also called *natural basis vector*) for some $i_k \in \{1, \ldots, d\}$, i.e., $\mathbf{e}_{i_k}$ is a $d \times 1$ vector whose i_kth element is equal to one and others are equal to zero.

Example 6.1 Given a vector $\mathbf{w}_k = [w_{k1}, \ldots, w_{kd}]^T \in \mathbb{R}^d$. If the function $f(\mathbf{w}_k) = \frac{1}{2}\|\mathbf{w}_k\|_2^2 = \frac{1}{2}(w_{k1}^2 + \cdots + w_{id}^2)$, then

$$\nabla_{i_k} f(\mathbf{w}_k) = w_{i_k}, \quad i_k \in \{1, \ldots, d\},$$

and the ith element of $\mathbf{w}_{k+1}$ is given by

$$w_{k+1,i} = \begin{cases} w_{k,i} - \alpha_k w_i, & i = i_k; \\ w_{k,i}, & \text{otherwise}; \end{cases}$$

for $i = 1, \ldots, d; i_k \in \{1, \ldots, k\}$.

That is to say, the solution estimates $\mathbf{w}_{k+1}$ and $\mathbf{w}_k$ differ only in their i_kth element as a result of a move in the i_kth coordinate from $\mathbf{w}_k$.

According to the choice of i_k, the coordinate descent methods can be divided into two categories.

1. *Cyclic coordinate descent:* Cyclic coordinate search through $\{1, \ldots, d\}$;
2. *Stochastic coordinate descent:* Random coordinate descent search [31]:
   - Cyclic random coordinate search through a random reordering of these indices (with the indices reordered after each set of d steps);
   - Select simply an index randomly with replacement in each iteration.

Stochastic coordinate descent algorithms are almost identical to cyclic coordinate descent algorithms except for selecting coordinates in a random manner.

Coordinate descent methods for solving $\min_\mathbf{x} f(\mathbf{x})$ consist of random choice step (R) and update step (U) at each iteration [166]:

R: Randomly choose an index $i_k \in \{1, \ldots, n\}$, read $\mathbf{x}$, and evaluate $\nabla_{i_k} f(\mathbf{x})$;
U: Update component i_k of the shared $\mathbf{x}_k$ as $\mathbf{x}_{k+1} \leftarrow \mathbf{x}_k - \alpha_k \nabla_{i_k} f(\mathbf{x}_k) \mathbf{e}_{i_k}$.

In the basic coordinate descent methods, a single coordinate is only updated at each iteration. In the following we consider how to update a random subset of coordinates at each iteration,

Consider the unconstrained minimization problem

$$\min_{\mathbf{x} \in \mathbb{R}^N} f(\mathbf{x}), \quad \mathbf{x} = [x_1, \ldots, x_N]^T \in \mathbb{R}^N, \quad (6.2.27)$$

where the objective function $f(\mathbf{x})$ is convex and differentiable on $\mathbb{R}^N$. Under the assumption that decision variables are separable, to decompose the decision vector $\mathbf{x} \in \mathbb{R}^N$ into n non-overlapping blocks in which each block contains N_i decision variables:

$$\mathbf{x} = \begin{bmatrix} \mathbf{x}^{(1)} \\ \vdots \\ \mathbf{x}^{(n)} \end{bmatrix}, \quad \text{where } \mathbf{x}^{(i)} = [x_{i_1}, \ldots, x_{i_{N_i}}]^T \in \mathbb{R}^{N_i} \quad (6.2.28)$$

such that

$$\mathbb{R}^N = \mathbb{R}^{N_1} \times \cdots \times \mathbb{R}^{N_n}, \quad N = \sum_{i=1}^n N_i. \quad (6.2.29)$$

Define the corresponding partition of the unit matrix $\mathbf{I}_{N \times N}$ as

$$\mathbf{I} = [\mathbf{U}_1, \ldots, \mathbf{U}_n] \in \mathbb{R}^{N \times N}, \quad \mathbf{U}_i \in \mathbb{R}^{N \times N_i}, \ i = 1, \ldots, n. \quad (6.2.30)$$

If letting $f_i(\mathbf{x})$ be differentiable convex functions such that $f_i(\mathbf{x})$ depends on blocks $\mathbf{x}^{(i)}$ for $i \in S_i = \{i_1, \ldots, i_{N_i}\}$ only, then the partial derivative of function $f(\mathbf{x})$ with respect to the block-vector $\mathbf{x}^{(i)}$ is defined as

$$\nabla_i f(\mathbf{x}) = \frac{\partial f(\mathbf{x})}{\partial \mathbf{x}^{(i)}} = \mathbf{U}_i^T \nabla f(\mathbf{x}) \in \mathbb{R}^{N_i}. \quad (6.2.31)$$

Proposition 6.1 (Block Decomposition [212]) *Any vector $\mathbf{x} \in \mathbb{R}^N$ can be written uniquely as*

$$\mathbf{x} = \sum_{i=1}^n \mathbf{U}_i \mathbf{x}^{(i)}, \quad (6.2.32)$$

where $\mathbf{x}^{(i)} \in \mathbb{R}^{N_i}$. Moreover, $\mathbf{x}^{(i)} = \mathbf{U}_i^T \mathbf{x}$.

6.2 Optimization in Machine Learning

Example 6.2 When $S_1 = \{1, 2\}$, $\mathbf{U}_1 = [\mathbf{e}_1, \mathbf{e}_2]$, where $\mathbf{e}_i$ is the ith unit coordinate vector. Then we have $\mathbf{x}^{(1)} = [x_1, x_2]^T$ and $f_i(\mathbf{x}) = \mathbf{U}_i^T f(\mathbf{x}) = [f(x_1), f(x_2)]^T$, which gives

$$\nabla_1 f(\mathbf{x}) = \nabla f_1(\mathbf{x}) = \nabla [f(x_1), f(x_2)]^T = \begin{bmatrix} f'(x_1) \\ f'(x_2) \end{bmatrix},$$

or

$$\nabla_1 f(\mathbf{x}) = [\mathbf{e}_1, \mathbf{e}_2]^T \nabla f(\mathbf{x}) = \begin{bmatrix} 1 & 0 & 0 & \cdots & 0 \\ 0 & 1 & 0 & \cdots & 0 \end{bmatrix} \begin{bmatrix} f'(x_1) \\ f'(x_2) \\ \vdots \\ f'(x_N) \end{bmatrix} = \begin{bmatrix} f'(x_1) \\ f'(x_2) \end{bmatrix},$$

where $f'(x_j) = \frac{\partial f(\mathbf{x})}{\partial x_j}$, $j = 1, 2$.

The *randomized coordinate descent method* (RCDM) [188] consists of the following two basic steps:

- Choose at random $i_k \in S_i = \{i_1, \ldots, i_{N_i}\}$;
- Update $\mathbf{x}_{k+1} = \mathbf{x}_k - \alpha_k \mathbf{U}_{i_k}^T \nabla_{i_k} f(\mathbf{x}_k)$.

To accelerate the randomized coordinate descent method, Nesterov [188] in 2012 proposed a computationally inefficient but theoretically interesting accelerated variant of RCDM, called *accelerated coordinate descent method (ACDM)*, for convex minimization without constraints.

Consider the regularized optimization problem with separable decision variables:

$$\min_{\mathbf{x} \in \mathbb{R}^N} \{F(\mathbf{x}) = f(\mathbf{x}) + h(\mathbf{x})\}$$

$$\text{subject to} \quad \mathbf{x} = \sum_{i=1}^n \mathbf{U}_i \mathbf{x}^{(i)} \in \mathbb{R}^{N_1} \times \cdots \times \mathbb{R}^{N_n} = \mathbb{R}^N, \quad (6.2.33)$$

where $f(\mathbf{x})$ is a smooth convex objective function that is separable in the blocks $\mathbf{x}^{(i)}$:

$$f(\mathbf{x}) = \sum_{i=1}^n f_i(\mathbf{x}), \quad \text{where} \quad f_i(\mathbf{x}) = \mathbf{U}_i^T f(\mathbf{x}), \quad (6.2.34)$$

and $h(\mathbf{x})$ is a (possibly nonsmooth) convex regularizer that is also separable in the blocks $\mathbf{x}^{(i)}$:

$$h(\mathbf{x}) = \sum_{i=1}^n h_i(\mathbf{x}), \quad \text{where} \quad h_i(\mathbf{x}) = \mathbf{U}_i^T h(\mathbf{x}). \quad (6.2.35)$$

Fercoq and Richtárk [91] in 2015 proposed the first randomized block coordinate descent method which is simultaneously accelerated, parallel, and proxima, called simply APPROX, for solving the optimization problem (6.2.33).

Let $\hat{S}$ be a random subset of $\mathcal{N} = \{1, \ldots, N\}$. Algorithm 6.4 shows the *APPROX coordinate descent method* in a form facilitating efficient implementation. APPROX is a remarkably versatile method.

Algorithm 6.4 APPROX coordinate descent method [91]
---
1. **input:**
2. **initialization:** Pick $\tilde{\mathbf{z}}_0 \in \mathbb{R}^N$ and set $\theta_0 = \frac{\tau}{n}$, $\mathbf{u}_0 = \mathbf{0}$.
3. **for** $k = 0, 1, \ldots$ **do**
4. Generate a random set of blocks $S_k \sim \hat{S}$.
5. $\quad \mathbf{u}_{k+1} \leftarrow \mathbf{u}_k$, $\tilde{\mathbf{z}}_{k+1} \leftarrow \tilde{\mathbf{z}}_k$.
6. $\quad$ **for** $i \in S_k$ **do**
7. $\quad\quad \mathbf{t}_k^{(i)} = \arg\max_{\mathbf{t} \in \mathbb{R}^{N_i}} \left\{ \langle \nabla_i f(\theta^2 \mathbf{u}_k + \tilde{\mathbf{z}}_k), \mathbf{t} \rangle + \frac{n\theta_k v_i}{2\tau} \|\mathbf{t}\|_{(i)}^2 + h_i(\tilde{\mathbf{z}}_k^{(i)}) \right\}$.
8. $\quad\quad \tilde{\mathbf{z}}_{k+1}^{(i)} \leftarrow \tilde{\mathbf{z}}_k^{(i)} + \mathbf{t}_k^{(i)}$.
9. $\quad\quad \mathbf{u}_{k+1}^{(i)} \leftarrow \mathbf{u}_k^{(i)} - \frac{1 - \frac{n}{\tau}\theta_k}{\theta_k^2} \mathbf{t}_k^{(i)}$.
10. $\quad$ **end for**
11. $\quad \theta_{k+1} = \frac{\sqrt{\theta_k^4 + 4\theta_k^2} - \theta_k^2}{2}$.
12. **end for**
13. **output:** $\mathbf{x} = \theta_k^2 \mathbf{u}_{k+1} + \tilde{\mathbf{z}}_{k+1}$.

6.2.4 Benchmark Functions for Single-Objective Optimization

The test of reliability, efficiency, and validation of an optimization algorithm is frequently carried out by using a chosen set of common *benchmark* or *test functions* [129]. There have been many benchmark or test functions reported in the literature. Ideally, benchmark functions should have diverse properties when testing new algorithms in an unbiased way. On the other hand, there is a serious phenomenon called "curse of dimensionality" in many optimization methods [21]. This implies that the optimization performance deteriorates quickly as the dimensionality of the search space increases. The reasons for this phenomenon appear to be two fold [240]:

- The solution space of a problem often increases exponentially with the problem dimension [21] and more efficient search strategies are required to explore all promising regions within a given time budget.
- The characteristics of a problem may change with the scale. Because an increase in scale may result in worsening of the features of an optimization problem, a previously successful search strategy may no longer be capable of finding the optimal solution.

To solve large-scale problems, it is necessary to provide a suite of benchmark functions for large-scale numerical optimization.

6.2 Optimization in Machine Learning

Definition 6.1 (Separable Function) Given $\mathbf{x} = [x_1, \ldots, x_n]^T$. A function $f(\mathbf{x})$ is a *separable function* if and only if

$$\arg\min_{(x_1,\ldots,x_n)} f(x_1, \ldots, x_n) = \left(\arg\min_{x_1} f(x_1, \ldots), \ldots, \arg\min_{x_n} f(\ldots, x_n)\right), \tag{6.2.36}$$

where $f(\ldots, x_i, \ldots)$ denotes that its decision variable is just x_i and other variables are constants with respect to x_i.

That is to say, a function of n decision variables is separable if it can be rewritten as a sum of n functions of just one variable, i.e., $f(x_1, \ldots, x_n) = f(x_1) + \cdots + f(x_n)$. If a function $f(\mathbf{x})$ is separable, its parameters x_i are said to be independent.

A separable function can be easily optimized, as it can be decomposed into a number of subproblems, each of which involving only one decision variable while treating all others as constants.

Definition 6.2 (Nonseparable Function) A *nonseparable function* $f(\mathbf{x})$ is called m-nonseparable function if at most m (where $m < n$) of its decision variables x_i are not independent. A nonseparable function $f(\mathbf{x})$ is known as a *fully nonseparable function* if any two of its parameters x_i are not independent.

A nonseparable function has at least partly coupled or dependent decision variables.

The definitions of separability provide us a measure of the difficulty of different optimization problems based on which a spectrum of benchmark problems can be designed. In general, separable problems are considered to be easiest, while the fully nonseparable ones usually are most difficult.

A benchmark problem for large-scale optimization is to provide test functions to researchers or users for testing the optimization algorithms' performance when applied to test problems. These test functions are as same as efficient in practical scenarios.

Jamil and Yang [129] presented a collection of 175 unconstrained optimization test problems which can be used to validate the performance of optimization algorithms. Some benchmark functions are described by classification as follows.

1. *Continuous, Differentiable, Separable, Multimodal Functions*

   - Giunta Function [179]

$$f(\mathbf{x}) = 0.6 + \sum_{i=1}^{2}\left[\sin\left(\frac{16}{15}x_i - 1\right) + \sin^2\left(\frac{16}{15}x_i - 1\right)\right.$$
$$\left. + \frac{1}{50}\sin\left(4\left(\frac{16}{15}x_i - 1\right)\right)\right] \tag{6.2.37}$$

subject to $-1 \le x_i \le 1$. The global minimum is $f(\mathbf{x}^*) = 0.060447$ located at $\mathbf{x}^* = (0.45834282, 0.45834282)$.

- Parsopoulos Function

$$f(\mathbf{x}) = \cos(x_1)^2 + \sin(x_2)^2 \qquad (6.2.38)$$

subject to $-5 \le x_i \le 5$, where $(x_1, x_2) \in \mathbb{R}^2$. This function has an infinite number of global minima in $\mathbb{R}^2$, at points $(\kappa \frac{2\pi}{2}, \lambda\pi)$, where $\kappa = \pm 1, \pm 3, \ldots$ and $\lambda = 0, \pm 1, \pm 2, \ldots$. In the given domain problem, the function has 12 global minima all equal to zero.

2. *Continuous, Differentiable, Nonseparable, Multimodal Functions*

- El-Attar-Vidyasagar-Dutta Function [85]

$$f(\mathbf{x}) = \left(x_1^2 + x_2 - 10\right)^2 + \left(x_1 + x_2^2 - 7\right)^2 + \left(x_1^2 + x_2^3 - 1\right)^2 \qquad (6.2.39)$$

subject to $-500 \le x_i \le 500$. The global minimum is $f(\mathbf{x}^*) = 0.470427$ located at $\mathbf{x}^* = (2.842503, 1.920175)$.

- Egg Holder Function

$$f(\mathbf{x}) = \sum_{i=1}^{n-1} \left[-(x_{i+1} + 47) \sin\sqrt{|x_{i+1} + x_i/2 + 47|} - x_i \sin\sqrt{|x_i - (x_{i+1} + 47)|} \right]$$

subject to $-512 \le x_i \le 512$. The global minimum for $n = 2$ is $f(\mathbf{x}^*) \approx -959.64$ located at $\mathbf{x}^* = (512, 404.2319)$.

- Exponential Function [209]

$$f(\mathbf{x}) = -\exp\left(-0.5 \sum_{i=1}^{D} x_i^2\right), \qquad (6.2.40)$$

subject to $-1 \le x_i \le 1$. The global minima is $f(\mathbf{x}^*) = 1$ located at $\mathbf{x}^* = (0, \ldots, 0)$.

- Langerman-5 Function [23]

$$f(\mathbf{x}) = -\sum_{i=1}^{m} c_i \exp\left(-\frac{1}{\pi} \sum_{j=1}^{D} (x_j - a_{ij})^2\right) \cos\left(\pi \sum_{j=1}^{D} (x_j - a_{ij})^2\right)$$

(6.2.41)

subject to $0 \le x_j \le 10$, where $j \in [1, D]$ and $m = 5$. It has a global minimum value of $f(\mathbf{x}^*) = -1.4$. The matrix $\mathbf{A} = [a_{ij}]$ and column vector $\mathbf{c} = [c_i]$ are given as

$$\mathbf{A} = \begin{bmatrix} 9.681 & 0.667 & 4.783 & 9.095 & 3.517 & 9.325 & 6.544 & 0.211 & 5.122 & 2.020 \\ 9.400 & 2.041 & 3.788 & 7.931 & 2.882 & 2.672 & 3.568 & 1.284 & 7.033 & 7.374 \\ 8.025 & 9.152 & 5.114 & 7.621 & 4.564 & 4.711 & 2.996 & 6.126 & 0.734 & 4.982 \\ 2.196 & 0.415 & 5.649 & 6.979 & 9.510 & 9.166 & 6.304 & 6.054 & 9.377 & 1.426 \\ 8.074 & 8.777 & 3.467 & 1.863 & 6.708 & 6.349 & 4.534 & 0.276 & 7.633 & 1.567 \end{bmatrix}$$

and $\mathbf{c} = [0.806, 0.517, 1.500, 0.908, 0.965]^T$.
- Mishra Function 1 [180]

$$f(\mathbf{x}) = \left(1 + D - \sum_{i=1}^{n-1} 0.5(x_i + x_{i+1})\right)^{n - \sum_{i=1}^{n-1} 0.5(x_i + x_{i+1})} \quad (6.2.42)$$

subject to $0 \leq x_i \leq 1$. The global minimum is $f(\mathbf{x}^*) = 2$.
- Mishra Function 2 [180]

$$f(\mathbf{x}) = -\ln\left[\sin^2\left((\cos(x_1) + \cos(x_2))^2\right) + \cos^2\left((\sin(x_1) + \sin(x_2))^2\right) + x_1\right]^2 \\ + 0.01((x_1 - 1)^2 + (x_2 - 1)^2). \quad (6.2.43)$$

The global minimum $f(\mathbf{x}^*) = -2.28395$ is located at $\mathbf{x}^* = (2.88631, 1.82326)$.

3. *Continuous, Non-Differentiable, Separable, Multimodal Functions*
    - Price Function [208]

    $$f(\mathbf{x}) = (|x_1| - 5)^2 + (|x_2| - 5)^2 \quad (6.2.44)$$

    subject to $-500 \leq x_i \leq 500$. The global minima $f(\mathbf{x}^*) = 0$ are located at $\mathbf{x}^* = (-5, -5), (-5, 5), (5, -5), (5, 5)$.
    - Schwefel Function [223]

    $$f(\mathbf{x}) = -\sum_{i=1}^{n} |x_i| \quad (6.2.45)$$

    subject to $-100 \leq x_i \leq 100$. The global minimum $f(\mathbf{x}^*) = 0$ is located at $\mathbf{x}^* = (0, \ldots, 0)$.

4. *Continuous, Non-Differentiable, Nonseparable, Multimodal Functions*
    - Bartels Conn Function

    $$f(\mathbf{x}) = \left|x_1^2 + x_2^2 + x_1 x_2\right| + |\sin(x_1)| + |\cos(x_2)| \quad (6.2.46)$$

subject to $-500 \leq x_i \leq 500$. The global minimum $f(\mathbf{x}^*) = 1$ is located at $\mathbf{x}^* = (0, 0)$.
- Bukin Function

$$f(\mathbf{x}) = 100\sqrt{|x_2 - 0.01x_1^2|} + 0.01|x_1 + 10| \tag{6.2.47}$$

subject to $-15 \leq x_1 \leq -5$ and $-13 \leq x_2 \leq -3$. The global minimum $F(\mathbf{x}^*) = 0$ is located at $\mathbf{x}^* = (-10, 1)$.

5. *Continuous, Differentiable, Partially Separable, Unimodal Functions*
   - Chung–Reynolds Function [54]

$$f(\mathbf{x}) = \left(\sum_{i=1}^{D} x_i^2\right)^2 \tag{6.2.48}$$

subject to $-100 \leq x_i \leq 100$. The global minimum $f(\mathbf{x}^*) = 0$ is located at $\mathbf{x}^* = (0, \ldots, 0)$.
   - Schwefel Function [223]

$$f(\mathbf{x}) = \left(\sum_{i=1}^{D} x_i^2\right)^\alpha \tag{6.2.49}$$

subject to $-100 \leq x_i \leq 100$, where $\alpha \geq 0$. The global minimum $f(\mathbf{x}^*) = 0$ is located at $\mathbf{x}^* = (0, \ldots, 0)$.

6. *Discontinuous, Non-Differentiable Functions*
   - Corana Function [61]

$$f(\mathbf{x}) = \begin{cases} 0.15 \left(z_i - 0.05 \, \text{sign}(z_i)^2\right) d_i, & \text{if } |v_i| < A, \\ d_i x_i^2, & \text{otherwise,} \end{cases} \tag{6.2.50}$$

where

$$v_i = |x_i - z_i|, \ A = 0.05,$$

$$z_i = 0.2 \left\lfloor \left|\frac{x_i}{0.2}\right| + 0.49999 \right\rfloor \text{sign}(x_i)$$

$$d_i = (1, 1000, 10, 100)$$

subject to $-500 \leq x_i \leq 500$. The global minimum $f(\mathbf{x}^*) = 0$ is located at $\mathbf{x}^* = (0, 0, 0, 0)$.

- Cosine Mixture Function [4]

$$f(\mathbf{x}) = -0.1 \sum_{i=1}^{n} \cos(5\pi x_i) - \sum_{i=1}^{n} x_i^2 \qquad (6.2.51)$$

subject to $-1 \leq x_i \leq 1$. The global minimum is located at $\mathbf{x}^* = (0, 0)$ or $\mathbf{x}^* = (0, 0, 0, 0)$, $f(\mathbf{x}^*) = 0.2$ or 0.4 for $n = 2$ and 4, respectively.

In this section we have focused on the single-objective optimization problems in machine learning. In a lot of applications in artificial intelligence, we will be faced with multi-objective optimization problems. Because these problems are closely related to evolutionary computation, we will deal with multi-objective optimization theory and methods in Chap. 9.

6.3 Majorization-Minimization Algorithms

In the era of big data, a fast development in data acquisition techniques and a growth of computing power, from an optimization perspective, can result in large-scale problems due to the tremendous amount of data and variables, which cause challenges to traditional algorithms [87]. By Hunter and Lange [125], the general principle behind *majorization-minimization (MM)* algorithms was first enunciated by the numerical analysts in 1970s [197] in the context of line search methods. MM algorithms are a set of analytic procedures for tackling difficult optimization problems. The basic idea of MM algorithms is to modify their objective functions so that solution spaces of the modified objective functions are easier to explore.

A successful MM algorithm substitutes a simple optimization problem for a difficult optimization problem. By [125], simplicity can be attained by

- avoiding large matrix inversions,
- linearizing an optimization problem,
- separating the parameters of an optimization problem,
- dealing with equality and inequality constraints gracefully, or
- turning a non-differentiable problem into a smooth problem.

MM has a long history that dates back to the 1970s [197], and is closely related to the famous expectation-maximization (EM) algorithm [278] intensively used in computational statistics.

6.3.1 MM Algorithm Framework

Consider the minimization problem

$$\hat{\boldsymbol{\theta}} = \arg\min_{\boldsymbol{\theta}\in\Theta} g(\boldsymbol{\theta}) \qquad (6.3.1)$$

for some difficult to manipulate objective function $g(\boldsymbol{\theta})$. For example, $g(\boldsymbol{\theta})$ with $\boldsymbol{\theta} \in \Theta$ is non-differential and/or non-convex, where Θ is a subset of some Euclidean space.

Let $\boldsymbol{\theta}^{(m)}$ represent a fixed value of the parameter $\boldsymbol{\theta}$ (where m is the mth iteration), and let $g(\boldsymbol{\theta}|\boldsymbol{\theta}^{(m)})$ denote a real-valued function of $\boldsymbol{\theta}$ whose form depends on $\boldsymbol{\theta}^{(m)}$.

Instead of operating on $g(\boldsymbol{\theta})$, we can consider an easier *surrogate function* $h(\boldsymbol{\theta}|\boldsymbol{\theta}^{(m)})$ at some point $\boldsymbol{\theta}^{(m)}$ instead.

Definition 6.3 (Majorizer [125, 191]) The surrogate function $h(\boldsymbol{\theta}|\boldsymbol{\theta}^{(m)})$ is said to be a *majorizer* of objective $g(\boldsymbol{\theta})$ provided

$$h(\boldsymbol{\theta}|\boldsymbol{\theta}^{(m)}) \geq g(\boldsymbol{\theta}) \quad \text{for all } \boldsymbol{\theta}, \qquad (6.3.2)$$

$$h(\boldsymbol{\theta}^{(m)}|\boldsymbol{\theta}^{(m)}) = g(\boldsymbol{\theta}^{(m)}). \qquad (6.3.3)$$

To find the majorizer $h(\boldsymbol{\theta}|\boldsymbol{\theta}^{(m)})$ of objective $g(\boldsymbol{\theta})$ constitutes the first step of MM algorithms, called *majorization step*.

The function $h(\boldsymbol{\theta}|\boldsymbol{\theta}^{(m)})$ satisfying the conditions (6.3.2) and (6.3.3) is said to majorize a real-valued function $g(\boldsymbol{\theta})$ at the point $\boldsymbol{\theta}^{(m)}$. In other words, the surface $\boldsymbol{\theta} \to h(\boldsymbol{\theta}|\boldsymbol{\theta}^{(m)})$ lies above the surface $g(\boldsymbol{\theta})$ and is tangent to it at the point $\boldsymbol{\theta} = \boldsymbol{\theta}^{(m)}$. In this sense, $h(\boldsymbol{\theta}|\boldsymbol{\theta}^{(m)})$ is called the majorization function. Moreover, if $\boldsymbol{\theta}^{(m+1)}$ is a minimizer of $h(\boldsymbol{\theta}|\boldsymbol{\theta}^{(m)})$, then both (6.3.2) and (6.3.3) further imply [288] that

$$g(\boldsymbol{\theta}^{(m)}) = h(\boldsymbol{\theta}^{(m)}|\boldsymbol{\theta}^{(m)}) \geq h(\boldsymbol{\theta}^{(m+1)}|\boldsymbol{\theta}^{(m)}) \geq g(\boldsymbol{\theta}^{(m+1)}). \qquad (6.3.4)$$

This is an important property of the MM algorithm, which means that the sequence $\{g(\boldsymbol{\theta}^{(m)})\}$, $m = \{1, 2, \ldots\}$ is non-increasing, so the iteration procedure $\boldsymbol{\theta}^{(m)}$ pushes $g(\boldsymbol{\theta})$ toward its minimum.

The function $h(\boldsymbol{\theta}|\boldsymbol{\theta}^{(m)})$ is said to minorize $g(\boldsymbol{\theta})$ at $\boldsymbol{\theta}^{(m)}$ if $-h(\boldsymbol{\theta}|\boldsymbol{\theta}^{(m)})$ majorizes $-g(\boldsymbol{\theta})$ at $\boldsymbol{\theta}^{(m)}$.

The second step, known as *minimization step*, is to minimize the surrogate function, namely update $\boldsymbol{\theta}$ as

$$\boldsymbol{\theta}^{(m+1)} = \arg\min_{\boldsymbol{\theta}} h(\boldsymbol{\theta}|\boldsymbol{\theta}^{(m)}). \qquad (6.3.5)$$

The descent property (6.3.2) lends an MM algorithm remarkable numerical stability. When minimizing a difficult loss function $g(\boldsymbol{\theta})$, we relax this loss function to a surrogate function $h(\boldsymbol{\theta}|\boldsymbol{\theta}^{(m)})$ that is easily minimized.

6.3 Majorization-Minimization Algorithms

Example 6.3 Given a matrix pencil $(\mathbf{A}, \mathbf{B})$, consider its generalized eigenvalue (GEV) problem $\mathbf{A}\mathbf{x} = \lambda \mathbf{B}\mathbf{x}$. From $\lambda = \mathbf{x}^T \mathbf{A}\mathbf{x}/(\mathbf{x}^T \mathbf{B}\mathbf{x})$ it is known that the variational formulation for the GEV problem in $\mathbf{A}\mathbf{x} = \lambda \mathbf{B}\mathbf{x}$ is given by

$$\lambda_{\max}(\mathbf{A}, \mathbf{B}) = \max_{\mathbf{x}} \{\mathbf{x}^T \mathbf{A}\mathbf{x}\} \quad \text{subject to} \quad \mathbf{x}^T \mathbf{B}\mathbf{x} = 1, \tag{6.3.6}$$

where $\lambda_{\max}(\mathbf{A}, \mathbf{B})$ is the maximum generalized eigenvalue associated with the matrix pencil $(\mathbf{A}, \mathbf{B})$. Then, the sparse GEV problem can be written as

$$\max_{\mathbf{x}} \{\mathbf{x}^T \mathbf{A}\mathbf{x}^T\} \quad \text{subject to} \quad \mathbf{x}^T \mathbf{B}\mathbf{x} = 1, \; \|\mathbf{x}\|_0 \leq k. \tag{6.3.7}$$

Consider the regularized (penalized) version of (6.3.7) given by [237]:

$$\max_{\mathbf{x}} \left\{ \mathbf{x}^T \mathbf{A}\mathbf{x} - \rho \|\mathbf{x}\|_0 \right\} \quad \text{subject to} \quad \mathbf{x}^T \mathbf{B}\mathbf{x} \leq 1, \tag{6.3.8}$$

where $\rho > 0$ is the regularization (penalization) parameter. The constrained equality condition $\mathbf{x}^T \mathbf{B}\mathbf{x} = 1$ in (6.3.7) has been relaxed to the inequality constraint $\mathbf{x}^T \mathbf{B}\mathbf{x} \leq 1$. To relax further the non-convex ℓ_0-norm $\|\mathbf{x}\|_0$, consider using

$$\|\mathbf{x}\|_0 = \sum_{i=1}^{n} \mathbb{1}\{|\mathbf{x}_i| \neq 0\} = \lim_{\epsilon \to 0} \sum_{i=1}^{n} \frac{\log(1 + |x_i|/\epsilon)}{\log(1 + 1/\epsilon)} \tag{6.3.9}$$

in (6.3.8) to rewrite as the equivalent form

$$\max_{\mathbf{x}} \left\{ \mathbf{x}^T \mathbf{A}\mathbf{x} - \rho \lim_{\epsilon \to 0} \sum_{i=1}^{n} \frac{\log(1 + |x_i|/\epsilon)}{\log(1 + 1/\epsilon)} \right\} \quad \text{subject to} \quad \mathbf{x}^T \mathbf{B}\mathbf{x} \leq 1. \tag{6.3.10}$$

The above program is approximated by the following approximate sparse GEV program by neglecting the limit in (6.3.10) and choosing $\epsilon > 0$:

$$\max_{\mathbf{x}} \left\{ \mathbf{x}^T \mathbf{A}\mathbf{x} - \rho \sum_{i=1}^{n} \frac{\log(1 + |x_i|/\epsilon)}{\log(1 + 1/\epsilon)} \right\} \quad \text{subject to} \quad \mathbf{x}^T \mathbf{B}\mathbf{x} \leq 1, \tag{6.3.11}$$

which is finally equivalent to [237]

$$\max_{\mathbf{x}} \left\{ \mathbf{x}^T \mathbf{A}\mathbf{x} - \rho_\epsilon \sum_{i=1}^{n} \log(1 + |x_i|/\epsilon) \right\} \quad \text{subject to} \quad \mathbf{x}^T \mathbf{B}\mathbf{x} \leq 1, \tag{6.3.12}$$

where $\rho_\epsilon = \rho/\log(1 + 1/\epsilon)$. Hence, instead of a non-convex sparse optimization problem (6.3.7), we can manipulate an easier surrogate convex optimization problem (6.3.12) instead by using the MM algorithm.

In addition to the typical examples above, there are a lot of applications of MM algorithms to machine learning, statistical estimation, and signal processing problems, see [191] for a list of 26 applications, which includes fully visible Boltzmann machine estimation, linear mixed model estimation, Markov random field estimation, matrix completion and imputation, quantile regression estimation, SVM estimation, and so on.

As pointed out by Nguyen [191], the MM algorithm framework is a popular tool for deriving useful algorithms for problems in machine learning, statistical estimation, and signal processing.

6.3.2 Examples of Majorization-Minimization Algorithms

Let $\boldsymbol{\theta}^{(0)}$ be some initial value and $\boldsymbol{\theta}^{(m)}$ be a sequence of iterates (in m) for the minimization problem (6.3.1). Definition 6.3 suggests the following scheme, referred to as an MM algorithm.

Definition 6.4 (Majorization-Minimization (MM) Algorithm [191]) Let $\boldsymbol{\theta}^{(0)}$ be some initial value and $\boldsymbol{\theta}^{(m)}$ be the mth iterate. Then, $\boldsymbol{\theta}^{(m+1)}$ is said to be the $(m+1)$th iterate of an MM algorithm if it satisfies

$$\boldsymbol{\theta}^{(m+1)} = \arg\min_{\boldsymbol{\theta} \in \Theta} h\left(\boldsymbol{\theta}|\boldsymbol{\theta}^{(m)}\right). \qquad (6.3.13)$$

From Definitions 6.3 and 6.4 it can be deduced that all MM algorithms have the monotonicity property. That is, if $\boldsymbol{\theta}^{(m)}$ is a sequence of MM algorithm iterates, the objective sequence $g(\boldsymbol{\theta})$ is monotonically decreasing in m.

Proposition 6.2 (Monotonicity Decreasing [125, 191]) *If $h(\boldsymbol{\theta}|\boldsymbol{\theta}^{(m)})$ is a majorizer of the objective $g(\boldsymbol{\theta})$ and $\boldsymbol{\theta}^{(m)}$ is a sequence of MM algorithm iterates, then MM algorithms have the following monotonicity decreasing:*

$$g(\boldsymbol{\theta}^{(m+1)}) \leq h(\boldsymbol{\theta}|\boldsymbol{\theta}^{(m+1)}) \leq h(\boldsymbol{\theta}^{(m)}|\boldsymbol{\theta}^{(m)}) \leq g(\boldsymbol{\theta}^{(m)}). \qquad (6.3.14)$$

Remarks It is notable that an algorithm need not be an MM algorithm in the strict sense of Definition 6.4 in order for (6.3.14) to hold. In fact, any algorithm with the $(m+1)$th iterate satisfying

$$\boldsymbol{\theta}^{(m+1)} \in \left\{\boldsymbol{\theta} \in \Theta : h(\boldsymbol{\theta}|\boldsymbol{\theta}^{(m)}) \leq h(\boldsymbol{\theta}^{(m)}|\boldsymbol{\theta}^{(m)})\right\} \qquad (6.3.15)$$

will generate a monotonically decreasing sequence of objective evaluates. Such an algorithm can be thought of as a generalized MM algorithm.

6.3 Majorization-Minimization Algorithms

Proposition 6.3 (Stationary Point [191]) *Starting from some initial value $\boldsymbol{\theta}^{(0)}$, if $\boldsymbol{\theta}^{(\infty)}$ is the limit point of an MM algorithm sequence of iterates $\boldsymbol{\theta}^{(m)}$ (i.e., satisfying Definition 6.4), then $\boldsymbol{\theta}^{(\infty)}$ is a stationary point of the minimization problem (6.3.1).*

Remarks Proposition 6.3 only guarantees the convergence of MM algorithm iterates to a stationary point and not a global, or even a local minimum. As such, for problems over difficult objective functions, multiple or good initial values are required in order to ensure that the obtained solution is of a high quality. Furthermore, Proposition 6.3 only guarantees convergence to a stationary point of (6.3.1) if a limit point exists for the chosen starting value. If a limit point does not exist, then the MM algorithm objective sequence may diverge.

The following result is useful when constructing the majorization function.

Lemma 6.1 ([236]) *Let $\mathbf{L}$ be an Hermitian matrix and $\mathbf{M}$ be another Hermitian matrix such that $\mathbf{M} \geq \mathbf{L}$ ($M_{ij} \geq L_{ij}$). Then for any point $\mathbf{x}_0 \in \mathbb{C}^n$, the quadratic function $\boldsymbol{\theta}^H \mathbf{L} \boldsymbol{\theta}$ is majorized by $\boldsymbol{\theta}^H \mathbf{M} \boldsymbol{\theta} + 2\,Re\bigl(\boldsymbol{\theta}^H (\mathbf{L} - \mathbf{M}) \boldsymbol{\theta}_0\bigr) + \boldsymbol{\theta}_0^H (\mathbf{M} - \mathbf{L}) \boldsymbol{\theta}_0$ at $\boldsymbol{\theta}_0$.*

To minimize over $g(\boldsymbol{\theta})$, the main steps of the majorization-minimization scheme are as follows [236]:

1. Find a feasible point $\boldsymbol{\theta}^{(0)}$ and set $m = 0$.
2. Construct a majorization function $h(\boldsymbol{\theta}|\boldsymbol{\theta}^{(m)})$ of $g(\boldsymbol{\theta})$ at the point $\boldsymbol{\theta}^{(m)}$.
3. Solve $\boldsymbol{\theta}^{(m+1)} = \arg\min_{\boldsymbol{\theta} \in \Theta} h(\boldsymbol{\theta}|\boldsymbol{\theta}^{(m)})$.
4. If some convergence criterion is met then exit; otherwise, set $k = k + 1$ and go to Step 2.

Consider the "weighted" ℓ_0 minimization problem [43]:

$$\min_{\mathbf{x} \in \mathbb{R}^n} \|\mathbf{W}\mathbf{x}\|_0 \quad \text{subject to} \quad \mathbf{y} = \boldsymbol{\Phi}\mathbf{x} \tag{6.3.16}$$

and the "weighted" ℓ_1 minimization problem

$$\min_{\mathbf{x} \in \mathbb{R}^n} \left\{ \sum_{i=1}^p w_i |x_i| \right\} \quad \text{subject to} \quad \mathbf{y} = \boldsymbol{\Phi}\mathbf{x}. \tag{6.3.17}$$

The weighted ℓ_1 minimization can be viewed as a relaxation of a weighted ℓ_0 minimization problem. To establish the connection between the weighted ℓ_1 minimization and the weighted ℓ_0 minimization, we consider the surrogate function

$$\frac{|x_i|}{|x_i^{(m)}| + \epsilon} \begin{cases} \approx 1, & x_i^{(m)} = x_i; \\ = 0, & x_i^{(m)} = x_i = 0. \end{cases} \tag{6.3.18}$$

Here ϵ is a small positive value. This shows that if updating

$$\mathbf{x}^{(m+1)} = \arg\min_{\mathbf{x}\in\mathbb{R}^n} \left\{ \sum_{i=1}^n \frac{|x_i|}{x_i^{(m)} + \epsilon} \right\}, \tag{6.3.19}$$

then $\sum_{i=1}^n \frac{|x_i|}{x_i^{(m)}+\epsilon}$ gives approximately $\|\mathbf{x}\|_0$ when $x_i^{(m)} \to x_i$ for all $i = 1, \ldots, n$ or $\mathbf{x}^{(m)} \to \mathbf{x}$.

Interestingly, (6.3.19) can be viewed as a form of the weighted ℓ_1 minimization (6.3.17), where $w_i^{(m+1)} = \frac{1}{|x_i^{(m)}|+\epsilon}$. Therefore, the MM algorithm for solving the weighted ℓ_1 minimization problem (6.3.17) consists of the following iterative steps [43]:

1. Set the iteration count m to zero and $w_i^{(0)} = 1$, $i = 1, \ldots, n$.
2. Solve the weighted ℓ_1 minimization problem

$$\mathbf{x}^{(m)} = \arg\min_{\mathbf{x}\in\mathbb{R}^n} \left\{ \sum_{i=1}^n w_i |x_i| \right\} \quad \text{subject to } \mathbf{y} = \mathbf{\Phi}\mathbf{x}. \tag{6.3.20}$$

3. Update the weights: for each $i = 1, \ldots, n$,

$$w_i^{(m+1)} = \frac{1}{|x_i^{(m)}| + \epsilon}. \tag{6.3.21}$$

4. Terminate on convergence or when m attains a specified maximum number of iterations $m_{\max}$. Otherwise, increment m by 1 and go to Step 2.

For a two-dimensional array $(x_{i,j})$, $1 \leq i, j \leq n$, let $(\mathbf{D}\mathbf{x})_{i,j}$ be the two-dimensional vector of forward differences $(\mathbf{D}\mathbf{x})_{i,j} = [x_{i+1,j} - x_{i,j}, x_{i,j+1} - x_{i,j}]$. The total-variation (TV) norm of $\mathbf{D}\mathbf{x}_{ij}$ is defined as

$$\|\mathbf{x}\|_{\text{TV}} = \sum_{1 \leq i,j \leq n-1} \|\mathbf{D}\mathbf{x}_{i,j}\|. \tag{6.3.22}$$

Because many natural images have a sparse or nearly sparse gradient, it is interesting to search for the reconstruction with minimal weighted TV norm, i.e.,

$$\min \left\{ \sum_{1 \leq i,j \leq n-1} w_{i,j}^{(m)} \|(\mathbf{D}\mathbf{x})_{i,j}\| \right\}, \quad \text{subject to } \mathbf{y} = \mathbf{\Phi}\mathbf{x}, \tag{6.3.23}$$

where $w_{i,j}^{(m)} = \frac{1}{\|(\mathbf{D}\mathbf{x})_{i,j}\|+\epsilon}$ by mimicking the derivation of (6.3.21).

The MM algorithm for minimizing a sequence of weighted TV norms is as follows [43]:

1. Set $m = 0$ and $w_{i,j}^{(0)}$, $1 \leq i, j \leq n-1$.
2. Solve the weighted TV minimization problem

$$\mathbf{x}^{(m)} = \arg\min_{\mathbf{x} \in \mathbb{R}^n} \left\{ \sum_{1 \leq i,j \leq n-1} w_{i,j}^{(m)} \|(D\mathbf{x})_{i,j}\| \right\}, \quad \text{subject to } \mathbf{y} = \mathbf{\Phi}\mathbf{x} \quad (6.3.24)$$

3. Update the weights

$$w_{i,j}^{(m)} = \frac{1}{\|(D\mathbf{x})_{i,j}\| + \epsilon} \quad (6.3.25)$$

for each (i, j), $1 \leq i, j \leq n-1$.
4. Terminate on convergence or when m attains a specified maximum number of iterations $m_{\max}$. Otherwise, increment m by 1 and go to Step 2.

6.4 Boosting and Probably Approximately Correct Learning

Valiant and Kearns [140, 254] put forward the concepts of weak learning and strong learning. A learning algorithm with error rate less than $1/2$ (i.e., its accuracy is only slightly higher than random guess) is called a *weak learning algorithm*, and the learning algorithm whose accuracy is very high and can be completed in polynomial time is called a *strong learning algorithm*. At the same time, Valiant and Kearns proposed the equivalence problem of weak learning algorithm and strong learning algorithm in probably approximately correct (PAC) learning model, that is, if any given weak learning algorithm is only slightly better than random guessing, can it be upgraded to a strong learning algorithm? In 1990, Schapire [221] first constructed a polynomial-level algorithm to prove that a weak learning algorithm which is slightly better than random guessing can be upgraded to a strong learning algorithm instead of having to find a strong learning algorithm that is too hard to get. This is the original boosting algorithm. In 1995, Freund and Schapire [96] improved Boosting algorithm and proposed AdaBoost (Adaptive Boosting) algorithm.

"*Boosting*" is a way of combining the performance of many "weak" classifiers to produce a powerful "committee." Boosting was proposed in the computational learning theory literature [96, 97, 221] and has since received much attention [100].

6.4.1 Boosting for Weak Learners

Boosting is a framework algorithm for obtaining a sample subset by an operation on the sample set. It then trains the weak classification algorithm on the sample subset to generate a series of base classifiers.

Let x_i be an instance and $X = \{x_1, \ldots, x_N\}$ be a set called a *domain* (also referred to as the *instance space*). If D_s is any fixed probability distribution over the instance space X, and D_t is a distribution over the set of training example (labeled instances) $X \times Y$, then D_s and D_t will be referred to as the *instance distribution* (or *source distribution*) and *target distribution*, respectively. In most of machine learning, it is assumed that $D_s = D_{tX}$, i.e., the instance distribution and the target distribution are the same over the instance space.

A *concept* c over X is some subset of the instance space X with unknown labels. A concept can be equivalently defined to be a Boolean mapping $c : X \to \{0, 1\}$, with $c(x) = 1$ indicating that x is a positive example of c and $c(x) = 0$ indicating that x is a negative example.

The Boolean mapping or concept of an instance or domain point $x \in X$ is sometimes known as a representation h of x.

Definition 6.5 (Agree, Consistent [140]) A representation h and an example (x, y) *agree* if $h(x) = y$; otherwise, they *disagree*. Let $S = \{(x_1, y_1), \ldots, (x_N, y_N)\}$ be any labeled set of instances, where each $x_i \in X$ and $y_i \in \{0, 1\}$. If c is a concept over X, and $c(x_i) = y_i$ for all $1 \leq i \leq N$, then the concept or representation c is *consistent* with a sample S (or equivalently, S is consistent with c). Otherwise, they are *inconsistent*.

For M target classes where each class is represented by $C_i, \forall i \in \{1, \ldots, M\}$, the task of a classifier is to assign the input sample x to one of the $(M + 1)$ classes, with the $(M + 1)$th class denoting that the classifier rejects x. Let $h_1, \ldots, h_K$ be K classifiers in which each gives its own classification result $h_i(x)$. The problem is how to produce a combined result $H(x) = j$, $j \in \{1, \ldots, M, M + 1\}$ from all K predictions $h_1(x), \ldots, h_K(x)$. The most common method to combine the outputs is the *majority voting* method.

Majority voting consists of the following three steps [281].

1. Define a binary function to represent the number of votes:

$$V_k(x \in C_i) = \begin{cases} 1, & \text{if } h_k(x) = i, \ i \in \{1, \ldots, M\}; \\ 0, & \text{otherwise.} \end{cases} \quad (6.4.1)$$

2. Sum the votes from all K classifiers for each C_i:

$$V_H(x \in C_i) = \sum_{k=1}^{K} V_k(x \in C_i), \quad i = 1, \ldots, M. \quad (6.4.2)$$

3. The combined result $H(\mathbf{x})$ is determined by

$$H(\mathbf{x}) = \begin{cases} i, & \text{if } \max_{i \in \{1,\ldots,M\}} V_E(\mathbf{x} \in C_i) \text{ and } V_E(\mathbf{x} \in C_i) \geq \alpha \cdot K; \\ M+1, & \text{otherwise.} \end{cases}$$

(6.4.3)

Here α is a user-defined threshold that controls the confidence in the final decision.

For a learning algorithm, we are primarily interested in its computational efficiency.

Definition 6.6 (Polynomially Evaluatable [140]) Let C be a representation class over X. C is said to be *polynomially evaluatable* if there is a polynomial-time (possibly randomized) evaluation algorithm A that, on input a representation $c \in C$ and a domain point $\mathbf{x} \in X$, runs in time polynomial in $|c|$ and $|x|$ and decides if $\mathbf{x} \in c(\mathbf{x})$.

A machine learning algorithm is expected to be a strong learning algorithm that is polynomially evaluatable and has a producing representation h that is consistent with a given sample S. However, it is difficult, even impractical, to design directly a strong learning algorithm. A simple and effective approach is to use a boosting framework to a weak machine learning algorithm.

A machine learner for producing a two-class classifier is a weak learner if it is a coinflip algorithm based completely on random guessing. Schapire [221] showed that a weak learner could always improve its performance by training two additional classifiers on filtered versions of the input data stream. After learning an initial classifier h_1 on the first N training points, if two additional classifiers are produced as follows [100, 221]:

1. h_2 is learned on a new sample of N points, half of which are misclassified by h_1,
2. h_3 is learned on N points for which h_1 and h_2 disagree, and
3. the boosted classifier is $h_B = $ Majority Vote(h_1, h_2, h_3),

then a boosted weak learner can produce a two-class classifier with performance guaranteed (with high probability) to be significantly better than a coinflip.

An *additive logistic regression model* is defined as

$$\frac{\log P(Y=1|\mathbf{x})}{\log P(Y=-1|\mathbf{x})} = \beta_0 + \beta_1 \mathbf{x}_1 + \cdots + \beta_p \mathbf{x}_p,$$

(6.4.4)

where $P(x) = \frac{1}{1+e^{-x}}$ is called the *logistic function*, commonly called the sigmoid function.

Let $F(\mathbf{x})$ be an additive function of the form

$$F(\mathbf{x}) = \sum_{m=1}^{M} f_m(\mathbf{x}),$$

(6.4.5)

and consider a problem for minimizing the exponential criterion

$$J(F) = E\{e^{-yF(\mathbf{x})}\} \tag{6.4.6}$$

for estimation of $F(\mathbf{x})$.

Lemma 6.2 ([100]) $E\{e^{-yF(\mathbf{x})}\}$ *is minimized at*

$$F(\mathbf{x}) = \frac{1}{2} \log \frac{P(y=1|\mathbf{x})}{P(y=-1|\mathbf{x})}. \tag{6.4.7}$$

Hence

$$P(y=1|\mathbf{x}) = \frac{e^{F(\mathbf{x})}}{e^{-F(\mathbf{x})} + e^{F(\mathbf{x})}}, \tag{6.4.8}$$

$$P(y=-1|\mathbf{x}) = \frac{e^{-F(\mathbf{x})}}{e^{F(\mathbf{x})} + e^{-F(\mathbf{x})}}. \tag{6.4.9}$$

Lemma 6.2 shows that for an additive function $F(\mathbf{x})$ with the form (6.4.5), minimizing the exponential criterion $J(F)$ in (6.4.6) is an additive logistic regression problem in two classes. Friedman, Hastie, and Tibshirani [100] developed a boosting algorithm for additive logistic regression, called simply *LogitBoost*, as shown in Algorithm 6.5.

Algorithm 6.5 LogitBoost (two classes) [100]

1. **input:** weights $w_i = 1/N$, $i = 1, \ldots, N$, $F(\mathbf{x}) = 0$ and probability estimates $p(x_i) = 1/2$.
2. **for** $m = 1$ to M **do**
3.   Compute the working response and weight
 $z_i = \frac{y_i^* - p(x_i)}{p(x_i)(1-p(x_i))}$,
 $w_i = p(x_i)(1 - p(x_i))$.
4. Fit the function $f_m(\mathbf{x})$ by a weighted least squares regression of z_i to x_i using weights w_i.
5. Update $F(\mathbf{x}) \leftarrow F(\mathbf{x}) + \frac{1}{2} f_m(\mathbf{x})$ and $p(\mathbf{x}) \leftarrow (e^{F(\mathbf{x})})/(e^{F(\mathbf{x})} + e^{-F(\mathbf{x})})$.
6. **end for**
7. **output:** the classifier $\operatorname{sign}[F(\mathbf{x})] = \operatorname{sign}[\sum_{m=1}^{M} f_m(\mathbf{x})]$.

6.4.2 Probably Approximately Correct Learning

We are given a set of N training examples $X = \{(\mathbf{x}_1, y_1), \ldots, (\mathbf{x}_N, y_N)\}$ drawn randomly from $X \times Y$ according to distribution D_s, where Y is the label set consisting of just two possible labels $Y = \{0, 1\}$ in two-classification or C possible labels $Y \in \{1, \ldots, C\}$ in multi-classification. Machine learner is to find a hypothesis or representation h_f which is consistent with most of the sample (i.e., $h_f(\mathbf{x}_i) = y_i$ for most $1 \leq i \leq N$). In general, a hypothesis which is accurate on the training

6.4 Boosting and Probably Approximately Correct Learning

set might not be accurate on examples outside the training set; this problem is sometimes referred to as *"overfitting"* [97]. To avoid overfitting is one of the important problems that must be considered in machine learning.

Definition 6.7 (Probably Approximately Correct (PAC) Model [141]) Let $\mathcal{C}$ be a concept class over X. $\mathcal{C}$ is said to be *probably approximately correct* (PAC) *learnable* if there exists an algorithm L with the property: given an *error parameter* ϵ and a *confidence parameter* δ, for every concept $c \in \mathcal{C}$, for every distribution D_s on X, and for all $0 \leq \epsilon \leq 1/2$ and $0 \leq \delta \leq 1/2$, if L, after some amount of time, outputs a hypothesis concept $h \in \mathcal{C}$ satisfying error$(h) \leq \epsilon$ with probability $1 - \delta$.

If L runs in time polynomial in $1/\epsilon$ and $1/\delta$, then c is said to be *efficiently PAC learnable*. The hypothesis $h \in \mathcal{C}$ of the PAC learning algorithm is "approximately correct" with high probability, hence is named as *"Probably Approximately Correct"* (PAC) *learning* [141].

In machine learning, it is usually difficult and even impracticable to find a learner whose concept over X, $c(\mathbf{x}_i) = y_i$ for all labels $1 \leq i \leq N$. Hence, the goal of PAC learning is to find a hypothesis $h : X \rightarrow \{0, 1\}$ which is consistent with *most* (rather than *all*) of the sample (i.e., $h(\mathbf{x}_i) = y_i$ for most $1 \leq i \leq N$). After some amount of time, the learner must output a hypothesis $h : X \rightarrow \{0, 1\}$. The value $h(\mathbf{x})$ can be interpreted as a randomized prediction of the label of $\mathbf{x}$ that is 1 with probability $h(\mathbf{x})$ and 0 with probability $1 - h(\mathbf{x})$.

Definition 6.8 (Strong PAC-Learning Algorithm [97]) A *strong PAC-learning algorithm* is an algorithm that, given an accuracy $\epsilon > 0$ and a reliability parameter $\delta > 0$ and access to random examples, outputs with probability $1 - \delta$ a hypothesis with error at most ϵ.

Definition 6.9 (Weak PAC-Learning Algorithm [97]) A *weak PAC-learning algorithm* is an algorithm that, given an accuracy $\epsilon > 0$ and access to random examples, outputs with probability $1 - \delta$ a hypothesis with error at least $1/2 - \gamma$, where $\gamma > 0$ is either a constant or decreases as $1/p$ where p is a polynomial in the relevant parameters.

In strong PAC learning, the learner is required to generate a hypothesis whose error is smaller than the required accuracy ϵ. On the other hand, in weak PAC learning, the accuracy of the hypothesis is required to be just slightly better than $1/2$, which is the accuracy of a completely random guess. When learning with respect to a given distribution over the instances, weak and strong learning are not equivalent.

For a great assortment of learning problems, the *"boosting algorithm"* can convert a "weak" PAC learning algorithm that performs just slightly better than random guessing into one with arbitrarily high accuracy [97].

There are two frameworks in which boosting can be applied: boosting by filtering and boosting by sampling [96].

Adaptive boosting (AdaBoost) of Freund and Schapire [97] is a popular boosting algorithm by sampling, which has been used in conjunction with a wide range of other machine learning algorithms to enhance their performance.

Finding multiple weak classification algorithms with low recognition rate is much easier than finding a strong classification algorithm with high recognition rate. AdaBoost aims at boosting the accuracy of a weak learner by carefully adjusting the weights of training instances and learn a classifier accordingly. After T such iterations, the final hypothesis h_f is output. The hypothesis h_f combines the outputs of the T weak hypotheses using a weighted majority vote. Algorithm 6.6 gives the adaptive boosting (AdaBoost) algorithm in [97].

Algorithm 6.6 Adaptive boosting (AdaBoost) algorithm [97]
---
1. **input:**
   1.1 sequence of N labeled examples $\{(\mathbf{x}_1, y_1), \ldots, (\mathbf{x}_N, y_N)\}$,
   1.2 distribution D over the N examples,
   1.3 weak learning algorithm **WeakLearn**,
   1.4 integer T specifying number of iterations.
2. **initialization:** the weight vector $w_i^1 = D(i)$ for $i = 1, \ldots, N$.
3. **for** $t = 1$ to T **do**
4. Set $\mathbf{p}^t = \frac{\mathbf{w}^t}{\sum_{i=1}^N w_i^t}$.
5. Call **WeakLearn**, providing it with the distribution $\mathbf{p}_t$; get back a hypothesis $h_t : X \to \{0, 1\}$.
6. Calculate the error of $\epsilon_t = \sum_{i=1}^N p_i^t |h_t(\mathbf{x}_i) - y_i|$.
7. Set $\beta_t = \epsilon_t/(1 - \epsilon_t)$.
8. Set the new weights vector to be $w_i^{t+1} = w_i^t \beta_t^{1-|h_t(\mathbf{x}_i) - y_i|}$.
9. **output:** the hypothesis
$$h_f(\mathbf{x}) = \begin{cases} 1, & \text{if } \sum_{t=1}^T (\log 1/\beta_t) h_t(\mathbf{x}) \geq \frac{1}{2} \sum_{t=1}^T \log(1/\epsilon_t); \\ 0, & \text{otherwise.} \end{cases}$$

Friedman et al. [100] analyze the AdaBoost procedures from a statistical perspective: AdaBoost can be rederived as a method for fitting an additive model $\sum_m f_m(\mathbf{x})$ in a forward stagewise manner, which largely explains why it tends to outperform a single base learner. By fitting an additive model of different and potentially simple functions, it expands the class of functions that can be approximated.

Using Newton stepping rather than exact optimization at each step, Friedman *et al.* [100] proposed a modified version of the AdaBoost algorithm, called "Gentle AdaBoost" procedure, which instead takes adaptive Newton steps much like the LogitBoost algorithm, see Algorithm 6.7.

Algorithm 6.7 Gentle AdaBoost [100]
---
1. **input:** weights $w_i = 1/N$, $i = 1, \ldots, N$, $F(\mathbf{x}) = 0$.
2. **for** $m = 1$ to M **do**
3. Fit the regression function $f_m(\mathbf{x})$ by weighted least squares of y_i to $\mathbf{x}_i$ with weights w_i.
4. Update $F(\mathbf{x}) \leftarrow F(\mathbf{x}) + f_m(\mathbf{x})$.
5. Update $w_i \leftarrow w_i \exp(-y_i f_m(\mathbf{x}_i))$ and renormalize.
6. **endfor**
7. **output:** the classifier $\text{sign}[F(\mathbf{x})] = \text{sign}[\sum_{m=1}^M f_m(\mathbf{x})]$.

The application of adaptive boosting in transfer learning will be discussed in Sect. 6.20.

6.5 Basic Theory of Machine Learning

The pioneer of machine learning, Arthur Samuel, defined machine learning as a "field of study that gives computers the ability to learn without being explicitly programmed."

6.5.1 Learning Machine

Consider a learning machine whose task is to learn a mapping $x_i \to y_i$. The machine is actually defined by a set of possible mappings $\mathbf{x} \to f(\mathbf{x}, \boldsymbol{\alpha})$ with the functions $f(\mathbf{x}, \boldsymbol{\alpha})$ labeled by the adjustable parameter vector $\boldsymbol{\alpha}$.

The machine is usually assumed to be deterministic: for a given input vector $\mathbf{x}$ and choice of $\boldsymbol{\alpha}$, it will always give the same output $f(\mathbf{x}, \boldsymbol{\alpha})$. A particular choice of $\boldsymbol{\alpha}$ generates a "*trained machine*," and hence a neural network with fixed architecture and containing $\boldsymbol{\alpha}$ corresponding to the weights and biases is called a *learning machine* in this sense [39].

The expectation of the test error for a trained machine is defined as

$$R(\boldsymbol{\alpha}) = \int \frac{1}{2}|y - f(\mathbf{x}, \boldsymbol{\alpha})| \mathrm{d}P(\mathbf{x}, y), \tag{6.5.1}$$

where $P(\mathbf{x}, \boldsymbol{\alpha})$ is some unknown cumulative probability distribution from which data $\mathbf{x}$ and y are drawn, i.e., the data are assumed to be independently and identically distributed (i.i.d.). When a density of cumulative probability distributions, $P(\mathbf{x}, y)$, exists, $\mathrm{d}P(\mathbf{x}, y)$ may be written as $p(\mathbf{x}, y)\mathrm{d}\mathbf{x}\mathrm{d}y$.

The quantity $R(\boldsymbol{\alpha})$ is called the expected risk, or just the risk for choice of $\boldsymbol{\alpha}$. The "*empirical risk*" $R_{\text{emp}}(\boldsymbol{\alpha})$ is defined to be just the measured mean error rate on the given training set:

$$R_{\text{emp}}(\boldsymbol{\alpha}) = \frac{1}{2N} \sum_{i=1}^{N} |y_i - f(\mathbf{x}_i, \boldsymbol{\alpha})|, \tag{6.5.2}$$

where no probability distribution appears.

The quantity $\frac{1}{2}|y_i - f(\mathbf{x}_i, \boldsymbol{\alpha})|$ is called the loss. $R_{\text{emp}}(\boldsymbol{\alpha})$ is a fixed number for a particular choice of $\boldsymbol{\alpha}$ and for a particular training set $\{\mathbf{x}_i, y_i\}$.

Another name for a family of functions $f(\mathbf{x}, \boldsymbol{\alpha})$ is "learning machine." Designing a good learning machine depends on choice of $\boldsymbol{\alpha}$. A popular method for building $\boldsymbol{\alpha}$

is *empirical risk minimization* (ERM). By ERM, we mean $R_{\text{emp}}(\alpha)$ is minimized over all possible choices of α, results in a risk $R(\alpha^*)$ which is close to its minimum.

The minimization of different loss functions results in different machine learning methods. In reality, most machine learning methods should have three phases rather than two: training, validation, and testing. After the training is complete, there maybe several models (e.g., artificial neural networks) available, and one needs to decide which one to have a good estimation of the error it will achieve on a test set. For this end, there should be a third separate data set: the validation data set.

6.5.2 Machine Learning Methods

There are many different machine learning methods for modeling the data to the underlying problem. These machine learning methods can be divided to the following types [38].

1. **Machine Learning Methods based on Network Structure.**

    - *Artificial Neural Networks (ANNs):* ANNs are inspired by the brain and composed of interconnected artificial neurons capable of certain computations on their inputs [120]. The input data activate the neurons in the first layer of the network whose output is the input to the second layer of neurons in the network. Similarly, each layer passes its output to the next layer and the last layer outputs the result. Layers in between the input and output layers are referred to as hidden layers. When an ANN is used as a classifier, the output layer generates the final classification category.
    - *Bayesian Network:* A Bayesian network is a probabilistic graphical model that represents the variables and the relationships between them [98, 120, 130]. The network is constructed with nodes as the discrete or continuous random variables and directed edges as the relationships between them, establishing a directed acyclic graph. Each node maintains the states of the random variable and the conditional probability form. Bayesian networks are built by using expert knowledge or efficient algorithms that perform inference.

2. **Machine Learning Methods based on Statistical Analysis.**

    - *Association Rules:* The concept of association rules was popularized particularly due to the article of Agrawal et al. in 1993 [3]. The goal of association rule mining is to discover previously unknown association rules from the data. An association rule describes a relationship $X \Rightarrow Y$ between an itemset X and a single item Y. Association rules have two metrics that tell how often a given relationship occurs in the data: the *support* is an indication of how frequently the itemset appears in the data set, and the *confidence* is an indication of how often the rule has been found to be true.

6.5 Basic Theory of Machine Learning

- *Clustering:* Clustering [126] is a set of techniques for finding patterns in high-dimensional unlabeled data. It is an unsupervised pattern discovery approach where the data are grouped together based on some similarity measure.
- *Ensemble Learning:* Supervised learning algorithms, in general, search the hypothesis space to determine the right hypothesis that will make good predictions for a given problem. Although good hypotheses might exist, it may be hard to find one. Ensemble methods combine multiple learning algorithms to obtain the better predictive performance than the constituent learning algorithms alone. Often, ensemble methods use multiple weak learners to build a strong learner [206].
- *Hidden Markov Models (HMMs):* An HMM is a statistical Markov model where the system being modeled is assumed to be a Markov process with unobserved (i.e., hidden) states [17]. The main challenge is to determine the hidden parameters from the observable parameters. The states of an HMM represent unobservable conditions being modeled. By having different output probability distributions in each state and allowing the system to change states over time, the model is capable of representing non-stationary sequences.
- *Inductive Learning:* Deduction and induction are two basic techniques of inferring information from data. Inductive learning is traditional supervised learning, and aims at learning a model from labeled examples, and trying to predict the labels of examples we have not seen or known about. In inductive learning, from specific observations one begins to detect patterns and regularities, formulates some tentative hypotheses to be explored, and lastly ends up developing some general conclusions or theories. Several ML algorithms are inductive, but by inductive learning, we usually mean "repeated incremental pruning to produce error reduction" (RIPPER) [57] and the quasi-optimal (AQ) algorithm [177].
- *Naive Bayes:* It is well known [98] that a surprisingly simple Bayesian classifier with strong assumptions of independence among features, called naive Bayes, is competitive with state-of-the-art classifiers such as C4.5. In general, the input features are assumed to be independent, whereas in practice this is seldom true. Naive Bayes classifiers [98, 270] can handle an arbitrary number of independent features whether continuous or categorical by reducing a high-dimensional density estimation task to a one-dimensional kernel density estimation, under the assumption that the features are independent. Although the Naive Bayes classifier has some limitations, it is an optimal classifier if the features are conditionally independent given the true class. One of the biggest advantages of the Naive Bayes classifier is that it is an online algorithm and its training can be completed in linear time.

3. **Machine Learning Methods based on Evolution.**

    - *Evolutionary Computation:* In computer science, evolutionary computation is a family of algorithms for global optimization inspired by biological evolution, and the subfield of artificial intelligence and soft computing studying these algorithms (by Wikipedia). The term evolutionary computation encompasses genetic algorithms (GA) [107], genetic programming (GP) [152], evolution strategies [26], particle Swarm optimization [142], ant colony optimization [78], and artificial immune systems [89], which is the topics in Chap. 9 Evolutionary Computations.

6.5.3 Expected Performance of Machine Learning Algorithms

A machine learning algorithm is expected to have the following *expected performance* [145].

- *Scalability:* This parameter can be defined as the ability for an algorithm to be able to handle an increase in its scale, such as feeding more data to the system, adding more features to the input data or adding more layers in a neural network, without it limitlessly increasing its complexity [5].
- *Training Time:* This is the amount of time that a machine learning algorithm takes to be fully trained and form the ability to make its predictions.
- *Response Time:* This parameter is related to the agility of a machine learning system, and represents the time that an algorithm takes, after it has been trained, to make a prediction for the desired self-organizing networks (SON) function.
- *Training Data:* This metric of machine learning algorithm is the amount and type of training data an algorithm needs. Algorithms supported by more training data usually have better accuracy, but they also take more time to be trained.
- *Complexity:* Complexity of a system can be defined as the amount of mathematical operations that it performs in order to achieve a desired solution. This parameter can determine if certain algorithms are more suitable to be deployed at the user side.
- *Accuracy:* Future networks are expected to be much more intelligent and quicker, enabling highly different types of applications and user requirements. Deploying algorithms that have high accuracy is critical to guarantee a good operability of certain self-organizing network functions.
- *Convergence Time:* This metric of an algorithm, different from the response time, relates to how fast it agrees that the solution found for that particular problem is the optimal solution at that time.
- *Convergence Reliability:* This parameter represents the susceptibility of some algorithm to be stuck at local minima and how initial conditions can affect its performance.

6.6 Classification and Regression

In essence, machine learning is to learn the given training samples for solving one of two basic problems: regression (for continuous outputs) or classification (for discrete outputs). *Classification* is closely related to pattern recognition, its goal

6.6 Classification and Regression

is to design a classifier thought learning a "training" set of input data for making recognition or classification for unknown samples. By regression, it means to design a regressor or predictor based on machine learning results of a set of training data for making a prediction for unknown continuous samples.

6.6.1 Pattern Recognition and Classification

Pattern recognition is the field devoted to the study of methods designed to categorize data into distinct classes. By Watanabe [262], a *pattern* is defined "as opposite of a chaos; it is an entity, vaguely defined, that could be given a name."

Pattern recognition is widely applied in recognitions of human biometrics (such as a person's face, fingerprint, iris) and various targets (such as aircraft, ships). In these applications, the extraction of object features is crucial. For example, when a target is considered as a linear system, the target parameter is a feature of the target signal.

Given a pattern, its recognition/classification may consist of one of the following two tasks [262]:

- *Supervised classification* (e.g., discriminant analysis) in which the input pattern is identified as a member of a predefined class;
- *Unsupervised classification* (e.g., clustering) in which the pattern is assigned to a hitherto unknown class.

The tasks of pattern recognition/classification are customarily divided into four distinct blocks:

1. *Data representation (acquisition and pre-processing).*
2. *Feature selection or extraction.*
3. *Clustering.*
4. *Classification.*

The four best known approaches for pattern recognition are [128]: (1) template matching, (2) statistical classification, (3) syntactic or structural matching, and (4) neural networks.

Data Representation

Data representation is mostly problem-specific. In matrix algebra, signal processing, pattern recognition, etc., data pre-processing most commonly involves the zero-mean normalization (i.e., *data centering*): for each data vector $\mathbf{x}_n$,

$$x_{i,j}^{\text{cent}} = x_{i,j} - \bar{x}, \quad \bar{x} = \frac{1}{N}\sum_{j=1}^{N} x_{i,j}, \quad \text{for } i = 1, \ldots, N \tag{6.6.1}$$

to remove the useless direct current (DC) components in data, where $x_{i,j}$ is the jth entry of the vector $\mathbf{x}_i$. The data centering is also referred to the data zero-meaning.

In order to prevent drastic changes in data amplitude, the input data are also required to be scaled to similar ranges:

$$x_{i,j}^{\text{scal}} = \frac{x_{i,j}}{s_i}, \quad s_i = \sqrt{\sum_{k=1}^{n} x_k^2}, \quad i = 1, \ldots, N; \; j = 1, \ldots, n. \tag{6.6.2}$$

This pre-processing is called the *data scaling*.

Feature Selection

In many applications, data vectors must become low-dimensional vectors via some transformation/processing method. These low-dimensional vectors are called the *pattern vectors* or *feature vectors* owing to the fact that they extract the features of the original data vectors, and are directly used for pattern clustering and classification. For example, colors of clouds and parameters of voice tones are pattern or feature vectors in weather forecasting and voice classification, respectively.

Due to high dimensionality, the original data vectors are not directly available for pattern recognition. Hence, one should try to find invariant features in data that describe the differences in classes as best as possible.

According to the given training set, feature selection can be divided into supervised and unsupervised methods.

- *Supervised feature selection:* Let $X_l = \{(\mathbf{x}_1, y_1), \ldots, (\mathbf{x}_N, y_N)\}$ be the labeled set of data, where $\mathbf{x}_i \in \mathbb{R}^n$ is the ith data vectors and $y_i = k$ indicates the kth class to which the data vector $\mathbf{x}_i$ belongs, where $k \in \{1, \ldots, M\}$ is the label of M target classes. The pattern recognition uses supervised machine learning. Let l_p be the number of $k = p$ and $l_1 + \cdots + l_M = M$, we can construct the data matrix in the pth class of targets:

$$\mathbf{x}^{(p)} = \left[\mathbf{x}_1^{(p)}, \ldots, \mathbf{x}_{l_p}^{(p)}\right]^T \in \mathbb{R}^{N \times l_p}, \quad p = 1, \ldots, M. \tag{6.6.3}$$

Thus we can select feature vectors in data by using matrix algebra methods such as principal component analysis, etc. This selection, as a supervised machine learning, is also known as *feature selection* or *dimensionality reduction* since the dimension of feature vector is much less than the dimension of original data vectors.

- *Unsupervised feature selection:* When data are unlabeled, i.e., only data vectors $\mathbf{x}_i, \ldots, \mathbf{x}_N$ are given, the pattern recognition is unsupervised machine learning which is more difficult than supervised patter recognition. In this case, we need to use mapping or signal processing for transforming the data into invariant features such as short time Fourier transform, bispectrum, wavelet, etc.

6.6 Classification and Regression

Feature selection is to select the most significant features of the data, and attempt to reduce the dimensionality (i.e., the number of features) for the remaining steps of the task.

We will discuss supervised and unsupervised feature extractions later in details.

Clustering

Let **w** be the weight vector of a cluster which aims at patterns within a cluster being more similar to each other rather than are patterns belonging to other clusters. Clustering methods are used in order to find the actual mapping between patterns and labels (or targets).

This categorization has two machine learning methods: supervised learning (which makes distinct labeling of the data) and unsupervised learning (division of the data into classes), or a combination of more than one of these tasks.

The above three stages (data representation, feature selection, and clustering) belong to the *training phase*. Having such a cluster, one may make a classification in the next testing phase.

Classification

Classification is one of the most frequently encountered decision making tasks of human activity. A classification problem occurs when an object needs to be assigned into a predefined group or classes related to that object after clustering a number of groups or classes based on a set of observed attributes. In other words, one needs to decide which class among a number of classes a given testing sample belongs to. This step is called the *testing phase*, compared with the previous training stage. Clearly, this is a form of supervised learning.

6.6.2 Regression

In statistical modeling and machine learning, *regression* is to fit a statistical process for estimating the relationships among variables. It focuses on the relationship between a dependent variable **x** and one or more independent variables (named as "*predictors*") β. More specifically, regression is to analyze how the typical value of the dependent variable changes when anyone of the independent variables is varied, while the other independent variables are fixed.

Machine learning is widely used for prediction and forecasting, where its use has substantial overlap with the field of statistical regression analysis.

Consider the more general multiple regression model with p independent variables:

$$y_i = \beta_1 x_{i1} + \beta_2 x_{i2} + \cdots + \beta_p x_{ip} + \varepsilon_i, \qquad (6.6.4)$$

where x_{ij} is the ith observation on the jth independent variable $\mathbf{x}_j = [x_{1j}, \ldots, x_{pj}]^T$. If the first independent variable takes the value 1 for all i, $x_{i1} = 1$, then the regression model reduces to

$$y_i = \beta_1 + \beta_2 x_{i2} + \cdots + \beta_p x_{ip} + \varepsilon_i$$

and β_1 is called the *regression intercept*.

The *regression residual* is defined as

$$\varepsilon_i = y_i - \hat{\beta}_1 x_{i1} - \cdots - \hat{\beta}_p x_{ip}. \tag{6.6.5}$$

Hence, the normal equations on regression are given by

$$\sum_{i=1}^{n} \sum_{k=1}^{p} x_{ij} x_{ik} \hat{\beta}_k = \sum_{i=1}^{n} x_{ij} y_i, \quad j = 1, \ldots, p. \tag{6.6.6}$$

In matrix notation, the above normal equations can be written as

$$(\mathbf{X}^T \mathbf{X}) \hat{\boldsymbol{\beta}} = \mathbf{X}^T \mathbf{y}, \tag{6.6.7}$$

where $\mathbf{X} = [x_{ij}]_{i=1,j=1}^{n,p}$ is an $n \times p$ matrix, $\mathbf{y} = [y_i]_{i=1}^{n}$ is an $n \times 1$ vector, and $\hat{\boldsymbol{\beta}} = [\hat{\beta}_j]_{j=1}^{p}$ is a $p \times 1$ vector. Finally, the solution of regression problem is given by

$$\hat{\boldsymbol{\beta}} = (\mathbf{X}^T \mathbf{X})^{-1} \mathbf{X}^T \mathbf{y}. \tag{6.6.8}$$

6.7 Feature Selection

The performance of machine learning methods generally depends on the choice of data representation. The *representation choice* heavily depends on *representation learning*. Representation learning is also known as *feature learning*.

In machine learning, representation learning [22] is a set of learning techniques that allows a system to automatically discover the representations needed for feature detection or classification from raw data.

When mining large data sets, both in dimension and size, we are given a set of original variables (e.g., in gene expression) or a set of extracted features (such as pre-processing via principal or minor component analysis) in data representation. An important problem is how to select a subset of the original variables or a subset of the original features. The former is called the *variable selection*, and the latter is known as the *feature selection or representation choice*.

Feature selection is a fundamental research topic in data mining and machine learning with a long history since the 1970s. With the advance of science and technology, more and more real-world data sets in data mining and machine learning often involve a large number of features. But, not all features are essential since many of them are redundant or even irrelevant, which may significantly degrade the accuracy of learned models as well as reduce the learning speed and ability of the models. By removing irrelevant and redundant features, feature selection can reduce the dimensionality of the data, speed up the learning process, simplify the learned model, and/or improve the performance [69, 112, 284].

A smaller set of representative variables or features, retaining the optimal salient characteristics of the data, not only decreases the processing complexity and time, but also overcomes the risk of "overfitting" and leads to more compactness of the model learned and better generalization. When the class labels of the data are given, we are faced with supervised variable or feature selection, otherwise we should use unsupervised variable or feature selection.

Since the learning methods used in both the variable selection and the feature selection are essentially the same, we take the feature selection as the subject of this section.

6.7.1 Supervised Feature Selection

An important problem in machine learning is to reduce the dimensionality D of the feature space F to overcome the risk of "overfitting." Data overfitting arises when the number n of features is large and the number m of training patterns is comparatively small. In such a situation, one may easily find a decision function that separates the training data but gives a poor result for test data.

Feature selection has become the focus of much research in areas of application (e.g., text processing of internet documents, gene expression array analysis, and combinatorial chemistry) for high-dimensional data. The objective of *feature selection* is three-fold [112]:

- improving the prediction performance of the predictors,
- providing faster and more cost-effective predictors, and
- providing a better understanding of the underlying process that generated the data.

Many feature selection algorithms are based on *variable ranking*. As a pre-processing step of feature selection, variable ranking is a filter method: it is independent of the choice of the predictor. For example, in the gene selection problem, the variables are gene expression coefficients corresponding to the abundance of mRNA in a sample for a number of patients. A typical classification task is to separate healthy patients from cancer patients, based on their gene expression "profile" [112].

The feature selection methods can be divided into two categories: the exhaustive enumeration method and the greedy method.

- *Exhaustive enumeration method* selects the best subset of features satisfying a given "model selection" criterion by exhaustive enumeration of all subsets of features. However, this method is impractical for large numbers of features.
- *Greedy method* is available particularly to the feature selection in large dimensional input spaces. Among various possible methods *feature ranking* techniques are very attractive. A number of top ranked features may be selected for further analysis or to design a classifier. Alternatively, a threshold can be set on the ranking criterion. Only the features whose criterion exceeds the threshold are retained.

Consider a set of m examples $\{\mathbf{x}_k, y_k\}(k = 1, \ldots, m)$, where $\mathbf{x}_k = [x_{k,1}, \ldots, x_{k,n}]^T$ and one output variable y_k. Let $\bar{x}_i = \frac{1}{m}\sum_{k=1}^{m} x_{k,i}$ and $\bar{y} = \frac{1}{m}\sum_{k=1}^{m} y_k$ stand for an average over the index k. Variable ranking makes use of a scoring function $S(i)$ computed from the values $x_{k,i}$ and y_k, $k = 1, \ldots, m$. By convention, a high score is indicative of a valuable variable (e.g., gene) and variables are sorted in decreasing order of $S(i)$.

A good feature selection method selects first a few of features that individually classify best the training data, and then increases the number of features. Classical feature selection methods include correlation methods and expression ratio methods.

A typical *correlation method* is based on the *Pearson correlation coefficients* defined as

$$R(i) = \frac{\sum_{k=1}^{m}(x_{k,i} - \bar{x}_i)(y_k - \bar{y})}{\sqrt{\sum_{k=1}^{m}(x_{k,i} - \bar{x}_i)^2 \sum_{k=1}^{m}(y_k - \bar{y})^2}}, \qquad (6.7.1)$$

where the numerator denotes the *between-class variance* and the denominator represents the *within-class variance*. In linear regression, using the square of $R(i)$ as a variable ranking criterion enforces a ranking according to goodness of linear fit of individual variables.

For two-class data $\{\mathbf{x}_k, y_k\}(k = 1, \ldots, m)$, where $\mathbf{x}_k = [x_{k,1}, \ldots, x_{k,n}]^T$ consists of n input variables $x_{k,i}(i = 1, \ldots, n)$ and $y_k \in \{+1, -1\}$ (e.g., certain disease vs. normal), let

$$\mu_i^+ = \frac{1}{m^+}\sum_{k=1}^{m^+} x_{k,i}, \quad \mu_i^- = \frac{1}{m^-}\sum_{k=1}^{m^-} x_{k,i}, \qquad (6.7.2)$$

$$\sigma_i^+ = \sum_{k=1}^{m^+}(x_{k,i} - \mu_i^+)^2, \quad \sigma_i^- = \sum_{k=1}^{m^-}(x_{k,i} - \mu_i^-)^2, \qquad (6.7.3)$$

where m^+ and m^- are the numbers of the ith variable $x_{k,i}$ corresponding to $y_i = +1$ and $y_i = -1$, respectively; μ_i^+ and μ_i^- are the means of $x_{k,i}$ over the index k associated with $y_i = +1$ and $y_i = -1$, respectively; and σ_i^+ and σ_i^- are, respectively, the standard deviations of $x_{k,i}$ over the index k associated with $y_i = +1$ and $y_i = -1$.

6.7 Feature Selection

A well-known *expression ratio method* uses the ratio [103, 109]:

$$F(x_i) = \left| \frac{\mu_i^+ - \mu_i^-}{\sigma_i^+ + \sigma_i^-} \right|. \tag{6.7.4}$$

This method selects highest ratios $F(x_i)$ differing most on average in the two classes and having small deviations in the scores in the respective classes as top features.

The *expression ratio criterion* (6.7.4) is similar to Fisher's discriminant criterion [83].

It should be noted that the variable dependencies cannot be ignored in feature selection as follows [112].

- Noise reduction and consequently better class separation may be obtained by adding variables that are presumably redundant.
- Perfectly correlated variables are truly redundant in the sense that no additional information is gained by adding them.
- Very high variable correlation (or anti-correlation) does not mean absence of variable complementarity.
- A variable that is completely useless by itself can provide a significant performance improvement when taken with others.
- Two variables that are useless by themselves can be useful together.

One possible use of feature ranking is the design of a class predictor (or classifier) based on a pre-selected subset of features. The weighted voting scheme yields a particular linear discriminant classifier:

$$D(\mathbf{x}) = \langle \mathbf{w}, \mathbf{x} - \boldsymbol{\mu} \rangle = \mathbf{w}^T (\mathbf{x} - \boldsymbol{\mu}), \tag{6.7.5}$$

where $\boldsymbol{\mu} = \frac{1}{2}(\boldsymbol{\mu}^+ + \boldsymbol{\mu}^-)$ with

$$\boldsymbol{\mu}^+ = \frac{1}{n^+} \sum_{\mathbf{x}_i \in X^+} \mathbf{x}_i, \tag{6.7.6}$$

$$\boldsymbol{\mu}^- = \frac{1}{n^-} \sum_{\mathbf{x}_i \in X^-} \mathbf{x}_i, \tag{6.7.7}$$

where $X^+ = \{(\mathbf{x}_i, y_i = +1)\}$, $X^- = \{(\mathbf{x}_i, y_i = -1)\}$, and n^+ and n^- are the numbers of the training data vectors $\mathbf{x}_i$ belonging to the classes (+) and (−), respectively.

Define the $n \times n$ within-class scatter matrix

$$\mathbf{S}_w = \sum_{\mathbf{x}_i \in X^+} (\mathbf{x}_i - \boldsymbol{\mu}^+)(\mathbf{x}_i - \boldsymbol{\mu}^+)^T + \sum_{\mathbf{x}_i \in X^-} (\mathbf{x}_i - \boldsymbol{\mu}^-)(\mathbf{x}_i - \boldsymbol{\mu}^-)^T. \tag{6.7.8}$$

By Fisher's linear discriminant, the classifier is given by [113]

$$\mathbf{w} = \mathbf{S}_w^{-1}(\boldsymbol{\mu}^+ - \boldsymbol{\mu}^-). \tag{6.7.9}$$

Once the classifier $\mathbf{w}$ is designed using the training data $(\mathbf{x}_i, y_i), i = 1, \ldots, n$ (where $y_i \in \{+1, -1\}$), for any given data vector $\mathbf{x}$, the new pattern can be classified according to the sign of the decision function [113]:

$$D(\mathbf{x}) = \mathbf{w}^T \mathbf{x} > 0 \Rightarrow \mathbf{x} \in \text{class } (+),$$

$$D(\mathbf{x}) = \mathbf{w}^T \mathbf{x} < 0 \Rightarrow \mathbf{x} \in \text{class } (-),$$

$$D(\mathbf{x}) = \mathbf{w}^T \mathbf{x} = 0, \quad \text{decision boundary}.$$

6.7.2 Unsupervised Feature Selection

Consider the unsupervised feature selection for given data vectors $\mathbf{x}_i, i = 1, \ldots, n$ without labeled class index $y_i, i = 1, \ldots, n$. In this case, there are two common similarity measures between two random vectors $\mathbf{x}$ and $\mathbf{y}$, as stated below [182].

1. *Correlation coefficient:* The correlation coefficient $\rho(\mathbf{x}, \mathbf{y})$ between two random vectors $\mathbf{x}$ and $\mathbf{y}$ is defined as

$$\rho(\mathbf{x}, \mathbf{y}) = \frac{\text{cov}(\mathbf{x}, \mathbf{y})}{\sqrt{\text{var}(\mathbf{x})\text{var}(\mathbf{y})}}, \tag{6.7.10}$$

where var($\mathbf{x}$) denotes the variance of the random vector $\mathbf{x}$ and cov($\mathbf{x}, \mathbf{y}$) denotes the covariance between $\mathbf{x}$ and $\mathbf{y}$. The correlation coefficient has the following properties.
   (a) $0 \leq 1 - |\rho(\mathbf{x}, \mathbf{y})| \leq 1$.
   (b) $1 - |\rho(\mathbf{x}, \mathbf{y})| = 0$ if and only if $\mathbf{x}$ and $\mathbf{y}$ are linearly dependent.
   (c) *Symmetric:* $1 - |\rho(\mathbf{x}, \mathbf{y})| = 1 - |\rho(\mathbf{y}, \mathbf{x})|$.
   (d) *Invariant to scaling and translation:* if $\mathbf{u} = \frac{\mathbf{x}-\mathbf{a}}{c}$ and $\mathbf{v} = \frac{\mathbf{y}-\mathbf{b}}{d}$ for some constant vectors $\mathbf{a}, \mathbf{b}$ and constants c, d, then $1 - |\rho(\mathbf{x}, \mathbf{y})| = 1 - |\rho(\mathbf{u}, \mathbf{v})|$.
   (e) *Sensitive to rotation:* If $(\mathbf{u}, \mathbf{v})$ is some rotation of $(\mathbf{x}, \mathbf{y})$, then $|\rho(\mathbf{x}, \mathbf{y})| \neq |\rho(\mathbf{u}, \mathbf{v})|$.

2. *Least square regression error:* Let $\mathbf{y} = a\mathbf{1} + b\mathbf{x}$ be the linear regression of $\mathbf{x}$ with the regression coefficients a and b. Least square regression error is defined as

$$e^2(\mathbf{x}, \mathbf{y}) = \frac{1}{n} \sum_{i=1}^{n} (e(\mathbf{x}, \mathbf{y})_i)^2, \tag{6.7.11}$$

6.7 Feature Selection

where $e(\mathbf{x}, \mathbf{y})_i = y_i - a - bx_i$. The regression coefficients are given by $a = \frac{1}{n} \sum_{i=1}^{n} y_i$, $b = \frac{\text{cov}(\mathbf{x},\mathbf{y})}{\text{var}(\mathbf{x})}$ and the mean square error $e(\mathbf{x}, \mathbf{y}) = \text{var}(\mathbf{y})(1 - \rho(\mathbf{x}, \mathbf{y})^2)$. If $\mathbf{x}$ and $\mathbf{y}$ are linearly correlated, then $e(\mathbf{x}, \mathbf{y}) = 0$, and if $\mathbf{x}$ and $\mathbf{y}$ are completely uncorrelated, then $e(\mathbf{x}, \mathbf{y}) = \text{var}(\mathbf{y})$. The measure e^2 is also called the *residual variance*. The least square regression error measure has the following properties.

(a) $0 \leq e(\mathbf{x}, \mathbf{y}) \leq \text{var}(\mathbf{y})$.
(b) $e(\mathbf{x}, \mathbf{y}) = 0$ if and only if $\mathbf{x}$ and $\mathbf{y}$ are linearly correlated.
(c) *Unsymmetric*: $e(\mathbf{x}, \mathbf{y}) \neq e(\mathbf{y}, \mathbf{x})$.
(d) *Sensitive to scaling*: If $\mathbf{u} = \mathbf{x}/c$ and $\mathbf{v} = \mathbf{y}/d$ for some constants c and d, then $e(\mathbf{x}, \mathbf{y}) = d^2 e(\mathbf{u}, \mathbf{v})$. However, e is invariant to translation of the variables.
(e) The measure e is sensitive to the rotation of the scatter diagram in x–y plane.

A feature similarity measure is expected to be symmetric, sensitive to scaling and translation of variables, and invariant to their rotation. Hence, the above two similarity measures are not expected, because the correlation coefficient has the unexpected properties (d) and (e), and the least square regression is not symmetric (property (c)) and is sensitive to the rotation (property (d)).

In order to have all the desired properties, Mitra et al. [182] proposed a *maximal information compression index* λ_2 defined as

$$2\lambda_2(\mathbf{x}, \mathbf{y}) = \text{var}(\mathbf{x}) + \text{var}(\mathbf{y})$$
$$- \sqrt{(\text{var}(\mathbf{x}) + \text{var}(\mathbf{y}))^2 - 4\text{var}(\mathbf{x})\text{var}(\mathbf{y})(1 - \rho(\mathbf{x}, \mathbf{y})^2)}. \quad (6.7.12)$$

Here, $\lambda_2(\mathbf{x}, \mathbf{y})$ denotes the smallest eigenvalue of the covariance matrix Σ of the random variables $\mathbf{x}$ and $\mathbf{y}$.

The feature similarity measure $\lambda_2(\mathbf{x}, \mathbf{y})$ has the following properties [182].

1. $0 \leq \lambda_2(\mathbf{x}, \mathbf{y}) \leq 0.5(\text{var}(\mathbf{x}) + \text{var}(\mathbf{y}))$.
2. $\lambda_2(\mathbf{x}, \mathbf{y}) = 0$ if and only if $\mathbf{x}$ and $\mathbf{y}$ are linearly correlated.
3. *Symmetric*: $\lambda_2(\mathbf{x}, \mathbf{y}) = \lambda_2(\mathbf{y}, \mathbf{x})$.
4. *Sensitive to scaling*: If $\mathbf{u} = \mathbf{x}/c$ and $\mathbf{v} = \mathbf{y}/d$ for some nonzero constants c and d, then $\lambda_2(\mathbf{x}, \mathbf{y}) \neq \lambda_2(\mathbf{u}, \mathbf{v})$. Since the expression of $\lambda_2(\mathbf{x}, \mathbf{y})$ does not contain the mean but only the variance and covariance terms, it is invariant to translation of $\mathbf{x}$ and $\mathbf{y}$.
5. λ_2 is *invariant to rotation* of the variables about the origin.

Let the original number of features be D, and the original feature set be $O = \{F_i, i = 1, \ldots, D\}$. Apply the measures of linear dependency (e.g., $\rho(F_i, F_j), e(F_i, F_j)$, and/or $\lambda_2(F_i, F_j)$) to compute the dissimilarity between F_i and F_j, denoted by $d(F_i, F_j)$. Smaller the value of $d(F_i, F_j)$ is, the more similar are the features F_i and F_j. Let r_i^k denote the dissimilarity between feature F_i and its nearest neighbor feature in R, and $\inf_{F_i \in R} r_i^k$ represent the infimum of r_i^k for $F_i \in R$.

Algorithm 6.8 is the feature clustering algorithm of Mitra et al. [182].

Algorithm 6.8 Feature clustering algorithm [182]
1. **input:** the original feature set $O = \{F_1, \ldots, F_D\}$.
2. **initialization:** Choose an initial value of $k \leq d_1$ and set $R \leftarrow O$.
3. For each $F_i \in R$ compute r_i^k.
4. Find feature $F_{i'}$ for which $r_{i'}^k$ is minimum. Retain this feature in R and discard k nearest features of $F_{i'}$.
5. Let $\epsilon = r_{i'}^k$.
6. If $k > |R| - 1$ then $k = |R| - 1$.
7. If $k = 1$ then goto Step 14.
8. **While** $r_{i'}^k > \epsilon$ **do**
9.   (a) $k = k - 1$.
10. $r_{i'}^k = \inf_{F_i \in R} r_i^k$.
    ("k is decremented by 1, until the "kth nearest neighbor of at least one of features in R is less than ϵ-dissimilar with the feature.)
11.   (b) If $k = 1$ then goto Step 14.
    (if no feature in R has less than ϵ-similar the "nearest neighbor" select all the remaining feature in R.)
12. **end while**
13. Goto Step 3.
14. **output:** return feature set R as the reduced feature set.

6.7.3 Nonlinear Joint Unsupervised Feature Selection

Let $\mathbf{X} = [\mathbf{x}_1, \ldots, \mathbf{x}_n]$ be the original feature matrix of n data samples, where $\mathbf{x}_i \in \mathbb{R}^D$ and x_{ip} denote the value of the pth ($p = 1, \ldots, D$) feature of $\mathbf{x}_i$. Consider selecting d ($d \leq D$) high-quality features. To this end, use $\mathbf{s} \in \{0, 1\}^D$ as the selection indicator vector, where $s_p = 1$ indicates the pth feature is selected and $s_p = 0$ indicates not selected.

Given a feature mapping $\boldsymbol{\phi}(\mathbf{x}) : X \rightarrow F$, $\boldsymbol{\phi}(\mathbf{x}) - E\{\boldsymbol{\phi}(\mathbf{x})\}$ is called the centered feature mapping, where $E\{\boldsymbol{\phi}(\mathbf{x})\}$ is its expectation.

Definition 6.10 (Centered Kernel Function [267]) The kernel function $K \in \mathbb{R}^{n \times n}$ after centering is given by

$$K_c(\mathbf{x}_i, \mathbf{x}_j) = (\boldsymbol{\phi}(\mathbf{x}_i) - E\{\boldsymbol{\phi}(\mathbf{x}_i)\})^T (\boldsymbol{\phi}(\mathbf{x}_i) - E\{\boldsymbol{\phi}(\mathbf{x}_i)\}). \quad (6.7.13)$$

For a finite number of samples, a feature vector is centered by subtracting its empirical expectation, i.e., $\boldsymbol{\phi}(\mathbf{x}_i) - \bar{\boldsymbol{\phi}}$, where $\bar{\boldsymbol{\phi}} = \frac{1}{n} \sum_{j=1}^{n} \boldsymbol{\phi}(\mathbf{x}_j)$.

Definition 6.11 (Centered Kernel Matrix [267]) The kernel matrix $\mathbf{K} \in \mathbb{R}^{n \times n}$ after centering is defined by its entry as follows:

$$(K_c)_{ij} = K_{ij} - \frac{1}{n} \sum_{i=1}^{n} K_{ij} - \frac{1}{n} \sum_{j=1}^{n} K_{ij} + \frac{1}{n^2} \sum_{i=1}^{n} \sum_{j=1}^{n} K_{ij}. \quad (6.7.14)$$

6.7 Feature Selection

Denoting $\mathbf{H} = \mathbf{I} - \frac{1}{n}\mathbf{11}^T$, then the kernel matrix $\mathbf{K} \in \mathbb{R}^{n \times n}$ after centering can be expressed as $\mathbf{K}_c = \mathbf{HKH}$. It is easy to know that $\mathbf{H}$ is an idempotent matrix, i.e., $\mathbf{HH} = \mathbf{H}$.

Definition 6.12 (Kernel Alignment [65, 267]) For two kernel matrices $\mathbf{K}_1 \in \mathbb{R}^{n \times n}$ and $\mathbf{K}_2 \in \mathbb{R}^{n \times n}$ (assume $\|\mathbf{K}_1\|_F > 0$ and $\|\mathbf{K}_2\|_F > 0$), the alignment between $\mathbf{K}_1$ and $\mathbf{K}_2$ is defined as the form of matrix trace: $\rho(\mathbf{K}_1, \mathbf{K}_2) = \text{tr}(\mathbf{K}_1 \mathbf{K}_2)$.

Assume $\mathbf{L} \in \mathbb{R}^{n \times n}$ is a kernel matrix computed from original feature matrix $\mathbf{X}$, $\mathbf{K} \in \mathbb{R}^{n \times n}$ is a kernel matrix computed from selected features.

Denoting

$$\mathbf{\Phi}(\mathbf{X}) = [\boldsymbol{\phi}(\mathbf{x}_1), \ldots, \boldsymbol{\phi}(\mathbf{x}_n)] \in \mathbb{R}^{D \times n}, \tag{6.7.15}$$

$$\mathbf{\Phi}(\mathbf{X}_s) = [\boldsymbol{\phi}(\mathbf{Diag}(\mathbf{s})\mathbf{x}_1), \ldots, \boldsymbol{\phi}(\mathbf{Diag}(\mathbf{s})\mathbf{x}_n)], \tag{6.7.16}$$

then

$$\mathbf{L} = (\mathbf{\Phi}(\mathbf{X}))^T \mathbf{\Phi}(\mathbf{X}) \in \mathbb{R}^{n \times n}, \tag{6.7.17}$$

$$\mathbf{K} = (\mathbf{\Phi}(\mathbf{X}_s))^T \mathbf{\Phi}(\mathbf{X}_s) \in \mathbb{R}^{n \times n}. \tag{6.7.18}$$

By Definition 6.12 and using $\text{tr}(\mathbf{AB}) = \text{tr}(\mathbf{BA})$ and $\mathbf{HH} = \mathbf{H}$, the kernel alignment between $\mathbf{L}_c$ and $\mathbf{K}_c$ can be represented as

$$\rho(\mathbf{L}_c, \mathbf{K}_c) = \text{tr}(\mathbf{HLHHKH}) = \text{tr}(\mathbf{HHLHHK}) = \text{tr}(\mathbf{HLHK}). \tag{6.7.19}$$

The core issue with nonlinear joint unsupervised feature selection is to search for an optimal selection vector $\mathbf{s}$ so that the selected d high-quality features from D original features can be the solution to the following constrained optimization problem:

$$\mathbf{s} = \arg\min_{\mathbf{s}} \ \{f(\mathbf{K}) = -\text{tr}(\mathbf{HLHK})\} \tag{6.7.20}$$

$$\text{subject to } \sum_{p=1}^{D} s_p = d, \quad s_p \in \{1, 0\}, \ \forall \ p = 1, \ldots, D, \tag{6.7.21}$$

or

$$\mathbf{s} = \arg\min_{\mathbf{s}} \ \{f(\mathbf{K}) = -\text{tr}(\mathbf{HLHK}) + \lambda \|\mathbf{s}\|_1\} \tag{6.7.22}$$

$$\text{subject to } s_p \in \{1, 0\}, \ \forall \ p = 1, \ldots, D. \tag{6.7.23}$$

For Gaussian kernel $K_{ij} \sim N(0, \sigma^2)$, its gradient with respect to the selection vector $\mathbf{s}_p$ is given by

$$\frac{\partial K_{ij}}{\partial \mathbf{s}_p} = -K_{ij} \frac{2(x_{iP} - x_{jp})^2 \mathbf{s}_p}{\sigma^2}. \quad (6.7.24)$$

Hence, the gradient of the objective $f(\mathbf{K})$ with respect to the selection vector $\mathbf{s}_p$ is given by [267]

$$\frac{\partial f(\mathbf{K})}{\partial \mathbf{s}_p} = -\sum_{i=1}^{n} \sum_{j=1}^{n} \left(\mathbf{HLH}_{ij} \cdot \frac{\partial K_{ij}}{\partial \mathbf{s}_p} \right) + \lambda$$

$$= \sum_{i=1}^{n} \sum_{j=1}^{n} \left(((\mathbf{HLH})_{ij} \odot \mathbf{K})_{ij} (x_{ip} - x_{jp})^2 \right) \frac{2\mathbf{s}_p}{\sigma^2} + \lambda. \quad (6.7.25)$$

This problem can be solved by using spectral projected gradient (SPG), see Algorithm 6.9.

Algorithm 6.9 Spectral projected gradient (SPG) algorithm for nonlinear joint unsupervised feature selection [267]

1. **input:** $\mathbf{s}_0 = \mathbf{1}$.
2. **initialization:** Set step length α_0, step length bound $\alpha_{\min} = 10^{-10}$, $\alpha_{\max} = 10^{10}$, history $h = 10, t = 0$.
3. **while** not converged **do**
4. $\alpha_t = \min\{\alpha_{\max}, \max(\alpha_{\min}, \alpha_0)\}$;
5. $\mathbf{d}_t = \text{Proj}_{[0,1]}(\mathbf{s}_t - \alpha_t \Delta f(\mathbf{s}_t)) - \mathbf{s}_t$;
6.   bound $f_b = \max\{f(\mathbf{s}_t), f(\mathbf{s}_{t-1}), \ldots, f(\mathbf{s}_{t-h})\}$.
7.   set $\alpha = 1$;
8.   **if** $f(\mathbf{s}_t + \alpha \mathbf{d}_t) > f_b + \eta \alpha (\Delta f(\mathbf{s}_t))^T \mathbf{d}_t$ **then**
9.   Choose $\alpha \in (0, \alpha)$;
10.   **end if**
11. $\mathbf{s}_{t+1} = \mathbf{s}_t + \alpha \mathbf{d}_t$;
12. $\mathbf{u}_t = \mathbf{s}_{t+1} - \mathbf{s}_t$;
13. $\mathbf{y}_t = \Delta f(\mathbf{s}_{t+1}) - \Delta f(\mathbf{s}_t)$;
14. $\alpha_0 = \mathbf{y}_t^T \mathbf{y}_t / \mathbf{u}_t^T \mathbf{y}_t$;
15. $t = t + 1$;
16. **end while**
17. **output:** the selected features with corresponding entry in $\mathbf{s}$ equal to 1.

6.8 Principal Component Analysis

Principal component analysis (PCA) is a powerful data processing and dimension reduction technique for extracting structure from possibly high-dimensional data set [134]. PCA has several important variants or generalizations: robust principal component analysis, kernel principal component analysis, sparse principal component analysis, etc. The PCA technique is widely applied in engineering, biology, and social science, such as in face recognition, handwritten zip code classification, gene expression data analysis, etc.

6.8.1 Principal Component Analysis Basis

Given a high-dimensional data vector $\mathbf{x} = [x_1, \ldots, x_N]^T \in \mathbb{R}^N$, where N is large. We use an N-dimensional feature extraction vector $\mathbf{a}_i = [a_{i1}, \ldots, a_{iN}]^T$ to extract the ith feature of the data vector $\mathbf{x}$ as $\tilde{x}_i = \mathbf{a}_i^T \mathbf{x} = \sum_{j=1}^{N} a_{ij} x_j$. If we hope to use an $K \times N$ feature extract matrix $\mathbf{A} = [\mathbf{a}_1, \ldots, \mathbf{a}_K] \in \mathbb{R}^{N \times K}$ for extracting the total K features of $\mathbf{x}$, then the K-dimensional feature vector $\tilde{\mathbf{x}} \in \mathbb{R}^K$ is given by

$$\tilde{\mathbf{x}} = \mathbf{A}^T \mathbf{x} = \begin{bmatrix} \mathbf{a}_1^T \mathbf{x} \\ \vdots \\ \mathbf{a}_K^T \mathbf{x} \end{bmatrix} = [\tilde{x}_1, \ldots, \tilde{x}_K]^T \in \mathbb{R}^K. \tag{6.8.1}$$

For a data matrix $\mathbf{X} = [\mathbf{x}_1, \ldots, \mathbf{x}_P] \in \mathbb{R}^{N \times P}$, its feature matrix $\tilde{\mathbf{X}}$ is given by

$$\tilde{\mathbf{X}} = \mathbf{A}^T \mathbf{X} = \mathbf{A}^T [\mathbf{x}_1, \ldots, \mathbf{x}_P] = \begin{bmatrix} \mathbf{a}_1^T \mathbf{x}_1 & \cdots & \mathbf{a}_1^T \mathbf{x}_P \\ \vdots & \ddots & \vdots \\ \mathbf{a}_K^T \mathbf{x}_1 & \cdots & \mathbf{a}_K^T \mathbf{x}_P \end{bmatrix} = [\tilde{\mathbf{x}}_1, \ldots, \tilde{\mathbf{x}}_P] \in \mathbb{R}^{K \times P}, \tag{6.8.2}$$

where

$$\tilde{\mathbf{x}}_i = \mathbf{A}^T \mathbf{x}_i = [\tilde{x}_{i1}, \ldots, \tilde{x}_{iK}]^T \in \mathbb{R}^K, \quad i = 1, \ldots, P. \tag{6.8.3}$$

The feature vector $\tilde{\mathbf{x}}_i$, $i = 1, \ldots, P$ should satisfy the following requirements.

1. *Dimensionality reduction:* Due to correlation of data vectors, there is redundant information among the N components of each data vector $\mathbf{x}_i = [x_{i1}, \ldots, x_{iN}]^T$ for $i = 1, \ldots, P$. We hope to construct a new set of lower-dimensional feature vectors $\tilde{\mathbf{x}}_i = [\tilde{x}_{i1}, \ldots, \tilde{x}_{iK}]^T \in \mathbb{R}^K$, where $K \ll N$. Such a process is known as *feature extraction* or *dimension reduction*.
2. *Orthogonalization:* In order to avoid redundant information, P feature vectors should be orthogonal to each other, i.e., $\tilde{\mathbf{x}}_i^T \tilde{\mathbf{x}}_j = \delta_{ij}$ for all $i, j \in \{1, \ldots, P\}$.

3. *Power maximization:* Under the assumption of $P < N$, let $\sigma_1^2 \geq \cdots \geq \sigma_K^2$ be K leading eigenvalues of the $P \times P$ covariance matrix $\mathbf{X}^T\mathbf{X}$. Then we have

$$E_{\tilde{x}_i} = E\{|\tilde{x}_i|_2^2\} = \mathbf{v}_i^T(\mathbf{X}^T\mathbf{X})\mathbf{v}_i = \sigma_i^2,$$

where σ_i are the i leading singular values of $\mathbf{X} = \mathbf{U}\boldsymbol{\Sigma}\mathbf{V}^T$ or $\lambda_i = \sigma_i^2$ are the ith leading eigenvalues of $\mathbf{X}^T\mathbf{X} = \mathbf{V}\boldsymbol{\Lambda}\mathbf{V}^T$ with $\boldsymbol{\Sigma} = \mathbf{Diag}(\sigma_1, \ldots, \sigma_P)$ and $\boldsymbol{\Lambda} = \mathbf{Diag}(\lambda_1, \ldots, \lambda_P)$. Because the eigenvalues σ_i^2 are arranged in nondescending order and thus $E_{\tilde{x}_1} \geq E_{\tilde{x}_2} \geq \cdots \geq E_{\tilde{x}_P}$. Therefore, $\tilde{\mathbf{x}}_1$ is often referred to as the first principal component feature of $\mathbf{X}$, $\tilde{\mathbf{x}}_2$ as the second principal component feature of $\mathbf{X}$, and so on. Hence, we have

$$E\{\tilde{\mathbf{x}}_1\} + \cdots + E\{\tilde{\mathbf{x}}_P\} = \sigma_1^2 + \cdots + \sigma_K^2 \approx E\{|\mathbf{x}_1|_2^2\} + \cdots + E\{|\mathbf{x}_P|_2^2\}. \quad (6.8.4)$$

A direct choice satisfying the above three requirements is to take $\mathbf{A} = \mathbf{U}_1$ in (6.8.2), giving the principal component feature matrix

$$\tilde{\mathbf{X}}_1 = [\tilde{\mathbf{x}}_1, \ldots, \tilde{\mathbf{x}}_P] = \mathbf{U}_1^T\mathbf{X} = \boldsymbol{\Sigma}_1\mathbf{V}_1^T = \begin{bmatrix} \sigma_1\mathbf{v}_1^T \\ \vdots \\ \sigma_K\mathbf{v}_K^T \end{bmatrix} \in \mathbb{R}^{K \times P} \quad (6.8.5)$$

because

$$\mathbf{X} = \mathbf{U}\boldsymbol{\Sigma}\mathbf{V}^T = [\mathbf{U}_1, \mathbf{U}_2]\begin{bmatrix} \boldsymbol{\Sigma}_1 & \mathbf{O}_{K \times (P-K)} \\ \mathbf{O}_{(P-K) \times K} & \boldsymbol{\Sigma}_2 \end{bmatrix}\begin{bmatrix} \mathbf{V}_1^T \\ \mathbf{V}_2^T \end{bmatrix} = \mathbf{U}_1\boldsymbol{\Sigma}_1\mathbf{V}_1^T + \mathbf{U}_2\boldsymbol{\Sigma}_2\mathbf{V}_2^T$$

with $\mathbf{U}_1 = [\mathbf{u}_1, \ldots, \mathbf{u}_K] \in \mathbb{R}^{N \times K}$, $\mathbf{U}_2 = [\mathbf{u}_{K+1}, \ldots, \mathbf{u}_P] \in \mathbb{R}^{N \times (P-K)}$, $\mathbf{V}_1 = [\mathbf{v}_1, \ldots, \mathbf{v}_K] \in \mathbb{R}^{P \times K}$, $\mathbf{V}_2 = [\mathbf{v}_{K+1}, \ldots, \mathbf{v}_P] \in \mathbb{R}^{P \times (P-K)}$, while $\boldsymbol{\Sigma}_1 = \mathbf{Diag}(\sigma_1, \ldots, \sigma_K)$ and $\boldsymbol{\Sigma}_2 = \mathbf{Diag}(\sigma_{K+1}, \ldots, \sigma_P)$.

The feature extraction method with the choice $\mathbf{A}_1 = \mathbf{U}_1$ is known as the *principal component analysis* (PCA). PCA has the following important properties:

- Principal components sequentially capture the maximum variability among the columns of $\mathbf{X}$, thus guaranteeing minimal information loss;
- Principal components are uncorrelated, so we can talk about one principal component without referring to others.

6.8.2 Minor Component Analysis

Definition 6.13 (Minor Components [169, 282]) Let $\mathbf{X}^T\mathbf{X}$ be the covariance matrix of an $N \times P$ data matrix $\mathbf{X}$ with K principal eigenvalues and $P - K$ minor eigenvalues (i.e., small eigenvalues). The $P - K$ eigenvectors $\mathbf{v}_{K+1}, \ldots, \mathbf{v}_P$ corresponding to these minor eigenvalues are called the *minor components* of the data matrix $\mathbf{X}$.

Data or signal analysis based on the $(P - K)$ minor components is known as *minor component analysis* (MCA). Unlike the PCA, the choice for the MCA is to take $\mathbf{A} = \mathbf{V}_2$ in (6.8.2), so that the $(P - K) \times P$ minor component feature matrix is determined by

$$\tilde{\mathbf{X}}_2 = \mathbf{U}_2 \mathbf{X} = \mathbf{\Sigma}_2 \mathbf{V}_2^T = \begin{bmatrix} \sigma_{K+1} \mathbf{v}_{K+1}^T \\ \vdots \\ \sigma_P \mathbf{v}_P^T \end{bmatrix} \in \mathbb{R}^{(P-K) \times P}. \quad (6.8.6)$$

Interestingly, the *contribution ratio* of the ith principal singular vector $\mathbf{v}_i$ in PCA is $\sigma_i/(\sigma_1^2 + \cdots + \sigma_K^2)^{1/2}$, $i = 1, \ldots, K$, while this ratio of the jth minor singular vector $\mathbf{v}_j$ in MCA is $\sigma_j/(\sigma_{K+1}^2 + \cdots + \sigma_P^2)^{1/2}$, $j = K+1, \ldots, P$.

In signal/data processing, pattern recognition (such as face, fingerprints, irises recognition, etc.), image processing, principal components correspond to the key body of a signal or image background, since they describe the global energy of a signal or an image. In contrast, minor components correspond to details of the signal or image, which are usually dictated by the moving objects in the image background.

6.8.3 Principal Subspace Analysis

Principal subspace analysis (PSA) and *minor subspace analysis* (MSA), in which PCA and MCA are special cases, are important for many applications. The goal of PSA or MSA is to extract, from a stationary random process $\mathbf{x}(k) \in \mathbb{R}^N$ with covariance $\mathbf{C} = E\{\mathbf{x}(k)\mathbf{x}^T(k)\}$, the subspaces spanned by the m principal or minor eigenvectors, respectively [79].

For the matrix $\mathbf{X} \in \mathbb{R}^{N \times P}$ with K principal component vectors $\mathbf{V}_s = [\mathbf{v}_1, \ldots, \mathbf{v}_K]$, the *principal subspace* (also known as *signal subspace*) of $\mathbf{X}$ is defined as

$$\mathcal{S} = \mathrm{span}(\mathbf{V}_s) = \mathrm{span}\{\mathbf{v}_1, \ldots, \mathbf{v}_K\}, \quad (6.8.7)$$

and the orthogonal projector (i.e., *signal projection matrix*) is given by

$$\mathbf{P}_\mathcal{S} = \mathbf{V}_s (\mathbf{V}_s^T \mathbf{V}_s)^{-1} \mathbf{V}_s^T = \mathbf{V}_s \mathbf{V}_s^T, \quad (6.8.8)$$

where $\mathbf{V}_s = [\mathbf{v}_1, \ldots, \mathbf{v}_K]$ is a $P \times K$ semi-orthogonal matrix such that $\mathbf{V}_s^T \mathbf{V}_s = \mathbf{I}_{K \times K}$.

Similarly, the *minor subspace* (also known as *noise subspace*) of $\mathbf{X}$ is defined as

$$\mathcal{N} = \mathrm{span}(\mathbf{V}_n) = \mathrm{span}\{\mathbf{v}_{K+1}, \ldots, \mathbf{v}_P\}, \quad (6.8.9)$$

and the orthogonal projector (i.e., *noise projection matrix*) is given by

$$\mathbf{P}_\mathcal{N} = \mathbf{V}_n(\mathbf{V}_n^T\mathbf{V}_n)^{-1}\mathbf{V}_n^T = \mathbf{V}_n\mathbf{V}_n^T, \tag{6.8.10}$$

where $\mathbf{V}_n = [\mathbf{v}_{K+1}, \ldots, \mathbf{v}_P]$ is a $P \times (P - K)$ semi-orthogonal matrix such that $\mathbf{V}_n^T\mathbf{V}_n = \mathbf{I}_{(P-K)\times(P-K)}$.

It is noticed that the principal subspace and the minor subspace are orthogonal, i.e., $\mathcal{S} \perp \mathcal{N}$, because

$$\mathbf{P}_\mathcal{S}^T\mathbf{P}_\mathcal{N} = \mathbf{V}_s\mathbf{V}_s^T\mathbf{V}_n\mathbf{V}_n^T = \mathbf{O}_{P\times P} \quad \text{(null matrix)}.$$

Moreover, as

$$\mathbf{P}_\mathcal{S} + \mathbf{P}_\mathcal{N} = \mathbf{V}_s\mathbf{V}_s^T + \mathbf{V}_n\mathbf{V}_n^T = [\mathbf{V}_s, \mathbf{V}_n]\begin{bmatrix}\mathbf{V}_s^T\\\mathbf{V}_n^T\end{bmatrix} = \mathbf{V}_{P\times P}\mathbf{V}_{P\times P}^T = \mathbf{I}_{P\times P},$$

there is the following relationship between the principal and minor subspaces:

$$\mathbf{P}_\mathcal{S} = \mathbf{I} - \mathbf{P}_\mathcal{N} \quad \text{or} \quad \mathbf{V}_s\mathbf{V}_s = \mathbf{I} - \mathbf{V}_n\mathbf{V}_n^T. \tag{6.8.11}$$

Therefore, the PSA seeks to find the principal subspace $\mathbf{V}_s\mathbf{V}_s^T$ instead of the principal components $\mathbf{V}_s = [\mathbf{v}_1, \ldots, \mathbf{v}_K]$ in the PCA, while the MSA finds the minor subspace $\mathbf{V}_n\mathbf{V}_n^T$ instead of the minor components $\mathbf{V}_n = [\mathbf{v}_{K+1}, \ldots, \mathbf{v}_P]$ in the MCA.

In order to deduce the PSA and MSA updating algorithms, we consider the minimization of the objective function $J(\mathbf{W})$, where $\mathbf{W}$ is an $n \times r$ matrix. The common constraints on $\mathbf{W}$ are of two types, as follows.

- *Orthogonality constraint:* $\mathbf{W}$ is required to satisfy an orthogonality condition, either $\mathbf{W}^H\mathbf{W} = \mathbf{I}_r$ ($n \geq r$) or $\mathbf{W}\mathbf{W}^H = \mathbf{I}_n$ ($n < r$). The matrix $\mathbf{W}$ satisfying such a condition is called *semi-orthogonal matrix*.
- *Homogeneity constraint:* It is required that $J(\mathbf{W}) = J(\mathbf{WQ})$, where $\mathbf{Q}$ is an $r \times r$ orthogonal matrix.

For an unconstrained optimization problem $\min J(\mathbf{W}_{n\times r})$ with $n > r$, its solution is a single matrix $\mathbf{W}$. However, for an optimization problem subject to the semi-orthogonality constraint $\mathbf{W}^H\mathbf{W} = \mathbf{I}_r$ and the homogeneity constraint $J(\mathbf{W}) = J(\mathbf{WQ})$, where $\mathbf{Q}$ is an $r \times r$ orthogonal matrix, i.e.,

$$\min J(\mathbf{W}) \quad \text{subject to} \quad \mathbf{W}^H\mathbf{W} = \mathbf{I}_r, \ J(\mathbf{W}) = J(\mathbf{WQ}), \tag{6.8.12}$$

its solution is any member of the set of semi-orthogonal matrices defined by

$$Gr(n, r) = \left\{\mathbf{W} \in \mathbb{C}^{n\times r} \big| \mathbf{W}^H\mathbf{W} = \mathbf{I}_r, \ \mathbf{W}_1\mathbf{W}_1^H = \mathbf{W}_2\mathbf{W}_2^H\right\}. \tag{6.8.13}$$

6.8 Principal Component Analysis

This set is known as the *Grassmann manifold* of the semi-orthogonal matrices $\mathbf{W}_{n \times r}$ such that $\mathbf{W}^T\mathbf{W} = \mathbf{I}$ but $\mathbf{W}\mathbf{W}^T$ is an $n \times n$ singular matrix.

For two given semi-orthogonal matrices $\mathbf{W}_1$ and $\mathbf{W}_2$, their subspaces are called *equivalent subspaces* if $\mathbf{W}_1\mathbf{W}_1^T = \mathbf{W}_2\mathbf{W}_2^T$. The feature extraction methods based on equivalent subspaces are *subspace analysis methods*.

Therefore, the goals of the PSA and MSA are to find the equivalent principal and minor subspaces, respectively.

Subspace analysis methods have the following characteristics [287].

- The signal (or principal) subspace method and the noise (or minor) subspace method need only a few singular vectors or eigenvectors.
- In many applications, it is not necessary to know singular values or eigenvalues; it is sufficient to know the rank of the matrix and its singular vectors or eigenvectors.
- In most cases, it is not necessary to know the singular vectors or eigenvectors exactly; it is sufficient to know the basis vectors spanning the principal subspace or minor subspace.
- Conversion between the principal subspace $\mathbf{V}_s\mathbf{V}_s^H$ and the minor subspace $\mathbf{V}_n\mathbf{V}_n^H$ can be made by using $\mathbf{V}_n\mathbf{V}_n^H = \mathbf{I} - \mathbf{V}_s\mathbf{V}_s^H$.

On the more basic theory and applications of subspace analysis, the reader can refer to [294].

In the following we deduce the PSA algorithm. Let $\mathbf{C} = E\{\mathbf{x}\mathbf{x}^T\}$ denote the covariance matrix of an $N \times 1$ random vector $\mathbf{x} \in \mathbb{R}^N$, and consider the following constrained optimization problem:

$$\min_{\mathbf{W}} E\{\|\mathbf{x} - \mathbf{P}_W\mathbf{x}\|_2^2\} \quad \text{subject to} \quad \mathbf{W}^T\mathbf{W} = \mathbf{I}, \tag{6.8.14}$$

where $\mathbf{P}_W = \mathbf{W}(\mathbf{W}^T\mathbf{W})^{-1}\mathbf{W}^T$ is the projector of the weighting matrix $\mathbf{W}_{N \times P}$ with $N > P$, while $\mathbf{P}_W\mathbf{x}$ denotes the feature vector of $\mathbf{x}$ extracted by $\mathbf{W}$. Clearly, $\mathbf{x} - \mathbf{P}_W\mathbf{x}$ denotes the error between the original input data vector and its extracted feature vector, and thus (6.8.14) is a minimum squares error (MSE) criterion.

Interestingly, the constrained optimization problem (6.8.14) can be equivalently written as the following unconstrained optimization problem:

$$\min_{\mathbf{W}} \quad E\{\|\mathbf{x} - \mathbf{W}\mathbf{W}^T\mathbf{x}\|_2^2\} \tag{6.8.15}$$

whose objective function is given by [287]:

$$J(\mathbf{W}) = E\{\|\mathbf{x} - \mathbf{W}\mathbf{W}^T\mathbf{x}\|_2^2\}$$
$$= E\{(\mathbf{x} - \mathbf{W}\mathbf{W}^T\mathbf{x})^T(\mathbf{x} - \mathbf{W}\mathbf{W}^T\mathbf{x})\}$$
$$= E\{\mathbf{x}^T\mathbf{x}\} - 2E\{\mathbf{x}^T\mathbf{W}\mathbf{W}^T\mathbf{x}\} + E\{\mathbf{x}^T\mathbf{W}\mathbf{W}^T\mathbf{W}\mathbf{W}^T\mathbf{x}\}. \tag{6.8.16}$$

Because

$$E\{\mathbf{x}^T\mathbf{x}\} = \sum_{i=1}^{n} E\{|x_i|^2\} = \operatorname{tr}\left(E\{\mathbf{x}\mathbf{x}^T\}\right) = \operatorname{tr}(\mathbf{C}),$$

$$E\{\mathbf{x}^T\mathbf{W}\mathbf{W}^T\mathbf{x}\} = \operatorname{tr}\left(E\{\mathbf{W}^T\mathbf{x}\mathbf{x}^T\mathbf{W}\}\right) = \operatorname{tr}(\mathbf{W}^T\mathbf{C}\mathbf{W}),$$

$$E\{\mathbf{x}^T\mathbf{W}\mathbf{W}^T\mathbf{W}\mathbf{W}^T\mathbf{x}\} = \operatorname{tr}\left(E\{\mathbf{W}^T\mathbf{x}\mathbf{x}^T\mathbf{W}\mathbf{W}^T\mathbf{W}\}\right)$$
$$= \operatorname{tr}(\mathbf{W}^T\mathbf{C}\mathbf{W}\mathbf{W}^T\mathbf{W}),$$

the objective function in (6.8.16) can be represented as the trace form

$$J(\mathbf{W}) = \operatorname{tr}(\mathbf{C}) - 2\operatorname{tr}(\mathbf{W}^T\mathbf{C}\mathbf{W}) + \operatorname{tr}(\mathbf{W}^T\mathbf{C}\mathbf{W}\mathbf{W}^T\mathbf{W}), \tag{6.8.17}$$

where $\mathbf{W}$ is an $N \times P$ matrix whose rank is assumed to be K.

From Eq. (6.8.17) it can be seen that in the time-varying case the matrix differential of the objective function $J(\mathbf{W}(t))$ is given by

$$\mathrm{d}J(\mathbf{W}(t)) = -2\operatorname{tr}\left(\mathbf{W}^T(t)\mathbf{C}(t)\mathrm{d}\mathbf{W}(t) + (\mathbf{C}(t)\mathbf{W}(t))^T\mathrm{d}\mathbf{W}^*(t)\right)$$
$$+ \operatorname{tr}\Big((\mathbf{W}^T(t)\mathbf{W}(t)\mathbf{W}^T(t)\mathbf{C}(t) + \mathbf{W}(t)\mathbf{C}(t)\mathbf{W}(t)\mathbf{W}^T(t))\mathrm{d}\mathbf{W}(t)$$
$$+ (\mathbf{C}(t)\mathbf{W}(t)\mathbf{W}^T(t)\mathbf{W}(t) + \mathbf{W}(t)\mathbf{W}^T(t)\mathbf{C}(t)\mathbf{W}(t))^T \mathrm{d}\mathbf{W}^*(t)\Big),$$

which yields the gradient matrix

$$\nabla_\mathbf{W} J(\mathbf{W}(t)) = -2\mathbf{C}(t)\mathbf{W}(t) + \mathbf{C}(t)\mathbf{W}(t)\mathbf{W}^T(t)\mathbf{W}(t) + \mathbf{W}(t)\mathbf{W}^T(t)\mathbf{C}(t)\mathbf{W}(t)$$
$$= \mathbf{W}(t)\mathbf{W}^T(t)\mathbf{C}(t)\mathbf{W}(t) - \mathbf{C}(t)\mathbf{W}(t),$$

where the semi-orthogonality constraint $\mathbf{W}^T(t)\mathbf{W}(t) = \mathbf{I}$ has been used.

Substituting $\mathbf{C}(t) = \mathbf{x}(t)\mathbf{x}^T(t)$ into the above gradient matrix formula, one obtains a gradient descent algorithm for solving the minimization problem $\mathbf{W}(t) = \mathbf{W}(t-1) - \mu \nabla_\mathbf{W} J(\mathbf{W}(t))$, as follows:

$$\mathbf{y}(t) = \mathbf{W}^T(t)\mathbf{x}(t), \tag{6.8.18}$$

$$\mathbf{W}(t+1) = \mathbf{W}(t) + \beta(t)\big(\mathbf{y}^T(t)\mathbf{x}(t) - \mathbf{W}(t)\mathbf{y}(t)\mathbf{y}^T(t)\big). \tag{6.8.19}$$

Here, $\beta(t) > 0$ is a learning parameter. This algorithm is called the *projection approximation subspace tracking* (PAST) that was developed by Yang in 1995 [287] from the viewpoint of subspace.

On the PAST algorithm, the following two theorems are true.

6.8 Principal Component Analysis

Theorem 6.1 ([287]) *The matrix $\mathbf{W}$ is a stationary point of $J(\mathbf{W})$ if and only if $\mathbf{W} = \mathbf{V}_r \mathbf{Q}$, where $\mathbf{V}_r \in \mathbb{R}^{P \times r}$ consists of r eigenvectors of the covariance matrix $\mathbf{C} = \mathbf{V}_r \mathbf{D}_r \mathbf{V}_r^T$ and $\mathbf{Q} \in \mathbb{R}^{r \times r}$ is any orthogonal matrix. At every stationary point, the value of the objective function $J(\mathbf{W})$ is equal to the sum of the eigenvalues whose corresponding eigenvectors are not in $\mathbf{V}_r$.*

Theorem 6.2 ([287]) *All its stationary points are saddle points of the objective function $J(\mathbf{W})$ unless $r = K$ (rank of $\mathbf{C}$), i.e., $\mathbf{V}_r = \mathbf{V}_s$ consists of K principal eigenvectors of the covariance matrix $\mathbf{C}$. In this particular case $J(\mathbf{W})$ reaches a global minimum.*

Theorems 6.1 and 6.2 show the following facts.

1. In the unconstrained minimization of the objective function $J(\mathbf{W})$ in (6.8.17), it is not required that the columns of $\mathbf{W}$ are orthogonal, but the two theorems show that minimization of the objective function $J(\mathbf{W})$ in (6.8.17) will automatically result in a semi-orthogonal matrix $\mathbf{W}$ such that $\mathbf{W}^T\mathbf{W} = \mathbf{I}$.
2. Theorem 6.2 shows that when the column space of $\mathbf{W}$ is equal to the signal or principal subspace: $\text{Col}(\mathbf{W}) = \text{Span}(\mathbf{V}_s)$, the objective function $J(\mathbf{W})$ reaches a global minimum, and it has no other local minimum.
3. From the definition formula (6.8.17) it is easy to see that $J(\mathbf{W}) = J(\mathbf{W}\mathbf{Q})$ holds for all $K \times K$ unitary matrices $\mathbf{Q}$, i.e., the objective function automatically satisfies the homogeneity constraint.
4. Because the objective function defined by (6.8.17) automatically satisfies the homogeneity constraint, and its minimization automatically has the result that $\mathbf{W}$ satisfies the orthogonality constraint $\mathbf{W}^H\mathbf{W} = \mathbf{I}$, the solution $\mathbf{W}$ is not uniquely determined but is a point on the Grassmann manifold. This greatly relaxes the solution for the optimization problem (6.8.17).
5. The projection matrix $\mathbf{P} = \mathbf{W}(\mathbf{W}^T\mathbf{W})^{-1}\mathbf{W}^T = \mathbf{W}\mathbf{W}^T = \mathbf{V}_s\mathbf{V}_s^T$ is uniquely determined. That is to say, the different solutions $\mathbf{W}$ span the same column space.
6. When $r = 1$, i.e., the objective function is a real function of a vector $\mathbf{w}$, the solution $\mathbf{w}$ of the minimization of $J(\mathbf{w})$ is the eigenvector corresponding to the largest eigenvalue of the covariance matrix $\mathbf{C}$.

Therefore, Eq. (6.8.19) gives the global solution to the optimization problem (6.8.16). It is worth emphasizing that (6.8.19) is just the PSA rule of Oja [194].

On the other hand, the two MSA algorithms are [49]

$$\mathbf{W}(t+1) = \mathbf{W}(t) + \beta(t)\left(\mathbf{W}(t)\mathbf{W}^T(t)\mathbf{y}(t)\mathbf{x}(t) - \mathbf{y}(t)\mathbf{y}^T(t)\mathbf{W}(t)\right), \quad (6.8.20)$$

and [79]

$$\mathbf{W}(t+1) = \mathbf{W}(t) - \beta(t)\Big(\mathbf{W}(t)\mathbf{W}^T(t)\mathbf{W}(t)\mathbf{W}^T(t)\mathbf{y}(t)\mathbf{x}^T(t) \\ - \mathbf{y}(t)\mathbf{y}^T(t)\mathbf{W}(t)\Big). \quad (6.8.21)$$

The Douglas algorithm (6.8.21) is a self-stabilizing in that the vectors do not need to be periodically normalized to unit modulus.

6.8.4 Robust Principal Component Analysis

Given an observation or data matrix $\mathbf{D} = \mathbf{A} + \mathbf{E}$, where $\mathbf{A}$ and $\mathbf{E}$ are unknown, but $\mathbf{A}$ is known to be a low- rank true data matrix and $\mathbf{E}$ is known to be an additive sparse noise matrix, the aim is to recover $\mathbf{A}$. Because the low-rank matrix $\mathbf{A}$ can be regarded as the principal component of the data matrix $\mathbf{D}$, and the sparse matrix $\mathbf{E}$ may have a few gross errors or outlying observations, the SVD-based principal component analysis (PCA) usually breaks down. Hence, this problem is a *robust principal component analysis* problem [164, 272].

So-called robust PCA is able to correctly recover underlying low-rank structure in the data matrix, even in the presence of gross errors or outlying observations. This problem is also called the *principal component pursuit* (PCP) problem [46] since it tracks down the principal components of the data matrix $\mathbf{D}$ by minimizing a combination of the nuclear norm $\|\mathbf{A}\|_*$ and the weighted ℓ_1-norm $\mu\|\mathbf{E}\|_1$:

$$\min_{\mathbf{A},\mathbf{E}} \left\{ \|\mathbf{A}\|_* + \mu \|\mathbf{E}\|_1 \right\} \quad \text{subject to} \quad \mathbf{D} = \mathbf{A} + \mathbf{E}. \tag{6.8.22}$$

Here, $\|\mathbf{A}\|_* = \sum_i^{\min\{m,n\}} \sigma_i(\mathbf{A})$ represents the nuclear norm of $\mathbf{A}$, i.e., the sum of all singular values, which reflects the cost of the low-rank matrix $\mathbf{A}$, while $\|\mathbf{E}\|_1 = \sum_{i=1}^m \sum_{j=1}^n |E_{ij}|$ is the sum of the absolute values of all entries of the additive error matrix $\mathbf{E}$, where some of its entries may be arbitrarily large. The role of the constant $\mu > 0$ is to balance the contradictory requirements of low rank and sparsity.

By using the minimization in (6.8.22), an initially unknown low-rank matrix $\mathbf{A}$ and unknown sparse matrix $\mathbf{E}$ can be recovered from the data matrix $\mathbf{D} \in \mathbb{R}^{m \times n}$. That is to say, the unconstrained minimization of the robust PCA or the PCP problem can be expressed as

$$\min_{\mathbf{A},\mathbf{E}} \left\{ \|\mathbf{A}\|_* + \mu \|\mathbf{E}\|_1 + \frac{1}{2}(\|\mathbf{A} + \mathbf{E} - \mathbf{D})\|_F^2 \right\}. \tag{6.8.23}$$

It can be shown [44] that, under rather weak assumptions, the PCP estimate can exactly recover the low-rank matrix $\mathbf{A}$ from the data matrix $\mathbf{D} = \mathbf{A} + \mathbf{E}$ with gross but sparse error matrix $\mathbf{E}$.

In order to solve the robust PCA or the PCP problem (6.8.23), consider a more general family of optimization problems of the form

$$F(\mathbf{X}) = f(\mathbf{X}) + \mu h(\mathbf{X}), \tag{6.8.24}$$

6.8 Principal Component Analysis

where $f(\mathbf{X})$ is a convex, smooth (i.e., differentiable), and L-Lipschitz function and $h(\mathbf{X})$ is a convex but nonsmooth function (such as $\|\mathbf{X}\|_1$, $\|\mathbf{X}\|_*$, and so on).

Instead of directly minimizing the composite function $F(\mathbf{X})$, we minimize its separable quadratic approximation $Q(\mathbf{X}, \mathbf{Y})$, formed at specially chosen points $\mathbf{Y}$:

$$Q(\mathbf{X}, \mathbf{Y}) = f(\mathbf{Y}) + \langle \nabla f(\mathbf{Y}), \mathbf{X} - \mathbf{Y} \rangle + \frac{1}{2}(\mathbf{X} - \mathbf{Y})^T \nabla^2 f(\mathbf{Y})(\mathbf{X} - \mathbf{Y}) + \mu h(\mathbf{X})$$

$$= f(\mathbf{Y}) + \langle \nabla f(\mathbf{Y}), \mathbf{X} - \mathbf{Y} \rangle + \frac{L}{2}\|\mathbf{X} - \mathbf{Y}\|_F^2 + \mu h(\mathbf{X}), \qquad (6.8.25)$$

where $\nabla^2 f(\mathbf{Y})$ is approximated by $L\mathbf{I}$.

When minimizing $Q(\mathbf{X}, \mathbf{Y})$ with respect to $\mathbf{X}$, the function term $f(\mathbf{Y})$ may be regarded as a constant term that is negligible. Hence, we have

$$\mathbf{X}_{k+1} = \arg\min_{\mathbf{X}} \left\{ \mu h(\mathbf{X}) + \frac{L}{2}\left\|\mathbf{X} - \mathbf{Y}_k + \frac{1}{L}\nabla f(\mathbf{Y}_k)\right\|_F^2 \right\}$$

$$= \mathbf{prox}_{\mu L^{-1} h}\left(\mathbf{Y}_k - \frac{1}{L}\nabla f(\mathbf{Y}_k)\right). \qquad (6.8.26)$$

If we let $f(\mathbf{Y}) = \frac{1}{2}\|\mathbf{A} + \mathbf{E} - \mathbf{D}\|_F^2$ and $h(\mathbf{X}) = \|\mathbf{A}\|_* + \lambda\|\mathbf{E}\|_1$, then the Lipschitz constant $L = 2$ and $\nabla f(\mathbf{Y}) = \mathbf{A} + \mathbf{E} - \mathbf{D}$. Hence

$$Q(\mathbf{X}, \mathbf{Y}) = \mu h(\mathbf{X}) + f(\mathbf{Y}) = (\mu\|\mathbf{A}\|_* + \mu\lambda\|\mathbf{E}\|_1) + \frac{1}{2}\|\mathbf{A} + \mathbf{E} - \mathbf{D}\|_F^2$$

reduces to the quadratic approximation of the robust PCA problem.

By Theorem 4.5, we have

$$\mathbf{A}_{k+1} = \mathbf{prox}_{\mu/2\|\cdot\|_*}\left(\mathbf{Y}_k^A - \frac{1}{2}(\mathbf{A}_k + \mathbf{E}_k - \mathbf{D})\right) = \mathbf{U}\text{soft}(\mathbf{\Sigma}, \lambda)\mathbf{V}^T,$$

$$\mathbf{E}_{k+1} = \mathbf{prox}_{\mu\lambda/2\|\cdot\|_1}\left(\mathbf{Y}_k^E - \frac{1}{2}(\mathbf{A}_k + \mathbf{E}_k - \mathbf{D})\right) = \text{soft}\left(\mathbf{W}_k^E, \frac{\mu\lambda}{2}\right),$$

where $\mathbf{U}\mathbf{\Sigma}\mathbf{V}^T$ is the SVD of $\mathbf{W}_k^A$ and

$$\mathbf{W}_k^A = \mathbf{Y}_k^A - \frac{1}{2}(\mathbf{A}_k + \mathbf{E}_k - \mathbf{D}), \qquad (6.8.27)$$

$$\mathbf{W}_k^E = \mathbf{Y}_k^E - \frac{1}{2}(\mathbf{A}_k + \mathbf{E}_k - \mathbf{D}). \qquad (6.8.28)$$

The robust PCA shown in Algorithm 6.10 has the following convergence.

Algorithm 6.10 Robust PCA via accelerated proximal gradient [164, 272]

1. **input:** Data matrix $\mathbf{D} \in \mathbb{R}^{m \times n}$, λ, allowed tolerance ϵ.
2. **initialization:** $\mathbf{A}_0, \mathbf{A}_{-1} \leftarrow \mathbf{O}$; $\mathbf{E}_0, \mathbf{E}_{-1} \leftarrow \mathbf{O}$; $t_0, t_{-1} \leftarrow 1$; $\bar{\mu} \leftarrow \delta \mu_0$.
3. **repeat**
4. $\quad \mathbf{Y}_k^A \leftarrow \mathbf{A}_k + \dfrac{t_{k-1} - 1}{t_k} (\mathbf{A}_k - \mathbf{A}_{k-1})$.
5. $\quad \mathbf{Y}_k^E \leftarrow \mathbf{E}_k + \dfrac{t_{k-1} - 1}{t_k} (\mathbf{E}_k - \mathbf{E}_{k-1})$.
6. $\quad \mathbf{W}_k^A \leftarrow \mathbf{Y}_k^A - \dfrac{1}{2}(\mathbf{A}_k + \mathbf{E}_k - \mathbf{D})$.
7. $\quad (\mathbf{U}, \boldsymbol{\Sigma}, \mathbf{V}) = \text{svd}(\mathbf{W}_k^A)$.
8. $\quad r = \max \left\{ j : \sigma_j > \dfrac{\mu_k}{2} \right\}$.
9. $\quad \mathbf{A}_{k+1} = \sum_{i=1}^{r} \left(\sigma_i - \dfrac{\mu_k}{2} \right) \mathbf{u}_i \mathbf{v}_i^T$.
10. $\quad \mathbf{W}_k^E \leftarrow \mathbf{Y}_k^E - \dfrac{1}{2}(\mathbf{A}_k + \mathbf{E}_k - \mathbf{D})$.
11. $\quad \mathbf{E}_{k+1} = \text{soft}\left(\mathbf{W}_k^E, \dfrac{\lambda \mu_k}{2} \right)$.
12. $\quad t_{k+1} \leftarrow \dfrac{1 + \sqrt{4 t_k^2 + 1}}{2}$.
13. $\quad \mu_{k+1} \leftarrow \max(\eta \mu_k, \bar{\mu})$.
14. $\quad \mathbf{S}_{k+1}^A = 2(\mathbf{Y}_k^A - \mathbf{A}_{k+1}) + (\mathbf{A}_{k+1} + \mathbf{E}_{k+1} - \mathbf{Y}_k^A - \mathbf{Y}_k^E)$.
15. $\quad \mathbf{S}_{k+1}^E = 2(\mathbf{Y}_k^E - \mathbf{E}_{k+1}) + (\mathbf{A}_{k+1} + \mathbf{E}_{k+1} - \mathbf{Y}_k^A - \mathbf{Y}_k^E)$.
16. $\quad$ **exit if** $\|\mathbf{S}_{k+1}\|_F^2 = \|\mathbf{S}_{k+1}^A\|_F^2 + \|\mathbf{S}_{k+1}^E\|_F^2 \leq \epsilon$.
17. **return** $k \leftarrow k + 1$.
18. **output:** $\mathbf{A} \leftarrow \mathbf{A}_k$, $\mathbf{E} \leftarrow \mathbf{E}_k$.

Theorem 6.3 ([164]) *Let* $F(\mathbf{X}) = F(\mathbf{A}, \mathbf{E}) = \mu \|\mathbf{A}\|_* + \mu \lambda \|\mathbf{E}\|_1 + \frac{1}{2} \|\mathbf{A} + \mathbf{E} - \mathbf{D}\|_F^2$. *Then, for all* $k > k_0 = C_1 / \log(1/n)$ *with* $C_1 = \log(\mu_0 / \mu)$, *one has*

$$F(\mathbf{X}) - F(\mathbf{X}^*) \leq \frac{4 \|\mathbf{X}_{k_0} - \mathbf{X}^*\|_F^2}{(k - k_0 + 1)^2}, \qquad (6.8.29)$$

where $\mathbf{X}^*$ *is a solution to the robust PCA problem* $\min_{\mathbf{X}} F(\mathbf{X})$.

6.8.5 Sparse Principal Component Analysis

Consider a *linear regression* problem: given an observed response vector $\mathbf{y} = [y_1, \ldots, y_n]^T \in \mathbb{R}^n$ and an observed data matrix (or predictors) $\mathbf{X} = [\mathbf{x}_1, \ldots, \mathbf{x}_p] \in \mathbb{R}^{n \times p}$, find a fitting coefficient vector $\boldsymbol{\beta} = [\beta_1, \ldots, \beta_p]^T \in \mathbb{R}^p$ such that

$$\hat{\boldsymbol{\beta}} = \arg\min_{\boldsymbol{\beta}} \|\mathbf{y} - \mathbf{X}\boldsymbol{\beta}\|_2^2 = \left\| \mathbf{y} - \sum_{i=1}^{p} \mathbf{x}_i \beta_i \right\|_2^2, \qquad (6.8.30)$$

6.8 Principal Component Analysis

where **y** and $\mathbf{x}_i$, $i = 1, \ldots, p$ are assumed to be centered, respectively.

Minimizing the above squared error usually leads to sensitive solutions. Many regularization methods have been proposed to decrease this sensitivity. Among them, Tikhonov regularization [247] and Lasso [84, 244] are two widely known and cited algorithms, as pointed out in [283].

The Lasso (*least absolute shrinkage and selection operator*) is a penalized least squares method:

$$\hat{\boldsymbol{\beta}}_{\text{Lasso}} = \arg\min_{\boldsymbol{\beta}} \left\| \mathbf{y} - \sum_{i=1}^{p} \mathbf{x}_i \beta_i \right\|_2^2 + \lambda \|\boldsymbol{\beta}\|_1, \tag{6.8.31}$$

where $\|\boldsymbol{\beta}\|_1 = \sum_{i=1}^{p} |\beta_i|$ is the ℓ_1-norm and λ is nonnegative.

The Lasso can provide a sparse prediction coefficient solution vector due to the nature of the ℓ_1-norm penalty through continuously shrinking the prediction coefficients toward zero, and achieving its prediction accuracy via the bias variance trade-off.

Although the Lasso has shown success in many situations, it has some limitations [307].

- If $p > n$, then the Lasso selects at most n variables before it saturates due to the nature of the convex optimization problem. This seems to be a limiting feature for a variable selection method. Moreover, the Lasso is not well defined unless the bound on the ℓ_1-norm of $\boldsymbol{\beta}$ is smaller than a certain value.
- If there is a group of variables among which the pairwise correlations are very high, then the Lasso tends to select only one variable from the group and does not care which one is selected.
- For usual $n > p$ cases, if there are high correlations between predictors $\mathbf{x}_i$, it has been empirically observed that the prediction performance of the Lasso is dominated by ridge regression [244].

To overcome these drawbacks, an *elastic net* was proposed by Zou and Hastie [307] through generalizing the Lasso:

$$\hat{\boldsymbol{\beta}}_{\text{en}} = (1 + \beta_2) \left(\arg\min_{\boldsymbol{\beta}} \left\| \mathbf{y} - \sum_{i=1}^{p} \mathbf{x}_i \beta_i \right\|_2^2 + \lambda_1 \sum_{i=1}^{p} |\beta_i| + \lambda_2 \sum_{i=1}^{p} |\lambda_i|^2 \right)$$
(6.8.32)

for any nonnegative λ_1 and λ_2. The elastic net penalty is a convex combination of the ridge and Lasso penalties. If $\lambda_2 = 0$, then the elastic net reduces to the Lasso.

Theorem 6.4 ([308]) *Let* $\mathbf{X} = \mathbf{U}\mathbf{D}\mathbf{V}^T$ *be the rank-K SVD of* $\mathbf{X}$. *For each* i, *denote by* $\mathbf{z}_i = d_{ii}\mathbf{u}_i$ *the* ith *principal component. Consider a positive* λ *and the ridge* *estimates* $\hat{\boldsymbol{\beta}}_{\text{ridge}}$ *given by*

$$\hat{\boldsymbol{\beta}}_{\text{ridge}} = \arg\min_{\boldsymbol{\beta}} \|\mathbf{z}_i - \mathbf{X}\boldsymbol{\beta}\|_2^2 + \lambda \|\boldsymbol{\beta}\|_2^2. \tag{6.8.33}$$

Letting $\hat{\mathbf{v}} = \dfrac{\hat{\boldsymbol{\beta}}_{\text{ridge}}}{\|\hat{\boldsymbol{\beta}}_{\text{ridge}}\|}$, *then* $\hat{\mathbf{v}} = \mathbf{v}_i$.

This theorem establishes the connection between PCA and a regression method.

Theorem 6.5 ([308]) *Let* $\mathbf{A}_{p\times k} = [\boldsymbol{\alpha}_1, \ldots, \boldsymbol{\alpha}_k]$ *and* $\mathbf{B}_{p\times k} = [\boldsymbol{\beta}_1, \ldots, \boldsymbol{\beta}_k]$. *For any* $\lambda > 0$, *consider the constrained optimization problem*

$$(\hat{\mathbf{A}}, \hat{\mathbf{B}}) = \arg\min_{\mathbf{A}, \mathbf{B}} \left\|\mathbf{X} - \mathbf{X}\mathbf{B}\mathbf{A}^T\right\|_F^2 + \lambda \sum_{j=1}^k \|\boldsymbol{\beta}_j\|_2^2 \tag{6.8.34}$$

subject to $\mathbf{A}^T\mathbf{A} = \mathbf{I}_{k\times k}$. *Then,* $\hat{\boldsymbol{\beta}}_j \propto \mathbf{v}_j$ *for* $j = 1, \ldots, k$.

If adding the Lasso penalty, then the criterion (6.8.34) becomes the following optimization problem:

$$(\hat{\mathbf{A}}, \hat{\mathbf{B}}) = \arg\min_{\mathbf{A}, \mathbf{B}} \left\|\mathbf{X} - \mathbf{X}\mathbf{A}\mathbf{B}^T\right\|_F^2 + \lambda \sum_{j=1}^k \|\boldsymbol{\beta}_j\|_2^2 + \sum_{j=1}^k \lambda_{1,j}\|\boldsymbol{\beta}_j\|_1 \tag{6.8.35}$$

subject to $\mathbf{A}^T\mathbf{A} = \mathbf{I}_{k\times k}$. Due to $\|\mathbf{A}\|_F^2 = \|\mathbf{A}^T\mathbf{A}\|_2 = 1$, one has

$$\|\mathbf{X} - \mathbf{X}\mathbf{B}\mathbf{A}^T\|_F^2 = \|\mathbf{X}\mathbf{A} - \mathbf{X}\mathbf{B}\|_F^2 = (\mathbf{A} - \mathbf{B})^T\mathbf{X}^T\mathbf{X}(\mathbf{A} - \mathbf{B}).$$

Substitute this result into (6.8.35) to get

$$(\hat{\boldsymbol{\alpha}}_j, \hat{\boldsymbol{\beta}}_j) = \arg\min_{\boldsymbol{\alpha}_j, \boldsymbol{\beta}_j}(\boldsymbol{\alpha}_j - \boldsymbol{\beta}_j)^T\mathbf{X}^T\mathbf{X}(\boldsymbol{\alpha}_j - \boldsymbol{\beta}_j) + \lambda\|\boldsymbol{\beta}_j\|_2^2 + \lambda_{1,j}\|\boldsymbol{\beta}_j\|_1. \tag{6.8.36}$$

This coupled optimization problem can be solved by using the well-known alternating method.

By summarizing the above discussion, the *sparse principal component analysis (SPCA) algorithm* developed by Zou et al. [308] can be described by Algorithm 6.11.

Algorithm 6.11 Sparse principal component analysis (SPCA) algorithm [308]
1. **input:** The data matrix $\mathbf{X} \in \mathbb{R}^{n \times p}$ and the response vector $\mathbf{y} \in \mathbb{R}^n$.
2. **initialization:**
    2.1 Make the truncated SVD $\mathbf{X} = \sum_{i=1}^{k} d_i \mathbf{u}_i \mathbf{v}_i^T$.
    2.2 Let $\boldsymbol{\alpha}_j = \mathbf{v}_j$, $j = 1, \ldots, k$.
    2.3 Select $\lambda > 0$ and $\lambda_{1,i} > 0$, $i = 1, \ldots, k$.
3. Given a fixed $\mathbf{A} = [\boldsymbol{\alpha}_1, \ldots, \boldsymbol{\alpha}_k]$, solve the following elastic net problem for $j = 1, \ldots, k$:
    $$\boldsymbol{\beta}_j = \arg\min_{\boldsymbol{\beta}} \ (\boldsymbol{\alpha}_j - \boldsymbol{\beta})^T \mathbf{X}^T \mathbf{X}(\boldsymbol{\alpha}_j - \boldsymbol{\beta}) + \lambda \|\boldsymbol{\beta}\|_2^2 + \lambda_{1,j} \|\boldsymbol{\beta}\|_1.$$
4. For a fixed $\mathbf{B} = [\boldsymbol{\beta}_1, \ldots, \boldsymbol{\beta}_k]$, compute the SVD of $\mathbf{X}^T \mathbf{X} \mathbf{B} = \mathbf{U} \mathbf{D} \mathbf{V}^T$, then update $\mathbf{A} = \mathbf{U} \mathbf{V}^T$.
5. Repeat Steps 3-4, until convergence.
6. Normalization: $\boldsymbol{\beta}_j \leftarrow \frac{\boldsymbol{\beta}_j}{\|\boldsymbol{\beta}_j\|_2}$ for $j = 1, \ldots, k$.
7. **output:** normalized fitting coefficient vectors $\boldsymbol{\beta}_j$, $j = 1, \ldots, k$.

6.9 Supervised Learning Regression

Supervised learning in the form of regression (for continuous outputs) and classification (for discrete outputs) is an important field of statistics and machine learning.

The aim of machine learning is to establish a mapping relationship between inputs and outputs via learning the correspondence between the sample and the label. As the name implies, supervised learning is based on a supervisor (data labeling expert): a type of machine learning of input (or training) data with labels.

In supervised learning, we are given a "training" set of input vectors $\{\mathbf{x}_n\}_{n=1}^{N}$ along with corresponding targets $\{y_n\}_{n=1}^{N}$. The data set $\{\mathbf{x}_n, y_n\}_{n=1}^{N}$ is known as the labeled data set.

Let the domain of input instances be X, and the domain of labels be Y, and let $P(x, y)$ be an (unknown) joint probability distribution on instances and labels $X \times Y$. Given an independent identically distributed (i.i.d.) training data $\{\mathbf{x}_i\}_{i=1}^{N}$ sampled from $P(\mathbf{x})$, denoted $\mathbf{x}_i \overset{\text{i.i.d.}}{\sim} P(\mathbf{x})$, supervised learning trains a function $f : X \to Y$ in some function family F in order to make $f(\mathbf{x})$ predict the true label y on future data $\mathbf{x}$, where $\mathbf{x} \overset{\text{i.i.d.}}{\sim} P(\mathbf{x})$ as well.

Depending on the target y, supervised learning problems are further divided into two of the most common supervised learning forms: classification and regression.

- *Classification* is the supervised learning problem with discrete class labels y, for example, $y = \{+1, -1\}$ for two classes. The function f is called a classifier.
- *Regression* is the supervised learning problem with real values y. The function f is called a regression function.

From this training set we wish to learn a model of the dependency of the targets on the inputs with the objective of making accurate predictions of y for previously unseen values of $\mathbf{x}$.

Goals in model selection include [116]:

- accurate predictions,
- *interpretable models:* determining which predictors are meaningful,

- *stability:* small changes in the data should not result in large changes in either the subset of predictors used, the associated coefficients, or the predictions,
- avoiding bias in hypothesis tests during or after variable selection.

6.9.1 Principle Component Regression

A task to approximate real-valued functions from a training sample is variously referred to as prediction, regression, interpolation, or function approximation.

In statistical modeling, regression analysis is a set of statistical processes for estimating the relationships between a dependent variable (called also "criterion variable") X and one or more independent variables (or predictors) Y. More specifically, regression analysis helps us understand how the typical value of the dependent variable changes when any one of the independent variables is varied, while the other independent variables are held fixed.

Most commonly, regression analysis estimates the conditional expectation of the dependent variable given the independent variables.

In all cases, a function of the independent variables called the regression function is to be estimated.

Given a training set $X = \{\mathbf{x}_1, \ldots, \mathbf{x}_n\}$ with $\mathbf{x}_i = [x_{i1}, \ldots, x_{ip}]^T$ for all $i \in [n] \stackrel{\text{def}}{=} \{1, \cdots, n\}$. Denote $\mathbf{X} = [\mathbf{x}_1, \ldots, \mathbf{x}_n]$, and consider the supervised learning problem: to learn a function $f(\mathbf{X}) : \mathbf{X} \to \mathbf{y}$ so that $f(\mathbf{X})$ is a good predictor for the corresponding value of $\mathbf{y}$, namely minimize $\frac{1}{2}\|\mathbf{y} - f(\mathbf{X})\|_2^2$. For historical reasons, this function f is called a *hypothesis*.

The regression models involve the following parameters and variables:

- The *unknown parameters*, denoted as $\boldsymbol{\beta} = [\beta_1, \ldots, \beta_p]^T$, which is usually a vector.
- The independent variable matrix $\mathbf{X} = [x_{ij}]_{i=1, j=2}^{n, p}$.
- The dependent variable vector, denoted as $\mathbf{y} = [y_1, \ldots, y_n]^T$.

A regression model uses the function of $\mathbf{X}$ and $\boldsymbol{\beta}$

$$\mathbf{y} \approx f(\mathbf{X}, \boldsymbol{\beta}) \tag{6.9.1}$$

as an approximation of $E(\mathbf{y}|\mathbf{X}) = f(\mathbf{X}, \boldsymbol{\beta})$.

Standard Linear Regression Method

Consider the more general multiple regression

$$y_i = \beta_1 x_{i1} + \beta_2 x_{i2} + \cdots + \beta_p x_{ip} + \varepsilon_i. \tag{6.9.2}$$

As data pre-processing, it is usually assumed that $\mathbf{y}$ and each of the p columns of $\mathbf{X}$ have already been centered so that all of them have zero empirical means.

6.9 Supervised Learning Regression

After centering, the standard Gauss–Markov linear regression model for $\mathbf{y}$ on $\mathbf{X}$ can be represented as:

$$\mathbf{y} = \mathbf{X}\boldsymbol{\beta} + \boldsymbol{\varepsilon}, \tag{6.9.3}$$

where $\boldsymbol{\beta} \in \mathbb{R}^p$ denotes the unknown parameter vector of regression coefficients and $\boldsymbol{\varepsilon}$ denotes the vector of random errors with $E(\boldsymbol{\varepsilon}) = \mathbf{0}$ and $\text{var}(\boldsymbol{\varepsilon}) = \sigma^2 \mathbf{I}_{n \times n}$ for some unknown variance parameter $\sigma^2 > 0$.

The LS estimate of regression parameter vector $\boldsymbol{\beta}$ is given by

$$\hat{\boldsymbol{\beta}}_{\text{LS}} = (\mathbf{X}^T \mathbf{X})^{-1} \mathbf{X}^T \mathbf{y} \tag{6.9.4}$$

for the data matrix $\mathbf{X}$ with full column rank.

Put the regularized cost function

$$L(\boldsymbol{\beta}) = \frac{1}{2} \|\mathbf{y} - \mathbf{X}\boldsymbol{\beta}\|_2^2 + \lambda \|\boldsymbol{\beta}\|_2^2, \tag{6.9.5}$$

then the Tikhonov regularization LS solution is given by

$$\hat{\boldsymbol{\beta}}_{\text{Tik}} = (\mathbf{X}^H \mathbf{X} + \lambda \mathbf{I})^{-1} \mathbf{X}^H \mathbf{y} \tag{6.9.6}$$

for the data matrix $\mathbf{X}$ with deficit column rank.

The total least squares (TLS) estimate of regression parameter vector $\boldsymbol{\beta}$ is given by

$$\hat{\boldsymbol{\beta}}_{\text{TLS}} = (\mathbf{X}^H \mathbf{X} - \lambda \mathbf{I})^{-1} \mathbf{X}^H \mathbf{y} \tag{6.9.7}$$

for the data matrix $\mathbf{X}$ with full column rank and observation errors.

Principle Component Regression Method

Principle component regression (PCR) [175] is another technique for estimating $\boldsymbol{\beta}$, based on principle component analysis (PCA).

Let $\mathbf{R} = \mathbf{X}^T \mathbf{X}$ be the sample autocorrelation matrix and its SVD be $\mathbf{R} = \mathbf{V}\boldsymbol{\Lambda}\mathbf{V}^T$, where $\mathbf{V} = [\mathbf{v}_1, \ldots, \mathbf{v}_p]$ is the principal loading matrix. Then, $\mathbf{X}\mathbf{V} = [\mathbf{X}\mathbf{v}_1, \ldots, \mathbf{X}\mathbf{v}_p]$ is the principal components of $\mathbf{X}$.

The idea of PCR is simple [175]: use principal components $\mathbf{W} = \mathbf{X}\mathbf{V}$ of $\mathbf{X}$ instead of the original data matrix $\mathbf{X}$ in standard linear regression $\mathbf{y} = \mathbf{X}\boldsymbol{\beta}$ to form the regression equation:

$$\mathbf{y} = \mathbf{W}\boldsymbol{\gamma} = \mathbf{X}\mathbf{V}\boldsymbol{\gamma}. \tag{6.9.8}$$

Comparing the PCR $\mathbf{y} = \mathbf{X}\mathbf{V}\boldsymbol{\gamma}$ with the linear regression $\mathbf{y} = \mathbf{X}\boldsymbol{\beta}$, we immediately have $\boldsymbol{\beta} = \mathbf{V}\boldsymbol{\gamma}$.

Then, the procedure of PCR consists of the following steps.

- *Principal Components Extraction:* PCR starts by performing an SVD on the autocorrelation matrix of centered data matrix $\mathbf{X}$: $\mathbf{R} = \mathbf{X}^T \mathbf{X} = \mathbf{V}\boldsymbol{\Lambda}\mathbf{V}^T$, where

$\mathbf{\Lambda}_{p \times p} = \mathbf{Diag}\left(\lambda_1^2, \ldots, \lambda_p^2\right)$ with $\lambda_1 \geq \cdots \geq \lambda_p \geq 0$ denoting the nonnegative eigenvalues (also called the principal eigenvalues) of $\mathbf{X}^T\mathbf{X}$. Then, $\mathbf{v}_j$ denotes the jth principal component direction (or known as PCA loading) corresponding to the jth largest principal value λ_j for each $j \in \{1, \ldots, p\}$. Hence, $\mathbf{V}$ is called the PCA loading matrix of $\mathbf{X}^T\mathbf{X}$.

- *PCR model:* For any $k \in \{1, \ldots, p\}$, let $\mathbf{V}_k$ denote the $p \times k$ matrix with orthonormal columns consisting of the first k columns of $\mathbf{V}$. Let $\mathbf{W}_k = \mathbf{X}\mathbf{V}_k = [\mathbf{X}\mathbf{v}_1, \ldots, \mathbf{X}\mathbf{v}_k]$ denote the $n \times k$ matrix having the first k principal components as its columns. Then, $\mathbf{W}$ may be viewed as the data matrix obtained by using the transformed covariants $\mathbf{x}_i^k = \mathbf{V}_k^T\mathbf{x}_i \in \mathbb{R}^k$ instead of using the original covariants $\mathbf{x}_i \in \mathbb{R}^p$, $\forall i = 1, \ldots, n$. Hence, if the standard Gauss–Markov linear regression model $\mathbf{y} = \mathbf{X}\boldsymbol{\beta}$ is viewed as a full component regression, then by using the principal component matrix $\mathbf{W}_k = \mathbf{X}\mathbf{V}_k$ instead of the original data matrix $\mathbf{X}$, we have the PCR model

$$\mathbf{y}_k = \mathbf{W}_k \boldsymbol{\gamma}_k. \tag{6.9.9}$$

- *PCR Estimator $\hat{\boldsymbol{\gamma}}_k$:* The solution of the above PCR equation is given by $\hat{\boldsymbol{\gamma}}_k = \left(\mathbf{W}_k^T\mathbf{W}_k\right)^{-1}\mathbf{W}_k^T\mathbf{y}_k \in \mathbb{R}^k$, which denotes the vector of estimated regression coefficients obtained by ordinary least squares regression of the response vector $\mathbf{y}$ on the data matrix $\mathbf{W}_k$.
- *PCR Estimator $\hat{\boldsymbol{\beta}}_k$:* Substitute $\mathbf{W}_k = \mathbf{X}\mathbf{V}_k$ into Eq. (6.9.9) to get $\mathbf{y}_k = \mathbf{X}\mathbf{V}_k\boldsymbol{\gamma}_k$. By comparing $\mathbf{y}_k = \mathbf{X}\mathbf{V}_k\boldsymbol{\gamma}_k$ with $\mathbf{y}_k = \mathbf{X}\boldsymbol{\beta}_k$, it is easily known that for any $k \in \{1, \ldots, p\}$, the final PCR estimator of $\boldsymbol{\beta}$ based on using the first k principal components is given by $\hat{\boldsymbol{\beta}}_k = \mathbf{V}_k\hat{\boldsymbol{\gamma}}_k \in \mathbb{R}^p$, $k \in \{1, \ldots, p\}$.

Algorithm 6.12 shows the principle component regression algorithm.

Algorithm 6.12 Principal component regression algorithm [175]

1. **input:**
   1.1 The data matrix $\mathbf{X} \in \mathbb{R}^{n \times p}$ of observed covariants;
   1.2 The vector $\mathbf{y} \in \mathbb{R}^n$ of observed outcomes, where $n \geq p$.
2. **initialization:** $x_{ij} \leftarrow x_{ij} - \bar{x}_j$ with $\bar{x}_j = \frac{1}{n}\sum_{i=1}^n x_{ij}$, $j = 1, \ldots, p$, and $y_i \leftarrow y_i - \bar{y}$, $\bar{y} = \frac{1}{n}\sum_{i=1}^n y_i$.
3. Compute the SVD: $\mathbf{X} = \mathbf{U}\boldsymbol{\Sigma}\mathbf{V}^T$, and save the singular values $\sigma_1, \ldots, \sigma_p$ and the right singular-vector matrix $\mathbf{V}$.
4. Determine k largest principal eigenvalues $\lambda_i = \sigma_i^2$, $i = 1, \ldots, k$ with $k \in \{1, \ldots, p\}$, and denote $\mathbf{V}_k = [\mathbf{v}_1, \ldots, \mathbf{v}_k]$.
5. Construct the transformed data matrix $\mathbf{W}_k = [\mathbf{X}\mathbf{v}_1, \ldots, \mathbf{X}\mathbf{v}_k]$, $k \in \{1, \ldots, p\}$.
6. Compute the PCR estimate $\hat{\boldsymbol{\gamma}}_k = \left(\mathbf{W}_k^T\mathbf{W}_k\right)^{-1}\mathbf{W}_k^T\mathbf{y} \in \mathbb{R}^k$, $k \in \{1, \ldots, p\}$.
7. The regression parameter estimate based on k principal components is given by
   $\hat{\boldsymbol{\beta}}_k = \mathbf{V}_k\hat{\boldsymbol{\gamma}}_k \in \mathbb{R}^p$, $k \in \{1, \ldots, p\}$.
8. **output:** $\hat{\boldsymbol{\beta}}_k$, $k \in \{1, \ldots, p\}$.

6.9.2 Partial Least Squares Regression

Let $\mathcal{X} \subset \mathbb{R}^N$ be an N-dimensional space of variables representing the first block and $\mathcal{Y} \subset \mathbb{R}^M$ be a space representing the second block of variables. The general setting of a linear *partial least squares* (PLS) algorithm is to model the relations between these two data sets (blocks of variables) by means of score vectors.

The PLS model can be considered as consisting of *outer relations* ($\mathcal{X}$- and $\mathcal{Y}$-blocks individually) and an *inner relation* (linking both blocks).

Let p denote the number of components in model, N be the number of objects. The main symbols in PLS model are given as follows:

X : $N \times K$ matrix of predictor variables,
Y : $N \times M$ matrix of response variables,
E : $N \times K$ residual matrix for **X**,
F* : $N \times M$ residual matrix for **Y**,
B : $K \times M$ matrix of PLS regression coefficients,
W : $K \times p$ matrix of the PLS weights for **X**,
C : $M \times p$ matrix of the PLS weights for **Y**,
P : $K \times p$ matrix of the PLS loadings for **X**,
Q : $M \times p$ matrix of the PLS loadings for **Y**,
T : $N \times p$ matrix of the PLS scores for **X**,
U : $N \times p$ matrix of the PLS scores for **Y**.

The outer relations for the $\mathcal{X}$- and $\mathcal{Y}$-blocks are given by [105]

$$\mathbf{X} = \mathbf{TP}^T + \mathbf{E} = \sum_{h=1}^{p} \mathbf{t}_h \mathbf{p}_h^T + \mathbf{E}, \qquad (6.9.10)$$

$$\mathbf{Y} = \mathbf{UQ}^T + \mathbf{F}^* = \sum_{h=1}^{p} \mathbf{u}_h \mathbf{q}_h^T + \mathbf{F}^*, \qquad (6.9.11)$$

where $\mathbf{T} = [\mathbf{t}_1, \ldots, \mathbf{t}_p] \in \mathbb{R}^{K \times p}$ and $\mathbf{U} = [\mathbf{u}_1, \ldots, \mathbf{u}_p] \in \mathbb{R}^{M \times p}$ with p score vectors (components, latent vectors) extracted from **X** and **Y**, respectively; the $K \times p$ matrix $\mathbf{P} = [\mathbf{p}_1, \ldots, \mathbf{p}_p]$ and the $M \times p$ matrix $\mathbf{Q} = [\mathbf{q}_1, \ldots, \mathbf{q}_p]$ represent matrices of loadings, and both $\mathbf{E} \in \mathbb{R}^{N \times K}$ and $\mathbf{F}^* \in \mathbb{R}^{N \times M}$ are the matrices of residuals.

Let $\mathbf{t} = \mathbf{Xw}$ be the $\mathcal{X}$-space score vector (simply $\mathcal{X}$-score) and $\mathbf{u} = \mathbf{Yc}$ be the $\mathcal{Y}$-space score vector (simply $\mathcal{Y}$-score). For the user of PLS, it is necessary to know the following main properties of the PLS factors [105, 121].

1. The PLS weight vectors $\mathbf{w}_i$ for **X** are mutually orthogonal:

$$\langle \mathbf{w}_i, \mathbf{w}_j \rangle = \mathbf{w}_i^T \mathbf{w}_j = 0, \text{ for } i \neq j. \qquad (6.9.12)$$

2. The PLS score vectors $\mathbf{t}_i$ are mutually orthogonal:

$$\langle \mathbf{t}_i, \mathbf{t}_j \rangle = \mathbf{t}_i^T \mathbf{t}_j = 0, \text{ for } i \neq j. \quad (6.9.13)$$

3. The PLS weight vectors $\mathbf{w}_i$ are orthogonal to the PLS loading vectors $\mathbf{p}_j$ for $\mathbf{X}$:

$$\langle \mathbf{w}_i, \mathbf{p}_j \rangle = \mathbf{w}_i^T \mathbf{p}_j = 0, \quad \text{for } i < j. \quad (6.9.14)$$

4. The PLS loading vectors $\mathbf{p}_i$ are orthogonal in the kernel space of $\mathbf{X}$:

$$\langle \mathbf{p}_i, \mathbf{p}_j \rangle_X = \mathbf{p}_i^T (\mathbf{X}^T \mathbf{X})^\dagger \mathbf{p}_j = 0, \quad \text{for } i \neq j. \quad (6.9.15)$$

5. Both the PLS loading vectors $\mathbf{p}_i$ for $\mathbf{X}$ and the PLS loading vectors $\mathbf{q}_i$ for $\mathbf{Y}$ have unit length for each i: $\|\mathbf{p}_i\| = \|\mathbf{q}_i\| = 1$.
6. Both the PLS score vector $\mathbf{t}_h$ for $\mathbf{X}$ and the PLS score vector $\mathbf{u}_h$ for $\mathbf{Y}$ have the zero means for each h: $\sum_{i=1}^{K} t_{hi} = 0$ and $\sum_{i=1}^{M} u_{hi} = 0$.

The classical form of RLS is to find weight vectors $\mathbf{w}$ and $\mathbf{c}$ such that the inner relation linking $\mathcal{X}$-block and $\mathcal{Y}$-block satisfies [121]:

$$[\text{cov}(\mathbf{t}, \mathbf{u})]^2 = [\text{cov}(\mathbf{Xw}, \mathbf{Yc})]^2 = \max_{\|\mathbf{r}\|=\|\mathbf{s}\|=1} [\text{cov}(\mathbf{Xr}, \mathbf{Ys})]^2, \quad (6.9.16)$$

where $\text{cov}(\mathbf{t}, \mathbf{u}) = \mathbf{t}^T \mathbf{u}/N$ denotes the sample covariance between the score vectors $\mathbf{t}$ and $\mathbf{u}$. This classical RLS algorithm is called the *nonlinear iterative partial least squares* (NIPALS) *algorithm*, proposed by Wold in 1975 [274], as shown in Algorithm 6.13.

Algorithm 6.13 Nonlinear iterative partial least squares (NIPALS) algorithm [274]

1. **input:** Data blocks $\mathbf{X} = [\mathbf{x}_1, \ldots, \mathbf{x}_K]$ and $\mathbf{Y} = [\mathbf{y}_1, \ldots, \mathbf{y}_M]$.
2. **initialization:** randomly generate $\mathcal{Y}$-space score vector $\mathbf{u}$.
3. **repeat**
    // $\mathcal{X}$-space score vector update:
4. $\quad \mathbf{w} = \mathbf{X}^T \mathbf{u}/(\mathbf{u}^T \mathbf{u})$,
5. $\quad \|\mathbf{w}\| \to 1$,
6. $\quad \mathbf{t} = \mathbf{Xw}$.
    // $\mathcal{Y}$-space score vector update:
7. $\quad \mathbf{c} = \mathbf{Y}^T \mathbf{t}/(\mathbf{t}^T \mathbf{t})$,
8. $\quad \|\mathbf{c}\| \to 1$,
9. $\quad \mathbf{u} = \mathbf{Yc}$.
10. **exit if** both $\mathbf{t}$ and $\mathbf{u}$ converge.
11. **return**
12. **output:** $\mathcal{X}$-space score vector $\mathbf{t}$ and $\mathcal{Y}$-space score vector $\mathbf{u}$.

Here are the characteristics of PLS [105].

- There are outer relations of the form $\mathbf{X} = \mathbf{TP}^T + \mathbf{E}$ and $\mathbf{Y} = \mathbf{UQ}^T + \mathbf{F}^*$.

6.9 Supervised Learning Regression

- There is an inner relation $\mathbf{u}_h = b_h \mathbf{t}_h$.
- The mixed relation is $\mathbf{Y} = \mathbf{TBQ}^T + \mathbf{F}^*$, where $\|\mathbf{F}^*\|$ is to be minimized.
- In the iterative algorithm, the blocks get each other's scores, this gives a better inner relation.
- In order to obtain orthogonal $\mathcal{X}$-scores, as in the PCA, it is necessary to introduce weights.

There are the interesting connections between PLS, principal component analysis (PCA), and canonical correlation analysis (CCA) as follows [215].

1. The optimization criterion of PCA can be written as

$$\max_{\|\mathbf{r}\|=1} \{\mathrm{var}(\mathbf{Xr})\}, \qquad (6.9.17)$$

where $\mathrm{var}(\mathbf{t}) = \mathbf{t}^T\mathbf{t}/N$ denotes the sample variance.

2. CCA finds the direction of maximal correlation by solving the optimization problem

$$\max_{\|\mathbf{r}\|=\|\mathbf{s}\|=1} \left\{[\mathrm{corr}(\mathbf{Xr}, \mathbf{Ys})]^2\right\}, \qquad (6.9.18)$$

where $[\mathrm{corr}(\mathbf{t}, \mathbf{u})]^2 = [\mathrm{cov}(\mathbf{t}, \mathbf{u})]^2/(\mathrm{var}(\mathbf{t})\mathrm{var}(\mathbf{u}))$ denotes the sample squared correlation.

3. The PLS optimization criterion (6.9.18) can be rewritten as

$$\max_{\|\mathbf{r}\|=\|\mathbf{s}\|=1} \left\{[\mathrm{cov}(\mathbf{Xr}, \mathbf{Ys})]^2\right\} = \max_{\|\mathbf{r}\|=\|\mathbf{s}\|=1} \left\{\mathrm{var}(\mathbf{Xr})[\mathrm{corr}(\mathbf{Xr}, \mathbf{Ys})]^2 \mathrm{var}(\mathbf{Ys})\right\}. \qquad (6.9.19)$$

This represents that PLS is a form of CCA where the criterion of maximal correlation is balanced with the requirement to explain as much variance as possible in both $\mathcal{X}$- and $\mathcal{Y}$-spaces. Note that only the $\mathcal{X}$-space variance is involved in the case of a one-dimensional $\mathcal{Y}$-space.

The regression based on PLS-model equations (6.9.10) and (6.9.11) is called *partial least squares* (PLS) *regression*. To combine (6.9.10) and (6.9.11) into the standard model of PLS regression, two assumptions are made:

- the score vectors $\{\mathbf{t}_i\}_{i=1}^p$ are good predictors of $\mathbf{Y}$, where p denotes the number of extracted $\mathcal{X}$-score vectors;
- a linear inner relation between the score's vectors $\mathbf{t}$ and $\mathbf{u}$ exists; that is, $\mathbf{u}_i = \alpha_i \mathbf{t}_i + \mathbf{h}_i$, where $\mathbf{h}_i$ denotes a residual vector. Therefore, $\mathbf{U} = [\mathbf{u}_1, \ldots, \mathbf{u}_p]$ and the $K \times M$ matrix of PLS regression coefficients $\mathbf{T} = [\mathbf{t}_1, \ldots, \mathbf{t}_p]$ are related by

$$\mathbf{U} = \mathbf{TD} + \mathbf{H}, \qquad (6.9.20)$$

where $\mathbf{D} = \mathbf{Diag}(\alpha_1, \ldots, \alpha_p)$ is $p \times p$ diagonal matrix and $\mathbf{H} = [\mathbf{h}_1, \ldots, \mathbf{h}_p]$ is the matrix of residuals.

Letting $\mathbf{W}$ be a weighting matrix such that $\mathbf{EW} = \mathbf{O}$ (zero matrix), and post-multiplying (6.9.10) by $\mathbf{W}$, then

$$\mathbf{XW} = \mathbf{TP}^T\mathbf{W} \quad \Rightarrow \quad \mathbf{T} = \mathbf{XW}(\mathbf{P}^T\mathbf{W})^{-1}. \tag{6.9.21}$$

On the other hand, substitute (6.9.20) into (6.9.11) to give

$$\mathbf{Y} = \mathbf{TDQ}^T + \mathbf{HQ}^T + \mathbf{F}^* \quad \Rightarrow \quad \mathbf{Y} = \mathbf{TC}^T + \mathbf{F}, \tag{6.9.22}$$

where $\mathbf{C} = \mathbf{QD}$ denotes the $N \times M$ matrix of regression coefficients and $\mathbf{F} = \mathbf{HQ}^T + \mathbf{F}^*$ is the residual matrix. Then, substituting (6.9.21) into (6.9.22), we have $\mathbf{Y} = \mathbf{XW}(\mathbf{P}^T\mathbf{W})^{-1}\mathbf{C}^T + \mathbf{F}^*$ or can write the PLS regression model as

$$\mathbf{Y} = \mathbf{XB}_{\text{PLS}} + \mathbf{F} \quad \text{and} \quad \mathbf{B}_{\text{PLS}} = \mathbf{W}(\mathbf{P}^T\mathbf{W})^{-1}\mathbf{C}^T \tag{6.9.23}$$

represents the matrix of PLS regression coefficients.

In NIPALS algorithm, $\mathbf{w} = \mathbf{X}^T\mathbf{u}/(\mathbf{u}^T\mathbf{u})$ and $\mathbf{c} = \mathbf{Y}^T\mathbf{t}/(\mathbf{t}^T\mathbf{t})$ are generated, which imply that

$$\mathbf{W} = \mathbf{X}^T\mathbf{U} \quad \text{and} \quad \mathbf{C} = \mathbf{Y}^T\mathbf{T}. \tag{6.9.24}$$

Moreover, pre-multiplying (6.9.21) by $\mathbf{T}^T$ and using $\mathbf{T}^T\mathbf{T} = \mathbf{I}$, then

$$(\mathbf{P}^T\mathbf{W})^{-1} = (\mathbf{T}^T\mathbf{XX}^T\mathbf{U})^{-1}. \tag{6.9.25}$$

Substitute (6.9.24) and (6.9.25) into (6.9.23) to give

$$\mathbf{B}_{\text{PLS}} = \mathbf{X}^T\mathbf{U}(\mathbf{T}^T\mathbf{XX}^T\mathbf{U})^{-1}\mathbf{T}^T\mathbf{Y}. \tag{6.9.26}$$

The NIPALS regression is an iterative process; i.e., after extraction of one component $\mathbf{t}_1$ the algorithm starts again using the deflated matrices $\mathbf{X}$ and $\mathbf{Y}$ to extract the next component $\mathbf{t}_2$. This process is repeated until the deflated matrix $\mathbf{X}$ is a null matrix.

Algorithm 6.14 shows a simple nonlinear iterative partial least squares (NIPALS) regression algorithm developed by Wold et al. [275].

The basic idea of the NIPALS regression can be summarized below.

- Find the first $\mathcal{X}$-score $\mathbf{t}_1 = \mathbf{Xw}$ and the first $\mathcal{Y}$-score $\mathbf{u}_1 = \mathbf{Yc}/(\mathbf{c}^T\mathbf{c})$ from the original data blocks $\mathbf{X}$ and $\mathbf{Y}$.
- Use contractive mapping to construct the residual matrices $\mathbf{X} = \mathbf{X} - \mathbf{tp}^T$ and $\mathbf{Y} = \mathbf{Y} - b\mathbf{tc}^T$.

6.9 Supervised Learning Regression

Algorithm 6.14 Simple nonlinear iterative partial least squares regression algorithm [275]

1. **input:** Optionally transformed, scaled, and centered data, $\mathbf{X} = \mathbf{X}_0$ and $\mathbf{Y} = \mathbf{Y}_0$.
2. **initialization:** A starting vector of $\mathbf{u}$, usually one of columns of $\mathbf{Y}$. With a single $\mathbf{y}$, take $\mathbf{u} = \mathbf{y}$. Put $k = 0$
3. **repeat**
4. Compute the $\mathcal{X}$-weights: $\mathbf{w} = \mathbf{X}^T \mathbf{u}/(\mathbf{u}^T \mathbf{u})$.
5. Calculate the $\mathcal{X}$-scores $\mathbf{t} = \mathbf{Xw}$.
6. Calculate the $\mathcal{Y}$-weights $\mathbf{c} = \mathbf{Y}^T \mathbf{t}/(\mathbf{t}^T \mathbf{t})$.
7. Update a set of $\mathcal{Y}$-scores $\mathbf{u}$ as $\mathbf{u} = \mathbf{Yc}/(\mathbf{c}^T \mathbf{c})$.
8. **exit if** the convergence $\|\mathbf{t}_{\text{old}} - \mathbf{t}_{\text{new}}\|/\|\mathbf{t}_{\text{new}}\| < \varepsilon$ has been reached, where ε is "small" (e.g., 10^{-6} or 10^{-8}).
9. **return**
10. Remove (deflate, peel off) the present component from $\mathbf{X}$ and $\mathbf{Y}$, and use these deflated matrices as $\mathbf{X}$ and $\mathbf{Y}$ in the next component. Here the deflation of $\mathbf{Y}$ is optional; the results are equivalent whether $\mathbf{Y}$ is deflated or not.
11. Compute $\mathcal{X}$-loadings: $\mathbf{p} = \mathbf{X}^T \mathbf{t}/(\mathbf{t}^T \mathbf{t})$.
12. Compute $\mathcal{Y}$-loadings: $\mathbf{q} = \mathbf{Y}^T \mathbf{u}/(\mathbf{u}^T \mathbf{u})$.
13. Regression ($\mathbf{u}$ on $\mathbf{t}$): $b = \mathbf{u}^T \mathbf{t}/(\mathbf{t}^T \mathbf{t})$.
14. Deflate $\mathbf{X}, \mathbf{Y}$ matrices: $\mathbf{X} \leftarrow \mathbf{X} - \mathbf{tp}^T$ and $\mathbf{Y} \leftarrow \mathbf{Y} - b\mathbf{tc}^T$.
15. $k \leftarrow k + 1$.
16. $\mathbf{u}_k = \mathbf{u}$ and $\mathbf{t}_k = \mathbf{t}$.
17. The next set of iterations starts with the new $\mathbf{X}$ and $\mathbf{Y}$ matrices as the residual matrices from the previous iteration. The iterations continue until a stopping criteria is used or $\mathbf{X}$ becomes the zero matrix.
18. $\mathbf{U} = [\mathbf{u}_1, \ldots, \mathbf{u}_p]$ and $\mathbf{T} = [\mathbf{t}_1, \ldots, \mathbf{t}_p]$.
19. $\mathbf{B}_{\text{PLS}} = \mathbf{X}^T \mathbf{U}(\mathbf{T}^T \mathbf{X}\mathbf{X}^T \mathbf{U})^{-1} \mathbf{T}^T \mathbf{Y}$.
20. **output:** the matrix of PLS regression coefficients $\mathbf{B}_{\text{PLS}}$.

- Find the second $\mathcal{X}$-score $\mathbf{t}_2$ and $\mathcal{Y}$-score $\mathbf{u}_2$ from the residual matrices $\mathbf{X} = \mathbf{X} - \mathbf{tp}^T$ and $\mathbf{Y} = \mathbf{Y} - b\mathbf{tc}^T$.
- Construct the new residual matrices $\mathbf{X}$ and $\mathbf{Y}$ and find the third $\mathcal{X}$- and $\mathcal{Y}$-scores $\mathbf{t}_3$ and $\mathbf{u}_3$. Repeat this process until the residual matrix $\mathbf{X}$ becomes a null matrix.
- Use (6.9.26) to compute the matrix of regression coefficients $\mathbf{B}$, where $\mathbf{U} = [\mathbf{u}_1, \ldots, \mathbf{u}_p]$ and $\mathbf{T} = [\mathbf{t}_1, \ldots, \mathbf{t}_p]$ consist of p $\mathcal{X}$- and $\mathcal{Y}$-scores, respectively.

Once the matrix of PLS regression coefficients $\mathbf{B}_{\text{PLS}}$ is found by using Algorithm 6.14, for a given new data block $\mathbf{X}_{\text{new}}$, then unknown $\mathcal{Y}$-values can be predicted by Eq. (6.9.23) as

$$\hat{\mathbf{Y}}_{\text{new}} = \mathbf{X}_{\text{new}} \mathbf{B}_{\text{PLS}}. \tag{6.9.27}$$

6.9.3 Penalized Regression

For linear regression problems $\mathbf{y} = \mathbf{X}\boldsymbol{\beta} \in \mathbb{R}^n$ with $\mathbf{X} \in \mathbb{R}^{n \times p}$, $\boldsymbol{\beta} \in \mathbb{R}^p$, ordinary least squares (OLS) regression minimizes squared regression error $\|\mathbf{y} - \mathbf{X}\boldsymbol{\beta}\|_2^2 =$

$(\mathbf{y} - \mathbf{X}\boldsymbol{\beta})^T(\mathbf{y} - \mathbf{X}\boldsymbol{\beta})$, and yields a $p \times 1$ unbiased estimator $\hat{\boldsymbol{\beta}}^{\text{OLS}} = (\mathbf{X}^T\mathbf{X})^{-1}\mathbf{X}^T\mathbf{y}$. Although the OLS estimator is simple and unbiased, if the design matrix $\mathbf{X}$ is not of full-rank, then $\hat{\boldsymbol{\beta}}$ is not unique and its variance $\text{var}(\hat{\boldsymbol{\beta}}) = (\mathbf{X}^T\mathbf{X})^{-1}\sigma^2$ (σ^2 is the variance of i.i.d. regression error) may be very large.

To achieve better prediction, Hoerl and Kennard [117, 118] introduced *ridge regression*

$$\hat{\boldsymbol{\beta}} = \arg\max_{\boldsymbol{\beta}} \left\{ \frac{1}{2}\|\mathbf{y} - \mathbf{X}\boldsymbol{\beta}\|_2^2 \right\} \quad \text{subject to} \quad \sum_{j=1}^{p} |\beta_j|^\gamma \leq t, \tag{6.9.28}$$

where $\gamma \geq 1$ and $t \geq 0$. The equivalent form of ridge regression is the following *penalized regression* [101]:

$$\hat{\boldsymbol{\beta}} = \arg\max_{\boldsymbol{\beta}} \left\{ \frac{1}{2}\|\mathbf{y} - \mathbf{X}\boldsymbol{\beta}\|_2^2 + \lambda \sum_{j=1}^{p} |\beta_j|^\gamma \right\}, \tag{6.9.29}$$

where $\gamma \geq 1$ and $\lambda \geq 0$ are a positive scalar; $\lambda = 0$ corresponds to ordinary least squares regression.

When γ takes different values, the penalized regression has different forms. The most well-known penalized regressions are ℓ_2 penalized regression with $\gamma = 2$ (often called the ridge regression):

$$\hat{\boldsymbol{\beta}}^{\text{ridge}} = \arg\min_{\boldsymbol{\beta}} \left\{ \frac{1}{2}\|\mathbf{y} - \mathbf{X}\boldsymbol{\beta}\|_2^2 + \lambda\|\boldsymbol{\beta}\|_2^2 \right\}, \tag{6.9.30}$$

and the ℓ_1 penalized regression with $\gamma = 1$, called *least absolute shrinkage and selection operator* (Lasso), given by

$$\hat{\boldsymbol{\beta}}^{\text{Lasso}} = \arg\min_{\boldsymbol{\beta}} \left\{ \frac{1}{2}\|\mathbf{y} - \mathbf{X}\boldsymbol{\beta}\|_2^2 + \lambda\|\boldsymbol{\beta}\|_1 \right\}, \tag{6.9.31}$$

where $\|\boldsymbol{\beta}\|_1 = \sum_{i=1}^{p} |\beta_i|$ is ℓ_1-norm.

The solution of the ridge regression is given by

$$\hat{\boldsymbol{\beta}}^{\text{ridge}} = (\mathbf{X}^T\mathbf{X} + \lambda\mathbf{I})^{-1}\mathbf{X}^T\mathbf{y} \tag{6.9.32}$$

that is just the Tikhonov regularized solution [246, 247].

6.9 Supervised Learning Regression

The Lasso solution is, in a soft-thresholded version [77], given by [99]

$$\hat{\beta}_i^{\text{Lasso}}(\gamma) = S(\hat{\beta}_i, \gamma) = \text{sign}(\hat{\beta}_i)(|\hat{\beta}_i| - \gamma)_+ \quad (6.9.33)$$

$$= \begin{cases} \hat{\beta}_i - \gamma, & \text{if } \hat{\beta}_i > 0 \text{ and } \gamma < |\hat{\beta}_i|; \\ \hat{\beta}_i + \gamma, & \text{if } \hat{\beta}_i < 0 \text{ and } \gamma < |\hat{\beta}_i|; \\ 0, & \text{if } \gamma \geq |\hat{\beta}_i|; \end{cases} \quad (6.9.34)$$

with

$$x_+ = \begin{cases} x, & \text{if } x > 0, \\ 0, & \text{if } x \leq 0. \end{cases} \quad (6.9.35)$$

Important differences between the ridge regression and the Lasso regression are twofold.

- The ℓ_1 penalty in Lasso results in variable selection, as variables with coefficients of zero are effectively omitted from the model, but the ℓ_2 penalty in ridge regression does not perform variable selection.
- An ℓ_2 penalty $\lambda \sum_{j=1}^{n} \beta_j^2$ pushes β_j toward zero with a force proportional to the value of the coefficient, whereas an ℓ_1 penalty $\lambda \sum_{j=1}^{n} |\beta_j|$ exerts the same force on all nonzero coefficients. Hence for variables that are most valuable in the model and where shrinkage toward zero is less desirable, an ℓ_1 penalty shrinks less [116].

Typically, the optimization problems with inequality constraints are carried out using a standard quadratic programming algorithm. An alternative is to explore "one-at-a-time" *coordinate-wise descent algorithms* for these problems [99].

For large optimization problems, coordinate descent can deliver a path of solutions efficiently, and can be applied to many other convex statistical problems such as the large Lasso, the garotte, elastic net, and so on [99].

The following are a few of typical versions of the Lasso and their respective solutions [99].

- *Nonnegative garotte:* This method, developed by Breiman [35], is a precursor to the Lasso, and solves

$$\min_{c} \left\{ \frac{1}{2} \sum_{i=1}^{n} \left(y_i - \sum_{j=1}^{p} x_{ij} c_j \hat{\beta}_j \right)^2 + \gamma \sum_{j=1}^{p} c_j \right\} \quad \text{subject to } c_j \geq 0, \quad (6.9.36)$$

where $\hat{\beta}_j$ is the usual least squares estimates ($p \leq n$ is assumed). The coordinate-wise update is given by

$$c_j \leftarrow \left(\frac{\tilde{\beta}_j \hat{\beta}_j - \lambda}{\hat{\beta}_j^2}\right)_+, \qquad (6.9.37)$$

where $\tilde{\beta}_j = \sum_{i=1}^n x_{ij}\left(y_i - \tilde{y}_i^{(j)}\right)$ and $\tilde{y}_i^{(j)} = \sum_{k \neq j} x_{ik} c_k \hat{\beta}_k$.

- *Elastic net:* This method [307] adds another constraint $\lambda_2 \sum_{j=1}^p \beta_j^2/2$ to Lasso to solve

$$\min_{\beta} \left\{\frac{1}{2}\sum_{i=1}^n \left(y_i - \sum_{j=1}^p x_{ij}\beta_j\right) + \lambda_1 \sum_{j=1}^p |\beta_j| + \lambda_2 \sum_{j=1}^p \beta_j^2/2\right\}. \qquad (6.9.38)$$

The coordinate-wise update has the form

$$\tilde{\beta}_j \leftarrow \frac{S\left(\sum_{i=1}^n x_{ij}\left(y_i - \tilde{y}_i^{(j)}\right), \lambda_1\right)_+}{1 + \lambda_2}. \qquad (6.9.39)$$

- *Grouped Lasso:* Let $\mathbf{X}_j$ be an $N \times p_j$ orthonormal matrix that represents the jth group of p_j variables, $j = 1, \ldots, m$, and β_j the corresponding coefficient vector. The grouped Lasso [292] solves

$$\min_{\beta} \left\{\left\|\mathbf{y} - \sum_{j=1}^m \mathbf{X}_j \boldsymbol{\beta}_j\right\|_2^2 + \sum_{j=1}^m \lambda_j \|\boldsymbol{\beta}_j\|_2\right\}, \qquad (6.9.40)$$

where $\lambda_j = \lambda \sqrt{p_j}$. The coordinate-wise update is

$$\tilde{\boldsymbol{\beta}}_j \leftarrow (\|\mathbf{s}_j\|_2 - \lambda_j)_+ \frac{\mathbf{s}_j}{\|\mathbf{s}_j\|_2}, \qquad (6.9.41)$$

here $\mathbf{s}_j = \mathbf{X}_j^T \left(\mathbf{y} - \tilde{\mathbf{y}}^{(j)}\right)$ with $\tilde{\mathbf{y}}^{(j)} = \sum_{k \neq j} \mathbf{X}_k \tilde{\boldsymbol{\beta}}_k$.

- *"Berhu" penalty:* This method [198] is a robust hybrid of Lasso and ridge regression, and its Lagrange form is

$$\min_{\beta} \left\{\frac{1}{2}\sum_{i=1}^n \left(y_i - \sum_{j=1}^p x_{ij}\beta_j\right) \right.$$
$$\left. + \lambda \sum_{j=1}^p \left(|\beta_j| \cdot I(|\beta_j| < \delta) + \frac{(\beta_j^2 + \delta^2)}{2\delta} \cdot I(|\beta_j| \geq \delta)\right)\right\}. \qquad (6.9.42)$$

Berhu penalty is the reverse of a "Huber" function. The coordinate-wise update has the form

$$\tilde{\beta}_j \leftarrow \begin{cases} S\left(\sum_{i=1}^n x_{ij}(y_i - \tilde{y}^{(j)}), \lambda\right), & \text{if } |\beta_j| < \delta; \\ \sum_{i=1}^n x_{ij}(y_i - \tilde{y}^{(j)})/(1 + \lambda/\delta), & \text{if } |\beta_j| \geq \delta. \end{cases} \quad (6.9.43)$$

This is a robust hybrid consisting of Lasso-style soft-thresholding for values less than δ and ridge-style beyond δ.

6.9.4 Gradient Projection for Sparse Reconstruction

As the general formulation of Lasso problem, we consider the unconstrained convex optimization problem

$$\min_{\mathbf{x}} \left\{ \frac{1}{2} \|\mathbf{y} - \mathbf{A}\mathbf{x}\|_2^2 + \tau \|\mathbf{x}\|_1 \right\}, \quad (6.9.44)$$

where $\mathbf{x} \in \mathbb{R}^n, \mathbf{y} \in \mathbb{R}^k, \mathbf{A} \in \mathbb{R}^{k \times n}$, and τ is a nonnegative parameter. When the variable $\mathbf{x}$ is a sparse vector, the above optimization is also called the *sparse reconstruction* of $\mathbf{x}$.

The optimization problem (6.9.44) is closely related to the following constrained convex optimization problems: *quadratically constrained linear program* (QCLP)

$$\min_{\mathbf{x}} \|\mathbf{x}\|_1 \quad \text{subject to} \quad \|\mathbf{y} - \mathbf{A}\mathbf{x}\|_2^2 \leq \epsilon \quad (6.9.45)$$

and *quadratic program* (QP)

$$\min_{\mathbf{x}} \|\mathbf{y} - \mathbf{A}\mathbf{x}\|_2^2 \quad \text{subject to} \quad \|\mathbf{x}\|_1 \leq \alpha, \quad (6.9.46)$$

where ϵ and α are nonnegative real parameters.

By various enhancements to the basic gradient projection (GP) algorithm applied to a quadratic programming formulation of (6.9.44), Figueiredo et al. [92] proposed a GPSR (*gradient projection for sparse reconstruction*). This GPSR approach, as in [102], splits the variable $\mathbf{x}$ into its positive and negative parts:

$$\mathbf{x} = \mathbf{u} - \mathbf{v}, \quad \mathbf{u} \geq \mathbf{0}, \ \mathbf{v} \geq \mathbf{0}, \quad (6.9.47)$$

where $u_i = (x_i)_+$ and $v_i = (-x_i)_+$ for all $i = 1, \ldots, n$ with $(x)_+ = \max\{0, x\}$. Then, (6.9.44) can be rewritten as the following *bound-constrained quadratic program* (BCQP):

$$\min_{\mathbf{u},\mathbf{v}} \left\{ \frac{1}{2} \|\mathbf{y} - \mathbf{A}(\mathbf{u} - \mathbf{v})\|_2^2 + \tau \mathbf{1}_n^T \mathbf{u} + \tau \mathbf{1}_n^T \mathbf{v} \right\},$$

subject to $\mathbf{u} \geq \mathbf{0}, \mathbf{v} \geq \mathbf{0}$. (6.9.48)

Equation (6.9.48) can be written in more standard BCQP form

$$\min_{\mathbf{z}} \left\{ F(\mathbf{z}) = \mathbf{c}^T \mathbf{z} + \frac{1}{2} \mathbf{z}^T \mathbf{B} \mathbf{z} \right\} \quad \text{subject to} \quad \mathbf{z} \geq \mathbf{0}, \quad (6.9.49)$$

where

$$\mathbf{z} = \begin{bmatrix} \mathbf{u} \\ \mathbf{v} \end{bmatrix}, \quad \mathbf{b} = \mathbf{A}^T \mathbf{y}, \quad \mathbf{c} = \tau \mathbf{1}_{2n} + \begin{bmatrix} -\mathbf{b} \\ \mathbf{b} \end{bmatrix}, \quad (6.9.50)$$

and

$$\mathbf{B} = \begin{bmatrix} \mathbf{A}^T \mathbf{A} & -\mathbf{A}^T \mathbf{A} \\ -\mathbf{A}^T \mathbf{A} & \mathbf{A}^T \mathbf{A} \end{bmatrix}. \quad (6.9.51)$$

When the solution of (6.9.44) is known in advance to be nonnegative, we have

$$\frac{1}{2} \|\mathbf{y} - \mathbf{A}\mathbf{x}\|_2^2 = \frac{1}{2} \mathbf{y}^T \mathbf{y} - \mathbf{y}^T \mathbf{A}\mathbf{x} + \frac{1}{2} \mathbf{x}^T \mathbf{A}^T \mathbf{A} \mathbf{x},$$

since $\mathbf{x}^T \mathbf{A}^T \mathbf{y} = (\mathbf{x}^T \mathbf{A}^T \mathbf{y})^T = \mathbf{y}^T \mathbf{A} \mathbf{x}$. It is noted that $\mathbf{y}^T \mathbf{y}$ is a constant independent of the variable $\mathbf{x}$ and $\tau \|\mathbf{x}\|_1 = \tau \mathbf{1}_n^T \mathbf{x}$ under the assumption that $\mathbf{x}$ is nonnegative, thus Eq. (6.9.44) can be equivalently written as

$$\min_{\mathbf{x}} \left\{ \left(\tau \mathbf{1}_n - \mathbf{A}^T \mathbf{y} \right)^T \mathbf{x} + \frac{1}{2} \mathbf{x}^T \mathbf{A}^T \mathbf{A} \mathbf{x} \right\} \quad \text{subject to} \quad \mathbf{x} \geq \mathbf{0} \quad (6.9.52)$$

which has the same form as (6.9.49).

Gradient projection is an efficient method for solving general minimization problems over a convex set C in unconstrained minimization form

$$\min_{\mathbf{x}} \{f(\mathbf{x}) + h(\mathbf{x})\}, \quad (6.9.53)$$

where $f(\mathbf{x})$ is continuously differentiable on C and $h(\mathbf{x})$ is non-smoothing on C. Following the first-order optimality condition, we have

$$\mathbf{x}^* \in \arg\min_{\mathbf{x} \in C} \{f(\mathbf{x}) + h(\mathbf{x})\} \Leftrightarrow \mathbf{0} \in \nabla f(\mathbf{x}^*) + \partial h(\mathbf{x}^*). \quad (6.9.54)$$

6.9 Supervised Learning Regression

where $\nabla f(\mathbf{x})$ is the gradient of the continuously differentiable function f and $\partial h(\mathbf{x})$ is the subdifferential of the non-smoothing function h.

Especially, when $f(\mathbf{x}) = \frac{1}{2}\|\mathbf{x}-\mathbf{z}\|_2^2$, we have $\nabla f(\mathbf{x}) = \mathbf{x} - \mathbf{z}$, and thus the above first-order condition becomes

$$\mathbf{0} \in (\mathbf{x} - \mathbf{z}) + \partial h(\mathbf{x}) \quad \Leftrightarrow \quad \mathbf{z} \in \mathbf{x} + \partial h(\mathbf{x}). \tag{6.9.55}$$

If taking $\partial h(\mathbf{x}) = -\nabla f(\mathbf{x})$ in Eq. (6.9.54), then Eq. (6.9.55) gives the following result:

$$\mathbf{z} = \mathbf{x} - \nabla f(\mathbf{x}). \tag{6.9.56}$$

This is just the *gradient projection* of $\mathbf{x}$ onto the convex set C. Then, the basic gradient projection updating formula is given by

$$\mathbf{x}_{k+1} = \mathbf{x}_k - \mu_k \nabla f(\mathbf{x}_k), \tag{6.9.57}$$

where μ_k is the step length at the kth update.

The basic gradient projection approach for solving (6.9.49) consists of the following two steps [92].

- First, choose some scalar parameter $\alpha_k > 0$ and set

$$\mathbf{w}_k = (\mathbf{z}_k - \alpha_k \nabla F(\mathbf{z}_k))_+. \tag{6.9.58}$$

- Then, choose the second scalar $\lambda_k \in [0, 1]$ and set

$$\mathbf{z}_{k+1} = \mathbf{z}_k + \lambda_k (\mathbf{w}_k - \mathbf{z}_k). \tag{6.9.59}$$

In the basic approach, each iterate $\mathbf{z}_k$ is searched along the negative gradient $-\nabla F(\mathbf{z}_k)$, projecting onto the nonnegative orthant, and performing a *backtracking line search* until a sufficient decrease is attained in F [92].

Figueiredo et al. [92] proposed two gradient projection algorithms for sparse reconstruction: a basic gradient projection (GPSR-Basic) algorithm and a Barzilai–Borwein gradient projection (GPSR-BB) algorithm.

The GPSR-BB algorithm, as shown in Algorithm 6.15, is based on the Barzilai–Borwein (BB) method, using $\mathbf{H}_k = \eta^{(k)}\mathbf{I}$ as an approximation of the Hessian matrix (MATLAB implementations are available at http://www.lx.it.pt/~mtf/GPSR).

Algorithm 6.15 GPSR-BB algorithm [92]

1. **input:** The data vector $\mathbf{y} \in \mathbb{R}^m$ and the input matrix $\mathbf{A} \in \mathbb{R}^{m \times n}$.
2. **initialization:** Randomly generate a nonnegative vector $\mathbf{z}^{(0)} \in \mathbb{R}_+^{2n}$.
3. Choose nonnegative parameters $\tau, \beta \in (0, 1)$, $\mu \in (0, 1/2)$ and $\alpha_{\min}, \alpha_{\max}, \alpha_0 \in [\alpha_{\min}, \alpha_{\max}]$.
4. Compute $\mathbf{b} = \mathbf{A}^T \mathbf{y}$, $\mathbf{c} = \tau \mathbf{1}_{2n} + \begin{bmatrix} -\mathbf{b} \\ \mathbf{b} \end{bmatrix}$ and $\mathbf{B} = \begin{bmatrix} \mathbf{A}^T \mathbf{A} & -\mathbf{A}^T \mathbf{A} \\ -\mathbf{A}^T \mathbf{A} & \mathbf{A}^T \mathbf{A} \end{bmatrix}$.
5. Set $k = 0$.
6. **repeat**
7. Calculate the gradient $\nabla F(\mathbf{z}^{(k)}) = \mathbf{c} + \mathbf{B}\mathbf{z}^{(k)}$.
8. Compute step: $\boldsymbol{\delta}^{(k)} = \left(\mathbf{z}^{(k)} - \alpha^{(k)} \nabla F(\mathbf{z}^{(k)})\right)_+ - \mathbf{z}^{(k)}$.
9. (**line search**): Find the scalar $\lambda^{(k)}$ that minimizes $F(\mathbf{z}^{(k)} + \lambda^{(k)} \boldsymbol{\delta}^{(k)})$ on the interval $\lambda^{(k)} \in [0, 1]$.
10. Set $\mathbf{z}^{(k+1)} = \mathbf{z}^{(k)} + \lambda^{(k)} \boldsymbol{\delta}^{(k)}$.
11. (**update** α): compute

$$\gamma^{(k)} = \left(\boldsymbol{\delta}^{(k)}\right)^T \mathbf{B} \boldsymbol{\delta}^{(k)};$$

if $\gamma^{(k)} = 0$ then let $\alpha^{(k)} = \alpha_{\max}$, otherwise

$$\alpha^{(k)} = \text{mid}\left\{\alpha_{\min}, \frac{\|\boldsymbol{\delta}^{(k)}\|_2^2}{\gamma^{(k)}}, \alpha_{\max}\right\}.$$

12. If $\left\|\mathbf{z}^{(k)} - \left(\mathbf{z}^{(k)} - \bar{\alpha} \nabla F(\mathbf{z})\right)_+\right\| \leq \text{tolP}$, where tolP is a small parameter and $\bar{\alpha}$ is a positive constant, then terminate; otherwise set $k \leftarrow k + 1$ return to Step 6.
13. **output:** $\mathbf{z}^{(k+1)}$.

6.10 Supervised Learning Classification

Linear classification is a useful tool in machine learning and data mining. Unlike nonlinear classifiers (such as kernel methods) mapping data to a higher-dimensional space, linear classifiers directly work on data in the original input space. For some data in a rich dimensional space, the performance (i.e., testing accuracy) of linear classifiers has shown to be close to that of nonlinear classifiers [293].

6.10.1 Binary Linear Classifiers

Definition 6.14 (Instance [305]) An *instance* $\mathbf{x}$ represents a specific object. The instance is often represented by a D-dimensional feature vector $\mathbf{x} = [x_1, \ldots, x_D]^T \in \mathbb{R}^D$, where each dimension is called a feature. The length D of the feature vector is known as the dimensionality of the feature vector.

It is assumed that these instances are sampled independently from an underlying distribution $P(\mathbf{x})$, which is unknown to us and is denoted by $\{\mathbf{x}_i\}_{i=1}^N \overset{\text{i.i.d.}}{\sim} P(\mathbf{x})$, where i.i.d. stands for *independent and identically distributed*.

Definition 6.15 (Training Sample [305]) A *training sample* is a collection of instances $\{\mathbf{x}_i\}_{i=1}^N = \{\mathbf{x}_1, \ldots, \mathbf{x}_N\}$, which acts as the input to the learning process.

6.10 Supervised Learning Classification

Definition 6.16 (Label [305]) The desired prediction y on an instance $\mathbf{x}$ is known as its *label*.

Definition 6.17 (Supervised Learning Classification) Let $X = \{\mathbf{x}_i\}$ be the domain of instances and $Y = \{y_i\}$ be the domain of labels. Let $P(\mathbf{x}, y)$ be an (unknown) joint probability distribution on instances and labels $X \times Y$. Given a set of training samples $\{(\mathbf{x}_i, y_i)\}_{i=1}^{N} \overset{\text{i.i.d.}}{\sim} P(\mathbf{x}, y)$, *supervised learning classification* trains a function $f : X \to Y$ in some function family F in order to make $f(\mathbf{x})$ predict the true label y on future testing data $\mathbf{x}$, where $(\mathbf{x}, y) \overset{\text{i.i.d.}}{\sim} P(\mathbf{x}, y)$ as well.

In binary classification, we are given training data $\{\mathbf{x}_i, y_i\} \in \mathbb{R}^D \times \{-1, +1\}$ for $i = 1, \ldots, N$. Linear classification methods construct the following decision function:

$$f(\mathbf{x}) = \langle \mathbf{w}, \mathbf{x} \rangle + b = \mathbf{w} \cdot \mathbf{x} + b = \mathbf{w}^T \mathbf{x} + b, \qquad (6.10.1)$$

where $\mathbf{w}$ is called the weight vector and b is an *intercept*, or called the *bias*.

To generate a decision function $f(\mathbf{x}) = \mathbf{w}^T \mathbf{x}$, linear classification involves the following risk minimization problem [293]:

$$\min_{\mathbf{w}} \left\{ f(\mathbf{w}) = R(\mathbf{w}) + C \sum_{i=1}^{N} \xi(\mathbf{w}^T \mathbf{x}_i, y_i) \right\}, \qquad (6.10.2)$$

where

- $\mathbf{w}$ is a weight vector consisting of linear classifier parameters;
- $R(\mathbf{w})$ is a regularization function that prevents the overfitting parameters;
- $\xi(\mathbf{w}^T \mathbf{x}; y_i)$ is a loss function measuring the discrepancy between the classifier's prediction and the true output y_i for the ith training example;
- $C > 0$ is a scalar constant (pre-specified by the user of the learning algorithm) that controls the balance between the regularization $R(\mathbf{w})$ and the loss function $\xi(\mathbf{w}^T \mathbf{x}; y_i)$.

The output of linear classifier, $d(\mathbf{w}) = \mathbf{w}^T \mathbf{x}$, is called the *decision function* of the classifier, since the classification decision (i.e., the label y) is made by $\hat{y} = \text{sign}(\mathbf{w}^T \mathbf{x})$.

The aim of the regularization function $R(\mathbf{w})$ is to prevent our model from overfitting the observations found in the training set. The following regularization items are commonly applied:

$$R_1(\mathbf{w}) = \|\mathbf{w}\|_1 = \sum_{k=1}^{N} |w_k|, \qquad (6.10.3)$$

or

$$R_2(\mathbf{w}) = \frac{1}{2}\|\mathbf{w}\|_2^2 = \frac{1}{2}\sum_{k=1}^{N} w_k^2. \qquad (6.10.4)$$

The regularization $R_1(\mathbf{w}) = \|\mathbf{w}\|_1$ has the following limitations:

- It is not strictly convex, so the solution may not be unique.
- For two highly correlated features, the solution obtained by R_1 regularization may select only one of these features. Consequently, R_1 regularization may discard the group effect of variables with high correlation [307].

To overcome the above two limitations of R_1 regularization, a convex combination of R_1 and R_2 regularizations forms the following elastic net [307]:

$$R_e(\mathbf{w}) = \lambda\|\mathbf{w}\|_2^2 + (1-\lambda)\|\mathbf{w}\|_1, \qquad (6.10.5)$$

where $\lambda \in (0, 1)$.

Popular loss functions $\xi(\mathbf{w}^T\mathbf{x}_i, y_i)$ include the hinge loss (i.e., interior penalty function)

$$\xi_1\left(\mathbf{w}^T\mathbf{x}_i, y_i\right) = \max\left(0, 1 - y_i\mathbf{w}^T\mathbf{x}_i\right), \qquad (6.10.6)$$

$$\xi_2\left(\mathbf{w}^T\mathbf{x}_i, y_i\right) = \max\left(0, 1 - y_i\mathbf{w}^T\mathbf{x}_i\right)^2, \qquad (6.10.7)$$

for linear support vector machines and the log loss (or log penalty term)

$$\xi_3\left(\mathbf{w}^T\mathbf{x}_i, y_i\right) = \log\left(1 + e^{-y_i\mathbf{w}^T\mathbf{x}_i}\right) \qquad (6.10.8)$$

for linear logistic regression [293].

Clearly, if all of the classifier's outputs $\mathbf{w}^T\mathbf{x}_i$ and the corresponding labels y_i have the same sign (i.e., correct decision making), then all terms $y_i\mathbf{w}^T\mathbf{x}_i > 0$ for $i = 1, \ldots, N$, thus resulting in any of the above loss functions evaluating to zero; otherwise, there is at least one of $\xi(\mathbf{w}^T\mathbf{x}_i, y_i) > 0$ for some i, so the loss function is punished.

6.10.2 Multiclass Linear Classifiers

In multiclass classification it is necessary to determine which of a finite set of classes the test input **x** belongs to.

Let $S = \{(\mathbf{x}_1, y_1), \ldots, (\mathbf{x}_N, y_N)\}$ be a set of N training examples, where each example $\mathbf{x}_i$ is drawn from a domain $\mathcal{X} \subseteq \mathbb{R}^D$, and each label y_i is an integer from the set $\mathcal{Y} = \{1, \ldots, k\}$.

As a concrete example of a supervised learning method, *k-nearest neighbor* (kNN) is a simple classification algorithm, as shown in Algorithm 6.16.

Being a D-dimensional feature vector, the test instance **x** can be viewed as a point in D-dimensional feature space. A classifier assigns a label to each point in

6.10 Supervised Learning Classification

Algorithm 6.16 k-nearest neighbor (kNN)

1. **input:** Training data $(\mathbf{x}_1, y_1), \ldots, (\mathbf{x}_N, y_N)$, distance function $d(\mathbf{a}, \mathbf{b})$, number of neighbors k, testing instance $\mathbf{x}$.
2. **initialization:** let the ith neighbor's data subset $L_i = \{\mathbf{x}_1^{(i)}, \ldots, \mathbf{x}_{n_i}^{(i)}\}$ for $y_1, \ldots, y_N \in \{i\}$, where $n_1 + \cdots + n_k = N$.
3. Compute centers of k neighbors $\mathbf{m}^{(i)} = \frac{1}{N_i} \sum_{j=1}^{n_i} \mathbf{x}_j^{(i)}$ for $i = 1, \ldots, k$.
4. Find the distances between $\mathbf{x}$ and the center $\mathbf{m}_i$ of the ith neighbor, denoted $d(\mathbf{x}, \mathbf{m}_i)$, for $i = 1, \ldots, k$.
5. Find the nearest neighbor of $\mathbf{x}$ using label of $y \leftarrow \min_i \{d(\mathbf{x}, \mathbf{m}^{(1)}), \ldots, d(\mathbf{x}, \mathbf{m}^{(k)})\}$.
6. **output:** label of y as the majority class of $y_{i_1}, \ldots, y_{i_k}$.

---

the feature space. This divides the feature space into decision regions within which points have the same label. The boundary separating these regions is called the *decision boundary* induced by the classifier.

A *multiclass classifier* is a function $H : \mathcal{X} \to \mathcal{Y}$ that maps an instance $\mathbf{x} \in \mathcal{X}$ to an element y of $\mathcal{Y}$.

Consider k-class classifiers of the form [64]

$$H_{\mathbf{W}}(\mathbf{x}) = \arg\max_{m=1,\ldots,k} \mathbf{w}_m^T \mathbf{x}, \quad (6.10.9)$$

where $\mathbf{W} = [\mathbf{w}_1, \ldots, \mathbf{w}_k]$ is an $N \times k$ weight matrix, while the value of the inner-product of the mth column of $\mathbf{W}$ with the instance $\mathbf{x}$ is called the *confidence* or the *similarity score* for the mth class. By this definition, the predicted label of the new instance $\mathbf{x}$ is the index of the column attaining the highest similarity score with $\mathbf{x}$.

For the case of $k = 2$, linear binary classifiers predict that the label of an instance $\mathbf{x}$ is 1 if $\mathbf{w}_1^T \mathbf{x} > 0$ and -1 (or 2) if $\mathbf{w}_2^T \mathbf{x} \leq 0$. In this case, the weight matrix $\mathbf{W} = [\mathbf{w}, -\mathbf{w}]$ is an $N \times 2$ matrix. By using the parsimonious model (6.10.9) when $k \geq 3$, Crammer and Singer [64] set the label of a new input instance by choosing the index of the most similar column of $\mathbf{W}$.

The multiclass classifier of Crammer and Singer [64] is to solve the following primal optimization problem with "soft" constraints:

$$\min_{\mathbf{W},\xi} \left\{ \frac{1}{2}\beta\|\mathbf{W}\|_2^2 + \sum_{i=1}^N \xi_i \right\} \quad (6.10.10)$$

subject to $\quad \mathbf{w}_{y_i}^T \mathbf{x}_i + \delta_{y_i,m} - \mathbf{w}_m^T \mathbf{x}_i \geq 1 - \xi_i, \ \forall i = 1, \ldots, N; \ m = 1, \ldots, k,$

where $\beta > 0$ is a regularization constant and for $m = y_i$ the above inequality constraints become $\xi_i \geq 0$, while $\delta_{p,q}$ is equal to 1 if $p = q$ and 0 otherwise.

By the Lagrange multiplier method, the dual optimization problem of (6.10.10) is given by

$$\min_{\mathbf{W}, \xi, \eta} \left\{ \mathcal{L}(\mathbf{W}, \xi, \eta) = \frac{1}{2} B \sum_{m=1}^{k} \|\mathbf{w}_m\|_2^2 + \sum_{i=1}^{N} \xi_i \right.$$

$$\left. + \sum_{i=1}^{N} \sum_{m=1}^{k} \eta_{i,m} \left[(\mathbf{w}_m - \mathbf{w}_{y_i})^T \mathbf{x}_i - \delta_{y_i, m} + 1 - \xi_i \right] \right\} \quad (6.10.11)$$

subject to $\eta_{i,m} \geq 0, \quad \forall i = 1, \ldots, N; \; m = 1, \ldots, k,$

where $\eta_{i,m}$ are the Lagrange multipliers.

From $\frac{\partial \mathcal{L}}{\partial \xi_i} = 0$ and $\frac{\partial \mathcal{L}}{\partial \mathbf{w}_m} = \mathbf{0}$ one has, respectively,

$$\sum_{m=1}^{k} \eta_{i,m} = 1, \quad \forall i = 1, \ldots, N \quad (6.10.12)$$

and

$$\mathbf{w}_m = \beta^{-1} \left(\sum_{i=1}^{N} (\delta_{y_i, m} - \eta_{i,m}) \mathbf{x}_i \right), \quad m = 1, \ldots, r, \quad (6.10.13)$$

which can be rewritten as

$$\mathbf{w}_m = \beta^{-1} \left(\sum_{i: y_i = m} (1 - \eta_{i,m} \mathbf{x}_i) + \sum_{i: y_i \neq m} (-\eta_{i,m}) \mathbf{x}_i \right) \quad (6.10.14)$$

for $m = 1, \ldots, r$.

Equations (6.10.12) and (6.10.14) show the following important performance of the linear multiclass classifier of Crammer and Singer [64]:

- Since the set of Lagrange multipliers, $\{\eta_{i,1}, \ldots, \eta_{i,k}\}$, satisfies the constraints $\eta_{i,1} \ldots, \eta_{i,k} \geq 0$ and $\sum_m \eta_{i,m} = 1$ in (6.10.12) for each pattern $\mathbf{x}_i$, each set can be viewed as a probability distribution over the labels $\{1, \ldots, k\}$. Under this probabilistic interpretation an example $\mathbf{x}_i$ is a support pattern if and only if its corresponding distribution is not concentrated on the correct label y_i. Therefore, the classifier is constructed using patterns whose labels are uncertain; the rest of the input patterns are ignored.
- The first sum in (6.10.14) is over all patterns that belong to the mth class. This shows that an example $\mathbf{x}_i$ labeled $y_i = m$ is a support pattern only if $\eta_{i,m} = \eta_{i,y_i} < 1$.
- The second sum in (6.10.14) is over the rest of the patterns whose labels are different from m. In this case, an example $\mathbf{x}_i$ is a support pattern only if $\eta_{i,m} > 0$.

6.11 Supervised Tensor Learning (STL)

In many disciplines, more and more problems need three or more subscripts for describing the data. Data with three or more subscripts are referred to as multi-channel data, whose representation is the *tensor*.

With development of modern applications, multi-way higher-order data models have been successfully applied across many fields, which include, among others, social network analysis/Web mining (e.g., [1, 150, 238]), computer vision (e.g., [111, 242, 258]), hyperspectral image [47, 157, 204], medicine and neuroscience [86], multilinear image analysis of face recognition (tensor face) [256], epilepsy tensors [2], Chemistry [36], and so on.

In this section, we focus on supervised tensor learning and tensor learning for regression.

6.11.1 Tensor Algebra Basics

Tensors will be represented by mathcal symbols, such as $\mathcal{T}, \mathcal{A}, \mathcal{X}$, and so on. A tensor represented in an N-way array is called an Nth-order tensor, and is defined as a multilinear function on an N-dimensional Cartesian product vector space, denoted $\mathcal{T} \in \mathbb{K}^{I_1 \times I_2 \times \cdots \times I_N}$, where $\mathbb{K}$ denotes either the real field $\mathbb{R}$ or the complex field $\mathbb{C}$ and where I_n is the number of entries or the dimension in the nth "direction" or mode. For example, for a third-order tensor $\mathcal{T} \in \mathbb{K}^{4 \times 2 \times 5}$, the first, second, and third modes have dimensions, respectively, equal to 4, 2, and 5. A scalar is a zero-order tensor, a vector is a first-order tensor, and a matrix is a second-order tensor. An Nth-order tensor is a multilinear mapping

$$\mathcal{T} : \mathbb{K}^{I_1} \times \mathbb{K}^{I_2} \times \cdots \times \mathbb{K}^{I_N} \to \mathbb{K}^{I_1 \times I_2 \times \cdots \times I_N}. \tag{6.11.1}$$

Because $\mathcal{T} \in \mathbb{K}^{I_1 \times \cdots \times I_N}$ can be regarded as an Nth-order matrix, *tensor algebra* is essentially *higher-order matrix algebra*.

The following figure shows, respectively, an example of third-order and firth-order tensors (Fig. 6.2).

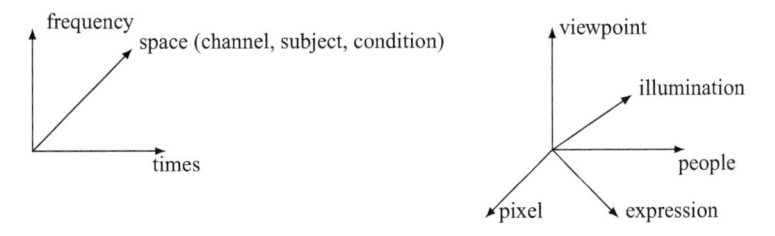

Fig. 6.2 Three-way array (left) and a tensor modeling a face (right)

In linear algebra (i.e., matrix algebra in a finite-dimensional vector space) a linear operator is defined in a finite-dimensional vector space. Because a matrix expresses a second-order tensor, linear algebra can also be regarded as the algebra of second-order tensors. In multiple linear algebra (i.e., higher-order tensor algebra) a multiple linear operator is also defined in a finite-dimensional vector space.

A matrix $\mathbf{A} \in \mathbb{K}^{m \times n}$ is represented by its entries and the symbol $[\cdot]$ as $\mathbf{A} = [a_{ij}]_{i,j=1}^{m,n}$. Similarly, an nth-order tensor $\mathcal{A} \in \mathbb{K}^{I_1 \times \cdots \times I_n}$ is, using a dual matrix symbol $[\![\cdot]\!]$, represented as $\mathcal{A} = [\![a_{i_1 \cdots i_n}]\!]_{i_1,\ldots,i_n=1}^{I_1,\ldots,I_n}$, where $a_{i_1 \cdots i_n}$ is the $(i_1, \ldots, i_n)$th entry of the tensor. An nth-order tensor is sometimes known as an *n-dimensional hypermatrix* [60]. The set of all $(I_1 \times \cdots \times I_n)$-dimensional tensors is denoted as $\mathcal{T}(I_1, \ldots, I_n)$.

The most common tensor is a *third-order tensor*, $\mathcal{A} = [\![a_{ijk}]\!]_{i,j,k}^{I,J,K} \in \mathbb{K}^{I \times J \times K}$. A third-order tensor is sometimes called a three-dimensional matrix. A square third-order tensor $\mathcal{X} \in \mathbb{K}^{I \times I \times I}$ is said to be *cubical*. In particular, a cubical tensor is called *supersymmetric tensor* [60], if its entries have the following symmetry:

$$x_{ijk} = x_{ikj} = x_{jik} = x_{jki} = x_{kij} = x_{kji}, \quad \forall i,j,k = 1, \ldots, I.$$

For a tensor $\mathcal{A} = [\![a_{i_1 \cdots i_m}]\!] \in \mathbb{R}^{N \times \cdots \times N}$, the line connecting $a_{11 \cdots 1}$ to $a_{NN \cdots N}$ is called the *superdiagonal line* of the tensor $\mathcal{A}$. A tensor $\mathcal{A}$ is known as *unit tensor*, if all its entries on the superdiagonal line are equal to 1 and all other entries are equal to zero, i.e.,

$$a_{i_1 \cdots i_m} = \delta_{i_1 \cdots i_m} = \begin{cases} 1, & \text{if } i_1 = \cdots = i_m; \\ 0, & \text{otherwise}. \end{cases} \quad (6.11.2)$$

A third-order unit tensor is denoted as $\mathcal{I} \in \mathbb{R}^{N \times N \times N}$.

In tensor algebra, it is convenient to view a third-order tensor as a set of vectors or matrices. Suppose that the ith entry of a vector $\mathbf{a}$ is denoted by a_i and that a matrix $\mathbf{A} = [a_{ij}] \in \mathbb{K}^{I \times J}$ has I row vectors $\mathbf{a}_{i:}, i = 1, \ldots, I$ and J column vectors $\mathbf{a}_{:j}, j = 1, \ldots, J$; the ith row vector is denoted by $\mathbf{a}_{i:} = [a_{i1}, \ldots, a_{iJ}]$, and the jth column vector is denoted by $\mathbf{a}_{:j} = [a_{1j}, \ldots, a_{Ij}]^T$. However, the concepts of row vectors and column vectors are no longer directly applicable for a higher-order tensor.

The three-way arrays of a third-order tensor are not known as row vectors and column vectors, but are renamed *tensor fibers*. A fiber is a one-way array with one subscript that is variable and the other subscripts fixed. The fibers of a third-order tensor are *vertical fibers*, *horizontal fibers*, and *"depth" fiber*. The vertical fibers of the third-order tensor $\mathcal{A} \in \mathbb{K}^{I \times J \times K}$ are also called *column fibers*, denoted $\mathbf{a}_{:jk}$; the horizontal fibers are also known as the *row fibers*, denoted $\mathbf{a}_{i:k}$; the depth fibers are also called the *tube fiber*, denoted $\mathbf{a}_{ij:}$.

6.11 Supervised Tensor Learning (STL)

Definition 6.18 (Tensor Vectorization) The *tensor vectorization* of an Nth-order tensor $\mathcal{A} = [\![\mathcal{A}_{i_1,\ldots,i_N}]\!] \in \mathbb{K}^{I_1 \times \cdots \times I_N}$ is denoted by $\mathbf{a} = \text{vec}(\mathcal{A})$ whose entries are given by

$$a_l = \mathcal{A}_{i_1,i_2,\ldots,i_N}, \quad l = i_1 + \sum_{n=2}^{N}\left((i_n - 1)\prod_{k=1}^{n-1} I_k\right). \qquad (6.11.3)$$

For example, $a_1 = \mathcal{A}_{1,1,\ldots,1}$, $a_{I_1} = \mathcal{A}_{I_1,1,\ldots,1}$, $a_{I_2} = \mathcal{A}_{1,I_2,1,\ldots,1}$, $a_{I_1 I_2 \cdots I_N} = \mathcal{A}_{I_1,I_2,\ldots,I_N}$.

The operation for transforming a tensor into a matrix is known as *tensor matrixing* or *tensor unfolding*.

Definition 6.19 (Tensor Unfolding) Given an Nth-order tensor $\mathcal{A} \in \mathbb{K}^{I_1 \times \cdots \times I_N}$, the mappings $\mathcal{A} \to \mathbf{A}_{(n)}$ and $\mathcal{A} \to \mathbf{A}^{(n)}$ are, respectively, said to be the *horizontal unfolding* and *longitudinal unfolding* of the tensor $\mathcal{A}$, where the matrix $\mathbf{A}_{(n)} \in \mathbb{K}^{I_n \times (I_1 \cdots I_{n-1} I_{n+1} \cdots I_N)}$ is known as the (mode-n) *horizontal unfolding matrix* of the tensor $\mathcal{A}$; and the matrix $\mathbf{A}^{(n)} = \mathbb{K}^{(I_1 \cdots I_{n-1} I_{n+1} \cdots I_N) \times I_n}$ is called the (mode-n) *longitudinal unfolding matrix* of $\mathcal{A}$.

There are three horizontal unfolding methods.

1. *Kiers horizontal unfolding method:* Given an Nth-order tensor $\mathcal{A} \in \mathbb{K}^{I_1 \times \cdots \times I_N}$, the Kiers horizontal unfolding method, presented by Kiers [143] in 2000, maps its entry $a_{i_1 i_2 \cdots i_N}$ to the (i_n, j)th entry of the matrix $\mathbf{A}_{(n)}^{\text{Kiers}} \in \mathbb{K}^{I_n \times (I_1 \cdots I_{n-1} I_{n+1} \cdots I_N)}$, i.e.,

$$\mathbf{A}_{(n)}^{\text{Kiers}}(i_n, j) = a_{i_n,j}^{\text{Kiers}} = a_{i_1,i_2,\ldots,i_N}, \qquad (6.11.4)$$

where $i_n = 1, \ldots, I_n$ and

$$j = \sum_{p=1}^{N-2}\left((i_{N+n-p} - 1)\prod_{q=n+1}^{N+n-p-1} I_q\right) + i_{n+1}, \quad n = 1, \ldots, N \qquad (6.11.5)$$

with $I_{N+m} = I_m$ and $i_{N+m} = i_m$ ($m > 0$).

2. *LMV horizontal unfolding method:* This method, introduced by Lathauwer, Moor, and Vanderwalle [154] in 2000, maps the entry $a_{i_1,\ldots,i_N}$ of an Nth-order tensor $\mathcal{A} \in \mathbb{K}^{I_1 \times \cdots \times I_N}$ to the entry $a_{i_n,j}^{\text{LMV}}$ of the matrix $\mathbf{A}_{(n)}^{\text{LMV}} \in \mathbb{K}^{I_n \times (I_1 \cdots I_{n-1} I_{n+1} \cdots I_N)}$, where

$$j = i_{n-1} + \sum_{k=1}^{n-2}\left((i_k - 1)\prod_{m=k+1}^{n-1} I_m\right) + \prod_{p=1}^{n-1} I_p \left(\sum_{k=n+1}^{N}\left((i_k - 1)\prod_{q=k}^{N} I_{q+1}\right)\right), \qquad (6.11.6)$$

where $I_q = 1$ if $q > N$.

3. *Kolda horizontal unfolding method:* This unfolding, introduced by Kolda [149] in 2006, maps the entry $a_{i_1,\ldots,i_N}$ of an Nth-order tensor to the entry $a_{i_n,j}^{\text{Kolda}}$ of the matrix $\mathbf{A}_{(n)}^{\text{Kolda}} \in \mathbb{K}^{I_n \times (I_1 \cdots I_{n-1} I_{n+1} \cdots I_N)}$, where

$$j = 1 + \sum_{k=1, k \neq n}^{N} \left((i_k - 1) \prod_{m=1, m \neq n}^{k-1} I_m \right). \quad (6.11.7)$$

Similar to the above three horizontal unfolding methods, there are three longitudinal unfolding methods for $\mathbf{A}^{(n)} = \mathbb{K}^{(I_1 \cdots I_{n-1} I_{n+1} \cdots I_N) \times I_n}$. The relationships between the longitudinal unfolding and the horizontal unfolding of a third-order tensor are

$$\mathbf{A}_{\text{Kiers}}^{(n)} = \left(\mathbf{A}_{(n)}^{\text{Kiers}} \right)^T, \quad \mathbf{A}_{\text{LMV}}^{(n)} = \left(\mathbf{A}_{(n)}^{\text{LMV}} \right)^T, \quad \mathbf{A}_{\text{Kolda}}^{(n)} = \left(\mathbf{A}_{(n)}^{\text{Kolda}} \right)^T. \quad (6.11.8)$$

Definition 6.20 (Tensor Inner Product) For two tensors $\mathcal{A}, \mathcal{B} \in \mathbb{K}^{I_1 \times \cdots \times I_N}$, their *tensor inner product* $\langle \mathcal{A}, \mathcal{B} \rangle$ is a scalar and is defined as the inner product of the column vectorizations of the two tensors, i.e.,

$$\langle \mathcal{A}, \mathcal{B} \rangle \stackrel{\text{def}}{=} \langle \text{vec}(\mathcal{A}), \text{vec}(\mathcal{B}) \rangle = (\text{vec}(\mathcal{A}))^H \text{vec}(\mathcal{B})$$

$$= \sum_{i_1=1}^{I_1} \sum_{i_2=1}^{I_2} \cdots \sum_{i_n=1}^{I_N} a_{i_1 i_2 \cdots i_n}^* b_{i_1 i_2 \cdots i_n}. \quad (6.11.9)$$

The tensor inner product is also called the *tensor dot product*, denoted by $\mathcal{A} \cdot \mathcal{B}$, namely $\mathcal{A} \cdot \mathcal{B} = \langle \mathcal{A}, \mathcal{B} \rangle$.

Definition 6.21 (Tensor Outer Product) Given two tensors $\mathcal{A} \in \mathbb{K}^{I_1 \times \cdots \times I_P}$ and $\mathcal{B} \in \mathbb{K}^{J_1 \times \cdots \times J_Q}$, their *tensor outer product* is denoted $\mathcal{A} \circ \mathcal{B} \in \mathbb{K}^{I_1 \times \cdots \times I_P \times J_1 \times \cdots \times J_Q}$, and is defined as

$$(\mathcal{A} \circ \mathcal{B})_{i_1 \cdots i_P j_1 \cdots j_Q} = a_{i_1 \cdots i_P} b_{j_1 \cdots j_Q}, \quad \forall i_1, \ldots, i_P, j_1, \ldots, j_Q. \quad (6.11.10)$$

Definition 6.22 (Tensor Frobenius Norm) The *Frobenius norm* of a tensor $\mathcal{A}$ is defined as

$$\|\mathcal{A}\|_F = \sqrt{\langle \mathcal{A}, \mathcal{A} \rangle} \stackrel{\text{def}}{=} \left(\sum_{i_1=1}^{I_1} \sum_{i_2=1}^{I_2} \cdots \sum_{i_n=1}^{I_N} |a_{i_1 i_2 \cdots i_n}|^2 \right)^{1/2}. \quad (6.11.11)$$

6.11 Supervised Tensor Learning (STL)

Definition 6.23 (Tucker Operator) Given an Nth-order tensor $\mathcal{G} \in \mathbb{K}^{J_1 \times J_2 \times \cdots \times J_N}$ and matrices $\mathbf{U}^{(n)} \in \mathbb{K}^{I_n \times J_n}$, where $n \in \{1, \ldots, N\}$, the *Tucker operator* is defined as [149]:

$$[\![\mathcal{G}; \mathbf{U}^{(1)}, \mathbf{U}^{(2)}, \ldots, \mathbf{U}^{(N)}]\!] \stackrel{\text{def}}{=} \mathcal{G} \times_1 \mathbf{U}^{(1)} \times_2 \mathbf{U}^{(2)} \times_3 \cdots \times_N \mathbf{U}^{(N)}. \quad (6.11.12)$$

The result is an Nth-order $I_1 \times I_2 \times \cdots \times I_N$ tensor.

The mode-n product of a tensor $\mathcal{X} \in \mathbb{R}^{I_1 \times \cdots \times I_N}$ with a matrix $\mathbf{U} \in \mathbb{R}^{J \times I_n}$, denoted as $\mathcal{X} \times_n \mathbf{U}$, is a tensor of size $I_1 \times \cdots \times I_{n-1} \times J \times I_{n+1} \times \cdots \times I_N$, its elements are defined as

$$(\mathcal{X} \times_n \mathbf{U})_{i_1, \ldots, i_{n-1}, j, i_{n+1}, \ldots, i_N} = \sum_{i_n=1}^{I_n} x_{i_1, \ldots, i_N} u_{j i_n}. \quad (6.11.13)$$

The equivalent unfolded expression is

$$\mathcal{Y} = \mathcal{X} \times_n \mathbf{U} \Leftrightarrow \mathbf{Y}_{(n)} = \mathbf{U} \mathbf{X}_{(n)}. \quad (6.11.14)$$

Especially, for a three-order tensor $\mathcal{X} \in \mathbb{K}^{I_1 \times I_2 \times I_3}$ and the matrices $\mathbf{A} \in \mathbb{K}^{J_1 \times I_1}, \mathbf{B} \in \mathbb{K}^{J_2 \times I_2}, \mathbf{C} \in \mathbb{K}^{J_3 \times I_3}$, the *Tucker mode-1 product* $\mathcal{X} \times_1 \mathbf{A}$, the *Tucker mode-2 product* $\mathcal{X} \times_2 \mathbf{B}$, and the *Tucker mode-3 product* $\mathcal{X} \times_3 \mathbf{C}$ are, respectively, defined as [155]:

$$(\mathcal{X} \times_1 \mathbf{A})_{j_1 i_2 i_3} = \sum_{i_1=1}^{I_1} x_{i_1 i_2 i_3} a_{j_1 i_1}, \; \forall j_1, i_2, i_3, \quad (6.11.15)$$

$$(\mathcal{X} \times_2 \mathbf{B})_{i_1 j_2 i_3} = \sum_{i_2=1}^{I_2} x_{i_1 i_2 i_3} b_{j_2 i_2}, \; \forall i_1, j_2, i_3, \quad (6.11.16)$$

$$(\mathcal{X} \times_3 \mathbf{C})_{i_1 i_2 j_3} = \sum_{i_3=1}^{I_3} x_{i_1 i_2 i_3} c_{j_3 i_3}, \; \forall i_1, i_2, j_3. \quad (6.11.17)$$

Decompositions of a third-order tensor have two basic forms.

1. *Tucker decomposition:*

$$\mathcal{X} = \mathcal{G} \times_1 \mathbf{A} \times_2 \mathbf{B} \times_3 \mathbf{C} \Leftrightarrow x_{ijk} = \sum_{p=1}^{P} \sum_{q=1}^{Q} \sum_{r=1}^{R} g_{pqr} a_{ip} b_{jq} c_{kr}. \quad (6.11.18)$$

Here, $\mathcal{G}$ is called the *core tensor* of tensor decomposition.

2. *Canonical decomposition* (CANDECOM)/*Parallel factors decomposition* (PAR-FAC), simply known as CP decomposition:

$$\mathcal{X} = \mathcal{I} \times_1 \mathbf{A} \times_2 \mathbf{B} \times_3 \mathbf{C} \Leftrightarrow x_{ijk} = \sum_{p=1}^{P} \sum_{q=1}^{Q} \sum_{r=1}^{R} a_{ip} b_{jq} c_{kr}, \quad (6.11.19)$$

where $\mathcal{I}$ is a unit tensor.

It can be easily seen that there are the following differences between the Tucker and CP decompositions:

- In the Tucker decomposition, the entry g_{pqr} of the core tensor $\mathcal{G}$ shows that there are interaction forms involving the entry a_{ip} of the mode-A vector $\mathbf{a}_i = [a_{i1}, \ldots, a_{iP}]^T$, the entry b_{jq} of the mode-B vector $\mathbf{b}_j = [b_{j1}, \ldots, b_{jQ}]^T$, and the entry c_{kr} of the mode-C vector $\mathbf{c}_k = [c_{k1}, \ldots, c_{kR}]^T$.
- In the CP decomposition, the core tensor is a unit tensor. Because the entries on the superdiagonal $p = q = r \in \{1, \ldots, R\}$ are equal to 1, and all the other entries are zero, there are interactions only between the rth factor a_{ir} of the mode-A vector $\mathbf{a}_i$, the rth factor b_{jr} of the mode-B vector $\mathbf{b}_j$, and the rth factor c_{kr} of the mode-C vector $\mathbf{c}_k$. This means that the mode-A, mode-B and mode-C vectors have the same number R of factors, i.e., all modes extract the same number of factors.

Therefore, the CP decomposition can be understood as a canonical Tucker decomposition.

The generalization of the Tucker decomposition to higher-order tensors is called the *higher-order singular value decomposition*.

Theorem 6.6 (Higher-Order SVD [154]) *Every* $I_1 \times I_2 \times \cdots \times I_N$ *real tensor* $\mathcal{X}$ *can be decomposed into the mode-n product*

$$\mathcal{X} = \mathcal{G} \times_1 \mathbf{U}^{(1)} \times_2 \mathbf{U}^{(2)} \times_3 \cdots \times_N \mathbf{U}^{(N)} = [\![\mathcal{G}; \mathbf{U}^{(1)}, \mathbf{U}^{(2)}, \ldots, \mathbf{U}^{(N)}]\!] \quad (6.11.20)$$

with entries

$$x_{i_1 i_2 \cdots i_N} = \sum_{j_1=1}^{J_1} \sum_{j_2=1}^{J_2} \cdots \sum_{j_N=1}^{J_N} g_{i_1 i_2 \cdots i_N} u_{i_1 j_1}^{(1)} u_{i_2 j_2}^{(2)} \cdots u_{i_N j_N}^{(N)}, \quad (6.11.21)$$

where $\mathbf{U}^{(n)} = [\mathbf{u}_1^{(n)}, \ldots, \mathbf{u}_{J_n}^{(n)}]$ *is an* $I_n \times J_n$ *semi-orthogonal matrix:* $(\mathbf{U}^{(n)})^T \mathbf{U}^{(n)} = \mathbf{I}_{J_n}$ *with* $J_n \leq I_n$, *the core tensor* $\mathcal{G}$ *is a* $J_1 \times J_2 \times \cdots \times J_N$ *tensor and the subtensor* $\mathcal{G}_{j_n=\alpha}$ *is the tensor* $\mathcal{X}$ *with the fixed index* $j_n = \alpha$. *The subtensors have the following properties.*

6.11 Supervised Tensor Learning (STL)

- All-orthogonality: two subtensors $\mathcal{G}_{j_n=\alpha}$ and $\mathcal{G}_{j_n=\beta}$, for $\alpha \neq \beta$, are orthogonal:

$$\langle \mathcal{G}_{j_n=\alpha}, \mathcal{G}_{j_n=\beta} \rangle = 0, \quad \forall \alpha \neq \beta, \, n = 1, \ldots, N. \tag{6.11.22}$$

- Ordering:

$$\|\mathcal{G}_{i_n=1}\|_F \geq \|\mathcal{G}_{i_n=2}\|_F \geq \cdots \geq \|\mathcal{G}_{i_n=N}\|_F. \tag{6.11.23}$$

The mode-n product is simply denoted as

$$\mathcal{X} \times_1 \mathbf{U}^{(1)} \times_2 \mathbf{U}^{(2)} \times_3 \cdots \times_N \mathbf{U}^{(N)} = \mathcal{X} \prod_{n=1}^{N} \times_n \mathbf{U}^{(n)}. \tag{6.11.24}$$

Note that the typical CP decomposition factorizes usually an N-order tensor $\mathcal{X} \in \mathbb{R}^{I_1 \times I_2 \times \cdots \times I_N}$ into a linear combination of the R rank-one tensors:

$$\mathcal{X} \approx \sum_{r=1}^{R} \mathbf{u}_r^{(1)} \circ \mathbf{u}_r^{(2)} \circ \cdots \circ \mathbf{u}_r^{(N)}$$

$$= [\![\mathbf{U}^{(1)}, \mathbf{U}^{(2)}, \ldots, \mathbf{U}^{(N)}]\!]. \tag{6.11.25}$$

This operation is called the *rank-one decomposition* of tensor. Here, the operator "$\circ$" denotes the outer product of vectors of the factor matrices $\mathbf{U}^{(n)} = [\mathbf{u}_1^{(n)}, \ldots, \mathbf{u}_R^{(n)}] \in \mathbb{R}^{I_n \times R}$, $n = 1, \ldots, N$.

On tensor analysis, readers can further refer to [294].

6.11.2 Supervised Tensor Learning Problems

Given N training tensor-scalar data $(\mathcal{X}_i, y_i)$, $i = 1, \ldots, N$, where the tensor $\mathcal{X}_i \in \mathbb{R}^{I_1, \ldots, I_M}$. The *supervised tensor learning* is to train a parameter/weight tensor $\mathcal{W}$ with rank-one decomposition $\mathcal{W} = \mathbf{w}_1 \circ \mathbf{w}_2 \circ \cdots \circ \mathbf{w}_M$ such that

$$(\mathbf{w}_1, \ldots, \mathbf{w}_M; b, \mathbf{x}) = \arg\min_{\mathbf{w}_1, \ldots, \mathbf{w}_M; b, \mathbf{x}} f(\mathbf{w}_1, \ldots, \mathbf{w}_M; b, \mathbf{x})$$

$$\text{subject to } y_i c_i \left(\mathcal{X}_i \prod_{k=1}^{M} \times_k \mathbf{w}_k + b \right) \geq \xi_i, \quad i = 1, \ldots, N. \tag{6.11.26}$$

If $y_i \in \{+1, -1\}$, then the weight tensor $\mathcal{W}$ is called the *tensor classifier* and if $y_i \in \mathbb{R}$, then $\mathcal{W}$ is known as the *tensor predictor*.

The Lagrangian for supervised tensor learning defined in (6.11.26) is given by

$$L(\mathbf{w}_1, \ldots, \mathbf{w}_M; b, \mathbf{x})$$
$$= f(\mathbf{w}_1, \ldots, \mathbf{w}_M; b, \mathbf{x}) - \sum_{i=1}^{N} \lambda_i \left[y_i c_i \left(\mathcal{X}_i \prod_{k=1}^{M} \times_k \mathbf{w}_k + b \right) - \xi_i \right]$$
$$= f(\mathbf{w}_1, \ldots, \mathbf{w}_M; b, \mathbf{x}) - \sum_{i=1}^{N} \lambda_i y_i c_i \left(\mathcal{X}_i \prod_{k=1}^{M} \times_k \mathbf{w}_k + b \right) - \boldsymbol{\lambda}^T \mathbf{x}, \quad (6.11.27)$$

where $\boldsymbol{\lambda} = [\lambda_1, \ldots, \lambda_N]^T \geq \mathbf{0}$ is a nonnegative Lagrangian multiplier vector and $\mathbf{x} = [\xi_1, \ldots, \xi_N]^T$ is a slack variable vector.

The partial derivative of $L(\mathbf{w}_1, \ldots, \mathbf{w}_M; b, \mathbf{x})$ with respect to $\mathbf{w}_j$ is given by

$$\frac{\partial L}{\partial \mathbf{w}_j} = \frac{\partial f}{\partial \mathbf{w}_j} - \sum_{i=1}^{N} \lambda_i y_i \frac{\partial c_i}{\partial \mathbf{w}_j} \frac{\partial}{\partial \mathbf{w}_j} \left(\mathcal{X}_i \prod_{k=1}^{M} \times_k \mathbf{w}_k + b \right)$$
$$= \frac{\partial f}{\partial \mathbf{w}_j} - \sum_{i=1}^{N} \lambda_i y_i \frac{\mathrm{d} c_i}{\mathrm{d} \mathbf{z}} (\mathcal{X}_i \bar{\times}_j \mathbf{w}_j), \quad (6.11.28)$$

where $\mathbf{z} = \mathcal{X}_i \prod_{k=1}^{M} \times_k \mathbf{w}_k + b$ and

$$\mathcal{X}_i \bar{\times}_j \mathbf{w}_j = \mathcal{X}_i \circ \mathbf{w}_1 \circ \cdots \circ \mathbf{w}_{j-1} \circ \mathbf{w}_{j+1} \circ \cdots \circ \mathbf{w}_M. \quad (6.11.29)$$

Similarly, the partial derivative of $L(\mathbf{w}_1, \ldots, \mathbf{w}_M; b, \mathbf{x})$ with respect to the bias b is

$$\frac{\partial L}{\partial b} = \frac{\partial f}{\partial b} - \sum_{i=1}^{N} \lambda_i y_i \frac{\partial c_i}{\partial b} \frac{\partial}{\partial b} \left(\mathcal{X}_i \prod_{k=1}^{M} \times_k \mathbf{w}_k + b \right)$$
$$= \frac{\partial f}{\partial b} - \sum_{i=1}^{N} \lambda_i y_i \frac{\mathrm{d} c_i}{\mathrm{d} \mathbf{z}} \frac{\partial \mathbf{z}}{\partial b} \frac{\partial}{\partial b} \left(\mathcal{X}_i \prod_{k=1}^{M} \times_k \mathbf{w}_k + b \right)$$
$$= \frac{\partial f}{\partial b} - \sum_{i=1}^{N} \lambda_i y_i \frac{\mathrm{d} c_i}{\mathrm{d} \mathbf{z}}. \quad (6.11.30)$$

Let $\frac{\partial L}{\partial \mathbf{w}_j} = \mathbf{0}$ and $\frac{\partial L}{\partial b} = \mathbf{0}$ then from (6.11.28) and (6.11.30) it is known that

$$\frac{\partial f}{\partial \mathbf{w}_j} = \sum_{i=1}^{N} \lambda_i y_i \frac{\mathrm{d} c_i}{\mathrm{d} \mathbf{z}} (\mathcal{X}_i \bar{\times}_j \mathbf{w}_j), \quad (6.11.31)$$

6.11 Supervised Tensor Learning (STL)

$$\frac{\partial f}{\partial b} = \sum_{i=1}^{N} \lambda_i y_i \frac{dc_i}{d\mathbf{z}}. \qquad (6.11.32)$$

Hence, one has the following update formulae [242]:

$$w_{j,t} \leftarrow \mathbf{w}_{j,(t-1)} - \eta_{1,t} \sum_{i=1}^{N} \lambda_i y_i \frac{dc_i}{d\mathbf{z}} (\mathcal{X}_i \bar{\times}_j \mathbf{w}_{j,t-1}), \qquad (6.11.33)$$

$$b_t \leftarrow b_{t-1} - \eta_{2,t} \sum_{i=1}^{N} \lambda_i y_i \frac{dc_i}{d\mathbf{z}}. \qquad (6.11.34)$$

Algorithm 6.17 shows an alternating projection algorithm for the supervised tensor learning [242].

Algorithm 6.17 Alternating projection algorithm for the supervised tensor learning [242]

1. **input:** Training data $(\mathcal{X}_i, y_i)$, where $\mathcal{X}_i \in \mathbb{R}^{I_1 \times \cdots \times I_N}$ and $y_i \in \mathbb{R}$ for $i = 1, \ldots, N$.
2. **initialization:** Set $\mathbf{w}_k$ equal to random unit vector in $\mathbb{R}^{I_k}$ for $k = 1, \ldots, M$.
3. **repeat**
4.   **for** $j = 1$ to M
5. $w_j^{(t)} \leftarrow \mathbf{w}_j^{(t-1)} - \eta_1 \sum_{i=1}^{N} \lambda_i y_i \frac{dc_i}{d\mathbf{z}} (\mathcal{X}_i \bar{\times}_j \mathbf{w}_j)$
6. $b^{(t)} \leftarrow b^{(t-1)} - \eta_2 \sum_{i=1}^{N} \lambda_i y_i \frac{dc_i}{d\mathbf{z}}$
7.   **end for**
8.   **if** $\sum_{i=k}^{M} \left(|\mathbf{w}_{k,t}^T \mathbf{w}_{k,t-1}| / \|\mathbf{w}_{k,t}\|_F^2 - 1 \right) \leq \epsilon$, **then** goto Step 11.
9. $t \leftarrow t + 1$
10. **return**
11. **output:** the parameters in classification tensorplane $\mathbf{w}_1, \ldots, \mathbf{w}_M$ and b.

6.11.3 Tensor Fisher Discriminant analysis

Fisher discriminant analysis (FDA) [94] is a widely applied method for classification. Suppose there are N training data $(\mathbf{x}_i \in \mathbb{R}^I, 1 \leq i \leq N)$ associated with the class labels $y_i \in \{+1, -1\}$.

Let N_+ and N_- be, respectively, the numbers of the positive training measurements $(\mathbf{x}_i, y_i = +1)$ and the negative training measurements $(\mathbf{x}_i, y_i = -1)$. Denote

$$\mathbb{1}(y_i = +1) = \begin{cases} 1, & y_i = +1; \\ 0, & \text{otherwise;} \end{cases} \qquad (6.11.35)$$

$$\mathbb{1}(y_i = -1) = \begin{cases} 1, & y_i = -1; \\ 0, & \text{otherwise.} \end{cases} \quad (6.11.36)$$

For N_+ positive training measurements $(\mathbf{x}_i, y_i = +1)$, their mean vector is $\mathbf{m}_+ = (1/N_+)\mathbb{1}(y_i = +1)\mathbf{x}_i$; and for N_- negative training measurements $(\mathbf{x}_i, y_i = -1)$, their mean vector can be calculated as $\mathbf{m}_- = (1/N_-)\mathbb{1}(y_i = -1)\mathbf{x}_i$, while the mean vector and the covariance matrix of all training measurements are $\mathbf{m} = (1/N)\sum_{i=1}^{N} \mathbf{x}_i$ and $\mathbf{\Sigma} = \sum_{i=1}^{N}(\mathbf{x}_i - \mathbf{m})(\mathbf{x}_i - \mathbf{m})^T$.

The *between-class scatter matrix* $\mathbf{S}_b$ and *within-class scatter matrix* $\mathbf{S}_w$ of two classes of targets $(\mathbf{x}_i, y_i = +1)$ and $(\mathbf{x}_i, y_i = -1)$ are, respectively, defined as

$$\mathbf{S}_b = (\mathbf{m}_+ - \mathbf{m}_-)(\mathbf{m}_+ - \mathbf{m}_-)^T, \quad (6.11.37)$$

$$\mathbf{S}_w = \sum_{i=1}^{N}(\mathbf{x}_i - \mathbf{m})(\mathbf{x}_i - \mathbf{m})^T = N\mathbf{\Sigma}. \quad (6.11.38)$$

The FDA criterion is to design the classifier $\mathbf{w}$ such that

$$\mathbf{w} = \arg\max_{\mathbf{w}} \left\{ J_{\text{FDA}} = \frac{\mathbf{w}^T \mathbf{S}_b \mathbf{w}}{\mathbf{w}^T \mathbf{S}_w \mathbf{w}} \right\}, \quad (6.11.39)$$

or equivalently

$$\mathbf{w} = \arg\max_{\mathbf{w}} \left\{ J_{\text{FDA}} = \frac{\|\mathbf{m}_+ - \mathbf{m}_-\|}{\sqrt{\mathbf{w}^T \mathbf{\Sigma} \mathbf{w}}} \right\}. \quad (6.11.40)$$

The tensor extension of FDA, called *tensor Fisher discriminant analysis* (TFDA), was proposed by Tao et al. [242], and is a combination of Fisher discriminant analysis and supervised tensor learning.

Given the training tensor measurements $\mathcal{X}_i \in \mathbb{R}^{I_1 \times \cdots \times I_M} (1 = 1, \ldots, N)$ and their corresponding class labels $y_i \in \{+1, -1\}$. The *mean tensor* of the training positive measurements is $\mathcal{M}_+ = (1/N_+)\sum_{i=1}^{N} \mathbb{1}(y_i = +1)\mathcal{X}_i$; the mean tensor of the training negative measurements is given by $\mathcal{M}_- = (1/N_-)\sum_{i=1}^{N} \mathbb{1}(y_i = -1)\mathcal{X}_i$; the mean tensor of all training measurements is $\mathcal{M} = (1/N)\sum_{i=1}^{N} \mathcal{X}_i$.

The TFDA criterion is to design a tensor classifier $\mathcal{W}$ such that

$$(\mathbf{w}_1, \ldots, \mathbf{w}_M) = \arg\max_{\mathbf{w}_1, \ldots, \mathbf{w}_M} \left\{ J_{\text{TFDA}} = \frac{\left\|(\mathcal{M}_+ - \mathcal{M}_-)\prod_{k=1}^{M} \times_k \mathbf{w}_k\right\|_2^2}{\sum_{i=1}^{N} \left\|(\mathcal{X}_i - \mathcal{M})\prod_{k=1}^{M} \times_k \mathbf{w}_k\right\|_2^2} \right\}. \quad (6.11.41)$$

On more applications of supervised tensor learning, see [242].

6.11.4 Tensor Learning for Regression

Given a set of labeled training set $\{\mathcal{X}_i, y_i\}_{i=1}^N$, where $\mathcal{X}_i \in \mathbb{R}^{I_1 \times \cdots \times I_M}$ is an M-mode tensor and y_i are the associated scalar targets.

A classic linear predictor in the vector space, $y = \langle \mathbf{x}, \mathbf{w} \rangle + b$, can be extended from the vector space to the tensor space as

$$\hat{y}_i = \langle \mathcal{X}_i, \mathcal{W} \rangle + b, \tag{6.11.42}$$

where $\mathcal{W} \in \mathbb{R}^{I_1 \times \cdots \times I_M}$ is the weight tensor and the scalar b is the bias.

The *tensor regression* seeks to design a weight tensor $\mathcal{W}$ to give the regression output $\hat{y}_i = \langle \mathcal{X}_i, \mathcal{W} \rangle + b$. To this end, let the weight tensor $\mathcal{W}$ is a sum of R rank-one tensors:

$$\mathcal{W} = \sum_{r=1}^R \mathbf{u}_r^{(1)} \circ \mathbf{u}_r^{(2)} \circ \cdots \circ \mathbf{u}_r^{(M)} \triangleq [\![\mathbf{U}^{(1)}, \mathbf{U}^{(2)}, \ldots, \mathbf{U}^{(M)}]\!], \tag{6.11.43}$$

where $\mathbf{U}^{(m)} = [\mathbf{u}_1^{(m)}, \ldots, \mathbf{u}_R^{(m)}]$, $m = 1, \ldots, M$. Then, the tensor regression can be rewritten as

$$\hat{y}_i = \langle \mathcal{X}_i, \mathcal{W} \rangle + b = \langle \mathcal{X}_i, [\![\mathbf{U}^{(1)}, \mathbf{U}^{(2)}, \ldots, \mathbf{U}^{(M)}]\!] \rangle + b. \tag{6.11.44}$$

Therefore, the *higher rank tensor ridge regression* (hrTRR) is to minimize the loss function [111]:

$$L(\mathbf{U}^{(1)}, \ldots, \mathbf{U}^{(M)}; b) = \frac{1}{2} \sum_{i=1}^N \left(y_i - \langle \mathcal{X}_i, [\![\mathbf{U}^{(1)}, \ldots, \mathbf{U}^{(M)}]\!] \rangle - b \right)^2$$

$$+ \frac{\lambda}{2} \| [\![\mathbf{U}^{(1)}, \ldots, \mathbf{U}^{(M)}]\!] \|_F^2. \tag{6.11.45}$$

The closed form solution of the optimization $\arg\min_{\mathbf{U}^{(1)},\ldots,\mathbf{U}^{(M)};b} L(\mathbf{U}^{(1)}, \ldots, \mathbf{U}^{(M)}; b)$ can be derived as [111]:

$$b = \sum_{i=1}^N \left(y_i - \langle \mathcal{X}_i, [\![\mathbf{U}^{(1)}, \ldots, \mathbf{U}^{(M)}]\!] \rangle \right), \tag{6.11.46}$$

and

$$\hat{\mathbf{u}}^{(m)} = \left(\mathbf{\Phi}_{(m)}^T \mathbf{\Phi}_{(m)} + \lambda \mathbf{I} \right)^{-1} \mathbf{\Phi}_{(m)}^T \mathbf{y}, \quad m = 1, \ldots, M, \tag{6.11.47}$$

where $\hat{\mathbf{u}}^{(m)} = [\text{vec}(\hat{\mathbf{U}}^{(m)})^T, b]^T$ is the vector of unknowns, $\mathbf{y} = [y_1, \ldots, y_N]^T$ are targets, and the ith row of the matrix $\mathbf{\Phi}_{(m)}$ is $[\text{vec}(\tilde{\mathbf{X}}_{i(m)}), 1]$. Here, the matrix $\tilde{\mathbf{X}}_{i(m)}$ can be constructed as follows:

$$\mathbf{U}^{(-m)} = \mathbf{U}^{(M)} \odot \cdots \odot \mathbf{U}^{(m+1)} \odot \mathbf{U}^{(m-1)} \odot \cdots \odot \mathbf{U}^{(1)}, \qquad (6.11.48)$$

$$\mathbf{B}_{(m)} = \mathbf{U}^{(-m)T} \mathbf{U}^{(-m)}, \qquad (6.11.49)$$

$$\tilde{\mathbf{X}}_{i(m)} = \mathbf{X}_{i(m)} \mathbf{U}^{(-m)} \mathbf{B}_{(m)}^{1/2}, \qquad (6.11.50)$$

where $\mathbf{X}_{i(m)} \in \mathbb{R}^{I_m \times (I_1 \cdots I_{m-1} I_{m+1} \cdots I_M)}$ is the horizontal unfolding matrix based on any of the Kiers method, the LMV method, and the Kolda method.

In the above hrTRR, the Frobenius norm regularization is used, requiring the a priori selection of the tensor rank R. Instead of the Frobenius norm regularization, consider the group sparsity norm regularization [111]:

$$\psi(\mathbf{W}) = \sum_{r=1}^{R} \left(\sum_{m=1}^{M} \|\mathbf{U}_{:,r}^{(m)}\|_2^2 \right)^{1/2}, \qquad (6.11.51)$$

where $\mathbf{U}_{:,r}^{(m)}, m = 1, \ldots, M$ denote the rth column of the matrix $\mathbf{U}^{(m)}$.

Lemma 6.3 ([111]) *The group sparsity norm regularization can be equivalently written as*

$$\psi(\mathbf{W}) = \sum_{r=1}^{R} \left(\sum_{m=1}^{M} \|\mathbf{U}_{:,r}^{(m)}\|_2^2 \right)^{1/2}$$

$$= \min_{\eta \in \mathbb{R}} \left\{ \frac{1}{2} \sum_{r=1}^{R} \frac{\sum_{m=1}^{M} \|\mathbf{U}_{:,r}^{(m)}\|_2^2}{\eta_r} + \frac{1}{2} \|\boldsymbol{\eta}\|_1 \right\}. \qquad (6.11.52)$$

The minimum is obtained for

$$\eta_r = \left(\sum_{m=1}^{M} \|\mathbf{U}_{:,r}^{(m)}\|_2^2 \right)^{1/2}, \quad \forall r = 1, \ldots, R. \qquad (6.11.53)$$

Since minimizing $\psi(\mathbf{W})$ contains the minimization of $\|\boldsymbol{\eta}\|_1$, the optimal rank R can be obtained from the sparsity of $\boldsymbol{\eta}$. Then, the *optimal rank tensor ridge regression* (orTRR) $(\hat{\mathbf{U}}^{(m)}, b) = \arg\min_{\mathbf{U}^{(m)}, b} L_m(\mathbf{U}^{(m)}, b)$ can be represented as

$$(\hat{\mathbf{U}}^{(m)}, b) = \arg\min_{\mathbf{U}^{(m)}, b} \left\{ \frac{1}{2} \sum_{i=1}^{N} \left(y_i - \text{tr}\left(\mathbf{U}^{(m)} \mathbf{U}^{(-m)T} \mathbf{X}_{i(m)}^T \right) - b \right)^2 + \frac{\lambda}{2} \text{tr}\left(\mathbf{U}^{(m)} \mathbf{\Lambda} \mathbf{U}^{(m)T} \right) \right\}, \qquad (6.11.54)$$

6.11 Supervised Tensor Learning (STL)

where $\mathbf{\Lambda} = \mathbf{Diag}\left(\frac{1}{\eta_1}, \ldots, \frac{1}{\eta_R}\right)$.

The closed form solution of the orTRR is given by

$$\hat{b} = \sum_{i=1}^{N} y_i - \mathrm{tr}\left(\mathbf{U}^{(m)}\mathbf{U}^{(-m)T}\mathbf{X}_{i(m)}^T\right), \quad (6.11.55)$$

$$\hat{\mathbf{u}}^{(m)} = \left(\mathbf{\Phi}^T\mathbf{\Phi} + \frac{\lambda}{2}\tilde{\mathbf{\Lambda}}\right)^{-1}\mathbf{\Phi}^T\mathbf{y}, \quad (6.11.56)$$

where $\tilde{\mathbf{\Lambda}} = \mathbf{\Lambda} \otimes \mathbf{I}_{I_m \times I_m}$.

Guo et al. [111] developed a tensor learning algorithm for regression, as shown in Algorithm 6.18.

Algorithm 6.18 Tensor learning algorithm for regression [111]

1. **input:** The set of training tensors and their corresponding targets, that is $\{\mathcal{X}_i, y_i\}_{i=1}^N$.
2. **initialization:** Construct randomly $\{\mathbf{U}_0^{(1)}, \ldots, \mathbf{U}_0^{(M)}\}$, unfolding $\mathcal{X}_i$ to the matrix $\mathbf{X}_{i(m)}$, and let $t = 0$.
3. **repeat**
4.   **for** $k = 1$ to M **do**
5.   for high rank tensor ridge regression (hrTRR), calculate
6. $\mathbf{U}_t^{(-j)} = \mathbf{U}_t^{(M)} \odot \cdots \odot \mathbf{U}_t^{(j+1)} \odot \mathbf{U}_t^{(j-1)} \odot \cdots \odot \mathbf{U}_t^{(1)}$,
7. $\mathbf{B}_t = \mathbf{U}_t^{(-j)T}\mathbf{U}_t^{(-j)}$,
8. $\tilde{\mathbf{X}}_{i(j)} = \mathbf{X}_{i(j)}\mathbf{U}_t^{(-j)}\mathbf{B}_t^{1/2}$,
9.   The ith row of the matrix $\mathbf{\Phi}_{(m)}$ is given by $[\mathrm{vec}(\tilde{\mathbf{X}}_{i(m)}), 1]$.
10.   Calculate $b = \sum_{i=1}^N \left(y_i - \langle \mathcal{X}_i, [\![\mathbf{U}^{(1)}, \ldots, \mathbf{U}^{(M)}]\!]\rangle\right)$.
11.   Calculate $\hat{\mathbf{u}}^{(j)} = (\mathbf{\Phi}_{(m)}^T\mathbf{\Phi}_{(m)} + \lambda\mathbf{I})^{-1}\mathbf{\Phi}_{(m)}^T\mathbf{y}$;
12.   for optimal rank tensor ridge regression (orTRR), calculate
13. $\eta_r = \left(\sum_{m=1}^M \|\mathbf{U}_{:,r}^{(m)}\|_2^2\right)^{1/2}, \forall r = 1, \ldots, R$.
14. $\mathbf{\Lambda} = \mathbf{Diag}\left(\frac{1}{\eta_1}, \ldots, \frac{1}{\eta_R}\right)$ and $\tilde{\mathbf{\Lambda}} = \mathbf{\Lambda} \otimes \mathbf{I}_{I_j \times I_j}$.
15.   Calculate $b = \sum_{i=1}^N y_i - \mathrm{tr}\left(\mathbf{U}^{(m)}\mathbf{U}^{(-m)T}\mathbf{X}_{i(m)}^T\right)$.
16.   Calculate $\hat{\mathbf{u}}^{(j)} = (\mathbf{\Phi}^T\mathbf{\Phi} + \frac{\lambda}{2}\tilde{\mathbf{\Lambda}})^{-1}\mathbf{\Phi}^T\mathbf{y}$,
17.   **end for**
18.   Update the parameter η given by $\eta_r = (\sum_{m=1}^M \|\mathbf{U}_{:,r}^{(m)}\|_2^2)^{1/2}, r = 1, \ldots, R$.
19.   Prune the columns $\mathbf{U}_{:,r}^{(m)}$ of factor matrices for $m \in \{1, \ldots, M\}$ and $r \in \{j | \eta_j \leq \epsilon, j = 1, \ldots, R\}$.
20. $\mathcal{W}^t \leftarrow [\![\mathbf{U}^{(1)}, \mathbf{U}^{(2)}, \ldots, \mathbf{U}^{(M)}]\!]$.
21. $t \leftarrow t + 1$
22. **until** $\|\mathcal{W}^{(t)} - \mathcal{W}^{(t-1)}\|/\|\mathcal{W}^{(t-1)}\| \leq \epsilon$ or $t \geq T_{\max}$.
23. **end repeat**
24. **output:** the weights $\{\mathbf{U}^{(1)}, \ldots, \mathbf{U}^{(M)}\}$ and the bias term $b \in \mathbb{R}$ that minimize the objective function.

6.11.5 Tensor K-Means Clustering

For one multi-modal object **x** with three modalities, i.e., image modality, text modality, and audio modality, there are three feature vectors: image feature vector **a**, text feature vector **b**, and audio feature vector **c**, after feature learning. Hence, we can build a three-order feature tensor $\mathcal{T} = \mathbf{a} \circ \mathbf{b} \circ \mathbf{c}$ for the object **x**.

One feature tensor is built for each heterogeneous object after feature fusion. However, the conventional K-means algorithm could cluster the objects represented by vectors, and hence it is necessary to extend the K-means clustering algorithm from feature vectors to feature tensors.

Given two feature tensors $\mathcal{X}, \mathcal{Y} \in \mathbb{R}^{I_1 \times I_1 \times \cdots \times I_n}$, the *tensor distance* (TD) between $\mathcal{X}$ and $\mathcal{Y}$, denoted as $d_{\text{TD}}(\mathcal{X}, \mathcal{Y})$, is defined as [37]:

$$d_{\text{TD}}(\mathcal{X}, \mathcal{Y}) = \sqrt{(\mathbf{x} - \mathbf{y})^T \mathbf{G} (\mathbf{x} - \mathbf{y})}, \qquad (6.11.57)$$

where $\mathbf{x} = \text{vec}(\mathcal{X})$ and $\mathbf{y} = \text{vec}(\mathcal{Y})$ are, respectively, the vectorizations of $\mathcal{X}$ and $\mathcal{Y}$, and **G** denotes the coefficient matrix used to reveal the location correlation between $\mathcal{X}$ and $\mathcal{Y}$ in the tensor space.

Therefore, the *tensor K-means clustering* algorithm based on the tensor distance is outlined in the following four steps [37].

1. Randomly select K objects as the clustering centers.
2. Use Eq. (6.11.57) to compute the tensor distance between each object and every clustering center, and assign each object to the nearest clustering center.
3. Recompute each clustering center.
4. If convergent, stop the algorithm, otherwise, return to Step 2.

The key step of the tensor K-means algorithm is to compute the tensor distance between each object and every clustering center. Hence the tensor K-means algorithm has a computational complexity of $O(tnk)$, where t, n, k denote the numbers of iterations, objects, and clustering centers, respectively.

6.12 Unsupervised Clustering

The main purpose of unsupervised learning methods is to extract generally useful features from unlabeled data, to detect and remove input redundancies, and to preserve only essential aspects of the data in robust and discriminative representations [174].

Since no or very little information about the underlying distribution is available, the exploration of complex data sets fundamentally relies on the identification of "natural" group structures in the data. The analysis based on natural group structures in the data is called "*cluster analysis*."

6.12 Unsupervised Clustering

Clustering is one of the most widely used techniques for exploratory data analysis, with applications ranging from statistics, computer science, information science and engineering, biology, medicine to social sciences or psychology.

Clustering criteria can be broadly divided into three fundamental categories [114]:

1. *Compactness:* This concept is generally implemented by keeping the intra-cluster variation small. The resulting methods tend to be very effective for spherical or well-separated clusters, but they may fail to detect more complicated cluster structures.
2. *Connectedness:* Its basic idea is that neighboring data items should share the same cluster. They are well-suited for the detection of arbitrarily shaped clusters, but can lack robustness when there is little spatial separation between the clusters.
3. *Spatial separation:* This is a criterion that gives little guidance during the clustering process and can easily lead to trivial solutions. It is therefore usually combined with other objectives, most notably measures of compactness or balance of cluster sizes.

In every scientific and engineering field dealing with empirical data, one attempts to get a first impression on their data by trying to identify groups of "similar behavior" in their data.

Unsupervised learning includes clustering algorithms, which partition input vectors (a data set covering various dimensions) into clusters satisfying certain criteria. The most popular modern clustering algorithms are hierarchical clustering, spectral clustering, K-means clustering, etc.

In unsupervised learning, we only have the objects and do not have their labels. That is to say, given a set of inputs, an unsupervised learning algorithm is to correctly infer the outputs without having a supervisor providing the correct answers or the degree of error for each observation.

A validation step is needed due to the following two issues that arise when using clustering algorithms [114]:

- Bias of clustering algorithms towards particular cluster properties is inevitable due to their own clustering criterion.
- Unsupervised classification relies on the existence of a distinct structure within the data. However, most clustering algorithms return a clustering even in the absence of actual structure, leaving it to the user to detect the lack of significance of the results returned.

Data clustering has been used for the following three main purposes [127].

- *Underlying structure:* It gains insight into data for generating hypotheses, detecting anomalies, and identifying salient features.
- *Natural classification:* It identifies the degree of similarity among forms or organisms (phylogenetic relationship).
- *Compression:* It is a method for organizing the data and summarizing it through cluster prototypes.

6.12.1 Similarity Measures

One of the main mathematical tools for unsupervised clustering and classification is the distance measure.

The distance between two vectors **p** and **g**, denoted by $D(\mathbf{p}\|\mathbf{g})$, is a *measure* if it has the following properties.

- *Nonnegativity and positiveness:* $D(\mathbf{p}\|\mathbf{g}) \geq 0$, and equality holds if and only if $\mathbf{p} = \mathbf{g}$.
- *Symmetry:* $D(\mathbf{p}\|\mathbf{g}) = D(\mathbf{g}\|\mathbf{p})$.
- *Triangle inequality:* $D(\mathbf{p}\|\mathbf{z}) \leq D(\mathbf{p}\|\mathbf{g}) + D(\mathbf{g}\|\mathbf{z})$.

The basic rule of unsupervised clustering and classification is to adopt some distance metric to measure the similarity of two feature vectors. As the name suggests, this *similarity* is a measure of the degree of similarity between vectors.

Consider a clustering problem in pattern recognition (e.g., template matching). For simplicity, suppose that there are N pattern vectors $\mathbf{s}_1, \ldots, \mathbf{s}_N$, but the number of classes is unknown. Our problem is to cluster N pattern vectors into a number of classes. For this purpose, we need to compare these pattern vectors in order to be clustered into a number of classes such that pattern vectors belonging to the same class are more similar than those belonging to other classes. On the basis of a similarity comparison, we can obtain the unsupervised pattern clustering.

A quantity known as the *dissimilarity* is used to make a reverse measurement of the similarity between vectors: two vectors with a smaller dissimilarity are more similar.

Let $D(\mathbf{s}_i, \mathbf{s}_j)$ ($i, j = 1, \ldots, N$, but $j \neq i$) be the dissimilarities between the ith and jth pattern vectors $\mathbf{s}_i$ and $\mathbf{s}_j$. If

$$D(\mathbf{s}_i, \mathbf{s}_{j_1}) < D(\mathbf{s}_i, \mathbf{s}_{j_2}), \tag{6.12.1}$$

then we say the pattern vector $\mathbf{s}_{j_1}$ is more similar to $\mathbf{s}_i$ than $\mathbf{s}_{j_2}$.

The simplest and most intuitive dissimilarity parameter is the Euclidean distance between vectors. The regularized Euclidean distance between the ith and the jth known pattern vectors $(\mathbf{s}_i, \mathbf{s}_j)$, denoted $D_E(\mathbf{s}_i, \mathbf{s}_j)$, is defined as

$$D_E(\mathbf{s}_i, \mathbf{s}_j) = \frac{\|\mathbf{s}_i - \mathbf{s}_j\|_2}{\sqrt{\|\mathbf{s}_i\|^2 + \|\mathbf{s}_j\|^2}} = \frac{\sqrt{(\mathbf{s}_i - \mathbf{s}_j)^T(\mathbf{s}_i - \mathbf{s}_j)}}{\sqrt{\|\mathbf{s}_i\|^2 + \|\mathbf{s}_j\|^2}}. \tag{6.12.2}$$

Two extreme values imply that two vectors are completely similar and completely dissimilar, respectively, namely

$$D_E(\mathbf{x}, \mathbf{y}) = \begin{cases} 0 & \Leftrightarrow \quad \mathbf{x} \text{ and } \mathbf{y} \text{ are completely similar,} \\ 1 & \Leftrightarrow \quad \mathbf{x} \text{ and } \mathbf{y} \text{ are completely dissimilar.} \end{cases} \tag{6.12.3}$$

6.12 Unsupervised Clustering

If

$$D_E(\mathbf{s}_1, \mathbf{s}_i) = \min_k D_E(\mathbf{s}_1, \mathbf{s}_k), \quad k = 1, \ldots, N, \, i \neq 1, \tag{6.12.4}$$

then $\mathbf{s}_i$ is said to be a *nearest neighbor* to $\mathbf{s}_1$.

A widely used classification method is *nearest neighbor classification*, which judges $\mathbf{s}_1$ and $\mathbf{s}_i$ to belong in the same model type.

All pattern vectors satisfying

$$D_E(\mathbf{s}_1, \mathbf{s}_j) \approx D_E(\mathbf{s}_1, \mathbf{s}_i), \quad j = 2, \ldots, N, \text{ but } j \neq i, \tag{6.12.5}$$

are said to be the nearest neighbors to $\mathbf{s}_1$ as well and thus these pattern vectors are judged to belong to Class 1. Mimicking this process for other unclustered pattern vectors until all vectors are clustered, we can cluster all other possible classes (Class 2, Class 3, and so on) and determine the number of classes. This is the basic idea of the unsupervised clustering approach.

Another frequently used distance function is the *Mahalanobis distance*, proposed by Mahalanobis in 1936 [172]. The Mahalanobis distance from the vector $\mathbf{x}$ to its mean $\boldsymbol{\mu}$ is given by

$$D_M(\mathbf{x}, \boldsymbol{\mu}) = \sqrt{(\mathbf{x} - \boldsymbol{\mu})^T \mathbf{C}_x^{-1} (\mathbf{x} - \boldsymbol{\mu})}, \tag{6.12.6}$$

where $\mathbf{C}_x = \text{cov}(\mathbf{x}, \mathbf{x}) = E\{(\mathbf{x} - \boldsymbol{\mu})(\mathbf{x} - \boldsymbol{\mu})^T\}$ is the autocovariance matrix of the vector $\mathbf{x}$.

The Mahalanobis distance between vectors $\mathbf{x} \in \mathbb{R}^n$ and $\mathbf{y} \in \mathbb{R}^n$ is denoted by $D_M(\mathbf{x}, \mathbf{y})$, and is defined as [172]

$$D_M(\mathbf{x}, \mathbf{y}) = \sqrt{(\mathbf{x} - \mathbf{y})^T \mathbf{C}_{xy}^{-1} (\mathbf{x} - \mathbf{y})}, \tag{6.12.7}$$

where $\mathbf{C}_{xy} = \text{cov}(\mathbf{x}, \mathbf{y}) = E\{(\mathbf{x} - \boldsymbol{\mu}_x)(\mathbf{y} - \boldsymbol{\mu}_y)^T\}$ is the cross-covariance matrix of $\mathbf{x}$ and $\mathbf{y}$, while $\boldsymbol{\mu}_x$ and $\boldsymbol{\mu}_y$ are the means of $\mathbf{x}$ and $\mathbf{y}$, respectively.

Clearly, if the covariance matrix is the identity matrix, i.e., $\mathbf{C} = \mathbf{I}$, then the Mahalanobis distance reduces to the Euclidean distance. If the covariance matrix takes a diagonal form, then the corresponding Mahalanobis distance is called the *normalized Euclidean distance* and is given by

$$D_M(\mathbf{x}, \mathbf{y}) = \sqrt{\sum_{i=1}^n \frac{(x_i - y_i)^2}{\sigma_i^2}}, \tag{6.12.8}$$

in which σ_i is the standard deviation of x_i and y_i in the whole sample set.

Let

$$\mu = \frac{1}{N}\sum_{i=1}^{N}\mathbf{s}_i, \quad \mathbf{C} = \sum_{i=1}^{N}\sum_{j=1}^{N}(\mathbf{s}_i - \mu)(\mathbf{s}_j - \mu)^T \quad (6.12.9)$$

be the sample mean vector of N known pattern vectors $\mathbf{s}_i$ and the sample cross-covariance matrix. Then the Mahalanobis distance from the ith pattern vector $\mathbf{s}_i$ to the jth pattern vector $\mathbf{s}_j$ is defined as

$$D_M(\mathbf{s}_i, \mathbf{s}_j) = \sqrt{(\mathbf{s}_i - \mathbf{s}_j)^T \mathbf{C}^{-1}(\mathbf{s}_i - \mathbf{s}_j)}. \quad (6.12.10)$$

By the nearest neighbor classification method, if

$$D_M(\mathbf{s}_1, \mathbf{s}_i) = \min_k D_M(\mathbf{s}_1, \mathbf{s}_k), \quad k = 1, \ldots, M, \quad (6.12.11)$$

then the pattern vector $\mathbf{s}_i$ is recognized as being in the pattern type to which $\mathbf{s}_1$ belongs.

The measure of dissimilarity between vectors is not necessarily limited to distance functions. The cosine function of the acute angle between two vectors,

$$D(\mathbf{x}, \mathbf{s}_i) = \cos\theta_i = \frac{\mathbf{x}^T \mathbf{s}_i}{\|\mathbf{x}\|_2 \|\mathbf{s}_i\|_2}, \quad (6.12.12)$$

is an effective measure of dissimilarities as well.

If $\cos\theta_i < \cos\theta_j, \forall j \neq i$, holds, then the unknown pattern vector $\mathbf{x}$ is said to be most similar to the known pattern vector $\mathbf{s}_i$. The variant of Eq. (6.12.12)

$$D(\mathbf{x}, \mathbf{s}_i) = \frac{\mathbf{x}^T \mathbf{s}_i}{\mathbf{x}^T \mathbf{x} + \mathbf{s}_i^T \mathbf{s}_i + \mathbf{x}^T \mathbf{s}_i} \quad (6.12.13)$$

is referred to as the *Tanimoto measure* [251] and is widely used in information retrieval, the classification of diseases, animal and plant classifications, etc.

Given a testing pattern vector $\mathbf{x}$ and the number m of clustered classes. Define the mean vector of the ith class as

$$\bar{\mathbf{s}}^{(i)} = \frac{1}{N_i}\sum_{k=1}^{N_i}\mathbf{s}_k^{(i)}, \quad (6.12.14)$$

where N_i is the number of pattern vectors in the ith class such that $N_1 + \cdots + N_m = N$ and $\mathbf{s}_k^{(i)}$ is the kth pattern vector in the ith class.

6.12 Unsupervised Clustering

By the nearest neighbor classification, we decide that **x** belongs to Class i, if

$$D\left(\mathbf{x}, \bar{\mathbf{s}}^{(i)}\right) = \min_{k=1,\ldots,m} D\left(\mathbf{x}, \bar{\mathbf{s}}^{(k)}\right). \tag{6.12.15}$$

The similarity measures used in clustering depend on application problems. In addition to the Euclidean distance, the Mahalanobis distance, and the Tanimoto measure described above, there are various measures. For two objects **x** and **y**, the sum of frequencies with which the object **x** was understood as the object **y** and **y** was understood as **x** is defined as [132, 178]

$$s(\mathbf{x}, \mathbf{y}) = \frac{f(\mathbf{x}, \mathbf{y})}{f(\mathbf{x}, \mathbf{x})} + \frac{f(\mathbf{y}, \mathbf{x})}{f(\mathbf{y}, \mathbf{y})}. \tag{6.12.16}$$

This is a useful similarity measure in various applications such as speech analysis, fault diagnosis, etc., where $f(\mathbf{x}, \mathbf{y})$ and $f(\mathbf{y}, \mathbf{x})$, for example, are the frequencies with which the consonant **x** was heard as the consonant **y** and the consonant **y** was heard as the consonant **x**, respectively.

A similarity value is also called *similarity strength*.

Given N objects $\mathbf{x}_1, \ldots, \mathbf{x}_N$. The matrix whose entries are pairwise similarities s_{ij} is called the *similarity matrix*, and is denoted by $\mathbf{S} = [s_{ij}]_{i=1, j=1}^{N,N}$, where

$$s_{ij} = s(\mathbf{x}_i, \mathbf{x}_j), \quad i = 1, \ldots, N; \; j = 1, \ldots, N, \tag{6.12.17}$$

with $s_{ii} = s(\mathbf{x}_i, \mathbf{x}_i) = 0$.

Similarly, when using the pairwise distances d_{ij} in clustering, the matrix **D** with entries

$$d_{ij} = d(\mathbf{x}_i, \mathbf{x}_j), \quad i = 1, \ldots, N; \; j = 1, \ldots, N, \tag{6.12.18}$$

is known as the *distance matrix*, where $d_{ii} = d(\mathbf{x}_i, \mathbf{x}_i) = 0$.

The similarity matrix is a symmetric and nonnegative matrix:

$$\mathbf{S}^T = \mathbf{S} \quad \text{and} \quad \mathbf{S} \geq 0. \tag{6.12.19}$$

This is because of the similarity symmetry $s(\mathbf{x}_i, \mathbf{x}_j) = s(\mathbf{x}_j, \mathbf{x}_i)$ and the nonnegativity of similarity, $s(\mathbf{x}_i, \mathbf{x}_j) \geq 0$, $\forall i, j$.

Similarly, the distance matrix **D** is also symmetric and nonnegative.

Similarity assessment usually depends on practical applications. For example, there are the following similarity assessment methods in clinical text recognition [50]:

- *Word similarity:* A vector of words weighted by the term frequency-inverse document frequency (TF-IDF) weighting scheme is used to represent each

sentence. Then the cosine similarity between two vectors is calculated as the similarity between the two sentences.
- *Syntax similarity:* Each sentence is parsed by the Stanford parser and the dependency relations derived from the parse tree are used to form the vector. Each dependency relation in the vector is weighted by using the TF-IDF weight scheme based on their counts in the sentence and the corpus. Finally, cosine similarity is computed for each pair of sentences, similar to the method of word similarity.
- *Semantic similarity:* This method calculates semantic similarity between two sentences based on concept similarity: (1) extraction of clinical concepts in each sentence so that each sentence can be represented using a vector of union concepts from the two sentences; and (2) calculation of the similarity between the two sentence vectors of concepts, by measuring similarity scores between any two concepts and computing the cosine similarity of two sentence vectors.
- *Combined similarity:* This method combines all word, syntactic, and semantic information for similarity calculation. First combine words and dependency relations for the same sentence into one vector, and then compute the cosine similarity for each pair of sentences based on the new vectors. The final combined similarity between the two sentences is the average similarity for both the newly computed cosine similarity between word/dependency vectors and the semantic similarity, which can be extracted from task-specific vocabularies such as the Unified Medical Language System (UMLS).

6.12.2 Hierarchical Clustering

In unsupervised clustering based on similarity measures, if the number of objects is large, since there is one similarity value for each pair of objects, the resulting array of similarity measures can be so enormous that the underlying pattern or structure is not evident.

If similarity measures are arranged into different rows according to their "values" or "strength," we have an array of similarity measures, and thus construct a hierarchical system of clustering representations, ranging from the first clustering (top row) to the last clustering (bottom line). The first clustering is the "weak" clustering (i.e., each of N objects is represented as a separate cluster), and the last clustering is the "strong" clustering (i.e., all N objects are grouped together as a single cluster). This is the basic idea of the *hierarchical clustering scheme* (HCS).

To introduce the general notion of a hierarchical clustering scheme, we are given N objects, represented by the integers 1 through N. Moreover, we are also given a sequence of $m + 1$ clusterings, $c_0, c_1, \ldots, c_m$ such that $C = c_1 \cup c_2 \cup \cdots \cup c_m$ and $|C| = m + 1$, and each clustering c_i corresponds to its value of similarity measure, $\alpha_i \in [0, 1]$. Let c_0 be the weak clustering of the N objects, with dissimilarity measure $\alpha_0 = D(\mathbf{x}_i, \mathbf{x}_j) = 0$ for any $i \neq j$ (i.e., any pair of N objects cannot be completely similar), and let c_m be the strong clustering with the smallest dissimilarity value. We require also that the numbers α_i increase, i.e., $\alpha_{i+1} \geq \alpha_i$ for

6.12 Unsupervised Clustering

$i = 1, 2, \ldots, m$, and the clusters "increase" also, where $c_{i+1} > c_i$ means that every cluster in c_{i+1} is the merging (or union) of clusters in c_i. This general arrangement is referred to as a hierarchical clustering scheme (HCS) [132].

The steps of hierarchical clustering methodology are given below [185]:

1. Define decision variables, feasible set, and objective functions.
2. Choose and apply a Pareto optimization algorithm, e.g., NSGA-II.
3. Clustering analysis:
   - *Clustering tendency:* By visual inspection or data projections verify that a hierarchical cluster structure is a reasonable model for the data.
   - *Data scaling:* Remove implicit variable weightings due to relative scales using range scaling.
   - *Proximity:* Select and apply an appropriate similarity measure for the data, here, Euclidean distance.
   - *Choice of algorithm(s):* Consider the assumptions and characteristics of clustering algorithms and select the most suitable algorithm for the application, here, group average linkage.
   - *Application of algorithm:* Apply the selected algorithm and obtain a dendrogram.
   - *Validation:* Examine the results based on application subject matter knowledge, assess the fit to the input data and stability of the cluster structure, and compare the results of multiple algorithms, if used.
4. Represent and use the clusters and structure: if the clustering is reasonable and valid, then examine the divisions in the hierarchy for trade-offs and other information to aid decision making.

Example 6.4 Six objects have the following similarity values:

$$0.04 : \quad (3, 6) \tag{6.12.20}$$

$$0.08 : \quad (3, 5), (5, 6) \tag{6.12.21}$$

$$0.22 : \quad (1, 3), (1, 5), (1, 6), (2, 4) \tag{6.12.22}$$

$$0.35 : \quad (1, 2), (1, 4) \tag{6.12.23}$$

$$> 0.35 : \quad (2, 3), (2, 4), (2, 5), (3, 4), (4, 5), (4, 6). \tag{6.12.24}$$

The hierarchical clustering results are

$$0.00 : \quad [1], [3], [5], [6], [4], [2] \tag{6.12.25}$$

$$0.04 : \quad [1], [3, 6], [2], [4], [5] \tag{6.12.26}$$

$$0.08 : \quad [3, 5, 6], [1], [2], [4] \tag{6.12.27}$$

$$0.22 : \quad [1, 3, 5, 6], [2, 4] \tag{6.12.28}$$

$$0.35 : \quad [1, 2, 3, 4, 5, 6]. \tag{6.12.29}$$

Table 6.1 Similarity values and clustering results

Values	Object number					
	1	3	6	5	2	4
0.00	•	•	•	•	•	•
0.04	•	XXXXXXXX		•	•	•
0.08	•	XXXXXXXXXXXXXXX			•	•
0.22	XXXXXXXXXXXXXXXXXXXXXXXX				•	•
0.35	XXXXXXXXXXXXXXXXXXXXXXXX				XXXXXXXX	

Table 6.1 shows similarity values and corresponding hierarchical clustering results.

In hierarchical clustering, similarity measures have the following properties [132]:

- $D(\mathbf{x}, \mathbf{y}) = 0$ if and only if $\mathbf{x} = \mathbf{y}$.
- $D(\mathbf{x}, \mathbf{y}) = D(\mathbf{y}, \mathbf{x})$ for all objects $\mathbf{x}$ and $\mathbf{y}$.
- The ultrametric inequality:

$$D(\mathbf{x}, \mathbf{z}) \leq \max\{D(\mathbf{x}, \mathbf{y}), D(\mathbf{y}, \mathbf{z})\}. \tag{6.12.30}$$

- The triangle inequality:

$$D(\mathbf{x}, \mathbf{z}) \leq D(\mathbf{x}, \mathbf{y}) + D(\mathbf{y}, \mathbf{z}). \tag{6.12.31}$$

Due to the symmetry $D(\mathbf{x}, \mathbf{y}) = D(\mathbf{y}, \mathbf{x})$, so for N objects the hierarchical clustering scheme method needs only computing $N(N-1)/2$ dissimilarities.

The (i, j)th entry of the similarity (or distance) matrix $\mathbf{S}$ is defined by the similarity between the ith and jth objects as follows:

$$s_{ij} \stackrel{\text{def}}{=} D(\mathbf{x}_i, \mathbf{x}_j). \tag{6.12.32}$$

It should be noted that if all objects are clustered into one class at the mth hierarchical clustering with similarity $\alpha^{(m)}$, then the (i, j)th entries of similarity matrix are given by

$$s_{ij} = \begin{cases} s_{ij}, & s_{ij} \leq \alpha^{(m)}; \\ \alpha^{(m)}, & s_{ij} > \alpha^{(m)}. \end{cases} \tag{6.12.33}$$

In hierarchical clustering, there are different distance matrices corresponding to different clusterings.

Table 6.2 gives the distance matrix before clustering in Example 6.4.

The distance matrix after the first clustering [1], [2], [3, 6], [4], [5] is given in Table 6.3.

6.12 Unsupervised Clustering

Table 6.2 Distance matrix in Example 6.4

d_{ij}	1	2	3	4	5	6
1	0.00	0.35	0.22	0.35	0.22	0.22
2	0.35	0.00	0.35	0.22	0.35	0.35
3	0.22	0.35	0.00	0.35	0.08	0.04
4	0.35	0.22	0.35	0.00	0.35	0.35
5	0.22	0.35	0.08	0.35	0.00	0.08
6	0.22	0.35	0.04	0.35	0.08	0.00

Table 6.3 Distance matrix after first clustering in Example 6.4

d_{ij}	1	2	[3, 6]	4	5
1	0.00	0.35	0.22	0.35	0.22
2	0.35	0.00	0.35	0.22	0.35
[3, 6]	0.22	0.35	0.00	0.35	0.08
4	0.35	0.22	0.35	0.00	0.35
5	0.22	0.35	0.08	0.35	0.00

Table 6.4 Distance matrix after second clustering in Example 6.4

d_{ij}	1	2	[3, 5, 6]	4
1	0.00	0.35	0.22	0.35
2	0.35	0.00	0.35	0.22
[3, 5, 6]	0.22	0.35	0.00	0.35
4	0.35	0.22	0.35	0.00

Table 6.4 shows the distance matrix after the second clustering [1], [2], [3, 5, 6], [4] in Example 6.4.

Mimicking the above processes, we can find the distance matrices after third clustering etc.

Given N objects and a similarity metric $d = D(\mathbf{x}, \mathbf{y})$ satisfying the ultrametric inequality (6.12.30), the minimum method for hierarchical clustering is as follows [132].

1. Make the weak clustering C_0 with $d = 0$.
2. For the given clustering C_{i-1} with the distance matrix defined for all objects or clusters in C_{i-1}, let α_i be the smallest nonzero entry in the distance matrix. Merge the pair of objects and/or clusters with distance α_i, to create C_i of value α_i.
3. Create a new similarity function for C_i in the following manner: if $\mathbf{x}$ and $\mathbf{y}$ are clustered in C_i and not in C_{i-1} (i.e., $d(\mathbf{x}, \mathbf{y}) = \alpha_i$), then define the distance from the cluster $[\mathbf{x}, \mathbf{y}]$ to any third object or cluster, $\mathbf{z}$, by $d([\mathbf{x}, \mathbf{y}], \mathbf{z}) = \min\{d(\mathbf{x}, \mathbf{z}), d(\mathbf{y}, \mathbf{z})\}$. If $\mathbf{x}$ and $\mathbf{y}$ are objects and/or clusters in C_{i-1} not clustered in C_i, then $d(\mathbf{x}, \mathbf{y})$ remains the same. Thus, we obtain a new similarity function d for C_i in this way.
4. Repeat Steps 2 and 3 above until the strong clustering is finally obtained.

If $d([\mathbf{x},\mathbf{y}],\mathbf{z}) = \min\{d(\mathbf{x},\mathbf{z}), d(\mathbf{y},\mathbf{z})\}$ in Step 3 above is replaced by $d([\mathbf{x},\mathbf{y}],\mathbf{z}) = \max\{d(\mathbf{x},\mathbf{z}), d(\mathbf{y},\mathbf{z})\}$, the resulting method is called the maximum method for hierarchical clustering.

6.12.3 Fisher Discriminant Analysis (FDA)

In Sect. 6.11.3, we have presented Fisher discriminant analysis (FDA) for supervised binary classification and its extension to tensor data. In this subsection, we deal with the FDA for unsupervised multiple clustering.

The divergence is a measure of the distance or dissimilarity between two signals and is often used in feature discrimination and evaluation of the effectiveness of class discrimination.

Let Q denote the common dimension of the signal-feature vectors extracted by various methods. Assume that there are c classes of signals; the Fisher class separability measure, called simply the *Fisher measure*, between the ith and the jth classes is used to determine the ranking of feature vectors. Consider the class discrimination of c classes of signals. Let $\mathbf{v}_k^{(l)}$ denote the kth sample feature vector of the lth class of signals, where $l = 1, \ldots, c$ and $k = 1, \ldots, l_K$, where l_K is the number of sample feature vectors in the lth signal class. Under the assumption that the prior probabilities of the random vectors $\mathbf{v}^{(l)} = \mathbf{v}_k^{(l)}$ are the same for $k = 1, \ldots, l_K$ (i.e., equal probabilities), the Fisher measure is defined as

$$m^{(i,j)} = \frac{\sum_{l=i,j}\left[\mathrm{mean}_k\left(\mathbf{v}_k^{(l)}\right) - \mathrm{mean}_l\left(\mathrm{mean}\left(\mathbf{v}_k^{(l)}\right)\right)\right]^2}{\sum_{l=i,j} \mathrm{var}_k\left(\mathbf{v}_k^{(l)}\right)}, \qquad (6.12.34)$$

where

- $\mathrm{mean}_k\left(\mathbf{v}_k^{(l)}\right)$ is the mean (centroid) of all signal-feature vectors of the lth class;
- $\mathrm{var}\left(\mathbf{v}_k^{(l)}\right)$ is the variance of all signal-feature vectors of the lth class;
- $\mathrm{mean}_l\left(\mathrm{mean}_k\left(\mathbf{v}_k^{(l)}\right)\right)$ is the total sample centroid of all classes.

As an extension of the Fisher measure, consider the projection of all $Q \times 1$ feature vectors onto the $(c-1)$th dimension of class discrimination space.

Put $N = N_1 + \cdots + N_c$, where N_i represents the number of feature vectors of the ith class signal extracted in the training phase. Suppose that

$$\mathbf{s}_{i,k} = [s_{i,k}(1), \ldots, s_{i,k}(Q)]^T$$

6.12 Unsupervised Clustering

denotes the ith ($Q \times 1$) feature vector obtained by the kth group of observation data of the ith signal in the training phase, while

$$\mathbf{m}_i = [m_i(1), \ldots, m_i(Q)]^T$$

is the sample mean of the ith signal-feature vector, where

$$m_i(q) = \frac{1}{N_i} \sum_{k=1}^{N_i} s_{i,k}(q), \quad i = 1, \ldots, c, \; q = 1, \ldots, Q.$$

Similarly, let

$$\mathbf{m} = [m(1), \ldots, m(Q)]^T$$

denote the ensemble mean of all the signal-feature vectors obtained from all the observation data, where

$$m(q) = \frac{1}{c} \sum_{i=1}^{c} m_i(q), \quad q = 1, \ldots, Q.$$

Using the above vectors, one can define a $Q \times Q$ within-class scatter matrix [83]

$$\mathbf{S}_w \stackrel{\text{def}}{=} \frac{1}{c} \sum_{i=1}^{c} \left(\frac{1}{N_i} \sum_{k=1}^{N_i} (\mathbf{s}_{i,k} - \mathbf{m}_i)(\mathbf{s}_{i,k} - \mathbf{m}_i)^T \right) \qquad (6.12.35)$$

and a $Q \times Q$ between-class scatter matrix [83]

$$\mathbf{S}_b \stackrel{\text{def}}{=} \frac{1}{c} \sum_{i=1}^{c} (\mathbf{m}_i - \mathbf{m})(\mathbf{m}_i - \mathbf{m})^T. \qquad (6.12.36)$$

For the optimal class discrimination matrix $\mathbf{U}$ to be determined, define a criterion function

$$J(\mathbf{U}) \stackrel{\text{def}}{=} \frac{\prod_{\text{diag}} \mathbf{U}^T \mathbf{S}_b \mathbf{U}}{\prod_{\text{diag}} \mathbf{U}^T \mathbf{S}_w \mathbf{U}}, \qquad (6.12.37)$$

where $\prod_{\text{diag}} \mathbf{A}$ denotes the product of diagonal entries of the matrix $\mathbf{A}$. As it is a measure of class discrimination ability, J should be maximized. We refer to Span($\mathbf{U}$) as the *class discrimination space*, if

$$\mathbf{U} = \arg\max_{\mathbf{U} \in \mathbb{R}^{Q \times Q}} J(\mathbf{U}) = \frac{\prod_{\text{diag}} \mathbf{U}^T \mathbf{S}_b \mathbf{U}}{\prod_{\text{diag}} \mathbf{U}^T \mathbf{S}_w \mathbf{U}}. \tag{6.12.38}$$

This optimization problem can be equivalently written as

$$[\mathbf{u}_1, \ldots, \mathbf{u}_Q] = \arg\max_{\mathbf{u}_i \in \mathbb{R}^Q} \left\{ \frac{\prod_{i=1}^Q \mathbf{u}_i^T \mathbf{S}_b \mathbf{u}_i}{\prod_{i=1}^Q \mathbf{u}_i^T \mathbf{S}_w \mathbf{u}_i} = \prod_{i=1}^Q \frac{\mathbf{u}_i^T \mathbf{S}_b \mathbf{u}_i}{\mathbf{u}_i^T \mathbf{S}_w \mathbf{u}_i} \right\} \tag{6.12.39}$$

whose solution is given by

$$\mathbf{u}_i = \arg\max_{\mathbf{u}_i \in \mathbb{R}^Q} \frac{\mathbf{u}_i^T \mathbf{S}_b \mathbf{u}_i}{\mathbf{u}_i^T \mathbf{S}_w \mathbf{u}_i}, \quad i = 1, \ldots, Q. \tag{6.12.40}$$

This is just the maximization of the generalized Rayleigh quotient. The above equation has a clear physical meaning: the column vectors of the matrix $\mathbf{U}$ constituting the optimal class discrimination subspace should maximize the between-class scatter and minimize the within-class scatter, namely these vectors should maximize the generalized Rayleigh quotient.

For c classes of signals, the optimal class discrimination subspace is $(c-1)$-dimensional. Therefore Eq. (6.12.40) is maximized only for $c-1$ generalized Rayleigh quotients. In other words, it is necessary to solve the generalized eigenvalue problem

$$\mathbf{S}_b \mathbf{u}_i = \lambda_i \mathbf{S}_w \mathbf{u}_i, \quad i = 1, 2, \ldots, c-1, \tag{6.12.41}$$

for $c-1$ generalized eigenvectors $\mathbf{u}_1, \ldots, \mathbf{u}_{c-1}$. These generalized eigenvectors constitute the $Q \times (c-1)$ matrix

$$\mathbf{U}_{c-1} = [\mathbf{u}_1, \ldots, \mathbf{u}_{c-1}] \tag{6.12.42}$$

whose columns span the optimal class discrimination subspace.

After obtaining the $Q \times (c-1)$ matrix $\mathbf{U}_{c-1}$, we can find its projection onto the optimal class discrimination subspace,

$$\mathbf{y}_{i,k} = \mathbf{U}_{c-1}^T \mathbf{s}_{i,k}, \quad i = 1, \ldots, c, \; k = 1, \ldots, N_i, \tag{6.12.43}$$

for every signal-feature vector obtained in the training phase.

When there are only three classes of signals ($c = 3$), the optimal class discrimination subspace is a plane, and the projection of every feature vector onto the optimal class discrimination subspace is a point. These projection points directly reflect the discrimination ability of different feature vectors in signal classification.

6.12.4 K-Means Clustering

Clustering technique may broadly be divided into two categories: hierarchical and non-hierarchical [6]. As one of the more widely used non-hierarchical clustering techniques, the K-means [251] aims to minimize the sum of squared Euclidean distance between patters and cluster centers via an iterative hill climbing algorithm.

Given a set of N data points in d-dimensional vector space $\mathbb{R}^d$, consider partitioning these points into a number (say $K \leq N$) of groups (or clusters) based on some similarity/dissimilarity metric which establishes a rule for assigning patterns to the domain of a particular cluster center.

Let the set of N data be $S = \{\mathbf{x}_1, \ldots, \mathbf{x}_N\}$ and the K clusters be $C_1, \ldots, C_K$. Then the clustering is to make K clusters satisfy the following conditions [186]:

$$C_i \not\subseteq \emptyset, \quad \text{for } i = 1, \ldots, K;$$

$$C_i \cap C_j = \emptyset, \quad \text{for } i = 1, \ldots, K; j = 1, \ldots, K \text{ and } i \neq j;$$

$$\bigcup_{i=1}^{K} C_i = S.$$

Here each observation vector belongs to the cluster with the nearest mean, serving as a prototype of the cluster.

Given an open set $S \in \mathbb{R}^d$, the set $\{V_i\}_{i=1}^{K}$ is called a *tessellation* of S if $V_i \cap V_j = \emptyset$ for $i \neq j$ and $\bigcup_{i=1}^{K} V_i = S$. When a set of points $\{\mathbf{z}_i\}_{i=1}^{K}$ is given, the sets $\hat{V}_i$ corresponding to the point $\mathbf{z}_i$ are defined by

$$\hat{V}_i = \left\{ \mathbf{x} \in S \big| \, \|\mathbf{x} - \mathbf{z}_i\| \leq \|\mathbf{x} - \mathbf{z}_j\| \text{ for } j = 1, \ldots, K, j \neq i \right\}. \quad (6.12.44)$$

The points $\{\mathbf{z}_i\}_{i=1}^{K}$ are called *generators*. The set $\{\hat{V}_i\}_{i=1}^{K}$ is a *Voronoi tessellation* or *Voronoi diagram* of S, and each $\hat{V}_i$ is known as the *Voronoi set* corresponding to the point $\mathbf{z}_i$.

The K-means algorithm requires three user-specified parameters: number of clusters K, cluster initialization, and distance metric [127, 245].

- *Number of clusters K:* The most critical parameter is K. As there are no perfect mathematical criterion available, the K-means algorithm is usually run independently for different values of K. The value of K is then selected using the partition that appears the most meaningful to the domain expert.

- *Cluster initialization:* To overcome the local minima of K-means clustering, the K-means algorithm selects K initial cluster centers $\mathbf{m}_1, \ldots, \mathbf{m}_K$ randomly from the N data points $\{\mathbf{x}_1, \ldots, \mathbf{x}_N\}$.
- *Distance metric:* Let d_{ik} denote the distance between data $\mathbf{x}_i = [x_{i1}, \ldots, x_{iN}]^T$ and $\mathbf{x}_k = [x_{k1}, \ldots, x_{kN}]^T$. K-means is typically used with the Euclidean distance $d_{ik} = \sum_j (x_{ik} - x_{jk})^2$ or Mahalanobis distance metric for computing the distance between points $\mathbf{x}_i, \mathbf{x}_k$ and cluster centers.

The most common algorithm uses an iterative refinement technique for finding a locally minimal solution without computing the Voronoi set $\hat{V}_i$. Due to its ubiquity it is often called the *K-means algorithm*. However, the K-means algorithm, as a greedy algorithm, can only converge to a local minimum.

Given any set Z of K centers, for each center $\mathbf{z} \in Z$, let $V(\mathbf{z})$ denote its neighborhood, namely the set of data points for which $\mathbf{z}$ is the nearest neighbor. Based on an elegant random sequential sampling method, the random K-means algorithm, or simply the K-means algorithm, does not require the calculation of Voronoi sets.

Algorithm 6.19 shows the *random K-means algorithm* [15, 81].

Algorithm 6.19 Random K-means algorithm [15, 81]

1. **input:** Dataset $S = \{\mathbf{x}_1, \ldots, \mathbf{x}_N\}$ with $\mathbf{x}_i \in \mathbb{R}^d$, a positive integer K.
2. **initialization:**
   2.1 Select an initial set of K centers $Z = \{\mathbf{z}_1, \ldots, \mathbf{z}_K\}$, e.g., by using a Monte Carlo method;
   2.2 Set $j_i = 1$ for $i = 1, \ldots, K$.
3. **repeat**
4.   Select a data point $\mathbf{x} \in S$ at random, according to the probability density function $\rho(x)$.
5.   Find the $\mathbf{z}_i \in \mathbb{R}^d$ that is closest to $\mathbf{x}$, for example, use
$$\mathbf{z}_i = \arg\min_{\mathbf{z}_1, \ldots, \mathbf{z}_N}\{\|\mathbf{x} - \mathbf{z}_1\|^2, \ldots, \|\mathbf{x} - \mathbf{z}_K\|^2\},$$
denote the index of that $\mathbf{z}_i$ by i^*. Ties are resolved arbitrarily.
6.   Set
$$\mathbf{z}_{i^*} \leftarrow \frac{j_{i^*}\mathbf{z}_{i^*} + \mathbf{x}}{j_{i^*} + 1} \quad \text{and} \quad j_{i^*} \leftarrow j_{i^*} + 1,$$
this new $\mathbf{z}_{i^*}$, along with the unchanged $\mathbf{z}_i, i \neq i^*$, forms the new set of cluster centers $\mathbf{z}_1, \ldots, \mathbf{z}_K$.
7.   Assign point $\mathbf{x}_i, i = 1, \ldots, n$ to cluster C_j, $j \in \{1, \ldots, K\}$ if and only if $\|\mathbf{x}_i - \mathbf{z}_j\| \leq \|\mathbf{x}_i - \mathbf{z}_p\|$, $p = 1, \ldots, K$; and $j \neq p$.
8. **exit if** K cluster centers $\mathbf{z}_1, \ldots, \mathbf{z}_K$ are converged.
9. **return**
10. **output:** K cluster centroids $\mathbf{z}_1, \ldots, \mathbf{z}_K$.

Once the K cluster centers $\mathbf{z}_1, \ldots, \mathbf{z}_K$ are obtained, we can perform the classification of the new testing data point $\mathbf{y} \in \mathbb{R}^d$ by

$$\text{class of } \mathbf{y} = \text{order number of } \mathbf{z}_k = \min\{d(\mathbf{y}, \mathbf{z}_1), \ldots, d(\mathbf{y}, \mathbf{z}_K)\} \quad (6.12.45)$$

where $d(\mathbf{y}, \mathbf{z})$ is the distance metric.

6.12 Unsupervised Clustering

The classic K-means clustering algorithm finds cluster centers that minimize the distance between data points and the nearest center. K-means can be viewed as a way of constructing a "dictionary" $\mathbf{D} \in \mathbb{R}^{n \times k}$ of k vectors so that a data vector $\mathbf{x}_i \in \mathbb{R}^n, i = 1, \ldots, m$ can be mapped to a code vector $\mathbf{s}_i$ that minimizes the error in reconstruction [56].

It is known [56] that initialization and pre-processing of the inputs are useful and necessary for K-means clustering.

1. Though it is common to initialize K-means to randomly chosen examples drawn from the data, this has been found to be a poor choice. Instead, it is better to randomly initialize the centroids from a normal distribution and then normalize them to unit length.
2. The centroids tend to be biased by the correlated data, and thought whitening input data, obtained centroids are more orthogonal.

After initializing and whitening inputs, a modified version of K-means, called *gain shape vector quantization* [290] or *spherical K-means* [75], is applied to find the dictionary $\mathbf{D}$.

Including the initialization and pre-processing, the full K-means training are as follows.

1. Normalize inputs:

$$\mathbf{x}_i \equiv \frac{\mathbf{x}_i - \text{mean}(\mathbf{x}_i)}{\text{var}(\mathbf{x}_i) + \epsilon_{\text{norm}}}, \quad \forall\, i, \qquad (6.12.46)$$

where mean($\mathbf{x}_i$) and var($\mathbf{x}_i$) are the mean and variance of the elements of $\mathbf{x}_i$, respectively.

2. Whiten inputs:

$$\text{cov}(\mathbf{x}) = \mathbf{V}\mathbf{D}\mathbf{V}^T \qquad (6.12.47)$$

$$\mathbf{x}_i = \mathbf{V}(\mathbf{D} + \epsilon)^{-1/2}\mathbf{V}^T \mathbf{x}_i, \quad \forall\, i. \qquad (6.12.48)$$

3. Loop until convergence (typically 10 iterations is enough):

$$s_i(j) = \begin{cases} \mathbf{d}_j^T \mathbf{x}_i, & \text{if } j = \arg\max_l |\mathbf{d}_l^T \mathbf{x}_i|; \\ 0, & \text{otherwise}. \end{cases} \quad (\forall\, j, i) \qquad (6.12.49)$$

$$\mathbf{D} = \mathbf{X}\mathbf{S}^T + \mathbf{D}, \qquad (6.12.50)$$

$$\mathbf{d}_j = \mathbf{d}_j / \|\mathbf{d}_j\|_2, \quad \forall\, j. \qquad (6.12.51)$$

The columns $\mathbf{d}_j$ of the dictionary $\mathbf{D}$ give the centroids, and K-means tends to learn low-frequency edge-like centroids.

This procedure of normalization, whitening, and K-means clustering is an effective "off the shelf" unsupervised learning module that can serve in many feature-learning roles [56].

6.13 Spectral Clustering

Spectral clustering is one of the most popular modern clustering algorithms, and has been extensively studied in the image processing, data mining, and machine learning communities, see, e.g., [190, 230, 257].

In multivariate statistics and the clustering of data, spectral clustering techniques make use of the spectrum (eigenvalues) of the similarity matrix of the data to perform dimensionality reduction before clustering in fewer dimensions. The similarity matrix is provided as an input and consists of a quantitative assessment of the relative similarity of each pair of points in the data set.

6.13.1 Spectral Clustering Algorithms

It is assumed that data consists of N "points" $\mathbf{x}_1, \ldots, \mathbf{x}_N$ which can be arbitrary objects. Let their pairwise similarities $s_{ij} = s(\mathbf{x}_i, \mathbf{x}_j)$ be measured by using some similarity function which is symmetric and nonnegative, and denote the corresponding similarity matrix by $\mathbf{S} = [s_{ij}]_{i,j=1}^{N,N}$.

Two popular spectral clustering algorithms [190, 230] are based on spectral graph partitioning. A promising alternative is to use matrix algebra approach.

Let $\mathbf{h}_k$ be a cluster that partitions the mapped data $\mathbf{\Phi} = [\boldsymbol{\phi}(\mathbf{x}_1), \ldots, \boldsymbol{\phi}(\mathbf{x}_N)]$, producing the output $\mathbf{\Phi}\mathbf{h}_k$. Maximize the output energy to get

$$\max \left\{ \|\mathbf{\Phi}\mathbf{h}_k\|_2^2 = \mathbf{h}_k^T \mathbf{\Phi}^T \mathbf{\Phi} \mathbf{h}_k = \mathbf{h}_k^T \mathbf{W} \mathbf{h}_k \right\},$$

where

$$\mathbf{W} = \mathbf{\Phi}^T \mathbf{\Phi} = [K(\mathbf{x}_i, \mathbf{x}_j)]_{i,j=1}^{N,N} = \left[\boldsymbol{\phi}^T(\mathbf{x}_i)\boldsymbol{\phi}(\mathbf{x}_j)\right]_{i,j=1}^{N,N}. \quad (6.13.1)$$

In spectral clustering, due to $K(\mathbf{x}_i, \mathbf{x}_j) = s(\mathbf{x}_i, \mathbf{x}_j)$, the affinity matrix $\mathbf{W}$ becomes the similarity matrix $\mathbf{S}$, i.e., $\mathbf{W} = \mathbf{S}$.

For K clusters we maximize the total output energy

$$\max_{\mathbf{h}_1, \ldots, \mathbf{h}_K} \left\{ \sum_{k=1}^{K} \mathbf{h}_k^T \mathbf{W} \mathbf{h}_k \right\} \quad (6.13.2)$$

subject to $\quad \mathbf{h}_k^T \mathbf{h}_k = 1, \ \forall k = 1, \ldots, K, \quad (6.13.3)$

6.13 Spectral Clustering

to get the primal constrained optimization problem:

$$\max_{\mathbf{H}=[\mathbf{h}_1,\ldots,\mathbf{h}_K]} \text{tr}(\mathbf{H}^T\mathbf{W}\mathbf{H}) \qquad (6.13.4)$$

$$\text{subject to} \quad \mathbf{H}^T\mathbf{H} = \mathbf{I}. \qquad (6.13.5)$$

By using the Lagrange multiplier method, the above constrained optimization can be rewritten as the dual unconstrained optimization:

$$\max_{\mathbf{H}} \left\{ \mathcal{L}(\mathbf{H}) = \text{tr}(\mathbf{H}^T\mathbf{W}\mathbf{H}) + \lambda\left(1 - \|\mathbf{H}\|_2^2\right) \right\}. \qquad (6.13.6)$$

From the first-order optimality condition we have

$$\frac{\partial \mathcal{L}}{\partial \mathbf{H}} = \mathbf{O} \quad \Rightarrow \quad 2\mathbf{W}\mathbf{H} - 2\lambda\mathbf{H} = \mathbf{O} \quad \Rightarrow \quad \mathbf{W}\mathbf{H} = \lambda\mathbf{H} \qquad (6.13.7)$$

due to the symmetry of the affinity matrix $\mathbf{W} = \mathbf{S}$. That is, $\mathbf{H} = [\mathbf{h}_1, \ldots, \mathbf{h}_K]$ consists of the K leading eigenvectors (with the largest eigenvalues) of $\mathbf{W}$ as columns.

If the affinity matrix $\mathbf{W}$ is replaced by the normalized Laplacian matrix $\mathbf{L}_{\text{sys}} = \mathbf{I} - \mathbf{D}^{-1/2}\mathbf{W}\mathbf{D}^{-1/2}$, then the K spectral clusters $\mathbf{h}_1, \ldots, \mathbf{h}_K$ are given directly by the K leading eigenvectors. This is just the key point of the normalized spectral clustering algorithm developed by Ng et al. [190], see Algorithm 6.20.

Algorithm 6.20 Normalized spectral clustering algorithm of Ng et al. [190]

1. **input:** Dataset $V = \{\mathbf{x}_1, \ldots, \mathbf{x}_N\}$ with $\mathbf{x}_i \in \mathbb{R}^n$, number k of clusters, kernel function $K(\mathbf{x}_i, \mathbf{x}_j) : \mathbb{R}^n \times \mathbb{R}^n \to \mathbb{R}$.
2. Construct the $N \times N$ weighted adjacency matrix $\mathbf{W} = [\delta_{ij} K(\mathbf{x}_i, \mathbf{x}_j)]_{i,j=1}^{N,N}$.
3. Compute the $N \times N$ diagonal matrix $[\mathbf{D}]_{ij} = \delta_{ij} \sum_{j=1}^{N} w_{ij}$.
4. Compute the $N \times N$ normalized Laplacian matrix $\mathbf{L}_{\text{sys}} = \mathbf{I} - \mathbf{D}^{-1/2}\mathbf{W}\mathbf{D}^{-1/2}$.
5. Compute the first k eigenvectors $\mathbf{u}_1, \ldots, \mathbf{u}_k$ of the eigenvalue problem $\mathbf{L}_{\text{sys}}\mathbf{u} = \lambda\mathbf{u}$.
6. Let the matrix $\mathbf{U} = [\mathbf{u}_1, \ldots, \mathbf{u}_k] \in \mathbb{R}^{N \times k}$.
7. Compute the (i, j)th entries of the $N \times k$ matrix $\mathbf{T} = [t_{ij}]_{i,j=1}^{N,k}$ as $t_{ij} = u_{ij}/(\sum_{m=1}^{k} u_{im}^2)^{1/2}$. Let $\mathbf{y}_i = [t_{i1}, \ldots, t_{ik}]^T \in \mathbb{R}^k$, $i = 1, \ldots, N$.
8. Cluster the points $\mathbf{y}_1, \ldots, \mathbf{y}_N$ via the K-means algorithm into clusters $C_1, \ldots, C_k$.
9. **output:** clusters $A_1, \ldots, A_k$ with $A_i = \{j | \mathbf{y}_j \in C_i\}$.

Interestingly, if letting $\tilde{\mathbf{H}} = \mathbf{D}^{1/2}\mathbf{H}$ and $\tilde{\mathbf{W}} = \mathbf{D}^{-1/2}\mathbf{W}\mathbf{D}^{-1/2}$, then the solution to the optimization problem

$$\max_{\mathbf{H}} \left\{ \mathcal{L}(\tilde{\mathbf{H}}) = \text{tr}(\tilde{\mathbf{H}}^T\tilde{\mathbf{W}}\tilde{\mathbf{H}}) + \lambda\left(1 - \|\tilde{\mathbf{H}}\|_2^2\right) \right\} \qquad (6.13.8)$$

is given by the eigenproblem $\tilde{\mathbf{W}}\tilde{\mathbf{H}} = \lambda \tilde{\mathbf{H}}$ which, by using $\tilde{\mathbf{H}} = \mathbf{D}^{1/2}\mathbf{H}$ and $\tilde{\mathbf{W}} = \mathbf{D}^{-1/2}\mathbf{W}\mathbf{D}^{-1/2}$, becomes the generalized eigenproblem $\mathbf{WH} = \lambda \mathbf{DH}$. Therefore, if the affinity matrix $\mathbf{W}$ is replaced by the normalized Laplacian matrix $\mathbf{L} = \mathbf{D} - \mathbf{W}$, then the K spectral clusters $\mathbf{h}_1, \ldots, \mathbf{h}_K$ consist of the K leading generalized eigenvectors (corresponding to the largest generalized eigenvalues) of the matrix pencil $(\mathbf{L}_{\text{sys}}, \mathbf{D})$. Therefore, we get the normalized spectral clustering algorithm of Shi and Malik [230], as shown in Algorithm 6.21.

Algorithm 6.21 Normalized spectral clustering algorithm of Shi and Malik [230]

1. **input:** Dataset $V = \{\mathbf{x}_1, \ldots, \mathbf{x}_N\}$ with $\mathbf{x}_i \in \mathbb{R}^n$, number k of clusters, kernel function $K(\mathbf{x}_i, \mathbf{x}_j) : \mathbb{R}^n \times \mathbb{R}^n \to \mathbb{R}$.
2. Construct the $N \times N$ weighted adjacency matrix $\mathbf{W} = [\delta_{ij} K(\mathbf{x}_i, \mathbf{x}_j)]_{i,j=1}^{N,N}$.
3. Compute the $N \times N$ diagonal matrix $[\mathbf{D}]_{ij} = \delta_{ij} \sum_{j=1}^{N} w_{ij}$.
4. Compute the $N \times N$ normalized Laplacian matrix $\mathbf{L}_{\text{sys}} = \mathbf{D}^{-1/2}(\mathbf{D} - \mathbf{W})\mathbf{D}^{-1/2}$.
5. Compute the first k generalized eigenvectors $\mathbf{u}_1, \ldots, \mathbf{u}_k$ of the generalized eigenproblem $\mathbf{L}_{\text{sys}} \mathbf{u} = \lambda \mathbf{D} \mathbf{u}$.
6. Let the matrix $\mathbf{U} = [\mathbf{u}_1, \ldots, \mathbf{u}_k] \in \mathbb{R}^{N \times k}$.
7. Compute the (i, j)th entries of the $N \times k$ matrix $\mathbf{T} = [t_{ij}]_{i,j=1}^{N,k}$ as $t_{ij} = u_{ij}/(\sum_{m=1}^{k} u_{im}^2)^{1/2}$. Let $\mathbf{y}_i = [t_{i1}, \ldots, t_{ik}]^T \in \mathbb{R}^k$, $i = 1, \ldots, N$.
8. Cluster the points $\mathbf{y}_1, \ldots, \mathbf{y}_N$ via the K-means algorithm into clusters $C_1, \ldots, C_k$.
9. **output:** clusters $A_1, \ldots, A_k$ with $A_i = \{j | \mathbf{y}_j \in C_i\}$.

Remark 1 Both Algorithms 6.20 and 6.21 are normalized spectral clustering algorithms, but different normalized Laplacian matrices are used: $\mathbf{L}_{\text{sys}} = \mathbf{D}^{-1/2}(\mathbf{D} - \mathbf{W})\mathbf{D}^{-1/2}$ in Algorithm 6.21, while $\mathbf{L}_{\text{sys}} = \mathbf{I} - \mathbf{D}^{-1/2}\mathbf{W}\mathbf{D}^{-1/2}$ in Algorithm 6.20.

Remark 2 If using the unnormalized Laplacian matrix $\mathbf{L} = \mathbf{D} - \mathbf{W}$ instead of the normalized Laplacian matrices $\mathbf{L}_{\text{sys}}$ in Step 3 in the algorithms above described, then we get the unnormalized spectral clustering algorithms.

Remark 3 In practical computation of the "ideal" weighted adjacency matrix $\mathbf{W}$, there is a perturbation to $\mathbf{A}$ that results in estimation errors of either generalized eigenvectors in Algorithm 6.21 or eigenvectors in Algorithm 6.20, i.e., there is a perturbed version $\hat{\mathbf{T}} = \mathbf{T} + \mathbf{E}$, where $\mathbf{E}$ is a perturbation matrix to the "ideal" principal (generalized) eigenvector matrix $\mathbf{T}$.

Spectral clustering techniques are widely used, due to their simplicity and empirical performance advantages compared to other clustering methods, such as hierarchical clustering and K-means clustering and so on.

In recent years, spectral clustering has gained new promotion and development. In the next two subsections, we introduce the constrained spectral clustering and the fast spectral clustering, respectively.

6.13.2 Constrained Spectral Clustering

Both K-means and hierarchical clustering are constrained clustering, but an intractable problem is how to satisfy many constraints in these algorithmic settings. Alternative to encode many constraints in clustering is to use spectral clustering.

To help spectral clustering recover from an undesirable partition, side information in various forms can be introduced, in either small or large amounts [259]:

- *Pairwise constraints:* In some cases a pair of instances must be in the same cluster (*Must-Link*), while a pair of instances cannot be in the same cluster (*Cannot-Link*).
- *Partial labeling:* There can be labels on some of the instances, which are neither complete nor exhaustive.
- *Alternative weak distance metrics:* In some situations there may be more than one distance metrics available.
- *Transfer of knowledge:* When the graph Laplacian is treated as the target domain, it can be viewed as the source domain to transfer knowledge from a different but related graph.

Consider encoding side information, in k-class clustering, with an $N \times N$ constraint matrix $\mathbf{Q}$ defined as [259]:

$$Q_{ij} = Q_{ji} = \begin{cases} +1, & \text{if nodes } \mathbf{x}_i, \mathbf{x}_j \text{ belong to same cluster;} \\ -1, & \text{if nodes } \mathbf{x}_i, \mathbf{x}_j \text{ belong to different clusters;} \\ 0, & \text{no side information available.} \end{cases} \quad (6.13.9)$$

Define the *cluster indicator vector* $\mathbf{u} = [u_1, \ldots, u_N]^T$ as

$$u_i = \begin{cases} +1, & \text{if } \mathbf{x}_i \text{ belongs to cluster } i; \\ -1, & \text{if } \mathbf{x}_i \text{ does not belong to cluster } i; \end{cases} \quad (6.13.10)$$

for $i = 1, \ldots, N$. Therefore,

$$\mathbf{u}^T \mathbf{Q} \mathbf{u} = \sum_{i=1}^{N} \sum_{j=1}^{N} u_i u_j Q_{ij} \quad (6.13.11)$$

becomes a real-valued measure of how well the constraints in $\mathbf{Q}$ are satisfied in the relaxed sense. The larger $\mathbf{u}^T \mathbf{Q} \mathbf{u}$ is, the better the cluster assignment $\mathbf{u}$ conforms to the given constraints in $\mathbf{Q}$.

Let constant $\alpha \in \mathbb{R}$ be a lower-bound of $\mathbf{u}^T\mathbf{Q}\mathbf{u}$, i.e., $\mathbf{u}^T\mathbf{Q}\mathbf{u} \geq \alpha$. If substituting $\mathbf{u}$ with $\mathbf{D}^{-1/2}\mathbf{v}$ (where $\mathbf{D}$ is a degree matrix), then this inequality constraint becomes

$$\mathbf{v}^T\bar{\mathbf{Q}}\mathbf{v} \geq \alpha, \quad (6.13.12)$$

where $\bar{\mathbf{Q}} = \mathbf{D}^{-1/2}\mathbf{Q}\mathbf{D}^{-1/2}$ is the *normalized constraint matrix*.

Therefore, the *constrained spectral clustering* can be written as a constrained optimization problem [259]:

$$\min_{\mathbf{v} \in \mathbb{R}^N} \mathbf{v}^T\bar{\mathbf{L}}\mathbf{v} \quad \text{subject to } \mathbf{v}^T\bar{\mathbf{Q}}\mathbf{v} \geq \alpha, \ \mathbf{v}^T\mathbf{v} = \text{vol}, \ \mathbf{v} \neq \mathbf{D}^{1/2}\mathbf{1}, \quad (6.13.13)$$

where $\bar{\mathbf{L}} = \mathbf{D}^{-1/2}\mathbf{L}\mathbf{D}^{-1/2}$ and $\text{vol} = \sum_{i=1}^{N} D_{ii}$ is the volume of graph $G(V, E)$. The second constraint $\mathbf{v}^T\mathbf{v} = \text{vol}$ normalizes $\mathbf{v}$; the third constraint $\mathbf{v} \neq \mathbf{D}^{1/2}\mathbf{1}$ rules out the trivial solution $\mathbf{D}^{1/2}\mathbf{1}$. By the Lagrange multiplier method, one gets the dual spectral clustering problem:

$$\min_{\mathbf{v} \in \mathbb{R}^N} \left\{ \mathcal{L}_D = \mathbf{v}^T\bar{\mathbf{L}}\mathbf{v} + \lambda\left(\alpha - \mathbf{v}^T\bar{\mathbf{Q}}\mathbf{v}\right) + \mu\left(\text{vol} - \mathbf{v}^T\mathbf{v}\right) \right\}. \quad (6.13.14)$$

From the stationary point condition, we have

$$\frac{\partial \mathcal{L}_D}{\partial \mathbf{v}} = \mathbf{0} \quad \Rightarrow \quad \bar{\mathbf{L}}\mathbf{v} - \lambda\bar{\mathbf{Q}}\mathbf{v} - \mu\mathbf{v} = \mathbf{0}. \quad (6.13.15)$$

Letting the auxiliary variable $\beta = -\frac{\mu}{\lambda}\text{vol}$ then the stationary point condition is

$$\bar{\mathbf{L}}\mathbf{v} - \lambda\bar{\mathbf{Q}}\mathbf{v} + \frac{\lambda\beta}{\text{vol}}\mathbf{v} = \mathbf{0} \quad \text{or} \quad \bar{\mathbf{L}}\mathbf{v} = \lambda\left(\bar{\mathbf{Q}} - \frac{\beta}{\text{vol}}\mathbf{I}\right)\mathbf{v}. \quad (6.13.16)$$

That is, $\mathbf{v}$ is the generalized eigenvector of the matrix pencil $\left(\bar{\mathbf{L}}, \bar{\mathbf{Q}} - \frac{\lambda\beta}{\text{vol}}\mathbf{I}\right)$ with the largest generalized eigenvalue $\lambda_{\max}$.

If there is one largest generalized eigenvalue $\lambda_{\max}$, then the associated generalized eigenvector $\mathbf{u}$ is the unique solution of the generalized eigenproblem. However, if there are multiple (say, p) largest generalized eigenvalues, then p associated generalized eigenvectors $\mathbf{v}_1, \ldots, \mathbf{v}_p$ are solutions of the generalized eigenproblem. In order to find the unique solution, let $\mathbf{V}_1 = [\mathbf{v}_1, \ldots, \mathbf{v}_p] \in \mathbb{R}^{N \times p}$. If using the Household transform matrix $\mathbf{H} \in \mathbb{R}^{p \times p}$ such that

$$\mathbf{V}_1\mathbf{H} = \begin{bmatrix} \gamma & 0 & \cdots & 0 \\ \vdots & & & \\ \mathbf{z} & & \times & \end{bmatrix}, \quad (6.13.17)$$

6.13 Spectral Clustering

where $\gamma \neq 0$ and $\times$ may be any values that are not interesting, then the unique solution of the generalized eigenproblem is given by $\mathbf{v} = \begin{bmatrix} \gamma \\ \mathbf{z} \end{bmatrix}$. Then, make $\mathbf{v} \leftarrow \frac{\mathbf{v}}{\|\mathbf{v}\|}\sqrt{\text{vol}}$.

Algorithm 6.22 shows the constrained spectral clustering method.

Algorithm 6.22 Constrained spectral clustering for two-way partition [259]

1. **input:** Affinity matrix $\mathbf{A}$, constraint matrix $\mathbf{Q}$, constant β.
2. $\text{vol} \leftarrow \sum_{i=1}^{N}\sum_{j=1}^{N} A_{ij}$, $\mathbf{D} \leftarrow \text{Diag}\left(\sum_{j=1}^{N} A_{1j}, \ldots, \sum_{j=1}^{N} A_{Nj}\right)$.
3. $\bar{\mathbf{L}} \leftarrow \mathbf{D}^{-1/2}\mathbf{L}\mathbf{D}^{-1/2}$, $\bar{\mathbf{Q}} \leftarrow \mathbf{D}^{-1/2}\mathbf{Q}\mathbf{D}^{-1/2}$.
4. $\lambda_{\max} \leftarrow$ the largest eigenvalue of $\bar{\mathbf{Q}}$.
5. **if** $\beta \geq \lambda_{\max} \cdot \text{vol}$, **then**
6.   return $\mathbf{u} = \mathbf{0}$;
7. **else**
8.   Solve the generalized eigenproblem $\bar{\mathbf{L}}\mathbf{v} = \lambda\left(\bar{\mathbf{Q}} - \frac{\beta}{\text{vol}}\mathbf{I}\right)\mathbf{v}$ for $(\lambda_i, \mathbf{v}_i)$
9.   **if** $\lambda_{\max} = \lambda_1$ is unique, **then**
10.   return $\mathbf{v} \leftarrow \frac{\mathbf{v}}{\|\mathbf{v}\|}\sqrt{\text{vol}}$.
11.   **else**
12. $\mathbf{V}_1 \leftarrow [\mathbf{v}_1, \ldots, \mathbf{v}_p]$ for $\lambda_{\max} = \lambda_1 \approx \cdots \approx \lambda_p$.
13.   Perform Household transform $\mathbf{V}_1\mathbf{H}$ in (6.13.17) to get $\mathbf{v} = \begin{bmatrix} \gamma \\ \mathbf{z} \end{bmatrix}$.
14.   return $\mathbf{v} \leftarrow \frac{\mathbf{v}}{\|\mathbf{v}\|}\sqrt{\text{vol}}$.
15. $\mathbf{v}^* \leftarrow \arg\min_{\mathbf{v}} \mathbf{v}^T\bar{\mathbf{L}}\mathbf{v}$, where $\mathbf{v}$ is among the feasible eigenvectors generated in the previous step;
16.   **end if**
17. **end if**
18. **output:** the optimal (relaxed) cluster indicator $\mathbf{u}^* \leftarrow \mathbf{D}^{-1/2}\mathbf{v}^*$.

For K-way partition, given all the feasible eigenvectors, the top $K-1$ eigenvectors are picked in terms of minimizing $\mathbf{v}^T\bar{\mathbf{L}}\mathbf{v}$. Instead of only using the optimal feasible eigenvector $\mathbf{u}^*$ in Algorithm 6.22, one needs to preserve top $(K-1)$ eigenvectors associated with positive eigenvalues, and perform K-means algorithm based on that embedding. Letting the $K-1$ eigenvectors form the columns of $\mathbf{V} \in \mathbb{R}^{N \times (K-1)}$, then K-means clustering is performed on the rows of $\mathbf{V}$ to get the final clustering. See Algorithm 6.23.

6.13.3 Fast Spectral Clustering

For large data sets, a serious computational challenge of spectral clustering is computation of an affinity matrix between pairs of data points for large data sets.

By combining the spectral clustering algorithm with the Nyström approximation to the graph Laplacian, a fast spectral clustering was developed by Choromanska et al. in 2013 [52]. The Nyström r-rank approximation for any symmetric positive semi-definite matrix $\mathbf{L} \in \mathbb{R}^{N \times N}$ with large N is below.

Algorithm 6.23 Constrained spectral clustering for K-way partition [259]
1. **input:** Affinity matrix $\mathbf{A}$, constraint matrix $\mathbf{Q}$, constant β, number of classes K.
2. vol $\leftarrow \sum_{i=1}^{N}\sum_{j=1}^{N} A_{ij}$, $\mathbf{D} \leftarrow \text{Diag}(\sum_{j=1}^{N} A_{1j}, \ldots, \sum_{j=1}^{N} A_{Nj})$.
3. $\bar{\mathbf{L}} \leftarrow \mathbf{D}^{-1/2}\mathbf{L}\mathbf{D}^{-1/2}$, $\bar{\mathbf{Q}} \leftarrow \mathbf{D}^{-1/2}\mathbf{Q}\mathbf{D}^{-1/2}$.
4. $\lambda_{\max} \leftarrow$ the largest eigenvalue of $\bar{\mathbf{Q}}$.
5. **if** $\beta \geq \lambda_{K-1}\text{vol}$, **then**
6.   **return** $\mathbf{u} = \mathbf{0}$
7. **else**
8.   Solve $\bar{\mathbf{L}}\mathbf{v}\left(\bar{\mathbf{L}} - \frac{\beta}{\text{vol}}\mathbf{I}\right)\mathbf{v}$ for $\mathbf{v}_1, \ldots, \mathbf{v}_{K-1}$.
9. $\mathbf{v} \leftarrow \frac{\mathbf{v}}{\|\mathbf{v}\|}\sqrt{\text{vol}}$.
10. $\mathbf{V}^* \leftarrow \arg\min_{\mathbf{V}\in\mathbb{R}^{N\times(K-1)}} \text{tr}(\mathbf{V}^T\bar{\mathbf{L}}\mathbf{V})$, where the columns of $\mathbf{V}$ are a subset of the feasible eigenvectors generated in the previous step;.
11.   **return** $\mathbf{u}^* \leftarrow \text{kmeans}(\mathbf{D}^{-1/2}\mathbf{V}^*, K)$.
12. **end if**

1. Perform uniform sampling of $\mathbf{L}$ (without replacement schemes) to create matrix $\mathbf{C} \in \mathbb{R}^{N\times l}$ from the sampled columns, and let $\mathcal{L}$ be indices of l columns sampled.
2. Form matrix $\mathbf{W} \in \mathbb{R}^{l\times l}$ from $\mathbf{C}$ by sampling its l rows via the indices in $\mathcal{L}$.
3. Compute the eigenvalue decomposition (EVD) $\mathbf{W} = \mathbf{U}\mathbf{\Sigma}\mathbf{U}^T$, where $\mathbf{U}$ is orthonormal and $\mathbf{\Sigma} = \mathbf{Diag}(\sigma_1, \ldots, \sigma_l)$ is an $l \times l$ real diagonal matrix with the diagonal sorted in decreasing order.
4. Calculate the best rank-r approximation to $\mathbf{W}$, denoted as $\mathbf{W}_r = \sum_{p=1}^{r} \sigma_p \mathbf{u}_p \mathbf{u}^p$, and its Moore–Penrose inverse $\mathbf{W}_r^{\dagger} = \sum_{p=1}^{r} \sigma_p^{-1} \mathbf{u}_p \mathbf{u}^p$, where $\mathbf{u}_p$ and $\mathbf{u}^p$ are the pth column and row of $\mathbf{U}$, respectively.
5. The Nyström r-rank approximation $\mathbf{L}_r$ of $\mathbf{L}$ can be obtained as $\mathbf{L}_r = \mathbf{C}\mathbf{W}_r^{\dagger}\mathbf{C}$.
6. Perform the EVD $\mathbf{W}_r = \mathbf{U}_{W_r}\mathbf{\Sigma}_{W_r}\mathbf{U}_{W_r}$.
7. The EVD of the Nyström r-rank approximation $\mathbf{L}_r$ is given by $\mathbf{L}_r = \mathbf{U}_{L_r}\mathbf{\Sigma}_{L_r}\mathbf{U}_{L_r}^T$, where $\mathbf{\Sigma}_{L_r} = \frac{N}{l}\mathbf{\Sigma}_{W_r}$ and $\mathbf{U}_{L_r} = N^{-1/2}\mathbf{C}\mathbf{U}_{W_r}\mathbf{\Sigma}_{W_r}^{-1}$.

Algorithm 6.24 shows the Nyström method for matrix approximation.

Algorithm 6.24 Nyström method for matrix approximation [52]
1. **input:** $N \times N$ symmetric matrix $\mathbf{L}$, number l of columns sampled, rank r of matrix approximation ($r \leq l \ll N$).
2. $\mathcal{L} \leftarrow$ indices of l columns sampled.
3. $\mathbf{C} \leftarrow \mathbf{L}(:, \mathcal{L})$.
4. $\mathbf{W} \leftarrow \mathbf{C}(\mathcal{L}, :)$.
5. $\mathbf{W} = \mathbf{U}\mathbf{\Sigma}\mathbf{U}^T$ and $\mathbf{W}_r = \sum_{p=1}^{r} \sigma_p \mathbf{u}_p \mathbf{u}^p$, where $\mathbf{u}_p$ and $\mathbf{u}^p$ are the pth column and row of $\mathbf{U}$.
6. $\mathbf{W}_r = \mathbf{U}_{W_r}\mathbf{\Sigma}_{W_r}\mathbf{U}_{W_r}$.
7. $\mathbf{\Sigma}_{L_r} = \frac{N}{l}\mathbf{\Sigma}_{W_r}$ and $\mathbf{U}_{L_r} = \sqrt{\frac{l}{N}}\mathbf{C}\mathbf{U}_{W_r}\mathbf{\Sigma}_{W_r}^{-1}$.
8. **output:** Nyström approximation of $\mathbf{L}$ is given by $\mathbf{L}_r = \mathbf{U}_{L_r}\mathbf{\Sigma}_{L_r}\mathbf{U}_{L_r}^T$.

Once the Nyström approximation of $\mathbf{L}$ is obtained, one can get the fast spectral clustering algorithm of Choromanska et al. [52]. See Algorithm 6.25.

Algorithm 6.25 Fast spectral clustering algorithm [52]
---
1. **input:** $N \times d$ dataset $S = \{\mathbf{x}_1, \ldots, \mathbf{x}_N\} \in \mathbb{R}^d$, number k of clusters, number l of columns sampled, rank r of approximation ($k \leq r \leq l \ll N$).
2. **initialization:** $\mathbf{A} \in \mathbb{R}^{N \times N}$ with $a_{ij} = [\mathbf{A}]_{ij} = \delta_{ij} K(\mathbf{x}_i, \mathbf{x}_j)$ if $i \neq j$ and 0 otherwise.
3. $\mathcal{L} \leftarrow$ indices of l columns sampled (uniformly without replacement).
4. $\hat{\mathbf{A}} = [\hat{a}_{ij}]_{i,j=1}^{N,l} \leftarrow \mathbf{A}(:, \mathcal{L})$.
5. $\mathbf{D} \in \mathbb{R}^{N \times N}$ with $d_{ij} = \delta_{ij} 1/\sqrt{\sum_{j=1}^{l} \hat{a}_{ij}}$.
6. $\mathbf{\Delta} \in \mathbb{R}^{l \times l}$ with $\Delta_{ij} = \delta_{ij} 1/\sqrt{\sum_{i=1}^{N} \hat{a}_{ij}}$.
7. $\mathbf{C} \leftarrow \hat{\mathbf{I}} - \sqrt{\frac{l}{N}} \mathbf{D}\hat{\mathbf{A}}\mathbf{\Delta}$, where $\hat{\mathbf{I}}$ is an $N \times l$ matrix consisting of columns of the identity matrix $\mathbf{I}_{N \times N}$ indexed by $\mathcal{L}$.
8. $\mathbf{W} \leftarrow \mathbf{C}(\mathcal{L}, :)$.
9. $\mathbf{W} = \mathbf{U}\mathbf{\Sigma}\mathbf{U}^T$ and $\mathbf{W}_r = \sum_{p=1}^{r} \sigma_p \mathbf{u}_p \mathbf{u}^p$, where $\mathbf{u}_p$ and $\mathbf{u}^p$ are the pth column and row of $\mathbf{U}$.
10. $\mathbf{W}_r = \mathbf{U}_{W_r} \mathbf{\Sigma}_{W_r} \mathbf{U}_{W_r}$.
11. $\mathbf{\Sigma}_{L_r} = \frac{N}{l} \mathbf{\Sigma}_{W_r}$ and $\mathbf{U}_{L_r} = \sqrt{\frac{l}{N}} \mathbf{C} \mathbf{U}_{W_r} \mathbf{\Sigma}_{W_r}^{-1}$.
12. $\mathbf{Z} = [z_{im}] \in \mathbb{R}^{N \times k} \leftarrow k$ eigenvectors of $\mathbf{U}_{L_r}$ with k smallest eigenvalues.
13. $[\mathbf{Y}]_{im} = y_{im} = z_{im}/(\sum_m z_{im}^2)^{1/2}$ for $i = 1, \ldots, N$; $j = 1, \ldots, k$.
14. Use any k-clustering algorithm (e.g., K-means) to N rows of $\mathbf{Y} \in \mathbb{R}^{N \times k}$ to get k clusters $S_i = \{\mathbf{y}_{i_1}^r, \ldots, \mathbf{y}_{i_{m_i}}^r\}$, $i = 1, \ldots, k$ with $m_1 + \cdots + m_k = N$, where $\mathbf{y}_p^r$ is the pth row of $\mathbf{Y}$.
15. **output:** k spectral clustering results of S are given by $S_i = \{\mathbf{x}_{i_1}, \ldots, \mathbf{x}_{i_{m_i}}\}$, $i = 1, \ldots, k$.

6.14 Semi-Supervised Learning Algorithms

The significance of *semi-supervised learning* (SSL) is mainly reflected in the following three aspects [20].

- From an engineering standpoint, due to the difficulty of collecting labeled data, an approach to pattern recognition that is able to make better use of unlabeled data to improve recognition performance is of potentially great practical significance.
- Arguably, most natural (human or animal) learning occurs in the semi-supervised regime. In many cases, a small amount of feedback is sufficient to allow the child to master the acoustic-to-phonetic mapping of any language.
- In most pattern recognition tasks, humans have access only to a small number of labeled examples. The ability of humans to learn unsupervised concepts (e.g., learning clusters and categories of objects) suggests that the success of human learning in this small sample regime is plausibly due to effective utilization of the large amounts of unlabeled data to extract information that is useful for generalization.

The semi-supervised learning problem has recently drawn a large amount of attention in the machine learning community, mainly due to considerable improvement in learning accuracy when unlabeled data is used in conjunction with a small amount of labeled data.

6.14.1 Semi-Supervised Inductive/Transductive Learning

Typically, semi-supervised learning uses a small amount of labeled data with a large amount of unlabeled data. Hence, as the name suggests, semi-supervised learning falls between unsupervised learning (without any labeled training data) and supervised learning (with completely labeled training data).

In supervised classification, one is always interested in classifying future test data by using fully labeled training data. In semi-supervised classification, however, the training samples contain both labeled and unlabeled data.

Given a training set $D_{\text{train}} = (X_{\text{train}}, Y_{\text{train}})$ that consists of the labeled set $(X_{\text{labeled}}, Y_{\text{labeled}}) = \{(\mathbf{x}_1, y_1), \ldots (\mathbf{x}_l, y_l)\}$ and the disjoint unlabeled set $X_{\text{unlabeled}} = \{\mathbf{x}_{l+1}, \ldots, \mathbf{x}_{l+u}\}$. Depending on different training sets and goals, there are two slightly different types of semi-supervised learning: semi-supervised inductive learning and semi-supervised transductive learning.

Definition 6.24 (Inductive Learning) If training set is $\mathcal{D}_{\text{train}} = (X_{\text{train}}, Y_{\text{train}}) = (X_{\text{labeled}}, Y_{\text{labeled}}) = \{(\mathbf{x}_1, y_1), \ldots, (\mathbf{x}_l, y_l)\}$, and test set is $\mathcal{D}_{\text{test}} = X_{\text{test}}$ (unlabeled) that is not appeared in training set, then this setting is called *inductive learning*.

Definition 6.25 (Semi-Supervised Inductive Learning [110, 234]) Let training set consist of a large amount of unlabeled data $X_{\text{unlabeled}} = \{\mathbf{x}_{l+1}, \ldots, \mathbf{x}_{l+u}\}$ and a small amount of the auxiliary labeled samples $X_{\text{auxiliary}} = (X_{\text{labeled}}, Y_{\text{labeled}}) = \{(\mathbf{x}_1, y_1), \ldots, (\mathbf{x}_l, y_l)\}$ with $l \ll u$. *Semi-supervised inductive learning* builds a function f so that f is expected to be a good predictor on future data set X_{test}, beyond $\{\mathbf{x}_j\}_{j=l+1}^{l+u}$.

If we do not care about X_{test}, but only want to know the effect of machine learning on $X_{\text{unlabeled}}$, this setting is known as semi-supervised transductive learning, because the unlabeled set $X_{\text{unlabeled}}$ was seen in training.

Definition 6.26 (Semi-Supervised Transductive Learning) Let training set consist of a large amount of unlabeled data $X_{\text{unlabeled}} = \{\mathbf{x}_{l+1}, \ldots, \mathbf{x}_{l+u}\}$ and a small amount of the auxiliary labeled samples $(X_{\text{labeled}}, Y_{\text{labeled}}) = \{(\mathbf{x}_1, y_1), \ldots, (\mathbf{x}_l, y_l)\}$ with $l \ll u$. *Semi-supervised transductive learning* uses all labeled data $(X_{\text{labeled}}, Y_{\text{labeled}})$ and unlabeled data $X_{\text{unlabeled}}$ to give the prediction or classification labels of $X_{\text{unlabeled}}$ or its some interesting subset without building the function f.

6.14 Semi-Supervised Learning Algorithms

To put it simply, there are the following differences between inductive learning, semi-supervised learning, semi-supervised inductive learning, and semi-supervised transductive learning.

- Differences in training set: Inductive learning includes only all labeled data (X_{labled}, Y_{labled}), while semi-supervised learning, semi-super inductive learning, and semi-supervised transductive learning use the same training set including both labeled data (X_{labeled}, Y_{labeled}) = $\{(\mathbf{x}_1, y_1), \ldots, (\mathbf{x}_l, y_l)\}$ and unlabeled data $X_{\text{unlabeled}} = \{\mathbf{x}_{l+1}, \ldots, x_{l+u}\}$.
- Differences in testing data: Unlabeled data X_{unlabled} is not the testing data in inductive learning, semi-supervised learning, and semi-supervised inductive learning, i.e., the samples we want to predict are not seen (or used) in training, whereas unlabeled data X_{unlabled} is the testing data in transductive learning, i.e., the samples we want to predict have been seen (or used) in training: $X_{\text{test}} = X_{\text{unlabeled}}$.
- Differences in prediction functions: In inductive learning, semi-supervised learning, and semi-supervised inductive learning, one needs to build the prediction function on unseen (or used) data (i.e., outside the training data), $f : D_{\text{train}} \to D_{\text{test}}$. However, this function does not need to be built in semi-supervised transductive learning, which only needs to give the predictions for some subset, rather than all, of the unlabeled data seen in training.

The semi-supervised transductive learning is simply called the transductive learning. In semi-supervised learning, semi-supervised inductive learning, and semi-supervised transductive learning, a small amount of labeled data set $\{(\mathbf{x}_1, y_1), \ldots, (\mathbf{x}_l, y_l)\}$ is usually viewed as an *auxiliary set*, denoted as $D_{\text{auxiliary}} = (X_{\text{labeled}}, Y_{\text{labeled}})$. It has been noted that a small amount of auxiliary labeled data $X_{\text{auxiliary}}$ typically helps boost the performance of the classifier or predictor significantly.

In fact, most semi-supervised learning strategies are based on extending either unsupervised or supervised learning to include additional information typical of the other learning paradigm. Specifically, semi-supervised learning encompasses the following different settings [305]:

- *Semi-supervised classification:* It is an extension to the supervised classification, and its goal is to train a classifier f from both the labeled and unlabeled data, such that it is better than the supervised classifier trained on a small amount of labeled data alone.
- *Constrained clustering:* This is an extension to unsupervised clustering. Some "supervised information" can be so-called must-link constraints: two instances $\mathbf{x}_i, \mathbf{x}_j$ must be in the same cluster; and cannot-link constraints: $\mathbf{x}_i, \mathbf{x}_j$ cannot be in the same cluster. One can also constrain the size of the clusters. The goal of constrained clustering is to obtain better clustering than the clustering from unlabeled data alone.

The acquisition of labeled data for a learning problem often is expensive since it requires a skilled human agent or a physical experiment. The cost associated with the labeling process thus may render a fully labeled training set infeasible, while acquisition of unlabeled data is relatively inexpensive. In such situations, semi-supervised learning can be particularly useful.

For example, in the process industry, it is a difficult task to assign the type of detected faulty data samples. This may require process knowledge and experiences of process engineers, and will be costly and may need a lot of time. As a result, there are only a small number of faulty samples that are assigned to their corresponding types, while most faulty samples are unlabeled. Since these unlabeled samples still contain important information of the process, they can be used to improve greatly the efficiency of the fault classification system. In this case, as compared to the model that only depends on a small part of labeled data samples, semi-supervised learning, with both labeled samples and unlabeled samples, will provide an improved fault classification model [104].

To achieve higher accuracy of machine learning with fewer training labels and a large pool of unlabeled data, several approaches have been proposed for semi-supervised learning. In the following, we summarize the common semi-supervised learning algorithms: self-training, cluster-then-label, and co-training.

6.14.2 Self-Training

Self-training is a commonly used technique for semi-supervised learning, and is the learning process that uses its own predictions to teach itself. For this reason, it is also known as *self-teaching* or *bootstrapping*.

In self-training a classifier is first found with the small amount of labeled data. Then, the classifier is used to classify the unlabeled data. Typically the most confident unlabeled points, together with their predicted labels, are added to the training set. The classifier is retrained and the procedure repeated until all of unlabeled data are classified. See Algorithm 6.26 for self-training.

Algorithm 6.26 Self-training algorithm [305]

---
1. **input:** labeled data $\{(\mathbf{x}_i, y_i)\}_{i=1}^{l}$, unlabeled data $\{\mathbf{x}_j\}_{j=l+1}^{l+u}$.
2. **initialization:** $L = \{(\mathbf{x}_i, y_i)\}_{i=1}^{l}$ and $U = \{\mathbf{x}_j\}_{j=l+1}^{l+u}$. Let $k = l + 1$.
3. **repeat**
4.   Train f from L using supervised learning.
5.   Apply f to the unlabeled instance $\mathbf{x}_k$ in U.
6.   Remove a subset S of learning success from U; and add $\{(\mathbf{x}_k, f(\mathbf{x}_k)) | \mathbf{x}_k \in S\}$ to L.
7.   **exist** if $k = l + u$.
8. $k \leftarrow k + 1$.
9. **return**
10. **output:** f.

On self-training, one has the following comments and considerations [305].

Remark 1 For self-training, the algorithm's own predictions, assuming high confidence, tend to be correct. This is likely to be the case when the classes form well-separated clusters.

Remark 2 Implementation for self-training is simple. Importantly, it is also a wrapper method, which means that the choice of learner for f in Step 4 is left completely open: the learner can be a simple kNN algorithm, or a very complicated classifier. The self-training procedure "wraps" around the learner without changing its inner workings. This is important for many real applications like natural language processing, where the learners can be complicated black boxes not amenable to changes.

Self-training is available for a larger labeled data set L. If L is small, an early mistake made by f can reinforce itself by generating incorrectly labeled data. Retraining with this data will lead to an even worse f in the next iteration.

The heuristics for *semi-supervised clustering* can alleviate the above problem, see Algorithm 6.27.

Algorithm 6.27 Propagating 1-nearest neighbor clustering algorithm [305]

1. **input:** labeled data $\{(\mathbf{x}_i, y_i)\}_{i=1}^{l}$, unlabeled data $\{\mathbf{x}_j\}_{j=l+1}^{l+u}$, distance function $d(\cdot, \cdot)$.
2. **initialization:** $L = \{(\mathbf{x}_i, y_i)\}_{i=1}^{l}$ and $U = \{\mathbf{x}_j\}_{j=l+1}^{l+u} = \{\mathbf{z}_j\}_{j=l+1}^{l+u}$.
3. **repeat**
4.   Select $\mathbf{z} = \arg\min_{\mathbf{z} \in U, \mathbf{x} \in L} d(\mathbf{z}, \mathbf{x})$.
5.   Let k be the label of $\mathbf{x} \in L$ nearest to $\mathbf{z}$, and add the new labeled data $(\mathbf{z}, y_k)$ to L.
6.   Remove $\mathbf{z}$ from U.
7.   **exit** if U is empty.
8. **return**
9. **output:** semi-supervised learning dataset $L = \{(\mathbf{x}_i, y_i)\}_{i=1}^{l+u}$.

6.14.3 Co-training

The traditional machine learning algorithms including self-training and the propagating 1-nearest neighbor clustering are based on the assumption that the instance has one attribute or one kind of information. However, in many real-world applications there are *two-views data* and *co-occurring data*. These data can be described using two different attributes or two different "kinds" of information. For example, a web-page contains two different kinds of information [30]:

- One kind of information about a web-page is the text appearing on the document itself.
- Another kind of information is the anchor text attached to hyperlinks pointing to this page, from other pages on the web.

Therefore, analyzing two views and co-occurring data is an important challenge in machine learning, pattern recognition, and signal processing communities, e.g., automatic annotation of text, audio, music, image, and video (see, e.g., [30, 80, 144, 193]) and sensor data mining [62].

More generally, the *co-training setting* [30] applies when a data set has a natural division of its features. Co-training setting assumes usually:

- features can be split into two sets;
- each sub-feature set is sufficient to train a good classifier;
- the two sets are conditionally independent given the class.

Traditional machine learning algorithms that learn over these domains ignore this division and pool all features together.

Definition 6.27 (Co-training Algorithm [30, 193]) *Co-training algorithm* is a multi-view weakly supervised algorithm using the co-training setting, and may learn separate classifiers over each of the feature sets, and combine their predictions to decrease classification error.

Given a small set of labeled two-view data $\mathcal{L} = \{(\mathbf{x}_i, \mathbf{y}_i), z_i\}_{i=1}^{l}$ and a large set of unlabeled two-view data $\mathcal{U} = \{(\mathbf{x}_j, \mathbf{y}_j)\}_{j=l+1}^{l+u}$, where $(\mathbf{x}_i, \mathbf{y}_i)$ is called the data bag and z_i is the class label associated labeled two-view data bag $(\mathbf{x}_i, \mathbf{y}_i)$, $i = 1, \ldots, l$.

Two classifiers $\mathbf{h}^{(1)}$ and $\mathbf{h}^{(2)}$ work in co-training fashion: $\mathbf{h}^{(1)}$ is a view-1 classifier that is only based the first view $\mathbf{x}^{(1)}$ (say $\mathbf{x}_i$) and ignores the second view $\mathbf{x}^{(2)}$ (say $\mathbf{y}_i$), while $\mathbf{h}^{(2)}$ is a view-2 classifier based on the second view $\mathbf{x}^{(2)}$ and ignores the first view $\mathbf{x}^{(1)}$.

The working process of co-training is [189, 304]: two separate classifiers are initially trained with the labeled data, on the two sub-feature sets, respectively. Then, each classifier uses one view of the data and selects most confident predictions from the pool and adds the corresponding instances with their predicted labels to the labeled data while maintaining the class distribution in the labeled data. One classifier "teaches" the other classifier with the few unlabeled examples and provides their feeling most confident unlabeled-data predictions as the training data for the other. Each classifier is then retrained with the additional training examples given by the other classifier, and the process repeats. In co-training process, the unlabeled data is eventually exhausted.

Co-training is a wrapper method. That is, it does not matter what the learning algorithms are for the two classifiers $\mathbf{h}^{(1)}$ and $\mathbf{h}^{(2)}$. The only requirement is that the two classifiers can assign a confidence score to their predictions. The confidence score is used to select which unlabeled instances to turn into additional training data for the other view. Being a wrapper method, co-training is widely applicable to many tasks.

As semi-supervised learning for two-view data and co-occurring data, the most popular algorithm is the standard co-training developed by Blum and Mitchell [30], as shown in Algorithm 6.28.

The following are two variants of co-training [304].

Algorithm 6.28 Co-training algorithm [30]

1. **input:** labeled data set $L = \{\mathbf{x}_i, y_i\}_{i=1}^{l}$, unlabeled data set $U = \{\mathbf{x}_j\}_{j=l+1}^{l+u}$. Each instance has two views $\mathbf{x}_i = [\mathbf{x}_i^{(1)}, \mathbf{x}_i^{(2)}]$.
2. **initialization:** create a pool U' of examples by choosing u examples at random from U.
3. **repeat** until unlabeled data is used up:
4.   Train a view-1 classifier $\mathbf{h}^{(1)}$ from $\{\mathbf{x}_i^{(1)}, y_i\}$ in L.
5.   Train a view-2 classifier $\mathbf{h}^{(2)}$ from $\{\mathbf{x}_i^{(2)}, y_i\}$ in L.
6.   Allow $\mathbf{h}^{(1)}$ to label p positive and n negative examples from U'.
7.   Allow $\mathbf{h}^{(2)}$ to label p positive and n negative examples from U'.
8.   Add these self-labeled examples to L, and remove these from the unlabeled data U.
9.   Randomly choose $2p + 2n$ examples from U to replenish U'.
10. **return**
11. **output:** $\{\mathbf{x}_i, y_i\}_{i=1}^{l+u}$.

1. *Democratic co-learning* [299]. Let $\mathcal{L}$ be the set of labeled data and $\mathcal{U}$ be the set of unlabeled data, and $A_1, \ldots, A_n$ (for $n \geq 3$) the provided supervised learning algorithms. Democratic co-learning begins by training all learners on the original labeled data set $\mathcal{L}$. For every example $\mathbf{x}_u \in \mathcal{U}$ in the unlabeled data set $\mathcal{U}$, each learner predicts a label. If a majority of learners confidently agree on the class of an unlabeled point $\mathbf{x}_u$, that classification is used as the label of $\mathbf{x}_u$. Then $\mathbf{x}_u$ and its label are added to the training data. All learners are retrained on the updated training set. The final prediction is made with a variant of a weighted majority vote among the n learners.
2. *Tri-training* [300]. If two of tri-training learners agree on the classification of an unlabeled point, the classification is used to teach the third classifier. This approach thus avoids the need of explicitly measuring label confidence of any learner. It can be applied to data sets without different views, or different types of classifiers.

6.15 Canonical Correlation Analysis

As the previous section illustrated, for a process described by two sets of variables corresponding to two different aspects or views, analyzing the relations between these two views helps improve the understanding of the underlying system. An alternative to two-view data analysis is canonical correlation analysis.

Canonical correlation analysis (CCA) [7, 122] is a standard two-view multivariate statistical method of correlating linear relationships between two random vectors such that they are maximally correlated, and hence is a powerful tool for analyzing multi-dimensional paired sample data. Paired samples are chosen in a random manner, and thus are random samples.

6.15.1 Canonical Correlation Analysis Algorithm

In the case of CCA, the variables of an observation can be partitioned into two sets that can be seen as the two views of the data. Let the two-view matrices $\mathbf{X} = [\mathbf{x}_1, \ldots, \mathbf{x}_N] \in \mathbb{R}^{N \times p}$ and $\mathbf{Y} = [\mathbf{y}_1, \ldots, \mathbf{y}_N] \in \mathbb{R}^{N \times q}$ have covariance matrices ($\mathbf{C}_{xx}, \mathbf{C}_{yy}$) and cross-covariance matrices ($\mathbf{C}_{xy}, \mathbf{C}_{yx}$), respectively. The row vectors $\mathbf{x}_n \in \mathbb{R}^{1 \times p}$ and $\mathbf{y}_n \in \mathbb{R}^{1 \times q}$ ($n = 1, \ldots, N$) denote the sets of empirical multivariate observations in $\mathbf{X}$ and $\mathbf{Y}$, respectively.

The aim of CCA is to extract the linear relations between the variables of $\mathbf{X}$ and $\mathbf{Y}$. For this end, consider the following transformations:

$$\mathbf{X}\mathbf{w}_x = \mathbf{z}_x \quad \text{and} \quad \mathbf{Y}\mathbf{w}_y = \mathbf{z}_y, \tag{6.15.1}$$

where $\mathbf{w}_x \in \mathbb{R}^p$ and $\mathbf{w}_y \in \mathbb{R}^q$ are, respectively, the projection vectors associated with the data matrices $\mathbf{X}$ and $\mathbf{Y}$, while $\mathbf{z}_x \in \mathbb{R}^N$ and $\mathbf{z}_y \in \mathbb{R}^N$ are, respectively, the images of the positions $\mathbf{w}_x$ and $\mathbf{w}_y$. The positions $\mathbf{w}_x$ and $\mathbf{w}_y$ are often referred to as *canonical weight vectors*, and the images $\mathbf{z}_x$ and $\mathbf{z}_y$ are termed as *canonical variants* or *score variants* [253, 265]. Hence, the data matrices $\mathbf{X}$ and $\mathbf{Y}$ represent linear transformations of the positions $\mathbf{w}_x$ and $\mathbf{w}_y$ onto the images $\mathbf{z}_x$ and $\mathbf{z}_y$ in the space $\mathbb{R}^N$, respectively.

Define the cosine of the angle between the images $\mathbf{z}_x$ and $\mathbf{z}_y$:

$$\cos(\mathbf{z}_x, \mathbf{z}_y) = \frac{\langle \mathbf{z}_x, \mathbf{z}_y \rangle}{\|\mathbf{z}_x\| \|\mathbf{z}_y\|}, \tag{6.15.2}$$

which is referred to as the *canonical correlation*.

The constraints of CCA on the mappings are given below:

- the position vectors of the images $\mathbf{z}_x$ and $\mathbf{z}_y$ are unit norm vectors;
- the enclosing angle $\theta \in [0, \frac{\pi}{2}]$ [108], between $\mathbf{z}_x$ and $\mathbf{z}_y$, is minimized.

The principle behind CCA is to find two positions in the two data spaces, respectively, that have images on a unit ball such that the angle between them is minimized and consequently the canonical correlation is maximized:

$$\cos \theta = \max_{\mathbf{z}_x, \mathbf{z}_y \in \mathbb{R}^N} \langle \mathbf{z}_x, \mathbf{z}_y \rangle \tag{6.15.3}$$

subject to $\|\mathbf{z}_x\|_2 = 1$ and $\|\mathbf{z}_y\|_2 = 1$.

For solving the above optimization problem, define the sample cross-covariance matrix

$$\mathbf{C}_{xy} = \mathbf{X}^T \mathbf{Y}, \quad \mathbf{C}_{yx} = \mathbf{Y}^T \mathbf{X} = \mathbf{C}_{xy}^T \tag{6.15.4}$$

and the empirical variance matrices

$$\mathbf{C}_{xx} = \mathbf{X}^T \mathbf{X}, \quad \mathbf{C}_{yy} = \mathbf{Y}^T \mathbf{Y}. \tag{6.15.5}$$

6.15 Canonical Correlation Analysis

Hence, the constraints $\|\mathbf{z}_x\|_2 = 1$ and $\|\mathbf{z}_y\|_2 = 1$ can be rewritten as follows:

$$\|\mathbf{z}_x\|_2^2 = \mathbf{z}_x^T \mathbf{z}_x = \mathbf{w}_x^T \mathbf{C}_{xx} \mathbf{w}_x = 1, \qquad (6.15.6)$$

$$\|\mathbf{z}_y\|_2^2 = \mathbf{z}_y^T \mathbf{z}_y = \mathbf{w}_y^T \mathbf{C}_{yy} \mathbf{w}_y = 1. \qquad (6.15.7)$$

In this case, the covariance matrix between $\mathbf{z}_x$ and $\mathbf{z}_y$ can be also written as

$$\mathbf{z}_x^T \mathbf{z}_y = \mathbf{w}_x^T \mathbf{X}^T \mathbf{Y} \mathbf{w}_y = \mathbf{w}_x^T \mathbf{C}_{xy} \mathbf{w}_y. \qquad (6.15.8)$$

Then, the optimization problem (6.15.3) becomes

$$\cos\theta = \max_{\mathbf{z}_x, \mathbf{z}_y \in \mathbb{R}^N} \langle \mathbf{z}_x, \mathbf{z}_y \rangle = \max_{w_x \in \mathbb{R}^p, w_y \in \mathbb{R}^q} \mathbf{w}_x^T \mathbf{C}_{xy} \mathbf{w}_y \qquad (6.15.9)$$

subject to $\|\mathbf{z}_x\|_2^2 = \mathbf{w}_x^T \mathbf{C}_{xx} \mathbf{w}_x = 1$ and $\|\mathbf{z}_y\|_2^2 = \mathbf{w}_y^T \mathbf{C}_{yy} \mathbf{w}_y = 1$.

In order to solve this constrained optimization problem, define the Lagrange objective function

$$\mathcal{L}(\mathbf{w}_x, \mathbf{w}_y) = \mathbf{w}_x^T \mathbf{C}_{xy} \mathbf{w}_y - \lambda_1 \left(\mathbf{w}_x^T \mathbf{C}_{xx} \mathbf{w}_x - 1 \right) - \lambda_2 \left(\mathbf{w}_y^T \mathbf{C}_{yy} \mathbf{w}_y - 1 \right), \qquad (6.15.10)$$

where λ_1 and λ_2 are the Lagrange multipliers. We then have

$$\frac{\partial \mathcal{L}}{\partial \mathbf{w}_x} = \mathbf{C}_{xy} w_y - \lambda_1 \mathbf{C}_{xx} \mathbf{w}_x = 0 \Rightarrow \mathbf{w}_x^T \mathbf{C}_{xy} w_y - \lambda_1 \mathbf{w}_x^T \mathbf{C}_{xx} \mathbf{w}_x = 0, \qquad (6.15.11)$$

$$\frac{\partial \mathcal{L}}{\partial \mathbf{w}_y} = \mathbf{C}_{yx} w_x - \lambda_2 \mathbf{C}_{yy} \mathbf{w}_y = 0 \Rightarrow \mathbf{w}_y^T \mathbf{C}_{yx} w_x - \lambda_2 \mathbf{w}_y^T \mathbf{C}_{yy} \mathbf{w}_y = 0. \qquad (6.15.12)$$

Since $\mathbf{w}_x^T \mathbf{C}_{xx} \mathbf{w}_x = 1$, $\mathbf{w}_y^T \mathbf{C}_{yy} \mathbf{w}_y = 1$ and $\cos(\mathbf{z}_x, \mathbf{z}_y) = \mathbf{w}_x^T \mathbf{C}_{xy} w_y = \cos(\mathbf{z}_y, \mathbf{z}_x) = \mathbf{w}_y^T \mathbf{C}_{yx} w_x$, we get

$$\lambda_1 = \lambda_2 = \lambda. \qquad (6.15.13)$$

Substitute (6.15.13) into (6.15.11) and (6.15.12), respectively, we obtain

$$\mathbf{C}_{xy} \mathbf{w}_y = \lambda \mathbf{C}_{xx} \mathbf{w}_x, \qquad (6.15.14)$$

$$\mathbf{C}_{yx} \mathbf{w}_x = \lambda \mathbf{C}_{yy} \mathbf{w}_y, \qquad (6.15.15)$$

which can be combined into the generalized eigenvalue decomposition problem [13, 115]:

$$\begin{bmatrix} \mathbf{O} & \mathbf{C}_{xy} \\ \mathbf{C}_{yx} & \mathbf{O} \end{bmatrix} \begin{bmatrix} \mathbf{w}_x \\ \mathbf{w}_y \end{bmatrix} = \lambda \begin{bmatrix} \mathbf{C}_{xx} & \mathbf{O} \\ \mathbf{O} & \mathbf{C}_{yy} \end{bmatrix} \begin{bmatrix} \mathbf{w}_x \\ \mathbf{w}_y \end{bmatrix}, \qquad (6.15.16)$$

where $\mathbf{O}$ denotes the null matrix.

Equation (6.15.16) shows that λ and $\begin{bmatrix} \mathbf{w}_x \\ \mathbf{w}_y \end{bmatrix}$ are, respectively, the generalized eigenvalue and the corresponding generalized eigenvector of the matrix pencil

$$\left(\begin{bmatrix} \mathbf{O} & \mathbf{C}_{xy} \\ \mathbf{C}_{yx} & \mathbf{O} \end{bmatrix}, \begin{bmatrix} \mathbf{C}_{xx} & \mathbf{O} \\ \mathbf{O} & \mathbf{C}_{yy} \end{bmatrix} \right). \qquad (6.15.17)$$

The generalized eigenvalue decomposition problem (6.15.16) can also solved by using the SVDs of the data matrices $\mathbf{X}$ and $\mathbf{Y}$.

Let $\mathbf{X} = \mathbf{U}_x \mathbf{D}_x \mathbf{V}_x^T$ and $\mathbf{Y} = \mathbf{U}_y \mathbf{D}_y \mathbf{V}_y^T$ be the SVDs of $\mathbf{X}$ and $\mathbf{Y}$. Then, we have

$$\mathbf{C}_{xx} = \mathbf{X}^T \mathbf{X} = \mathbf{V}_x \mathbf{D}_x^2 \mathbf{V}_x^T, \qquad (6.15.18)$$

$$\mathbf{C}_{yy} = \mathbf{Y}^T \mathbf{Y} = \mathbf{V}_y \mathbf{D}_y^2 \mathbf{V}_y^T, \qquad (6.15.19)$$

$$\mathbf{C}_{xy} = \mathbf{X}^T \mathbf{Y} = \mathbf{V}_x \mathbf{D}_x \mathbf{U}_x^T \mathbf{U}_y \mathbf{D}_y \mathbf{V}_y^T = \mathbf{C}_{yx}^T. \qquad (6.15.20)$$

From (6.15.14) it is known that $\mathbf{w}_x = \frac{1}{\lambda} \mathbf{C}_{xx}^{-1} \mathbf{C}_{xy} \mathbf{w}_y$. Substitute this result into (6.15.15) to get

$$\left(\mathbf{C}_{yx} \mathbf{C}_{xx}^{-1} \mathbf{C}_{xy} - \lambda^2 \mathbf{C}_{yy} \right) \mathbf{w}_y = \mathbf{0}. \qquad (6.15.21)$$

Substituting $\mathbf{C}_{xx}^{-1} = \mathbf{V}_x \mathbf{D}_x^{-2} \mathbf{V}_x^T$ into the above equation, we obtain

$$\left(\mathbf{V}_y \mathbf{D}_y \mathbf{U}_y^T \mathbf{U}_x \mathbf{V}_x^T \mathbf{V}_x \mathbf{D}_x \mathbf{U}_x^T \mathbf{U}_y \mathbf{D}_y \mathbf{V}_y^T - \lambda^2 \mathbf{V}_y \mathbf{D}_y^2 \mathbf{V}_y^T \right) = \mathbf{0}$$

By using $\mathbf{V}_x \mathbf{V}_x = \mathbf{I}$, the above equation can be rewritten as

$$\left(\mathbf{V}_y \mathbf{D}_y \mathbf{U}_y^T \mathbf{U}_x \mathbf{U}_y \mathbf{D}_y \mathbf{V}_y^T - \lambda^2 \mathbf{V}_y \mathbf{D}_y^2 \mathbf{V}_y^T \right) = \mathbf{0}. \qquad (6.15.22)$$

Premultiplying $\mathbf{D}_y^{-1} \mathbf{V}_y^T$, then the above equation reduces to

$$\left(\mathbf{U}_y^T \mathbf{U}_x \mathbf{U}_x^T \mathbf{U}_y \mathbf{V}_y^T - \lambda^2 \right) \mathbf{D}_y \mathbf{V}_y^T \mathbf{w}_y = \mathbf{0}. \qquad (6.15.23)$$

Let $\mathbf{U}_x^T\mathbf{U}_y = \mathbf{U}\mathbf{D}\mathbf{V}^T$ be the SVD of $\mathbf{U}_x^T\mathbf{U}_y$, then

$$\left(\mathbf{V}\mathbf{D}^2\mathbf{V}^T - \lambda^2\mathbf{I}\right)\mathbf{D}_y\mathbf{V}_y^T\mathbf{w}_y = \mathbf{0}. \tag{6.15.24}$$

or

$$\mathbf{V}\left(\mathbf{D}^2 - \lambda^2\mathbf{I}\right)\mathbf{V}^T\mathbf{D}_y\mathbf{V}_y^T\mathbf{w}_y = \mathbf{0}, \tag{6.15.25}$$

Premultiplying $\mathbf{V}^T$ and noting that $\mathbf{V}^T\mathbf{V} = \mathbf{I}$, the above equation can be simplified to [265]:

$$\left(\mathbf{D}^2 - \lambda^2\mathbf{I}\right)\mathbf{V}^T\mathbf{D}_y\mathbf{V}_y^T\mathbf{w}_y = \mathbf{0}. \tag{6.15.26}$$

Clearly, when

$$\mathbf{w}_y = \mathbf{V}_y\mathbf{D}_y^{-1}\mathbf{V}, \tag{6.15.27}$$

Equation (6.15.26) yields the result

$$\left(\mathbf{D}^2 - \lambda^2\mathbf{I}\right) = \mathbf{0} \quad \text{or} \quad |\mathbf{D}^2 - \lambda^2\mathbf{I}| = 0. \tag{6.15.28}$$

By mimicking this process, we immediately have that when

$$\mathbf{w}_x = \mathbf{V}_x\mathbf{D}_x^{-1}\mathbf{V}, \tag{6.15.29}$$

then

$$\left(\mathbf{D}^2 - \lambda^2\mathbf{I}\right)\mathbf{V}^T\mathbf{D}_x\mathbf{V}_x^T\mathbf{w}_x = \mathbf{0}. \tag{6.15.30}$$

The above analysis shows that the generalized eigenvalues λ in the generalized eigenvalue problem (6.15.16) is given by the singular value matrix $\mathbf{D}$ of the matrix $\mathbf{U}_x^T\mathbf{U}_y$, as shown in (6.15.28), and the corresponding generalized eigenvectors (i.e., canonical variants) are given by $\mathbf{w}_x = \mathbf{V}_x\mathbf{D}_x^{-1}\mathbf{V}$ in (6.15.29) and $\mathbf{w}_y = \mathbf{V}_y\mathbf{D}_y^{-1}\mathbf{V}$ in (6.15.27), respectively.

Algorithm 6.29 summarizes the CCA algorithm.

6.15.2 Kernel Canonical Correlation Analysis

Let X and Y denote two views (i.e., two attribute sets describing the data), and $(\mathbf{x}, \mathbf{y}, c)$ be a labeled example where $\mathbf{x} \in X$ and $\mathbf{y} \in Y$ are the two portions of the example, and c is their label. For simplicity, assume that $c \in \{0, 1\}$ where 0 and 1 denote negative and positive classes, respectively. Assume that there exist

---

Algorithm 6.29 Canonical correlation analysis (CCA) algorithm [13, 115]

1. **input:** The two-view data vectors $\{(\mathbf{x}_n, \mathbf{y}_n)\}_{n=1}^{N}$ with $\mathbf{x}_n \in \mathbb{R}^{1\times p}$, $\mathbf{y}_n \in \mathbb{R}^{1\times q}$.
2. **initialization:** Normalize $\mathbf{x}_n \leftarrow \mathbf{x}_n - \frac{1}{p}\sum_{i=1}^{p} x_n(i)\mathbf{1}_{p\times 1}^{T}$ and $\mathbf{y}_n \leftarrow \mathbf{y}_n - \frac{1}{q}\sum_{i=1}^{q} y_n(i)\mathbf{1}_{q\times 1}^{T}$.
3. Let $\mathbf{X} = [\mathbf{x}_1, \ldots, \mathbf{x}_N]$ and $\mathbf{Y} = [\mathbf{y}_1, \ldots, \mathbf{y}_N]$.
4. Make the SVD $\mathbf{X} = \mathbf{U}_x \mathbf{D}_x \mathbf{V}_x^T$, $\mathbf{Y} = \mathbf{U}_y \mathbf{D}_y \mathbf{V}_y^T$.
5. Calculate the SVD $\mathbf{U}_x^T \mathbf{U}_y = \mathbf{U}\mathbf{D}\mathbf{V}^T$.
6. Calculate $\mathbf{w}_x = \mathbf{V}_x \mathbf{D}_x^{-1} \mathbf{V}$ and $\mathbf{w}_y = \mathbf{V}_y \mathbf{D}_y^{-1} \mathbf{V}$.
7. **output:** the canonical variants $(\mathbf{w}_x, \mathbf{w}_y)$ of $(\mathbf{X}, \mathbf{Y})$.

---

two decision functions f_X and f_Y over X and Y, respectively, such that $f_X(\mathbf{x}) = f_Y(\mathbf{y}) = c$. Intuitively, this means that every example is associated with two views each contains sufficient information for determining the label of the example.

Given one labeled example $((\mathbf{x}_0, \mathbf{y}_0), 1)$ and a large number of unlabeled examples $U = \{\mathbf{x}_i, \mathbf{y}_i\}_{i=1}^{N-1}$, the task of semi-supervised learning is to train a classifier for classifying unlabeled examples $U = \{\mathbf{x}_i, \mathbf{y}_i\}_{i=1}^{N-1}$, i.e., determining unknown labels $c_i, i = 1, \ldots, N-1$.

For the data described by two sufficient views, some projections in these two views should have strong correlation.

Let $\mathbf{X} = [\mathbf{x}_0, \mathbf{x}_1, \ldots, \mathbf{x}_{N-1}]$ and $\mathbf{Y} = [\mathbf{y}_0, \mathbf{y}_1, \ldots, \mathbf{y}_{N-1}]$ be two-view data matrices consisting of one labeled data $((\mathbf{x}_0, \mathbf{y}_0), 1)$ and $N-1$ unlabeled data $\{\mathbf{x}_i, \mathbf{y}_i\}_{i=1}^{N-1}$. CCA finds two projector vectors $\mathbf{w}_x$ and $\mathbf{w}_y$ such that the projections $\mathbf{X}^T \mathbf{w}_x$ and $\mathbf{Y}^T \mathbf{w}_y$ are strongly correlated as soon as possible, namely their correlation coefficient is maximized:

$$(\mathbf{w}_x, \mathbf{w}_y) = \arg\max_{\mathbf{w}_x, \mathbf{w}_y} \frac{\langle \mathbf{X}^T \mathbf{w}_x, \mathbf{Y}^T \mathbf{w}_y \rangle}{\|\mathbf{X}^T \mathbf{w}_x\| \cdot \|\mathbf{Y}^T \mathbf{w}_y\|}$$

$$= \arg\max_{\mathbf{w}_x, \mathbf{w}_y} \left(\frac{\mathbf{w}_x^T \mathbf{C}_{xy} \mathbf{w}_y}{\sqrt{\mathbf{w}_x^T \mathbf{C}_{xx} \mathbf{w}_x \cdot \mathbf{w}_y^T \mathbf{C}_{yy} \mathbf{w}_y}} \right) \quad (6.15.31)$$

$$\text{subject to} \begin{cases} \mathbf{w}_x^T \mathbf{C}_{xx} \mathbf{w}_x = 1, \\ \mathbf{w}_y^T \mathbf{C}_{yy} \mathbf{w}_y = 1, \end{cases} \quad (6.15.32)$$

where $\mathbf{C}_{xy} = \frac{1}{N} \mathbf{X}\mathbf{Y}^T$ is the between-sets covariance matrix of $\mathbf{X}$ and $\mathbf{Y}$, while $\mathbf{C}_{xx} = \frac{1}{N} \mathbf{X}\mathbf{X}^T$ and $\mathbf{C}_{yy} = \frac{1}{N} \mathbf{Y}\mathbf{Y}^T$ are the within-sets covariance matrices of $\mathbf{X}$ and $\mathbf{Y}$, respectively.

6.15 Canonical Correlation Analysis

By the Lagrange multiplier method, the objective of the above primal constrained optimization can be rewritten as the objective of the following dual unconstrained optimization:

$$\mathcal{L}_D(\mathbf{w}_x, \mathbf{w}_y) = \mathbf{w}_x^T \mathbf{C}_{xy} \mathbf{w}_y - \frac{\lambda_x}{2}\left(\mathbf{w}_x^T \mathbf{C}_{xx} \mathbf{w}_x - 1\right) + \frac{\lambda_y}{2}\left(\mathbf{w}_y^T \mathbf{C}_{yy} \mathbf{w}_y - 1\right). \tag{6.15.33}$$

From the first-order optimality conditions we have

$$\frac{\partial \mathcal{L}_D}{\partial \mathbf{w}_x} = \mathbf{0} \quad \Rightarrow \quad \mathbf{C}_{xy} \mathbf{w}_y = \lambda_x \mathbf{C}_{xx} \mathbf{w}_x, \tag{6.15.34}$$

$$\frac{\partial \mathcal{L}_D}{\partial \mathbf{w}_y} = \mathbf{0} \quad \Rightarrow \quad \mathbf{C}_{yx} \mathbf{w}_x = \lambda_y \mathbf{C}_{yy} \mathbf{w}_y, \tag{6.15.35}$$

which can be combined into

$$0 = \mathbf{w}_x^T \mathbf{C}_{xy} \mathbf{w}_y - \lambda_x \mathbf{w}_x^T \mathbf{C}_{xx} \mathbf{w}_x - \mathbf{w}_y^T \mathbf{C}_{yx} \mathbf{w}_x + \lambda_y \mathbf{w}_y^T \mathbf{C}_{yy} \mathbf{w}_y$$
$$= \lambda_y \mathbf{w}_y^T \mathbf{C}_{yy} \mathbf{w}_y - \lambda_x \mathbf{w}_x^T \mathbf{C}_{xx} \mathbf{w}_x,$$
$$= \lambda_y - \lambda_x.$$

Letting $\lambda = \lambda_x = \lambda_y$ and assuming that $\mathbf{C}_{yy}$ is invertible, then (6.15.35) becomes

$$\mathbf{w}_y = \frac{1}{\lambda} \mathbf{C}_{yy}^{-1} \mathbf{C}_{yx} \mathbf{w}_x. \tag{6.15.36}$$

Substituting (6.15.36) into (6.15.34) yields

$$\mathbf{C}_{xy} \mathbf{C}_{yy}^{-1} \mathbf{C}_{yx} \mathbf{w}_x = \lambda^2 \mathbf{C}_{xx} \mathbf{w}_x. \tag{6.15.37}$$

This implies that the projection vector $\mathbf{w}_x$ is the generalized eigenvector corresponding to the largest generalized eigenvalue $\lambda_{\max}^2$ of the matrix pencil $(\mathbf{C}_{xy} \mathbf{C}_{yy}^{-1} \mathbf{C}_{yx}, \mathbf{C}_{xx})$.

In order to identify nonlinearly correlated projections between the two views, we can apply the kernel extensions of canonical correlation analysis, called simply *Kernel canonical correlation analysis* (kernel CCD) [115]. The kernel CCA maps the instances $\mathbf{x}$ and $\mathbf{y}$ to the higher-dimensional kernel instances $\boldsymbol{\phi}_x(\mathbf{x})$ and $\boldsymbol{\phi}_y(\mathbf{y})$, respectively.

Letting

$$\mathbf{S}_x = [\boldsymbol{\phi}_x(\mathbf{x}_0), \boldsymbol{\phi}_x(\mathbf{x}_1), \ldots, \boldsymbol{\phi}_x(\mathbf{x}_{N-1})], \tag{6.15.38}$$

$$\mathbf{S}_y = [\boldsymbol{\phi}_y(\mathbf{y}_0), \boldsymbol{\phi}_y(\mathbf{y}_1), \ldots, \boldsymbol{\phi}_y(\mathbf{y}_{N-1})], \tag{6.15.39}$$

then the projection vectors $\boldsymbol{\phi}_x(\mathbf{x}_i)$ and $\boldsymbol{\phi}_y(\mathbf{y}_i)$ in higher-dimensional kernel space can be rewritten as $\mathbf{w}_x^\phi = \mathbf{S}_x \boldsymbol{\alpha}$ and $\mathbf{w}_y^\phi = \mathbf{S}_y \boldsymbol{\beta}$, where $\boldsymbol{\alpha}, \boldsymbol{\beta} \in \mathbb{R}^N$.

Therefore, the objective function (6.15.33) for linearly correlated projections becomes the objective function for nonlinearly correlated projections below [303]:

$$L(\boldsymbol{\alpha}, \boldsymbol{\beta}) = \frac{\langle \mathbf{S}_x^T \mathbf{w}_x^\phi, \mathbf{S}_y^T \mathbf{w}_y^\phi \rangle}{\|\mathbf{S}_x^T \mathbf{w}_x^\phi\| \cdot \|\mathbf{S}_y^T \mathbf{w}_y^\phi\|} = \frac{\boldsymbol{\alpha}^T \mathbf{S}_x^T \mathbf{S}_x \mathbf{S}_y^T \mathbf{S}_y \boldsymbol{\beta}}{\|\boldsymbol{\alpha}^T \mathbf{S}_x^T \mathbf{S}_x\| \cdot \|\boldsymbol{\beta}^T \mathbf{S}_y^T \mathbf{S}_y\|}. \tag{6.15.40}$$

Let

$$K_x(\mathbf{x}_i, \mathbf{x}_j) = \langle \boldsymbol{\phi}_x(\mathbf{x}_i), \boldsymbol{\phi}_x(\mathbf{x}_j) \rangle = \boldsymbol{\phi}_x^T(\mathbf{x}_i) \boldsymbol{\phi}_x(\mathbf{x}_j), \ i, j = 0, 1, \ldots, N-1, \tag{6.15.41}$$

$$K_y(\mathbf{y}_i, \mathbf{y}_j) = \langle \boldsymbol{\phi}_y(\mathbf{y}_i), \boldsymbol{\phi}_y(\mathbf{y}_j) \rangle = \boldsymbol{\phi}_y^T(\mathbf{y}_i) \boldsymbol{\phi}_y(\mathbf{y}_j), \ i, j = 0, 1, \ldots, N-1 \tag{6.15.42}$$

be the kernel functions on the two views, respectively.

Defining the kernel matrices

$$\mathbf{K}_x = \mathbf{S}_x^T \mathbf{S}_x = \begin{bmatrix} \boldsymbol{\phi}_x^T(\mathbf{x}_0) \\ \boldsymbol{\phi}_x^T(\mathbf{x}_1) \\ \vdots \\ \boldsymbol{\phi}_x^T(\mathbf{x}_{N-1}) \end{bmatrix} [\boldsymbol{\phi}_x(\mathbf{x}_0), \boldsymbol{\phi}_x(\mathbf{x}_1), \ldots, \boldsymbol{\phi}_x(\mathbf{x}_{N-1})]$$

$$= [K_x(\mathbf{x}_i, \mathbf{x}_j)]_{i,j=0}^{N-1, N-1}, \tag{6.15.43}$$

and

$$\mathbf{K}_y = \mathbf{S}_y^T \mathbf{S}_y = \begin{bmatrix} \boldsymbol{\phi}_y^T(\mathbf{y}_0) \\ \boldsymbol{\phi}_y^T(\mathbf{y}_1) \\ \vdots \\ \boldsymbol{\phi}_y^T(\mathbf{y}_{N-1}) \end{bmatrix} [\boldsymbol{\phi}_y(\mathbf{y}_0), \boldsymbol{\phi}_y(\mathbf{y}_1), \ldots, \boldsymbol{\phi}_y(\mathbf{y}_{N-1})]$$

$$= [K_y(\mathbf{y}_i, \mathbf{y}_j)]_{i,j=0}^{N-1, N-1}, \tag{6.15.44}$$

then the objective function in (6.15.40) can be simplified to

$$L(\boldsymbol{\alpha}, \boldsymbol{\beta}) = \frac{\langle \mathbf{K}_x \boldsymbol{\alpha}, \mathbf{K}_y \boldsymbol{\beta} \rangle}{\|\mathbf{K}_x \boldsymbol{\alpha}\| \cdot \|\mathbf{K}_y \boldsymbol{\beta}\|} = \frac{\boldsymbol{\alpha}^T \mathbf{K}_x^T \mathbf{K}_y \boldsymbol{\beta}}{\sqrt{\boldsymbol{\alpha}^T \mathbf{K}_x^T \mathbf{K}_x \boldsymbol{\alpha} \cdot \boldsymbol{\beta}^T \mathbf{K}_y^T \mathbf{K}_y \boldsymbol{\beta}}}, \tag{6.15.45}$$

6.15 Canonical Correlation Analysis

$$\text{subject to } \begin{cases} \boldsymbol{\alpha}^T \mathbf{K}_x^T \mathbf{K}_x \boldsymbol{\alpha} = 1, \\ \boldsymbol{\beta}^T \mathbf{K}_y^T \mathbf{K}_y \boldsymbol{\beta} = 1. \end{cases} \quad (6.15.46)$$

By the Lagrange multiplier method and the regularization method, we have the following objective function for dual optimization:

$$\mathcal{L}_D(\boldsymbol{\alpha}, \boldsymbol{\beta}) = \boldsymbol{\alpha}^T \mathbf{K}_x^T \mathbf{K}_y \boldsymbol{\beta} - \frac{\lambda_x}{2}\left(\boldsymbol{\alpha}^T \mathbf{K}_x^T \mathbf{K}_x \boldsymbol{\alpha} - 1\right) - \frac{\nu_x}{2}\|\mathbf{S}_x^T \boldsymbol{\alpha}\|^2$$
$$- \frac{\lambda_y}{2}\left(\boldsymbol{\beta}^T \mathbf{K}_y^T \mathbf{K}_y \boldsymbol{\beta} - 1\right) - \frac{\nu_y}{2}\|\mathbf{S}_y^T \boldsymbol{\beta}\|^2. \quad (6.15.47)$$

The first-order optimality conditions give the results:

$$\frac{\partial \mathcal{L}_D}{\partial \boldsymbol{\alpha}} = \mathbf{0} \Rightarrow \mathbf{K}_x^T \mathbf{K}_y \boldsymbol{\beta} = \lambda_x \mathbf{K}_x^T \mathbf{K}_x \boldsymbol{\alpha} + \nu_x \mathbf{K}_x^T \boldsymbol{\alpha}$$
$$\Rightarrow \mathbf{K}_y \boldsymbol{\beta} = \lambda_x \mathbf{K}_x \boldsymbol{\alpha} + \nu_x \boldsymbol{\alpha}, \quad (6.15.48)$$

$$\frac{\partial \mathcal{L}_D}{\partial \boldsymbol{\beta}} = \mathbf{0} \Rightarrow \mathbf{K}_y^T \mathbf{K}_x \boldsymbol{\alpha} = \lambda_y \mathbf{K}_y^T \mathbf{K}_y \boldsymbol{\beta} + \nu_y \mathbf{K}_y^T \boldsymbol{\beta}$$
$$\Rightarrow \mathbf{K}_x \boldsymbol{\alpha} = \lambda_y \mathbf{K}_y \boldsymbol{\beta} + \nu_y \boldsymbol{\beta}. \quad (6.15.49)$$

For simplicity, let $\lambda = \lambda_1 = \lambda_2$, $\nu = \nu_1 = \nu_2$, and $\kappa = \frac{\nu}{\lambda}$. Then, from (6.15.49) we get

$$\boldsymbol{\beta} = \frac{1}{\lambda}(\mathbf{K}_y + \kappa \mathbf{I})^{-1} \mathbf{K}_x \boldsymbol{\alpha}. \quad (6.15.50)$$

Substituting this result into (6.15.48), we obtain

$$(\mathbf{K}_x + \kappa \mathbf{I})^{-1} \mathbf{K}_y (\mathbf{K}_y + \kappa \mathbf{I})^{-1} \mathbf{K}_x \boldsymbol{\alpha} = \lambda^2 \boldsymbol{\alpha}, \quad (6.15.51)$$

or

$$\mathbf{K}_y (\mathbf{K}_y + \kappa \mathbf{I})^{-1} \mathbf{K}_x \boldsymbol{\alpha} = \lambda^2 (\mathbf{K}_x + \kappa \mathbf{I}) \boldsymbol{\alpha}. \quad (6.15.52)$$

This implies that a number of $\boldsymbol{\alpha}$ (and corresponding λ) can be found by solving the eigenvalue problem (6.15.51) or the generalized eigenvalue problem (6.15.52).

Once $\boldsymbol{\alpha}$ is found, Eq. (6.15.50) can be used to find the unique $\boldsymbol{\beta}$ for each $\boldsymbol{\alpha}$. That is to say, in addition to the most strongly correlated pair of projections, one can also identify the correlations between other pairs of projections and the strength of the correlations can be measured by the values of λ.

In the jth projection, the similarity between an original unlabeled instance $(\mathbf{x}_i, \mathbf{y}_i)$, $i = 1, 2, \ldots, N - 1$ and the original labeled instance $(\mathbf{x}_0, \mathbf{y}_0)$ can be measured by [303]

$$\text{sim}_{i,j} = \exp(-d^2(P_j(\mathbf{x}_i), P_j(\mathbf{x}_0))) + \exp(-d^2(P_j(\mathbf{y}_i), P_j(\mathbf{y}_0))),$$
$$i = 1, \ldots, N-1; \; j = 1, \ldots, m, \quad (6.15.53)$$

where $d(\mathbf{a}, \mathbf{b})$ is the Euclidean distance between $\mathbf{a}$ and $\mathbf{b}$, m is the number of unlabeled instances $(\mathbf{x}_i, \mathbf{y}_i)$ correlating with the labeled instance $(\mathbf{x}_0, \mathbf{y}_0)$, while $(P_j(\mathbf{x}_i), P_j(\mathbf{y}_i))$ is the projection of the instance $(\mathbf{x}_i, \mathbf{y}_i)$ on higher-dimensional kernel instances, $\mathbf{K}_x\boldsymbol{\alpha}$ and $\mathbf{K}_y\boldsymbol{\beta}$.

6.15.3 Penalized Canonical Correlation Analysis

Given two-view data matrices $(\mathbf{X}, \mathbf{Y})$, the standard CCA seeks to find $(\mathbf{u}, \mathbf{v})$ that maximize $\text{cor}(\mathbf{Xu}, \mathbf{Yv}) = \mathbf{u}^T \mathbf{XYv}$:

$$\max_{\mathbf{u},\mathbf{v}} \mathbf{u}^T \mathbf{XYv} \quad \text{subject to } \|\mathbf{Xu}\|_2^2 \leq 1 \text{ and } \|\mathbf{Yv}\|_2^2 \leq 1. \quad (6.15.54)$$

If including penalties $p_1(\mathbf{u}) \leq c_1$ and $p_2(\mathbf{v}) \leq c_2$ in the CCA then one obtains a *penalized canonical correlation analysis* (penalized CCA) in [271]:

$$\max_{\mathbf{u},\mathbf{v}} \mathbf{u}^T \mathbf{XYv}$$

$$\text{subject to } \|\mathbf{Xu}\|_2^2 \leq 1, \; \|\mathbf{Yv}\|_2^2 \leq 1, \; p_1(\mathbf{u}) \leq c_1, \; p_2(\mathbf{v}) \leq c_2. \quad (6.15.55)$$

If pre-normalizing the column vectors of the data matrices $\mathbf{X}, \mathbf{Y} \in \mathbb{R}^{N \times P}$ such that they become the semi-orthogonal, i.e., $\mathbf{X}^T\mathbf{X} = \mathbf{I}_{P \times P}$ and $\mathbf{Y}^T\mathbf{Y} = \mathbf{I}_{P \times P}$, then the penalized CCA is simplified to [271]

$$\max_{\mathbf{u},\mathbf{v}} \mathbf{u}^T \mathbf{XYv}$$

$$\text{subject to } \|\mathbf{u}\|_2^2 \leq 1, \; \|\mathbf{v}\|_2^2 \leq 1, \; p_1(\mathbf{u}) \leq c_1, \; p_2(\mathbf{v}) \leq c_2 \quad (6.15.56)$$

because $\|\mathbf{Xu}\|_2^2 = \mathbf{u}^T\mathbf{X}^T\mathbf{Xu} = \|\mathbf{u}\|_2^2$ and $\|\mathbf{Yv}\|_2^2 = \|\mathbf{v}\|_2^2$.

Equation (6.15.56) is called "*diagonal penalized CCA.*" If penalties $p_1(\mathbf{u}) = \|\mathbf{u}\|_1$ and $p_2(\mathbf{v}) = \|\mathbf{v}\|_1$, then the diagonal penalized CCA become the *sparse CCA* that gives the *sparse canonical variants* $\mathbf{u}$ and $\mathbf{v}$.

Let soft denote the soft thresholding operator defined by

$$\text{soft}(a, c) = \text{sign}(a)(|a| - c)_+, \quad (6.15.57)$$

6.15 Canonical Correlation Analysis

where $c > 0$ is a constant and

$$x_+ = \begin{cases} x, & \text{if } x > 0, \\ 0, & \text{if } x \leq 0. \end{cases} \qquad (6.15.58)$$

Lemma 6.4 ([271]) *For the optimization problem*

$$\max_{\mathbf{u}} \mathbf{u}^T \mathbf{a} \quad \text{subject to } \|\mathbf{u}\|_2^2 \leq 1, \ \|\mathbf{u}\|_1 \leq c, \qquad (6.15.59)$$

its solution is given by

$$\mathbf{u} = \frac{\mathbf{s}(\mathbf{a}, \Delta)}{\|\mathbf{s}(\mathbf{a}, \Delta)\|_2}, \qquad (6.15.60)$$

where $\mathbf{s}(\mathbf{a}, \Delta) = [\text{soft}(a_1, \Delta), \ldots, \text{soft}(a_p, \Delta)]^T$, *and* $\Delta = 0$ *if it results in* $\|\mathbf{u}\|_1 \leq c$, *otherwise* Δ *is chosen to be a positive constant such that* $\|\mathbf{u}\|_1 = c$.

For the penalized (sparse) CCA problem

$$(\mathbf{u}, \mathbf{v}) = \arg\max_{\mathbf{u}, \mathbf{v}} \mathbf{u}^T \mathbf{X} \mathbf{v}$$

$$\text{subject to } \|\mathbf{u}\|_2^2 \leq 1, \ \|\mathbf{v}\|_2^2 \leq 1, \ \|\mathbf{u}\|_1 \leq c_1, \ \|\mathbf{v}\|_1 \leq c_2. \qquad (6.15.61)$$

Algorithm 6.30 [271] gives the sparse canonical variants $\mathbf{u}_i, \mathbf{v}_i$, $i = 1, \ldots, K$.

Algorithm 6.30 Penalized (sparse) canonical component analysis (CCA) algorithm

1. **input:** The matrix $\mathbf{X}$.
2. **initialization:** Make the truncated SVD $\mathbf{X} = \sum_{i=1}^{K} d_i \mathbf{u}_i \mathbf{v}_i$. Put $\mathbf{X}_1 = \mathbf{X}$.
3. **for** $k = 1$ to K **do**
4. $\mathbf{u}_k \leftarrow \frac{\mathbf{s}(\mathbf{X}_k \mathbf{v}_k, \Delta_1)}{\|\mathbf{s}(\mathbf{X}_k \mathbf{v}_k, \Delta_1)\|_2}$, where $\Delta_1 = 0$ if this results in $\|\mathbf{u}\|_1 \leq c_1$; otherwise, Δ_1 is chosen to be a positive constant such that $\|\mathbf{u}\|_1 = c_1$.
5. $\mathbf{u}_k \leftarrow \mathbf{u}_k / \|\mathbf{u}_k\|_2$.
6. $\mathbf{v}_k \leftarrow \frac{\mathbf{s}(\mathbf{X}_k^T \mathbf{u}_k, \Delta_2)}{\|\mathbf{s}(\mathbf{X}_k^T \mathbf{u}_k, \Delta_2)\|_2}$, where $\Delta_2 = 0$ if this results in $\|\mathbf{v}\|_1 \leq c_2$; otherwise, Δ_2 is chosen to be a positive constant such that $\|\mathbf{v}\|_1 = c_2$.
7. $\mathbf{v}_k \leftarrow \mathbf{v}_k / \|\mathbf{v}_k\|_2$.
8.   **if** $\mathbf{u}_k, \mathbf{v}_k$ are converged **then do**
9. $d_k = \mathbf{u}_k^T \mathbf{X}_k \mathbf{v}_k$,
10.   **else goto** Step 4.
11.   **end if**
12.   **exit if** $k = K$.
13. $\mathbf{X}_{k+1} \leftarrow \mathbf{X}_k - d_k \mathbf{u}_k \mathbf{v}_k^T$.
14. **end for**
15. **output:** $\mathbf{u}_k, \mathbf{v}_k$, $k = 1, \ldots, K$.

6.16 Graph Machine Learning

Many practical applications in machine learning build on irregular or *non-Euclidean structures* rather than regular or *Euclidean structures*. Non-Euclidean structures are also called *graph structures* because this structure is a topological map in abstract sense of graph theory. Prominent examples of underlying *graph-structured data* are social networks, information networks, gene data on biological regulatory networks, text documents on word embeddings [73], log data on telecommunication networks, genetic regulatory networks, functional networks of the brain, 3D shapes represented as discrete manifolds, and so on [158]. Moreover, the signal values associated with the vertices of the graph carry the information of interest in observations or physical measurements. Numerous examples can be found in real world applications, such as temperatures within a geographical area, transportation capacities at hubs in a transportation network, or human behaviors in a social network [76].

When signal values are defined on the vertex set of a weighted and undirected graph, structured data are referred to as *graph signals*, where the vertices of the graph represent the entities and the edge weights reflect the pairwise relationships or similarities between these entities.

Graphs can encode complex geometric structures and can be studied with strong mathematical tools such as *spectral graph theory* [53]. One of the main goals in spectral graph theory is to deduce the principal properties and structure of a graph from its graph spectrum. Spectral graph theory contains three basic aspects: similarity graph, graph Laplacian matrices, and graph spectrum.

This section focuses on graph-based machine learning and contains the following two parts:

- *Spectral graph theory:* Similarity graph, graph Laplacian matrices, and graph spectrum.
- *Graph-based learning:* Learning the structure of a graph, called also graph construction, from training samples in semi-supervised and unsupervised learning cases.

6.16.1 Graphs

The standard *Euclidean structure* refers to a space structure on n-dimensional vector space $\mathbb{R}^n$ equipped with the following four *Euclidean measures*:

1. the standard inner product (also known as the dot product) on $\mathbb{R}^m$,

$$\langle \mathbf{x}, \mathbf{y} \rangle = \mathbf{x} \cdot \mathbf{y} = \mathbf{x}^T \mathbf{y} = x_1 y_1 + \cdots + x_m y_m;$$

2. the Euclidean length of a vector **x** on $\mathbb{R}^m$,

$$\|\mathbf{x}\| = \sqrt{\langle \mathbf{x}, \mathbf{y} \rangle} = \sqrt{x_1^2 + \cdots + x_m^2};$$

3. the Euclidean distance between **x** and **y** on $\mathbb{R}^m$,

$$d(\mathbf{x}, \mathbf{y}) = \|\mathbf{x} - \mathbf{y}\| = \sqrt{(x_1 - y_1)^2 + \cdots + (x_m - y_m)^2};$$

4. the (nonreflex) angle θ ($0° \leq \theta \leq 180°$) between vectors **x** and **y** on $\mathbb{R}^m$,

$$\theta = \arccos\left(\frac{\langle \mathbf{x}, \mathbf{y} \rangle}{\|\mathbf{x}\|\|\mathbf{y}\|}\right) = \arccos\left(\frac{\mathbf{x}^T \mathbf{y}}{\|\mathbf{x}\|\|\mathbf{y}\|}\right).$$

A space structure with no Euclidean measure above is known as the *non-Euclidean structure*.

Given a set of data points $\mathbf{x}_1, \ldots, \mathbf{x}_N$ with $\mathbf{x}_i \in \mathbb{R}^d$ and some kernel function $K(\mathbf{x}_i, \mathbf{x}_j)$ between all pairs of data points $\mathbf{x}_i$ and $\mathbf{x}_j$.

The kernel function $K(\mathbf{x}_i, \mathbf{x}_j)$ must be symmetric and nonnegative:

$$K(\mathbf{x}_i, \mathbf{x}_j) = K(\mathbf{x}_j, \mathbf{x}_i) \quad \text{and} \quad K(\mathbf{x}_i, \mathbf{x}_j) \geq 0, \tag{6.16.1}$$

with $K(\mathbf{x}_i, \mathbf{x}_i) = 0$ for $i = 1, \ldots, N$.

The kernel functions satisfying the above conditions may be any distance measure $d(\mathbf{x}_i, \mathbf{x}_j)$, any similarity measure $s(\mathbf{x}_i, \mathbf{x}_j)$ or Gaussian similarity function $s(\mathbf{x}_i, \mathbf{x}_j) = \exp\left(-\|\mathbf{x}_i - \mathbf{x}_j\|^2/(2\sigma^2)\right)$ (where the parameter σ controls the width of the neighborhoods), depending on application objects.

If we do not have more information than similarities between data points, a nice way of data representing is in the form of a weighted undirected *similarity graph* or (called simply *graph*) $G = (V, E, \mathbf{W})$ or simply $G = (V, E)$.

Definition 6.28 (Graph [241, 261]) A *graph* $G(V, E, \mathbf{W})$ or simply $G(V, E)$ is a collection of a *vertex set* (or *node set*) $V = \{v_1, \ldots, v_N\}$ and an *edge set* $E = \{e_{ij}\}_{i,j=1}^N$. A graph G contains nonnegative *edge weights* associated with each edge: $w_{ij} \geq 0$. If v_i and v_j are not connected to each other, then $w_{ij} = 0$. For an *undirected graph* G, $e_{ij} = e_{ji}$ and $w_{ij} = w_{ji}$; if G is a *directed graph*, then $e_{ij} \neq e_{ji}$ and $w_{ij} \neq w_{ji}$.

Clearly, any graph is of non-Euclidean structure.

The edge weight w_{ij} is generally treated as a measure of similarity between the nodes v_i and v_j. The higher the edge weight, the more similar the two nodes are expected to be.

Definition 6.29 (Adjacency Matrix) The adjacency matrix of a graph, denoted as $\mathbf{W} = [w_{ij}]_{i,j=1}^{N,N}$, refers to the matrix used to represent the connection of nodes in the graph. The adjacency matrix **W** can be either binary or weighted. For undirected graphs with N nodes, the adjacency matrix is an $N \times N$ real symmetric matrix.

The adjacency matrix is also called the *affinity matrix* of the graph.

Definition 6.30 (Degree Matrix) The *degree* (or degree function) of node j in the graph $G(V, E, \mathbf{W})$, denoted as $d_j : V \to \mathbb{R}$, represents the number of edges connected to node j:

$$d_j = \sum_{i \sim j} w_{ij}, \qquad (6.16.2)$$

where $i \sim j$ denotes all vertices i connected to j by the edges $(i, j) \in E$. The *degree matrix* of a graph is a diagonal matrix $\mathbf{D} = \mathbf{Diag}(D_{11}, \ldots, D_{NN}) = \mathbf{Diag}(d_1, \ldots, d_N)$ whose ith diagonal element $D_{ii} = d_i$ is used to describe the degree of node i in the graph.

That is, the degree function d_i is defined by the sum of all entries of the ith row of the affinity matrix **W**.

Each vertex (or node) of the graph corresponds to a datum, and the edges encode the pairwise relationships or similarities among the data. For example, the vertices of the web are just the web pages, and the edges denote the hyperlinks; in market basket analysis, the items also form a graph by connecting any two items which have appeared in the same shopping basket [301].

A signal $\mathbf{x} : V \to \mathbb{R}$ defined on the nodes of the graph may be regarded as a vector $\mathbf{x} \in \mathbb{R}^d$. Let the vertex $\mathbf{v}_i \in V = \{\mathbf{v}_1, \ldots, \mathbf{v}_N\}$ represent a data point $\mathbf{x}_i$ or a document doc_i. Then, each edge $e(i, j) \in E$ is assigned an affinity score w_{ij} forming matrix **W** which reflects the similarity between vertex $\mathbf{v}_i$ and vertex $\mathbf{v}_j$.

The nodes of the graph are the points in the feature space, and an edge is formed between every pair of nodes. The nonnegative weight on each edge, $w_{ij} \geq 0$, is a function of the similarity between nodes i and j. If the *edge weighting functions* $w_{ij} = 0$, then the vertices $\mathbf{v}_i$ and $\mathbf{v}_j$ are not connected by an edge.

Clustering seeks to partition the vertex set V into disjoint sets $V_1, \ldots, V_m$, where by some measure the similarity among the vertices in a set V_i is high, and across different sets (V_i, V_j) is low, where $V_i = \{\mathbf{x}_{i1}, \ldots, \mathbf{x}_{i,m_i}\}$ is the ith clustered vertex subset such that $m_1 + \cdots + m_k = N$. Disjoint subsets $V_1, \ldots, V_m$ of the vertex set V imply that $V_1 \cup V_2 \cup \cdots \cup V_m = V$ and $V_i \cap V_j = \emptyset$ for all $i \neq j$.

Due to the assumption of undirected graph $G = (V, E)$, we require $w_{ij} = w_{ji}$. The affinity matrix or *weighted adjacency matrix* of the graph is the matrix $\mathbf{W} = [w_{ij}]_{i,j=1}^{N,N}$ with *edge weights* w_{ij} as entries

$$w_{ij} = \begin{cases} K(\mathbf{x}_i, \mathbf{x}_j), & i \neq j; \\ 0, & i = j. \end{cases} \qquad (6.16.3)$$

6.16 Graph Machine Learning

Definition 6.31 (Weighed Graph [301]) A graph is weighted if it is associated with a function $w : V \times V \to \mathbb{R}$ satisfying

$$w_{ij} > 0, \quad \text{if } (i, j) \in E, \tag{6.16.4}$$

and

$$w_{ij} = w_{ji}. \tag{6.16.5}$$

Consider a graph with N vertices where each vertex corresponds to a data point. For two data points $\mathbf{x}_i$ and $\mathbf{x}_j$, the weight w_{ij} of an edge connecting vertices i and j has four most commonly used choices as follows [41, 232].

1. *0–1 weighting:* $w_{ij} = 1$ if and only if nodes i and j are connected by an edge; otherwise $w_{ij} = 0$. This is the simplest weighting method and is very easy to compute.
2. *Heat kernel weighting:* If nodes i and j are connected, then

$$w_{ij} = e^{-\frac{\|\mathbf{x}_i - \mathbf{x}_j\|^2}{\sigma}} \tag{6.16.6}$$

is called the *heat kernel weighting*. Here σ is some pre-selected parameter.
3. *Thresholded Gaussian kernel weighting:* The weight of an edge connecting vertices i and j is defined, via a thresholded Gaussian kernel weighting function, as

$$w_{ij} = \begin{cases} \exp\left(-\frac{[\text{dist}(i,j)]^2}{2\theta^2}\right), & \text{dist}(i, j) \leq \kappa, \\ 0, & \text{otherwise,} \end{cases} \tag{6.16.7}$$

for some parameters θ and κ. Here, $\text{dist}(i, j)$ may represent a physical distance between vertices i and j or the Euclidean distance between two feature vectors describing i and j, the latter of which is especially common in graph-based semi-supervised learning methods.
4. *Dot-product weighting:* If nodes i and j are connected, then

$$w_{ij} = \mathbf{x}_i \cdot \mathbf{x}_j = \mathbf{x}_i^T \mathbf{x}_j \tag{6.16.8}$$

is known as the *dot-product weighting*. Note that if $\mathbf{x}$ is normalized to have unit norm, then the dot product of two vectors is equivalent to the cosine similarity of the two vectors.

There are different popular similarity graphs, depending on transforming a given set $\mathbf{x}_1, \ldots, \mathbf{x}_N$ of data points with pairwise similarities s_{ij} or pairwise distances d_{ij} into a graph. The following are three popular methods for constructing similarity graphs [257].

1. *ϵ-neighborhood graph:* All points are connected with pairwise distances smaller than ϵ. As the distances between all connected points are roughly of the same scale (at most ϵ), weighting the edges would not incorporate more information about the data to the graph. Hence, the ϵ-neighborhood graph is usually considered as an unweighted graph.
2. *K-nearest neighbor graph:* Connect vertex $\mathbf{v}_i$ with vertex $\mathbf{v}_j$ if $\mathbf{v}_j$ is among the K-nearest neighbors of $\mathbf{v}_i$. Due to the nonsymmetric neighborhood relationship, this definition leads to a directed graph. There are two methods for making this graph undirected:

   - One can simply ignore the directions of the edges, that is, $\mathbf{v}_i$ and $\mathbf{v}_j$ are connected with an undirected edge if $\mathbf{v}_i$ is among the K-nearest neighbors of $\mathbf{v}_j$ or if $\mathbf{v}_j$ is among the K-nearest neighbors of $\mathbf{v}_i$. The resulting graph is called the K-nearest neighbor graph.
   - Vertices $\mathbf{v}_i$ and $\mathbf{v}_j$ are connected if $\mathbf{v}_i$ is among the K-nearest neighbors of $\mathbf{v}_j$ and $\mathbf{v}_j$ is among the K-nearest neighbors of $\mathbf{v}_i$, which results in a *mutual K-nearest neighbor graph*. In these cases, after connecting the appropriate vertices, the edges are weighted by the similarity of their endpoints.

3. *Fully connected graph:* Connect simply all points with positive similarity with each other, and weight all edges by s_{ij}. Since the graph should represent the local neighborhood relationships, this construction is only useful if using the similarity function itself to model local neighborhoods. The Gaussian similarity function $s(\mathbf{x}_i, \mathbf{x}_j) = \exp\left(-\|\mathbf{x}_i - \mathbf{x}_j\|^2/(2\sigma^2)\right)$ is an example of such a positive similarity function. This parameter plays a similar role as the parameter ϵ in case of the ϵ-neighborhood graph.

The main differences between the three graphs above-mentioned are as follows.

- The ϵ-neighborhood graph uses the distance measure smaller than a threshold ϵ, and is an unweighted graph.
- The K-nearest neighbor graph considers K-nearest neighbors of vertices, and the edges are weighted by the similarity of their endpoints.
- The fully connected graph uses the positive (rather than nonnegative) similarity.

6.16.2 Graph Laplacian Matrices

The key problem of learning graph topology can be casted as a problem of learning the so-called *graph Laplacian matrices* as it uniquely characterizes a graph. In other words, graph Laplacian matrices are an essential operator in spectral graph theory [53].

6.16 Graph Machine Learning

Definition 6.32 (Graph Laplacian Matrix) The Laplacian matrix of graph $G(V, E, \mathbf{W})$ is defined as $\mathbf{L} = \mathbf{D} - \mathbf{W}$ with elements

$$L(u, v) = \begin{cases} d_v, & \text{if } u = v; \\ -1, & \text{if } (u, v) \in E; \\ 0, & \text{otherwise.} \end{cases} \quad (6.16.9)$$

Here, d_v denotes the degree of node i. The corresponding normalized Laplacian matrix is given by [53]

$$\mathcal{L}(u, v) = \begin{cases} 1, & \text{if } u = v \text{ and } d_v \neq 0; \\ -\frac{1}{\sqrt{d_u}\sqrt{d_v}}, & (u, v) \in E; \\ 0, & \text{otherwise.} \end{cases} \quad (6.16.10)$$

The unnormalized graph Laplacian $\mathbf{L}$ is also called *combinatorial graph Laplacian*.

There are two common normalized graph Laplacian matrices [257]:

- *Symmetric normalized graph Laplacian:*

$$\mathbf{L}_{\text{sym}} = \mathbf{D}^{-1/2} \mathbf{L} \mathbf{D}^{-1/2} = \mathbf{I} - \mathbf{D}^{-1/2} \mathbf{W} \mathbf{D}^{-1/2}. \quad (6.16.11)$$

- *Random walk normalized graph Laplacian:*

$$\mathbf{L}_{\text{rw}} = \mathbf{D}^{-1} \mathbf{L} = \mathbf{D}^{-1} \mathbf{W}. \quad (6.16.12)$$

The quadratic form of Laplacian matrices is given by

$$\mathbf{u}^T \mathbf{L} \mathbf{u} = \mathbf{u}^T \mathbf{D} \mathbf{u} - \mathbf{u}^T \mathbf{W} \mathbf{u} = \sum_{i=1}^{N} d_i u_i^2 - \sum_{j=1}^{N} \left(\sum_{i=1}^{N} u_i w_{ij} \right) u_j$$

$$= \sum_{i=1}^{N} d_i u_i^2 - \sum_{i=1}^{N} \sum_{j=1}^{N} u_i u_j w_{ij}$$

$$= \frac{1}{2} \left(\sum_{i=1}^{N} d_i u_i^2 - 2 \sum_{i=1}^{N} \sum_{j=1}^{N} u_i u_j w_{ij} + \sum_{j=1}^{N} d_j u_j^2 \right)$$

$$= \frac{1}{2} \left(\sum_{i=1}^{N} u_i^2 \sum_{j=1}^{N} w_{ij} - 2 \sum_{i=1}^{N} \sum_{j=1}^{N} u_i u_j w_{ij} + \sum_{j=1}^{N} u_j^2 \sum_{i=1}^{N} w_{ij} \right),$$

namely

$$\mathbf{u}^T \mathbf{L} \mathbf{u} = \frac{1}{2} \sum_{i=1}^{N} \sum_{j=1}^{N} w_{ij}(u_i - u_j)^2 \geq 0, \qquad (6.16.13)$$

the equality holds if and only if $u_i = u_j, \forall i, j \in \{1, \ldots, N\}$, i.e., $\mathbf{u} = \mathbf{1}_N = [1, \ldots, 1]^T \in \mathbb{R}^N$.

If $\mathbf{u} = \mathbf{1}_N$, then

$$[\mathbf{L}\mathbf{u}]_i = [(\mathbf{D} - \mathbf{W})\mathbf{u}]_i = \sum_{j=1}^{N} d_{ij} u_j - \sum_{j=1}^{N} w_{ij} u_j = d_i - d_i = 0$$

because $u_j \equiv 1$, $\sum_{j=1}^{N} d_{ij} = d_i$, and $\sum_{j=1}^{N} w_{ij} = d_i$. Therefore, we have

$$\mathbf{L}\mathbf{u} = \mathbf{0} \overset{\mathbf{L}\mathbf{u}=\lambda\mathbf{u}}{\Longleftrightarrow} \lambda = 0 \quad \text{for } \mathbf{u} = \mathbf{1}_N. \qquad (6.16.14)$$

Equations (6.16.13) and (6.16.14) give the following important properties of Laplacian matrices.

- The Laplacian matrix $\mathbf{L} = \mathbf{D} - \mathbf{W}$ is a positive semi-definite matrix.
- The minimum eigenvalue of the Laplacian matrix $\mathbf{L}$ is 0, and the eigenvector corresponding to the minimum eigenvalue is a vector with all values of 1.

6.16.3 Graph Spectrum

As the combinatorial graph Laplacian and the two normalized graph Laplacians are all real symmetric positive semi-definite matrices, they admit an eigendecomposition $\mathbf{L} = \mathbf{U}\boldsymbol{\Lambda}\mathbf{U}^T$ with $\mathbf{U} = [\mathbf{u}_0, \mathbf{u}_1, \ldots, \mathbf{u}_{n-1}] \in \mathbb{R}^{n \times n}$ and $\boldsymbol{\Lambda} = \mathbf{Diag}(\lambda_0, \lambda_1, \ldots, \lambda_{n-1}) \in \mathbb{R}^{n \times n}$. A complete set of orthonormal eigenvectors $\{\mathbf{u}_0, \mathbf{u}_1, \ldots, \mathbf{u}_{n-1}\}$ is known as the *graph Fourier basis*. The multiplicity of a zero Laplacian eigenvalue is equal to the number of connected components of the graph, and hence the real nonnegative eigenvalues are ordered as $0 = \lambda_0 \leq \lambda_1 \leq \lambda_2 \leq \cdots \leq \lambda_{n-1} = \lambda_{\max}$. The set of the graph Laplacian eigenvalues, $\sigma(L) = \{\lambda_0, \lambda_1, \ldots, \lambda_{n-1}\}$, is usually referred to as the *graph spectrum* of $\mathbf{L}$ (or the spectrum of the associated graph G).

Therefore, in graph signal processing and machine learning, eigenvalues of the Laplacian $\mathbf{L}$ are usually ordered increasingly, respecting multiplicities. By "the first k eigenvectors" we refer to the eigenvectors corresponding to the k smallest eigenvalues [257].

6.16 Graph Machine Learning

Proposition 6.4 ([53, 184]) *The graph Laplacian* **L** *satisfies the following properties:*

(1) For every vector $\mathbf{f} \in \mathbb{R}^n$ *and any graph Laplacian among* $\mathbf{L}, \mathbf{L}_{sys}, \mathbf{L}_{rw}$, *we have*

$$\mathbf{f}^T \mathbf{L} \mathbf{f} = \frac{1}{2} \sum_{i=1}^{n} \sum_{j=1}^{n} w_{ij}(f_i - f_j)^2.$$

(2) Any graph Laplacian among $\mathbf{L}, \mathbf{L}_{sys}, \mathbf{L}_{rw}$ *is symmetric and positive semi-definite.*
(3) The smallest eigenvalue of either $\mathbf{L}$ *or* $\mathbf{L}_{rw}$ *is* 0, *and* **1** *is an eigenvector associated with eigenvalue 0 (here* **1** *is a constant vector whose entries are all* 1). *Although* 0 *is also an eigenvalue of* $\mathbf{L}_{norm}$, *its associated eigenvector is* $\mathbf{D}^{1/2}\mathbf{1}$.
(4) If a graph Laplacian has k eigenvalues 0, then it has $N - k$ *positive, real-valued eigenvalues.*

The irrelevant eigenvectors are the ones that correspond to the several smallest eigenvalues of the Laplacian matrix **L** except for the smallest eigenvalue equal to 0. For computational efficiency, we thus often focus on calculating the eigenvectors corresponding to the largest several eigenvalues of the Laplacian matrix **L**. These results imply the following graph-based methods.

1. *Graph principal component analysis:* For given data vectors $\mathbf{x}_1, \ldots, \mathbf{x}_N$ and a graph $G(V, E, \mathbf{W})$ with Laplacian **L**, if p is the number of eigenvalues equal to 0 and q is the number of the smallest eigenvalues not equal to zero, then the Laplacian **L** has $N - (p + q)$ dominant eigenvalues. The corresponding eigenvectors of **L** give the graph principal component analysis (GPCA) after principal components in the standard PCA (see Sect. 6.8) are replaced by principal components of the Laplacian **L**.
2. *Graph minor component analysis:* If the minor components in standard minor component analysis (MCA) (see Sect. 6.8) are replaced by the relevant eigenvectors corresponding to the q smallest eigenvalues of **L** not equal to zero, then the standard MCA is extended to the *graph minor component analysis* (GMCA).
3. *Graph K-means clustering:* If there are K distinct clustered regions within the N data samples, then there are $K = N - (p+q)$ dominant nonnegative eigenvalues that provide a means of estimating the possible number of clusters within the data samples in graph K-means clustering.

Definition 6.33 (Edge Derivative [301]) Let $e = (i, j)$ denote the edge between vertices i and j. The *edge derivative* of function f with respect to e at the vertex i is defined as

$$\left.\frac{\partial f}{\partial e}\right|_i = \sqrt{\frac{w_{ij}}{d_i}} f_i - \sqrt{\frac{w_{ij}}{d_j}} f_j. \qquad (6.16.15)$$

The local variation of f at each vertex j is then defined to be

$$\|\nabla_j f\| = \sqrt{\sum_{e \vdash j} \left(\left.\frac{\partial f}{\partial e}\right|_j\right)^2}, \quad (6.16.16)$$

where $e \vdash j$ denotes the set of the edges incident with vertex j.

The smoothness of f is then naturally measured by the sum of the local variations at each vertex:

$$S(f) = \frac{1}{2} \sum_j \|\nabla_j f\|^2. \quad (6.16.17)$$

The graph Laplacian $\mathbf{L} = \mathbf{D} - \mathbf{W}$ is a difference operator, as, for any signal $\mathbf{f} \in \mathbb{R}^n$, it satisfies

$$(\mathbf{Lf})(i) = \sum_{j \in \mathcal{N}_i} w_{ij}[f_i - f_j], \quad (6.16.18)$$

where the neighborhood $\mathcal{N}_i$ is the set of vertices connected to vertex i by an edge.

The classical Fourier transform

$$\hat{f}(\xi) = \langle f, e^{j2\pi\xi t} \rangle = \int_{\mathbb{R}} f(t) e^{-j2\pi\xi t} dt \quad (6.16.19)$$

is the expansion of a function f in terms of the complex exponentials, which are the eigenfunctions of the one-dimensional (1-D) Laplace operator

$$-L(e^{j2\pi\xi t}) = -\frac{\partial^2}{\partial t^2} e^{j2\pi\xi t} = (2\pi\xi)^2 e^{j2\pi\xi t}. \quad (6.16.20)$$

Similarly, the *graph Fourier transform* $\hat{f}(\lambda)$ of any function $\mathbf{f} = [f(0), f(1), \ldots, f(n-1)]^T \in \mathbb{R}^n$ on the vertices of G is defined as the expansion of $\mathbf{f}$ in terms of the eigenvectors $\mathbf{u}_l = [u_l(0), u_l(1), \ldots, u_l(n-1)]^T$ of the graph Laplacian:

$$\hat{f}(\lambda_l) = \langle \mathbf{f}, \mathbf{u}_l \rangle = \sum_{i=0}^{n-1} f_i u_l^*(i), \quad l = 0, 1, \ldots, n-1, \quad (6.16.21)$$

or written as matrix-vector form

$$\hat{\mathbf{f}} = \mathbf{U}^H \mathbf{f} \in \mathbb{R}^n, \text{ where } \mathbf{U} = [\mathbf{u}_0, \mathbf{u}_1, \ldots, \mathbf{u}_{n-1}]. \quad (6.16.22)$$

The *graph inverse Fourier transform* is then given by

$$f_i = \sum_{l=0}^{n-1} \hat{f}(\lambda_l) u_l(i), \quad i = 0, 1, \ldots, n-1 \quad \text{or} \quad \mathbf{f} = \mathbf{U}\hat{\mathbf{f}}. \tag{6.16.23}$$

In classical Fourier analysis, the eigenvalues $\{(2\pi\xi)^2\}_{\xi \in \mathbb{R}}$ carry a specific notion of frequency: for ξ close to zero (low frequencies), the associated complex exponential eigenfunctions are smooth, slowly oscillating functions. On the contrary, if ξ is far from zero (high frequencies), then the associated complex exponential eigenfunctions oscillate much more rapidly. In the graph setting, the graph Laplacian eigenvalues and eigenvectors provide a similar notion of frequency [232]: the graph Laplacian eigenvectors associated with low frequencies λ_l, vary slowly across the graph, i.e., if two vertices are connected by an edge with a large weight, then the eigenvectors at those locations are likely to be similar. The eigenvectors associated with larger eigenvalues oscillate more rapidly and are more likely to have dissimilar values on vertices connected by an edge with high weight.

6.16.4 Graph Signal Processing

In this subsection, we focus on some basic operations in graph signal processing.

Graph Filtering
In classical signal processing, given an input time signal $f(t)$ and a time-domain filter $h(t)$, the frequency filtering is defined as

$$\hat{f}_{\text{out}}(\xi) = \hat{f}_{\text{in}}(\xi)\hat{h}(\xi), \tag{6.16.24}$$

where $\hat{f}_{\text{in}}(\xi)$ and $\hat{f}_{\text{out}}(\xi)$ are the spectrum of the input and output signals, respectively; while $\hat{h}(\xi)$ is the transfer function of the filter. Taking an inverse Fourier transform of (6.16.24), we have the time filtering given by

$$f_{\text{out}}(t) = \int_{\mathbb{R}} \hat{f}_{\text{in}}(\xi)\hat{h}(\xi) e^{j2\pi\xi t} d\xi = \int_{\mathbb{R}} f_{\text{in}}(\tau) h(t-\tau) \, d\tau = (f * h)(t), \tag{6.16.25}$$

where

$$(f * h)(t) = \int_{\mathbb{R}} f_{\text{in}}(\tau) h(t-\tau) d\tau \tag{6.16.26}$$

denotes the convolution product of the continue signal $f(t)$ and the continue filter $h(t)$.

For discrete signal $\mathbf{f} = [f_1, \ldots, f_N]^T$, (6.16.25) gives the filtering result of discrete signals

$$f_{\text{out}}(i) = (f * h)(i) = \sum_{k=0}^{N-1} f(k)h(i-k) = \sum_{k=0}^{N-1} h(k)f(i-k). \quad (6.16.27)$$

Similar to (6.16.24), the graph spectral filtering can be defined as

$$\hat{f}_{\text{out}}(\lambda_\ell) = \hat{f}_{\text{in}}(\lambda_\ell)\hat{h}(\lambda_\ell), \quad (6.16.28)$$

where λ_ℓ is the ℓth eigenvalue of the Laplacian matrix $\mathbf{L}$. Taking the inverse Fourier transform of the above equation, we have the discrete-time graph filtering

$$f_{\text{out}}(i) = \sum_{\ell=0}^{N-1} \hat{f}_{\text{in}}(\lambda_\ell)\hat{h}(\lambda_\ell)u_\ell(i), \quad (6.16.29)$$

where $u_\ell(i)$ is the ith element of $N \times 1$ eigenvector $\mathbf{u}_\ell = [u_\ell(1), \ldots, u_\ell(N)]^T$ corresponding to the eigenvalue λ_ℓ.

Graph Convolution

Because the graph signal is discrete, the graph filtering (6.16.29) can be represented using the same convolution as discrete signals, i.e.,

$$f_{\text{out}}(i) = (f_{\text{in}} * h)_G(i), \quad (6.16.30)$$

But, the graph convolution does not have the standard form of classical convolution $(f * h)(i) = \sum_{k=0}^{N-1} f(k)h(i-k) = \sum_{k=0}^{N-1} h(k)f(i-k)$, because there is no delayed form $f(i-k)$ for any graphic signal $f(i)$.

By comparing (6.16.30) with (6.16.25), we get the convolution product of graph signals as follows:

$$(f * h)_G(i) = \sum_{\ell=0}^{N-1} \hat{f}_{\text{in}}(\lambda_\ell)\hat{h}(\lambda_\ell)u_\ell(i). \quad (6.16.31)$$

When using the Laplacian matrix $\mathbf{L}$ to represent the graph convolution, we denote

$$\mathbf{f}_{\text{in}} = [f_{\text{in}}(1), \ldots, f_{\text{in}}(N)]^T, \quad (6.16.32)$$

$$\mathbf{f}_{\text{out}} = [f_{\text{out}}(1), \ldots, f_{\text{out}}(N)]^T, \quad (6.16.33)$$

$$\hat{h}(\mathbf{L}) = \mathbf{U} \begin{bmatrix} \lambda_0 & & 0 \\ & \ddots & \\ 0 & & \lambda_{N-1} \end{bmatrix} \mathbf{U}^H. \quad (6.16.34)$$

Therefore, the graph convolution (6.16.31) can be rewritten in matrix-vector form as [232]

$$(\mathbf{f} * \mathbf{h})_G = \hat{h}(\mathbf{L})\mathbf{f}_{\text{in}}. \quad (6.16.35)$$

That is to say, the graph filtering can be represented, in Laplacian form, as follows:

$$\mathbf{f}_{\text{out}} = \hat{h}(\mathbf{L})\mathbf{f}_{\text{in}}. \quad (6.16.36)$$

p-Dirichlet Norm

Consider the approximation problem of discrete functions on a graph G.

A lot of learning algorithms operate on input spaces $\mathcal{X}$ other than $\mathbb{R}^n$, specifically, discrete input spaces, such as strings, graphs, trees, automata, etc. Such an input space is known as the manifold. In Euclidean space, the smoothness of a vector function $\mathbf{f}$ is usually represented in Euclidean norm as $\|\mathbf{f}\|_2^2 = \sum_{i=1}^{N} \sum_{j=1}^{N} (f_j - f_i)^2$. However, this representation is available for graph signals in manifold.

Consider an undirected unweighted graph $G(V, E, \mathbf{W})$ consisting of a set of vertices V numbered 1 to n, and a set of edges E (i.e., pairs (i, j)) where $i, j \in V$ and $(i, j) \in E \Leftrightarrow (j, i) \in E$. For convenience, we will sometimes use $i \sim j$ to denote that i and j are neighbors, i.e., $(i, j) \in E$. The adjacency matrix of G is an $n \times n$ real matrix $\mathbf{W}$ whose entries $w_{ij} = 1$ if $i \sim j$, and 0 otherwise. By construction, $\mathbf{W}$ is symmetric and its diagonal entries are zero.

To approximate a function on a graph G, with the weight matrix $\mathbf{W}$, we need finding a "good" function that does not make too many "jumps" for two connected nodes i and j. This good function can be formalized by the smoothness functional $S(\mathbf{f}) = \sum_{i \sim j} w_{ij}(f_i - f_j)^2$ that should be minimized. More generally, we have the following definition of smoothness function on graphs.

Definition 6.34 ([232]) The discrete p-Dirichlet norm of the graph signal $\mathbf{f} = [f_1, \ldots, f_N]^T$ is defined by

$$S_p(\mathbf{f}) = \frac{1}{p} \sum_{i \in V} \left[\sum_{j \in \mathcal{N}_i} w_{ij}(f_j - f_i)^2 \right]^{p/2}. \quad (6.16.37)$$

The following are two common p-Dirichlet norms of the graph signal $\mathbf{f}$.

- When $p = 1$, the 1-Dirichlet norm

$$S_1(\mathbf{f}) = \sum_{i \in V} \left[\sum_{j \in \mathcal{N}_i} w_{ij}(f_j - f_i)^2 \right]^{1/2} \quad (6.16.38)$$

is the total variation of the signal with respect to the graph.

- When $p = 2$, the 2-Dirichlet norm

$$S_2(\mathbf{f}) = \frac{1}{2} \sum_{i \in V} \sum_{j \in \mathcal{N}_i} w_{ij}(f_j - f_i)^2$$

$$= \sum_{i,j \in E} w_{ij}(f_j - f_i)^2 = \mathbf{f}^T \mathbf{L} \mathbf{f} \qquad (6.16.39)$$

is a vector norm of the graph signal vector $\mathbf{f}$ weighted by the Laplacian matrix $\mathbf{L}$.

As compared with Euclidean norm, the 2-Dirichlet norm can be reviewed as the Euclidean norm weighted by adjacency w_{ij} of two nodes.

Graph Tikhonov Regularization

Given a noisy graph signal $\mathbf{y} = \mathbf{f}_0 + \mathbf{e}$, where $\mathbf{f}_0$ is a graph signal vector and $\mathbf{e}$ is uncorrelated additive Gaussian noise. In order to recover $\mathbf{f}_0$, if using the 2-Dirichlet form $S_2(\mathbf{f}) = \mathbf{f}^T \mathbf{L} \mathbf{f}$ of the graph signal $\mathbf{f}$ instead of its regularization term $\|\mathbf{f}\|_2^2$ in the Tikhonov regularization method, we have that the Tikhonov regularization on the graph signal is given by

$$\arg\min_{\mathbf{f}} \left\{ \|\mathbf{f} - \mathbf{y}\|_2^2 + \lambda \mathbf{f}^T \mathbf{L} \mathbf{f} \right\}. \qquad (6.16.40)$$

Proposition 6.5 ([231]) *The Tikhonov solution $\mathbf{f}_*$ to (6.16.40) is given by*

$$\mathbf{f}_*(i) = \sum_{\ell=0}^{N-1} \left[\frac{1}{1 + \gamma \lambda_\ell} \right] \hat{y}(\lambda_\ell) u_\ell(i), \qquad (6.16.41)$$

for all $i = \{1, 2, \ldots, N\}$.

Graph Kerners

Kernel-based algorithms capture the structure of $\mathcal{X}$ via the kernel $K : \mathcal{X} \times \mathcal{X} \to \mathbb{R}$. One of the most general representations of discrete metric spaces are graphs. Regularization on graphs is an important step for manifold learning.

For a Mercer kernel $K : \mathcal{X} \times \mathcal{X} \to \mathbb{R}$, there is an associated reproducing kernel Hilbert space (RKHS) $\mathcal{H}_K$ of functions $\mathcal{X} \to \mathbb{R}$ with the corresponding norm $\|\cdot\|_K$. Given a set of labeled examples $(\mathbf{x}_i, y_i)$, $i = 1, \ldots, l$ and a set of unlabeled examples $\{\mathbf{x}_j\}_{j=l+1}^{l+u}$, we consider the following optimization problem:

$$\mathbf{f}^* = \arg\min_{\mathbf{f} \in \mathcal{H}_K} \left\{ \frac{1}{l} \sum_{i=1}^{l} V(\mathbf{x}_i, y_i, \mathbf{f}) + \gamma_A \|\mathbf{f}\|_K^2 + \frac{\gamma_I}{l+u} \hat{\mathbf{f}}^T \mathbf{L} \hat{\mathbf{f}} \right\}, \qquad (6.16.42)$$

where $\hat{\mathbf{f}} = [f(\mathbf{x}_1), \ldots, f(\mathbf{x}_{l+u})]^T$ as $n = l + u$ and $f_i = f(\mathbf{x}_i)$ for $i = 1, \ldots, n$, and $\mathbf{L}$ is the Laplacian matrix $\mathbf{L} = \mathbf{D} - \mathbf{W}$ where w_{ij} are the edge weights in the data adjacency graph. Here, the diagonal matrix $\mathbf{D}$ is given by $\mathbf{D}_{l+u}$ with

$D_{ii} = \sum_{j=1}^{l+u} w_{ij}$. The normalizing coefficient $\frac{1}{(l+u)^2}$ is the natural scale factor for the empirical estimate of the Laplace operator. On a sparse adjacency graph it may be replaced by $\sum_{i=1}^{l+u} \sum_{j=1}^{l+u} w_{ij}$.

Corollary 6.1 ([235]) *Denote by* $\mathbf{P} = r(\tilde{\mathbf{L}})$ *a regularization matrix, then the corresponding kernel matrix is given by the inverse* $\mathbf{K} = r^{-1}(\tilde{\mathbf{L}})$ *or the pseudo-inverse* $\mathbf{K} = r^{\dagger}(\tilde{\mathbf{L}})$ *wherever necessary. More specifically, if* $\{(\lambda_i, \mathbf{v}_i)\}$ *constitute the eigensystem of* $\tilde{\mathbf{L}}$, *we have*

$$\mathbf{K} = \sum_{i=1}^{m} r^{-1}(\lambda_i) \mathbf{v}_i \mathbf{v}_i^T, \qquad (6.16.43)$$

where $0^{-1} = 0$.

In the context of spectral graph theory and segmentation, the following graph kernel matrices are of particular interest [235]:

$\mathbf{K} = (\mathbf{I} + \sigma^2 \tilde{\mathbf{L}})^{-1}$ (Regularized Laplacian)

$\mathbf{K} = \exp(\sigma^2/2\tilde{\mathbf{L}})$ (Diffusion Process)

$\mathbf{K} = (a\mathbf{I} - \tilde{\mathbf{L}})^p$ with $a \geq 2$ (One-Step Random Walk)

$\mathbf{K} = \cos(\tilde{\mathbf{L}}\pi/4)$ (Inverse Cosine).

Theorem 6.7 ([20]) *The minimizer of optimization problem (6.16.42) admits an expansion*

$$f^*(\mathbf{x}) = \sum_{i=1}^{l+u} \alpha_i K(\mathbf{x}_i, \mathbf{x}) \qquad (6.16.44)$$

in terms of the labeled examples $(\mathbf{x}_i, y_i)_{i=1}^{l}$ *and unlabeled examples* $\{\mathbf{x}_j\}_{j=1}^{l+u}$.

If taking squared loss function in (6.16.42) as

$$V(\mathbf{x}_i, y_i, f) = (y_i - f(\mathbf{x}_i))^2, \qquad (6.16.45)$$

then the minimizer of the $(l+u)$-dimensional expansion coefficient vector $\boldsymbol{\alpha} = [\alpha_1, \ldots, \alpha_{l+u}]^T$ in (6.16.44) is given by [20]

$$\boldsymbol{\alpha}^* = \left(\mathbf{JK} + \gamma_A l \mathbf{I} + \frac{\gamma_I l}{(l+u)^2} \mathbf{LK} \right)^{-1} \mathbf{y}, \qquad (6.16.46)$$

where $\mathbf{K}$ is the $(l+u) \times (l+u)$ Gram matrix over labeled and unlabeled points; $\mathbf{y}$ is an $(l+u)$ dimensional label vector given by $\mathbf{y} = [y_1, \ldots, y_l, 0, \ldots, 0]^T$; and

$\mathbf{J} = \mathbf{Diag}(1, \ldots, 1, 0, \ldots, 0)$ is an $(l+u) \times (l+u)$ diagonal matrix with the first l diagonal entries as 1 and the others as 0.

Algorithm 6.31 shows the manifold regularization algorithm mentioned above.

Algorithm 6.31 Manifold regularization algorithms [20]

1. **input:** l labeled examples $\{\mathbf{x}_i, y_i\}_{i=1}^{l}$ and u unlabeled examples $\{\mathbf{x}_j\}_{j=1}^{l+u}$
2. Construct data adjacency graph with $(l+u)$ nodes using, e.g., k-nearest neighbors. Choose edge weights $w_{ij} = \exp\left(-\|\mathbf{x}_i - \mathbf{x}_j\|^2/(4l)\right)$
3. Choose a kernel function $K(x, y)$ and compute the Gram matrix $K_{ij} = K(\mathbf{x}_i, \mathbf{x}_j)$
4. Compute graph Laplacian matrix $\mathbf{L} = \mathbf{D} - \mathbf{W}$ where $\mathbf{D} = \mathbf{Diag}(W_{1,1}, \ldots, W_{l+u,l+u})$
5. Choose γ_A and γ_I
6. Compute $\boldsymbol{\alpha}^* = \left(\mathbf{JK} + \gamma_A l \mathbf{I} + \frac{\gamma_I l}{(l+u)^2}\mathbf{LK}\right)^{-1}\mathbf{y}$
7. **output:** $f^*(\mathbf{x}) = \sum_{i=1}^{l+u} \alpha_i K(\mathbf{x}_i, \mathbf{x})$

When $\gamma_I = 0$, Eq. (6.16.42) gives zero coefficients over unlabeled data, so the coefficients over labeled data are exactly those for standard RLS.

6.16.5 Semi-Supervised Graph Learning: Harmonic Function Method

In practical applications of graph signal processing, graph pattern recognition, graph machine learning (spectral clustering, graph principal component analysis, graph K-means clustering, etc.), and so on, even graph-structured data (such as social networks, information networks, text documents, etc.) are usually given in convenient sampled data rather than the graph form. To apply the spectral graph theory to the sampled data, we must learn these data and transform them to the graph form, i.e., the weighted adjacency matrix $\mathbf{W}$ or the graph Laplacian matrix $\mathbf{L}$.

The graph learning problems can be generally thought of as looking for a function $\mathbf{f}$ which is smooth and simultaneously close to another given function $\mathbf{y}$. This view can be formalized as the following optimization problem [301]:

$$\mathbf{f} = \arg\min_{\mathbf{f} \in \ell^2(V)} \left\{ S(\mathbf{f}) + \frac{\mu}{2}\|\mathbf{f} - \mathbf{y}\|_2^2 \right\}. \tag{6.16.47}$$

The first term $S(\mathbf{f})$ measures the smoothness of the function $\mathbf{f}$, and the second term $\frac{\mu}{2}\|\mathbf{f} - \mathbf{y}\|_2^2$ measures its closeness to the given function $\mathbf{y}$. The trade-off between these two terms is captured by a nonnegative parameter μ.

We are given l labeled points $\{(\mathbf{x}_1, y_1), \ldots, (\mathbf{x}_l, y_l)\}$ and u unlabeled points $\{\mathbf{x}_{l+1}, \ldots, \mathbf{x}_{l+u}\}$ with $\mathbf{x}_i \in \mathbb{R}^d$, typically $l \ll u$. Let $n = l + u$ be the total number of data points.

Consider a connected graph $G(V, E, \mathbf{W})$ with node set V corresponding to the n data points, with nodes $L = \{1, \ldots, l\}$ corresponding to the labeled points with class labels $y_1, \ldots, y_l$, and nodes $U = \{l+1, \ldots, l+u\}$ corresponding to the unlabeled points. The task of semi-supervised graph labeling is to first estimate a real-valued function $f : V \to \mathbb{R}$ on the graph G such that f satisfies following two conditions at the same time [304]:

- f_i should be close to the given labels y_i on the labeled nodes,
- **f** should be smooth on the whole graph.

The function f is usually constrained to take values

$$f_i = f_i^{(l)} = f^{(l)}(\mathbf{x}_i) \equiv y_i \tag{6.16.48}$$

on the labeled data $\{(\mathbf{x}_1, y_1), \ldots, (\mathbf{x}_l, y_l)\}$. At the same time, the function $\mathbf{f}_u = [f_{l+1}, \ldots, f_{l+u}]^T$ assigns the labels y_{l+i} to the unlabeled instances $\mathbf{x}_{l+1}, \ldots, \mathbf{x}_{l+u}$ via $\hat{y}_{l+i} = \text{sign}(f_{l+i})$ in binary classification $y_i \in \{-1, +1\}$.

An approach to graph-based semi-supervised labeling was proposed by Zhu et al. [306]. In this approach, labeled and unlabeled data are represented as vertices in a weighted graph, with edge weights encoding the similarity between instances, and the semi-supervised labeling problem is then formulated in terms of a Gaussian random field on this graph, where the mean of the field is characterized in terms of harmonic functions, and is efficiently obtained using matrix algebra methods.

It is assumed that an $n \times n$ symmetric weight matrix $\mathbf{W}$ on the edges of the graph is given. For example, when $\mathbf{x}_i \in \mathbb{R}^d$, the (i, j)th entries of the weight matrix can be

$$w_{ij} = \exp\left(-\sum_{k=1}^d \frac{(x_{ik} - x_{jk})^2}{\sigma_k}\right), \tag{6.16.49}$$

where x_{ik} is the kth component of instance $\mathbf{x}_i \in \mathbb{R}^d$ and $\sigma_1, \ldots, \sigma_d$ are length scale hyperparameters for each dimension. Thus, the weight matrix can be decomposed into the block form [306]:

$$\mathbf{W} = \begin{bmatrix} \mathbf{W}_{ll} & \mathbf{W}_{lu} \\ \mathbf{W}_{ul} & \mathbf{W}_{uu} \end{bmatrix}, \tag{6.16.50}$$

where

$W_{ll}(i, j) = w_{ij}, i, j = 1, \ldots, l; \quad W_{lu}(i, j) = w_{i,l+j}, i = 1, \ldots, l; j = 1, \ldots, u;$

$W_{ul}(i, j) = w_{l+i,j}, i = 1, \ldots, u; j = 1, \ldots, l;$

$W_{uu}(i, j) = w_{l+i,l+j}, i, j = 1, \ldots, u.$

The Gaussian function (6.16.49) shows that nearby points in Euclidean space are assigned large edge weight or other more appropriate weight values.

Let the real-valued function vector

$$\mathbf{f} = \begin{bmatrix} \mathbf{f}_l \\ \mathbf{f}_u \end{bmatrix} = [f(1), \ldots, f(l), f(l+1), \ldots, f(l+u)]^T, \qquad (6.16.51)$$

where

$$\mathbf{f}_l = [f_l(1), \ldots, f_l(l)]^T \quad \text{or} \quad f_l(i) = f(i) \equiv y_i, \ i = 1, \ldots, l; \qquad (6.16.52)$$

$$\mathbf{f}_u = [f_u(1), \ldots, f_u(u)]^T \quad \text{or} \quad f_u(i) = f(l+i), \ i = 1, \ldots, u. \qquad (6.16.53)$$

Since $f_l(i) \equiv y_i, i = 1, \ldots, l$ are known, $\mathbf{f}_u$ is the $u \times 1$ vector to be found. Once $\mathbf{f}_u$ can be found by using the graph, then the labels to unlabeled data $\mathbf{x}_{l+1}, \ldots, \mathbf{x}_{l+u}$ can be assigned: $\hat{y}_{l+i} = \text{sign}(f_u(i)), i = 1, \ldots, u$. Intuitively, unlabeled points that are nearby in the graph should have similar labels. This results in the quadratic energy function

$$E(f) = \frac{1}{2} \sum_{i,j} w_{ij} (f_i - f_j)^2. \qquad (6.16.54)$$

By [306], the minimum energy function $\mathbf{f} = \arg\min_{\mathbf{f}|_L = \mathbf{f}^{(l)}} E(f)$ is harmonic; namely, it satisfies $\Delta \mathbf{f} = \mathbf{0}$ on unlabeled data points U, and is equal to $\mathbf{f}_l$ on the labeled data points L, where $\Delta = \mathbf{D} - \mathbf{W}$ is the Laplacian of the graph G.

From $\Delta \mathbf{f} = \mathbf{0}$ or

$$\left(\begin{bmatrix} \mathbf{D}_{ll} & \\ & \mathbf{D}_{uu} \end{bmatrix} - \begin{bmatrix} \mathbf{W}_{ll} & \mathbf{W}_{lu} \\ \mathbf{W}_{ul} & \mathbf{W}_{uu} \end{bmatrix} \right) \begin{bmatrix} \mathbf{f}_l \\ \mathbf{f}_u \end{bmatrix} = \begin{bmatrix} \mathbf{0}_{l \times 1} \\ \mathbf{0}_{u \times 1} \end{bmatrix}$$

it follows that the harmonic solution is given by [306]:

$$\mathbf{f}_u = (\mathbf{D}_{uu} - \mathbf{W}_{uu})^{-1} \mathbf{W}_{ul} \mathbf{f}_l = (\mathbf{I} - \mathbf{P}_{uu})^{-1} \mathbf{P}_{ul} \mathbf{f}_l, \qquad (6.16.55)$$

where

$$\mathbf{P} = \mathbf{D}^{-1} \mathbf{W} = \begin{bmatrix} \mathbf{D}_{ll}^{-1} \mathbf{W}_{ll} & \mathbf{D}_{ll}^{-1} \mathbf{W}_{lu} \\ \mathbf{D}_{uu}^{-1} \mathbf{W}_{ul} & \mathbf{D}_{uu}^{-1} \mathbf{W}_{uu} \end{bmatrix}$$

$$= \begin{bmatrix} \mathbf{P}_{ll} & \mathbf{P}_{lu} \\ \mathbf{P}_{ul} & \mathbf{P}_{uu} \end{bmatrix}. \qquad (6.16.56)$$

The above discussions can be summarized into the harmonic function algorithm for semi-supervised graph learning, as shown in Algorithm 6.32.

Semi-supervised labeling is closely related to important applications. For example, in electric networks the edges of a graph G can be imagined to be resistors with conductance W, where nodes labeled 1 are connected to a positive voltage source,

6.16 Graph Machine Learning

Algorithm 6.32 Harmonic function algorithm for semi-supervised graph learning [306]

1. **input**: Labeled data $(\mathbf{x}_1, y_1), \ldots, (\mathbf{x}_l, y_l)\}$ and unlabeled data $\{\mathbf{x}_{l+1}, \ldots, \mathbf{x}_{l+u}\}$ with $l \ll u$.
2. Choice a Gaussian weight function w_{ij} on the graph $G(E, V)$, where $i, j = 1, \ldots, n$ and $n = l + u$.
3. Compute the degree $d_j = \sum_{i=1} w_{ij}, j = 1, \ldots, n$.
4. Compute $\mathbf{D}_{uu} = \text{Diag}(d_{l+1}, \ldots, d_{l+u})$, $\mathbf{W}_{uu} = [w_{l+i,l+j}]_{i,j=1}^{u,u}$ and $\mathbf{W}_{ul} = [w_{l+i,j}]_{i,j=1}^{u,l}$.
5. $\mathbf{P}_{uu} \leftarrow \mathbf{D}_{uu}^{-1} \mathbf{W}_{uu}$, $\mathbf{P}_{ul} \leftarrow \mathbf{D}_{uu}^{-1} \mathbf{W}_{ul}$.
6. $\mathbf{f}_u \leftarrow (\mathbf{I} - \mathbf{P}_{uu})^{-1} \mathbf{P}_{ul} \mathbf{f}_l$.
7. **output**: the labels $\hat{y}_{l+i} = \text{sign}(f_u(i))$, $i = 1, \ldots, u$.

---

and points labeled 0 to ground [306]. Then $f_u(i)$ is the voltage in the resulting electric network on the ith unlabeled node.

Once the labels $\hat{y}_{l+1}$ are estimated by applying Algorithm 6.32, the semi-supervised graph classification or regression problem becomes a supervised one.

6.16.6 Semi-Supervised Graph Learning: Min-Cut Method

Semi-supervised graph learning aims at constructing a graph from the labeled data $\{(\mathbf{x}_1, y_1), \ldots, (\mathbf{x}_l, y_l)\}$ and unlabeled data $\{\mathbf{x}_{l+1}, \ldots, \mathbf{x}_{l+u}\}$ with $\mathbf{x}_i \in \mathbb{R}^d$, $i = 1, \ldots, l + u$. These data are drawn independently at random from a probability density function $p(x)$ on a domain $M \subseteq \mathbb{R}^d$.

The question is how to partition the nodes in a graph into disjoint subsets S and T such that the source s is in S and the sink t is in T. This problem is called s/t cut problem. In combinatorial optimization, the cost of an s/t cut, $C = \{S, T\}$, is defined as the sum of the costs of "boundary" edges (p, q), where $p \in S$ and $q \in T$. The *minimum cut* (min-cut) problem on a graph is to find a cut that has the minimum cost among all cuts.

Min-cut

Let $X = \{\mathbf{x}_1, \ldots, \mathbf{x}_{l+u}\}$ be a set of $l + u$ points, and S be a smooth hypersurface that is separated into two parts S_1 and S_2 such that $X_1 = X \cap S_1 = \{\mathbf{x}_1, \ldots, \mathbf{x}_l\}$ and $X_2 = X \cap S_2 = \{\mathbf{x}_{l+1}, \ldots, \mathbf{x}_{l+u}\}$ are the data subsets which land in S_1 and S_2, respectively.

We are given two disjoint subsets of nodes $L = \{1, \ldots, l\}$, $S = L_+$, and $T = L_-$, corresponding to the l labeled points with labels $y_1, \ldots, y_l$, i.e., $L = S \cup T$ and $S \cap T = \emptyset$. The degree of dissimilarity between these two pieces S and T can be computed as total weight of the edges that have been removed. In graph theoretic language, it is called the *cut* that is denoted by $\text{cut}(S, T)$ and is defined as [230]:

$$\text{cut}(S, T) = \sum_{i \in S, j \in T} w_{ij}. \tag{6.16.57}$$

The optimal binary classification of a graph is to minimize this cut value, resulting in a most popular method for finding the minimum cut of a graph.

Normalized Cut

However, the minimum cut criteria favors cutting small sets of isolated nodes in the graph [279]. To avoid this unnatural bias for partitioning out small sets of points, Shi and Malik [230] proposed the normalized cut (Ncut) as the disassociation measure to compute the cut cost as a fraction of the total edge connections to all the nodes in the graph, instead of looking at the value of total edge weight connecting the two partitions (S, T).

Definition 6.35 (Normalized Cut [230]) The *normalized cut* (Ncut) of two disjointed partitions $S \cup T = L$ and $S \cap T = \emptyset$ is denoted by Ncut(S, T) and is defined as

$$\text{Ncut}(S, T) = \frac{\text{cut}(S, T)}{\text{assoc}(S, L)} + \frac{\text{cut}(S, T)}{\text{assoc}(T, L)}, \quad (6.16.58)$$

where

$$\text{assoc}(S, L) = \sum_{i \in S, j \in L} w_{ij} \quad \text{and} \quad \text{assoc}(T, L) = \sum_{i \in T, j \in L} w_{ij} \quad (6.16.59)$$

are, respectively, the total connections from nodes in S and T to all nodes in the graph.

In the same spirit, a measure for total normalized association within groups for a given partition can be defined as [230]

$$\text{Nassoc}(S, T) = \frac{\text{assoc}(S, S)}{\text{assoc}(S, L)} + \frac{\text{assoc}(T, T)}{\text{assoc}(T, L)}, \quad (6.16.60)$$

where $\text{assoc}(S, S)$ and $\text{assoc}(T, T)$ are total weights of edges connecting nodes within S and T, respectively.

The association and disassociation of a partition are naturally related:

$$\begin{aligned}
\text{Ncut}(S, T) &= \frac{\text{cut}(S, T)}{\text{assoc}(S, L)} + \frac{\text{cut}(S, T)}{\text{assoc}(T, L)} \\
&= \frac{\text{assoc}(S, L) - \text{assoc}(S, S)}{\text{assoc}(S, L)} + \frac{\text{assoc}(T, L) - \text{assoc}(T, T)}{\text{assoc}(T, L)} \\
&= 2 - \left(\frac{\text{assoc}(S, S)}{\text{assoc}(S, L)} + \frac{\text{assoc}(T, T)}{\text{assoc}(T, L)} \right) \\
&= 2 - \text{Nassoc}(S, T). \quad (6.16.61)
\end{aligned}$$

This means that the two partition criteria, attempting to minimize Ncut(S, T) (disassociation between the groups (S, T)) and attempting to maximize Nassoc(S, T) (association within the groups), are in fact identical.

Graph Mincut Learning

Consider the complete graph whose vertices are associated with the points $\mathbf{x}_i, \mathbf{x}_j$, and where the weight of the edge between $\mathbf{x}_i$ and $\mathbf{x}_j$, $i \neq j$ is given.

Let $\mathbf{f} = [f_1, \ldots, f_N]^T$ be the indicator vector for X_1:

$$f_i = \begin{cases} 1, & \text{if } x_i \in X_1, \\ 0, & \text{otherwise.} \end{cases} \tag{6.16.62}$$

There are two quantities of interest [187]:

- $\int_S p(s) ds$, which measures the quality of the partition S in accordance with the weighted volume of the boundary.
- $\mathbf{f}^T \mathbf{L} \mathbf{f}$, which measures the quality of the empirical partition in terms of its cut size.

When making graph constructions, it is necessary to consider choosing which graph to use from the following three graph types [19, 173].

1. *k-nearest neighbor* (kNN) *graph:* The samples $\mathbf{x}_i$ and $\mathbf{x}_j$ are considered as neighbors if $\mathbf{x}_i$ is among the k-nearest neighbors of $\mathbf{x}_j$ or $\mathbf{x}_j$ is among the k-nearest neighbors of $\mathbf{x}_i$, where k is a positive integer and the k-nearest neighbors are measured by the usual Euclidean distance.
2. *r-ball neighborhood graph:* The samples $\mathbf{x}_i$ and $\mathbf{x}_j$ are considered as neighbors if and only if $\|\mathbf{x}_i - \mathbf{x}_j\| < r$, where the norm is the usual Euclidean norm, r is a fixed radius and two points are connected if their distance does not exceed the threshold radius r. Note that due to the symmetry of the distance we do not have to distinguish between directed and undirected graphs. The r-ball neighborhood graph is simply called the "r-neighborhood" as well.
3. *Complete weighted graph:* There is an edge $e = (i, j)$ between each pair of distinct nodes $\mathbf{x}_i$ and $\mathbf{x}_j$ (but no loops). This graph is not considered as a neighborhood graph in general, but if the weight function is chosen in such a way that the weights of edges between nearby nodes are high and the weights between points far away from each other are almost negligible, then the behavior of this graph should be similar to that of a neighborhood graph. One such weight function is the Gaussian weight function $w_{ij} = \exp(-\|\mathbf{x}_i - \mathbf{x}_j\|^2/(2\sigma^2))$.

The k-nearest neighbor graph is a directed graph, while the r-ball neighborhood graph is an undirected graph.

The weights used on neighborhood graphs usually depend on the distance of the vertices of the edge and are non-increasing. In other words, the weight w_{ij} of an edge $(\mathbf{x}_i, \mathbf{x}_j)$ is given by $w_{ij} = f(\text{dist}(\mathbf{x}_i, \mathbf{x}_j))$ with a non-increasing weight function f. The weight functions to be considered here are the unit weight function

$f \equiv 1$, which results in the unweighted graph, and the Gaussian weight function

$$f(u) = \frac{1}{(2\pi\sigma^2)^{d/2}} \exp\left(-\frac{1}{2}\frac{u^2}{\sigma^2}\right), \quad (6.16.63)$$

where the parameter $\sigma > 0$ defines the bandwidth of the Gaussian function.

It is usually assumed that for a measurable set $S \subseteq \mathbb{R}^d$, the surface integral along a decision boundary S, $\mu(S) = \int_S p(s) ds$, is the measure on $\mathbb{R}^d$ induced by probability distribution p. This measure is called *"weighted boundary volume."*

Narayanan et al. [187] prove that the weighted boundary volume is approximated by $\frac{\sqrt{\pi}}{N\sqrt{\sigma}} \mathbf{f}^T \mathbf{L} \mathbf{f}$:

$$\int_S p(s) ds \approx \frac{\sqrt{\pi}}{N\sqrt{\sigma}} \mathbf{f}^T \mathbf{L} \mathbf{f}, \quad (6.16.64)$$

where $\mathbf{L}$ is the normalized graph Laplacian, $\mathbf{f} = [f_1, \ldots, f_{l+u}]^T$ is the weight vector function, and σ is the bandwidth of the edge weight Gaussian function.

For the classification problem, we are given the training sample data as $\mathbf{X} = [\mathbf{x}_1, \ldots, \mathbf{x}_N]$, where $\mathbf{x}_i \in \mathbb{R}^d$ and N is the total number of training samples. Traditional graph construction methods typically decompose the graph construction process into two steps: graph adjacency construction and graph weight calculation [286].

1. *Graph adjacency construction:* There are two main construction methods [19]: ϵ-ball neighborhood and k-nearest neighbors.
2. *Three Graph weight calculation approaches* [286]

   - *Heat Kernel* [19]

   $$w_{ij} = \begin{cases} \frac{e^{-\|\mathbf{x}_i - \mathbf{x}_j\|^2}}{t}, & \text{if } \mathbf{x}_i \text{ and } \mathbf{x}_j \text{ are neighbors;} \\ 0, & \text{otherwise,} \end{cases} \quad (6.16.65)$$

   where t is the heat kernel parameter. Note when $t \to \infty$, the heat kernel will produce binary weight and the graph constructed will be a binary graph, i.e.,

   $$w_{ij} = \begin{cases} 1, & \text{if } \mathbf{x}_i \text{ and } \mathbf{x}_j \text{ are neighbors;} \\ 0, & \text{otherwise.} \end{cases} \quad (6.16.66)$$

   - *Inverse Euclidean Distance* [63]

   $$w_{ij} = \|\mathbf{x}_i - \mathbf{x}_j\|_2^{-1}, \quad (6.16.67)$$

   where $\|\mathbf{x}_i - \mathbf{x}_j\|_2$ is the Euclidean distance between $\mathbf{x}_i$ and $\mathbf{x}_j$.

- *Local linear reconstruction coefficient* [216]

$$\xi(\mathbf{W}) = \sum_i \left\| \mathbf{x}_i - \sum_j w_{ij} \mathbf{x}_j \right\|^2, \quad \text{subject to} \sum_j w_{ij} = 1, \ \forall i, \quad (6.16.68)$$

where $w_{ij} = 0$ if sample $\mathbf{x}_i$ and $\mathbf{x}_j$ are not neighbors.

If there are N samples in the original data, then *leave-one-out cross-validation* (LOOCV) uses each sample as a testing set separately, and the other $N - 1$ samples are used as a training set. Therefore, LOOCV will get N models, and the average of the classification accuracy of the final testing set of N models will be used as the performance index of LOOCV classifier under this condition.

LOOCV error is defined as the count of $\mathbf{x} \in L \cup U$ with label of $\mathbf{x}$ different from label of its nearest neighbor. This is also the value of the cut.

Let $nn_{\mathbf{uv}}$ be the indicator of "$\mathbf{v}$ is the nearest neighbor of $\mathbf{u}$" for K-nearest neighborhood graph, and $w_{\mathbf{uv}}$ be the weight of the label of $\mathbf{v}$ when classifying $\mathbf{u}$ for K-averaged nearest neighborhood.

Theorem 6.8 ([29]) *Suppose edge weights between example nodes are defined as follows: for each pair of nodes $\mathbf{u}$ and $\mathbf{v}$, define $nn_{\mathbf{uv}} = 1$ if $\mathbf{v}$ is the nearest neighbor of $\mathbf{u}$, and $nn_{\mathbf{uv}} = 0$ otherwise. If defining $w(\mathbf{u}, \mathbf{v}) = nn_{\mathbf{uv}} + nn_{\mathbf{vu}}$ for K-nearest neighborhood graph or $w(\mathbf{u}, \mathbf{v}) = nn_{\mathbf{uv}} + nn_{\mathbf{vu}}$ for K-averaged nearest neighborhood graph, then for any binary labeling of the examples $\mathbf{u} \in U$, the cost of the associated cut is equal to the number of LOOCV mistakes made by 1-nearest neighbor on $L \cup U$.*

The above theorem implies that min-cut for K-nearest neighborhood graph or K-averaged nearest neighborhood graph is identical to minimizing the LOOCV error on $L \cup U$.

The following summarize the steps in *Graph Mincut Learning Algorithm* [29].

1. Construct a weighted graph $G = (V, E, \mathbf{W})$, where $V = L \cup U \cup \{v_+, v_-\}$, and $E \subseteq V \times V$. Associated with each edge $e \in E$ is a weight $w(e)$. The vertices v_+ and v_- are called the *classification vertices*, and all other vertices the *example vertices*.
2. $w(v, v_+) = \infty$ for all $v \in L_+$ and $w(v, v_-) = \infty$ for all $v \in L_-$.
3. The function assigning weights to edges between example nodes $\mathbf{x}_i, \mathbf{x}_j$ are referred to as the edge weighting function w_{ij}.
4. v_+ is viewed as the source, v_- is viewed as the sink, and the edge weights are treated as capacities. Removing the edges in the cut partitions the graph into two sets of vertices: V_+ with $v_+ \in V_+$ and V_- with $v_- \in V_-$. For concreteness, if there are multiple minimum cuts, then the algorithm chooses the one such that V_+ is smallest (this is always well defined and easy to obtain from the flow).
5. Assign a positive label to all unlabeled examples in the set V_+ and a negative label to all unlabeled examples in the set V_-.

Minimum Normalized Cut

If a node in $S = L_+$ is viewed as a source $s \in S$ and a node in $T = L_-$ is viewed as a sink $t \in T$, then minimizing $\text{Ncut}(S, T)$ is a minimum s/t cut problem. One of the fundamental results in combinatorial optimization is [34] that the minimum s/t cut problem can be solved by finding a maximum flow from the source s to the sink t. Loosely speaking, maximum flow is the maximum "amount of water" that can be sent from the source to the sink by interpreting graph edges as directed "pipes" with capacities equal to edge weights. The theorem of Ford and Fulkerson [95] states that a maximum flow from s to t saturates a set of edges in the graph dividing the nodes into two disjoint parts $\{S, T\}$ corresponding to a minimum cut. Thus, min-cut and max-flow problems are equivalent. In this sense, min-cut, max-flow, and *min-cut/max-flow methods* refer to as the same method.

Let **f** be an $N = |L|$ dimensional indicator vector, $f_i = 1$ if node i is in L_+ and -1, otherwise. Under the assumption that $d_i = \sum_{j=1} w_{ij}$ is the total connection from node i to all other nodes, $\text{Ncut}(S, T)$ can be rewritten as

$$\text{Ncut}(S, T) = \frac{\text{cut}(S, T)}{\text{assoc}(S, L)} + \frac{\text{cut}(S, T)}{\text{assoc}(T, L)}$$

$$= \frac{\sum_{(f_i>0, f_j<0)} -w_{ij} f_i f_j}{\sum_{f_i>0} d_i} + \frac{\sum_{(f_i<0, f_j>0)} -w_{ij} f_i f_j}{\sum_{f_i<0} d_i}$$

$$= \text{Ncut}(\mathbf{f}). \tag{6.16.69}$$

It was shown [230] that the minimization of normalized cut $\text{Ncut}(S, T) = \text{Ncut}(\mathbf{f})$ is identical to the minimization of Rayleigh quotient $R(\mathbf{f})$:

$$\min_{\mathbf{f}} \text{Ncut}(\mathbf{f}) = \min_{\mathbf{u}} \left\{ R(\mathbf{u}) = \frac{\mathbf{u}^T (\mathbf{D} - \mathbf{W}) \mathbf{u}}{\mathbf{u}^T \mathbf{D} \mathbf{u}} \right\} \tag{6.16.70}$$

subject to the normalization constraint $\mathbf{f}^T \mathbf{1} = 0$ and $\|\mathbf{f}\| = 1$.

Proposition 6.6 ([230]) *The constrained minimization of Rayleigh quotient (6.16.70) can be formulated as a generalized eigenvalue problem* $\mathbf{Lf} = \lambda \mathbf{Df}$: *the solution is the second smallest generalized eigenvector of the matrix pencil* $(\mathbf{L}, \mathbf{D})$, *where* $\mathbf{L} = \mathbf{D} - \mathbf{W}$ *is the Laplacian in the graph G.*

In image segmentation, a key problem is how to partition the domain I of an image into subsets, and pick the "right" one. Based on the graph theoretic formulation of grouping, a minimum normalized cuts based grouping approach was proposed by Shi and Malik [230]. In this grouping, the set of vertices is partitioned into disjoint sets $V_1, V_2, \ldots, V_m$, where by some measure the similarity among the vertices in a set V_i is high and, across different sets V_i, V_j is low.

Using the relation between minimum normalized cut and the generalized eigenvalue decomposition of the matrix pencil $(\mathbf{D} - \mathbf{W}, \mathbf{D})$, a normalized cuts based algorithm for image segmentation is shown below [230].

1. Given an image or image sequence, set up a weighted graph $G(V, E, \mathbf{W})$ and set the weight on the edge connecting two nodes to be a measure of the similarity between the two nodes.
2. Solve $(\mathbf{D} - \mathbf{W})\mathbf{u} = \lambda \mathbf{D}\mathbf{u}$ for eigenvectors with the smallest eigenvalues.
3. Use the eigenvector with the second smallest eigenvalue to bipartition the graph.
4. Decide if the current partition should be subdivided and recursively repartition the segmented parts if necessary.

6.16.7 Unsupervised Graph Learning: Sparse Coding Method

In Sects. 6.16.5 and 6.16.6 we have discussed the harmonic function method and the min-cut method for semi-supervised graph learning or graph construction. Now we turn to graph construction from only unlabeled training examples.

Given a set of N d-dimension training examples $\mathbf{x}_1, \ldots, \mathbf{x}_N$ that constitute an $N \times d$ training matrix $\mathbf{X} = [\mathbf{x}_1, \ldots, \mathbf{x}_N]$, typically $d \ll N$. Let $\mathbf{y} = [y_1, \ldots, y_d]^T$ be the target signal and $\mathbf{a} = [a_1, \ldots, a_N]^T$ be a coefficient vector such that

$$\mathbf{e} = \mathbf{y} - \mathbf{X}\mathbf{a}. \tag{6.16.71}$$

The coefficient vector $\mathbf{a}$ can be estimated via the optimization $\min \|\mathbf{y} - \mathbf{X}\mathbf{a}\|_2^2$. The corresponding matrix equation is

$$[\mathbf{X}, \mathbf{I}_d] \begin{bmatrix} \mathbf{a} \\ \mathbf{e} \end{bmatrix} = \mathbf{y} \quad \text{or} \quad \mathbf{B}\boldsymbol{\alpha} = \mathbf{y}, \tag{6.16.72}$$

where $\mathbf{B} = [\mathbf{X}, \mathbf{I}_d] \in \mathbb{R}^{d \times (N+d)}$ with the $d \times d$ identity matrix $\mathbf{I}_d$, and $\boldsymbol{\alpha} = \begin{bmatrix} \mathbf{a} \\ \mathbf{e} \end{bmatrix} \in \mathbb{R}^{d+N}$. However, due to $d \ll N$, the matrix equation $\mathbf{B}\boldsymbol{\alpha} = \mathbf{y}$ is under-determined, so it has infinite solutions $\boldsymbol{\alpha}$.

Using the sparse coding method, the unique solution to the under-determined equation (6.16.72) can be obtained by minimizing both the reconstruction error $\|\mathbf{e}\|_2^2$ and the ℓ_1-norm of the coefficient vector $\boldsymbol{\alpha}$, and turned to solve [273]:

$$\hat{\boldsymbol{\alpha}} = \arg\min_{\boldsymbol{\alpha}} \|\boldsymbol{\alpha}\|_1$$

$$\text{subject to } \mathbf{B}\boldsymbol{\alpha} = \mathbf{y}. \tag{6.16.73}$$

In unsupervised cases, the target $\mathbf{y}$ is not available, and one can think of the ith training sample $\mathbf{x}_i$ as the target $\mathbf{y}$ to transform the ℓ_1 optimization problem (6.16.73) to [286]:

$$\hat{\boldsymbol{\alpha}}_i = \arg\min_{\boldsymbol{\alpha}} \|\boldsymbol{\alpha}\|_1$$

$$\text{subject to } \mathbf{B}\boldsymbol{\alpha} = \mathbf{x}_i, \tag{6.16.74}$$

where

$$\boldsymbol{\alpha}_i = \left[a_1^{(i)}, \ldots, a_{N-1}^{(i)}, e_1^{(i)}, \ldots, e_d^{(i)} \right]^T \in \mathbb{R}^{N-1+d}, \tag{6.16.75}$$

$$\mathbf{B} = [\mathbf{X} \setminus \mathbf{x}_i, \mathbf{I}_d] \in \mathbb{R}^{d \times (N-1+d)}, \; i = 1, \ldots, N \tag{6.16.76}$$

with $[\mathbf{X} \setminus \mathbf{x}_i] = [\mathbf{x}_1, \ldots, \mathbf{x}_{i-1}, \mathbf{x}_{i+1}, \ldots, \mathbf{x}_N]$ for $i = 1, \ldots, N$.

Furthermore, graph weights are given by $w_{ij} = |a_j^{(i)}|$. This graph obtained by using sparse coding from the training samples $\mathbf{x}_1, \ldots, \mathbf{x}_N$ is called ℓ_1 *graph* [286].

Algorithm 6.33 shows an ℓ_1 directed graph construction algorithm.

Algorithm 6.33 ℓ_1 directed graph construction algorithm [286]

1. **input:** Column sample matrix $\mathbf{X} = [\mathbf{x}_1, \ldots \mathbf{x}_N]$.
2. **initialization:** Normalize the training samples to have unit ℓ_2 norm.
3. **for** $i = 1 : N$ **do**
4.   Set $[\mathbf{X} \setminus \mathbf{x}_k] = [\mathbf{x}_1, \ldots, \mathbf{x}_{k-1}, \mathbf{x}_{k+1}, \ldots, \mathbf{x}_N] \in \mathbb{R}^{d \times (N-1)}$, and $\mathbf{B} = [\mathbf{X} \setminus \mathbf{x}_k, \mathbf{I}_d] \in \mathbb{R}^{d \times (d+N-1)}$.
5.   Solve the ℓ_1 optimization problem $\hat{\boldsymbol{\alpha}} = \arg\min_{\boldsymbol{\alpha}} \|\boldsymbol{\alpha}\|_1$ subject to $\mathbf{B}\boldsymbol{\alpha} = \mathbf{x}_k$.
6.   Let $\mathbf{a} = \hat{\boldsymbol{\alpha}}_{1:N}$ and $\mathbf{e} = \hat{\boldsymbol{\alpha}}_{N+1:N+d-1}$.
7.   **for** $j = 1 : N - 1$ **do**
8.   if $j < i$ then set $w_{ij} = |a_j|$,
9.   else set $w_{ij} = |a_{j-1}|$.
10.   **end for**
11. **end for**
12. **output:** $\mathbf{W} = [w_{ij}]_{i,j=1}^{N,N}$.

6.17 Active Learning

Machine learning can be divided into "passive" learning and "active" learning from the perspective of the relationship between learners and supervisors.

- *Passive learning* is instructor/supervisor-centered, where the learners/users receive information from the instructor/supervisor and internalize it to make classification or prediction (or regression).
- *Active learning* (AL) is generally used to refer to a learning problem or system where the learner has some active or participatory role in determining on what data it will be trained. This is in contrast to passive learning, where the learner is simply presented with a training set over which it has no control [58].

Active learning problems have been studied for many decades under the rubric of experimental design from 1970s ([51, 90]).

Active learning and semi-supervised learning face the same issue: data without class labels are quite massive, and labeled data is rather scarce and hard to obtain. In these cases, we can make the learning algorithm take the initiative to propose which data to be labeled, and then add these new-labeled data to the training sample set to train the learning algorithm. This process is usually referred to as "*active learning.*" Hence, active learning is a special case of semi-supervised machine learning in which a learning algorithm is able to interactively query the user (or some other information source) to obtain the desired outputs at new data points [40]. It is quite natural to combine active learning and semi-supervised learning to address this issue from both ends.

The basic difference between active learning and semi-supervised learning is that the learner has the ability or need to influence or select its own training data in active learning, while semi-supervised learning has neither this ability nor need.

6.17.1 Active Learning: Background

In many machine learning problems, the training data are treated as a fixed and given part of the problem definition. In practice, however, the training data are often not fixed beforehand. Rather, as an active participant in the training process, the learner has an opportunity to play a role in deciding what data will be acquired for training. In active learning, the learner must take actions to gain information, and must decide what actions will give him/her the information that will best minimize future loss [58].

In many applications, unlabeled data are usually abundant but manual labeling is costly. If more sophisticated supervised (and semi-supervised) learning algorithms are used in these applications, labeled instances are very difficult, time-consuming, or costly to obtain. A few examples are given below [40]:

- *Information extraction.* Good information extraction systems must be trained by using labeled documents with detailed annotations. Locating entities and relations can take half an hour for even simple newswire stories [224]. Annotations for other knowledge domains may require additional expertise, e.g., annotating gene and disease mentions for biomedical information extraction often requires Ph.D-level biologists.
- *Classification and filtering.* Learning to classify documents (e.g., articles or web pages) or any other kind of media (e.g., image, audio, and video files) requires users to label each document or media file with particular labels, like "relevant" or "not relevant" to the query. It can be tedious and even redundant for one to have to annotate thousands of these instances .
- *Clinical named entity recognition.* Identification of clinical concepts or clinical named entity recognition (NER) is an important task for building clinical natural language processing (NLP) systems. Chen et al. [50] reported that for the annotated NER corpus from the 2010 i2b2/VA NLP challenge that contained 349 clinical documents with 20,423 unique sentences, it is very difficult to simulate experiments using an existing clinical NER corpus with annotated medical problems, treatments, and lab tests in clinical notes.

Effective management of digital video resources for human action retrieval is a difficult task, as action retrieval is more challenging than action recognition.

Relevance feedback is one common technique used for mitigating the effect of the lack of training in human action retrieval systems. In order to improve accuracy of a retrieval system, by relevance feedback the user can feedback each result returned to the retrieval system for marking them either as relevant or irrelevant to their query. Then, the retrieval system will perform an initial query and return the top T most relevant results to the user for labeling as many "relevant" or as few "irrelevant" as desired. Next, the user run the query again to generate improved results until intraclass variability is better represented in the query.

Even with relevance feedback, however, retrieval results can still be poor [135]: the amount of feedback is often very small, as users have limited patience to provide feedback to improve the query; with only few training samples the retrieval results will often still be unstable and unreliable.

Active learning is a better choice than relevance feedback for retrieval systems. Similar to relevance feedback, "*Active learning*" (also called query learning) also requires the user to label several database items according to their relevance to the query. These labels are then used to update the query and improve the retrieval results.

The key difference between relevance feedback and active learning is which database items the user provides labels for [135]:

- In relevance feedback, the user himself/herself labels the top T most relevant results returned by the retrieval system.
- In active learning, the user can (actively) query the teacher/expert for labels in order to achieve higher accuracy.

With more informative feedback, active learning has a faster learning rate, and typically performs better than relevance feedback.

Active learning is a subfield of machine learning and, more generally, artificial intelligence [40]. The key idea behind active learning is that if a machine learning algorithm is allowed to choose the data from which it learns, it will perform better with less training.

6.17.2 Statistical Active Learning

When active learning is used for classification or regression, there are three most important paradigms: membership query learning, pool-based active learning, and stream-based active learning:

- *Membership query learning [9]:* It is also called *constructive active learning.* In membership query learning, the learner is allowed to create or select unlabeled instances for the human expert to label [42].

6.17 Active Learning

- *Pool-based active learning [159, 176]:* It is popular for many applications such as text classification and speech recognition where unlabeled data are plentiful and cheap, but labels are expensive and slow to acquire. In pool-based active learning, the learner may not propose arbitrary points to label, but instead has access to a set of unlabeled examples, and is allowed to select which of them to request labels for.
- *Stream-based active learning [11]:* It is also called *sequential active learning* or *selective sampling*. Stream-based active learning resembles pool-based learning in many ways, except that the learner must decide whether to query or discard the unlabeled instances as a stream.

The choice of examples to label can be seen as a dilemma between the exploration and the exploitation over the data space representation [33]. Active learning aims at involving the best method to choose the data points for $\mathcal{T}_{C,i}$ (a subset chosen to be labeled) via some query strategy.

Query strategies for determining which data points should be labeled can be organized into a number of different categories [40]:

- *Uncertainty sampling* [159] is one of the most popular query criterions, and can be used to select the most uncertain instances of the unlabeled data.
- *Query by committee (QBC) method* [226] is an effective active learning method. Given a committee $C = \{\theta^{(1)}, \ldots, \theta^{(C)}\}$ of models, these models are all trained on the current labeled set $\mathcal{L}$, and are query candidates. The QBC method directly measures the voting differences of committee members, and selects samples with inconsistent voting by class labels as training data, while the most informative query is considered to be the instance about which they most disagree.
- *Expected model change method* [225] is to query the instance that would impart the greatest change to the current model if we knew its label.
- *Variance reduction method* [59] labels those points that would minimize output variance, which is one of the components of error.
- *Balance exploration and exploitation:* in this setting, the choice of examples to label is seen as a dilemma between the exploration and the exploitation over the data space representation. This strategy manages this compromise by modeling the active learning problem as a contextual conflict problem. The commonly used algorithms have the greedy algorithm, the Softmax algorithm, Bayes Bandit algorithm, etc.
- *Least certainty* selects subset $\mathcal{N}_a$ whose elements are in the bottom of the rank queue Q.
- *Middle certainty* selects subset $\mathcal{N}_a$ whose elements are in the middle of the rank queue Q.
- *Exponentiated gradient exploration* [32] can improve any active learning algorithm by an optimal random exploration.

Relying on the learner's ability to statistically model its own uncertainty, the objective of statistical active learning is usually to find model parameters that minimize some form of expected loss. The process of statistical active learning is usually as follows [58]:

1. Begin by requesting labels for a small random sub-sample of the examples and fit a statistical model to the labeled data.
2. For any **x**, a statistical model is used to estimate both the conditional expectation of its label $y = y(\mathbf{x})$ and the variance of that expectation, $\sigma^2_{\hat{y}(\mathbf{x})}$.
3. Consider a candidate point $\tilde{\mathbf{x}}$, and ask what reduction in loss we would obtain if we had labeled it $\tilde{y}$.
4. Given the ability to estimate the expected effect of obtaining label $\tilde{y}$ for candidate point $\tilde{\mathbf{x}}$, repeat this computation for a sample of M candidates, and then request a label for the candidate with the largest expected decrease in loss. Add the newly labeled example to the training set, retrain, and begin looking at candidate points to add on the next iteration.

6.17.3 Active Learning Algorithms

Given a large pool $\mathcal{U}$ of unlabeled samples and a small set of labeled samples $\mathcal{L}$, with $|\mathcal{L}| \ll |\mathcal{U}|$ and $\mathcal{X} = \mathcal{L} \cup \mathcal{U}$, where $|\cdot|$ represents the cardinality of a set. Typically, *pool-based active learning* (PAL) uses $\mathcal{L}$ to train a classifier **h**. Then, a query set $\mathcal{S}$ of unlabeled samples is determined for labeling based on a selection strategy Q, which uses the "knowledge" contained in **h** for sample selection, and presented to the oracle $\mathcal{O}$. The labeled samples are then added to $\mathcal{L}$ and the classifier **h** is updated.

The pool-based active learning framework can be summarized as follows [50]:

1. *Initial model generation:* At the beginning, a small number of samples are queried for annotation to build the initial model. Two common initial sampling strategies are random sampling or application-oriented sampling (e.g., longest sentence sampling for named entity recognition in clinical text).
2. *Querying:* The unannotated samples are then ranked based on the querying algorithm. Different algorithms use different query strategies for ranking.
3. *Training:* The selected unlabeled subset is retrained on the updated annotated set.
4. *Iteration:* Steps 2 and 3 are repeated until the stop criterion is met.

Pool-based active learning has been widely used in many real-world applications, e.g., text categorization [119, 250], video search [266], image classification [160], action retrieval [135], and so on.

Algorithm 6.34 is a traditional active learning approach based on the least (or middle) certainty query strategy [295].

Active learning is easily extended to the case of two-view instances. This extended active learning is called *co-active learning*, sketched in Algorithm 6.35.

Algorithm 6.34 Active learning [295]

1. **input:** labeled dataset $\mathcal{L} = \{\mathbf{x}_i, y_i\}_{i=1}^{l}$ and unlabeled dataset $\mathcal{U} = \{\mathbf{x}_i\}_{i=l+1}^{l+u}$.
2. **repeat**
3. Upsample the labeled dataset $\mathcal{L}$ to obtain even class distribution $\mathcal{L}_D$;
4. Use $\mathcal{L}/\mathcal{L}_D$ to train a classifier $\mathbf{h}$, and then classify the unlabeled dataset $\mathcal{U}$;
5. Rank the data based on the prediction confidence value C and store them in queue Q.
6. Select subset $\mathcal{N}_a$ whose elements are in the bottom of the rank queue Q (least certainty); or select $\mathcal{N}_a$ whose elements are in the middle of the rank queue Q (middle certainty);
7. Submit the selected subset $\mathcal{N}_a$ to human annotation;
8. Remove $\mathcal{N}_a$ from the unlabeled dataset to get $\mathcal{U} = \mathcal{U} \setminus \mathcal{N}_a$;
9. Add $\mathcal{N}_a$ to the labeled dataset $\mathcal{L}$, i.e., $\mathcal{L} = \mathcal{L} \cup \mathcal{N}_a$;
10. **until** unlabeled data is used up.
11. **output:** $\{\mathbf{x}_i, y_i\}_{i=1}^{l+u}$.

Algorithm 6.35 Co-active learning [295]

1. **input:** a learning domain with features $V = \{\mathbf{x}_i, y_i\}_{i=1}^{l} \cup \{\mathbf{x}_i\}_{i=l+1}^{l+u}$.
2. **repeat**
3. Split the domain V into two "views": $V_1 = \{\mathbf{x}_i^{(1)}\}_{i=1}^{l}$ and $V_2 = \{\mathbf{x}_i^{(2)}\}_{i=1}^{l}$, where $V_1 \cap V_2 = \emptyset$;
4. Upsample "views" V_1 and V_2 to obtain even class distributions V_{D_1} and V_{D_2}, respectively;
5. Use V/V_{D_1} and V/V_{D_2} to train classifiers $\mathbf{h}_1$ and $\mathbf{h}_2$, respectively; and then use $\mathbf{h}_1$ and $\mathbf{h}_2$ to classify the unlabeled dataset $\mathcal{U}$, respectively;
6. Rank the data based on the prediction confidence value C and store them in queue Q;
7. Select subset $\mathcal{N}_a$ whose elements are in the middle of the rank queue Q (middle certainty);
8. Submit the selected subset $\mathcal{N}_{ca} = \mathcal{U}_{a1} \cup \mathcal{U}_{a2}$ to human annotation;
9. Remove $\mathcal{N}_{ca}$ from the unlabeled dataset $\mathcal{U}: \mathcal{U} = \mathcal{U} \setminus \mathcal{N}_{ca}$;
10. Add $\mathcal{N}_{ca}$ to the labeled dataset $\mathcal{L}$, i.e., $\mathcal{L} = \mathcal{L} \cup \mathcal{N}_{ca}$;
11. **until** unlabeled data is used up.
12. **output:** $\{\mathbf{x}_i, y_i\}_{i=1}^{l+u}$.

6.17.4 Active Learning Based Binary Linear Classifiers

Consider binary classification with two classes "+" and "−." We are given a small labeled instance set $\mathcal{L} = \{\mathbf{x}_i, y_i\}_{i=1}^{l}$ and a large unlabeled instance set $\mathcal{U} = \{\mathbf{x}_i\}_{i=l+1}^{l+u}$, where $\mathbf{x}_i \in \mathbb{R}^d$, $y_i \in \{+1, -1\}$. Our task is that given any testing instance $\mathbf{x}$, we want to make a decision on which class $\mathbf{x}$ belongs to by using pool-based active learning.

To this end, we need to find an opposite pair $\{\mathbf{x}_+, \mathbf{x}_-\}$ close to the separating hyperplane.

Let C_+ and C_- be the centroid of positive and negative labeled instances, respectively:

$$\mathbf{c}_+ = \frac{1}{|\mathcal{L}_+|} \sum_{i \in \mathcal{L}_+} \mathbf{x}_i, \quad \mathbf{c}_- = \frac{1}{|\mathcal{L}_-|} \sum_{i \in \mathcal{L}_-} \mathbf{x}_i. \qquad (6.17.1)$$

When the labeled set $\mathcal{L}$ is small, we cannot use $\mathbf{x}_+ = \mathbf{c}_+$ and $\mathbf{x}_- = \mathbf{c}_-$ as the estimation results of $\mathbf{x}_+$ and $\mathbf{x}_-$, respectively, as these estimates are usually very poor.

By pool-based active learning, we synthesize the first instance $\mathbf{x}_s = \frac{1}{2}(\mathbf{x}_+ + \mathbf{x}_-)$ and query its label. Since this is a pseudo-instance, it might not be recognized by the human annotator, especially in vision-based tasks [260]. Instead, its nearest neighbor (denoted by $\mathbf{x}_q$) should be searched from the unlabeled set $\mathcal{U}$ and its label queried. If the label is positive, we use the midpoint of $\mathbf{x}_q$ and $\mathbf{c}_-$ as a new synthesized instance. Repeating this process until we can find an opposite pair $\{\mathbf{x}_+, \mathbf{x}_-\}$ close to the classification boundary.

Algorithm 6.36 shows a pool-based active learning for finding opposite pair close to separating hyperplane, which was introduced by Wang et al. [260].

Algorithm 6.36 Active learning for finding opposite pair close to separating hyperplane [260]

---

1. **input:** labeled data set $\mathcal{L}^0 = \{\mathbf{x}_i, y_i\}_{i=1}^l$, unlabeled data set $\mathcal{U}^0 = \{\mathbf{x}\}_{i=l+1}^{l+u}$.
2. **initialization:** let $\mathbf{x}_+^0 = \frac{1}{|\mathcal{L}_+|} \sum_{i \in \mathcal{L}_+} \mathbf{x}_i$ and $\mathbf{x}_-^0 = \frac{1}{|\mathcal{L}_-|} \sum_{i \in \mathcal{L}_-} \mathbf{x}_i$.
3. **repeat**
4.   Synthesize an instance $\mathbf{x}_s = \frac{1}{2}(\mathbf{x}_+ + \mathbf{x}_-)$;
5.   Use the nearest neighbor method to compute the instance $\mathbf{x}_q$ in $\mathcal{U}$ for query, e.g.,
6. $\mathbf{x}_q \leftarrow \arg\min_{\mathbf{x}_i \in \mathcal{U}} d(\mathbf{x}_s, \mathbf{x}_i)$;
7.   Query $\mathbf{x}_q$;
8. $\mathcal{L} = \mathcal{L} \cup \mathbf{x}_q$ and $\mathcal{U} = \mathcal{U} \setminus \mathbf{x}_q$;
9.   **if** $\mathbf{x}_q$ is positive, **then**
10. $\mathbf{x}_+ \leftarrow \mathbf{x}_q$.
11.   **else**
12. $\mathbf{x}_- \leftarrow \mathbf{x}_q$.
13.   **end if**
14. **until** $\mathbf{x}_+$ and $\mathbf{x}_-$ are converged, respectively; or the number of binary search being run.
15. **output:** $(\{\mathbf{x}_+, \mathbf{x}_-\}, \mathcal{L}, \mathcal{U}) = findPair(\{\mathbf{x}_+^0, \mathbf{x}_-^0\}, \mathcal{L}^0, \mathcal{U}^0)$.

In the above algorithm, $d(\mathbf{a}, \mathbf{b})$ is a distance function available for practical application.

Once the enhanced labeled data set $\mathcal{L}$ and the remained unlabeled data set $\mathcal{U}$ are found by Algorithm 6.36, we can apply any semi-supervised binary classification/regression method to find the binary classifier or regressor $\mathbf{h}$. Then, for any testing instance $\mathbf{x}$, the classification $\text{sign}(\mathbf{h}^T \mathbf{x})$ or the regression $\hat{y} = \mathbf{h}^T \mathbf{x}$ can be obtained.

6.17.5 Active Learning Using Extreme Learning Machine

Suppose we are given N arbitrary distinct training instances $\{\mathbf{x}_i, \mathbf{y}_i\}_{i=1}^N$, where $\mathbf{x}_i \in \mathbb{R}^n$ and $\mathbf{y}_i \in \mathbb{R}^m$.

6.17 Active Learning

Consider a single-hidden layer feedforward network (SLFN) with L hidden nodes that approximates these N samples with zero error: there exist $\boldsymbol{\beta}_i, a_i$, and b_i such that

$$f_L(\mathbf{x}_j) = \sum_{i=1}^{L} \boldsymbol{\beta}_i G(a_i, b_i, \mathbf{x}_j) = \mathbf{y}_j, \quad j = 1, \ldots, N, \tag{6.17.2}$$

where a_i and b_i denote the ith weight and bias of the hidden layer, respectively, and $\boldsymbol{\beta}_i$ is the weight vector connecting the ith hidden node to the output nodes.

Letting

$$\mathbf{H} = \begin{bmatrix} G(a_1, b_1, \mathbf{x}_1) & \cdots & G(a_L, b_L, \mathbf{x}_1) \\ \vdots & \ddots & \vdots \\ G(a_1, b_1, \mathbf{x}_N) & \cdots & G(a_L, b_L, \mathbf{x}_N) \end{bmatrix} \in \mathbb{R}^{N \times L}, \tag{6.17.3}$$

$$\mathbf{B} = \begin{bmatrix} \boldsymbol{\beta}_1^T \\ \vdots \\ \boldsymbol{\beta}_L^T \end{bmatrix} \in \mathbb{R}^{L \times m}, \tag{6.17.4}$$

$$\mathbf{Y} = \begin{bmatrix} \mathbf{y}_1^T \\ \vdots \\ \mathbf{y}_N^T \end{bmatrix} \in \mathbb{R}^{N \times m}, \tag{6.17.5}$$

then (6.17.2) can be written as the following compact matrix form:

$$\mathbf{HB} = \mathbf{Y}. \tag{6.17.6}$$

Here, $G(a_i, b_i, \mathbf{x}_j)$ denotes the activation function used to calculate the output of the ith hidden node on the jth training instance, $\mathbf{H}$ is called the hidden layer output matrix, where its ith column denotes the ith hidden node's output vector with respect to inputs $\mathbf{x}_1, \ldots, \mathbf{x}_N$ and its jth row represents the output vector of the hidden layer with respect to the input $\mathbf{x}_j$. The solution of Eq. (6.17.6) is given by $\hat{\mathbf{B}} = \mathbf{H}^\dagger \mathbf{Y}$, where $\mathbf{H}^\dagger$ is the Moore–Penrose inverse of the hidden layer output matrix $\mathbf{H}$, and $\mathbf{H}^\dagger = (\mathbf{H}^T \mathbf{H})^{-1} \mathbf{H}^T$ if $\mathbf{H}$ is of full column rank or $\mathbf{H}^\dagger = \mathbf{H}^T (\mathbf{H} \mathbf{H}^T)^{-1}$ if $\mathbf{H}$ is of full row rank. The solution $\hat{\mathbf{H}}$ of the overdetermined matrix equation (6.17.6) is the minimum norm least squares solution, i.e., $\hat{\mathbf{B}} = \mathbf{H}^\dagger \mathbf{Y}$ has the minimum norm $\|\hat{\mathbf{H}}\| \leq \|\mathbf{Z}\|$ and the least variance $\text{var}(\hat{\mathbf{H}}) \leq \text{var}(\mathbf{Z})$ for all other positive solutions $\mathbf{Z}$. Because the norm of output weights $\|\mathbf{B}\|$ is closely related with the generalization ability of the neural network [16], extreme learning machine should minimize both $\|\mathbf{B}\|$ and $\|\mathbf{HB} - \mathbf{Y}\|$ simultaneously, namely

$$\min_{\mathbf{B}} \left\{ L_{\text{ELM}} = \frac{1}{2} \|\mathbf{B}\|^2 + \frac{C}{2} \|\mathbf{HB} - \mathbf{Y}\|^2 \right\}. \tag{6.17.7}$$

By the first-order optimality condition, we have

$$\frac{\partial L_{\text{ELM}}}{\partial \mathbf{B}} = \mathbf{O} \quad \Rightarrow \quad \mathbf{B} + C(\mathbf{H}^T \mathbf{H} \mathbf{B} - \mathbf{H}^T \mathbf{Y}) = \mathbf{O}. \tag{6.17.8}$$

Letting $\lambda = 1/C$, then the solution of the optimization problem (6.17.7) is given by

$$\mathbf{B} = \begin{cases} (\mathbf{H}^T \mathbf{H} + \lambda \mathbf{I})^{-1} \mathbf{H}^T \mathbf{Y}, & \text{if } \mathbf{H}^T \mathbf{H} \text{ is nonsingular;} \\ \mathbf{H}^T (\mathbf{H} \mathbf{H}^T + \lambda \mathbf{I})^{-1} \mathbf{Y}, & \text{if } \mathbf{H} \mathbf{H}^T \text{ is nonsingular.} \end{cases} \tag{6.17.9}$$

In other words, the solution of the optimization problem (6.17.7) is equivalent to the Tikhonov regularization solution of the overdetermined matrix equation (6.17.6).

Define the output function (row) vector $\mathbf{f}(\mathbf{x}) = [f_1(\mathbf{x}), \ldots, f_m(\mathbf{x})] \in \mathbb{R}^{1 \times m}$ and the active function (row) vector $\mathbf{h}(\mathbf{x}) = [G(a_1, b_1, \mathbf{x}), \ldots, G(a_L, b_L, \mathbf{x})] \in \mathbb{R}^{1 \times L}$, where $f_i(\mathbf{x})$ denotes the actual output of the ith output node corresponding to the instance $\mathbf{x}$.

From (6.17.6) it follows $\mathbf{f}(\mathbf{x}) = \mathbf{h}(\mathbf{x})\mathbf{B} = \mathbf{y}$.

In extreme learning machine, an instance is always labeled as the category having the maximum output [123, 124]. Once the weight matrix $\mathbf{B}$ is found using (6.17.9) and a new testing instance $\mathbf{x}$ is given, one can make the binary classification label$(\mathbf{x}) = \arg\max_i f_i(\mathbf{x})$, $i = [1, \ldots, m]$ or regression $\hat{\mathbf{y}} = \mathbf{h}(\mathbf{x})\hat{\mathbf{B}}$.

By defining the posteriori probabilities of the instance $\mathbf{x}$ defined from the actual output $f(\mathbf{x})$ in extreme learning machine as

$$P(y = 1 | f_i(\mathbf{x})) = \frac{1}{1 + \exp(-f_i(\mathbf{x}))} \tag{6.17.10}$$

and the normalized posteriori probabilities as

$$\bar{P}(y = 1 | f_i(\mathbf{x})) = \frac{P(y = 1 | f_i(\mathbf{x}))}{\sum_{j=1}^{m} P(y = 1 | f_j(\mathbf{x}))}, \tag{6.17.11}$$

an active learning-extreme learning machine (AL-ELM) algorithm was developed by Yu et al. [291], as shown in Algorithm 6.37.

Algorithm 6.37 Active learning-extreme learning machine (AL-ELM) algorithm [291]

1. **input**: a small labeled set $\mathcal{L} = \{\mathbf{x}_i, y_i\}_{i=1}^{l}$, $\mathbf{x}_i \in \mathbb{R}^n$, $y_i \in \mathbb{R}^m$ and a large unlabeled set $\mathcal{U} = \{\mathbf{x}_i\}_{i=1}^{l+u}$ activation function $G(\mathbf{x})$, penalty factor C and the number of hidden nodes L.
2. **initialization**: let $k = 0$.
   2.1 Assign random hidden nodes by randomly generating parameters (a_i, b_i) according to any continuous sampling distribution, where $i = 1, \ldots, L$;
   2.2 Calculate the initial hidden layer output matrix $\mathbf{H}_0$ by using (6.17.3) from the labeled instances $\mathbf{x}_i$ in set $\mathcal{L}$;
   2.3 Use (6.17.9) to find the solution $\mathbf{B}^{(0)}$ of the matrix equation (6.17.8).
3. **repeat**
4. **for** $k = 1$ **to** M **do**
5. Use $\mathbf{B}^{(k)}$ to calculate the actual outputs of each instance in the unlabeled set $\mathcal{U}$;
6. Calculate converted posteriori probabilities by (6.17.10). For multiclass problems, they need to be further converted to the normalized posteriori probabilities by (6.17.11);
7. Calculate uncertainty level of each instance in $\mathcal{U}$ by margin sampling strategy;
8. Sort instances based on the uncertainty levels in ascending order;
9. Extract some instances corresponding to the smallest uncertainty levels and remove them from the unlabeled set **U**;
10. Label the extracted unlabeled instances manually by the human annotators, and then put them into $\mathcal{L}$;
11. Update the hidden layer output matrix, getting $\mathbf{H}_{k+1}$ by using new set $\mathcal{L}$;
12. Update the output weights, getting $\mathbf{B}^{(k+1)}$ by using (6.17.9).
13. **end for**
14. **until** the pre-designed stopping criterion being satisfied.
15. **output**: $\mathbf{h}(\mathbf{x}) = [G(a_1, b_1, \mathbf{x}), \ldots, G(a_L, b_L, \mathbf{x})]$.

6.18 Reinforcement Learning

One of the primary goals of artificial intelligence (AI) is to produce fully autonomous agents that are able to interact with their environments through trial-and-error way in order to learn optimal behaviors. This experience-driven autonomous learning is known as *"reinforcement learning"* (RL) [269].

6.18.1 Basic Concepts and Theory

The reinforcement learning machine learns to achieve a goal by trial-and-error interaction within a dynamic environment. At each discrete-time step $t \in \{0, 1, 2, \ldots\}$, the agent observes the state description vector $\mathbf{s}_t$ of the environment, and decides and takes an action $\mathbf{a}_t \in A(\mathbf{s}_t)$ based on the state $\mathbf{s}_t$. After receiving a reward signal from the environment, $r_t \in \mathbb{R}$, the agent decides the action to be performed via a function called a policy $\pi(\mathbf{s}) = \mathbf{a}$, which could also be stochastic $\pi(\mathbf{a}|\mathbf{s}) = \Pr\{\mathbf{a}_t = \mathbf{a}|\mathbf{s}_t = \mathbf{s}\}$. A policy that maximizes the cumulative reward is denoted as π^*. The policy gives probabilities of selecting actions for a certain state. Then the environment state moves to the next state $\mathbf{s}_{t+1}$ and thus the agent obtains a new reward r_{t+1}, where the state $\mathbf{s}_{t+1}$ is drawn from transition probability with respect to the previous state $\mathbf{s}_t$

and the action $\mathbf{a}_t$. The sequence of the steps from an initial state to a terminal state is called as *episode*.

By [137], a standard reinforcement learning model usually consists of

- an agent,
- a discrete set of environment states $\mathbf{s}_t$, $S = \{\mathbf{s}_t\}$,
- a discrete set of agent actions $\mathbf{a}_t$, $A = \{\mathbf{a}_t\}$; and
- a reward function $R : S \times A \to R$.

Definition 6.36 (Agent [276]) An *agent* refers to a hardware or (more usually) software-based computer system that enjoy the following properties:

- *Autonomy:* agents operate without the direct intervention of humans or others, and have some kind of control over their actions and internal state.
- *Social ability:* agents interact with other agents (and possibly humans) via some kind of agent-communication language.
- *Reactivity:* agents operate on their environment (which may be the physical world, a user of these combined), and respond in a timely fashion to changes that occur in it.
- *Pro-activeness:* agents do not simply act in response to their environment, they are able to exhibit goal-directed behavior by taking the initiative.

Definition 6.37 (State [220]) State refers to all variables (discrete or continuous) that are necessary to model a system. All states of a system are compactly denoted as a vector $\mathbf{s}$.

A state $\mathbf{s}$ contains all relevant information about the current situation to predict future states (or observables) of a system.

Definition 6.38 (State-Action Space [220]) The mathematical space spanned by actions $\mathbf{a}$ and states $\mathbf{s}$ jointly is known as *state-action space*. Solving a movement task can be thought of as finding a path between two points in this state, the initial state and the goal state.

Reinforcement learning is based on the interaction between an agent and its environment. The agent selects the action $\mathbf{a}_t$, resulting in a change in state $\mathbf{s}_t$ and a reward r_t returned by the environment.

Definition 6.39 (Action [220]) An *action* refers to all variables (discrete or continuous) that can actively change the state of a system. Usually, actions are motor commands, abbreviated as a vector $\mathbf{a}$.

Definition 6.40 (Reward Function [220]) A function r_t is called a *reward function*, if it provides a scalar (discrete or continuous) value about the goodness of an action $\mathbf{a}$ in a state $\mathbf{s}$.

There are three different types of commonly used reward functions [147].

- Rewards depending only on the current state $R = R(\mathbf{s})$.
- Rewards depending on the current state and action $R = R(\mathbf{s}, \mathbf{a})$.

6.18 Reinforcement Learning

- Rewards including the state transitions $R = R(\mathbf{s}', \mathbf{a}, \mathbf{s})$, where $\mathbf{s}'$ is the next state and $\mathbf{s}$ is the previous state.

Reinforcement learning usually requires an unambiguous representation of states and actions of the movement system and the existence of a scalar reward function.

Reinforcement learning proceeds by trying actions in a particular state and updating an evaluation function that assigns expected rewards to possible actions. To deal with possibly delayed rewards, often many iterations are performed. After learning, the action $\mathbf{a}$ with the highest expected value in each state is chosen to achieve the task goal [220].

The goal of reinforcement learning is to find a mapping from states $\mathbf{s}_t$ to actions $\mathbf{a}_t$, called policy π, that picks actions $\mathbf{a}$ in given states $\mathbf{s}$ maximizing the cumulative expected reward.

Definition 6.41 (Control Policy [220]) A function is known as a *control policy*, if it maps the state $\mathbf{s}$ of a movement system and its environment into an appropriate action $\mathbf{a}$ for a particular task, i.e., $\mathbf{a} = \pi(\mathbf{s}, t, \boldsymbol{\alpha})$. As indicated, the function π can directly depend on the time t, and the vector of open parameters, $\boldsymbol{\alpha}$, that may be useful to adjust the policy for a particular task goal.

The policy has the following two types [147].

- *Deterministic policy:* π always uses the exact same action for a given state in the form $\mathbf{a} = \pi(\mathbf{s})$,
- *Probabilistic policy:* π draws a sample from a distribution over actions when it encounters a state, i.e., $\mathbf{a} \sim \pi(\mathbf{s}, \mathbf{a}) = P(\mathbf{a}|\mathbf{s})$.

Because an action taken does not necessarily have an immediate impact on the reward, but can also affect a reward in the distant future, the goal of the agency is to minimize its average long-term penalty. That is, the agent needs to deduce a "good" (or ideally optimal) policy (i.e., strategy or controller) π, i.e., a good mapping between the states and the actions.

The essence of reinforcement learning is to perform learning through interaction with its environment. The reinforcement learning agent starts at receiving a scalar value called a *payoff* for transitions from one state to another. Then the agent determines a control policy which maximizes the *return*. "Return" here refers to the expected future discounted sum of payoffs received. Upon observing the consequences of its actions, a reinforcement learning agent can learn to alter its own behavior in response to *rewards* received.

Reinforcement learning differs from the more widely studied problem of supervised learning in several ways [138]:

- The biggest difference is that there is no presentation of input/output pairs in reinforcement learning. Instead, after choosing an action the agent is given the immediate reward and the subsequent state, but is not told which action would have been in its best long-term interests. It is necessary for the agent to gather useful experience about the possible system states, actions, transitions, and rewards actively to act optimally.

- Reinforcement learning uses the Markov decision process as the model, while the supervised learning uses the linear or nonlinear classification/regression model.
- In supervised learning, expected future predictive accuracy or statistical efficiency are the prime concerns.
- In supervised learning, an agent is directly presented a sequence of independent examples of correct predictions to make in different circumstances. In imitation learning, an agent is provided demonstrations of actions of a good strategy to follow in given situations [147].

By Kaelbling et al. in 1998 [138], reinforcement learning is the problem faced by an agent that must learn behavior through trial-and-error interactions with a dynamic environment. Thus it is appropriately thought of as a class of problems, rather than as a set of techniques.

Reinforcement learning with function approximation is a popular framework for approximate policy evaluation and dynamic programming.

Definition 6.42 (Function Approximation [213]) This is a family of mathematical and statistical techniques used to represent a function of interest when it is computationally or information-theoretically intractable to represent the function exactly or explicitly (e.g., in tabular form).

Typically, in reinforcement learning the function approximation is based on sample data collected during interaction with the environment. Function approximation is critical in nearly every reinforcement learning problem, and becomes inevitable in continuous state ones [147].

There are three common *optimal behavior models* [137].

1. *Finite-horizon model:* At a given moment in time, the agent should optimize its expected reward for the next h steps:

$$E\left\{\sum_{t=0}^{h} r_t\right\}, \qquad (6.18.1)$$

where r_t represents the scalar reward received t steps into the future. The finite-horizon model is not always appropriate. In many cases we may not know the precise length of the agent's life in advance.

2. *Infinite-horizon discounted model:* It takes the long-run reward of the agent into account, but rewards that are received in the future are geometrically discounted according to discount factor γ (where $0 \leq \gamma < 1$):

$$E\left\{\sum_{t=0}^{\infty} \gamma^t r_t\right\}. \qquad (6.18.2)$$

Here, γ can be seen as an interest rate, a probability of living another step, or as a mathematical trick to bound the infinite sum. The discounted model is more mathematically tractable than the finite-horizon model. This is a dominant reason for the wide attention this model has received.

3. *Average-reward model:* In this model, the agent is supposed to take actions that optimize its long-run average reward:

$$\lim_{h \to \infty} E \left\{ \sum_{t=0}^{\infty} r_t \right\} \quad (6.18.3)$$

which can be seen as the limiting case of the infinite-horizon discounted model (6.18.2) as the discount factor γ approaches 1 [24]. Such a policy is referred to as a *gain optimal policy*.

6.18.2 Markov Decision Process (MDP)

A *Markov decision process* (MDP) is discrete-time stochastic control process and is used to model the synchronous interaction between agent and environment [138], as shown in Fig. 6.3.

The Markov decision process setting corresponds to the standard reinforcement learning model: the environment at time step t can be represented by a finite discrete set of state description vectors, $S = \{\mathbf{s}_1, \ldots, \mathbf{s}_n\}$, with a finite discrete set of actions in each state performed by the agent, $A = \{\mathbf{a}_1, \ldots, \mathbf{a}_n\}$. Associated with each action in each state is a *state transition function* $P(\mathbf{s}_{t+1}|\mathbf{s}_t, \mathbf{a}_t) : S \times A \times S \to \mathbb{R}$ which determine the transition probability distribution of ending in state $\mathbf{s}_{t+1}$ given environment state $\mathbf{s}_t$ and agent action $\mathbf{a}_t$. Therefore, in the standard reinforcement learning framework [138, 239], a Markov decision process can be described as a five-tuple (S, A, P, R, γ).

1. $S = \{\mathbf{s}_1, \ldots, \mathbf{s}_n\}$ is a finite set of states of the environment, called the *state space*.
2. $A = \{\mathbf{a}_1, \ldots, \mathbf{a}_n\}$ is a finite set of actions that can be performed by the agent, called the *action space*.
3. The state transition function $P(\mathbf{s}_{t+1}|\mathbf{s}_t, \mathbf{a}_t) : S \times A \times S \to \mathbb{R}$ gives for environment state $\mathbf{s}_t$ and agent action $\mathbf{a}_t$, a probability distribution of ending in state $\mathbf{s}_{t+1}$.

Fig. 6.3 An MDP models the synchronous interaction between agent and environment

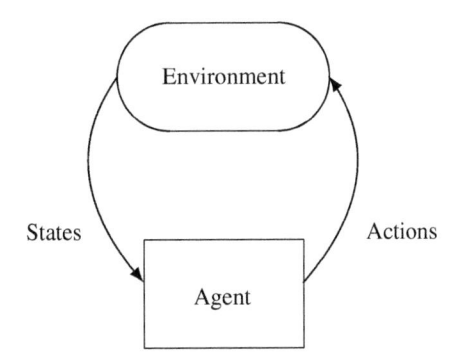

4. $R(\mathbf{s}_t, \mathbf{a}_t) : S \times A \to \mathbb{R}$ is a *reward function*, and denotes an immediate/instantaneous reward function generating an *immediate reward* $r_{t+1} = R(\mathbf{s}_t, \mathbf{a}_t, \mathbf{s}_{t+1})$ obtained by doing action $\mathbf{a}_t$ at state transition from state $\mathbf{s} \in S$ to $\mathbf{s}_{t+1} \in S$, and taking the expected rewards $R_s^a = E\{r_{t+1}|\mathbf{s}_t = \mathbf{s}, \mathbf{a}_t = \mathbf{a}\}$.
5. $\gamma \in [0, 1]$ is a predefined *discount factor* that represents the difference in importance between future rewards and present rewards, in which lower values place more emphasis on immediate rewards $r_t = R(\mathbf{s}_t, \mathbf{a}_t, \mathbf{s}_{t+1})$.

In other words, an agent is connected to its environment through perception and action as follows.

- On each step of interaction the agent receives as input i some indication of the current state $\mathbf{s}_t$ of the environment.
- The agent chooses an action $\mathbf{a}_t$ to generate an output.
- The action changes the state of the environment, and the value of this state transition is communicated to the agent through a scalar "reinforcement signal" r_t.
- The agent's behavior B chooses actions that tend to increase the long-run sum of values of the reinforcement signal.

A common way to solve the reinforcement learning problem is to describe the environment as a Markov decision problem with a set of state value function pairs, a set of action-values, a policy, and a reward function.

When action $\mathbf{a}_t$ is triggered in state $\mathbf{s}_t$, the system emits a reinforcement signal $R(\mathbf{s}_t, \mathbf{a}_t)$. The action choice in each state, i.e., the learner's behavior B is determined by its policy. The policy is a mapping from states to actions, and is denoted by $\pi = [\pi(\mathbf{s}_1), \ldots, \pi(\mathbf{s}_{|S|})]$, where $|S|$ is the cardinal of state set S. Following policy π, whenever the learner is in state $\mathbf{s}_t$, it applies action $\mathbf{a}_t = \pi(\mathbf{s}_t)$ (in the stationary policies). Each policy has an evaluation function

$$V^\pi(\mathbf{s}) = E_\pi \left\{ \sum_{t=0}^{\infty} \gamma^t R(\mathbf{s}_t, \pi(\mathbf{s}_t)) | \mathbf{s}_0 = \mathbf{s} \right\} \qquad (6.18.4)$$

qualifying the corresponding learner's behavior with respect to the reinforcement function. Here, E_π denotes the expected value under the assumptions that policy π is used and $\mathbf{s}$ is the starting state. The discount factor $\gamma \in [0, 1]$ is used to weight primary reinforcements with respect to time. $\gamma = 0$ means that evaluation functions represent primary reinforcements. Otherwise, they denote the expected discounted return over an infinite number of time steps.

Definition 6.43 (Improvement [136]) Let π and π' be two polices; π' is said to be an *improvement* over π if $V^{\pi'}(\mathbf{s}) \geq V^\pi(\mathbf{s})$, $\forall \mathbf{s} \in S$ with strict inequality holding for at least one state.

The general framework of reinforcement learning encompasses a broad variety of problems ranging from various forms of function optimization at one extreme to learning control at the other [269]. In the following, we discuss reinforcement learning from the view of point of learning control [14].

6.18 Reinforcement Learning

Reinforcement learning is considered as one of the most crucial and effective online control strategies. At the beginning, selection is based on the trial-and-error model. However, after a certain time horizon is passed, the algorithm is trained to apply suitable control actions. For each state in this method, a different control action is applied. For instance, if $\mathbf{s}_t$ and $\mathbf{a}_t$ denote the system state and system control at time t, then the state at time $t+1$ is given by [21]

$$\mathbf{s}_{t+1} = \mathbf{f}(\mathbf{s}_t, \mathbf{a}_t), \quad \mathbf{a}_t \in A, \; \forall t > 0. \tag{6.18.5}$$

For each control action $\mathbf{a}_t$, a reward or a punishment is presumed. In other words, if this action improves the system error, then a reward will be anticipated for this action; otherwise a punishment will be applied. The selection of reward or punishment is specified by comparison between errors in t and $t+1$. Thus, the immediate reward function at time t for each action $\mathbf{a}_t$ is given by

$$R(\mathbf{s}_t, \mathbf{a}_t) = \sum_{t=0}^{\infty} \gamma^t r(\mathbf{s}_t, \mathbf{a}_t), \tag{6.18.6}$$

where $0 < \gamma < 1$ and $r(\mathbf{s}_t, \mathbf{a}_t) \leq B$ in which B is the maximum value of the reward and is selected through trial-and-error.

The *state value function* is defined as the maximum value of the immediate reward functions:

$$V(\mathbf{s}) = \max_{\mathbf{a}_t \in A} R(\mathbf{s}_t, \mathbf{a}_t). \tag{6.18.7}$$

The state value function is described by the *Bellman equation* [21]

$$V(\mathbf{s}) = \max_{\mathbf{a} \in A} \{r(\mathbf{s}, \mathbf{a}) + \gamma V(\mathbf{f}(\mathbf{s}, \mathbf{a}))\}. \tag{6.18.8}$$

A scalar value called a *payoff* is received by the control system for transitions from one state to another. The aim of the system is to find a control policy which maximizes the expected future discount of the payoff received, known as the "*return*" [218].

The state value function $V(\mathbf{s}_t)$ represents the discounted sum of payoff received from time step t onwards, and this sum will depend on the sequence of actions taken. The taken sequence of actions can be determined by the *policy*.

Hence, the *optimal control policy* can be expressed as

$$\mathbf{a}^*(\mathbf{s}) = \arg\max_{\mathbf{a} \in A} \{r(\mathbf{s}, \mathbf{a}) + \gamma V(\mathbf{f}(\mathbf{s}, \mathbf{a}))\}, \tag{6.18.9}$$

from which a reward or punishment function, called *action-value function* or simply *Q-function*, can be obtained as

$$Q(\mathbf{s}, \mathbf{a}) = r(\mathbf{s}, \mathbf{a}) + \gamma V(\mathbf{f}(\mathbf{s}, \mathbf{a})). \tag{6.18.10}$$

In terms of Q-function, the state value function $V(\mathbf{s})$ can be reexpressed as

$$V(\mathbf{s}) = \max_{\mathbf{a} \in A} Q(\mathbf{s}, \mathbf{a}), \qquad (6.18.11)$$

and thus the optimal control policy is given by

$$\mathbf{a}^*(\mathbf{s}) = \arg\max_{\mathbf{a} \in A} Q(\mathbf{s}, \mathbf{a}). \qquad (6.18.12)$$

The state value function $V(\mathbf{s})$ or action-value function $Q(\mathbf{s}, \mathbf{a})$ is updated by a certain policy π, where π is a function that maps states $\mathbf{s} \in S$ to actions $\mathbf{a} \in A$, i.e., $\pi : S \to A \Rightarrow \mathbf{a} = \pi(\mathbf{s})$.

6.19 Q-Learning

Q-learning, proposed by Watkins in 1989 [263], is a popular model-free reinforcement learning algorithm, and can be used to optimally solve Markov decision processes (MDPs). Moreover, Q-learning is a technology based on action-value function to evaluate which action to take, and action-value function determines the value of taking an action in a certain state.

6.19.1 Basic Q-Learning

As an important model-free asynchronous dynamic programming (DP) technique, Q-learning provides agents with the capability of learning to act optimally in Markovian domains by experiencing the consequences of actions, without requiring them to build maps of the domains.

As the name implies, *Q-learning* is how to learn a value (i.e., Q) function. There are two methods for learning the *Q-function* [248].

- *On-policy methods* are namely Sarsa for on-policy control [218]:

$$\Delta_{\text{Sarsa}} \leftarrow \left[r_{t+1} + \gamma Q(\mathbf{s}_{t+1}, \mathbf{a}_{t+1}) - Q(\mathbf{s}_t, \mathbf{a}_t)\right], \qquad (6.19.1)$$

$$Q(\mathbf{s}_t, \mathbf{a}_t) \leftarrow Q(\mathbf{s}_t, \mathbf{a}_t) + \alpha \Delta_{\text{Sarsa}}. \qquad (6.19.2)$$

- *Off-policy methods* are Q-learning for off-policy control [263]:

$$\mathbf{b}^* \leftarrow \arg\max_{\mathbf{b} \in A(\mathbf{s}_{t+1})} Q(\mathbf{s}_{t+1}, \mathbf{b}), \qquad (6.19.3)$$

$$\Delta_{\text{Qlearning}} \leftarrow \left[r_{t+1} + \gamma Q(\mathbf{s}_{t+1}, \mathbf{b}^*) - Q(\mathbf{s}_t, \mathbf{a}_t)\right], \qquad (6.19.4)$$

$$Q(\mathbf{s}_t, \mathbf{a}_t) \leftarrow Q(\mathbf{s}_t, \mathbf{a}_t) + \alpha \Delta_{\text{Qlearning}}, \qquad (6.19.5)$$

where α is a stepsize parameter [106].

6.19 Q-Learning

The basic framework of Q-learning is as follows. Under the assumption that the number of system states and the number of agent actions must be finite, the learning system consists of a finite set of environment states $S = \{\mathbf{s}_1, \ldots, \mathbf{s}_n\}$ and a finite set of agent actions $A = \{\mathbf{a}_1, \ldots, \mathbf{a}_n\}$. At each step of interaction with the environment, the agent observes the environment state $\mathbf{s}_t$ and issues an action $\mathbf{a}_t$ based on the system state. By performing the action $\mathbf{a}_t$, the system moves from one state $\mathbf{s}_t$ to another $\mathbf{s}_{t+1}$. The new state $\mathbf{s}_{t+1}$ gives the agent a penalty $p_t = p(\mathbf{s}_t)$ which indicates the value of the state transition $\mathbf{s}_t \to \mathbf{s}_{t+1}$. The agent keeps a value function (named Q-function) $Q^\pi(\mathbf{s}, \mathbf{a})$ for each state-action pair $(\mathbf{s}, \mathbf{a})$, which represents the expected long-term penalty or reward if the system starts from state $\mathbf{s}$, taking action $\mathbf{a}$, and then following policy π. Based on this Q-function, the agent decides which action should be taken in current state to achieve the minimum long-term penalties or the maximum rewards.

Therefore, the core of the Q-learning algorithm is a value iteration update of the Q-value function. The Q-value for each state-action pair is initially chosen by the designer and then it will be updated each time an action is issued and a penalty is received based on the following expression [229]:

$$Q(\mathbf{s}_{t+1}, \mathbf{a}_{t+1})$$
$$= \underbrace{Q(\mathbf{s}_t, \mathbf{a}_t)}_{\text{old value}} + \underbrace{\eta_t(\mathbf{s}_t, \mathbf{a}_t)}_{\text{learning rate}} \times \left(\underbrace{p_{t+1}}_{\text{pernalty}} + \overbrace{\underbrace{\gamma}_{\text{discount factor}} \underbrace{\min_{\mathbf{a}} Q(\mathbf{s}_{t+1}, \mathbf{a})}_{\text{min future value}}}^{\text{expected discount pernalty}} - \underbrace{Q(\mathbf{s}_t, \mathbf{a}_t)}_{\text{old value}} \right). \tag{6.19.6}$$

Here, $\mathbf{s}_t$, $\mathbf{a}_t$, and p_t are the state visited, action taken, and penalty received at time t, respectively; and $\eta_t(\mathbf{s}, \mathbf{a}) \in (0, 1)$ is the learning rate. The discount factor γ is a value between 0 and 1 which gives more weight to the penalties in the near future than the far future.

Interestingly, if we let $Q(\mathbf{s}_t, \mathbf{a}_t) = x_i$, $\eta_t(\mathbf{s}_t, \mathbf{a}_t) = \alpha$, $\gamma \min_\mathbf{a} Q(\mathbf{s}_{t+1}, \mathbf{a}) = F_i(\mathbf{x})$ and $p_{t+1} = n_i$, then Q-learning (6.19.6) has the same general structure as stochastic approximation algorithms [252]:

$$x_i = x_i + \alpha \big(F_i(\mathbf{x}) - x_i + n_i \big), \quad i = 1, \ldots, n, \tag{6.19.7}$$

where $\mathbf{x} = [x_1, \ldots, x_n]^T \in \mathbb{R}^n$, $F_1, \ldots, F_n$ are mappings from $\mathbb{R}^n$ into $\mathbb{R}$, i.e., $F_i : \mathbb{R}^n \to \mathbb{R}$, n_i is a random noise term and α is a small, usually decreasing, stepsize.

On the mapping vector function $\mathbf{f}(\mathbf{x}) = [F_1(\mathbf{x}), \ldots, F_n(\mathbf{x})]^T \in \mathbb{R}^n$, one makes the following assumptions [252].

(a) The mapping $\mathbf{f}$ is monotone; that is, if $\mathbf{x} \leq \mathbf{y}$, then $\mathbf{f}(\mathbf{x}) \leq \mathbf{f}(\mathbf{y})$.
(b) The mapping $\mathbf{f}$ is continuous.
(c) The mapping $\mathbf{f}(\mathbf{x})$ has a unique fixed point $\mathbf{x}^*$.

(d) If $\mathbf{1} \in \mathbb{R}^n$ is the summing vector with all components equal to 1, and r is a positive scalar, then

$$\mathbf{f(x)} - r\mathbf{1} \leq \mathbf{f(x} - r\mathbf{1}) \leq \mathbf{f(x} + r\mathbf{1}) \leq \mathbf{f(x)} + r\mathbf{1}. \tag{6.19.8}$$

In basic Q-learning, the agent's experience consists of a sequence of distinct stages or *episodes*. In the tth episode, the agent [264]:

- observes its current state $\mathbf{s}_t$,
- selects and performs an action $\mathbf{a}_t$,
- observes the subsequent state $\mathbf{s}_{t+1}$,
- receives an immediate payoff r_t, and
- adjust its Q-function value using a learning factor $\eta_t(\mathbf{s}_t, \mathbf{a}_t)$, according to Eq. (6.19.6).

Algorithm 6.38 shows a basic Q-learning algorithm.

Algorithm 6.38 Q-learning [296]
---
1. **initialization:** $Q, \mathbf{s}$
2. **Repeat** (for each step of episode)
3.   Choose action $\mathbf{a}$ from state $\mathbf{s}$ based on Q and some exploration strategy (e.g., ϵ-greedy)
4.   Take action $\mathbf{a}$, observe $r, \mathbf{s}'$
5. $\mathbf{a}^* \leftarrow \arg\max_{\mathbf{a}} Q(\mathbf{s}', \mathbf{a})$
6. $\delta \leftarrow r + \gamma Q(\mathbf{s}', \mathbf{a}^*) - Q(\mathbf{s}, \mathbf{a})$
7. $Q(\mathbf{s}, \mathbf{a}) \leftarrow Q(\mathbf{s}, \mathbf{a}) + \alpha(\mathbf{s}, \mathbf{a})\delta$
8. $\mathbf{s} \leftarrow \mathbf{s}'$
9. **until end**
10. **output:** $\mathbf{s}$.

Q-learning has four important variants: double Q-learning, weighted double Q-learning, online connectionist Q-learning, and Q-learning with experience replay. They are the topics in the next three subsections.

6.19.2 Double Q-Learning and Weighted Double Q-Learning

Although Q-learning can be used to optimally solve Markov decision processes (MDPs), its performance may be poor in stochastic MDPs because of large overestimation of the action-values [255].

To cope with overestimation caused by Q-learning, *double Q-learning* that was developed in [255] uses the *double estimator* method due to using two sets of estimators: $\mu^A = \{\boldsymbol{\mu}_1^A, \ldots, \boldsymbol{\mu}_M^A\}$ and $\mu^B = \{\boldsymbol{\mu}_1^B, \ldots, \boldsymbol{\mu}_M^B\}$ to estimate the maximum expected value of the Q-function for the next state, $\max_{\mathbf{a}'} E\{Q(\mathbf{s}', \mathbf{a}')\}$. Both sets of estimators are updated with a subset of the samples drawn such that $S = S^A \cup S^B$ and $S_A \cap S_B = \emptyset$. Like the single estimator $\boldsymbol{\mu}_i$, both $\boldsymbol{\mu}_i^A$ and $\boldsymbol{\mu}_i^B$ are unbiased if one assumes that the samples are split in a proper manner, for instance, randomly, over the two sets of estimators.

Suppose we are given a set of M random variables $X = \{\mathbf{x}_1, \ldots, \mathbf{x}_M\}$. In many problems, one is interested in finding the maximum expected value $\max_i E\{\mathbf{x}_i\}$. However, it is impossible to determine $\max_i E\{\mathbf{x}_i\}$ exactly if there is no knowledge of the functional form and parameters of the underlying distributions of the variables in X. Therefore, this maximum expected value $\max_i \{\mathbf{x}_i\}$ is usually approximated by constructing approximations for $E\{\mathbf{x}_i\}$ for all i.

Let $S = \bigcup_{i=1}^{M} S_i$ be a set of samples, where S_i is the subset containing samples for the variable $\mathbf{x}_i$. It is assumed that the samples in S_i are independent and identically distributed (i.i.d.). Then, unbiased estimates for the expected values can be obtained by computing the sample average for each variable: $E\{\mathbf{x}_i\} = E\{\boldsymbol{\mu}_i\} \approx \boldsymbol{\mu}_i(S) \stackrel{\text{def}}{=} \frac{1}{|S_i|} \sum_{\mathbf{s} \in S_i} \mathbf{s}$, where $\boldsymbol{\mu}_i$ is a mean estimator for variables $\mathbf{x}_i$. This approximation is unbiased since every sample $\mathbf{s} \in S_i$ is an unbiased estimate for the value of $E\{\mathbf{x}_i\}$.

As shown in Algorithm 6.39, the double Q-learning stores two Q-functions, Q^A and Q^B, and uses two separate subsets of experience samples to learn them. The action selected is calculated based on the average of the two Q-values for each action and the ϵ-greedy exploration strategy. With the same probability, each Q-function is updated using a value from the other Q-function for the next state. Then, the action $\mathbf{a}^*$ is the maximizing action in state $\mathbf{s}'$ based on the Q_1 value.

Algorithm 6.39 Double Q-learning [255]

1. **input:** $X = \{\mathbf{x}_1, \ldots, \mathbf{x}_M\}$
2. **initialize** $Q^A, Q^B, \mathbf{s}$
3. **repeat**
4.   Choose $\mathbf{a}$ from $\mathbf{s}$ based on $Q^A(\mathbf{s}, \cdot)$ and $Q^B(\mathbf{s}, \cdot)$ (e.g., Q-greedy in $Q^A + Q^B$)
5.   Take action $\mathbf{a}$, observe $r, \mathbf{s}'$
6.   With 0.5 probability to Choose whether to update Q^A or Q^B
7.   **if** update Q^A **then**
8. $\mathbf{a}^* \leftarrow \arg\max_{\mathbf{a}} Q^A(\mathbf{s}', \mathbf{a})$
9. $\alpha^A(\mathbf{s}, \mathbf{a}) \leftarrow r + \gamma Q^B(\mathbf{s}', \mathbf{a}^*) - Q^A(\mathbf{s}, \mathbf{a})$
10. $\delta^A \leftarrow r + \gamma Q^B(\mathbf{s}', \mathbf{a}^*) - Q^A(\mathbf{s}, \mathbf{a})$
11. $Q^A(\mathbf{s}, \mathbf{a}) \leftarrow Q^A(\mathbf{s}, \mathbf{a}) + \alpha^A(\mathbf{s}, \mathbf{a})\delta^A$
12.   **else if** update Q^B **then**
13. $\mathbf{a}^* \leftarrow \arg\max_{\mathbf{a}} Q^B(\mathbf{s}', \mathbf{a})$
14. $\alpha^B(\mathbf{s}, \mathbf{a}) \leftarrow r + \gamma Q^A(\mathbf{s}', \mathbf{a}^*) - Q^B(\mathbf{s}, \mathbf{a})$
15. $\delta^B \leftarrow r + \gamma Q^A(\mathbf{s}', \mathbf{a}^*) - Q^B(\mathbf{s}, \mathbf{a})$
16. $Q^B(\mathbf{s}, \mathbf{a}) \leftarrow Q^B(\mathbf{s}, \mathbf{a}) + \alpha^B(\mathbf{s}, \mathbf{a})\delta^B$
17. $\mathbf{s} \leftarrow \mathbf{s}'$
18.   **end if**
19. **until end**
20. **output:** $\hat{\mathbf{x}}_i \approx \mu_i(S) = \frac{1}{|S_i|} \sum_{\mathbf{s} \in S_i} \mathbf{s}$.

As opposed to Q-learning, the double Q-learning uses the value $Q^B(\mathbf{s}', \mathbf{a}^*)$ rather than the value $Q^A(\mathbf{s}', \mathbf{a}^*)$ to update Q^A.

Since Q^B is updated with a different set of experience samples, $Q^B(\mathbf{s}', \mathbf{a}^*)$ is an unbiased estimate for the value of the action $\mathbf{a}^*$ in the sense that $E\{Q^B(\mathbf{s}', \mathbf{a}^*)\} = E\{Q(\mathbf{s}', \mathbf{a}^*)\}$. Similarly, the value of $Q^A(\mathbf{s}', \mathbf{a}^*)$ is also unbiased. However, since $E\{Q^B(\mathbf{s}', \mathbf{a}^*)\} \leq \max_{\mathbf{a}} E\{Q^B(\mathbf{s}', \mathbf{a})\}$ and $E\{Q^A(\mathbf{s}', \mathbf{a}^*)\} \leq \max_{\mathbf{a}} E\{Q^A(\mathbf{s}', \mathbf{a})\}$, both estimators $Q^B(\mathbf{s}', \mathbf{a}^*)$ and $Q^A(\mathbf{s}', \mathbf{a}^*)$ sometime have negative biases. This results in double Q-learning underestimating action values in some stochastic environments.

Weighted double Q-learning [296] combines Q-learning and double Q-learning for avoiding the overestimation in Q-learning and the underestimation in double Q-learning, as shown in Algorithm 6.40.

Algorithm 6.40 Weighted double Q-learning [296]

1. **input:** $X = \{\mathbf{x}_1, \ldots, \mathbf{x}_M\}$
2. **initialize** Q^A, Q^B, $\mathbf{s}$
3. **repeat**
4.   Choose $\mathbf{a}$ from $\mathbf{s}$ based on Q^A and Q^B (e.g., ϵ-greedy in $Q^A + Q^B$)
5.   Take action $\mathbf{a}$, observe r, $\mathbf{s}'$
6.   With 0.5 probability to Choose whether to update Q^A or Q^B
7.   **if** chose to update Q^A **then**
8. $\mathbf{a}^* \leftarrow \arg\max_{\mathbf{a}} Q^A(\mathbf{s}', \mathbf{a})$
9. $\mathbf{a}_L \leftarrow \arg\min_{\mathbf{a}} Q^A(\mathbf{s}', \mathbf{a})$
10. $\beta^A \leftarrow \frac{|Q^B(\mathbf{s}', \mathbf{a}^*) - Q^B(\mathbf{s}', \mathbf{a}_L)|}{c + |Q^B(\mathbf{s}', \mathbf{a}^*) - Q^B(\mathbf{s}', \mathbf{a}_L)|}$
11. $\delta^A \leftarrow r + \gamma[\beta^A Q^A(\mathbf{s}', \mathbf{a}^*) + (1 - \beta^A) Q_B(\mathbf{s}', \mathbf{a}^*)] - Q^A(\mathbf{s}, \mathbf{a})$
12. $\alpha^A(\mathbf{s}, \mathbf{a}) \leftarrow r + \gamma Q^B(\mathbf{s}', \mathbf{a}^*) - Q^A(\mathbf{s}, \mathbf{a})$
13. $Q^A(\mathbf{s}, \mathbf{a}) \leftarrow Q^A(\mathbf{s}, \mathbf{a}) + \alpha^A(\mathbf{s}, \mathbf{a}) \delta^A$
14.   **else if** chose to update Q^B **then**
15. $\mathbf{a}^* \leftarrow \arg\max_{\mathbf{a}} Q^B(\mathbf{s}', \mathbf{a})$
16. $\mathbf{a}_L \leftarrow \arg\min_{\mathbf{a}} Q^B(\mathbf{s}', \mathbf{a})$
17. $\beta^B \leftarrow \frac{|Q^A(\mathbf{s}', \mathbf{a}^*) - Q^A(\mathbf{s}', \mathbf{a}_L)|}{c + |Q^A(\mathbf{s}', \mathbf{a}^*) - Q^A(\mathbf{s}', \mathbf{a}_L)|}$
18. $\delta^B \leftarrow r + \gamma[\beta^B Q^B(\mathbf{s}', \mathbf{a}^*) + (1 - \beta^B) Q^A(\mathbf{s}', \mathbf{a}^*)] - Q^B(\mathbf{s}, \mathbf{a})$
19. $\alpha^B(\mathbf{s}, \mathbf{a}) \leftarrow r + \gamma Q^A(\mathbf{s}', \mathbf{a}^*) - Q^B(\mathbf{s}, \mathbf{a})$
20. $Q^B(\mathbf{s}, \mathbf{a}) \leftarrow Q^B(\mathbf{s}, \mathbf{a}) + \alpha^B(\mathbf{s}, \mathbf{a}) \delta^B$
21. $\mathbf{s} \leftarrow \mathbf{s}'$
22.   **end if**
23. **until end**
24. **output:** $\hat{\mathbf{x}}_i \approx \mu_i(S) = \frac{1}{|S_i|} \sum_{\mathbf{s} \in S_i} \mathbf{s}$.

The following are the comparison among the Q-learning, the double Q-learning, and the weighted double Q-learning.

1. The Q-learning is a single estimator for $Q(\mathbf{s}, \mathbf{a})$.
2. The double Q-learning is a double estimator for both $Q^A(\mathbf{s}, \mathbf{a})$ and $Q^B(\mathbf{s}, \mathbf{a})$.
3. The weighted double Q-learning is a weighted double estimator for both $Q^A(\mathbf{s}, \mathbf{a})$ and $Q^B(\mathbf{s}, \mathbf{a})$. The main differences between the double Q-learning

Algorithm 6.39 and the weighted double Q-learning Algorithm 6.40 are:

- Algorithm 6.40 uses the weighted term $\beta^A Q^A(\mathbf{s}', \mathbf{a}^*) + (1 - \beta^A) Q^B(\mathbf{s}', \mathbf{a}^*)$ to replace $Q^B(\mathbf{s}', \mathbf{a}^*)$ when updating δ^A in Algorithm 6.39;
- Algorithm 6.40 uses the weighted term $\beta^B Q^B(\mathbf{s}', \mathbf{a}^*) + (1 - \beta^B) Q^A(\mathbf{s}', \mathbf{a}^*)$ to replace $Q^A(\mathbf{s}', \mathbf{a}^*)$ when updating δ^B in Algorithm 6.39.

These changes result in the name "weighted double Q-learning."

6.19.3 Online Connectionist Q-Learning Algorithm

In order to represent the Q-function, Lin [163] chose to use one neural network for each action, which avoids the hidden nodes receiving conflicting error signals from different outputs.

The neural network weights are updated by [163]

$$\mathbf{w}_{t+1} = \mathbf{w}_t + \eta \left(r_t + \gamma \max_{\mathbf{a} \in A} Q_{t+1} - Q_t \right) \nabla_w Q_t, \qquad (6.19.9)$$

where η is the learning constant, r_t is the payoff received for the transition from state $\mathbf{x}_t$ to $\mathbf{x}_{t+1}$, $Q_t = Q(\mathbf{x}_t, \mathbf{a}_t)$, and $\nabla_w Q_t$ is a vector of the output gradients $\partial Q_t / \partial \mathbf{w}_t$ calculated by backpropagation.

Rummery and Niranjan [218] proposed an alternative update algorithm, called *modified connectionist Q-learning*. This algorithm is based more strongly on temporal difference, and uses the following update rule:

$$\Delta \mathbf{w}_t = \eta \left(r_t + \gamma Q_{t+1} - Q_t \right) \sum_{k=0}^{t} (\gamma \lambda)^{t-k} \nabla_w Q_k. \qquad (6.19.10)$$

This update rule is different from the normal Q-learning in the use of the Q_{t+1} associated with the action selected instead of the greedy $\max_{\mathbf{a} \in A} Q_{t+1}$.

In the following, we discuss how to calculate the gradient $\nabla_w Q_t$. For this end, define a backpropagation neural network as a collection of interconnected units arranged in layers. Let i, j, and k refer to the different neurons in the network; with error signal propagating from right to left, as shown in Fig. 6.4.

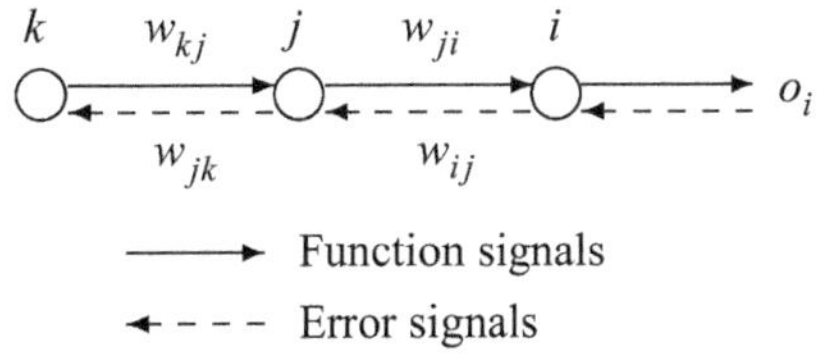

Fig. 6.4 One neural network as a collection of interconnected units, where i, j, and k are neurons in output layer, hidden layer, and input layer, respectively

Neuron j lies in a layer to the left of neuron i, and neuron k lies to the left of the neuron j when neuron j is a hidden unit.

Letting w_{ij} be a weight on a connection from layer i to j, then the output from layer i is given by [218]:

$$o_i = f(\sigma_i), \qquad (6.19.11)$$

where $\sigma_i = \sum_j w_{ij} o_j$ and $f(\sigma)$ is a sigmoid function.

The output gradient with respect to the output layer weights w_{ij} is defined as

$$\frac{\partial o_i}{\partial w_{ij}} = \frac{\partial f(\sigma_i)}{\partial w_{ij}} = \frac{\partial f(\sigma_i)}{\partial \sigma_i} \cdot \frac{\partial \sigma_i}{\partial w_{ij}} = f'(\sigma_i) o_j, \qquad (6.19.12)$$

where $f'(\sigma_i) = \frac{\partial f(\sigma_i)}{\partial \sigma_i}$ is the first-order differential of the sigmoid function $f(\sigma_i)$.

The gradient of the output o_i with respect to the hidden layer weights w_{jk} is given by

$$\frac{\partial o_i}{\partial w_{jk}} = \frac{\partial f(\sigma_i)}{\partial w_{jk}} = \frac{\partial f(\sigma_i)}{\partial \sigma_i} \cdot \frac{\partial \sigma_i}{\partial o_j} \cdot \frac{\partial o_j}{\partial \sigma_j} \cdot \frac{\partial \sigma_j}{\partial w_{jk}}$$
$$= f'(\sigma_i) \cdot w_{ij} \cdot f'(\sigma_j) \cdot o_k. \qquad (6.19.13)$$

Modified connectionist Q-learning of Rummery and Niranjan [218] is shown in Algorithm 6.41.

Algorithm 6.41 Modified connectionist Q-learning algorithm [218]

1. **initialization:** Reset all eligibilities $\mathbf{e}_0 = \mathbf{0}$ and put $t = 0$.
2. **repeat**
3.   Select action $\mathbf{a}_t$;
4.   If $t > 0$,
5. $\mathbf{w}_t = \mathbf{w}_{t-1} + \eta(r_{t-1} + Q_t - Q_{t-1})\mathbf{e}_{t-1}$.
6.   Calculate $\nabla_w Q_t$ with respect to selected action $\mathbf{a}_t$ only.
7. $\mathbf{e}_t = \nabla_w Q_t + \gamma \lambda \mathbf{e}_{t-1}$.
8.   Perform action $\mathbf{a}_t$, and receive the payoff r_t.
9.   If trial has not ended, $t \leftarrow t + 1$ and go to Step 3.
10. **until end**
11. **output:** Q_t.

6.19.4 Q-Learning with Experience Replay

Reinforcement learning is known to be unstable and can even diverge when a nonlinear function approximator such as a neural network is used to represent the

6.19 Q-Learning

action-value, i.e., Q function. This instability has a few causes [183]:

- The correlations present in the sequence of observations.
- Small updates to Q may significantly change the policy and therefore change the data distribution.
- The correlation between the action-value Q and the target value $r + \max_{a'} Q(s', a')$.

To deal with these instabilities, a novel variant of Q-learning was developed by Mnih et al. [183], which uses two key ideas.

1. A biologically inspired mechanism termed experience replay is used, which randomizes over the data, thereby removing correlations in the observation sequence and smoothing over changes in the data distribution.
2. An iterative update is used for adjusting the action-values Q towards target values that are only periodically updated, thereby reducing correlations with the target.

To generate the experience replay, an approximate value function $Q(s, a; \theta_i)$ is first parameterized, where θ_i are the parameters (that is, weights) of the Q-network at iteration i. Then, the agent's experiences $e_t = (s_t, a_t, r_t, s_{t+1})$ are stored at each time-step t in a data set $D_t = \{e_1, \ldots, e_t\}$ to perform experience replay.

Algorithm 6.42 shows the deep Q-learning with experience replay developed by Mnih et al. [183].

Algorithm 6.42 Deep Q-learning with experience replay [183]

1. **initialization:**
    1.1 replay memory D to capacity N,
    1.2 action-value function Q with random weights θ,
    1.3 target action-value function $\hat{Q}$ with weights $\theta^- = \theta$.
2. **for** episode $= 1, M$ **do**
3.   Initialize sequence $s_1 = \{x_1\}$ and preprocessed sequence $\phi_1 = \phi(s_1)$.
4.   **for** $t = 1, T$ **do**
5.   With probability ϵ select a random action a_t,
6.   otherwise select $a_t = \arg\max_a Q(\phi(s_t), a; \theta)$.
7.   Execute action a_t in emulator and observe reward r_t and image x_{t+1}.
8.   Set $s_{t+1} = s_t, a_t, x_{t+1}$ and preprocess $\phi_{t+1} = \phi(s_{t+1})$.
9.   Store transition $(\phi_t, a_t, r_t, \phi_{t+1})$ in D.
10.   Sample random mini batch of transitions $(\phi_j, a_j, r_j, \phi_{j+1})$ from D.
11.   Set $y_j = \begin{cases} r_j, & \text{if episode terminates at step } j+1; \\ r_j + \max_{a'} \hat{Q}(\phi_{j+1}, a'; \theta^-), & \text{otherwise.} \end{cases}$
12.   Perform a gradient descent step on $(y_j - Q(\phi_j, a_j; \theta))^2$ with respect to the network parameters θ.
13.   Every C steps reset $\hat{Q} = Q$.
14.   **end for**
15. **end for**

During learning, Q-learning updates are applied on samples (or mini-batches) of experience $(s, a, r, s') \sim U(D)$, drawn uniformly at random from the pool of stored samples. The Q-learning update at iteration i uses the following loss function:

$$L_i(\theta_i) = E_{(\mathbf{s},\mathbf{a},r,\mathbf{s}')\sim U(D)}\left\{r + \gamma \max_{\mathbf{a}'} Q(\mathbf{s}', \mathbf{a}'; \theta_i^-) - Q(\mathbf{s}, \mathbf{a}; \theta_i)\right\}, \qquad (6.19.14)$$

where γ is the discount factor determining the agent's horizon, θ_i are the parameters of the Q-network at iteration i and θ_i^- are the network parameters used to compute the target at iteration i. The target network parameters θ_i^- are only updated with the Q-network parameters θ_i every C steps and are held fixed between individual updates.

6.20 Transfer Learning

Traditional machine learning/data mining works well only when both training set and test set come from the same feature space and have the same distribution. This implies that whenever the data is changed, the model needs to be retrained, which may be too troublesome because

- it is very expensive and difficult to obtain new training data for new data sets,
- some data sets are easily outdated, that is, the data distribution in different periods will be different.

Therefore, in many real-world applications, when the distribution changes or data are outdated, most statistical models are no longer applicable, and need to be rebuilt using newly collected training data. It is expensive or impossible to recollect the needed training data and rebuild the models. In such cases, knowledge transfer or transfer learning between tasks and domains would be desirable.

Unlike traditional machine learning, transfer learning allows the domains, tasks, and distributions used in training and testing to be different.

On the other hand, we get easily a large number of unlabeled images (or audio samples, or text documents) randomly downloaded from the Internet, and it is desirable to use such unlabeled images for improving performance on a given image (or audio, or text) classification task. Clearly, in such cases it is not assumed that the unlabeled data follows the same class labels or generative distribution as the labeled data. Using unlabeled data in supervised classification tasks is known as self-taught learning (transfer learning from unlabeled data) [210]. Therefore, transfer learning or self-taught learning is widely applicable for typical semi-supervised or transfer learning settings in many practical learning problems.

Transfer learning can transfer previously knowledge to a new task, helping the learning of a new task.

Transfer learning has attracted more and more attention since 1995 in different names: learning to learn, life-long learning, knowledge transfer, inductive learning, multitask learning, knowledge consolidation, context-sensitive learning, knowledge-based inductive bias, metalearning, incremental/cumulative learning, and self-taught learning [199, 210, 243].

6.20.1 Notations and Definitions

Definition 6.44 (Domain [199]) A *domain* $\mathcal{D}$ consists of two parts: a feature space $\mathcal{X}$ and a marginal probability distribution $P(X)$, denoted as $\mathcal{D} = \{\mathcal{X}, P(X)\}$, where $X = \{\mathbf{x}_1, \ldots, \mathbf{x}_n\} \in \mathcal{X}$ is a particular learning sample (set).

For document classification with a bag-of-words representation, $\mathcal{X}$ is the space of all document representations, $\mathbf{x}_i$ is the ith term vector corresponding to some document and X is the sample set of documents used for training.

In supervised learning, unsupervised learning, semi-supervised learning, inductive learning and semi-supervised transductive learning discussed before, the source domain and the target domain are assumed to have the same feature space or the same marginal probability distribution. In these cases, we denote the labeled set $D^{(l)}$, the unlabeled set $D^{(u)}$ and the test set $D^{(t)}$ as

$$D^{(l)} = \left\{(\mathbf{x}_{l_1}, y_{l_1}), \ldots, (\mathbf{x}_{l_{n_l}}, y_{l_{n_l}})\right\}, \qquad (6.20.1)$$

$$D^{(u)} = \{\mathbf{x}_{u_1}, \ldots, \mathbf{x}_{u_{n_u}}\}, \qquad (6.20.2)$$

$$D^{(t)} = \{\mathbf{x}_{t_1}, \ldots, \mathbf{x}_{t_{n_t}}\}, \qquad (6.20.3)$$

respectively.

There are two domains in machine learning: *source domain* $\mathcal{D}_S = \{\mathcal{X}_S, P(X_S)\}$ and *target domain* $\mathcal{D}_T = \{\mathcal{X}_T, P(X_T)\}$. These two domains are generally different: they may have different feature spaces $\mathcal{X}_S \neq \mathcal{X}_T$ or different marginal probability distributions $P(X_S) \neq P(X_T)$.

Let $X_S = \{\mathbf{x}_{S_1}, \ldots, \mathbf{x}_{S_{n_s}}\}$ and $Y_S = \{y_{S_1}, \ldots, y_{S_{n_s}}\}$ be, respectively, the sets of all labeled instances of source-domain $\mathcal{D}_S$ and all corresponding labels of source-domain. Similarly, denote $X_T = \{\mathbf{x}_{T_1}, \ldots, \mathbf{x}_{T_{n_t}}\}$ and $Y_T = \{y_{T_1}, \ldots, y_{T_{n_t}}\}$. The *source-domain data* and the *target-domain data* are, respectively, represented as

$$D_S = (X_S, Y_S) = X_S \cup Y_S = \left\{(\mathbf{x}_{S_1}, y_{S_1}), \ldots, (\mathbf{x}_{S_{n_s}}, y_{S_{n_s}})\right\} \in \mathcal{D}_S, \quad (6.20.4)$$

$$D_T = (X_T, Y_T) = X_T \cup Y_T = \left\{(\mathbf{x}_{T_1}, y_{T_1}), \ldots, (\mathbf{x}_{T_{n_t}}, y_{T_{n_t}})\right\} \in \mathcal{D}_T, \quad (6.20.5)$$

where

- $\mathbf{x}_{S_i} \in \mathcal{X}_S$ is the ith data instance of source-domain $\mathcal{D}_S$ and $y_{S_i} \in \mathcal{Y}_S$ is the corresponding class label for $\mathbf{x}_{S_i}$. For example, in document classification, the label set is a binary set containing true and false, i.e., $y_{S_i} \in \{\text{true, false}\}$ takes on a value of true or false, and $f(\mathbf{x})$ is the learner that predicts the label value for $\mathbf{x}$.
- $\mathbf{x}_{T_i} \in \mathcal{X}_T$ is the ith data instance of target-domain $\mathcal{D}_T$ and $y_{T_i} \in \mathcal{Y}_T$ is the corresponding class label for $\mathbf{x}_{T_i}$.
- n_s and n_t are the numbers of source-domain data and target-domain data, respectively. In most cases, $0 \leq n_t \ll n_s$.

Let $D_S^{(l)}$ and $D_S^{(u)}$ be the labeled set and unlabeled set of data drawn from source domain $\mathcal{D}_S$, respectively. Similarly, $D_T^{(l)}$ and $D_T^{(u)}$ denote the labeled set and unlabeled set drawn from $\mathcal{D}_T$, respectively. Then we have the following notations:

$$D_S^{(l)} = \left\{ \left(\mathbf{x}_{S_1}^{(l)}, y_{S_1}^{(l)}\right), \ldots, \left(\mathbf{x}_{S_{n_l}}^{(l)}, y_{S_{n_l}}^{(l)}\right) \right\} \quad \text{and} \quad D_S^{(u)} = \left\{ \mathbf{x}_{S_1}^{(u)}, \ldots, \mathbf{x}_{S_{n_u}}^{(u)} \right\},$$

$$D_T^{(l)} = \left\{ \left(\mathbf{x}_{T_1}^{(l)}, y_{T_1}^{(l)}\right), \ldots, \left(\mathbf{x}_{T_{n_l}}^{(l)}, y_{T_{n_l}}^{(l)}\right) \right\} \quad \text{and} \quad D_T^{(u)} = \left\{ \mathbf{x}_{T_1}^{(u)}, \ldots, \mathbf{x}_{T_{n_u}}^{(u)} \right\},$$

where

- $\mathbf{x}_{S_i}^{(l)}, \mathbf{x}_{S_i}^{(u)} \in \mathcal{D}_S$ are the S_i-th labeled and unlabeled data vectors drawn from source domain, respectively; $y_{S_i}^{(l)} \in \mathcal{Y}_S$ is the class label corresponding to $\mathbf{x}_{S_i}^{(l)}$ drawn from source domain;
- $\mathbf{x}_{T_i}^{(l)}, \mathbf{x}_{T_i}^{(u)} \in \mathcal{D}_T$ are the T_i-th labeled and unlabeled data vectors drawn from target domain, respectively; $y_{T_i}^{(l)} \in \mathcal{Y}_T$ is the class label corresponding to $\mathbf{x}_{T_i}^{(l)}$ drawn from target domain;
- n_l and n_u are the numbers of labeled and unlabeled data in source (or target) domain.

It should be noted that $D_S^{(l)}$ and $D_S^{(u)}$ are disjoint: $D_S^{(l)} \cap D_S^{(u)} = \emptyset$. Similarly, we have also $D^{(l)} \cap D^{(u)} = \emptyset$, as required in supervised learning, unsupervised learning, semi-supervised learning, inductive learning and semi-supervised transductive learning discussed before.

When discussing machine learning in two different domains $\mathcal{D}_S$ and $\mathcal{D}_T$, we have two possible training sets D_S^{train} drawn from source domain $\mathcal{D}_S$ and D_T^{train} drawn from target domain $\mathcal{D}_T$ as follows:

$$D_S^{\text{train}} = D_S^{(l)} \cup D_S^{(u)} \in \mathcal{D}_S, \tag{6.20.6}$$

$$D_T^{\text{train}} = D_T^{(l)} \cup D_T^{(u)} \in \mathcal{D}_T, \tag{6.20.7}$$

where one of $D_S^{(l)}$ and $D_S^{(u)}$ may be empty, while $D_T^{(l)}$ or $D_T^{(u)}$ may be empty also, which corresponds to unsupervised and supervised learning cases, respectively.

Definition 6.45 (Task [199]) For a given (source or target) domain $\mathcal{D} = \{\mathcal{X}, P(X)\}$, a *task* $\mathcal{T}$ consists of two parts: a label space $\mathcal{Y}$ and an objective predictive function $f(\cdot)$, denoted by $\mathcal{T} = \{\mathcal{Y}, f(\cdot)\}$. The objective prediction function $f(\cdot)$ is not observed but can be typically learned from the feature vector and label pairs $(\mathbf{x}_i, y_i)$, where $\mathbf{x}_i \in X$ and $y_i \in \mathcal{Y}$.

The function $f(\cdot)$ can be used to predict the corresponding label $f(\mathbf{x}_i)$ of a new instance $\mathbf{x}_i$. From a probabilistic viewpoint, $f(\mathbf{x}_i)$ is usually written as $P(y_i|\mathbf{x}_i)$.

If the given specific domain is the source domain, i.e., $\mathcal{D} = \mathcal{D}_S$, then learning task is called the *source learning task*, denoted as $\mathcal{T}_S = \{\mathcal{Y}_S, f_S(\cdot)\}$. Similarly,

6.20 Transfer Learning

learning task in target domain is known as the *target learning task*, denoted as $\mathcal{T}_T = \{\mathcal{Y}_T, f_T(\cdot)\}$.

For two different but relevant domains or tasks, if we transfer the model parameters trained in one domain (or task) to the new model in another domain (or task), it helps clearly to train the new data set because this transfer can share the learned parameters with the new model so as to speed up and optimize the learning of the new model without learning from zero as before. Such a machine learning is called transfer learning.

The following is a unified definition of transfer learning.

Definition 6.46 (Transfer Learning [199]) Given a source domain $\mathcal{D}_S$ and source learning task $\mathcal{T}_S$, a target domain $\mathcal{D}_T$ and target learning task $\mathcal{T}_T$. *Transfer learning emphasizes the transfer of knowledge across domains, tasks, and distributions that are similar but not the same and aims to help improve the learning of the target predictive function $f_T(\cdot)$ in $\mathcal{D}_T$ using the knowledge in $\mathcal{D}_S$ and $\mathcal{T}_S$, where $\mathcal{D}_S \neq \mathcal{D}_T$ or $\mathcal{T}_S \neq \mathcal{T}_T$.*

This definition shows that transfer learning consists of three parts [302]:

- define source and target domains;
- learn on the source domain; and
- predict or generalize on the target domain.

In transfer learning the knowledge gained in solving the source (or target) task in the source (or target) domain is stored and then applied to solving another problem of interest. Figure 6.5 shows this transfer learning setup.

In most cases, there is only a single-loop transfer learning (shown by arrows with solid line). However, in machine translation of two languages there are two-loop transfer learning setups, as shown in Fig. 6.5.

In transfer learning, there are the following three main research issues [199, 268].

- *What to transfer:* Which part of knowledge can be transferred across domains or tasks? If some knowledge is specific for individual domains or tasks, then there is no need to transfer it. Conversely, when some knowledge may be common

Fig. 6.5 Transfer learning setup: storing knowledge gained from solving one problem and applying it to a different but related problem

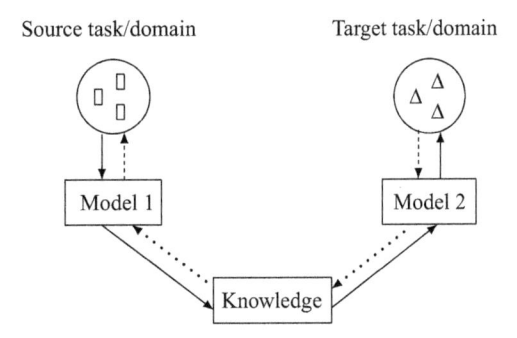

between different domains, it is necessary to transfer this part of knowledge because they may help improve performance for the target domain or task.
- *How to transfer:* How to train a suitable model? After discovering which knowledge can be transferred, learning algorithms need to be developed to transfer the knowledge.
- *When to transfer:* The high-level concept of transfer learning is to improve a target learner by using data from a related source domain. But if the source domain is not well-related to the target, then the target learner can be negatively impacted by this weak relation, which is referred to as *negative transfer*.

In Definition 6.46 on transfer learning, a domain is defined as a pair $\mathcal{D} = \{\mathcal{X}, P(X)\}$, and a task is a pair $\mathcal{T} = \{\mathcal{Y}, P(Y|X)\}$. Hence, the condition on two different domains $\mathcal{D}_S \neq \mathcal{D}_T$ has two possible scenarios: $\mathcal{X}_S \neq \mathcal{X}_T$ and/or $P(X_S) \neq P(X_T)$, while the condition on two different tasks $\mathcal{T}_S \neq \mathcal{T}_T$ has also two possible scenarios: $\mathcal{Y}_S \neq \mathcal{Y}_T$ and/or $P(Y_S|X_S) \neq P(Y_T|X_T)$. In other words, there are the following four transfer learning scenarios:

1. $\mathcal{X}_S \neq \mathcal{X}_T$ means that the feature spaces between the source and target domains are different. For example, in the context of natural language processing, the documents written in two different languages belong to this scenario, which is generally referred to as *cross-lingual adaptation*.
2. $P(X_S) \neq P(X_T)$ implies that the marginal probability distributions between source and target domains are different, for example, when the documents focus on different topics. This scenario is generally known as *domain adaptation*.
3. $\mathcal{Y}_S \neq \mathcal{Y}_T$ means that the label spaces between the source and target domains are different. The case of $\mathcal{Y}_S \neq \mathcal{Y}_T$ refers to a mismatch in the class space. An example of this case is when the source software project has a binary label space of true for defect prone and false for not defect prone, and the target domain has a label space that defines five levels of fault prone modules [268].
4. $P(Y_S|X_S) \neq P(Y_T|X_T)$ implies that conditional probability distributions between the source and target domains are different. For example, source and target documents are unbalanced with regard to their classes. This scenario is quite common in practice.

6.20.2 Categorization of Transfer Learning

Feature-based transfer learning approaches are categorized in two ways [268], as shown in Fig. 6.6.

- *Asymmetric transformation* transforms the features of the source via reweighting to more closely match the target domain [201], as depicted in Fig. 6.6b.
- *Symmetric transformation* discovers underlying meaningful structures between the domains, shown in Fig. 6.6a.

6.20 Transfer Learning

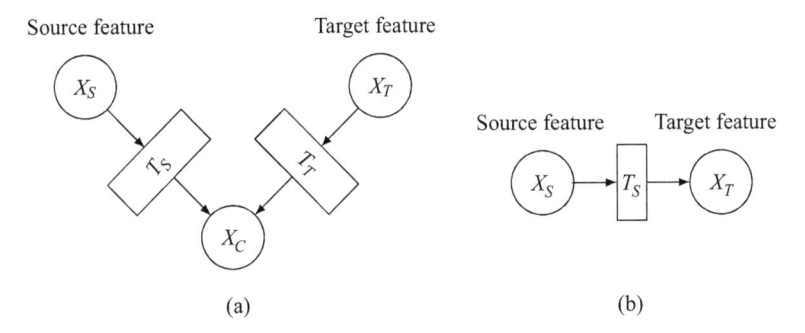

Fig. 6.6 The transformation mapping [268]. (**a**) The symmetric transformation (T_S and T_T) of the source-domain feature set $X_S = \{\mathbf{x}_{S_i}\}$ and target-domain feature set $X_T = \{\mathbf{x}_{T_i}\}$ into a common latent feature set $X_C = \{\mathbf{x}_{C_i}\}$. (**b**) The asymmetric transformation T_T of the source-domain feature set X_S to the target-domain feature set X_T

According to different perspectives, there are different categorization methods in transfer learning.

1. According to feature spaces, transfer learning can be categorized to two classes [268].

   - *Homogeneous transfer learning:* Source domain and target domain have the same feature space, i.e., $\mathcal{X}_S = \mathcal{X}_T$, including instance-based transfer learning [48], asymmetric feature-based transfer learning [70], domain adaptation [203], parameter-based transfer learning [249], relational-based transfer learning [161], hybrid-based (instance and parameter) transfer learning [280] and so on.
   - *Heterogeneous transfer learning:* Source domain and target domain have different feature spaces, i.e., $\mathcal{X}_S \neq \mathcal{X}_T$, including symmetric feature-based transfer learning in [207], asymmetric feature-based transfer learning in [153], and so on.

2. According to computational intelligence, transfer learning can be categorized as follows [167].

   - *Transfer learning using neural network* including transfer learning using deep neural network [55], transfer learning using convolutional neural networks [196], transfer learning using multiple task neural network [45, 233], and transfer learning using radial basis function neural networks [285].
   - *Transfer learning using Bayes* including transfer learning using naive Bayes [170, 217], transfer learning using Bayesian network [168, 192], and transfer learning using hierarchical Bayesian model [93].
   - *Transfer learning using fuzzy system and genetic algorithm* including fuzzy-based transductive transfer learning [18], framework of fuzzy transfer learning to form a prediction model in intelligent environments [227, 228], generalized hidden-mapping ridge regression [74], genetic transfer learning [148], and so on.

3. According to tasks, transfer learning can be categorized into the following three classes [199]:

   - inductive transfer learning,
   - transductive transfer learning [10], and
   - unsupervised transfer learning.

Definition 6.47 (Inductive Transfer Learning [199]) Let the target task $\mathcal{T}_T$ be different from the source task $\mathcal{T}_S$, i.e., $\mathcal{T}_S \neq \mathcal{T}_T$, no matter when the source and target domains are the same or not. *Inductive transfer learning* uses some labeled data in the target domain to induce an objective predictive model $f_T(\cdot)$ for use in the target domain.

Definition 6.48 (Transductive Transfer Learning [10, 199]) Given a source domain $\mathcal{D}_S$ and a corresponding learning task $\mathcal{T}_S$, a target domain $\mathcal{D}_T$ and a corresponding learning task $\mathcal{T}_T$, *transductive transfer learning* aims to improve the learning of the target predictive function $f_T(\cdot)$ using the knowledge in $\mathcal{D}_S$ and $\mathcal{T}_S$, where $\mathcal{D}_S \neq \mathcal{D}_T$ and $\mathcal{T}_S = \mathcal{T}_T$. In addition, some unlabeled target-domain data must be available at training time.

Definition 6.49 (Unsupervised Transfer Learning [199]) When $\mathcal{D}_S \neq \mathcal{D}_T$ and $\mathcal{T}_S \neq \mathcal{T}_T$, and labeled data in source domain and target domain are not available, *unsupervised transfer learning* is to help to improve the learning of the target predictive function $f_T(\cdot)$ in $\mathcal{D}_T$ using the knowledge in $\mathcal{D}_S$ and $\mathcal{T}_S$.

From the above three definitions, we can compare the three transfer learning settings as follows.

1. *Common point:* The inductive transfer learning, the transductive transfer learning, and the unsupervised transfer learning do not assume that the unlabeled data $\mathbf{x}_j^{(u)}$ was drawn from the same distribution as, nor that it can be associated with the same class labels as, the labeled data. This is the common point in any type of transfer learning, and is the basic difference from the traditional machine learning.
2. *Comparison of assumptions:* The transductive transfer learning requires the source task and the target task must be the same, i.e., $\mathcal{T}_S = \mathcal{T}_T$, but both the inductive transfer learning and the unsupervised transfer learning require $\mathcal{T}_S \neq \mathcal{T}_T$.
3. *Comparison of data used:* Different from the unsupervised transfer learning that uses no labeled data, the inductive transfer learning uses some labeled data in the target domain, both transductive transfer learning and self-taught learning use some unlabeled data available at training time, while unsupervised transfer learning does not use any labeled data.
4. *Comparison of tasks:* The inductive transfer learning induces the target predictive model $f_T(\cdot)$, while both the transductive transfer learning and the unsupervised transfer learning aim to improve the learning of the target predictive function $f_T(\cdot)$ using the knowledge in $\mathcal{D}_S$ and $\mathcal{T}_S$, but the inductive and unsupervised

6.20 Transfer Learning

transfer learning require $\mathcal{T}_S \neq \mathcal{T}_T$, while the transductive transfer learning requires $\mathcal{T}_S = \mathcal{T}_T$.

According to different situations of labeled and unlabeled data in the source domain, the inductive transfer learning setting can be categorized further into two cases [199].

- *Multitask learning:* a lot of labeled data in the source domain are available. In this case, the inductive transfer learning setting is similar to the multitask learning setting. However, the inductive transfer learning setting only aims at achieving high performance in the target task by transferring knowledge from the source task, while the multitask learning tries to learn the target and source tasks simultaneously.
- *Self-taught transfer learning:* no labeled data in the source domain are available. In this case, the inductive transfer learning setting is similar to the self-taught transfer learning setting proposed by Raina et al. [210].

Definition 6.50 (Multitask Learning) For multiple related tasks, *multitask learning* is defined as an inductive transfer learning method based on shared representation, in which multiple related tasks are learned together by using labeled data for source domain. What is learned for each task can help other tasks to learn better.

This definition shows that multitask learning consists of three parts [302]:

- model the task relatedness;
- learn all tasks simultaneously;
- tasks may have different data/features.

Definition 6.51 (Self-Taught Transfer Learning [210]) Given a labeled training set of m examples $\{(\mathbf{x}_1^{(l)}, y_1^{(l)}), \ldots, (\mathbf{x}_m^{(l)}, y_m^{(l)})\}$ drawn i.i.d. from some distribution D, where each $\mathbf{x}_i^{(l)} \in \mathbb{R}^n$ denotes an input feature vector (the subscript "l" indicates that it is a labeled example), and $y_i^{(l)} \in \{1, \ldots, C\}$ is the corresponding class label in the classification problems. In addition, we are given a set of k unlabeled examples $\mathbf{x}_1^{(u)}, \ldots, \mathbf{x}_k^{(u)} \in \mathbb{R}^n$. *Self-taught transfer learning* aims to determine if $\mathbf{x}_i^{(l)}$ and $\mathbf{x}_j^{(u)}$ come from the same input "type" or "modality" through transfer learning from unlabeled data.

In the self-taught learning setting, the label spaces between the source and target domains may be different, which implies the side information of the source domain cannot be used directly. Thus, it is similar to the inductive transfer learning setting where the labeled data in the source domain are unavailable.

In the traditional machine learning setting, the transductive learning [131] refers to the situation where all test data are required to be seen at training time, and that the learned model cannot be reused for future data. In contrast, the term "transductive" in Definition 6.48 emphasizes the concept in this type of transfer learning, the tasks must be the same and there must be some unlabeled data available in the target domain. Thus, when some new test data arrive, they must be classified together with all existing data.

In the transductive transfer learning setting, the source and target tasks are required to be the same, while the source and target domains are different. In this situation, no labeled data in the target domain are available, while a lot of labeled data in the source domain are available. According to different situations between the source and target domains, the transductive transfer learning can be further categorized into two cases.

- The feature spaces between the source and target domains are different, namely $\mathcal{X}_S \neq \mathcal{X}_T$. This is a so-called heterogeneous transfer learning.
- The feature spaces between domains are the same, i.e., $\mathcal{X}_S = \mathcal{X}_T$, but the marginal probability distributions of the input data are different: $P(X_S) \neq P(X_T)$. This special case is closely related to "domain adaptation".

Definition 6.52 (Domain Adaptation) Given two different domains $\mathcal{D}_S \neq \mathcal{D}_T$ and a single task $\mathcal{T}_S = \mathcal{T}_T$, *domain adaptation* aims to improve the learning of the target predictive function $f_T(\cdot)$ using labeled source-domain data and unlabeled or few labeled target-domain data.

There are two main existing approaches to transfer learning.

1. *Instance-based approach* learns different weights to rank training examples in a source domain for better learning in a target domain. For example, boosting for transfer learning of Dai et al. [66].
2. *Feature-based approach* tries to learn a common feature structure from different domains that can bridge the two domains for knowledge transfer. For example, multitask learning of Ando and Zhang [8]; multi-domain learning of Blitzer et al. [27]; self-taught learning of Raina et al. [210]; and transfer learning via dimensionality reduction of Pan et al. [201].

6.20.3 Boosting for Transfer Learning

Consider transfer learning in which there are only labeled data from a similar old domain, when a task from one new domain comes. In these cases, labeling the new data can be costly and it would also be a waste to throw away all the old data. A natural question to ask is whether it is possible to construct a high-quality classification model using only a tiny amount of new data and a large amount of old data, even when the new data are not sufficient to train a model alone. For this end, as an extension of adaptive boosting (AdaBoost) [97], an AdaBoost tailored for transfer learning purpose, called *TrAdaBoost*, was proposed by Dai et al. [66].

For the more or less outdated training data, there are certain parts of the data that can still be reused. In other words, knowledge learned from this part of the data can still be used in training a classifier for the new data. The aim of TrAdaBoost is to iteratively reduce low quality source domain data and to remain the reusable training data.

Assume that there are two types of training data [66].

- *Same-distribution training data:* a part of the labeled training data has the same distribution as the test data. The quantity of the same-distribution training data is often inadequate to train a good classifier for the test data, but access to this data is helpful for voting on the usefulness of each of the old data instances.
- *Diff-distribution training data:* the training data has a different distribution from the test data, perhaps because they are outdated. These data are assumed to be abundant, and the classifiers learned from these data cannot classify the test data well due to different data distributions.

The TrAdaBoost model uses boosting to filter out the different-distribution training data which are very different from the same-distribution data by automatically adjusting the weights of training instances, so that the remaining different-distribution data are treated as the additional training data which greatly boost the confidence of the learned model even when the same-distribution training data are scarce.

Let X_s be the same-distribution instance space, X_d be the diff-distribution instance space, and $Y = \{0, 1\}$ be the set of category labels. A concept is a Boolean function c mapping from X to Y, where $X = X_s \cup X_d$. The test data set is denoted by $S = \{(\mathbf{x}_i^t)\}$, where $\mathbf{x}_i^t \in X_s$, $i = 1, \ldots, k$. Here, k is the size of the unlabeled test set S. The training data set $T \subseteq \{X \times Y\}$ is partitioned into two labeled sets T_d and T_s, where T_d represents the diff-distribution training data that $T_d = \{(\mathbf{x}_i^d, c(\mathbf{x}_i^d))\}$ with $\mathbf{x}_i^d \in X_d$, $i = 1, \ldots, n$. T_s represents the same-distribution training data that $T_s = \{(\mathbf{x}_j^s, c(\mathbf{x}_j^s))\}$, where $\mathbf{x}_j^s \in X_s$, $j = 1, \ldots, m$, and n and m are the sizes of T_d and T_s, respectively. The combined training set $T = \{(\mathbf{x}_i, c(\mathbf{x}_i))\}$ is defined as

$$\mathbf{x}_i = \begin{cases} \mathbf{x}_i^d, & \text{if } i = 1, \ldots, n; \\ \mathbf{x}_i^s, & \text{if } i = n+1, \ldots, n+m. \end{cases} \quad (6.20.8)$$

Algorithm 6.43 gives a *transfer AdaBoost learning framework* (TrAdaBoost) developed by Dai et al. [66].

6.20.4 Multitask Learning

The standard supervised learning task is to seek a predictor that maps an input vector $\mathbf{x} \in \mathcal{X}$ to the corresponding output $y \in \mathcal{Y}$. Consider m learning problems indexed by $\ell \in \{1, \ldots, m\}$. Suppose we are given a finite set of training examples $\{(\mathbf{x}_1^\ell, y_1^\ell), \ldots, (\mathbf{x}_{n_\ell}^\ell, y_{n_\ell}^\ell)\}$ that are independently generated according to some unknown probability distribution $\mathcal{D}$, where $\ell = 1, \ldots, m$. Based on this finite set of training examples, the predictor is selected from a set $\mathcal{H}$ of functions. The set $\mathcal{H}$, called the *hypothesis space*, consists of functions mapping from $\mathcal{X}$ to $\mathcal{Y}$ that can be used to predict the output in $\mathcal{Y}$ of an input datum in $\mathcal{X}$.

Algorithm 6.43 TrAdaBoost algorithm [66]

1. **input:** two labeled data sets T_d and T_s, the unlabeled data set S, a base learning algorithm **Learner**, and the maximum number of iterations N.
2. **initialization:** the initial weight vector $\mathbf{w}^1 = [w_1^1, \ldots, w_{n+m}^1]^T$, or the initial values is specified for $\mathbf{w}^1$.
3. **for** $t = 1$ to N **do**
4. Set $\mathbf{p}^t = \mathbf{w}^t / (\sum_{i=1}^{n+m} \mathbf{w}_i^t)$.
5. Call **Learner**, providing it the combined training set T with the distribution $\mathbf{p}^t$ over T and the unlabeled data set S. Then, get back a hypothesis $h_t : X \to Y$ (or [0, 1] by confidence).
6. Calculate the error of h_t on T_s:
    $$\epsilon_t = \sum_{i=n+1}^{n+m} \frac{w_i^t \cdot |h_t(\mathbf{x}_i) - c(\mathbf{x}_i)|}{\sum_{i=n+1}^{n+m} w_i^t}.$$
7. Set $\beta_t = \epsilon_t / (1 - \epsilon_t)$ and $\beta = 1/(1 + \sqrt{2 \ln n / N})$. Note that $\epsilon_t \leq 1/2$ is required.
8. Update the new weight vector:
    $$w_i^{t+1} = \begin{cases} w_i^t \beta^{|h_t(\mathbf{x}_i) - c(\mathbf{x}_i)|}, & \text{for } 1 \leq i \leq n; \\ w_i^t \beta_t^{|h_t(\mathbf{x}_i) - c(\mathbf{x}_i)|}, & \text{for } n+1 \leq i \leq n+m. \end{cases}$$
9. **end for**
10. **output:** the hypothesis:
    $$h_f(\mathbf{x}) = \begin{cases} 1, & \text{if } \prod_{t=\lceil N/2 \rceil} \beta_t^{-h_t(\mathbf{x})} \geq \prod_{t=\lceil N/2 \rceil} \beta_t^{-1/2}; \\ 0, & \text{otherwise}. \end{cases}$$

In the multitask scenario the goal of a learning system is to solve all multiple supervised learning tasks (such as recognizing objects or predicting attributes) by sharing information between them.

Definition 6.53 (Structural Learning [8]) *Structural learning* aims to learn some underlying predictive functional structures (smooth function classes) that can characterize what good predictors are like. In other words, its goal is to find a predictor f so that its error with respect to $\mathcal{D}$ is as small as possible.

Given the input space $\mathcal{X}$, a linear predictor is not necessarily linear on the original space $\mathcal{X}$, but rather can be regarded as a linear functional on a high-dimensional feature space $\mathcal{F}$. Assume there is a known feature map $\Phi : \mathcal{X} \to \mathcal{F}$. Therefore, a known high-dimensional feature map is given by $\mathbf{w}^T \boldsymbol{\phi}(\mathbf{x})$, where $\mathbf{w}$ is a weight vector on high-dimensional feature map $\boldsymbol{\phi}(\mathbf{x})$. In order to apply the structural learning framework, consider a parameterized low-dimensional feature map given by $\mathbf{v}^T \boldsymbol{\psi}_\theta(\mathbf{x})$, where $\mathbf{v}$ is a weight vector on low-dimensional feature map $\boldsymbol{\psi}(\mathbf{x})$, θ denotes the common structure parameter shared by all m learning problems. That is, the linear predictor $f(\mathbf{x})$ in structural learning framework has a form [8]:

$$f(\mathbf{x}) = \mathbf{w}^T \boldsymbol{\phi}(\mathbf{x}) + \mathbf{v}^T \boldsymbol{\psi}_\theta(\mathbf{x}). \tag{6.20.9}$$

In order to simplify numerical computation, Ando and Zhang [8] propose using a simple linear form of feature map, $\boldsymbol{\psi}_\theta(\mathbf{x}) = \Theta \boldsymbol{\psi}(\mathbf{x})$, where Θ is an $h \times p$ matrix, and $\boldsymbol{\psi}$ is a known p-dimensional vector function. Then, the linear predictor can be reformulated as [8]:

$$f_\Theta(\mathbf{w}, \mathbf{v}; \mathbf{x}) = \mathbf{w}^T \boldsymbol{\phi}(\mathbf{x}) + \mathbf{v}^T \Theta \boldsymbol{\psi}(\mathbf{x}). \tag{6.20.10}$$

Letting $\phi(\mathbf{x}) = \psi(\mathbf{x}) = \mathbf{x}_i^\ell \in \mathbb{R}^p$, then

$$f_\ell\left(\mathbf{w}_\ell, \mathbf{v}_\ell; \mathbf{x}_i^\ell\right) = \left(\mathbf{w}_\ell^T + \mathbf{v}_\ell^T \mathbf{\Theta}\right)\mathbf{x}_i^\ell, \quad \ell = 1, \ldots, m, \tag{6.20.11}$$

and the empirical error $L(f_\ell(\mathbf{w}_\ell, \mathbf{v}_\ell; \mathbf{\Theta}), y_i^\ell)$ on training data $\{(\mathbf{x}_1^\ell, y_1^\ell), \ldots, (\mathbf{x}_{n_\ell}^\ell, y_{n_\ell}^\ell)\}$ can be taken as the modified Huber loss function [8]:

$$L(p, y) = \begin{cases} \max(0, 1 - py)^2, & \text{if } py \geq -1; \\ -4py, & \text{otherwise.} \end{cases} \tag{6.20.12}$$

Hence, the optimization problem for multitask learning can be written as

$$[\{\hat{\mathbf{w}}_\ell, \hat{\mathbf{v}}_\ell\}; \hat{\mathbf{\Theta}}] = \arg\min_{\{\mathbf{w}, \mathbf{v}\}; \mathbf{\Theta}} \left\{ \frac{1}{n_\ell} \sum_{i=1}^{n_\ell} L\left(f_\ell(\mathbf{w}_\ell, \mathbf{v}_\ell; \mathbf{\Theta}), y_i^\ell\right) + \lambda_\ell \|\mathbf{w}\|_2^2 \right\}. \tag{6.20.13}$$

Algorithm 6.44 shows the SVD-based alternative LS solution for $\mathbf{w}_\ell$.

Algorithm 6.44 SVD-based alternating structure optimization algorithm [8]

1. **input:** training data $\{(\mathbf{x}_i^\ell, y_i^\ell)\}, l = 1, \ldots, m; i = 1, \ldots, n_\ell$; parameters: h and $\lambda_1, \ldots, \lambda_m$.
2. **initialization:** $\mathbf{u}_\ell = \mathbf{0}$ ($\ell = 1, \ldots, m$) and arbitrary matrix $\mathbf{\Theta}$.
3. **iterate**
4.   **for** $\ell = 1$ to m **do**
5.   With fixed $\mathbf{\Theta}$ and $\mathbf{v}_\ell = \mathbf{\Theta}\mathbf{u}_\ell$, approximately solve for $\hat{\mathbf{w}}_\ell$:
$$\hat{\mathbf{w}}_\ell = \arg\min_{\mathbf{w}_\ell} \left\{ \frac{1}{n_\ell} \sum_{i=1}^{n_\ell} L(\mathbf{w}_\ell^T \mathbf{x}_i^\ell + (\mathbf{v}_\ell^T \mathbf{\Theta})\mathbf{x}_i^\ell, y_i^\ell) + \lambda_\ell \|\mathbf{w}_\ell\|_2^2 \right\}$$
6.   Let $\mathbf{u}_\ell = \hat{\mathbf{w}}_\ell + \mathbf{\Theta}^T \mathbf{v}_\ell$.
7.   **end for**
8.   Compute the SVD of $\mathbf{U} = [\sqrt{\lambda_1}\mathbf{u}_1, \ldots, \sqrt{\lambda_m}\mathbf{u}_m]$:
$$\mathbf{U} = \mathbf{V}_1 \mathbf{D} \mathbf{V}_2^T \text{ (with diagonals of } \mathbf{D} \text{ in descending order).}$$
9.   Let the rows of $\mathbf{\Theta}$ be the first h rows of $\mathbf{V}_1^T$.
10. **until converge**
11. **output:** $h \times p$ dimensional matrix $\mathbf{\Theta}$.

For many natural language processing (NLP) tasks, due to new domains in which labeled data is scarce or non-existent, one seeks to adapt existing models from a resource-rich source domain to a resource-poor target domain. For this end, a structural correspondence learning is introduced by Blitzer et al. [27] to automatically induce correspondences among features from different domains.

Definition 6.54 (Structural Correspondence Learning [27]) *Structural correspondence learning* (SCL) is to identify correspondences among features from different domains by modeling their correlations with *pivot features*. Pivot features are features which behave in the same way for discriminative learning in both domains. *Non-pivot features* from different domains which are correlated with many of the same pivot features are assumed to correspond, and they are treated similarly in a discriminative learner.

The structural correspondence learning method consists of the following three steps.

- Find pivot features, e.g., using one of the feature selection methods presented in Sect. 6.7. The pivot predictors are the key element in structural correspondence learning.
- Use the weigh vectors $\hat{\mathbf{w}}_\ell$ to encode the covariance of the non-pivot features with the pivot features in order to learn a mapping matrix Θ from the original feature spaces of both domains to a shared, low-dimensional real-valued feature space. This is the core of structural correspondence learning.
- For each source $\mathcal{D}_\ell$, $\ell = 1, \ldots, m$, construct its linear predictor on target domain:

$$f_\ell(\mathbf{x}) = \text{sign}\left(\hat{\mathbf{w}}_\ell^T \mathbf{x}\right), \quad \ell = 1, \ldots, m. \tag{6.20.14}$$

Algorithm 6.45 summarizes the structural correspondence learning algorithm developed by Blitzer et al. [27].

Algorithm 6.45 Structural correspondence learning algorithm [27]

1. **input:** labeled source data $\{(\mathbf{x}_t, y_t)_{t=1}^T\}$ and unlabeled data $\{\mathbf{x}_j\}$ from both domains.
2. Choose m pivot features $\tilde{\mathbf{x}}_i$, $i = 1, \ldots, m$ using feature selection method.
3. **for** $\ell = 1$ to m **do**
4. $\quad \hat{\mathbf{w}}_\ell = \arg\min\limits_{\mathbf{w}} \left\{ \sum_j L(\mathbf{w}^T \mathbf{x}_j, p_\ell(\mathbf{x}_j)) + \lambda \|\mathbf{w}\|_2^2 \right\}$,
   where $L(p, q)$ is the modified Huber loss function given in (6.20.12).
5. **end for**
6. Construct the matrix $\mathbf{W} = [\hat{\mathbf{w}}_1, \ldots, \hat{\mathbf{w}}_m]$.
7. Compute the SVD $[\mathbf{U}, \mathbf{D}, \mathbf{V}^T]$ = SVD of $\mathbf{W}$.
8. Put $\Theta = \mathbf{U}_{1:h,:}^T$.
9. **output:** predictor $f_\ell : X_\ell \to Y_T$ is given by $f_\ell(\mathbf{x}) = \text{sign}(\hat{\mathbf{w}}_\ell^T \mathbf{x})$, where $\ell = 1, \ldots, m$.

6.20.5 EigenTransfer

In transfer learning, we are given two target data sets: a target training data set $\mathcal{X}_t = \{\mathbf{x}_i^{(t)}\}_{i=1}^n$ with labels, and a target test data set $\mathcal{X}_u = \{\mathbf{x}_i^{(u)}\}_{i=1}^k$ to be predicted. Different from traditional machine learning, we are also given an auxiliary data set $\mathcal{X}_a = \{\mathbf{x}_i^{(t)}\}_{i=1}^m$ to help the target learning.

In spectral learning [53], the input is a weighted graph $G = (V, E)$ to represent the target task, where $V = \{v_i\}_{i=1}^n$ and $E = \{e_{ij}\}_{i,j=1}^{n,n}$ represent the node set and the edge set in the graph, respectively.

6.20 Transfer Learning

To apply spectral learning to transfer learning, construct a task graph $G(V, E)$ in the following way [68].

- *Node construction:* the nodes V in the graph G represent features $\{f^{(i)}\}_{i=1}^{s}$, instances $\{\mathbf{x}_i(t)\}_{i=1}^{n}$, $\{\mathbf{x}_i^{(a)}\}_{i=1}^{m}$, $\{\mathbf{x}_i^{(u)}\}_{i=1}^{k}$ or class labels $c(\mathbf{x}_i)$.
- *Edge construction:* the edges E represent the relations between these nodes, where the weights of the edges are based on the number of co-occurrences between the end nodes in the target and auxiliary data.

Consequently, the task graph $G(V, E)$ contains almost all the information for the transfer learning task, including the target data and the auxiliary data. Usually, the task graph G is sparse, symmetric, real, and positive semi-definite. Thus, the spectra of the graph G can be calculated very efficiently.

The following are two steps for a unified framework for graph transfer learning [68].

1. Construct the weight graph $G(V, E)$ for three types of transfer learning.

    - *Cross-domain learning:* a graph $G(V, E)$ for representing the problem of cross-domain learning is given by

    $$V = \mathcal{X}_t \cup \mathcal{X}_a \cup \mathcal{X}_u \cup \mathcal{F} \cup \mathcal{C}, \tag{6.20.15}$$

    $$e_{ij} = \begin{cases} \phi_{v_i, v_j}, & \text{if } v_i \in \mathcal{X}_t \cup \mathcal{X}_a \cup \mathcal{X}_u \wedge v_j \in \mathcal{F}; \\ \phi_{v_j, v_i}, & \text{if } v_i \in \mathcal{F} \wedge v_j \in \mathcal{X}_t \cup \mathcal{X}_a \cup \mathcal{X}_u; \\ 1, & \text{if } v_i \in \mathcal{X}_t \wedge v_j \in \mathcal{C} \wedge \mathcal{C}(v_i) = v_j; \\ 1, & \text{if } v_i \in \mathcal{C} \wedge v_j \in \mathcal{X}_t \wedge \mathcal{C}(v_j) = v_i; \\ 1, & \text{if } v_i \in \mathcal{X}_a \wedge v_j \in \mathcal{C} \wedge \mathcal{C}(v_i) = v_j; \\ 1, & \text{if } v_i \in \mathcal{C} \wedge v_j \in \mathcal{X}_a \wedge \mathcal{C}(v_j) = v_i; \\ 0, & \text{otherwise.} \end{cases} \tag{6.20.16}$$

    - *Cross-category learning:* a graph $G(V, E)$ for representing the problem of cross-category learning is given by

    $$V = \mathcal{X}_t \cup \mathcal{X}_a \cup \mathcal{X}_u \cup \mathcal{F} \cup \mathcal{C}_t \cup \mathcal{C}_a, \tag{6.20.17}$$

    $$e_{ij} = \begin{cases} \phi_{v_i, v_j}, & \text{if } v_i \in \mathcal{X}_t \cup \mathcal{X}_a \cup \mathcal{X}_u \wedge v_j \in \mathcal{F}; \\ \phi_{v_j, v_i}, & \text{if } v_i \in \mathcal{F} \wedge v_j \in \mathcal{X}_t \cup \mathcal{X}_a \cup \mathcal{X}_u; \\ 1, & \text{if } v_i \in \mathcal{X}_t \wedge v_j \in \mathcal{C}_t \wedge \mathcal{C}(v_i) = v_j; \\ 1, & \text{if } v_i \in \mathcal{C}_t \wedge v_j \in \mathcal{X}_t \wedge \mathcal{C}(v_j) = v_i; \\ 1, & \text{if } v_i \in \mathcal{X}_a \wedge v_j \in \mathcal{C}_a \wedge \mathcal{C}(v_i) = v_j; \\ 1, & \text{if } v_i \in \mathcal{C}_a \wedge v_j \in \mathcal{X}_a \wedge \mathcal{C}(v_j) = v_i; \\ 0, & \text{otherwise.} \end{cases} \tag{6.20.18}$$

- *Self-taught learning:* $G(V, E)$ is given by

$$V = \mathcal{X}_t \cup \mathcal{X}_a \cup \mathcal{X}_u \cup \mathcal{F} \cup \mathcal{C}_t, \qquad (6.20.19)$$

$$e_{ij} = \begin{cases} \phi_{v_i, v_j}, & \text{if } v_i \in \mathcal{X}_t \cup \mathcal{X}_a \cup \mathcal{X}_u \wedge v_j \in \mathcal{F}; \\ \phi_{v_j, v_i}, & \text{if } v_i \in \mathcal{F} \wedge v_j \in \mathcal{X}_t \cup \mathcal{X}_a \cup \mathcal{X}_u; \\ 1, & \text{if } v_i \in \mathcal{X}_t \wedge v_j \in \mathcal{C}_t \wedge \mathcal{C}(v_i) = v_j; \\ 1, & \text{if } v_i \in \mathcal{C}_t \wedge v_j \in \mathcal{X}_t \wedge \mathcal{C}(v_j) = v_i; \\ 0, & \text{otherwise.} \end{cases} \qquad (6.20.20)$$

In the above equations, $\phi_{x,f}$ denotes the importance of the feature $f \in \mathcal{F}$ appearing in the instance $\mathbf{x} \in \mathcal{X}_t \cup \mathcal{X}_a \cup \mathcal{X}_u$, and $\mathcal{C}(\mathbf{x})$ means the true label of the instance $\mathbf{x}$.

2. Learning graph spectra.

    - Construct an adjacency matrix $\mathbf{W} \in \mathbb{R}^{n \times n}$ with respect to the graph $G(V, E)$:

    $$\mathbf{W} = \begin{bmatrix} w_{11} & \cdots & w_{1n} \\ \vdots & \ddots & \vdots \\ w_{n1} & \cdots & w_{nn} \end{bmatrix} \quad (w_{ij} = e_{ij}). \qquad (6.20.21)$$

    - Construct the diagonal matrix $\mathbf{D} = [d_{ij}]_{i,j=1}^{n,n}$:

    $$d_{ij} = \begin{cases} \sum_{t=1}^{n} w_{it}, & \text{if } i = j, \\ 0, & \text{if } i \neq j. \end{cases} \qquad (6.20.22)$$

    - Calculate the Laplacian matrix of G: $\mathbf{L} = \mathbf{D} - \mathbf{W}$.
    - Use the normalized cut technique [230] to learn the new feature representation to enable transfer learning:
        (a) calculate the first m generalized eigenvectors $\mathbf{v}_1, \ldots, \mathbf{v}_m$ of the generalized eigenproblem $\mathbf{L}\mathbf{v} = \lambda \mathbf{D}\mathbf{v}$.
        (b) use the first m generalized eigenvectors to construct the feature representation for transfer learning: $\mathbf{U} = [\mathbf{v}_1, \ldots, \mathbf{v}_m]$.

The above graph transfer learning based on spectrum features is called *EigenTransfer* [68].

EigenTransfer can model a variety of existing transfer learning problems and solutions. In this framework, a task graph is first constructed to represent the transfer learning task. Then, an eigen feature representation is learned from the task graph based on spectral learning theory.

Under the new feature representation, the knowledge from the auxiliary data tends to be transferred to help the target learning.

Algorithm 6.46 gives an *EigenCluster* algorithm.

Algorithm 6.46 EigenCluster: a unified framework for transfer learning [68]
---
1. **input:** the target data set $\mathcal{X}_t = \{\mathbf{x}_i^{(t)}\}_{i=1}^n$, the auxiliary data set $\mathcal{X}_a = \{\mathbf{x}_i^{(a)}\}_{i=1}^m$, and the test data set $\mathcal{X}_u = \{\mathbf{x}_i^{(u)}\}_{i=1}^k$.
2. Construct the task graph $G(V, E)$ based on the target clustering task (c.f. (6.20.15)–(6.20.19)).
3. Construct the $n \times n$ adjacent matrix $\mathbf{W}$ whose entries $w_{ij} = e_{ij}$ are based on the task graph $G(V, E)$.
4. Use (6.20.22) to calculate the diagonal matrix $\mathbf{D}$.
5. Construct the Laplacian matrix $\mathbf{L} = \mathbf{D} - \mathbf{W}$.
6. Calculate the first N generalized eigenvectors $\mathbf{v}_1, \ldots, \mathbf{v}_N$ of $\mathbf{Lv} = \lambda \mathbf{Dv}$.
7. Let $\mathbf{U} = [\mathbf{v}_1, \ldots, \mathbf{v}_N]$.
8. **for** each $\mathbf{x}_i^{(t)} \in \mathcal{X}_t$ **do**
9.   Let $y_i^{(t)}$ be the corresponding row in $\mathbf{U}$ with respect to $\mathbf{x}_i^{(t)}$.
10. **end for**
11. Train a classifier based on $\mathcal{Y}^{(t)} = \{y_i^{(i)}\}_{i=1}^n$ instead of $\mathcal{X}_t = \{\mathbf{x}_i^{(t)}\}_{i=1}^n$ using a traditional classification algorithm, and then classify the test data $\mathcal{X}_u = \{\mathbf{x}_i^{(u)}\}_{i=1}^k$.
12. **output:** classification result on $\mathcal{X}_u$.

In the next section, we focus on domain adaptation.

6.21 Domain Adaptation

The task of *domain adaptation* is to develop learning algorithms that can be easily ported from one domain to another. This problem is particularly interesting and important when we have a large collection of labeled data in one "source" domain but truly desire a model that performs well in a second "target" domain.

Let $\mathcal{S} = \{\mathbf{x}_i^{(s)}, y_i^{(s)}\}_{i=1}^{N_s}$, where $\mathbf{x}_i^{(s)} \in \mathbb{R}^N$ denotes the labeled data from the source domain and is referred to as an observation, and $y_i^{(s)}$ is the corresponding class label. Labeled data from the target domain is denoted by $\mathcal{T}_l = \{\mathbf{x}_i^{(t)}, y_i^{(t)}\}_{i=1}^{N_{t_l}}$, where $\mathbf{x}_i^{(t)} \in \mathbb{R}^M$. Similarly, unlabeled data in the target domain is denoted by $\mathcal{T}_u = \{\mathbf{x}_i^{(u)}\}_{i=1}^{N_{t_u}}$, where $\mathbf{x} \in \mathbb{R}^M$. It is usually assumed that $N = M$. Denoting $\mathcal{T} = \mathcal{T}_l \cup \mathcal{T}_u$, then $N_t = N_{t_l} + N_{t_u}$ denotes the total number of samples in the target domain.

Let $\mathbf{S} = [\mathbf{x}_1^{(s)}, \ldots, \mathbf{x}_{N_s}^{(s)}]$ be the matrix of N_s data points from the source domain $\mathcal{S}$. Similarly, for target domain $\mathcal{T}$, let $\mathbf{T}_l = [\mathbf{x}_1^{(t)}, \ldots, \mathbf{x}_{N_{t_l}}^{(t)}]$ be the matrix of N_{t_l} data from $\mathcal{T}_l$, $\mathbf{T}_u = [\mathbf{x}_1^{(u)}, \ldots, \mathbf{x}_{N_{t_u}}^{(u)}]$ be the matrix of N_{t_u} data from $\mathcal{T}_u$, and $\mathbf{T} = [\mathbf{T}_l, \mathbf{T}_u] = [\mathbf{x}_1^{(t)}, \ldots, \mathbf{x}_{N_t}^{(t)}]$ be the matrix of N_t data from $\mathcal{T}$.

The goal of domain adaptation is to learn a function $f(\cdot)$ that predicts the class label of a new test sample from the target domain. Depending on the availability of the source and target-domain data, the domain adaptation problem can be defined in many different ways [205].

- In semi-supervised domain adaptation, the function $f(\cdot)$ is learned using the knowledge in both $\mathcal{S}$ and $\mathcal{T}_l$.
- In unsupervised domain adaptation, the function $f(\cdot)$ is learned using the knowledge in both $\mathcal{S}$ and $\mathcal{T}_u$.
- In multi-source domain adaptation, $f(\cdot)$ is learned from more than one domain in $\mathcal{S}$ accompanying each of the first two cases.
- In the heterogeneous domain adaptation, the dimensions of features in the source and target domains are assumed to be different. In other words, $N \neq M$.

In what follows, we present a number of domain adaptation approaches.

6.21.1 Feature Augmentation Method

One of the simplest domain adaptation approaches is the *feature augmentation* of Daumé [70].

For each given feature $\mathbf{x} \in \mathbb{R}^F$ in the original problem, make its three versions: a general version, a source-specific version, and a target-specific version. The augmented source data $\boldsymbol{\phi}_s(\mathbf{x}) : \mathbb{R}^F \to \mathbb{R}^{3F}$ will contain only general and source-specific versions, and the augmented target data $\boldsymbol{\phi}_t(\mathbf{x}) : \mathbb{R}^F \to \mathbb{R}^{3F}$ contains general and target-specific versions as follows:

$$\boldsymbol{\phi}_s(\mathbf{x}) = \begin{bmatrix} \mathbf{x} \\ \mathbf{x} \\ \mathbf{0} \end{bmatrix}, \quad \boldsymbol{\phi}_t(\mathbf{x}) = \begin{bmatrix} \mathbf{x} \\ \mathbf{0} \\ \mathbf{x} \end{bmatrix}, \quad (6.21.1)$$

where $\mathbf{0}$ is an $F \times 1$ zero vector.

Consider the *heterogeneous domain adaptation* (HDA) problem where the dimensions of data from the source and target domains are different. By introducing a common subspace for the source and target data, both the source and target data of dimension N and M are, respectively, projected onto a latent domain of dimension l using two projection matrices $\mathbf{W}_1 \in \mathbb{R}^{l+N}$ and $\mathbf{W}_2 \in \mathbb{R}^{l+M}$. The augmented feature maps for the source and target domains in the common space are then defined as [162]:

$$\boldsymbol{\phi}_s\left(\mathbf{x}_i^{(s)}\right) = \begin{bmatrix} \mathbf{W}_1 \mathbf{x}_i^{(s)} \\ \mathbf{x}_i^{(s)} \\ \mathbf{0}_M \end{bmatrix} \in \mathbb{R}^{l+N+M}, \quad (6.21.2)$$

6.21 Domain Adaptation

$$\phi_t\left(\mathbf{x}_i^{(l)}\right) = \begin{bmatrix} \mathbf{W}_2 \mathbf{x}_i^{(t)} \\ \mathbf{0}_N \\ \mathbf{x}_i^{(t)} \end{bmatrix} \in \mathbb{R}^{l+N+M}, \quad (6.21.3)$$

where $\mathbf{x}_i^{(s)} \in \mathcal{S}$, $\mathbf{x}_i^{(t)} \in \mathcal{T}_l$ and $\mathbf{0}_M$ is an $M \times 1$ zero vector.

After introducing $\mathbf{W}_1$ and $\mathbf{W}_2$, the data from two domains can be readily compared in the common subspace. Once the data from both domains are transformed onto a common space, *heterogeneous feature augmentation* (HFA) [162] can be readily formulated as follows:

$$\min_{\mathbf{W}_1, \mathbf{W}_2} \min_{\mathbf{w}, b, \xi_i^{(s)}, \xi_i^{(t)}} \left\{ \frac{1}{2} \|\mathbf{w}\|^2 + C \left(\sum_{i=1}^{N_s} \xi_i^{(s)} + \sum_{i=1}^{N_t} \xi_i^{(t)} \right) \right\}, \quad (6.21.4)$$

subject to $\quad y_i^{(s)} \left(\mathbf{w}^T \phi_s\left(\mathbf{x}_i^{(s)}\right) + b \right) \geq 1 - \xi_i^{(s)}, \; \xi_i^{(s)} \geq 0, \quad (6.21.5)$

$$y_i^{(t)} \left(\mathbf{w}^T \phi_t\left(\mathbf{x}_i^{(t)}\right) + b \right) \geq 1 - \xi_i^{(t)}, \; \xi_i^{(t)} \geq 0, \quad (6.21.6)$$

$$\|\mathbf{W}_1\|_F^2 \leq \lambda_{w1}, \; \|\mathbf{W}_2\|_F^2 \leq \lambda_{w2}, \quad (6.21.7)$$

where $C > 0$ is a tradeoff parameter which balances the model complexity and the empirical losses on the training samples from two domains, and λ_{w1} and $\lambda_{w2} > 0$ are predefined parameters to control the complexities of $\mathbf{W}_1$ and $\mathbf{W}_2$, respectively.

Denote the kernel on the source domain samples as $\mathbf{K}_s = \mathbf{\Phi}_s^T \mathbf{\Phi}_s \in \mathbb{R}^{n_s \times n_s}$, where $\mathbf{\Phi}_s = [\phi_s(\mathbf{x}_1^{(s)}), \ldots, \phi_s(\mathbf{x}_{n_s}^{(s)})]$ and $\phi_s(\cdot)$ is the nonlinear feature mapping function induced by $\mathbf{K}_s$. Similarly, the kernel on the target-domain samples is denoted as $\mathbf{K}_t = \mathbf{\Phi}_t^T \mathbf{\Phi}_t \in \mathbb{R}^{n_t \times n_t}$ where $\mathbf{\Phi}_t = [\phi_t(\mathbf{x}_1^{(t)}), \ldots, \phi_t(\mathbf{x}_{n_t}^{(t)})]$ and $\phi_t(\cdot)$ is the nonlinear feature mapping function induced by $\mathbf{K}_t$.

Define the feature weight vector $\mathbf{w} = [\mathbf{w}_c, \mathbf{w}_s, \mathbf{w}_t] \in \mathbb{R}^{d_c + d_s + d_t}$ for the augmented feature space, where $\mathbf{w}_c \in \mathbb{R}^{d_c}$, $\mathbf{w}_s \in \mathbb{R}^{d_s}$ and $\mathbf{w}_t \in \mathbb{R}^{d_t}$ are also the weight vectors defined for the common subspace, the source domain, and the target domain, respectively. Introduce a *transformation metric*

$$\mathbf{H} = [\mathbf{W}_1, \mathbf{W}_2]^T [\mathbf{W}_1, \mathbf{W}_2] \in \mathbb{R}^{(d_s + d_t) \times (d_s + d_t)} \quad (6.21.8)$$

which is positive semi-definite, i.e., $\mathbf{H} \succeq 0$.

With $\mathbf{H}$, the optimization problem can be reformulated as follows [162]:

$$\min_{\mathbf{H} \succeq 0} \max_{\boldsymbol{\alpha}} \left\{ J(\boldsymbol{\alpha}) = \mathbf{1}^T \boldsymbol{\alpha} - \frac{1}{2} (\boldsymbol{\alpha} \odot \mathbf{y})^T \mathbf{K}_\mathbf{H} (\boldsymbol{\alpha} \odot \mathbf{y}) \right\}, \quad (6.21.9)$$

subject to $\quad \mathbf{y}^T \boldsymbol{\alpha} = 0, \; \mathbf{0} \leq \boldsymbol{\alpha} \leq C\mathbf{1}, \quad \mathrm{tr}(\mathbf{H}) \leq \lambda, \quad (6.21.10)$

where $\mathbf{a} \odot \mathbf{b}$ denotes the elementwise product of two vectors $\mathbf{a}$ and $\mathbf{b}$, and

- $\boldsymbol{\alpha} = [\alpha_1^{(s)}, \ldots, \alpha_{n_s}^{(s)}, \alpha_1^{(t)}, \ldots, \alpha_{n_t}^{(t)}]^T$ is the vector of dual variables;
- $\mathbf{y} = [y_1^{(s)}, \ldots, y_{n_s}^{(s)}, y_1^{(t)}, \ldots, y_{n_t}^{(t)}]^T$ is the label vector of all training instances;
- $\mathbf{K}_H = \boldsymbol{\Phi}^T (\mathbf{H} + \mathbf{I}) \boldsymbol{\Phi}$ is the derived kernel matrix for the samples from source and target domains, where $\boldsymbol{\Phi} = \begin{bmatrix} \boldsymbol{\Phi}_s & \mathbf{O}_{d_s \times n_t} \\ \mathbf{O}_{d_t \times n_s} & \boldsymbol{\Phi}_t \end{bmatrix} \in \mathbb{R}^{(d_s + d_t) \times (n_s + n_t)}$;
- $\lambda = \lambda_{w1} + \lambda_{w2}$.

Let transformation metric $\mathbf{H}$ consist of the linear combination of rank-one normalized positive semi-definite (PSD) matrices:

$$\mathbf{H}_{\boldsymbol{\theta}} = \sum_{r=1}^{\infty} \theta_r \mathbf{M}_r, \quad \mathbf{M}_r \in \mathcal{M}, \tag{6.21.11}$$

where θ_r is the linear combination coefficients, and $\mathcal{M} = \{\mathbf{M}_r|_{r=1}^{\infty}\}$ is the set of rank-one normalized PSD matrices $\mathbf{M}_r = \mathbf{h}_r \mathbf{h}_r^T$ with $\mathbf{h}_r \in \mathbb{R}^{n_s + n_t}$ and $\mathbf{h}_r^T \mathbf{h}_r = 1$. If denoting $\mathbf{K} = \boldsymbol{\Phi}^T \boldsymbol{\Phi} = \mathbf{K}^{1/2} \mathbf{K}^{1/2}$, then the optimization problem of heterogeneous feature augmentation in Eqs. (6.21.9)–(6.21.10) can be reformulated as [162]:

$$\min_{\boldsymbol{\theta}} \max_{\boldsymbol{\alpha} \in \mathcal{A}} \left\{ \mathbf{1}^T \boldsymbol{\alpha} - \frac{1}{2} (\boldsymbol{\alpha} \odot \mathbf{y})^T \mathbf{K}^{1/2} (\mathbf{H}_{\boldsymbol{\theta}} + \mathbf{I}) \mathbf{K}^{1/2} (\boldsymbol{\alpha} \odot \mathbf{y}) \right\}, \tag{6.21.12}$$

subject to $\quad \mathbf{H}_{\boldsymbol{\theta}} = \sum_{r=1}^{\infty} \theta_r \mathbf{M}_r, \ \mathbf{M}_r \in \mathcal{M}, \ \mathbf{1}^T \boldsymbol{\theta} \leq \lambda, \ \boldsymbol{\theta} \geq 0. \tag{6.21.13}$

By setting $\boldsymbol{\theta} \leftarrow \frac{1}{\lambda} \boldsymbol{\theta}$, then the above heterogeneous feature augmentation can be further rewritten as [162]:

$$\min_{\boldsymbol{\theta} \in \mathcal{D}_{\boldsymbol{\theta}}} \max_{\boldsymbol{\alpha} \in \mathcal{A}} \left\{ \mathbf{1}^T \boldsymbol{\alpha} - \frac{1}{2} (\boldsymbol{\alpha} \odot \mathbf{y})^T \left(\sum_{r=1}^{\infty} \theta_r \mathbf{K}_r \right) (\boldsymbol{\alpha} \odot \mathbf{y}) \right\}, \tag{6.21.14}$$

where $\mathcal{A} = \{\boldsymbol{\alpha} | \boldsymbol{\alpha}^T \mathbf{y} = 0, \mathbf{0}_n \leq \boldsymbol{\alpha} \leq C \mathbf{1}_n\}$ is the feasible set of the dual variables $\boldsymbol{\alpha}$, and $\mathbf{K}_r = [k_r(\mathbf{x}_i, \mathbf{x}_j)]_{i,j=1}^{n,n} = [\boldsymbol{\phi}_k^T(\mathbf{x}_i) \boldsymbol{\phi}_k(\mathbf{x}_j)]_{i,j=1}^{n,n} \in \mathbb{R}^{n \times n}$ is the kth base kernel matrix of the labeled patterns, while the linear combination coefficient vector $\boldsymbol{\theta} = [\theta_1, \ldots, \theta_{\infty}]^T$ is nonnegative, i.e., $\boldsymbol{\theta} \geq 0$.

The optimization problem in (6.21.14) is an *infinite kernel learning* (IKL) problem with each base kernel as $\mathbf{K}_r$, which can be readily solved with the multiple kernel learning solver.

The ordinary machine learning and support vector machine operate on a single kernel matrix, representing information from a single data source. However, many machine learning problems are often better characterized by incorporating data from more than one source of data. By combining multiple kernel matrices into a single classifier, multiple kernel learning realizes the representation of multiple feature sets, which is conducive to the processing of multi-source data.

6.21 Domain Adaptation

Definition 6.55 (Multiple Kernel Learning) *Multiple kernel learning* (MKL) is a machine learning that aims at the unified representation of multiple domain data with multiple kernels and consists of the following three parts.

- Use kernel matrices to serve as the unified representation of multi-source data.
- Transform data under each feature representation to a kernel matrix $\mathbf{K}$.
- M kinds of features will lead to M kernels $\{k_m(\mathbf{x}_i, \mathbf{x}_j)\}_{m=1}^{M}$. That is, the ensemble kernel for unified representation is the linear combination of base ones given by

$$K_{ij} = K(\mathbf{x}_i, \mathbf{x}_j) = \sum_{m=1}^{M} \theta_m k_m(\mathbf{x}_i, \mathbf{x}_j). \tag{6.21.15}$$

The (i, j)th entry of the kernel matrix $\mathbf{K}$ may be the linear kernel function $K_{ij} = K(\mathbf{x}_i, \mathbf{x}_j) = \mathbf{x}_i^T \mathbf{x}_j$, the polynomial kernel function $K_{ij} = K(\mathbf{x}_i, \mathbf{x}_j) = (\gamma \mathbf{x}_i^T \mathbf{x}_j + b)^d$, $\gamma > 0$, the Gaussian kernel function $K_{ij} = K(\mathbf{x}_i, \mathbf{x}_j) = \exp(-\gamma d^2(\mathbf{x}_i, \mathbf{x}_j))$, and so on. Here, $d(\mathbf{x}_i, \mathbf{x}_j)$ is the dissimilarity or distance between $\mathbf{x}_i$ and $\mathbf{x}_j$.

With the training data $\{\mathbf{x}_i, y_i\}_{i=1}^{N}$ where $y_i \in \{-1, +1\}$ and base kernels $\{k_m\}_{m=1}^{M}$, the learned model is of the form

$$h(\mathbf{x}) = \sum_{i=1}^{N} \alpha_i y_i k(\mathbf{x}_i, \mathbf{x}) + b$$

$$= \sum_{i=1}^{N} \alpha_i y_i \sum_{m=1}^{M} \theta_m k_m(\mathbf{x}_i, \mathbf{x}) + b, \tag{6.21.16}$$

where linear combination coefficient θ_m is the importance of the mth base kernel $k_m(\mathbf{x}_i, \mathbf{x}_j)$, and b is a learning bias.

Then, task of multiple kernel learning is to optimize both $\{\alpha_i\}_{i=1}^{N}$ and $\{\theta_m\}_{m=1}^{M}$ by using all the training data $\{(\mathbf{x}_n, y_n)\}_{n=1}^{N} = \{(\mathbf{x}_i^{(s)}, y_i^{(s)})\}_{i=1}^{N_s} \cup \{(\mathbf{x}_j^{(t)}, y_j^{(t)})\}_{j=1}^{N_t}$ with $N = N_s + N_t$.

Algorithm 6.47 shows the ℓ_p-norm multiple kernel learning algorithm for interleaved optimization of $\boldsymbol{\theta}$ and $\boldsymbol{\alpha}$ in (6.21.9).

Algorithm 6.48 is the heterogeneous feature augmentation algorithm [162].

6.21.2 Cross-Domain Transform Method

Let $\mathbf{W} : \mathbf{x}^{(t)} \to \mathbf{x}^{(s)}$ denote a linear transformation matrix from the target domain $\mathbf{x}^{(t)} \in \mathcal{D}_T$ to the source domain $\mathbf{x}^{(s)} \in \mathcal{D}_S$. Then, the inner product similarity function between $\mathbf{x}^{(s)}$ and the transformed $\mathbf{W}\mathbf{x}^{(t)}$ can be described as

$$\text{sim}(\mathbf{W}) = \langle \mathbf{x}^{(s)}, \mathbf{W}\mathbf{x}^{(t)} \rangle = (\mathbf{x}^{(s)})^T \mathbf{W}\mathbf{x}^{(t)}. \tag{6.21.17}$$

Algorithm 6.47 ℓ_p-norm multiple kernel learning algorithm [146]

1. **input:** the accuracy parameter ϵ and the subproblem size Q.
2. **initialization:** $g_{m,i} = \hat{g}_i = \alpha_i = 0, \forall i = 1, \ldots, N; L = S = -\infty; \theta_m = \sqrt[p]{1/M}$, $\forall m = 1, \ldots, M$.
3. **iterate**
4.   Calculate the gradient $\hat{g}$ of $J(\boldsymbol{\alpha})$ in (6.21.9) w.r.t. $\boldsymbol{\alpha}$;
    $$\hat{\mathbf{g}} = \frac{\partial J(\boldsymbol{\alpha})}{\partial \boldsymbol{\alpha}} = \mathbf{1} - \left(\sum_{m=1}^{M} \theta_m \mathbf{K}_m\right)(\boldsymbol{\alpha} - \mathbf{y}).$$
5.   Select Q variables $\alpha_{i_1}, \ldots, \alpha_{i_Q}$ based on the gradient $\hat{\mathbf{g}}$.
6.   Store $\boldsymbol{\alpha}^{\text{old}} = \boldsymbol{\alpha}$ and then update $\boldsymbol{\alpha} \leftarrow \boldsymbol{\alpha} - \mu \hat{\mathbf{g}}$.
7.   Update gradient $g_{m,i} \leftarrow g_{m,i} + \sum_{q=1}^{Q}(\alpha_{i_q} - \alpha_{i_q}^{\text{old}}) k_m(x_{i_q}, x_i), \forall m = 1, \ldots, M$; $i = 1, \ldots, N$.
8.   Compute the quadratic terms $S_m = \frac{1}{2}\sum_i g_{m,i}\alpha_i$, $q_m = 2\theta_m^2 S_m, \forall m = 1, \ldots, M$.
9. $L_{\text{old}} = L$, $L = \sum_i y_i \alpha_i$, $S_{\text{old}} = S$, $S = \sum_m \theta_m S_m$.
10. **if** $\left|1 - \frac{L-S}{L_{\text{old}} - S_{\text{old}}}\right| \geq \epsilon$ **then**
    $$\theta_m = (q_m)^{1/(p+1)} \bigg/ \left(\sum_{m'=1}^{M}(q_{m'})^{p/(p+1)}\right)^{1/p}, \forall m = 1, \ldots, M.$$
11. **else**
12.   **break**
13. **end if**
14. $\hat{g}_i = \sum_m \theta_m g_{m,i}, \forall i = 1, \ldots, N$.
15. **output:** $\boldsymbol{\theta} = [\theta_1, \ldots, \theta_M]^T$ and $\boldsymbol{\alpha} = [\alpha_1, \ldots, \alpha_N]^T$.

Algorithm 6.48 Heterogeneous feature augmentation [162]

1. **input:** labeled source samples $\{(\mathbf{x}_i^{(s)}, y_i^{(s)})\}_{i=1}^{N_s}$ and labeled target samples $\{(\mathbf{x}_j^{(t)}, y_j^{(t)})\}_{j=1}^{N_t}$.
2. **initialization:** $\mathcal{M}_1 = \{\mathbf{M}_1\}$ with $\mathbf{M}_1 = \mathbf{h}_1 \mathbf{h}_1^T$ and $\mathbf{h}_1 = \frac{1}{\sqrt{N_s + N_t}} \mathbf{1}_{N_s + N_t}$.
3. **repeat**
4.   Use Algorithm 6.47 to solve $\boldsymbol{\theta}$ and $\boldsymbol{\alpha}$ in (6.21.9) based on $\mathbf{M}_r$.
5.   Obtain a rank-one PSD matrix $\mathbf{M}_{r+1} = \mathbf{h}\mathbf{h}^T \in \mathbb{R}^{(N_s + N_t) \times (N_s + N_t)}$ with $\mathbf{h} = \frac{\mathbf{K}^{1/2}(\boldsymbol{\alpha} \odot \mathbf{y})}{\|\mathbf{K}^{1/2}(\boldsymbol{\alpha} \odot \mathbf{y})\|}$.
6.   Set $\mathcal{M}_{r+1} = \mathcal{M}_r \cup \{\mathbf{M}_{r+1}\}$, and $r = r + 1$.
7. **exit** if the objective converges.
8. **output:** $\boldsymbol{\alpha}$ and $\mathbf{H} = \lambda \sum_{r=1}^{Q} \theta_r \mathbf{M}_r$.

The goal of cross-domain transform-based domain adaptation [219] is to learn the linear transformation $\mathbf{W}$ given some form of supervision, and then to utilize the learned similarity function in a classification or clustering algorithm.

Given a set of N points $\{\mathbf{x}_1, \ldots, \mathbf{x}_N\} \in \mathbb{R}^d$, we seek a positive definite Mahalanobis matrix $\mathbf{W}$ which parameterizes the (squared) Mahalanobis distance:

$$d_W(\mathbf{x}_i, \mathbf{x}_j) = (\mathbf{x}_i - \mathbf{x}_j)^T \mathbf{W}(\mathbf{x}_i - \mathbf{x}_j). \tag{6.21.18}$$

Sample a random pair consisting of a labeled source domain sample $(\mathbf{x}_i^{(s)}, y_i^{(s)})$ and a labeled target-domain sample $(\mathbf{x}_j^{(t)}, y_j^{(t)})$, and then create two constraints described by

$$d_W\left(\mathbf{x}_i^{(s)}, \mathbf{x}_j^{(t)}\right) \leq u, \quad \text{if } y_i^{(s)} = y_j^{(t)}; \tag{6.21.19}$$

$$d_W\left(\mathbf{x}_i^{(s)}, \mathbf{x}_j^{(t)}\right) \geq \ell, \quad \text{if } y_i^{(s)} \neq y_j^{(t)}, \tag{6.21.20}$$

where u is a relatively small value and ℓ is a sufficiently large value. The above two constraints can be rewritten as

$$\mathbf{x}_i^{(s)} \text{ similar to } \mathbf{x}_j^{(t)}, \quad \text{if } d_W\left(\mathbf{x}_i^{(s)}, \mathbf{x}_j^{(t)}\right) \leq u; \tag{6.21.21}$$

$$\mathbf{x}_i^{(s)} \text{ dissimilar to } \mathbf{x}_j^{(t)}, \quad \text{if } d_W\left(\mathbf{x}_i^{(s)}, \mathbf{x}_j^{(t)}\right) \geq \ell. \tag{6.21.22}$$

Learning a Mahalanobis distance can be stated as follows.

1. Given N source feature points $\{\mathbf{x}_1^{(s)}, \ldots, \mathbf{x}_N^{(s)}\} \in \mathbb{R}^d$ and N target feature points $\{\mathbf{x}_1^{(t)}, \ldots, \mathbf{x}_N^{(t)}\} \in \mathbb{R}^d$.
2. Given inequality constraints relating pairs of feature points.
   - Similarity constraints: $d_W(\mathbf{x}_i^{(s)}, \mathbf{x}_j^{(t)}) \leq u$;
   - Dissimilarity constraints: $d_W(\mathbf{x}_i^{(s)}, \mathbf{x}_j^{(t)}) \geq \ell$.
3. Problem: Learn a positive semi-definite Mahalanobis matrix $\mathbf{W}$ such that the Mahalanobis distance $d_W(\mathbf{x}_i^{(s)}, \mathbf{x}_j^{(t)}) = (\mathbf{x}_i^{(s)} - \mathbf{x}_j^{(t)})^T \mathbf{W}(\mathbf{x}_i^{(s)} - \mathbf{x}_j^{(t)})$ satisfies the above similarity and dissimilarity constraints.

This problem is known as a *Mahalanobis metric learning* problem which can be formulated as the constrained optimization problem [219]:

$$\min_{\mathbf{W} \succeq 0} \{\text{tr}(\mathbf{W}) - \log \det(\mathbf{W})\}, \tag{6.21.23}$$

$$\text{subject to} \quad d_W\left(\mathbf{x}_i^{(s)}, \mathbf{x}_j^{(t)}\right) \leq u, \ (i,j) \in S, \tag{6.21.24}$$

$$d_W\left(\mathbf{x}_i^{(s)}, \mathbf{x}_j^{(t)}\right) \geq \ell, \ (i,j) \in D. \tag{6.21.25}$$

Here, the regularizer $\text{tr}(\mathbf{W}) - \log \det(\mathbf{W})$ is defined only between positive semi-definite matrices, and S and D are the set of similar pairs and the set of dissimilar pairs, respectively.

The constrained optimization (6.21.23) provides a Mahalanobis metric learning method and is called *information theoretic metric learning* (ITML), as shown in Algorithm 6.49.

6.21.3 Transfer Component Analysis Method

Let $\mathcal{P}(X_S)$ and $\mathcal{Q}(X_T)$ be the marginal distributions of $X_S = \{\mathbf{x}_{S_i}\}$ and $X_T = \{\mathbf{x}_{T_i}\}$ from the source and target domains, respectively; where $\mathcal{P}$ and $\mathcal{Q}$ are different.

Algorithm 6.49 Information-theoretic metric learning [71]
1. **input:** $d \times n$ matrix $\mathbf{X} = [\mathbf{x}_1, \ldots, \mathbf{x}_n]$, set of similar pairs S, set of dissimilar pairs D, distance threshold u for similarity constraint, distance threshold ℓ for dissimilarity constrain, Mahalanobis matrix $\mathbf{W}_0$, slack parameter γ, constraint index function $c(\cdot)$.
2. **initialization:** $\mathbf{W} \leftarrow \mathbf{W}_0$, $\lambda_{ij} \leftarrow 0$, $\forall i, j$.
3. Let $\xi_{c(i,j)} = \begin{cases} u, & \text{if } (i,j) \in S; \\ \ell, & \text{otherwise}. \end{cases}$
4. **repeat**
5.   Pick a constraint $(i,j) \in S$ or $(i,j) \in D$.
6.   Update $p \leftarrow (\mathbf{x}_i - \mathbf{x}_j)^T \mathbf{W}(\mathbf{x}_i - \mathbf{x}_j)$.
7. $\delta = \begin{cases} 1, & \text{if } (i,j) \in S; \\ -1, & \text{otherwise}. \end{cases}$
8. $\alpha \leftarrow \min\left\{\lambda_{ij}, \frac{\delta}{2}\left(\frac{1}{p} - \gamma \xi_{c(i,j)}\right)\right\}$.
9. $\beta \leftarrow \delta\alpha/(1 - \delta\alpha p)$.
10. $\xi_{c(i,j)} \leftarrow \gamma \xi_{c(i,j)}/(\gamma + \delta\alpha \xi_{c(i,j)})$.
11. $\lambda_{ij} \leftarrow \lambda_{ij} - \alpha$.
12. $\mathbf{W} \leftarrow \mathbf{W} + \beta \mathbf{W}(\mathbf{x}_i - \mathbf{x}_j)(\mathbf{x}_i - \mathbf{x}_j)^T \mathbf{W}$.
13. **until convergence**
14. **output:** Mahalanobis matrix $\mathbf{W}$.

Most domain adaptation methods are based on a key assumption: $\mathcal{P} \neq \mathcal{Q}$, but $P(Y_S|X_S) = P(Y_T|X_T)$. However, in many real-world applications, the conditional probability $P(Y|X)$ may also change across domains due to noisy or dynamic factors underlying the observed data.

Let $\boldsymbol{\phi}$ be a transformation vector such that $\boldsymbol{\phi}(X_S) = [\phi(\mathbf{x}_{S_1}), \cdots, \phi(\mathbf{x}_{S_{n_s}})] \in \mathbb{R}^{1 \times n_s}$ and $\boldsymbol{\phi}(X_T) = [\phi(\mathbf{x}_{T_1}), \ldots, \phi(\mathbf{x}_{T_{n_t}})] \in \mathbb{R}^{1 \times n_t}$, where n_s and n_t are the numbers of the source and target features. Consider domain adaptation under weak assumption that $\mathcal{P} \neq \mathcal{Q}$ but there exists a transformation vector $\boldsymbol{\phi}$ such that $P(\boldsymbol{\phi}(X_S)) \approx P(\boldsymbol{\phi}(X_T))$ and $P(Y_S|\boldsymbol{\phi}(X_S)) \approx P(Y_T|\boldsymbol{\phi}(X_T))$.

A key problem is how to find this transformation $\boldsymbol{\phi}$. Since there are no labeled data in the target domain, $\boldsymbol{\phi}$ cannot be learned by directly minimizing the distance between $P(Y_S|\boldsymbol{\phi}(X_S))$ and $P(Y_T|\boldsymbol{\phi}(X_T))$. Pan et al. [203] propose to learn $\boldsymbol{\phi}$ such that:

- the distance between the marginal distributions $P(\boldsymbol{\phi}(X_S))$ and $P(\boldsymbol{\phi}(X_T))$ is small,
- $\boldsymbol{\phi}(X_S)$ and $\boldsymbol{\phi}(X_T)$ preserve important properties of X_S and X_T, respectively.

For finding a transformation $\boldsymbol{\phi}$ satisfying $P(Y_S|\boldsymbol{\phi}(X_S)) \approx P(Y_T|\boldsymbol{\phi}(X_T))$, assume that $\boldsymbol{\phi}$ is the feature mapping vector induced by a universal kernel. *Maximum mean discrepancy embedding* (MMDE) [201] embeds both the source and target-domain data into a shared low-dimensional common latent space using a nonlinear mapping $\boldsymbol{\phi}$, which is a symmetric feature transformation with $\phi = T_S = T_T$ shown in Fig. 6.6b.

6.21 Domain Adaptation

Define the Gram matrices defined on the source domain (S, S), target domain (T, T), and cross-domain (S, T) in the embedded space as follows:

$$\mathbf{K}_{S,S} = \boldsymbol{\phi}^T(X_S)\boldsymbol{\phi}(X_S) = \begin{bmatrix} \phi(\mathbf{x}_{S_1}) \\ \vdots \\ \phi(\mathbf{x}_{S_{n_s}}) \end{bmatrix} \begin{bmatrix} \phi(\mathbf{x}_{S_1}) & \cdots & \phi(\mathbf{x}_{S_{n_s}}) \end{bmatrix} \in \mathbb{R}^{n_s \times n_s}, \tag{6.21.26}$$

$$\mathbf{K}_{S,T} = \boldsymbol{\phi}^T(X_S)\boldsymbol{\phi}(X_T) = \begin{bmatrix} \phi(\mathbf{x}_{S_1}) \\ \vdots \\ \phi(\mathbf{x}_{S_{n_s}}) \end{bmatrix} \begin{bmatrix} \phi(\mathbf{x}_{T_1}) & \cdots & \phi(\mathbf{x}_{T_{n_t}}) \end{bmatrix} \in \mathbb{R}^{n_s \times n_t}, \tag{6.21.27}$$

and

$$\mathbf{K}_{T,S} = \boldsymbol{\phi}^T(X_T)\boldsymbol{\phi}(X_S) = \mathbf{K}_{S,T}^T \in \mathbb{R}^{n_t \times n_s}, \tag{6.21.28}$$

$$\mathbf{K}_{T,T} = \boldsymbol{\phi}^T(X_T)\boldsymbol{\phi}(X_T) = \begin{bmatrix} \phi(\mathbf{x}_{T_1}) \\ \vdots \\ \phi(\mathbf{x}_{T_{n_t}}) \end{bmatrix} \begin{bmatrix} \phi(\mathbf{x}_{T_1}) & \cdots & \phi(\mathbf{x}_{T_{n_t}}) \end{bmatrix} \in \mathbb{R}^{n_t \times n_t}. \tag{6.21.29}$$

To learn the symmetric kernel matrix

$$\mathbf{K} = \begin{bmatrix} \mathbf{K}_{S,S} & \mathbf{K}_{S,T} \\ \mathbf{K}_{T,S} & \mathbf{K}_{T,T} \end{bmatrix} \in \mathbb{R}^{(n_s+n_t) \times (n_s+n_t)}, \tag{6.21.30}$$

construct a matrix $\mathbf{L}$ with entries

$$L_{ij} = \begin{cases} 1/n_s^2, & \text{if } \mathbf{x}_i, \mathbf{x}_j \in X_S; \\ 1/n_t^2, & \text{if } \mathbf{x}_i, \mathbf{x}_j \in X_T; \\ -1/(n_s n_t), & \text{otherwise.} \end{cases} \tag{6.21.31}$$

Then, the objective function of MMDE can be written as [203]:

$$\max_{\mathbf{K} \geq 0} \{\text{tr}(\mathbf{K}\mathbf{L}) - \lambda \text{tr}(\mathbf{K})\} \quad \text{subject to constraints on } \mathbf{K}, \tag{6.21.32}$$

where the first term in the objective minimizes the distance between distributions, while the second term maximizes the variance in the feature space, and $\lambda \geq 0$ is a trade-off parameter. However, $\mathbf{K}$ is a high-dimensional kernel matrix due to the fact that n_s and/or n_t are large. As such it is necessary to use a semi-orthogonal transformation matrix $\bar{\mathbf{W}} \in \mathbb{R}^{(n_s+n_t) \times m}$ such that $\bar{\mathbf{W}}\bar{\mathbf{W}}^T = \mathbf{I}_m$ to transform $\mathbf{K}^{-1/2}$

to an m-dimensional matrix $\mathbf{W} = \mathbf{K}^{-1/2}\bar{\mathbf{W}}$ (where $m \ll n_s + n_t$). Then, the distance between distributions, $\text{dist}(X'_S, X'_T) = \text{tr}(\mathbf{KL})$, becomes

$$\text{tr}(\mathbf{KL}) = \text{tr}(\mathbf{KK}^{-1/2}\bar{\mathbf{W}}(\mathbf{K}^{-1/2}\bar{\mathbf{W}})^T\mathbf{KL}) = \text{tr}(\mathbf{KWW}^T\mathbf{KL}) = \text{tr}(\mathbf{W}^T\mathbf{KLKW}).$$

Therefore, the objective function of MMDE in (6.21.32) can be rewritten as [203]:

$$\min_{\mathbf{W}} \left\{ \text{tr}(\mathbf{W}^T\mathbf{KLKW}) + \mu \text{tr}(\mathbf{W}^T\mathbf{W}) \right\}, \tag{6.21.33}$$

$$\text{subject to} \quad \mathbf{W}^T\mathbf{KHKW} = \mathbf{I}_m, \tag{6.21.34}$$

where $\mathbf{H}$ is the centering matrix given by

$$\mathbf{H} = \mathbf{I}_{n_s+n_t} - \frac{1}{n_t + n_t}\mathbf{1}\mathbf{1}^T. \tag{6.21.35}$$

The constrained optimization problem (6.21.33) and (6.21.34) can be reformulated as the unconstrained optimization problem [203]:

$$\max_{\mathbf{W}} \left\{ J(\mathbf{W}) = \text{tr}\left(\left(\mathbf{W}^T(\mathbf{KLK} + \mu\mathbf{I})\mathbf{W}\right)^{-1} \mathbf{W}^T\mathbf{KHKW} \right) \right\}. \tag{6.21.36}$$

Under the condition that the column vectors of $\mathbf{W} = [\mathbf{w}_1, \ldots, \mathbf{w}_m]$ are orthogonal to each other, the above optimization problem can be divided into m subproblems:

$$\mathbf{w}_i = \arg\max_{\mathbf{w}_i} \left\{ \text{tr}\left(\left(\mathbf{w}_i^T(\mathbf{KLK} + \mu\mathbf{I})\mathbf{w}_i\right)^{-1} \mathbf{w}_i^T\mathbf{KHKw}_i \right) \right\}$$

$$= \arg\max_{\mathbf{w}_i} \frac{\mathbf{w}_i^T\mathbf{KHKw}_i}{\mathbf{w}_i^T(\mathbf{KLK} + \mu\mathbf{I})\mathbf{w}_i}, \quad i = 1, \ldots, m. \tag{6.21.37}$$

This is a typical generalized Rayleigh quotient, and thus $\mathbf{w}_i$ is the ith generalized eigenvector corresponding to the ith maximum generalized eigenvalue of the matrix pencil $(\mathbf{KHK}, \mathbf{KLK} + \mu\mathbf{I})$ or the ith eigenvector corresponding to the ith maximum eigenvalue of the matrix $(\mathbf{KLK} + \mu\mathbf{I})^{-1}\mathbf{KLK}$. Moreover, m is the number of leading generalized eigenvalues of $(\mathbf{KHK}, \mathbf{KLK} + \mu\mathbf{I})$ or leading eigenvalues of $(\mathbf{KLK} + \mu\mathbf{I})^{-1}\mathbf{KLK}$. This is the unsupervised *transfer component analysis* (TCA) for domain adaptation, which was proposed by Pan et al. [203].

Transfer learning has been applied to many real-world applications [199, 268].

- Natural language processing [27] including sentiment classification [28], cross-language classification [165], text classification [67], etc.
- Image classification [277], WiFi localization classification [200, 202, 289, 297, 298], video concept detection [82], and so on.

- Pattern recognition including face recognition [139], head pose classification [211], disease prediction [195], atmospheric dust aerosol particle classification [171], sign language recognition [88], image recognition [205], etc.

Brief Summary of This Chapter

- This chapter presents the machine learning tree.
- Machine learning aims at using learning machines to establish the mapping of input data to output data.
- The main tasks of machine learning are classification (for discrete data) and prediction (for continue data).
- This chapter highlights selected topics and advances in machine learning: graph machine learning, reinforcement learning, Q-learning, and transfer learning, etc.

References

1. Acar, E., Camtepe, S.A., Krishnamoorthy, M., Yener, B.: Modeling and multiway analysis of chatroom tensors. In: Proceedings of the IEEE International Conference on Intelligence and Security Informatics, pp. 256–268. Springer, Berlin (2005)
2. Acar, E., Aykut-Bingo, C., Bingo, H., Bro, R., Yener, B.: Multiway analysis of epilepsy tensors. Bioinformatics **23**, i10–i18 (2007)
3. Agrawal, R., Imielinski, T., Swami, A.N.: Mining association rules between sets of items in large databases. In: Proceedings of the ACM SIGMOD International Conference on Management of Data, pp. 207–216 (1992)
4. Ali, M.M., Khompatraporn, C., Zabinsky, Z.B.: A numerical evaluation of several stochastic algorithms on selected continuous global optimization on test problems. J. Global Optim. **31**, 635–672 (2005)
5. Aliu, O.G., Imran, A., Imran, M.A., Evans, B.: A survey of self organisation in future cellular networks. IEEE Commun. Surveys Tutorials. **15**(1), 336–361 (2013)
6. Anderberg, M.R.: Cluster Analysis for Application. Academic, New York (1973)
7. Anderson, T.W.: An Introduction to Multivariate Statistical Analysis, 2nd edn. Wiley, New York (1984)
8. Ando, R.K., Zhang, T.: A framework for learning predictive structures from multiple tasks and unlabeled data. J. Mach. Learn. Res. **6**, 1817–1853 (2005)
9. Angluin D.: Queries and concept learning. Mach. Learn. **2**(4), 319–342 (1988)
10. Arnold, A., Nallapati, R., Cohen, W.W.: A comparative study of methods for transductive transfer learning. In: Proceedings of the Seventh IEEE International Conference on Data Mining Workshops, pp. 77–82 (2007)
11. Atlas, L., Cohn, D., Ladner, R., El-Sharkawi, M.A., Marks II, R.J.: Training connectionist networks with queries and selective sampling. In: Advances in Neural Information Processing Systems 2, Morgan Kaufmann, pp. 566–573 (1990)
12. Auslender, A.: Optimisation Méthodes Numériques. Masson, Paris (1976)
13. Bach, F.R., Jordan, M.I.: Kernel independent component analysis. J. Mach. Learn. Res. **3**, 1–48 (2002)
14. Bagheri, M., Nurmanova, V., Abedinia, O., Naderi, M.S.: Enhancing power quality in microgrids with a new online control Strategy for DSTATCOM using reinforcement learning algorithm. IEEE Access **6**, 38986–38996 (2018)

15. Bandyopdhyay, S., Maulik, U.: An evolutionary technique based on K-means algorithm for optimal clustering in $\mathbb{R}^N$. Inform. Sci. **146**(1–4), 221–237 (2002)
16. Bartlett, P.L.: The sample complexity of pattern classification with neural networks: the size of the weights is more important than the size of the network. IEEE Trans. Inf. Theory. **44**(2), 525–536 (1998)
17. Baum, L.E., Eagon, J.A.: An inequality with applications to statistical estimation for probabilistic functions of Markov processes and to a model for ecology. Bull. Amer. Math. Soc. **73**(3), 360 (1967)
18. Behbood, V., Lu, J., Zhang, G.: Fuzzy bridged refinement domain adaptation: long-term bank failure prediction. Int. J. Comput Intell. Appl. **12**(1), Art. no. 1350003 (2013)
19. Belkin, M., Niyogi, P.: Laplacian eigenmaps for dimensionality reduction and data representation. Neural Comput. **15**(6), 1373–1396 (2003)
20. Belkin, M., Niyogi, P., Sindhwani, V.: Manifold regularization: a geometric framework for learning from labeled and unlabeled examples. J. Mach. Learn. Res. **7**, 2399–2434 (2006)
21. Bellman, R.: Dynamic Programming. Princeton University Press, Princeton (1957)
22. Bengio, Y., Courville, A., Vincent, P.: Representation learning: a review and new perspectives. IEEE Trans. Pattern Anal. Mach. Intell. **35**(8), 1798–1828 (2013)
23. Bersini, H., Dorigo, M., Langerman, S.: Results of the first international contest on evolutionary optimization. In: Proceedings of IEEE International Conference on Evolutionary Computation, Nagoya, pp. 611–615 (1996)
24. Bertsekas, D.P.: Dynamic Programming and Optimal Sequence of States of the Markov Decision Process. Control, vol. 11. Athena Scientific, Nashua (1995)
25. Bertsekas, D.P.: Nonlinear Programming, 2nd edn. Athena Scientific, Nashua (1999)
26. Beyer, H.G., Schwefel, H.P.: Evolution strategies: a comprehensive introduction. J. Nat. Comput. **1**(1), 3–52 (2002)
27. Blitzer, J., McDonald, R., Pereira, F.: Domain adaptation with structural correspondence learning. In: Proceedings of the 2006 Conference on Empirical Methods in Natural Language Processing, pp. 120–128 (2006)
28. Blitzer, J., Dredze, M., Pereira, F.: Biographies, Bollywood, Boom-Boxes and Blenders: Domain adaptation for sentiment classification. In: Proceedings of the 45th Annual Meeting of the Association of Computational Linguistics, pp. 432–439 (2007)
29. Blum, A., Chawla, S.: Learning from labeled and unlabeled data using graph mincuts. In: Proceedings of the 18th International Conference on Machine Learning (2001)
30. Blum, A., Mitchell, T.: Combining labeled and unlabeled data with co-training. In: Proceedings of the 11th Annual Conference on Computational Learning Theorem (COLT 98), pp. 92–100 (1998)
31. Bottou, L., Curtis, F.E., Nocedal, J.: Optimization methods for large-scale machine learning. SIAM Rev. **60**(2), 223–311 (2018)
32. Bouneffouf, D.: Exponentiated gradient exploration for active learning. Computers **5**(1), 1–12 (2016)
33. Bouneffouf, D., Laroche, R., Urvoy, T., Fèraud, R., Allesiardo, R.: Contextual bandit for active learning: Active Thompson sampling. In: Proceedings of the 21st International Conference on Neural Information Processing, ICONIP (2014)
34. Boykov, Y., Kolmogorov, V.: An experimental comparison of min-cut/max-flow algorithms for energy minimization in vision. IEEE Trans. Pattern Analy. Mach. Intell. **26**(9), 1124–1137 (2004)
35. Breiman, L.: Better subset selection using the nonnegative garrote. Technometrics **37**, 738–754 (1995)
36. Bro, R.: PARAFAC: tutorial and applications. Chemome. Intell. Lab. Syst. **38**, 149–171 (1997)
37. Bu, F.: A high-order clustering algorithm based on dropout deep learning for heterogeneous data in Cyber-Physical-Social systems. IEEE Access **6**, 11687–11693 (2018)
38. Buczak, A.L., Guven, E.: A survey of data mining and machine learning methods for cyber security intrusion detection. IEEE Commun. Surv. Tut. **18**(2), 1153–1176 (2016)

39. Burges, C.J.C.: A tutorial on support vector machines for pattern recognition. Data Min. Knowl. Disc. **2**, 121–167 (1998)
40. Burr, S.: Active Learning Literature Survey. Computer Sciences Technical Report 1648, University of Wisconsin-Madison, Retrieved 2014-11-18 (2010)
41. Cai, D., Zhang, C., He, S.: Unsupervised feature selection for multi-cluster data. In: Proceedings of the 16th ACM SIGKDD, July 25–28, Washington, pp. 333–342 (2010)
42. Campbell, C., Cristianini, N., Smola, A.: Query learning with large margin classifiers. In: Proceedings of the International Conference on Machine Learning (ICML) (2000)
43. Candès, E.J., Wakin, M.B., Boyd, S.P.: Enhancing sparsity by reweighted ℓ_1 minimization. J. Fourier Analy. Appl. **14**(5–6), 877–905 (2008)
44. Candès, E.J., Li, X., Ma, Y., Wright, J.: Robust principal component analysis? J. ACM **58**(3), 1–37 (2011)
45. Caruana, R.A.: Multitask learning. Mach. Learn. **28**, 41–75 (1997)
46. Chandrasekaran, V., Sanghavi, S., Parrilo, P.A., Wilisky, A.S.: Rank-sparsity incoherence for matrix decomposition. SIAM J. Optim. **21**(2), 572–596 (2011)
47. Chang, C.I., Du, Q.: Estimation of number of spectrally distinct signal sources in hyperspectral imagery. IEEE Trans. Geosci. Remote Sens. **42**(3), 608–619 (2004)
48. Chattopadhyay, R., Sun, Q., Fan, W., Davidson, I., Panchanathan, S., Ye, J.: Multisource domain adaptation and its application to early detection of fatigue. ACM Trans. Knowl. Discov. From Data **6**(4), 1–26 (2012)
49. Chen, T., Amari, S., Lin, Q.: A unified algorithm for principal and minor components extraction. Neural Netw. **11**, 385–390 (1998)
50. Chen, Y., Lasko, T.A., Mei, Q., Denny, J.C, Xu, H.: A study of active learning methods for named entity recognition in clinical text. J. Biomed. Inform. **58**, 11–18 (2015)
51. Chernoff, H.: Sequential analysis and optimal design. In: CBMS-NSF Regional Conference Series in Applied Mathematics, vol. 8. SIAM, Philadelphia (1972)
52. Choromanska, A., Jebara, T., Kim, H., Mohan, M., Monteleoni, C.: Fast spectral clustering via the Nyström method. In: International Conference on Algorithmic Learning Theory ALT 2013, pp. 367–381 (2013)
53. Chung, F.R.K.: Spectral graph theory. In: CBMS Regional Conference Series, vol.92. Conference Board of the Mathematical Sciences, Washington (1997)
54. Chung, C.J., Reynolds, R.G.: CAEP: An evolution-based tool for real-valued function optimization using cultural algorithms. Int. J. Artif. Intell. Tool **7**(3), 239–291 (1998)
55. Ciresan, D.C., Meier, U., Schmidhuber, J.: Transfer learning for Latin and Chinese characters with deep neural networks. In: Proceedings of the International Joint Conference on Neural Networks (IJCNN), Brisbane, pp. 1–6 (2012)
56. Coates, A., Ng, A.Y.: Learning feature representations with K-means. In: Montavon, G., Orr, G.B., Müller, K.-R. (eds.) Neural Networks: Tricks of the Trade, 2nd edn., pp. 561–580. Springer, Berlin (2012)
57. Cohen, W.W.: Fast effective rule induction. In: Proceedings of the 12th International Conference on International Conference on Machine Learning, Lake Tahoe, pp. 115–123 (1995)
58. Cohn, D.: Active learning. In: Sammut, C., Webb, G.I. (eds.) Encyclopedia of Machine Learning, pp. 10–14 (2011)
59. Cohn, D., Ghahramani, Z., Jordan, M.I.: Active learning with statistical models. J. Artific. Intell. Res. **4**, 129–145 (1996)
60. Comon, P., Golub, G., Lim, L.H., Mourrain, B.: Symmetric tensors and symmetric tensor rank. SIAM J. Matrix Anal. Appl. **30**(3), 1254–1279 (2008)
61. Corana, A., Marchesi, M., Martini, C., Ridella, S.: Minimizing multimodal functions of continuous variables with simulated annealing algorithms. ACM Trans. Math. Softw. **13**(3), 262–280 (1987)
62. Correa, N.M., Adali, T., Li, Y.Q., Calhoun, V.D.: Canonical correlation analysis for data fusion and group inferences. IEEE Signal Proc. Mag. **27**(4), 39–50 (2010)

63. Cortes, C., Mohri, M.: On transductive regression. In: Proceedings of the Neural Information Processing Systems (NIPS), pp. 305–312 (2006)
64. Crammer, K., Singer, Y.: On the algorithmic implementation of multiclass kernel-based vector machines. J. Mach. Learn. Res. **2**, 265–292 (2001)
65. Cristianini, N., Shawe-Taylor, J., Eliseeff, A., Kandola, J.S.: On kernel-target alignment. In: NIPS'01 Proceedings of the 14th International Conference on Neural Information Processing Systems: Natural and Synthetic, pp. 367–373 (2001)
66. Dai, W., Yang, Q., Xue, G.R., Yu, Y.: Boosting for transfer learning. In: Proceedings of the 24th International Conference on Machine Learning, pp. 193–200 (2007)
67. Dai, W., Xue, G., Yang, Q., Yu, Y.: Transferring naive Bayes classifiers for text classification. In: Proc. 22nd Association for the Advancement of Artificial Intelligence (AAAI) Conference on Artificial Intelligence, pp. 540–545 (2007)
68. Dai, W., Jin, O., Xue, G.-R., Yang, Q., Yu, Y.: EigenTransfer: A unified framework for transfer learning. In: Proceedings of the the 26th International Conference on Machine Learning, Montreal, pp. 193–200 (2009)
69. Dash, M., Liu, H.: Feature selection for classification. Intell. Data Anal. **1**(1–4), 131–156 (1997)
70. Daumé III, H.: Frustratingly easy domain adaptation. In: Proceedings of the 45th Annual Meeting of the Association of Computational Linguistics, pp. 256–263 (2007)
71. Davis, J.V., Kulis, B., Jain, P., Sra, S., Dhillon, I.S.: Information-theoretic metric learning. In: Proceedings of the international Conference on Machine Learning, pp. 209–216 (2007)
72. Defazio, A., Bach, F., Lacoste-Julien, S.: SAGA: A fast incremental gradient method with support for non-strongly convex composite objectives. In: Advances in Neural Information Processing Systems, vol. 27, pp. 1646–1654 (2014)
73. Defferrard, M., Bresson, X., Vandergheynst, P.: Convolutional neural networks on graphs with fast localized spectral filtering. In: Proceedings of the 30th Conference on Neural Information Processing Systems (NIPS 2016), Barcelona, pp. 3837–3845 (2016)
74. Deng, Z., Choi, K., Jiang, Y.: Generalized hidden-mapping ridge regression, knowledge-leveraged inductive transfer learning for neural networks, fuzzy systems and kernel method. IEEE Trans. Cybern. **44**(12), 2585–2599 (2014)
75. Dhillon, I.S., Modha, D.M.: Concept decompositions for large sparse text data using clustering. Mach. Learn. **42**(1), 143–175 (2001)
76. Dong, X., Thanou, D., Frossard, P., Vandergheynst, P.: Learning Laplacian matrix in smooth graph signal representations. IEEE Trans. Sign. Proc. **64**(23), 6160–6173 (2016)
77. Donoho, D.L., Johnstone, I.: Adapting to unknown smoothness via wavelet shrinkage. J. Amer. Statist. Assoc. **90**, 1200–1224 (1995)
78. Dorigo, M., Gambardella, L.M.: Ant colony system: A cooperative learning approach to the traveling salesman problem. IEEE Trans. Evol. Comput. **1**(1), 53–66 (1997)
79. Douglas, S.C., Kung, S.-Y., Amari, S.: A self-stabilized minor subspace rule. IEEE Sign. Proc. Lett. **5**(12), 328–330 (1998)
80. Downie, J.S.: A window into music information retrieval research. Acoust. Sci. Technol. **29**(4), 247–255 (2008)
81. Du, Q., Faber, V., Gunzburger, M.: Centroidal Voronoi tessellations: applications and algorithms. SIAM Rev. **41**, 637–676 (1999)
82. Duan, L., Tsang, I.W., Xu, D., Maybank, S.J.: Domain transfer SVM for video concept detection. In: Proceedings of the 2009 IEEE Conference on Computer Vision and Pattern Recognition (CVPR), pp. 1375–1381 (2009)

83. Duda, R.O., Hart, P.E.: Pattern Classification and Scene Analysis. Wiley, New York (1973)
84. Efron, B., Hastie, T., Johnstone, I., Tibshirani, R.: Least angle regression. Ann. Statist. **32**, 407–499 (2004)
85. El-Attar, R.A., Vidyasagar, M., Dutta, S.R.K.: An algorithm for ll-norm minimization with application to nonlinear ll-approximation. SIAM J. Numer. Anal. **16**(1), 70–86 (1979)
86. Estienne, F., Matthijs, N., Massart, D.L., Ricoux, P., Leibovici, D.: Multi-way modeling of high-dimensionality electroencephalographic data. Chemometr. Intell. Lab. Syst. **58**(1), 59–72 (2001)
87. Fan, J., Han, F., Liu, H.: Challenges of big data analysis. Nat. Sci. Rev. **1**(2), 293–314 (2014)
88. Farhadi, A., Forsyth, D., White, R.: Transfer learning in sign language. In: Proceedings of the IEEE 2007 Conference on Computer Vision and Pattern Recognition, pp. 1–8 (2007)
89. Farmer, J., Packard, N., Perelson, A.: The immune system, adaptation and machine learning. Phys. D: Nonlinear Phenom. **2**, 187–204 (1986)
90. Fedorov, V.V.: Theory of Optimal Experiments. (Trans. by Studden, W.J., Klimko, E.M.). Academic, New York (1972)
91. Fercoq, O., Richtárk, P.: Accelerated, parallel, and proximal coordinate descent. SIAM J. Optim. **25**(4), 1997–2023 (2015)
92. Figueiredo, M.A.T., Nowak, R.D., Wright, S.J.: Gradient projection for sparse reconstruction: application to compressed sensing and other inverse problems. IEEE J. Sel. Top. Signa. Proc. **1**(4), 586–597 (2007)
93. Finkel, J.R., Manning, C.D.: Hierarchical Bayesian domain adaptation. In: Proceedings of the Annual Conference of the North American Chapter of the Association for Computational Linguistics, Los Angeles, pp. 602–610 (2009)
94. Fisher, R.A.: The statistical utilization of multiple measurements. Ann. Eugenic. **8**, 376–386 (1938)
95. Ford, L., Fulkerson, D.: Flows in Networks. Princeton University Press, Princeton (1962)
96. Freund, Y.: Boosting a weak learning algorithm by majority. Inform. Comput. **12**(2), 256–285 (1995)
97. Freund, Y., Schapire, R.E.: A decision-theoretic generalization of on-line learning and an application to boosting. J. Comput. Syst. Sci. **55**, 119–139 (1997)
98. Friedman, N., Geiger, D., Goldszmidt, M.: Bayesian network classifiers. Mach. Learn. **29**, 131–163 (1997)
99. Friedman, J., Hastie, T., Höeling, H., Tibshirani, R.: Pathwise coordinate optimization. Ann. Appl. Stat. **1**(2), 302–332 (2007)
100. Friedman, J., Hastie, T., Tibshirani, R.: Additive logistic regression: a statistical view of boosting. Ann. Stat. **28**(2), 337–407 (2000)
101. Fu, W.J.: Penalized regressions: the bridge versus the Lasso. J. Comput. Graph. Stat. **7**(3), 397–416 (1998)
102. Fuchs, J.J.: Multipath time-delay detection and estimation. IEEE Trans. Signal Process. **47**(1), 237–243 (1999)
103. Furey, T.S., Cristianini, N., Duffy, N., Bednarski, D.W., Schummer, M., Haussler, D.: Support vector machine classification and validation of cancer tissue samples using microarray expression data. Bioinformatics **16**(10), 906–914 (2000)
104. Ge, Z., Song, Z., Ding, S.X., Huang, B.: Data mining and analytics in the process industry: the role of machine learning. IEEE Access **5**, 20590–20616 (2017)
105. Geladi, P., Kowalski, B.R.: Partial least squares regression: a tutorial. Anal. Chim. Acta **186**, 1–17 (1986)
106. George, A.P., Powell, W.B.: Adaptive stepsizes for recursive estimation with applications in approximate dynamic programming. Mach. Learn. **65**(1), 167–198 (2006)
107. Goldberg, D.E., Holland, J.H.: Genetic algorithms and machine learning. Mach. Learn. **3**(2), 95–99 (1988)
108. Golub, G.H., Zha, H.: The canonical correlations of matrix pairs and their numerical computation. In: Linear Algebra for Signal Processing, pp. 27–49. Springer, Berlin (1995)

109. Golub, T.R., Slonim, D.K., Tamayo, P., Huard, C., Gaasenbeek, M., Mesirov, J.P., Coller, H., Loh, M.L., Downing, J.R., Caligiuri, M.A., Bloomfield, C.D., Lander, E.S.: Molecular classification of cancer: class discovery and class prediction by gene expression monitoring. Science **286**, 531–537 (1999)
110. Grandvalet, Y., Bengio, Y.: Semi-supervised learning by entropy minimization. In: Advances in Neural Information Processing Systems, vol. 17, pp. 529–536 (2005)
111. Guo, W., Kotsia, I., Ioannis, P.: Tensor learning for regression. IEEE Trans. Image Process. **21**(2), 816–827 (2012)
112. Guyon, I., Elisseeff, A.: An introduction to variable and feature selection. J. Mach. Learn. Res. **3**, 1157–1182 (2003)
113. Guyon, I., Weston, J., Barnhill, S., Vapnik, V.: Gene selection for cancer classification using support vector machines. Mach. Learn. **46**, 389–422 (2002)
114. Handl, J., Knowles, J., Kell, D.B.: Computational cluster validation in post-genomic data analysis. Bioinformatics **21**(15), 3201–3212 (2005)
115. Hardoon, D.R., Szedmak, S., Shawe-Taylor, J.: Canonical correlation analysis: an overview with application to learning methods. Neural Comput. **16**(12), 2639–2664 (2004)
116. Hesterberg, T., Choi, N.H., Meier, L., Fraley, C.: Least angle and ℓ_1 penalized regression: a review. Stat. Surv. **2**, 61–93 (2008)
117. Hoerl, A.E., Kennard, R.W.: Ridge regression: biased estimates for non-orthogonal problems. Technometrics **12**, 55–67 (1970)
118. Hoerl, A.E., Kennard, R.W.: Ridge regression: applications to nonorthogonal problems. Technometrics **12**, 69–82 (1970)
119. Hoi, S.C.H., Jin, R., Lyu, M.R.: Batch mode active learning with applications to text categorization and image retrieval. IEEE Trans. Knowl. Data Eng. **21**(9), 1233–1247 (2009)
120. Hornik, K., Stinchcombe, M., White, H.: Multilayer feedforward networks are universal approximators. Neural Netw. **2**(5), 359–366 (1989)
121. Höskuldsson, A.: PLS regression methods. J. Chemometr. **2**, 211–228 (1988)
122. Hotelling, H.: Relations between two sets of variants. Biometrika **28**(3/4), 321–377 (1936)
123. Huang, G.-B., Zhu, Q.-Y., Siew, C.-K.: Extreme learning machine: theory and applications. Neurocomputing **70**(1–3), 489–501 (2006)
124. Huang, G.-B., Zhou, H., Ding, X., Zhang, R.: Extreme learning machine for regression and multiclass classification. IEEE Trans. Syst. Man Cybern. B Cybern. **42**(2), 513–529 (2012)
125. Hunter, D.R., Lange, K.: A tutorial on MM algorithms. Amer. Statist. **58**, 30–37 (2004)
126. Jain, K., Dubes, R.C.: Algorithms for Clustering Data. Prentice-Hall, Englewood Cliffs (1988)
127. Jain, A.K.: Data clustering: 50 years beyond K-means. Pattern Recogn. Lett. **31**, 651–666 (2010)
128. Jain, A.K., Duin, R.P.W., Mao, J.: Statistical pattern recognition: a review. IEEE Trans. Pattern Anal. Mach. Intell. **22**(1), 4–37 (2000)
129. Jamil, M., Yang, X.-S.: A literature survey of benchmark functions for global optimization problems. Int. J. Math. Modell. Numer. Optim. **4**(2), 150–194 (2013)
130. Jensen, F.V.: Bayesian Networks and Decision Graphs. Springer, New York (2001)
131. Joachims, T.: Transductive inference for text classification using support vector machines. In: Proceedings of the 16th International Conference on Machine Learning, pp. 200–209 (1999)
132. Johnson, S.C.: Hierarchical clustering schemes. Psycioietrika **32**(3), 241–254 (1967)
133. Johnson, R., Zhang, T.: Accelerating stochastic gradient descent using predictive variance reduction. In: Advances in Neural Information Processing Systems, vol. 26, pp. 315–323 (2013)
134. Jolliffe, I.: Principal Component Analysis. Springer, New York (1986)
135. Jonesb, S., Shaoa, L., Dub, K.: Active learning for human action retrieval using query pool selection. Neurocomputing **124**, 89–96 (2014)
136. Jouffe, L.: Fuzzy inference system learning by reinforcement methods. IEEE Trans. Syst. Man Cybern. Part C **28**(3), 338–355 (1998)

137. Kaelbling, L.P., Littman, M.L., Moore, A.W.: Reinforcement learning: a survey. J. Artif. Intell. Res. **4**, 237–285 (1996)
138. Kaelbling, L.P., Littman, M.L., Cassandra, A.R.: Planning and acting in partially observable stochastic domains. Artif. Intell. **101**(1), 99–134 (1998)
139. Kan, M., Wu, J., Shan, S., Chen, X.: Domain adaptation for face recognition: targetize source domain bridged by common subspace. Int. J. Comput. Vis. **109**(1–2), 94–109 (2014)
140. Kearns, M., Valiant, L.: Crytographic limitations on learning Boolean formulae and finite automata. In: Proceedings of the Twenty-first Annual ACM Symposium on Theory of Computing, pp. 433–444 (1989); See J. ACM **41**(1), 67–95 (1994)
141. Kearns, M.J., Vazirani, U.V.: An Introduction to Computational Learning Theory. MIT Press, Cambridge (1994)
142. Kennedy, J., Eberhart, R.: Particle swarm optimization. In: Proceedings of the IEEE International Conference on Neural Networks (ICNN), vol. IV, pp. 1942–1948 (1995)
143. Kiers, H.A.L.: Towards a standardized notation and terminology in multiway analysis. J. Chemometr. **14**, 105–122 (2000)
144. Kimura, A., Kameoka, H., Sugiyama, M., Nakano, T., Maeda, E., Sakano, H., Ishiguro, K.: SemiCCA: Efficient semi-supervised learning of canonical correlations. Inform. Media Technol. **8**(2), 311–318 (2013)
145. Klaine, P.V., Imran, M.A., Souza, R.D., Onireti, O.: A survey of machine learning techniques applied to self-organizing cellular networks. IEEE Commun. Surv. Tut. **19**(4), 2392–2431 (2017)
146. Kloft, M., Brefeld, U., Sonnenburg, S., and Zien, A.: ℓ_p-norm multiple kernel learning. J. Mach. Learn. Res. **12**, 953–997 (2011)
147. Kober, J., Bangell, J., Peters, J.: Reinforcement learning in robotics: a survey. Int. J. Robustics Res. **32**(11), 1238–1274 (2013)
148. Kocer, B., Arslan, A.: Genetic transfer learning. Expert Syst. Appl. **37**(10), 6997–7002 (2010)
149. Kolda, T.G.: Multilinear operators for higher-order decompositions. Sandia Report SAND2006-2081, California (2006)
150. Kolda, T.G., Bader, B.W., Kenny, J.P.: Higher-order web link analysis using multilinear algebra. In: Proceedings of the 5th IEEE International Conference on Data Mining, pp. 242–249 (2005)
151. Konečný J., Liu, J., Richtárik, P., Takáč, M.: Mini-batch semi-stochastic gradient descent in the proximal setting. IEEE J. Sel. Top. Signa. Process. **10**(2), 242–255 (2016)
152. Koza, J.R.: Genetic Programming: On the Programming of Computers by Means of Natural Selection. MIT Press, Cambridge (1992)
153. Kulis, B., Saenko, K., Darrell, T.: What you saw is not what you get: domain adaptation using asymmetric kernel transforms. In: Proceedings of the IEEE 2011 Conference on Computer Vision and Pattern Recognition, pp. 1785–1292 (2011)
154. Lathauwer, L.D., Moor, B.D., Vandewalle, J.: A multilinear singular value decomposition. SIAM J. Matrix Anal. Appl. **21**, 1253–1278 (2000)
155. Lathauwer, L.D., Nion, D.: Decompositions of a higher-order tensor in block terms—part III: alternating least squares algorithms. SIAM J. Matrix Anal. Appl. **30**(3), 1067–1083 (2008)
156. Le Roux, N., Schmidt, M., Bach, F.R.: A stochastic gradient method with an exponential convergence rate for finite training sets. In: Advances in Neural Information Processing Systems, vol. 25, pp. 2663–2671 (2012)
157. Letexier, D., Bourennane, S., Blanc-Talon, J.: Nonorthogonal tensor matricization for hyperspectral image filtering. IEEE Geosci. Remote Sensing. Lett. **5**(1), 3–7 (2008)
158. Levie, R., Monti, F., Bresson, X., Bronstein, M.M.: CayleyNets: Graph convolutional neural networks with complex rational spectral filters (2018). Available at: https://arXiv:1705.07664v2
159. Lewis, D., Gale, W.: A sequential algorithm for training text classifiers. In Proceedings of the ACM SIGIR Conference on Research and Development in Information Retrieval, pp. 3–12. ACM/Springer, New York/Berlin (1994)

160. Li, X., Guo, Y.: Adaptive active learning for image classification. In: Proceedings of the 26th IEEE Conference on Computer Vision and Pattern Recognition, pp. 1–8 (2013)
161. Li, F., Pan, S.J., Jin, O., Yang, Q., Zhu, X.: Cross-domain co-extraction of sentiment and topic lexicons. In: Proceedings of the 50th annual meeting of the association for computational linguistics long papers, vol. 1, pp. 410–419 (2012)
162. Li, W., Duan, L., Xu, D., Tsang, I.: Learning with augmented features for supervised and semi-supervised heterogeneous domain adaptation. IEEE Trans. Pattern Anal. Mach. Intell. **36**(6), 1134–1148 (2014)
163. Lin, L.: Self-improving reactive agents based on reinforcement learning, planning and teaching. Mach. Learn. **8**, 293–321 (1992)
164. Lin, Z., Chen, M., Ma, Y.: The augmented Lagrange multiplier method for exact recovery of corrupted low-rank matrices. Technical Report UILU-ENG-09-2215 (2009)
165. Ling, X., G.-R. Xue, G. -R., Dai, W., Jiang, Y., Yang, Q., Yu, Y.: Can Chinese Web pages be classified with English data source? In: Proceedings of the 17th International Conference on World Wide Web, pp. 969–978 (2008)
166. Liu, J., Wright, S.J., Re, C., Bittorf, V., Sridhar, S.: An asynchronous parallel stochastic coordinate descent algorithm. J. Mach. Learn. Res., **16**, 285–322 (2015)
167. Lu, J., Behbood, V., Hao, P., Zuo, H., Xue, S., Zhang, G.: Transfer learning using computational intelligence: a survey. Knowl. Based Syst. **80**, 14–23 (2015)
168. Luis, R., Sucar, L.E., Morales, E.F.: Inductive transfer for learning Bayesian networks. Mach. Learn. **79**(1–2), 227–255 (2010)
169. Luo, F.L., Unbehauen, R., Cichock, R.: A minor component analysis algorithm. Neural Netw. **10**(2), 291–297 (1997)
170. Ma, Y., Luo, G., Zeng, X., Chen, A.: Transfer learning for cross-company software defect prediction. Inform. Softw. Technol. **54**(3), 248–256 (2012)
171. Ma, Y., Gong, W., Mao, F.: Transfer learning used to analyze the dynamic evolution of the dust aerosol. J. Quant. Spectrosc. Radiat. Transf. **153**, 119–130 (2015)
172. Mahalanobis, P.C.: On the generalised distance in statistics. Proc. Natl. Inst. Sci. India **2**(1), 49–55 (1936)
173. Maier, M., von Luxburg, U., Hein, M.: How the result of graph clustering methods depends on the construction of the graph. ESAIM: Probab. Stat. **17**, 370–418 (2013)
174. Masci, J., Meier, U., Ciresan, D., Schmidhuber, J.: Stacked convolutional auto-encoders for hierarchical feature extraction. In: Proceedings of the 21st International Conference on Artificial Neural Networks, Part I, Espoo, pp. 52–59 (2011)
175. Massy, W.F.: Principal components regression in exploratory statistical research. J. Am. Stat. Assoc. **60**(309), 234–256 (1965)
176. McCallum, A., Nigam, K.: Employing EM and pool-based active learning for text classification. In: ICML '98: Proceedings of the Fifteenth International Conference on Machine Learning, pp. 359–367 (1998)
177. Michalski, R.: A theory and methodology of inductive learning. Mach. Learn. **1**, 83–134 (1983)
178. Miller, G.A., Nicely, P.E.: An analysis of perceptual confusions among some English consonants. J. Acoust. Soc. Am. **27**, 338–352 (1955)
179. Mishra, S.K.: Global optimization by differential evolution and particle swarm methods: Evaluation on some benchmark functions. Munich Research Papers in Economics (2006). Available at: https://mpra.ub.uni-muenchen.de/1005/
180. Mishra, S.K.: Performance of differential evolution and particle swarm methods on some relatively Harder multi-modal benchmark functions (2006). Available at: https://mpra.ub.uni-muenchen.de/449/
181. Mitchell, T.M.: Machine Learning, vol. 45. McGraw Hill, Burr Ridge (1997)
182. Mitra, P., Murthu, C.A., Pal, S.K.: Unsupervised feature selection using feature similarity. IEEE Trans. Pattern Anal. Mach. Intell. **24**(3), 301–312 (2002)
183. Mnih, V., et al.: Human-level control through deep reinforcement learning. Nature **518**, 529–533 (2015)

184. Mohar, B.: Some applications of Laplace eigenvalues of graphs. In: Hahn, G., Sabidussi, G. (eds.) Graph Symmetry: Algebraic Methods and Applications. NATO Science Series C, vol.497, pp. 225–275. Kluwer, Dordrecht (1997)
185. Moulton, C.M., Roberts, S.A., Calatn, P.H.: Hierarchical clustering of multiobjective optimization results to inform land-use decision making. URISA J. **21**(2), 25–38 (2009)
186. Murthy, C.A., Chowdhury, N.: In search of optimal clusters using genetic algorithms. Pattern Recog. Lett. **17**, 825–832 (1996)
187. Narayanan, H., Belkin, M., Niyogi, P.: On the relation between low density separation, spectral clustering and graph cuts. In: Schölkopf, B., Platt, J., Hoffman, T. (eds.) Advances in Neural Information Processing Systems, vol. 19, pp. 1025–1032. MIT Press, Cambridge (2007)
188. Nesterov, Y.: Efficiency of coordinate descent methods on huge-scale optimization problems. SIAM J. Optim. **22**(2), 341–362 (2012)
189. Ng, V., Vardie, C.: Weakly supervised natural language learning without redundant views. In: Proceedings of the Human Language Technology/Conference of the North American Chapter of the Association for Computational Linguistics (HLT-NAACL), Main Papers, pp. 94–101 (2003)
190. Ng, A., Jordan, M., Weiss, Y.: On spectral clustering: analysis and an algorithm. In: Dietterich, T., Becker, S., Ghahramani, Z. (eds.) Advances in Neural Information Processing Systems, vol. 14, pp. 849–856. MIT Press, Cambridge (2002)
191. Nguyen, H.D.: An introduction to Majorization-minimization algorithms for machine learning and statistical estimation. WIREs Data Min. Knowl. Discovery **7**(2), e1198 (2017)
192. Niculescu-Mizil, A., Caruana, R.: Inductive transfer for Bayesian network structure learning. In: Proceedings of the 11th International Conference on Artificial Intelligence and Statistics (AISTATS), San Juan (2007)
193. Nigam, K., Ghani, R.: Analyzing the effectiveness and applicability of co-training. In: Proceedings of the International Conference on Information and Knowledge Management (CIKM), pp. 86–93 (2000)
194. Oja, E., Karhunen, J.: On stochastic approximation of the eigenvectors and eigenvalues of the expectation of a random matrix. J. Math Anal. Appl. **106**, 69–84 (1985)
195. Ogoe, H.A., Visweswaran, S., Lu, X., Gopalakrishnan, V.: Knowledge transfer via classification rules using functional mapping for integrative modeling of gene expression data. BMC Bioinform. **16**, 1–15 (2015)
196. Oquab, M., Bottou, L., Laptev, I.: Learning and transferring mid-level image representations using convolutional neural networks. In: Proceedings of the IEEE Conference on Computer Vision and Pattern Recognition, pp. 1717–1724 (2014)
197. Ortega, J.M., Rheinboldt, W.C.: Iterative Solutions of Nonlinear Equations in Several Variables, pp. 253–255. Academic, New York (1970)
198. Owen, A.B.: A robust hybrid of lasso and ridge regression. Prediction and Discovery (Contemp. Math.), **443**, 59–71 (2007)
199. Pan, S.J., Yang, Q.: A survey on transfer learning. IEEE Trans. Knowl. Data Eng. **22**(10), 1345–1359 (2010)
200. Pan, S.J., Kwok, J.T., Yang, Q., Pan, J.J.: Adaptive localization in a dynamic WiFi environment through multi-view learning. In: Proceedings of the 22nd Association for the Advancement of Artificial Intelligence (AAAI) Conference Artificial Intelligence, pp. 1108–1113 (2007)
201. Pan, S.J., Kwok, J.T., Yang, Q.: Transfer learning via dimensionality reduction. In: Proceedings of the 23rd National Conference on Artificial Intelligence, vol. 2, pp. 677–682 (2008)
202. Pan, S.J., Shen, D., Yang, Q., Kwok, J.T.: Transferring localization models across space. In: Proceedings of the 23rd Association for the Advancement of Artificial Intelligence (AAAI) Conference on Artificial Intelligence, pp. 1383–1388 (2008)
203. Pan, S.J., Tsang, I.W., Kwok, J.T, Yang, Q.: Domain adaptation via transfer component analysis. IEEE Trans. Neural Netw. **22**(2), 199–210 (2011)

204. Parra, L., Spence, C., Sajda, P., Ziehe, A., Muller, K.: Unmixing hyperspectral data. In: Advances in Neural Information Processing Systems, vol. 12, pp. 942–948. MIT Press, Cambridge (2000)
205. Patel, V.M, Gopalan, R., Li, R., Chellappa, R.: Visual domain adaptation: a survey of recent advances. IEEE Signal Process. Mag. **32**(3), 53–69 (2015)
206. Polikar, R.: Ensemble based systems in decision making. IEEE Circ. Syst. Mag. **6**(3), 21–45 (2006)
207. Prettenhofer, P., Stein, B.: Cross-language text classification using structural correspondence learning. In: Proceedings of the 48th Annual Meeting of the Association for Computational Linguistics, pp. 1118–1127 (2010)
208. Price, W.L.: A controlled random search procedure for global optimisation. Comput. J. **20**(4), 367–370 (1977)
209. Rahnamayan, S., Tizhoosh, H.R., Salama, N.M.M.: Opposition-based differential evolution. IEEE Trans. Evol. Comput. **12**(1), 64–79 (2008)
210. Raina, R., Battle, A., Lee, H., Packer, B., Ng, A.Y.: Self-taught learning: Transfer learning from unlabeled data. In: Proceedings of the 24th International Conference on Machine Learning, Corvallis, pp. 759–766 (2007)
211. Rajagopal, A.N., Subramanian, R., Ricci, E., Vieriu, R.L., Lanz, O., Ramak-rishnan, K.R., Sebe, N.: Exploring transfer learning approaches for head pose classification from multi-view surveillance images. Int. J. Comput. Vis. **109**(1–2), 146–167 (2014)
212. Richtárik, P., Takáč M.: Parallel coordinate descent methods for big data optimization. Math. Program. Ser. A **156**, 433–484 (2016)
213. Rivli, J.: An Introduction to the Approximation of Functions. Courier Dover Publications, New York (1969)
214. Robbins, H., Monro, S.: A stochastic approximation method. Ann. Math. Stat. **22**, 400–407 (1951)
215. Rosipal, R., Krämer, N.: Overview and recent advances in partial least squares. In: Proceedings of the Workshop on Subspace, Latent Structure and Feature Selection (SLSFS) 2005, pp. 34–51 (2006)
216. Roweis, S., Saul, L.: Nonlinear dimensionality reduction by locally linear embedding. Science **290**(5500), 2323–2326 (2000)
217. Roy, D.M., Kaelbling, L.P.: Efficient Bayesian task-level transfer learning. In: Proceedings of the 20th International Joint Conference on Artificial Intelligence, Hyderabad, pp. 2599–2604 (2007)
218. Rummery, G.A., Niranjan, M.: On-line Q-learning using connectionist systems. Technical Report CUED/F-INFENG/TR 166, Cambridge University (1994)
219. Saenko, K., Kulis, B., Fritz, M., Darrell, T.: Adapting visual category models to new domains. In: Proceedings of the European Conference on Computer Vision, vol. 6314, pp. 213–226 (2010)
220. Schaal, S.: Is imitation learning the route to humanoid robots? Trends Cogn. Sci. **3**(6), 233–242 (1999)
221. Schapire, R.E.: The strength of weak learnability. Mach. Learn. **5**, 197–227 (1990)
222. Schmidt, M., Le Roux, N., Bach, F.: Minimizing finite sums with the stochastic average gradient. Technical Report, INRIA, hal-0086005 (2013). See also Math. Program. **162**, 83–112 (2017)
223. Schwefel, H.P.: Numerical Optimization of Computer Models. Wiley, Hoboken (1981)
224. Settles, B., Craven, M., Friedland, L.: Active learning with real annotation costs. In: Proceedings of the NIPS Workshop on Cost-Sensitive Learning, pp. 1–10 (2008)
225. Settles, B., Craven, M., Ray, S.: Multiple-instance active learning. In: Advances in Neural Information Processing Systems (NIPS), vol.20, pp. 1289–1296, MIT Press, Cambridge (2008)
226. Seung, H.S., Opper, M., Sompolinsky, H.: Query by committee. In: Proceedings of the ACM Workshop on Computational Learning Theory, pp. 287–294 (1992)

227. Shell, J., Coupland, S.: Towards fuzzy transfer learning for intelligent environments. Ambient. Intell. Lect. Notes Comput. Sci. **7683**, 145–160 (2012)
228. Shell, J., Coupland, S.: Fuzzy transfer learning: Methodology and application. Inform. Sci. **293**, 59–79 (2015)
229. Shen, H., Tan, Y., Lu, J., Wu, Q., Qiu, Q.: Achieving autonomous power management using reinforcement learning. ACM Trans. Des. Autom. Electron. Syst. **18**(2), 24:1–24:32 (2013)
230. Shi, J., Malik, J.: Normalized cuts and image segmentation. IEEE Trans. Pattern Anal. Mach. Intell. **22**(8), 888–905 (2000)
231. Shuman, D.I., Vandergheynst, P., Frossard, P.: Chebyshev polynomial approximation for distributed signal processing. In: Proceedings of the International Conference on Distributed Computing in Sensor Systems, Barcelona, pp. 1–8 (2011)
232. Shuman, D.I., Narang, S.K, Frossard, P., Ortega, A., Vandergheynst, P.: Extending high-dimensional data analysis to networks and other irregular domains. IEEE Signal Process. Mag. **30**(3), 83–98 (2013)
233. Silver, D.L., Mercer, R.E.: The parallel transfer of task knowledge using dynamic learning rates based on a measure of relatedness. In: Thrun, S., Pratt, L.Y. (eds.) Learning to Learn, pp. 213–233. Kluwer Academic, Boston (1997)
234. Sindhwani, V., Niyogi, P., Belkin, M.: Beyond the point cloud: From transductive to semi-supervised learning. In: Proceedings of the 22nd International Conference on Machine Learning (ICML), pp. 824–831. ACM, New York (2005)
235. Smola, J., Kondor, R.: Kernels and regularization on graphs. In: Learning Theory and Kernel Machines, pp. 144–158. Springer, Berlin (2003)
236. Song, J., Babu, P., Palomar, D.P.: Optimization methods for designing sequences with low autocorrelation sidelobes. IEEE Trans. Signal Process. **63**(15), 3998–4009 (2015)
237. Sriperumbudur, B.K., Torres, D.A., Lanckriet, G.R.G.: A majorization-minimization approach to the sparse generalized eigenvalue problem. Mach. Learn. **85**, 3–39 (2011)
238. Sun, J., Zeng, H., Liu, H., Lu, Y., Chen, Z.: CubeSVD: a novel approach to personalized web search. In: Proceedings of the 14th International Conference on World Wide Web, pp. 652–662 (2005)
239. Sutton, R.S., Barto, A.G.: Reinforcement Learning: An Introduction. Adaptive Computation and Machine Learning Series. MIT Press, Cambridge (1998)
240. Tang, K., Li, X., Suganthan, P.N., Yang, Z., Weise, T.: Benchmark functions for the CEC'2010 special session and competition on large-scale global optimization. Technical Report, 2009. Available at: https://www.researchgate.net/publication/228932005
241. Tang, J., Qu, M., Wang, M., Zhang, M., Yan, J.,, Mei, Q.: LINE: Large-scale information network embedding. In: Proceedings of the International World Wide Web Conference Committee (IW3C2), Florence, pp. 1067–1077 (2015)
242. Tao, D., Li, X., Wu, X., Hu, W., Maybank, S.J.: Supervised tensor learning. Knowl. Inform. Syst. **13**, 1–42 (2007)
243. Thrun, S., Pratt, L. (eds.): Learning to Learn. Kluwer Academic, Dordrecht (1998)
244. Tibshirani, R.: Regression shrinkage and selection via the lasso. J. R. Statist. Soc. B **58**, 267–288 (1996)
245. Tibshirani, R., Walther, G., Hastie, T.: Estimating the number of clusters in a data set via the gap statistic. J. Roy. Statist. Soc. B **63**(2), 411–423 (2001)
246. Tikhonov, A.: Solution of incorrectly formulated problems and the regularization method. Soviet Math. Dokl., **4**, 1035–1038 (1963)
247. Tikhonov, A.N., Arsenin, V.Y.: Solutions of Ill-Posed Problems. Wiley, New York (1977)
248. Tokic, M., Palm, G.: Value-difference based exploration: Adaptive control between epsilon-greedy and softmax. In: KI 2011: Advances in Artificial Intelligence, pp. 335–346 (2011)
249. Tommasi, T., Orabona, F., Caputo, B.: Safety in numbers: learning categories from few examples with multi model knowledge transfer. In: Proceedings of the IEEE Conference on Computer Vision Pattern Recognition 2010, pp. 3081–3088 (2010)
250. Tong, S., Koller, D.: Support vector machine active learning with applications to text classification. J. Mach. Learn. Res. **3**, 45–66 (2001)

251. Tou, J.T., Gonzalez, R.C.: Pattern Recognition Principles. Addison-Wesley, London (1974)
252. Tsitsiklis, J.N.: Asynchronous stochastic approximation and Q-Learning. Mach. Learn. **16**, 185–202 (1994)
253. Uurtio, V., Monteiro, J.M., Kandola, J., Shawe-Taylor, J., Fernandez-Reyes, D., Rousu, J.: A tutorial on canonical correlation methods. ACM Comput. Surv. **50**(95), 14–38 (2017)
254. Valiant, L.G.: A theory of the learnable. Commun. ACM **27**, 1134–1142 (1984)
255. van Hasselt, H.: Double Q-learning. In: Proceedings of the Advances in Neural Information Processing Systems (NIPS), pp. 2613–2621 (2010)
256. Vasilescu, M.A.O., Terzopoulos, D.: Multilinear analysis of image ensembles: TensorFaces. In: Proceedings of the European Conference on Computer Vision, Copenhagen, pp. 447–460 (2002)
257. von Luxburg, U.: A tutorial on spectral clustering. Stat. Comput. **17**(4), 395–416 (2007)
258. Wang, H., Ahuja, N.: Compact representation of multidimensional data using tensor rank-one decomposition. In: Proceedings of the International Conference on Pattern Recognition, vol. 1, pp. 44–47 (2004)
259. Wang, X., Qian, B., Davidson, I.: On constrained spectral clustering and its applications. Data Min. Knowl. Disc. **28**, 1–30 (2014)
260. Wang, L., Hua, X., Yuan, B., Lu, J.: Active learning via query synthesis and nearest neighbour search. Neurocomputing **147**, 426–434 (2015)
261. Wang, D., Cui, P., Zhu, W.: Structural deep network embedding. In: Proceedings of the 22nd International Conference on Knowledge Discovery and Data Mining, pp. 1225–1234. ACM, New York (2016)
262. Watanabe, S.: Pattern Recognition: Human and Mechanical. Wiley, New York (1985)
263. Watldns, C.J.C.H.: Learning from delayed rewards. PhD Thesis, University of Cambridge, England (1989)
264. Watkins, C.J.C.H., Dayan, R.: Q-learning. Mach. Learn. **8**, 279–292 (1992)
265. Weenink, D.: Canonical correlation analysis. IFA Proc. **25**, 81–99 (2003)
266. Wei, X.-Y., Yang, Z.-Q.: Coached active learning for interactive video search. In: Proceedings of the ACM International Conference on Multimedia, pp. 443–452 (2011)
267. Wei, X., Cao, B. Yu, P.S.: Nonlinear joint unsupervised feature selection. In: Proceedings of the 2016 SIAM International Conference on Data Mining, pp. 414–422 (2016)
268. Weiss, K., Khoshgoftaar, T.M., Wang, D.D.: A survey of transfer learning. J. Big Data **3**(9), 1–40 (2016)
269. Williams, R.J.: Simple statistical gradient-following algorithms for connectionist reinforcement learning. Mach. Learn. **8**, 229–256 (1992)
270. Witten, I.H., Frank, E.: Data Mining: Practical Machine Learning Tools and Techniques, 3rd edn. Morgan Kaufmann, San Mateo (2011)
271. Witten, D.M., Tibshirani, R., Hastie, T.: A penalized matrix decomposition, with applications to sparse principal components and canonical correlation analysis. Biostatistics **10**(3), 515–534 (2009)
272. Wright, J., Ganesh, A., Rao, S., Peng, Y., Ma, Y.: Robust principal component analysis: exact recovery of corrupted low-rank matrices via convex optimization. In: Proceedings of the Advances in Neural Information Processing Systems, vol. 87, pp. 20:3–20:56 (2009)
273. Wright, J., Ganesh, A., Yang, A.Y., Ganesh, A., Sastry, S., Ma, Y.: Robust face recognition via sparse representation. IEEE Trans. Pattern Reconginit. Mach. Intell. **31**(2), 210–227 (2009)
274. Wold, H.: Path models with latent variables: The NIPALS approach. In: Blalock, H.M., et al. (eds.) Quantitative Sociology: International Perspectives on Mathematical and Statistical Model Building, pp. 307–357. Academic, Cambridge (1975)
275. Wold, S., Sjöström, M., Eriksson, L.: PLS-regression: a basic tool of chemometrics. Chemom. Intell. Lab. Syst. **58**(2), 109–130 (2001)
276. Wooldridge, M.J., Jennings, N.R.: Intelligent agent: theory and practice. Knowl. Eng. Rev. **10**(2), 115–152 (1995)
277. Wu, P., Dietterich, T.G.: Improving SVM accuracy by training on auxiliary data sources. In: Proceedings of the Twenty-First International Conference on Machine Learning, pp. 871–878 (2004)

278. Wu, T.T., Lange, K.: The MM alternative to EM. Statist. Sci. **25**(4), 492–505 (2010)
279. Wu, Z., Leahy, R.: An optimal graph theoretic approach to data clustering: Theory and its application to image segmentation. IEEE Trans. Pattern Anal. Mach. Intell. **15**(11), 1101–1113 (1993)
280. Xia, R., Zong, C., Hu, X., Cambria, E.: Feature ensemble plus sample selection: domain adaptation for sentiment classification. IEEE Intell. Syst. **28**(3), 10–18 (2013)
281. Xu, L., Krzyzak, A., Suen, C.Y.: Methods of combining multiple classifiers and their applications to handwriting recognition. IEEE Trans. Syst. Man Cybern. **22**, 418–435 (1992)
282. Xu, L., Oja, E., Suen, C.: Modified Hebbian learning for curve and surface fitting. Neural Netw. **5**, 441–457 (1992)
283. Xu, H., Caramanis, C., Mannor, S.: Robust regression and Lasso. IEEE Trans. Inform. Theory **56**(7), 3561–3574 (2010)
284. Xue, B., Zhang, M., Browne, W.N., Yao, X.: A survey on evolutionary computation approaches to feature selection. IEEE Trans. Evol. Comput. **20**(4), 606–626 (2016)
285. Yamauchi, K.: Covariate shift and incremental learning. In: Advances in Neuro-Information Processing, pp. 1154–1162. Springer, Berlin (2009)
286. Yan, S., Wang, H.: Semi-supervised Learning by sparse representation. In: Proceedings of the SIAM International Conference on Data Mining, Philadelphia, pp. 792–801 (2009)
287. Yang, B.: Projection approximation subspace tracking. IEEE Trans. Signal Process. **43**, 95–107 (1995)
288. Yen, T.-J.: A majorization-minimization approach to variable selection using spike and slab priors. Ann. Stat. **39**(3), 1748–1775 (2011)
289. Yin, J., Yang, Q., Ni, L.M.: Adaptive temporal radio maps for indoor location estimation. In: Proceedings of the Third IEEE International Conference on Pervasive Computing and Communications (2005)
290. Yu, K., Zhang, T., Gong, Y.: Nonlinear learning using local coordinate coding. In: Advances in Neural Information Processing Systems, vol. 22, pp. 2223–2231 (2009)
291. Yu, H., Sun, C., Yang, W., Yang, X., Zuo, X.: AL-ELM: One uncertainty-based active learning algorithm using extreme learning machine. Neurocomputing **166**(20), 140–150 (2015)
292. Yuan, M., Lin, Y.: Model selection and estimation in regression with grouped variables. J. Roy. Stat. Soc. Ser. B **68**, 49–67 (2006)
293. Yuan, G.-X., Ho, C.-H., Lin, C.-J.: Recent advances of large-scale linear classification. Proc. IEEE **100**(9), 2584–2603 (2012)
294. Zhang, X.D.: Matrix Analysis and Applications. Cambridge University Press, Cambridge (2017)
295. Zhang, Z., Coutinho, E., Deng, J., Schuller, B.: Cooperative learning and its application to emotion recognition from speech. IEEE Trans. Audio Speech Lang. Process. **23**(1), 115–126 (2015)
296. Zhang, Z., Pan, Z., Kochenderfer, M.J.: Weighted Double Q-learning. In: Proceedings of the Twenty-Sixth International Joint Conference on Artificial Intelligence (IJCAI-17), pp. 3455–3461 (2017)
297. Zheng, V.W., Pan, S.J., Yang, Q., Pan, J.J.: Transferring multi-device localization models using latent multi-task learning. In: Proceedings of the 23rd Association for the Advancement of Artificial Intelligence (AAAI) Conference on Artificial Intelligence, pp. 1427–1432 (2008)
298. Zheng, V.W., Yang, Q., Xiang, W., Shen, D.: Transferring localization models over time. In: Proceedings of the 23rd Association for the Advancement of Artificial Intelligence (AAAI) Conference on Artificial Intelligence, pp. 1421–1426 (2008)
299. Zhou, Y., Goldman, S.: Democratic co-learning. In: Proceedings of the 16th IEEE International Conference on Tools with Artificial Intelligence (ICTAI), pp. 594–602 (2004)
300. Zhou, Z.-H., Li, M.: Tri-training: exploiting unlabeled data using three classifiers. IEEE Trans. Knowl. Data Eng. **17**, 1529–1541 (2005)
301. Zhou, D., Schölkopf, B.: A regularization framework for learning from graph data. In: Proceedings of the ICML Workshop on Statistical Relational Learning, pp. 132–137 (2004)

302. Zhou, J., Chen, J., Ye, J.: Multi-task learning: Theory, algorithms, and applications (2012). Available at: https://archive.siam.org/meetings/sdm12/zhou_-chen_-ye.pdf
303. Zhou, Z.-H., Zhan, D.-C., Yang, Q.: Semi-supervised learning with very few labeled training examples. In: Proceedings of the Twenty-Second AAAI Conference on Artificial Intelligence (AAAI-07) (2007)
304. Zhu, X.: Semi-Supervised Learning Literature Survey. Computer Sciences TR 1530, University of Wisconsin, Madison, (2005)
305. Zhu, X., Goldberg, A.B.: Introduction to Semi-Supervised Learning. In: Brachman, R.J., Dietterich, T. (eds.) Synthesis Lectures on Artificial Intelligence and Machine Learning. Morgan & Claypoo, San Rafael (2009)
306. Zhu, X., Ghahramani, Z., Laffer, J.: Semi-supervised learning using Gaussian fields and harmonic functions. In: Proceedings of the Twentieth International Conference on Machine Learning (ICML-2003), Washington (2003)
307. Zou, H., Hastie, T.: Regularization and variable selection via the elastic net. J. Roy. Stat. Soc. B, **67**(2), 301–320 (2005)
308. Zou, H., Hastie,, T., Tibshirani, R.: Sparse principal component analysis. J. Comput. Graph. Stat. **15**(2), 265–286 (2006)

Chapter 7
Neural Networks

Neural networks can be viewed as a kind of cognitive intelligence. This chapter presents a neural network tree, and then deals with the optimization problem in neural networks, activation functions, and basic neural networks from the perspective of matrix algebra. The topical subjects of this chapter are selected topics and advances in neural networks: convolutional neural networks (CNNs), dropout learning, autoencoder, extreme learning machine (ELM), graph embedding, manifold learning, network embedding, graph neural networks (GNNs), batch normalization networks, and generative adversarial networks (GANs).

7.1 Neural Network Tree

A *neural network* is viewed as an adaptive machine, and its good definition given by Haykin [52] is:

> A neural network is a massively parallel distributed processor that has a natural propensity for storing experimental knowledge and making it available for use. It resembles the brain in two respects:
>
> 1. Knowledge is acquired by the network through a learning process.
> 2. Interneuron connection strengths known as synaptic weights are used to store the knowledge.

Machine learning learns how to associate an input with some output, given a training set of examples of inputs and outputs. In Chap. 6 we focused on using algorithm forms to implement machine learning. In this chapter we will discuss how to use neural networks with layered structures for implementing machine learning.

For a neural network, the most important thing is how to get the model parameters of each layer according to the data provided in the training set, so as to minimize the loss. Because of its strong nonlinear fitting ability, it has important applications in various fields.

As described in Chap. 6, according to data structure, machine learning can be divided into two broad categories: "regular or Euclidean structured data learning" and "irregular or non-Euclidean structured data learning." Similarly, neural networks (NNs) can be divided into two broad categories: NNs for regular or Euclidean structured model learning and NNs for irregular or non-Euclidean structured model learning.

1. *Neural networks for Euclidean structured model learning*

    - *Neural networks for stochastic (probabilistic) model learning*

        – *Bayesian neural networks:* Their core idea is to minimize KL divergence. Bayesian neural networks can be easily learned from small data sets. Bayesian methods provide further uncertainty estimation in the form of a probability distribution through its parameters. At the same time, the prior probability distribution is used to integrate the parameters to provide a regularization effect for the network, so as to prevent overfitting.
        – *Boltzmann machine:* The basic idea of the popular Boltzmann machine and restricted Boltzmann machine is that with the introduction of hidden units in the Hopfield network, the model can be upgraded from the initial ideas of the Hopfield network and evolve to replace it. The model of Boltzmann machine splits into two parts: visible units and hidden units.
        – *Generative adversarial networks (GANs):* GANs can be used for unsupervised learning of rich feature representations for arbitrary data distributions. GANs learn reusable feature representations from large unlabeled data sets. GANs have emerged as a powerful framework for learning generative models of arbitrarily complex data distributions. The GAN plays an adversarial game with two linked models: *Generative Model* and *Discriminative Model*.

    - *Neural networks for deterministic model learning*

        – *Multilayer Perceptrons (MLPs)*, also known as *Feedforward Neural Networks*, aim at working as an approximator of the mapping function from inputs to outputs. The approximated function is usually built by stacking together several hidden layers that are activated in chain to obtain the desired output.
        – *Recurrent Neural Networks (RNNs)*, in contrast to MLPs, are models in which the output is a function of not only the current inputs but also of the previous outputs, which are encoded into a hidden state **h**. Due to having memory of the previous outputs, RNNs therefore can encode the information present in the sequence itself, something that MLPs cannot do. As a consequence, this type of model can be very useful to learn from sequential data.
        – *Convolutional Neural Networks (CNNs)* are a specific type of models conceived to accept two-dimensional input data, such as images or time series data. The output of the convolution operation is usually run through

a nonlinear activation function and then further modified by means of a pooling function, which replaces the output in a certain location with a value obtained from nearby outputs.
- *Dropout networks* are a form of regularization for fully connected neural network layers and is a very efficient version of model combination through setting the output of each hidden neuron to zero with probability q while training. Moreover, overfitting is a serious problem in deep neural networks with a large number of parameter to be learned. This overfitting can be greatly reduced by using dropout.
- *Autoenconder* is an MLP with symmetric structure and used to learn an efficient representation (encoding) for a set of data in an unsupervised manner. The aim of an autoencoder is to make the output as similar to the input as possible. Autoencoders are unsupervised learning models.

2. *Neural networks for non-Euclidean structured model learning*

- *Graph embedding:* Embedding a graph or an information network into a low-dimensional space is useful in a variety of applications. Given the inputs of a graph $G = (V, E)$, and a predefined dimensionality of the embedding d ($d \ll |V|$), the graph embedding is to convert G into a d-dimensional space $\mathbb{R}^d$. In this space, the graph properties (such as the first-order, second-order, and higher-order proximities) are preserved as much as possible. The graph is represented as either a d-dimensional vector (for a whole graph) or a set of d-dimensional vectors with each vector representing the embedding of part of the graph (e.g., node, edge, substructure).
- *Network embedding:* Given a graph denoted as $G = (V, E)$, *network embedding* aims to learn a mapping function $f : \mathbf{v}_i \to \mathbf{y}_i \in \mathbb{R}^d$, where $d \ll |V|$. The objective of the function is to make the similarity between $\mathbf{y}_i$ and $\mathbf{y}_j$ explicitly preserve the first-order, second-order, and higher-order proximities of $\mathbf{v}_i$ and $\mathbf{v}_j$.
- *Neural networks on graph:* Neural networks in graph domains are divided into semi-supervised networks (Graph neural networks (GNNs) and graph convolutional networks (GCNs)) and unsupervised networks (graph autoencoders).

    - *Graph neural networks (GNNs)* are a kind of neural network which processes directly the data represented in graph domain. GNNs can capture the dependence of graphs and rich relation information among elements via message passing between nodes of a graph.
    - *Graph convolutional networks (GCNs)*, introduced by Kipf and Welling [82], are a machine learning method of graph structured data through extracting spatial features. Because the standard convolution for image or text cannot be directly applied to graphs without a grid structure, it is necessary to transform graphs into another spectral domain with a grid structure.
    - *Graph autoencoders (GAEs):* This is a TensorFlow implementation of the (Variational) graph autoencoder model [81].

```
                                              ┌ Bayesian neural networks
                         ┌ NNs for stochastic ┤ Boltzmann machine
                         │ model learning     └ Generative adversarial networks
         ┌ NNs with Euclidean
         │ structured model    ┌ Multilayer perceptrons
         │           ┌ NNs for │ Recurrent neural networks
         │           │ deterministic ┤ Convolutional neural networks
Neural  ─┤           │ model learning│ Dropout networks
networks │           └               └ Autoencoders
         │                      ┌ Graph embedding
         │                      │ Network embedding
         └ NNs with non-Euclidean ┤
           structured model     │                  ┌ Graph neural networks
                                │ Neural Networks  ┤ Graph convolutional networks
                                └ on graph         └ Graph autoencoders
```

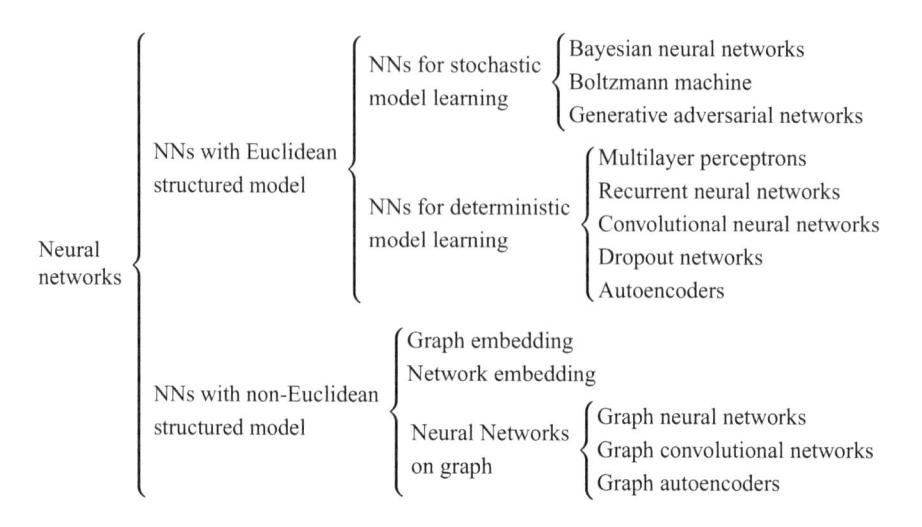

Fig. 7.1 Neural network tree

The above classification of neural networks can be vividly represented by a neural network tree, as shown in Fig. 7.1.

7.2 From Modern Neural Networks to Deep Learning

A standard neural network (NN) consists of many simple, connected neurons, each producing a sequence of real-valued activations. Modern artificial neural networks are divided into shallow and deep neural networks. Shallow neural networks typically consists of three layers (one input layer, one or a few hidden layers, and one output layer), while deep neural networks contain many hidden layers. Shallow neural networks can only capture shallow features in data, while deep neural networks can provide a deep understanding of features in data.

The following are major milestones of modern neural network and deep learning research [130, 155]:

- McCulloch and Pitts in 1943 [105] introduced MCP Model, which is considered to be the ancestor of Artificial Neural Model.
- Donald Hebb in 1949 [57] is considered to be the father of neural networks, as he introduced Hebbian Learning Rule, which lays the foundation for modern neural networks.
- Frank Rosenblatt in 1958 [121] introduced the first perceptron, which highly resembles the modern perceptron.

- Paul Werbos in 1975 [158] mentioned the possibility of applying Backpropagation to artificial neural networks.
- Kunihiko Fukushima in 1980 [32] introduced a Neocognitron, which inspired the Convolutional Neural Network, and Teuvo Kohonen in 1981 [83] introduced the Self Organizing Map principal.
- John Hopfield in 1982 [65] introduced the Hopfield Network.
- Ackley et al. in 1985 [1] invented the Boltzmann Machine.
- In 1986 Paul Smolensky [139] introduced Harmonium, which was later known as the Restricted Boltzmann Machine, and Michael I. Jordan [75] defined and introduced the Recurrent Neural Network.
- Le Cun in 1990 introduced LeNet, showing the possibility of deep neural networks in practice.
- In 1991 Hochreiter in his diploma thesis represented a milestone of explicit Deep Learning research.
- In 1997 Schuster and Paliwal [132] introduced the Bidirectional Recurrent Neural Network, and Hochreiter and Schmidhuber [63] introduced the LSTM, which solved the problem of vanishing gradients in recurrent neural networks.
- Geoffrey Hinton in 2006 [61] introduced Deep Belief Networks and the layer-wise pretraining technique, opening the path to the current deep learning era.
- Salakhutdinov and Hinton in 2009 [124] introduced Deep Boltzmann Machines.
- In 2012 Geoffrey Hinton [62] introduced Dropout, an efficient way of training neural networks, and Krizhevsky et al. [84] invented AlexNet, starting the era of CNN used for ImageNet classification.
- In 2016 Goodfellow, Bergio, and Courville published the comprehensive book on deep learning [39].

The universal approximation properties of shallow neural networks come at a price of exponentially many neurons and therefore are not realistic.

7.3 Optimization of Neural Networks

Consider a (time) sequence of input data vectors $X = \{\mathbf{x}_1, \ldots, \mathbf{x}_N\}$ and a sequence of corresponding output data vectors $Y = \{\mathbf{y}_1, \ldots, \mathbf{y}_N\}$ with neighboring data-pairs (in time) being somehow statistically dependent. Given time sequences X and Y as training data, the aim of a neural network is to learn the rules to predict the output data $\mathbf{y}$ given the input data $\mathbf{x}$. Inputs and outputs can, in general, be continuous and/or categorical variables. When outputs are continuous, the problem is known as a *regression problem*, and when they are categorical (class labels), the problem is known as a *classification problem*. In the context of neural networks, the term prediction is usually used as a general term including regression and classification.

The convex optimization in machine learning problems are offline convex optimizations: Given a set of labeled data $(\mathbf{x}_i, y_i), i = 1, \ldots, N$ or unlabeled data $\mathbf{x}_1, \ldots, \mathbf{x}_N$, solve the optimization problem $\min_{\mathbf{x} \in F} f(\mathbf{x})$ through a fixed regularization function such as ℓ_2-squared and modify it only via a single time-dependent parameter. However, in many practical applications in science and engineering, we are faced with online convex optimization problems that require optimization algorithms to adaptively choose its regularization function based on the loss functions observed so far.

7.3.1 Online Optimization Problems

Given a convex set $F \subseteq \mathbb{R}^n$ and a convex cost function $f : F \to \mathbb{R}$. Convex optimization involves finding a point $\mathbf{x}^*$ in F which minimizes f. In *online convex optimization*, the convex set F is known in advance, but in each step of some repeated optimization problem, one must select a point in F before seeing the cost function for that step. Let $\mathbf{x} \in F$ be the feasible decision vector lie in the convex feasible set F.

Definition 7.1 (Convex Optimization Problem [177]) Convex optimization consists of a convex feasible set F and a convex cost function $f(\mathbf{x}) : F \to \mathbb{R}$. The optimal solution is the solution that minimizes the cost.

Definition 7.2 (Online Convex Optimization Problem [177]) Online convex optimization consists of a feasible set $F \subseteq \mathbb{R}^n$ and an infinite sequence $\{f_1, f_2, \ldots\}$, where each $f_t : F \to \mathbb{R}$ is a convex function at the time step t.

At each time step t, an online convex optimization algorithm selects a vector $\mathbf{x}_t \in F$. After the vector is selected, it receives the cost function f_t.

Definition 7.3 (Projection) For the distance function $d(\mathbf{x}, \mathbf{y}) = \|\mathbf{x} - \mathbf{y}\|$ between $\mathbf{x} \in F$ and $\mathbf{y}$, the projection of a point $\mathbf{y}$ onto F is defined as [177]:

$$P(\mathbf{y}) = \arg\min_{\mathbf{x} \in F} d(\mathbf{x}, \mathbf{y}) = \arg\min_{\mathbf{x} \in F} \|\mathbf{x} - \mathbf{y}\|. \tag{7.3.1}$$

Especially, standard subgradient algorithms move the predictor $\mathbf{x}_t$ in the opposite direction of the subgradient vector $\mathbf{g}_t$, while maintaining $\mathbf{x}_{t+1} \in F$ via the *projected gradient update* [27, 177]:

$$\mathbf{x}_{t+1} = P(\mathbf{x}_t - \eta \mathbf{g}_t) = \arg\min_{\mathbf{x} \in F} \|\mathbf{x} - (\mathbf{x}_t - \eta \mathbf{g}_t)\|_2^2. \tag{7.3.2}$$

If taking the Mahalanobis norm $\|\mathbf{x}\|_A = \langle \mathbf{x}, \mathbf{A}\mathbf{x}\rangle^{1/2}$ and denoting, by $P_A(\mathbf{y})$, the projection of a point $\mathbf{y}$ onto F according to $\mathbf{A}$, then the greedy projection algorithm becomes

$$P_A(\mathbf{y}) = \arg\min_{\mathbf{x} \in F} \langle \mathbf{x} - \mathbf{y}, \mathbf{A}(\mathbf{x} - \mathbf{y})\rangle. \tag{7.3.3}$$

7.3 Optimization of Neural Networks

Select an arbitrary $\mathbf{x}_1 \in F$ as an initial value and a sequence of learning rates $\eta_1, \eta_2, \ldots \in \mathbb{R}_+$ (nonnegative quadrant). In time step t, after receiving a cost function $f_t(\mathbf{x}_t)$, the *greedy projection algorithm* selects the next vector $\mathbf{x}_{t+1}$ as follows [177]:

$$\mathbf{x}_{t+1} = P\left(\mathbf{x}_t - \eta_t \partial f_t(\mathbf{x}_t)\right) = \arg\min_{\mathbf{x} \in F} \|\mathbf{x} - (\mathbf{x}_t - \eta_t \partial f_t(\mathbf{x}_t))\|, \quad (7.3.4)$$

or

$$\mathbf{x}_{t+1} = P_A\left(\mathbf{x}_t - \eta_t \partial f_t(\mathbf{x}_t)\right)$$
$$= \arg\min_{\mathbf{x} \in F} \langle \mathbf{x} - (\mathbf{x}_t - \eta_t \partial f_t(\mathbf{x}_t)), \mathbf{A}(\mathbf{x} - (\mathbf{x}_t - \eta_t \partial f_t(\mathbf{x}_t)))\rangle. \quad (7.3.5)$$

Definition 7.4 (Cost, Regret [177]) Given an algorithm A and a convex optimization problem $(F, \{f_1, f_2, \ldots\})$, if $\{\mathbf{x}_1, \mathbf{x}_2, \ldots\}$ are the vectors selected by A, then the *cost* of A until time T is

$$C_A(T) = \sum_{t=1}^{T} f_t(\mathbf{x}_t). \quad (7.3.6)$$

The cost of a static feasible solution $\mathbf{x} \in F$ until time T is given by

$$C_x(T) = \sum_{t=1}^{T} f_t(\mathbf{x}). \quad (7.3.7)$$

The *regret* of algorithm A until time T is denoted by $R_A(T)$ and is defined as

$$R_A(T) = C_A(T) - \min_{\mathbf{x} \in F} C_x(T) = \sum_{t=1}^{T} f_t(\mathbf{x}_t) - \min_{\mathbf{x} \in F} \sum_{t=1}^{T} f_t(\mathbf{x}). \quad (7.3.8)$$

The goal of online optimization is to minimize the regret via adaptively choosing the regularization function based on the loss functions observed so far.

7.3.2 Adaptive Gradient Algorithm

In many applications of online and stochastic learning, the input data are of very high dimension, yet within any particular high-dimensional data only a few features are nonzero. Standard stochastic subgradient methods largely follow a predetermined procedural scheme that is oblivious to the characteristics of the data being observed.

Consider the minimization of composite functions [27]:

$$\min_{\mathbf{x} \in F} \{\phi_t(\mathbf{x}) = f_t(\mathbf{x}) + \varphi(\mathbf{x})\}, \qquad (7.3.9)$$

where f_t and φ are (closed) convex functions. In machine learning setting, f_t is either an instantaneous loss or a stochastic estimate of the objective function in an optimization task. The function φ serves as a fixed regularization function and is typically used to control the complexity of $\mathbf{x}$. At each round the algorithm makes a prediction $\mathbf{x}_t \in F$ and then receives the function f_t.

Define the regret with respect to the fixed (optimal) predictor $\mathbf{x}^*$ as [27]

$$R_\phi(T) \triangleq \sum_{t=1}^{T} \left[\phi_t(\mathbf{x}_t) - \phi_t(\mathbf{x}_t^*) \right]$$

$$= \sum_{t=1}^{T} \left[f_t(\mathbf{x}_t) + \varphi(\mathbf{x}_t) - f_t(\mathbf{x}^*) - \varphi(\mathbf{x}^*) \right]. \qquad (7.3.10)$$

The goal of online optimization is to minimize the regret $R_\phi(T)$.

The following are two methods for minimizing the regret (7.3.10):

1. *Primal-Dual Subgradient Method* [27, 107, 162]: In this method, the algorithm uses the average gradient $\bar{\mathbf{g}}_t = \frac{1}{t} \sum_{\tau=1}^{t} \mathbf{g}_\tau$ to make a prediction $\mathbf{x}_t$ on round t through encompassing a trade-off between a gradient-dependent linear term $\langle \bar{\mathbf{g}}_t, \mathbf{x} \rangle$, the regularizer $\varphi(\mathbf{x})$, and a proximal term $h_t(\mathbf{x})$ for well-conditioned predictions. The update amounts to solving [27]:

$$\mathbf{x}_{t+1} = \arg\min_{\mathbf{x} \in F} \left\{ \eta \langle \bar{\mathbf{g}}_t, \mathbf{x} \rangle + \eta \varphi(\mathbf{x}) + \frac{1}{t} h_t(\mathbf{x}) \right\}, \qquad (7.3.11)$$

where η is a fixed stepsize and $\mathbf{x}_1 = \arg\min_{\mathbf{x} \in F} \varphi(\mathbf{x})$.

2. *Composite Mirror Descent Method:* This method is also known as proximal gradient, forward-backward splitting [26, 147]. The update formula is given by

$$\mathbf{x}_{t+1} = \arg\min_{\mathbf{x} \in F} \left\{ \eta \langle \mathbf{g}_t, \mathbf{x} \rangle + \eta \varphi(\mathbf{x}) + B_{h_t}(\mathbf{x}, \mathbf{x}_t) \right\}, \qquad (7.3.12)$$

where $B_h(\mathbf{x}, \mathbf{y})$ is the Bregman divergence associated with strongly convex and differentiable function $h(\mathbf{x})$, and is defined as

$$B_h(\mathbf{x}, \mathbf{y}) = h(\mathbf{x}) - h(\mathbf{y}) - \langle \nabla h(\mathbf{y}), \mathbf{x} - \mathbf{y} \rangle. \qquad (7.3.13)$$

7.3 Optimization of Neural Networks

Two typical regularization functions are $\varphi(\mathbf{x}) = \|\mathbf{x}\|_1$ and $\varphi(\mathbf{x}) = \|\mathbf{x}\|_2$, the proximal function often takes the squared Mahalanobis norms $h_t(\mathbf{x}) = \langle \mathbf{x}, \mathbf{H}_t \mathbf{x} \rangle = \mathbf{x}^T \mathbf{H}_t \mathbf{x}$ for a symmetric nonnegative matrix $\mathbf{H}_t \geq 0$. For example,

$$\mathbf{H}_t = \delta \mathbf{I} + \mathbf{Diag}(\mathbf{G}_t)^{1/2} \quad \text{(Diagonal)} \tag{7.3.14}$$

and

$$\mathbf{H}_t = \delta \mathbf{I} + (\mathbf{G}_t)^{1/2} \quad \text{(Full)}. \tag{7.3.15}$$

where $\mathbf{G}_t = \sum_{\tau=1}^{t} \mathbf{g}_\tau \mathbf{g}_\tau^T$ is the outer product matrix of the subgradient vector $\mathbf{g}_t$, and $(\mathbf{G}_t)^{1/2}$ is the square-root of $\mathbf{G}_t$.

If taking $\mathbf{u} = \eta \bar{\mathbf{g}}_t$, $\eta \varphi(\mathbf{x}) = \lambda \|\mathbf{x}\|_2$, and $h_t(\mathbf{x}_t) = \frac{t}{2} \|\mathbf{x}_t\|_H = \langle \mathbf{x}_t, \mathbf{H}_t \mathbf{x}_t \rangle$, then the primal-dual subgradient optimization (7.3.11) becomes the following online optimization of composite functions:

$$\mathbf{x}_{t+1} = \arg\min_{\mathbf{x}_t \in F} \left\{ \langle \mathbf{u}, \mathbf{x}_t \rangle + \lambda \|\mathbf{x}_t\|_2 + \frac{1}{2} \langle \mathbf{x}_t, \mathbf{H}_t \mathbf{x}_t \rangle \right\}, \tag{7.3.16}$$

the subgradient $\mathbf{g}_t = \mathbf{u} + \lambda \mathbf{x}_t + \mathbf{H}_t \mathbf{x}_t$ and $\mathbf{G}_t = \sum_{\tau=1}^{t} \mathbf{g}_\tau \mathbf{g}_\tau^T$.

Introducing a variable $\mathbf{z} = \mathbf{x}$, then the equivalent problem of (7.3.16) becomes the constrained optimization problem:

$$\min \left\{ L(\mathbf{x}, \mathbf{z}, \boldsymbol{\alpha}) = \langle \mathbf{u}, \mathbf{x} \rangle + \frac{1}{2} \langle \mathbf{x}, \mathbf{H} \mathbf{x} \rangle + \lambda \|\mathbf{z}\|_2 \right\}$$

subject to $\mathbf{x} = \mathbf{z}$. With Lagrange multiplier vector $\boldsymbol{\alpha}$ for the equality constraint, the Lagrangian is given by

$$L(\mathbf{x}, \mathbf{z}, \boldsymbol{\alpha}) = \langle \mathbf{u}, \mathbf{x} \rangle + \frac{1}{2} \langle \mathbf{x}, \mathbf{H} \mathbf{x} \rangle + \lambda \|\mathbf{z}\|_2 + \langle \boldsymbol{\alpha}, \mathbf{x} - \mathbf{z} \rangle. \tag{7.3.17}$$

From the first-order optimized condition $\frac{\partial L(\mathbf{x}, \mathbf{z}, \boldsymbol{\alpha})}{\partial \mathbf{x}} = \mathbf{u} + \mathbf{H}\mathbf{x} + \boldsymbol{\alpha} = \mathbf{0}$, the infimum of $L(\mathbf{x}, \mathbf{z}, \boldsymbol{\alpha})$ is attained at $\mathbf{x} = -\mathbf{H}^{-1}(\mathbf{u} + \boldsymbol{\alpha})$.

Based on the above analysis, the *adaptive gradient* (ADAGRAD) algorithm was developed in [27] for minimizing the composite functions $\langle \mathbf{u}, \mathbf{x} \rangle + \frac{1}{2} \langle \mathbf{x}, \mathbf{H}\mathbf{x} \rangle + \lambda \|\mathbf{x}\|_2$, see Algorithm 7.1.

7.3.3 Adaptive Moment Estimation

In many fields of science and engineering, many problems can be cast as the optimization (minimization or maximization) of some scalar parameterized objective function with respect to its parameters. Objective functions are often stochastic. In

Algorithm 7.1 ADAGRAD algorithm for minimizing $\{\langle \mathbf{u}, \mathbf{x} \rangle + \frac{1}{2}\langle \mathbf{x}, \mathbf{Hx} \rangle + \lambda \|\mathbf{x}\|_2\}$ [27]

1. **input:** $\mathbf{u} \in \mathbb{R}^d, \lambda > 0$.
2. **initialization:** Generate randomly a nonnegative matrix $\mathbf{H}_0$ and put $\mathbf{x}_0 = \mathbf{0}$.
3. **for** $t = 1, 2, \ldots$ **do**
4. $\mathbf{g}_t = \mathbf{u} + (\mathbf{H}_{t-1} + \lambda \mathbf{I})\mathbf{x}_{t-1}$.
5. $\mathbf{G}_t = \sum_{\tau=1}^{t} \mathbf{g}_\tau \mathbf{g}_\tau^T$.
6. $\mathbf{H}_t = \delta \mathbf{I} + \mathbf{diag}(\mathbf{G}_t)^{1/2}$ or $\mathbf{H}_t = \delta \mathbf{I} + (\mathbf{G}_t)^{1/2}$.
7.   Make the SVD $\mathbf{H}_t = \mathbf{U\Sigma V}^T$ and denote the minimal and maximal singular values by $\sigma_{\min}(\mathbf{H}_t)$ and $\sigma_{\max}(\mathbf{H}_t)$, respectively.
8. $\mathbf{v}_t = \mathbf{H}_t^{-1}\mathbf{u} = \mathbf{V\Sigma}^{-1}\mathbf{U}^T\mathbf{u}$.
9. $\theta_{\max} = \|\mathbf{v}_t\|_2/\lambda - 1/\sigma_{\min}(\mathbf{H}_t)$.
10. $\theta_{\min} = \|\mathbf{v}_t\|_2/\lambda - 1/\sigma_{\max}(\mathbf{H}_t)$.
11.   **while** $\theta_{\max} - \theta_{\min} > \epsilon$ **do**
12. $\theta = (\theta_{\max} + \theta_{\min})/2$.
13. $\boldsymbol{\alpha}_t = -(\mathbf{H}_t^{-1} + \theta \mathbf{I})^{-1}\mathbf{v}_t$.
14.   **if** $\|\boldsymbol{\alpha}_t\|_2 > \lambda$ **do**
15. $\theta_{\min} = \theta$,
16.   **else**
17. $\theta_{\max} = \theta$.
18.   **end if**
19. $\mathbf{x}_t = -\mathbf{H}^{-1}\big(\mathbf{u} + \boldsymbol{\alpha}_t(\theta)\big)$.
20.   **end while**
21.   **exit:** if $\mathbf{x}_t$ converges.
22. **end for**
23. **output:** $\mathbf{x}_t = -\mathbf{H}^{-1}\big(\mathbf{u} + \boldsymbol{\alpha}_t(\theta)\big)$.

this case optimization can be made more efficient by taking gradient steps with respect to individual subfunctions, i.e., stochastic gradient descent (SGD) or ascent.

Let $f(\boldsymbol{\theta})$ be a noisy objective function, i.e., a stochastic scalar function that is differentiable with respect to parameters $\boldsymbol{\theta}$. We consider how to minimize the expected value of this function, $E\{f(\boldsymbol{\theta})\}$, with respect to its parameters $\boldsymbol{\theta}$. Let $f_1(\boldsymbol{\theta}), \ldots, f_T(\boldsymbol{\theta})$ be the realizations of the stochastic function at subsequent time steps $1, \ldots, T$. The stochasticity might come from the evaluation at random subsamples (mini-batches) of data points, or arise from inherent function noise. Use $g_t = \nabla_{\boldsymbol{\theta}} f_t(\boldsymbol{\theta})$ to denote the gradient, i.e., the vector of partial derivatives of f_t, with respect to $\boldsymbol{\theta}$ evaluated at time step t.

Given an arbitrary, unknown sequence of convex cost functions $f_1(\boldsymbol{\theta}), \ldots, f_T(\boldsymbol{\theta})$. At each time t, the goal of an online optimization is to predict the parameter $\boldsymbol{\theta}_t$ and evaluate it on a previously unknown cost function f_t. Since the nature of the sequence is unknown in advance, we design an online optimization algorithm using the regret, that is the sum of all the previous differences between the online prediction $f_t(\boldsymbol{\theta}_t)$ and the best fixed point parameter $f_t(\boldsymbol{\theta}^*)$ from a feasible set F

7.3 Optimization of Neural Networks

for all the previous steps. Concretely, the regret is defined as

$$R(T) = \sum_{t=1}^{T}[f_t(\boldsymbol{\theta}_t) - f_t(\boldsymbol{\theta}^*)], \tag{7.3.18}$$

where $\boldsymbol{\theta}^* = \arg\min_{\boldsymbol{\theta} \in F} \sum_{t=1}^{T} f_t(\boldsymbol{\theta})$.

ADAM (*Ada*ptive *M*oment) of Kingma and Ba [80] is a first-order gradient-based optimization algorithm, which uses estimates of the first moment vector **m** and the second moment vector **v** to optimize a stochastic objective function. The first and second moment vectors are updated as

$$\mathbf{m}_{t+1} = \beta_1 \mathbf{m}_t + (1 - \beta_1)\mathbf{g}_{t+1}, \tag{7.3.19}$$

$$\mathbf{v}_{t+1} = \beta_2 \mathbf{v}_t + (1 - \beta_2)\mathbf{g}_{t+1}^2, \tag{7.3.20}$$

where $\mathbf{g}_t$ is the gradient of loss function, and $\mathbf{g}_t^2$ is an elementwise square function. The exponential decay rates for the moment estimates are recommended to be $\beta_1 = 0.9$ and $\beta_2 = 0.999$. The bias correction of first and second moment estimates are

$$\hat{\mathbf{m}}_{t+1} = \mathbf{m}_t / \left(1 - \beta_1^{t+1}\right), \tag{7.3.21}$$

$$\hat{\mathbf{v}}_{t+1} = \mathbf{v}_t / \left(1 - \beta_2^{t+1}\right). \tag{7.3.22}$$

Algorithm 7.2 gives the ADAM algorithm for stochastic optimization.

Algorithm 7.2 ADAM algorithm for stochastic optimization [80]

---
1. **input:** Stepsize α (e.g., $\alpha = 0.001$), exponential decay rates $\beta_1 = 0.9$, $\beta_2 = 0.999$, $f(\boldsymbol{\theta})$, and put $\epsilon = 10^{-8}$.
2. **initialization:**
    2.1 Initial parameter vector $\boldsymbol{\theta}_0$.
    2.2 $\mathbf{m}_0 \leftarrow \mathbf{0}$ (Initialize 1st moment vector)
    2.3 $\mathbf{v}_0 \leftarrow \mathbf{0}$ (Initialize 2nd moment vector)
    2.4 $t \leftarrow 0$ (Initialize timestep)
3. **while** $\boldsymbol{\theta}$ not converged **do**
4. $t \leftarrow t + 1$,
5. $g_t \leftarrow \nabla_\theta f_t(\boldsymbol{\theta}_{t-1})$,
6. $\mathbf{m}_t \leftarrow \beta_1 \mathbf{m}_{t-1} + (1 - \beta_1) \cdot g_t$,
7. $\mathbf{v}_t \leftarrow \beta_2 \mathbf{v}_{t-1} + (1 - \beta_2) \cdot g_t^2$,
8. $\hat{\mathbf{m}}_t \leftarrow \mathbf{m}_t / (1 - \beta_1^t)$,
9. $\hat{\mathbf{v}}_t \leftarrow \mathbf{v}_t / (1 - \beta_1^t)$,
10. $\alpha_t = \alpha \cdot \sqrt{1 - \beta_2^t}/(1 - \beta_1^t)$,
11. $\boldsymbol{\theta}_t \leftarrow \boldsymbol{\theta}_{t-1} + \alpha_t \cdot \hat{\mathbf{m}}_t/(\sqrt{\hat{\mathbf{v}}_t} + \epsilon)$.
12. **end while**
13. **output:** $\boldsymbol{\theta}_t$.

7.4 Activation Functions

In neural networks, a unit's or node's expected value is referred as its activation, i.e., the *activation function* of a unit or node (or neuron) defines the output of that unit (or node or neuron) given an input or set of inputs.

7.4.1 Logistic Regression and Sigmoid Function

Neural networks typically employ a sigmoidal nonlinearity function as the activation function.

Given T training samples $\{(\mathbf{x}_1, y_1), \ldots, (\mathbf{x}_T, y_T)\}$, where the feature vector $\mathbf{x}_t \in \mathbb{R}^n$. Consider the binary classification problem in which $y_t \in \{0, 1\}$.

A logistic *regression* problem can be described as follows: for the given training samples, design a weight vector $\mathbf{w} \in \mathbb{R}^n$ to minimize the *logistic loss function* (also called the *cross-entropy loss function*)

$$J(\mathbf{w}) = \frac{1}{T} \sum_{t=1}^{T} H(p_t, q_t) = -\frac{1}{T} \sum_{t=1}^{T} \left[y_t \log \hat{y}_t + (1 - y_t) \log(1 - \hat{y}_t) \right] \quad (7.4.1)$$

where $H(p_t, q_t)$, the cross-entropy between p_t and q_t, will be defined later. Here, $\hat{y}_t \equiv g(\mathbf{w}^T \mathbf{x}_t)$, and is the *logistic regressor* with an attached *sigmoid function* (also known as a *logistic function* or *soft-step function*) [64]:

$$g(z) = 1/(1 + e^{-z}) \in (0, 1), \quad (7.4.2)$$

whose derivative is given by [11]

$$g'(z) = \frac{\partial}{\partial z}\left(\frac{1}{1+e^{-z}}\right) = \frac{e^{-z}}{(1+e^{-z})^2} = \frac{1}{1+e^{-z}}\left(1 - \frac{1}{1+e^{-z}}\right),$$

i.e.,

$$g'(z) = g(z)(1 - g(z)). \quad (7.4.3)$$

In information theory, for two discrete distributions p and q with K states, the *Kullback–Leibler (KL) divergence* of q from p is defined as

$$\mathrm{KL}(p \| q) \triangleq \sum_{k=1}^{K} p_k \log \frac{p_k}{q_k} \quad (7.4.4)$$

7.4 Activation Functions

or equivalently is expressed as

$$\mathrm{KL}(p\|q) = \sum_{k=1}^{K} p_k \log p_k - \sum_{k=1}^{K} p_k \log q_k = -H(p) + H(p,q) \qquad (7.4.5)$$

or

$$H(p,q) = H(p) + \mathrm{KL}(p\|q), \qquad (7.4.6)$$

where

$$H(p) = -\sum_{k=1}^{K} p_k \log p_k \quad \text{and} \quad H(p,q) = -\sum_{k=1}^{K} p_k \log q_k \qquad (7.4.7)$$

are, respectively, the entropy of p and the *cross-entropy* between two probability distributions p and q over a given set.

The KL divergence $\mathrm{KL}(p\|q)$ is also known as the *relative entropy* of p with respect to q.

In logistic regression, there are only two states $y_t = 0$ and $y_t = 1$, i.e., $K = 2$, the probability of finding the output $y = 0$ (corresponding to $k = 1$) is given by

$$p_{y=0} = 1 - y \quad \text{and} \quad q_{y=0} = 1 - \hat{y}, \qquad (7.4.8)$$

while the probability of finding the output $y = 1$ (corresponding to $k = 2$) is given by

$$p_{y=1} = y \quad \text{and} \quad q_{y=1} = \hat{y}. \qquad (7.4.9)$$

Substituting (7.4.8) and (7.4.9) into (7.4.4), then the KL divergence in logistic regression is given by

$$\mathrm{KL}(p\|q) = (1-y)\log\frac{1-y}{1-\hat{y}} + y\log\frac{y}{\hat{y}}. \qquad (7.4.10)$$

Similarly, substituting (7.4.8) and (7.4.9) into (7.4.7) to yield the cross-entropy

$$H(p,q) = -(1-y)\log(1-\hat{y}) - y\log\hat{y}. \qquad (7.4.11)$$

The cross-entropy $H(p,q)$ denotes the information loss produced when an "unnatural" probability distribution q is used to fit the "true" distribution p. Because the entropy of the true distribution value p is invariable, it is known from (7.4.6) that the cross-entropy and thus the KL divergence also describes the similarity between the predicted results $\hat{y}_t$ and the real results y_t, which can be used as a loss function to ensure that the predicted value $\hat{y}_t$ conforms to the true value y_t.

Therefore, the logistic loss is given by

$$J(\mathbf{w}) = \frac{1}{T}\sum_{t=1}^{T} H(p_t, q_t) = -\frac{1}{T}\sum_{t=1}^{T}\left[y_t \log \hat{y}_t + (1-y_t)\log(1-\hat{y}_t)\right]. \quad (7.4.12)$$

From $\hat{y}_t = g(\mathbf{w}^T\mathbf{x}_t) = 1/(1+e^{-\mathbf{w}^T\mathbf{x}_t})$ and (7.4.11), the above logistic loss becomes

$$J(\mathbf{w}) = \frac{1}{T}\sum_{t=1}^{T}\left[y_t \log\left(1 + e^{-\mathbf{w}^T\mathbf{x}_t}\right) + (1-y_t)\mathbf{w}^T\mathbf{x}_t\right]. \quad (7.4.13)$$

It is easily known that the gradient of $J(\mathbf{w})$ with respect to $\mathbf{w}$ is given by

$$\frac{\partial J(\mathbf{w})}{\partial \mathbf{w}} = \frac{1}{T}\sum_{t=1}^{T}\left(1 - \frac{y_t}{1 + e^{-\mathbf{w}^T\mathbf{x}_t}}\right)\mathbf{x}_t. \quad (7.4.14)$$

In logistic regression, the weight vector $\mathbf{w}$ can be updated by the gradient descent algorithm

$$\mathbf{w}_{k+1} = \mathbf{w}_k - \eta\frac{1}{T}\sum_{t=1}^{T}\left(1 - \frac{y_t}{1 + e^{-\mathbf{w}_k^T\mathbf{x}_t}}\right)\mathbf{x}_t \quad (7.4.15)$$

until $\mathbf{w}_k$ is converged.

For a layer with a sigmoid activation function $\mathbf{g}(\mathbf{z}) = \mathbf{g}(\mathbf{Wx} + \mathbf{b})$, where $\mathbf{x}$ is the layer input, the weight matrix $\mathbf{W}$, and the bias vector $\mathbf{b}$ are the layer parameters to be learned, and $g(z_i) = 1/(1 + \exp(-z_i))$. As some $|z_i|$ increases, $g'(z_i) = g(z_i)(1 - g(z_i))$ tends to zero. This means that the corresponding gradient flowing $-g'(z_i)$ will vanish and thus the model will train slowly. For a deep neural network (DNN) with multiple hidden layers, a layer with slow training will affect convergence of all the layers below. Because this effect is amplified as the network depth increases, the sigmoid function is not available for activation function of hidden layers in DNN.

7.4.2 Softmax Regression and Softmax Function

The logistic regression is only available for binary classification problems, but there are many multiple classification problems in science and engineering. Let the class labels be k different values that are assumed to be $\{1, \ldots, k\}$ without generality.

7.4 Activation Functions

Suppose we are given a set of training samples $\{(\mathbf{x}_1, y_1), \ldots, (\mathbf{x}_T, y_T)\}$, where $\mathbf{x}_t \in \mathbb{R}^n$, $y_t \in \{1, \ldots, k\}$. In this case, we design a weight matrix $\mathbf{W} = [\mathbf{w}_1, \ldots, \mathbf{w}_k] \in \mathbb{R}^{p \times n}$ such that the hypothetical function has the following vector form:

$$\mathbf{h_W}(\mathbf{x}_t) = \begin{bmatrix} p(y_t = 1 | \mathbf{x}_t; \mathbf{W}) \\ p(y_t = 2 | \mathbf{x}_t; \mathbf{W}) \\ \vdots \\ p(y_t = k | \mathbf{x}_t; \mathbf{W}) \end{bmatrix} = \frac{1}{\sum_{j=1}^{k} e^{\mathbf{w}_j^T \mathbf{x}_t}} \begin{bmatrix} e^{\mathbf{w}_1^T \mathbf{x}_t} \\ e^{\mathbf{w}_2^T \mathbf{x}_t} \\ \vdots \\ e^{\mathbf{w}_k^T \mathbf{x}_t} \end{bmatrix}. \qquad (7.4.16)$$

On the other hand, for an RNN network, its output vector $\mathbf{g}(\mathbf{W}\mathbf{x}_t)$ is directly used as the hypothetical function, i.e., $\mathbf{h_W}(\mathbf{x}_t) = \mathbf{g}(\mathbf{W}\mathbf{x}_t)$. Letting $\mathbf{z} = \mathbf{W}\mathbf{x}_t$ or $z_i = \mathbf{w}_i^T \mathbf{x}_t$, $i = 1, \ldots, k$, then we apply naturally the *softmax function*

$$\text{softmax}_i(\mathbf{z}) = \frac{e^{z_i}}{\sum_{j=1}^{k} e^{z_j}} = \frac{e^{\mathbf{w}_i^T \mathbf{x}_t}}{\sum_{j=1}^{k} e^{\mathbf{w}_j^T \mathbf{x}_t}}, \quad i = 1, \ldots, k, \qquad (7.4.17)$$

or

$$\mathbf{softmax}(\mathbf{z}) = \frac{1}{\sum_{j=1}^{k} e^{\mathbf{w}_j^T \mathbf{x}_t}} \begin{bmatrix} e^{\mathbf{w}_1^T \mathbf{x}_t} \\ \vdots \\ e^{\mathbf{w}_k \mathbf{w}_t} \end{bmatrix} \in (0, 1) \qquad (7.4.18)$$

to express the hypothetical function

$$\mathbf{h}_W(\mathbf{x}_t) = \mathbf{softmax}(\mathbf{W}\mathbf{x}_t). \qquad (7.4.19)$$

Due to $\frac{\partial z_i}{\partial z_j} = \delta_{ij}$ the derivative of the softmax function is given by

$$\frac{\partial \text{softmax}_i(\mathbf{z})}{\partial z_j} = \frac{\partial}{\partial z_j} \left(\frac{e^{z_i}}{\sum_{j=1}^{k} e^{z_j}} \right) = \frac{e^{z_i} \delta_{ij}}{\sum_{j=1}^{k} e^{z_j}} + e^{z_i} \cdot \frac{-e^{z_j}}{\left(\sum_{j=1}^{k} e^{z_j}\right)^2},$$

namely

$$\frac{\partial \text{softmax}_i(\mathbf{z})}{\partial z_j} = \text{softmax}_i(\mathbf{z}) \left(\delta_{ij} - \text{softmax}_j(\mathbf{z}) \right) \in (0, 0.25), \qquad (7.4.20)$$

because the range of $\text{softmax}(\mathbf{z})$ is $(0, 1)$.

The common (hard) max function $\max\{z_1, \ldots, z_k\}$ takes only a maximum value among $z_1, \ldots, z_k$. Unlike the max function, the softmax function takes k values in probability form. Different from the normalized function, the softmax function has no negative probability.

Example 7.1 If $\mathbf{z} = [3, 5, -4]^T$, then the max function $\max\{3, 5, -4\} = 5$, the normalized function gives the normalized vector $[0.4243, 0.7071, -0.5657]^T$ and **softmax(z)** $= [0.1192, 0.8807, 0.0001]^T$.

It is interesting to compare the softmax function with the max function and the normalized function.

- The max function is a single-valued function and gives only one result while rejecting all others.
- The normalized function is a multivalued function and may retain the negative value(s), and cannot be used as the probability representation.
- The softmax function is also a multivalued function but provides the probability of z_i belonging to the class i, and thus it can transform a matrix into a nonnegative even sparse matrix.

Therefore, the softmax function is especially available for multiple classification problems.

7.4.3 Other Activation Functions

Recently, there is increasing evidence that other types of nonlinearities can improve the performance of DNNs.

The hyperbolic *tangent (tanh) function* serves as a point-wise nonlinearity applied to all hidden units of a DNN, and is defined by

$$f(z) = \tanh(z) = \frac{e^z - e^{-z}}{e^z + e^{-z}} \in (-1, +1), \tag{7.4.21}$$

$$f'(z) = \frac{\partial \tanh(z)}{\partial z} = 1 - \tanh^2(z). \tag{7.4.22}$$

As an alternative to the tanh function, the softsign function is given by [11]

$$\text{softsign}(z) = \frac{z}{1 + |z|} \in (-1, 1), \tag{7.4.23}$$

$$\text{softsign}'(z) = \frac{\partial \text{softsign}(z)}{\partial z} = \frac{1}{(1 + |z|)^2}. \tag{7.4.24}$$

Clearly, when $\tanh(z) \to -1$ or $+1$, the tanh function has a vanishing gradient, as shown in (7.4.22). But, it is seen from (7.4.24) that the softsign function has no such trouble. As a matter of fact, the ratio of their derivatives has the following

7.4 Activation Functions

asymptotic limit [11]:

$$\lim_{z \to \infty} \frac{\frac{\partial \tanh(z)}{\partial z}}{\partial \left(\frac{1}{1+z}\right)/\partial z} = \lim_{z \to \infty} \frac{\frac{\partial}{\partial z}\left(\frac{2}{1+\exp(-z)}\right)}{\frac{\partial}{\partial z}\left(\frac{z}{1+z}\right)} = \lim_{z \to \infty} \exp(-z)z = 0.$$

Rectified Linear Unit (ReLU) function [106] is defined as

$$f(z) = \text{ReLU}(z) = \max\{z, 0\} = \begin{cases} 0, & z < 0; \\ z, & z \geq 0, \end{cases} \quad (7.4.25)$$

with range $[0, \infty)$. The derivative is given by

$$f'(z) = \frac{\partial \text{ReLU}(z)}{\partial z} = \begin{cases} 0, & z < 0; \\ 1, & z \geq 0. \end{cases} \quad (7.4.26)$$

Softplus function is defined as

$$\text{softplus}(z) = \log(1 + e^z). \quad (7.4.27)$$

This function can be seen as ReLU's smoothing. According to neuroscientists' research, softplus and ReLU are similar to the activation frequency function of brain neurons. That is to say, compared with the early activation functions, softplus and ReLU are closer to the activation model of brain neurons, while neural networks are based on the development of brain neuroscience. It could be said that the application of these two activation functions has contributed to a new wave of neural network research.

As a comparison with the sigmoid and tanh functions, the ReLU has the following two advantages:

- When the input is positive, there is no gradient saturation problem.
- The computation speed is much faster than sigmoid and tanh.

However, there are the following two disadvantages in the ReLU:

- When the input is negative, the ReLU is completely inactive. This is not a problem for the forward transmission, but in the backpropagation process, the negative input may make the gradient be completely equal to zero, this is the same problem as sigmoid function and tanh function.
- The output of the ReLU function is either 0 or positive, which means that the ReLU function is not a zero-centric function.

In order to improve the ReLU function, the following variants were developed:

1. *Leaky rectified linear unit* (Leaky ReLU) [102]

$$f(z) = \text{LReLU}(z) = \max\{z, 0\} = \begin{cases} 0.01z, & z < 0; \\ z, & z \geq 0, \end{cases} \in (-\infty, +\infty), \quad (7.4.28)$$

$$f'(z) = \frac{\partial \text{LReLU}(z)}{\partial z} = \begin{cases} 0.01, & z < 0; \\ 1, & z \geq 0. \end{cases} \quad (7.4.29)$$

2. *Parametric rectified linear unit (PReLU)* [55]

$$f(z) = \text{PReLU}(z) = \max\{z, 0\} = \begin{cases} \alpha z, & z < 0; \\ z, & z \geq 0, \end{cases} \in (-\infty, +\infty) \quad (7.4.30)$$

$$f'(z) = \frac{\partial \text{PReLU}(z)}{\partial z} = \begin{cases} \alpha, & z < 0; \\ 1, & z \geq 0. \end{cases} \quad (7.4.31)$$

Clearly, when $\alpha = 0.01$ is taken, the PReLU reduces to the Leaky ReLU. Note that the randomized leaky rectified linear unit (RLReLU) was independently proposed in [164] and both the RLReLU and the PReLU have the same form.

Deep learning methods aim at learning feature hierarchies with features from higher levels of the hierarchy formed by composing lower level features. Training deep neural networks is complicated by the fact that the distribution of each layer's inputs changes during training, as the parameters of the previous layers change [74].

Given values of z over a mini-batch $B = \{z_1, \ldots, z_m\}$. Let

$$\mu_B = \frac{1}{m} \sum_{i=1}^{m} z_i \quad \text{and} \quad \sigma_B^2 = \frac{1}{m} \sum_{i=1}^{m} (z_i - \mu_B)^2 \quad (7.4.32)$$

are, respectively, the *mini-batch mean* and the *mini-batch variance*. Then, the *mini-batch normalization* is given by

$$\hat{z}_i = \frac{z_i - E\{z_i\}}{\sqrt{\text{var}(z_i)}} = \frac{z_i - \mu_B}{\sqrt{\sigma_B^2 + \epsilon}} \quad \text{or} \quad \hat{\mathbf{z}} = \frac{\mathbf{z} - \mu_B \mathbf{1}}{\sigma_B^2 + \epsilon}, \quad (7.4.33)$$

where ϵ is a small constant added to the mini-batch variance for numerical stability.

7.4 Activation Functions

Batch normalization is defined as [74]:

$$\text{BN}_{\gamma_i, \beta_i}(z_i) = \gamma_i \hat{z}_i + \beta_i \quad \text{or} \quad \mathbf{BN}_{\boldsymbol{\gamma}, \boldsymbol{\beta}}(\mathbf{z}) = \boldsymbol{\gamma} \odot \hat{\mathbf{z}} + \boldsymbol{\beta} \qquad (7.4.34)$$

with $\boldsymbol{\gamma} = [\gamma_1, \ldots, \gamma_m]^T$ and $\boldsymbol{\beta} = [\beta_1, \ldots, \beta_m]^T$, where γ_i and β_i are two *batch normalization parameters* to be learned, and $\odot$ denotes the elementwise product of two vectors, i.e., $[\boldsymbol{\gamma} \odot \hat{\mathbf{z}}]_i = \gamma_i \hat{z}_i$.

As shown in [90], such a normalization speeds up convergence, even when the features are not decorrelated.

Batch normalization can be applied to any set of activations in the network. For example, for an affine transformation followed by an elementwise nonlinearity:

$$\mathbf{z} = g(\mathbf{W}\mathbf{u} + \mathbf{b}), \qquad (7.4.35)$$

where $\mathbf{W}$ and $\mathbf{b}$ are learned parameters of the model, if the nonlinear activation function $g(\cdot)$ such as sigmoid or ReLU is replaced with

$$\mathbf{z} = g(\mathbf{BN}(\mathbf{W}\mathbf{u})), \qquad (7.4.36)$$

then the bias $\mathbf{b}$ can be ignored since its effect will be canceled by the subsequent mean subtraction.

It is noted that $\mathbf{z}$ of a neuron in the some layer (say the lth layer) is not the original input, that is, not the output of each neuron in the $(l-1)$th layer, but the linear activation $\mathbf{z} = \mathbf{W}\mathbf{u} + \mathbf{b}$ of the neuron in the lth layer, here $\mathbf{u}$ is the output of the neuron in the $(l-1)$th layer. Therefore, the original activation $\mathbf{z}$ of a neuron is converted by subtracting the mini-batch mean $E\{\mathbf{z}\}$ and dividing it by the mini-batch variance $\text{var}(\mathbf{z})$.

By normalizing activations throughout the network, it prevents small changes to the parameters from amplifying into larger and suboptimal changes in activations in gradients. Hence, the batch normalization helps to address the following issues in traditional deep networks: too-high learning rate may result in the gradients that explode or vanish, as well as getting stuck in poor local minima [74].

Another advantage of batch normalization is that backpropagation through a layer is unaffected by the scale of its parameters, because for a scaled batch normalization $\mathbf{BN}(\mathbf{W}\mathbf{u}) = \mathbf{BN}((\alpha \mathbf{W})\mathbf{u})$ with a scalar α, its derivatives with respect to $\mathbf{u}$ and $\mathbf{W}$ are, respectively, given by [74]

$$\frac{\partial \mathbf{BN}((\alpha \mathbf{W})\mathbf{u})}{\partial \mathbf{u}} = \frac{\partial \mathbf{BN}(\mathbf{W}\mathbf{u})}{\partial \mathbf{u}}, \qquad (7.4.37)$$

$$\frac{\partial \mathbf{BN}((\alpha \mathbf{W})\mathbf{u})}{\partial (\alpha \mathbf{W})} = \frac{1}{\alpha} \cdot \frac{\partial \mathbf{BN}(\mathbf{W}\mathbf{u})}{\partial \mathbf{W}}. \qquad (7.4.38)$$

That is to say, the scale does not affect the layer Jacobian nor, consequently, the gradient propagation. Moreover, larger weights lead to smaller gradients, and batch normalization will stabilize the parameter growth.

On batch normalization networks, we will discuss further in Sect. 7.15.

7.5 Recurrent Neural Networks

A *recurrent neural network* (RNN), proposed by Rumelhartin et al. in 1986 [123], is a neural network model for modeling time series. RNNs provide a very powerful way of dealing with (time) sequential data. In traditional neural network models, it is operated from the input layer to hidden layer to output layer. These layers are fully connected, and there is no connection between nodes of each layer. For tasks that involve sequential inputs, such as speech and language, it is often better to use RNNs.

RNNs are "fuzzy" in the sense that they do not use exact templates from the training data to make predictions, but, like other neural networks, use their internal representation to perform a high-dimensional interpolation between training examples [44].

7.5.1 Conventional Recurrent Neural Networks

The modern definition of "recurrent" was initially introduced by Jordan [75]:

> If a network has one or more cycles, that is, if it is possible to follow a path from a unit back to itself, then the network is referred to as recurrent. A nonrecurrent network has no cycles.

Let $x_i(t)$ denote the activation of input unit i at time t, $h_j(t)$ denote the activation of hidden unit j at time t, and $y_k(t)$ denote the activation of output unit k at time t.

The variables and parameters that need to be used and learned during training are denoted as follows:

- $\mathbf{x}_t \in \mathbb{R}^n$: input vector to the hidden layer,
- $\mathbf{h}_t \in \mathbb{R}^l$: output of hidden layer and denotes hidden state,
- $\hat{\mathbf{y}}_t \in \mathbb{R}^m$: output produced by RNN, $\mathbf{y} \in \mathbb{R}^m$: desired output,
- $\mathbf{U}_{hx} = [U_{ji}] \in \mathbb{R}^{l \times n}$: the input-hidden weight matrix from the ith input node to the jth hidden state,
- $\mathbf{W}_{hy} = [W_{jk}] \in \mathbb{R}^{l \times m}$: output-hidden weight matrix from the kth output node to the jth hidden state,
- $\mathbf{V}_{yh} = [V_{kj}] \in \mathbb{R}^{m \times l}$: hidden-output weight matrix from the jth hidden state to the kth output node,

7.5 Recurrent Neural Networks

- $\mathbf{b}_t \in \mathbb{R}^l$: bias vectors (called also intercept terms) of hidden layers,
- $\mathbf{c}_t \in \mathbb{R}^m$: bias vectors (called also intercept terms) of output layers,
- $f(h_i) \in \mathbb{R}$: elementwise activation function of hidden layer,
- $g(y_i) \in \mathbb{R}$: elementwise activation function of output layer.

The dynamical system of a feedforward network (FFN) is given by

– *Hidden states for FFN:*

$$h_j(t) = f(\text{net}_j), \quad \text{for } j = 1, \ldots, l, \tag{7.5.1}$$

$$\text{net}_j = \sum_{i=1}^{n} U_{ji} x_i(t) + b_j(t), \quad \text{for } j = 1, \ldots, l. \tag{7.5.2}$$

– *Output nodes for FFN:*

$$y_k(t) = g(\text{net}_k), \quad \text{for } k = 1, \ldots, m, \tag{7.5.3}$$

$$\text{net}_k = \sum_{j=1}^{l} V_{kj} h_j(t) + c_j(t), \quad \text{for } k = 1, \ldots, m. \tag{7.5.4}$$

The dynamical system for FFNs can be written as the matrix-vector form:

$$\mathbf{h}_t = f(\mathbf{U}_{hx}\mathbf{x}_t + \mathbf{b}_t), \tag{7.5.5}$$

$$\mathbf{y}_t = g(\mathbf{V}_{yh}\mathbf{h}_t + \mathbf{c}_t). \tag{7.5.6}$$

An RNN is a neural network that simulates a discrete-time dynamical system with an input $\mathbf{x}_t = [x_1(t), \ldots, x_n(t)]^T$, an output $\mathbf{y}_t = [y_1(t), \ldots, y_m(t)]^T$, and a hidden state vector $\mathbf{h}_t = [h_1(t), \ldots, h_l(t)]^T$, where the subscript t represents the time step.

RNNs have two recurrence forms. Figure 7.2 shows the general structure of a three-layer RNN with hidden state recurrence.

The RNN shown in Fig. 7.2 can be described by the following dynamical system:

- *Hidden states for hidden state recurrence:*

$$h_j(t) = f(\text{net}_j), \quad \text{for } j = 1, \ldots, l, \tag{7.5.7}$$

$$\text{net}_j = \sum_{i=1}^{n} U_{ji} x_i(t) + \sum_{\tilde{j}=1}^{l} W_{j\tilde{j}} h_{\tilde{j}}(t-1) + b_j, \quad \text{for } j = 1, \ldots, l. \tag{7.5.8}$$

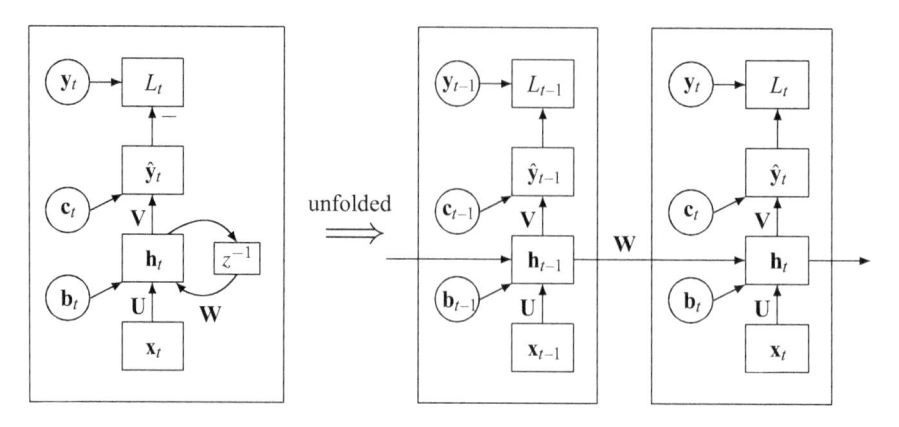

Fig. 7.2 General structure of a regular unidirectional three-layer RNN with hidden state recurrence, where z^{-1} denotes a delay line. Left is RNN with a delay line, and right is RNN unfolded in time for two time steps

- *Output nodes for hidden state recurrence:*

$$\hat{y}_k(t) = g(\text{net}_k), \quad \text{for } k = 1, \ldots, m, \tag{7.5.9}$$

$$\text{net}_k = \sum_{j=1}^{l} V_{kj} h_j(t) + c_j(t), \quad \text{for } k = 1, \ldots, m. \tag{7.5.10}$$

The dynamical system of the RNN with hidden state recurrence can be written in matrix-vector form as

$$\left. \begin{array}{l} \mathbf{h}_t = f(\mathbf{U}_{hx}\mathbf{x}_t + \mathbf{W}_{hh}\mathbf{h}_{t-1} + \mathbf{b}_t), \\ \hat{\mathbf{y}}_t = g(\mathbf{V}_{yh}\mathbf{h}_t + \mathbf{c}_t), \\ L_t(\mathbf{y}_t, \hat{\mathbf{y}}_t) = \mathbf{y}_t - \hat{\mathbf{y}}_t. \end{array} \right\} \tag{7.5.11}$$

Figure 7.3 shows the general structure of a three-layer RNN with output recurrence.

The RNN with output recurrence can be described by the following equations:

- *Hidden states for output recurrence:*

$$h_j(t) = f(\text{net}_j), \quad \text{for } j = 1, \ldots, l \tag{7.5.12}$$

$$\text{net}_j = \sum_{i=1}^{n} U_{ji} x_i(t) + \sum_{k=1}^{m} W_{jk} \hat{y}_k(t-1) + b_j, \quad \text{for } j = 1, \ldots, l. \tag{7.5.13}$$

7.5 Recurrent Neural Networks

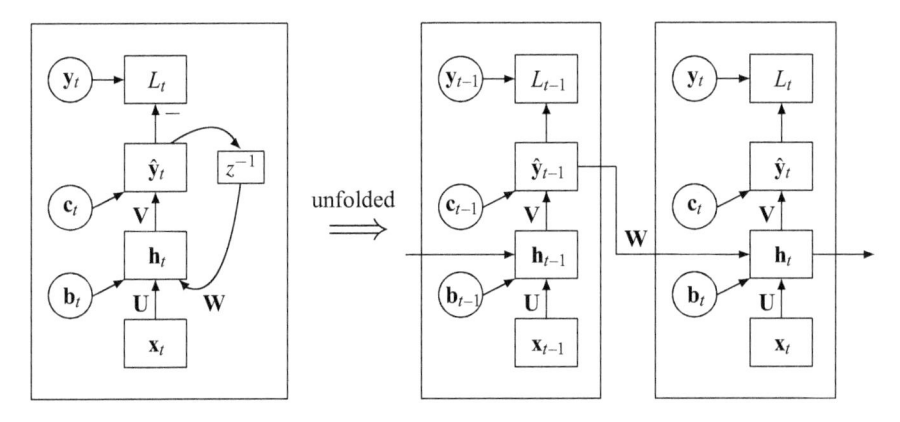

Fig. 7.3 General structure of a regular unidirectional three-layer RNN with output recurrence, where z^{-1} denotes a delay line. Left is RNN with a delay line, and right is RNN unfolded in time for two time steps

- *Output nodes for output recurrence:*

$$\hat{y}_k(t) = g(\text{net}_k), \quad \text{for } k = 1, \ldots, m, \tag{7.5.14}$$

$$\text{net}_k = \sum_{j=1}^{l} V_{kj} h_j(t) + c_j(t), \quad \text{for } k = 1, \ldots, m. \tag{7.5.15}$$

The dynamical system for an RNN with output recurrence can be written as the following matrix-vector form:

$$\left. \begin{array}{l} \mathbf{h}_t = f(\mathbf{U}_{hx}\mathbf{x}_t + \mathbf{W}_{hy}\hat{\mathbf{y}}_{t-1} + \mathbf{b}_t), \\ \hat{\mathbf{y}}_t = g(\mathbf{V}_{yh}\mathbf{h}_t + \mathbf{c}_t), \\ L_t(\mathbf{y}_t, \hat{\mathbf{y}}_t) = \mathbf{y}_t - \hat{\mathbf{y}}_t. \end{array} \right\} \tag{7.5.16}$$

In the following we focus on RNNs with output recurrence.

7.5.2 Backpropagation Through Time (BPTT)

An input sequence $\mathbf{x} \in \mathbb{R}^m$ of length T is denoted by $(\mathbb{R}^m)^T$ and an output sequence $\mathbf{y} \in \mathbb{R}^n$ is denoted by $(\mathbb{R}^n)^T$. Then, a recurrent neural network with m inputs, n outputs, and weight matrix $\mathbf{W}$ such that $\mathbf{y} = \mathbf{W}\mathbf{x}$ can be viewed as a continuous map $(\mathbb{R}^m)^T \to (\mathbb{R}^n)^T$.

Suppose we are given a set of N training sequences $D = \left\{ \left(\mathbf{x}_t^{(n)}, \mathbf{y}_t^{(n)}\right)_{t=1}^T \right\}_{n=1}^N$, where $\left(\mathbf{x}_t^{(n)}, \mathbf{y}_t^{(n)}\right)$ denotes the tth pair of samples in the nth subset of the training data set. Then, the parameters $\boldsymbol{\theta} = (\mathbf{W}, \mathbf{U}, \mathbf{V})$ of an RNN can be estimated by minimizing the cost function

$$J(\boldsymbol{\theta}) = \frac{1}{N} \sum_{n=1}^{N} \sum_{t=1}^{T} d\left(\mathbf{y}_t^{(n)}, \hat{\mathbf{y}}_t^{(n)}\right), \tag{7.5.17}$$

where $d(\mathbf{a}, \mathbf{b})$ denotes a predefined divergence measure between $\mathbf{a}$ and $\mathbf{b}$, such as Euclidean distance $d(\mathbf{a}, \mathbf{b}) = \|\mathbf{a} - \mathbf{b}\|_2 = \langle \mathbf{a} - \mathbf{b}, \mathbf{a} - \mathbf{b} \rangle^{1/2}$ or Mahalanobis distance $d_H(\mathbf{a}, \mathbf{b}) = \|\mathbf{a} - \mathbf{b}\|_H = \langle \mathbf{a} - \mathbf{b}, \mathbf{H}(\mathbf{a} - \mathbf{b}) \rangle^{1/2}$ (where $\mathbf{H}$ is a nonnegative matrix) or the KL divergence $\mathrm{KL}(\mathbf{a}\|\mathbf{b})$.

The most frequently used cost function is the summed squared error (SSE). Suppose that there are P patterns or presentations in the training set and all K output units in an RNN. Then, the *loss* or *error* function at time step t is defined as

$$E_t(y_{pk}(t), \hat{y}_{pk}(t)) = \frac{1}{2} \sum_{p=1}^{P} \sum_{k=1}^{m} \left(y_{pk}(t) - \hat{y}_{pk}(t)\right)^2, \tag{7.5.18}$$

where $\mathbf{y}_{pk}(t)$ are the desired output vectors (or given sample vectors) associated with the pth pattern at the kth output unit at time step t, and $\hat{\mathbf{y}}_{pk}(t)$ are the estimated or predicted output vectors at time step t.

Define the total loss or error as

$$E(\mathbf{y}, \hat{\mathbf{y}}) = \sum_{t=1}^{T} E_t\left(y_{pk}(t), \hat{y}_{pk}(t)\right). \tag{7.5.19}$$

By gradient descent, each weight change $\Delta \mathbf{W}$ in the network should be proportional to the negative gradient of the cost with respect to the specific weight matrix $\mathbf{W}$:

$$\Delta \mathbf{W} = -\eta \frac{\partial E}{\partial \mathbf{W}}, \tag{7.5.20}$$

where η is a learning step size or rate.

7.5 Recurrent Neural Networks

The design goal of an RNN is to calculate the gradient of the error with respect to parameters **U**, **V**, and **W**, and then learn good parameters using stochastic gradient descent (SGD). By mimicking the sum of the errors, the gradients at each time step are also summed up for one training example to get

$$\frac{\partial E}{\partial \mathbf{W}} = \sum_{t=1}^{T} \frac{\partial E_t}{\partial \mathbf{W}}, \tag{7.5.21}$$

$$\frac{\partial E}{\partial \mathbf{V}} = \sum_{t=1}^{T} \frac{\partial E_t}{\partial \mathbf{V}}, \tag{7.5.22}$$

$$\frac{\partial E}{\partial \mathbf{U}} = \sum_{t=1}^{T} \frac{\partial E_t}{\partial \mathbf{U}}. \tag{7.5.23}$$

The *backpropagation through time* (BPTT) algorithm [159] plays an important role in designing RNNs. The basic idea of this algorithm is to apply the chain rule of differentiation to calculate the above gradients backwards starting from the error.

1. Calculate the gradient $\frac{\partial E_t}{\partial V_{kj}(t)}$:

$$\begin{aligned}\frac{\partial E_t}{\partial V_{kj}(t)} &= \frac{\partial E_t}{\partial \hat{y}_{pk}(t)} \cdot \frac{\partial \hat{y}_{pk}(t)}{\partial \text{net}_{pk}(t)} \cdot \frac{\partial \text{net}_{pk}(t)}{\partial V_{kj}(t)} \\ &= -\left(y_{pk}(t) - \hat{y}_{pk}(t)\right) g'\left(\hat{y}_{pk}(t)\right) \sum_{j=1}^{l} h_{pj}(t) \\ &= -\delta_{pk}(t) \cdot \sum_{j=1}^{l} h_{pj}(t), \end{aligned} \tag{7.5.24}$$

where $\delta_{pk}(t)$ is called the *error for output nodes* at time step t and is defined as

$$\begin{aligned}\delta_{pk}(t) &= -\frac{\partial E_t}{\partial \hat{y}_{pk}(t)} \cdot \frac{\partial \hat{y}_{pk}(t)}{\partial \text{net}_{pk}(t)} \\ &= \left(y_{pk}(t) - \hat{y}_{pk}(t)\right) g'\left(\hat{y}_{pk}(t)\right). \end{aligned} \tag{7.5.25}$$

2. Calculate the gradient $\frac{\partial E_t}{\partial W_{jk}(t)}$:

$$\frac{\partial E_t}{\partial W_{jk}(t)} = \left[\sum_{k=1}^{m}\left(\frac{\partial E_t}{\partial \hat{y}_{pk}(t)} \cdot \frac{\partial \hat{y}_{pk}(t)}{\partial \text{net}_{pk}(t)} \cdot \frac{\partial \text{net}_{pk}(t)}{\partial h_{pj}(t)}\right)\right] \cdot \frac{\partial h_{pj}(t)}{\partial \text{net}_{pj}(t)} \cdot \frac{\partial \text{net}_{pj}(t)}{\partial W_{jk}(t)}$$

$$= -\delta_{pj}(t) \cdot \frac{\partial \text{net}_{pj}(t)}{\partial W_{jk}(t)} = -\delta_{pj}(t) \cdot \sum_{k=1}^{m} \hat{y}_{pk}(t-1), \quad (7.5.26)$$

where $\delta_{pj}(t)$ is known as the *error for hidden nodes* given by

$$\delta_{pj}(t) = -\left(\sum_{k=1}^{m} \frac{\partial E_t}{\partial \hat{y}_{pk}(t)} \cdot \frac{\partial \hat{y}_{pk}(t)}{\partial \text{net}_{pk}} \cdot \frac{\partial \text{net}_{pk}}{\partial h_{pj}(t)}\right) \cdot \frac{\partial h_{pj}(t)}{\partial \text{net}_{pj}}$$

$$= \left(\sum_{k=1}^{m} \delta_{pk} \cdot V_{kj}\right) f'(h_{pj}). \quad (7.5.27)$$

3. Calculate the gradient $\frac{\partial E_t}{\partial U_{ji}(t)}$:

$$\frac{\partial E_t}{\partial U_{ji}(t)} = \left[\sum_{k=1}^{m}\left(\frac{\partial E_t}{\partial \hat{y}_{pk}(t)} \cdot \frac{\partial \hat{y}_{pk}(t)}{\partial \text{net}_{pk}(t)} \cdot \frac{\partial \text{net}_{pk}(t)}{\partial h_{pj}(t)}\right)\right] \cdot \frac{\partial h_{pj}(t)}{\partial \text{net}_{pj}(t)} \cdot \frac{\partial \text{net}_{pj}(t)}{\partial U_{ji}(t)}$$

$$= -\delta_{pj}(t) \cdot \sum_{i=1}^{n} x_{pi}(t). \quad (7.5.28)$$

The above discussion can be summarized into the following formulae:

$$\delta_{pk}(t) = (y_{pk}(t) - \hat{y}_{pk}(t))g'(\hat{y}_{pk}(t)), \quad (7.5.29)$$

$$\delta_{pj}(t) = \left(\sum_{k=1}^{m}(\delta_{pk}(t) \cdot V_{jk}(t))\right) \cdot f'(h_{pj}(t)), \quad (7.5.30)$$

$$V_{kj}(t+1) = V_{kj}(t) + \Delta V_{kj}(t) = V_{kj}(t) + \eta \sum_{p=1}^{P} \delta_{pk}\left(\sum_{j=1}^{l} h_{pj}(t)\right), \quad (7.5.31)$$

$$W_{jk}(t+1) = W_{jk}(t) + \Delta W_{jk}(t) = W_{jk}(t) + \eta \sum_{p=1}^{P} \delta_{pj}\left(\sum_{k=1}^{m} \hat{y}_{pk}(t-1)\right), \quad (7.5.32)$$

$$U_{ji}(t+1) = U_{ji}(t) + \Delta U_{ji}(t) = U_{ji}(t) + \eta \sum_{p=1}^{P} \delta_{pj}\left(\sum_{i=1}^{n} x_{pk}(t)\right). \quad (7.5.33)$$

7.5.3 Jordan Network and Elman Network

After discussing RNNs with output recurrence, we now focus on RNNs with the forget gate. The key equations for this type of RNNs are given by [34, 47]

$$\mathbf{i}_t = \sigma\left(\mathbf{W}_{ix}\mathbf{x}_t + \mathbf{W}_{ih}\mathbf{h}_{t-1} + \mathbf{W}_{ic}\mathbf{c}_{t-1} + \mathbf{b}_i\right), \tag{7.5.34}$$

$$\mathbf{f}_t = \sigma\left(\mathbf{W}_{fx}\mathbf{x}_t + \mathbf{W}_{fh}\mathbf{h}_{t-1} + \mathbf{W}_{fc}\mathbf{c}_{t-1} + \mathbf{b}_f\right), \tag{7.5.35}$$

$$\mathbf{c}_t = \mathbf{f}_t \odot \mathbf{c}_{t-1} + \mathbf{i}_t \odot \tanh\left(\mathbf{W}_{cx}\mathbf{x}_t + \mathbf{W}_{ch}\mathbf{h}_{t-1} + \mathbf{b}_c\right), \tag{7.5.36}$$

$$\mathbf{y}_t = \sigma\left(\mathbf{W}_{ox}\mathbf{x}_t + \mathbf{W}_{oh}\mathbf{h}_{t-1} + \mathbf{W}_{oc}\mathbf{c}_t + \mathbf{b}_o\right), \tag{7.5.37}$$

$$\mathbf{h}_t = \mathbf{y}_t \odot \tanh(\mathbf{c}_t), \tag{7.5.38}$$

where $\odot$ denotes Hadamard product or elementwise product of two vectors, and variables that need to be learned during training are as follows:

- $\mathbf{x}_t \in \mathbb{R}^d$: input vector to the network unit,
- $\mathbf{f}_t \in \mathbb{R}^h$: *forget gate*'s activation vector,
- $\mathbf{i}_t \in \mathbb{R}^h$: *input gate*'s activation vector,
- $\mathbf{y}_t \in \mathbb{R}^h$: *output gate*'s activation vector,
- $\mathbf{h}_t \in \mathbb{R}^h$: output vector of the neural unit,
- $\mathbf{c}_t \in \mathbb{R}^h$: *cell state* vector,
- $\mathbf{W}_{ix}, \mathbf{W}_{fx}, \mathbf{W}_{ox}, \mathbf{W}_{cx} \in \mathbb{R}^{h \times d}$: weight matrices from input vector to input gate, forget gate to input gate, output gate to input gate and cell to input gate, respectively,
- $\mathbf{W}_{ih}, \mathbf{W}_{fh}, \mathbf{W}_{oh}, \mathbf{W}_{ch} \in \mathbb{R}^{h \times h}$: weight matrices from input gate to hidden gate, forget gate to hidden gate, output gate to hidden gate and cell to hidden gate, respectively,
- $\mathbf{b} \in \mathbb{R}^h$: bias vector.

Here the superscripts d and h refer to the number of input features and number of hidden units, respectively, and $\sigma(\cdot)$ is the logistic sigmoid function.

At each step t, the standard RNNs compute the hidden vector $\mathbf{h}_t$ from the current input vector $\mathbf{x}_t$ and the previous hidden vector $\mathbf{h}_{t-1}$ using the nonlinear transformation function ϕ : $\mathbf{h}_t = \phi(\mathbf{x}_t, \mathbf{h}_{t-1})$. This computation is known as *recurrent computation* and can be done via three different directional mechanisms [110]:

1. *Forward mechanism:* It recurs from 1 to n and generate the forward hidden vector sequence: $R(\mathbf{x}_1, \ldots, \mathbf{x}_n) = \overrightarrow{\mathbf{h}}_1, \ldots, \overrightarrow{\mathbf{h}}_n$;
2. *Backward mechanism:* RNNs are performed from n to 1 and provide in the backward hidden vector sequence $R(\mathbf{x}_n, \ldots, \mathbf{x}_1) = \overleftarrow{\mathbf{h}}_n, \ldots, \overleftarrow{\mathbf{h}}_1$;
3. *Bidirectional mechanism:* It runs RNNs in both directions to produce the forward and backward hidden vector sequences, and then concatenate them at each

position to generate the new hidden vector sequence $\mathbf{h}_1^b, \ldots, \mathbf{h}_n^b$ with $\mathbf{h}_i^b = [\overrightarrow{\mathbf{h}}_i, \overleftarrow{\mathbf{h}}_i]$.

The RNN dynamics can be described using deterministic transitions from previous to current hidden states. The deterministic state transition is a function described as

$$\text{RNN}: \quad h_t^{l-1}, h_{t-1}^l \rightarrow h_t^l. \tag{7.5.39}$$

For classical RNNs, this function is given by

$$h_t^l = f\left(T_{n,n} h_t^{l-1} + T_{n,n} h_{t-1}^l\right), \text{ where } f \in \{\text{sigm, tanh}\}. \tag{7.5.40}$$

The RNN also has a special initial bias $\mathbf{b}_h^{\text{init}} \in \mathbb{R}^m$ which replaces the formally undefined expression $\mathbf{W}_{hh} \mathbf{h}_0$ at time $t = 1$.

The RNNs have the following two networks:

- *Jordan network:* For a simple neural network with one hidden layer, if the input matrix is denoted as $\mathbf{X} = [\mathbf{x}_1, \ldots, \mathbf{x}_T]$, weights of hidden layer are denoted as $\mathbf{W}_h$, weights of output layer are denoted as $\mathbf{W}_y$, weights of recurrent computation are denoted as $\mathbf{W}_r$, the hidden representation is denoted as $\mathbf{h}$, and the output is denoted as $\mathbf{y}$, then Jordan Network can be formulated as

$$\mathbf{h}_t = \sigma(\mathbf{W}_h \mathbf{X} + \mathbf{W}_r \mathbf{y}_{t-1}), \quad \mathbf{y} = \sigma(\mathbf{W}_y \mathbf{h}_t).$$

- *Elman network:* It was introduced by Elman in 1990 [29] and is described by

$$\mathbf{h}_t = \sigma(\mathbf{W}_h \mathbf{X} + \mathbf{W}_r \mathbf{h}_{t-1}), \quad \mathbf{y} = \sigma(\mathbf{W}_y \mathbf{h}_t).$$

Figures 7.4 and 7.5 show the difference between recurrent structures of Jordan network and Elman network.

7.5.4 Bidirectional Recurrent Neural Networks

If unfolding an RNN, one can get the structure of a feedforward neural network with infinite depth. A key question is: What the recurrent structures that correspond to the infinite layer of bidirectional models are. The answer lies in the *Bidirectional Recurrent Neural Network (BRNN)* [132].

BRNN was invented by Schuster and Paliwal in 1997 [132] with the goal to introduce a structure that was unfolded to be a bidirectional neural network and can be trained using available input information in the past and future of a specific time frame.

7.5 Recurrent Neural Networks

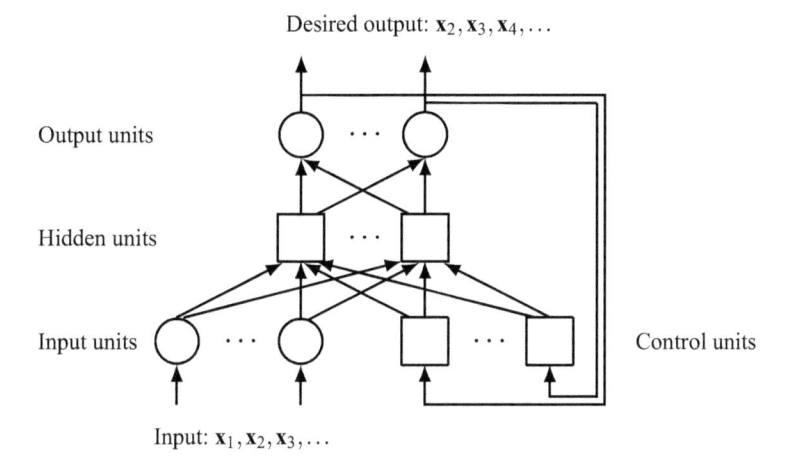

Fig. 7.4 Illustration of Jordan network

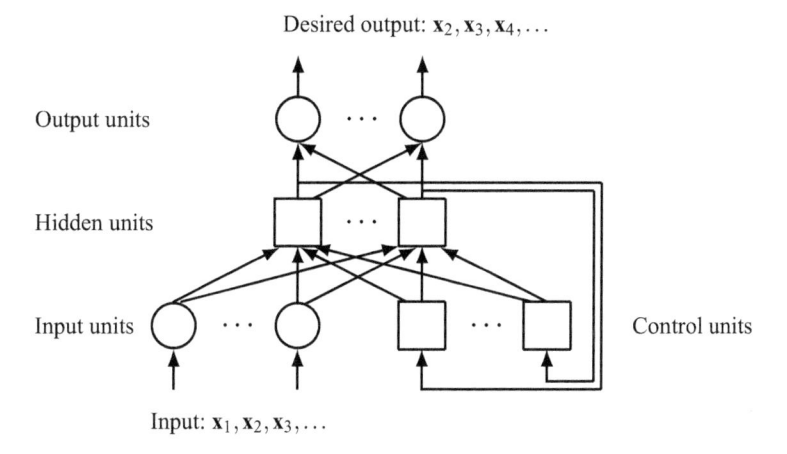

Fig. 7.5 Illustration of Elman network [97]

As shown in Fig. 7.6, a BRNN connects two hidden layers running in opposite directions to a single output, allowing them to receive information from both past and future states.

Unlike standard RNNs, the idea of BRNN is to split the state neurons of a regular RNN into one part that is responsible for the positive time direction (forward states) and another part for the negative time direction (backward states). Neither of these output states are connected to inputs of the opposite directions. With these two time directions considered simultaneously, input data from the past and future of the current time frame can be directly used to minimize the objective function.

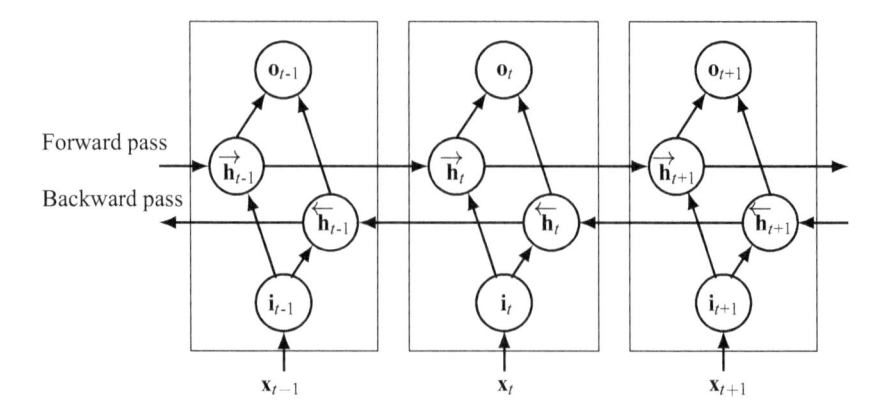

Fig. 7.6 Illustration of the Bidirectional Recurrent Neural Network (BRNN) unfolded in time for three time steps. **Upper:** output units, **Middle:** forward and backward hidden units, **Lower:** input units

By Graves [42], bidirectional RNNs are preferred because each output vector depends on the whole input sequence (rather than on the previous inputs only, as is the case with normal RNNs).

In general training, forward and backward states are processed first in the "forward" pass, then output neurons are passed. For the backward pass, the opposite takes place: output neurons are processed first, then forward and backward states are passed next. Weights are updated only after the forward and backward passes are complete.

For BRNN, the training procedure for the unfolded bidirectional network over time can be summarized as follows [42, 132]:

1. *Forward pass:* Run all input data for one time slice $1 \leq t \leq T$ through the BRNN and determine all predicted outputs.

    - Do forward pass just for forward states (from $t = 1$ to $t = T$) and backward states (from $t = T$ to $t = 1$).
    - Do forward pass for output neurons.

2. *Backward pass:* Calculate the part of the objective function derivative for the time slice $1 \leq t \leq T$ used in the forward pass.

    - Do backward pass for output neurons.
    - Do backward pass just for forward states (from $t = 1$ to $t = T$) and backward states (from $t = T$ to $t = 1$).

7.5 Recurrent Neural Networks

Given an input sequence of length T, $(\mathbf{x}_1, \ldots, \mathbf{x}_T)$, a BRNN computes the forward hidden sequence $(\overrightarrow{h}_1, \ldots, \overrightarrow{h}_T)$, the backward hidden sequence $(\overleftarrow{h}_1, \ldots, \overleftarrow{h}_T)$, and the transcription sequence $(f_1, \ldots, f_T)$ by first iterating the backward layer from $t = T$ to 1 to compute the backward hidden sequence [42]:

$$\overleftarrow{\mathbf{h}}_t = f_h\left(\mathbf{W}_{\overleftarrow{h}i}\mathbf{x}_t + \mathbf{W}_{\overleftarrow{h}\overleftarrow{h}}\overleftarrow{\mathbf{h}}_{t-1} + \mathbf{b}_{\overleftarrow{\mathbf{h}}}\right). \tag{7.5.41}$$

Then iterating the forward and output layers from $t = 1$ to T to compute the backward hidden sequence and the output sequence:

$$\overrightarrow{\mathbf{h}}_t = f_h\left(\mathbf{W}_{\overrightarrow{h}i}\mathbf{x}_t + \mathbf{W}_{\overrightarrow{h}\overrightarrow{h}}\overrightarrow{\mathbf{h}}_{t-1} + \mathbf{b}_{\overrightarrow{h}}\right), \tag{7.5.42}$$

$$\mathbf{y}_t = \mathbf{W}_{o\overrightarrow{h}}\overrightarrow{\mathbf{h}}_t + \mathbf{W}_{o\overleftarrow{h}}\overleftarrow{\mathbf{h}}_t + \mathbf{b}_o. \tag{7.5.43}$$

7.5.5 Long Short-Term Memory (LSTM)

Standard RNNs [113] have been plagued by two major practical problems.

- RNNs have short-term memory problems: short-term memory has a greater impact, but the long-term impact is small, so the gradient of the total output error with respect to previous inputs quickly vanishes as the time lags between relevant inputs, and errors increase. Therefore, long time lags are inaccessible to existing architectures [33].
- Training RNN requires a lot of cost.

Long Short-Term Memory (LSTM) [63] is an RNN architecture designed to be better at storing and accessing information than standard RNNs. LSTM is essentially a recurrent network architecture in conjunction with an appropriate gradient-based learning algorithm.

For general-purpose sequence modeling, it is widely recognized (see, e.g., [43, 44, 63, 112, 135, 143]) that LSTM, as a special RNN structure, has proven to be stable and powerful for modeling long-range dependencies. The major innovation of LSTM is its memory cell $\mathbf{c}_t$ which essentially acts as an accumulator of the state information. The cell is accessed, written, and cleared by several self-parameterized controlling gates. At every time, a new input comes, its information will be accumulated to the cell if the input gate is activated. Moreover, the past cell status $\mathbf{c}_{t-1}$ could be "forgotten" in this process if the forget gate $\mathbf{f}_t$ is on [135].

An RNN composed of LSTM units is often called an LSTM network. A common LSTM unit is composed of a cell, an input gate, an output gate, and a forget gate.

An LSTM layer consists of a set of recurrently connected blocks, known as *memory blocks*. These blocks can be thought of as a differentiable version of the memory chips in a digital computer. Each block contains one or more recurrently connected memory cells and three multiplicative units: the *input gate* (denoted as **i**), *output gate* (denoted as **y**), and *forget gate* (denoted as **f**). Each of these gates can be viewed as a "standard" neuron in a feedforward (or multilayer) neural network that compute an activation (using an activation function) of a weighted sum. The vectors $\mathbf{i}_t, \mathbf{y}_t$, and $\mathbf{f}_t$ represent the activations of the input, output, and forget gates, at time step t, respectively.

The cell output y_c is calculated via the current cell state and four sources of inputs: net_c is input to the cell itself, while net_i, net_f, and net_o are inputs to the input gates, forget gates, and output gates, respectively.

Let j denote the indexes of memory blocks; v be the indexes of memory cells in block j (with S_j cells), such that c_j^v denotes the vth cell of the jth memory block; and w_{lm} be the weight on the connection from unit m to unit l. For the gates f_l, $l \in \{i, o, f\}$ is a logistic sigmoid with range [0, 1].

In standard LSTM, a typical LSTM cell is made of input, forget, and output gates and a cell activation component. These units receive the activation signals from different sources and control the activation of the cell by the designed multipliers, as described below [126].

Introduce the following notation.

- S denotes the input sequence over which the training takes place, and it runs from time τ_0 to τ_1.
- i, f, o denote, respectively, the input gate, forget gate, and output gate.
- c denotes cell, and refers to an element of the set of cells C.
- $x_k(\tau)$ is the network input to unit k at time τ, and $y_k(\tau)$ denotes its activation.
- $s_c(\tau)$ is the state value of cell c at time τ, i.e., its value after the input and forget gates have been applied.
- $t_k(\tau)$ denotes the training target for output unit k at time τ.
- $E(\tau)$ refers to the (scalar) output error of the net at time τ.
- $g(x_c)$ and $h(s_c)$ are, respectively, the cell input and state squashing functions.
- $\sigma(x_i), \sigma(x_f)$, and $\sigma(x_o)$: the squashing function of the input gate, the forget gate, and the output gate, respectively,
- w_{ij} is the weight from unit j to unit i.

The following are pseudocode for full gradient LSTM layer in a multilayer net [45]:

1. *Forward pass*

    - Reset all activations to 0.
    - Running forwards from time τ_0 to time τ_1, feed in the inputs and update the activations. Store all hidden layer and output activations at every time step.

7.5 Recurrent Neural Networks

- For each LSTM block, the activations are updated as follows:
  Input gates:
  $$x_i(\tau) = \sum_{j \in N} w_{ij} y_j(\tau - 1) + \sum_{c \in C} w_{ic} s_c(\tau - 1),$$
  $$y_i(\tau) = \sigma(x_i(\tau)).$$
  Forget gates:
  $$x_f(\tau) = \sum_{j \in N} w_{fj} y_j(\tau - 1) + \sum_{c \in C} w_{fc} s_c(\tau - 1),$$
  $$y_f(\tau) = \sigma(x_f(\tau)).$$

  Cells:
  $$x_c(\tau) = \sum_{j \in N} w_{cj} y_j(\tau - 1), \forall c \in C,$$
  $$s_c(\tau) = y_f s_c(\tau - 1) + y_i g(x_c(\tau)).$$

  Output gates:
  $$x_o(\tau) = \sum_{j \in N} w_{oj} y_j(\tau - 1) + \sum_{c \in C} w_{oc} s_c(\tau),$$
  $$y_o(\tau) = \sigma(x_o(\tau)).$$
  Cell outputs:
  $$y_c(\tau) = y_o(\tau) h(s_c(\tau)), \forall c \in C.$$

2. *Backward pass*

   - Reset all partial derivatives to 0.
   - Starting at time τ_1, propagate the output errors backwards through the unfolded net, using the standard BPTT equations for a softmax output layer and the crossentropy error function:
   $$\delta_k(\tau) = \frac{\partial E(\tau)}{\partial x_k(\tau)} = y_k(\tau) - t_k(\tau), \ k \in \text{output units}.$$

   - For each LSTM block, from time $\tau_1 - 1$ to time τ_0 the δ's are calculated as follows:
     Cell Outputs:
     $$\epsilon_c(\tau) = \sum_{j \in N} w_{jc} \delta_j(\tau + 1), \ \forall c \in C.$$
     Output Gates:
     $$\delta_o(\tau) = \sigma'(x_o(\tau)) \sum_{c \in C} \epsilon_c(\tau) h(s_c(\tau)).$$
     States:
     $$\frac{\partial E(\tau)}{\partial s_c(\tau)} = \epsilon_c(\tau) y_o(\tau) h'(s_c(\tau)) + \frac{\partial E(\tau + 1)}{\partial s_c(\tau + 1)} y_o(\tau + 1)$$
     $$+ \delta_i(\tau + 1) w_{ic} + \delta_f(\tau + 1) w_{fc} + \delta_o(\tau + 1) w_{oc}$$
     Cells:
     $$\delta_c(\tau) = y_i(\tau) g'(x_c(\tau)) \frac{\partial E(\tau)}{\partial s_c(\tau)}.$$
     Forget Gates:
     $$\delta_f(\tau) = \sigma'(x_f(\tau)) \sum_{c \in C} \frac{\partial E(\tau - 1)}{\partial s_c(\tau - 1)}.$$

Input Gates:

$$\delta_i(\tau) = \sigma'(x_i(\tau)) \sum_{c \in C} \frac{\partial E(\tau)}{\partial s_c(\tau)} g(x_c(\tau)).$$

- Using the standard BPTT equation, accumulate the δ's to get the partial derivatives of the cumulative sequence error:

$$\text{define } E_{\text{total}}(S) = \sum_{\tau=\tau_0}^{\tau_1} E(\tau),$$

$$\text{define } \nabla_{ij}(S) = \frac{E_{\text{total}}(S)}{\partial w_{ij}},$$

$$\Rightarrow \nabla_{ij}(S) = \sum_{\tau=\tau_0+1}^{\tau_1} \delta_i(\tau) y_j(\tau-1).$$

3. *Update Weights:* After the presentation of sequence S, with learning rate α and momentum m, update all weights with the standard equation for gradient descent with momentum:

$$\Delta w_{ij}(S) = -\alpha \nabla_{ij}(S) + m \Delta w_{ij}(S-1).$$

Activation functions may be taken as follows:

- σ_f: sigmoid function.
- σ_i: hyperbolic tangent function.
- σ_o: hyperbolic tangent function or, as the peephole LSTM suggests, $\sigma_h(x) = x$.

7.5.6 Improvement of Long Short-Term Memory

RNNs are powerful sequence learners, but they have following limitations:

- They require pre-segmented training data, and post-processing to transform their outputs into label sequences, their applicability for labeling unsegmented sequence data has limited [46].
- They may fail to understand input structures more complicated than a sequence [175].

To avoid the pre-segmented training data and post-processing into label sequences, Graves et al. [46] in 2006 proposed a method for training RNNs to label unsegmented sequences directly, thereby solving both problems. This method is called the *Connectionist Temporal Classification* (CTC).

7.5 Recurrent Neural Networks

Let S be a set of training examples drawn from a fixed distribution $\mathcal{D}_{\mathcal{X} \times \mathcal{Z}}$. Each example in S consists of a pair of sequences $(\mathbf{x}, \mathbf{z})$, where the target sequence $\mathbf{z} = (z_1, \ldots, z_U)$ is at most as long as the input sequence $\mathbf{x} = (x_1, \ldots, x_T)$, i.e., $U \leq T$. These input and target sequences cannot be aligned a priori due to having not the same length.

The CTC method uses a softmax layer to define a separate output distribution $P(k|t)$ at every step t along the input sequence. This distribution covers the K phonemes plus an extra blank symbol Ø which represents a non-output (the softmax layer is therefore size $K + 1$).

RNN strained with CTC are generally bidirectional, to ensure that every $P(k|t)$ depends on the entire input sequence, and not just the inputs up to t. For example, for deep bidirectional networks, with $P(k|t)$ defined by [47]

$$\mathbf{y}_t = \mathbf{W}_{\vec{h}^N \mathbf{y}} \vec{\mathbf{h}}_t^N + \mathbf{W}_{\overleftarrow{h}^N \mathbf{y}} \overleftarrow{\mathbf{h}}_t^N + \mathbf{b}_y, \tag{7.5.44}$$

$$P(k|t) = \frac{\exp(y_t[k])}{\sum_{k'=1}^{K} \exp(y_t[k'])}, \tag{7.5.45}$$

where $y_t[k]$ is the kth element of the length $K+1$ unnormalized output vector $\mathbf{y}_t$, N denotes the number of bidirectional levels, and $\mathbf{W}_{\vec{h}^N \mathbf{y}}$ and $\mathbf{W}_{\overleftarrow{h}^N \mathbf{y}}$ denote the weight matrix from forward and backward hidden states to the output gate, respectively.

The LSTM can learn longer sequence correlation, but it may fail to understand input structures more complicated than a sequence. To overcome the gradient vanishing problem in LSTM and learn longer term dependencies from input, Zhu et al. [175] developed a model of multiple child cells or multiple descendant cells for LSTM, called S-LSTM network. S-LSTM is made of S-LSTM memory blocks and works based on a hierarchical structure.

Each S-LSTM memory block contains one input gate and one output gate, but different from LSTM, S-LSTM has two or more forget gates. The number of forget gates depends on the number of children of a node. For two children, their hidden vectors are denoted as $\mathbf{h}_{t-1}^L$ for the left child and $\mathbf{h}_{t-1}^R$ for the right child. These hidden vectors are taken in as the inputs of the current block.

The forward computation of a S-LSTM memory block is specified as follows [175]:

1. *Input gate:* The input gate $\mathbf{i}_t$ contains four resources of information: the hidden vectors ($\mathbf{h}_{t-1}^L$ and $\mathbf{h}_{t-1}^R$) and cell vectors ($\mathbf{c}_{t-1}^L$ and $\mathbf{c}_{t-1}^R$) of its two children, i.e.,

$$\mathbf{i}_t = \sigma\big(\mathbf{W}_{hi}^L \mathbf{h}_{t-1}^L + \mathbf{W}_{hi}^R \mathbf{h}_{t-1}^R + \mathbf{W}_{ci}^L \mathbf{c}_{t-1}^L + \mathbf{W}_{ci}^R \mathbf{c}_{t-1}^R + \mathbf{b}_i\big), \tag{7.5.46}$$

where σ is the elementwise logistic function used to confine the gating signals to be in the range of $[0, 1]$.

2. *Forget gate:* The above four sources of information are also used to form the gating signals for the left forget gate $\mathbf{f}^L_{t-1}$ and right forget gate $\mathbf{f}^R_{t-1}$ via different weight matrices:

$$\mathbf{f}^L_t = \sigma\left(\mathbf{W}^L_{hf_l}\mathbf{h}^L_{t-1} + \mathbf{W}^R_{hf_l}\mathbf{h}^R_{t-1} + \mathbf{W}^L_{cf_l}\mathbf{c}^L_{t-1} + \mathbf{W}^R_{cf_l}\mathbf{c}^R_{t-1} + \mathbf{b}_{f_l}\right), \quad (7.5.47)$$

$$\mathbf{f}^R_t = \sigma\left(\mathbf{W}^L_{hf_r}\mathbf{h}^L_{t-1} + \mathbf{W}^R_{hf_r}\mathbf{h}^R_{t-1} + \mathbf{W}^L_{cf_r}\mathbf{c}^L_{t-1} + \mathbf{W}^R_{cf_r}\mathbf{c}^R_{t-1} + \mathbf{b}_{f_r}\right). \quad (7.5.48)$$

3. *Cell gate:* The cell here considers the copies from both children's cell vectors ($\mathbf{c}^L_{t-1}, \mathbf{c}^R_{t-1}$), gated with separated forget gates. The left and right forget gates can be controlled independently, allowing the pass-through of information from children's cell vectors:

$$\mathbf{x}_t = \mathbf{W}^L_{hx}\mathbf{h}^L_{t-1} + \mathbf{W}^R_{hx}\mathbf{h}^R_{t-1} + \mathbf{b}_x, \quad (7.5.49)$$

$$\mathbf{c}_t = \mathbf{f}^L_t \odot \mathbf{c}^L_{t-1} + \mathbf{f}^R_t \odot \mathbf{c}^R_{t-1} + \mathbf{i}_t \odot \tanh(\mathbf{x}_t). \quad (7.5.50)$$

4. *Output gate:* The output gate $\mathbf{o}_t$ considers the hidden vectors from the children and the current cell vector:

$$\mathbf{o}_t = \sigma\left(\mathbf{W}^L_{ho}\mathbf{h}^L_{t-1} + \mathbf{W}^R_{ho}\mathbf{h}^R_{t-1} + \mathbf{W}_{co}\mathbf{c}_t + \mathbf{b}_o\right). \quad (7.5.51)$$

5. *Hidden state:* The hidden vector $\mathbf{h}_t$ and the cell vector $\mathbf{c}_t$ of the current block are passed to the parent and are used depending on if the current block is a left or right child of its parent:

$$\mathbf{h}_t = \mathbf{o}_t \odot \tanh(\mathbf{c}_t). \quad (7.5.52)$$

The backward computation of a S-LSTM memory block uses backpropagation over structures [175]:

$$\epsilon^h_t = \frac{\partial o}{\partial \mathbf{h}_t}, \quad (7.5.53)$$

$$\frac{\partial o_t}{\partial \mathbf{x}} = \epsilon^h_t \odot \tanh(\mathbf{c}_t) \odot \sigma'(o_t), \quad (7.5.54)$$

$$\frac{\partial f^l_t}{\partial \mathbf{x}} = \epsilon^c_t \odot \mathbf{c}^L_{t-1} \odot \sigma'\left(f^L_t\right). \quad (7.5.55)$$

In convenient LSTM each gate receives connections from the input units and the outputs of all cells, but there is no direct connection from the *"Constant Error Carousel"* (CEC) that provides short-term memory storage for extended time periods. A simple but effective remedy is to add weighted "peephole" connections from the CEC to the gates of the same memory block [34]. The peephole connections allow all input gate $\mathbf{i}$, output gate $\mathbf{y}$, and forget gate $\mathbf{f}$ to inspect the current cell state even when the output gate is closed. This information can be essential for finding well-working network solutions.

The peephole connections from the memory cell **c** to the three gates **i**, **y**, and **f** actually denote the contributions of the activation of the memory cell **c** at time step $t-1$, i.e., the contribution of $\mathbf{c}_{t-1}$ rather than $\mathbf{c}_t$.

LSTM with peephole connections is called the *peephole LSTM* that is proposed by Gers et al. [33, 34]. The key update equations of peephole LSTM are as follows:

$$\mathbf{f}_t = \sigma_g(\mathbf{W}_f \mathbf{x}_t + \mathbf{U}_f \mathbf{c}_{t-1} + \mathbf{b}_f), \tag{7.5.56}$$

$$\mathbf{i}_t = \sigma_g(\mathbf{W}_i \mathbf{x}_t + \mathbf{U}_i \mathbf{c}_{t-1} + \mathbf{b}_i), \tag{7.5.57}$$

$$\mathbf{y}_t = \sigma_g(\mathbf{W}_o \mathbf{x}_t + \mathbf{U}_o \mathbf{c}_{t-1} + \mathbf{b}_o), \tag{7.5.58}$$

$$\mathbf{c}_t = \mathbf{f}_t \odot \mathbf{c}_{t-1} + \mathbf{i}_t \odot \sigma_c(\mathbf{W}_c \mathbf{x}_t + \mathbf{U}_c \mathbf{c}_{t-1} + \mathbf{b}_c), \tag{7.5.59}$$

$$\mathbf{h}_t = \mathbf{y}_t \odot \sigma_h(\mathbf{c}_t). \tag{7.5.60}$$

7.6 Boltzmann Machines

The *Boltzmann machine*, invented by Ackley et al. in 1985 [1], is another important type of neural networks that relies on a stochastic (probabilistic) form of learning. The Boltzmann machine is closely related to the Hopfield network.

7.6.1 Hopfield Network and Boltzmann Machines

Energy-based models are a special class of neural networks. The simplest energy model is the *Hopfield Network*, first introduced by Hopfield in 1982 [65], which is usually viewed as a form of recurrent neural network.

The Hopfield network is a fully connected neural network with binary thresholding neural units whose values are either 0 or 1. These units are fully connected in a "recurrent" way in which the connection between weights and neurons are bidirectional.

With this setting, the energy of a Hopfield network is defined as

$$E = -\sum_i s_i b_i - \sum_{i,j} s_i s_j w_{ij}, \tag{7.6.1}$$

where s_i is the state of unit i, b_i denotes its bias; w_{ij} denotes the bidirectional weights connecting units i and j.

There are two inference procedures of Hopfield network [155].

(a) *Asynchronous update:* For a state of data, the network tests if inverting the state of one units, will cause the energy to decrease. If so, the network will invert the state and proceed to test the next unit.
(b) *Synchronous update:* The network first tests for all the units and then inverts all the units-to-invert simultaneously.

The main disadvantages of Hopfield network are twofold [155].

- Both the asynchronous and synchronous update methods may lead to a local optimum. The synchronous update may even result in an increasing of energy and may converge to an oscillation or loop of states.
- The Hopfield network cannot keep the memory very efficient because a network of N units can only store memory up to $0.15N^2$ bits, but the number of bits required to store N units are $N^2 \log(2M + 1)$ (M is instances of data) [65].

Interestingly and importantly, with the introduction of hidden units in the Hopfield network, the model can be upgraded from the initial ideas of the Hopfield network and evolve to replace it. This is the basic idea of the popular Boltzmann machine and restricted Boltzmann machine.

The stochastic neurons of a Boltzmann machine partition into two functional groups: hidden units and visible units. Figure 7.7 shows a comparison between the Hopfield network and the Boltzmann machine [155].

In a Boltzmann machine, only visible units are connected with data and hidden units are used to assist visible units to describe the distribution of data, but it still maintains a fully connected network among these units.

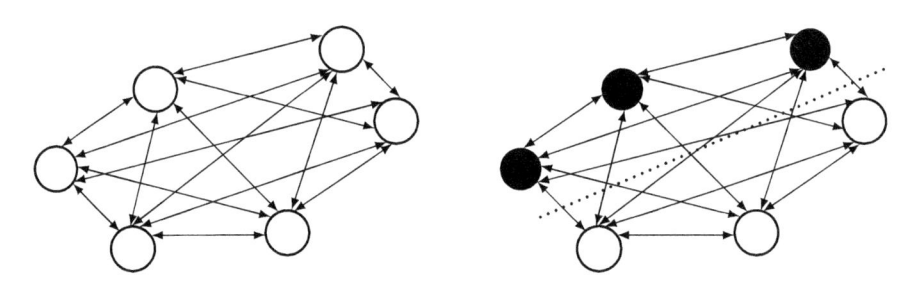

Fig. 7.7 A comparison between Hopfield network and Boltzmann machine. **Left**: is Hopfield network in which a fully connected network of six binary thresholding neural units. Every unit is connected with data. **Right**: is Boltzmann machine whose model splits into two parts: visible units and hidden units (shaded nodes). The dashed line is used to highlight the model separation

7.6 Boltzmann Machines

The Boltzmann machine and the Hopfield network share the following common features [52]:

- their processing units have binary values (say $+1$ and -1) for their states;
- all the synaptic connections between their units are symmetric;
- the units are picked at random and one at a time for updating;
- they have no self-feedback.

There are three important differences between the Boltzmann machine and the Hopfield network [52].

(a) The Boltzmann machine permits the use of hidden neurons, while no such neurons exist in the Hopfield network.
(b) The Boltzmann machine uses stochastic neurons with a probabilistic firing mechanism, whereas the standard Hopfield network uses neurons with a deterministic firing mechanism.
(c) The Boltzmann machine operates in a supervised manner, while the Hopfield network operates in an unsupervised manner.

The above common features and important differences help in providing a better understanding of the following discussion on the Boltzmann machine.

The Boltzmann machine has the same definition of energy functions as that of the Hopfield network, except the Boltzmann machine splits the energy function according to hidden units and visible units [65]:

$$E(\mathbf{x}, \mathbf{h}) = -\mathbf{b}^T \mathbf{x} - \mathbf{c}^T \mathbf{h} - \mathbf{h}^T \mathbf{W} \mathbf{x} - \mathbf{x}^T \mathbf{U} \mathbf{x} - \mathbf{h}^T \mathbf{V} \mathbf{h}, \qquad (7.6.2)$$

where $\mathbf{b}$ and $\mathbf{c}$ are the offsets associated with the input $\mathbf{x}$ (visible vector) and the hidden output $\mathbf{h}$ (hidden vector), respectively, while $\mathbf{W}$, $\mathbf{U}$, and $\mathbf{V}$ are the weight matrices of hidden-visible units, visible-visible units, and hidden-hidden units, respectively.

If considering only one observed part (denoted by $\mathbf{x}$) and a hidden part $\mathbf{h}$ then the energy-based probability distribution can be defined as [8, 65]:

$$P(\mathbf{x}, \mathbf{h}) = \frac{e^{-E(\mathbf{x},\mathbf{h})}}{Z}, \qquad (7.6.3)$$

where the normalizing factor with a sum running over the visible and hidden spaces given by

$$Z = \sum_{\mathbf{x},\mathbf{h}} e^{-E(\mathbf{x},\mathbf{h})} \qquad (7.6.4)$$

is called the *partition function* by analogy with physical systems.

Because only **x** is observed, it only needs to care about the marginal

$$p(\mathbf{x}) = \frac{1}{Z} \sum_{\mathbf{h}} e^{-E(\mathbf{x},\mathbf{h})}. \tag{7.6.5}$$

Substituting (7.6.4) into (7.6.5) yields

$$p(\mathbf{x}) = \frac{\sum_{\mathbf{h}} e^{-E(\mathbf{x},\mathbf{h})}}{\sum_{\bar{\mathbf{x}},\mathbf{h}} e^{-E(\bar{\mathbf{x}},\mathbf{h})}} \tag{7.6.6}$$

whose log-likelihood form is given by

$$\log p(\mathbf{x}) = \log \sum_{\mathbf{h}} e^{-E(\mathbf{x},\mathbf{h})} - \log \sum_{\bar{\mathbf{x}},\mathbf{h}} e^{-E(\bar{\mathbf{x}},\mathbf{h})}. \tag{7.6.7}$$

Then, by letting $\theta = (\mathbf{b}, \mathbf{c}, \mathbf{W}, \mathbf{U}, \mathbf{V})$ denote the parameters of the Boltzmann model, one has [8]

$$\begin{aligned}
\frac{\partial \log p(\mathbf{x})}{\partial \theta} &= \frac{\partial \log \sum_{\mathbf{h}} e^{-E(\mathbf{x},\mathbf{h})}}{\partial \theta} - \frac{\partial \log \sum_{\bar{\mathbf{x}},\mathbf{h}} e^{-E(\bar{\mathbf{x}},\mathbf{h})}}{\partial \theta} \\
&= -\frac{1}{\sum_{\mathbf{h}} e^{-E(\mathbf{x},\mathbf{h})}} \sum_{\mathbf{h}} e^{-E(\mathbf{x},\mathbf{h})} \frac{\partial E(\mathbf{x},\mathbf{h})}{\partial \theta} \\
&\quad + \frac{1}{\sum_{\bar{\mathbf{x}},\mathbf{h}} e^{-E(\bar{\mathbf{x}},\mathbf{h})}} \sum_{\bar{\mathbf{x}},\mathbf{h}} e^{-E(\bar{\mathbf{x}},\mathbf{h})} \frac{\partial e^{-E(\bar{\mathbf{x}},\mathbf{h})}}{\partial \theta} \\
&= -\sum_{\mathbf{h}} p(\mathbf{h}|\mathbf{x}) \frac{\partial E(\mathbf{x},\mathbf{h})}{\partial \theta} + \sum_{\bar{\mathbf{x}},\mathbf{h}} p(\bar{\mathbf{x}},\mathbf{h}) \frac{\partial E(\bar{\mathbf{x}},\mathbf{h})}{\partial \theta}.
\end{aligned} \tag{7.6.8}$$

Let $\mathbf{s} = \begin{bmatrix} \mathbf{x} \\ \mathbf{h} \end{bmatrix}$ denote all the units in the Boltzmann machine. Then (7.6.2) can be rewritten as

$$E(\mathbf{s}) = -[\mathbf{b}^T, \mathbf{c}^T] \begin{bmatrix} \mathbf{x} \\ \mathbf{h} \end{bmatrix} - [\mathbf{x}^T, \mathbf{h}^T] \begin{bmatrix} \mathbf{U} & \mathbf{O} \\ \mathbf{W} & \mathbf{V} \end{bmatrix} \begin{bmatrix} \mathbf{x} \\ \mathbf{h} \end{bmatrix} = -\mathbf{d}^T \mathbf{s} - \mathbf{s}^T \mathbf{A} \mathbf{s}, \tag{7.6.9}$$

where

$$\mathbf{d} = \begin{bmatrix} \mathbf{b} \\ \mathbf{c} \end{bmatrix}, \quad \mathbf{A} = \begin{bmatrix} \mathbf{U} & \mathbf{O} \\ \mathbf{W} & \mathbf{V} \end{bmatrix}. \tag{7.6.10}$$

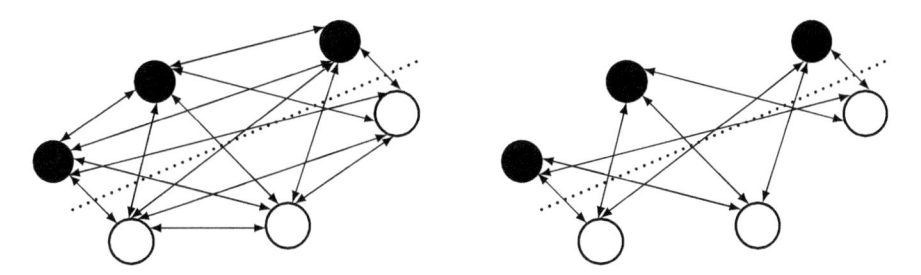

Fig. 7.8 A comparison between **Left** Boltzmann machine and **Right** restricted Boltzmann machine. In the restricted Boltzmann machine there are no connections between hidden units (shaded nodes) and no connections between visible units (unshaded nodes)

7.6.2 Restricted Boltzmann Machine

A *Restricted Boltzmann machine* (RBM) is a version of the Boltzmann machine with an added restriction: there should be no connections either between visible units or between hidden units. The RBM was originally known as Harmonium when invented by Smolensky in 1986 [139].

Figure 7.8 shows a comparison between the restricted Boltzmann machine and the Boltzmann machine [155].

The RBM is a two layer, bipartite, undirected graphical model with a set of binary hidden units **h**, a set of (binary or real-valued) visible units **x**, and symmetric connections between these two layers represented by a weight matrix **W**. The probabilistic semantics for an RBM is denoted by $p(\mathbf{x}; \mathbf{h})$, and is defined by its energy function:

$$p(\mathbf{x}, \mathbf{h}) = \frac{e^{-E(\mathbf{x},\mathbf{h})}}{Z(\mathbf{h})}, \qquad (7.6.11)$$

where $Z(\mathbf{h}) = \sum_{\bar{\mathbf{h}}} e^{-E(\mathbf{x},\bar{\mathbf{h}})}$ is known as the *partition function* for hidden units.

Consider a training set of binary vectors which are assumed to be binary images. The training set can be modeled using a two-layer network called a RBM in which stochastic, binary pixels are connected to stochastic, binary feature detectors using symmetrically weighted connections. Because states of the pixels are observed, these pixels correspond to "visible" units of the RBM, while the feature detectors correspond to "hidden" units. Without connections either between visible units or between hidden units, **U** and **V** in (7.6.2) are two null weight matrices.

Let $x_i, i = 1, \ldots, m$ and $h_j, j = 1, \ldots, F$ be the observed visible variables and the binary values of hidden (latent) variables, respectively. Then the energy function (7.6.2) of the Boltzmann machine reduces to that of the restricted Boltzmann machine [65]:

$$E(\mathbf{x}, \mathbf{h}) = -\sum_{i=1}^{m} b_i x_i - \sum_{j=1}^{F} c_j h_j - \sum_{i,j} x_i W_{ij} h_j$$
$$= -\mathbf{x}^T \mathbf{W} \mathbf{h} - \mathbf{b}^T \mathbf{x} - \mathbf{c}^T \mathbf{h}, \tag{7.6.12}$$

where x_i and h_j are the binary states of visible unit i and hidden unit j, b_i and c_j are their biases, and W_{ij} is the weight between them, while $\mathbf{b} = [b_i]$ is visible unit bias vector and $\mathbf{c} = [c_j]$ is hidden unit bias vector.

In regression problems, the visible units $\mathbf{x}$ are real-valued, and the energy function is defined as

$$E(\mathbf{x}, \mathbf{h}) = \frac{1}{2}\sum_{i=1}^{m} x_i^2 - \sum_{i=1}^{m}\sum_{j=1}^{F} x_i W_{ij} h_j - \sum_{i=1}^{m} b_i x_i - \sum_{j=1}^{F} c_j h_j$$
$$= \frac{1}{2}\mathbf{x}^T \mathbf{x} - \mathbf{x}^T \mathbf{W} \mathbf{h} - \mathbf{b}^T \mathbf{x} - \mathbf{c}^T \mathbf{h}. \tag{7.6.13}$$

Clearly, the hidden units are conditionally independent of one another given the visible layer, and vice versa. In particular, the units of a binary layer (conditioned on the other layer) are independent Bernoulli random variables, and if the visible layer is real-valued then the visible units (conditioned on the hidden layer) are Gaussian with diagonal covariance [92].

Therefore, a tractable expression for the conditional probability for RBM can be readily obtained as [8]:

$$p(\mathbf{h}|\mathbf{x}) = \frac{p(\mathbf{x})}{p(\mathbf{x}, \mathbf{h})}$$
$$= \frac{\exp\left(\mathbf{b}^T \mathbf{x} + \mathbf{c}^T \mathbf{h} + \mathbf{x}^T \mathbf{W} \mathbf{h}\right)}{\sum_{\bar{\mathbf{h}}} \exp\left(\mathbf{b}^T \mathbf{x} + \mathbf{c}^T \bar{\mathbf{h}} + \mathbf{x}^T \mathbf{W} \bar{\mathbf{h}}\right)}$$
$$= \frac{\prod_{j=1}^{F} \exp\left(c_j h_j + h_j \mathbf{w}_j^T \mathbf{x}\right)}{\prod_{j=1}^{F} \sum_{j'=1}^{F} \exp\left(c_j h_{j'} + h_{j'} \mathbf{w}_j^T \mathbf{x}\right)}$$
$$= \prod_{j=1}^{F} \frac{\exp\left(h_j(c_j + \mathbf{w}_j^T \mathbf{x})\right)}{\sum_{j'=1}^{F} \exp\left(h_{j'}(c_j + \mathbf{w}_j^T \mathbf{x})\right)}$$
$$= \prod_{j=1}^{F} P(h_j|\mathbf{x}), \tag{7.6.14}$$

where $\mathbf{w}_j$ is the jth column of the weight matrix $\mathbf{W} = [\mathbf{w}_1, \ldots, \mathbf{w}_F] \in \mathbb{R}^{m \times F}$.

7.6 Boltzmann Machines

By the special structure of RBM (that is, there is connection between layers and no connection within layers) it is known that the activation states of hidden units are conditionally independent given the state of visible units. At this time, the activation probability of the jth hidden unit is given by

$$p(h_j = 1|\mathbf{x}) = \frac{e^{c_j + \mathbf{w}_j^T \mathbf{x}}}{1 + e^{c_j + \mathbf{w}_j^T \mathbf{x}}} = \sigma\left(c_j + \mathbf{w}_j^T \mathbf{x}\right), \quad (7.6.15)$$

where $\sigma(z)$ is the logistic sigmoid function $1/(1 + \exp(z))$.

Since $\mathbf{x}$ and $\mathbf{h}$ play a symmetric role in the energy function, a similar derivation gives the conditional probability of $\mathbf{x}$ given $\mathbf{h}$:

$$p(\mathbf{x}|\mathbf{h}) = \prod_{i=1}^{m} p(x_i|\mathbf{h}). \quad (7.6.16)$$

Because the structure of RBM is symmetrical, when the state of the hidden unit is given, the activation state of each visible unit is conditionally independent, that is, the activation probability of the ith visible unit is given by

$$p(x_i = 1|\mathbf{h}) = \sigma\left(b_i + \tilde{\mathbf{w}}_i \mathbf{h}\right), \quad (7.6.17)$$

where $\tilde{\mathbf{w}}_i$ is the ith row of $\mathbf{W} \in \mathbb{R}^{m \times F}$.

Consider a probability distribution over a vector $\mathbf{x}$ and with parameters $\mathbf{W}$ [15]:

$$p(\mathbf{x}; \mathbf{W}) = \frac{1}{Z(\mathbf{W})} e^{-E(\mathbf{x}; \mathbf{W})}, \quad (7.6.18)$$

where $Z(\mathbf{W}) = \sum_{\mathbf{x}} e^{-(\mathbf{x}; \mathbf{W})}$ is a normalization constant and $E(\mathbf{x}; \mathbf{W})$ is an energy function.

Maximum-likelihood (ML) learning of the parameters $\mathbf{W}$ given an i.i.d. sample $\mathcal{X} = \{\mathbf{x}_n\}_{n=1}^{N}$ can be updated by gradient ascent:

$$\mathbf{W}^{(t+1)} = \mathbf{W}^{(t)} + \eta \left.\frac{\partial L(\mathbf{W}; \mathcal{X})}{\partial \mathbf{W}}\right|_{\mathbf{W}^{(t)}}, \quad (7.6.19)$$

where the learning rate η needs not be constant, and the average log-likelihood is

$$L(\mathbf{W}; \mathcal{X}) = \frac{1}{N} \sum_{n=1}^{N} \log p(\mathbf{x}_n; \mathbf{W})$$
$$= \langle \log p(\mathbf{x}; \mathbf{W}) \rangle_0$$
$$= -\langle E(\mathbf{x}; \mathbf{W}) \rangle_0 - \log Z(\mathbf{W}), \quad (7.6.20)$$

where $\langle \cdot \rangle_0$ denotes an average with respect to the data distribution, i.e., $P_0(\mathbf{x}) = \langle E(\mathbf{x}; \mathbf{W}) \rangle_0 = \frac{1}{N} \sum_{n=1}^{N} \delta(\mathbf{x} - \mathbf{x}_n)$. A well-known difficulty arises in the computation of the gradient

$$\frac{\partial L(\mathbf{W}; \mathcal{X})}{\partial \mathbf{W}} = -\left\langle \frac{\partial E(\mathbf{x}; \mathbf{W})}{\partial \mathbf{W}} \right\rangle_0 + \left\langle \frac{\partial E(\mathbf{x}; \mathbf{W})}{\partial \mathbf{W}} \right\rangle_\infty, \qquad (7.6.21)$$

where $\langle \cdot \rangle_\infty$ denotes an average with respect to the model distribution $p_\infty(\mathbf{x}; \mathbf{W}) = p(\mathbf{x}; \mathbf{W})$. The average $\langle \cdot \rangle_0$ is readily computed using the sample data $\mathcal{X} = \{\mathbf{x}_n\}_{n=1}^{N}$, but the average $\langle \cdot \rangle_\infty$ involves the normalization constant $Z(\mathbf{W})$, which cannot generally be computed efficiently (being a sum of an exponential number of terms).

7.6.3 Contrastive Divergence Learning

In the RBM framework, there are no direct interactions between the hidden units and no direct interactions between the visible units that represent the pixels. In this case, there is a simple and efficient method called "*contrastive divergence*" (CD) [58] to learn a good set of feature detectors from a set of training images.

The RBM parameters are optimized by performing stochastic gradient ascent on the log-likelihood of training data. Unfortunately, computing the exact gradient of the log-likelihood is usually intractable. Instead of computing the exact gradient of the log-likelihood, one can use the contrastive divergence method to approximate the exact gradient, which has been shown to work well in practice.

Consider a probability distribution over a vector $\mathbf{x}$ weighted by a parameter matrix $\mathbf{W}$:

$$p(\mathbf{x}; \mathbf{W}) = \frac{1}{Z(\mathbf{W})} e^{-E(\mathbf{x}; \mathbf{W})}, \qquad (7.6.22)$$

where $E(\mathbf{x}; \mathbf{W})$ is an energy function and

$$Z(\mathbf{W}) = \sum_{\mathbf{x}} e^{-E(\mathbf{x}; \mathbf{W})} \qquad (7.6.23)$$

is a normalization constant. The log-likelihood of $p(\mathbf{x}; \mathbf{W})$ is given by

$$L(\mathbf{x}; \mathbf{W}) = \log p(\mathbf{x}; \mathbf{W}) = \log e^{-E(\mathbf{x}; \mathbf{W})} - \log Z(\mathbf{W})$$
$$= -E(\mathbf{x}; \mathbf{W}) + \sum_{\mathbf{x}} E(\mathbf{x}; \mathbf{W}). \qquad (7.6.24)$$

7.6 Boltzmann Machines

Given an independent identically distributed (i.i.d.) sample set $X = \{\mathbf{x}_n\}_{n=1}^N$, the average log-likelihood of $p(\{\mathbf{x}_n\}_{n=1}^N; \mathbf{W})$, denoted by $L(X; \mathbf{W})$, is defined as

$$L(X; \mathbf{W}) = \frac{1}{N}\sum_{n=1}^N L(\mathbf{x}_n; \mathbf{W}) = -\frac{1}{N}\sum_{n=1}^N E(\mathbf{x}_n; \mathbf{W}) + \sum_{\mathbf{x}} \frac{1}{N}\sum_{n=1}^N E(\mathbf{x}_n; \mathbf{W})$$

$$= -\langle E(\mathbf{x}; \mathbf{W})\rangle_0 + \langle E(\mathbf{x}; \mathbf{W})\rangle_\infty, \qquad (7.6.25)$$

where

$$\langle E(\mathbf{x}; \mathbf{W})\rangle_0 = \frac{1}{N}\sum_{n=1}^N E(\mathbf{x}; \mathbf{W})|_{\mathbf{x}=\mathbf{x}_n}, \qquad (7.6.26)$$

$$\langle E(\mathbf{x}; \mathbf{W})\rangle_\infty = \sum_{\mathbf{x}} \frac{1}{N}\sum_{n=1}^N E(\mathbf{x}; \mathbf{W})|_{\mathbf{x}=\mathbf{x}_n} \qquad (7.6.27)$$

denote an average with respect to the data distribution $p_0(\mathbf{x}) = \sum_{n=1}^N \delta(\mathbf{x} - \mathbf{x}_n)$ and an average with respect to the model distribution $p_\infty(\mathbf{x}; \mathbf{W}) = P(\mathbf{x}; \mathbf{W})$, respectively.

Therefore, maximum likelihood (ML) learning of the weight matrix $\mathbf{W}$ at the time t can be updated by

$$\mathbf{W}^{(t+1)} = \mathbf{W}^{(t)} - \eta \left.\frac{\partial L(\mathbf{W}; X)}{\partial \mathbf{W}}\right|_{\mathbf{W}^{(t)}}, \qquad (7.6.28)$$

where η is a learning rate that may allow to be nonconstant, and the gradient is given by

$$\frac{\partial L(\mathbf{W}; X)}{\partial \mathbf{W}} = -\left\langle \frac{\partial E(\mathbf{x}; \mathbf{W})}{\partial \mathbf{W}}\right\rangle_0 + \left\langle \frac{\partial E(\mathbf{x}; \mathbf{W})}{\partial \mathbf{W}}\right\rangle_\infty. \qquad (7.6.29)$$

Although the first term $\langle\cdot\rangle_0$ is readily computed, the second term $\langle\cdot\rangle_\infty$ involves a sum of an exponential number of terms, which cannot generally be computed efficiently.

To avoid the difficulty in computing the log-likelihood gradient, Hinton [58] proposed the contrastive divergence (CD) method which approximately follows the gradient of a different function. ML learning minimizes the Kullback–Leibler divergence

$$\mathrm{KL}(p_0\|p_\infty) = \sum_{\mathbf{x}} p_0(\mathbf{x}) \log \frac{p_0(\mathbf{x})}{p(\mathbf{x}; \mathbf{W})}. \qquad (7.6.30)$$

CD learning approximately follows the gradient of the difference of two divergences [58]:

$$\text{CD}_n = \text{KL}(p_0 \| p_\infty) - \text{KL}(p_n \| p_\infty). \tag{7.6.31}$$

For fully visible Boltzmann machines (VBMs) without hidden units (i.e., $\mathbf{h} = \mathbf{0}$) with the input $\mathbf{x} = [x_1, \ldots, x_d]^T \in \mathbb{R}^{d \times d}$, the energy function $E(\mathbf{x}; \mathbf{W}) = -\frac{1}{2}\mathbf{x}^T \mathbf{W}\mathbf{x}$, where $\mathbf{W} = [w_{ij}]$ is a symmetric $d \times d$ matrix of real-valued weights. In VBMs, the log-likelihood has a unique optimum because its Hessian is negative definite. Since $\partial E/\partial w_{ij} = -x_i x_j$, ML learning takes the form [15]:

$$w_{ij}(t+1) = w_{ij}(t) + \eta\big(\langle x_i x_j \rangle_0 - \langle x_i x_j \rangle_\infty\big), \tag{7.6.32}$$

while CD_n learning takes the form

$$w_{ij}(t+1) = w_{ij}(t) + \eta\big(\langle x_i x_j \rangle_0 - \langle x_i x_j \rangle_n\big), \tag{7.6.33}$$

$$b_i(t+1) = b_i(t) + \eta\big(\langle x_i \rangle_0 - \langle x_i \rangle_n\big), \tag{7.6.34}$$

$$c_j(t+1) = c_j(t) + \eta\big(\langle h_j \rangle_0 - \langle h_j \rangle_n\big). \tag{7.6.35}$$

Based on the symmetrical structure of RBM model and the conditional independence of neuron states, we can use Gibbs sampling to obtain random samples of distribution defined by RBM. The specific algorithm of k-step Gibbs sampling in RBM is to use a training sample (or any randomized state of the visible layer) for initializing the state $\mathbf{x}_0$ of the visible layer and alternately sample as follows:

$$\mathbf{h}_0 \sim P(\mathbf{h}|\mathbf{x}_0), \quad \mathbf{x}_1 \sim P(\mathbf{x}|\mathbf{h}_0),$$
$$\mathbf{h}_1 \sim P(\mathbf{h}|\mathbf{x}_1), \quad \mathbf{x}_2 \sim P(\mathbf{x}|\mathbf{h}_1),$$
$$\vdots \quad \quad \vdots$$
$$\mathbf{h}_2 \sim P(\mathbf{h}|\mathbf{x}_k), \quad \mathbf{x}_{k+1} \sim P(\mathbf{x}|\mathbf{h}_k).$$

Unlike Gibbs sampling, when using contrastive divergence method to train data, we only need to use $k = 1$ step Gibbs sampling to get good enough approximation $(\mathbf{x}_1, \mathbf{h}_1)$ and $(\mathbf{x}_2, \mathbf{h}_2)$. $k = 1$ Gibbs sampling based contrastive divergence (CD) learning is called CD_1 fast RBM learning algorithm. In this case, CD_1 learning forms (7.6.33)–(7.6.35) can be rewritten as the matrix-vector forms:

$$\mathbf{W} \leftarrow \mathbf{W} + \eta\bigg(P(\mathbf{h}_1 = \mathbf{1}|\mathbf{x}_1)\mathbf{x}_1^T - P(\mathbf{h}_2 = \mathbf{1}|\mathbf{x}_2)\mathbf{x}_2^T\bigg), \tag{7.6.36}$$

$$\mathbf{b} \leftarrow \mathbf{b} + \eta(\mathbf{x}_1 - \mathbf{x}_2), \tag{7.6.37}$$

$$\mathbf{c} \leftarrow \mathbf{c} + \eta\big(P(\mathbf{h}_1 = \mathbf{1}|\mathbf{x}_1) - P(\mathbf{h}_2 = \mathbf{1}|\mathbf{x}_2)\big). \tag{7.6.38}$$

Algorithm 7.3 summarizes the steps of CD_1 fast RBM learning algorithm.

Algorithm 7.3 CD_1 fast RBM learning algorithm
---
1. **input:** a training sample x_0, the number of hidden layer units, m, learning rate η, maximum training period T.
2. **initialization:** initial state of visible layer unit $x_1 = x_0$; select randomly W, b, c.
3. **for** $t = 1$ to T **do**
4.   **for** $j = 1$ to m **do** (for all hidden units)
5.   calculate $P(h_{1j} = 1|\mathbf{x}_1) = \text{sigmoid}\left(c_j + \sum_i W_{ij} x_{1i}\right)$
6.   **end for**
7.   construct $P(\mathbf{h}_1 = \mathbf{1}|\mathbf{x}_1) = [P(h_{11} = 1|\mathbf{x}_1), \ldots, P(h_{1m} = 1|\mathbf{x}_1)]^T$
8.   **for** $i = 1$ to n (for all visible units)
9.   Calculate $P(x_{2i} = 1|\mathbf{h}_1) = \text{sigmoid}\left(b_i + \sum_j W_{ij} h_{1j}\right)$
10.   **end for**
11.   **for** $j = 1$ to m **do**
12.   calculate $P(h_{2j} = 1|\mathbf{x}_2) = \text{sigmoid}\left(c_j + \sum_i w_{ij} x_{2i}\right)$
13.   **end for**
14.   construct $P(\mathbf{h}_2 = \mathbf{1}|\mathbf{x}_2) = [P(h_{21} = 1|\mathbf{x}_2), \ldots, P(h_{2m} = 1|\mathbf{x}_2)]^T$.
15. $W \leftarrow W + \eta\left(P(\mathbf{h}_1 = \mathbf{1}|\mathbf{x}_1)\mathbf{x}_1^T - P(\mathbf{h}_2 = \mathbf{1}|\mathbf{x}_2)\mathbf{x}_2^T\right)$
16. $b \leftarrow b + \eta(\mathbf{x}_1 - \mathbf{x}_2)$
17. $c \leftarrow c + \eta(P(\mathbf{h}_1 = \mathbf{1}|\mathbf{x}_1) - P(\mathbf{h}_2 = \mathbf{1}|\mathbf{x}_2))$
18. **end for**
19. **output:** W, b, c.

7.6.4 Multiple Restricted Boltzmann Machines

In the above discussion, the observed "visible" unit is denoted by the binary vector $\mathbf{x} = [x_1, \ldots, x_m]^T$. Now, consider the case where we have a set of m user rated movies or commodities. Support the user rated movie i as k, where $k \in \{1, \ldots, K\}$. Then the observed *"visible" binary rating matrix* is defined as

$$\mathbf{X} = \left[x_i^k\right]_{k=1, i=1}^{K,m} = [\mathbf{x}_1, \ldots, \mathbf{x}_m] = \begin{bmatrix} x_1^1 & x_2^1 & \cdots & x_m^1 \\ x_1^2 & x_2^2 & \cdots & x_m^2 \\ \vdots & \vdots & \ddots & \vdots \\ x_1^K & x_2^K & \cdots & x_m^K \end{bmatrix} \in \mathbb{R}^{K \times m}, \quad (7.6.39)$$

where the (k, i)th entry is given by

$$x_i^k = \begin{cases} 1, & \text{if the user rated movie } i \text{ as } k; \\ 0, & \text{otherwise.} \end{cases} \quad (7.6.40)$$

Therefore, we have K restricted Boltzmann machines with binary hidden units and softmax visible units. For each user, the RBM only includes softmax units for the movies that user has rated [125]. All of the K RBMs have the same binary values of hidden (latent) variables **h**.

If using a conditional multinomial distribution (a "softmax") for modeling each column of the observed visible binary rating matrix **X** and a conditional Bernoulli distribution for modeling hidden user features **h**, then [125]

$$p\left(x_i^k = 1 | \mathbf{h}\right) = \frac{\exp\left(b_i^k + \sum_{j=1} h_j W_{ij}^k\right)}{\sum_{l=1} \exp\left(b_i^l + \sum_{j=1} h_j W_{ij}^l\right)}, \tag{7.6.41}$$

$$p(h_j = 1 | \mathbf{X}) = \sigma\left(c_j + \sum_{i=1}^m \sum_{k=1}^K x_i^k W_{ij}^k\right), \tag{7.6.42}$$

where W_{ij}^k is a symmetric interaction parameter between feature j and rating k of movie i, b_i^k is the bias of rating k for movie i, and c_j is the bias of feature j. Note that the b_i^k can be initialized to the logs of their respective base rates over all users.

The marginal distribution over the visible ratings **X** is [125]

$$p(\mathbf{X}) = \sum_{\mathbf{h}} \frac{\exp(-E(\mathbf{X}, \mathbf{h}))}{\sum_{\mathbf{X}'} \sum_{\mathbf{h}'} \exp(-E(\mathbf{X}', \mathbf{h}'))} \tag{7.6.43}$$

with an "energy" term given by

$$E(\mathbf{X}, \mathbf{h}) = -\sum_{i=1}^m \sum_{j=1}^F \sum_{k=1}^K W_{ij}^k h_j x_i^k + \log(Z_i) - \sum_{i=1}^m \sum_{k=1}^K x_i^k b_i^k - \sum_{j=1}^F h_j c_j, \tag{7.6.44}$$

where $Z_i = \sum_{l=1}^K \exp(b_i^l + \sum_{j=1}^F h_j W_{ij}^l)$ is the normalization term that ensures that $\sum_{l=1}^K p(x_i^l = 1 | \mathbf{h}) = 1$. The movies with missing ratings do not make any contribution to the energy function.

The symmetric interaction matrix **W** is updated in gradient ascent form as $W_{ij} \leftarrow W_{ij} - \eta \Delta W_{ij}$, where

$$\Delta W_{ij} = \frac{\partial \log p(\mathbf{X})}{\partial W_{ij}^k} = \left\langle x_i^k h_j \right\rangle_{\text{data}} - \left\langle x_i^k h_j \right\rangle_{\text{model}}, \tag{7.6.45}$$

where the expectation $\langle x_i^k h_j \rangle_{\text{data}}$ defines the frequency with which movie i with rating k and feature j are on together when the feature detectors are being driven by the observed user-rating data from the training set using Eq. (7.6.42), and

$\langle x_i^k h_j \rangle_{\text{model}}$ is the corresponding frequency when the hidden units are being driven by reconstructed images.

To avoid computing $\langle \cdot \rangle_{\text{model}}$, Salakhutdinov et al. [125] proposed to follow an approximation to the gradient of a different objective function called "contrastive divergence" (CD) [58]:

$$\Delta W_{ij}^k = \langle x_i^k h_j \rangle_{\text{data}} - \langle x_i^k h_j \rangle_T. \tag{7.6.46}$$

The expectation $\langle \cdot \rangle_T$ represents a distribution of samples from running the Gibbs sampler (Eqs. (7.6.41) and (7.6.42)), initialized at the data, for T full steps. T is typically set to one at the beginning of learning and increased as the learning converges.

Restricted Boltzmann machines discussed above are assumed to use binary visible and hidden units, but many other types of unit can also be used. The main use of other types of unit is for dealing with data that is not well-modeled by binary (or logistic) visible units.

The following are two typical units that can be used in restricted Boltzmann machines [59, 157]:

1. *Softmax and multinomial units:* For a binary unit, the probability of turning on is given by the logistic sigmoid function of its total input **x**:

$$p = \sigma(x) = \frac{1}{1 + e^x} = \frac{e^x}{e^x + e^0}. \tag{7.6.47}$$

The energy contributed by the unit is $-x$ if it is on and 0 if it is off. This logistic sigmoid function of two states can be generalized to K alternative states, i.e.,

$$p_j = \frac{e^{x_j}}{\sum_{i=1} e^{x_i}}, \tag{7.6.48}$$

which is often called a "softmax" unit. A further generalization of the softmax unit is to sample N times (with replacement) from the probability distribution instead of just sampling once. The K different states can then have integer values bigger than 1, but the values must add to N. This is called a multinomial unit and the learning rule is again unchanged.

2. *Gaussian visible units:* The binary visible units are replaced by linear units with independent Gaussian noise. In this case, the energy function becomes

$$E(\mathbf{x}, \mathbf{h}) = \sum_{j \in \text{visible}} \frac{(x_j - b_j)^2}{2\sigma_j^2} - \sum_{i \in \text{hidden}} c_i h_i - \sum_{i,j} h_i \frac{x_j}{\sigma_j} w_{ij}, \tag{7.6.49}$$

where σ_i is the standard deviation of the Gaussian noise for visible unit i.

7.7 Bayesian Neural Networks

Consider the task of classification in a Bayesian learning framework. Assume that the data was generated by a parametric model, and use training data to calculate Bayesian optimal estimates of the model parameters. Then, equipped with these estimates, the classifier classifies new test data using Bayes' rule for turning the generative model around and calculate the posterior probability that a class would have generated the test data in question. Classification then becomes a simple matter of selecting the most probable class.

7.7.1 Naive Bayesian Classification

A widely used framework for classification is provided by a simple theorem of probability known as *Bayes' rule* (called also Bayes' theorem or Bayes' formula) [94, 104]:

$$p(c_k|\mathbf{x}) = p(c_k) \times \frac{p(\mathbf{x}|c_k)}{p(\mathbf{x})} \quad (7.7.1)$$

with

$$p(\mathbf{x}) = \sum_{i=1}^{M} p(\mathbf{x}|c_i) p(c_i), \quad (7.7.2)$$

where M is the number of classes or groups, $\mathbf{x} = [x_1, \ldots, x_d]^T$ is a vector of feature values, and $p(c_k|\mathbf{x})$ is the conditional probability that a given vector $\mathbf{x}$ belongs to the class c_k. It is assumed that all possible events fall into exactly one of M classes or groups $\{c_1, \ldots, c_M\}$.

Naive Bayesian classification is to assign the given feature vector $\mathbf{x}$ to the class c_k with the highest conditional probability $p(c_k|\mathbf{x})$. Of course, the conditional probabilities $p(c_1|\mathbf{x}), \ldots, p(c_M|\mathbf{x})$ are not known; they must be estimated from data. This estimation is difficult to do directly.

Bayes' rule suggests first estimating $p(\mathbf{x}|c_k)$, $p(c_k)$, and $p(\mathbf{x})$ and then using these estimates to get an estimate of $p(c_k|\mathbf{x})$, but estimating $p(\mathbf{x}|c_k)$ is still a problem, as $\mathbf{x}$ contains d components $x_1, \ldots, x_d$. A common strategy is to assume that the distribution of $\mathbf{x}$ conditional on c_k can be decomposed in the following fashion for all c_k [94]:

$$p(\mathbf{x}|c_k) = \prod_{i=1}^{d} p(x_i|c_k). \quad (7.7.3)$$

7.7 Bayesian Neural Networks

This assumption implies that the occurrence of a particular value of x_i is statistically independent of the occurrence of any other x_j, $j \neq i$, given a data of class c_k.

Under the assumption in (7.7.3), Eq. (7.7.1) becomes

$$p(c_k|\mathbf{x}) = p(c_k) \times \frac{\prod_{i=1}^{d} p(x_i|c_k)}{p(\mathbf{x})} \tag{7.7.4}$$

from which it follows that

$$p(\mathbf{x}) = \sum_{l=1}^{M} p(c_l) \times \prod_{i=1}^{d} p(x_i|c_l). \tag{7.7.5}$$

Then, the estimation formula is given by [94]

$$\hat{p}(c_k|\mathbf{x}) = \frac{\hat{p}(c_k) \times \prod_{i=1}^{d} \hat{p}(x_i|c_k)}{\hat{p}(\mathbf{x})}. \tag{7.7.6}$$

This estimate can then be used for classification.

7.7.2 Bayesian Classification Theory

Bayesian decision theory provides the fundamental probability model for well-known classification procedures.

Consider a general M-group classification problem in which each object has an associated attribute vector of dimension d. Let $x \in R^d$ be an associated attribute vector, and w_j denote the membership variable that takes value 1 if an object belongs to group j. Define $p(\omega_j)$ as the prior probability of group j and $f(\mathbf{x}|\omega_j)$ as the probability density function. According to the Bayes rule, we have

$$p(\omega_j|\mathbf{x}) = \frac{f(\mathbf{x}|\omega_j)p(\omega_j)}{f(\mathbf{x})}, \tag{7.7.7}$$

where $p(\omega_j|\mathbf{x})$ is the posterior probability of group j and $f(\mathbf{x})$ is the probability density function: $f(\mathbf{x}) = \sum_{j=1}^{M} f(\mathbf{x}|\omega_j) p(\omega_j)$.

Suppose that an object with a particular feature vector $\mathbf{x}$ is observed. If the group membership of $\mathbf{x}$ is decided to be j, then the probability of classification error is given by [170]

$$p(\text{Error}|\mathbf{x}) = \sum_{i \neq j} p(\omega_i|\mathbf{x}) = 1 - p(\omega_j|\mathbf{x}). \tag{7.7.8}$$

The purpose of Bayesian classification rule is to minimize the probability of total classification error (*misclassification rate*):

$$\text{Decide } \omega_k \text{ if } p(\omega_k|\mathbf{x}) = \max\{p(\omega_1|\mathbf{x}), \ldots, p(\omega_M|\mathbf{x})\}. \tag{7.7.9}$$

Let c_{ij} be the cost of misclassifying to group i when it actually belongs to group j. The expected cost associated with assigning to group i is

$$C_i(\mathbf{x}) = \sum_{j=1}^{M} c_{ij} p(\omega_j|\mathbf{x}), \quad i = 1, \ldots, M. \tag{7.7.10}$$

C_i is also known as the *conditional risk function*. The optimal Bayesian decision rule that minimizes the overall expected cost can be represented as

$$\text{Decide } \omega_k \text{ for } \mathbf{x} \text{ if } C_k(\mathbf{x}) = \min\{C_1(\mathbf{x}), \ldots, C_M(\mathbf{x})\}. \tag{7.7.11}$$

For binary classification with the two classes of ω_1 and ω_2, we should assign to class $+1$ if

$$c_{12}(\mathbf{x}) p(\omega_2) f(\mathbf{x}|\omega_2) < c_{21}(\mathbf{x}) p(\omega_1) f(\mathbf{x}|\omega_1) \tag{7.7.12}$$

or

$$\frac{f(\mathbf{x}|\omega_1)}{f(\mathbf{x}|\omega_2)} > \frac{c_{21}(\mathbf{x}) p(\omega_2)}{c_{12}(\mathbf{x}) p(\omega_1)}. \tag{7.7.13}$$

Otherwise, we should assign to class -1.

At this point, we discuss the relationship between Bayes classification and neural network classification. To this end, let C be a random variable which specifies the class membership, i.e., $C = c_k$ denotes membership in the kth class. The goal of classification is then to determine C given $\mathbf{x} \in X$. To minimize the probability of classification error, the optimal decision rule corresponds to choosing the class c_k which maximizes the posterior probability $p(c_k|\mathbf{x})$.

Define $\mathbf{y} = [y_1, \ldots, y_m]^T \in \mathbb{R}^M$, where M is the number of classes. Let $\mathbf{e}_k = [0, \ldots, 0, 1, 0, \ldots, 0]^T \in \mathbb{R}^M$ be the $M \times 1$ basic vector whose kth element equals 1 and others equal zero. Hence, if $\mathbf{y} = \mathbf{e}_k$ then $\mathbf{x}$ belongs to class k. This implies that $\mathbf{y}$ and c_k relate the same information if and only if $\mathbf{y} = \mathbf{e}_k$. The advantage of this representation in [151] is: the kth component of the least squares estimate $\hat{\mathbf{y}} = E\{\mathbf{y}|\mathbf{x}\}$ can be written as

$$\hat{y}_k = E\{y_k|\mathbf{x}\} = \sum_{y_k \in \{0,1\}} y_k p(y_k|\mathbf{x}) = p(y_k = 1|\mathbf{x}) = p(c_k|\mathbf{x}),$$

7.7 Bayesian Neural Networks

namely

$$E\{y_k|\mathbf{x}\} = p(c_k|\mathbf{x}). \tag{7.7.14}$$

This result provides a theoretical interpretation for neural network classifiers that are also trained by minimizing the mean squared errors. A neural network can be considered as a non-parametric technique for estimating posterior probability distributions.

7.7.3 Sparse Bayesian Learning

Suppose we are given a sample of N "training" pairs $\{\mathbf{x}_n, \mathbf{t}_n\}_{n=1}^{N}$, where $\mathbf{t} = [t_1, \ldots, t_N]^T$ is a target vector expressed as the sum of an approximation vector $\mathbf{y} = [y(x_1), \ldots, y(\mathbf{x}_N)]^T$ and an "error" vector $\boldsymbol{\epsilon} = [\epsilon_1, \ldots, \epsilon_N]^T$:

$$\mathbf{t} = \mathbf{y} + \boldsymbol{\epsilon} = \boldsymbol{\Phi}\mathbf{w} + \boldsymbol{\epsilon}, \tag{7.7.15}$$

where $\mathbf{w}$ is the parameter vector and $\boldsymbol{\Phi} = [\boldsymbol{\phi}_1, \ldots, \boldsymbol{\phi}_M]$ is the $N \times M$ "design" matrix whose columns comprise the complete set of M "basis vectors."

In the sparse Bayesian framework, the errors are assumed to be modeled probabilistically as independent zero-mean Gaussians, with variance σ^2, i.e.,

$$p(\boldsymbol{\epsilon}) = \prod_{n=1}^{N} N(\epsilon_n|0, \sigma^2). \tag{7.7.16}$$

The above error model implies a multivariate Gaussian likelihood for the target vector $\mathbf{t}$:

$$p(\mathbf{t}|\mathbf{w}, \sigma^2) = \frac{(2\pi)^{-N/2}}{\sigma^N} \exp\left\{-\frac{\|\mathbf{t} - \mathbf{y}\|^2}{2\sigma^2}\right\}. \tag{7.7.17}$$

By [146], given hyperparameters $\boldsymbol{\alpha} = [\alpha_1, \ldots, \alpha_M]^T$, the posterior parameter distribution conditioned on the data is given by combining the likelihood and prior within Bayes' rule:

$$p(\mathbf{w}|\mathbf{t}, \boldsymbol{\alpha}, \sigma^2) = p(\mathbf{t}|\mathbf{w}, \sigma^2) p(\mathbf{w}|\boldsymbol{\alpha}) / p(\mathbf{t}|\boldsymbol{\alpha}, \sigma^2), \tag{7.7.18}$$

and is Gaussian $N(\boldsymbol{\mu}, \boldsymbol{\Sigma})$ with

$$\boldsymbol{\Sigma} = \left(\mathbf{A} + \sigma^{-2} - \boldsymbol{\Phi}^T \boldsymbol{\Phi}\right)^{-1}, \qquad (7.7.19)$$

$$\boldsymbol{\mu} = \sigma^{-2} \boldsymbol{\Sigma} \boldsymbol{\Phi}^T \hat{\mathbf{t}}, \qquad (7.7.20)$$

where $\mathbf{A} = \mathbf{Diag}(\alpha_1, \ldots, \alpha_M)$ is an $M \times M$ diagonal matrix.

To find the solution vector $\boldsymbol{\alpha} = \boldsymbol{\alpha}_{MP}$, Tipping et al. [146] proposed to use sparse Bayesian learning for formulating the (local) maximization with respect to $\boldsymbol{\alpha}$ of its logarithm

$$\begin{aligned}\mathcal{L}(\boldsymbol{\alpha}) &= \log p\left(\mathbf{t} | \boldsymbol{\alpha}, \sigma^2\right) \\ &= \log \int_{-\infty}^{\infty} p\left(\mathbf{t} | \mathbf{w}, \sigma^2\right) p(\mathbf{w}|\boldsymbol{\alpha}) d\mathbf{w} \\ &= -\frac{1}{2}\left(N \log 2\pi + \log |\mathbf{C}| + \mathbf{t}^T \mathbf{C}^{-1} \mathbf{t}\right),\end{aligned} \qquad (7.7.21)$$

with

$$\mathbf{C} = \sigma^2 \mathbf{I} + \boldsymbol{\Phi} \mathbf{A}^{-1} \boldsymbol{\Phi}^T. \qquad (7.7.22)$$

Considering the dependence of $\mathcal{L}(\boldsymbol{\alpha})$ on a single hyperparameter α_i, $i \in \{1, \ldots, M\}$, then $\mathbf{C}$ in (7.7.22) can be decomposed as

$$\begin{aligned}\mathbf{C} &= \sigma^2 \mathbf{I} + \sum_{m \neq i} \alpha_m^{-1} \boldsymbol{\phi}_m \boldsymbol{\phi}_m^T + \alpha_i^{-1} \boldsymbol{\phi}_i \boldsymbol{\phi}_i^T, \\ &= \mathbf{C}_{-i} + \alpha_i^{-1} \boldsymbol{\phi}_i \boldsymbol{\phi}_i^T,\end{aligned} \qquad (7.7.23)$$

where $\mathbf{C}_{-i}$ is $\mathbf{C}$ with the contribution of basis vector i removed.

Then, the matrix determinant $|\mathbf{C}|$ and the inverse $\mathbf{C}^{-1}$ in the loss $\mathcal{L}(\boldsymbol{\alpha})$ can be obtained as [146]:

$$|\mathbf{C}| = |\mathbf{C}_{-i}| \cdot \left|1 + \alpha_i^{-1} \boldsymbol{\phi}_i^T \mathbf{C}_{-i}^{-1} \boldsymbol{\phi}_i\right| \qquad (7.7.24)$$

and

$$\mathbf{C}^{-1} = \mathbf{C}_{-i}^{-1} - \frac{\mathbf{C}_{-i}^{-1} \boldsymbol{\phi}_i \boldsymbol{\phi}_i^T \mathbf{C}_{-i}^{-1}}{\alpha_i + \boldsymbol{\phi}_i^T \mathbf{C}_{-i}^{-1} \boldsymbol{\phi}_i}. \qquad (7.7.25)$$

7.7 Bayesian Neural Networks

Hence, we have

$$\mathcal{L}(\boldsymbol{\alpha}) = -\frac{1}{2}\bigg(N\log(2\pi) + \log|\mathbf{C}_{-i}| + \mathbf{t}^T\mathbf{C}_{-i}^{-1}\mathbf{t}$$

$$-\log\alpha_i + \log\left(\alpha_i + \boldsymbol{\phi}_i^T\mathbf{C}_{-i}^{-1}\boldsymbol{\phi}_i\right) - \frac{(\boldsymbol{\phi}_i^T\mathbf{C}_{-i}^{-1}\mathbf{t})^2}{\alpha_i + \boldsymbol{\phi}_i^T\mathbf{C}_{-i}^{-1}\boldsymbol{\phi}_i}\bigg)$$

$$= \mathcal{L}(\boldsymbol{\alpha}_{-i}) + \frac{1}{2}\left(\log\alpha_i - \log(\alpha_i + s_i) + \frac{q_i^2}{\alpha_i + s_i}\right),$$

$$= \mathcal{L}(\boldsymbol{\alpha}_{-i}) + \ell(\alpha_i), \tag{7.7.26}$$

where

$$s_i = \boldsymbol{\phi}_i^T \mathbf{C}_{-i}^{-1} \boldsymbol{\phi}_i \quad \text{and} \quad q_i = \boldsymbol{\phi}_i^T \mathbf{C}_{-i}^{-1} \mathbf{t}. \tag{7.7.27}$$

- *Sparsity factor* s_i can be seen to be a measure of the extent that basis vector $\boldsymbol{\phi}_i$ "overlaps" those already present in the model.
- *Quality factor* q_i can be written as $q_i = \sigma^{-2}\boldsymbol{\phi}_i^T(\mathbf{t} - \mathbf{y}_{-i})$, and is thus a measure of the alignment of $\boldsymbol{\phi}_i$ with the error of the model with that vector excluded.

Analysis of sparse Bayesian learning [30] shows that $\mathcal{L}(\boldsymbol{\alpha})$ has a unique maximum with respect to α_i:

$$\alpha_i = \begin{cases} s_i^2 q_i^2 - s_i, & \text{if } q_i^2 > s_i; \\ \infty, & \text{if } q_i^2 \le s_i. \end{cases} \tag{7.7.28}$$

This result implies the following:

- If $\boldsymbol{\phi}_i$ is "in the model" (i.e., $\alpha_i < \infty$) yet $q_i^2 \le s_i$, then $\boldsymbol{\phi}_i$ may be deleted (i.e., set $\alpha_i = \infty$),
- If $\boldsymbol{\phi}_i$ is excluded from the model (i.e., $\alpha_i = \infty$) and $q_i^2 > s_i$, $\boldsymbol{\phi}_i$ may be added (i.e., set $\alpha_i = s_i^2 q_i^2 - s_i$).

Through the above analysis, Tipping et al. [146] proposed the following sequential sparse Bayesian learning algorithm:

1. If regression initialize σ^2 to some sensible value (e.g., $\sigma^2 = \text{var}(\mathbf{t}) \times 0.1$).
2. Initialize with a single basis vector $\boldsymbol{\phi}_i$, setting

$$\alpha_i = \frac{\|\boldsymbol{\phi}_i\|^2}{\|\boldsymbol{\phi}_i^T\mathbf{t}\|^2/\|\boldsymbol{\phi}_i\|^2 - \sigma^2}. \tag{7.7.29}$$

All other α_m are notionally set to infinity.

3. Explicitly compute Σ and μ (which are scalars initially), along with initial values of s_m and q_m for all M bases $\boldsymbol{\phi}_m$.
4. Recompute/update Σ, μ:

$$\Sigma = \left(\boldsymbol{\Phi}^T \mathbf{B} \boldsymbol{\Phi} + \mathbf{A}\right)^{-1}, \tag{7.7.30}$$

$$\hat{\mathbf{t}} = \left(\mathbf{I} - \boldsymbol{\Phi} \Sigma \boldsymbol{\Phi}^T \mathbf{B}\right)^{-1} (\mathbf{t} - \mathbf{y}), \tag{7.7.31}$$

$$\boldsymbol{\mu}_{\mathrm{MP}} = \Sigma \boldsymbol{\Phi}^T \mathbf{B} \hat{\mathbf{t}}, \tag{7.7.32}$$

where $\mathbf{A} = \mathbf{Diag}(\alpha_1, \ldots, \alpha_M)$ and $\mathbf{B} = \sigma^{-2}\mathbf{I}$.
5. Select a candidate basis vector $\boldsymbol{\phi}_i$ from the set of all M.
6. Compute $\theta_i = q_i^2 - s_i$.
7. If $\theta_i > 0$ and $\alpha_i < \infty$ (i.e., $\boldsymbol{\phi}_i$ is in the model), re-estimate α_i.
8. If $\theta_i > 0$ and $\alpha_i = \infty$, add $\boldsymbol{\phi}_i$ to the model with updated α_i.
9. If $\theta_i \leq 0$ and $\alpha_i < \infty$, then delete $\boldsymbol{\phi}_i$ from the model and set $\alpha_i = \infty$.
10. In regression and estimating the noise level, update $\sigma^2 = \|\mathbf{t} - \mathbf{y}\|^2/(N - M + \sum_m \alpha_m \Sigma_{mm})$, where $\mathbf{y} \approx \boldsymbol{\Phi}\boldsymbol{\mu}_{\mathrm{MP}}$ and Σ_{mm} is the (m,m)th diagonal element.
11. Compute

$$S_m = \boldsymbol{\phi}_m^T \mathbf{B} \boldsymbol{\phi}_m - \boldsymbol{\phi}_m^T \mathbf{B} \boldsymbol{\Phi} \Sigma \boldsymbol{\Phi}^T \mathbf{B} \boldsymbol{\phi}_m, \tag{7.7.33}$$

$$Q_m = \boldsymbol{\phi}_m^T \mathbf{B} \hat{\mathbf{t}} - \boldsymbol{\phi}_m^T \mathbf{B} \boldsymbol{\Phi} \Sigma \boldsymbol{\Phi}^T \mathbf{B} \hat{\mathbf{t}}, \tag{7.7.34}$$

where $\hat{\mathbf{t}} = \mathbf{t}$ in the regression case, and $\hat{\mathbf{t}} = \boldsymbol{\Phi}\boldsymbol{\mu}_{\mathrm{MP}} + \mathbf{B}^{-1}(\mathbf{t} - \mathbf{y})$ in the classification case. Here quantities $\boldsymbol{\Phi}$ and Σ contain only those basis functions that are currently included in the model, and computation thus scales in the cube of that measure which is typically only a very small fraction of the full M.
12. Compute all s_m and q_m:

$$s_m = \frac{\alpha_m S_m}{\alpha_m - S_m}, \quad q_m = \frac{\alpha_m Q_m}{\alpha_m - S_m}. \tag{7.7.35}$$

13. If converged terminate, otherwise go to Step 5.

Once a maximum posterior estimate $\boldsymbol{\mu}_{\mathrm{MP}}$ is obtained by evaluating (7.7.20) with $\boldsymbol{\alpha} = \boldsymbol{\alpha}_{\mathrm{MP}}$, then a final (posterior mean) approximator is given by $\mathbf{y} = \boldsymbol{\Phi}\mathbf{w} \approx \boldsymbol{\Phi}\boldsymbol{\mu}_{\mathrm{MP}}$ for sparse Bayesian regression.

7.8 Convolutional Neural Networks

A few classic neural networks were introduced in the previous sections. Starting from this section, we will focus on selected topics and advances in neural networks. Here, we first deal with convolutional neural networks.

7.8 Convolutional Neural Networks

Convolutional Neural Networks (CNNs) have been extremely successful in images, speech, audio, and video recognition tasks, where the coordinates of the underlying data representation have a grid structure (in 1, 2, and 3 dimensions), thanks to their ability to exploit translational equivariance/invariance with respect to this grid structure [13].

Gabor filter banks are a powerful method for extracting features. In this method, a bank of N Gabor filters are created. Then, each filter is convolved with the input image to produces N different images. Next, pixels of each image are pooled to extract information from each image. There are mainly two steps in the Gabor filter method including convolution and pooling.

The Neocognitron, proposed by Fukushima [32], is generally seen as the model that inspires CNNs on the computation side.

The first convolutional neural network was LeNet, invented by Le Cun et al. [88] for handwritten digit recognition and further made popular with LeCun et al. [89].

As compared with traditional fully connected neural networks, the main benefit of using CNNs is the reduced amount of parameters to be learned. A CNN has the following layers [31]:

1. *Input layer:* It feeds data to the network. Inputs can be either raw data (e.g., image pixels) or their transformations, whichever better emphasize some specific aspects of the data.
2. *Convolutional layers:* They contain a series of filters with fixed size used to perform convolutions on the data, for generating feature maps.
3. *Pooling layers:* These layers make the network focus only on the most important patterns, reducing the dimensionality of the feature maps used by the following layers. Pooling layers are known as *downsampling layers* as well.
4. *Rectified Linear Unit (ReLU):* ReLU layers are responsible for applying a nonlinear function to the output x of the previous layer, such as $f(x) = \max(0, x)$. By [84], they can be used for fast convergence in the training of CNNs, speeding-up the training.
5. *Fully connected layers:* They are used for the understanding of patterns generated by the previous layers. Neurons in this layer have full connections to all activations in the previous layer. They are also called the *inner product layers*. After trained, transfer learning approaches can extract features in these layers to train another classifier.
6. *Loss layers:* These layers specify how the network training penalizes the deviation between the predicted and true labels. Various loss functions appropriate for different tasks can be used: Softmax, Sigmoid, Cross-entropy, Euclidean loss, among others.

Convolutional layers are the most essential components of the CNN, and we will discuss them first.

7.8.1 Hankel Matrix and Convolution

Given an input signal $\mathbf{x} = [x(1), \ldots, x(n)]^T \in \mathbb{R}^n$, a *convolution* of filter (or kernel) $\mathbf{f} = [f(1), \ldots, f(d)]^T \in \mathbb{R}^d$ with input $\mathbf{x}$ has the following three forms:

$$y_i = (\mathbf{x} * \mathbf{f})_i = \sum_{j=1}^{n} x(j) f(i - j + 1), \quad i = 1, \ldots, 2d - 1, \quad (7.8.1)$$

or

$$y_i = (\mathbf{x} * \mathbf{f})_i = \sum_{j=1}^{d} x(i - j + 1) f(j), \quad i = 1, \ldots, 2d - 1, \quad (7.8.2)$$

or

$$y_i = (\mathbf{x} * \mathbf{f})_i = \sum_{j=1}^{d} x(i + j - 1) f(j), \quad i = 1, \ldots, n - d + 1. \quad (7.8.3)$$

As it provides the most outputs when $n \gg d$, Eq. (7.8.3) is commonly used as the convolution operation in CNNs.

The Matrix-vector form of (7.8.3) is given by

$$\mathbf{y} = (\mathbf{x} * \mathbf{f}) = \begin{bmatrix} x(1) & x(2) & \cdots & x(d) \\ x(2) & x(3) & \cdots & x(d+1) \\ \vdots & \vdots & \ddots & \vdots \\ x(n-d+1) & x(n-d+2) & \cdots & x(n) \end{bmatrix} \begin{bmatrix} f(1) \\ f(2) \\ \vdots \\ f(d) \end{bmatrix} \quad (7.8.4)$$

whose compact form is

$$\mathbf{y} = (\mathbf{x} * \mathbf{f}) = \mathbf{H}(\mathbf{x})\mathbf{f}, \quad (7.8.5)$$

where

$$\mathbf{H}(\mathbf{x}) = \begin{bmatrix} x(1) & x(2) & \cdots & x(d) \\ x(2) & x(3) & \cdots & x(d+1) \\ \vdots & \vdots & \ddots & \vdots \\ x(n-d+1) & x(n-d+2) & \cdots & x(n) \end{bmatrix} \in \mathbb{R}^{(n-d+1) \times d} \quad (7.8.6)$$

is called the *Hankel structured matrix* generated from an n-dimensional vector $\mathbf{x} = [x(1), \ldots, x(n)]^T \in \mathbb{R}^n$. Here, d is called a matrix pencil parameter. The space of these types of Hankel structured matrices is denoted as $\mathcal{H}(n, d)$. A Hankel structured matrix sometimes is simply called the *Hankel matrix*.

7.8 Convolutional Neural Networks

An $n \times d$ *wrap-around Hankel matrix* generated from an n-dimensional vector $\mathbf{x} = [x(1), \ldots, x(n)]^T \in \mathbb{R}^n$ is defined as [165]:

$$\mathbf{H}_d(\mathbf{x}) = \begin{bmatrix} x(1) & x(2) & \cdots & x(d) \\ x(2) & x(3) & \cdots & x(d+1) \\ \vdots & \vdots & \ddots & \vdots \\ x(n-d+1) & x(n-d+2) & \cdots & x(n) \\ x(n-d+2) & x(n-d+3) & \cdots & x(1) \\ \vdots & \vdots & \ddots & \vdots \\ x(n) & x(1) & \cdots & x(d-1) \end{bmatrix} \in \mathbb{R}^{n \times d}. \quad (7.8.7)$$

Clearly, an $n \times d$ wrap-around Hankel matrix generated by $\mathbf{x} \in \mathbb{R}^n$ can be considered as a Hankel matrix of the following elongated vector

$$\bar{\mathbf{x}} = \left[\mathbf{x}^T, x(1), x(2), \ldots, x(d-1) \right]^T \in \mathbb{R}^{n+d-1}. \quad (7.8.8)$$

Theorem 7.1 ([165]) *Let $r + 1$ denote the minimum length of the annihilating filters that annihilate the signal $\mathbf{x} = [x(1), \ldots, x(n)]^T$. Then, for a given Hankel structured matrix $\mathbf{H}_d(\mathbf{x}) \in \mathcal{H}(n, d)$ with $d > r$, its rank is given by*

$$\text{rank}(\mathbf{H}_d(\mathbf{x})) = r. \quad (7.8.9)$$

This theorem implies the following two results:

- If d is sufficiently large then the resulting Hankel matrix is low-rank.
- The rank of the Hankel matrix can be explicitly calculated as shown in the above theorem.

The convolutional operation can be extended to two-dimensional data. Given an input $I(m, n)$ and a kernel $K(a, b)$, their convolution operation is then given by [77]

$$s(t) = I(a, b) * K(a, b) = \sum_a \sum_b I(a, b) \cdot K(m - a, n - b) \quad (7.8.10)$$

or equivalently by

$$s(t) = I(a, b) * K(a, b) = \sum_a \sum_b I(m - a, n - b) \cdot K(a, b). \quad (7.8.11)$$

The 2-D convolutional operation can also be written as the cross-correlation form:

$$s(t) = I(a, b) * K(a, b) = \sum_a \sum_b I(m + a, n + b) \cdot K(a, b). \quad (7.8.12)$$

Suppose we are given a 2-D image $\mathbf{X} = [\mathbf{x}_1, \ldots, \mathbf{x}_p] \in \mathbb{R}^{n \times p}$ and a 2-D filter $\mathbf{\Phi} = [\boldsymbol{\phi}_1, \ldots, \boldsymbol{\phi}_q] \in \mathbb{R}^{d \times q}$, where $\mathbf{x}_j = [x_j(1), \ldots, x_j(n)]^T \in \mathbb{R}^n$, $j = 1, \ldots, p$ and $\boldsymbol{\phi}_i = [\phi_i(1), \ldots, \phi_i(d)]^T \in \mathbb{R}^d$, $i = 1, \ldots, q$. Then, the *2-D convolution* of the filter $\mathbf{\Phi}$ with the image $\mathbf{X}$ is given by

$$(\mathbf{X} * \mathbf{\Phi})_{m,k} = \sum_{i=1}^{d} \sum_{j=1}^{q} x_{m+i-1, k+j-1} \phi_{i,j},$$

$$m = 1, 2, \ldots, n - d + 1; \ k = 1, 2, \ldots, p - q + 1, \qquad (7.8.13)$$

or written as

$$\mathbf{Y} = (\mathbf{X} * \mathbf{\Phi}) = \begin{bmatrix} \mathbf{H}(\mathbf{x}_1)\boldsymbol{\phi}_1 & \cdots & \mathbf{H}(\mathbf{x}_1)\boldsymbol{\phi}_q \\ \vdots & \ddots & \vdots \\ \mathbf{H}(\mathbf{x}_p)\boldsymbol{\phi}_1 & \cdots & \mathbf{H}(\mathbf{x}_p)\boldsymbol{\phi}_q \end{bmatrix} = \mathbf{H}_{d,q}(\mathbf{X})\mathbf{\Phi}, \qquad (7.8.14)$$

where

$$\mathbf{Y} = [\mathbf{y}_1, \ldots, \mathbf{y}_p] = \begin{bmatrix} y_1(1) & \cdots & y_{p-q+1}(1) \\ \vdots & \ddots & \vdots \\ y_1(n-d+1) & \cdots & y_{p-q+1}(n-d+1) \end{bmatrix}, \qquad (7.8.15)$$

$$\mathbf{H}_{d,q}(\mathbf{X}) = \begin{bmatrix} \mathbf{H}(\mathbf{x}_1) \\ \vdots \\ \mathbf{H}(\mathbf{x}_p) \end{bmatrix}, \qquad (7.8.16)$$

$$\mathbf{H}(\mathbf{x}_j) = \begin{bmatrix} x_j(1) & x_j(2) & \cdots & x_j(d) \\ x_j(2) & x_j(3) & \cdots & x_j(d+1) \\ \vdots & \vdots & \ddots & \vdots \\ x_j(n-d+1) & x_j(n-d+2) & \cdots & x_j(n) \end{bmatrix}, \qquad (7.8.17)$$

for $j = 1, \ldots, p$. Here $x_j(i)$ is the ith element of the jth input channel $\mathbf{x}_j$.

Since 2-D convolution (7.8.13) is calculated for every m and k of $\mathbf{X}$, we say that the *stride of convolution* is equal to one. In some cases, it might be interesting to compute the convolution with a larger stride. For example, suppose we want to compute the convolution of alternating pixels. In this case, the stride of convolution is taken as two, leading to the equation [2, p.95]:

$$(\mathbf{X} * \mathbf{\Phi})_{m,k} = \sum_{i=1}^{d} \sum_{j=1}^{q} x_{m+i-1, k+j-1} \phi_{i,j},$$

$$m = 1, 3, 5, \ldots, n - d + 1; \ k = 1, 3, 5, \ldots, p - q + 1. \qquad (7.8.18)$$

7.8 Convolutional Neural Networks

Given a vector $\mathbf{v} = [v(1), \ldots, v(n)]^T \in \mathbb{R}^n$, the vector with reversed indices, $\overline{\mathbf{v}} = [v(n), \ldots, v(1)]^T \in \mathbb{R}^n$, is referred to as the *flipped vector* of $\mathbf{v}$.

Similarly, for a matrix

$$\boldsymbol{\Phi} = [\boldsymbol{\phi}_1, \ldots, \boldsymbol{\phi}_q] = \begin{bmatrix} \phi_1(1) & \phi_2(1) & \cdots & \phi_q(1) \\ \phi_1(2) & \phi_2(2) & \cdots & \phi_q(2) \\ \vdots & \vdots & \cdots & \vdots \\ \phi_1(d) & \phi_2(d) & \cdots & \phi_q(d) \end{bmatrix} \in \mathbb{R}^{d \times q}, \quad (7.8.19)$$

its *flipped matrix* is given by

$$\overline{\boldsymbol{\Phi}} = [\overline{\boldsymbol{\phi}}_1, \ldots, \overline{\boldsymbol{\phi}}_q] = \begin{bmatrix} \phi_1(d) & \phi_2(d) & \cdots & \phi_q(d) \\ \phi_1(d-1) & \phi_2(d-1) & \cdots & \phi_q(d-1) \\ \vdots & \vdots & \cdots & \vdots \\ \phi_1(1) & \phi_2(1) & \cdots & \phi_q(1) \end{bmatrix} \in \mathbb{R}^{d \times q}.$$

(7.8.20)

For a $pd \times q$ block-structured matrix

$$\boldsymbol{\Phi} = \left[\boldsymbol{\Phi}_1^T, \ldots, \boldsymbol{\Phi}_p^T\right]^T \in \mathbb{R}^{pd \times q}, \quad (7.8.21)$$

its *flipped block-structured matrix* has the form:

$$\overline{\boldsymbol{\Phi}} = \begin{bmatrix} \overline{\boldsymbol{\Phi}}_1 \\ \overline{\boldsymbol{\Phi}}_2 \\ \vdots \\ \overline{\boldsymbol{\Phi}}_p \end{bmatrix}, \quad (7.8.22)$$

where the sub-flipped matrix is defined as

$$\overline{\boldsymbol{\Phi}}_j = [\overline{\boldsymbol{\phi}}_1^j, \ldots, \overline{\boldsymbol{\phi}}_q^j] = \begin{bmatrix} \phi_1^j(d) & \phi_2^j(d) & \cdots & \phi_q^j(d) \\ \phi_1^j(d-1) & \phi_2^j(d-1) & \cdots & \phi_q^j(d-1) \\ \vdots & \vdots & \cdots & \vdots \\ \phi_1^j(1) & \phi_2^j(1) & \cdots & \phi_q^j(1) \end{bmatrix} \in \mathbb{R}^{d \times q},$$

(7.8.23)

for $j = 1, \ldots, p$.

The following are four *convolutions* in different cases [166]:

1. *Single-input single-output (SISO) convolution:* For the single-input channel $\mathbf{x} = [x(1), \ldots, x(n)]^T \in \mathbb{R}^n$ and the single-filter vector $\boldsymbol{\phi} = [\phi(1), \ldots, \phi(d)]^T \in \mathbb{R}^d$, the SISO convolution of the input vector $\mathbf{x}$ with the filter vector $\boldsymbol{\phi}$ can be

represented as matrix form:

$$\mathbf{y} = \mathbf{x} * \boldsymbol{\phi} = \mathbf{H}_d(\mathbf{x})\boldsymbol{\phi}, \qquad (7.8.24)$$

where $\mathbf{y} = [y_1, \ldots, y_d]^T$ and $\mathbf{H}_d(\mathbf{x})$ is a wrap-around Hankel matrix given by

$$\mathbf{H}_d(\mathbf{x}) = \begin{bmatrix} x(1) & x(2) & \cdots & x(d) \\ x(2) & x(3) & \cdots & x(d+1) \\ \vdots & \vdots & \ddots & \vdots \\ x(n) & x(1) & \cdots & x(d-1) \end{bmatrix} \in \mathbb{R}^{n \times d}. \qquad (7.8.25)$$

2. *Single-input multi-output (SIMO) convolution:* The convolution of the single-input vector $\mathbf{x}$ with q-filter vectors $\boldsymbol{\phi}_1, \ldots, \boldsymbol{\phi}_q$ can be represented by

$$\mathbf{Y} = \mathbf{x} * \boldsymbol{\Phi} = \mathbf{H}_d(\mathbf{x})\boldsymbol{\Phi}, \qquad (7.8.26)$$

where

$$\mathbf{Y} = [\mathbf{y}_1, \ldots, \mathbf{y}_q] \in \mathbb{R}^{n \times q} \quad \text{and} \quad \boldsymbol{\Phi} = [\boldsymbol{\phi}_1, \ldots, \boldsymbol{\phi}_q] \in \mathbb{R}^{d \times q}. \qquad (7.8.27)$$

3. *Multi-input multi-output (MIMO) convolution:* The convolution of p-channel input vectors $\mathbf{Z} = [\mathbf{z}_1, \ldots, \mathbf{z}_p] \in \mathbb{R}^{n \times p}$ with q-channel filters $\boldsymbol{\phi}^1, \ldots, \boldsymbol{\phi}^q$ can be represented as

$$\mathbf{y}_i = \sum_{j=1}^{p} \mathbf{z}_j * \boldsymbol{\phi}_i^j, \quad i = 1, \ldots, q, \qquad (7.8.28)$$

where $\boldsymbol{\phi}_i^j \in \mathbb{R}^d$ denotes the filter of length d with respect to the jth input channel and the ith output channel. If defining the MIMO filter kernel as

$$\boldsymbol{\Phi} = \begin{bmatrix} \boldsymbol{\Phi}_1 \\ \vdots \\ \boldsymbol{\Phi}_p \end{bmatrix}, \quad \text{where} \quad \boldsymbol{\Phi}_j = [\boldsymbol{\phi}_1^j, \ldots, \boldsymbol{\phi}_q^j] \in \mathbb{R}^{d \times q}, \qquad (7.8.29)$$

then the corresponding matrix representation of the MIMO convolution is given by

$$\mathbf{Y} = \sum_{j=1}^{p} \mathbf{H}_d(\mathbf{z}_j)\boldsymbol{\Phi}_j = \mathbf{H}_{d|p}(\mathbf{Z})\boldsymbol{\Phi}, \qquad (7.8.30)$$

7.8 Convolutional Neural Networks

where

$$\mathbf{H}_{d|p}(\mathbf{Z}) = [\mathbf{H}_d(\mathbf{z}_1), \ldots, \mathbf{H}_d(\mathbf{z}_p)] \qquad (7.8.31)$$

is an *extended Hankel matrix* by stacking p Hankel matrices side by side.

4. *Multi-input single-output (MISO) convolution:* If $q = 1$, then the MIMO convolution reduces to the MISO convolution

$$\mathbf{y} = \mathbf{H}_{d|p}(\mathbf{Z})\boldsymbol{\phi}, \qquad (7.8.32)$$

where $\boldsymbol{\phi} = [\phi_1, \ldots, \phi_p]^T$.

The above SISO, SIMO, MIMO, and MISO convolutional operations are easily extended to the multichannel 2-D convolution operation for an image domain convolutional neural network (CNN). For this end, the (extended) Hankel matrix is needed to be defined as a *block Hankel matrix*. For a 2-D input $\mathbf{X} = [\mathbf{x}_1, \ldots, \mathbf{x}_{n_2}] \in \mathbb{R}^{n_1 \times n_2}$ with $\mathbf{x}_i \in \mathbb{R}^{n_1}, i = 1, \ldots, n_2$, the block Hankel matrix associated with a $d_1 \times d_2$ filter is defined as [166]:

$$\mathbf{H}_{d_1,d_2}(\mathbf{X}) = \begin{bmatrix} \mathbf{H}_{d_1}(\mathbf{x}_1) & \mathbf{H}_{d_1}(\mathbf{x}_2) & \cdots & \mathbf{H}_{d_1}(\mathbf{x}_{d_2}) \\ \mathbf{H}_{d_1}(\mathbf{x}_2) & \mathbf{H}_{d_1}(\mathbf{x}_3) & \cdots & \mathbf{H}_{d_1}(\mathbf{x}_{d_2+1}) \\ \vdots & \vdots & \ddots & \vdots \\ \mathbf{H}_{d_1}(\mathbf{x}_{n_2}) & \mathbf{H}_{d_1}(\mathbf{x}_1) & \cdots & \mathbf{H}_{d_1}(\mathbf{x}_{d_2-1}) \end{bmatrix} \in \mathbb{R}^{n_1 n_2 \times d_1 d_2}. \qquad (7.8.33)$$

Then, the output $\mathbf{Y} = \mathbf{H}_{d_1,d_2}(\mathbf{X})\mathbf{K} \in \mathbb{R}^{n_1 \times n_2}$ from the 2-D SISO convolution for a given image $\mathbf{X} \in \mathbb{R}^{n_1 \times n_2}$ with the 2-D filter $\mathbf{K} \in \mathbb{R}^{d_1 \times d_2}$ can be represented in a matrix-vector form as follows:

$$\text{vec}(\mathbf{Y}) = \mathbf{H}_{d_1,d_2}(\mathbf{X})\text{vec}(\mathbf{K}), \qquad (7.8.34)$$

where $\text{vec}(\mathbf{Y})$ denotes the vectorization of the matrix $\mathbf{Y}$ by stacking its column vectors one by one.

Similarly, for a p-channel $n_1 \times n_2$ input image $\mathbf{X}^{(j)} = [\mathbf{x}_1^{(j)}, \ldots, \mathbf{x}_{n_2}^{(j)}], j = 1, \ldots, p$, its extended block Hankel matrix is defined as

$$\mathbf{H}_{d_1,d_2|p}\left([\mathbf{X}^{(1)}, \ldots, \mathbf{X}^{(p)}]\right) = \left[\mathbf{H}_{d_1,d_2}(\mathbf{X}^{(1)}), \ldots, \mathbf{H}_{d_1,d_2}(\mathbf{X}^{(p)})\right] \in \mathbb{R}^{n_1 n_2 \times d_1 d_2 p}. \qquad (7.8.35)$$

Then 2-D MIMO convolution for given p-input images $\mathbf{X}^{(j)} \in \mathbb{R}^{n_1 \times n_2}, j = 1, \ldots, p$ with 2-D filter $\overline{\mathbf{K}}_{(i)}^{(j)} \in \mathbb{R}^{d_1 \times d_2}$ associated with jth input channel and the ith output channel can be represented in a matrix-vector form as

$$\text{vec}\left(\mathbf{Y}^{(i)}\right) = \sum_{j=1}^{p} \mathbf{H}_{d_1,d_2}(\mathbf{X}^{(j)})\text{vec}\left(\mathbf{K}_{(i)}^{(j)}\right), \quad i = 1, \ldots, q. \qquad (7.8.36)$$

If defining

$$\mathcal{Y} = \left[\text{vec}(\mathbf{Y}^{(1)}), \ldots, \text{vec}(\mathbf{Y}^{(p)})\right], \tag{7.8.37}$$

$$\mathcal{K} = \begin{bmatrix} \text{vec}(\mathbf{K}_{(1)}^{(1)}) & \cdots & \text{vec}(\mathbf{K}_{(q)}^{(1)}) \\ \vdots & \ddots & \vdots \\ \text{vec}(\mathbf{K}_{(1)}^{(p)}) & \cdots & \text{vec}(\mathbf{K}_{(q)}^{(p)}) \end{bmatrix}, \tag{7.8.38}$$

then 2-D MIMO convolution can be rewritten as

$$\mathcal{Y} = \mathbf{H}_{d_1,d_2|p}\left(\left[\mathbf{X}^{(1)}, \ldots, \mathbf{X}^{(p)}\right]\right)\mathcal{K}. \tag{7.8.39}$$

7.8.2 Pooling Layer

We now deal with other layers of a CNN after discussing convolutional layers. The pooling layer is used to conduct a form of downsampling, i.e., to transform the raw data so that some specific aspects of the data are emphasized. In computer vision for example, sometimes the image needs to be converted to grayscale, removing color information from the analysis. Given an input RGB image I, with one pixel represented by $I(i, j) = (R(i, j), G(i, j), B(i, j))$, its conversion to graylevels happens using the following equation [31]:

$$C(i, j) = 0.2989 R(i, j) + 0.5870 G(i, j) + 0.1140 B(i, j), \tag{7.8.40}$$

where C is the image converted to grayscale and used as input to the network and R, G, and B are the color channels from the initial image.

Assume an 180×180 image which is connected to a convolution layer containing 50 filters of size 7×7. The output of the convolution layer will be an $174 \times 174 \times 50 = 1{,}563{,}300$ dimensional vector. Clearly, this number of dimensions is very high. To this end, a pooling layer is needed to reduce the dimensionality of feature maps, i.e., downsample the feature vector. The *pooling stride* is also called *downsampling factor*. Let s denote the pooling stride. For example, for the 12-dimensional vector $\mathbf{x} = [1, 10, 8, 2, 3, 6, 7, 0, 5, 4, 9, 2]$, downsampling $\mathbf{x}$ with stride $s = 2$ means we have to pick every alternate pixel starting from the element at index 0 which will generate the vector $[1, 8, 3, 7, 5, 9]$. By doing this, the dimensionality of $\mathbf{x}$ is divided by $s = 2$ and it becomes a six-dimensional vector.

Pooling is one of the most important operations in CNN. A pooling operator operates on an individual feature channel, aggregating data of a local region (e.g., a rectangle) and transforming them into one single value. Common choices include max pooling (using the maximum operator) and average pooling (using the average operator), both of which are hand-crafted. Traditional convolution with sliding strides larger than one pixel can also be regarded as a pooling operation. However, this kind of pooling operation does not deal with each input feature channel independently. Instead, it uses all the input feature channels to generate each output feature channel. Therefore, the convolutional pooling operation consumes many extra parameters.

On the pooling operation, the following points are to be noted [77]:

1. Pooling operates over a portion of the input and applies a function f over this input to produce the output.
2. The function f is commonly the max operation (leading to max pooling), but other variants such as average or ℓ_2 norm can be used as an alternative.
3. For a two-dimensional input, this is a rectangular portion.
4. The output produced as a result of pooling is much smaller in dimensionality as compared to the input.

Let $\mathbf{w}_k^l$ and b_k^l be the weight vector and bias term of the kth filter of the lth layer, respectively, and $\mathbf{x}_{i,j}^l$ be the input patch centered at location (i, j) of the lth layer. Then, the feature value at location (i, j) in the kth feature map of lth layer, $z_{i,j,k}^l$, is calculated by

$$z_{i,j,k}^l = \left(\mathbf{w}_k^l\right)^T \mathbf{x}_{i,j}^l + b_k^l. \tag{7.8.41}$$

The activation value $a_{i,j,k}^l$ of convolutional feature $z_{i,j,k}^l$ can be computed as

$$a_{i,j,k}^l = a\left(z_{i,j,k}^l\right), \tag{7.8.42}$$

where $a(\cdot)$ denotes an activation function.

The pooling layer aims to achieve shift-invariance by reducing the resolution of the feature maps. It is usually placed between two convolutional layers. Each feature map of a pooling layer is connected to its corresponding feature map of the preceding convolutional layer. Denoting the pooling function as $\text{pool}(\cdot)$ for each feature map $a_{i,j,k}^l$, then

$$y_{i,j,k}^l = \text{pool}\left(a_{m,n,k}^l\right), \quad \forall m, n \in \mathbb{R}_{ij}, \tag{7.8.43}$$

where $\mathbb{R}_{ij}$ is a local neighborhood around location (i, j).

The following are some pooling methods used in CNNs [48].

1. ℓ_p *pooling:* This pooling is a biologically inspired pooling process modeled on complex cells [72]. The summary statistic in ℓ_p pooling is the ℓ_p norm of the inputs into the pool. That is, if nodes (i, j) are in a pool k, then the output of the pool is

$$y_{i,j,k} = \left[\sum_{(m,n)\in\mathbb{R}_{ij}} (a_{m,n,k})^p\right]^{1/p}, \tag{7.8.44}$$

where $a_{m,n,k}$ is the feature value at location (m, n) within the pooling region $\mathbb{R}_{ij}$ in the kth feature map. The ℓ_p pooling contains the following two conventional choices:

- When $p = 1$, the ℓ_1 pooling gives the *sum pooling*:

$$y_{i,j,k} = \sum_{(m,n) \in \mathbb{R}_{ij}} a_{m,n,k}. \tag{7.8.45}$$

In sum pooling, all elements in a pooling region are considered. When combined with linear rectification nonlinearities, strong activations may be down-weighted since many zero elements are included in the sum. Even worse, with tanh($\cdot$) nonlinearities, strong positive and negative activations can cancel each other out, leading to small pooled responses.
- When $p \to \infty$, ℓ_p pooling corresponds to the *max pooling*

$$y_{i,j,k} = \max |a_{m,n,k}| \cdot \text{sign}(\text{argmax}|a_{m,n,k}|), \quad \text{for all } i, j. \tag{7.8.46}$$

A common form of downsampling is max pooling where the stride size of the pooling layer equals the pooling size. Max pooling serves to reduce the sizes of the feature vectors and the parameters of CNNs, which decreases training time and memory requirements, but this pooling easily overfits the training set in practice, making it hard to generalize well to test examples.

2. *Mixed pooling:* This pooling method is the combination of a max pooling and an average pooling, namely [167]:

$$y_{i,j,k} = \lambda \max_{(m,n) \in \mathbb{R}_{ij}} a_{m,n,k} + (1 - \lambda) \frac{1}{|\mathbb{R}_{ij}|} \sum_{(m,n) \in \mathbb{R}_{ij}} a_{m,n,k}, \tag{7.8.47}$$

where λ is a random value being either 0 or 1 which indicates the choice of either using average pooling or max pooling. During forward propagation process, λ is recorded and will be used for the backpropagation operation.
3. *Stochastic pooling:* It first computes the probabilities p for each region (i, j) by normalizing the activations within the region as [168]:

$$p_j = \frac{a_j}{\sum_{i \in R_j} a_i}, \tag{7.8.48}$$

then samples from the multinomial distribution based on p to pick a location j within the region. The pooled activation is then simply

$$y_j = a_l \quad \text{where } l \sim P(p_1, \ldots, p_{|R_j|}). \tag{7.8.49}$$

Max pooling only captures the strongest activation of the filter template with the input for each region. However, there may be additional activations in the same pooling region that should be taken into account when passing information up the network and stochastic pooling ensures that these non-maximal activations will also be utilized. Unlike the average pooling and the max pooling representing only one-modal distribution of activations within a region, the stochastic pooling, via selecting different l, can represent multi-modal distributions of activations within a region. Compared with max pooling, stochastic pooling can avoid overfitting due to the stochastic component.

4. *Spectral pooling:* The idea behind spectral pooling, presented by Rippel et al. in 2015 [120], stems from the observation that the frequency domain provides an ideal basis for inputs with spatial structure. Suppose we are given an input $\mathbf{x} \in \mathbb{R}^{M \times N}$, and some desired output map dimensionality $H \times W$. First, compute the discrete Fourier transform (DFT) of the input into the frequency domain as $\mathbf{y} = \mathcal{F}(\mathbf{x}) \in \mathbb{C}^{M \times N}$, and assume that the direct current component has been shifted to the center of the domain as is standard practice. Then, truncate $\mathbf{y}$ to $\hat{\mathbf{y}} \in \mathbb{C}^{H \times W}$. Finally, return its inverse DFT as $\hat{\mathbf{x}} = \mathcal{F}^{-1}(\hat{\mathbf{y}}) \in \mathbb{R}^{H \times W}$. The main advantages of spectral pooling are [138]: fast convolution computations, high parameter information retention (compression), flexibility in pooling output dimensions, and further computation savings by implementing spectral parameterization.

5. *Spatial pyramid pooling (SPP):* The last pooling layer (e.g., after the last convolutional layer) is replaced with a spatial pyramid pooling layer [56]. In each spatial bin with the number M of bins, the SPP using some pooling method (e.g., max pooling) generates a fixed-length representation regardless of the input sizes, resulting in the kM-dimensional output vector of the spatial pyramid pooling (k is the number of filters in the last convolutional layer). Then, fixed-dimensional vectors are the input to the fully connected layer.

Figure 7.9 shows a toy example using four different pooling techniques.

7.8.3 Activation Functions in CNNs

The following are conveniently used activation functions in CNNs [48].

1. *Rectified Linear Unit* (ReLU): The standard way to model a neuron's output f as a function of its input x is to use $f(x) = \tanh(x)$ or $f(x) = (1 + e^{-x})^{-1}$. However, these saturating nonlinear activation functions are much slower than the non-saturating nonlinear activation function $f(x) = \max(0, x)$ in gradient descent algorithm. Following Nair and Hinton [106], the neurons with the nonlinearity $f(x) = \max(0, x)$ are known as rectified linear units (ReLUs). The ReLU is one of the most notable non-saturated activation functions and is defined as

$$a_{i,j,k} = \max\{z_{i,j,k}, 0\}, \qquad (7.8.50)$$

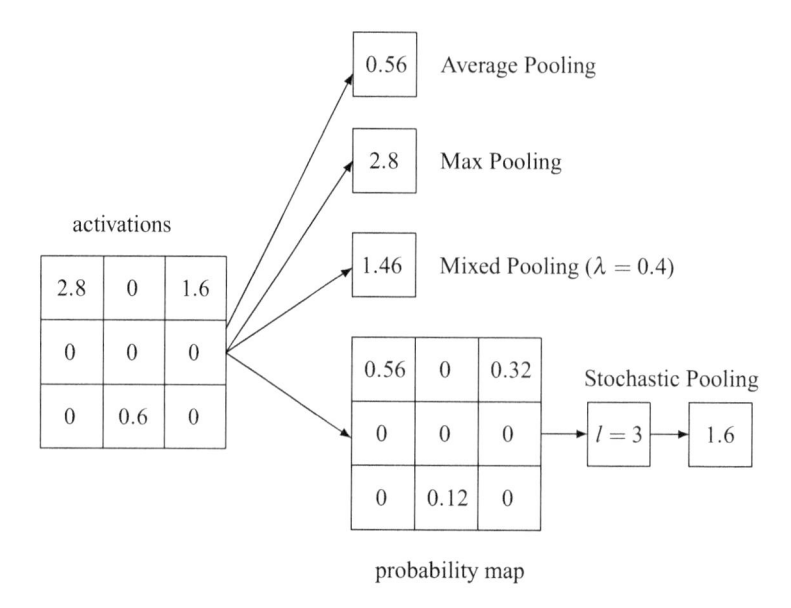

Fig. 7.9 A toy example using four different pooling techniques. **Left:** resulting activations within a given pooling region. **Right:** pooling results given by four different pooling techniques. If $\lambda = 0.4$ is taken then the mixed pooling result is 1.46. The mixed pooling and the stochastic pooling can represent multi-modal distributions of activations within a region

where $z_{i,j,k}$ is the input of the activation function at location (i, j) on the kth channel. A potential disadvantage of the ReLU unit is: it has zero gradient whenever the unit is not active, which may cause units that are not active initially to never become active since the gradient-based optimization will not adjust their weights. Moreover, it may slow down the training process due to the constant zero gradients.

2. *Noisy Rectified Linear Unit* (NReLU) [106]

$$a_{i,j,k} = \max\left\{z_{i,j,k} + N(0, \sigma^2)\right\}, \qquad (7.8.51)$$

where $N(0, \sigma^2)$ is a Gaussian noise with zero mean and variance σ^2. It was shown [106] that NReLUs work better than binary hidden units for recognizing objects and comparing faces, and can deal with large intensity variations much more naturally than binary units.

3. *Leaky Rectified Linear Unit* (Leaky ReLU) [102]

$$a_{i,j,k} = \max\left\{z_{i,j,k}, 0\right\} + \lambda \min\left\{z_{i,j,k}, 0\right\}, \qquad (7.8.52)$$

7.8 Convolutional Neural Networks

where λ is a predefined parameter in range $(0, 1)$. As compared with ReLU, Leaky ReLU compresses the negative part to a small, nonzero gradient when the unit is not active.

4. *Parametric Rectified Linear Unit* (PReLU) [55]

$$a_{i,j,k} = \max\{z_{i,j,k}, 0\} + \lambda \min\{z_{i,j,k}, 0\}. \quad (7.8.53)$$

5. *Randomized Rectified Linear Unit* (RReLU) [164]

$$a_{i,j,k}^{(n)} = \max\{z_{i,j,k}^{(n)}, 0\} + \lambda_k^{(n)} \min\{z_{i,j,k}^{(n)}, 0\}, \quad (7.8.54)$$

where $z_{i,j,k}^{(n)}$ denotes the input of activation function at location (i, j) on the kth channel of the nth example, $\lambda_k^{(n)}$ denotes its corresponding sampled parameter, and $a_{i,j,k}^{(n)}$ denotes its corresponding output.

6. *Exponential Linear Unit* (ELU) [16]

$$a_{i,j,k} = \max\{z_{i,j,k}, 0\} + \min\{\lambda(e^{z_{i,j,k}} - 1), 0\}, \quad (7.8.55)$$

where λ is a predefined parameter for controlling the value to which an ELU saturates for negative inputs.

7. *Maxout* [37]

$$a_{i,j,k} = \max_{k \in [1,K]} z_{i,j,k}, \quad (7.8.56)$$

where $z_{i,j,k}$ is the kth channel of the feature map. Maxout enjoys all the benefits of ReLU since ReLU is actually a special case of maxout, e.g., $\max\{\mathbf{w}_1^T \mathbf{x} + b_1, \mathbf{w}_2^T + b_2\}$, where $\mathbf{w}_1$ is a zero vector and b_1 is zero. Besides, maxout is particularly well suited for training with Dropout.

8. *Probout* [141] is a probabilistic variant of maxout, and first defines a probability for each of k linear units as:

$$\hat{p}_0 = 0.5, \quad (7.8.57)$$

$$\hat{p}_i = \frac{e^{\lambda z_i}}{\left(2 \sum_{j=1}^{k} e^{\lambda z_j}\right)}. \quad (7.8.58)$$

The activation function is then sampled as

$$a_i = \begin{cases} 0, & \text{if } i = 0; \\ z_i, & \text{otherwise,} \end{cases} \quad (7.8.59)$$

where $i \sim \text{multinomial}\{\hat{p}_0, \ldots, \hat{p}_k\}$. Probout can achieve the balance between preserving the desirable properties of maxout units and improving their invariance properties, but in testing process, probout is computationally expensive than maxout due to the additional probability calculations.

7.8.4 Loss Function

A key of CNNs is to learn the low-dimensional mapping: given a set of input vectors $\mathcal{I} = \{\mathbf{x}_1, \ldots, \mathbf{x}_N\}$, where $\mathbf{x}_i \in \mathbb{R}^D, \forall i = 1, \ldots, N$, find a parametric function $\mathbf{w} : \mathbb{R}^D \to \mathbb{R}^d$ with $d \ll D$, such that it has the following properties [49]:

- It only needs neighborhood relationships between training samples. These relationships could come from prior knowledge, or manual labeling, and be independent of any distance metric.
- It may learn functions that are invariant to complicated nonlinear transformations of the inputs such as lighting changes and geometric distortions.
- The learned function can be used to map new samples not seen during training, with no prior knowledge.

In order to find the parametric function $\mathbf{w}$ having the above three properties, the choice of a suitable *loss function* is an important problem. The commonly used loss functions in neural network designs are introduced below [48].

1. *Hinge loss:* It is usually used to train large-margin classifiers. Let $\mathbf{w}$ be the weight vector of a classifier and $\mathbf{y}^{(i)} \in \{1, \ldots, K\}$ indicate its correct class label among the K classes. Hinge loss of the classifier $\mathbf{w}$ is defined as

$$\mathcal{L}_{\text{hinge}} = \frac{1}{N} \sum_{i=1}^{N} \sum_{j=1}^{K} \left(\max \left(0, 1 - \delta(y^{(i)}, j) \mathbf{w}^T \mathbf{x}_i \right) \right)^p, \quad (7.8.60)$$

where

$$\delta(y^{(i)}, j) = \begin{cases} 1, & \text{if } y^{(i)} = j; \\ -1, & \text{otherwise.} \end{cases} \quad (7.8.61)$$

Two special examples of hinge loss are as follows:

- If $p = 1$, then the hinge loss is called the ℓ_1-*loss*:

$$\mathcal{L}_1 = \frac{1}{N} \sum_{i=1}^{N} \sum_{j=1}^{K} \max \left(0, 1 - \delta(y^{(i)}, j) \mathbf{w}^T \mathbf{x}_i \right). \quad (7.8.62)$$

7.8 Convolutional Neural Networks

- If $p = 2$, then the hinge loss is known as the *squared hinge loss* or ℓ_2-*loss* [171]:

$$\mathcal{L}_2 = \frac{1}{N} \sum_{i=1}^{N} \sum_{j=1}^{K} \left(\max\left(0, 1 - \delta(y^{(i)}, j) \mathbf{w}^T \mathbf{x}_i \right) \right)^2. \tag{7.8.63}$$

2. *Softmax loss:* Due to its simplicity and probabilistic interpretation, the softmax function is widely adopted in many CNNs (see, e.g., [55, 84]). Given a training set $(\mathbf{x}^{(i)}, y^{(i)})$ for $i \in \{1, \ldots, N\}$, $y^{(i)} \in \{1, \ldots, K\}$, where $\mathbf{x}^{(i)}$ is the ith input image patch, and $y^{(i)}$ is its target class label among the K classes. The prediction of the jth class for ith input is transformed with the softmax function:

$$p_j^{(i)} = \frac{e^{z_j^{(i)}}}{\sum_{l=1}^{K} e^{z_l^{(i)}}}, \tag{7.8.64}$$

where $z_j^{(i)}$ is usually the activations of a densely connected layer, so $z_j^{(i)}$ can be written as $z_j^{(i)} = \mathbf{w}_j^T a^{(i)} + b_j$. Softmax turns the predictions into nonnegative values and normalizes them to get a probability distribution over classes. These probabilistic predictions are used to compute the multinomial logistic loss, i.e., the softmax loss is defined by

$$\mathcal{L}_{\text{softmax}} = -\frac{1}{N} \sum_{i=1}^{N} \sum_{j=1}^{K} \mathbb{1}\{y^{(i)} = j\} \log p^{(i)}, \tag{7.8.65}$$

where $\mathbb{1}\{y^{(i)} = j\}$ denotes the indicator function returning 1 if $y^{(i)} = j$, and 0 otherwise. Despite its popularity, current softmax loss does not explicitly enjoy both within-class compactness and between-class-separability. In order to reinforce CNNs with more discriminative information, the learned features are expected to maximize simultaneously their within-class compactness and between-class separability. Inspired by such an idea, some improvements to softmax loss were developed to enforce extra within-class compactness and between-class separability: the contrastive loss [49], the triplet loss [131], and large-margin softmax loss [99].

3. *Contrastive loss:* Consider a weakly supervised scheme for learning a similarity measure from pairs of data instances labeled as matching or non-matching. Suppose we are given the ith pair of input vectors $(\mathbf{x}_1^{(i)}, \mathbf{x}_2^{(i)})$ and a binary label y assigned to this pair: $y = 0$ if $\mathbf{x}_1$ and $\mathbf{x}_2$ are deemed similar, and $y = 1$ if they are deemed dissimilar. Let $(\mathbf{z}_1^{(i,l)}, \mathbf{z}_2^{(i,l)})$ denotes its corresponding output pair of the lth ($l \in [1, \ldots, L]$) layer. The contrastive loss function is defined by [49]

$$\mathcal{L}_{\text{contrastive}} = \frac{1}{2N} \sum_{i=1}^{N} (1 - y) \cdot \left(d^{(i,L)}\right)^2 + y \cdot \max\left\{m - d^{(i,L)}, 0\right\}, \tag{7.8.66}$$

where $d^{(i,L)} = \|\mathbf{z}_1^{(i,L)} - \mathbf{z}_2^{(i,L)}\|_2^2$, and m is a margin parameter affecting non-matching pairs. However, Lin et al. [96] found from experiment results that the contrastive loss function would cause a dramatic drop in retrieval results when fine-turning the network on all pairs. As a solution, Lin et al. [96] propose a *double-contrastive loss* with an additional parameter affecting matching pairs:

$$\mathcal{L}_{\text{d-contrastive}} = \frac{1}{2N} \sum_{i=1}^{N} y \cdot \max\left\{d^{(i,L)} - m_1, 0\right\}$$
$$+ (1-y) \cdot \max\left\{m_2 - d^{(i,L)}, 0\right\}. \qquad (7.8.67)$$

4. *Triplet loss:* Consider the triplet units $(\mathbf{x}_a^{(i)}, \mathbf{x}_p^{(i)}, \mathbf{x}_n^{(i)})$, where $\mathbf{x}_a^{(i)}$ is an anchor instance, $\mathbf{x}_p^{(i)}$ is a positive instance from the same class of $\mathbf{x}_a^{(i)}$, and $\mathbf{x}_n^{(i)}$ is a negative instance. Let $(\mathbf{z}_a^{(i)}, \mathbf{z}_p^{(i)}, \mathbf{z}_n^{(i)})$ denote the feature representation of the triplet units, the triplet loss is defined as [131]:

$$\mathcal{L}_{\text{triplet}} = \frac{1}{N} \sum_{i=1}^{N} \max\left\{d_{(a,p)}^{(i)} - d_{(a,n)}^{(i)} + m, 0\right\}, \qquad (7.8.68)$$

where $d_{(a,p)}^{(i)} = \|\mathbf{z}_a^{(i)} - \mathbf{z}_p^{(i)}\|_2^2$ and $d_{(a,n)}^{(i)} = \|\mathbf{z}_a^{(i)} - \mathbf{z}_n^{(i)}\|_2^2$. The objective of triplet loss is to minimize the distance between the anchor and positive, and maximize the distance between the negative and the anchor. The *coupled clusters loss* function is defined as [98]:

$$\mathcal{L}_{\text{cc}} = \frac{1}{N_p} \sum_{i=1}^{N_p} \frac{1}{2} \max\left\{\|\mathbf{z}_p^{(i)} - \mathbf{c}_p\|_2^2 - \|\mathbf{z}_n^{(*)} - \mathbf{c}_p\|_2^2 + m, 0\right\}, \qquad (7.8.69)$$

where N_p is the number of samples per set, $\mathbf{z}_n^{(*)}$ is the feature representation of $\mathbf{x}_n^{(*)}$ which is the nearest negative sample to the estimated center point $\mathbf{c}_p = (\sum_{i=1}^{N_p} \mathbf{z}_p^{(i)})/N_p$.

5. *Large-margin softmax loss:* Consider a binary classification and a sample $\mathbf{x}$ given from class 1. The original softmax is to force $\mathbf{W}_1^T \mathbf{x} > \mathbf{W}_2^T \mathbf{x}$ (i.e., $\|\mathbf{W}_1\|\|\mathbf{x}\|\cos(\theta_1) > \|\mathbf{W}_2\|\|\mathbf{x}\|\cos(\theta_2)$) in order to classify $\mathbf{x}$ correctly. However, we want to make the classification more rigorous in order to produce a decision margin. To this end, consider instead requiring $\|\mathbf{W}_1\|\|\mathbf{x}\|\cos(m\theta_1) > \|\mathbf{W}_2\|\|\mathbf{x}\|\cos(\theta_2)$ ($0 \le \theta_1 \le \pi/m$), where m is a positive integer. Because the inequality holds

$$\|\mathbf{W}_1\|\|\mathbf{x}\|\cos(\theta_1) \ge \|\mathbf{W}_1\|\|\mathbf{x}\|\cos(m\theta_1) > \|\mathbf{W}_2\|\|\mathbf{x}\|\cos(\theta_2),$$

the new classification criteria is a stronger requirement to correctly classify **x**, producing a more rigorous decision boundary for class 1. By introducing an angular margin to the angle θ_j between input feature vector $a^{(i)}$ and the jth column $\mathbf{w}_j$ of weight matrix $\mathbf{W}$, Liu et al. [99] proposed the *large-margin softmax* (L-Softmax) *loss function*. The prediction $p_j^{(i)}$ for L-Softmax loss is defined as

$$p_j^{(i)} = \frac{e^{\|\mathbf{w}_j\|\|\mathbf{a}^{(i)}\|\psi(\theta_j)}}{e^{\|\mathbf{w}_j\|\|\mathbf{a}^{(i)}\|\psi(\theta_j)} + \sum_{l \neq j} e^{\|\mathbf{w}_l\|\|\mathbf{a}^{(i)}\|\cos(\theta_l)}}, \qquad (7.8.70)$$

where

$$\psi(\theta_j) = (-1)^k \cos\left(m\theta_j\right) - 2^k, \quad \theta_j \in [k\pi/m, (k+1)\pi/m]. \qquad (7.8.71)$$

Here $k \in [0, m-1]$ is an integer, m controls the margin among classes. As compared with the Softmax loss, the L-Softmax has the following performance [99]:

- In the training stage, the original softmax loss requires $\theta_1 < \theta_2$ to classify the sample **x** as class 1, while the L-Softmax loss requires $m\theta_1 < \theta_2$ to make the same decision. Hence, with bigger m (under the same training loss), the ideal margin between classes becomes larger and the learning difficulty is also increased. With $m = 1$, the L-Softmax loss becomes identical to the original softmax loss.
- The L-Softmax loss defines a relatively difficult learning objective with adjustable margin (difficulty). A difficult learning objective can effectively avoid overfitting and take full advantage of the strong learning ability from deep and wide architectures.
- The L-Softmax loss can be easily used as a replacement for standard loss, as well as used in tandem with other performance-boosting approaches and modules, including learning activation functions, data augmentation, pooling functions, or other modified network architectures.

6. *Kullback–Leibler (KL) divergence:* Given a discrete variable x and its two probability distributions $p(x)$ and $q(x)$. The KL divergence from $q(x)$ to $p(x)$ is defined as

$$\begin{aligned} \mathrm{KL}(q\|p) &= -H(p(x)) - E_p\{\log q(x)\} \\ &= \sum_x p(x) \log p(x) - \sum_x p(x) \log q(x) \\ &= \sum_x p(x) \log \frac{p(x)}{q(x)}, \end{aligned} \qquad (7.8.72)$$

where $H(p(x))$ is the Shannon entropy of $p(x)$, $E_p(\log q(x))$ is the cross-entropy between $p(x)$ and $q(x)$. The KL divergence does not have symmetry, i.e., $\mathrm{KL}(q\|p) \neq \mathrm{KL}(p\|q)$.

7. *Jensen-Shannon (JS) divergence:*

$$\begin{aligned}D_{\mathrm{JS}}(p\|q) = &\frac{1}{2}\mathrm{KL}\left(p(x)\left\|\frac{p(x)+q(x)}{2}\right.\right) \\ &+ \frac{1}{2}\mathrm{KL}\left(q(x)\left\|\frac{p(x)+q(x)}{2}\right.\right)\end{aligned} \quad (7.8.73)$$

is a symmetrical form of KL divergence. It measures the similarity between $p(x)$ and $q(x)$. By minimizing the JS divergence, the two distributions $p(x)$ and $q(x)$ are made as close as possible.

7.9 Dropout Learning

Combining the predictions of many different models is a very successful way to reduce test errors, but it appears to be too expensive for big neural networks. *Dropout* is a very efficient version of model combination through setting to zero the output of each hidden neuron with probability q. Moreover, overfitting is a serious problem in deep neural networks with a large number of parameter to be learned. This overfitting can be greatly reduced by using dropout.

Dropout, introduced by Hinton et al. in 2012 [62], is a form of regularization for fully connected neural network layers. Each element of a layer's output is kept with probability p, otherwise is randomly dropped with probability $(1 - p)$ out some of the units while training. This dropout procedure not only has good performance but also its implementation is simple.

The dropout procedure can be explained from two different perspectives.

- *Randomly omitting:* Each hidden unit is randomly omitted from the network with a probability $q = 1 - p$ (say $q = 0.5$) on each training case, the dropout can prevent complex co-adaptations on the training data.
- *Model averaging:* It is computationally expensive to train all neurons or nodes in many separate layers and then to apply each of these networks to the test data in the standard way. However, the dropout randomly performs model averaging with neural networks, and makes it possible for all of these networks to share the same weights for the hidden units that are present.

Dropout is effective for preventing the co-adaptation of feature detectors [5], and hence can prevent overfitting associated with their co-adaptation.

7.9.1 Dropout for Shallow and Deep Learning

Dropout is an intriguing algorithm for shallow and deep learning, which seems to be more effective than full-connection neural networks, but comes with little formal understanding and raises several interesting questions [5]:

1. What kind of model averaging is dropout implementing, exactly or in approximation, when applied to multiple layers?
2. How crucial are its parameters? For instance, is $q = 0.5$ necessary and what happens when other values are used? What happens when other transfer functions are used?
3. What are the effects of different values of q for different layers? What happens if dropout is applied to connections rather than units?
4. What are precisely the regularization and averaging properties of dropout?
5. What are the convergence properties of dropout?

The motivation for dropout may be better understood by thinking about common inference problems in machine learning or neural networks. Consider a scheme where an inference needs to be made by one big conspiracy of one hundred people M. With less time, a smaller conspiracy involving fifty people can make a first inference $\mathbf{y}^{(1)}$. If another conspiracy of a new group of fifty people make the second inference $\mathbf{y}^{(2)}$ based on $\mathbf{y}^{(l)}$, and so on, then a good inference probably is made by a few smaller conspiracies faster than a big conspiracy of one hundred people.

To model the dropout problem, we consider a neural network with L hidden layers. Let $l \in \{1, \ldots, L\}$ be the index of the lth hidden layer of the network, and $\mathbf{z}^{(l)}$ denote the input vector of layer l, $\mathbf{y}^{(l)}$ denote the output vector of layer l (where $\mathbf{y}^{(0)} = \mathbf{x}$ is the input). The feedforward operation of a standard full-connection neural network can be described as

$$\mathbf{z}^{(l+1)} = \mathbf{W}^{(l)} \mathbf{y}^{(l)} + \mathbf{b}^{(l)}, \tag{7.9.1}$$

$$\mathbf{y}^{(l+1)} = f\left(\mathbf{z}^{(l+1)}\right), \tag{7.9.2}$$

for $l \in \{0, \ldots, L-1\}$ and any hidden unit i. Here $\mathbf{W}^{(l)}$ and $\mathbf{b}^{(l)}$ are, respectively, the weight matrix and bias vector at layer l, while $f(\cdot)$ is any elementwise activation function, such as $f(z) = 1/(1 + \exp(-z))$ or $f(z) = \tanh(z)$.

If a dropout is applied to the outputs of a fully connected layer, then the above two equations can be rewritten as

$$\mathbf{z}^{(l+1)} = \left(\mathbf{M}^{(l)} \odot \mathbf{W}\right) \mathbf{y}^{(l)} + \mathbf{b}^{(l)}, \tag{7.9.3}$$

$$\mathbf{y}^{(l+1)} = f\left(\mathbf{z}^{(l+1)}\right), \tag{7.9.4}$$

Fig. 7.10 An example of a thinned network produced by applying dropout to a full-connection neural network. Crossed units have been dropped. The dropout probabilities of the first, second, and third layers are 0.4, 0.6, and 0.4, respectively,

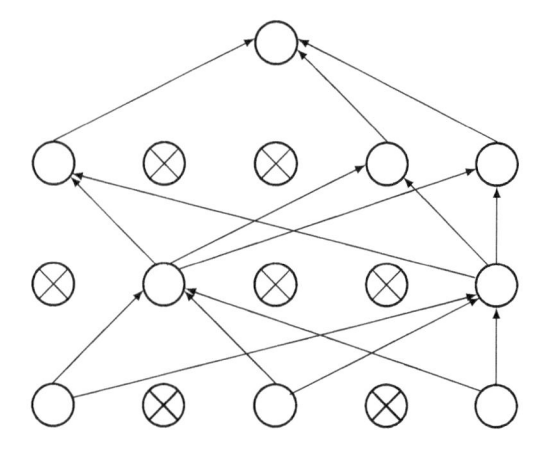

where $\mathbf{M} \odot \mathbf{W}$ denotes the elementwise product of two $d \times n$ matrices $\mathbf{M}$ and $\mathbf{W}$, and $\mathbf{M}^{(l)} \in \mathbb{R}^{d \times n}$ is a binary matrix encoding the connection information with elements

$$M_{ij}^{(l)} \sim \text{Bernoulli}(p). \tag{7.9.5}$$

Here, Bernoulli(p) denotes a *Bernoulli distribution* of binary random variable v:

$$P(v = 1) = p = 1 - P(v = 0) = 1 - q. \tag{7.9.6}$$

Figure 7.10 shows an example of dropout applying to a full-connection neural network.

Although dropout is available for hidden layers with different probability q, i.e., the number of dropped out units in each hidden layer may be different, this number is fixed to be same for each hidden layer in practical applications.

Let $\mathbf{y}^{(l)} = \left[y_1^{(l)}, \ldots, y_n^{(l)} \right]^T$ with $\mathbf{y}^{(0)} = \mathbf{x} \in \mathbb{R}^{n \times 1}$, $\mathbf{m}^{(l)} = \left[m_1^{(l)}, \ldots, m_n^{(l)} \right]^T$ be binary *Bernoulli vector* with elements 1 and 0, d be the number of all units in each hidden layer, and $\mathbf{w}_i \in \mathbb{R}^{n \times 1}$ be the ith row of the weight matrix $\mathbf{W} \in \mathbb{R}^{d \times n}$.

With dropout, the feedforward operations constitute the *dropout neural network model* [142]:

$$m_j^{(l)} \sim \text{Bernoulli}(p), \tag{7.9.7}$$

$$\tilde{\mathbf{y}}^{(l)} = \mathbf{m}^{(l)} \odot \mathbf{y}^{(l)}, \tag{7.9.8}$$

$$\mathbf{z}^{(l+1)} = \mathbf{W}^{(l+1)} \tilde{\mathbf{y}}^{(l)} + \mathbf{b}^{(l+1)}, \tag{7.9.9}$$

$$\mathbf{y}^{(l+1)} = f(\mathbf{z}^{(l+1)}), \tag{7.9.10}$$

7.9 Dropout Learning

for $l \in \{0, 1, \ldots, L-1\}$ and the hidden unit i. The vector $\mathbf{m}^{(l)}$ is sampled and multiplied elementwise with the outputs of the lth layer, $\mathbf{y}^{(l)}$, to create the thinned outputs $\tilde{\mathbf{y}}^{(l)}$. The thinned outputs $\tilde{\mathbf{y}}^{(l)}$ are then used as inputs to the next layer.

Consider a standard CNN composed of alternating convolutional and pooling layers, with fully connected layers on top. On each presentation of a training example, if layer l is followed by a pooling layer, the forward propagation without dropout can be described as

$$a_j^{(l+1)} = \text{pool}\left(a_1^{(l)}, \ldots, a_n^{(l)}\right), \quad i \in \mathbb{R}_j^{(l)}, \quad (7.9.11)$$

where $a_i^{(l)}$ with $i \in \mathbb{R}_j^{(l)}$ that is pooling region j at layer l and $a_i^{(l)}$ is the activity of each neuron within it, while $n = \left|R_j^{(l)}\right|$ is the number of units in $\mathbb{R}_j^{(l)}$.

With dropout, the forward propagation of max-pooling dropout at training time becomes [160]

$$\hat{\mathbf{a}}^{(l)} = \mathbf{m}^{(l)} \odot \mathbf{a}^{(l)}, \quad (7.9.12)$$

$$a_j^{(l+1)} = \text{pool}\left(\hat{a}_1^{(l)}, \ldots, \hat{a}_n^{(l)}\right), \quad i \in \mathbb{R}_j^{(l)}, \quad (7.9.13)$$

where the activations $\left(\hat{a}_1^{(l)}, \ldots, \hat{a}_n^{(l)}\right)$ in each pooling region j are reordered in nondecreasing order, i.e., $0 \leq \hat{a}_1^{(l)} \leq \ldots \leq \hat{a}_n^{(l)}$.

If layer l is followed by a convolutional layer, the forward propagation with dropout is formulated as [160]:

$$m_k^{(l)}(i) \sim \text{Bernoulli}(p), \quad (7.9.14)$$

$$\hat{\mathbf{a}}_k^{(l)} = \mathbf{a}_k^{(l)} \odot \mathbf{m}_k^{(l)}, \quad (7.9.15)$$

$$\mathbf{z}_j^{(l+1)} = \sum_{k=1}^{n^{(l)}} \text{conv}\left(\mathbf{W}_j^{(l+1)}, \hat{\mathbf{a}}_k^{(l)}\right), \quad (7.9.16)$$

$$\mathbf{a}_j^{(l+1)} = f\left(\mathbf{z}_j^{(l+1)}\right), \quad (7.9.17)$$

where $\mathbf{a}_k^{(l)}$ denotes the activations of feature map k ($k = 1, \ldots, n^{(l)}$) at layer l. The mask vector $\mathbf{m}_k^{(l)}$ consists of independent Bernoulli variables $m_k^{(l)}(i)$. Then, this mask is sampled and multiplied with activations in kth feature map at layer l in order to produce dropout-modified activations $\hat{\mathbf{a}}_k^{(l)}$. These modified activations are convolved with filter $W_k^{(l+1)}$ to produce convolved features $\mathbf{z}_j^{(l)}$. The function $f(\cdot)$ is an elementwise function applied to the convolved features to get the activations of convolutional layers, $a_j^{(l+1)}$.

7.9.2 Dropout Spherical K-Means

A major goal in machine learning or neural networks is to learn deep hierarchies of features for other tasks. Many algorithms are available to learn deep hierarchies of features from unlabeled data, especially text documents and images. It has been found [17, 18, 21] that using K-means clustering as the unsupervised learning module in these types of "feature learning" pipelines can lead to excellent results, often rivaling the state-of-the-art systems.

The classic K-means clustering algorithm finds cluster centroids that minimize the distance between data points and the nearest centroid. K-means is also called "*vector quantization*" (VQ) in the sense that K-means can be viewed as a way of constructing a "dictionary" $\mathbf{D} \in \mathbb{R}^{n \times K}$ of K column vectors $\mathbf{d}^{(j)}$, $j = 1, \ldots, K$ so that a data vector $\mathbf{x}^{(i)} \in \mathbb{R}^n$ ($i = 1, \ldots, m$) can be mapped to a code vector $\mathbf{s}^{(i)}$ that minimizes the error in reconstruction. A popular modified version of K-means is known as "*gain shape vector quantization*" [169] or "*spherical K-means*" [21].

Given m data vectors $\mathbf{x}^{(i)} \in \mathbb{R}^n$, $i = 1, \ldots, m$. The VQ of the data vector $\mathbf{x}^{(i)}$ can be described by the model

$$\mathbf{x}^{(i)} = \mathbf{C}\mathbf{s}^{(i)}, \ i = 1, \ldots, m, \tag{7.9.18}$$

where $\mathbf{s}^{(i)}$ is a quantization coefficient vector or "*code vector*" associated with the input $\mathbf{x}^{(i)}$, $\mathbf{D} \in \mathbb{R}^{K \times n}$ denotes the *codebook matrix* or "*dictionary*" whose jth column is denoted by $\mathbf{d}^{(j)}$. The goal of spherical K-means is equivalently to find $\mathbf{D}$ and $\mathbf{s}^{(i)}$ from the input $\mathbf{x}^{(i)}$ according to [18]:

$$\min_{\mathbf{D},\mathbf{s}} \left\{ \sum_{i=1}^{m} \left\| \mathbf{D}\mathbf{s}^{(i)} - \mathbf{x}^{(i)} \right\|_2^2 \right\}, \tag{7.9.19}$$

subject to $\|\mathbf{s}^{(i)}\|_0 \leq 1$, $\forall i = 1, \ldots, m$ and $\|\mathbf{d}^{(j)}\|_2 = 1$, $\forall j = 1, \ldots, K$.
(7.9.20)

In the above equation, the code vector $\mathbf{s}^{(i)}$ can be thought of as a "*feature representation*" of each example $\mathbf{x}^{(i)}$ that satisfies several criteria:

- Given $\mathbf{s}^{(i)}$ and $\mathbf{D}$, the original example $\mathbf{x}^{(i)}$ should be able to be reconstructed well.
- The first constraint $\|\mathbf{s}^{(i)}\|_0 \leq 1$ means that each $\mathbf{s}^{(i)}$ should have at most one nonzero entry (associated with vector quantization). Thus a new representation of $\mathbf{x}^{(i)}$ should preserve it as well as possible, but also be a very simple or parsimonious representation.
- The second constraint $\|\mathbf{d}^{(j)}\|_2 = 1$ requires that each dictionary column have unit length, preventing them from becoming arbitrarily large or small.

7.9 Dropout Learning

The full K-means algorithm for learning feature representation can be summarized below [18]:

1. *Normalize inputs:*

$$\mathbf{x}^{(i)} = \frac{\mathbf{x}^{(i)} - \mathrm{mean}(\mathbf{x}^{(i)})}{\sqrt{\mathrm{var}(\mathbf{x}^{(i)}) + \epsilon_{\mathrm{norm}}}}, \quad \forall i = 1, \ldots, m. \tag{7.9.21}$$

2. *Eigenvalue decomposition and estimate inputs:*

$$\mathrm{cov}(\mathbf{x}^{(i)}) = \mathbf{V}\mathbf{D}\mathbf{V}^T, \quad i = 1, \ldots, m, \tag{7.9.22}$$

$$\mathbf{x}^{(i)} = \mathbf{V}(\mathbf{D} + \epsilon_{\mathrm{norm}}\mathbf{I})^{-1/2}\mathbf{V}^T\mathbf{x}^{(i)}, \quad i = 1, \ldots, m. \tag{7.9.23}$$

3. *Loop until convergence (typically 10 iterations is enough):*

$$s_j^{(i)} = \begin{cases} \mathbf{d}^{(j)T}\mathbf{x}^{(i)}, & \text{if } j = \arg\max_l \|\mathbf{D}^{(l)}\mathbf{x}^{(i)}\|_1, \\ 0, & \text{otherwise,} \end{cases} \tag{7.9.24}$$

$$\mathbf{d}^{(j)} = \mathbf{X}\mathbf{s}_j + \mathbf{d}^{(j)}, \tag{7.9.25}$$

$$\mathbf{d}^{(j)} = \mathbf{d}^{(j)}/\|\mathbf{d}^{(j)}\|_2, \tag{7.9.26}$$

where $j = 1, \ldots, K$, $i = 1, \ldots, m$; while $\mathbf{X} = [\mathbf{x}^{(1)}, \ldots, \mathbf{x}^{(m)}] \in \mathbb{R}^{n \times m}$ and $\mathbf{s}_j = [s_j^{(1)}, \ldots, s_j^{(m)}]^T \in \mathbb{R}^{m \times 1}$ are the data (or example) matrix and the code vectors, respectively.

When dropout is applied to the outputs of a dictionary, the spherical K-means is extended to the *dropout spherical K-means* [173]. The optimization problem of the dropout spherical K-means can be described as

$$\min_{\mathbf{D},\mathbf{s}} \left\{ \sum_{i=1}^m \|(\mathbf{M} \odot \mathbf{D})\mathbf{s}^{(i)} - \mathbf{x}^{(i)}\|_2^2 \right\}, \tag{7.9.27}$$

subject to $\|\mathbf{s}^{(i)}\|_0 \leq 1$, $\forall i = 1, \ldots, m$ and $\|\mathbf{d}^{(j)}\|_2 = 1$, $\forall j = 1, \ldots, K$,
$\tag{7.9.28}$

where $\mathbf{M}$ is a binary mask matrix with the same size as $\mathbf{D}$, and each column is drawn independently from $\mathbf{M}_j \sim \mathrm{Bernoulli}(p)$ trials.

Given a dictionary $\mathbf{D}$ and a dropout mask matrix $\mathbf{M}$, then $\mathbf{M} \odot \mathbf{D}$ can be viewed as "*thinned dictionary*" $\mathbf{D}_{\text{thin}} = \mathbf{M} \odot \mathbf{D}$, where $\mathbf{D}_{\text{thin}} \in \mathbf{D}$. Hence, Eq. (7.9.24) becomes

$$s_j^{(i)} = \begin{cases} \mathbf{d}_{\text{thin}}^{(j)T} \mathbf{x}^{(i)}, & \text{if } j = \arg\max_l \|\mathbf{D}_{\text{thin}}^{(l)} \mathbf{x}^{(i)}\|_1, \\ 0, & \text{otherwise}, \end{cases} \quad (7.9.29)$$

for all $j = 1, \ldots, K$; $i = 1, \ldots, m$.

After the code vector $\mathbf{s}^{(i)}$ is obtained, new centroids can be computed according to

$$\begin{aligned} \mathbf{D}_{\text{new}} &= \arg\min_{\mathbf{D}_{\text{thin}}} \left\{ \|\mathbf{D}_{\text{thin}} \mathbf{S} - \mathbf{X}\|_2^2 + \|\mathbf{D}_{\text{thin}} - \mathbf{D}_{\text{old}}\|_2^2 \right\}, \\ &= (\mathbf{S}\mathbf{S}^T + \mathbf{I})^{-1} (\mathbf{X}\mathbf{S}^T + \mathbf{D}_{\text{thin}}), \\ &\propto \mathbf{X}\mathbf{S}^T + \mathbf{D}_{\text{thin}}. \end{aligned} \quad (7.9.30)$$

The full dropout K-means algorithm [173] for learning feature representation can be summarized below.

1. *Normalize inputs:* use (7.9.21) to compute the normalized input vectors $\mathbf{x}^{(i)}$.
2. *Eigenvalue decomposition and estimate inputs:* use (7.9.22) to compute the EVD of the variance matrix $\text{cov}(\mathbf{x}^{(i)})$, then estimate the inputs $\mathbf{x}^{(i)}$ via (7.9.23).
3. *Loop until convergence:* calculate the code vectors $\mathbf{s}^{(i)}$ by using (7.9.29), and update $\mathbf{D}_{\text{new}} = \mathbf{X}\mathbf{S}^T + \mathbf{D}_{\text{thin}}$. Finally, make the normalization $\mathbf{d}^{(j)} = \mathbf{d}^{(j)}/\|\mathbf{d}^{(j)}\|_2$.

7.9.3 DropConnect

As pointed out in [5], an important and interesting question is: what happens if dropout is applied to connections rather than units? Dropout applied to connections is called *DropConnect* which was developed by Wan et al. [152]. More specifically, DropConnect is the generalization of Dropout in which each connection, rather than each output unit, can be dropped out with probability $1 - p$ [152].

The following is a comparison between Dropout and DropConnect from the perspective of training networks:

- *Training network with Dropout:* Each element of a layer's output $\mathbf{r}$ is kept with probability p, otherwise set to 0 with probability $1 - p$:

$$\mathbf{r} = \mathbf{m} \odot a(\mathbf{W}\mathbf{v}) = a(\mathbf{m} \odot (\mathbf{W}\mathbf{v})), \quad (7.9.31)$$

7.9 Dropout Learning

Fig. 7.11 An example of the structure of DropConnect. The dotted lines show dropped connections

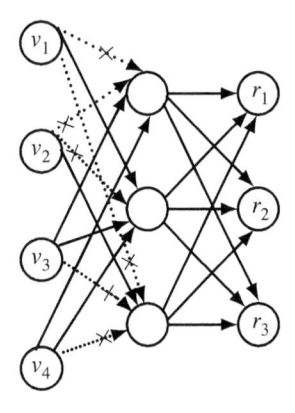

where **v** is the fully connected layer inputs (i.e., output feature vector of the input data vector **x**), **W** is the fully connected layer weights, $\odot$ denotes elementwise multiplication, and **m** is a *binary mask vector* of size d with each element j drawn independently from $m_j \sim$ Bernoulli(p). The activation function $a(\cdot)$ may be tanh and reLU functions with $a(0) = 0$.

- *Training network with DropConnect:* Generalization of Dropout in which each connection, rather than each output unit, can be dropped with probability $1 - p$:

$$\mathbf{r} = a\big((\mathbf{M} \odot \mathbf{W})\mathbf{v}\big), \tag{7.9.32}$$

where **M** is *weight mask matrix* with $M_{ij} \sim$ Bernoulli(p). In DropConnect, dropout is applied at the inputs to the activation function.

Figure 7.11 shows an example of DropConnect for a full-connection neural network.

From Figs. 7.10 and 7.11 we can see further the similarity and difference between Dropout and DropConnect.

- Both Dropout and DropConnect are suitable for fully connected layers only, and can introduce dynamic sparsity within the model.
- Dropout has sparsity on output units by dropping activations with probability $1 - p$, while DropConnect makes the weight matrix **W** become a sparse matrix by introducing zero-elements with probability $1 - p$, rather than modifying the output vectors of a layer. That is to say, the fully connected layer with DropConnect becomes a sparsely connected layer. However, this is not equivalent to setting **W** to be a fixed sparse matrix during training.
- Dropout is to remove the outputs of randomly dropped hidden units, while DropConnect is to set weight values connecting the input units and hidden units to be zero with a probability of $1 - p$. In other words, any dropped hidden unit has no outputs in all output directions, but a DropConnected layer has no outputs in DropConnect directions only.

Definition 7.5 (DropConnect Network [153]) Let $\{x_1, \ldots, x_l\}$ with labels $\{y_1, \ldots, y_l\}$ be the data set S of l entries. A *DropConnect network* is defined as a mixture model:

$$\mathbf{o} = \sum_m p(\mathbf{M}) f(\mathbf{x}; \boldsymbol{\theta}, \mathbf{M}) = E_m\{f(\mathbf{x}; \boldsymbol{\theta}, \mathbf{M})\}, \quad (7.9.33)$$

where m is the DropConnect layer mask, $\boldsymbol{\theta} = \{\mathbf{W}_s, \mathbf{W}, \mathbf{W}_g\}$ are network parameters: $\mathbf{W}_s$ are the softmax layer parameters, $\mathbf{W}$ are the DropConnect layer parameters and $\mathbf{W}_g$ are the feature extractor parameters. Each network $f(\mathbf{x}; \boldsymbol{\theta}, \mathbf{M})$ has weights $p(\mathbf{M})$ such that $M_{ij} \sim \text{Bernoulli}(p)$.

When each element of $\mathbf{M}$ has equal probability of being on and off ($p = 0.5$), the mixture model has equal weights for all sub-models $f(\mathbf{x}; \boldsymbol{\theta}, \mathbf{M})$, otherwise the mixture model has larger weights in some sub-models than others.

A standard DropConnect model architecture comprises the following basic steps [152].

1. *Feature Extractor:*

$$\mathbf{v} = g(\mathbf{x}; \mathbf{W}_g), \quad (7.9.34)$$

where $\mathbf{v} \in \mathbb{R}^{n \times 1}$ is the output feature vector, $\mathbf{x} \in \mathbb{R}^{I \times 1}$ is the input data vector to the overall model, and $\mathbf{W}_g \in \mathbb{R}^{n \times I}$ is the parameter matrix for the feature extractor. $g(\cdot)$ is chosen to be a multilayered convolutional neural network (CNN), with $\mathbf{W}_g$ being the convolutional filters (and biases) of the CNN.

2. *DropConnect Layer:*

$$\mathbf{r} = f(\mathbf{u}) = f((\mathbf{M} \odot \mathbf{W})\mathbf{v}) \in \mathbb{R}^{d \times 1}, \quad (7.9.35)$$

where $\mathbf{v}$ is the output of the feature extractor, $\mathbf{W} \in \mathbb{R}^{d \times n}$ is a fully connected weight matrix, f is a nonlinear activation function, and $\mathbf{M} \in \mathbb{R}^{d \times n}$ is the *binary mask matrix*.

3. *Softmax Classification Layer:*

$$o_i = \text{softmax}(\mathbf{r}; \mathbf{W}_s) = \frac{\exp\left(-\mathbf{w}_{s,i}^T \mathbf{r}\right)}{1 + \sum_j \exp\left(-\mathbf{w}_{s,j}^T \mathbf{r}\right)}, \quad (7.9.36)$$

where $\mathbf{W}_s \in \mathbb{R}^{k \times d}$ is the weight matrix of softmax classification layer which takes $\mathbf{r}$ as its input and uses $\mathbf{W}_s$ to map this input to a k-dimensional output vector $\mathbf{o}$ (k is the number of classes).

7.9 Dropout Learning

Definition 7.6 (Logistic Loss [153]) The following loss function defined on k-class classification is called the *logistic loss function*:

$$A(\mathbf{o}) = -\sum_i y_i \ln \frac{\exp(o_i)}{\sum_j \exp(o_j)} = -o_i + \ln \sum_j \exp(o_j), \quad (7.9.37)$$

where $\mathbf{y} = [y_1, \ldots, y_k]^T$ is a binary vector with the ith bit set to on.

Lemma 7.1 *The logistic loss function A has the following properties [153]:*

1. $A(0) = \ln k$, *i.e.*, $A(0)$ *depends on some constant related to the number of labels.*
2. $-1 \le A'(0) \le 1$ *with* $A'(0) = \frac{\partial A(\mathbf{o})}{\partial \mathbf{o}}\big|_{\mathbf{0}} = \mathbf{0}$. *That is to say, A is a Lipschitz function with $L = 1$.*
3. $A''(\mathbf{o}) \ge 0$, *i.e., A is a convex function with respect to* $\mathbf{x}$.

With DropConnect, the feedforward operation becomes the DropConnect network [152]:

$$M_{ij}^{(l)} \sim \text{Bernoulli}(p), \quad (7.9.38)$$

$$\tilde{\mathbf{W}}^{(l)} = \mathbf{W}^{(l)} \odot \mathbf{M}, \quad (7.9.39)$$

$$z_i^{(l+1)} = \sum_j \tilde{W}_{ij} y_j^{(l)} + b_i^{(l)}, \quad (7.9.40)$$

$$\mathbf{y}^{(l+1)} = f(\mathbf{z}^{(l+1)}). \quad (7.9.41)$$

Due to the weighted sum of Bernoulli variables M_{ij}, a single unit $u_i = \sum_j (W_{ij} v_j) M_{ij}$ can be approximated by a Gaussian $N(\mu_u, \sigma_u^2)$, i.e., $\mathbf{u} \sim N(\boldsymbol{\mu}_u, \sigma_u^2 \mathbf{I})$, where the mean vector and variance matrix are given by $\boldsymbol{\mu}_u = E\{\mathbf{u}\} = p\mathbf{W}\mathbf{v}$ and $\sigma_u^2 \mathbf{I} = E\{\mathbf{u}\mathbf{u}^T\} = p(1-p)(\mathbf{W} \odot \mathbf{W})(\mathbf{v} \odot \mathbf{v})$, respectively.

A comparison between Dropout and DropConnect network inferences is as follows [152]:

- *Dropout network inference* (mean-inference):

$$E_m\{a(\mathbf{m} \odot \mathbf{W}\mathbf{v})\} \approx a(E_m\{\mathbf{m} \odot \mathbf{W}\mathbf{v}\}) = a(p\mathbf{W}\mathbf{v}). \quad (7.9.42)$$

- *DropConnect Network Inference* (sampling):

$$E_M\{a((\mathbf{M} \odot \mathbf{W})\mathbf{v})\} \approx E_u\{a(\mathbf{u})\}, \quad (7.9.43)$$

where $\mathbf{u} = (\mathbf{M} \odot \mathbf{W})\mathbf{v}$ with entry $u_i = \sum_j (W_{ij} v_j) M_{ij}$ that is a weighted sum of Bernoulli variables M_{ij}. Each neuron u_i can be approximated by a Gaussian via moment matching to give

$$\mathbf{u} \sim N\big(p\mathbf{W}\mathbf{v},\ p(1-p)(\mathbf{W} \odot \mathbf{W})(\mathbf{v} \odot \mathbf{v})\big), \quad (7.9.44)$$

where the mean vector is $E_M\{\mathbf{u}\} = p\mathbf{W}\mathbf{v}$ and the variance matrix is $\text{var}_M(\mathbf{u}) = p(1-p)(\mathbf{W} \odot \mathbf{W})(\mathbf{v} \odot \mathbf{v})$.

The overall model $f(\mathbf{x}; \boldsymbol{\theta}, \mathbf{M})$ therefore maps input data $\mathbf{x}$ to an output $\mathbf{o}$ through a sequence of the above operations given the parameters $\boldsymbol{\theta} = \{\mathbf{W}_g, \mathbf{W}, \mathbf{W}_s\}$ and a randomly drawn mask $\mathbf{M}$. The correct value of $\mathbf{o}$ is obtained by summing out over all possible masks $\mathbf{M}$:

$$\mathbf{o} = \sum_M p(\mathbf{M}) f(\mathbf{x}; \boldsymbol{\theta}, \mathbf{M}). \tag{7.9.45}$$

If $p(\mathbf{M})$, the probability for a mask, is uniform over all masks $\mathbf{M}$, then

$$\mathbf{o} = \frac{1}{|\mathbf{M}|} \sum_M f(\mathbf{x}; \boldsymbol{\theta}, \mathbf{M})$$

$$= \frac{1}{|\mathbf{M}|} \sum_M \text{softmax}(f((\mathbf{M} \odot \mathbf{W})\mathbf{v}); \mathbf{W}_s). \tag{7.9.46}$$

The parameters throughout the model $\boldsymbol{\theta}$ then can be updated via stochastic gradient descent (SGD) by back-propagating gradients of the loss function with respect to the parameters, $A'_{\boldsymbol{\theta}}$, see [152].

7.10 Autoencoders

Multilayer perceptrons (MLPs) form a class of neural networks with a feedforward layered structure.

Consider a general MLP with one hidden layer. It consists of an input layer containing n_i units and two layers with computational units: the hidden layer with p units and the output layer with n_o units.

In the standard MLP, in order to achieve dimension reduction by auto-association, the input units are desired to communicate their values to the output units through a hidden layer acting as a limited capacity bottleneck which must optimally encode the input vectors. Thus, in this particular application, $n_i = n_o = n$ and $p < n$.

Figure 7.12 shows an MLP with one hidden layer for auto association.

When entering an n-dimensional real input vector $\mathbf{x}_k$ ($k = 1, \ldots, N$), the output values of the hidden units form a p-vector given by

$$\mathbf{h}_k = F\left(\mathbf{W}^{(1)}\mathbf{x}_k + \mathbf{b}^{(1)}\right), \tag{7.10.1}$$

where $\mathbf{W}^{(1)}$ is the (input-to-hidden) $p \times n$ weight matrix and $\mathbf{b}^{(1)}$ is a p-vector of biases.

Fig. 7.12 MLP with one hidden layer for auto association. The shape of the whole network is similar to an hourglass

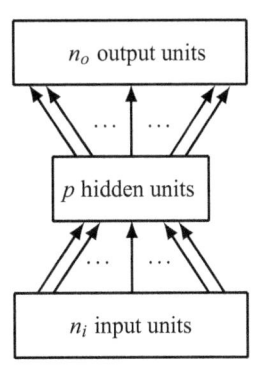

n_o output units

p hidden units

n_i input units

7.10.1 Basic Autoencoder

In the context of object recognition, a particularly interesting and challenging question is whether unsupervised learning can be used to learn hierarchies of invariant feature extractors [117]. Autoencoders provide an effective way to solve this problem. The basic idea of the autoencoder, proposed by Rumelhart et al. [123] in 1986, is the layer-by-layer training.

Features that are invariant to small distortions are called the *invariant features*. The key idea to *invariant feature learning* is to represent an input patch with two components [117]: the invariant feature vector (which represents what is in the image) and the transformation parameters (which encodes where each feature appears in the image).

The *Autoencoder* is an MLP with a symmetric structure, and is used to learn an efficient representation (encoding) for a set of data in an unsupervised manner. The aim of an autoencoder is to make the output as similar to the input as possible.

Architecturally, the simplest form of an autoencoder is a feedforward, non-recurrent neural network very similar to the many single-layer perceptrons in an MLP: having an input layer, an output layer (also known as reconstruction layer), and one or more hidden layers connecting them, but with the output layer having the same number of nodes as the input layer. The purpose of autoencoders is to reconstruct its own inputs instead of predicting the target value Y given inputs X. Therefore, autoencoders are unsupervised learning models.

An autoencoder is composed of two networks: an encoder network and a decoder network, as shown in Fig. 7.13.

The basic idea of an autoencoder is to encode information automatically (as compressed, not encrypted), and hence it gets its name. The hidden layer in the middle is smaller, and the input layer and output layer on both sides are larger. The autoencoder is always symmetrical, with the middle layer (comprising one or two layers, depending on whether the number of neural network layers is odd or even). The layer in front of the middle layer is the encoding layer, and the layer behind the middle layer is the decoding layer.

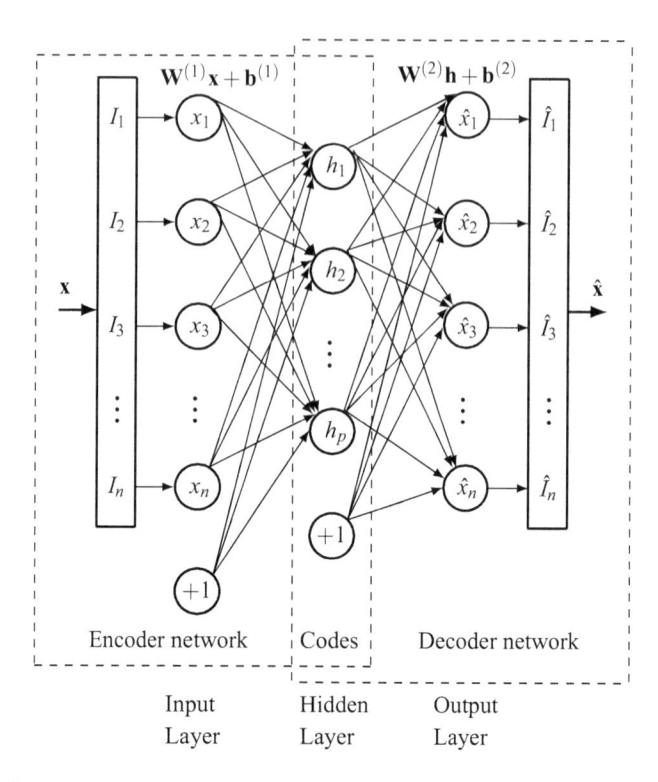

Fig. 7.13 Illustration of the architecture of the basic autoencoder with "encoder" and "decoder" networks for high-level feature learning. The leftmost layer of the whole network is called the input layer, and the rightmost layer the output layer. The middle layer of nodes is called the hidden layer, because its values are not observed in the training set. The circles labeled "+1" are called bias units, and correspond to the intercept term **b** [109]. I_i denotes the index i, $i = 1, \ldots, n$

- *Encoder network:* the encoder **f(x)** is a feature-extracting vector function in a specific parameterized closed form, and maps the original high-dimensional inputs (n-vector) $\mathbf{x} \in \mathbb{R}^n$ to a low-dimensional feature or intermediate (p-vector) representation $\mathbf{h} = \mathbf{f}(\mathbf{W}^{(1)}\mathbf{x}) \in \mathbb{R}^p$:

$$\mathbf{h} = \mathbf{f}(\mathbf{x}) = s_f\left(\mathbf{W}^{(1)}\mathbf{x} + \mathbf{b}_h\right) : \mathbb{R}^n \to \mathbb{R}^p \quad \text{with } n > p, \qquad (7.10.2)$$

where **h** is the *feature vector* or *representation* or *code* computed from **x**, $\mathbf{W}^{(1)} \in \mathbb{R}^{p \times n}$ is the encoding matrix, and s_f is an elementwise nonlinear activation function, typically a logistic function $s_f(z_i) = \text{sigmoid}(z_i) = \frac{1}{1+e^{-z_i}}$ for the argument vector $\mathbf{z} = [z_i]_{i=1}^p$ of $s_f(\mathbf{z})$. The encoder's parameters are denoted by a bias vector $\mathbf{b}_h \in \mathbb{R}^p$.
- *Decoder network:* the symmetric decoder **g(h)** is another closed form parameterized vector function that maps **h** back to an n-dimensional output vector, producing a *reconstruction* $\mathbf{y} = \mathbf{g}(\mathbf{h})$ that is desired to be as similar as possible

7.10 Autoencoders

to the original input vector **x**, i.e., $\mathbf{y} = \hat{\mathbf{x}}$:

$$\mathbf{y} = \mathbf{g}(\mathbf{h}) = s_g\left(\mathbf{W}^{(2)}\mathbf{h} + \mathbf{b}_y\right) : \mathbb{R}^p \to \mathbb{R}^n, \tag{7.10.3}$$

where $\mathbf{W}^{(2)} \in \mathbb{R}^{n \times p}$ is the decoding matrix and **g** is also an elementwise nonlinear function, and s_g is the decoder's activation function, typically either the identity (yielding linear reconstruction) or a sigmoid. The decoder's parameters are represented by a bias vector $\mathbf{b}_y \in \mathbb{R}^n$. In general, one only explores the *tied weights* case, in which $\mathbf{W}^{(2)} = (\mathbf{W}^{(1)})^T = \mathbf{W}^T$.

Because the input layer and the output layer have the same numbers of neurons n that is larger than the number p of neurons of the hidden layer, the middle hidden layer is the bottleneck of the whole autoencoder.

The autoencoder shown in Fig. 7.13 is a typical feedforward neural network, since the connectivity graph does not have any directed loops or cycles. Autoencoder training is to find parameters $\theta = \{\mathbf{W}, \mathbf{b}_h, \mathbf{b}_y\}$ that minimize the reconstruction error on a training set of examples D_n:

$$\min_{\theta} \left\{ J_{\text{AE}}(\theta) = \sum_{\mathbf{x} \in D_n} L(\mathbf{x}, \mathbf{g}(\mathbf{f}(\mathbf{x}))) \right\}, \tag{7.10.4}$$

where $L(\mathbf{x}, \mathbf{y})$ is the reconstruction error. Its typical choices include [119]:

- the squared error $L(\mathbf{x}, \mathbf{y}) = \|\mathbf{x} - \mathbf{y}\|^2$ and
- the *cross-entropy loss function* $L(\mathbf{x}, \mathbf{y}) = -\sum_{i=1}^{n} x_i \log(y_i) + (1 - x_i) \log(1 - y_i)$.

An Autoencoder training flowchart is shown in Fig. 7.14.

In the autoencoder framework [10, 12, 60], one starts by explicitly defining a feature-extracting function in a specific parameterized closed form. This function, denoted by $\mathbf{f}_\theta$, is called the encoder and will allow the forthright and efficient computation of a feature vector $\mathbf{h} = \mathbf{f}_{\theta_1}(\mathbf{x})$ from an input **x**. For each example $\mathbf{x}_t$ from a data set $\{\mathbf{x}_1, \ldots, \mathbf{x}_T\}$, define the hidden neurons and activations of the

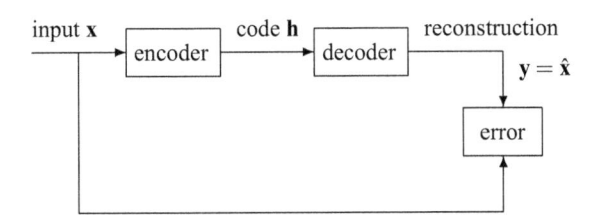

Fig. 7.14 Autoencoder training flowchart. Encoder f encodes the input **x** to $\mathbf{h} = f(\mathbf{x})$, and the decoder g decodes $\mathbf{h} = f(\mathbf{x})$ to the output $\mathbf{y} = \hat{\mathbf{x}} = g(f(\mathbf{x}))$ so that the reconstruction error $L(\mathbf{x}, \mathbf{y})$ is minimized

second layer (hidden layer) as

$$\text{net}_j^{(2)} = \sum_{i=1}^{n} w_{ji}^{(1)} x_i + b_j^{(1)}, \ j = 1, \ldots, m, \quad (7.10.5)$$

$$a_j^{(2)} = f\left(\text{net}_j^{(2)}\right), \ j = 1, \ldots, m. \quad (7.10.6)$$

Similarly, the hidden neurons and activations of the third layer (output layer) are, respectively, defined as

$$\text{net}_i^{(3)} = \sum_{j=1}^{m} w_{ij}^{(2)} a_j^{(2)} + b_i^{(2)}, \ i = 1, \ldots, n, \quad (7.10.7)$$

$$a_i^{(3)} = f\left(\text{net}_i^{(3)}\right), \ i = 1, \ldots, n, \quad (7.10.8)$$

where f is a nonlinear (typically sigmoid) function, i.e., $f(z) = 1/(1 + e^{-z})$.

The problem of the basic autoencoder is to find optimal weight matrices $\mathbf{W}^{(1)}$, $\mathbf{W}^{(2)}$ and the bias vectors $\mathbf{b}^{(1)}$, $\mathbf{b}^{(2)}$ for minimizing the total error energy of the system.

Denoting $\boldsymbol{\theta} = (\boldsymbol{\theta}_1; \boldsymbol{\theta}_2) = (\mathbf{W}^{(1)}, \mathbf{b}^{(1)}; \mathbf{W}^{(2)}, \mathbf{b}^{(2)})$, the energy of the system is the sum of two terms [115]:

$$J_{\text{AE}}(\boldsymbol{\theta}) = \frac{1}{2m}\left[\sum_{j=1}^{m} E_e\left(a_j^{(2)}, \mathbf{W}^{(1)}\right)\right] + \frac{1}{2n}\left[\sum_{i=1}^{n} E_d\left(a_i^{(3)}, \mathbf{W}^{(2)}\right)\right], \quad (7.10.9)$$

where

- $E_e\left(a_j^{(2)}, \mathbf{W}^{(1)}\right)$ is the *code prediction energy* which measures the discrepancy between the output of the encoder and the code vector $\mathbf{a}^{(2)} = [a_1^{(2)}, \ldots, a_m^{(2)}]^T$. A typical definition is given by

$$E_e\left(a_j^{(2)}, \mathbf{W}^{(1)}\right) = \frac{1}{2}\left(a_j^{(2)} - \text{Encode}(x_i, \mathbf{W}^{(1)})\right)^2$$

$$= \frac{1}{2}\left(a_j^{(2)} - w_{ji}^{(1)} x_i\right)^2. \quad (7.10.10)$$

- $E_d(a_i^{(3)}, \mathbf{W}^{(2)})$ is the *reconstruction energy* which measures the discrepancy between the reconstructed image patch produced by the decoder and the input image patch X. A common choice is

$$E_d\left(a_i^{(3)}, \mathbf{W}^{(2)}\right) = \frac{1}{2}\left(a_i^{(3)} - \text{Decode}(\bar{\mathbf{y}}_t, \mathbf{W}^{(2)})\right)^2$$

$$= \frac{1}{2}\left(a_i^{(3)} - x_i\right)^2. \quad (7.10.11)$$

7.10 Autoencoders

Therefore, the loss function (7.10.9) in the basic autoencoder optimization can be represented as

$$J_{\text{AE}}(\boldsymbol{\theta}) = \frac{1}{2m}\sum_{j=1}^{m}\left(a_j^{(2)} - w_{ji}^{(1)}x_i\right)^2 + \frac{1}{2n}\sum_{i=1}^{n}\left(a_i^{(3)} - x_i\right)^2. \quad (7.10.12)$$

Usually, the backpropagation algorithm is an effective method to train the parameters of an autoencoder with n input units and m hidden units. By the backpropagation algorithm, the partial derivative of the loss function $J_{\text{AE}}(\boldsymbol{\theta})$ with respect to weights $w_{ij}^{(2)}$ and $b_i^{(2)}$ are given using the chain rule twice, respectively:

$$\frac{\partial J_{\text{AE}}}{\partial w_{ij}^{(2)}} = \frac{\partial J_{\text{AE}}}{\partial a_i^{(3)}} \cdot \frac{\partial a_i^{(3)}}{\partial \text{net}_i^{(3)}} \cdot \frac{\partial \text{net}_i^{(3)}}{\partial w_{ij}^{(2)}}, \quad (7.10.13)$$

$$\frac{\partial J_{\text{AE}}}{\partial b_i^{(2)}} = \frac{\partial J_{\text{AE}}}{\partial a_i^{(3)}} \cdot \frac{\partial a_i^{(3)}}{\partial \text{net}_i^{(3)}} \cdot \frac{\partial \text{net}_i^{(3)}}{\partial b_i^{(2)}}, \quad (7.10.14)$$

where

$$\frac{\partial \text{net}_i^{(3)}}{\partial w_{ij}^{(2)}} = \frac{\partial}{\partial w_{ij}^{(2)}}\left(\sum_{j=1}^{m} w_{ij}^{(2)} a_j^{(2)} + b_i^{(2)}\right) = a_j^{(2)} \quad (7.10.15)$$

$$\frac{\partial \text{net}_i^{(3)}}{\partial b_i^{(2)}} = \frac{\partial}{\partial b_i^{(2)}}\left(\sum_{j=1}^{m} w_{ij}^{(2)} a_j^{(2)} + b_i^{(2)}\right) = 1, \quad (7.10.16)$$

$$\frac{\partial a_i^{(3)}}{\partial \text{net}_i^{(3)}} = f'\left(\text{net}_i^{(3)}\right) = f\left(\text{net}_i^{(3)}\right)\left(1 - f\left(\text{net}_i^{(3)}\right)\right) = a_i^{(3)}\left(1 - a_i^{(3)}\right), \quad (7.10.17)$$

$$\frac{\partial J_{\text{AE}}}{\partial a_i^{(3)}} = \frac{\partial}{\partial a_i^{(3)}} \frac{1}{2}\left(a_i^{(3)} - x_i\right)^2 = a_i^{(3)} - x_i. \quad (7.10.18)$$

Hence, we have

$$\sigma_i^{(3)} = \frac{\partial J_{\text{AE}}}{\partial a_i^{(3)}} \frac{\partial a_i^{(3)}}{\partial \text{net}_i^{(3)}} = \left(a_i^{(3)} - x_i\right)\left(a_i^{(3)}\left(1 - a_i^{(3)}\right)\right), \quad (7.10.19)$$

$$\Delta b_i^{(2)} = \Delta b_i^{(2)} + \sigma_i^{(3)}, \quad (7.10.20)$$

$$\Delta w_{ij}^{(2)} = \Delta w_{ij}^{(2)} + a_j^{(2)} \sigma_i^{(3)}, \quad (7.10.21)$$

for $i = 1, \ldots, n;\ j = 1, \ldots, m$.

Similarly, the partial derivative of the loss $J_{AE}(\theta)$ with respect to weights $w_{ji}^{(1)}$ and $b_j^{(1)}$ are given using the chain rule twice, respectively:

$$\frac{\partial J_{AE}}{\partial w_{ji}^{(1)}} = \frac{\partial J_{AE}}{\partial a_j^{(2)}} \cdot \frac{\partial a_j^{(2)}}{\partial \text{net}_j^{(2)}} \cdot \frac{\partial \text{net}_j^{(2)}}{\partial w_{ji}^{(1)}}, \quad (7.10.22)$$

$$\frac{\partial J_{AE}}{\partial b_j^{(1)}} = \frac{\partial J_{AE}}{\partial a_j^{(2)}} \cdot \frac{\partial a_j^{(2)}}{\partial \text{net}_j^{(2)}} \cdot \frac{\partial \text{net}_j^{(2)}}{\partial b_j^{(1)}}, \quad (7.10.23)$$

where

$$\frac{\partial \text{net}_j^{(2)}}{\partial w_{ji}^{(1)}} = \frac{\partial}{\partial w_{ji}^{(1)}} \left(\sum_{i=1}^n w_{ji}^{(1)} x_i + b_j^{(1)} \right) = x_i,$$

and

$$\frac{\partial \text{net}_j^{(2)}}{\partial b_j^{(1)}} = \frac{\partial}{\partial b_j^{(1)}} \left(\sum_{i=1}^n w_{ji}^{(1)} x_i + b_j^{(1)} \right) = 1,$$

$$\frac{\partial a_j^{(2)}}{\partial \text{net}_j^{(2)}} = f'\left(\text{net}_j^{(2)}\right) = f\left(\text{net}_j^{(2)}\right)\left(1 - f\left(\text{net}_j^{(2)}\right)\right) = a_j^{(2)}\left(1 - a_j^{(2)}\right),$$

$$\frac{\partial J_{AE}}{\partial a_j^{(2)}} = \sum_{i=1}^n \frac{\partial J_{AE}}{\partial a_i^{(3)}} \frac{\partial a_i^{(3)}}{\partial \text{net}_i^{(3)}} \frac{\partial \text{net}_i^{(3)}}{\partial a_j^{(2)}} = \sum_{i=1}^n \frac{\partial J_{AE}}{\partial a_i^{(3)}} \frac{\partial a_i^{(3)}}{\partial \text{net}_i^{(3)}} w_{ij}^{(2)} = \sum_{i=1}^n \sigma_i^{(3)} w_{ij}^{(2)}.$$

Thus we have [174]

$$\sigma_j^{(2)} = \frac{\partial J_{AE}}{\partial a_j^{(2)}} \frac{\partial a_j^{(2)}}{\partial \text{net}_j^{(2)}} = \left(\sum_{i=1}^n \sigma_i^{(3)} w_{ij}^{(2)} \right)\left(a_j^{(2)}(1 - a_j^{(2)}) \right), \quad (7.10.24)$$

$$\Delta b_j^{(2)} = \Delta b_j^{(2)} + \sigma_j^{(2)}, \quad (7.10.25)$$

$$\Delta w_{ji}^{(1)} = \Delta w_{ji}^{(1)} + x_i \sigma_j^{(2)}, \quad (7.10.26)$$

for $i = 1, \ldots, n; \ j = 1, \ldots, m$.

The Backpropagation algorithm for the basic autoencoder is summarized in Algorithm 7.4.

7.10 Autoencoders

Algorithm 7.4 Backpropagation algorithm for the basic autoencoder [174]

1. **input:** $\{(\mathbf{x}^{(j)}, \mathbf{y}^{(j)})\}_{j=1}^{m}$ with $\mathbf{x}^{(j)} = [x_1^{(j)}, \ldots, x_n^{(j)}]^T$, η and $threshold$.
2. **initialization:**
3. **for** $iteration = 1, 2, \ldots, iterater_{\max}$ **do**
4.   **for** $example = 1, 2, \ldots, N$ **do**
5.   **for** $j = 1, 2, \ldots, m$ **do**
6. $net_j^{(2)} = \sum_{i=1}^{n} w_{ji}^{(1)} x_i + b_j^{(1)}$;
7. $a_j^{(2)} = f(net_j^{(2)})$;
8.   **end for**
9.   **for** $i = 1, 2, \ldots, n$ **do**
10. $net_i^{(3)} = \sum_{j=1}^{m} w_{ij}^{(2)} a_j^{(2)} + b_i^{(2)}$;
11. $a_i^{(3)} = f(net_i^{(3)})$;
12.   **end for**
13.   **if** $J_{TAE}(\boldsymbol{\theta}) > threshold$ **then** compute
14. $\sigma_i^{(3)} = \left(a_i^{(3)} \cdot (1 - a_i^{(3)})\right) \cdot \left(a_i^{(3)} - x_i\right), i = 1, \ldots, n$;
15. $\sigma_j^{(2)} = \left(\sum_{i=1}^{n} w_{ij}^{(2)} \sigma_i^{(3)}\right)\left(a_j^{(2)}(1 - a_j^{(2)})\right), j = 1, \ldots, m$;
16. $\Delta b_i^{(2)} = \Delta b_i^{(2)} + \sigma_i^{(3)}, i = 1, \ldots, n$;
17. $\Delta w_{ij}^{(2)} = \Delta w_{ij}^{(2)} + a_j^{(2)} \cdot \sigma_i^{(3)}, i = 1, \ldots, n; j = 1, \ldots, m$;
18. $\Delta b_j^{(1)} = \Delta b_j^{(1)} + \sigma_j^{(2)}, j = 1, \ldots, m$;
19. $\Delta w_{ji}^{(1)} = \Delta w_{ji}^{(1)} + x_i \cdot \sigma_j^{(2)}, i = 1, \ldots, n; j = 1, \ldots, m$;
20. $\mathbf{W}^{(1)} = \mathbf{W}^{(1)} - \eta(\frac{1}{N}\Delta \mathbf{W}^{(1)})$;
21. $\mathbf{W}^{(2)} = \mathbf{W}^{(2)} - \eta(\frac{1}{N}\Delta \mathbf{W}^{(2)})$;
22. $\mathbf{b}^{(1)} = \mathbf{b}^{(1)} - \eta(\frac{1}{N}\Delta \mathbf{b}^{(1)})$;
23. $\mathbf{b}^{(2)} = \mathbf{b}^{(2)} - \eta(\frac{1}{N}\Delta \mathbf{b}^{(2)})$;
24.   **end if**
25.   **end for**
26. **end for**
27. **output:** $\boldsymbol{\theta} = \{\mathbf{W}^{(1)}, \mathbf{b}^{(1)}; \mathbf{W}^{(2)}, \mathbf{b}^{(2)}\}$.

For regularized autoencoders, the simplest form of regularization is *weight decay* (wd) which favors small weights by optimizing instead the following regularized objective:

$$J_{\text{AE+wd}}(\boldsymbol{\theta}) = \sum_{\mathbf{x} \in D_n} L(\mathbf{x}, \mathbf{g}(\mathbf{f}(\mathbf{x}))) + \lambda \sum_{i,j} W_{ij}^2, \quad (7.10.27)$$

where λ is the hyperparameter for controlling the strength of the regularization.

If taking the penalization term

$$\|\mathbf{J}_f(\mathbf{x})\|_F^2 = \sum_{i,j} \left(\frac{\partial h_j(\mathbf{x})}{\partial x_i}\right)^2 \quad (7.10.28)$$

as the regularization term, then the obtained regularized autoencoder is called the *contractive autoencoder* (CAE) that was proposed by Rifai et al. [119]. Hence, the objective function of the contractive autoencoder is given by

$$J_{\text{CAE}}(\boldsymbol{\theta}) = \sum_{\mathbf{x} \in D_n} L(\mathbf{x}, \mathbf{g}(\mathbf{f}(\mathbf{x}))) + \lambda \|\mathbf{J}_f(\mathbf{x})\|_F^2, \tag{7.10.29}$$

where λ is a hyperparameter controlling the strength of the regularization. Penalizing $\|\mathbf{J}_f\|_F^2$ encourages the mapping to the feature space to be contractive in the neighborhood of the training data, which gives the name "contractive autoencoder."

7.10.2 Stacked Sparse Autoencoder

Autoencoders have various major extensions: the stacked autoencoder, the sparse autoencoder, the stacked sparse autoencoder, the stacked denoising autoencoder, the stacked convolutional autoencoder, and the stacked denoising convolutional autoencoder.

The *stacked autoencoder* of Bengio et al. [9] is a multilayer neural network in which each layer is the hidden layer of the former autoencoder and the input layer of the next autoencoder. Hence, a stacked autoencoder can be viewed as a neural network composed of multiple autoencoders connected successively in a greedy layer-wise fashion.

Figure 7.15 gives an illustration of a stacked autoencoder with 2 hidden layers.

Letting the input in the dth stacked autoencoder be $\mathbf{x}_t^{(d)}$, and $\mathbf{x}_t^{(1)} = \mathbf{x}_t$, then the equations of the dth autoencoder are given by hidden layer as

$$\mathbf{h}_t^{(d)} = f\left(\mathbf{W}_1^{(d)} \mathbf{x}_t^{(d)} + \mathbf{b}_1^{(d)}\right), \tag{7.10.30}$$

$$\mathbf{y}_t^{(d)} = f\left(\mathbf{W}_2^{(d)} \mathbf{h}_t^{(d)} + \mathbf{b}_2^{(d)}\right), \tag{7.10.31}$$

where $\mathbf{h}_t^{(d)}$ and $\mathbf{y}_t^{(d)}$ denote, respectively, the feature representation at hidden layer and the output at the output layer associated with the dth autoencoder, and $\mathbf{W}_1^{(d)}$ and $\mathbf{W}_2^{(d)}$ stand, respectively, for weight matrices in the encoding step and the decoding step, and $\mathbf{b}_1^{(d)}$ and $\mathbf{b}_2^{(d)}$ are the biases of the hidden layer and the output layer. The activation function $f(\cdot)$ usually uses the sigmoid function $f(x) = 1/(1+\exp(-x))$.

7.10 Autoencoders

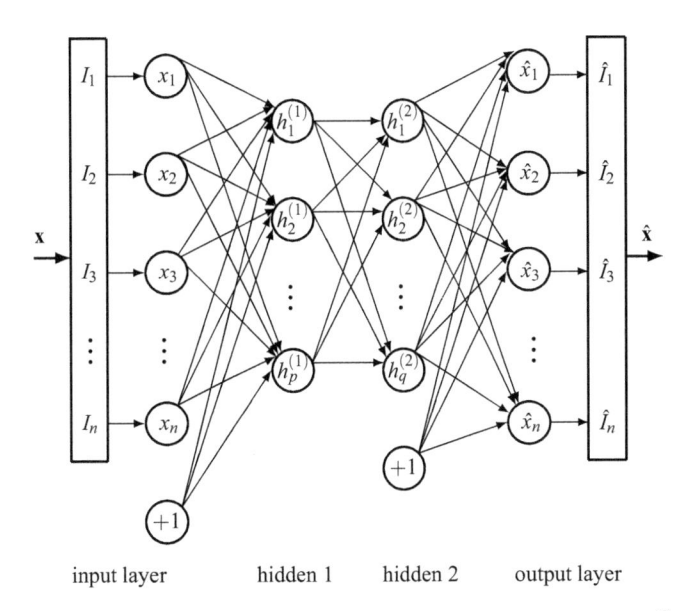

Fig. 7.15 Stacked autoencoder scheme with 2 hidden layers, where $\mathbf{h}^{(d)} = f(\mathbf{W}_1^{(d)} \mathbf{x}^{(d)} + \mathbf{b}_1^{(d)})$ and $\mathbf{y}^{(d)} = f(\mathbf{W}_2^{(d)} \mathbf{h}^{(d)} + \mathbf{b}_2^{(d)})$, $d = 1, 2$ with $\mathbf{x}^{(1)} = \mathbf{x}$, $\mathbf{x}^{(2)} = \mathbf{y}^{(1)}$ and $\hat{\mathbf{x}} = \mathbf{y}^{(2)}$

By taking the feature representation $\mathbf{y}^{(d-1)}$ of the $(d-1)$th layer as the input $\mathbf{x}^{(d)}$ of the dth layer, i.e., $\mathbf{x}^{(d)} = \mathbf{y}^{(d-1)}$, then the objective function is given by

$$J\left(\mathbf{W}_1^{(d)}, \mathbf{b}_1^{(d)}; \mathbf{W}_2^{(d)}, \mathbf{b}_2^{(d)}\right)$$
$$= \frac{1}{2T} \sum_{t=1}^{T} \left\| \mathbf{y}_t^{(d)} - \mathbf{x}_t^{(d)} \right\|_2^2 + \frac{\beta}{2} \left(\|\mathbf{W}_1^{(d)}\|_2^2 + \|\mathbf{W}_2^{(d)}\|_2^2 \right). \quad (7.10.32)$$

Sparse autoencoder [116, 117] is an autoencoder that introduces a sparse penalty term in its hidden layer, so that the autoencoder can obtain more concise and efficient low-dimensional data features under sparse constraints to better express the input data.

Suppose that the average activation of the neurons in the hidden layer is

$$\hat{\rho}_j = \frac{1}{T} \sum_{t=1}^{T} x_{t,j}, \quad j = 1, \ldots, n, \quad (7.10.33)$$

where $x_{t,j}$ is the jth element of training data $\mathbf{x}_t \in \mathbb{R}^n$.

To enforce sparsity, it is hoped that the average activation $\hat{\rho}_j$ approaches a constant ρ which is the sparsity parameter chosen to be a small positive number near 0. To this end, the Kullback–Leibler (KL) divergence between $\hat{\rho}_j$ and ρ, i.e.,

$$\mathrm{KL}(\rho\|\hat{\rho}) = \sum_{j=1}^{n} \rho \log \frac{\rho}{\hat{\rho}_j} + (1-\rho)\log\frac{1-\rho}{1-\hat{\rho}_j}, \quad j=1,\ldots,n, \qquad (7.10.34)$$

as a regularization term, is added to the error function $J_{\mathrm{AE}}(\boldsymbol{\theta})$ of the basic autoencoder. Here $\hat{\boldsymbol{\rho}} = [\hat{\rho}_1,\ldots,\hat{\rho}_n]^T$ is the vector of average hidden activities.

Denote $\mathbf{W} = (\mathbf{W}^{(1)}, \mathbf{W}^{(2)})$ and $\mathbf{b} = (\mathbf{b}^{(1)}, \mathbf{b}^{(2)})$. To prevent overfitting, a weight decay term is also added to the cost function of $J_{\mathrm{AE}}(\boldsymbol{\theta}) = J_{\mathrm{AE}}(\mathbf{W},\mathbf{b})$, giving the final cost function of the sparse autoencoder, denoted by $J_{\mathrm{SAE}}(\mathbf{W},\mathbf{b})$, as follows [67]:

$$J_{\mathrm{SAE}}(\mathbf{W},\mathbf{b}) = J_{\mathrm{AE}}(\mathbf{W},\mathbf{b}) + \beta J_{\mathrm{KL}}(\rho\|\hat{\rho}) + \frac{\lambda}{2}\sum_{l=1}^{2}\sum_{i=1}^{s_l}\sum_{j=1}^{s_{l+1}} \left(w_{ij}^{(l)}\right)^2, \qquad (7.10.35)$$

where β controls the sparsity penalty term, $J_{\mathrm{KL}}(\rho\|\hat{\rho}) = \mathrm{KL}(\rho\|\hat{\rho})$, λ controls the penalty term facilitating weight decay, and s_l and s_{l+1} are the sizes of adjacent layers.

Stacked sparse autoencoder (SSAE) is a neural network composed of multiple sparse autoencoders connected end to end. The output of the previous layer of each sparse autoencoder is used as the input of the next layer of the autoencoder, so that higher-level feature representations of the input data can be obtained. Hence, a stacked sparse autodecoder is a combination of stacked and sparse autoencoders. SSAE has a structure similar to stacked autoencoders by adding a sparse constraint.

The objective function of the dth layer of a stacked sparse autoencoder is described by [68]

$$J_{\mathrm{SSAE}}\left(\mathbf{W}_1^{(d)}, \mathbf{b}_1^{(d)}, \mathbf{W}_2^{(d)}, \mathbf{b}_2^{(d)}\right) = \frac{1}{2T}\sum_{t=1}^{T}\left\|\mathbf{y}_t^{(d)} - \mathbf{x}_t^{(d)}\right\|_2^2 + \beta\sum_{j=1}^{K_d}\mathrm{KL}(\rho\|\hat{\rho}_j)$$

$$+ \frac{\alpha}{2}\left(\|\mathbf{W}_1^{(d)}\|_2^2 + \|\mathbf{W}_2^{(d)}\|_2^2\right), \qquad (7.10.36)$$

where

- K_d denotes the number of hidden units in the dth layer with $d = 2,\ldots,D$, and D is the number of layers,
- T is the number of inputs,
- α means the weight decay parameter,
- β denotes the weight of the sparsity penalty,
- ρ is the sparsity parameter, usually set to a small value,
- $\hat{\rho}_j$ is the average activation of the jth hidden unit in the dth layer over the training set,
- $\mathrm{KL}(\rho,\hat{\rho}_j)$ denotes the Kullback–Leibler divergence between ρ and $\hat{\rho}_j$.

7.10 Autoencoders

The parameters $\mathbf{W}_1^{(d)}, \mathbf{W}_2^{(d)}, \mathbf{b}_1^{(d)}, \mathbf{b}_2^{(d)}$ are updated by the following stochastic gradient descent algorithm [68]:

$$\mathbf{W}_1^{(d)} \leftarrow \mathbf{W}_1^{(d)} - \eta \left(\frac{1}{T} \sum_{t=1}^{T} \mathbf{h}_i^{(d)} \delta_i^1 + \alpha \mathbf{W}_1^{(d)} \right), \tag{7.10.37}$$

$$\mathbf{W}_2^{(d)} \leftarrow \mathbf{W}_2^{(d)} - \eta \left(\frac{1}{T} \sum_{t=1}^{T} \mathbf{h}_i^{(d)} \delta_i^2 + \alpha \mathbf{W}_2^{(d)} \right), \tag{7.10.38}$$

$$\mathbf{b}_1^{(d)} \leftarrow \mathbf{b}_1^{(d)} - \frac{\eta}{T} \sum_{i=1}^{T} \delta_i^1, \tag{7.10.39}$$

$$\mathbf{b}_2^{(d)} \leftarrow \mathbf{b}_2^{(d)} - \frac{\eta}{T} \sum_{i=1}^{T} \delta_i^2, \tag{7.10.40}$$

where η is a learning rate, and

$$\delta_i^1 = \left(\mathbf{W}_2^{(d)} \delta_i^2 + \beta(-(\rho/\hat{\rho}_t) + (1-\rho)/(1-\hat{\rho}_t)) \right) f' \left(\mathbf{W}_1^{(d)} \mathbf{x}_i^{(d)} + \mathbf{b}_1^{(d)} \right), \tag{7.10.41}$$

$$\delta_i^2 = \left(\mathbf{y}_i^{(d)} - \mathbf{x}_i^{(d)} \right) f' \left(\mathbf{W}_2^{(d)} \mathbf{h}_i^{(d)} + \mathbf{b}_2^{(d)} \right). \tag{7.10.42}$$

7.10.3 Stacked Denoising Autoencoders

Without any additional constraints, conventional autoencoders learn only the identity mapping. This problem can be circumvented by using *denoising autoencoder* (DAE) [149] which tries to train an autoencoder which could automatically denoise the input data from a corrupted version that has been manually added with random noise, and thus generates better feature representations for the subsequent classification tasks.

Let the original input data be $\mathbf{x} \in \mathbb{R}^d$, where d denotes the dimension of data. DAE firstly produces a vector $\tilde{\mathbf{x}}$ by adding the Gaussian noise to $\mathbf{x}$ or a variable amount $\mathbf{v}$ of noise distributed according to the characteristics of the input image, where the parameter $\mathbf{v}$ represents the percentage of permissible corruption. DAE reconstructs the denoising input $\mathbf{x}$ from the corrupted input $\tilde{\mathbf{x}}$. The number of units in the input layer is d, which is equal to the dimension of the input data $\tilde{\mathbf{x}}$.

The DAE is trained to denoise the inputs by first finding the latent representation

$$\mathbf{h} = f_e(\mathbf{W}\tilde{\mathbf{x}} + \mathbf{b}), \tag{7.10.43}$$

where $\mathbf{h} \in \mathbb{R}^h$ is the output of the hidden layer and can also be called the feature representation or code, h is the number of units in the hidden layer, $\mathbf{W} \in \mathbb{R}^{h \times d}$ is the input-to-hidden weights, $\mathbf{b}$ denotes the bias, $\mathbf{W}\tilde{\mathbf{x}} + \mathbf{b}$ stands for the input of the hidden layer, and $f_e(\cdot)$ is called activation function of the hidden layer. When choosing the ReLU function as the activation function, then

$$\mathbf{h} = f_e(\mathbf{W}\tilde{\mathbf{x}} + \mathbf{b}) = \max(0, \mathbf{W}\tilde{\mathbf{x}} + \mathbf{b}). \tag{7.10.44}$$

Clearly, if the value of $\mathbf{W}\tilde{\mathbf{x}}+\mathbf{b}$ is smaller than zero, the output of the hidden layer will be zero, and hence the ReLU activation function is able to produce a sparse feature representation, which may have better separation capability. Moreover, ReLU can train the neural network for large-scale data faster and more effectively than the other activation functions.

The decoding or reconstruction of the original input is obtained by using a mapping function $f_d(\cdot)$:

$$\mathbf{z} = f_d(\mathbf{W}'\mathbf{y} + \mathbf{b}'), \tag{7.10.45}$$

where $\mathbf{z} \in \mathbb{R}^d$ is the output of DAE, which is also the reconstruction of original data $\mathbf{x}$. The output layer has the same number of nodes as the input layer. $\mathbf{W}' = \mathbf{W}^T$ is referred to as the *tied weight matrix*.

If $\mathbf{x}$ ranges from 0 to 1, the softplus function is selected as the decoding function $f_d(z_i) = \log(1+\mathrm{e}^{z_i})$; otherwise, we preprocess $\mathbf{x}$ by *zero-phase component analysis* (ZCA) whitening and use a linear function as the decoding function:

$$f_d(\mathbf{W}'\mathbf{y} + \mathbf{b}') = \begin{cases} \log\left(1 + \mathrm{e}^{\mathbf{W}'\mathbf{y}+\mathbf{b}'}\right), & \mathbf{x} \in [0, 1]^d; \\ \mathbf{W}'\mathbf{y} + \mathbf{b}', & \text{otherwise.} \end{cases} \tag{7.10.46}$$

DAE aims to train the network by requiring the output data $\mathbf{z}$ to reconstruct the input data $\mathbf{x}$, which is also called reconstruction-oriented training. Therefore, the reconstruction error should be used as the objective function or cost function, which is defined as follows [163]:

$$J(\mathbf{W}) = \begin{cases} -\frac{1}{m}\sum_{i=1}^{m}\sum_{j=1}^{d}\left[\mathbf{x}_j^{(i)} \log\left(\mathbf{z}_j^{(i)}\right) + \left(1 - \mathbf{x}_j^{(i)}\right)\log\left(1 - \mathbf{z}_j^{(i)}\right)\right] + \frac{\lambda}{2}\|\mathbf{W}\|_2^2, & \text{if } \mathbf{x} \in [0, 1]^d; \\ \frac{1}{m}\sum_{i=1}^{m}\left\|\mathbf{x}^{(i)} - \mathbf{z}^{(i)}\right\|_F^2 + \frac{\lambda}{2}\|\mathbf{W}\|_2^2, & \text{otherwise.} \end{cases} \tag{7.10.47}$$

Here $\mathbf{x}_j^{(i)}$ is the jth element of the ith sample, $\|\mathbf{W}\|_2^2$ is ℓ_2-regularization term, also called the weight decay term, and parameter λ controls the importance of the regularization term.

7.10 Autoencoders

DAE can be stacked to build a deep network which has more than one hidden layer. This type of stacked DAEs is simply called the SDAE. Typical SDAE structure includes two encoding layers and two decoding layers. In the encoding part, the output of the first encoding layer acts as the input data of the second encoding layer. Supposing there are L hidden layers in the encoding part, we have the activation function of the kth encoding layer:

$$\mathbf{y}^{(k+1)} = f_e\big(\mathbf{W}^{(k)}\mathbf{y}^{(k)} + \mathbf{b}^{(k)}\big), \quad k = 0, \ldots, L-1, \tag{7.10.48}$$

where the input $\mathbf{y}^{(0)}$ is the original data $\mathbf{x}$. The output $\mathbf{y}^{(L)}$ of the last encoding layer are the high-level features extracted by the SDAE network.

In the decoding part, the output of the first decoding layer is regarded as the input of the second decoding layer. The decoding function of the kth decode layer is given by

$$\mathbf{z}^{(k+1)} = f_e\bigg(\mathbf{W}^{(L-k)T}\mathbf{a}^{(k)} + \mathbf{b}^{'(k+1)}\bigg), \quad k = 0, \ldots, L-1, \tag{7.10.49}$$

where the input $\mathbf{z}^{(0)}$ of the first decoding layer is the output $\mathbf{y}^{(L)}$ of the last encoding layer. The output $\mathbf{z}^{(L)}$ of the last decoding layer is the reconstruction of the original data $\mathbf{x}$.

The training process of SDAE is stated as follows [163]:

1. Choose input data, which can be randomly selected from the hyperspectral images.
2. Train the first DAE, which includes the first encoding layer and the last decoding layer. Obtain the network weights $\mathbf{W}^{(1)}$, $\mathbf{b}^{(1)}$ and the features $\mathbf{y}^{(1)}$ which are the output of the first encoding layer.
3. Use $\mathbf{y}^{(k)}$ as the input data of the $(k+1)$th encoding layer. Train the $(k+1)$th DAE and obtain $\mathbf{W}^{(k+1)}$, $\mathbf{b}^{(k+1)}$ and the features $\mathbf{y}^{(k+1)}$, where $k = 1, \ldots, L-1$ and L is the number of hidden layers in the network.

This training of SDAE is called layer-wise training, since each DAE is trained independently.

Unlike the general SDAE network stacked by two DAE structures, the SDAE of Xing et al. [163] retains the encoding part for feature extraction to produce the initial features and adds a *logistic regression* (LR) part instead of the decoding part for fine-tuning and classification. The output layer of the whole network is also called the LR layer which uses the following sigmoid function as the activation function:

$$h(\mathbf{x}) = \frac{1}{1 + \exp(-\mathbf{W}\mathbf{x} - \mathbf{b})}, \tag{7.10.50}$$

where $\mathbf{x}$ is the output $\mathbf{y}^{(L)}$ of the last encoding layer.

The steps of SDAE-LR network training are as follows [163]:

1. SDAE is utilized to train the initial network weights.
2. Initial weights of the LR layer are randomly set.
3. Training data are used as input data, and their predicted classification results are produced with the initial weights of the whole network.
4. Network weights are iteratively tuned by minimizing the cross-entropy function

$$\text{Cost} = -\frac{1}{m}\left[\sum_{i=1}^{m} l^{(i)} \log\left(h\left(\mathbf{x}^{(i)}\right)\right) + \left(1 - l^{(i)}\right) \log\left(1 - h\left(\mathbf{x}^{(i)}\right)\right)\right] \tag{7.10.51}$$

using mini-batch stochastic gradient descent (MSGD) algorithm. Here $l^{(i)}$ denotes the label of the sample $\mathbf{x}^{(i)}$.

7.10.4 Convolutional Autoencoders (CAE)

Fully connected AEs and DAEs both ignore the 2D image structure, which introduces redundancy in the parameters, forcing each feature to be global. To discover localized features, *convolutional autoencoder* (CAE) is an alternative.

CAEs differ from conventional AEs as their weights are shared among all locations in the input, preserving spatial locality [103]. The CAE architecture is intuitively similar to the DAE except that the weights are shared.

For a mono-channel input $\mathbf{x}$, the latent representation of the kth feature map is given by [103]

$$\mathbf{h}^{(k)} = \sigma\left(\mathbf{x} * \mathbf{W}^{(k)} + \mathbf{b}^{(k)}\right), \tag{7.10.52}$$

where the bias is broadcasted to the whole map, $\sigma(\cdot)$ is an activation function, and $*$ denotes the 2D convolution. A single bias per latent map is used so that each filter specializes on features of the whole input (one bias per pixel would introduce too many degrees of freedom).

The reconstruction of the input $\mathbf{x}$ is given by

$$\mathbf{y}^{(k)} = \sigma\left(\sum_{k \in H} \mathbf{h}^{(k)} * \bar{\mathbf{W}}^{(k)} + \mathbf{c}\right), \tag{7.10.53}$$

where $\mathbf{c}$ is one bias per input channel; H identifies the group of latent feature maps; $\bar{\mathbf{W}}$ identifies the flip operation over both dimensions of the weights.

7.10 Autoencoders

The cost function of CAE is defined as the mean squared error (MSE):

$$E(\boldsymbol{\theta}) = \frac{1}{2n} \sum_{i=1}^{n} (x_i - y_i)^2. \quad (7.10.54)$$

It is easy to know by using a convolution operation that

$$\frac{\partial E(\boldsymbol{\theta})}{\partial \mathbf{W}^{(k)}} = \mathbf{x} * \delta \mathbf{h}^{(k)} + \bar{\mathbf{h}}^{(k)} \delta \mathbf{y}, \quad (7.10.55)$$

where $\delta \mathbf{h}$ and $\delta \mathbf{y}$ are the deltas of the hidden states and the reconstruction, respectively; and $\bar{\mathbf{h}}$ is the flip operation of $\mathbf{h}$. Hence, the weight matrix $\mathbf{W}$ can be updated as

$$\begin{aligned} \mathbf{W}^{(k+1)} &= \mathbf{W}^{(k)} - \eta \frac{\partial E(\boldsymbol{\theta})}{\partial \mathbf{W}^{(k)}} \\ &= \mathbf{W}^{(k)} - \eta \left(\mathbf{x} * \delta \mathbf{h}^{(k)} + \bar{\mathbf{h}}^{(k)} \delta \mathbf{y} \right). \end{aligned} \quad (7.10.56)$$

The network architecture of CAE consists of three basic building blocks: the convolutional layer, the max-pooling layer, and the classification layer.

7.10.5 Stacked Convolutional Denoising Autoencoder

The *stacked convolutional denoising autoencoder* (SCDAE), proposed by Du et al. [25], is an unsupervised deep network that stacks well-designed DAEs in a convolutional way to generate high-level feature representations. The overall architecture is optimized by layer-wise training.

The structure of the SCDAE with dropout is as follows [25]:

1. *Latent vector mapping:* A DAE corrupts the input vector $\mathbf{x}$ into vector $\tilde{\mathbf{x}}$ with a certain probability λ by means of a stochastic mapping

$$\tilde{\mathbf{x}} \sim D(\tilde{\mathbf{x}}|\mathbf{x}, \lambda), \quad (7.10.57)$$

where D is a type of distribution determined by the original distribution of $\mathbf{x}$ and the type of random noise added to $\mathbf{x}$. Then, $\tilde{\mathbf{x}}$ is mapped to a latent vector representation $\mathbf{h}$ using a deterministic function f:

$$\mathbf{h} = f_e(\mathbf{W}\tilde{\mathbf{x}} + \mathbf{b}). \quad (7.10.58)$$

2. *Dropout:* By using the dropout technique, the network optimizes the training process; neural units in the hidden layers are randomly omitted with a probability

q. The representation **h** is then transformed into a dropped representation $\tilde{\mathbf{h}}$ by a scalar product with a masking vector $\mathbf{m} \sim \text{Bernoulli}(1-q)$:

$$\tilde{\mathbf{h}} = \mathbf{m} \odot \mathbf{h}, \tag{7.10.59}$$

where $(\mathbf{m} \odot \mathbf{h})_i = m_i \cdot h_i$. Since a large neural network is updated iteratively, a unique network is trained in each iteration as a result of randomly dropping the neurons in the hidden layer. When the training process converges, the network gets an average representation of $2^{|\mathbf{m}|}$ networks.

3. *Decoding:* The dropped hidden feature vector $\tilde{\mathbf{h}}$ is then reversely mapped to a final feature **z** to reconstruct the original input **x** by another mapping function:

$$\mathbf{z} = f_d(\mathbf{W}'\tilde{\mathbf{h}} + \mathbf{b}'). \tag{7.10.60}$$

4. *Optimization:* The optimization function is given by

$$(\mathbf{W}, \mathbf{W}', \mathbf{b}, \mathbf{b}') = \underset{\mathbf{W},\mathbf{W}',\mathbf{b},\mathbf{b}'}{\arg\min} \|\mathbf{z} - \mathbf{x}\|_2^2 + \text{sparse}(\tilde{\mathbf{h}}), \tag{7.10.61}$$

where $\text{sparse}(\tilde{\mathbf{h}})$ denotes a type of sparse constraint, such as the KL distance.

5. *Convolutional layer:* The input of each convolutional layer is a 3D feature map, a mid-level representation of the input image. In each convolutional layer, the convolutional DAE transforms the input features into more robust and abstract feature maps through the learned denoising filters.

7.10.6 Nonnegative Sparse Autoencoder

Nonnegative sparse autoencoders are inspired by the idea of *nonnegative matrix factorization* (NMF) [91] and *sparse coding*.

1. *Nonnegative constraints:* The factorization $\mathbf{X} = \mathbf{AS}$ of a nonnegative matrix $\mathbf{X} \in \mathbb{R}_{+,0}^{M \times N}$ is called the nonnegative matrix factorization, if both the $M \times p$ factor matrix **A** and the $p \times N$ factor matrix **S** are nonnegative. Inspired by NMF, nonnegative constraints are added to weights of a neural network. There are the following key features of NMF [172]:

    - *Distributed nonnegative encoding:* NMF does not allow negative elements in the factor matrices **A** and **S**. Unlike the single constraint of the *vector quantization* (VQ), the nonnegative constraint allows the use of a combination of basis parts to represent a whole signal. Unlike the principal component analysis (PCA), the NMF allows only additive combinations; because the nonzero elements of **A** and **S** are all positive, the occurrence of any subtraction between basis images or signals can be avoided. In terms of optimization criteria, the VQ adopts the *"winner-take-all"* constraint and the PCA is based

7.10 Autoencoders

on the "*all-sharing*" constraint, but the NMF subjects to a "*group-sharing*" constraint together with a nonnegative constraint. From the encoding point of view, the NMF is a distributed nonnegative encoding, which often leads to sparse encoding.
- *Parts combination:* The NMF gives the intuitive impression that it is not a combination of all features; rather, just some of the features, (simply called the parts), are combined into a (target) whole. From the perspective of machine learning, the NMF is a kind of machine learning method based on the combination of parts which has the ability to extract the main features.
- *Multilinear data analysis capability:* The PCA uses a linear combination of all the basis vectors to represent the data and can extract only the linear structure of the data. In contrast, the NMF uses combinations of different numbers of and different labels of the basis vectors (parts) to represent the data and can extract its multilinear structure; thereby it has certain nonlinear data analysis capabilities.

2. *Sparse constraints:* Inspired by sparse coding, the sparse constraints imposed on the connecting weights can play the "dropout" role and optimize the whole autoencoder. Dropout is a technique applied to the fully connected layers to prevent overfitting.

Many signals in practice have the nonnegative values. The following are four important actual examples of nonnegative matrices [87]:

- Text documents are stored as vectors. Each element of a document vector is a count (possibly weighted) of the number of times a corresponding term appears in that document. Stacking document vectors one after the other creates a nonnegative term-by-document matrix that represents the entire document collection numerically.
- In image collections, each image is represented by a vector, and each element of the vector corresponds to a pixel. The intensity and color of a pixel is given by a nonnegative value, thereby creating a nonnegative pixel-by-image matrix.
- For itemsets or recommendation systems, the information for a purchase history of customers or ratings on a subset of items is stored in a nonnegative sparse matrix.
- In gene expression, gene-by-experiment matrices are formed from observations of the gene sequences produced under various experimental conditions.

To encourage nonnegativity in the connecting weights $\mathbf{W}$, the weight decay term $(w_{ij}^{(l)})^2$ in the sparse autoencoder (7.10.35) is replaced by a quadratic function $f(w_{ij}^{(l)})$, resulting in the cost function for *nonnegativity constrained autoencoder* (NCAE):

$$J_{\text{NCAE}}(\mathbf{W}, \mathbf{b}) = J_{\text{AE}}(\mathbf{W}, \mathbf{b}) + \beta J_{\text{KL}}(\rho \| \hat{\boldsymbol{\rho}}) + \frac{\alpha}{2} \sum_{l=1}^{2} \sum_{i=1}^{s_l} \sum_{j=1}^{s_{l+1}} f\left(w_{ij}^{(l)}\right), \quad (7.10.62)$$

where $\alpha \geq 0$ and

$$f(w_{ij}) = \begin{cases} w_{ij}^2, & w_{ij} < 0; \\ 0, & w_{ij} \geq 0. \end{cases} \tag{7.10.63}$$

Then, use the backpropagation algorithm to compute the gradients:

$$w_{ij}^{(l)} = w_{ij}^{(l)} - \eta \frac{\partial J_{\text{NCAE}}(\mathbf{W}, \mathbf{b})}{\partial w_{ij}^{(l)}}, \tag{7.10.64}$$

$$b_i^{(l)} = b_i^{(l)} - \eta \frac{\partial J_{\text{NCAE}}(\mathbf{W}, \mathbf{b})}{\partial b_i^{(l)}}, \tag{7.10.65}$$

where $\eta > 0$ is the learning rate, and

$$\frac{\partial J_{\text{NCAE}}(\mathbf{W}, \mathbf{b})}{\partial w_{ij}^{(l)}} = \frac{\partial J_{\text{AE}}(\mathbf{W}, \mathbf{b})}{\partial w_{ij}^{(l)}} + \beta \frac{\partial J_{\text{KL}}(\rho \| \hat{\boldsymbol{\rho}})}{\partial w_{ij}^{(l)}} + \alpha g\left(w_{ij}^{(l)}\right), \tag{7.10.66}$$

$$\frac{\partial J_{\text{NCAE}}(\mathbf{W}, \mathbf{b})}{\partial b_i^{(l)}} = \frac{\partial J_{\text{AE}}(\mathbf{W}, \mathbf{b})}{\partial b_i^{(l)}} + \beta \frac{\partial J_{\text{KL}}(\rho \| \hat{\boldsymbol{\rho}})}{\partial b_i^{(l)}}, \tag{7.10.67}$$

where

$$g(w_{ij}) = \begin{cases} w_{ij}, & w_{ij} < 0; \\ 0, & w_{ij} \geq 0. \end{cases} \tag{7.10.68}$$

Autoencoders with regularization constraints are known as *regularized autoencoders* [10]. The stacked sparse autoencoder and the nonnegative sparse autoencoder are two typical regularized autoencoders due to their sparsity regularization [115]. Importantly, regularized autoencoders can capture local structure of the density of signals [10].

7.11 Extreme Learning Machine

Extreme learning machine (ELM) is a machine learning algorithm for *single-hidden layer feedforward neural networks* (SLFNs) which randomly chooses hidden nodes and analytically determines the output weights of SLFNs. In theory, this algorithm tends to provide good generalization performance while retaining an extremely fast learning speed [70].

7.11 Extreme Learning Machine

For function approximation in a finite training set, SLFNs with at most N hidden nodes and almost any nonlinear activation function can exactly learn N distinct observations [69]. ELM is an effective solution for SLFNs.

Because ELM has a fast speed for classification, it is widely applied in data stream classification tasks.

7.11.1 Single-Hidden Layer Feedforward Networks with Random Hidden Nodes

First, we introduce the concept of random hidden nodes.

Definition 7.7 (Piecewise Continuous [150, p.334]) A function is said to be *piecewise continuous* if it has only a finite number of discontinuities in any interval and its left and right limits are defined (not necessarily equal) at each discontinuity.

Definition 7.8 (Randomly Generated) The function sequence $\{g_n = g(\langle \mathbf{w}_n, \mathbf{x} \rangle + b_n)\}$ or $\{g_n = g(\|\mathbf{w}_n - \mathbf{x}\|/b_n)\}$ is said to be randomly generated if the corresponding parameters are randomly generated from or based on a continuous sampling distribution probability.

Definition 7.9 (Random Node) A node is called a random node if its parameters $(\mathbf{w}, b)$ are randomly generated based on a continuous sampling distribution probability.

SLFNs have two main network architectures:

- the SLFNs with additive hidden nodes and
- the SLFNs with radial basis function (RBF) networks which apply RBF nodes in the hidden layer.

The network function of a SLFN with d hidden nodes can be represented by

$$f_d(\mathbf{x}) = \sum_{i=1}^{d} \beta_i g_i(\mathbf{x}), \quad \mathbf{x} \in \mathbb{R}^n, \ \beta_i \in \mathbb{R}, \tag{7.11.1}$$

where g_i denotes the ith hidden node output function.

Two commonly used ith hidden node output functions g_i are defined as

$$g_i(\mathbf{x}) = g\left(\langle \mathbf{w}_i, \mathbf{x} \rangle + b_i\right) = g(\mathbf{w}_i^T \mathbf{x} + b_i), \quad \mathbf{w}_i \in \mathbb{R}^n, \ b_i \in \mathbb{R} \tag{7.11.2}$$

for additive nodes, and

$$g_i(\mathbf{x}) = g\left(\frac{\|\mathbf{x} - \mathbf{a}_i\|}{b_i}\right), \quad \mathbf{a}_i \in \mathbb{R}^n, \ b_i \in \mathbb{R}_+ \tag{7.11.3}$$

for RBF nodes, where $\mathbf{a}_i$ and b_i are the center and impact factor of the ith RBF node and are the weights connecting the ith RBF hidden node to the output node, while $\mathbb{R}_+$ indicates the set of all positive real value.

In other words, the output function of an SLFN with d additive nodes and d RBF nodes can be, respectively, given by

$$f_d(\mathbf{x}) = \sum_{i=1}^{d} \beta_i g\left(\mathbf{w}_i^T \mathbf{x} + b_i\right) \in \mathbb{R} \quad (7.11.4)$$

and

$$f_d(\mathbf{x}) = \sum_{i=1}^{d} \beta_i g\left(\frac{\|\mathbf{x} - \mathbf{a}_i\|}{b_i}\right), \quad \mathbf{a}_i \in \mathbb{R}^n \in \mathbb{R}. \quad (7.11.5)$$

It was proved by Hornik [66] that if the activation function is continuous, bounded, and nonconstant, then continuous mappings can be approximated by SLFNs with additive hidden nodes over compact input sets. It is the more important [93] that SLFNs with additive hidden nodes and with a nonpolynomial activation function can approximate any continuous target functions.

Consider the SLFN $\mathbf{w}$ with the weights $\boldsymbol{\beta}_1, \ldots, \boldsymbol{\beta}_d$. If the SLFN produces N distinct samples $(\mathbf{x}_i, \mathbf{y}_i)$, $i = 1, \ldots, N$, where $\mathbf{x}_i = [x_{i1}, \ldots, x_{in}]^T \in \mathbb{R}^n$ and $\mathbf{y}_i = [y_{i1}, \ldots, y_{im}]^T \in \mathbb{R}^m$, then standard SLFNs with d hidden nodes and activation function $g(x)$ are mathematically modeled as

$$\begin{cases} \sum_{i=1}^{d} \boldsymbol{\beta}_i g\left(\mathbf{w}_i^T \mathbf{x}_1 + b_i\right) = \mathbf{y}_1, \\ \quad\quad\quad\quad \vdots \\ \sum_{i=1}^{d} \boldsymbol{\beta}_i g\left(\mathbf{w}_i^T \mathbf{x}_N + b_i\right) = \mathbf{y}_N, \end{cases} \quad (7.11.6)$$

where $\mathbf{w}_i = [w_{i1}, \ldots, w_{in}]^T \in \mathbb{R}^n$ is the weight vector connecting the ith hidden node and the input nodes, $\boldsymbol{\beta}_i = [\beta_{i1}, \ldots, \beta_{iN}]^T \in \mathbb{R}^N$ is the weight vector connecting the ith hidden node and the output nodes, and b_i is the threshold of the ith hidden node.

The N equations in (7.11.6) can be expressed in the following matrix form [70]:

$$\mathbf{HB} = \mathbf{Y}, \quad (7.11.7)$$

7.11 Extreme Learning Machine

where

$$\mathbf{H} = \begin{bmatrix} g\left(\mathbf{w}_1^T \mathbf{x}_1 + b_1\right) & \cdots & g\left(\mathbf{w}_d^T \mathbf{x}_1 + b_d\right) \\ \vdots & \ddots & \vdots \\ g(\mathbf{w}_1^T \mathbf{x}_N + b_1) & \cdots & g\left(\mathbf{w}_d^T \mathbf{x}_N + b_d\right) \end{bmatrix} \in \mathbb{R}^{N \times d}, \quad (7.11.8)$$

$$\mathbf{B} = \begin{bmatrix} \boldsymbol{\beta}_1^T \\ \vdots \\ \boldsymbol{\beta}_d^T \end{bmatrix} = \begin{bmatrix} \beta_{11} & \cdots & \beta_{1m} \\ \vdots & \ddots & \vdots \\ \beta_{d1} & \cdots & \beta_{dm} \end{bmatrix} \in \mathbb{R}^{d \times m}, \quad (7.11.9)$$

$$\mathbf{Y} = \begin{bmatrix} \mathbf{y}_1^T \\ \vdots \\ \mathbf{y}_N^T \end{bmatrix} = \begin{bmatrix} y_{11} & \cdots & y_{1m} \\ \vdots & \ddots & \vdots \\ y_{N1} & \cdots & y_{Nm} \end{bmatrix} \in \mathbb{R}^{N \times m}. \quad (7.11.10)$$

$\mathbf{H}$ is known as the hidden layer output matrix of the neural network [69], and the ith column of $\mathbf{H}$ is the ith hidden node output when inputting $\mathbf{x}_1, \ldots, \mathbf{x}_N$.

Theorem 7.2 ([70]) *Suppose we are given a standard SLFN with N hidden nodes and activation function $g : \mathbb{R} \to \mathbb{R}$ which is infinitely differentiable in any interval, for N arbitrary distinct samples $(\mathbf{x}_i, \mathbf{y}_i)$, $i = 1, \ldots, N$, where $\mathbf{x}_i \in \mathbb{R}^n$ and $\mathbf{y}_i \in \mathbb{R}^m$. Then, for any $\mathbf{w}_i$ and b_i randomly chosen, respectively, from any intervals of $\mathbb{R}^n$ and $\mathbb{R}$ according to any continuous probability distribution, the hidden layer output matrix $\mathbf{H}$ of the SLFN, with probability one, is invertible and $\|\mathbf{HB} - \mathbf{Y}\| = 0$.*

Theorem 7.3 ([70]) *If we are given any small positive value $\epsilon > 0$ and activation function $g : \mathbb{R} \to \mathbb{R}$ which is infinitely differentiable in any interval, and there exists $d \le N$ such that for N arbitrary distinct samples $\mathbf{x}_1, \ldots, \mathbf{x}_N$ with $\mathbf{x}_i \in \mathbb{R}^n$ and $\mathbf{y}_i \in \mathbb{R}^m$, for any $\mathbf{w}_i$ and b_i randomly chosen, respectively, from any intervals of $\mathbb{R}^n$ and $\mathbb{R}$ according to any continuous probability distribution, then with probability one $\|\mathbf{H}_{N \times d} \mathbf{B}_{d \times m} - \mathbf{Y}_{N \times m}\| < \epsilon$.*

Infinitely differentiable activation functions include the sigmoidal functions as well as the radial basis, sine, cosine, exponential, and many other nonregular functions, as shown by Huang and Babri [69].

7.11.2 Extreme Learning Machine Algorithm for Regression and Binary Classification

The offline solution of (7.11.7) is given by

$$\mathbf{B} = \mathbf{H}^\dagger \mathbf{Y}, \quad (7.11.11)$$

where $\mathbf{H}^\dagger$ is the Moore–Penrose generalized inverse of the hidden layer output matrix $\mathbf{H}$.

Based on the minimum norm LS solution $\mathbf{B} = \mathbf{H}^\dagger \mathbf{Y}$, Huang *et. al.* [70] proposed an extreme learning machine (ELM) as a simple learning algorithm for SLFNs, as shown in Algorithm 7.5.

Algorithm 7.5 Extreme learning machine (ELM) algorithm

1. **input:** A training set $\{(\mathbf{x}_i, \mathbf{y}_i) | \mathbf{x}_i \in \mathbb{R}^n, \mathbf{y}_i \in \mathbb{R}^m, i = 1, \ldots, N\}$, activation function $g(x)$ and hidden node number d.
2. **initialization:** Randomly assign input weight $\mathbf{w}_i$ and bias $b_i, i = 1, \ldots, d$.
3. **learning step:**
   3.1. Use (7.11.8) to calculate the hidden layer output matrix $\mathbf{H}$.

   3.2. Use (7.11.10) to construct the training output matrix $\mathbf{Y}$.

   3.3. Calculate the minimum norm least squares weight matrix $\mathbf{B} = \mathbf{H}^\dagger \mathbf{Y}$, and get $\boldsymbol{\beta}_i$ from $\mathbf{B}^T = [\boldsymbol{\beta}_1, \ldots, \boldsymbol{\beta}_d]$.
4. **testing step:** for given testing sample $\mathbf{x} \in \mathbb{R}^n$, the output of SLFNs is given by
$\mathbf{y} = \sum_{i=1}^d \boldsymbol{\beta}_i g\left(\mathbf{w}_i^T \mathbf{x} + b_i\right)$.

It is shown [71] that the ELM can use a wide type of feature mappings (hidden layer output functions), including random hidden nodes and kernels. With this extension, the unified ELM solution can be obtained for feedforward neural networks, RBF network, LS-SVM, and PSVM.

In ELM, the hidden layer need not be tuned. For one output node case, the output function of ELM for generalized SLFNs is given by

$$f_d(\mathbf{x}) = \sum_{j=1}^d \beta_j h_j(\mathbf{x}) = \mathbf{h}^T(\mathbf{x})\boldsymbol{\beta}, \qquad (7.11.12)$$

where $\mathbf{h}(\mathbf{x}) = [h_1(\mathbf{x}), \ldots, h_d(\mathbf{x})]^T$ is the output vector of the hidden layer with respect to the input $\mathbf{x}$, and $\boldsymbol{\beta} = [\beta_1, \ldots, \beta_d]^T$ is the vector of the output weights between the hidden layer of d nodes.

The output vector $\mathbf{h}(\mathbf{x})$ is a feature mapping: it actually maps the data from the n-dimensional input space to the d-dimensional hidden layer feature space (ELM feature space) $\mathcal{H}$. For the binary classification problem, the decision function of ELM is

$$f_d(\mathbf{x}) = \text{sign}\left(\mathbf{h}^T(\mathbf{x})\boldsymbol{\beta}\right). \qquad (7.11.13)$$

An ELM with a single-output node is used to generate N output samples:

$$\sum_{j=1}^d \beta_j h_j(\mathbf{x}_i) = y_i, \quad i = 1, \ldots, N \qquad (7.11.14)$$

7.11 Extreme Learning Machine

which can be rewritten as the form

$$\mathbf{h}^T(\mathbf{x}_i)\boldsymbol{\beta} = y_i, \quad i = 1, \ldots, N \quad \text{or} \quad \mathbf{H}\boldsymbol{\beta} = \mathbf{y}, \tag{7.11.15}$$

where

$$\mathbf{H} = \begin{bmatrix} h_1(\mathbf{x}_1) & \cdots & h_d(\mathbf{x}_1) \\ \vdots & \ddots & \vdots \\ h_1(\mathbf{x}_N) & \cdots & h_d(\mathbf{x}_N) \end{bmatrix}, \tag{7.11.16}$$

$$\boldsymbol{\beta} = \begin{bmatrix} \beta_1 \\ \vdots \\ \beta_d \end{bmatrix}, \tag{7.11.17}$$

$$\mathbf{y} = \begin{bmatrix} y_1 \\ \vdots \\ y_N \end{bmatrix} = \begin{bmatrix} \mathbf{h}^T(\mathbf{x}_1) \\ \vdots \\ \mathbf{h}^T(\mathbf{x}_N) \end{bmatrix}. \tag{7.11.18}$$

The constrained optimization problem for ELM regression and binary classification with a single-output node can be formulated as [71]:

$$\min_{\boldsymbol{\beta},\xi_i} \left\{ \mathcal{L}_{\text{PELM}} = \frac{1}{2}\|\boldsymbol{\beta}\|_2^2 + \frac{C}{2}\sum_{i=1}^N \xi_i^2 \right\}, \tag{7.11.19}$$

subject to $\quad \mathbf{h}^T(\mathbf{x}_i)\boldsymbol{\beta} = y_i - \xi_i, \quad i = 1, \ldots, N.$ (7.11.20)

The dual unconstrained optimization problem is

$$\min_{\boldsymbol{\beta},\xi_i,\alpha_i} \left\{ \mathcal{L}_{\text{DELM}}(\boldsymbol{\beta}, \xi_i, \alpha_i) = \frac{1}{2}\|\boldsymbol{\beta}\|_2^2 + \frac{C}{2}\sum_{i=1}^N \xi_i^2 \right.$$

$$\left. - \sum_{i=1}^N \alpha_i \left(\mathbf{h}^T(\mathbf{x}_i)\boldsymbol{\beta} - y_i + \xi_i \right) \right\} \tag{7.11.21}$$

with the Lagrange multipliers $\alpha_i \geq 0$, $i = 1, \ldots, N$.

From the optimality conditions one has

$$\frac{\partial \mathcal{L}_{\text{DELM}}}{\partial \boldsymbol{\beta}} = 0 \quad \Rightarrow \quad \boldsymbol{\beta} = \sum_{i=1}^N \alpha_i \mathbf{h}(\mathbf{x}_i) = \mathbf{H}^T \boldsymbol{\alpha} = \mathbf{H}^T \boldsymbol{\alpha}, \tag{7.11.22}$$

where $\boldsymbol{\alpha} = [\alpha_1, \ldots, \alpha_N]^T$.

Substituting (7.11.22) into (7.11.15) and using (7.11.18), we get the equation

$$\mathbf{H}^T \mathbf{H} \boldsymbol{\alpha} = \mathbf{y}, \qquad (7.11.23)$$

where

$$\mathbf{H}^T \mathbf{H} = \begin{bmatrix} \mathbf{h}^T(\mathbf{x}_1) \\ \vdots \\ \mathbf{h}^T(\mathbf{x}_N) \end{bmatrix} [\mathbf{h}(\mathbf{x}_1), \ldots, \mathbf{h}(\mathbf{x}_N)]$$

$$= \begin{bmatrix} \mathbf{h}^T(\mathbf{x}_1)\mathbf{h}(\mathbf{x}_1) & \cdots & \mathbf{h}^T(\mathbf{x}_1)\mathbf{h}(\mathbf{x}_N) \\ \vdots & \ddots & \vdots \\ \mathbf{h}^T(\mathbf{x}_N)\mathbf{h}(\mathbf{x}_1) & \cdots & \mathbf{h}^T(\mathbf{x}_N)\mathbf{h}(\mathbf{x}_N) \end{bmatrix}. \qquad (7.11.24)$$

The LS solution of Eq. (7.11.23) is given by

$$\boldsymbol{\alpha} = (\mathbf{H}^T \mathbf{H})^\dagger \mathbf{y}. \qquad (7.11.25)$$

If a feature mapping $\mathbf{h}(\mathbf{x})$ is unknown to users, one can apply Mercer's conditions on ELM, and hence a kernel matrix for ELM can be defined as follows [71]:

$$K(\mathbf{x}, \mathbf{x}_i) = \mathbf{h}^T(\mathbf{x})\mathbf{h}(\mathbf{x}_i). \qquad (7.11.26)$$

If using (7.11.26), then (7.11.24) can be rewritten as in the following kernel form:

$$\mathbf{H}^T \mathbf{H} = \begin{bmatrix} K(\mathbf{x}_1, \mathbf{x}_1) & \cdots & K(\mathbf{x}_1, \mathbf{x}_N) \\ \vdots & \ddots & \vdots \\ K(\mathbf{x}_N, \mathbf{x}_1) & \cdots & K(\mathbf{x}_N, \mathbf{x}_N) \end{bmatrix} \qquad (7.11.27)$$

whose (i,j)th entry is $\left[\mathbf{H}^T \mathbf{H}\right]_{ij} = K(\mathbf{x}_i, \mathbf{x}_j)$ for $i = 1, \ldots, N;\ j = 1, \ldots, N$.

This shows that the feature mapping $\mathbf{h}(\mathbf{x})$ need not be used; instead, it is enough to use the corresponding kernel $K(\mathbf{x}, \mathbf{x}_i)$, e.g., $K(\mathbf{x}, \mathbf{x}_i) = \exp(-\gamma \|\mathbf{x} - \mathbf{x}_i\|)$.

Finally, from (7.11.22) it follows that the EML regression function is

$$\hat{y} = \mathbf{h}^T(\mathbf{x})\boldsymbol{\beta} = \mathbf{h}^T(\mathbf{x}) \sum_{i=1}^{N} \alpha_i \mathbf{h}(\mathbf{x}_i) = \sum_{i=1}^{N} \alpha_i K(\mathbf{x}, \mathbf{x}_i), \qquad (7.11.28)$$

while the decision function for EML binary classification is

$$\text{class of } \mathbf{x} = \text{sign}\left(\sum_{i=1}^{N} \alpha_i K(\mathbf{x}, \mathbf{x}_i)\right). \qquad (7.11.29)$$

7.11 Extreme Learning Machine

Algorithm 7.6 shows the ELM algorithm for regression and binary classification.

Algorithm 7.6 ELM algorithm for regression and binary classification [71]

input: A training set $\{(\mathbf{x}_i, y_i) | \mathbf{x}_i \in \mathbb{R}^n, y_i \in \mathbb{R}, i = 1, \ldots, N\}$, hidden node number d and the kernel function $K(\mathbf{u}, \mathbf{v}) = \exp(-\gamma \|\mathbf{u} - \mathbf{v}\|)$.
initialization: $\mathbf{y} = [y_1, \ldots, y_N]^T$.
learning step:
  1. Use the kernel function to construct the matrix $[\mathbf{H}^T \mathbf{H}]_{ij} = K(\mathbf{x}_i, \mathbf{x}_j), i, j = 1, \ldots, N$.
  2. Calculate the minimum norm least squares solution $\boldsymbol{\alpha} = (\mathbf{H}^T \mathbf{H})^\dagger \mathbf{y}$.
testing step: for given testing sample $\mathbf{x} \in \mathbb{R}^n$, the ELM regression is $\sum_{i=1}^{N} \alpha_i K(\mathbf{x}, \mathbf{x}_i)$, while the ELM binary classification is class of $\mathbf{x} = \text{sign}\left(\sum_{i=1}^{N} \alpha_i K(\mathbf{x}, \mathbf{x}_i)\right)$.

7.11.3 Extreme Learning Machine Algorithm for Multiclass Classification

For multiclass applications, let ELM have multi-output nodes instead of a single-output node. For a k-class of classifier, we are given N training set $\{\mathbf{x}_i, \mathbf{y}_i | \mathbf{x}_i \in \mathbb{R}^n, \mathbf{y}_i \in \mathbb{R}^d\}, i = 1, \ldots, N$. If the original class label is p, then the jth entry of expected output vector of the d output nodes, $\mathbf{y}_i = [y_{i1}, \ldots, y_{id}]^T \in \mathbb{R}^d$, is denoted as

$$y_{ij} = \begin{cases} +1, & j = p, \\ -1, & j \neq p, \end{cases} \qquad (7.11.30)$$

where $p = \{1, \ldots, k\}$. That is, only the pth entry of $\mathbf{y}_i$ is $+1$, while the rest of the entries are set to -1.

The primal classification problem of the mth-class for ELM with multi-output nodes can be formulated as [71]:

$$\min \quad \mathcal{L}_{\text{PELM}}^{(m)} = \frac{1}{2}\left(\|\boldsymbol{\beta}_m\|_2^2 + b_m^2\right) + \frac{C}{2}\sum_{i=1}^{N} \xi_{m,i}^2 \qquad (7.11.31)$$

$$\text{subject to} \quad \begin{cases} \mathbf{h}_m^T(\mathbf{x}_1)\boldsymbol{\beta}_m + b_m = 1 - \xi_{m,1}, \\ \quad \vdots \\ \mathbf{h}_m^T(\mathbf{x}_N)\boldsymbol{\beta}_m + b_m = 1 - \xi_{m,N}. \end{cases} \qquad (7.11.32)$$

The corresponding dual unconstrained optimization problem is given by

$$\min\left\{\mathcal{L}_{\text{DELM}}^{(m)} = \frac{1}{2}\left(\|\boldsymbol{\beta}_m\|_2^2 + b_m^2\right) + \frac{C}{2}\sum_{i=1}^{N}\xi_{m,i}^2 \right.$$
$$\left. - \sum_{i=1}^{N}\alpha_{m,i}\left(y_i^{(m)}\left(\mathbf{h}_m^T(\mathbf{x}_i)\boldsymbol{\beta}_m + b_m\right) - 1 + \xi_{m,i}\right)\right\}. \quad (7.11.33)$$

From the optimality conditions one has

$$\frac{\partial \mathcal{L}_{\text{DELM}}^{(m)}}{\partial \boldsymbol{\beta}_m} = \mathbf{0} \Rightarrow \boldsymbol{\beta}_m = \sum_{i=1}^{N}\alpha_{m,i}y_i^{(m)}\mathbf{h}_m(\mathbf{x}_i), \quad (7.11.34)$$

$$\frac{\partial \mathcal{L}_{\text{DELM}}^{(m)}}{\partial b_m} = 0 \Rightarrow b_m = \sum_{i=1}^{N}\alpha_{m,i}y_i^{(m)}, \quad (7.11.35)$$

$$\frac{\partial \mathcal{L}_{\text{DELM}}^{(m)}}{\partial \xi_{m,i}} = 0 \Rightarrow \xi_{m,i} = C^{-1}\alpha_{m,i}, \quad (7.11.36)$$

$$\frac{\partial \mathcal{L}_{\text{DELM}}^{(m)}}{\partial \alpha_{m,i}} = 0 \Rightarrow y_i^{(m)}\left(\mathbf{h}_m^T(\mathbf{x}_i)\boldsymbol{\beta}_m + b_m\right) - 1 + \xi_{m,i} = 0, \quad (7.11.37)$$

for $i = 1, \ldots, N$.

Eliminating $\boldsymbol{\beta}_m$ and $\xi_{m,i}$ in the above equations yields the KKT equation

$$b_m = \sum_{i=1}^{N}\alpha_{m,i}y_i^{(m)} = \boldsymbol{\alpha}_m^T\mathbf{y}_m \quad (7.11.38)$$

and

$$\left(C^{-1}\mathbf{I} + \mathbf{H}_m^T\mathbf{H}_m + \mathbf{y}_m\mathbf{y}_m^T\right)\boldsymbol{\alpha}_m = \mathbf{1}, \quad (7.11.39)$$

where $\mathbf{I}$ is an $N \times N$ identity matrix, $\mathbf{1}$ is an $N \times 1$ summing vector with all entries equal to 1, and

$$\mathbf{H}_m = \left[y_1^{(m)}\mathbf{h}_m(\mathbf{x}_1), \ldots, y_N^{(m)}\mathbf{h}_m(\mathbf{x}_N)\right], \quad (7.11.40)$$

$$\mathbf{y}_m = \left[y_1^{(m)}, \ldots, y_N^{(m)}\right]^T, \quad (7.11.41)$$

$$\boldsymbol{\alpha}_m = [\alpha_{m,1}, \ldots, \alpha_{m,N}]^T. \quad (7.11.42)$$

It easily follows that the (i, j)th entry of the matrix $\mathbf{H}_m \mathbf{H}_m^T$ can be represented as

$$\left[\mathbf{H}_m^T \mathbf{H}_m\right]_{ij} = y_i y_j \mathbf{h}_m^T(\mathbf{x}_i) \mathbf{h}_m(\mathbf{x}_j) = y_i y_j K_m(\mathbf{x}_i, \mathbf{x}_j), \qquad (7.11.43)$$

where $K_m(\mathbf{x}_i, \mathbf{x}_j) = \mathbf{h}_m^T(\mathbf{x}_i) \mathbf{h}_m(\mathbf{x}_j)$ is the kernel function for the mth ELM classifier.

Algorithm 7.7 shows the ELM multiclass classification algorithm.

Algorithm 7.7 ELM multiclass classification algorithm [71]

1. **input:** A training set $\{(\mathbf{x}_i, \mathbf{y}_i) | \mathbf{x}_i \in \mathbb{R}^n, \mathbf{y}_i \in \mathbb{R}^d, i = 1, \ldots, N\}$, hidden node number d, the number k of classes and the kernel function $K_m(\mathbf{u}, \mathbf{v})$ for the mth classifier $m = 1, \ldots, k$, such as $K_m(\mathbf{u}, \mathbf{v}) = \exp(-\gamma \|\mathbf{u} - \mathbf{v}\|_2^2)$.
2. **initialization:** Reconstruct the mth class's output vector $\mathbf{y}^{(m)} = \left[y_1^{(m)}, \ldots, y_N^{(m)}\right]^T$ with
$$y_i^{(m)} = \begin{cases} +1, & y_i = m; \\ -1, & y_i \neq m; \end{cases} \text{ for } m = 1, \ldots, k; i = 1, \ldots, N.$$
3. **learning step:**
4. **while** $m = 1, \ldots, k$
5.   Use the kernel function $K_m(\mathbf{x}_i, \mathbf{x}_j)$ and the mth class's output vector $\mathbf{y}^{(m)}$ to construct the matrix $\left[\mathbf{H}_m^T \mathbf{H}_m\right]_{ij} = y_i^{(m)} y_j^{(m)} K_m(\mathbf{x}_i, \mathbf{x}_j), i, j = 1, \ldots, N$.
6.   Calculate the minimum norm least square solution $\boldsymbol{\alpha}_m = \left(\mathbf{H}_m^T \mathbf{H}_m + \mathbf{y}_m \mathbf{y}_m^T + C^{-1}\mathbf{I}\right)^\dagger \mathbf{1}$.
7.   Calculate $b_m = \boldsymbol{\alpha}_m^T \mathbf{y}_m$.
8. **endwhile**
9. **testing step:** For given testing sample $\mathbf{x} \in \mathbb{R}^n$, the decision function of the ELM multiclass classifier is given by
$$\text{class of } \mathbf{x} = \arg\max_{m=1,\ldots,k} \left(\sum_{i=1}^N \alpha_{m,i} y_i^{(m)} K_m(\mathbf{x}, \mathbf{x}_i) + b_m\right).$$

7.12 Graph Embedding

Graphs, such as social graph/diffusion graph in social media networks, word co-occurrence networks, user interest graph in electronic commerce area, knowledge graph, and communication networks etc., exist naturally in a wide diversity of real-world scenarios.

Analyzing or learning these graphs yields insight into the structure of society, language, and different patterns of communication. For example, in social network research, classification of users into meaningful social groups based on social graphs can lead to many useful practical applications such as user search, targeted advertising and recommendations.

Embedding a graph or an information network into a low-dimensional space is useful in a variety of applications. To conduct the embedding, the graph structures must be preserved. The first intuition is that the local graph structure, i.e., the local pairwise proximity between the vertices, must be preserved. Therefore, the two main tasks of graph embedding are dimension reduction and local graph structure preservation. The problem of dimensionality reduction arises in many fields of information processing: machine learning, data compression, neural computation, pattern recognition, scientific visualization, and so on.

7.12.1 Proximity Measures and Graph Embedding

We first introduce the definition of the basic concepts in graph embedding.

Suppose we are given a graph $G = (V, E)$, where $v \in V$ is a vertex or node and $e \in E$ is an edge. G is associated with a node type mapping function $f_v : V \to \mathcal{T}^v$ and an edge type mapping function $f_e : E \to \mathcal{T}^e$, where $\mathcal{T}^v$ and $\mathcal{T}^e$ denote the set of node types and edge types, respectively. Each node $v_i \in V$ belongs to one particular type, i.e., $f_v(v_i) \in \mathcal{T}^v$. Similarly, for $e_{ij} \in E$, $f_e(e_{ij}) \in \mathcal{T}^e$.

Graph learning is closely related to graph proximities and the graph embedding. Graph learning tasks can be broadly abstracted into the following four categories [41]:

- *Node classification* aims at determining the label of nodes (a.k.a. vertices) based on other labeled nodes and the topology of the network.
- *Link prediction* refers to the task of predicting missing links or links that are likely to occur in the future.
- *Clustering* is used to find subsets of similar nodes and group them together.
- *Visualization* helps in providing insights into the structure of the network.

The most basic measure for both dimension reduction and structure preservation of a graph is the graph proximity. Proximity measures are usually adopted to quantify the graph property to be preserved in the embedded space.

The microscopic structures of a graph can be described by its first-order proximity and second-order proximity. The first-order proximity between the vertices is their local pairwise similarity between only the nodes connected by edges.

Definition 7.10 (First-Order Proximity [144]) The *first-order proximity* is the observed pairwise proximity between two nodes v_i and v_j, denoted as $S_{ij}^{(1)} = s_{ij}$, where s_{ij} is the edge weight between the two nodes. If no edge is observed between nodes i and j, then their first-order proximity $S_{ij}^{(1)} = 0$.

The first-order proximity is the first and foremost measure of similarity between two nodes. The first-order proximity implies that two nodes in real-world networks are always similar if they are linked by an observed edge. For example, if a paper cites another paper, they should contain some common topic or keywords. However,

7.12 Graph Embedding

it is not sufficient to only capture the first-order proximity, and it is also necessary to introduce the second-order proximity to capture the global network structure.

Definition 7.11 (Second-Order Proximity [144]) Let $\mathbf{s}_i^{(1)} = \mathbf{s}_i = [S_{i,1}^{(1)}, \ldots, S_{i,n}^{(1)}]^T$ and $\mathbf{s}_i^{(2)} = [S_{i,1}^{(2)}, \ldots, S_{i,n}^{(2)}]^T$ be the first-order and the second-order proximity vectors between node i and other nodes, respectively. Then the *second-order proximity* $S_{ij}^{(2)}$ is determined by the similarity of $\mathbf{s}_i$ and $\mathbf{s}_j$. If no vertex is linked from/to both i and j, then the second-order proximity between v_i and v_j is zero, i.e., $S_{ij}^{(2)} = 0$.

As a matter of fact, the second-order proximity $S_{ij}^{(2)}$ between a pair of vertices (i, j) in a graph is the similarity between v_i's neighborhood $s_i^{(1)}$ and v_j's neighborhood $s_j^{(1)}$.

- The first-order proximity compares the similarity between the nodes i and j. The more similar two nodes are, the larger the first-order proximity value between them.
- The second-order proximity compares the similarity between the nodes' neighborhood structures. The more similar two nodes' neighborhoods are, the larger the second-order proximity value between them.

Similarly, we can define the higher-order proximity $S_{ij}^{(k)}$ (where $k \geq 3$) between a pair of vertices (i, j) in a graph.

Definition 7.12 (k-Order Proximity) Let $\mathbf{s}_i^{(k)} = [S_{i,1}^{(k)}, \ldots, S_{i,n}^{(k)}]^T$ be the kth-order proximity vector between node i and other nodes. Then the kth-order proximity $S_{ij}^{(k)}$ is determined by the similarity of $\mathbf{s}_i^{(k-1)}$ and $\mathbf{s}_j^{(k-1)}$.

In particular, when $k \geq 3$, the kth-order proximity is generally referred to as the *higher-order proximity*. The matrix $\mathbf{S}^{(k)} = [S_{ij}^{(k)}]$ is known as the k-order *proximity matrix*. When $k \geq 3$, $\mathbf{S}^{(k)}$ is called the *higher-order proximity matrix*. The higher-order proximity matrices are also defined using some other metrics, e.g., Katz Index, Rooted PageRank, Adamic-Adar, etc. that will be discussed in Sect. 7.13.3.

By Definitions 7.10, 7.11, and 7.12, the first-order, second-order, and third-order proximity matrices $\mathbf{S}^{(k)} = [S_{ij}^{(k)}] \in \mathbb{R}^{n \times n}$ (where $k = 1, 2, 3$) are nonnegative matrices, respectively.

If considering the cosine similarity as the k-order proximity, then for nodes v_i and v_j, we have the following results:

1. The first-order proximity $S_{ij}^{(1)} = s_{ij}$, where s_{ij} is the edge weight between the two nodes.

2. The second-order proximity

$$S_{ij}^{(2)} = \frac{\langle \mathbf{s}_i, \mathbf{s}_j \rangle}{\|\mathbf{s}_i\| \cdot \|\mathbf{s}_j\|}$$

$$= \frac{\sum_{l=1}^{n} S_{il}^{(1)} S_{jl}^{(1)}}{\sqrt{\sum_{l=1}^{n} |S_{il}^{(1)}|^2} \sqrt{\sum_{l=1}^{n} |S_{jl}^{(1)}|^2}}$$

$$= \frac{\sum_{l=1}^{n} s_{il} s_{jl}}{\sqrt{\sum_{l=1}^{n} |s_{il}|^2} \sqrt{\sum_{l=1}^{n} |s_{jl}|^2}}. \qquad (7.12.1)$$

In this way, the second-order proximity is between [0, 1].

3. The third-order proximity

$$S_{ij}^{(3)} = \frac{\sum_{l=1}^{n} S_{il}^{(2)} S_{jl}^{(2)}}{\sqrt{\sum_{l=1}^{n} |S_{il}^{(2)}|^2} \sqrt{\sum_{l=1}^{n} |S_{jl}^{(2)}|^2}}. \qquad (7.12.2)$$

Definition 7.13 (Graph Embedding [14, 41]) Given the inputs of a graph $G = (V, E)$, and a predefined dimensionality of the embedding d ($d \ll |V|$), the graph embedding is to convert G into a d-dimensional space $\mathbb{R}^d$. In this space, the graph properties (such as the first-order, second-order, and higher-order proximities) are preserved as much as possible. The graph is represented as either a d-dimensional vector (for a whole graph) or a set of d-dimensional vectors with each vector representing the embedding of part of the graph (e.g., node, edge, substructure).

Therefore, a graph embedding maps each node of graph $G(E, V)$ to a low-dimensional feature vector $\mathbf{y}_i$ and tries to preserve the connection strengths between vertices. For example, a graph embedding preserving first-order proximity might be obtained by minimizing $\sum_{i,j} s_{ij} \|\mathbf{y}_i - \mathbf{y}_j\|_2^2$.

Graph embedding is an important method for learning low-dimensional representations of vertices in networks, aiming to capture and preserve the network structure. Learning network representations faces the following great challenges [154]:

1. *High nonlinearity:* The underlying structure of a graph or network is highly nonlinear. Therefore, designing a model to capture the highly nonlinear structure is rather difficult.
2. *Topology structure-preserving:* To support applications in analyzing networks, network embedding is required to preserve the network structure. However, the underlying structure of the network is very complex. The similarity of vertices is dependent on both the local and global network structures. Therefore, how to simultaneously preserve the local and global structure is a tough problem.
3. *Sparsity:* Many real-world networks are often so sparse that only utilizing the very limited observed links is not enough to reach a satisfactory performance.

7.12 Graph Embedding

Definition 7.14 (Local Topology Preserving [101]) Suppose we are given a symmetric undirected graph G with edge weights $s_{ij} = w_{ij}$, and a corresponding embedding $(\mathbf{y}_1, \ldots, \mathbf{y}_n)$ for the n nodes of the graph. The embedding is said to be *local topology preserving* if the following condition holds:

$$\text{if } w_{ij} \geq w_{pq} \text{ then } \|\mathbf{y}_i - \mathbf{y}_j\|_2^2 \leq \|\mathbf{y}_p - \mathbf{y}_q\|_2^2, \ \forall i, j, p, q. \tag{7.12.3}$$

Roughly speaking, this definition says that if two node pairs (v_i, v_j) and (v_p, v_q) are associated with connections strengths such that $s_{ij} \geq s_{pq}$, then v_i and v_j will be mapped to points in the embedding space that will be closer to each other than the mapping of v_p and v_q. That is, for any pair of nodes (v_i, v_j), the more similar they are (the bigger the edge weight w_{ij} or the first-order proximity $S_{ij}^{(1)}$ is), the closer they should be embedded together (the smaller $\|\mathbf{y}_i - \mathbf{y}_j\|$ should be).

To describe topology structure of the graph or network, it is necessary to introduce the metric distance between two nodes. If n points $\mathbf{x}_i \in \mathbb{R}^d$, $i = 1, \ldots, n$ have (orthogonal) coordinates $(x_{i1}, \ldots, x_{id})$, then the *Euclidean* (or Pythagorean) *metric distance* from x_i to x_j is given by

$$d_{ij} = \left(\sum_{l=1}^{d} (x_{il} - x_{jl})^2 \right)^{1/2}. \tag{7.12.4}$$

Another commonly used metric distance is the *Minkowski p-metric distance* (or called ℓ_p-*metric distance*) given by

$$d_{ij} = \left(\sum_{l=1}^{d} (x_{il} - x_{jl})^p \right)^{1/p}, \quad p \geq 1. \tag{7.12.5}$$

Clearly, the Euclidean metric distance is a special example of the Minkowski p-metric distance when $p = 2$. For $p = 1$, the Minkowski metric becomes the so-called *city-block distance* or *Manhattan metric*:

$$d_{ij} = \sum_{l=1}^{d} |x_{il} - x_{jl}|. \tag{7.12.6}$$

For $p = \infty$, the Minkowski metric becomes a familiar metric

$$d_{ij} = \max_{l} |x_{il} - x_{jl}|. \tag{7.12.7}$$

It is well known that a local subspace of a nonlinear space can be linear. Similarly, a local structured subspace of a non-Euclidean space can be a locally Euclidean structured space as well. It should be noted that the Euclidean distance is available only for Euclidean structured space or locally Euclidean structured space in a graph or network, but the Minkowski p-metric distance ($p \neq 2$) can be used to described the distance between two nodes in non-Euclidean structured spaces.

The generic problem of linear dimensionality reduction is: given a set of points $\mathbf{x}_1, \ldots, \mathbf{x}_n$ in $\mathbb{R}^D$, find an $n \times d$ transformation matrix $\mathbf{W}$ that maps these n points to a set of points $\mathbf{y}_1, \ldots, \mathbf{y}_n$ in $\mathbb{R}^d (d \ll D)$, such that $\mathbf{y}_i = \mathbf{W}^T \mathbf{x}_i$ "represents" $\mathbf{x}_i$.

The two popular linear dimensionality reduction techniques are the principal component analysis (PCA) and *linear discriminant analysis* (LDA).

Let $\mathbf{C}_x = \frac{1}{n} \sum_{i=1}^{n} \mathbf{x}_i \mathbf{x}_i^T$ be the covariance matrix of the n data vectors $\mathbf{x}_1, \ldots, \mathbf{x}_n$ and $\mathbf{u}_1, \ldots, \mathbf{u}_p$ be p eigenvectors associated with principal eigenvalues of $\mathbf{C}$, where $p \ll D$. Then, the transformation matrix is given by $\mathbf{W} = [\mathbf{u}_1, \ldots, \mathbf{u}_p] \in \mathbb{R}^{n \times p}$ and the low-dimensional vectors $\mathbf{y}_i = \mathbf{W}^T \mathbf{x}_i \in \mathbb{R}^p$ are the good representation of the high-dimensional data vectors $\mathbf{x}_i$.

While PCA seeks directions that are efficient for data representation, LDA seeks directions that are efficient for data discrimination. Suppose the data points belong to c classes. To ensure the between-class separability and within-class compactness, the objective function of LDA is given by

$$\mathbf{w}_{\text{opt}} = \arg\max_{\mathbf{w}} \frac{\mathbf{w}^T \mathbf{S}_b \mathbf{w}}{\mathbf{w}^T \mathbf{S}_w \mathbf{w}}, \quad (7.12.8)$$

where the between-class scatter matrix $\mathbf{S}_b$ and the within-class scatter matrix $\mathbf{S}_w$ are computed as

$$\mathbf{S}_b = \sum_{i=1}^{c} n_i \left(\mathbf{m}^{(i)} - \mathbf{m}\right)\left(\mathbf{m}^{(i)} - \mathbf{m}\right)^T, \quad (7.12.9)$$

$$\mathbf{S}_w = \sum_{i=1}^{c} \left[\sum_{j=1}^{n_i} \left(\mathbf{x}_j^{(i)} - \mathbf{m}^{(i)}\right)\left(\mathbf{x}_j^{(i)} - \mathbf{m}^{(i)}\right)^T\right], \quad (7.12.10)$$

where $\mathbf{m}$ is the total sample mean vector, n_i is the number of samples in the ith class, $\mathbf{m}^{(i)}$ is the mean vector of the ith class, and $\mathbf{x}_j^{(i)}$ is the jth sample vector in the ith class.

The LDA solution $\mathbf{w}$ is the generalized eigenvector associated with the maximum generalized eigenvalue of the matrix pencil $(\mathbf{S}_b, \mathbf{S}_w)$. The transformation matrix $\mathbf{W}$ consists of p principal generalized eigenvectors associated with p principal generalized eigenvalues of $(\mathbf{S}_b, \mathbf{S}_w)$.

In the following, we focus on four graph embedding techniques: classical multidimensional scaling and three manifold learning techniques: isometric map, locally linear embedding, and Laplacian eigenmap.

7.12.2 Multidimensional Scaling

We are given n points $\mathbf{x}_1, \ldots, \mathbf{x}_n$ with $\mathbf{x}_i \in \mathbb{R}^D$. While a configuration is considered as n points in the model space, one may with equal validity consider it as a single point

$$\left(x_{11}, \ldots, x_{1D}, \ldots, x_{n1}, \ldots, x_{nD}\right) \tag{7.12.11}$$

in the configuration space.

Let d_{ij} denote the distance between nodes i and j, and s_{ij} be a measure of the similarity between them. For example, a measure of the psychological similarity of two stimuli in a paired-associate experiment is given by [133]

$$s_{ij} = \sqrt{\frac{p_{ij} p_{ji}}{p_{ii} p_{jj}}}, \tag{7.12.12}$$

where p_{ij} is the empirical estimate of the conditional probability of the response assigned to stimulus j when stimulus i is presented.

Another example of the similarity data s_{ij} is [134]

$$s_{ij} = \sum_{k=1}^{K} w_k p_{ik} p_{jk}, \tag{7.12.13}$$

where w_k are the positive psychological weights of the discrete properties that both stimuli have in common, and

$$p_{ik} = \begin{cases} 1, & \text{if object } i \text{ has property } k; \\ 0, & \text{otherwise.} \end{cases} \tag{7.12.14}$$

Define the *normalized stress* [85] by

$$S^* = \sum_{i<j} (d_{ij} - \hat{d}_{ij})^2, \tag{7.12.15}$$

$$T^* = \sum_{i<j} d_{ij}^2, \tag{7.12.16}$$

$$\text{Stress} = S = \sqrt{\frac{S^*}{T^*}} = \sqrt{\frac{\sum_{i<j} \left(d_{ij} - \hat{d}_{ij}\right)^2}{\sum_{i<j} d_{ij}^2}}. \tag{7.12.17}$$

Here, $\hat{d}_{ij}$ are numbers that are monotonic with the similarity data s_{ij} and minimize stress relative to the spatial distance d_{ij} at each iteration.

Suppose now that the values of s_{ij} are given. Then for any point in configuration space, that is, for any configuration, there is a definite stress value S. In other words, S is a function

$$S = S\left(x_{11}, \ldots, x_{1D}, \ldots, x_{n1}, \ldots, x_{nD}\right) \qquad (7.12.18)$$

defined on the points of configuration space. The *multidimensional scaling* (MDS) problem is to find that point which minimizes S. This is a standard problem of numerical analysis: to minimize a function of several variables.

The following is the verbal evaluation suggested in [85, 86].

Stress	Goodness of fit
20%	poor
10%	fair
5%	good
2.5%	excellent
0%	"perfect."

By "perfect" here, it means only that there is a perfect monotone relationship between dissimilarities and the distances.

In the method of steepest descent, the (negative) gradient is given by

$$\left(-\frac{\partial S}{\partial x_{11}}, \ldots, -\frac{\partial S}{\partial x_{1d}}, \ldots, -\frac{\partial S}{\partial x_{n1}}, \ldots, -\frac{\partial S}{\partial x_{nd}}\right). \qquad (7.12.19)$$

For Minkowski p-metric, the (negative) gradient terms $g_{kl} = -\frac{\partial S}{\partial x_{kl}}$ are given by [86]

$$g_{kl} = \sum_{i,j}(\delta_{ki} - \delta_{jk})\left(\frac{d_{ij} - \hat{d}_{ij}}{S^*} - \frac{d_{ij}}{T^*}\right)\frac{|x_{il} - x_{jl}|^{p-1}}{d_{ij}^{p-1}}\text{sign}(x_{il} - x_{jl}),$$
$$(7.12.20)$$

where

$$\text{sign}(x) = \begin{cases} +1, & x > 0; \\ -1, & x < 0; \\ 0, & x = 0. \end{cases} \qquad (7.12.21)$$

7.12.3 Manifold Learning: Isometric Map

Given n points $\mathbf{x}_1, \ldots, \mathbf{x}_n$ in a high-dimensional vector space $\mathbb{R}^D$, our aim is to find a set of points $\mathbf{y}_1, \ldots, \mathbf{y}_n$ in low-dimensional space $\mathbb{R}^d$ (where $d \ll D$) such that each low-dimensional point $\mathbf{y}_i$ can "represent" the corresponding high-dimensional data point $\mathbf{x}_i$. Here, we consider the special case where $\mathbf{x}_1, \ldots, \mathbf{x}_n \in \mathcal{M}$ and $\mathcal{M}$ is a manifold embedded in the high-dimensional vector space $\mathbb{R}^D$. To construct low-dimensional points $\mathbf{y}_1, \ldots, \mathbf{y}_n$, the structure of the manifold $\mathcal{M}$ must be learned.

Manifold learning is used to extract low-dimensional features $\mathbf{y}_i$ in the graph (dimensionality reduction) in order to represent the original high-dimensional data points $\mathbf{x}_i$. In this subsection, we focus on an *isometric map* (Isomap) based manifold learning algorithm, and other manifold learning algorithms will be discussed in the next subsections.

The Isomap, presented by Tenenbaum et al. [145], is an algorithm for computing a quasi-isometric, low-dimensional embedding of a set of high-dimensional data points. The Isomap algorithm defines the connectivity of each data point via its nearest Euclidean neighbors in the high-dimensional space.

The Isomap algorithm finds a low-dimensional representation that most faithfully preserves the pairwise geodesic distances between feature vectors in all scales as measured Isomap along the submanifold from which they were sampled. There are three steps for Isomap [145]:

1. *Construct neighborhood graph* of $G(V, E)$ of the observed data $\{\mathbf{x}\}$ in a suitable way.

    - *k-nearest neighborhood:* $G(V, E)$ might contain the edge $e = (i, j)$ if and only if $\mathbf{x}_j$ is one of the k-nearest neighbors of $\mathbf{x}_i$ (and vice versa). This will result in a directed graph and is called an *k-Isomap*.
    - *ϵ-neighborhood:* $G(V, E)$ might contain the edge $e = (i, j)$ if and only if $\|\mathbf{x}_i - \mathbf{x}_j\|^2 < \epsilon$ for some ϵ. This will result in an undirected graph and is known as *ϵ-Isomap*.

2. *Compute shortest paths* in the graph for all pairs of data points. Initialize

$$d_G(i, j) = \begin{cases} d_x(i, j), & \text{if } i, j \text{ are linked by an edge;} \\ \infty, & \text{otherwise.} \end{cases} \quad (7.12.22)$$

Then for each value of $k = 1, 2, \ldots, n$ in turn, update all entries

$$d_G(i, j) = \min\{d_G(i, j), d_G(i, k) + d_G(k, j)\}. \quad (7.12.23)$$

The matrix of final values $\mathbf{D}_G = [d_G(i, j)]$ will contain the shortest path distances between all pairs of points in $G(V, E)$.

3. *Construct d-dimensional embedding:* Let λ_p be the pth eigenvalue (in decreasing order) of the matrix $\mathbf{D}_G$, and $v_p(i)$ be the ith component of the pth eigenvector $\mathbf{v}_p$. Then set the pth component of the d-dimensional coordinate vector $\mathbf{y}_i$ equal to $\sqrt{\lambda_p} v_p(i)$, i.e., $y_i(p) = \sqrt{\lambda_p} v_p(i)$.

The final step applies classical multidimensional scaling (MDS) to the matrix of graph distances $\mathbf{D}_G = [d_G(i, j)]$ for constructing an embedding of the data in a d-dimensional Euclidean space Y that best preserves the manifold's estimated intrinsic geometry. The coordinate vectors $\mathbf{y}_i$ for points in Y are chosen to minimize the cost function

$$E = \|\mathbf{D}_G - \mathbf{D}_Y\|_F^2, \qquad (7.12.24)$$

where $\mathbf{D}_Y$ denotes the matrix of Euclidean distances with entries $d_Y(i, j) = \|\mathbf{y}_i - \mathbf{y}_j\|_2$. The global minimum of E is achieved by setting the coordinates $\mathbf{y}_i$ to the top d eigenvectors of the matrix $\mathbf{D}_G$.

Conformal Isomap (C-Isomap) [137] is an extension of Isomap which is capable of learning the structure of certain curved manifolds. In C-Isomap, k-nearest neighborhood is constructed in Step 1 and Step 2 is replaced by the following: Compute shortest paths in the graph for all pairs of data points. Each edge (i, j) in the graph is weighted by $\|\mathbf{x}_i - \mathbf{x}_j\|/\sqrt{M(i)M(j)}$, where $M(i)$ is the mean distance of $\mathbf{x}_i$ to its k-nearest neighbors and the rescaling factor $\sqrt{M(i)M(j)}$ is an asymptotically accurate approximation to the conformal scaling factor in the neighborhood of $\mathbf{x}_i$ and $\mathbf{x}_j$.

7.12.4 Manifold Learning: Locally Linear Embedding

By eliminating the estimation of pairwise distances between widely separated data points, the *locally linear embedding* (LLE) [122] recovers global nonlinear structure from locally linear fits. The LLE algorithm is based on simple geometric intuitions.

Unlike other methods, the LLE recovers global nonlinear structure from locally linear fits.

Let the data consist of N real-valued vectors $\mathbf{x}_i \in \mathbb{R}^D$, $i = 1, \ldots, N$, sampled from some underlying manifold. Suppose there is sufficient data (such that the manifold is well-sampled); it is expected that each data point and its neighbors lie on or close to a locally linear patch of the manifold. The local geometry of these patches is characterized by linear coefficients that reconstruct each data point from its neighbors, where reconstruction errors are measured by the cost function

$$E(\mathbf{W}) = \sum_{i=1}^{N} \left\| \mathbf{x}_i - \sum_{j=1}^{N} w_{ij} \mathbf{x}_j \right\|_2^2 \qquad (7.12.25)$$

7.12 Graph Embedding

which adds up the squared distances between all the data points $\mathbf{x}_1, \ldots, \mathbf{x}_N$ and their reconstructions $\sum_j w_{ij} \mathbf{x}_j$. The weights w_{ij} summarize the contribution of the jth data point $\mathbf{x}_j$ to the reconstruction of the ith data point $\mathbf{x}_i$.

The cost function is minimized subject to the following two constraints:

- Each data point $\mathbf{x}_i$ is reconstructed only from its neighbors to enforce $w_{ij} = 0$ if $\mathbf{x}_j$ does not belong to the set of neighbors of $\mathbf{x}_i$. The set of neighbors for each data point can be assigned in a variety of ways: by choosing the k nearest neighbors in Euclidean distance, by considering all data points within a ball of fixed radius, or by using prior knowledge. For fixed number of neighbors, the maximum number of embedding dimensions that LLE can be expected to recover is strictly less than the number of neighbors.
- The rows of the weight matrix sum to one: $\sum_j w_{ij} = 1$.

The embedding vectors $\mathbf{y}_i$ are found by minimizing the cost function

$$\phi(\mathbf{y}) = \sum_i \|\mathbf{y}_i - w_{ij}\mathbf{y}_j\|^2 \qquad (7.12.26)$$

over $\mathbf{y}_i$ with fixed weights w_{ij}.

The LLE algorithm can be described as follows [122, 128]:

1. Assign neighbors: For each data point $\mathbf{x}_i, i = 1, \ldots, N$, identify the indices corresponding to its k-nearest neighbors in Euclidean distance. Let $\mathcal{N}_i = \{\boldsymbol{\eta}_1, \ldots, \boldsymbol{\eta}_k\}$ (where $k = |\mathcal{N}_i|$) denote the collection of those neighbors.
2. Compute the weights w_{ij} that best reconstruct each data point $\mathbf{x}_i$ from its neighbors, minimizing the cost in Eq. (7.12.25) by constrained linear fits. This fits consist of the following three steps:

   - Evaluate inner products between neighbors to compute the neighborhood correlation matrix, $C_{jk} = \langle \boldsymbol{\eta}_j, \boldsymbol{\eta}_k \rangle = \boldsymbol{\eta}_j^T \boldsymbol{\eta}_k$ and its matrix inverse, $\mathbf{C}^{-1} = (\mathbf{C} + \sigma \mathbf{I})^{-1}$, where $\sigma > 0$ is a small value.
   - Compute the Lagrange multiplier, $\lambda = \alpha/\beta$, that enforces the sum-to-one constraint, where $\alpha = 1 - \sum_{j,k} C_{jk}^{-1} \mathbf{x}_i^T \boldsymbol{\eta}_k$ and $\beta = C_{jk}^{-1}$.
   - Compute the reconstruction weights associated with $\mathbf{x}_i$:

$$w_{ij} = w_j^{(i)} = \sum_k C_{jk}^{-1} \left(\mathbf{x}_i^T \boldsymbol{\eta}_k + \lambda \right). \qquad (7.12.27)$$

3. Compute the low-dimensional embedding vector $\mathbf{y}_i$ best reconstructed by w_{ij} by solving the constrained minimization

$$\mathbf{y} = \arg\min_{\mathbf{y}^T \mathbf{y} = 1} \phi(\mathbf{y}) = \sum_i \|\mathbf{y}_i - w_{ij} \mathbf{y}_j\|^2, \qquad (7.12.28)$$

or

$$\mathbf{y} = \arg\min_{\mathbf{y}} \mathcal{L}(\mathbf{y}) = \sum_i \|\mathbf{y}_i - w_{ij}\mathbf{y}_j\|^2 + \lambda\left(1 - \mathbf{y}_j^T\mathbf{y}_j\right), \quad (7.12.29)$$

where the constraint $\mathbf{y}^T\mathbf{y} = \sum_i y_i^2 = 1$ is used to avoid degenerate solutions $\mathbf{y} = \mathbf{0}$. From $\frac{\partial \mathcal{L}}{\partial \mathbf{y}_j} = \mathbf{0}$, we get

$$\frac{\partial \mathcal{L}}{\partial \mathbf{y}_j} = (\delta_{ij} - w_{ji})(\delta_{ij}y_j - w_{ij}y_j) - \lambda \mathbf{y}_j = \mathbf{0},$$

and thus we have the matrix equation

$$(\mathbf{I} - \mathbf{W})^T(\mathbf{I} - \mathbf{W})\mathbf{y}_j = \lambda \mathbf{y}_j. \quad (7.12.30)$$

That is, the embedding coordinates $\mathbf{y}_j$ can be found efficiently from the bottom $d+1$ eigenvectors (those corresponding to its smallest $d+1$ eigenvalues) of the matrix $(\mathbf{I} - \mathbf{W})^T(\mathbf{I} - \mathbf{W})$ without performing a full matrix diagonalization. Once $\mathbf{y}_j$ are obtained, we can make the coordinates be centered on the origin: $\mathbf{y}_j \leftarrow \mathbf{y}_j - \boldsymbol{\mu}$ ($\boldsymbol{\mu}$ is the mean vector of $\mathbf{y}_j$) in order to remove the degree of freedom of each embedding coordinate.

Neighborhood preserving embedding (NPE) in [54] is a linear approximation to the LLE algorithm.

Different from principal component analysis which preserves global structure, NPE preserves local manifold structure in which each data point can be represented as a linear combination of its neighbors.

Let $\mathbf{y} = \mathbf{X}^T\mathbf{a}$ be a linear transformation approximation to $\mathbf{y}$, and put

$$z_i = y_i - \sum_j w_{ij} y_j \quad \text{or} \quad \mathbf{z} = \mathbf{y} - \mathbf{W}\mathbf{y} = (\mathbf{I} - \mathbf{W})\mathbf{X}^T\mathbf{a}. \quad (7.12.31)$$

Then the cost function in (7.12.26) can be written as

$$\phi(\mathbf{a}) = \sum_i (z_i)^2 = \mathbf{z}^T\mathbf{z} = \mathbf{a}^T\mathbf{X}(\mathbf{I} - \mathbf{W})^T(\mathbf{I} - \mathbf{W})\mathbf{X}^T\mathbf{a}$$

$$= \mathbf{a}^T\mathbf{X}\mathbf{M}\mathbf{X}^T\mathbf{a}, \quad (7.12.32)$$

where $\mathbf{M} = (\mathbf{I} - \mathbf{W})^T(\mathbf{I} - \mathbf{W})$. Hence, the constrained LLE minimization (7.12.28) becomes the NPE minimization:

$$\mathbf{a} = \arg\min_{\mathbf{a}^T\mathbf{X}\mathbf{X}^T\mathbf{a}=1} \phi(\mathbf{a}) = \mathbf{a}^T\mathbf{X}\mathbf{M}\mathbf{X}^T\mathbf{a}, \quad (7.12.33)$$

7.12 Graph Embedding

which can be rewritten as the unconstrained minimization problem

$$\mathbf{a} = \arg\min_{\mathbf{a}} \left\{ \mathcal{L}(\mathbf{a}) = \mathbf{a}^T \mathbf{XMX}^T \mathbf{a} + \lambda \left(1 - \mathbf{a}^T \mathbf{XX}^T \mathbf{a}\right) \right\}. \tag{7.12.34}$$

The optimization condition $\frac{\partial \mathcal{L}}{\partial \mathbf{a}} = \mathbf{0}$ yields directly the optimal solution formula

$$\mathbf{XMX}^T \mathbf{a} = \lambda \mathbf{XX}^T \mathbf{a}. \tag{7.12.35}$$

That is, d linear transformation vectors $\mathbf{a}_i$ are given by the generalized eigenvectors of the matrix pencil ($\mathbf{XMX}^T, \mathbf{XX}^T$), ordered according to their generalized eigenvalues, $\lambda_0 < \ldots < \lambda_{d-1}$.

Donoho and Grimes [24] presented a Hessian-based locally linear embedding (HLLE) method, derived from a conceptual framework of local isometry in the manifold $\mathcal{M}$. Due to requiring estimation of second derivatives, the Hessian approach is known to be numerically noisy or difficult in very high-dimensional data samples.

The LLE algorithm requires the solution of N separate $k \times k$ eigenproblems and a single $N \times N$ sparse eigenproblem. The sparsity of this eigenproblem can confer a substantial advantage over the general nonsparse eigenproblem. This is an important factor distinguishing LLE techniques from the Isomap technique, which poses a completely dense $N \times N$ matrix for eigenanalysis [24].

7.12.5 Manifold Learning: Laplacian Eigenmap

Lapalacian eigenmap [6] finds the low-dimensional representation that most faithfully preserves local similarity structure in the feature space.

Given a data set $\{\mathbf{x}_1, \ldots, \mathbf{x}_n\}$, it is expected to construct a connected weighted graph $G = (V, E)$ with edges connecting nearby points to each other. Consider the problem of mapping the weighted graph $G(V, E)$ to a line so that connected points stay as close together as possible. Let $\mathbf{y} = [y_1, \ldots, y_n]^T$ be such a map.

In Laplacian embedding, the input data is a matrix $\mathbf{W}$ of pairwise similarities among n data objects. $\mathbf{W}$ is viewed as the edge weights on a graph with n nodes. The task is to embed the nodes of the graph into 1-D space with coordinates $(y_1, \ldots, y_n)$. The objective is that if i, j are similar (i.e., w_{ij} is large), they should be adjacent in embedded space, i.e., $(y_i - y_j)^2$ should be as small as possible. This can be achieved by minimizing the objective function [6, 7]

$$\min_{\mathbf{y}} \left\{ E(\mathbf{y}) = \frac{1}{2} \sum_{i,j} (y_i - y_j)^2 w_{ij} \right\}. \tag{7.12.36}$$

This unconstrained optimization can be rewritten as

$$\min_{\mathbf{y}} \{E(\mathbf{y}) = \mathbf{y}^T(\mathbf{D} - \mathbf{W})\mathbf{y}\}, \qquad (7.12.37)$$

because

$$\sum_{i,j}(y_i - y_j)^2 w_{ij} = \sum_i y_i^2 \sum_j w_{ij} - 2\sum_i \sum_j y_i y_j w_{ij} + \sum_j y_j^2 \sum_i w_{ij}$$

$$= \sum_i y_i^2 d_{ii} + \sum_j y_j^2 d_{jj} - 2\sum_{i,j} y_i y_j w_{ij}$$

$$= 2\mathbf{y}^T(\mathbf{D} - \mathbf{W})\mathbf{y},$$

where $\mathbf{D} = \mathbf{Diag}(d_{11}, \ldots, d_{nn})$ with $d_{ii} = \sum_{j=1}^n w_{ij}$, $i = 1, \ldots, n$.

However, the unconstrained optimization problem (7.12.37) has only the zero solution $\mathbf{y} = \mathbf{0}$ due to $\frac{\partial E(\mathbf{y})}{\partial \mathbf{y}} = 2(\mathbf{D} - \mathbf{W})\mathbf{y} = \mathbf{0}$.

To avoid the trivial solution $\mathbf{y} = \mathbf{0}$, the unconstrained minimization (7.12.37) should impose the constraint conditions: one is the normalization $\sum_i y_i^2 = 1$, another is the centralization $\sum_i y_i = 0$.

For removing an arbitrary scaling factor in the embedding, the normalization constraint $\sum_i y_i^2 = \mathbf{y}^T\mathbf{y} = 1$ can be relaxed to $\mathbf{y}^T\mathbf{D}\mathbf{y} = 1$. Then, the constrained form (7.12.37) becomes

$$\min_{\mathbf{y}} \{E(\mathbf{y}) = \mathbf{y}^T\mathbf{L}\mathbf{y}\}$$

subject to $\quad \mathbf{y}^T\mathbf{D}\mathbf{y} = 1, \qquad (7.12.38)$

where $\mathbf{L} = \mathbf{D} - \mathbf{W}$ is the graph Laplacian. The above constrained minimization can be rewritten as the Lagrange multiplier form:

$$\min_{\mathbf{y}} \ L(\mathbf{y}) = \mathbf{y}^T\mathbf{L}\mathbf{y} + \lambda(1 - \mathbf{y}^T\mathbf{D}\mathbf{y}). \qquad (7.12.39)$$

From $\frac{\partial L(\mathbf{y})}{\partial \mathbf{y}} = \mathbf{0}$, it follows that

$$\mathbf{L}\mathbf{y} = \lambda\mathbf{D}\mathbf{y}, \qquad (7.12.40)$$

or equivalent to the eigenvalue decomposition of the matrix $\mathbf{D}^{-1}\mathbf{L}$:

$$\mathbf{D}^{-1}\mathbf{L}\mathbf{y} = \lambda\mathbf{y}. \qquad (7.12.41)$$

Therefore, this Laplacian embedding method is called the *Laplacian eigenmap method* [7].

7.12 Graph Embedding

Moreover, another constraint condition $\sum_{i=1} y_i = \mathbf{y}^T \mathbf{1} = 0$ can be relaxed to $\mathbf{y}^T \mathbf{D} \mathbf{1} = 0$, which can be interpreted as removing a translation invariance in $\mathbf{y}$. But, in practical applications, we usually implement only the minimization problem with the constraint $\mathbf{y}^T \mathbf{D} \mathbf{y}$, because the constraint $\mathbf{y}^T \mathbf{1} = 0$ is easily implemented via $\mathbf{y} - \boldsymbol{\mu}$ to the found solution $\mathbf{y}$ (where $\boldsymbol{\mu}$ is the mean vector of $\mathbf{y}$). In other words, the Laplacian eigenmap becomes the following minimization problem:

$$\mathbf{y} = \underset{\mathbf{y}^T \mathbf{D} \mathbf{y} = 1}{\arg \min} \left\{ \mathbf{y}^T \mathbf{L} \mathbf{y} \right\}. \tag{7.12.42}$$

Laplacian Eigenmaps contain the following three steps [6]:

1. Construct neighborhood graph through ϵ-neighborhoods or n-nearest neighbors.

   - ϵ-neighborhoods: Nodes i and j are connected by an edge if $\|\mathbf{x}_i - \mathbf{x}_j\|_2^2 < \epsilon$.
   - n-nearest neighbors: Nodes i and j are connected by an edge if i is among n-nearest neighbors of j or j is among n-nearest neighbors of i.

2. Choose edge weights using the heat kernel or simply set edge weight to be 1 if connected and 0 otherwise.

   - *Heat kernel* is given by

$$w_{ij} = \begin{cases} \exp\left(-\frac{\|\mathbf{x}_i - \mathbf{x}_j\|_2^2}{t}\right), & \text{if nodes } i \text{ and } j \text{ are connected;} \\ 0, & \text{otherwise.} \end{cases} \tag{7.12.43}$$

   - Simple-minded: Put $w_{ij} = 1$ if and only if vertices i and j are connected by an edge. This simplification avoids the necessity of choosing t.

3. Make eigenmaps: For the graph $G(V, E)$, solve the generalized eigenvector problem $\mathbf{L}\mathbf{y} = \lambda \mathbf{D} \mathbf{y}$. Leave out the eigenvector $\mathbf{y}$ associated with zero eigenvalues and use the next d eigenvectors for embedding in d-dimensional Euclidean space: $\mathbf{x}_i \rightarrow (y_1(i), \ldots, y_d(i))$. Finally, make the centralization $\mathbf{y} \leftarrow \mathbf{y} - \boldsymbol{\mu}$.

Locality preserving projections (LPP) in [53] is a linear approximation to Laplacian eigenmaps in which the embedding coordinate vector $\mathbf{y}$ in Eq. (7.12.36) is approximated by the linear transformation $\mathbf{y} = \mathbf{X}^T \mathbf{a}$, where $\mathbf{X} = [\mathbf{x}_1, \ldots, \mathbf{x}_n]$ is the data matrix and $\mathbf{a}$ is a linear transformation vector. After using $\mathbf{X}^T \mathbf{a}$ instead of $\mathbf{y}$ in LLE optimization problem (7.12.42), the objective in LLP is given by

$$\mathbf{a} = \underset{\mathbf{a}^T \mathbf{X} \mathbf{D} \mathbf{X}^T \mathbf{a} = 1}{\arg \min} \ \mathbf{a}^T \mathbf{X} \mathbf{L} \mathbf{X}^T \mathbf{a}. \tag{7.12.44}$$

Put $\mathcal{L}(\mathbf{a}) = \mathbf{a}^T \mathbf{XLX}^T \mathbf{a} + \lambda(1 - \mathbf{a}^T \mathbf{XDX}^T \mathbf{a})$. Then, from $\frac{\partial \mathcal{L}}{\partial \mathbf{a}} = \mathbf{0}$ it is immediately known that the optimal linear transformation $\mathbf{a}$ is given by

$$\mathbf{XLX}^T \mathbf{a} = \lambda \mathbf{XDX}^T \mathbf{a}. \qquad (7.12.45)$$

Therefore the linear transformation vectors $\mathbf{a}_0, \ldots, \mathbf{a}_{d-1}$ are given by the generalized eigenvectors of the matrix pencil $(\mathbf{XLX}^T, \mathbf{XDX}^T)$, ordered according to their generalized eigenvalues, $\lambda_0 < \ldots < \lambda_{d-1}$.

The perceived notion of local topology preservation in Laplacian eigenmaps is in fact false in many cases due to the following aspects [101]:

- At large distances (small similarity), the quadratic function of the Laplacian embedding **emphasizes** the **large distance** pairs, which enforces node pair $(\mathbf{x}_i, \mathbf{x}_j)$ with small w_{ij} to be separated far away.
- At small distances (large similarity), the quadratic function of the Laplacian embedding **de-emphasizes** the **small distance** pairs, leading to many violations of local topology preservation at small distance pairs.

In other words, due to using a quadratic penalty function on the distance between embeddings, the objective function of Laplacian eigenmaps emphasizes preservation of dissimilarity between nodes more than their similarity. This may yield embeddings which do not preserve local topology, which can be defined as the equality between relative order of edge weights (w_{ij}) and inverse order of distances in the embedded space ($\|\mathbf{y}_i - \mathbf{y}_j\|$). In order to tackle this problem, Luo et al. [101] presented *Cauchy graph embedding* by replacing the quadratic function $\|\mathbf{y}_i - \mathbf{y}_j\|^2$ with $\|\mathbf{y}_i - \mathbf{y}_j\|^2 / (\|\mathbf{y}_i - \mathbf{y}_j\|^2 + \sigma^2)$. Upon rearrangement, Cauchy graph embedding becomes the constrained maximization problem:

$$\mathbf{y} = \arg\max_{\mathbf{y}^T \mathbf{y} = 1, \mathbf{y}^T \mathbf{1} = 0} \phi(\mathbf{y}) = \sum_{i,j} \frac{\|\mathbf{y}_i - \mathbf{y}_j\|^2}{\|\mathbf{y}_i - \mathbf{y}_j\|^2 + \sigma^2} \qquad (7.12.46)$$

for each i.

The above objective is an inverse function of distance and thus puts emphasis on similar nodes rather than dissimilar nodes, which enforces the local topology preserving in embedded space.

The following are global versus local approaches to graph embedding [137]:

- Local approaches (the LLE, the Laplacian eigenmaps, the Cauchy graph embedding) attempt to preserve the local geometry of the data; essentially, they seek to map nearby points on the manifold to nearby points in the low-dimensional representation.
- Global approaches (the Isomap) attempt to preserve geometry at all scales, mapping nearby points on the manifold to nearby points in low-dimensional space, and faraway points to faraway points.

Local approaches to low-dimensional embedding (LLE or Laplacian Eigenmaps, Cauchy graph embedding) have two principal advantages over a global approach (Isomap): they tolerate a certain amount of curvature and they lead naturally to a sparse eigenvalue problem. Local isometry assumption seems much more likely to hold in practice than the more restrictive global isometry assumption in Isomaps [24].

The advantage of a global approach over local approaches is: it will give a more faithful representation of the data's global structure.

7.13 Network Embedding

Many complex network systems, such as social networks, biological networks, and information networks, are usually represented as a graph $G(V, E)$, where V is a vertex set representing the nodes in a network, and E is an edge set representing the relationships among the nodes. For large networks with billions of nodes, the traditional network representation poses several challenges to network processing and analysis [19].

1. *High computational complexity:* The nodes in a network are related to each other to a certain degree, encoded by the edge set E in the traditional network representation. These relationships will result in high computational complexity: most of the network processing or analysis algorithms are either iterative or need combinatorial computation steps.
2. *Low parallelizability:* Network data represented in the traditional way causes servere difficulties in the design and implementation of parallel and distributed algorithms. The bottleneck is that nodes in a network are coupled to each other, which is explicitly reflected by the presence of E.
3. *Inapplicability of machine learning methods:* For network data represented in the traditional way, most of the off-the-shelf machine learning methods may not be applicable, because those methods usually assume that data samples can be represented by independent vectors in a vector space, while the samples in network data (i.e., the nodes) are dependent on each other to some degree determined by E.

Graph embedding cannot tackle the above challenges, because graph embedding mostly works on graphs constructed from feature represented data sets, where the proximity among nodes encoded by the edge weights are well defined in the original feature space. In contrast, in naturally formed networks, such as social networks, biology networks, and e-commerce networks, the proximities among nodes are not explicitly or directly defined.

7.13.1 Structure and Property Preserving Network Embedding

Rich structural information for network embedding from nodes and links is closely related to neighborhood structure [114], higher-order proximities of nodes [154], and community structures [155].

Definition 7.15 (Network Embedding [154]) Given a graph denoted as $G = (V, E)$, *network embedding* aims to learn a mapping function $f : \mathbf{v}_i \to \mathbf{y}_i \in \mathbb{R}^d$, where $d \ll |V|$. The objective of the function is to make the similarity between $\mathbf{y}_i$ and $\mathbf{y}_j$ explicitly preserve the first-order, second-order, and higher-order proximities of $\mathbf{v}_i$ and $\mathbf{v}_j$.

The microscopic structures of a network can be described by its first-order proximity and second-order proximity.

Network embedding usually has the following two goals [19]:

- *Network reconstruction:* To learn low-dimensional vector representations for network nodes, the relationships among the nodes, which were originally represented by edges or other higher-order topological measures in graphs, are captured by the distances between nodes in the vector space, and the topological and structural characteristics of a node are encoded into its embedding vector.
- *Network inference:* The learned embedding space can effectively support network inference, such as predicting unseen links, identifying important nodes, and inferring node labels.

The problem of large-scale information network embedding can be described by using first-order and second-order proximities.

Definition 7.16 (Information Network [144]) An *information network* is defined as $G = (V, E)$, where V is the set of vertices, each representing a data object and E is the set of edges between the vertices, each representing a relationship between two data objects. Each edge $e \in E$ is an ordered pair $e = (i, j)$ and is associated with a weight $w_{ij} > 0$, which indicates the strength of the relation. If $G(V, E)$ is undirected, then $(i, j) \equiv (j, i)$ and $w_{ij} \equiv w_{ji}$; and if $G(V, E)$ is directed, then $(i, j) \neq (j, i)$ and $w_{ij} \neq w_{ji}$.

Definition 7.17 (Large-Scale Information Network Embedding [144]) Given a large network $G = (V, E)$, the problem of *large-scale information network embedding* aims to represent each vertex $v \in V$ in a low-dimensional space $\mathbb{R}^d$, i.e., learning a function $f_G : V \to \mathbb{R}^r$, where $d \ll |V|$. In the space $\mathbb{R}^d$, both the first-order proximity and the second-order proximity between the vertices are preserved.

The first-order proximity can be measured by the joint probability distribution between two nodes v_i and v_j as

$$p_1(\mathbf{v}_i, \mathbf{v}_j) = \frac{1}{1 + \exp\left(-\mathbf{u}_i^T \mathbf{u}_j\right)}, \quad (7.13.1)$$

where $\mathbf{u}_i \in \mathbb{R}^d$ is the low-dimensional vector representation of vertex $\mathbf{v}_i$. A straightforward way to preserve the first-order proximity is to minimize the following objective function:

$$O_1 = -\sum_{(i,j)\in E} \log p_1(\mathbf{v}_i, \mathbf{v}_j). \quad (7.13.2)$$

The second-order proximity is modeled by the probability of the context node $\mathbf{x}_j$ being generated by node $\mathbf{x}_i$, that is,

$$p_2(\mathbf{v}_j|\mathbf{v}_i) = \frac{\exp\left(\bar{\mathbf{u}}_j^T \mathbf{u}_i\right)}{\sum_k^{|V|} \exp\left(\bar{\mathbf{u}}_k^T \mathbf{u}_i\right)}, \quad (7.13.3)$$

where $\mathbf{u}_i$ is the representation of $\mathbf{v}_i$ when it is treated as a vertex, while $\bar{\mathbf{u}}_i$ is the representation of $\mathbf{v}_i$ when it is treated as a specific "context." A straightforward way to preserve the second-order proximity is to minimize the following objective function:

$$O_2 = -\sum_{(i,j)\in E} \log p_2(\mathbf{v}_j|\mathbf{v}_i). \quad (7.13.4)$$

By learning $\{\mathbf{u}_i\}, i = 1, \ldots, |V|$ and $\{\bar{\mathbf{u}}_i\}, i = 1, \ldots, |V|$ that minimize this objective, we are able to represent every vertex $\mathbf{v}_i$ with a d-dimensional vector $\mathbf{u}_i$.

7.13.2 Community Preserving Network Embedding

Given a network $G(V, E)$ with $|V|$ nodes and $|E|$ edges, represented by an adjacency or similarity matrix $\mathbf{S} = [S_{ij}] \in \mathbb{R}^{|V|\times|V|}$. We want to determine whether there is any natural division of its vertices into nonoverlapping groups or communities, where these communities may be of any size. True community structure in a network corresponds to a statistically surprising arrangement of edges, can be quantified by using the measure known as modularity [108].

Suppose the given network contains n vertices. For a particular division of the network into two groups let the *community membership* $h_i = 1$ if vertex i belongs to group 1 and $h_i = -1$ if it belongs to group 2. And let the number of edges between vertices i and j be A_{ij}, which will normally be 0 or 1, although larger values are possible in networks where multiple edges are allowed. If edges are placed at random, then the expected number of edges between vertices i and j is $k_i k_j/2m$, where k_i and k_j are the degrees of the vertices and $m = \frac{1}{2}\sum_i k_i$ is the total number of edges in the network.

Definition 7.18 (Modularity [108]) The *modularity* is, up to a multiplicative constant, the number of edges falling within groups minus the expected number in an equivalent network with edges placed at random. Mathematically, the modularity Q can be expressed as

$$Q = \frac{1}{4m} \sum_{i,j} \left(A_{ij} - \frac{k_i k_j}{2m} \right) h_i h_j. \qquad (7.13.5)$$

Equation (7.13.5) can conveniently be written in matrix-vector form as

$$Q = \frac{1}{4m} \mathbf{h}^T \mathbf{B} \mathbf{h}, \qquad (7.13.6)$$

where **h** is known as the *community membership vector* (column vector) whose elements are the community membership h_i, and **B** is called the *modularity matrix* with elements

$$B_{ij} = A_{ij} - \frac{k_i k_j}{2m}. \qquad (7.13.7)$$

The modularity can be either positive or negative, where positive values indicate the possible presence of community structure.

Many networks, however, contain more than two communities. A straightforward approach is to divide repeatedly such a network into two: first to divide the network into two parts, then divide those parts, and so forth. A better approach is to write the additional contribution ΔQ to the modularity upon further dividing a group g of size n_g in two as [108]

$$\Delta Q = \frac{1}{2m} \left(\frac{1}{2} \sum_{i,j \in g} B_{ij}(h_i h_j + 1) - \sum_{i,j \in g} B_{ij} \right)$$

$$= \frac{1}{4m} \left(\sum_{i,j \in g} B_{ij} h_i h_j - \sum_{i,j \in g} B_{ij} \right)$$

$$= \frac{1}{4m} \left(\sum_{i,j \in g} B_{ij} - \delta_{ij} \sum_{k \in g} B_{ik} \right) h_i h_j$$

$$= \frac{1}{4m} \mathbf{h}^T \mathbf{B}^{(g)} \mathbf{h}, \qquad (7.13.8)$$

where $\delta_{ij} = 1$ if $i = j$ and 0 otherwise, and $\mathbf{B}^{(g)}$ is the $n_g \times n_g$ matrix with entries

$$B_{ij}^{(g)} = B_{ij} - \delta_{ij} \sum_{k \in g} B_{ik}. \qquad (7.13.9)$$

7.13 Network Embedding

To preserve both of the first-order and second-order proximities, define the final similarity matrix as $\mathbf{S} = \mathbf{S}^{(1)} + \eta \mathbf{S}^{(2)}$, where $\eta > 0$ is the weight of the second-order proximity. Because both $\mathbf{S}^{(1)}$ and $\mathbf{S}^{(2)}$ are nonnegative matrices, the similarity matrix $\mathbf{S}$ is nonnegative as well.

Let $\mathbf{M} \in \mathbb{R}^{n \times m}$ be a nonnegative basis matrix and $\mathbf{U} \in \mathbb{R}^{n \times m}$ be a nonnegative representation matrix, where m is the dimension of representation and the ith row $\mathbf{u}_i$ of $\mathbf{U}$ is the representation of node i.

Community preserving network embedding [156] finds the nonnegative matrices $\mathbf{M}$ and $\mathbf{U}$ via nonnegative matrix factorization (NMF)

$$(\mathbf{M}, \mathbf{U}) = \arg\min_{\mathbf{M}, \mathbf{U}} \ \|\mathbf{S} - \mathbf{M}\mathbf{U}\|_F^2 \tag{7.13.10}$$

$$\text{subject to } \mathbf{M} \geq 0, \ \mathbf{U} \geq 0. \tag{7.13.11}$$

Due to $\mathbf{S}$ containing the first-order and second-order proximity matrices, the above NMF preserves only the *microscopic structure* (pairwise node similarity).

To preserve the *mesoscopic structure* of a network, Wang et al. [156] proposed to introduce community structure of network with an auxiliary community representation matrix.

For a network with $k > 2$ communities, the community membership indicator is defined as $\mathbf{H} \in \mathbb{R}^{n \times k}$ with one column for each community. In each row of $\mathbf{H}$, only one element is 1 and all the others are 0, and thus the constraint over community is given by

$$\text{tr}(\mathbf{H}^T \mathbf{H}) = n. \tag{7.13.12}$$

After suppressing the constant which has no effect on the maximum of the modularity, we have

$$\mathbf{H} = \arg\min_{\mathbf{H}} \ \left\{ Q = \text{tr}(\mathbf{H}^T \mathbf{B} \mathbf{H}) \right\} \tag{7.13.13}$$

$$\text{subject to } \ \text{tr}(\mathbf{H}^T \mathbf{H}) = n. \tag{7.13.14}$$

In addition to the above two models (7.13.10) and (7.13.13), it is necessary to combine together the community structure to guide the learning process of representation matrix $\mathbf{U}$ as well. To this end, introduce an auxiliary nonnegative matrix $\mathbf{C} \in \mathbb{R}^{k \times m}$, namely *community representation matrix*, where the rth row $\mathbf{c}_r$ is the representation of community r. Therefore, the nonnegative community indicator matrix $\mathbf{H}$ must satisfy the following nonnegative matrix factorization [156]:

$$(\mathbf{U}, \mathbf{C}) = \arg\min_{\mathbf{U}, \mathbf{C}} \ \|\mathbf{H} - \mathbf{C}\mathbf{U}\|_F^2 \tag{7.13.15}$$

$$\text{subject to } \mathbf{U} \geq 0, \mathbf{C} \geq 0. \tag{7.13.16}$$

Combining Eqs. (7.13.10), (7.13.13), and (7.13.15), the objective of the community preserving network embedding is given by [156]

$$\underset{\mathbf{M},\mathbf{U},\mathbf{H},\mathbf{C}}{\arg\min} \quad (1-\alpha)\|\mathbf{S}-\mathbf{M}\mathbf{U}\|_F^2 + \alpha\|\mathbf{H}-\mathbf{C}\mathbf{U}\|_F^2 - \beta\mathrm{tr}(\mathbf{H}^T\mathbf{B}\mathbf{H}), \tag{7.13.17}$$

subject to $\quad \mathbf{M} \geq 0, \mathbf{U} \geq 0, \mathbf{H} \geq 0, \mathbf{C} \geq 0, \mathrm{tr}(\mathbf{H}^T\mathbf{H}) = n,$ (7.13.18)

where α and β are positive parameters for adjusting the contribution of corresponding terms.

For the non-convex objective function (7.13.17), its optimization can be solved by using two subproblems.

1. **H-subproblem:** Updating $\mathbf{H}$ with the other parameters in (7.13.17) fixed leads to the following optimization subproblem [156]:

$$\underset{\mathbf{H}}{\arg\min} \; \alpha\|\mathbf{H}-\mathbf{U}\mathbf{C}^T\|_F^2 - \beta\mathrm{tr}(\mathbf{H}^T\mathbf{B}\mathbf{H}), \tag{7.13.19}$$

subject to $\mathrm{tr}(\mathbf{H}^T\mathbf{H}) = n.$ (7.13.20)

The constrained condition $\mathrm{tr}(\mathbf{H}^T\mathbf{H}) = n$ can be relaxed to the regularization $\mathbf{H}^T\mathbf{H} = \mathbf{I}$, and thus the optimization subproblem becomes

$$\mathbf{H} = \underset{\mathbf{H}}{\arg\min} \; \alpha\|\mathbf{H}-\mathbf{U}\mathbf{C}^T\|_F^2 - \beta\mathrm{tr}(\mathbf{H}^T\mathbf{B}\mathbf{H}) + \lambda\|\mathbf{H}^T\mathbf{H}-\mathbf{I}\|_F^2, \tag{7.13.21}$$

where $\lambda > 0$ should be large enough to ensure the orthogonality is satisfied. By the Karush–Kuhn–Tucker (KKT) condition for the nonnegativity of $\mathbf{H}$, it can be shown that given an initial value of $\mathbf{H}$, its successive updating rule is given by

$$\mathbf{H} \leftarrow \mathbf{H} \odot \sqrt{\frac{-2\beta\mathbf{B}_1\mathbf{H}+\sqrt{\mathbf{\Delta}}}{8\lambda\mathbf{H}\mathbf{H}^T\mathbf{H}}}, \tag{7.13.22}$$

where $\odot$ denotes the Hadamard product of two matrices, and

$$\mathbf{\Delta} = 2\beta(\mathbf{B}_1\mathbf{H}) \odot 2\beta(\mathbf{B}_1\mathbf{H}) \\ + 16\lambda(\mathbf{H}\mathbf{H}^T\mathbf{H}) \odot (2\beta\mathbf{A}\mathbf{H} + 2\alpha\mathbf{U}\mathbf{C}^T + (4\lambda - 2\alpha)\mathbf{H}), \tag{7.13.23}$$

with the Hadamard division of two matrices given by

$$\left[\sqrt{\frac{-2\beta\mathbf{B}_1\mathbf{H}+\sqrt{\mathbf{\Delta}}}{8\lambda\mathbf{H}\mathbf{H}^T\mathbf{H}}}\right]_{ij} = \sqrt{\frac{-2\beta[\mathbf{B}_1\mathbf{H}]_{ij}+\sqrt{\mathbf{\Delta}_{ij}}}{8\lambda[\mathbf{H}\mathbf{H}^T\mathbf{H}]_{ij}}}. \tag{7.13.24}$$

7.13 Network Embedding

2. **Joint NMF subproblem:** Updating **M, U, C** with **H** in (7.13.17) fixed leads to the joint NMF problems [3]:

$$\arg\min_{\mathbf{M},\mathbf{U},\mathbf{C}} \quad (1-\alpha)\|\mathbf{S}-\mathbf{MU}\|_F^2 + \alpha\|\mathbf{H}-\mathbf{CU}\|_F^2 \tag{7.13.25}$$

$$\text{subject to} \quad \mathbf{M} \geq 0, \mathbf{U} \geq 0, \mathbf{C} \geq 0. \tag{7.13.26}$$

For the matrices of basis vectors **M** and **C**, the update rules immediately carry through and read:

$$\mathbf{M} = \mathbf{M} \odot \frac{\mathbf{SU}^T}{\mathbf{MUU}^T}, \tag{7.13.27}$$

$$\mathbf{C} = \mathbf{C} \odot \frac{\mathbf{HU}^T}{\mathbf{CUU}^T}. \tag{7.13.28}$$

Since the coefficient matrix **U** now couples two bases **M** and **C**, its update is slightly more involved. The simplified version of the fixed point iteration for **U** is given by [3]

$$\mathbf{U} = \mathbf{U} \odot \frac{(1-\lambda)\mathbf{M}^T\mathbf{S} + \lambda\mathbf{C}^T\mathbf{H}}{((1-\lambda)\mathbf{M}^T\mathbf{M} + \lambda\mathbf{C}^T\mathbf{C})\mathbf{U}}. \tag{7.13.29}$$

7.13.3 Higher-Order Proximity Preserved Network Embedding

Graph embedding algorithms aim to embed a graph into a vector space where the structure and the inherent properties of the graph are preserved without considering how to preserve its asymmetric transitivity, which is a critical property of directed graphs.

Transitivity is a common characteristic of undirected and directed graphs [111, 140], and plays a key role in graph inference and analysis tasks, such as calculating similarities between nodes and measuring the importance of nodes.

- In undirected graphs, if there is an edge between vertices **u** and **w**, and another between **w** and **v**, then **u** and **v** are likely connected by an edge. Transitivity is symmetric in undirected graphs.
- In directed graph, there is a directed path from **u** to **v**, but not from **v** to **u**. That is, transitivity is asymmetric in directed graphs.

Consider a directed graph $G = (V, E)$, where $V = \{\mathbf{v}_1, \ldots, \mathbf{v}_N\}$ is the vertex set, N is the number of vertexes. E is the directed edge set, i.e., $e_{ij} = (\mathbf{v}_i, \mathbf{v}_j) \in E$ represents a directed edge from $\mathbf{v}_i$ to $\mathbf{v}_j$. The adjacency matrix is denoted by **A**. If S_{ij} are the *higher-order proximities* between $\mathbf{v}_i$ and $\mathbf{v}_j$, then $\mathbf{S} = [S_{ij}]$ is known as a *higher-order proximity matrix*. Let $\mathbf{U} = [\mathbf{U}_s, \mathbf{U}_t]$ be the embedding matrix whose ith row $\mathbf{u}_i$ is the embedding vector of $\mathbf{v}_i$, and let $\mathbf{U}_s, \mathbf{U}_t \in \mathbb{R}^{N \times K}$ be the *source embedding vectors* and *target embedding vectors*, respectively, where K is the embedding dimensions.

The *higher-order proximity preserved embedding* (HOPE) in [111] can be stated as follows: Given a higher-order proximity matrix $\mathbf{S}$, find the source embedding matrix $\mathbf{U}_s$ and the target embedding matrix $\mathbf{U}_t$. The objective of this problem is defined as [111]:

$$\min \|\mathbf{S} - \mathbf{U}_s \mathbf{U}_t^T\|_F^2. \quad (7.13.30)$$

Let the singular value decomposition (SVD) of the higher-order proximity matrix $\mathbf{S}$ be given by

$$\mathbf{S} = \sum_{i=1}^{N} \sigma_i \mathbf{v}_i^s \left(\mathbf{v}_i^t\right)^T, \quad (7.13.31)$$

where $\sigma_1, \ldots, \sigma_N$ are the singular values sorted in decreasing order, and $\mathbf{v}_i^s$ and $\mathbf{v}_i^t$ are the left- and right-singular vectors associated with σ_i of $\mathbf{S}$.

By comparison of (7.13.31) with (7.13.30), then it is easily known that the source and target embedding matrices can be determined by

$$\mathbf{U}_s = \left[\sqrt{\sigma_1} \mathbf{v}_1^s, \ldots, \sqrt{\sigma_K} \mathbf{v}_K^s\right], \quad (7.13.32)$$

$$\mathbf{U}_t = \left[\sqrt{\sigma_1} \mathbf{v}_1^t, \ldots, \sqrt{\sigma_K} \mathbf{v}_K^t\right], \quad (7.13.33)$$

where K is the number of the largest singular values of $\mathbf{S}$, giving the estimate of the embedding dimension.

Many higher-order proximity measurements in graph can reflect the asymmetric transitivity. The High-order proximity matrix shares a general formulation:

$$\mathbf{S} = \mathbf{M}_g^{-1} \mathbf{M}_l, \quad (7.13.34)$$

where $\mathbf{M}_g$ and $\mathbf{M}_l$ are both polynomials of matrices.

The following are a few examples of higher-order proximity matrices [111]:

1. *Katz Index* [76]:

$$\mathbf{S}^{\text{Katz}} = \sum_{l=1}^{\infty} \beta \mathbf{A}_l = \beta \mathbf{A} \mathbf{S}^{\text{Katz}} + \beta \mathbf{A}, \quad (7.13.35)$$

7.13 Network Embedding

from which it follows that

$$\mathbf{S}^{\text{Katz}} = (\mathbf{I} - \beta \mathbf{A})^{-1} \beta \mathbf{A}, \quad (7.13.36)$$

where β is a decay parameter. β should be smaller than the spectral radius of adjacency matrix. Clearly, for Katz index, one has

$$\mathbf{M}_g = (\mathbf{I} - \beta \mathbf{A}) \quad \text{and} \quad \mathbf{M}_l = \beta \mathbf{A}. \quad (7.13.37)$$

2. *Rooted PageRank* (RPR):

$$\mathbf{S}^{\text{RPR}} = \alpha \mathbf{S}^{\text{RPR}} \mathbf{P} + (1-\alpha)\mathbf{I} \Rightarrow \mathbf{S}^{\text{RPR}} = (\mathbf{I} - \alpha \mathbf{P})^{-1}(1-\alpha)\mathbf{I}, \quad (7.13.38)$$

where $\alpha \in [0, 1)$ is the probability to randomly walk to a neighbor, and $\mathbf{P}$ is the probability transition matrix satisfying the condition $\sum_{i=1}^{N} P_{ij} = 1$. Clearly, for RPR,

$$\mathbf{M}_g = \mathbf{I} - \alpha \mathbf{P} \quad \text{and} \quad \mathbf{M}_l = (1-\alpha)\mathbf{I}. \quad (7.13.39)$$

3. *Common Neighbors* (CN): S_{ij}^{CN} counts the number of vertices connecting to both $\mathbf{v}_i$ and $\mathbf{v}_j$. For directed graphs, S_{ij}^{CN} is the number of vertices which are the target of an edge from $\mathbf{v}_i$ and the source of an edge to $\mathbf{v}_j$. Formally,

$$\mathbf{S}^{\text{CN}} = \mathbf{A}^2 \quad (7.13.40)$$

from which we get

$$\mathbf{M}_g = \mathbf{I} \quad \text{and} \quad \mathbf{M}_l = \mathbf{A}^2. \quad (7.13.41)$$

4. *Adamic-Adar* (AA): Adamic-Adar is a variant of common neighbors:

$$\mathbf{S}^{\text{AA}} = \mathbf{ADA} \quad (7.13.42)$$

which gives

$$\mathbf{M}_g = \mathbf{I} \quad \text{and} \quad \mathbf{M}_l = \mathbf{ADA}, \quad (7.13.43)$$

where

$$D_{ii} = \left(\sum_j (A_{ij} + A_{ji}) \right)^{-1}. \quad (7.13.44)$$

It needs the matrix inversion $\mathbf{M}_g^{-1}$ to compute the higher-order proximity matrix $\mathbf{S}$ from $\mathbf{M}_g$ and $\mathbf{M}_l$. To improve the numerical stability of the higher-order proximity preserved embedding, Ou et al. [111] suggested to compute the generalized singular value decomposition (GSVD) of the matrix pair $(\mathbf{M}_g, \mathbf{M}_l)$ instead of the SVD of $\mathbf{S}$:

$$\mathbf{V}_t^T \mathbf{M}_l^T \mathbf{X} = \mathbf{Diag}\left(\sigma_1^l, \ldots, \sigma_N^l\right), \qquad (7.13.45)$$

$$\mathbf{V}_s^T \mathbf{M}_g^T \mathbf{X} = \mathbf{Diag}\left(\sigma_1^g, \ldots, \sigma_N^g\right), \qquad (7.13.46)$$

where $\mathbf{X}$ is a nonsingular matrix, and

$$\sigma_1^l \geq \sigma_2^l \geq \ldots \geq \sigma_N^l \geq 0, \qquad (7.13.47)$$

$$0 \leq \sigma_1^g \leq \sigma_2^g \leq \ldots \leq \sigma_N^g, \qquad (7.13.48)$$

$$\left(\sigma_i^l\right)^2 + \left(\sigma_i^g\right)^2 = 1, \ \forall\, i. \qquad (7.13.49)$$

Most existing embedding methods focus on the static network, while neglecting the evolving characteristic of real-world networks. Recently, Zhu et al. [176] proposed a higher-order proximity preserved embedding for dynamic networks.

7.14 Neural Networks on Graphs

Lots of learning tasks deal with graph data. Graphs are a kind of irregular data structure which models a set of objects (nodes) and their relationships (edges) in graph domains, not Euclidean domains. In Sect. 6.16, we have discussed graph machine learning. This section focuses on neural networks that operate on graphs.

Neural networks in graph domains are divided into semi-supervised networks and unsupervised networks. The earliest semi-supervised neural networks on graph were introduced by Gori et al. [40] in 2005, called graph neural networks (GNNs). GNNs can capture the dependence of graphs and rich relation information among elements via message passing between nodes of a graph. Another kind of semi-supervised networks on graph is graph convolutional networks (GCNs). Unsupervised networks on graph are graph autoencoders (GAEs).

The original GNN framework in [40] and [129] requires the repeated application of contraction maps as propagation functions until node representations reach a stable fixed point. This restriction was later alleviated by Li et al. [95] in 2016 by introducing modern practices for recurrent neural network training to the original GNN framework. These alleviated GNNs are called *gated graph neural networks* (GG-NN). Duvenaud et al. [28] in 2015 introduced a convolution-like propagation rule on graphs, called *graph convolutional networks* (GCNs). In 2017, Gilmer et al.

[35] proposed a unified framework for several variants of GNNs, called *message passing neural networks* (MPNNs).

7.14.1 Graph Neural Networks (GNNs)

GNN is a kind of neural network which processes directly the data represented in a graph domain. A typical application of GNN is node classification. Essentially, each node in the graph is associated with a label. Our goal is to predict the label of a node without an associated ground-truth.

Let the vertex or node x_i ($i = 1, \ldots, n$) represent the ith data point (feature). The goal of GNNs is to learn a state embedding $\mathbf{h}_i \in \mathbb{R}^s$ which contains the information of the neighborhood for the ith node x_i.

GNNs are a general neural network architecture defined according to a graph structure $G = (V, E)$. Nodes $j \in V$ take unique values from $\{1, \ldots, |V|\}$, and edges are pairs $e = (i, j) \in V \times V$. In directed graphs, (i, j) represents a directed edge $i \to j$. The *node vector*, called also *node representation* or *node embedding*, for node j is denoted by $\mathbf{x}_j \in \mathbb{R}^D$. Graphs may also contain *node labels* $l_j \in \{1, \ldots, L_{|V|}\}$ for each node j and *edge labels* or *edge types* $l_e \in \{1, \ldots, L_{|E|}\}$ for each edge. Let $\mathbf{x}_S = \{\mathbf{x}_j | j \in S\}$ when S is a set of nodes, and $l_E = \{l_e | e \in E\}$ when E is a set of edges.

Let f_i be a parametric function (for the node i), called *local transition function*, that expresses the dependence of node i on its neighborhood and let g_i be the *local output function* (for the node i) that describes how the output is produced.

The set ne[n] stands for the neighbors of the vertex n, i.e., the nodes connected to n by an arc, while co[n] denotes the set of arcs having n as a vertex.

The *state vector* or *state embedding* $\mathbf{h}_i$ and the output $\mathbf{o}_i$ can be represented by [129]:

$$\mathbf{h}_i = f_i\left(\mathbf{x}_i, \mathbf{x}_{\text{co}[i]}, \mathbf{h}_{\text{ne}[i]}, \mathbf{x}_{\text{ne}[i]}\right), \tag{7.14.1}$$

$$\mathbf{o}_i = g_i(\mathbf{h}_i, \mathbf{x}_i), \tag{7.14.2}$$

where

- $\mathbf{x}_i$: the features of the node i,
- $\mathbf{x}_{\text{co}[i]}$: the features of edges connecting with i,
- $\mathbf{h}_{\text{ne}[i]}$: the embedding of the nodes in the neighborhood of i,
- $\mathbf{x}_{\text{ne}[i]}$: the features of the nodes in the neighborhood of i,
- f_i: a transition function that maps the above four inputs to d-dimensional space,
- g_i: an output function when the input is $\mathbf{x}_i$ and the transited state is $\mathbf{h}_i$.

Example 7.2 Let $\mathbf{x}_1, \ldots, \mathbf{x}_8$ be eight data points, where $\mathbf{x}_2, \mathbf{x}_3, \mathbf{x}_4, \mathbf{x}_6$ are the neighbors of $\mathbf{x}_1$. Then, $\mathbf{x}_{\text{co}[1]} = (e_{(1,2)}, e_{(3,1)}, e_{(1,4)}, e_{(6,1)})$, where $e_{(ij)}$ denotes the edge label connecting node i to node j, $\mathbf{x}_{\text{ne}[1]} = (\mathbf{x}_2, \mathbf{x}_3, \mathbf{x}_4, \mathbf{x}_6)$, and $\mathbf{h}_{\text{ne}[1]} =$

($\mathbf{h}_2, \mathbf{h}_3, \mathbf{h}_4, \mathbf{h}_6$). In other words,

$$\mathbf{h}_1 = f\Big(\mathbf{x}_1, \underbrace{(e_{(1,2)}, e_{(3,1)}, e_{(1,4)}, e_{(6,1)})}_{\mathbf{x}_{co[1]}}, \underbrace{(\mathbf{h}_2, \mathbf{h}_3, \mathbf{h}_4, \mathbf{h}_6)}_{\mathbf{h}_{ne[1]}}, \underbrace{(\mathbf{x}_2, \mathbf{x}_3, \mathbf{x}_4, \mathbf{x}_6)}_{\mathbf{x}_{ne[1]}}\Big),$$

which contains information on the neighborhood for the first node $\mathbf{x}_1$.

Let $\mathbf{h}, \mathbf{o}, \mathbf{x}$ and $\mathbf{x}_N$ be the vectors constructed by stacking all the states, all the outputs, all the features, and all the node features, respectively. Equations (7.14.1) and (7.14.2) can be rewritten in a compact form as [129]:

$$\mathbf{h} = \mathbf{f}(\mathbf{h}, \mathbf{x}), \tag{7.14.3}$$

$$\mathbf{o} = \mathbf{g}(\mathbf{h}, \mathbf{x}_N), \tag{7.14.4}$$

where $\mathbf{f} = [f_1, \ldots, f_N]^T$ and $\mathbf{g} = [g_1, \ldots, g_N]^T$ are stacked versions of local transition functions and local output functions corresponding to all N nodes, respectively; and are known as the *global transition function* and *global output function* for all nodes in a graph, respectively. The aim of GNNs is to learn the global transition function $\mathbf{f}$ and the global output function $\mathbf{g}$.

Let $\mathbf{t}_i = \mathbf{W}\mathbf{h}_i$ be the target information (for a specific node i) for the supervision, the loss can be written as follows:

$$\mathcal{L}(\mathbf{W}) = \frac{1}{2} \sum_{i=1}^{p} \|\mathbf{t}_i - \mathbf{o}_i\|_2^2, \tag{7.14.5}$$

where p is the number of supervised nodes.

By Banach's fixed point theorem [78], GNN uses the following classic iterative scheme for updating the state:

$$\mathbf{h}_{t+1} = \mathbf{f}(\mathbf{h}_t, \mathbf{x}), \tag{7.14.6}$$

where $\mathbf{h}_t$ denotes the tth iteration of $\mathbf{h}$. This updating converges exponentially fast to the solution of Eq. (7.14.3) for any initial value $\mathbf{h}_0$.

The dynamical systems based on the computations of $\mathbf{f}$ and $\mathbf{g}$ can be interpreted as feedforward neural networks. To learn the parameters of $\mathbf{f}$ and $\mathbf{g}$, given the target information $\mathbf{t}$ for the supervision, the loss in (7.14.5) can be rewritten as follows:

$$\mathcal{L}(\mathbf{W}) = \frac{1}{2}\|\mathbf{t} - \mathbf{o}\|_2^2 = \frac{1}{2}\|\mathbf{W}\mathbf{h} - \mathbf{o}\|_2^2. \tag{7.14.7}$$

7.14 Neural Networks on Graphs

The learning algorithm is based on a gradient descent strategy composed of the following steps [129]:

1. The states $\mathbf{h}_i^{(t)}$ are iteratively updated by Eq. (7.14.1) until a time T. They approach the fixed point solution to Eq. (7.14.3): $\mathbf{h}(T) \approx \mathbf{h}$.
2. Compute the gradient of weights

$$\nabla \mathcal{L}(\mathbf{W}_t) = \frac{\partial \mathcal{L}}{\partial \mathbf{W}_t} = (\mathbf{W}_t \mathbf{h} - \mathbf{o}_t)\mathbf{h}^T. \quad (7.14.8)$$

3. The weights are updated as $\mathbf{W}_{t+1} = \mathbf{W}_t - \mu_t \nabla \mathcal{L}(\mathbf{W}_t)$.
4. Return to Step 1 and repeat the above steps until $\mathbf{W}$ is converged.

7.14.2 DeepWalk and GraphSAGE

The original GNN has three main limitations.

- If the "fixed point" hypothesis is relaxed, a more stable representation can be learned using multilayer perceptrons and the iterative update process can be deleted.
- It does not handle edge information (for example, different edges in a knowledge graph may represent different relationships between nodes).
- Fixed points hinder the diversity of node distribution and are not suitable for learning good representations of nodes.

In view of the above limitations of the original GNN, its many variants are proposed. Two typical variants are DeepWalk and GraphSAGE.

DeepWalk

Social representations are expected to have the following characteristics [114]:

- *Adaptable:* Real social networks are constantly evolving; new social relations should not require repeating the learning process all over again.
- *Community aware:* The distance between latent dimensions should represent a metric for evaluating social similarity between the corresponding members of the network. This allows generalization in networks with homophily.
- *Low dimensional:* When labeled data is scarce, low-dimensional models generalize better, and speed up convergence and inference.
- *Continuous:* Latent representations are required to model partial community membership in continuous space. In addition to providing a nuanced view of community membership, a continuous representation has smooth decision boundaries between communities which allows more robust classification.

DeepWalk, introduced by Perozzi et al. [114], satisfies these requirements by learning representation for vertices from a stream of short random walks, using optimization techniques originally designed for language modeling.

1. *Random Walks:* Random walks are executed on nodes in graph to generate a sequence of nodes. A random walk rooted at vertex v_i is denoted as $\mathcal{W}_{v_i}$ which is a stochastic process with random variables $\mathcal{W}_{v_i}^1, \ldots, \mathcal{W}_{v_i}^k$ such that $\mathcal{W}_{v_i}^{k+1}$ is a vertex chosen at random from the neighbors of vertex v_k. Due to local structure, a stream of short random walks can be used as a basic tool for extracting information from a network. Moreover, using random walks has two other desirable properties [114].

   - Local exploration is easy to parallelize. Several random walkers (in different threads, processes, or machines) can simultaneously explore different parts of the same graph.
   - Relying on information obtained from short random walks make it possible to accommodate small changes in the graph structure without the need for global recomputation so that the learned model can be iteratively updated with new random walks from the changed region in time sub-linear to the entire graph.

2. *Language Modeling:* Run skip-gram to learn the embedding of each node according to the sequence of nodes generated in random walks. Language modeling can be generalized to explore the graph through a stream of short random walks. These random walks can be thought of as short sentences and phrases in a special language; the direct analog is to estimate the likelihood of observing vertex v_i given all the previous vertices visited so far in the random walk, i.e.,

$$\Pr\big(v_i | (v_1, v_2, \ldots, v_{i-1})\big). \tag{7.14.9}$$

Algorithm 7.8 shows the DeepWalk-gram and Algorithm 7.9 provides the skip-gram.

Algorithm 7.8 DeepWalk(G, w, d, γ, t) [114]

1. **input:** Graph $G(V, E)$; window size w; embedding size d; walks per vertex γ; walk length t.
2. **initialization:** Sample Φ from $\mathcal{U}^{|V| \times d}$.
3. Build a binary Tree T from V.
4. **for** $i = 0$ to γ **do**
5. $\mathcal{O} = \text{Shuffle}(V)$.
6.   **for** each $v_i \in \mathcal{O}$ **do**
7. $\mathcal{W}_{v_i} = \text{RandomWalk}(G, v_i, t)$.
8.   SkipGram$(\Phi, \mathcal{W}_{v_i}, w)$.
9.   **end for**
10. **end for**
11. **output:** matrix of vertex representations $\Phi \in \mathbb{R}^{|V| \times d}$.

7.14 Neural Networks on Graphs

Algorithm 7.9 SkipGram(Φ, $\mathcal{W}_{v_i}$, w) [114]
1. **for** each $v_j \in \mathcal{W}_{v_i}$ **do**
2.   **for** each $u_k \in \mathcal{W}_{v_i}[j - w : j + w]$ **do**
3. $J(\Phi) = -\log \Pr(u_k | \Phi(v_j))$
4. $\Phi = \Phi - \alpha \frac{\partial J}{\partial \Phi}$
5.   **end for**
6. **end for**

However, DeepWalk's main problem is lack of generalization ability. Whenever a new node appears, it must retrain the model to represent the node. Therefore, this GNN is not suitable for dynamic graphs with changing nodes.

GraphSAGE

Most graph embedding methods need all the nodes in the graph to participate in the training process, which is a property of transductive learning, and cannot directly generalize to nodes that have not been seen before.

Graph with SAmple and aggreGatE (simply called *GraphSAGE*), introduced by Hamilton et al. [50], uses an inductive node embedding for large-scale networks, which can generate embeddings quickly for new nodes without an additional training process. The framework of GraphSAGE is as follows [50]:

- *Embedding generation:* Unlike embedding approaches that are based on matrix factorization, GraphSAGE leverages node features (e.g., text attributes, node profile information, node degrees) in order to learn an embedding function that generalizes to unseen nodes.
- *Parameter learning:* By incorporating node features in the learning of parameters, GraphSAGE simultaneously learns the topological structure of each node's neighborhood as well as the distribution of node features in the neighborhood.
- *Aggregator architecture:* Instead of training a distinct embedding vector for each node, GraphSAGE trains a set of aggregate functions that learn to aggregate feature information from a node's local neighborhood. Each aggregate function aggregates information from a different number of hops, or search depth, away from a given node.

The *aggregate functions* applied in GraphSAGE must meet the following two basic requirements:

- The aggregate functions must operate over an unordered set of vectors because a node's neighbors have no natural ordering.
- An aggregate function should be symmetric (i.e., invariant to permutations of its inputs), but should still be trainable and maintain high representational capacity.

There are the following three aggregate functions satisfying the above two basic requirements [50].

1. *Mean aggregate function:* The basic operator simply takes the elementwise mean of the vectors in $\{\mathbf{h}_u^{k-1}, \forall u \in \mathcal{N}(v)\}$:

$$\text{AGG}_{\text{mean}} = \text{MEAN}(\{\mathbf{h}_u^{k-1}, \forall u \in \mathcal{N}(v)\}) = \sum_{u \in \mathcal{N}(v)} \frac{\mathbf{h}_u^{k-1}}{|\mathcal{N}(v)|}. \quad (7.14.10)$$

The embedding in the kth layer of the node v is given by

$$\mathbf{h}_v^k = \sigma \left(\mathbf{W}_k \sum_{u \in \mathcal{N}(v)} \frac{\mathbf{h}_v^{k-1}}{|\mathcal{N}(v)|} + \mathbf{B}_k \mathbf{h}_v^{k-1} \right), \quad \forall k > 0. \quad (7.14.11)$$

A better choice is the Graph Convolution Network (GCN) neighborhood aggregation defined as

$$\text{AGG}_{\text{GCN}} = \text{MEAN}(\{\mathbf{h}_v^{k-1}\} \cup \{\mathbf{h}_u^{k-1}, \forall u \in \mathcal{N}(v)\})$$

$$= \sum_{u \in \mathcal{N}(v) \cup v} \frac{\mathbf{h}_u^{k-1}}{\sqrt{|\mathcal{N}(u)| \cdot |\mathcal{N}(v)|}}, \quad (7.14.12)$$

$$\Rightarrow \mathbf{h}_v^k = \sigma \left(\mathbf{W}_k \sum_{u \in \mathcal{N}(v) \cup v} \frac{\mathbf{h}_v^{k-1}}{\sqrt{|\mathcal{N}(u)| \cdot |\mathcal{N}(v)|}} \right), \quad \forall k > 0. \quad (7.14.13)$$

2. *LSTM aggregate function:* This is a more complex aggregator based on an LSTM architecture [63]:

$$\text{AGG}_{\text{LSTM}} = \text{LSTM}(\{\mathbf{h}_v^{k-1}, \forall v \in \mathcal{N}(v)\}). \quad (7.14.14)$$

Compared to the mean aggregate function, the advantage of LSTM aggregate functions is their larger expressive capability. However, LSTMs are not inherently symmetric, it is necessary to adapt LSTMs for operating on an unordered set by simply applying the LSTMs to a random permutation of the node's neighbors.

3. *Pooling aggregate function:* This aggregator is symmetric and trainable. In this pooling approach, each neighbor's vector is independently fed through a fully connected neural network; following this transformation, an elementwise max-pooling operation is applied to aggregate information across the neighbor set:

$$\text{AGG}_{\text{pool}} = \gamma \left(\{\mathbf{h}_{u_i}^k + \mathbf{b}, \forall u_i \in \mathcal{N}(v)\} \right), \quad (7.14.15)$$

where γ takes usually elementwise mean/max function.

Algorithm 7.10 shows the GraphSAGE embedding generation algorithm [50].

7.14 Neural Networks on Graphs

Algorithm 7.10 GraphSAGE embedding generation algorithm [50]
1. **input:** Graph $G(V, E)$; input features $\{\mathbf{x}_v, \forall v \in V\}$; depth K; weight matrices $\mathbf{W}_k, k=1, \ldots, K$; nonlinearity σ; neighborhood function $\mathcal{N} : v \to 2^V$.
2. $\mathbf{h}_v^0 \leftarrow \mathbf{x}_v, \forall v \in V$;
3. **for** $k = 1, \ldots, K$ **do**
4.   **for** $v \in V$ **do**
5. $\mathbf{h}_v^k \leftarrow \sigma\left(\mathbf{W} \cdot \text{MEAN}(\{\mathbf{h}_u^{k-1}, \forall u \in \mathcal{N}(v)\})\right)$
      or $\mathbf{h}_v^k \leftarrow \sigma\left(\mathbf{W} \cdot \text{MEAN}(\{\mathbf{h}_v^{k-1}\}) \cup \{\mathbf{h}_u^{k-1}, \forall u \in \mathcal{N}(v)\})\right)$;
6.   **end for**
7. $\mathbf{h}_v^k \leftarrow \mathbf{h}_v^k / \|\mathbf{h}_v^k\|, \forall v \in V$.
8. **end for**
9. $\mathbf{z}_v \leftarrow \mathbf{h}_v^K, \forall v \in V$.
10. **output:** vector representations $\mathbf{z}_v$ for all $v \in V$.

GraphSAGE is not only applied to feature-rich graphs (e.g., citation data with text attributes, biological data with functional/molecular markers) for making use of structural features that are present in all graphs (e.g., node degrees), but also can be applied to graphs without node features.

7.14.3 Graph Convolutional Networks (GCNs)

There are two types of spatial information in graph data:

- *Node information:* each vertex or node has its own information or characteristic that is represented by the nodes themselves.
- *Structural information:* each node in the graph data has its own structural information, which is association information between nodes and is represented by edges connecting a node and other nodes.

Generally speaking, graph data should consider not only node information but also structure information. A graph convolution neural network can automatically learn not only node information, but also association information between nodes.

Graph convolutional networks (GCNs), introduced by Kipf and Welling [82], are machine learning methods that "learn" graph structured data through extracting spatial features. Because the standard convolution for image or text cannot be directly applied to graphs without a grid structure, a graph must necessarily be mapped onto another spectral domain that does have a grid structure.

Unlike the spatial convolution (such as GraphSAGE) which is a vertex domain (spatial domain) method based on convolutions in the vertex domain (spatial domain) defined directly by the connections relationships of each node, the spectral convolution is a frequency domain method.

Bruna et al. [13] first introduced a convolution for graph data from spectral domains using the graph Laplacian matrix **L**. This convolution on the spectral domain of a graph is called the *spectral convolution*.

The Laplacian matrix has many important properties. The following are the two points related to GCNs:

- The Laplacian matrix is a symmetric matrix and can perform eigenvalue decomposition (spectral decomposition), which corresponds to the spectral domain of GCNs.
- The Laplacian matrix has only nonzero elements at the center apex and the first-order connected vertices, and the rest are 0.

In classical signal processing, we have *convolution the theorem in the time domain*

$$f(t) * h(t) = \int_{-\infty}^{\infty} \hat{f}(\omega)\hat{h}(\omega)e^{-j\omega t} d\omega, \qquad (7.14.16)$$

and the *convolution theorem in the frequency domain*

$$\hat{f}(\omega) * \hat{h}(\omega) = \int_{-\infty}^{\infty} f(t)h(t)e^{j\omega t} dt. \qquad (7.14.17)$$

Because a graph signal **f** has no grid structure, the standard convolution cannot be directly applied to **f**. Compared with the graph signal **f** in the vertex domain, its spectral graph signal $\hat{\mathbf{f}}$ in the graph spectral domain has grid structure, and thus the convolution can be directly to $\hat{\mathbf{f}}$. The spectral signals $\hat{f}(\lambda_\ell)$ are referred to as kernels of the vertex signals $f(i)$.

To introduce the convolution on graph signals, we need to search for an orthonormal basis instead of the basis function $e^{\pm j\omega t}$ in the standard convolution on Euclidean structures.

Since both normalized and unnormalized Laplacians are symmetric and positive semi-definite matrices, they admit an eigenvalue decomposition $\mathbf{L} = \mathbf{U}\mathbf{\Lambda}\mathbf{U}^T$, where $\mathbf{U} = [\mathbf{u}_1, \ldots, \mathbf{u}_n]$ are the orthonormal eigenvectors and $\mathbf{\Lambda} = \mathbf{diag}(\lambda_1, \ldots, \lambda_n)$ is the diagonal matrix of the corresponding nonnegative eigenvalues (spectrum) $\lambda_1 \geq \lambda_2 \geq \ldots \geq \lambda_n = 0$. The eigenvectors play the role of Fourier atoms in classical harmonic analysis and the eigenvalues can be interpreted as (the square of) frequencies.

Consider signals defined on an undirected, connected, weighted graph $G(V, E)$ which consists of a finite set of vertices V with $|V| = N$, a set of edges E and a weighted adjacency matrix **A**. If there is an edge $e = (i, j)$ connecting vertices i and j, then the entry w_{ij} represents the weight of the edge; otherwise, $w_{ij} = 0$. If the graph $G(V, E)$ is not connected and has M connected components ($M > 1$), then $G(V, E)$ is separated into M subgraphs $G_1, \ldots, G_M$, and signals on $G(V, E)$ are separated into M pieces corresponding to the M connected components, and independently process the separated signals on each of the subgraphs.

A signal or function $f : V \to \mathbb{R}$ defined on the vertices of the graph may be represented as a vector $\mathbf{f} \in \mathbb{R}^N$, where the ith component of the vector **f** represents the function value at the ith vertex in V.

7.14 Neural Networks on Graphs

Suppose we are given two signals $\mathbf{f} = [f_1, \ldots, f_N]^T$ and $\mathbf{h} = [h_1, \ldots, h_N]^T$ on the vertices of graph $G(V, E)$. By replacing $e^{\pm j\omega t}$ with eigenvectors $\mathbf{u}_i$ and $\mathbf{u}_i^*$, one can define the graph Fourier transforms and the graph convolution as follows [136]:

1. *Graph Fourier transform* $\hat{\mathbf{f}}$ of any graph signal or function vector $\mathbf{f} \in \mathbb{R}^N$ on the vertices of $G(V, E)$ is defined as the expansion of $\mathbf{f}$ in terms of the eigenvectors $\mathbf{u}_1, \ldots, \mathbf{u}_N$ of the graph Laplacian $\mathbf{L}$, namely

$$\hat{f}(\lambda_\ell) = \sum_{i=1}^{N} f(i) u_\ell^*(i) \quad \text{or} \quad \hat{\mathbf{f}} = \mathbf{U}^H \mathbf{f}, \tag{7.14.18}$$

where $\mathbf{U} = [\mathbf{u}_1, \ldots, \mathbf{u}_N]^T$ is the eigenvector-matrix of the EVD $\mathbf{L} = \mathbf{U}\mathbf{\Lambda}\mathbf{U}^H$.

2. *Inverse graph Fourier transform* is given by

$$f(i) = \sum_{\ell=0}^{N-1} \hat{f}(\lambda_\ell) u_\ell(i) \quad \text{or} \quad \mathbf{f} = \mathbf{U}\hat{\mathbf{f}}. \tag{7.14.19}$$

3. *Convolution theorem in the time domain*, $f(t) * h(t) = \int_{-\infty}^{\infty} \hat{f}(\omega)\hat{h}(\omega) e^{-j\omega t} d\omega$, becomes the *graph convolution theorem*

$$f(i) *_G h(i) = \sum_{\ell=0}^{N-1} \hat{f}(\lambda_\ell)\hat{h}(\lambda_\ell) u_\ell(i) \tag{7.14.20}$$

and can be written in the spectral convolution form:

$$\mathbf{f} *_G \mathbf{h} = \mathbf{U}\text{Diag}(\hat{h}_1, \ldots, \hat{h}_N)\hat{\mathbf{f}} = \mathbf{U}\text{Diag}(\hat{h}_1, \ldots, \hat{h}_N)\mathbf{U}^T \mathbf{f} \tag{7.14.21}$$

which enforces the property that convolution in the vertex domain is equivalent to multiplication in the graph spectral domain.

Given a signal $\mathbf{x} \in \mathbb{R}^N$ and a filter $\mathbf{g}_\theta = \text{Diag}(\theta_1, \ldots, \theta_N)$ parameterized by $\boldsymbol{\theta} = [\theta_1, \ldots, \theta_N]^T \in \mathbb{R}^N$, the spectral convolutions on graphs in Eq. (7.14.21) becomes

$$\mathbf{x} * \mathbf{g}_\theta = \mathbf{U}\mathbf{g}_\theta \mathbf{U}^T \mathbf{x}, \tag{7.14.22}$$

where $\mathbf{U}$ is the matrix of eigenvectors of the normalized graph Laplacian $\mathbf{L} = \mathbf{I}_N - \mathbf{D}^{-1/2}\mathbf{A}\mathbf{D}^{-1/2} = \mathbf{U}\mathbf{\Lambda}\mathbf{U}^T$, with a diagonal matrix of its eigenvalues $\mathbf{\Lambda}$ and $\mathbf{U}^T \mathbf{x}$ being the graph Fourier transform of $\mathbf{x}$. $\mathbf{g}_\theta$ can be understood as a function of the eigenvalues of L, i.e., $\mathbf{g}_\theta(\mathbf{\Lambda})$. This eigenvalue function can be well-approximated by a truncated expansion in terms of Chebyshev polynomials $T_k(\mathbf{x})$ up to the Kth order [51, 82]:

$$\mathbf{g}_{\theta'}(\mathbf{\Lambda}) \approx \sum_{k=0}^{K} \theta'_k T_k(\tilde{\mathbf{\Lambda}}), \qquad (7.14.23)$$

where $\tilde{\mathbf{\Lambda}}' = \frac{2}{\lambda_{\max}} \mathbf{\Lambda} - \mathbf{I}_N$ with $\lambda_{\max}$ denoting the largest eigenvalue of L, $\boldsymbol{\theta}' \in \mathbb{R}^K$ is a vector of Chebyshev coefficients.

From Eqs. (7.14.22) and (7.14.23) it follows that the spectral convolution of a graph signal $\mathbf{x}$ with a filter $\mathbf{g}_{\theta'}$ is given by [82]

$$\mathbf{x} * \mathbf{g}_{\theta'} \approx \sum_{k=0}^{K} \theta'_k T_k(\tilde{\mathbf{L}})\mathbf{x}, \qquad (7.14.24)$$

where $\tilde{\mathbf{L}} = \frac{2}{\lambda_{\max}} \mathbf{L} - \mathbf{I}_N$ and $T_k(\tilde{\mathbf{L}}) = 2\tilde{\mathbf{L}} T_{k-1}(\tilde{\mathbf{L}}) - T_{k-2}(\tilde{\mathbf{L}})$ with $T_0(\tilde{\mathbf{L}}) = 1$ and $T_1(\tilde{\mathbf{L}}) = \tilde{\mathbf{L}}$.

Under the approximation $\lambda \approx 2$, Eq. (7.14.24) simplifies to

$$\mathbf{x} * \mathbf{g}_{\theta'} \approx \theta'_0 \mathbf{x} + \theta'_1 (\mathbf{L} - \mathbf{I}_N)\mathbf{x} = \theta'_0 \mathbf{x} - \mathbf{D}^{-1/2} \mathbf{A} \mathbf{D}^{-1/2} \mathbf{x}. \qquad (7.14.25)$$

If taking a single parameter $\theta = \theta'_0 = -\theta'_1$, then (7.14.25) gives the following expression [82]:

$$\mathbf{x} * \mathbf{g}_{\theta'} \approx \theta \left(\mathbf{I}_N + \mathbf{D}^{-1/2} \mathbf{A} \mathbf{D}^{-1/2} \right) \mathbf{x}, \qquad (7.14.26)$$

where $\mathbf{I}_N + \mathbf{D}^{-1/2} \mathbf{W} \mathbf{D}^{-1/2}$ has eigenvalues in the range $[0, 2]$.

Direct application of the iteration in Eq. (7.14.26) will lead to numerical instability, and may lead to exploding or vanishing gradients in deep neural network models. In order to alleviate this problem, the renormalization technique is necessary: $\mathbf{I}_N + \mathbf{D}^{-1/2} \mathbf{A} \mathbf{D}^{-1/2} \to \tilde{\mathbf{D}}^{-1/2} \tilde{\mathbf{A}} \tilde{\mathbf{D}}^{-1/2}$, where $\tilde{\mathbf{A}} = \mathbf{A} + \mathbf{I}_N$ and $\tilde{D}_{ii} = \sum_{j=1}^{N} \tilde{A}_{ij}$.

The above definition can be generalized to a signal $\mathbf{X} \in \mathbb{R}^{N \times C}$ with C input channels (i.e., a C-dimensional feature vector for every node) and F filters or feature maps, giving

$$\mathbf{Z} = \tilde{\mathbf{D}}^{-1/2} \tilde{\mathbf{A}} \tilde{\mathbf{D}}^{-1/2} \mathbf{X} \boldsymbol{\Theta}, \qquad (7.14.27)$$

where $\mathbf{Z} \in \mathbb{R}^{N \times F}$ is the convolved signal matrix and $\boldsymbol{\Theta} \in \mathbb{R}^{C \times F}$ is a matrix of filter parameters.

7.14 Neural Networks on Graphs

For semi-supervised multiclass classification, the loss is defined by the cross-entropy error over all labeled examples as follows:

$$\mathcal{L} = -\sum_{l \in \mathcal{Y}_L} \sum_{f=1}^{F} Y_{lf} \ln Z_{lf}, \tag{7.14.28}$$

where $\mathcal{Y}_L$ is the set of node indices that have labels.

For a two-layer GCN for semi-supervised node classification on a graph with a symmetric adjacency matrix **A** (binary or weighted), the forward model takes the simple form

$$\mathbf{Z} = f(\mathbf{X}, \mathbf{A}) = \text{softmax}\left(\hat{\mathbf{A}} \text{ReLU}(\hat{\mathbf{A}} \mathbf{X} \mathbf{W}^{(0)}) \mathbf{W}^{(1)}\right), \tag{7.14.29}$$

where

- $\hat{\mathbf{A}} = \tilde{\mathbf{D}}^{-1/2} \tilde{\mathbf{A}} \tilde{\mathbf{D}}^{-1/2}$ is calculated in the pre-processing step.
- $\mathbf{W}^{(0)} \in \mathbb{R}^{C \times H}$ is an input-to-hidden weight matrix for a hidden layer with H feature maps.
- $\mathbf{W}^{(1)} \in \mathbb{R}^{H \times F}$ is a hidden-to-output weight matrix.
- The softmax activation function is defined as $\text{softmax}(x_i) = \frac{1}{Q} \exp(x_i)$ with $Q = \sum_i \exp(x_i)$ and is applied row-wise.

GCNs have the following four characteristics:

1. GCNs are a natural extension of convolutional neural networks in graph domain. Graph convolution is widely applicable to nodes and graphs of any topological structure.
2. *Local characteristics:* GCN networks focus on information within the K-order neighborhood centered on a node, which is essentially different from GNNs.
3. *First-order characteristics:* After many approximations, a GCN becomes a first-order model. That is to say, a single-layer GCN can be used to process information on first-order neighbors in graphs, and a multilayer GCN can be used to process K-order neighbors.
4. *Parameter sharing:* The filter parameter W is shared over all nodes, which is one of the reasons the graph convolution network has its name.

The following is a comparison of the GCN neighborhood aggregation with the basic neighborhood aggregation in GraphSAGE:

- *Basic neighborhood aggregation:*

$$\mathbf{h}_v^k = \sigma\left(\mathbf{W}_k \sum_{u \in \mathcal{N}(v)} \frac{\mathbf{h}_u^{k-1}}{|\mathcal{N}(v)|} + \mathbf{B}_k \mathbf{h}_v^{k-1}\right), \tag{7.14.30}$$

where $\mathbf{h}_v^k$ is the embedding in the kth layer of the node v, $\sigma(\cdot)$ is a nonlinear activation function, the summing term denotes the embedding in the $(k-1)$th layer of neighbor nodes $u \in \mathcal{N}(v)$ of average node v, and the second term denotes the embedding in layer $k-1$ of the node v.
- *GCN neighborhood aggregation:*

$$\mathbf{h}_v^k = \sigma \left(\mathbf{W}_k \sum_{u \in \mathcal{N}(v) \cup v} \frac{\mathbf{h}_v^{k-1}}{\sqrt{|\mathcal{N}(u)| \cdot |\mathcal{N}(v)|}} \right), \qquad (7.14.31)$$

where $\mathbf{W}_k \sum_{u \in \mathcal{N}(v) \cup v}$ denotes the same matrix for the node v self and its neighbor embedding, and $\frac{\mathbf{h}_v^{k-1}}{\sqrt{|\mathcal{N}(u)| \cdot |\mathcal{N}(v)|}}$ denotes per-neighbor normalization.

7.15 Batch Normalization Networks

Batch normalization (BatchNorm) [74] is a milestone technique in the development of deep learning, enabling various networks to train. It is proved [127] that BatchNorm impacts network training in a fundamental way: it makes the landscape of the corresponding optimization problem significantly more smooth. This ensures, in particular, that the gradients are more predictive. This allows for use of a larger range of learning rates and leads to faster network convergence.

7.15.1 Batch Normalization

Consider a layer with D-dimensional input $\mathbf{x} = [x_1, \ldots, x_D]^T$ and a fully connected matrix $\mathbf{W}$ for extracting d-dimensional feature vectors $\mathbf{y} = \mathbf{W}\mathbf{x} = [y_1, \ldots, y_d]^T$, where $d \ll D$. Batch Normalization (BatchNorm or BN) addresses the problem of internal covariate shift by reducing the dependency of the distribution of the input activations of each layer on all the preceding layers. This is achieved by normalizing activation y_i for each feature.

Suppose we are given T feature vectors $\mathbf{y}_1, \ldots, \mathbf{y}_T$, where $\mathbf{y}_i = [y_{i1}, \ldots, y_{id}]^T$, $i = 1, \ldots, T$. Divide the training data set $\{\mathbf{y}_1, \ldots, \mathbf{y}_T\}$ into K mini-batch data sets of size N such that $T = KN$. The kth mini-batch $\mathcal{B}^{(k)}$ consists of the batch data $\mathbf{y}_1^{(k)}, \ldots, \mathbf{y}_N^{(k)}$:

$$\mathcal{B}^{(k)} = \left\{ \mathbf{y}_{1\ldots N}^{(k)} \right\} = \left\{ \mathbf{y}_1^{(k)}, \ldots, \mathbf{y}_N^{(k)} \right\}. \qquad (7.15.1)$$

For the sake of brevity, we withdraw the batch index k hereafter.

7.15 Batch Normalization Networks

Definition 7.19 (Batch Normalization [74]) Given a mini-batch feature vector $\mathbf{y}_{1...N} = \{\mathbf{y}_1, \ldots, \mathbf{y}_N\}$, let the normalized values of the mini-batch $\mathbf{y}_{1...N}$ be $\hat{\mathbf{y}}_{1...N}$, and their linear transformations be $\mathbf{z}_{1...N}$, where

$$\mathbf{z}_{1...N} = \gamma \hat{\mathbf{y}}_{1...N} + \boldsymbol{\beta} \tag{7.15.2}$$

with $\boldsymbol{\beta} = \beta \mathbf{1}$ and

$$\hat{\mathbf{y}}_{1...N} = \frac{\mathbf{y}_{1...N} - \mu_\mathcal{B} \mathbf{1}}{\sqrt{\sigma_\mathcal{B}^2 + \epsilon}}, \tag{7.15.3}$$

where

$$\mu_\mathcal{B} = \frac{1}{T} \sum_{n=1}^{N} y_t, \tag{7.15.4}$$

$$\sigma_\mathcal{B} = \sqrt{\frac{1}{N} \sum_{n=1}^{N} (y_n - \mu_\mathcal{B})^2}. \tag{7.15.5}$$

Then, $\hat{\mathbf{y}}_{1...N}$ is known as the *batch normalization* (BatchNorm) of the batch data $\mathbf{y}_{1...N}$, and the linear transform

$$\mathrm{BN}_{\gamma,\beta} : \mathbf{y}_{1...N} \to \hat{\mathbf{y}}_{1...N} \tag{7.15.6}$$

is referred to as the *batch normalizing transform*.

In the above definition, γ and β are, respectively, the scale and shift parameters, another parameter ϵ is a very small positive number for avoiding $\sigma_\mathcal{B} \approx 0$.

Equation (7.15.5) implies that the BatchNorm performs the normalization over a whole mini-batch. This is the reason why the normalization described by Eqs. (7.15.3) and (7.15.5) is named as the batch normalization.

Figure 7.16 shows a comparison between two networks (one has no BatchNorm layer and another has a BatchNorm layer).

Given feature vectors over a mini-batch: $\mathcal{B} = \{\mathbf{y}_1, \ldots, \mathbf{y}_N\}$. BatchNorm consists of the following steps [74]:

$$\mu_\mathcal{B} \leftarrow \frac{1}{N} \sum_{n=1}^{N} \mathbf{y}_n, \tag{7.15.7}$$

(a) Network

(b) Network + BatchNorm layer

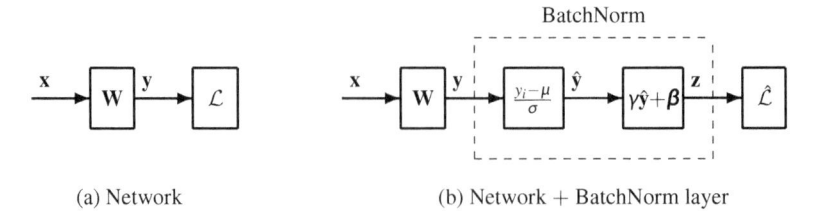

Fig. 7.16 Comparison between the two network architectures for a batch data: (**a**) the network with no BatchNorm layer; (**b**) the same network as in (**a**) with a BatchNorm layer inserted after the fully connected layer **W**. All the layer parameters have exactly the same value in both networks, and the two networks have the same loss function $\mathcal{L}$, i.e., $\hat{\mathcal{L}} = \mathcal{L}$

and

$$\sigma_{\mathcal{B}} \leftarrow \sqrt{\frac{\sum_{n=1}^{N}(\mathbf{y}_n - \boldsymbol{\mu}_{\mathcal{B}})^T(\mathbf{y}_n - \boldsymbol{\mu}_{\mathcal{B}})}{N}} + \epsilon, \tag{7.15.8}$$

$$\hat{\mathbf{y}}_n \leftarrow \frac{\mathbf{y}_n - \boldsymbol{\mu}_{\mathcal{B}}}{\sigma_{\mathcal{B}}}, \quad n = 1, \ldots, N, \tag{7.15.9}$$

$$\mathbf{z}_n \leftarrow \gamma \hat{\mathbf{y}}_n + \beta \mathbf{1} = \text{BN}(\mathbf{y}_n), \quad n = 1, \ldots, N. \tag{7.15.10}$$

Design the loss $\mathcal{L}(\mathbf{y}, \hat{\mathbf{y}}, \mathbf{z}, \gamma, \beta)$ corresponding to some application problem to be solved, where γ and β are the parameters to be learned in the batch normalization.

To learn two parameters γ and β, during training the gradient of loss $\mathcal{L}$ needs to backpropagate through linear transformation $\text{BN}_{\gamma,\beta} : \mathbf{y}_{1\ldots N} \to \hat{\mathbf{y}}_{1\ldots N}$ as follows [74]:

$$\frac{\partial \mathcal{L}}{\partial \hat{\mathbf{y}}_i} = \frac{\partial \mathcal{L}}{\partial \mathbf{z}_i} \cdot \gamma, \tag{7.15.11}$$

$$\frac{\partial \mathcal{L}}{\partial \sigma_{\mathcal{B}}^2} = \sum_{i=1}^{N} \frac{\partial \mathcal{L}}{\partial \hat{\mathbf{y}}_i}(\mathbf{y}_i - \boldsymbol{\mu}_{\mathcal{B}}) \cdot \frac{-1}{2}\left(\sigma_{\mathcal{B}}^2 + \epsilon\right)^{-3/2}, \tag{7.15.12}$$

$$\frac{\partial \mathcal{L}}{\partial \boldsymbol{\mu}_{\mathcal{B}}} = \left(\sum_{i=1}^{N} \frac{\partial \mathcal{L}}{\partial \hat{\mathbf{y}}_i} \cdot \frac{-1}{\sqrt{\sigma_{\mathcal{B}}^2 + \epsilon}}\right) + \frac{\partial \mathcal{L}}{\partial \sigma_{\mathcal{B}}^2} \cdot \frac{-\sum_{i=1}^{N} 2(\mathbf{y}_i - \boldsymbol{\mu}_{\mathcal{L}})}{N}, \tag{7.15.13}$$

$$\frac{\partial \mathcal{L}}{\partial \mathbf{y}_i} = \frac{\partial \mathcal{L}}{\partial \hat{\mathbf{y}}_i} \cdot \frac{1}{\sqrt{\sigma_{\mathcal{B}}^2 + \epsilon}} + \frac{\partial \mathcal{L}}{\partial \sigma_{\mathcal{B}}^2} \cdot \frac{2(\mathbf{y}_i - \boldsymbol{\mu}_{\mathcal{B}})}{N} + \frac{\partial \mathcal{L}}{\partial \boldsymbol{\mu}_{\mathcal{B}}} \cdot \frac{1}{N}, \tag{7.15.14}$$

$$\frac{\partial \mathcal{L}}{\partial \gamma} = \frac{1}{N}\sum_{i=1}^{N} \frac{\partial \mathcal{L}}{\partial \mathbf{z}_i} \cdot \hat{\mathbf{y}}_i = \frac{1}{N}\sum_{i=1}^{N}\left(\frac{\partial \mathcal{L}}{\partial \mathbf{z}_i}\right)^T \hat{\mathbf{y}}_i, \tag{7.15.15}$$

7.15 Batch Normalization Networks

and

$$\frac{\partial \mathcal{L}}{\partial \beta} = \frac{1}{N} \sum_{i=1}^{N} \left(\frac{\partial \mathcal{L}}{\partial \mathbf{z}_i}\right)^T \mathbf{1}. \tag{7.15.16}$$

During inference, the standard practice is to normalize the activations using the moving averages μ and σ^2 instead of mini-batch mean μ_B and variance σ_B^2 [74]:

$$\mathbf{z}_{\text{inference}} = \frac{\mathbf{y} - \mu}{\sigma} \cdot \gamma + \beta \mathbf{1} \tag{7.15.17}$$

which depends only on a single input example $\mathbf{y}$ rather than requiring a whole mini-batch.

Consider the optimization landscape with respect to the batch-normalized activations z_j.

- For the loss function $\mathcal{L}$ that is assumed to be Lipschitz continue, its Lipschitz constant L plays a crucial role in optimization, since it controls the amount by which the loss can change when taking a step.
- The gradient magnitude $\|\nabla_{z_j}\hat{\mathcal{L}}\|$ captures the Lipschitzness of the loss $\hat{\mathcal{L}}$.

Without any assumptions on the specific weights or the loss being used, Santurkar et al. [127] have showed that the optimization landscape with the batch-normalized activations z_j are more well-behaved, including favorable properties in Lipschitz-continuity, and predictability of the gradients.

The Effect of BatchNorm on the Lipschitzness

Theorem 7.4 *[[127]] For a BatchNorm network with loss $\hat{\mathcal{L}}$ and an identical non-BatchNorm network with (identical) loss $\mathcal{L}$, the following inequality is true:*

$$\|\nabla_{z_j}\hat{\mathcal{L}}\|^2 \leq \frac{\gamma^2}{\sigma_j^2} \left(\|\nabla_{z_j}\mathcal{L}\|^2 - \frac{1}{N}\langle \mathbf{1}, \nabla_{z_j}\mathcal{L}\rangle^2 - \frac{1}{\sqrt{N}}\langle \nabla_{z_j}\mathcal{L}, \hat{z}_j\rangle^2 \right). \tag{7.15.18}$$

The reduction of the gradient magnitude $\|\nabla_{z_i}\hat{\mathcal{L}}\| \leq \frac{\gamma^2}{\sigma_j^2}\|\nabla_{z_i}\mathcal{L}\|$ has an effect even when the scaling of BatchNorm is identical to the original layer scaling (i.e., even when $\gamma = \sigma_j$). Because the gradient magnitude $\|\nabla_{z_j}\hat{\mathcal{L}}\|$ captures the Lipschitzness of the loss $\hat{\mathcal{L}}$, BatchNorm exhibits a better Lipschitz constant of the loss $\mathcal{L}$.

The Effect of BatchNorm to Smoothness

Theorem 7.5 *[[127]] Let $\hat{\mathbf{g}}_j = \nabla_{z_j}\hat{\mathcal{L}}$ and $\mathbf{H}_{jj} = \frac{\partial \hat{\mathcal{L}}}{\partial \mathbf{z}_j \partial \mathbf{z}_j}$ be the gradient vector and Hessian matrix of the loss with respect to the layer outputs, respectively. Then*

$$(\nabla_{z_j}\hat{\mathcal{L}})^T \frac{\partial \hat{\mathcal{L}}}{\partial \mathbf{z}_j \partial \mathbf{z}_j} (\nabla_{z_j}\hat{\mathcal{L}}) \leq \frac{\gamma^2}{\sigma^2} (\nabla_{z_i}\hat{\mathcal{L}})^T \mathbf{H}_{jj} (\nabla_{z_j}\hat{\mathcal{L}}) - \frac{\gamma}{N\sigma^2} \langle \hat{\mathbf{g}}_j, \hat{\mathbf{z}}_j \rangle \left\| \frac{\partial \hat{\mathcal{L}}}{\partial \mathbf{z}_j} \right\|^2. \tag{7.15.19}$$

If the $\mathbf{H}_{jj}$ preserves also the relative norms of $\hat{\mathbf{g}}_j$ and $\nabla_{z_j}\hat{\mathcal{L}}$, then

$$(\nabla_{z_j}\hat{\mathcal{L}})^T \frac{\partial \hat{\mathcal{L}}}{\partial \mathbf{z}_j \partial \mathbf{z}_j} (\nabla_{z_j}\hat{\mathcal{L}}) \leq \frac{\gamma^2}{\sigma^2} \left(\hat{\mathbf{g}}_j^T \mathbf{H}_{jj} \hat{\mathbf{g}}_j - \frac{1}{N\gamma} \langle \hat{\mathbf{g}}_j, \hat{\mathbf{z}}_j \rangle \left\| \frac{\partial \hat{\mathcal{L}}}{\partial \mathbf{z}_j} \right\| \right). \tag{7.15.20}$$

The quadratic form of the loss Hessian matrix captures the second-order term of the Taylor expansion of the gradient around the current point. Therefore, if the quadratic forms involving the loss Hessian $\mathbf{H}_{jj}$ and the inner product $\langle \hat{\mathbf{y}}_j, \hat{\mathbf{g}}_j \rangle$ are nonnegative (both fairly mild assumptions), Theorem 7.5 implies that the quadratic form of the loss Hessian is reduced for a BatchNorm network than the standard networks, and thus the first-order term (gradient) is more predictive in BatchNorm networks.

BatchNorm Leads to a Favorable Initialization

Lemma 7.2 ([127]) *Let $\mathbf{W}^*$ and $\hat{\mathbf{W}}^*$ be the sets of local optima for the weights in the normal and BatchNorm networks, respectively. For any initialization $\mathbf{W}_0$, if $\langle \mathbf{W}^*, \mathbf{W}_0 \rangle > 0$, where $\hat{\mathbf{W}}^*$ and $\mathbf{W}^*$ are closet optima for BatchNorm and standard networks, respectively, then*

$$\|\mathbf{W}_0 - \hat{\mathbf{W}}^*\|^2 \leq \|\mathbf{W}_0 - \mathbf{W}^*\|^2 - \frac{1}{\|\mathbf{W}^*\|^2} \left(\|\mathbf{W}^*\|^2 - \langle \mathbf{W}^*, \mathbf{W}_0 \rangle \right)^2. \tag{7.15.21}$$

This lemma shows that $\|\mathbf{W}_0 - \hat{\mathbf{W}}^*\|^2 < \|\mathbf{W}_0 - \mathbf{W}^*\|^2$. That is, the effect of any initialization $\mathbf{W}_0$ to the closet optima $\hat{\mathbf{W}}^*$ for BatchNorm networks is smaller, as compared with standard networks. In other words, the initialization in optimization for BatchNorm networks is more favorable.

7.15.2 Variants and Extensions of Batch Normalization

As pointed by Ioffe [73], the dependence of the batch-normalized activations on the entire mini-batch makes BatchNorm powerful, but it is also the source of its drawbacks.

- When the training mini-batches are small, the estimates of the mean and variance become less accurate. These inaccuracies are compounded with depth, and will lead to its performance degradation.
- If the training mini-batches do not consist of independent samples, then different activations are produced between training and inference, which may lead to error in inference.

To address the above two problems, variants and extensions of batch normalizations were proposed. These variants include batch renormalization, layer normalization (LN), instance normalization (IN), group normalization (GN), and so on.

Batch Renormalization [73]

Given m mini-batches $\mathcal{B}_t, \ldots, \mathcal{B}_{t-m+1}$, where t denotes the current time. The moving averages of the mini-batch statistics $(\mu_\mathcal{B}, \sigma_\mathcal{B}^2)$, denoted as (μ, σ^2), are given by

$$\mu = \frac{1}{m} \sum_{i=1}^{m} \mu_{\mathcal{B}_{t-i+1}}, \tag{7.15.22}$$

$$\sigma = \frac{1}{m} \sum_{i=1}^{m} \sigma_{\mathcal{B}_{t-i+1}}. \tag{7.15.23}$$

If we use the moving averages (μ, σ^2) for normalizing a source data $\mathbf{y}$ to obtain its batch normalization $\hat{\mathbf{y}}$ and activation $\mathbf{z}$:

$$\hat{y}_i = \frac{y_i - \mu}{\sigma}, \quad z_i = \gamma \hat{y}_i + \beta, \tag{7.15.24}$$

there exists still a problem: when the source data and the target data have different distributions, the moving average normalization statistics of the source data still cannot represent the normalized statistics of the target (or testing) data for making inference of the testing data.

To address the above problem, consider using the mini-batch statistics (μ_B, σ_B^2) to make batch normalization and provide activation:

$$\hat{y}_i = \frac{y_i - \mu_B}{\sigma_B} \cdot r + d, \quad z_i = \gamma \hat{y}_i + \beta, \qquad (7.15.25)$$

where r and b are two hyperparameters to be determined.

Clearly, if selecting

$$r = \frac{\sigma_B}{\sigma}, \quad d = \frac{\mu_B - \mu}{\sigma}, \qquad (7.15.26)$$

then Eq. (7.15.25) is identical to Eq. (7.15.24).

To address the non-independently identical distributions and the mini-batch, Ioffe [73] proposed the parameters r and d in Eq. (7.15.25) determined by

$$r \leftarrow \text{stop\_gradient}\left(\text{clip}_{[1/r_{max}, r_{max}]}\left(\frac{\sigma_B}{\sigma}\right)\right), \qquad (7.15.27)$$

$$d \leftarrow \text{stop\_gradient}\left(\text{clip}_{[-d_{max}, d_{max}]}\left(\frac{\mu_B - \mu}{\sigma}\right)\right), \qquad (7.15.28)$$

where the values marked with stop_gradient are treated as constant for a given training step, the gradient is not propagated through them, and

$$\text{clip}_{[1/r_{max}, r_{max}]}\left(\frac{\sigma_B}{\sigma}\right) = \begin{cases} 1/r_{max}, & \text{if } \frac{\sigma_B}{\sigma} < 1/r_{max}; \\ \frac{\sigma_B}{\sigma}, & \text{if } 1/r_{max} \leq \frac{\sigma_B}{\sigma} \leq r_{max}; \\ r_{max}, & \text{if } \frac{\sigma_B}{\sigma} > r_{max}; \end{cases} \qquad (7.15.29)$$

$$\text{clip}_{[-d_{min}, d_{max}]}\left(\frac{\mu_B - \mu}{\sigma}\right) = \begin{cases} -d_{min}, & \text{if } \frac{\mu_B - \mu}{\sigma} < -d_{min}; \\ \frac{\mu_B - \mu}{\sigma}, & \text{if } -d_{min} \leq \frac{\mu_B - \mu}{\sigma} \leq d_{max}; \\ d_{max}, & \text{if } \frac{\mu_B - \mu}{\sigma} > d_{max}. \end{cases}$$

$$(7.15.30)$$

The batch normalization $\hat{y}_i$ based on r in (7.15.27) and d in (7.15.28), developed by Ioffe [73], is known as the *batch renormalization* of the feature data y_i.

Algorithm 7.11 shows the batch renormalization of Ioffe [73].

7.15 Batch Normalization Networks

Algorithm 7.11 Batch renormalization, applied to activation **y** over a mini-batch [73]

1. **input:** Feature vectors **y** over a training mini-batch $\mathcal{B} = \{\mathbf{y}_{1...m}\}$; parameters γ, β; current moving mean μ and standard deviation σ; moving average update rate ; maximum allowed correction r_{max}, d_{max}.
2. $\mu_\mathcal{B} \leftarrow \frac{1}{m}\sum_{i=1}^{m}\mathbf{y}_i$.
3. $\sigma_\mathcal{B} \leftarrow \sqrt{\epsilon + \frac{1}{m}\sum_{i=1}^{m}(\mathbf{y}_i - \mu_\mathcal{B})^T(\mathbf{y}_i - \mu_\mathcal{B})}$.
4. $r \leftarrow \text{stop\_gradient}\left(\text{clip}_{[1/r_{max},r_{max}]}\left(\frac{\sigma_\mathcal{B}}{\sigma}\right)\right)$.
5. $d \leftarrow \text{stop\_gradient}\left(\text{clip}_{[-d_{max},d_{max}]}\left(\frac{\mu_\mathcal{B}-\mu}{\sigma}\right)\right)$.
6. $\hat{\mathbf{y}}_i \leftarrow \frac{\mathbf{y}_i - \mu_\mathcal{B}}{\sigma_\mathcal{B}} \cdot r + d\mathbf{1}$.
7. $\mathbf{z}_i \leftarrow \gamma \hat{\mathbf{y}}_i + \beta \mathbf{1}$.
8. $\mu \leftarrow \mu + \alpha(\mu_\mathcal{B} - \mu)$.   // Update moving averages
9. $\sigma \leftarrow \sigma + \alpha(\sigma_\mathcal{B} - \sigma)$.
10. **output:** $\mathbf{z}_i = \text{BatchRenorm}(\mathbf{y}_i)$; updated μ, σ.
11. Inference: $\mathbf{z} \leftarrow \gamma \cdot \frac{\mathbf{y}-\mu}{\sigma} + \beta \mathbf{1}$.

This algorithm updates the exponentially decayed moving averages μ and σ, and optimizes the rest of the model using gradient optimization, with the gradients calculated via backpropagation:

$$\frac{\partial \mathcal{L}}{\partial \hat{\mathbf{y}}_i} = \frac{\partial \mathcal{L}}{\partial \mathbf{z}_i} \cdot \gamma, \tag{7.15.31}$$

$$\frac{\partial \mathcal{L}}{\partial \sigma_\mathcal{B}} = \left(\sum_{i=1}^{m} \frac{\partial \mathcal{L}}{\partial \hat{\mathbf{y}}_i}\right)^T (\mathbf{y}_i - \mu_\mathcal{B}) \cdot \frac{-r}{\sigma^2}, \tag{7.15.32}$$

$$\frac{\partial \mathcal{L}}{\partial \mu_\mathcal{B}} = \sum_{i=1}^{m} \frac{\partial \mathcal{L}}{\partial \hat{\mathbf{y}}_i} \cdot \frac{-r}{\sigma_\mathcal{B}}, \tag{7.15.33}$$

$$\frac{\partial \mathcal{L}}{\partial \mathbf{y}_i} = \frac{\partial \mathcal{L}}{\partial \hat{\mathbf{y}}_i} \cdot \frac{r}{\sigma_\mathcal{B}} + \frac{\partial \mathcal{L}}{\partial \sigma_\mathcal{B}} \cdot \frac{\mathbf{y}_i - \mu_\mathcal{B}}{m\sigma_\mathcal{B}} + \frac{\partial \mathcal{L}}{\partial \sigma_\mathcal{B}} \cdot \frac{1}{m}\mathbf{1}, \tag{7.15.34}$$

$$\frac{\partial \mathcal{L}}{\partial \gamma} = \left(\sum_{i=1}^{m} \frac{\partial \mathcal{L}}{\partial \mathbf{z}_i}\right)^T \hat{\mathbf{y}}_i, \tag{7.15.35}$$

$$\frac{\partial \mathcal{L}}{\partial \beta} = \left(\sum_{i=1}^{m} \frac{\partial \mathcal{L}}{\partial \mathbf{z}_i}\right)^T \mathbf{1}. \tag{7.15.36}$$

Consider batch normalization for convolutional neural networks, the input and output (activation) of a BatchNorm layer are four-dimensional tensors $\mathcal{Y}, \mathcal{Z} \in \mathbb{R}^{N \times C \times H \times W}$ with the elements y_{nijk} and z_{nijk}, respectively. Here, n is the index of image in mini-batch, i is the index of channel, j and k span the spatial height and width dimensions, respectively. For input images, the channels correspond to the RGB channels. Then, N is the numbers of input images in each batch, C is the number of feature channels, H and W are the spatial height and width of the activation map in the layer, respectively.

Layer Normalization (LN) [4]

Changes in the output of one layer will tend to cause highly correlated changes in the summed inputs to the next layer, especially with ReLU units whose outputs can change by a lot. To reduce this "covariate shift" problem, a simple but efficient way is to fix the mean and the variance of the summed inputs within each layer.

The normalization over all the hidden units in the same layer is called the *layer normalization* (LN). Two-layer normalization statistics are computed as follows:

$$\mu_{n,i} = \frac{1}{HW} \sum_{j=1}^{H} \sum_{k=1}^{W} y_{nijk}, \qquad (7.15.37)$$

$$\sigma_{n,i} = \sqrt{\frac{1}{HW} \sum_{j=1}^{H} \sum_{k=1}^{W} (y_{nijk} - \mu_i)^2} \qquad (7.15.38)$$

for $i = 1, \ldots, C$. This shows that the LN performs the normalization over C channels and is named as the layer normalization.

Instance Normalization (IN) [148]

For the nth image in a mini-batch, two normalization statistics are given by

$$\mu_n = \frac{1}{CHW} \sum_{i=1}^{C} \sum_{j=1}^{H} \sum_{k=1}^{W} y_{nijk}, \qquad (7.15.39)$$

$$\sigma_n = \sqrt{\frac{1}{CHW} \sum_{i=1}^{C} \sum_{j=1}^{H} \sum_{k=1}^{W} (y_{nijk} - \mu_n)^2}. \qquad (7.15.40)$$

This shows that IN performs the normalization of each image (i.e., instance), and is named as the *instance normalization* (IN).

The instance normalization is also known as "contrast normalization."

Group Normalization (GN) [161]

Different from BN, LN, and IN, *group normalization* (GN) [161] divides C channels into G groups, and each group contains C/G channels. GN computes the mean and variance within each group for normalization as follows:

$$\mu_n = \frac{1}{(C/G)HW} \sum_{i=1}^{C/G} \sum_{j=1}^{H} \sum_{k=1}^{W} y_{nijk}, \quad (7.15.41)$$

$$\sigma_n = \sqrt{\frac{1}{(C/G)HW} \sum_{i=1}^{C/G} \sum_{j=1}^{H} \sum_{k=1}^{W} (y_{nijk} - \mu_n)^2}. \quad (7.15.42)$$

The channels of visual representations are not entirely independent. Therefore, group-wise representations are widely used in visual representation, for example,

- the *Scale Invariant Feature Transform* (SIFT) [100] transforms image data into scale-invariant coordinates relative to local features.
- the grids of *Histograms of Oriented Gradient* (HOG) descriptors [20] significantly outperform existing feature sets for human detection.

In these applications, each group of channels is constructed by some kind of histogram. These features are often processed by group-wise normalization over each histogram or each orientation [161].

GN is related to LN and IN as follows:

- If $G = 1$, then Eqs. (7.15.41) and (7.15.42) reduce to Eqs. (7.15.39) and (7.15.40), i.e., the group normalization for $G = 1$ is identical to the instance normalization.
- When $G = C$, Eqs. (7.15.41) and (7.15.42) become Eqs. (7.15.37) and (7.15.38), i.e., the group normalization for $G = C$ is identical to the layer normalization.

A common characteristic of LN, IN, and GN is that their normalization is independent of the batch size N.

The normalization statistics in BN, LN, IN, and GN can be unified as

$$\mu_l = \frac{1}{m} \sum_{p \in S} y_{nijk}, \quad (7.15.43)$$

$$\sigma_l = \sqrt{\frac{1}{m} \sum_{p \in S} (y_{nijk} - \mu_l)^2}, \quad (7.15.44)$$

where S is some designed subset of $\{n, i, j, k\}$ and l is the index of $\{n, i, j, k\} \setminus S$.

- BN: $S = \{p\} = \{1, \ldots, N\}$ and $m = |S| = N$ are taken. In this case, $\sum_{p \in S} = \sum_{n=1}^{N}$ and the two normalization statistics in (7.15.43) and (7.15.44) are, respectively, given by

$$\mu_{\text{BN}}(i, j, k) = \frac{1}{N} \sum_{n=1}^{N} y_{nijk}, \tag{7.15.45}$$

$$\sigma_{\text{BN}}(i, j, k) = \sqrt{\frac{1}{N} \sum_{n=1}^{N} (y_{nijk} - \mu_{\text{BN}}(i, j, k))^2}, \tag{7.15.46}$$

which are just the two normalization statistics in batch normalization.
- LN: $S = \{p\} = (H, W)$ and $m = HW$. In this case, $\sum_{p \in S} = \sum_{j=1}^{H} \sum_{k=1}^{W}$ and the two normalization statistics in (7.15.43) and (7.15.44) are, respectively, given by Eqs. (7.15.37) and (7.15.38).
- IN: $S = \{p\} = (C, H, W)$ and $m = CHW$. In this case, $\sum_{p \in S} = \sum_{i=1}^{C} \sum_{j=1}^{H} \sum_{k=1}^{W}$ and the two normalization statistics in (7.15.43) and (7.15.44) are, respectively, given by Eqs. (7.15.39) and (7.15.40).
- GN: $S = \{p\} = (C/G, H, W)$ and $m = (C/G)HW$. In this case, $\sum_{p \in S} = \sum_{i=1}^{C/G} \sum_{j=1}^{H} \sum_{k=1}^{W}$ and the two normalization statistics in (7.15.43) and (7.15.44) are, respectively, given by Eqs. (7.15.41) and (7.15.42).

7.16 Generative Adversarial Networks (GANs)

In the context of pattern recognition (e.g., face recognition) and computer vision, there are the practically unlimited amount of reusable images and videos from large unlabeled data sets. It has been active research to learn reusable feature representations from large unlabeled data sets. *Generative Adversarial Networks* (GANs) [38] is such a research area.

7.16.1 Generative Adversarial Network Framework

As one type of deep learning model, GANs have emerged as a powerful framework for learning generative models of arbitrarily complex data distributions. GAN plays an adversarial game with two linked models: *Generative Model* and *Discriminative Model*. The loss function in discriminant models is easily defined because its output objective is relatively simple. But for generating models, the definition of loss function is not so easy. A simple and efficient choice for defining the loss function of generative models is to use a discriminant model as the feedback in the generative model. In this way, a generative model and a discriminative model are closely

7.16 Generative Adversarial Networks (GANs)

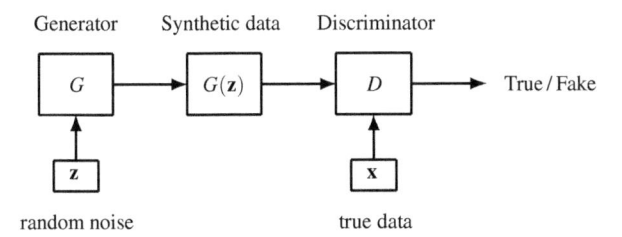

Fig. 7.17 The basic structure of generative adversarial networks (GANs). The generator G uses a random noise $\mathbf{z}$ to generate a synthetic data $\hat{\mathbf{x}} = G(\mathbf{z})$, and the discriminator D tries to identify whether the synthesized data $\hat{\mathbf{x}}$ is a real data $\mathbf{x}$, i.e., making a real or fake inference

connected. This is the basic idea of Goodfellow's GANs [36]. In other words, the basic idea of GANs is to set up a game between two players. One of them is called the *generator G*, and another player is the *adversarial discriminator D*.

Let a set of data $\{\mathbf{x}_1, \ldots, \mathbf{x}_m\}$ be sampled from the real distribution $P_{\text{data}}(\mathbf{x})$ (e.g., natural images), and let $P_{\mathbf{z}}(\mathbf{z})$ be a prior on input noise variables $\mathbf{z}$ (e.g., observed images). The basic structure of GANs is shown in Fig. 7.17.

The framework of GAN is as follows:

- *Generator* is simply defined by a differentiable function G that takes $\mathbf{z}$ as input and uses $\boldsymbol{\theta}_G$ as parameters. When $\mathbf{z} = (\mathbf{x}_i; \boldsymbol{\theta})$ is sampled from some simple prior distribution, $G(\mathbf{z}) = G(\mathbf{x}_i; \boldsymbol{\theta})$ yields the i.i.d. samples $\mathbf{x}_i$ drawn from $P_{\text{model}} = P_G$. The generator wishes to find the maximum likelihood estimate

$$\begin{aligned}
\boldsymbol{\theta}_G^* &= \arg\max_{\boldsymbol{\theta}} \log \prod_{i=1}^{m} P_G(\mathbf{x}_i; \boldsymbol{\theta}) \\
&= \arg\max_{\boldsymbol{\theta}} \sum_{i=1}^{m} \log P_G(\mathbf{x}_i; \boldsymbol{\theta}) \\
&\approx \arg\max_{\boldsymbol{\theta}} E_{\mathbf{x} \sim P_{\text{data}}}[\log P_G(\mathbf{x}_i; \boldsymbol{\theta})] \\
&= \arg\max_{\boldsymbol{\theta}} \int_{\mathbf{x}} P_{\text{data}}(\mathbf{x}) \log P_G(\mathbf{x}; \boldsymbol{\theta}) d\mathbf{x} - \arg\max_{\boldsymbol{\theta}} P_{\text{data}}(\mathbf{x}) \int_{\mathbf{x}} \log P_{\text{data}}(\mathbf{x}) d\mathbf{x} \\
&= \arg\max_{\boldsymbol{\theta}} \int_{\mathbf{x}} P_{\text{data}}(\mathbf{x}) [\log P_G(\mathbf{x}; \boldsymbol{\theta}) - \log P_{\text{data}}(\mathbf{x})] d\mathbf{x} \\
&= \arg\min_{\boldsymbol{\theta}} \int_{\mathbf{x}} P_{\text{data}}(\mathbf{x}) \log \frac{P_{\text{data}}(\mathbf{x})}{P_G(\mathbf{x}; \boldsymbol{\theta})} d\mathbf{x} \\
&= \arg\min_{\boldsymbol{\theta}} \text{KL}(P_{\text{data}}(\mathbf{x}) \| P_G(\mathbf{x}; \boldsymbol{\theta})). \quad (7.16.1)
\end{aligned}$$

The generator G tries to convince the discriminator that the generated samples come from the prior data distribution. The generator can be thought of as being like a counterfeiter, trying to make fake money.
- *Discriminator* is a function D that takes a sample $\mathbf{x}$ as input and uses $\boldsymbol{\theta}_D$ as parameters, and tries to discriminate whether an input sample comes from the data distribution or from the built generator, namely distinguish between real and generated samples as accurately as possible. The discriminator can be thought of as being like police, trying to allow legitimate money and catch counterfeit money.

The GAN framework learns a generator mapping samples from an arbitrary latent distribution to data, as well as an adversarial discriminator which tries to distinguish between real and generated samples as accurately as possible. The generator's goal is to "fool" the discriminator by producing samples which are as close to real data as possible [23].

Because each player's cost depends on the other player's parameters, but each player cannot control the other player's parameters, this scenario is most straightforward to describe as a game rather than as an optimization problem [36]. The solution to an optimization problem is a (local) minimum that is a point in parameter space where all neighboring points have greater or equal cost, while the solution to a game is a *Nash equilibrium*. In the GAN framework, the terminology of local differential Nash equilibria [118] is used. In this context, a Nash equilibrium is a tuple $(\boldsymbol{\theta}_D, \boldsymbol{\theta}_G)$ that is a local minimum of J_D with respect to $\boldsymbol{\theta}_D$ and a local minimum of J_G with respect to $\boldsymbol{\theta}_G$.

If defining a probability distribution $P_G(\mathbf{z})$ as the distribution of the samples $G(\mathbf{z})$ obtained when $\mathbf{z} \sim P_{\mathbf{z}}(\mathbf{z})$, then the probability distribution $P_G(\mathbf{z})$ becomes $P_G(\mathbf{x})$ when the generator G outputs $\mathbf{x}$ from the input $\mathbf{z}$, i.e., $G(\mathbf{z}) = \mathbf{x}$. In other words, $P_G(\mathbf{x})$ can be viewed as the probability distribution of $\mathbf{x}$ generated by the generator G, while $P_{\text{data}}(\mathbf{x})$ is defined as the data generating distribution.

On the optimal discriminator $D_G^*(\mathbf{x})$, Goodfellow et al. [38] proved the following proposition.

Proposition 7.1 *For G fixed, the optimal discriminator D is given by*

$$D_G^*(\mathbf{x}) = \frac{P_{\text{data}}(\mathbf{x})}{P_{\text{data}}(\mathbf{x}) + P_G(\mathbf{x}; \boldsymbol{\theta})}. \qquad (7.16.2)$$

By the definitions of both $P_G(\mathbf{x})$ and $P_{\text{data}}(\mathbf{x})$, we have the following results:
- $P_G(\mathbf{x}; \boldsymbol{\theta}) \leq P_{\text{data}}(\mathbf{x})$.
- $E_{\mathbf{x} \sim P_{\text{data}}(\mathbf{x})}\{\log D(\mathbf{x})\}$ denotes the log-probability of the discriminator D being correct.
- $E_{\mathbf{z} \sim P_{\text{data}}(\mathbf{z})}\{\log(1 - D(G(\mathbf{z})))\} = E_{\mathbf{x} \sim P_G(\mathbf{x}; \boldsymbol{\theta})}\{\log(1 - D(\mathbf{x}))\}$ represents the log-probability of discriminator D being mistaken.

In the GAN, the discriminator D is trained to maximize the log-probability $E_{\mathbf{x} \sim P_{\text{data}}(\mathbf{x})}\{\log D(\mathbf{x})\}$ and the generator G is simultaneously trained to minimize the log-probability $E_{\mathbf{z} \sim P_{\mathbf{z}}(\mathbf{z})}\{\log(1 - D(G(\mathbf{z})))\}$ of discriminator D being mistaken [38]:

$$\min_{G} \max_{D} V(D, G) = E_{\mathbf{x} \sim P_{\text{data}}(\mathbf{x})} \big\{ \log D(\mathbf{x}) \big\}$$
$$+ E_{\mathbf{z} \sim P_{\mathbf{z}}(\mathbf{z})} \big\{ \log(1 - D(G(\mathbf{z}))) \big\}. \quad (7.16.3)$$

That is to say, the discriminator D and the generator G play a two-player minimax game with value function $V(D, G)$, their solution involves minimization with respect to G in an outer loop and maximization with respect to D in an inner loop. Importantly, the minimax game is mostly of interest because it is easily amenable to theoretical analysis [36].

Mini-batch stochastic gradient descent training of GANs for optimizing (7.16.3) is shown in Algorithm 7.12.

Algorithm 7.12 Minibatch stochastic gradient descent training of GANs [38]

1. **input:** The number k of steps to apply to the discriminator.
2. **for** number of training iterations **do**
3.   **for** k steps **do**
4.   Sample mini-batch of m noise samples $\{\mathbf{z}_1, \ldots, \mathbf{z}_m\}$ from noise prior $P_g(\mathbf{z})$.
5.   Sample mini-batch of m examples $\{\mathbf{x}_1, \ldots, \mathbf{x}_m\}$ from data generating distribution $P_{\text{data}}(\mathbf{x})$.
6.   Update the discriminator by ascending its stochastic gradient:
   $$\nabla \theta_d \frac{1}{m} \sum_{i=1}^{m} \big[\log D(\mathbf{x}_i) + \log((1 - D)(G(\mathbf{z}_i))) \big].$$
7.   **end for**
8.   Sample mini-batch of m noise samples $\{\mathbf{z}_1, \ldots, \mathbf{z}_m\}$ from noise prior $P_g(\mathbf{z})$.
9.   Update the generator by descending its stochastic gradient:
   $$\nabla \theta_g \frac{1}{m} \sum_{i=1}^{m} \log(1 - D(G(\mathbf{z}_i))).$$
10. **end for**

The training objective for D can be interpreted as maximizing the log-likelihood for estimating the conditional probability $p(Y = y|\mathbf{x})$, where Y indicates whether $\mathbf{x}$ comes from $P_{\text{data}}(\mathbf{x})$ (with $y = 1$) or from $P_G(\mathbf{x}; \boldsymbol{\theta})$ (with $y = 0$). Denoting

$$C(G) = \max_{D} V(D, G)$$
$$= E_{\mathbf{x} \sim P_{\text{data}}(\mathbf{x})} \left\{ \log \frac{P_{\text{data}}(\mathbf{x})}{P_{\text{data}}(\mathbf{x}) + P_G(\mathbf{x}; \boldsymbol{\theta})} \right\}$$
$$+ E_{\mathbf{x} \sim P_G(\mathbf{x}; \boldsymbol{\theta})} \left\{ \log \frac{P_G(\mathbf{x}; \boldsymbol{\theta})}{P_{\text{data}}(\mathbf{x}) + P_G(\mathbf{x}; \boldsymbol{\theta})} \right\} \quad (7.16.4)$$

then the minimax game in Eq. (7.16.3) can be reformulated as [38]:

$$\min_{G} C(G) = \min_{G} \max_{D} V(D, G). \quad (7.16.5)$$

Theorem 7.6 ([38]) *The global minimum of the virtual training criterion $C(G)$ is achieved if and only if $P_G(\mathbf{x}; \boldsymbol{\theta}) = P_{\text{data}}(\mathbf{x})$. At that point, $C(G)$ achieves the value $-\log 4$.*

Consider generative models that work via the principle of maximum likelihood (ML). Here, the likelihood is referred to as the probability that the model assigns to the training data [36]: $\prod_{i=1}^{m} P_G(\mathbf{x}_i, \boldsymbol{\theta})$, for a data set containing m training examples $\mathbf{x}_i$. The principle of maximum likelihood simply says to choose the parameters $\boldsymbol{\theta}$ for the model that maximize the likelihood of the training data $\mathbf{x}_1, \ldots, \mathbf{x}_m$.

As shown in Eq. (7.16.1), the ML estimation is to minimize the KL divergence between the data generating distribution $P_{\text{data}}(\mathbf{x})$ and the model distribution $P_G(\mathbf{x}; \boldsymbol{\theta})$. If this ML estimate can be found and $P_{\text{data}}(\mathbf{x})$ lies within the family of distributions $P_G(\mathbf{x}; \boldsymbol{\theta})$, then the model would recover $P_{\text{data}}(\mathbf{x})$ exactly. However, $P_{\text{data}}(\mathbf{x})$ itself can only be estimated from a training set consisting of m samples. This estimate is denoted as $\hat{P}_{\text{data}}(\mathbf{x})$, and it is an empirical distribution for approximating $P_{\text{data}}(\mathbf{x})$. Minimizing the KL divergence between $\hat{P}_{\text{data}}(\mathbf{x})$ and $P_G(\mathbf{x}; \boldsymbol{\theta})$ is exactly equivalent to maximizing the log-likelihood of the training set.

Let $V(\boldsymbol{\theta}_D, \boldsymbol{\theta}_G)$ be a value function specifying the discriminator's payoff $J_D(\boldsymbol{\theta}_D, \boldsymbol{\theta}_G)$, i.e.,

$$V(\boldsymbol{\theta}_D, \boldsymbol{\theta}_G) = -J_D(\boldsymbol{\theta}_D, \boldsymbol{\theta}_G). \tag{7.16.6}$$

Then the solution of the generator G involves minimization with respect to $\boldsymbol{\theta}_G$ in an outer loop and maximization with respect to $\boldsymbol{\theta}_D$ in an inner loop [36]:

$$\boldsymbol{\theta}_G^* = \arg\min_{\boldsymbol{\theta}_G} \max_{\boldsymbol{\theta}_D} V(\boldsymbol{\theta}_D, \boldsymbol{\theta}_G). \tag{7.16.7}$$

In the original GAN theory, G and D are not neural networks, but only functions that can be generated and distinguished. But in practice, deep neural networks are generally used for G and D.

7.16.2 Bidirectional Generative Adversarial Networks

The GANs framework can learn generative models mapping from simple latent distributions to arbitrarily complex data distributions. However, in many applications, it is important as well for GANs to have the ability of learning the inverse mapping: projecting data back into the latent space. A natural question that arises when using the GAN framework is: can GANs be used for unsupervised learning of rich feature representations for arbitrary data distributions?

GANs with the forward mapping and the inverse mapping are known as *Bidirectional Generative Adversarial Networks* (BiGANs) and were developed by Donahue et al. [23].

7.16 Generative Adversarial Networks (GANs)

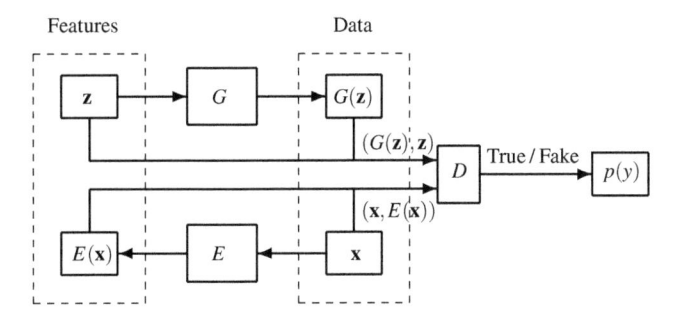

Fig. 7.18 Bidirectional generative adversarial network (BiGAN) is a combination of a standard GAN and an Encoder [23]. The generator G from the standard GAN framework maps latent samples $\mathbf{z}$ to generated data $G(\mathbf{z})$. An encoder E maps data $\mathbf{x}$ to the output $E(\mathbf{x})$. Both the generator tuple $(G(\mathbf{z}), \mathbf{z})$ and the encoder tuple $(\mathbf{x}, E(\mathbf{x}))$, act, in a bidirectional way, as inputs to the discriminator D

As shown in Fig. 7.18, the overall model of a BiGNN consists of two models:

- The generator G from the standard GAN framework [38] maps latent samples $\mathbf{z}$ to generated data $G(\mathbf{z})$. Then, the generative tuple $(G(\mathbf{z}), \mathbf{z})$, as one of two input sets, inputs to the discriminator D.
- An encoder E maps the data $\mathbf{x}$ to the output $E(\mathbf{x})$. The encoder tuple $(\mathbf{x}, E(\mathbf{x}))$, the other of the two input sets, inputs to the discriminator D. According to the two sets of inputs $(G(\mathbf{z}), \mathbf{z})$ and $(\mathbf{x}, E(\mathbf{x}))$, the discriminator D produces the inferring output on prior data distribution $p(y)$.

The BiGAN discriminator D discriminates not only in data space ($\mathbf{x}$ versus $G(\mathbf{z})$), but jointly in data and latent space (tuples $(\mathbf{x}, E(\mathbf{x}))$ versus $(G(\mathbf{z}), \mathbf{z})$), where the latent component is either an encoder output $E(\mathbf{x})$ or a generator input $\mathbf{z}$. When the BiGAN encoder E learns to invert the generator G, the two modules cannot directly "communicate" with one another: the encoder E never "sees" the generator's outputs $G(\mathbf{z})$, namely $E(G(\mathbf{z}))$ is not computed, and vice versa.

Let $\mathbf{x} \in \Omega_{\mathbf{x}}$ and $\mathbf{z} \in \Omega_{\mathbf{z}}$, where $\Omega_{\mathbf{x}}$ and $\Omega_{\mathbf{z}}$ are called the data space and the latent space, respectively. The BiGAN not only trains a generator, but additionally trains an encoder $E: \Omega_{\mathbf{x}} \to \Omega_{\mathbf{z}}$, and the training objective is defined as a minimax objective [23]:

$$\min_{G,E} \max_{D} V(G, E, D) = E_{\mathbf{x} \sim P_{\mathbf{x}}} \{ \underbrace{E_{\mathbf{z} \sim P_E(\cdot|\mathbf{x})} \{\log D(\mathbf{x}, \mathbf{z})\}}_{\log D(\mathbf{x}, E(\mathbf{x}))} \}$$
$$+ E_{\mathbf{z} \sim P_{\mathbf{z}}} \{ \underbrace{E_{\mathbf{x} \sim P_G(\cdot|\mathbf{z})} \{\log(1 - D(\mathbf{x}, \mathbf{z}))\}}_{\log(1 - D(G(\mathbf{z}), \mathbf{z}))} \}. \quad (7.16.8)$$

In visual feature learning, the encoder E may take higher resolution input, while the generator output and discriminator input remain low resolution. Hence, it is often useful to parametrize the output of the generator G and the input of the encoder E in different, usually smaller, spaces $\Omega'_\mathbf{x}$ and $\Omega'_\mathbf{z}$ rather than the original data space $\Omega_\mathbf{x}$ and the original latent space $\Omega_\mathbf{z}$.

Introduce

- two generalized functions $g(\mathbf{x}) : \Omega_\mathbf{x} \to \Omega'_\mathbf{x}$ and $g(\mathbf{z}) : \Omega_\mathbf{z} \to \Omega'_\mathbf{z}$,
- the generalized encoder $E : \Omega_\mathbf{x} \to \Omega'_\mathbf{z}$,
- the generalized generator $G : \Omega_\mathbf{z} \to \Omega'_\mathbf{x}$,
- the generalized discriminator $D : \Omega'_\mathbf{x} \times \Omega'_\mathbf{z} \to [0, 1]$.

Therefore, the BiGAN objective $V(G, E, D)$ can be generalized to [23]:

$$V(G, E, D) = E_{\mathbf{x} \sim P_\mathbf{x}} \Big\{ \underbrace{E_{\mathbf{z}' \sim P_E(\cdot|\mathbf{x})} \{\log D(g(\mathbf{x}), \mathbf{z}')\}}_{\log D(g(\mathbf{x}), E(\mathbf{x}))} \Big\}$$
$$+ E_{\mathbf{z} \sim P_\mathbf{z}} \Big\{ \underbrace{E_{\mathbf{x}' \sim P_G(\cdot|\mathbf{z})} \{\log(1 - D(\mathbf{x}', g(\mathbf{z})))\}}_{\log(1 - D(G(\mathbf{z}), g(\mathbf{z})))} \Big\}. \qquad (7.16.9)$$

Clearly, if $g(\mathbf{x}) = \mathbf{x}$ and $g(\mathbf{z}) = \mathbf{z}$ (and thus $\Omega'_\mathbf{x} = \Omega_\mathbf{x}$ and $\Omega'_\mathbf{z} = \Omega_\mathbf{z}$), then the above generalized BiGAN objective yields the original BiGAN objective in (7.16.8).

7.16.3 Variational Autoencoders

In the former two subsections we deal with two methods for generating datapoint $\mathbf{x}$ from a noise vector $\mathbf{z}$: standard GAN and BiGAN. Now, we focus on an alternative method: *Variational Autoencoders* (VAEs). They are called "autoencoders" only because the final training objective that derives from this setup does have an encoder and a decoder, and resembles a traditional autoencoder [22]. Figure 7.19 shows the construction of VAEs,

Fig. 7.19 A variational autoencoder (VAEs) connected with a standard generative adversarial network (GAN). Unlike the BiGAN in which the encoder's output $E(\mathbf{x})$ is one of two bidirectional inputs, the autoencoder E here executes variation $\mathbf{z} \sim E(\mathbf{x})$ that is used as the random noise in GAN

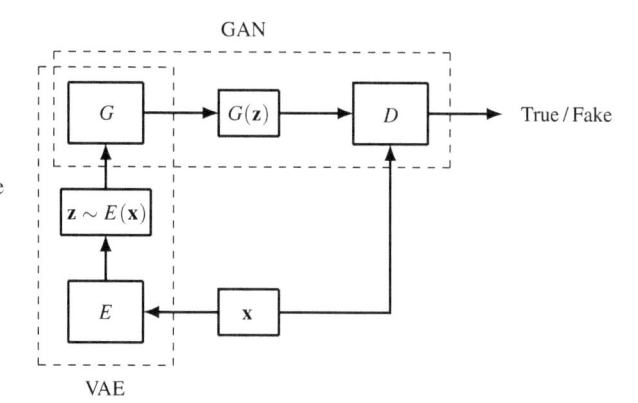

7.16 Generative Adversarial Networks (GANs)

"Generative modeling" is a broad area of machine learning which deals with models of distributions $p(\mathbf{x})$, where $\mathbf{x} \in X$ which is a potentially high-dimensional space. For images, a popular kind of data, each "datapoint" (image) has thousands or millions of dimensions (pixels), and the generative model has to somehow capture the dependencies between pixels, e.g., that nearby pixels have similar color, and are organized into objects [22].

Let $f_\theta(\mathbf{z})$ be a family of random functions in the generator G, parameterized by a fixed generative model parameter vector $\boldsymbol{\theta}$ in some space Θ. Consider how to use $f : Z \times \Theta \rightarrow X$ for generating a datapoint $\mathbf{x} = f_\theta(\mathbf{z}) \in X$ when $\mathbf{z}$ is a random noise vector in the space Z. Our aim is to generate a datapoint according to

$$\mathbf{x} = \arg\max_{\mathbf{x}} \left\{ P_\theta(\mathbf{x}) = \int_\mathbf{z} f_\theta(\mathbf{z}) P_\theta(\mathbf{z}) d\mathbf{z} \right\}. \tag{7.16.10}$$

Because the random function $f_\theta(\mathbf{z})$ is difficult to determine, a reasonable solution is to use a distribution $P_\theta(\mathbf{x}|\mathbf{z})$ instead of $f_\theta(\mathbf{z})$, which allows us to make the dependence of $\mathbf{x}$ on $\mathbf{z}$ explicit by using the law of total probability.

Then, the maximization in (7.16.10) becomes the maximization of the probability of each $\mathbf{x}$ in the training set under the entire generative process, according to [22]:

$$\mathbf{x} = \arg\max_{\mathbf{x}} \left\{ P_\theta(\mathbf{x}) = \int_\mathbf{z} P_\theta(\mathbf{x}|\mathbf{z}) P_\theta(\mathbf{z}) d\mathbf{z} \right\}. \tag{7.16.11}$$

To solve Eq. (7.16.11), samples of $\mathbf{z}$ can be drawn from a standard Gaussian distribution $N(\mathbf{0}, \mathbf{I})$.

Let the prior over the latent variables be the centered isotropic multivariate Gaussian $P_\theta(\mathbf{z}) = N(\mathbf{z}; \mathbf{0}, \mathbf{I})$, and $P_\theta(\mathbf{x}|\mathbf{z})$ be a multivariate Gaussian (in case of real-valued data) or Bernoulli (in case of binary data) whose distribution parameters are computed from $\mathbf{z}$ with a multilayered perceptron (MLP) (a fully connected neural network with a single-hidden layer) as follows [79]:

- *Bernoulli MLP as decoder:* a multivariate Bernoulli probability $P_\theta(\mathbf{x}|\mathbf{z})$ is computed from $\mathbf{z}$ with a fully connected neural network with a single-hidden layer:

$$\log P_\theta(\mathbf{x}|\mathbf{z}) = \sum_{i=1}^{D} x_i \log y_i + (1 - x_i) \log(1 - y_i), \tag{7.16.12}$$

where

$$\mathbf{y} = f_\sigma(\mathbf{W}_2 \tanh(\mathbf{W}_1 + \mathbf{z} + \mathbf{b}_1) + \mathbf{b}_2) \tag{7.16.13}$$

with f_σ being the elementwise sigmoid activation function, and where $\boldsymbol{\theta} = \{\mathbf{W}_1, \mathbf{W}_2, \mathbf{b}_1, \mathbf{b}_2\}$ are the weights and biases of the MLP.

- *Gaussian MLP as encoder or decoder:* the encoder or decoder is a multivariate Gaussian with a diagonal covariance structure:

$$\log P_\theta(\mathbf{x}|\mathbf{z}) = \log N(\mathbf{x}; \boldsymbol{\mu}, \sigma^2 \mathbf{I}), \quad (7.16.14)$$

where

$$\boldsymbol{\mu} = \mathbf{W}_4 \mathbf{h} + \mathbf{b}_4 \quad (7.16.15)$$

$$\log \sigma^2 = \|\mathbf{W}_5 \mathbf{h} + \mathbf{b}_5\|_2 \quad (7.16.16)$$

$$\mathbf{h} = \tanh(\mathbf{W}_3 \mathbf{z} + \mathbf{b}_3), \quad (7.16.17)$$

where $\{\mathbf{W}_3, \mathbf{W}_4, \mathbf{W}_5, \mathbf{b}_3, \mathbf{b}_4, \mathbf{b}_5\}$ are the weights and biases of the MLP and are a part of θ when used as a decoder. Note that when this network is used as an encoder $q_\phi(\mathbf{z}|\mathbf{x})$, then $\mathbf{z}$ and $\mathbf{x}$ are swapped, and the weights and biases are variational parameters ϕ.

However, when using samples $\mathbf{z} \sim N(\mathbf{z}; \mathbf{0}, \mathbf{I})$ to compute Eq. (7.16.11), we have the following two problems [22]:

- If a large number of $\mathbf{z}$ values $\{\mathbf{z}_1, \ldots, \mathbf{z}_n\}$ is sampled from $N(\mathbf{z}; \mathbf{0}, \mathbf{I})$, then $P_\theta(\mathbf{x})$ can be approximately computed by $P_\theta(\mathbf{x}) \approx \frac{1}{n} \sum_{i=1}^{n} P_\theta(\mathbf{x}|\mathbf{z}_i)$. A new problem here is that in high-dimensional spaces, n might need to be extremely large for an accurate estimate of $P_\theta(\mathbf{x})$.
- For most $\mathbf{z}$, $P_\theta(\mathbf{x}|\mathbf{z})$ will be nearly zero, and hence contribute almost nothing to the estimate of $P_\theta(\mathbf{x})$.

The key idea behind the variational autoencoder is to attempt to sample values of $\mathbf{z}$ that are likely to have produced $\mathbf{x}$, and compute $P_\theta(\mathbf{x})$ just from those. This means [22] that we need a new function $Q(\mathbf{z}|\mathbf{x})$ which can take a value of $\mathbf{x}$ and give us a distribution over $\mathbf{z}$ values that are likely to produce $\mathbf{x}$.

The relationship between $E_{\mathbf{z}\sim Q} P(\mathbf{x}|\mathbf{z})$ and $P(\mathbf{x})$ is one of the cornerstones of variational Bayesian methods. Define the Kullback–Leibler divergence between $P(\mathbf{z}|\mathbf{x})$ and $Q(\mathbf{z})$ for some arbitrary Q (which may or may not depend on $\mathbf{x}$ as):

$$\mathrm{KL}[Q(\mathbf{z})\|P(\mathbf{z}|\mathbf{x})] = E_{\mathbf{z}\sim Q}\{\log Q(\mathbf{z}) - \log P(\mathbf{z}|\mathbf{x})\}. \quad (7.16.18)$$

By applying Bayes rule $P(\mathbf{z}|\mathbf{x}) = \frac{P(\mathbf{x}|\mathbf{z})P(\mathbf{z})}{P(\mathbf{x})}$ and noticing that $\log P(\mathbf{x})$ does not depend on $\mathbf{z}$, the above equation becomes [22]

$$\mathrm{KL}[Q(\mathbf{z})\|P(\mathbf{z}|\mathbf{x})]$$
$$= E_{\mathbf{z}\sim Q}\{\log Q(\mathbf{z}) - \log P(\mathbf{x}|\mathbf{z}) - \log P(\mathbf{z})\} + \log P(\mathbf{x}), \quad (7.16.19)$$

or equivalently

$$\log P(\mathbf{x}) - \mathrm{KL}[Q(\mathbf{z}) \| P(\mathbf{z}|\mathbf{x})] = E_{\mathbf{z} \sim Q}\{\log P(\mathbf{x}|\mathbf{z}) - \mathrm{KL}[Q(\mathbf{z}) \| P(\mathbf{z})]\}. \qquad (7.16.20)$$

Since we are interested in inferring $P(\mathbf{x})$, it makes sense to construct a Q which does depend on $\mathbf{x}$, and in particular, one which makes $\mathrm{KL}[Q(\mathbf{z}) \| P(\mathbf{z}|\mathbf{x})]$ small:

$$\begin{aligned}&\log P(\mathbf{x}) - \mathrm{KL}[Q(\mathbf{z}|\mathbf{x}) \| P(\mathbf{z}|\mathbf{x})] \\ &= E_{\mathbf{z} \sim Q}\{\log P(\mathbf{x}|\mathbf{z}) - \mathrm{KL}[Q(\mathbf{z}|\mathbf{x}) \| P(\mathbf{z})]\}.\end{aligned} \qquad (7.16.21)$$

This equation serves as the core of the variational autoencoder [22].

- The left-hand side contains the quantity that should be maximized: $\log P(\mathbf{x})$ (plus an error term $\mathrm{KL}[Q(\mathbf{z}|\mathbf{x}) \| P(\mathbf{z}|\mathbf{x})]$, which makes Q produce $\mathbf{z}$'s that can reproduce a given $\mathbf{x}$; this term will become small if Q is high-capacity).
- The right-hand side should be optimized, i.e. via stochastic gradient descent, given the right choice of Q (although it may not be obvious yet how).

In conclusion, Q is "encoding" $\mathbf{x}$ into $\mathbf{z}$, and P is "decoding" $\mathbf{z}$ to reconstruct $\mathbf{x}$.

Brief Summary of This Chapter

- This chapter introduces the neural network tree.
- Neural networks aim at using networks with layered structure to implement machine learning.
- Neural networks (NNs) are divided into two broad categories: NNs for Euclidean structured model learning and NNs for non-Euclidean structured model learning.
- This chapter mainly introduces selected topics and advances in neural networks: convolutional neural networks (CNNs), dropout learning, autoencoders, extreme learning machine (ELM), graph embeddings, manifold learning, network embeddings, graph neural networks (GNNs), batch normalization networks, and generative adversarial networks (GANs).

References

1. Ackley, D.H., Hinton, G.E., Sejnowski, T.J.: A learning algorithm for Boltzmann machines. Cognitive Sci. **9**(1), 147–169 (1985)
2. Aghdam, H.H., Heravi, E.J.: Guide to Convolutional Neural Networks. Springer, Berlin (2017)
3. Akata, Z., Thurau, C., Bauckhage, C.: Non-negative matrix factorization in multimodality data for segmentation and label prediction. In: Proceedings of the 16th Computer Vision Winter Workshop (2011)

4. Ba, J.L., Kiros, J.R., Hinton, G.E.: Layer normalization. Available at: arXiv:1607.06450 (2016)
5. Baldi, P., Sadowski, P.: The dropout learning algorithm. Artif. Int. **210**, 78–122 (2014)
6. Belkin, M., Niyogi, P.: Laplacian eigenmaps and spectral techniques for embedding and clustering. In: Proceedings of the 14th International Conference on Neural Information Processing Systems: Natural and Synthetic, pp. 585–591 (2001)
7. Belkin, M., Niyogi, P.: Laplacian eigenmaps for dimensionality reduction and data representation. Neural Comput. **15**(6), 1373–1396 (2003)
8. Bengio, Y.: Learning deep architectures for AI. Found. Trends Mach. Learn. **2**(1), 1–127 (2009)
9. Bengio, Y., Lamblin, P., Popovici, D., Larochelle, H.: Greedy layer-wise training of deep networks. In: Conference on Advances in Neural Information Processing Systems (NIPS), vol. 19, pp. 153–160 (2007)
10. Bengio, Y., Courville, A., Vincent, P.: Representation learning: a review and new perspectives. IEEE Trans. Pattern Anal. Mach. Int. **35**(8), 1798–1828 (2013)
11. Bergstra, J., Guillaume, D., Pascal, L., Yoshua, B.: Quadratic polynomials learn better image features. Technical Report 1337. Département d'Informatique et de Recherche Opérationnelle, Université de Montréal (2009)
12. Bourlard, H., Kamp, Y.: Auto-association by multilayer perceptrons and singular value decomposition. Biol. Cybern. **59**, 291–294 (1988)
13. Bruna, J., Zaremba, W., Szlam, A., LeCun, Y.: Spectral networks and locally connected networks on graphs (2014). Available at: https://arXiv:1312.6203v3
14. Cai, H., Zheng, V.W., Chang, K.C.-C: A comprehensive survey of graph embedding: problems, techniques, and applications. IEEE Trans. Knowl. Data Eng. **30**(9), 1616–1677 (2018)
15. Carreira-Perpinan, M.A., Hinton, G.E.: On contrastive divergence learning. In: Cowell R.G., Ghahramani, Z. (eds.) Proceedings of the Tenth International Workshop on Artificial Intelligence and Statistics (AISTATS05). Society for Artificial Intelligence and Statistics, pp. 33–40 (2005)
16. Clevert, D.-A., Unterthiner, T., Hochreiter, S.: Fast and accurate deep network learning by exponential linear units (ELUs). In: Proceedings of the International Conference on Learning Representations (ICLR) (2016)
17. Coates, A., Lee, H., Ng, A.Y.: An analysis of single-layer networks in unsupervised feature learning. In: Proceedings of the 14th International Conference on AI and Statistics, pp. 215–223 (2011)
18. Coates, A., Ng, A.Y.: Learning feature representations with K-means. In: Montavon G., Orr G.B., Müller K.-R. (eds.) Neural Networks: Tricks of the Trade, 2nd edn., pp. 561–580. Springer, Berlin (2012)
19. Cui, P., Wang, X., Pei, J., Zhu, W.: A survey on network embedding. IEEE Trans. Knowl. Data Eng. (2018)
20. Dalal, N., Triggs, B.: Histograms of oriented gradients for human detection. In: Proceedings of the International Conference on Computer Vision and Pattern Recognition (CVPR), San Diego, pp. 886–893 (2005)
21. Dhillon, I.S., Modha, D.M.: Concept decompositions for large sparse text data using clustering. Mach. Learn. **42**(1), 143–175 (2001)
22. Doersch, C.: Tutorial on variational autoencoders (2016). Available at: arXiv:1606.05908v2
23. Donahue, J., Krähenbühl, P., Darrell, T.: Adversarial feature learning. In: Proceedings of the International Conference on Learning Representations ICLR (2017)
24. Donoho, D.L., Grimes, C.: Hessian eigenmaps: locally linear embedding techniques for high-dimensional data. Proc. Natl. Acad. Sci. **100**(10), 5591–5596 (2003)
25. Du, B., Xiong, W., Wu, J., Zhang, L.F., Zhang, L.P., Tao, D.: Stacked convolutional denoising auto-encoders for feature representation. IEEE Trans. Cybern. **47**(4), 1017–1027 (2017)
26. Duchi, J., Shalev-Shwartz, S., Singer, Y., Tewari, A.: Composite objective mirror descent. In: Proceedings of the Twenty Third Annual Conference on Computational Learning Theory (2010)

27. Duchi, J., Hazan, E., Singer, Y.: Adaptive subgradient methods for online learning and stochastic optimization. J. Mach. Learn. Res. **12**, 2121–2159 (2011)
28. Duvenaud, D.K., Maclaurin, D., Iparraguirre, J., Bombarell, R., Hirzel, T., Aspuru-Guzik, A., Adams, R.P.: Convolutional networks on graphs for learning molecular fingerprints. In: Proceedings of the Advances in Neural Information Processing Systems (NIPS), pp. 2224–2232 (2015)
29. Elman, J.L.: Finding structure in time. Cognitive Sci. **14**(2), 179–211 (1990)
30. Faul, A.C., Tipping, M.E.: Analysis of sparse Bayesian learning. In: Dietterich T.G., Becker S., Ghahramani Z. (eds.) Advances in Neural Information Processing (NIPS 14), pp. 383–389 (2002). Available at: http://citeseer.ist.psu.edu/faul01analysis.html
31. Ferreira, A., Giraldi, G.: Convolutional Neural Network approaches to granite tiles classification. Expert Syst. Appl. **84**, 1–11 (2017)
32. Fukushima, K.: Neocognitron: A self-organizing neural network model for a mechanism of pattern recognition unaffected by shift in position. Biol. Cybern. **36**(4), 193–202 (1980)
33. Gers, F.A., Schmidhuber, J.: LSTM recurrent networks learn simple context free and context sensitive languages. IEEE Trans. Neural Networ. **12**(6), 1333–1340 (2001)
34. Gers, F.A., Schraudolph, N., Schmidhuber, J.: Learning precise timing with LSTM recurrent networks. J. Mach. Learn. Res. **3**, 115–143 (2002)
35. Gilmer, J., Schoenholz, S.S., Riley, P.F., Vinyals, O., Dahl, G.E.: Neural message passing for quantum chemistry. In: Proceedings of the 34th International Conference on Machine Learning, Sydney, PMLR 70 (2017)
36. Goodfellow, I.: NIPS 2016 Tutorial: Generative Adversarial Networks (2017). Available at: arXiv:1701.00160v4 [cs.LG]
37. Goodfellow, I.J., Warde-Farley, D., Mirza, M., Courville, A., Bengio, Y.: Maxout networks. In: Proceedings of the International Conference on Machine Learning (ICML), pp. 1319–1327 (2013). Goodfellow16
38. Goodfellow, I., Pouget-Abadie, J., Mirza, M., Xu, B., Warde-Farley, D., Ozair, S., Courville, A., Bengio, Y.: Generative adversarial nets. In: Proceedings of the Advances in Neural Information Processing Systems (NIPS), pp. 2672–2680 (2014)
39. Goodfellow, I.J., Bengio, Y., Courville, A.: Deep Learning. MIT Press, Cambridge (2016)
40. Gori, M., Monfardini, G., Scarselli, F.: A new model for learning in graph domains. In: Proceedings of the 2005 IEEE International Joint Conference on Neural Networks., vol. 2, pp. 729–734 (2005)
41. Goyal, P., Ferrara, E.: Graph embedding techniques, applications, and performance: a survey. Knowl.-Based Syst. **151**, 78–94 (2018)
42. Graves, A.: Sequence transduction with recurrent neural networks. In: ICML Representation Learning Worksop (2012). Available at: https://arxiv.org/pdf/1211.3711.pdf
43. Graves, A.: Long short-term memory. In: Supervised Sequence Labelling with Recurrent Neural Networks, pp. 37–45. Springer, Berlin (2012)
44. Graves, A.: Generating sequences with recurrent neural networks (2013). Available at: https://arxiv.org/pdf/1308.0850v5.pdf
45. Graves, A., Schmidhuber, J.: Framewise phoneme classification with bidirectional LSTM and other neural network architectures. Neural Netw. **18**, 602–610 (2005)
46. Graves, A., Fernandez, S., Gomez, F., Schmidhuber, J.: Connectionist temporal classification: Labelling unsegmented sequence data with recurrent neural networks. In: Proceedings of the 23rd International Conference on Machine Learning (ICML), Pittsburgh, pp. 369–376 (2006)
47. Graves, A., Mohamed, A., Hinton, G.: Speech recognition with deep recurrent neural networks. In: Proceedings of the IEEE International Conference on Acoustics, Speech and Signal Processing ICASSP 13, pp. 6645–6649 (2013)
48. Gu, J., Wang, Z., Kuen, J., Ma, L., Shahroudy, A., Shuai, B., Liu, T., Wang, X., Wang, G., Cai, J., Chen, T.: Recent advances in convolutional neural networks. Pattern Recogn. **77**, 354–377 (2018)
49. Hadsell, R., Chopra, S., LeCun, Y.: Dimensionality reduction by learning an invariant mapping. In: Proceedings of the 2006 IEEE Computer Society Conference on Computer Vision and Pattern Recognition (IEEE CVPR'06), pp. 1735–1742 (2006)

50. Hamilton, W.I., Ying, R., Leskovec, J.: Inductive representation learning on large graphs. In: Proceedings of the 31st Conference on Neural Information Processing Systems (NIPS 2017), Long Beach (2017)
51. Hammond, D.K., Vandergheynst, P., Gribonval, R.: Wavelets on graphs via spectral graph theory. Appl. Comput. Harm. Anal. **30**(2), 129–150 (2011)
52. Haykin, S.: Neural Networks: A Comprehensive Foundation. Macmillan College, New York (1994)
53. He, X., Niyogi, P.: Locality preserving projections. In: NIPS'03 Proceedings of the 16th International Conference on Neural Information Processing Systems 16, Vancouver, British Columbia, pp. 153–160 (2003)
54. He, X., Cai, D., Yan, S., Zhang, H.-J.: Neighborhood preserving embedding. In: Proceedings of the Tenth IEEE International Conference on Computer Vision (ICCV'05), vol. 1, (2005)
55. He, K., Zhang, X., Ren, S., Sun, J.: Delving deep into rectifiers: Surpassing human-level performance on ImageNet classification. In: Proceedings of the International Conference on Computer Vision (ICCV), pp. 1026–1034 (2015)
56. He, K., Zhang, X., Ren, S., Sun, R.: Spatial pyramid pooling in deep convolutional networks for visual recognition. IEEE Trans. Pattern Anal. Mach. Intell. **37**(9), 1904–1916 (2015)
57. Hebb, D.O.: The Organization of Behavior: A Neuropsychological Theory. Psychology Press, London (1949)
58. Hinton, G.E.: Training products of experts by minimizing contrastive divergence. Neural Comput. **14**(8), 1771–1800 (2002)
59. Hinton, G.E.: A practical guide to training restricted Boltzmann machines (Version 1). Technical Report UTML TR 2010–003, pp. 1–20 (2010)
60. Hinton, G.E., Zemel, R.S.: Autoencoders, minimum description length, and Helmholtz free energy. In: Neural Information and Processing Systems (1993)
61. Hinton, G.E., Osindero, S., Teh, Y.-W.: A fast learning algorithm for deep belief nets. Neural Comput. **18**, 1527–1554 (2006)
62. Hinton, G.E., Srivastava, N., Krizhevsky, A., Sutskever, I., Salakhutdinov, R.R.: Improving neural networks by preventing co-adaptation of feature detectors. Preprint arXiv:1207.0580 (2012)
63. Hochreiter, S., Schmidhuber, J.: Long short-term memory. Neural Comput. **9**(8), 1735–1780 (1997)
64. Hodgkin, A.L., Huxley, A.F.: A quantitative description of membrane current and its application to conduction and excitation in nerve. J. Physiol. **117**(4), 500–544 (1952)
65. Hopfield, J.J.: Neural networks and physical systems with emergent collective computational abilities. Proc. Natl. Acad. Sci. **79**(8), 2554–2558 (1982)
66. Hornik, K.: Approximation capabilities of multilayer feedforward networks. Neural Netw. **4**, 251–257 (1991)
67. Hosseini-Asl, E., Zurada, J.M., Nasraoui, O.: Deep learning of part-based representation of data using sparse autoencoders with nonnegativity constraints. IEEE Trans. Neural Netw. Learn. Stst. **27**(12), 3486–3498 (2016)
68. Hu, Y., Fan, J., Wang, J.: Classification of PolSAR images based on adaptive nonlocal stacked sparse autoencoder. IEEE Geosci. Remote Sens. Lett. **15**(7), 1050–1054 (2018)
69. Huang, G.-B., Babri, H.A.: Upper bounds on the number of hidden neurons in feedforward networks with arbitrary bounded nonlinear activation functions. IEEE Trans. Neural Netw. **9**(1), 224–229 (1998)
70. Huang, G.-B., Zhu, Q.-Y., Siew, C.-K.: Extreme learning machine: theory and applications. Neurocomputing **70**(1–3), 489–501 (2006)
71. Huang, G.-B., Zhou, H., Ding, X., Zhang, R.: Extreme learning machine for regression and multiclass classification. IEEE Trans. Syst. Man Cybern. Part B: Cybern. **42**(2), 513–529 (2012)
72. Hyvärinen, A., Köster, U.: Complex cell pooling and the statistics of natural images. Network **18**(2), 81–100 (2007)

73. Ioffe, A.: Batch renormalization: Towards reducing minibatch dependence in batch-normalized models. In: Proceedings of the 31st Conference on Neural Information Processing Systems (NIPS 2017), Long Beach (2017)
74. Ioffe, A., Szegedy, C.: Banch normalization: Accelerating deep network training by reducing internal covariate shift. In: Proceedings of the 32nd International Conference on Machine Learning (ICML-15), pp. 448–456 (2015)
75. Jordan, M.I.: Serial order: a parallel distributed processing approach. Adv. Psychol. **121**, 471–495 (1986)
76. Katz, L.: A new status index derived from sociometric analysis. Psychometrika **18**(1), 39–43 (1953)
77. Ketkar, N.: Deep Learning with Python. Springer, Berlin (2017)
78. Khamsi, M.A., Kirk, W.A.: An Introduction to Metric Spaces and Fixed Point Theory, vol. 53. Wiley, Hoboken (2011)
79. Kingma, D.P., Welling, M.: Auto-encoding variational Bayes. Available at: arXiv:1312.6114v10 (2014)
80. Kingma, D.P., Mohamed, S., Rezende, D.J., Welling, M.: Semisupervised learning with deep generative models. In: Proceedings of the Advances in Neural Information Processing Systems, pp. 3581–3589 (2014)
81. Kipf, T.N., Welling, M.: Variational graph auto-encoders. In: NIPS Workshop on Bayesian Deep Learning (2016)
82. Kipf, T.N., Welling, M.: Semi-supervised classification with graph convolutional networks. In: Proceedings of the International Conference on Learning Representations (ICLR) (2017)
83. Kohonen, T.: Automatic formation of topological maps of patterns in a self-organizing system. In: Proceedings of the 2nd Scandinavian Conference on Image Analysis (Espoo, 1981), pp. 214–220 (1981)
84. Krizhevsky, A., Sutskever, I., Hinton, G.E.: ImageNet classification with deep convolutional neural networks. In: Proceedings of the Advances in Neural Information Processing Systems (NIPS), vol. 25, pp. 1097–1105 (2012)
85. Kruskal, J.B.: Multidimensional scaling by optimizing goodness of fit to a nonmetric hypothesis. Psychometrika **29**(1), 1–26 (1964)
86. Kruskal, J.B.: Nonmetric multidimensional scaling: a numerical method. Psychometrika **29**(2), 115–129 (1964)
87. Langville, A.N., Meyer, C.D., Albright, R., Cox, J., Duling, D.: Algorithms, initializations, convergence for the nonnegative matrix factorization. SAS Technical Report (2014). arXiv:1407.7299
88. LeCun, B.B., Denker, J.S., Henderson, D., Howard, R.E., Hubbard, W., Jackel, L.D.: Handwritten digit recognition with a back-propagation network. In: Proceedings of the Advances in Neural Information Processing Systems. Citeseer (1990)
89. LeCun, Y., Bottou, L., Bengio, Y., Haffner, P.: Gradient-based learning applied to document recognition. Proc. IEEE **86**(11), 2278–2324 (1998)
90. LeCun, Y., Bottou, L., Orr, G., Muller, K.: Efficient backprop. In: Orr G., Muller K. (eds.) Neural Networks: Tricks of the Trade. Springer, Berlin (1998)
91. Lee, D.D., Seung, H.S.: Learning the parts of objects by non-negative matrix factorization. Nature **401**, 788–791 (1999)
92. Lee, H., Grosse, R., Ranganath, R., Ng, A.Y.: Convolutional deep belief networks for scalable unsupervised learning of hierarchical representations. In: Proceedings of the 26th Annual International Conference on Machine Learning, pp. 609–616 (2009)
93. Leshno, M., Lin, V.Y., Pinkus, A., Schocken, S.: Multilayer feedforward networks with a nonpolynomial activation function can approximate any function. Neural Netw. **6**, 861–867 (1993)
94. Lewis, D.D.: Naive (Bayes) at forty: The independence assumption in information retrieval. In: Proceedings of the European Conference on Machine Learning (1998)
95. Li, Y., Tarlow, D., Brockschmidt, M., Zemel, R.: Gated graph sequence neural networks. In: Proceedings of the International Conference on Learning Representations (ICLR) (2016)

96. Lin, J., Morere, O., Chandrasekhar, V., Veillard, A., Goh, H.: Deephash: getting regularization, depth and fine-tuning right. In: Proceedings of the 2017 ACM on International Conference on Multimedia Retrieval (ICMR), pp. 133–141 (2017)
97. Liou, C.-Y., Cheng, W.-C., Liou, J.-W., Liou, D.-R.: Autoencoder for words. Neurocomputing **139**, 84–96 (2014)
98. Liu, H., Tian, Y., Yang, Y., Pang, L., Huang, T.: Deep relative distance learning: Tell the difference between similar vehicles. In: Proceedings of the IEEE Conference on Computer Vision and Pattern Recognition (CVPR), pp. 2167–2175 (2016)
99. Liu, W., Wen, Y., Yu, Z., Yang, M.: Large-margin softmax loss for convolutional neural networks. In: Proceedings of the 33rd International Conference on Machine Learning (ICML), New York, pp. 507–516 (2016)
100. Lowe, D.G.: Distinctive image features from scale-invariant keypoints. Int. J. Comput. Vis. **60**(2), 91–110 (2004)
101. Luo, D., Ding, C., Nie, F., Huang, H.: Cauchy graph embedding. In: Proceedings of the 28th International Conference on Machine Learning, Bellevue (2011)
102. Maas, A.L., Hannun, A.Y., Ng, A.Y.: Rectifier nonlinearities improve neural network acoustic models. In: Proceedings of the 30th International Conference on Machine Learning (ICML), vol. 30 (2013). Retrieved 2 January 2017
103. Masci, J., Meier, U., Ciresan, D., Schmidhuber, J.: Stacked convolutional auto-encoders for hierarchical feature extraction. In: Proceedings of the 21st International Conference on Artificial Neural Networks, Part I, Espoo, pp. 52–59 (2011)
104. McCallum, A., Nigam, K.: A comparison of event models for Naive Bayes text classification. In: Proceedings of the AAAI-98 Workshop on Learning for Text Categorization (1998)
105. McCulloch, W.S., Pitts, W.: A logical calculus of the ideas immanent in nervous activity. Bull. Math. Biophys. **5**(4), 115–133 (1943)
106. Nair, V., Hinton, G.E.: Rectified linear units improve restricted Boltzmann machines. In: Proceedings of the 27th International Conference on Machine Learning (ICML), pp. 807–814 (2010)
107. Nesterov, Y.: Primal-dual subgradient methods for convex problems. Math. Program. **120**(1), 221–259 (2009)
108. Newman, M.E.: Modularity and community structure in networks. Proc. Natl. Acad. Sci. **103**(23), 8577–8582 (2006)
109. Ng, A.: Sparse autoencoder. CS294A Lecture Notes (2011)
110. Nguyen, T.R., Grishman, R.: Combining neural networks and log-linear models to improve relation extraction (2015). Available at https://arXiv:1511.05926v1
111. Ou, M., Cui, P., Pei, J., Zhang, Z., Zhu, W.: Asymmetric transitivity preserving graph embedding. In: KDD '16 Proceedings of the 22nd ACM SIGKDD International Conference on Knowledge Discovery and Data Mining, pp. 1105–1114 (2016)
112. Pascanu, R., Mikolov, T., Bengio, Y.: On the difficulty of training recurrent neural networks. In: Proceedings of the 30th International Conference on Machine Learning (ICML), Atlanta, pp. 1310–1318 (2013)
113. Pearlmutter, B.A.: Gradient calculations for dynamic recurrent neural networks: a survey. IEEE Trans. Neural Netw. **6**(5), 1212–1228 (1995)
114. Perozzi, B., Al-Rfou, R., Skiena, S.: DeepWalk: Online learning of social representations. In: Proceedings of the 20th ACM SIGKDD International Conference on Knowledge Discovery and Data Mining, pp. 701–710 (2014)
115. Ranzato, M., Poultney, C., Chopra, S., LeCun, Y.: Efficient learning of sparse representations with an energy-based Model. In: Proceedings of the 19th International Conference on Neural Information and Processing Systems, pp. 1137–1144 (2006)
116. Ranzato, C.P.M., Chopra, S., LeCun, Y.: Efficient learning of sparse representations with an energy-based model. In: Proceedings of the 2007 Advances in Neural Information Processing Systems, 2007, pp. 1137–1144 (2007)

117. Ranzato, M., Huang, F.J., Boureau, Y.-L., LeCun, Y.: Unsupervised learning of invariant feature hierarchies with applications to object recognition. In: Proceedings of the 2007 Conference on Computer Vision and Pattern Recognition, pp. 1–8. IEEE, Piscataway (2007)
118. Ratliff, L.J., Burden, S.A., Sastry, S.S.: Characterization and computation of local Nash equilibria in continuous games. In 51st Annual Allerton Conference on Communication, Control, and Computing (Allerton), pp. 917–924 (2013)
119. Rifai, S., Vincent, P., Muller, X., Glorot, X., Bengio, Y.: Contractive auto-encoders: Explicit invariance during feature extraction. In: Proceedings of the International Conference on Machine Learning, pp. 833–840 (2011)
120. Rippel, O., Snoek, J., Adams, R.P.: Spectral representations for convolutional neural networks. In: Proceedings of the Advances in Neural Information Processing Systems (NIPS), pp. 2449–2457 (2015)
121. Rosenblatt, F.: The perceptron: a probabilistic model for information storage and organization in the brain. Psychol. Rev. **65**(6), 386 (1958)
122. Roweis, S., Saul, L.: Nonlinear dimensionality reduction by locally linear embedding. Science **290**(5500), 2323–2326 (2000)
123. Rumelhart, D.E., Hinton, G.E., Williams, R.J.: Learning representations by backpropagating errors. Nature **323**(6088), 533–536 (1986)
124. Salakhutdinov, R., Hinton, G.E.: Deep Boltzmann machines. In: Proceedings of the 13th International Conference on Artificial Intelligence and Statistics (AISTATS), vol. 1, p. 3 (2009)
125. Salakhutdinov, R., Mnih, A., Hinton, G.E.: Restricted Boltzmann machines for collaborative filtering. In: Proceedings of the 24th International Conference on Machine Learning, pp. 791–798 (2007)
126. Salehinejad, H., Sankar, S., Barfett, J., Colak, E., Valaee, S.: Recent advances in Recurrent Neural Networks (2018). Preprint arXiv:1801.01078v3 [cs.NE]
127. Santurkar, S., Tsipras, D., Ilyas, A., Madry, A.: How does batch normalization help optimization? In: Proceedings of the Advances in Neural Information Processing Systems 31 (NIPS 2018) (2018)
128. Saul, L.K., Roweis, S.T.: Think globally, fit locally: unsupervised learning of low dimensional manifolds. J. Mach. Learn. Res. **4**, 119–155 (2003)
129. Scarselli, F., Gori, M., Tsoi, A.C., Hagenbuchner, M., Monfardini, G.: The graph neural network model. IEEE Trans. Neural Netw. **20**(1), 61–80 (2009)
130. Schmidhuber, J.: Deep learning in neural networks: an overview. Neural Netw. **61**, 85–117 (2015)
131. Schroff, F., Kalenichenko, D., Philbin, J.: Facenet: a unified embedding for face recognition and clustering. In: Proceedings of the IEEE Conference on Computer Vision and Pattern Recognition (CVPR), pp. 815–823 (2015)
132. Schuster, M., Paliwal, K.K.: Bidirectional recurrent neural networks. IEEE Trans. Signal Process. **45**(11), 2673–2681 (1997)
133. Shepard, R.N.: Stimulus and response generalization: deduction of the generalization gradient from a trace model. Psychol. Rev. **55**, 242–256 (1958)
134. Shepard, R.N.: Multidimensional scaling, tree-fitting, and clustering. Science **210**(24), 390–398 (1980)
135. Shi, J., Chen, Z., Wang, H., Yeung, D.-U., Wong, W.-K., Woo, W.-C.: Convolutional LSTM network: A machine learning approach for precipitation nowcasting. In: Proceedings of the 28th International Conference on Neural Information Processing Systems, pp. 802–810 (2015)
136. Shuman, D.I., Ricaud, B., Vandergheynst, P.: A windowed graph Fourier transform. In: Proceedings of the 2012 IEEE Statistical Signal Processing Workshop (SSP), Ann Arbor, pp. 133–136 (2012)
137. Silva, V.D., Tenenbaum, J.B.: Global versus local methods in nonlinear dimensionality reduction. In: Advances in Neural Information Processing Systems 15, pp. 705–712. MIT Press, Cambridge (2003)

138. Smith, J., Wishum, M.: Review: spectral representations for convolutional neural networks. In: COMP 7970 Presentations. Springer, Berlin (2018)
139. Smolensky, P.: Information processing in dynamical systems: Foundations of harmony theory. Technical Report, DTIC Document (1986)
140. Snijders, T.A., Pattison, P.E., Robins, G.L., Handcock, M.S.: New specifications for exponential random graph models. Sociol. Methodol. **36**(1), 99–153 (2006)
141. Springenberg, J.T., Riedmiller, M.: Improving deep neural networks with probabilistic maxout units (2013). Preprint CoRR arXiv:abs/1312.6116
142. Srivastava, V., Hinton, G.E., Krizhevsky, A., Sutskever, I., Salakhutdinov, R.: Dropout: a simple way to prevent neural networks from overfitting. J. Mach. Learn. Res. **15**(1), 1929–1958 (2014)
143. Sutskever, T., Vinyals, O., Le, Q.V.: Sequence to sequence learning with neural networks. In Conference on Neural Information Processing Systems (NIPS), pp. 3104–3112 (2014)
144. Tang, J., Qu, M., Wang, M., Zhang, M., Yan, J., Mei, Q.: LINE: Large-scale information network embedding. In: Proceedings of the International World Wide Web Conference Committee (IW3C2), Florence, pp. 1067–1077 (2015)
145. Tenenbaum, J.B., de Silva, V., Langford3 J. C.: A global geometric framework for nonlinear dimensionality reduction. Science **290**(5500), 2319–2323 (2000)
146. Tipping, M.E., Faul, A.C.: Fast marginal likelihood maximization for sparse Bayesian models. In: Bishop C.M., Frey B.J.(eds.) Proceedings of the 9th International Workshop Artificial Intelligence and Statistics (2003). Available at: http://www.miketipping.com/papers.htm
147. Tseng, P.: On accelerated proximal gradient methods for convex-concave optimization. Technical Report, Department of Mathematics, University of Washington (2008)
148. Ulungu, E.L., Teghem, J., Fortemps, Ph., Tuyttens, D.: Mosa method: a tool for solving multiobjective combinatorial optimization problems. J. Multi-Criteria Decis. Anal. **8**, 221–236 (1999)
149. Vincent, P., Larochelle, H., Bengio, Y., Manzagol, P.A.: Extracting and composing robust features with denoising autoencoders. In: Proceedings of the 25th International Conference on Machine Learning, pp. 1096–1103 (2008)
150. Voxman, W.L., Roy, J., Goetschel, H.: Advanced Calculus: An Introduction to Modern Analysis. Dekker, New York (1981)
151. Wan, E.A.: Neural network classification: a Bayesian interpretation. IEEE Trans. Neural Netw. **1**(4), 303–305 (1990)
152. Wan, L., Zeiler, M., Zhang, S., LeCun, Y., Fergus, R.: Regularization of neural networks using DropConnect. J. Mach. Learn. Res. **28**(3), 1058–1066 (2013)
153. Wan, L., Zeiler, M., Zhang, S., LeCun, Y., Fergus, R.: Regularization of neural networks using DropConnect (Supplementary Material). In: Proceedings of the 30th International Conference on Machine Learning (ICML), vol. 30 (2013). Available at: https://cs.nyu.edu/~wanli/dropc/dropc$_-$supp.pdf
154. Wang, D., Cui, P., Zhu, W.: Structural deep network embedding. In: Proceedings of the 22nd International Conference on Knowledge Discovery and Data Mining, pp. 1225–1234. ACM, New York (2016)
155. Wang, H., Raj, B.: On the origin of deep learning (2017). Available at: https://arXiv:1702.07800v4
156. Wang, X., Cui, P., Wang, J., Pei, J., Zhu, W., Yang, S.: Community preserving network embedding. In: Proceedings of the Thirty-First AAAI Conference on Artificial Intelligence (AAAI-17), pp. 203–209 (2017)
157. Welling, M., Rosen-Zvi, M., Hinton, G.E.: Exponential family harmoniums with an application to information retrieval. In: Proceedings of the Advances in Neural Information Processing Systems, pp. 1481–1488. MIT Press, Cambridge (2005)
158. Werbos, P.J.: Beyond regression: New tools for prediction and analysis in the behavioral sciences. Ph.D. Dissertation, Harvard University (1975)
159. Werbos, P.J.: Backpropagation through time: what it does and how to do it. Proc. IEEE **78**(10), 1550–1560 (1990)

160. Wu, H., Gu, X.: Towards dropout training for convolutional neural networks. Neural Netw. **71**, 1–10 (2015)
161. Wu, Y., He, K.: Group normalization. In: Proceedings of the European Conference on Computer Vision (ECCV-2018), pp. 3–19 (2018)
162. Xiao, L.: Dual averaging methods for regularized stochastic learning and online optimization. J. Mach. Learn. Res. **11**(1), 2543–2596 (2010)
163. Xing, C., Ma, L., Yang, X.: Stacked denoise autoencoder based feature extraction and classification for hyperspectral images. J. Sensors **2016**, Article ID 3632943, 10 (2016)
164. Xu, B., Wang, N., Chen, T., Li, M.: Empirical evaluation of rectified activations in convolutional network. In: Proceedings of the International Conference on Machine Learning (ICML) Workshop (2015)
165. Ye, J.C., Kim, J.M., Jin, K.H., Lee, K.: Compressive sampling using annihilating filter-based low-rank interpolation. IEEE Trans. Inform. Theory **63**(2), 777–801 (2017)
166. Ye, J.C., Han, Y., Cha, E.: Deep convolutional framelets: a general deep learning framework for inverse problems. SIAM J. Imaging Sci. **11**(2), 991–1048 (2018)
167. Yu, D., Wang, H., Chen, P., Wei, Z.: Mixed pooling for convolutional neural networks. In: Proceedings of the Rough Sets and Knowledge Technology (RSKT), pp. 364–375 (2014)
168. Zeiler, M.D., Fergus, R.: Stochastic pooling for regularization of deep convolutional neural networks. In: Proceedings of the International Conference on Learning Representations (ICLR) (2013)
169. Zetzsche, C., Krieger, G., Wegmann, B.: The atoms of vision: Cartesian or polar? J. Opt. Soc. Am. **16**(7), 1554–1565 (1999)
170. Zhang, G.P.: Neural networks for classification: a survey. IEEE Trans. Syst. Man Cybern. Part C Appl. Rev. **30**(4), 451–462 (2000)
171. Zhang, T.: Solving large scale linear prediction problems using stochastic gradient descent algorithms. In: Proceedings of the 21st International Conference on Machine Learning (ICML), pp. 116–123 (2004)
172. Zhang, X.D.: Matrix Analysis and Applications. Cambridge University Press, Cambridge (2017)
173. Zhang, F., Du, B., Zhang, L.P., Zhang, L.F.: Hierarchical feature learning with dropout k-means for hyperspectral image classification. Neurocomputing **187**, 75–82 (2016)
174. Zhang, Z., Cui, P., Zhu, W.: Deep learning on graphs: A survey (2018). Available at: arXiv.1812.04202v1 [cs.LG]
175. Zhu, X., Sobhani, P., Guo, H.: Long short-term memory over recursive structures. In: Proceedings of the International Conference on Machine Learning, pp. 1604–1612 (2015)
176. Zhu, D., Cui, P., Zhang, Z., Pei, J., Zhu, W.: High-order proximity preserved embedding for dynamic networks. IEEE Trans. Knowl. Data Eng. **30**(11), 2134–2144 (2018)
177. Zinkevich, M.: Online convex programming and generalized infinitesimal gradient ascent. In: Proceedings of the Twentieth International Conference on Machine Learning (ICML-2003), Washington, pp. 928–936 (2003)

Chapter 8
Support Vector Machines

Supervised regression/classification methods learn a model of relation between the target vectors $\{y_i\}_{i=1}^{N}$ and the corresponding input vectors $\{\mathbf{x}_i\}_{i=1}^{N}$ consisting of N training samples and utilize this model to predict/classify target values for the previously unseen inputs.

In real-world data, the presence of noise (in regression) and class overlap (in classification) implies that the principal modeling challenge is to avoid "overfitting" of the training set, which is an important concern with this modeling.

Unlike traditional neural network approaches which suffer difficulties with generalization, producing models that may overfit the data, SVMs [43] are a popular machine learning approach for classification, regression, and other learning tasks, based on statistical learning theory (Vapnik–Chervonenkis or VC-theory).

In regression and classification applications, SVMs draw on two main practical observations [5, pp. 12–13]:

- At a sufficiently high dimension, patterns are orthogonal to each other, and thus it is easier to find a separating hyperplane for data in a high dimension space.
- Not all patterns are necessary for finding a separating hyperplane. In fact, it is sufficient to use only those points that are near the boundary between groups to construct the boundary.

As a kind of generalization intelligence, this chapter presents SVMs together with applications in target regression and classification.

8.1 Support Vector Machines: Basic Theory

Support vector machines comprise a new class of learning algorithms, originally developed for pattern recognition [3, 46], and motivated by results of statistical learning theory [43].

The problem of empirical data modeling is germane to many engineering applications. In empirical data modeling a process of induction is used to build up a model of the system under study, and it is hoped to deduce responses of the system that have yet to be observed.

The SVMs aim at minimizing an upper bound of the generalization error by maximizing the margin between the separating hyperplane[1] and the data. What makes SVMs attractive is the property of condensing information in the training data and providing a sparse representation by using a very small number of data points (support vectors, SVs) [17].

8.1.1 Statistical Learning Theory

This subsection is a very brief introduction to statistical learning theory [42, 44].

Statistical learning theory based modeling aims at choosing a model from the hypothesis space, which is closest (with respect to some error measure) to the underlying function in the target space.

Errors in SVM modeling arise from two cases [19]:

- *Approximation Error* is a consequence of a poor choice of the hypothesis space smaller than the target space, which will result in a large approximation error, and is referred to as *model mismatch*.
- *Estimation Error* is the error due to the learning procedure which results in a technique selecting the non-optimal model from the hypothesis space.

In order to choose the best available approximation to the supervisor's response, it is important to measure the loss or discrepancy $L(y, f(\mathbf{w}, \mathbf{x}))$ between the response y of the supervisor to a given input $\mathbf{x}$ and the response $f(\mathbf{w}, \mathbf{x})$ provided by the learning machine.

Given a training set of N independent observations $(\mathbf{x}_1, y_1), \ldots, (\mathbf{x}_N, y_N)$. We would like to find the function f such that the expected value of the loss, called the risk function, is minimized:

$$R(\mathbf{w}) = \int L(y, f(\mathbf{w}, \mathbf{x})) \mathrm{d}P(\mathbf{x}, y), \qquad (8.1.1)$$

where $P(\mathbf{x}, y)$ is the joint probability distribution. The goal of SVM is to minimize the risk function $R(\mathbf{w})$ over the class of functions $f(\mathbf{w}, \mathbf{x})$, $\mathbf{w} \in W$.

---

[1] In geometry a hyperplane is a subspace whose dimension is one less than that of its ambient space. For example, if a space is three-dimensional then its hyperplanes are the two-dimensional planes, while if the space is two-dimensional, its hyperplanes are the one-dimensional lines.

8.1 Support Vector Machines: Basic Theory

The problem is: the joint probability distribution $P(\mathbf{x}, y) = P(y|\mathbf{x})P(\mathbf{x})$ is unknown, and the only available information is contained in the training set $\{(\mathbf{x}_1, y_1), \ldots, (\mathbf{x}_N, y_N)\}$.

In order to solve this problem, the incalculable risk function $R(\mathbf{w})$ should be approximated by the empirical risk function

$$R_{\text{emp}}(\mathbf{w}) = \frac{1}{N} \sum_{i=1}^{N} L(y_i, f(\mathbf{w}, \mathbf{x}_i)) \tag{8.1.2}$$

which can be constructed by using the training set $\{(\mathbf{x}_1, y_1), \ldots, (\mathbf{x}_N, y_N)\}$. In the above equation, the Boolean function $f(\mathbf{w}, \mathbf{x}_i)$ on the input feature $\mathbf{x}_i$ and a set of Boolean functions on the input features $\{f(\mathbf{w}, \mathbf{x}_i), \mathbf{w} \in W, \mathbf{x}_i \in X\}$ are known as *hypothesis* and *hypothesis space*, respectively.

Empirical risk minimization (ERM) principle is to minimize $R_{\text{emp}}(\mathbf{w})$ over the set $\mathbf{w} \in W$, results in a risk $R(\mathbf{w}_i^*)$ which is close to its minimum.

The evaluation of the soundness of the ERM principle requires answers to the following two questions [42]:

1. Does the empirical risk $R_{\text{emp}}(\mathbf{w})$ converge uniformly to the actual risk $R(\mathbf{w})$ over the full set $f(\mathbf{w}, \mathbf{x}), \mathbf{w} \in W$, namely

$$\text{Prob}\left\{ \sup_{\mathbf{w} \in W} |R(\mathbf{w}) - R_{\text{emp}}(\mathbf{w})| > \varepsilon \right\} \to 0 \quad \text{as } N \to \infty. \tag{8.1.3}$$

2. What is the rate of convergence?

The theory of uniform convergence of empirical risk to actual risk includes necessary and sufficient conditions as well as bounds for the rate of convergence. These bounds independent of the distribution function $P(x, y)$ are based on the *Vapnik–Chervonekis (VC) dimension* of the set of functions $f(\mathbf{w}, \mathbf{x})$ implemented by the learning machine.

For a set of functions $\{f(\mathbf{w})\}$, $\mathbf{w}$ is a generic set of parameters: a choice of $\mathbf{w}$ specifies a particular function and can be defined for various classes of function f.

Let us consider only the functions corresponding to the two-class pattern recognition case, so that $f(\mathbf{w}, \mathbf{x}) \in \{-1, 1\}, \forall \mathbf{w}, \mathbf{x}$.

If a given set of N data points can be labeled in all possible 2^N ways, and for each labeling, a member of the set $\{f(\mathbf{w})\}$ with correctly assigned labels can be found, then the set of points is said to be shattered by the set of functions.

Suppose that the space in which the data live is $\mathbb{R}^2$, and the set $\{f(\mathbf{w})\}$ consists of oriented straight lines, so that for a given line, all points on one side are assigned to the class 1, and all points on the other side, the class -1, as shown in Fig. 8.1.

One can find at least one set of 3 points in 2D all of whose $2^3 = 8$ possible labeling can be separated by some hyperplane. The orientation is shown by an arrow, specifying on which side of the line points is to be assigned the label 1. Any set of 4 points, all of whose $2^4 = 16$ possible labeling, are not separable by hyperplanes.

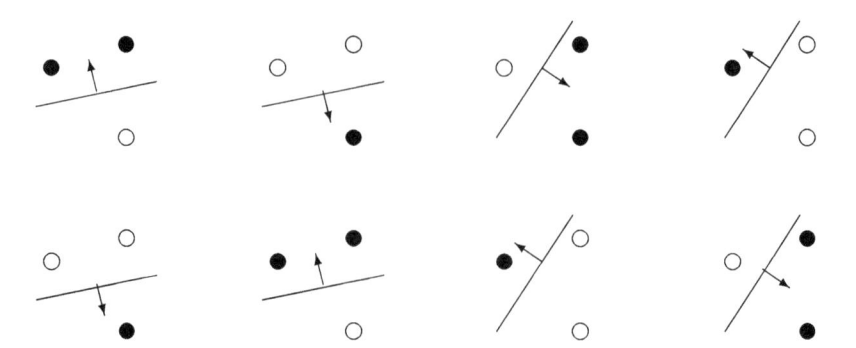

Fig. 8.1 Three points in $\mathbb{R}^2$, shattered by oriented lines. The arrow points to the side with the points labeled black

Suppose we have a data set containing N points. These N points can be labeled in $2N$ ways as positive "+" and negative "−." Therefore, $2N$ different learning problems can be defined by N data points. If for any of these problems, we can find a hypothesis $h \in \mathcal{H}$ that separates the positive examples from the negative, then we say $\mathcal{H}$ shatters N points. That is, any learning problem definable by N examples can be learned without error by a hypothesis drawn from $\mathcal{H}$.

Definition 8.1 (Indicator Function [12]) Let $\mathcal{D}$ be a $2n$ full factorial design with levels -1 and $+1$, and a fractional factorial design $\mathcal{F}$ be a subset of $\mathcal{D}$. The indicator function F of its fraction $\mathcal{F}$ is a function defined on $\mathcal{D}$ as follows:

$$F(\mathbf{x}) = \begin{cases} +1, & \text{if } \mathbf{x} \in \mathcal{F}; \\ -1, & \text{if } \mathbf{x} \in \mathcal{D} \setminus \mathcal{F}. \end{cases} \tag{8.1.4}$$

Definition 8.2 (VC-Dimension [42]) The *VC-dimension* of a set of indicator functions $f(\mathbf{w}, \mathbf{x}), \mathbf{w} \in W$ is the maximal number h of vectors which can be shattered in all possible $2h$ ways by $f(\mathbf{w}, \mathbf{x}), \mathbf{w} \in W$.

For example, $h = n + 1$ for linear decision rules in n-dimensional space, since they can shatter at most $n + 1$ points.

It is well-known [45] that the finiteness of the VC-dimension of the set of indicator functions implemented by the learning machine forms the necessary and sufficient condition for consistency of the ERM method independent of probability measure. Finiteness of VC-dimension also implies fast convergence.

Theorem 8.1 ([6]) *Consider some set of m points in $\mathbb{R}^n$. Choose any one of the points as origin. Then the m points can be shattered by oriented hyperplanes if and only if the position vectors of the remaining points are linearly independent.*

Corollary 8.1 ([6]) *The VC-dimension of the set of oriented hyperplanes in $\mathbb{R}^n$ is $n + 1$, since we can always choose $n + 1$ points, and then choose one of the points as origin, such that the position vectors of the remaining n points are linearly* independent, but can never choose $n + 2$ such points (since no $n + 1$ vectors in $\mathbb{R}^n$ can be linearly independent).

8.1.2 Linear Support Vector Machines

We discuss linear SVMs in two cases: the separable case and the nonseparable case.

The Separable Case

Consider first the simplest case: linear machines trained on separable data. We are given the labeled training data $\{\mathbf{x}_i, y_i\}$ with $i = 1, \ldots, N$, $y_i \in \{-1, +1\}$, $\mathbf{x}_i \in \mathbb{R}^n$. Suppose we have a "separating hyperplane" which separates the positive from the negative data samples. The points $\mathbf{x}$ on the hyperplane satisfy $\mathbf{w}^T \mathbf{x} + b = 0$, where $\mathbf{w}$ is normal to the hyperplane, $|b|/\|\mathbf{w}\|_2$ is the perpendicular distance from the hyperplane to the origin, and $\|\mathbf{w}\|_2$ is the Euclidean norm of $\mathbf{w}$.

Lemma 8.1 ([6]) *Two sets of points in $\mathbb{R}^n$ may be separated by a hyperplane if and only if the intersection of their convex hulls is empty.*

Let d_+ and d_- be the shortest distances from the separating hyperplane to the closest positive and negative data samples, respectively. Then, the "margin" of a separating hyperplane can be defined to be $d_+ + d_-$.

For the linearly separable case, the support vector algorithm aims at the separating hyperplane with the largest margin $d_+ + d_-$. For this end, all the training data need to satisfy the following constrains:

$$\mathbf{w}^T \mathbf{x}_i + b \geq +1, \text{ for } y_i = +1, \qquad (8.1.5)$$

$$\mathbf{w}^T \mathbf{x}_i + b \leq -1, \text{ for } y_i = -1. \qquad (8.1.6)$$

The above two constrains can be combined into one set of inequality constraints:

$$y_i(\mathbf{w}^T \mathbf{x}_i + b) - 1 \geq 0, \quad \forall\, i = 1, \ldots, N. \qquad (8.1.7)$$

Clearly, the points satisfying the equality in (8.1.5) lie on the hyperplane $H_1 : \mathbf{w}^T \mathbf{x}_i + b = 1$ with normal $\mathbf{w}$ and perpendicular distance from the origin $|1 - b|/\|\mathbf{w}\|_2$. Similarly, the points satisfying the equality in Eq. (8.1.6) lie on the hyperplane $H_2 : \mathbf{w}^T \mathbf{x} + b = -1$ with normal again $\mathbf{w}$ and perpendicular distance from the origin $|-1 - b|/\|\mathbf{w}\|_2$. Then $d_+ = d_- = 1/\|\mathbf{w}\|_2$ and hence the margin is simply $2/\|\mathbf{w}\|_2$.

Due to having the same normal, H_1 and H_2 are parallel, and no training points fall between them. Thus the pair of hyperplanes (H_1, H_2) which gives the maximum margin can be found by minimizing $\frac{1}{2}\|\mathbf{w}\|_2^2$ subject to constraints (8.1.7), namely we have the constrained optimization problem:

$$\min_{\mathbf{w},b} \frac{1}{2}\|\mathbf{w}\|_2^2, \tag{8.1.8}$$

$$\text{subject to } y_i(\mathbf{w}^T\mathbf{x}_i + b) - 1 \geq 0, \quad \forall\, i = 1, \ldots, N. \tag{8.1.9}$$

For the above equality constraints, the Lagrange multiplier method gives the following unconstrained primal optimization problem:

$$\min_{\mathbf{w},b,\boldsymbol{\alpha}} \mathcal{L}_P(\mathbf{w},b,\boldsymbol{\alpha}) = \frac{1}{2}\|\mathbf{w}\|_2^2 - \sum_{i=1}^{N} \alpha_i y_i(\mathbf{w}^T\mathbf{x}_i + b) + \sum_{i=1}^{N} \alpha_i \tag{8.1.10}$$

with nonnegative Lagrange multipliers $\alpha_i \geq 0,\ \forall\, i = 1, \ldots, N$.

From the first-order optimality conditions we have

$$\frac{\partial \mathcal{L}}{\partial \mathbf{w}} = 0 \ \Rightarrow\ \mathbf{w} = \sum_{i=1}^{N} \alpha_i y_i \mathbf{x}_i, \tag{8.1.11}$$

$$\frac{\partial \mathcal{L}_P}{\partial b} = 0 \ \Rightarrow\ \sum_{i=1}^{N} \alpha_i y_i = 0, \tag{8.1.12}$$

Substituting (8.1.11) and (8.1.12) into (8.1.10) can eliminate b, and thus yield the dual objective function

$$\mathcal{L}_D(\boldsymbol{\alpha}) = \sum_{i=1}^{N} \alpha_i - \frac{1}{2} \sum_{i=1}^{N} \sum_{j=1}^{N} \alpha_i \alpha_j y_i y_j \mathbf{x}_i^T \mathbf{x}_j. \tag{8.1.13}$$

Combining (8.1.13) with (8.1.12), we get the *Wolfe dual optimization problem* corresponding to (8.1.10) as follows:

$$\max_{\boldsymbol{\alpha}} \mathcal{W}(\boldsymbol{\alpha}) = \sum_{i=1}^{N} \alpha_i - \frac{1}{2} \sum_{i=1}^{N} \sum_{j=1}^{N} \alpha_i \alpha_j y_i y_j \mathbf{x}_i^T \mathbf{x}_j, \tag{8.1.14}$$

$$\text{subject to } \sum_{i=1}^{N} \alpha_i y_i = 0, \tag{8.1.15}$$

for the separable case.

8.1 Support Vector Machines: Basic Theory

The Nonseparable Case

When applied to nonseparable data, the above separable optimization algorithm will result into no feasible solution. Hence, the inequality constraints (8.1.5) and (8.1.6) must be relaxed by introducing positive slack variables $\xi_i, i = 1, \ldots, N$:

$$\mathbf{w}^T \mathbf{x}_i + b \geq +1 - \xi_i, \quad \text{for } y_i = +1, \tag{8.1.16}$$

and

$$\mathbf{w}^T \mathbf{x}_i + b \leq -1 + \xi_i, \quad \text{for } y_i = -1, \tag{8.1.17}$$

$$\xi_i \geq 0, \quad \forall i = 1, \ldots, N. \tag{8.1.18}$$

For an error to occur, the corresponding ξ_i must exceed unity, so $\sum_i \xi_i$ is an upper bound on the number of training errors. Hence, a natural way to assign an extra cost for errors is to change the objective function to be minimized from $\|\mathbf{w}\|_2^2/2$ to $\|\mathbf{w}\|_2^2/2 + C(\sum_i \xi_i)$, where C is a parameter to be chosen by the user, a larger C corresponding to assigning a higher penalty to errors.

Consider the primal minimization problem

$$\min_{\mathbf{w},b,\boldsymbol{\alpha},C} \mathcal{L}_P(\mathbf{w}, b, \boldsymbol{\alpha}, C) = \frac{1}{2}\|\mathbf{w}\|_2^2 - \sum_{i=1}^{N} \alpha_i y_i (\mathbf{w}^T \mathbf{x}_i + b)$$

$$+ C \sum_{i=1}^{N} \xi_i + \sum_{i=1}^{N} \alpha_i, \tag{8.1.19}$$

$$\text{subject to} \quad 0 \leq \alpha_i \leq C, \quad i = 1, \ldots, N. \tag{8.1.20}$$

By the first-order optimality conditions it follows that

$$\frac{\partial \mathcal{L}_P}{\partial \mathbf{w}} = 0 \quad \Rightarrow \quad \mathbf{w} = \sum_{i=1}^{N} \alpha_i y_i \mathbf{x}_i, \tag{8.1.21}$$

$$\frac{\partial \mathcal{L}_P}{\partial C} = 0 \quad \Rightarrow \quad \sum_{i=1}^{N} \xi_i = 0, \tag{8.1.22}$$

$$\frac{\partial \mathcal{L}_P}{\partial b} = 0 \quad \Rightarrow \quad \sum_{i=1}^{N} \alpha_i y_i = 0. \tag{8.1.23}$$

By substituting (8.1.21) and (8.1.22) into (8.1.19), and using the constraints (8.1.20) and (8.1.23), we can eliminate ξ_i and C in the loss to get

$$\max_{\boldsymbol{\alpha}} \left\{ \sum_{i=1}^{N} \alpha_i - \sum_{i=1}^{N}\sum_{j=1}^{N} \alpha_i \alpha_j y_i y_j \mathbf{x}_i^T \mathbf{x}_j \right\}, \qquad (8.1.24)$$

$$\text{subject to} \quad 0 \leq \alpha_i \leq C, \quad i = 1, \ldots, N, \qquad (8.1.25)$$

$$\sum_{i=1}^{N} \alpha_i y_i = 0, \qquad (8.1.26)$$

which is called the Wolfe dual optimization problem for finding $\boldsymbol{\alpha}$.

8.2 Kernel Regression Methods

Following the development of support vector machines, positive definite kernels have recently attracted considerable attention in the machine learning community.

8.2.1 Reproducing Kernel and Mercer Kernel

Consider how a linear SVM is generalized to the case where the decision function $f(\mathbf{x})$, whose sign represents the class assigned to data point $\mathbf{x}$, is not a linear function of the data $\mathbf{x}$?

Definition 8.3 (Reproducing Kernel Hilbert Space [47]) A *reproducing kernel Hilbert space* (RKHS) is a Hilbert space $\mathcal{H}$ of real-valued function on a compact domain X with the *reproducing property* that for $f \in \mathcal{H}$, and each $\mathbf{x} \in X$, there exists M_x, not depending on f such that

$$\mathcal{F}_x[f] = |f(\mathbf{x})| = \langle f(\cdot), K_x(\cdot) \rangle \leq M_x \|f\|, \quad \forall f \in \mathcal{H}, \qquad (8.2.1)$$

where $\mathcal{F}_x[f]$ is called the *evaluation functional* of f. That is to say, $\mathcal{F}_x[f]$ is a bounded linear functional.

The reproducing property above states that the inner product of a function f and a kernel function K_x, $\langle f(\cdot), K_x(\cdot) \rangle$, is $f(\mathbf{x})$ in the RKHS. Because K_x must ensure reproducing the function f, the function $K_x(\cdot)$ is called the *reproducing kernel* of f for the RKHS H.

8.2 Kernel Regression Methods

A natural question to ask is: how to choose a function K_x which is a reproducing kernel? To answer this question, we define K_x as a two-variable function

$$K(\mathbf{x}, \mathbf{y}) = K_x(\mathbf{y}). \qquad (8.2.2)$$

Definition 8.4 (Gram Matrix [35]) Given a kernel $K(\mathbf{x}, \mathbf{y})$ and patterns $\mathbf{x}_1, \ldots, \mathbf{x}_N \in X$, where X is a closed subset of $\mathbb{R}^n$. The $N \times N$ matrix $\mathbf{K}$ with entries

$$K_{ij} = K(\mathbf{x}_i, \mathbf{x}_j) \quad \text{and} \quad K_{ij} = K_{ji} \qquad (8.2.3)$$

is called the *Gram matrix* of the kernel K with respect to $\mathbf{x}_1, \ldots, \mathbf{x}_N$.

Definition 8.5 (Positive Definite Kernel [47]) Let X be a closed subset of $\mathbb{R}^n$, and $K : X \times X \to \mathbb{R}$ a symmetric function satisfying: for any finite set of points $\{\mathbf{x}_i\}_{i=1}^N$ in X and real numbers $\{a_i\}_{i=1}^N$,

$$\sum_{i=1}^{N} \sum_{j=1}^{N} a_i a_j K(\mathbf{x}_i, \mathbf{x}_j) \geq 0. \qquad (8.2.4)$$

Then K is said to be a *positive definite kernel* on X.

Denote kernels $K_i : X \times X \to \mathbb{R}$, then positive definition (pd) kernels have the following general properties:

1. If $K_1, \ldots, K_n$ are p.d. kernels and $\lambda_1, \ldots, \lambda_n \geq 0$, then the sum $\sum_{i=1}^{n} \lambda_i K_i$ is p.d.
2. If $K_1, \ldots, K_n$ are p.d. kernels and $\alpha_1, \ldots, \alpha_n \in \{1, \ldots, N\}$, then their product $\prod_{i=1}^{n} K_i^{\alpha_i}$ is p.d.
3. For a sequence of p.d. kernels, the limit $K = \lim_{n \to \infty} K_n$ is p.d., if the limit exists.
4. If $X_0 \subseteq X$, then $K_0 : X_0 \times X_0 \to \mathbb{R}$ is also a p.d. kernel.
5. If $K_i : X_i \times X_i \to \mathbb{R}$ is a sequence of p.d. kernels, then

    - $K((\mathbf{x}_1, \ldots, \mathbf{x}_n), (\mathbf{y}_1, \ldots, \mathbf{y}_n)) = \prod_{i=1}^{n} K_i(\mathbf{x}_i, \mathbf{y}_i)$ is p.d. kernel on $X_1 \times \cdots \times X_n$.
    - $K((\mathbf{x}_1, \ldots, \mathbf{x}_n), (\mathbf{y}_1, \ldots, \mathbf{y}_n)) = \sum_{i=1}^{n} K_i(\mathbf{x}_i, \mathbf{y}_i)$ is p.d. kernel on $X_1 \times \cdots \times X_n$.

Definition 8.6 (Reproducing Kernel [1]) Let F be a class of functions $f(\mathbf{x})$ defined in a set E, forming a Hilbert space (complex or real). The function $K(\mathbf{x}, \mathbf{y})$ of $\mathbf{x}$ and $\mathbf{y}$ in E is called a *reproducing kernel* of F if

1. For every $\mathbf{y}$, $K(\mathbf{x}, \mathbf{y})$ as function of $\mathbf{x}$ belongs to F.

2. The *reproducing property*: for every $\mathbf{y} \in E$ and every $f \in F$,

$$f(\mathbf{y}) = \langle f(\mathbf{x}), K(\mathbf{x}, \mathbf{y}) \rangle_{x^*}. \quad (8.2.5)$$

Here the subscript x^* by the scalar product indicates that the scalar product applies to functions of $\mathbf{x}$.

Theorem 8.2 (Moore–Aronszajn Theorem [1]) *Let* $\mathbf{x}, \mathbf{y} \in X$ *and* $K(\mathbf{x}, \mathbf{y}) : X \times X \to \mathbb{R}$ *be positive defined. Then, there must be a unique RKHS* $\mathcal{H} \subset \mathbb{R}^X$ *whose reproducing kernel is* $K(\mathbf{x}, \mathbf{y})$. *Moreover, if the space* $\mathcal{H}_0 = \text{span}(\{K(\cdot, \mathbf{x})\}_{\mathbf{x} \in X})$ *has the inner product*

$$\langle f, g \rangle_{\mathcal{H}_0} = \sum_{i=1}^{n} \sum_{j=1}^{n} \alpha_i \beta_j K(\mathbf{x}_i, \mathbf{x}_j), \quad (8.2.6)$$

where $f = \sum_{i=1}^{n} \alpha_i K(\cdot, \mathbf{x}_i)$ *and* $g = \sum_{j=1}^{n} \beta_j K(\cdot, \mathbf{x}_j)$, *then* $\mathcal{H}_0$ *is an effective RKHS.*

Moore–Aronszajn theorem states that there is one-to-one correspondence between positive definite kernel function $K(\mathbf{x}, \mathbf{y})$ and RKHS $\mathcal{H}$. This is why the kernel function is limited to the reproducing kernel Hilbert space $\mathcal{H}$.

The reproducing kernel has the following basic properties [1]:

- *Uniqueness*. If a reproducing kernel $K(\mathbf{x}, \mathbf{y})$ exists, then it is unique.
- *Existence*. For the existence of a reproducing kernel $K(\mathbf{x}, \mathbf{y})$ it is necessary and sufficient that for every $\mathbf{y}$ of the set E, $f(\mathbf{y})$ be a continuous functional of f running through the Hilbert space F.
- *Positiveness*. $\mathbf{K} = [K(\mathbf{x}_i, \mathbf{y}_j)]_{i,j=1}^{n,n}$ is a *positive matrix* in the sense of E.
- *One-to-one correspondence*. To every positive matrix $\mathbf{K} = [K(\mathbf{x}_i, \mathbf{y}_j)]_{i,j=1}^{n,n}$, there corresponds one and only one class of functions with a uniquely determined quadratic form in it, forming a Hilbert space and admitting $K(\mathbf{x}, \mathbf{y})$ as a reproducing kernel.
- *Convergence*. If the class F possesses a reproducing kernel $K(\mathbf{x}, \mathbf{y})$, every sequence of functions $\{f_n\}$ which converges strongly to a function f in the Hilbert space F, converges also at every point in the ordinary sense, $\lim f_n(\mathbf{x}) = f(\mathbf{x})$.

The following Mercer's theorem provides a construction method of the nonlinear reproducing kernel function.

Theorem 8.3 (Mercer's Theorem) *Each positive definite kernel* $K(\mathbf{x}, \mathbf{y})$ *defined on a compact domain* $X \times X$ *can be written in the form*

$$K(\mathbf{x}, \mathbf{y}) = \sum_{i=1}^{M} \lambda_i \phi_i(\mathbf{x}) \phi_i(\mathbf{y}), \quad M \leq \infty. \quad (8.2.7)$$

8.2 Kernel Regression Methods

The kernel satisfying Mercer's theorem is called the *Mercer kernel*. According to Mercer's theorem, the nonlinear kernel function is usually constructed by

$$K(\mathbf{x}_i, \mathbf{y}_i) = \boldsymbol{\phi}^T(\mathbf{x}_i)\boldsymbol{\phi}(\mathbf{y}_i,) \tag{8.2.8}$$

where $\boldsymbol{\phi}(\mathbf{x}) = [\phi_1(\mathbf{x}), \ldots, \phi_M(\mathbf{x})]^T$ is a nonlinear function.

The widely used Mercer kernel is the *Gaussian radial basis function (GRBF) kernel*:

$$K(\mathbf{x}_i, \mathbf{x}_j) = \exp\left(-\frac{\|\mathbf{x}_i - \mathbf{x}_j\|_2^2}{2\sigma^2}\right) \tag{8.2.9}$$

or the *exponential radial basis function kernel*

$$K(\mathbf{x}_i, \mathbf{x}_j) = \exp\left(-\frac{\|\mathbf{x}_i - \mathbf{x}_j\|_2}{2\sigma^2}\right). \tag{8.2.10}$$

For a radial basis function (RBF) kernel one has the following approximation [16]:

$$\int_{\mathbf{x}} p(\mathbf{x})^2 d\mathbf{x} \approx \frac{1}{N^2} \sum_{i=1}^{N} \sum_{j=1}^{N} K_{ij} = \mathbf{1}_N^T \mathbf{K} \mathbf{1}_N, \tag{8.2.11}$$

where $\mathbf{1}_N$ is an $N \times 1$ summation vector with all entries equal to 1.

If the $N \times N$ kernel matrix $\mathbf{K}$ has the eigenvalue decomposition $\mathbf{K} = \mathbf{U}\boldsymbol{\Lambda}\mathbf{U}^T$, then

$$\mathbf{1}_N^T \mathbf{K} \mathbf{1}_N = \mathbf{1}_N^T \left(\sum_{i=1} \lambda_i \mathbf{u}_i \mathbf{u}_i^T\right) \mathbf{1}_N = \sum_{i=1}^{N} \lambda_i \left(\mathbf{1}_N^T \mathbf{u}_i\right)^2. \tag{8.2.12}$$

This result has two important applications.

1. *Kernel PCA:* For given data vectors $\mathbf{x}_1, \ldots, \mathbf{x}_N$ and an RBF kernel, K dominant eigenvalues and their corresponding eigenvectors of the kernel matrix $\mathbf{K}$ give the *kernel PCA* after the principal components in standard PCA are replaced by the principal components of the kernel matrix $\mathbf{K}$. The kernel PCA was originally proposed by Schölkopf et al. [33].
2. *Kernel K-means clustering:* If there are K distinct clustered regions within the N data samples, then there will be K dominant terms $\lambda_i \left(\mathbf{1}_N^T \mathbf{u}_i\right)$ that provide a means of estimating the possible number of clusters within the data sample in kernel based K-means clustering [16].

Kernel PCA is a nonlinear generalization of PCA in the sense that it is performing PCA in feature spaces of arbitrarily large (possibly infinite) dimensionality. If the kernels $K(\mathbf{x}_i, \mathbf{x}_j) = \mathbf{x}_i^T \mathbf{x}_j$ are used, then the kernel PCA reduces to standard PCA. Compared to the standard PCA, kernel PCA has the main advantage that no nonlinear optimization is involved; it is essentially linear algebra, as simple as standard PCA.

8.2.2 Representer Theorem and Kernel Regression

We now focus on applications of nonlinear kernel functions in SVMs. The following are widely used types of *kernel functions* in SVMs [19, 36]:

1. *Linear SVM* uses the linear kernel function

$$K(\mathbf{x}, \mathbf{x}_k) = \langle \mathbf{x}, \mathbf{x}_k \rangle = \mathbf{x}^T \mathbf{x}_k. \tag{8.2.13}$$

2. *Polynomial SVM* uses the polynomial kernel of degree d

$$K(\mathbf{x}, \mathbf{x}_k) = (\langle \mathbf{x}, \mathbf{x}_k \rangle + 1)^d. \tag{8.2.14}$$

3. *Radial basis function (RBF) SVM* consists of the *Gaussian radial basis function*

$$K(\mathbf{x}, \mathbf{x}_k) = \exp\left(\frac{\|\mathbf{x} - \mathbf{x}_k\|^2}{2\sigma^2}\right) \tag{8.2.15}$$

or *exponential radial basis function*

$$K(\mathbf{x}, \mathbf{x}_k) = \exp\left(\frac{\|\mathbf{x} - \mathbf{x}_k\|}{2\sigma^2}\right). \tag{8.2.16}$$

4. *Multilayer SVM* uses the *multilayer perceptron kernel function*

$$K(\mathbf{x}, \mathbf{x}_k) = \tanh(\rho \langle \mathbf{x}, \mathbf{x}_k \rangle + \vartheta) = \tanh(\rho \mathbf{x}^T \mathbf{x}_k + \vartheta) \tag{8.2.17}$$

for certain values of the scale ρ and offset ϑ parameters. Here the support vector (SV) corresponds to the first layer and the Lagrange multipliers to the weights.

5. *B-splines SVM* consists of B-splines kernel function of order $2M + 1$:

$$K(\mathbf{x}, \mathbf{x}_k) = B_{2M+1}(\|\mathbf{x} - \mathbf{x}_k\|) \tag{8.2.18}$$

with

$$B_k(x) = \sum_{r=0}^{k+1} \frac{(-1)^r}{k!} \binom{k+1}{r} \left(x + \frac{k+1}{2} - r\right)_+. \tag{8.2.19}$$

8.2 Kernel Regression Methods

6. *Summing SVM* uses *additive kernel function*

$$K(\mathbf{x}, \mathbf{x}_k) = \sum_i K_i(\mathbf{x}, \mathbf{x}_k) \tag{8.2.20}$$

which is obtained by forming summing kernels, since the sum of two positive definite functions is positive definite.

To address the poor generalization properties of SVMs, regularization networks, Gaussian processes, and spline methods, a popular technique is to solve regularized optimization problem in a reproducing kernel Hilbert space $\mathcal{H}$:

$$\mathbf{f}^* = \arg\min_{\mathbf{f} \in \mathcal{H}} \frac{1}{l} \sum_{i=1}^{l} V(\mathbf{x}_i, y_i, \mathbf{f}) + \frac{1}{2}\gamma \|\mathbf{f}\|_{\mathcal{H}}^2, \tag{8.2.21}$$

where $\mathbf{f}$ is regressor or classifier.

The solution to the problem (8.2.21) was given by Kimeldorf and Wahba in 1971 [24], known as the *representer theorem*.

Theorem 8.4 (Representer Theorem for Supervised Learning [24, 47]) *Given a set of l labeled examples* $\{(\mathbf{x}_i, y_i)\}_{i=1}^{l}$. *Any solution to the optimization problem (8.2.21) for SVMs has a representation of the form*

$$f(\mathbf{x}) = \sum_{j=1}^{l} \alpha_j K(\mathbf{x}, \mathbf{x}_j), \tag{8.2.22}$$

where $\{\alpha_j\}_{j=1}^{l} \in \mathbb{R}$.

Denoting $f_i = f(\mathbf{x}_i) = \sum_{j=1}^{l} \alpha_j K(\mathbf{x}_i, \mathbf{x}_j)$ then we have

$$\mathbf{f} = [f_1, \ldots, f_l]^T = \begin{bmatrix} \sum_{j=1}^{l} \alpha_j K(\mathbf{x}_1, \mathbf{x}_j) \\ \vdots \\ \sum_{j=1}^{l} \alpha_j K(\mathbf{x}_l, \mathbf{x}_j) \end{bmatrix} \Rightarrow \|\mathbf{f}\|_{\mathcal{H}}^2 = \mathbf{f}^T \mathbf{f} = \boldsymbol{\alpha}^T \mathbf{K} \boldsymbol{\alpha}. \tag{8.2.23}$$

Letting $\sum_{i=1}^{l} V(\mathbf{x}_i, y_i, \mathbf{f}) = \sum_{i=1}^{l}(y_i - \hat{f}_i)^2 = \|\mathbf{y} - \mathbf{K}\boldsymbol{\alpha}\|_2^2$, substituting (8.2.22) into (8.2.21), and using (8.2.23), we get the loss expressed by $\boldsymbol{\alpha}$:

$$L(\boldsymbol{\alpha}) = (\mathbf{y} - \mathbf{K}\boldsymbol{\alpha})^T (\mathbf{y} - \mathbf{K}\boldsymbol{\alpha}) + \frac{1}{2}\gamma_A \boldsymbol{\alpha}^T \mathbf{K} \boldsymbol{\alpha}. \tag{8.2.24}$$

From $\frac{\partial L(\alpha)}{\partial \alpha} = \mathbf{0}$ it follows that $(-\mathbf{K})(\mathbf{y} - \mathbf{K}\alpha) + \gamma_A \mathbf{K}\alpha = \mathbf{0}$, which yields the optimal solution

$$\alpha^* = (\mathbf{K} + \gamma_A \mathbf{I})^{-1} \mathbf{y}. \tag{8.2.25}$$

This is just the Tikhonov regularization least squares solution.

8.2.3 Semi-Supervised and Graph Regression

Representer theorem for supervised learning (Theorem 8.4) can be extended to semi-supervised learning and/or graph signals.

Theorem 8.5 (Representer Theorem for Semi-Supervised Learning [2]) *Given a set of l labeled examples $\{(\mathbf{x}_i, y_i)\}_{i=1}^{l}$, a set of u unlabeled examples $\{\mathbf{x}_j\}_{j=l+1}^{l+u}$, and the graph Laplacian $\mathbf{L}$. The minimizer of semi-supervised/graph optimization problem*

$$\mathbf{f}^* = \arg\min_{\mathbf{f}} \left\{ \frac{1}{l} \sum_{i=1}^{l} V(\mathbf{x}_i, y_i, \mathbf{f}) + \frac{1}{2}\gamma_A \|\mathbf{f}\|_{\mathcal{H}}^2 + \frac{\gamma_I}{(u+l)^2} \mathbf{f}^T \mathbf{L} \mathbf{f} \right\}, \tag{8.2.26}$$

admits an expansion

$$\mathbf{f}^*(\mathbf{x}) = \sum_{i=1}^{l+u} \alpha_i K(\mathbf{x}, \mathbf{x}_i) \tag{8.2.27}$$

in terms of the labeled and unlabeled examples.

Note that the above representer theorem [2] is different from the generalized representer theorem in [35]. When the graph Laplacian $\mathbf{L} = \mathbf{I}$, Theorem 8.5 reduces to the representer theorem for semi-supervised learning. If letting further $u = 0$, then Theorem 8.5 reduces to the representer theorem 8.4 for supervised learning.

Theorem 8.5 shows that the optimization problem (8.2.26) is equivalent to finding the optimal solution α^*.

Letting the loss $V(\alpha) = \sum_{i=1}^{l}(y_i - \hat{f}_i(\alpha))^2 = \|\mathbf{y} - \mathbf{JK}\alpha\|_2^2$ and using (8.2.23), then the loss in (8.2.26) can be represented as [2]:

$$L(\alpha) = \frac{1}{l}(\mathbf{y} - \mathbf{JK}\alpha)^T (\mathbf{y} - \mathbf{JK}\alpha) + \frac{1}{2}\gamma_A \alpha^T \mathbf{K}\alpha + \frac{\gamma_I}{2(u+l)^2} \alpha^T \mathbf{KLK}\alpha, \tag{8.2.28}$$

where $\mathbf{K}$ is the $(l+u) \times (l+u)$ Gram matrix $K_{ij} = K(\mathbf{x}_i, \mathbf{x}_j)$, $\mathbf{y} = [y_1, \ldots, y_l, 0, \ldots, 0]^T$ is the $(l+u)$-dimensional label vector, and $\mathbf{J} =$

8.2 Kernel Regression Methods

Diag$(1, \ldots, 1, 0, \ldots, 0)$ is an $(l + u) \times (l + u)$ diagonal matrix with the first l diagonal entries 1 and the rest 0.

From $\frac{\partial L(\alpha)}{\partial \alpha} = \mathbf{0}$, one has the first-order optimization condition

$$\frac{1}{l}\left(-(\mathbf{JK})^T\right)(\mathbf{y} - \mathbf{JK}\alpha) + \left(\gamma_A \mathbf{K} + \frac{\gamma_I}{(u+l)^2}\mathbf{KLK}\right)\alpha = \mathbf{0}. \tag{8.2.29}$$

Due to $\mathbf{K}^T\mathbf{J}^T = \mathbf{KJ}$, $\mathbf{JJ} = \mathbf{J}$, and $\mathbf{Jy} - \mathbf{JJK}\alpha = \mathbf{Jy} - \mathbf{JK}\alpha$, the optimization condition (8.2.29) gives the optimal solution

$$\alpha^* = \left(\mathbf{JK} + \gamma_A l\mathbf{I} + \frac{\gamma_I l}{(u+l)^2}\mathbf{LK}\right)^{-1}\mathbf{Jy}. \tag{8.2.30}$$

This solution is known as *Laplacian regularized least squares (LapRLS) solution* [2].

Interestingly, LapRLS is also available for graph supervised regression and classification. In this setting, the number u of unlabeled samples is equal to zero, resulting to the $l \times l$ matrix $\mathbf{J} = \mathbf{I}$, and thus (8.2.30) becomes

$$\alpha^* = \left(\mathbf{K} + \gamma_A l\mathbf{I} + \frac{\gamma_I}{l}\mathbf{LK}\right)^{-1}\mathbf{y}. \tag{8.2.31}$$

This is just LapRLS solution for supervised regression and classification.

Another interesting fact is that LapRLS contains the regularized least squares (RLS) as a special example. The regularized least squares algorithm for non-graph signals is a fully supervised method where the optimal problem is to find the solution

$$\mathbf{f}^* = \arg\min_{\mathbf{f}} \frac{1}{l}\sum_{i=1}^{l}(y_i - f(\mathbf{x}_i))^2 + \gamma_A\|\mathbf{f}\|_K^2. \tag{8.2.32}$$

By the classical representer theorem, this solution is given by

$$\mathbf{f}^*(\mathbf{x}) = \sum_{i=1}^{l}\alpha_i^* K(\mathbf{x}, \mathbf{x}_i). \tag{8.2.33}$$

Substituting (8.2.33) into (8.2.32) then

$$\alpha^* = \arg\min_{\alpha}\left\{\frac{1}{l}(\mathbf{y} - \mathbf{K}\alpha)^T(\mathbf{y} - \mathbf{K}\alpha) + \gamma_A\alpha^T\mathbf{K}\alpha\right\}. \tag{8.2.34}$$

From the first-order optimization condition $\frac{\partial V(\alpha)}{\partial \alpha} = \mathbf{0}$, it is easy to get

$$\alpha^* = (\mathbf{K} + \gamma_A l\mathbf{I})^{-1}\mathbf{y}, \tag{8.2.35}$$

which is just the Tikhonov regularization least squares (RLS) solution. Clearly, the Tikhonov RLS solution is a special example of LapRLS solution (8.2.31) for supervised machine learning when the graph Laplacian $\mathbf{L}$ is a null matrix.

8.2.4 Kernel Partial Least Squares Regression

Given two data blocks $\mathbf{X}$ and $\mathbf{Y}$, consider kernel methods for partial least squares (PLS) regression. This type of methods is known as the *kernel partial least squares regression*.

Kernel PLS regression is a natural extension of PLS regression discussed in Sect. 6.9.2.

The key steps of PLS regression are given by

1. $\mathbf{w} = \mathbf{X}^T \mathbf{u}/(\mathbf{u}^T \mathbf{u})$.
2. $\mathbf{t} = \mathbf{X}\mathbf{w}$.
3. $\mathbf{c} = \mathbf{Y}^T \mathbf{t}/(\mathbf{t}^T \mathbf{t})$.
4. $\mathbf{u} = \mathbf{Y}^T \mathbf{c}/(\mathbf{c}^T \mathbf{c})$.
5. $\mathbf{p} = \mathbf{X}^T \mathbf{t}/(\mathbf{t}^T \mathbf{t})$.
6. $\mathbf{q} = \mathbf{Y}^T \mathbf{u}/(\mathbf{u}^T \mathbf{u})$.
7. $\mathbf{X} = \mathbf{X} - \mathbf{t}\mathbf{p}^T$.
8. $\mathbf{Y} = \mathbf{Y} - \mathbf{t}\mathbf{c}^T$.

Then, we have

$$\mathbf{t} = \mathbf{X}\mathbf{X}^T \mathbf{u}/(\mathbf{u}^T \mathbf{u}), \tag{8.2.36}$$

$$\mathbf{c} = \mathbf{Y}^T \mathbf{t}, \tag{8.2.37}$$

$$\mathbf{u} = \mathbf{Y}^T \mathbf{u}/(\mathbf{u}^T \mathbf{u}), \tag{8.2.38}$$

$$\mathbf{X} = \mathbf{X} - \mathbf{t}\mathbf{t}^T \mathbf{X}, \tag{8.2.39}$$

$$\mathbf{Y} = \mathbf{Y} - \mathbf{t}\mathbf{c}^T. \tag{8.2.40}$$

Using $\boldsymbol{\Phi} = \boldsymbol{\Phi}(\mathbf{X})$ instead of $\mathbf{X}$, then Eqs. (8.2.36) and (8.2.39) become

$$\mathbf{t} = \boldsymbol{\Phi}\boldsymbol{\Phi}^T \mathbf{u}/(\mathbf{u}^T \mathbf{u}) \quad \text{and} \quad \boldsymbol{\Phi} = \boldsymbol{\Phi} - \mathbf{t}\mathbf{t}^T \boldsymbol{\Phi}. \tag{8.2.41}$$

Therefore, the key steps of kernel *nonlinear iterative partial least squares* (NIPALS) regression are as follows [28, 32]:

Given $\boldsymbol{\Phi}_0 = \boldsymbol{\Phi}$ and the data block $\mathbf{Y}_0 = \mathbf{Y}$.

1. Randomly initialize $\mathbf{u}$.
2. $\mathbf{t} = \boldsymbol{\Phi}\boldsymbol{\Phi}^T \mathbf{u}$, $\mathbf{t} \leftarrow \mathbf{t}/(\mathbf{t}^T \mathbf{t})$.
3. $\mathbf{c} = \mathbf{Y}^T \mathbf{t}$.
4. $\mathbf{u} = \mathbf{Y}^T \mathbf{u}$, $\mathbf{u} \leftarrow \mathbf{u}/(\mathbf{u}^T \mathbf{u})$.

8.2 Kernel Regression Methods

5. Repeat Steps 2–4 until convergence of **t**.
6. Deflate the matrix: $\mathbf{\Phi \Phi}^T = (\mathbf{I} - \mathbf{tt}^T)\mathbf{\Phi \Phi}^T(\mathbf{I} - \mathbf{tt}^T)^T$.
7. Deflate the matrix: $\mathbf{Y} = \mathbf{Y} - \mathbf{tc}^T$.

The kernel NIPALS regression is an iterative process: after extraction of the first component $\mathbf{t}_1$ the algorithm starts again using the deflated matrices $\mathbf{\Phi \Phi}^T$ and $\mathbf{Y}$ computed in Step 6 and Step 7, and repeat Steps 2–7 until the deflated matrix $\mathbf{\Phi \Phi}^T$ or $\mathbf{Y}$ becomes a null matrix.

Once two matrices $\mathbf{T} = [\mathbf{t}_1, \ldots, \mathbf{t}_p]$ and $\mathbf{U} = [\mathbf{u}_1, \ldots, \mathbf{u}_p]$ are found by using the NIPALS regression algorithm, then the matrix regression coefficients **B** can be computed in the form similar to (6.9.26):

$$\mathbf{B} = \mathbf{\Phi}_0^T \mathbf{U}(\mathbf{T}^T \mathbf{\Phi}_0 \mathbf{\Phi}_0^T \mathbf{U})^{-1} \mathbf{T}^T \mathbf{Y}_0. \tag{8.2.42}$$

Then for a given new data block $\mathbf{X}_{\text{new}}$ and $\mathbf{\Phi}_{\text{new}} = \mathbf{\Phi}(\mathbf{X}_{\text{new}})$, then unknown $\mathcal{Y}$-values can be predicted as

$$\hat{\mathbf{Y}}_{\text{new}} = \mathbf{\Phi}_{\text{new}} \mathbf{B}. \tag{8.2.43}$$

MATLIB code for Kernel RLS algorithm can be found in [28, Appendix III].

8.2.5 Laplacian Support Vector Machines

Consider support vector machines for graph signals. Given a set of l labeled graph examples $\{(\mathbf{x}_i, y_i)\}_{i=1}^l$ and a set of u unlabeled graph examples $\{\mathbf{x}_j\}_{j=l+1}^{l+u}$. Let **f** be a semi-supervised SVM for graph training samples. The graph or *Laplacian support vector machines* are extended by solving the following optimization problem [2]:

$$\mathbf{f}^* = \arg\min_{\mathbf{f}} \left\{ \frac{1}{l} \sum_{i=1}^{l} \left(1 - y_i f(\mathbf{x}_i)\right)_+ + \gamma_A \|\mathbf{f}\|_K^2 + \frac{\gamma_I}{(u+l)^2} \mathbf{f}^T \mathbf{L} \mathbf{f} \right\}. \tag{8.2.44}$$

By the extended representer theorem (Theorem 8.5), the solution to the problem (8.2.44) is given by

$$\mathbf{f}^* = \sum_{i=1}^{l+u} \alpha_i^* K(\mathbf{x}, \mathbf{x}_i). \tag{8.2.45}$$

Substituting (8.2.45) into (8.2.44) and adding an unregularized bias term b in $\mathbf{f}^*$, then the primal problem for optimizing $\boldsymbol{\alpha}$ can be written as

$$\arg\min_{\boldsymbol{\alpha} \in \mathbb{R}^{l+u}, \boldsymbol{\xi} \in \mathbb{R}^l} \left\{ \frac{1}{l} \sum_{i=1}^{l} \xi_i + \gamma_A \boldsymbol{\alpha}^T \mathbf{K} \boldsymbol{\alpha} + \frac{\gamma_I}{(u+l)^2} \boldsymbol{\alpha}^T \mathbf{K} \mathbf{L} \mathbf{K} \boldsymbol{\alpha} \right\}, \tag{8.2.46}$$

subject to $y_i \left(\sum_{j=1}^{l+u} \alpha_j K(\mathbf{x}_i, \mathbf{x}_j) + b \right) \geq 1 - \xi_i, \quad i = 1, \ldots, l,$ (8.2.47)

$$\xi_i \geq 0, \quad i = 1, \ldots, l.$$ (8.2.48)

By using the augmented Lagrange multiplier method, the Lagrange function for dual unconstrained optimization is then given by

$$L(\boldsymbol{\alpha}, \boldsymbol{\xi}, b, \boldsymbol{\beta}, \boldsymbol{\zeta}) = \frac{1}{l} \sum_{i=1}^{l} \xi_i + \frac{1}{2} \boldsymbol{\alpha}^T \left(2\gamma_A \mathbf{K} + 2 \frac{\gamma_I}{(l+u)^2} \mathbf{KLK} \right) \boldsymbol{\alpha}$$

$$- \sum_{i=1}^{l} \beta_i \left[y_i \left(\sum_{j=1}^{l+u} \alpha_j K(\mathbf{x}_i, \mathbf{x}_j) + b \right) - 1 + \xi_i \right] - \sum_{i=1}^{l} \zeta_i \xi_i.$$
(8.2.49)

From the first-order optimization conditions

$$\frac{\partial L}{\partial b} = 0 \Rightarrow \sum_{i=1}^{l} \beta_i y_i = 0,$$

$$\frac{\partial L}{\partial \xi_i} = 0 \Rightarrow \frac{1}{l} - \beta_i - \zeta_i = 0,$$

$$\Rightarrow 0 \leq \beta_i \leq \frac{1}{l} \quad (\zeta_i, \xi_i \text{ are nonnegative}).$$

By using above identities, the Lagrange function in (8.2.49) can be reduced to

$$L^R(\boldsymbol{\alpha}, \boldsymbol{\beta}) = \frac{1}{2} \boldsymbol{\alpha}^T \left(2\gamma_A \mathbf{K} + 2 \frac{\gamma_I}{(u+l)^2} \mathbf{KLK} \right) \boldsymbol{\alpha} - \sum_{i=1}^{l} \beta_i \left(y_i \sum_{j=1}^{l+u} \alpha_j K(\mathbf{x}_i, \mathbf{x}_j) - 1 \right)$$

$$= \frac{1}{2} \boldsymbol{\alpha}^T \left(2\gamma_A \mathbf{K} + 2 \frac{\gamma_I}{(u+l)^2} \mathbf{KLK} \right) \boldsymbol{\alpha} - \boldsymbol{\alpha}^T \mathbf{KJ}^T \mathbf{Y} \boldsymbol{\beta} + \sum_{i=1}^{l} \beta_i,$$ (8.2.50)

where $\mathbf{J} = [1, \ldots, 1, 0, \ldots, 0]$ is an $1 \times (l+u)$ matrix with the first l entries as 1 and the rest as 0, and $\mathbf{Y} = \mathbf{Diag}(y_1, \ldots, y_l)$.

From the first-order optimization condition

$$\frac{\partial L^R}{\partial \boldsymbol{\alpha}} = \left(2\gamma_A \mathbf{K} + 2 \frac{\gamma_I}{(u+l)^2} \mathbf{KLK} \right) \boldsymbol{\alpha} - \mathbf{KJ}^T \mathbf{Y} \boldsymbol{\beta} = 0$$ (8.2.51)

it follows that

$$\boldsymbol{\alpha}^* = \left(2\gamma_A \mathbf{I} + 2\frac{\gamma_l}{(u+l)^2}\mathbf{LK}\right)^{-1}\mathbf{J}^T\mathbf{Y}\boldsymbol{\beta}^*. \qquad (8.2.52)$$

8.3 Support Vector Machine Regression

The support vector machine regression is a binary regressor algorithm that looks for an optimal hyperplane as a decision function in a high-dimensional space.

8.3.1 Support Vector Machine Regressor

Suppose we are given a training set of N data points $\{\mathbf{x}_k, y_k\}, k = 1, \ldots, N$, where $\mathbf{x}_k \in \mathbb{R}^n$ is the kth input pattern and $y_k \in \mathbb{R}$ is the kth associated "truth".

Let $\boldsymbol{\phi} : I \subseteq \mathbb{R}^n \to F \subseteq \mathbb{R}^N$ be a mapping from the input space $I \subseteq \mathbb{R}^n$ to the feature space F. Here, $\boldsymbol{\phi}(\mathbf{x}_i)$ is the extracted feature of the input $\mathbf{x}_i$.

The SVM learning algorithm aims at finding a hyperplane $(\mathbf{w}, b)$ through the constrained optimization

$$\min_{\mathbf{w}, b} \left\{ f(\mathbf{w}, b) = \|\mathbf{w}\|_2^2 \right\}, \qquad (8.3.1)$$

$$\text{subject to } \sum_{i=1}^{N} y_i \left[\mathbf{w}^T \boldsymbol{\phi}(\mathbf{x}_i) - b\right] \geq 0, \qquad (8.3.2)$$

where the quality $(\langle \mathbf{w}, \boldsymbol{\phi}(\mathbf{x})\rangle - b)$ corresponds to the distance between the point $\mathbf{x}_i$ and the decision boundary, and the quality

$$\gamma = \sum_{i=1}^{N} y_i [\langle \mathbf{w}, \boldsymbol{\phi}(\mathbf{x}_i)\rangle - b] \qquad (8.3.3)$$

is called the *margin*.

The constrained optimization problem (8.3.1) can be rewritten as an unconstrained optimization in Lagrangian form:

$$\min_{\mathbf{w}, b} \left\{ L(\mathbf{w}, b) = \|\mathbf{w}\|_2^2 - \sum_{i=1}^{N} \alpha_i y_i \left[\mathbf{w}^T \boldsymbol{\phi}(\mathbf{x}_i) - b\right] \right\}, \qquad (8.3.4)$$

where the Lagrangian multiplies α_i are nonnegative.

From the optimization conditions, we have

$$\frac{\partial L(\mathbf{w}, b)}{\partial \mathbf{w}} = \mathbf{w} - \sum_{i=1}^{N} \alpha_i y_i \boldsymbol{\phi}(\mathbf{x}_i) = 0 \Rightarrow \mathbf{w} = \sum_{i=1}^{N} \alpha_i y_i \boldsymbol{\phi}(\mathbf{x}_i), \quad (8.3.5)$$

$$\frac{\partial L(\mathbf{w}, b)}{\partial b} = \sum_{i=1}^{N} \alpha_i y_i = 0. \quad (8.3.6)$$

Substituting these two results into (8.3.1) to give the following constrained optimization with respect to $\boldsymbol{\alpha}$:

$$\min_{\boldsymbol{\alpha}} \left\{ J_1(\boldsymbol{\alpha}) = \sum_{i=1}^{N} \sum_{j=1}^{N} \alpha_i \alpha_j y_i y_j K(\mathbf{x}_i, \mathbf{x}_j) - \sum_{i=1}^{N} \alpha_i \right\}, \quad (8.3.7)$$

or

$$\max_{\boldsymbol{\alpha}} \left\{ J_2(\boldsymbol{\alpha}) = \sum_{i=1}^{N} \alpha_i - \sum_{i=1}^{N} \sum_{j=1}^{N} \alpha_i \alpha_j y_i y_j K(\mathbf{x}_i, \mathbf{x}_j) \right\} \quad (8.3.8)$$

subject to

$$\sum_{i=1}^{N} \alpha_i y_i = 0 \quad \text{and} \quad \alpha_i > 0, \quad i = 1, \ldots, N, \quad (8.3.9)$$

where $K(\mathbf{x}_i, \mathbf{x}_j) = \langle \boldsymbol{\phi}(\mathbf{x}_i), \boldsymbol{\phi}(\mathbf{x}_j) \rangle = \boldsymbol{\phi}^T(\mathbf{x}_i) \boldsymbol{\phi}(\mathbf{x}_j)$.

The support vector machine regression aims at designing the following machine learning algorithms:

- solve the maximization problem (8.3.8) and (8.3.9) for the Lagrangian multiplies $\alpha_i, i = 1, \ldots, N$;
- update the bias via $b \leftarrow b - \eta \sum_{i=1}^{N} \alpha_i y_i$;
- use (8.3.5) to calculate the support vector regressor $\mathbf{w}$.

8.3.2 ϵ-Support Vector Regression

The basic idea of ϵ-support vector (SV) regression is to find a function $f(\mathbf{x}) = \mathbf{w}^T \boldsymbol{\phi}(\mathbf{x}) + b$ that has at most ϵ deviation from the actually obtained targets y_i for all the training data $\mathbf{x}_1, \ldots, \mathbf{x}_N$, namely $|f(\mathbf{x}_i) - y_i| \leq \epsilon$ for $i = 1, \ldots, N$, and at the same time $\mathbf{w}$ is as flat as possible. In other words, we do not count errors as long as they are less than ϵ, but will not accept any deviation larger than this [36].

8.3 Support Vector Machine Regression

One way to ensure the flatness of $\mathbf{w}$ is to minimize its norm $\|\mathbf{w}\|_2^2 = \langle \mathbf{w}, \mathbf{w}\rangle$. Hence, the basic form of ϵ-support vector regression can be written as a convex optimization problem [36, 43]:

$$\min_{\mathbf{w}} \quad \frac{1}{2}\|\mathbf{w}\|_2^2, \tag{8.3.10}$$

$$\text{subject to} \quad \begin{cases} y_k - \langle \mathbf{w}, \boldsymbol{\phi}(\mathbf{x}_k)\rangle - b \leq \epsilon, \\ \langle \mathbf{w}, \boldsymbol{\phi}(\mathbf{x}_k)\rangle + b - y_k \leq \epsilon; \end{cases} \tag{8.3.11}$$

where $\epsilon > 0$ is a regression error.

In order to avoid a violation of constrained conditions in (8.3.11), one can introduce slackness parameters (ξ_k, ξ_k^*), for given parameters $C > 0$ and $\epsilon > 0$. Hence we have the standard form of support vector regression given by Vapnik [44]

$$\min_{\mathbf{w},b,\xi_k,\xi_k^*} \left\{ \frac{1}{2}\langle \mathbf{w}, \mathbf{w}\rangle + C \sum_{k=1}^{N}(\xi_k + \xi_k^*) \right\} \tag{8.3.12}$$

$$\text{subject to} \quad y_k - (\langle \mathbf{w}, \boldsymbol{\phi}(\mathbf{x}_k)\rangle + b) \leq \epsilon + \xi_k, \tag{8.3.13}$$

$$(\langle \mathbf{w}, \boldsymbol{\phi}(\mathbf{x}_k)\rangle + b) - y_k \leq \epsilon + \xi_k^*, \tag{8.3.14}$$

$$\xi_k, \xi_k^* \geq 0, \quad k = 1, \ldots, N. \tag{8.3.15}$$

Here $C > 0$ is the regularization parameter that denotes SVM misclassification tolerance and is a pre-specified value. ξ_k represents the upper training error, and ξ_k^* is the lower training error subject to ϵ-insensitive tube $|y_k - (\langle \mathbf{w}, \boldsymbol{\phi}(\mathbf{x}_k)\rangle + b)| \leq \epsilon$.

To solve the above constrained optimization problem, one can define the Lagrange function (or Lagrangian) as follows [36]:

$$\mathcal{L} = \mathcal{L}(\mathbf{w}, b, \xi_k, \xi_k^*)$$

$$= \frac{1}{2}\|\mathbf{w}\|_2^2 + C\sum_{k=1}^{N}(\xi_k + \xi_k^*) - \sum_{k=1}^{N}(\eta_k\xi_k + \eta_k^*\xi_k^*)$$

$$- \sum_{k=1}^{N}\alpha_k\left(\epsilon + \xi_k - y_k + \langle \mathbf{w}, \boldsymbol{\phi}(\mathbf{x}_k)\rangle + b\right)$$

$$- \sum_{k=1}^{N}\alpha_k^*\left(\epsilon + \xi_k^* + y_k - \langle \mathbf{w}, \boldsymbol{\phi}(\mathbf{x}_k)\rangle - b\right), \tag{8.3.16}$$

where $\eta_k, \eta_k^*, \alpha_k, \alpha_k^*$ are Lagrange multipliers which have to satisfy the positivity constraints:

$$\eta_k^{(*)}, \alpha_k^{(*)} \geq 0, \tag{8.3.17}$$

where $\eta_k^{(*)} = \{\eta_k, \eta_k^*\}$ and $\alpha_k^{(*)} = \{\alpha_k, \alpha_k^*\}$.

From the first-order optimality conditions it follows that

$$\frac{\partial \mathcal{L}}{\partial \mathbf{w}} = 0 \Rightarrow \mathbf{w} = \sum_{k=1}^{N}(\alpha_k - \alpha_k^*)\boldsymbol{\phi}(\mathbf{x}_k), \tag{8.3.18}$$

$$\frac{\partial \mathcal{L}}{\partial b} = 0 \Rightarrow \sum_{k=1}^{N}(\alpha_k - \alpha_k^*) = 0, \tag{8.3.19}$$

$$\frac{\partial \mathcal{L}}{\partial \xi_k} = 0 \Rightarrow \eta_k + \alpha_k = C, \tag{8.3.20}$$

$$\frac{\partial \mathcal{L}}{\partial \xi_k^*} = 0 \Rightarrow \eta_k^* + \alpha_k^* = C. \tag{8.3.21}$$

Equation (8.3.18) is the so-called *support vector expansion*, i.e., $\mathbf{w}$ can be completely described as a linear combination of the training patterns $\mathbf{x}_i$.

Substituting (8.3.18) and (8.3.19) into (8.3.16) can eliminate the dual variables η_i, η_i^* and gives the dual function

$$\mathcal{L}_D(\boldsymbol{\alpha}, \boldsymbol{\alpha}^*) = -\frac{1}{2}\sum_{i=1}^{N}\sum_{j=1}^{N}(\alpha_i - \alpha_i^*)(\alpha_j - \alpha_j^*)\langle\boldsymbol{\phi}(\mathbf{x}_i), \boldsymbol{\phi}(\mathbf{x}_j)\rangle$$

$$- \epsilon\sum_{i=1}^{N}(\alpha_i + \alpha_i^*) + \sum_{i=1}^{N}y_i(\alpha_i - \alpha_j^*). \tag{8.3.22}$$

On the other side, from (8.3.17), (8.3.20), and (8.3.21) it is known that

$$0 \leq \alpha_i, \alpha_i^* \leq C, \quad i = 1, \ldots, N. \tag{8.3.23}$$

Equations (8.3.22) together with constraints (8.3.19) and (8.3.23) constitute the Wolfe dual ϵ-support vector machine regression problem:

$$\max_{\boldsymbol{\alpha}, \boldsymbol{\alpha}^*}\left\{-\frac{1}{2}(\boldsymbol{\alpha} - \boldsymbol{\alpha}^*)^T\mathbf{Q}(\boldsymbol{\alpha} - \boldsymbol{\alpha}^*) - \epsilon\sum_{i=1}^{N}(\alpha_i + \alpha_i^*) + \sum_{i=1}^{N}y_i(\alpha_i - \alpha_j^*)\right\}, \tag{8.3.24}$$

$$\text{subject to } \sum_{k=1}^{N}(\alpha_k - \alpha_k^*) = 0, \tag{8.3.25}$$

$$0 \leq \alpha_i, \alpha_i^* \leq C, \quad i = 1, \ldots, N, \tag{8.3.26}$$

8.3 Support Vector Machine Regression

where $\mathbf{Q} = [K(\mathbf{x}_i, \mathbf{x}_j)]_{i,j=1}^{N,N}$ is an $N \times N$ positive semi-definite matrix with $K(\mathbf{x}_i, \mathbf{x}_j) = \langle \boldsymbol{\phi}(\mathbf{x}_i), \boldsymbol{\phi}(\mathbf{x}_j) \rangle$ being the kernel function of SVM.

Then the regression function is given by

$$f(\mathbf{x}) = \mathbf{w}^T \boldsymbol{\phi}(\mathbf{x}) + b = \sum_{i=1}^{N} (\alpha_i - \alpha_i^*) K(\mathbf{x}_i, \mathbf{x}) + b, \qquad (8.3.27)$$

where $\mathbf{w}$ is described by the support vector expansion Eq. (8.3.18).

The KKT conditions for the dual constrained optimization problem (8.3.24) are given by [36]

$$\alpha_i (\epsilon + \xi_i - y_i + \langle \mathbf{w}, \mathbf{x}_i \rangle + b) = 0, \qquad (8.3.28)$$

$$\alpha_i^* (\epsilon + \xi_i^* - y_i + \langle \mathbf{w}, \mathbf{x}_i \rangle + b) = 0, \qquad (8.3.29)$$

$$(C - \alpha_i)\xi_i = 0. \qquad (8.3.30)$$

$$(C - \alpha_i^*)\xi_i^* = 0. \qquad (8.3.31)$$

From the above conditions it follows that

$$\epsilon - y_i + \langle \mathbf{w}, \mathbf{x}_i \rangle + b \geq 0 \quad \text{and} \quad \xi_i = 0 \quad \text{if } \alpha_i < C, \qquad (8.3.32)$$

$$\epsilon - y_i + \langle \mathbf{w}, \mathbf{x}_i \rangle + b \leq 0 \quad \text{if } \alpha_i > 0, \qquad (8.3.33)$$

$$\epsilon - y_i + \langle \mathbf{w}, \mathbf{x}_i \rangle + b \geq 0 \quad \text{and} \quad \xi_i^* = 0 \quad \text{if } \alpha_i^* < C, \qquad (8.3.34)$$

$$\epsilon - y_i + \langle \mathbf{w}, \mathbf{x}_i \rangle + b \leq 0 \quad \text{if } \alpha_i^* > 0. \qquad (8.3.35)$$

Then the computation of b is given by Smola and Schölkopf [36]:

$$\max\{-\epsilon + y_i - \langle \mathbf{w}, \mathbf{x}_i \rangle | \alpha_i < C \text{ or } \alpha_i^* > 0\} \leq b \leq$$
$$\min\{-\epsilon + y_i - \langle \mathbf{w}, \mathbf{x}_i \rangle | \alpha_i > 0 \text{ or } \alpha_i^* < C\}. \qquad (8.3.36)$$

If some α_i or $\alpha_i^* \in (0, C)$, then the inequalities become equalities.

8.3.3 ν-Support Vector Machine Regression

In the standard support vector machine regression described above, an error of ϵ is allowed at each point $\mathbf{x}_i$. By introducing a constant $\nu \geq 0$, the size of ϵ can be traded off against model complexity and slack variables. This is the basic idea of the following ν-support vector machine regression developed by Schölkopf et al. [34]:

$$\min_{\mathbf{w}, \xi_i, \xi_i^*, \epsilon} \left\{ \frac{1}{2} \|\mathbf{w}\|_2^2 + C \left(\nu\epsilon + \frac{1}{N} \sum_{i=1}^{N} (\xi_i + \xi_i^*) \right) \right\} \qquad (8.3.37)$$

subject to $(\mathbf{w}^T\mathbf{x}_i + b) - y_i \leq \epsilon + \xi_i,$ \hfill (8.3.38)

$$y_i - (\mathbf{w}^T\mathbf{x}_i + b) \leq \epsilon + \xi_i^*,$$ \hfill (8.3.39)

$$\xi_i \geq 0, \xi_i^* \geq 0, \quad i = 1, \ldots, N.$$ \hfill (8.3.40)

By introducing the Lagrange multipliers $\alpha_i^* . \eta_i^*, \beta \geq 0$, one has the Lagrange function [34]:

$$\mathcal{L}\left(\mathbf{w}, b, \boldsymbol{\xi}^*, \epsilon, \boldsymbol{\alpha}^*, \beta, \boldsymbol{\eta}^*\right)$$

$$= \frac{1}{2}\|\mathbf{w}\|_2^2 + C\nu\epsilon + \frac{C}{N}\sum_{i=1}^{N}(\xi_i + \xi_i^*) - \beta\epsilon - \sum_{i=1}^{N}(\eta_i\xi_i + \eta_i^*\xi_i^*)$$

$$- \sum_{i=1}^{N}\alpha_i\left(\xi_i + y_i - \mathbf{w}^T\mathbf{x}_i - b + \epsilon\right)$$

$$- \sum_{i=1}^{N}\alpha_i^*\left(\xi_i^* + \mathbf{w}^T\mathbf{x}_i + b - y_i + \epsilon\right).$$ \hfill (8.3.41)

The first-order optimality conditions give the results

$$\frac{\partial \mathcal{L}}{\partial \mathbf{w}} = 0 \quad \Rightarrow \quad \mathbf{w} = \sum_{i=1}^{N}(\alpha_i^* - \alpha_i)\mathbf{x}_i,$$ \hfill (8.3.42)

$$\frac{\partial \mathcal{L}}{\partial \epsilon} = 0 \quad \Rightarrow \quad C\nu - \sum_{i=1}^{N}(\alpha_i + \alpha_i^*) - \beta = 0,$$ \hfill (8.3.43)

$$\frac{\partial \mathcal{L}}{\partial b} = 0 \quad \Rightarrow \quad \sum_{i=1}^{N}(\alpha_i - \alpha_i^*) = 0,$$ \hfill (8.3.44)

$$\frac{\partial \mathcal{L}}{\partial \xi_i^*} = 0 \quad \Rightarrow \quad \frac{C}{N} - \alpha_i^* - \eta_i^* = 0, \quad i = 1, \ldots, N;$$ \hfill (8.3.45)

for minimizing over the primal variables $\mathbf{w}, \epsilon, b, \xi_i^*$, while maximizing over the dual variables $\alpha_i^*, \eta_i^*, \beta$ yields

$$\frac{\partial \mathcal{L}}{\partial \alpha_i^*} = 0 \quad \Rightarrow \quad \xi_i^* + \mathbf{w}^T\mathbf{x}_i + b - y_i + \epsilon = 0,$$

and

$$\frac{\partial \mathcal{L}}{\partial \eta_i^*} = 0 \quad \Rightarrow \quad \epsilon = 0, \quad \frac{\partial \mathcal{L}}{\partial \beta} = 0 \quad \Rightarrow \quad \xi_i^* = 0.$$

Hence, the regression result is given by

$$y_i = \mathbf{w}^T \mathbf{x}_i + b. \tag{8.3.46}$$

Substituting the four constraints in (8.3.42)–(8.3.45) into the Lagrange function $\mathcal{L}$ defined in (8.3.41) leads to the optimization problem called the Wolfe dual. The Wolfe dual ν-support vector machine regression problem can be written as [34]

$$\max \left\{ \mathcal{W}(\boldsymbol{\alpha}^*) = \sum_{i=1}^{N} (\alpha_i^* - \alpha_i) y_i - \frac{1}{2} \sum_{i=1}^{N} \sum_{j=1}^{N} (\alpha_i^* - \alpha_i)(\alpha_j^* - \alpha_j) K(\mathbf{x}_i, \mathbf{x}_j) \right\} \tag{8.3.47}$$

$$\text{subject to } \sum_{i=1}^{N} (\alpha_i - \alpha_i^*) = 0, \quad \alpha_i^* \in \left[0, \frac{C}{N}\right], \quad \sum_{i=1}^{N} (\alpha_i + \alpha_i^*) \le C\nu. \tag{8.3.48}$$

After solving the Wolfe dual problem, the regression value of the new instance $\mathbf{x}$ is given by

$$f(\mathbf{x}) = \sum_{i=1}^{N} (\alpha_i^* - \alpha_i) K(\mathbf{x}, \mathbf{x}_i) + b. \tag{8.3.49}$$

Proposition 8.1 ([34]) *Suppose ν-SVR is applied to some data set, and the resulting ϵ is nonzero. The following statements hold:*

- *ν is an upper bound on the fraction of errors.*
- *ν is a lower bound on the fraction of SVs.*

The following function is a popular choice to illustrate SVR for regression in the literature:

$$y(x) = \begin{cases} \sin(x)/x, & x \ne 0; \\ 0, & x = 0. \end{cases} \tag{8.3.50}$$

8.4 Support Vector Machine Binary Classification

A basic task in data analysis and pattern recognition is classification that requires the construction of a classifier, namely a function that assigns a class label to instances described by a set of attributes.

In this section we discuss SVM binary classification problems with two classes.

8.4.1 Support Vector Machine Binary Classifier

Given a training set of N data points $\{\mathbf{x}_k, y_k\}, k = 1, \ldots, N$, where $\mathbf{x}_k$ is the kth input pattern and y_k is the kth output pattern. Denote the set of data

$$\mathcal{D} = \{(\mathbf{x}_1, y_1), \ldots, (\mathbf{x}_N, y_N)\}, \quad \mathbf{x}_k \in \mathbb{R}^n, \; y_k \in \{-1, +1\} \tag{8.4.1}$$

A SVM classifier is a classifier which finds the hyperplane that separates the data with the largest distance between the hyperplane and the closest data point (called margin).

A linear separating hyperplane is a decision function in the form

$$f(\mathbf{x}) = \text{sign}(\mathbf{w}^T \mathbf{x} + b), \tag{8.4.2}$$

where $\mathbf{x}$ is the input pattern, $\mathbf{w}$ is the weight vector, and b is a bias term. Hence, we have

$$\mathbf{w}^T \mathbf{x} + b \begin{cases} > 0, & \text{then } \mathbf{x} \in \text{class } S_+; \\ < 0, & \text{then } \mathbf{x} \in \text{class } S_-; \\ = 0, & \text{then } \mathbf{x} \in \text{class } S_+ \text{ or } \mathbf{x} \in \text{class } S_-. \end{cases} \tag{8.4.3}$$

Similarly, for a nonlinear classifier $\mathbf{w}$ with the nonlinear mapping $\boldsymbol{\phi} : \mathbf{x} \to \boldsymbol{\phi}(\mathbf{x})$, its testing output $y = f(\mathbf{x}) = \mathbf{w}^T \boldsymbol{\phi}(\mathbf{x}) + b$, where b denotes a bias term. Then, we have

$$\mathbf{w}^T \boldsymbol{\phi}(\mathbf{x}) + b \begin{cases} > 0, & \text{then } \mathbf{x} \in \text{class } S_+; \\ < 0, & \text{then } \mathbf{x} \in \text{class } S_-; \\ = 0, & \text{then } \mathbf{x} \in \text{class } S_+ \text{ or } \mathbf{x} \in \text{class } S_-. \end{cases} \tag{8.4.4}$$

Therefore the decision function of a nonlinear classifier is given by

$$\text{class of } \mathbf{x} = \text{sign}\left(\mathbf{w}^T \boldsymbol{\phi}(\mathbf{x}) + b\right). \tag{8.4.5}$$

For a SVM with the kernel function $K(\mathbf{x}, \mathbf{x}_i) = \boldsymbol{\phi}^T(\mathbf{x})\boldsymbol{\phi}(\mathbf{x}_i)$, the SVM classifier $\mathbf{w}$ is usually designed as

$$\mathbf{w} = \sum_{i=1}^{N} \alpha_i y_i \boldsymbol{\phi}(\mathbf{x}_i). \tag{8.4.6}$$

Substituting (8.4.6) into (8.4.5), we directly get the decision function of any SVM classifier as follows:

$$\text{class of } \mathbf{x} = \text{sign}\left(\sum_{i=1}^{N} \alpha_i y_i K(\mathbf{x}, \mathbf{x}_i) + b\right). \tag{8.4.7}$$

8.4 Support Vector Machine Binary Classification

Hence, when designing any SVM classifier, its weighting vector $\mathbf{w}$ must have the form in (8.4.6).

The set of vectors $\{\mathbf{x}_1, \ldots, \mathbf{x}_N\}$ is said to be optimally separated by the hyperplane if it is separated without error and the distance between the closest vector to the hyperplane is maximal.

For designing the classifier $\mathbf{w}$, we assume that

$$\mathbf{w}^T \boldsymbol{\phi}(\mathbf{x}_k) + b \geq 1, \quad \text{if } y_k = +1, \tag{8.4.8}$$

$$\mathbf{w}^T \boldsymbol{\phi}(\mathbf{x}_k) + b \leq -1, \text{ if } y_k = -1. \tag{8.4.9}$$

That is to say, if the data is to be classified correctly, this hyperplane should ensure that

$$y_k \left(\mathbf{w}^T \boldsymbol{\phi}(\mathbf{x}_k) + b \right) > 0, \text{ for all } k = 1, \ldots, N, \tag{8.4.10}$$

assuming that $y \in \{-1, +1\}$. Here $\boldsymbol{\phi}(\mathbf{x}_k)$ is a nonlinear vector function mapping the input space $\mathbb{R}^n$ into a higher-dimensional space, but this function is not explicitly constructed.

The distance of a point $\mathbf{x}_k$ from the hyperplane, denoted as $d(\mathbf{w}, b; \mathbf{x}_k)$, is defined as

$$d(\mathbf{w}, b; \mathbf{x}_k) = \frac{|\langle \mathbf{w}, \boldsymbol{\phi}(\mathbf{x}_k) \rangle + b|}{\|\mathbf{w}\|}. \tag{8.4.11}$$

By the constraint condition $y_k (\langle \mathbf{w}, \boldsymbol{\phi}(\mathbf{x}_k) \rangle + b) \geq 1$ in (8.4.10), the margin of a classifier is given by

$$\begin{aligned}
\rho(\mathbf{w}, b) &= \min_{\mathbf{x}_k : y_k = -1} d(\mathbf{w}, b; \mathbf{x}_k) + \min_{\mathbf{x}_k : y_k = 1} d(\mathbf{w}, b; \mathbf{x}_k) \\
&= \min_{\mathbf{x}_k : y_k = -1} \frac{|\langle \mathbf{w}, \boldsymbol{\phi}(\mathbf{x}_k) \rangle + b|}{\|\mathbf{w}\|} + \min_{\mathbf{x}_k : y_k = 1} \frac{|\langle \mathbf{w}, \boldsymbol{\phi}(\mathbf{x}_k) \rangle + b|}{\|\mathbf{w}\|} \\
&= \frac{1}{\|\mathbf{w}\|} \left(\min_{\mathbf{x}_k : y_k = -1} |\langle \mathbf{w}, \boldsymbol{\phi}(\mathbf{x}_k) \rangle + b| + \min_{\mathbf{x}_k : y_k = 1} |\langle \mathbf{w}, \boldsymbol{\phi}(\mathbf{x}_k) \rangle + b| \right) \\
&= \frac{2}{\|\mathbf{w}\|}.
\end{aligned} \tag{8.4.12}$$

The optimal hyperplane $\mathbf{w}_{\text{opt}}$ is given by maximizing the above margin, namely

$$\mathbf{w} = \arg\max_{\mathbf{w}} \left\{ \rho(\mathbf{w}, b) = \frac{2}{\|\mathbf{w}\|} \right\} \tag{8.4.13}$$

which is equivalent to

$$\mathbf{w} = \arg\min_{\mathbf{w}} \frac{1}{2}\|\mathbf{w}\|. \qquad (8.4.14)$$

In order to avoid the possibility to violate (8.4.10), variables ξ_k need to be introduced such that

$$y_k\left(\mathbf{w}^T\boldsymbol{\phi}(\mathbf{x}_k) + b\right) \geq 1 - \xi_k, \quad k = 1, \ldots, N, \qquad (8.4.15)$$

$$\xi_k \geq 0, \quad k = 1, \ldots, N. \qquad (8.4.16)$$

According to the structural risk minimization principle, the risk bound is minimized by formulating the optimization problem

$$\min\left\{\frac{1}{2}\|\mathbf{w}\|^2 + C\sum_{k=1}^{N}\xi_k\right\} \qquad (8.4.17)$$

subject to (8.4.15), where C is a user-specified parameter for providing a trade-off between the distance of the separating margin and the training error.

Therefore the primary problem for SVM binary classifier is a constrained optimization problem:

$$\min\left\{\mathcal{L}_{\text{P}_{\text{SVM}}} = \frac{1}{2}\|\mathbf{w}\|_2^2 + C\sum_{i=1}^{N}\xi_i\right\} \qquad (8.4.18)$$

subject to $y_i\left(\mathbf{w}^T\boldsymbol{\phi}(\mathbf{x}_i) + b\right) \geq 1 - \xi_i, \ \xi_i \geq 0 \ (i = 1, \ldots, N).$ \qquad (8.4.19)

The dual form of the above primary optimization problem is given by

$$\min\left\{\mathcal{L}_{\text{D}_{\text{SVM}}} = \frac{1}{2}\sum_{i=1}^{N}\sum_{j=1}^{N}y_i y_j \alpha_i \alpha_j \langle\boldsymbol{\phi}(\mathbf{x}_i), \boldsymbol{\phi}(\mathbf{x}_j)\rangle - \sum_{i=1}^{N}\alpha_i\right\} \qquad (8.4.20)$$

subject to $\sum_{i=1}^{N}\alpha_i y_i = 0, \ 0 \leq \alpha_i \leq C \ (i = 1, \ldots, N),$ \qquad (8.4.21)

where α_i is the Lagrange multiplier corresponding to ith training sample $(\mathbf{x}_i, y_i)$, and vectors $\mathbf{x}_i$ satisfying $y_i(\mathbf{w}^T\boldsymbol{\phi}(\mathbf{x}_i) + b) = 1$ are termed *support vectors*.

In SVM learning algorithms, kernel functions $K(\mathbf{u}, \mathbf{v}) = \langle \boldsymbol{\phi}(\mathbf{u}), \boldsymbol{\phi}(\mathbf{v}) \rangle$ are usually used, and the dual optimization problem for SVM binary classifier is represented as

$$\min \left\{ \mathcal{L}_{\text{DSVM}} = \frac{1}{2} \sum_{i=1}^{N} \sum_{j=1}^{N} y_i y_j \alpha_i \alpha_j K(\mathbf{x}_i, \mathbf{x}_j) - \sum_{i=1}^{N} \alpha_i \right\} \quad (8.4.22)$$

$$\text{subject to } \sum_{i=1}^{N} \alpha_i y_i = 0, \ 0 \leq \alpha_i \leq C \ (i = 1, \ldots, N). \quad (8.4.23)$$

By Bousquet et al. [5, pp. 14–15], several important practical points should be taken into account when designing classifiers.

1. In order to reduce the likelihood of overfitting the classifier to the training data, the ratio of the number of training examples to the number of features should be at least 10:1. For the same reason the ratio of the number of training examples to the number of unknown parameters should be at least 10:1.
2. Importantly, proper error-estimation methods should be used, especially when selecting parameters for the classifier.
3. Some algorithms require the input features to be scaled to similar ranges, such as some kind of a weighted average of the inputs.
4. There is no single best classification algorithm!

8.4.2 ν-Support Vector Machine Binary Classifier

Similar to ν-SVR, by introducing a new parameter $\nu \in (0, 1]$ in SVM classification, Schölkopf et al. [34] presented ν-support vector classifier (ν-SVC).

The primal optimization problem of ν-SVC is described by

$$\min_{\mathbf{w}, \boldsymbol{\xi}, \rho} \left\{ \frac{1}{2} \|\mathbf{w}\|_2^2 - \nu\rho + \sum_{i=1}^{N} \xi_i \right\} \quad (8.4.24)$$

$$\text{subject to } y_i(\mathbf{w}^T \boldsymbol{\phi}(\mathbf{x}_i) + b) \geq \epsilon - \xi_i, \ \xi_i \geq 0 \ (i = 1, \ldots, N), \ \rho \geq 0. \quad (8.4.25)$$

Let Lagrange multipliers $\alpha_i, \beta_i, \delta \geq 0$, and consider the Lagrange function

$$\mathcal{L}(\mathbf{w}, \boldsymbol{\xi}, b, \rho, \boldsymbol{\alpha}, \boldsymbol{\beta}, \delta) = \frac{1}{2} \|\mathbf{w}\|_2^2 - \nu\rho + \frac{1}{N} \sum_{i=1}^{N} \xi_i - \delta\rho$$

$$- \sum_{i=1}^{N} \left(\alpha_i \left[y_i(\mathbf{w}^T \mathbf{x}_i + b) - \rho + \xi_i \right] + \beta_i \xi_i \right). \quad (8.4.26)$$

This function needs to be minimized with respect to the primal variables $\mathbf{w}, \boldsymbol{\xi}, b, \rho$ and maximized over the dual variables $\boldsymbol{\alpha}, \boldsymbol{\beta}, \delta$.

The first-order optimization conditions for minimization give the following results:

$$\frac{\partial \mathcal{L}}{\partial \mathbf{w}} = 0 \quad \Rightarrow \quad \mathbf{w} = \sum_{i=1}^{N} \alpha_i y_i \mathbf{x}_i, \tag{8.4.27}$$

$$\frac{\partial \mathcal{L}}{\partial \xi_i} = 0 \quad \Rightarrow \quad \alpha_i + \beta_i = 1/N, \quad i = 1, \ldots, N, \tag{8.4.28}$$

$$\frac{\partial \mathcal{L}}{\partial b} = 0 \quad \Rightarrow \quad \sum_{i=1}^{N} \alpha_i y_i = 0, \tag{8.4.29}$$

$$\frac{\partial \mathcal{L}}{\partial \rho} = 0 \quad \Rightarrow \quad \sum_{i=1}^{N} \alpha_i - \delta = \nu. \tag{8.4.30}$$

If substituting (8.4.27)–(8.4.30) into the Lagrange function $\mathcal{L}$, using $\alpha_i, \beta_i, \delta \geq 0$ and incorporating the kernel function $K(\mathbf{x}, \mathbf{x}_i)$ instead of $\mathbf{x}_i$ in dot product, then one obtains the Wolfe dual optimization problem for ν-SVC as follows:

$$\max_{\boldsymbol{\alpha}} \left\{ \mathcal{W}(\boldsymbol{\alpha}) = -\frac{1}{2} \sum_{i=1}^{N} \sum_{j=1}^{N} \alpha_i \alpha_j y_i y_j K(\mathbf{x}_i, \mathbf{x}_j) \right\} \tag{8.4.31}$$

subject to $0 \leq \alpha_i \leq 1/N$ $(i = 1, \ldots, N)$; $\sum_{i=1}^{N} \alpha_i y_i = 0$; $\sum_{i=1}^{N} \alpha_i \geq \nu.$

$$\tag{8.4.32}$$

To compute parameters b and ρ in the primal ν-SVC optimization problem, consider two sets S_+ and S_-, containing support vectors $\mathbf{x}_i$ with $0 \leq \alpha_i < 1$ and $y_i = \pm 1$.

Let

$$s_1 = |S_+| = |\{i \,|\, 0 < \alpha_i < 1, y_i = 1\}|, \tag{8.4.33}$$

$$s_2 = |S_-| = |\{i \,|\, 0 < \alpha_i < 1, y_i = -1\}| \tag{8.4.34}$$

be the sizes of the sets S_+ and S_-, respectively.

If defining

$$r_1 = \frac{1}{s_1} \sum_{\mathbf{x} \in S_+} \sum_{j=1}^{N} \alpha_j y_j K(\mathbf{x}, \mathbf{x}_j), \tag{8.4.35}$$

8.4 Support Vector Machine Binary Classification

$$r_2 = -\frac{1}{s_2} \sum_{\mathbf{x} \in S_-} \sum_{j=1}^{N} \alpha_j y_j K(\mathbf{x}, \mathbf{x}_j), \quad (8.4.36)$$

then one has [7, 34]

$$b = -\frac{r_1 - r_2}{2} \quad \text{and} \quad \rho = \frac{r_1 + r_2}{2}. \quad (8.4.37)$$

8.4.3 Least Squares SVM Binary Classifier

Standard SVMs are powerful tools for data classification by assigning them to one of two disjoint halfspaces in either the original input space for linear classifiers, or in a higher-dimensional feature space for nonlinear classifiers. The least squares SVM (LS-SVMs) developed in [37] and the proximal SVMs (PSVMs) presented in [14, 15] are two much simpler classifiers, in which each class of points is assigned to the closest of two parallel planes (in input or feature space) such that they are pushed apart as far as possible.

The LS-SVMs formulate the binary classification problem as

$$\min_{\mathbf{w},b,e} \left\{ \frac{1}{2} \|\mathbf{w}\|_2^2 + \frac{C}{2} \sum_{k=1}^{N} e_k^2 \right\} \quad (8.4.38)$$

$$\text{subject to } y_k \left(\mathbf{w}^T \boldsymbol{\phi}(\mathbf{x}_k) + b \right) = 1 - e_k, \quad k = 1, \ldots, N, \quad (8.4.39)$$

for the binary classification $y_k \in \{-1, 1\}$.

Define the Lagrange function (Lagrangian)

$$\mathcal{L} = \mathcal{L}(\mathbf{w}, b, \mathbf{e}; \boldsymbol{\alpha})$$

$$= \frac{1}{2} \|\mathbf{w}\|_2^2 + \frac{C}{2} \sum_{k=1}^{N} e_k^2 - \sum_{k=1}^{N} \alpha_k \left[y_k \left(\mathbf{w}^T \boldsymbol{\phi}(\mathbf{x}_k) + b \right) - 1 + e_k \right], \quad (8.4.40)$$

where α_k are Lagrange multipliers. Different from Lagrange multipliers in SVM with inequality constraints, in LS-SVM, Lagrange multipliers α_k can be either positive or negative due to the equality constraints used.

Based on the KKT conditions, one has the optimality conditions of (8.4.40) as follows [11]:

$$\nabla_{\mathbf{w}} \mathcal{L} = \frac{\partial \mathcal{L}}{\partial \mathbf{w}} = \mathbf{0} \Rightarrow \mathbf{w} = \sum_{k=1}^{N} \alpha_k y_k \boldsymbol{\phi}(\mathbf{x}_k) \Rightarrow \mathbf{w} = \mathbf{Z}\boldsymbol{\alpha}, \quad (8.4.41)$$

$$\nabla_b \mathcal{L} = \frac{\partial \mathcal{L}}{\partial b} = 0 \Rightarrow \sum_{k=1}^{N} \alpha_k y_k = 0 \Rightarrow \mathbf{y}^T \boldsymbol{\alpha} = 0, \tag{8.4.42}$$

$$\nabla_e \mathcal{L} = \frac{\partial \mathcal{L}}{\partial e_k} = 0 \Rightarrow \alpha_k = C e_k, k = 1, \ldots, N \Rightarrow \boldsymbol{\alpha} = C\mathbf{e}, \tag{8.4.43}$$

$$\nabla_{\alpha_k} \mathcal{L} = \frac{\partial \mathcal{L}}{\partial \alpha_k} = 0 \Rightarrow y_k \left(\mathbf{w}^T \boldsymbol{\phi}(\mathbf{x}_k) + b \right) - 1 + e_k = 0, k = 1, \ldots, N,$$

$$\Rightarrow \mathbf{Z}^T \mathbf{w} + b\mathbf{y} + \mathbf{e} = \mathbf{1}, \tag{8.4.44}$$

where $\mathbf{Z} = [y_1 \boldsymbol{\phi}(\mathbf{x}_1), \ldots, y_N \boldsymbol{\phi}(\mathbf{x}_N)] \in \mathbb{R}^{m \times N}$, $\mathbf{y} = [y_1, \ldots, y_N]^T$, $\mathbf{1} = [1, \ldots, 1]^T$, $\mathbf{e} = [e_1, \ldots, e_N]^T$ and $\boldsymbol{\alpha} = [\alpha_1, \ldots, \alpha_N]^T$.

The KKT equations (8.4.41)–(8.4.44) can be written as the following matrix equation form [11]:

$$\begin{bmatrix} \mathbf{I} & \mathbf{0} & \mathbf{0} & -\mathbf{Z} \\ \mathbf{0} & 0 & \mathbf{0} & -\mathbf{y}^T \\ \mathbf{0} & \mathbf{0} & C\mathbf{I} & -\mathbf{I} \\ \mathbf{Z}^T & \mathbf{y} & \mathbf{I} & \mathbf{0} \end{bmatrix} \begin{bmatrix} \mathbf{w} \\ b \\ \mathbf{e} \\ \boldsymbol{\alpha} \end{bmatrix} = \begin{bmatrix} \mathbf{0} \\ 0 \\ \mathbf{0} \\ \mathbf{1} \end{bmatrix}. \tag{8.4.45}$$

By eliminating $\mathbf{w}$ and $\mathbf{e}$, the above KKT equations are simplified as

$$\begin{bmatrix} 0 & \mathbf{y}^T \\ \mathbf{y} & \mathbf{Z}^T \mathbf{Z} + C^{-1}\mathbf{I} \end{bmatrix} \begin{bmatrix} b \\ \boldsymbol{\alpha} \end{bmatrix} = \begin{bmatrix} 0 \\ \mathbf{1} \end{bmatrix}. \tag{8.4.46}$$

Mercer's condition can be applied again to the matrix $\mathbf{Z}^T \mathbf{Z}$ to get [37]

$$[\mathbf{Z}^T \mathbf{Z}]_{ij} = y_i y_j \boldsymbol{\phi}^T(\mathbf{x}_i) \boldsymbol{\phi}(\mathbf{x}_j) = y_i y_j K(\mathbf{x}_i, \mathbf{x}_j). \tag{8.4.47}$$

Given a training set $\{(\mathbf{x}_i, y_i) | \mathbf{x}_i \in \mathbb{R}^n, y_i \in \{-1, 1\}, i = 1, \ldots, N\}$, constant $C > 0$, and the kernel function $K(\mathbf{x}_i, \mathbf{x}_j)$, the LS-SVM binary classification algorithm performs the following learning step:

- Construct the $N \times N$ matrix $[\mathbf{Z}^T \mathbf{Z}]_{ij} = y_i y_j \boldsymbol{\phi}^T(\mathbf{x}_i) \boldsymbol{\phi}(\mathbf{x}_j) = y_i y_j K(\mathbf{x}_i, \mathbf{x}_j)$.
- Solve the KKT matrix equation (8.4.46) for $\boldsymbol{\alpha} = [\alpha_1, \ldots, \alpha_N]^T$ and b.

In testing step, for given testing sample $\mathbf{x} \in \mathbb{R}^n$, its decision is given by

$$\text{class of } \mathbf{x} = \text{sign} \left(\sum_{j=1}^{N} \alpha_j y_j K(\mathbf{x}, \mathbf{x}_j) + b \right). \tag{8.4.48}$$

8.4.4 Proximal Support Vector Machine Binary Classifier

Due to the simplicity of their implementations, least square SVM (LS-SVM) and proximal support vector machine (PSVM) have been widely used in binary classification applications.

In the standard SVM binary classifier a plane midway between two parallel bounding planes is used to bound two disjoint halfspaces each of which contains points mostly of class 1 or 2. Similar to the LS-SVM, the key idea of PSVM [14] is that the separation hyperplanes are "proximal" planes rather than bounded planes anymore. Proximal planes classify data points depending on proximity to either one of the two separation planes. We aim to push away proximal planes as far apart as possible. Different from LS-SVM, PSVM uses $(\|\mathbf{w}\|_2^2 + b^2)$ instead of $\|\mathbf{w}\|_2^2$ as the objective function, making the optimization problem strongly convex and has little or no effect on the original optimization problem [22].

The primal optimization problem of linear PSVM is described by

$$\min \left\{ \mathcal{L}_{\text{PPSVM}}(\mathbf{w}, b, \xi_i) = \frac{1}{2}(\|\mathbf{w}\|_2^2 + b^2) + \frac{C}{2}\sum_{i=1}^{N} \xi_i^2 \right\}, \quad (8.4.49)$$

subject to $y_i(\mathbf{w}^T \mathbf{x}_i + b) = 1 - \xi_i, \quad i = 1, \ldots, N.$ $\quad (8.4.50)$

By Fung and Mangasarian [15], the PSVM formulation (8.4.49) can be also interpreted as a regularized least squares solution [38] of the system of linear equations $y_i(\mathbf{x}_i^T \mathbf{w} + b) = 1, i = 1, \ldots, N$, that is, finding an approximate solution $(\mathbf{w}, b)$ to $y_i(\mathbf{x}_i^T \mathbf{w} + b) = 1$ with least 2-norm $\left\| \begin{matrix} \mathbf{w} \\ b \end{matrix} \right\|_2^2 = \|\mathbf{w}\|_2^2 + b^2$.

The corresponding dual unconstrained optimization problem is given by

$$\min \ \mathcal{L}_{\text{DPSVM}}(\mathbf{w}, b, \xi_i, \alpha_i)$$

$$= \frac{1}{2}(\|\mathbf{w}\|_2^2 + b^2) + \frac{C}{2}\sum_{i=1}^{N} \xi_i^2 - \sum_{i=1}^{N} \alpha_i \left(y_i(\mathbf{w}^T \mathbf{x}_i + b) - 1 + \xi_i \right).$$

$$(8.4.51)$$

Using the first-order optimality conditions $\frac{\partial \mathcal{L}}{\partial \mathbf{w}} = \mathbf{0}$, $\frac{\partial \mathcal{L}}{\partial b} = 0$, $\frac{\partial \mathcal{L}}{\partial \xi_i} = 0$, and $\frac{\partial \mathcal{L}}{\partial \alpha_i} = 0$, and eliminating $\mathbf{w}$ and ξ_i, one gets the KKT equation of linear PSVM classifier in matrix form:

$$\left(C^{-1}\mathbf{I} + \mathbf{Z}\mathbf{Z}^T + \mathbf{y}\mathbf{y}^T \right) \boldsymbol{\alpha} = \mathbf{1}, \quad (8.4.52)$$

and

$$b = \sum_{i=1}^{N} \alpha_i y_i. \tag{8.4.53}$$

Here

$$\mathbf{Z} = [y_1 \mathbf{x}_1, \ldots, y_N \mathbf{x}_N], \tag{8.4.54}$$

$$\mathbf{y} = [y_1, \ldots, y_N]^T, \tag{8.4.55}$$

$$\boldsymbol{\alpha} = [\alpha_1, \ldots, \alpha_N]^T. \tag{8.4.56}$$

Similar to LS-SVM, the training data $\mathbf{x}$ can be mapped from the input space $\mathbb{R}^n$ into the feature space $\boldsymbol{\phi} : \mathbf{x} \to \boldsymbol{\phi}(\mathbf{x})$. Hence, the nonlinear PSVM classifier has still the KKT equation (8.4.52) with $\mathbf{Z} = [y_1 \boldsymbol{\phi}(\mathbf{x}_1), \ldots, y_N \boldsymbol{\phi}(\mathbf{x}_N)]$ in Eq. (8.4.54).

Algorithm 8.1 shows the PSVM binary classification algorithm.

Algorithm 8.1 PSVM binary classification algorithm
---
input: Training set $\{(\mathbf{x}_i, y_i) | \mathbf{x}_i \in \mathbb{R}^n, \ y_i \in \{-1, 1\}, \ i = 1, \ldots, N\}$, constant $C > 0$ and the kernel function $K(\mathbf{x}, \mathbf{x}_i)$.
initialization: $\mathbf{y} = [y_1, \ldots, y_N]^T$.
learning step:
 1. Construct the $N \times N$ matrix $[\mathbf{Z}^T \mathbf{Z}]_{ij} = y_i y_j \boldsymbol{\phi}^T(\mathbf{x}_i) \boldsymbol{\phi}(\mathbf{x}_j) = y_i y_j K(\mathbf{x}_i, \mathbf{x}_j)$.
 2. Solve the KKT matrix equation (8.4.52) for $\boldsymbol{\alpha} = [\alpha_1, \ldots, \alpha_N]^T$.
 3. Compute $b = \sum_{i=1}^{N} \alpha_i y_i$.
testing step: for given testing sample $\mathbf{x} \in \mathbb{R}^n$, its decision is given by
 class of $\mathbf{x} = \text{sign}\left(\sum_{i=1}^{N} \alpha_i y_i K(\mathbf{x}, \mathbf{x}_i) + b\right)$.

The decision functions of binary SVM, LS-SVM, and PSVM classifiers have the same form

$$f(\mathbf{x}) = \text{sign}\left(\sum_{i=1}^{N} \alpha_i y_i K(\mathbf{x}, \mathbf{x}_i) + b\right), \tag{8.4.57}$$

where y_i is the corresponding target class label of the training data $\mathbf{x}_i$, α_i is the Lagrange multiplier to be computed by the learning machines, and $K(\mathbf{x}, \mathbf{x}_i)$ is a suitable kernel function to be given by users.

The followings are the comparisons of SVM, LS-SVM, and PSVM.

- SVM, LS-SVM, and PSVM are originally proposed for binary classification.
- LS-SVM and PSVM provide fast implementations of the traditional SVM. Both LS-SVM and PSVM use equality optimization constraints instead of inequalities from the traditional SVM, which results in a direct least square solution by avoiding quadratic programming.

8.4 Support Vector Machine Binary Classification

It should be noted [22] that the Lagrange multipliers α_i are proportional to the training errors ξ_i in LS-SVM, while in the conventional SVM, many Lagrange multipliers α_i are typically equal to zero. As compared to the conventional SVM, sparsity is lost in LS-SVM; this is true to PSVM as well.

8.4.5 SVM-Recursive Feature Elimination

Decision functions that are simple weighted sums of the training patterns plus a bias are called *linear discriminant functions* [9]:

$$D(\mathbf{x}) = \langle \mathbf{w}, \mathbf{x} \rangle + b, \tag{8.4.58}$$

where $\mathbf{w}$ is the weight vector and b is a bias value.

Construct the optimal hyperplane $(\mathbf{w}, b)$ such that

$$\mathbf{w}_{\text{opt}}^T \mathbf{x} + b_{\text{opt}} = 0 \tag{8.4.59}$$

which separates a set of training data $(\mathbf{x}_1, y_1), \ldots, (\mathbf{x}_n, y_n)$. Vectors $\mathbf{x}_i$ such that $y_i(\mathbf{w}^T \mathbf{x}_i + b) = 1$ is termed *support vectors*. Hence, the constrained optimization problem for finding the optimal weight vector $\mathbf{w}$ can be described as

$$\min_{\mathbf{w},b} \left\{ f(\mathbf{w}, b) = \frac{1}{2} \|\mathbf{w}\|_2^2 \right\}, \tag{8.4.60}$$

$$\text{subject to } y_i(\mathbf{x}_i^T \mathbf{w} + b) \geq 1, \quad i = 1, \ldots, n, \tag{8.4.61}$$

where the inequality constraint is for ensuring $\mathbf{x}$ to be support vectors.

The above constrained optimization can be rewritten as the unconstrained optimization in Lagrangian form:

$$\min_{\mathbf{w},b,\boldsymbol{\alpha}} \left\{ L(\mathbf{w}, b, \boldsymbol{\alpha}) = \frac{1}{2} \|\mathbf{w}\|_2 - \sum_{i=1}^{n} \alpha_i [y_i(\mathbf{x}_i^T \mathbf{w} + b) - 1] \right\}, \tag{8.4.62}$$

where $\boldsymbol{\alpha} = [\alpha_1, \ldots, \alpha_n]^T$ is the vector of nonnegative Lagrangian multiplies $\alpha_i \geq 0, i = 1, \ldots, n$.

The optimization conditions with respect to $\mathbf{w}$ and b are given by

$$\frac{\partial L(\mathbf{w}, b, \boldsymbol{\alpha})}{\partial \mathbf{w}} = \left(\mathbf{w} - \sum_{i=1}^{n} \alpha_i y_i \mathbf{x}_i \right) = 0 \Rightarrow \mathbf{w} = \sum_{i=1}^{n} \alpha_i y_i \mathbf{x}_i, \tag{8.4.63}$$

$$\frac{\partial L(\mathbf{x}, b, \boldsymbol{\alpha})}{\partial b} = \sum_{i=1}^{n} \alpha_i y_i = 0, \tag{8.4.64}$$

Substituting the above two equations into (8.4.62) to yield [8]

$$\min_{\alpha} \left\{ J(\alpha) = \frac{1}{2} \sum_{i=1}^{n} \sum_{j=1}^{n} \alpha_i \alpha_j y_i y_j \mathbf{x}_i^T \mathbf{x}_j - \sum_{i=1}^{n} \alpha_i \right\}$$

subject to $\mathbf{0} \leq \boldsymbol{\alpha} \leq C\mathbf{I}$ and $\boldsymbol{\alpha}^T \mathbf{y} = 0$. (8.4.65)

In conclusion, when used for classification, SVMs separate a given set of binary labeled training data $(\mathbf{x}_i, y_i)$ with hyperplane $(\mathbf{w}, b)$ that is maximally distant from them. Such a hyperplane is known as "the *maximal margin hyperplane.*"

In nonlinear binary classification cases, Eq. (8.4.65) becomes

$$\min_{\alpha} \left\{ \frac{1}{2} \sum_{i=1}^{n} \sum_{j=1}^{n} \alpha_i \alpha_j y_i y_j \boldsymbol{\phi}^T(\mathbf{x}_i) \boldsymbol{\phi}(\mathbf{x}_j) - \sum_{i=1}^{n} \alpha_i \right\}$$

subject to $\mathbf{0} \leq \boldsymbol{\alpha} \leq C\mathbf{I}$ and $\boldsymbol{\alpha}^T \mathbf{y} = 0$. (8.4.66)

The goal of the *SVM-recursive feature elimination* (SVM-RFE) algorithm, proposed by Guyon et al. [20], is to find a subset of size r among d variables ($r < d$) which maximizes the performance of the predictor, based on a backward sequential selection. Starting all the features, one feature is removed at a time. The removed ith feature is the one whose removal minimizes the variation of $\|\mathbf{w}\|_2^2$ [31]:

$$\left| \|\mathbf{w}\|_2^2 - \|\mathbf{w}^{(i)}\|_2^2 \right|$$
$$= \frac{1}{2} \left| \sum_{j=1}^{d} \sum_{k=1}^{d} \alpha_j \alpha_k y_j y_k K(\mathbf{x}_j, \mathbf{x}_k) - \sum_{j=1}^{d} \sum_{k=1}^{d} \alpha_j^{(i)} \alpha_k^{(i)} y_j y_k K^{(i)}(\mathbf{x}_j, \mathbf{x}_k) \right|,$$
(8.4.67)

where $[\mathbf{K}^{(i)}]_{jk} = K_{jk}^{(i)} = \langle \boldsymbol{\phi}(\mathbf{x}_j^{(i)}), \boldsymbol{\phi}(\mathbf{x}_k^{(i)}) \rangle$ is the (j,k)th entry of the Gram matrix $\mathbf{K}^{(i)}$ of the training data when variable i is removed, and $\alpha_j^{(i)}$ is the corresponding solution of (8.4.66). For the sake of simplicity, $\alpha_j^{(i)} = \alpha_j$ is usually taken even if a variable has been removed in order to reduce the computational complexity of the SVM-RFE algorithm.

Algorithm 8.2 shows the SVM-RFE algorithm that is an application of RFE using the weight magnitude as ranking criterion.

Algorithm 8.2 SVM-recursive feature elimination (SVM-RFE) algorithm [20]

1. **input:** Training data $X = \{\mathbf{x}_1, \ldots, \mathbf{x}_d\}$, class labels $Y = \{y_1, \ldots, y_d\}$ and expected feature number r.
2. **initialization:** Index subset of surviving features $S = \{1, \ldots, d\}$.
3. **repeat**
4.   Restrict training examples to good feature indices (X, Y).
5.   Solve (8.4.65) or (8.4.66) for the classifier $\boldsymbol{\alpha}$.
6.   Compute the weight vector of dimension $m = \text{length}(S)$ as $\mathbf{w} = \sum_{k=1}^{m} \alpha_k y_k \mathbf{x}_k$.
7.   Compute the ranking criteria $c_i = (w_i)^2$ for all $i = 1, \ldots, m$.
8.   Find the feature index with smallest ranking criterion using
       index $i = \arg \min\{c_1, \ldots, c_m\}$.
9.   Eliminate the variable i with smallest ranking criterion and update $X \leftarrow X \setminus \mathbf{x}_i, Y \leftarrow Y \setminus y_i$ and $S \leftarrow S \setminus i$.
10. **until** length$(S) = r$.
11. **output:** feature ranked list X.

8.5 Support Vector Machine Multiclass Classification

A multiclass classifier is a function $H : \mathcal{X} \to \mathcal{Y}$ that maps an instance $\mathbf{x} \in \mathcal{X}$ (for example, $\mathcal{X} = \mathbb{R}^n$) into an element y of $\mathcal{Y}$ (for example, $y \in \{1, \ldots, k\}$).

8.5.1 Decomposition Methods for Multiclass Classification

A popular way to solve a k-class problem is to decompose it to a set of L binary classification problems. One-versus-one, one-versus-rest, and directed acyclic graph SVM methods are three of the most common decomposition approaches [21, 41, 48].

One-Against-All Method
The standard method for k-class SVM classification is to construct $L = k$ binary classifiers $\mathcal{C}_m, m = 1, \ldots, k$. The ith SVM will be trained with all of the examples in the ith class with positive labels, and all other examples with negative labels. Then, the weight vector $\mathbf{w}_m$ for the mth model can be generated by any linear classifier. Such an SVM classification method is referred to as *one-against-all* (OAA) or *one-versus-rest* (OVR) method [4, 27].

Let $S = \{(\mathbf{x}_1, y_1), \ldots, (\mathbf{x}_N, y_N)\}$ be a set of N training examples, and each example $\mathbf{x}_i$ be drawn from a domain $\mathcal{X} \subseteq \mathbb{R}^n$ and $y_i \in \{1, \ldots, k\}$ is the class of $\mathbf{x}_i$.

The mth one-against-all classifier solves the following constrained optimization problem:

$$\min_{\mathbf{w}_m, b_m, \xi_m} \left\{ \frac{1}{2} \|\mathbf{w}_m\|_2^2 + C \sum_{i=1}^{N} \xi_{m,i} \right\}, \tag{8.5.1}$$

$$\text{subject to} \quad \mathbf{w}_m^T \boldsymbol{\phi}(\mathbf{x}_i) + b_m \geq 1 - \xi_{m,i}, \quad \text{if } y_i = m,$$

$$\mathbf{w}_m^T \boldsymbol{\phi}(\mathbf{x}_i) + b_m \leq -1 + \xi_{m,i}, \quad \text{if } y_i \neq m,$$

$$\xi_{m,i} \geq 0, \quad i = 1, \ldots, N, \tag{8.5.2}$$

where $\mathbf{w}_m = [w_{m,1}, \cdots, w_{m,n}]^T$, $m = 1, \ldots, k$ is the weight vector of the mth classifier, C is the penalty parameter, and $\xi_{m,i}$ is the training error corresponding to the mth class and the ith training data $\mathbf{x}_i$ that is mapped to a higher-dimensional space by the function $\boldsymbol{\phi}(\mathbf{x}_i)$.

Minimizing $\frac{1}{2}\|\mathbf{w}_m\|_2^2$ means maximizing the margin $2/\|\mathbf{w}_m\|$ between two groups of data. When data are not linear separable, there is a penalty term $C \sum_{i=1}^{N} \xi_{m,i}$ which can reduce the training errors in the mth classifier. The basic concept behind one-against-all SVM is to search for a balance between the regularization term $\frac{1}{2}\|\mathbf{w}_m\|_2^2$ and the training error $C \sum_{i=1}^{N} \xi_{m,i}$ for the m classifiers.

By "*decision function*" it means a function $f(\mathbf{x})$ whose sign represents the class assigned to data point $\mathbf{x}$.

The solution of (8.5.1) gives k decision functions $\mathbf{w}_m^T \boldsymbol{\phi}(\mathbf{x}) + b_m$, $m = 1, \ldots, k$. Hence, a new testing instance $\mathbf{x}$ is said to be in the class which has the largest value among k decision functions:

$$\text{class of } \mathbf{x} = \arg\max_{m=1,\ldots,k} \left\{ \mathbf{w}_m^T \boldsymbol{\phi}(\mathbf{x}) + b_m \right\}. \tag{8.5.3}$$

One-Against-One Method

The *one-against-one* (OAO) method [13, 25] is also known as one-versus-one (OVO) method. This method constructs $L = k(k-1)/2$ binary classifiers by solving $k(k-1)/2$ binary classification problems [25]. Each binary classifier constructs a model with data from one class as positive and another class as negative. Since there is $k(k-1)/2$ combinations of two classes, one needs to construct $k(k-1)/2$ weight vectors: $\mathbf{w}_{1,2}, \ldots, \mathbf{w}_{1,k}; \mathbf{w}_{2,3}, \ldots, \mathbf{w}_{2,k}; \ldots; \mathbf{w}_{k-1,k}$.

For training data from the ith and the jth classes, one solves the following binary classification problem:

$$\min_{\mathbf{w}^{ij}, b^{ij}, \xi^{ij}} \left\{ \frac{1}{2} (\mathbf{w}^{ij})^T \mathbf{w}^{ij} + C \sum_{n=1}^{N} \xi_n^{ij} (\mathbf{w}^{ij})^T \boldsymbol{\phi}(\mathbf{x}_n) \right\}, \tag{8.5.4}$$

$$\text{subject to} \quad (\mathbf{w}^{ij})^T \boldsymbol{\phi}(\mathbf{x}_n) + b^{ij} \geq 1 - \xi_n^{ij}, \quad \text{if } y_n = i, \tag{8.5.5}$$

$$(\mathbf{w}^{ij})^T \boldsymbol{\phi}(\mathbf{x}_n) + b^{ij} \leq -1 + \xi_n^{ij}, \quad \text{if } y_n = j, \tag{8.5.6}$$

$$\xi_n^{ij} \geq 0, \quad n = 1, \ldots, N, \tag{8.5.7}$$

where $\mathbf{w}^{ij}$ is the binary classifier for the ith and the jth classes, b^{ij} is a real parameter, and ξ_{ij} is the training error corresponding to the ith and the jth classes.

8.5 Support Vector Machine Multiclass Classification

After all $k(k-1)/2$ binary classifiers are constructed by solving the above binary problems for $i, j = 1, \ldots, k$, the following voting strategy suggested in [25] can be used: if the decision function $\text{sign}((\mathbf{w}^{ij})^T \boldsymbol{\phi}(\mathbf{x}) + b^{ij})$ determines that $\mathbf{x}$ is in ith class, then vote for the ith class is added by one. Otherwise, the vote for jth is added by one. Finally, $\mathbf{x}$ is predicted to be in the class with the largest vote. This voting approach is also called the "Max Wins" strategy. In case that two classes have identical votes, thought it may not be a good strategy, we simply select the one with the smaller index [21].

DAGSVM Method

Directed acyclic graph SVM method [27] is simply called DAGSVM method whose training phase is the same as the one-against-one method by solving binary SVMs. However, in the testing phase, it uses a rooted binary directed acyclic graph which has internal nodes and leaves.

The training phase of the DAGSVM method is the same as the one-against-one method by solving $k(k-1)/2$ binary SVM classification problems, but uses one different voting strategy, called a rooted binary directed acyclic graph which has $k(k-1)/2$ internal nodes and k leaves, in the testing phase. Each node is a binary SVM classifier of ith and jth classes.

A directed acyclic graph (DAG) is a graph whose edges have an orientation and no cycles.

Definition 8.7 (Decision Directed Acyclic Graph [27]) Given a space X and a set of Boolean functions $\mathcal{F} = \{f : X \to \{0, 1\}\}$, the *decision directed acyclic graphs* (DDAGs) on k classes over $\mathcal{F}$ are functions which can be implemented using a rooted binary DAG with k leaves labeled by the classes where each of the $L = k(k-1)/2$ internal nodes is labeled with an element of $\mathcal{F}$. The nodes are arranged in a triangle with the single root node at the top, two nodes in the second layer, and so on until the final layer of k leaves. The ith node in layer $j < k$ is connected to the ith and $(i+1)$th node in the $(j+1)$st layer.

For k-class classification, a rooted binary directed acyclic graph has k layers: the top layer has a single root node, the second layer has two nodes, and so on until the kth (i.e., final) layer has k nodes, so a rooted binary DAG has k leaves and $1 + 2 + \cdots + k = k(k-1)/2$ internal nodes. Each node is a binary SVM of ith and jth classes.

The DDAG is equivalent to operating on a list, where each node eliminates one class from the list. The list is initialized with a list of all classes. A test point is evaluated against the decision node that corresponds to the first and last elements of the list. If the node prefers one of the two classes, the other class is eliminated from the list, and the DDAG proceeds to test the first and last elements of the new list. The DDAG terminates when only one class remains in the list. Thus, for a problem with k classes, $k - 1$ decision nodes will be evaluated in order to derive an answer.

Example 8.1 Given N test samples $\{\mathbf{x}_i, y_i\}, i = 1, \ldots, N$ with $\mathbf{x}_i \in \mathbb{R}^n$ and $y_i \in \{1, 2, 3, 4\}$. The DDAG is equivalent to operating on a list $\{1, 2, 3, 4\}$, where each

Fig. 8.2 The decision directed acyclic graphs (DDAG) for finding the best class out of four classes

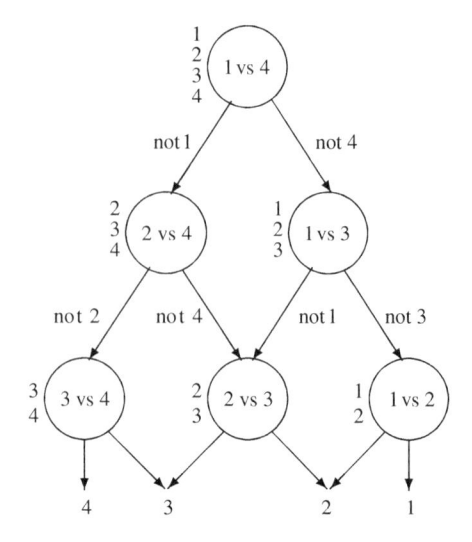

node eliminates one class from the list. The list is starting at the root node 1 versus 4 at the top layer, the binary decision function is evaluated. If the output value is not Class 1, then it is eliminated from the list to get a new list {2, 3, 4} and thus we make the binary decision of two classes 2 versus 4. If the root node prefers Class 1, then Class 4 at the root node is removed from the list to yield a new list {1, 2, 3} and the binary decision is made for two classes 1 versus 3. Then, the second layer has two nodes (2 vs 4) and (1 vs 3). Therefore, we go through a path until DDAG terminates when only one class remains in the list.

Figure 8.2 shows the decision directed acyclic graphs (DDAG) of the above four classes [27].

DDAGs naturally generalize the class of Decision Trees, allowing for a more efficient representation of redundancies and repetitions that can occur in different branches of the tree, by allowing the merging of different decision paths [27].

8.5.2 Least Squares SVM Multiclass Classifier

The previous subsection discussed the support vector machine binary classifier. In this subsection we discuss multi-class classification cases.

For the multiclass case with k labels, LS-SVM uses k output nodes in order to encode multiclasses, where $y_{i,j}$ denotes the output value of the jth output node for the training data $\mathbf{x}_i$ [41]. The k outputs can be used to encode up to 2^k different

8.5 Support Vector Machine Multiclass Classification

classes. For multiclass case, the primal optimization problem of LS-SVM can be represented as [41]

$$\min\left\{\mathcal{L}_{\text{LS-SVM}} = \frac{1}{2}\sum_{m=1}^{k}\|\mathbf{w}_m\|^2 + \frac{C}{2}\sum_{i=1}^{N}\sum_{m=1}^{k}\xi_{m,i}^2\right\}$$

$$\text{subject to } \begin{cases} y_i^{(1)}(\mathbf{w}_1^T\boldsymbol{\phi}_1(\mathbf{x}_i) + b_1) = 1 - \xi_{1,i} \\ \quad\vdots \\ y_i^{(k)}(\mathbf{w}_k^T\boldsymbol{\phi}_k(\mathbf{x}_i) + b_k) = 1 - \xi_{k,i} \end{cases} \quad (8.5.8)$$

for $i = 1, \ldots, N$.

The Lagrange function in dual LS-SVM multiclass classifier is given by

$$\mathcal{L}_D = \frac{1}{2}\sum_{m=1}^{k}\|\mathbf{w}_m\|^2 + \frac{C}{2}\sum_{i=1}^{N}\sum_{m=1}^{k}\xi_{m,i}^2 \quad (8.5.9)$$

$$-\sum_{i=1}^{N}\sum_{m=1}^{k}\alpha_{m,i}\left[y_i\left(\mathbf{w}_m^T\boldsymbol{\phi}_m(\mathbf{x}_i) + b_m\right) - 1 + \xi_{m,i}\right] \quad (8.5.10)$$

Similar to the LS-SVM solution to the binary classification, the conditions for optimality are given by

$$\frac{\partial \mathcal{L}_D}{\partial \mathbf{w}_m} = \mathbf{0} \Rightarrow \mathbf{w}_m = \sum_{i=1}^{N}\alpha_{m,i}y_i^{(m)}\boldsymbol{\phi}_m(\mathbf{x}_i) \Rightarrow \mathbf{w}_m = \mathbf{Z}_m\boldsymbol{\alpha}_m, \quad (8.5.11)$$

$$\frac{\partial \mathcal{L}_D}{\partial b_m} = 0 \Rightarrow \sum_{i=1}^{N}\alpha_{m,i}y_i^{(m)} = 0 \Rightarrow \boldsymbol{\alpha}_m^T\mathbf{y}^{(m)} = 0, \quad (8.5.12)$$

$$\frac{\partial \mathcal{L}_D}{\partial \xi_{m,i}} = 0 \Rightarrow \alpha_{m,i} = C\xi_{m,i} \Rightarrow \boldsymbol{\alpha}_m = C\boldsymbol{\xi}_m, \quad (8.5.13)$$

$$\frac{\partial \mathcal{L}_D}{\partial \alpha_{m,i}} = 0 \Rightarrow y_i^{(m)}\left(\mathbf{w}_m^T\boldsymbol{\phi}_m(\mathbf{x}_i) + b_m\right) - 1 + \xi_{m,i} = 0,$$

$$\Rightarrow \mathbf{Z}_m^T\mathbf{w}_m + b_m\mathbf{y}^{(m)} + \boldsymbol{\xi}_m = \mathbf{1}. \quad (8.5.14)$$

In the above equations:

$$\mathbf{Z}_m = [y_1^{(m)}\boldsymbol{\phi}_m(\mathbf{x}_1), \ldots, y_N^{(m)}\boldsymbol{\phi}_m(\mathbf{x}_N)], \quad (8.5.15)$$

$$\mathbf{w}_m = [w_{m,1}, \ldots, w_{m,N}]^T, \quad (8.5.16)$$

$$\mathbf{y}^{(m)} = [y_1^{(m)}, \ldots, y_N^{(m)}]^T, \quad (8.5.17)$$

$$\boldsymbol{\alpha}_m = [\alpha_{m,1}, \ldots, \alpha_{m,N}]^T, \qquad (8.5.18)$$

$$\boldsymbol{\xi}_m = [\xi_{m,1}, \ldots, \xi_{m,N}]^T. \qquad (8.5.19)$$

Due to each classifier being binary, for given $y_i \in \{1, \ldots, k\}$, one has

$$y_i^{(m)} = \begin{cases} +1, & y_i = m; \\ -1, & y_i \neq m. \end{cases} \quad (m = 1, \ldots, k; i = 1, \ldots, N) \qquad (8.5.20)$$

Equations (8.5.15)–(8.5.19) can be rewritten as the KKT equation in matrix form:

$$\begin{bmatrix} 0 & (\mathbf{y}^{(m)})^T \\ \mathbf{y}^{(m)} & \boldsymbol{\Omega}^{(m)} \end{bmatrix} \begin{bmatrix} b_m \\ \boldsymbol{\alpha}_m \end{bmatrix} = \begin{bmatrix} 0 \\ \mathbf{1} \end{bmatrix}, \quad m = 1, \ldots, k, \qquad (8.5.21)$$

where $\boldsymbol{\Omega}^{(m)} = \left(\mathbf{Z}_m^T \mathbf{Z}_m + C^{-1}\mathbf{I}\right)$ with the (i, j)th entries

$$\Omega_{ij}^{(m)} = y_i^{(m)} y_j^{(m)} K_m(\mathbf{x}_i, \mathbf{x}_j) + C^{-1}\delta_{ij}, \quad i, j = 1, \ldots, N, \qquad (8.5.22)$$

where $\delta_{ij} = 1$ (if $i = j$) or 0 (otherwise); and $K_m(\mathbf{x}_i, \mathbf{x}_j) = \boldsymbol{\phi}_m^T(\mathbf{x}_i)\boldsymbol{\phi}_m(\mathbf{x}_j)$ is the kernel function of the mth SVM for multiclass classification.

Algorithm 8.3 shows the LS-SVM multiclass classification algorithm.

Algorithm 8.3 LS-SVM multiclass classification algorithm [41]

input: Training set $\{(\mathbf{x}_i, y_i) | \mathbf{x}_i \in \mathbb{R}^n, y_i \in \{1, \ldots, k\}, i = 1, \ldots, N\}$, constant $C > 0$ and the kernel
function $K_m(\mathbf{x}, \mathbf{x}_i), m = 1, \ldots, k$.

initialization: $y_i^{(m)} = \begin{cases} +1, & y_i = m; \\ -1, & y_i \neq m; \end{cases}$ for $m = 1, \ldots, k; i = 1, \ldots, N$.

learning step:
  while $m = 1, \ldots, k$
  1. Construct the $N \times 1$ vector $\mathbf{y}^{(m)} = \left[y_1^{(m)}, \ldots, y_N^{(m)}\right]^T$.
  2. Use (8.5.22) to construct all (i, j)th entries of the $N \times N$ matrix $\boldsymbol{\Omega}^{(m)}$.
  3. Solve the KKT matrix equation (8.5.21) for b_m and $\boldsymbol{\alpha}_m = [\alpha_{m,i}]_{i=1}^N$.
  end while

testing step: for given testing sample $\mathbf{x} \in \mathbb{R}^n$, its decision is given by
$$\text{class of } \mathbf{x} = \arg\max_{m=1,\ldots,k} \left(\sum_{i=1}^N \alpha_{m,i} y_i^{(m)} K_m(\mathbf{x}, \mathbf{x}_i) + b_m\right).$$

8.5.3 Proximal Support Vector Machine Multiclass Classifier

The primal constrained optimization problem of PSVM multiclass classifier is described by

$$\min\left\{\mathcal{L}_{\text{PPSVM}}(m)(\mathbf{w}_m, b_m, \xi_{m,i}) = \frac{1}{2}(\|\mathbf{w}_m\|_2^2 + b_m^2) + \frac{C}{2}\sum_{i=1}^{N}\xi_{m,i}^2\right\} \quad (8.5.23)$$

subject to
$$\begin{cases} y_i^{(1)}(\mathbf{w}_1^T \boldsymbol{\phi}_m(\mathbf{x}_1) + b_1) = 1 - \xi_{1,i}, \\ \quad \vdots \\ y_i^{(k)}(\mathbf{w}_k^T \boldsymbol{\phi}_m(\mathbf{x}_i) + b_k) = 1 - \xi_{k,i}. \end{cases} \quad (8.5.24)$$

The corresponding dual unconstrained optimization problem is given by

$$\min \quad \mathcal{L}_{\text{DPSVM}}^{(m)} = \frac{1}{2}(\|\mathbf{w}_m\|_2^2 + b_m^2) + \frac{C}{2}\sum_{i=1}^{N}\xi_{m,i}^2$$

$$-\sum_{i=1}^{N}\alpha_{m,i}\left(y_i^{(m)}(\mathbf{w}_m^T \boldsymbol{\phi}_m(\mathbf{x}_i) + b_m) - 1 + \xi_{m,i}\right). \quad (8.5.25)$$

From the optimality conditions we have

$$\frac{\partial \mathcal{L}_{\text{DPSVM}}^{(m)}}{\partial \mathbf{w}_m} = \mathbf{0} \quad \Rightarrow \quad \mathbf{w}_m = \sum_{i=1}^{N}\alpha_{m,i} y_i^{(m)} \boldsymbol{\phi}_m(\mathbf{x}_i), \quad (8.5.26)$$

$$\frac{\partial \mathcal{L}_{\text{DPSVM}}^{(m)}}{\partial b_m} = 0 \quad \Rightarrow \quad b_m = \sum_{i=1}^{N}\alpha_{m,i} y_i^{(m)}, \quad (8.5.27)$$

$$\frac{\partial \mathcal{L}_{\text{DPSVM}}^{(m)}}{\partial \xi_{m,i}} = 0 \quad \Rightarrow \quad \xi_{m,i} = C^{-1}\alpha_{m,i}, \quad (8.5.28)$$

$$\frac{\partial \mathcal{L}_{\text{DPSVM}}^{(m)}}{\partial \alpha_{m,i}} = 0 \quad \Rightarrow \quad y_i^{(m)}(\mathbf{w}_m^T \boldsymbol{\phi}_m(\mathbf{x}_i) + b_m) - 1 + \xi_{m,i} = 0, \quad (8.5.29)$$

for $m = 1, \ldots, k$.

Eliminating $\mathbf{w}_m$ and $\xi_{m,i}$ in the above equations yields the KKT equations

$$b_m = \sum_{i=1}^{N}\alpha_{m,i} y_i^{(m)} = \boldsymbol{\alpha}_m^T \mathbf{y}_m, \quad (8.5.30)$$

and

$$(C^{-1}\mathbf{I} + \mathbf{Z}_m^T\mathbf{Z}_m + \mathbf{y}_m\mathbf{y}_m^T)\boldsymbol{\alpha}_m = \mathbf{1}, \tag{8.5.31}$$

where

$$\mathbf{Z}_m = \left[y_1^{(m)}\boldsymbol{\phi}_m(\mathbf{x}_1), \ldots, y_N^{(m)}\boldsymbol{\phi}_m(\mathbf{x}_N)\right], \tag{8.5.32}$$

$$\mathbf{y}_m = [y_1^{(m)}, \ldots, y_N^{(m)}]^T, \tag{8.5.33}$$

$$\boldsymbol{\alpha}_m = [\alpha_{m,1}, \ldots, \alpha_{m,N}]^T. \tag{8.5.34}$$

Algorithm 8.4 shows the PSVM multiclass classification algorithm.

Algorithm 8.4 PSVM multiclass classification algorithm

1. **input:** Training set $\{(\mathbf{x}_i, y_i) | \mathbf{x}_i \in \mathbb{R}^n, y_i \in \{1, \ldots, k\}, i = 1, \ldots, N\}$, constant $C > 0$ and the kernel function $K_m(\mathbf{x}, \mathbf{x}_j), m = 1, \ldots, k$.
2. **initialization:** $y_i^{(m)} = \begin{cases} +1, & y_i = m; \\ -1, & y_i \neq m; \end{cases}$ for $m = 1, \ldots, k; i = 1, \ldots, N$.
3. **learning step:**
   while $m = 1, \ldots, k$
   3.1. Construct the $N \times N$ matrix $[\mathbf{Z}_m^T\mathbf{Z}_m]_{ij} = y_i^{(m)} y_j^{(m)} K_m(\mathbf{x}_i, \mathbf{x}_j)$.
   3.2. Solve the KKT matrix equation (8.5.31) for $\boldsymbol{\alpha}_m = (C^{-1}\mathbf{I} + \mathbf{Z}_m^T\mathbf{Z}_m + \mathbf{y}_m\mathbf{y}_m^T)^\dagger \mathbf{1}$.
   3.3. Compute $b_m = \boldsymbol{\alpha}_m^T \mathbf{y}_m$.
   endwhile
4. **testing step:** for given testing sample $\mathbf{x} \in \mathbb{R}^n$, its decision is given by
   $$\text{class of } \mathbf{x} = \arg\max_{m=1,\ldots,k} \left(\sum_{i=1}^N \alpha_{m,i} y_i^{(m)} K_m(\mathbf{x}, \mathbf{x}_i) + b_m\right).$$

8.6 Gaussian Process for Regression and Classification

In Chap. 6 we discussed traditionally parametric models based machine learning. The parametric models have a possible advantage in ease of interpretability, but for complex data sets, simple parametric models may lack expressive power, and their more complex counterparts (such as feedforward neural networks) may not be easy to work with in practice [29, 30]. The advent of kernel machines, such as Gaussian processes, sparse Bayesian learning, and relevance vector machine, has opened the possibility of flexible models which are practical to work with.

In this section we deal with Gaussian process methods for regression and classification problems.

8.6.1 Joint, Marginal, and Conditional Probabilities

Let the n (discrete or continuous) random variables $y_1, \ldots, y_n$ have a joint probability $p(y_1, \ldots, y_n)$, or $p(\mathbf{y})$ for short. Technically, one ought to distinguish between probabilities (for discrete variables) and probability densities for continuous variables. Throughout the book we commonly use the term "probability" to refer to both. Let us partition the variables in $\mathbf{y}$ into two groups, $\mathbf{y}_A$ and $\mathbf{y}_B$, where $A \cup B = \{1, \ldots, n\}$ and $A \cap B = \emptyset$, so that $p(\mathbf{y}) = p(\mathbf{y}_A, \mathbf{y}_B)$. Each group may contain one or more variables. The marginal probability of $\mathbf{y}_A$, denoted by $p(\mathbf{y}_A)$, is given by marginal probability

$$p(\mathbf{y}_A) = \int p(\mathbf{y}_A, \mathbf{y}_B) d\mathbf{y}_B. \tag{8.6.1}$$

The integral is replaced by a sum if the variables are discrete valued. Notice that if the set A contains more than one variable, then the marginal probability is itself a joint probability of these variables. If the joint distribution is equal to the product of the marginals, then the variables are said to be independent, otherwise they are dependent.

The *conditional probability function* is defined as conditional probability

$$p(\mathbf{y}_A | \mathbf{y}_B) = \frac{p(\mathbf{y}_A, \mathbf{y}_B)}{p(\mathbf{y}_B)} \tag{8.6.2}$$

defined for $p(\mathbf{y}_B) > 0$, as it is not meaningful to condition on an impossible event. If $\mathbf{y}_A$ and $\mathbf{y}_B$ are independent, then the marginal $p(\mathbf{y}_A)$ and the conditional $p(\mathbf{y}_A|\mathbf{y}_B)$ are equal.

Using the definitions of both $p(\mathbf{y}_A|\mathbf{y}_B)$ and $p(\mathbf{y}_B|\mathbf{y}_A)$ we obtain Bayes theorem

$$p(\mathbf{y}_A|\mathbf{y}_B) = \frac{p(\mathbf{y}_A) p(\mathbf{y}_B|\mathbf{y}_A)}{p(\mathbf{y}_B)}. \tag{8.6.3}$$

8.6.2 Gaussian Process

A random variable x with the normal distribution $p(x) = (2\pi\sigma^2)^{-1} \exp\left(\frac{|x-\mu|^2}{2\sigma^2}\right)$ is called the univariate Gaussian distribution, and is denoted as $x \sim N(\mu, \sigma^2)$, where $\mu = E\{x\}$ and $\sigma^2 = \text{var}(x)$ are its mean and variance, respectively.

A multivariate Gaussian (or normal) distribution has a joint probability density given by

$$p(\mathbf{x}|\boldsymbol{\mu}, \boldsymbol{\Sigma}) = (2\pi)^{-N/2} |\boldsymbol{\Sigma}|^{-1/2} \exp\left(-\frac{1}{2}(\mathbf{x} - \boldsymbol{\mu})^T \boldsymbol{\Sigma}^{-1} (\mathbf{x} - \boldsymbol{\mu})\right), \tag{8.6.4}$$

where $\boldsymbol{\mu} = [\mu_1, \ldots, \mu_N]^T$ with $\mu_i = E\{x_i\}$ is the mean vector (of length N) of $\mathbf{x}$ and $\boldsymbol{\Sigma}$ is the $N \times N$ (symmetric, positive definite) covariance matrix of $\mathbf{x}$. The multivariate Gaussian distribution is denoted simply as $\mathbf{x} \sim N(\boldsymbol{\mu}, \boldsymbol{\Sigma})$.

If letting $\mathbf{x}$ and $\mathbf{y}$ be jointly Gaussian random vectors

$$\begin{bmatrix} \mathbf{x} \\ \mathbf{y} \end{bmatrix} \sim N\left(\begin{bmatrix} \boldsymbol{\mu}_x \\ \boldsymbol{\mu}_y \end{bmatrix}, \begin{bmatrix} \mathbf{A} & \mathbf{C} \\ \mathbf{C}^T & \mathbf{B} \end{bmatrix} \right) = N\left(\begin{bmatrix} \boldsymbol{\mu}_x \\ \boldsymbol{\mu}_y \end{bmatrix}, \begin{bmatrix} \bar{\mathbf{A}} & \bar{\mathbf{C}} \\ \bar{\mathbf{C}}^T & \bar{\mathbf{B}} \end{bmatrix}^{-1} \right), \tag{8.6.5}$$

then the marginal distribution of $\mathbf{x}$ and the conditional distribution of $\mathbf{x}$ given marginalizing $\mathbf{y}$ are

$$\begin{bmatrix} \mathbf{x} \\ \mathbf{y} \end{bmatrix} \sim N\left(\begin{bmatrix} \boldsymbol{\mu}_x \\ \boldsymbol{\mu}_y \end{bmatrix}, \begin{bmatrix} \mathbf{A} & \mathbf{C} \\ \mathbf{C}^T & \mathbf{B} \end{bmatrix} \right) \Rightarrow \mathbf{y}|\mathbf{x} \sim N\left(\boldsymbol{\mu}_y + \mathbf{C}^T \mathbf{A}^{-1}(\mathbf{x} - \boldsymbol{\mu}_x), \mathbf{B} - \mathbf{C}^T \mathbf{A}^{-1} \mathbf{C} \right)$$
$$\tag{8.6.6}$$

or

$$\begin{bmatrix} \mathbf{x} \\ \mathbf{y} \end{bmatrix} \sim N\left(\begin{bmatrix} \boldsymbol{\mu}_x \\ \boldsymbol{\mu}_y \end{bmatrix}, \begin{bmatrix} \bar{\mathbf{A}} & \bar{\mathbf{C}} \\ \bar{\mathbf{C}}^T & \bar{\mathbf{B}} \end{bmatrix}^{-1} \right) \Rightarrow \mathbf{y}|\mathbf{x} \sim N\left(\boldsymbol{\mu}_y - \bar{\mathbf{B}}^{-1} \bar{\mathbf{C}}^T (\mathbf{x} - \boldsymbol{\mu}_x), \bar{\mathbf{B}}^{-1} \right). \tag{8.6.7}$$

A Gaussian process is a natural generalization of multivariate Gaussian distribution.

Definition 8.8 (Gaussian Process [29, 30]) A *Gaussian Process* $f(\mathbf{x})$ is a collection of random variables $\mathbf{x}$, any finite number of which have (consistent) joint Gaussian distributions.

Clearly, the Gaussian distribution is over vectors, whereas the Gaussian process is over functions $f(\mathbf{x})$, where $\mathbf{x}$ is a Gaussian vector.

A scalar Gaussian process $f(x)$ can be represented as

$$f \sim \text{GP}(\mu, K), \tag{8.6.8}$$

where $\mu = E\{x\}$ is the mean of x and $K(x, x')$ is the covariance function of x. Equation (8.6.8) means [29]: "the function f is distributed as a GP with mean function μ and covariance function K."

A vector-valued Gaussian process $\mathbf{f}(\mathbf{x})$ is completely specified by its mean function and covariance function

$$\boldsymbol{\mu}(\mathbf{x}) = E\{\mathbf{f}(\mathbf{x})\}, \tag{8.6.9}$$

$$K(\mathbf{x}_i, \mathbf{x}_j) = E\{(\mathbf{f}(\mathbf{x}_i) - \boldsymbol{\mu}(\mathbf{x}_i))^T (\mathbf{f}(\mathbf{x}_j) - \boldsymbol{\mu}(\mathbf{x}_j))\}, \tag{8.6.10}$$

and the Gaussian process is expressed as

$$\mathbf{f}(\mathbf{x}) \sim GP(\boldsymbol{\mu}, \boldsymbol{\Sigma}), \tag{8.6.11}$$

where the (i, j)th elements of the covariance matrix $\boldsymbol{\Sigma}$ are defined as $\Sigma_{ij} = K(\mathbf{x}_i, \mathbf{x}_j)$.

The covariance function $K(\mathbf{x}_i, \mathbf{x}_j)$ (also known as kernel, kernel function, or covariance kernel) is the driving factor in Gaussian processes for regression and/or classification. Actually, the kernel represents the particular structure present in the data $\mathbf{x}_1, \ldots, \mathbf{x}_N$ being modeled. One of the main difficulties in applying Gaussian processes is to construct such a kernel.

According to Mercer's theorem (Theorem 8.3), the nonlinear kernel function is usually constructed by

$$K(\mathbf{x}_i, \mathbf{x}_j) = \boldsymbol{\phi}^T(\mathbf{x}_i)\boldsymbol{\phi}(\mathbf{x}_j). \tag{8.6.12}$$

8.6.3 Gaussian Process Regression

Let $\{(\mathbf{x}_n, f_n) | n = 1, \ldots, N\}$ be the training samples, where $f_n, n = 1, \ldots, N$ are the noise-free training outputs of $\mathbf{x}_n$, typically continuous functions for regression or discrete function for classification.

Denote the noise-free training output vector $\mathbf{f} = [f_1, \ldots, f_N]^T$ and the test output vector $\mathbf{f}_* = [f_1^*, \ldots, f_{N_*}^*]^T$ given the test samples $\mathbf{x}_1^*, \ldots, \mathbf{x}_{N_*}^*$. Under the assumption of Gaussian distributions $\mathbf{f} \sim N(\mathbf{0}, \mathbf{K}(\mathbf{X}, \mathbf{X}))$ and $\mathbf{f}_* \sim N(\mathbf{0}, \mathbf{K}(\mathbf{X}_*, \mathbf{X}_*))$, the joint distribution of $\mathbf{f}$ and $\mathbf{f}_*$ is given by

$$\begin{bmatrix} \mathbf{f} \\ \mathbf{f}_* \end{bmatrix} \sim N\left(\mathbf{0}, \begin{bmatrix} \mathbf{K}(\mathbf{X}, \mathbf{X}) & \mathbf{K}(\mathbf{X}, \mathbf{X}_*) \\ \mathbf{K}(\mathbf{X}_*, \mathbf{X}) & \mathbf{K}(\mathbf{X}_*, \mathbf{X}_*) \end{bmatrix}\right), \tag{8.6.13}$$

where $\mathbf{X} = [\mathbf{x}_1, \ldots, \mathbf{x}_N]$ and $\mathbf{X}_* = [\mathbf{x}_1^*, \ldots, \mathbf{x}_{N_*}^*]$ are the $N \times N$ training sample matrix and the $N_* \times N_*$ test sample matrix, respectively; and $\mathbf{K}(\mathbf{X}, \mathbf{X}) = \text{cov}(\mathbf{X}) \in \mathbb{R}^{N \times N}$, $\mathbf{K}(\mathbf{X}, \mathbf{X}_*) = \text{cov}(\mathbf{X}, \mathbf{X}_*) \in \mathbb{R}^{N \times N_*}$, $\mathbf{K}(\mathbf{X}_*, \mathbf{X}) = \text{cov}(\mathbf{X}_*, \mathbf{X}) = \mathbf{K}^T(\mathbf{X}, \mathbf{X}_*) \in \mathbb{R}^{N_* \times N}$, and $\mathbf{K}(\mathbf{X}_*, \mathbf{X}_*) = \text{cov}(\mathbf{X}_*) \in \mathbb{R}^{N_* \times N_*}$ are the covariance function matrices, respectively.

From Eq. (8.6.6) it is known that the conditional distribution of $\mathbf{f}_*$ given $\mathbf{f}$ can be expressed as

$$\mathbf{f}_* | \mathbf{f} \sim N\big(\mathbf{K}(\mathbf{X}_*, \mathbf{X})\mathbf{K}^{-1}(\mathbf{X}, \mathbf{X})\mathbf{f}, \mathbf{K}(\mathbf{X}_*, \mathbf{X}_*) - \mathbf{K}(\mathbf{X}_*, \mathbf{X})\mathbf{K}^{-1}(\mathbf{X}, \mathbf{X})\mathbf{K}(\mathbf{X}, \mathbf{X}_*)\big). \tag{8.6.14}$$

This is just the posterior distribution for a specific set of test cases.

The goal of Gaussian processes regression is to find $\mathbf{f}_*$ via maximizing the posterior probability $\mathbf{f}_*|\mathbf{f}$. However, $\mathbf{f}$ is unobservable. In practical applications, the noisy training output vector $\mathbf{y} = \mathbf{f} + \mathbf{e}$ is observed in the additive Gaussian white noise $\mathbf{e} \sim N(\mathbf{0}, \sigma_n^2 \mathbf{I})$. In this case, $\mathbf{y} \sim N(\mathbf{0}, \mathrm{cov}(\mathbf{y}))$ with

$$\mathrm{cov}(y_i, y_j) = K(y_i, y_j) + \sigma_n^2 \delta_{ij} \quad \text{or} \quad \mathbf{y} \sim N\bigl(\mathbf{0}, \mathbf{K}(\mathbf{X}, \mathbf{X}) + \sigma_n^2 \mathbf{I}\bigr). \tag{8.6.15}$$

Then, the joint distribution of $\mathbf{y}$ and $\mathbf{f}_*$ according to the prior is given by

$$\begin{bmatrix} \mathbf{y} \\ \mathbf{f}_* \end{bmatrix} \sim N\left(\mathbf{0}, \begin{bmatrix} \mathbf{K}(\mathbf{X},\mathbf{X}) + \sigma_n^2 \mathbf{I} & \mathbf{K}(\mathbf{X}, \mathbf{X}_*) \\ \mathbf{K}(\mathbf{X}_*, \mathbf{X}) & \mathbf{K}(\mathbf{X}_*, \mathbf{X}_*) \end{bmatrix}\right), \tag{8.6.16}$$

From Eq. (8.6.6) we get the conditional distribution of $\mathbf{f}_*$ given $\mathbf{y}$ below:

$$\mathbf{f}_*|\mathbf{y} \sim N\bigl(\bar{\mathbf{f}}_*, \mathrm{cov}(\mathbf{f}_*)\bigr), \tag{8.6.17}$$

where

$$\bar{\mathbf{f}}_* = E\{\mathbf{f}_*|\mathbf{y}\} = \mathbf{K}(\mathbf{X}_*, \mathbf{X})\bigl[\mathbf{K}(\mathbf{X}, \mathbf{X}) + \sigma_n^2 \mathbf{I}\bigr]^{-1} \mathbf{y}, \tag{8.6.18}$$

$$\mathrm{cov}(\mathbf{f}_*) = \mathbf{K}(\mathbf{X}_*, \mathbf{X}_*) - \mathbf{K}(\mathbf{X}_*, \mathbf{X})\bigl[\mathbf{K}(\mathbf{X}, \mathbf{X}) + \sigma_n^2 \mathbf{I}\bigr]^{-1} \mathbf{K}(\mathbf{X}, \mathbf{X}_*). \tag{8.6.19}$$

Here, $\bar{\mathbf{f}}_*$ is the mean vector over N_* test points, i.e., $\bar{\mathbf{f}}_* = \bar{\mathbf{f}}(\mathbf{x}_1^*, \ldots, \mathbf{x}_{N_*}^*)$.
If there is only one test point $\mathbf{x}_*$, i.e., $\mathbf{X}_* = \mathbf{x}_*$, then $\mathbf{K}(\mathbf{X}, \mathbf{X}_*)$ reduce to $\mathbf{k}_* = \mathbf{k}(\mathbf{x}_*) = \mathbf{k}(\mathbf{x}, \mathbf{x}_*)$ which denotes the vector of covariances between the test point $\mathbf{x}_*$ and the n training points $\mathbf{x}_1, \ldots, \mathbf{x}_N$ in $\mathbf{X}$. In this case, Eqs. (8.6.18) and (8.6.19) reduce to

$$\bar{\mathbf{f}}_* = \mathbf{k}_*^T (\mathbf{K} + \sigma_n^2 \mathbf{I})^{-1} \mathbf{y}, \tag{8.6.20}$$

$$\mathrm{cov}(\mathbf{f}_*) = K(\mathbf{x}_*, \mathbf{x}_*) - \mathbf{k}_*^T (\mathbf{K} + \sigma_n^* \mathbf{I})^{-1} \mathbf{k}_*. \tag{8.6.21}$$

Letting $\boldsymbol{\alpha} = (\mathbf{K} + \sigma_n^2 \mathbf{I})^{-1} \mathbf{y}$, then the mean vector centered on a test point $\mathbf{x}_*$, denoted by $\bar{\mathbf{f}}_* = \bar{\mathbf{f}}(\mathbf{x}_*)$, is given by

$$\bar{\mathbf{f}}(\mathbf{x}_*) = \sum_{i=1}^{N} \alpha_i K(\mathbf{x}_i, \mathbf{x}_*) = \mathbf{k}_*^T \boldsymbol{\alpha}. \tag{8.6.22}$$

When using Eqs. (8.6.20) and (8.6.21) to compute directly $\bar{\mathbf{f}}_*$ and $\mathrm{var}(\mathbf{f}_*)$, it needs to invert the matrix $\mathbf{K} + \sigma_n^2 \mathbf{I}$. A faster and numerically more stable computation method is to use Cholesky decomposition.

8.6 Gaussian Process for Regression and Classification

Let the Cholesky decomposition of $\mathbf{K}+\sigma_n^2\mathbf{I}$ be given by $(\mathbf{K}+\sigma_n^2\mathbf{I}) = \mathbf{L}\mathbf{L}^T$, where $\mathbf{L}$ is a lower triangular matrix. Then, $\boldsymbol{\alpha} = (\mathbf{K}+\sigma_n^2\mathbf{I})^{-1}\mathbf{y}$ becomes $\boldsymbol{\alpha} = (\mathbf{L}^T)^{-1}\mathbf{L}^{-1}\mathbf{y}$. Denoting $\mathbf{z} = \mathbf{L}^{-1}\mathbf{y}$ and $\boldsymbol{\alpha} = (\mathbf{L}^T)^{-1}\mathbf{z}$, then

$$\mathbf{z} = \mathbf{L}^{-1}\mathbf{y} \Leftrightarrow \mathbf{L}\mathbf{z} = \mathbf{y}, \tag{8.6.23}$$

$$\boldsymbol{\alpha} = (\mathbf{L}^T)^{-1}\mathbf{z} \Leftrightarrow \mathbf{L}^T\boldsymbol{\alpha} = \mathbf{z}. \tag{8.6.24}$$

The above two equations suggest an efficient algorithm for finding $\boldsymbol{\alpha}$ given $\mathbf{K}$ and $\mathbf{y}$ as follows:

1. Make Cholesky decomposition: $(\mathbf{K} + \sigma_n^2\mathbf{I}) = \mathbf{L}\mathbf{L}^T$.
2. Solve the triangular system $\mathbf{L}\mathbf{z} = \mathbf{y}$ for $\mathbf{z}$ by forward substitution.
3. Solve the triangular system $\mathbf{L}^T\boldsymbol{\alpha} = \mathbf{z}$ for $\boldsymbol{\alpha}$ by back substitution.

Similarly, we have $\mathbf{k}_*^T(\mathbf{K} + \sigma_n^2\mathbf{I})^{-1}\mathbf{k}_* = \mathbf{k}_*^T(\mathbf{L}^T)^{-1}\mathbf{L}^{-1}\mathbf{k}_* = \mathbf{v}^T\mathbf{v}$, where $\mathbf{v} = \mathbf{L}^{-1}\mathbf{k}_* \Leftrightarrow \mathbf{L}\mathbf{v} = \mathbf{k}_*$. This implies $\mathbf{v}$ is the solution for the triangular system $\mathbf{L}\mathbf{v} = \mathbf{k}_*$.

Once $\boldsymbol{\alpha}$ and $\mathbf{v}$ are found, one can use Eqs. (8.6.20) and (8.6.21) to get $\bar{\mathbf{f}}(\mathbf{x}_*) = \mathbf{k}_*^T\boldsymbol{\alpha}$ and $\text{var}(\mathbf{f}_*) = k(\mathbf{x}_*, \mathbf{x}_*) - \mathbf{v}^T\mathbf{v}$, respectively.

The marginal likelihood is the integral marginal of the likelihood times the prior

$$p(\mathbf{y}|\mathbf{X}) = \int p(\mathbf{y}|\mathbf{f}, \mathbf{X})p(\mathbf{f}|\mathbf{X})d\mathbf{f}. \tag{8.6.25}$$

By the term "marginal likelihood," it refers to the marginalization over the function values $\mathbf{f}$. Under the Gaussian process model, the prior is Gaussian, $\mathbf{f}|\mathbf{X} \sim N(\mathbf{0}, \mathbf{K}(\mathbf{X}, \mathbf{X}))$, or

$$\log p(\mathbf{f}|\mathbf{X}) = -\frac{1}{2}\mathbf{f}^T\mathbf{K}^{-1}\mathbf{f} - \frac{1}{2}\log|\mathbf{K}| - \frac{N}{2}\log(2\pi). \tag{8.6.26}$$

Because of $\mathbf{y} = \mathbf{f} + \mathbf{e}$ with $\mathbf{e} \sim N(\mathbf{0}, \sigma_n^2\mathbf{I})$, the marginal likelihood

$$\log p(\mathbf{y}|\mathbf{X}) = -\frac{1}{2}\mathbf{y}^T[\mathbf{K} + \sigma_n^2\mathbf{I}]^{-1}\mathbf{y} - \frac{1}{2}\log|\mathbf{K} + \sigma_n^2\mathbf{I}| - \frac{N}{2}\log(2\pi). \tag{8.6.27}$$

Here, the first term is a data fitting term as it is the only one involving the observed targets $\mathbf{y}$. The second term $\log|\mathbf{K} + \sigma_n^2\mathbf{I}|/2$ is the model complexity depending only on the covariance function and the inputs. The last term $\log(2\pi)/2$ is just a normalization constant.

Algorithm 8.5 shows the Gaussian process regression (GPR) in [30].

Algorithm 8.5 Gaussian process regression algorithm [30]

1. **input:** $\mathbf{X}$ (inputs), $\mathbf{y}$ (targets), k (covariance function), σ_n^2 (noise level), $\mathbf{x}_*$ (test input).
2. Construct the matrix $\mathbf{K}$ whose (i, j) elements $K_{ij} = k(\mathbf{x}_i, \mathbf{x}_j), i, j = 1, \ldots, N$.
3. Make Cholesky decomposition $\mathbf{K} + \sigma_n^2 \mathbf{I} = \mathbf{L}\mathbf{L}^T$.
4. Solve the triangular system $\mathbf{L}\mathbf{z} = \mathbf{y}$ for $\mathbf{z}$ by forward substitution.
5. Solve the triangular system $\mathbf{L}^T \boldsymbol{\alpha} = \mathbf{z}$ for $\boldsymbol{\alpha}$ by back substitution.
6. $\bar{\mathbf{f}}_* = \mathbf{k}_*^T \boldsymbol{\alpha}$.
7. Solve the triangular system $\mathbf{L}\mathbf{v} = \mathbf{k}_*$ for $\mathbf{v}$ by forward substitution.
8. $\text{var}(\mathbf{f}_*) = k(\mathbf{x}_*, \mathbf{x}_*) - \mathbf{v}^T \mathbf{v}$.
9. $\log p(\mathbf{y}|\mathbf{X}) = -\frac{1}{2}\mathbf{y}^T \boldsymbol{\alpha} - \sum_{i=1}^N L_{ii} - \frac{N}{2}\log(2\pi)$.
10. **output:** $\bar{\mathbf{f}}_*$ (mean), $\text{var}(\mathbf{f}_*)$ (variance), $\log(\mathbf{y}|\mathbf{X})$ (log marginal likelihood).

8.6.4 Gaussian Process Classification

Both regression and classification can be viewed as function approximation problems, but their solutions are rather different. This is because the targets in regression are continuous functions in which the likelihood function is Gaussian; a Gaussian process prior combined with a Gaussian likelihood gives rise to a posterior Gaussian process over functions, and everything remains analytically tractable [30]. But, in classification models, the targets are discrete class labels, so the Gaussian likelihood is no longer appropriate. Hence, approximate inference is only available for classification, where exact inference is not feasible.

Inference is naturally divided into two steps [30]:

1. Computing the distribution of the latent variable corresponding to a test case:

$$p(\mathbf{f}_*|\mathbf{X}, \mathbf{y}, \mathbf{x}_*) = \int p(\mathbf{f}_*|\mathbf{X}, \mathbf{x}_*, \mathbf{f}) p(\mathbf{f}|\mathbf{X}, \mathbf{y}) d\mathbf{f}, \qquad (8.6.28)$$

where $p(\mathbf{f}|\mathbf{X}, \mathbf{y}) = p(\mathbf{y}|\mathbf{f}) p(\mathbf{f}|\mathbf{X}) / p(\mathbf{y}|\mathbf{X})$ is the posterior over the latent variables.

2. Using this distribution over the latent $\mathbf{f}_*$ to produce a probabilistic prediction

$$\bar{\pi}_* = p(\mathbf{y}_* = +1|\mathbf{X}, \mathbf{y}, \mathbf{x}_*) = \int \sigma(\mathbf{f}_*) p(\mathbf{f}_*|\mathbf{X}, \mathbf{y}, \mathbf{x}_*) d\mathbf{f}_*. \qquad (8.6.29)$$

In classification due to discrete class labels the posterior $p(\mathbf{f}|\mathbf{X}, \mathbf{y})$ in (8.6.28) is non-Gaussian, which makes the likelihood in (8.6.28) is non-Gaussian, which makes its integral analytically intractable. Similarly, Eq. (8.6.29) can be intractable analytically for certain sigmoid functions.

The non-Gaussian joint posterior $p(\mathbf{f}|\mathbf{X}, \mathbf{y})$ can be approximated analytically with a Gaussian one. To this end, consider *Laplace's method* that utilizes a *Gaussian approximation*

$$q(\mathbf{f}|\mathbf{X}, \mathbf{y}) = N(\mathbf{f}|\hat{\mathbf{f}}, \mathbf{A}^{-1}) \propto \exp\left(\frac{1}{2}(\mathbf{f} - \hat{\mathbf{f}})^T \mathbf{A}(\mathbf{f} - \hat{\mathbf{f}})\right) \qquad (8.6.30)$$

8.7 Relevance Vector Machine

to the non-Gaussian posterior $p(\mathbf{f}|\mathbf{X}, \mathbf{y})$ in the integral (8.6.28). In the above equation, $\hat{\mathbf{f}} = \arg\min_\mathbf{f} p(\mathbf{f}|\mathbf{X}, \mathbf{y})$, and $\mathbf{A} = -\nabla^2 \log p(\mathbf{f}|\mathbf{X}, \mathbf{y})|_{\mathbf{f}=\hat{\mathbf{f}}}$ is the Hessian matrix of the negative log posterior at that point.

By Bayes' rule the posterior over the latent variables is given by $p(\mathbf{f}|\mathbf{X}, \mathbf{y}) = p(\mathbf{y}|\mathbf{f})p(\mathbf{f}|\mathbf{X})/p(\mathbf{y}|\mathbf{X})$, but as $p(\mathbf{y}|\mathbf{X})$ is independent of $\mathbf{f}$, we need to consider only the unnormalized posterior when maximizing with respect to $\mathbf{f}$, namely consider only $p(\mathbf{f}|\mathbf{X}, \mathbf{y}) = p(\mathbf{y}|\mathbf{f})p(\mathbf{f}|\mathbf{X})$. Taking its logarithm and using Eq. (8.6.26), then

$$\Psi(\mathbf{f}) = \log p(\mathbf{y}|\mathbf{f}) + \log p(\mathbf{f}|\mathbf{X})$$
$$= \log p(\mathbf{y}|\mathbf{f}) - \frac{1}{2}\mathbf{f}^T \mathbf{K}^{-1} \mathbf{f} - \frac{1}{2} \log |\mathbf{K}| - \frac{N}{2} \log(2\pi) \quad (8.6.31)$$

whose first and second derivatives with respect to $\mathbf{f}$ are, respectively, given by

$$\nabla \Psi(\mathbf{f}) = \nabla \log p(\mathbf{y}|\mathbf{f}) - \mathbf{K}^{-1} \mathbf{f},$$
$$\nabla^2 \Psi(\mathbf{f}) = \nabla^2 \log p(\mathbf{y}|\mathbf{f}) - \mathbf{K}^{-1}.$$

From $\nabla \Psi(\mathbf{f}) = \mathbf{0}$ it follows that

$$\mathbf{f} = \mathbf{K}\mathbf{a}, \quad \mathbf{a} = \nabla \log p(\mathbf{y}|\mathbf{f}). \quad (8.6.32)$$

The Hessian matrix can be written as

$$\nabla^2 \Psi(\mathbf{f}) = -\mathbf{W} - \mathbf{K}^{-1}, \quad \mathbf{W} = -\nabla^2 \log p(\mathbf{y}|\mathbf{X}). \quad (8.6.33)$$

Here, $\mathbf{W}$ is a diagonal matrix, since the distribution for y_i depends only on f_i, not on $f_{j \neq i}$.

Gaussian processes are a main mathematical tool for sparse Bayesian learning and hence the relevance vector machine that will be discussed in the next section.

8.7 Relevance Vector Machine

In real-world data, the presence of noise (in regression) and class overlap (in classification) implies that the principal modeling challenge is to avoid "overfitting" of the training set. In order to avoid the overfitting of SVMs (because of too many support vectors), Tipping [39] proposed a relevant vectors based learning machine, called the *relevance vector machine* (RVM). "*Sparse Bayesian learning*" is the basis for the RVM.

8.7.1 Sparse Bayesian Regression

We are given a set of data $\{\mathbf{x}_n, t_n\}_{n=1}^{N}$, where the "target" samples $t_n = y(\mathbf{x}_n) + \epsilon_n$ are conventionally considered to be realizations of a deterministic function y that is corrupted by some additive noise process $\{\epsilon_n\}$. This function f is usually modeled by a linearly weighted sum of M fixed basis functions $\{\phi_m(\mathbf{x})\}_{m=1}^{M}$:

$$\hat{y}(\mathbf{x}) = \sum_{m=1}^{M} w_m \phi_m(\mathbf{x}). \tag{8.7.1}$$

Our objective is to infer values of the parameters/weights $w_m, m = 1, \ldots, M$ such that $\hat{y}(\mathbf{x})$ is a "good" approximation of $y(\mathbf{x})$.

The function approximation has two important benefits: accuracy and "sparsity." By sparsity, it means that a sparse learning algorithm can set significant numbers of the parameters w_m to zero.

The support vector machine (SVM) makes predictions based on the function:

$$y(\mathbf{x}; \mathbf{w}) = \sum_{i=1}^{N} w_i K(\mathbf{x}; \mathbf{x}_i) + \mathbf{w}_0, \tag{8.7.2}$$

where $K(\mathbf{x}; \mathbf{x}_i)$ is a kernel function, effectively defining one basis function for each example in the training set.

There are two key features of the SVM:

- This can avoid overfitting, which leads to good generalization.
- This furthermore results in a sparse model dependent only on a subset of kernel functions.

However, there are a number of significant and practical disadvantages of the support vector learning methodology [39]:

1. Although relatively sparse, SVMs make unnecessarily liberal use of basis functions since the number of support vectors required typically grows linearly with the size of the training set. In order to reduce computational complexity, some form of post-processing is often required.
2. Predictions are not probabilistic. The SVM outputs a point estimate in regression, and a "hard" binary decision in classification. Ideally, it is desired to estimate the conditional distribution $p(t|\mathbf{x})$ in order to capture uncertainty in prediction. Although posterior probability estimates can be coerced from SVMs via post-processing, these estimates are unreliable.
3. Owing to satisfying Mercer's condition, the kernel function $K(\mathbf{x}; \mathbf{x}_i)$ must be the continuous symmetric kernel of a positive integral operator.

8.7 Relevance Vector Machine

To avoid any of the above limitations, Tipping [39] proposed to use the "relevance vector machine" (RVM) instead of the SVM as a Bayesian learning of (8.7.2):

$$y(\mathbf{x}; \mathbf{w}) = \sum_{i=1}^{m} w_i \phi_i(\mathbf{x}) = \mathbf{w}^T \boldsymbol{\phi}(\mathbf{x}), \tag{8.7.3}$$

where the output is a linearly weighted sum of M generally nonlinear and fixed basis functions $\boldsymbol{\phi}(\mathbf{x}) = [\phi_1(\mathbf{x}), \ldots, \phi_m(\mathbf{x})]^T$. The regression process is to determine (or learn) the adjustable parameters (or "weights") $\mathbf{w} = [w_1, \ldots, w_m]^T$ by using a Bayesian learning framework.

The key feature of this learning approach is that in order to offer good generalization performance, the inferred predictors are exceedingly *sparse* in that they contain relatively few nonzero w_i parameters. Since the majority of parameters are automatically set to zero during the learning process, this learning approach is known as the *sparse Bayesian learning* for regression [39].

The following are the framework of sparse Bayesian learning for regression.

Model Specification

Given a data set of input-target pairs $\{\mathbf{x}_n, t_n\}_{n=1}^{N}$, where $\mathbf{x}_n \in \mathbb{R}^d$, $t_n \in \mathbb{R}$. It is usually assumed that the target values are samples from the model with additive noise ϵ_n. Under this assumption, the standard linear regression model with additive Gaussian noise is given by

$$t_n = y_n + \epsilon_n = y(\mathbf{x}_n; \mathbf{w}) + \epsilon_n, \tag{8.7.4}$$

where $y_n = y(\mathbf{x}_n; \mathbf{w})$ are approximation signals of the target signal t_n, and ϵ_n are independent samples from some Gaussian noise $N(0, \sigma^2)$ with mean-zero and variance σ^2. If letting $\mathbf{t} = [t_1, \ldots, t_N]^T$ be the target vector and $\mathbf{y} = [y_1, \ldots, y_N]^T$ be the approximation vector, then under the assumption of independence of the target vector $\mathbf{t}$ and the Gaussian noise $\mathbf{e} = \mathbf{t} - \mathbf{y}$, the likelihood of the complete data set can be written as

$$p(\mathbf{t}|\mathbf{w}, \sigma^2) = (2\pi)^{-N/2} \sigma^{-N} \exp\left(-\frac{1}{2\sigma^2} \|\mathbf{t} - \mathbf{y}\|_2^2\right) \tag{8.7.5}$$

In different regression cases, the above likelihood has different forms.

- The linear neural network makes predictions based on the function

$$y_n = y(\mathbf{x}_n; \mathbf{w}) = \mathbf{x}_n^T \mathbf{w} \quad \text{or} \quad \mathbf{y} = \mathbf{X}^T \mathbf{w}, \tag{8.7.6}$$

where $\mathbf{w} = [w_1, \ldots, w_N]^T$ is the weight vector, $\mathbf{X} = [\mathbf{x}_1, \ldots, \mathbf{x}_N]$ is the data matrix, and thus the likelihood (because of the independence assumption) is given by

$$p(\mathbf{t}|\mathbf{w}, \sigma^2) = \prod_{i=1}^{N} p(t_i|\mathbf{w}, \sigma^2)$$

$$= \prod_{i=1}^{N} \frac{1}{\sqrt{2\pi}\sigma} \exp\left(-\frac{\|t_i - \mathbf{x}_i^T \mathbf{w}\|_2^2}{2\sigma^2}\right)$$

$$= (2\pi)^{-N/2} \sigma^{-N} \exp\left(-\frac{1}{2\sigma^2}\|\mathbf{t} - \mathbf{X}^T \mathbf{w}\|_2^2\right). \quad (8.7.7)$$

- The SVM makes predictions

$$y_n = \sum_{i=1}^{N} w_i K(\mathbf{x}_n; \mathbf{x}_i) + w_0 \quad \text{or} \quad \mathbf{y} = \mathbf{K}\mathbf{w}, \quad (8.7.8)$$

where $\mathbf{K} = [\mathbf{k}(\mathbf{x}_1), \ldots, \mathbf{k}(\mathbf{x}_N)]^T$ is the $N \times (N+1)$ kernel matrix with $\mathbf{k}(\mathbf{x}_n) = [1, K(\mathbf{x}_n, \mathbf{x}_1), \ldots, K(\mathbf{x}_n, \mathbf{x}_N)]^T$, and $\mathbf{w} = [w_0, w_1, \ldots, w_N]^T$. So under the independence assumption, the likelihood is given by

$$p(\mathbf{t}|\mathbf{w}, \sigma^2) = (2\pi)^{-N/2} \sigma^{-N} \exp\left(-\frac{1}{2\sigma^2}\|\mathbf{t} - \mathbf{K}\mathbf{w}\|_2^2\right). \quad (8.7.9)$$

- The RVM makes predictions

$$y_n = \sum_{i=1}^{N} \boldsymbol{\phi}^T(\mathbf{x}_i)\mathbf{w} + w_0 \quad \text{or} \quad \mathbf{y} = \boldsymbol{\Phi}\mathbf{w}, \quad (8.7.10)$$

where $\boldsymbol{\Phi} = [\boldsymbol{\phi}(\mathbf{x}_1), \ldots, \boldsymbol{\phi}(\mathbf{x}_N)]^T$ is the $N \times (N+1)$ "design" matrix with $\boldsymbol{\phi}(\mathbf{x}_n) = [1, \phi_1(\mathbf{x}_n), \ldots, \phi_N(\mathbf{x}_n)]^T$, and $\mathbf{w} = [w_0, w_1, \ldots, w_N]^T$. For example, $\boldsymbol{\phi}(\mathbf{x}_n) = [1, K(\mathbf{x}_n, \mathbf{x}_1), K(\mathbf{x}_n, \mathbf{x}_2), \ldots, K(\mathbf{x}_n, \mathbf{x}_N)]^T$. Therefore, the likelihood (because of the independence assumption) is given by

$$p(\mathbf{t}|\mathbf{w}, \sigma^2) = (2\pi)^{-N/2} \sigma^{-N} \exp\left(-\frac{1}{2\sigma^2}\|\mathbf{t} - \boldsymbol{\Phi}\mathbf{w}\|_2^2\right). \quad (8.7.11)$$

Parameter Estimation

Inference in the Bayesian linear model is based on the posterior distribution over the weights, computed by Bayes' rule [30]:

$$\text{posterior} = \frac{\text{likelihood} \times \text{prior}}{\text{marginal likelihood}}. \quad (8.7.12)$$

8.7 Relevance Vector Machine

Hence, the posterior distribution over the weight vector **w** is given by Tipping [39]

$$p(\mathbf{w}|\mathbf{t}, \boldsymbol{\alpha}, \sigma^2) = \frac{p(\mathbf{t}|\mathbf{w}, \sigma^2) p(\mathbf{w}, \boldsymbol{\alpha})}{p(\mathbf{t}|\boldsymbol{\alpha}, \sigma^2)}$$

$$= (2\pi)^{-(N+1)/2} |\boldsymbol{\Sigma}|^{-1/2} \exp\left(-\frac{1}{2}(\mathbf{w}-\boldsymbol{\mu})^T \boldsymbol{\Sigma}^{-1}(\mathbf{w}-\boldsymbol{\mu})\right),$$
(8.7.13)

where the posterior covariance and mean are, respectively, as follows [26]:

$$\boldsymbol{\Sigma} = (\sigma^{-2} \boldsymbol{\Phi}^T \boldsymbol{\Phi} + \mathbf{A})^{-1}, \quad (8.7.14)$$

$$\boldsymbol{\mu} = \sigma^{-2} \boldsymbol{\Sigma} \boldsymbol{\Phi}^T \mathbf{t}, \quad (8.7.15)$$

with $\mathbf{A} = \mathbf{Diag}(\alpha_0, \alpha_1, \ldots, \alpha_N)$.

Sparse Bayesian "Learning"

The procedure which chooses a model with the maximum marginal likelihood is called the *type II maximum likelihood* procedure by Good [18]. By this procedure, sparse Bayesian learning is formulated as the (local) maximization with respect to $\boldsymbol{\alpha}$ of the marginal likelihood, or equivalently, its logarithm marginal likelihood is given by Tipping [39]:

$$\mathcal{L}(\boldsymbol{\alpha}) = \log p(\mathbf{t}|\boldsymbol{\alpha}, \sigma^2) = \log \int_{-\infty}^{\infty} p(\mathbf{t}|\mathbf{w}, \sigma^2) p(\mathbf{w}|\boldsymbol{\alpha}) d\mathbf{w}$$

$$= -\frac{1}{2}\left(N \log(2\pi) + \log|\mathbf{C}| + \mathbf{t}^T \mathbf{C}^{-1} \mathbf{t}\right) \quad (8.7.16)$$

with

$$\mathbf{C} = \sigma^2 \mathbf{I} + \boldsymbol{\Phi} \mathbf{A}^{-1} \boldsymbol{\Phi}^T. \quad (8.7.17)$$

Here, by the term "marginal" it emphasizes a non-parametric model.

The Bayesian learning problem, in the context of the RVM, thus becomes the search for the hyperparameters $\boldsymbol{\alpha}$. In the RVM, these hyperparameters are estimated from minimizing $\mathcal{L}(\boldsymbol{\alpha})$. This estimation problem will be discussed in Sect. 8.7.3.

As a typical application of RVMs, consider the *compressive sensing* (CS) measurements **g** represented as [23]:

$$\mathbf{g} = \boldsymbol{\Phi} \mathbf{w}, \quad (8.7.18)$$

where $\boldsymbol{\Phi} = [\mathbf{r}_1, \ldots, \mathbf{r}_K]$ is a $K \times N$ matrix, assuming random CS measurements are made.

A typical means of solving the CS problem is ℓ_1 regularization:

$$\mathbf{w}^* = \arg\min_{\mathbf{w}} \left\{ \|\mathbf{g} - \mathbf{\Phi}\mathbf{w}\|_2^2 + \gamma\|\mathbf{w}\|_1 \right\}. \tag{8.7.19}$$

Under the assumption that the additive noise $\mathbf{n}$ is mean-mean Gaussian distribution $N(\mathbf{0}, \sigma^2\mathbf{I})$, the Gaussian likelihood model is described as

$$p(\mathbf{g}|\mathbf{w}, \sigma^2) = (2\pi)^{-K/2}\sigma^{-K}\exp\left(-\frac{\|\mathbf{g} - \mathbf{\Phi}\mathbf{w}\|_2^2}{2\sigma^2}\right). \tag{8.7.20}$$

Assuming the hyperparameters $\boldsymbol{\alpha}$ and α_0 are known, given the CS measurements $\mathbf{g}$ and the projection matrix $\mathbf{\Phi}$, the posterior for $\mathbf{w}$ can be expressed analytically as a multivariate Gaussian distribution with mean and covariance

$$\boldsymbol{\mu} = \alpha_0 \mathbf{\Sigma}\mathbf{\Phi}^T\mathbf{g} \quad \text{and} \quad \mathbf{\Sigma} = (\alpha_0 \mathbf{\Phi}^T\mathbf{\Phi} + \mathbf{A})^{-1}. \tag{8.7.21}$$

Then, Bayesian compressive sensing problem becomes sparse Bayesian regression one.

8.7.2 Sparse Bayesian Classification

Sparse Bayesian classification follows an essentially identical framework for regression above, but using a Bernoulli likelihood and a sigmoidal link function to account for the change in the target quantities [39].

For an input vector $\mathbf{x}$, an RVM classifier models the probability distribution of its class label using logistic sigmoid link function $\sigma(y) = 1/(1 + e^{-y})$ as

$$p(d = 1|\mathbf{x}) = \frac{1}{1 + \exp(-f_{\text{RVM}}(\mathbf{x}))}, \tag{8.7.22}$$

where $f_{\text{RVM}}(\mathbf{x})$, called the RVM classifier function, is given by

$$f_{\text{RVM}}(\mathbf{x}) = \sum_{i=1}^{N} \alpha_i K(\mathbf{x}, \mathbf{x}_i), \tag{8.7.23}$$

where $K(\mathbf{x}, \mathbf{x}_i)$ is a kernel function, and $\mathbf{x}_i, i = 1, \ldots, N$ are the training samples.

By adopting the Bernoulli distribution for $P(\mathbf{t}|\mathbf{x})$, the likelihood can be written as

$$P(\mathbf{t}|\mathbf{w}) = \prod_{n=1}^{N} \sigma\big(y(\mathbf{x}_n, \mathbf{w})\big)^{t_n} \big(1 - \sigma\big(y(\mathbf{x}_n, \mathbf{w})\big)\big)^{1-t_n} \tag{8.7.24}$$

where the target $t_n \in \{0, 1\}$.

8.7 Relevance Vector Machine

In sparse Bayesian regression, the weights $\mathbf{w}$ can be integrated out analytically. Unlike the regression case, sparse Bayesian classification cannot be integrated out analytically, precluding closed form expressions for either the weight posterior $p(\mathbf{w}|\mathbf{t}, \boldsymbol{\alpha})$ or the marginal likelihood $P(\mathbf{t}|\boldsymbol{\alpha})$. Thus, it utilizes the Laplace approximation procedure.

In the statistics and machine leaning literature, the *Laplace approximation* refers to the evaluation of the marginal likelihood or free energy using Laplace's method. This is equivalent to a local Gaussian approximation of $P(\mathbf{t}|\mathbf{w})$ around a maximum a posteriori (MAP) estimate [40].

8.7.3 Fast Marginal Likelihood Maximization

Consider the dependence of $\mathcal{L}(\boldsymbol{\alpha})$ on a single hyperparameter α_m, $m \in \{1, \ldots, M\}$, the matrix $\mathbf{C}$ in the Eq. (8.7.16) can be decomposed as

$$\begin{aligned}\mathbf{C} &= \sigma^2 \mathbf{I} + \sum_{i \neq m} \alpha_i^{-1} \boldsymbol{\phi}_i \boldsymbol{\phi}_i^T + \alpha_m^{-1} \boldsymbol{\phi}_m \boldsymbol{\phi}_m^T \\ &= \mathbf{C}_{-m} + \alpha_m^{-1} \boldsymbol{\phi}_m \boldsymbol{\phi}_m^T,\end{aligned} \qquad (8.7.25)$$

where $\mathbf{C}_{-m} = [\mathbf{C} \setminus \mathbf{c}_m]$ is $\mathbf{C}$ with the mth column $\mathbf{c}_m$ removed. Thus, by applying the determinant identity and matrix inverse lemma, the terms of interest in $\mathcal{L}$ can be written as

$$|\mathbf{C}| = |\mathbf{C}_m| \cdot |1 + \alpha_m^{-1} \boldsymbol{\phi}_m^T \mathbf{C}_{-m}^{-1} \boldsymbol{\phi}_m|, \qquad (8.7.26)$$

$$\mathbf{C}^{-1} = \mathbf{C}_{-m}^{-1} - \frac{\mathbf{C}_{-m}^{-1} \boldsymbol{\phi}_m \boldsymbol{\phi}_m^T \mathbf{C}_{-m}^{-1}}{\alpha_m + \boldsymbol{\phi}_m^T \mathbf{C}_{-m}^{-1} \boldsymbol{\phi}_m}. \qquad (8.7.27)$$

Then, $\mathcal{L}(\boldsymbol{\alpha})$ can be rewritten as

$$\begin{aligned}\mathcal{L}(\boldsymbol{\alpha}) &= -\frac{1}{2} \Bigg(N \log(2\pi) + \log |\mathbf{C}_{-m}| + \mathbf{t}^T \mathbf{C}_{-m}^{-1} \mathbf{t} \\ &\quad - \log \alpha_m + \log(\alpha_m + \boldsymbol{\phi}_m^T \mathbf{C}_{-m}^{-1} \boldsymbol{\phi}_m) - \frac{(\boldsymbol{\phi}_m^T \mathbf{C}_{-m}^{-1} \mathbf{t})^2}{\alpha_m + \boldsymbol{\phi}_m^T \mathbf{C}_{-m}^{-1} \boldsymbol{\phi}_m} \Bigg) \\ &= \mathcal{L}(\boldsymbol{\alpha}_{-m}) + \frac{1}{2} \left(\log \alpha_m - \log(\alpha_m + s_m) + \frac{q_m^2}{\alpha_m + s_m} \right) \\ &= \mathcal{L}(\boldsymbol{\alpha}_{-m}) + \ell(\alpha_m), \end{aligned} \qquad (8.7.28)$$

where

$$s_m = \boldsymbol{\phi}_m^T \mathbf{C}_{-m}^{-1} \boldsymbol{\phi}_m \quad \text{and} \quad q_m = \boldsymbol{\phi}_m^T \mathbf{C}_{-m}^{-1} \mathbf{t}, \quad \text{for } m = 1, \ldots, M. \tag{8.7.29}$$

The "sparsity factor" s_m can be seen to be a measure of the extent that basis vector $\boldsymbol{\phi}_m$ "overlaps" those already present in the model, while the "quality factor" q_m is a measure of the alignment of $\boldsymbol{\phi}_m$ with the error of the model with that vector excluded.

From (8.7.28) and the optimization condition $\frac{\partial \mathcal{L}(\boldsymbol{\alpha})}{\partial \alpha_m} = \frac{\partial \ell(\alpha_m)}{\partial \alpha_m} = 0$ we have the result

$$\frac{1}{2}\left(\frac{1}{\alpha_m} - \frac{1}{\alpha_m + s_m} - \frac{q_m^2}{(\alpha_m + s_m)^2}\right) = 0 \quad \text{or} \quad \alpha_m = \frac{s_m^2}{q_m^2 - s_m}. \tag{8.7.30}$$

Hence, $\mathcal{L}(\boldsymbol{\alpha})$ has a unique maximum with respect to α_m [10]:

$$\alpha_m = \begin{cases} \frac{s_m^2}{q_m^2 - s_m}, & \text{if } q_m^2 > s_m, \\ \infty, & \text{otherwise}, \end{cases} \tag{8.7.31}$$

for $m = 1, \ldots, M$.

Equation (8.7.31) shows that:

- If $\boldsymbol{\phi}_m$ is "in the model" (i.e., $\alpha_m < \infty$) yet $q_m^2 \leq s_m$, then $\boldsymbol{\phi}_m$ may be deleted (i.e., α_m set to ∞).
- If $\boldsymbol{\phi}_m$ is excluded from the model ($\alpha_m = \infty$) and $q_m^2 > s_m$, then $\boldsymbol{\phi}_m$ may be added (i.e., α_m is set to some optimal finite value).

It is obvious that as compared with $\mathbf{C}_{-m}^{-1}, m = 1, \ldots, M$, it is easier to maintain and compute value of $\mathbf{C}^{-1}$. Let

$$S_m = \boldsymbol{\phi}_m^T \mathbf{C}^{-1} \boldsymbol{\phi}_m \quad \text{and} \quad Q_m = \boldsymbol{\phi}_m^T \mathbf{C}^{-1} \mathbf{t}, \quad m = 1, \ldots, M. \tag{8.7.32}$$

By premultiplying $\boldsymbol{\phi}_m^T$, postmultiplying $\boldsymbol{\phi}_m$, and using (8.7.29), Eq. (8.7.27) yields

$$S_m = \boldsymbol{\phi}_m^T \mathbf{C}^{-1} \boldsymbol{\phi}_m = \boldsymbol{\phi}_m^T \left(\mathbf{C}_{-m}^{-1} - \frac{\mathbf{C}_{-m}^{-1} \boldsymbol{\phi}_m \boldsymbol{\phi}_m^T \mathbf{C}_{-m}^{-1}}{\alpha_m + \boldsymbol{\phi}_m^T \mathbf{C}_{-m}^{-1} \boldsymbol{\phi}_m} \right) \boldsymbol{\phi}_m$$

$$= s_m - \frac{s_m^2}{\alpha_m + s_m}, \quad m = 1, \ldots, M. \tag{8.7.33}$$

8.7 Relevance Vector Machine

Similarly, we have

$$Q_m = \boldsymbol{\phi}_m^T \mathbf{C}^{-1} \mathbf{t} = \boldsymbol{\phi}_m^T \left(\mathbf{C}_{-m}^{-1} - \frac{\mathbf{C}_{-m}^{-1} \boldsymbol{\phi}_m \boldsymbol{\phi}_m^T \mathbf{C}_{-m}^{-1}}{\alpha_m + \boldsymbol{\phi}_m^T \mathbf{C}_{-m}^{-1} \boldsymbol{\phi}_m} \right) \mathbf{t}$$

$$= q_m - \frac{s_m q_m}{\alpha_m + s_m}, \quad m = 1, \ldots, M. \tag{8.7.34}$$

From (8.7.33) and (8.7.34) it follows that

$$s_m = \frac{\alpha_m S_m}{\alpha_m - S_m} \quad \text{and} \quad q_m = \frac{\alpha_m Q_m}{\alpha_m - S_m} \tag{8.7.35}$$

for $m = 1, \ldots, M$. Note that when $\alpha_m = \infty$, $s_m = S_m$ and $q_m = Q_m$. By Duncan–Guttman inversion formula (1.6.5)

$$(\mathbf{A} + \mathbf{U}\mathbf{D}^{-1}\mathbf{V})^{-1} = \mathbf{A}^{-1} - \mathbf{A}^{-1}\mathbf{U}(\mathbf{D} + \mathbf{V}\mathbf{A}^{-1}\mathbf{U})^{-1}\mathbf{V}\mathbf{A}^{-1}, \tag{8.7.36}$$

we have

$$\mathbf{C}^{-1} = (\sigma^2 \mathbf{I} + \boldsymbol{\Phi} \mathbf{A}^{-1} \boldsymbol{\Phi}^T)^{-1}$$

$$= \mathbf{B} - \mathbf{B}\boldsymbol{\Phi}(\mathbf{A} + \boldsymbol{\Phi}^T \mathbf{B} \boldsymbol{\Phi})^{-1} \boldsymbol{\Phi}^T \mathbf{B}$$

$$= \mathbf{B} - \mathbf{B}\boldsymbol{\Phi}\boldsymbol{\Sigma}\boldsymbol{\Phi}^T \mathbf{B}, \tag{8.7.37}$$

where

$$\mathbf{B} = \sigma^{-2}\mathbf{I} \quad \text{and} \quad \boldsymbol{\Sigma} = (\mathbf{A} + \boldsymbol{\Phi}^T \mathbf{B} \boldsymbol{\Phi})^{-1}. \tag{8.7.38}$$

Substituting (8.7.37) into (8.7.32) to yield

$$S_m = \boldsymbol{\phi}_m^T \mathbf{B} \boldsymbol{\phi}_m - \boldsymbol{\phi}_m \mathbf{B} \boldsymbol{\Phi} \boldsymbol{\Sigma} \boldsymbol{\Phi}^T \mathbf{B} \boldsymbol{\phi}_m^T, \tag{8.7.39}$$

$$Q_m = \boldsymbol{\phi}_m^T \mathbf{B} \hat{\mathbf{t}} - \boldsymbol{\phi}_m \mathbf{B} \boldsymbol{\Phi} \boldsymbol{\Sigma} \boldsymbol{\Phi}^T \mathbf{B} \hat{\mathbf{t}}. \tag{8.7.40}$$

The following is the marginal likelihood maximization algorithm for sequential sparse Bayesian learning proposed by Tipping and Faul [40]:

1. Initialize σ^2 to some sensible value (e.g., var$[t] \times 0.1$).
2. Initialize with a single basis vector $\boldsymbol{\phi}_i$, setting, from (8.7.31):

$$\alpha_i = \frac{\|\boldsymbol{\phi}_i\|_2^2}{\|\boldsymbol{\phi}_i^T \mathbf{t}\|_2^2/\| - \sigma^2}. \tag{8.7.41}$$

All other α_m are notionally set to infinity.

3. Explicitly compute Σ and μ (which are scalars initially), along with initial values of s_m and q_m for all M bases $\boldsymbol{\phi}_m$.
4. Select a candidate basis vector $\boldsymbol{\phi}_i$ from the set of all M.
5. Compute $\theta_i = q_i^2 - s_i$.
6. If $\theta_i > 0$ and $\alpha_i < \infty$ (i.e., $\boldsymbol{\phi}_i$ is in the model), re-estimate α_i. Defining $\kappa_j = (\Sigma_{jj} + (\bar{\alpha}_i - \alpha_i)^{-1})^{-1}$ and $\boldsymbol{\Sigma}_j$ as the jth column of $\boldsymbol{\Sigma}$:

$$2\Delta \mathcal{L} = \frac{Q_i^2}{S_i + (\bar{\alpha}_i^{-1} - \alpha_i^{-1})^{-1}} - \log\left(1 + S_i(\bar{\alpha}_i^{-1} - \alpha_i^{-1})\right), \qquad (8.7.42)$$

$$\bar{\boldsymbol{\Sigma}} = \boldsymbol{\Sigma} - \kappa_j \boldsymbol{\Sigma} \boldsymbol{\Sigma}_j^T, \qquad (8.7.43)$$

$$\bar{\boldsymbol{\mu}} = \boldsymbol{\mu} - \kappa_j \mu_j \boldsymbol{\Sigma}_j, \qquad (8.7.44)$$

$$\bar{S}_m = S_m + \kappa_j (\beta \boldsymbol{\Sigma}_j^T \boldsymbol{\Phi}^T \boldsymbol{\phi}_m)^2, \qquad (8.7.45)$$

$$\bar{Q}_m = Q_m + \kappa_j \mu_j (\beta \boldsymbol{\Sigma}_j^T \boldsymbol{\Phi}^T \boldsymbol{\phi}_m). \qquad (8.7.46)$$

7. If $\theta_i > 0$ and $\alpha_i = \infty$, add $\boldsymbol{\phi}_i$ to the model with updated α_i:

$$2\Delta \mathcal{L} = \frac{Q_i^2 - S_i}{S_i} + \log \frac{S_i}{Q_i^2}, \qquad (8.7.47)$$

$$\bar{\boldsymbol{\Sigma}} = \begin{bmatrix} \boldsymbol{\Sigma} + \beta^2 \Sigma_{ii} \boldsymbol{\Sigma} \boldsymbol{\Phi}^T \boldsymbol{\phi}_i \boldsymbol{\phi}_i^T \boldsymbol{\Phi} \boldsymbol{\Sigma} & -\beta^2 \Sigma_{ii} \boldsymbol{\Sigma} \boldsymbol{\Phi}^T \boldsymbol{\phi}_i \\ -\beta^2 \Sigma_{ii} (\boldsymbol{\Sigma} \boldsymbol{\Phi}^T \boldsymbol{\phi}_i)^T & \Sigma_{ii} \end{bmatrix}, \qquad (8.7.48)$$

$$\bar{\boldsymbol{\mu}} = \begin{bmatrix} \boldsymbol{\mu} - \mu_i \beta \boldsymbol{\Sigma} \boldsymbol{\Phi}^T \boldsymbol{\phi}_i \\ \mu_i \end{bmatrix}, \qquad (8.7.49)$$

$$\bar{S}_m = S_m - \Sigma_{ii} (\beta \boldsymbol{\phi}_m^T \mathbf{e}_i)^2, \qquad (8.7.50)$$

$$\bar{Q}_m = Q_m - \mu_i (\beta \boldsymbol{\phi}_m^T \mathbf{e}_i), \qquad (8.7.51)$$

where $\Sigma_{ii} = (\alpha_i + S_i)^{-1}$, $\mu_i = \Sigma_{ii} Q_i$, and $\mathbf{e}_i = \boldsymbol{\phi}_i - \beta \boldsymbol{\Phi} \boldsymbol{\Sigma} \boldsymbol{\Phi}^T \boldsymbol{\phi}_i$.

8. If $\theta_i \leq 0$ and $\alpha_i < \infty$, then delete $\boldsymbol{\phi}_i$ from the model and set $\alpha_i = \infty$:

$$2\Delta \mathcal{L} = \frac{Q_i^2}{S_i - \alpha_i} - \log\left(1 - \frac{S_i}{\alpha_i}\right), \qquad (8.7.52)$$

$$\bar{\boldsymbol{\Sigma}} = \boldsymbol{\Sigma} - \frac{1}{\Sigma_{jj}} \boldsymbol{\Sigma}_j \boldsymbol{\Sigma}_j^T, \qquad (8.7.53)$$

$$\bar{\boldsymbol{\mu}} = \boldsymbol{\mu} - \frac{\mu_j}{\Sigma_{jj}} \boldsymbol{\Sigma}_j, \qquad (8.7.54)$$

$$\bar{S}_m = S_m + \frac{1}{\Sigma_{jj}}(\beta \Sigma_j^T \Phi^T \phi_m)^2, \qquad (8.7.55)$$

$$\bar{Q}_m = Q_m + \frac{\mu_j}{\Sigma_{jj}}(\beta \Sigma_j^T \Phi^T \phi_m) \qquad (8.7.56)$$

Following updates (8.7.53) and (8.7.54), the appropriate row and/or column j is removed from $\bar{\Sigma}$ and $\bar{\mu}$.

9. If it is a regression model and estimating the noise level, update $\sigma^2 = \|\mathbf{t}-\mathbf{y}\|_2^2/(N - M + \sum_m \alpha_m \Sigma_{mm})$.
10. Recompute/update Σ, μ (using the Laplace approximation procedure in classification) and all s_m and q_m using Eqs. (8.7.35)–(8.7.40).
11. If converged terminate, otherwise goto 4.

Brief Summary of This Chapter

- SVMs have high classification accuracy due to introducing maximum interval.
- When the sample size is small, SVMs can also classify accurately and have good generalization ability.
- The kernel function can be used to solve nonlinear problems.
- SVMs can solve the problem of classification and regression with high-dimensional features.

References

1. Aronszajn, N.: Theory of reproducing kernels. Trans. Am. Math. Soc. **68**, 337–404 (1950)
2. Belkin, M., Niyogi, P., Sindhwani, V.: Manifold regularization: a geometric framework for learning from labeled and unlabeled examples. J. Mach. Learn. Res. **7**, 2399–2434 (2006)
3. Boser, B.E., Guyon, I.M., Vapnik, V.N.: A training algorithm for optimal margin classifiers. In: Haussler D. (ed.) Proceedings of the 5th Annual ACM Workshop on Computational Learning Theory, pp. 144–152. ACM Press, Pittsburgh (1992)
4. Bottou, L., Cortes, C., Denker, J., Drucker, H., Guyon, I., Jackel, L., LeCun, Y., Muller, U., Sackinger, E., Simard, P., Vapnik, V.: Comparison of classifier methods: a case study in handwriting digit recognition. In: Proceedings of International Conference on Pattern Recognition, pp. 77–87 (1994)
5. Bousquet, O., von Luxburg, U., Rätsch, G. (eds.): Advanced Lectures on Machine Learning. Springer, Berlin (2004)
6. Burges, C.J.C.: A tutorial on support vector machines for pattern recognition. Data Min. Knowl. Disc. **2**, 121–167 (1998)
7. Chang, C.-C., Lin, C.-J.: LIBSVM: a library for support vector machines. ACM Trans. Intell. Syst. Technol. **2**(3), 27 (2011)
8. Cortes, C., Vapnik, V.: Support vector networks. Mach. Learn. **20**(3), 273–297 (1995)
9. Duda, R.O., Hart, P.E., Stork, D.G.: Pattern Classification and Scene Analysis. Wiley, New York (1973)

10. Faul, A.C., Tipping, M.E.: Analysis of sparse Bayesian learning. In: Dietterich, T.G., Becker, S., Ghahramani, Z. (eds.) Advances in Neural Information Processing, vol. 14, pp. 383–389 (2002). http://citeseer.ist.psu.edu/faul01analysis.html
11. Fletcher, R.: Practical Methods of Optimization. Wiley, Chichester (1987)
12. Fontana, R., Pistone, G., Rogantin, M.P.: Classification of two-level factorial fractions. J. Stat. Plann. Inference **87**, 149–172 (2000)
13. Friedman, J.: Another approach to polychotomous classification. Department of Statistics, Stanford University, Stanford (1996). https://www-stat.stanford.edu/reports/friedman/poly.ps.Z
14. Fung, G.M., Mangasarian, O.L.: Proximal support vector machine classifiers. In: Proceedings of International Conference on Knowledge Discovery and Data Mining, San Francisco, pp. 77–86 (2001)
15. Fung, G.M., Mangasarian, O.L.: Multicategory proximal support vector machine classifiers. Mach. Learn. **59**(1–2), 77–97 (2005)
16. Girolami, M.: Mercer kernel-based clustering in feature space. IEEE Trans. Neural Netw. **13**(3), 780–784 (2002)
17. Girosi, F.: An equivalence between sparse approximation and support vector machines. Neural Comput. **20**, 1455–1480 (1998)
18. Good, I.J.: The Estimation of Probabilities. MIT Press, Cambridge (1965)
19. Gunn, S.R.: Support vector machines for classification and regression. Technical Report, Faculty of Engineering, Science and Mathematics School of Electronics and Computer Science, University of Southampton (1998)
20. Guyon, I., Weston, J., Barnhill, S., Vapnik, V.: Gene selection for cancer classification using support vector machines. Mach. Learn. **46**, 389–422 (2002)
21. Hsu, C.-W., Lin C.-J.: A comparison of methods for multiclass support vector machines. IEEE Trans. Neural Netw. **13**(2), 415–425 (2002)
22. Huang, G.-B., Zhou, H., Ding, X., Zhang, R.: Extreme learning machine for regression and multiclass classification. IEEE Trans. Syst. Man Cybern. B Cybern. **42**(2), 513–529 (2012)
23. Ji, S., Xue, Y., Carin, L.: Bayesian compressive sensing. IEEE Trans. Signal Process. **56**(6), 2346–2356 (2008)
24. Kimeldorf, G., Wahba, G.: Some results on Tchebycheffian spline functions. J. Math. Anal. Appl. **33**, 82–95 (1971)
25. Knerr, S., Personnaz, L., Dreyfus, G.: Single-layer learning revisited: a stepwise procedure for building and training a neural network. In: Fogelman, J. (ed.) Neurocomputing: Algorithms, Architectures and Applications. Springer, New York (1990)
26. MacKay, D.J.C.: Bayesian interpolation. Neural Comput. **4**(3), 415–447 (1992)
27. Platt, J.C., Cristianini, N., Shawe-Taylor, J.: Large margin DAGs for multiclass classification. In: Advances in Neural Information Processing Systems, vol.12, pp. 547–553. MIT Press, Cambridge (2000)
28. Rännar, S., Lindegren, F., Geladi, P., Wold, S.: A PLS kernel algorithm for data sets with many variables and fewer objects. Part 1: theory and algorithm. J. Chemometr. **8**, 111–125 (1994)
29. Rasmussen, C.E.: Gaussian processes in machine learning. In: Bousquet, O. et al. (ed.) Machine Learning 2003. Lecture Notes in Artificial Intelligence, vol. 3176, pp. 63–71. Springer, Berlin (2004)
30. Rasmussen, C.E., Williams, C.K.I.: Gaussian Processes for Machine Learning. MIT Press, Cambridge (2005)
31. Rokotomamonjy, A.: Variable selection using SVM-based criteria. J. Mach. Learn. Res. **3**, 1357–1370 (2003)
32. Rosipal, R., Trejo, L.J.: Kernel partial least squares regression in reproducing kernel Hilbert space. J. Mach. Learn. Res. **2**, 97–123 (2001)
33. Schölkopf, B., Smola, A., Müller, K.-R.: Nonlinear component analysis as a kernel eigenvalue problem. Neural Comput. **10**, 1299–1319 (1998)
34. Schölkopf, B., Smola, A.J., Williamson, R.C., Bartlett, P.L.: New support vector algorithms. Neural Comput. **12**, 1207–1245 (2000)

35. Schölkopf, B., Herbrich, R., Smola, A.J.: A generalized representer theorem. In: Proceedings of the International Conference on Computational Learning Theory (COLT2001), pp. 416–426 (2001)
36. Smola, A.J., Schölkopf, B.: A tutorial on support vector regression. Stat. Comput. **14**, 199–222 (2004)
37. Suykens, J.A.K., Vandewalle, J.: Least squares support vector machine classifiers. Neural Process. Lett. **9**, 293–300 (1999)
38. Tikhonov, A.N., Arsenin, V.Y.: Solutions of Ill-Posed Problems. Wiley, New York (1977)
39. Tipping, M.E.: Sparse Bayesian learning and the relevance vector machine. J. Mach. Learn. Res. **1**, 211–244 (2001)
40. Tipping, M.E., Faul, A.C.: Fast marginal likelihood maximization for sparse Bayesian models. In: Bishop, C.M., Frey, B.J. (eds.) Proceedings of the 9th International Workshop Artificial Intelligence and Statistics (2003). http://www.miketipping.com/papers.htm
41. Van Gestel, T., Suykens, J.A.K., Lanckriet, G., Lambrechts, A., De Moor, B., Vandewalle, J.: Multiclass LS-SVMs: moderated outputs and coding-decoding schemes. Neural Process. Lett. **15**(1), 48–58 (2002)
42. Vapnik, V.N.: Principles of risk minimization for learning theory. Adv. Neural Inf. Proces. Syst. **4**, 831–838 (1992)
43. Vapnik, V.N.: The Nature of Statistical Learning Theory. Springer, New York (1995)
44. Vapnik, V.N.: Statistical Learning Theory, vol. 1. Wiley, New York (1998)
45. Vapnik, V.N.: An overview of statistical learning theory. IEEE Trans. Neural Netw. **10**(5), 988–999 (1999)
46. Vapnik, V.N., Chervonenkis, A.: Theory of Pattern Recognition (in Russian). Moscow, Nauka (1974). (German Translation: Wapnik W., Tscherwonenkis A.: Theorie der Zeichenerkennung), Akademie-Verlag, Berlin (1979)
47. Wahba, G.: Support vector machines, reproducing Kernel Hilbert spaces, and randomized GACV. In: Schölkopf, B., Burges, C.J.C., Smola, A.J. (eds.) Advances in Kernel Methods – Support Vector Learning, pp. 69–88. The MIT Press, Cambridge (1999)
48. Yuan, G.-X., Ho, C.-H., Lin C.-J.: Recent advances of large-scale linear classification. Proc. IEEE **100**(9), 2584–2603 (2012)

Chapter 9
Evolutionary Computation

From the perspective of artificial intelligence, evolutionary computation belongs to computation intelligence. The origins of evolutionary computation can be traced back to the late 1950s (see, e.g., the influencing works [12, 15, 46, 47]), and has started to receive significant attention during the 1970s (see, e.g., [41, 68, 133]).

The first issue of the Evolutionary Computation in 1993 and the first issue of the IEEE Transactions on Evolutionary Computation in 1997 mark two important milestones in the history of the rapidly growing field of evolutionary computation.

This chapter is focused primarily on basic theory and methods of evolutionary computation, including multiobjective optimization, multiobjective simulated annealing, multiobjective genetic algorithms, multiobjective evolutionary algorithms, evolutionary programming, differential evolution together with ant colony optimization, artificial bee colony algorithms, and particle swarm optimization. In particular, this chapter also highlights selected topics and advances in evolutionary computation: Pareto optimization theory, noisy multiobjective optimization, and opposition-based evolutionary computation.

9.1 Evolutionary Computation Tree

Evolutionary computation can be broadly classified into three major categories.

1. *Physic-inspired evolutionary computation:* Its typical representative is simulated annealing [100] inspired by heating and cooling of materials.
2. *Darwin's evolution theory inspired evolutionary computation*

    - *Genetic Algorithm* (GA) is inspired by the process of natural selections. A GA is commonly used to generate high-quality solutions to optimization and search problems by relying on bio-inspired operators such as mutation, crossover, and selection.

- *Evolutionary Algorithms* (EAs) are inspired by Darwin's evolution theory which is based on survival of fittest candidate for a given environment. These algorithms begin with a population (set of solutions) which tries to survive in an environment (defined with fitness evaluation). The parent population shares their properties of adaptation to the environment to the children with various mechanisms of evolution such as genetic crossover and mutation.
- *Evolutionary Strategy* (ES) was developed by Schwefel in 1981 [143]. Similar to the GA, the ES uses evolutionary theory to optimize, that is to say, genetic information is used to inherit and mutate the survival of the fittest generation by generation, so as to get the optimal solution. The difference lies in: (a) the DNA sequence in ES is encoded by real number instead of 0-1 binary code in GA; (b) In the ES the mutation intensity is added for each real number value on the DNA sequence in order to make a mutation.
- *Evolutionary Programming* (EP) [43]: The EP and the ES adopt the same coding (digital string) to the optimization problem and the same mutation operation mode, but they adopt the different mutation expressions and survival selections. Moreover, the crossover operator in ES is optional, while there is no crossover operator in EP. In the aspect of parent selection, the ES adopts the method of probability selection to form the parent, and each individual of the parent can be selected with the same probability, while the EP adopts a deterministic approach, that is, each parent in the current population must undergo mutation to generate offspring.
- *Differential Evolution* (DE) was developed by Storn and Price in 1997 [147]. Similar to the GA, the main process includes three steps: mutation, crossover, and selection. The difference is that the GA controls the parent's crossover according to the fitness value, while the mutation vector of the DE is generated by the difference vector of the parent generation, and crossover with the individual vector of the parent generation to generate a new individual vector, which directly selects with the individual of the parent generation. Therefore, the approximation effect of DE is more significant than that of the GA.

3. *Swarm intelligence inspired evolutionary computation:* Swarm intelligence includes two important concepts: (a) The concept of a swarm means multiplicity, stochasticity, randomness, and messiness. (b) The concept of intelligence implies a kind of problem-solving ability through the interactions of simple information-processing units. This "collective intelligence" is built up through a population of homogeneous agents interacting with each other and with their environment. The major swarm intelligence inspired evolutionary computation includes:

   - *Ant Colony Optimization (ACO)* by Dorigo in 1992 [28]; a swarm of ants can solve very complex problems such as finding the shortest path between their nest and the food source. If finding the shortest path is looked upon as an optimization problem, then each path between the starting point (their nest) and the terminal (the food source) can be viewed as a feasible solution.
   - *Artificial Bee Colony (ABC)* algorithm, proposed by Karaboga in 2005 [81], is inspired by the intelligent foraging behavior of the honey bee swarm. In the

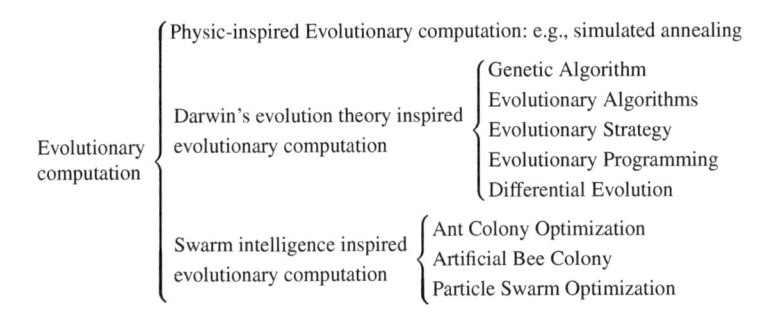

9.2 Multiobjective Optimization

$$
\text{Evolutionary computation} \begin{cases} \text{Physic-inspired Evolutionary computation: e.g., simulated annealing} \\ \text{Darwin's evolution theory inspired evolutionary computation} \begin{cases} \text{Genetic Algorithm} \\ \text{Evolutionary Algorithms} \\ \text{Evolutionary Strategy} \\ \text{Evolutionary Programming} \\ \text{Differential Evolution} \end{cases} \\ \text{Swarm intelligence inspired evolutionary computation} \begin{cases} \text{Ant Colony Optimization} \\ \text{Artificial Bee Colony} \\ \text{Particle Swarm Optimization} \end{cases} \end{cases}
$$

Fig. 9.1 Evolutionary computation tree

ABC algorithm, the position of a food source represents a possible solution of the optimization problem and the nectar amount of a food source corresponds to the quality (fitness) of the associated solution.

- *Particle Swarm Optimization* (PSO), developed by Kennedy and Eberhart in 1995 [86], is essentially an optimization technique based on swarm intelligence, and adopts a population-based stochastic algorithm for finding optimal regions of complex search spaces through the interaction of individuals in a population of particles. Different from evolutionary algorithms, the particle swarm does not use selection operation; while interactions of all population members result in iterative improvement of the quality of problem solutions from the beginning of a trial until the end.

The above classification of evolutionary computation can be represented by evolutionary computation tree, as shown in Fig. 9.1.

9.2 Multiobjective Optimization

In science and engineering applications, a lot of optimization problems involve with multiple objectives. Different from single-objective problems, objectives in multiobjective problems are usually conflict. For example, in the design of a complex hardware/software system, an optimal design might be an architecture that minimizes cost and power consumption but maximizing the overall performance. This structural contradiction leads to multiobjective optimization theories and methods different from single-objective optimization. The need for solving multiobjective optimization problems has spawned various evolutionary computation methods.

This section discusses two multiobjective optimizations: multiobjective combinatorial optimization problems and multiobjective optimization problems.

9.2.1 Multiobjective Combinatorial Optimization

Combinatorial optimization is a topic that consists of finding an optimal object from a finite set of objects.

The general framework of *multiobjective combinatorial optimization* (MOCO) is described as [152]

$$\min_{\mathbf{x} \in \Omega} \left\{ \mathbf{c}_1^T \mathbf{x} = z_1(\mathbf{x}), \ldots, \mathbf{c}_K^T \mathbf{x} = z_K(\mathbf{x}) \right\}, \tag{9.2.1}$$

$$\text{subject to } D = \{\mathbf{x} : \mathbf{x} \in LD, \mathbf{x} \in B^n\}, \quad LD = \{\mathbf{x} : \mathbf{Ax} \leq \mathbf{b}\}, \tag{9.2.2}$$

where $\mathbf{c}_k \in \mathbb{R}^{n \times 1}$, the solution $\mathbf{x} = [x_1, \ldots, x_n]^T \in \mathbb{R}^n$ is a vector of discrete decision variables, $\mathbf{A} \in \mathbb{R}^{m \times n}, \mathbf{b} \in \mathbb{R}^m$ and $B = \{0, 1\}$, and $\mathbf{Ax} \leq \mathbf{b}$ denotes the elementwise inequality $(\mathbf{Ax})_i \leq b_i$ for all $i = 1, \ldots, m$.

Most of multiobjective combinatorial optimization are related to assignment problems (APs), transportation problems (TPs) and network flow problems (NFPs).

1. *Multiobjective assignment problems [17]:*

$$\min z_k(\mathbf{x}) = \sum_{i=1}^{n} \sum_{j=1}^{n} c_{ij}^k x_{ij}, \quad k = 1, \ldots K, \tag{9.2.3}$$

$$\sum_{j=1}^{n} x_{ij} = 1, \qquad i = 1, \ldots, n, \tag{9.2.4}$$

$$\sum_{i=1}^{n} x_{ij} = 1, \qquad j = 1, \ldots, n, \tag{9.2.5}$$

$$x_{ij} = (0, 1), \qquad i, j = 1, \ldots, n. \tag{9.2.6}$$

2. *Multiobjective transportation problems [152]:*

$$\min z_k(\mathbf{x}) = \sum_{i=1}^{r} \sum_{j=1}^{s} c_{ij}^k x_{ij}, \quad k = 1, \ldots K, \tag{9.2.7}$$

$$\sum_{j=1}^{s} x_{ij} = a_i, \qquad i = 1, \ldots, n, \tag{9.2.8}$$

$$\sum_{i=1}^{r} x_{ij} = b_j, \qquad j = 1, \ldots, n, \tag{9.2.9}$$

$$x_{ij} \geq 0 \text{ and integer}, \qquad i = 1, \ldots, r; \ j = 1, \ldots, q. \tag{9.2.10}$$

9.2 Multiobjective Optimization

Here, it is not restrictive to suppose equality constraints with $\sum_{i=1}^{r} a_i = \sum_{j=1}^{q} b_j$.

3. *Multiobjective network flow* or *transhipment problems [152]:* Let $G(N, A)$ be a network flow with a node set N and an arc set A; the model can be stated as

$$\min z_k(\mathbf{x}) = \sum_{(i,j) \in A} c_{ij}^k x_{ij}, \quad k = 1, \ldots, K, \tag{9.2.11}$$

$$\sum_{j \in A \setminus i} x_{ij} - \sum_{j \in A \setminus i} x_{ji} = 0, \quad \forall i \in \mathcal{N}, \tag{9.2.12}$$

$$l_{ij} \le x_{ij} \le u_{ij}, \quad \forall (i,j) \in \mathcal{A}, \tag{9.2.13}$$

$$x_{ij} \ge 0 \text{ and integer}, \tag{9.2.14}$$

where x_{ij} is the flow through arc (i, j), c_{ij}^k is the linear transhipment "cost" for arc (i, j) in objective k, and l_{ij} and u_{ij} are lower and upper bounds on x_{ij}, respectively.

4. *Multiobjective Min Sum Location Problems [134]:*

$$\min \sum_{i=1}^{m} f_i^k z_i + \sum_{i=1}^{m} \sum_{j=1}^{n} c_{ij}^k x_{ij}, \quad k = 1, \ldots, K, \tag{9.2.15}$$

$$\sum_{i=1}^{m} x_{ij} = 1, \quad j = 1, \ldots, n, \tag{9.2.16}$$

$$a_i z_i \le \sum_{j=1}^{n} d_{ij} x_{ij} \le b_i z_i, \quad i = 1, \ldots, m, \tag{9.2.17}$$

$$x_{ij} = (0, 1), \quad i = 1, \ldots, m; \ j = 1, \ldots, n, \tag{9.2.18}$$

$$z_i = (0, 1), \quad i = 1, \ldots, m, \tag{9.2.19}$$

where $z_i = 1$ if a facility is established at site i; $x_{ii} = 1$ if customer j is assigned to the facility at site i; d_{ij} is a certain usage of the facility at site i by customer j if he/she is assigned to that facility; a_i and b_i are possible limitations on the total customer usage permitted at the facility i; c_{ij}^k is, for objective k, a variable cost (or distance, etc.) if customer j is assigned to facility i; f_i^k is, for objective k, a fixed cost associated with the facility at site i.

An *unconstrained combinatorial optimization problem* $P = (S, f)$ is an optimization problem [120], where S is a finite set of solutions (called search space), and $f : S \to \mathbb{R}_+$ is an objective function that assigns a positive cost value to each solution vector $\mathbf{s} \in S$. The goal is either to find a solution of minimum cost value or a good enough solution in a reasonable amount of time in the case of approximate solution techniques.

The difficulty of multiobjective combinatorial optimization problems comes from the following two factors [22].

- It requires intensive co-operation with the decision maker for solving a multiobjective combinatorial optimization problem; this results in especially high requirements for effective tools used to generate efficient solutions.
- Many combinatorial problems are hard even in single-objective versions; their multiple objective versions are frequently more difficult.

An multiobjective decision making (MODM) problem is ill-posed from the mathematical point of view because, except in trivial cases, it has no optimal solution. The goal of MODM methods is to find a solution most consistent with the decision maker's preferences, i.e., the best compromise. Under very weak assumptions about decision maker's preferences the best compromise solution belongs to the set of efficient solutions [136].

9.2.2 Multiobjective Optimization Problems

Consider optimization problems with multiple objectives that may usually be incompatible.

Definition 9.1 (Multiobjective Optimization Problem [78, 150]) A *multiobjective optimization problem* (MOP) consists of a set of n parameters (i.e., decision variables), a set of m objective functions, and a set of p inequality constraints and q equality constraints, and can be mathematically formulated as

$$\text{minimize/maximize} \quad \left\{ \mathbf{y} = \mathbf{f}(\mathbf{x}) = [f_1(\mathbf{x}), \ldots, f_m(\mathbf{x})]^T \right\}, \quad (9.2.20)$$

$$\text{subject to} \quad g_i(\mathbf{x}) \leq 0 \ (i = 1, \ldots, p) \ \text{or} \ \mathbf{g}(\mathbf{x}) \leq \mathbf{0}, \quad (9.2.21)$$

$$h_j(\mathbf{x}) = 0 \ (j = 1, \ldots, q) \ \text{or} \ \mathbf{h}(\mathbf{x}) = \mathbf{0}. \quad (9.2.22)$$

Here the integer $m \geq 2$ is the number of objectives; $\mathbf{x} = [x_1, \ldots, x_n]^T \in X$ is a *decision vector*, $\mathbf{y} = [y_1, \ldots, y_m]^T$ is the *objective vector*; $f_i : \mathbb{R}^n \to \mathbb{R}$, $i = 1, \ldots, m$ are the objective functions, and $g_i, h_j : \mathbb{R}^n \to \mathbb{R}$, $i = 1, \ldots, m$; $j = 1, \ldots, q$ are p inequality constraints and q equality constraints of the problem, respectively; while $X = \{\mathbf{x} \in \mathbb{R}^n\}$ is called the *decision space*, and $Y = \{\mathbf{y} \in \mathbb{R}^m | \mathbf{y} = \mathbf{f}(\mathbf{x}), \mathbf{x} \in X\}$ is known as the *objective space*.

The decision space and the objective space are two Euclidean spaces in MOPs:

- The n-dimensional decision space $X \in \mathbb{R}^n$ in which each coordinate axis corresponds to a component (called decision variable) of decision vector $\mathbf{x}$.

9.2 Multiobjective Optimization

- The m-dimensional objective space $Y \in \mathbb{R}^m$ in which each coordinate axis corresponds to a component (i.e., a scalar objective function) of objective function vector $\mathbf{y} = \mathbf{f}(\mathbf{x}) = [f_1(\mathbf{x}), \ldots, f_m(\mathbf{x})]^T$.

The MOP's evaluation function $\mathbf{f} : \mathbf{x} \to \mathbf{y}$ maps decision vector $\mathbf{x} = [x_1, \ldots, x_n]^T$ to objective vector $\mathbf{y} = [y_1, \ldots, y_m]^T = [f_1(\mathbf{x}), \ldots, f_m(\mathbf{x})]^T$.

If some objective function f_i is to be maximized (or minimized), the original problem is written as the minimization (or maximization) of the negative objective $-f_i$. Thereafter, we consider only the multiobjective minimization (or maximization) problems, rather than optimization problems mixing both minimization and maximization.

The constraints $\mathbf{g}(\mathbf{x}) \leq \mathbf{0}$ and $\mathbf{h}(\mathbf{x}) = \mathbf{0}$ determine the set of feasible solutions.

Definition 9.2 (Feasible Set) The feasible set X_f is defined as the set of decision vectors $\mathbf{x}$ that satisfy the inequality constraint $\mathbf{g}(\mathbf{x}) \leq \mathbf{0}$ and the equality constraint $\mathbf{h}(\mathbf{x}) = \mathbf{0}$, i.e.,

$$X_f = \{\mathbf{x} \in X \,|\, \mathbf{g}(\mathbf{x}) \leq \mathbf{0};\, \mathbf{h}(\mathbf{x}) = \mathbf{0}\}. \tag{9.2.23}$$

Example 9.1 Dynamic economic emission dispatch problem is a typical multiobjective optimization which attempts to minimize both cost function and emission function simultaneously while satisfying equality and inequality constraints [8].

1. *Objectives:*
   - *Cost function:* The fuel cost function of each thermal generator, considering the valve-point effect, is expressed as the sum of a quadratic and a sinusoidal function. The total fuel cost in terms of real power can be expressed as

$$f_1 = \sum_{m=1}^{M} \sum_{i=1}^{N} \left(a_i + b_i P_{im} + c_i P_{im}^2 + d_i \left| \sin\left(e_i(P_i^{\min} - P_{im})\right) \right| \right), \tag{9.2.24}$$

where N is number of generating units, M is number of hours in the time horizon; a_i, b_i, c_i, d_i, e_i are cost coefficients of the ith unit; P_{im} is power output of ith unit at time m, and $P_i^{\min}$ is lower generation limits for ith unit.
   - *Emission function:* The atmospheric pollutants such as sulfur oxides (SOx) and nitrogen oxides (NOx) caused by fossil-fueled generating units can be modeled separately. However, for comparison purpose, the total emission of these pollutants which is the sum of a quadratic and an exponential function can be expressed as

$$f_2 = \sum_{m=1}^{M} \sum_{i=1}^{N} \left(\alpha_i + \beta_i P_{im} + \gamma_i P_{im}^2 + \eta_i \exp(\delta_i P_{im}) \right), \tag{9.2.25}$$

where $\alpha_i, \beta_i, \gamma_i, \eta_i, \delta_i$ are emission coefficients of the ith unit.

2. *Constraints:*

- *Real power balance constraints:* The total real power generation must balance the predicted power demand plus the real power losses in the transmission lines, at each time interval over the scheduling horizon, i.e.,

$$\sum_{i=1}^{N} P_{im} - P_{Dm} - P_{Lm} = 0, \quad m = 1, \ldots, M, \quad (9.2.26)$$

  where P_{Dm} is load demand at time m, P_{Lm} is transmission line losses at time m.

- *Real power operating limits:*

$$P_i^{\min} \leq P_{im} \leq P_i^{\max}, \quad i = 1, \ldots, N, \, m = 1, \ldots, M, \quad (9.2.27)$$

  where $P_i^{\max}$ is upper generation limits for the ith unit.

- *Generating unit ramp rate limits:*

$$P_{im} - P_{i(m-1)} \leq UR_i, \quad i = 1, \ldots, N, \, m = 1, \ldots, M, \quad (9.2.28)$$

$$P_{i(m-1)} - P_{im} \leq DR_i, \quad i = 1, \ldots, N, \, m = 1, \ldots, M, \quad (9.2.29)$$

  where UR_i and DR_i are ramp-up and ramp-down rate limits of the ith unit, respectively.

Definition 9.3 (Ideal Vector [20, p. 8]) Let $\mathbf{x}^{\mathrm{opt},k} = [x_1^{\mathrm{opt},k}, \ldots, x_n^{\mathrm{opt},k}]^T$ be a vector of variables which optimizes (either minimizes or maximizes) the kth objective function $f_k(\mathbf{x})$. In other words, if the vector $\mathbf{x}^{\mathrm{opt},k} \in X_f$ is such that

$$f_k^{\mathrm{opt},k} = \operatorname*{opt}_{\mathbf{x} \in X} f_k(\mathbf{x}), \quad k \in \{1, \ldots, m\}, \quad (9.2.30)$$

then the objective vector $\mathbf{f}^{\mathrm{opt}} = [f_1^{\mathrm{opt}}, \ldots, f_m^{\mathrm{opt}}]^T$ (where f_k^{opt} denotes the optimum of the kth objective function) is ideal for an MOP, and the point in $\mathbb{R}^n$ which determined this vector is the *ideal solution*, and is consequently called the *ideal vector*.

However, due to objective conflict and/or interdependence among the n decision variables of $\mathbf{x}$, such an ideal solution vector is impossible for MOPs: none of the feasible solutions allows simultaneous optimal solutions for all objectives [63].

In other words, individual optimal solutions for each objective are usually different. Thus, a mathematically most favorable solution should offer the least objective conflict.

Many real-life problems can be described as multiobjective optimization problems. The most typical multiobjective optimization problem is traveling salesman problem.

9.2 Multiobjective Optimization

The *traveling salesman problem* (TSP) can be stated as (see, e.g., [152])

$$\min \sum_{i \in \rho(i)} c_{i,\rho(i)}^{(k)}, \quad k = 1, \ldots, K, \quad (9.2.31)$$

where $\rho(i)$ is a tour of $\{1, \ldots, n\}$ and k denotes trip k.

Let $C = \{c_1, \ldots, c_{N_c}\}$ be a set of cities, where c_i is the ith city and N_c is the number of all cities; $A = \{(r, s) : r, s \in C\}$ be the edge set, and $d(r, s)$ be a cost measure associated with edge $(r, s) \in A$.

For a set of N_c cities, the TSP problem involves finding the shortest length closed tour visiting each city only once. In other words, the TSP problem is to find the shortest-route to visit each city once, ending up back at the starting city.

The following are the different forms of TSP [105]:

- *Euclidean TSP:* If cities $r \in C$ are given by their coordinates (x_r, y_r) and $d(r, s)$ is the Euclidean distance between city r and s, then TSP is an Euclidean TSP.
- *Symmetric TSP:* If $d(r, s) = d(s, r)$ for all (r, s), then the TSP becomes a symmetric TSP (STSP).
- *Asymmetric TSP:* If $d(r, s) \neq d(s, r)$ for at least some (r, s), then the TSP is an asymmetric TSP (ATSP).
- *Dynamic TSP:* The dynamic TSP (DTSP) is a TSP in which cities can be added or removed at run time.

The goal of TSP is to find the shortest closed tour which visits all the cities in a given set as early as possible after each and every iteration.

Another typical example of multiobjective optimizations is multiobjective data mining.

One of the most important questions in data mining problems is how to evaluate a candidate model which depends on the type of data mining task at hand. Data mining involves discovering interesting and potentially useful patterns of different types, such as associations, summaries, rules, changes, outliers, and significant structures [110]. Data mining tasks can broadly be classified into two categories [6, 62]:

- Predictive or supervised techniques learn from the current data in order to make predictions about the behavior of new data sets.
- Descriptive or unsupervised techniques provide a summary of the data.

The most commonly used data mining tasks include feature selection, classification, regression, clustering, association rule mining, deviation detection, and so on [110].

1. *Feature Selection:* It deals with selection of an optimum relevant set of features or attributes that are necessary for the recognition process (classification or clustering). In general, the feature selection problem (Ω, P) can formally be defined as an optimization problem: determine the feature set F^* for which

$$P(F^*) = \min_{F \in \Omega} P(F, X), \quad (9.2.32)$$

where Ω is the set of possible feature subsets, F refers to a feature subset, and $P : \Omega \times \Psi \to \mathbb{R}$ denotes a criterion to measure the quality of a feature subset with respect to its utility in classifying/clustering the set of points $X \in \Psi$.

2. *Classification:* The problem of supervised classification can formally be stated as follows. Given an unknown function $g : X \to Y$ (the ground truth) that maps input instances $\mathbf{x} \in X$ to output class labels $y \in Y$, and a training data set $D = \{(\mathbf{x}_1, y_1), \ldots, (\mathbf{x}_n, y_n)\}$ which is assumed to represent accurate examples of the mapping g, produce a function $h : X \to Y$ that approximates the correct mapping g as closely as possible.

3. *Clustering:* Clustering is an important unsupervised classification technique where a set of n patterns $\mathbf{x}_i \in \mathbb{R}^d$ are grouped into clusters $\{C_1, \ldots, C_K\}$ in such a way that patterns in the same cluster are similar in some sense and patterns in different clusters are dissimilar in the same sense. The main objective of any clustering technique is to produce a $K \times n$ partition matrix $\mathbf{U} = [u_{kj}]$, $k = 1, \ldots, K$; $j = 1, \ldots, n$, where u_{kj} is the membership of pattern $\mathbf{x}_j$ to cluster C_k:

$$u_{kj} = \begin{cases} 1, & \text{if } \mathbf{x}_j \in C_k; \\ 0, & \text{if } \mathbf{x}_j \notin C_k; \end{cases} \tag{9.2.33}$$

for hard or crisp partitioning of the data, and

$$\forall k \in \{1, \ldots, K\}, \quad 0 < \sum_{j=1}^{n} u_{kj} < n, \tag{9.2.34}$$

$$\forall j \in \{1, \ldots, n\}, \quad \sum_{k=1}^{K} u_{kj} = 1, \tag{9.2.35}$$

with $\sum_{k=1}^{K} \sum_{j=1}^{n} u_{kj} = n$ for probabilistic fuzzy partitioning of the data.

4. *Association Rule Mining:* The principle of association rule mining (ARM) [1] lies in the market basket or transaction data analysis. Association analysis is the discovery of rules showing attribute-value associations that occur frequently. The objective of ARM is to find all rules of the form $X \Rightarrow Y$, $X \cap Y = \emptyset$ with probability $c\%$, indicating that if itemset X occurs in a transaction, the itemset Y also occurs with probability $c\%$.

Most of the data mining problems can be thought of as optimization problems, while the majority of data mining problems have multiple criteria to be optimized. For example, a feature selection problem may try to maximize the classification accuracy while minimizing the size of the feature subset. Similarly, a rule mining problem may optimize several rule interestingness measures such as support, confidence, comprehensibility, and lift at the same time [110, 148].

9.3 Pareto Optimization Theory

Many real-world problems in artificial intelligence involve simultaneous optimization of several incommensurable and often competing objectives. In these cases, there may exist solutions in which the performance on one objective cannot be improved without reducing performance on at least one other. In other words, there is often no single optimal solution with respect to all objectives. In practice, there could be a number of optimal solutions in multiobjective optimization problems (MOPs) and the suitability of one solution depends on a number of factors including designer's choice and problem environment. These solutions are optimal in the wider sense that no other solutions in the search space are superior to them when all objectives are considered. In these problems, the Pareto optimization approach is a natural choice. Solutions given by the Pareto optimization approach are called *Pareto-optimal solutions* that are closely related to *Pareto concepts*.

9.3.1 Pareto Concepts

Pareto concepts are named after Vilfredo Pareto (1848–1923). These concepts constitute the *Pareto optimization theory* for multiple objectives.

Consider the multiobjective optimization problems

$$\max/\min_{\mathbf{x} \in S} \left\{ \mathbf{f}(\mathbf{x}) = [f_1(\mathbf{x}), \ldots, f_m(\mathbf{x})]^T \right\} \quad (9.3.1)$$

in which the objectives (or criterion functions) $f_1, \ldots, f_m$ are real-valued, continuous, and quasi-concave (for maximization) or quasi-convex (for minimization) on a closed convex set S.

Three approaches can be taken to find the solution(s) to the above problems [10].

- *Reformulation approach:* This approach entails reformulating the problem as a single objective problem. To do so, additional information is required from the decision makers, such as the relative importance or weights of the objectives, goal levels for the objectives, value functions, etc.
- *Decision making approach:* This approach requires that the decision makers interact with the optimization procedure typically by specifying preferences between pairs of presented solutions.
- *Pareto optimization approach:* It finds a representative set of nondominated solutions approximating the Pareto front. Pareto optimization methods, such as evolutionary multiobjective optimization algorithms, allow decision makers to investigate the potential solutions without a priori judgments regarding to the relative importance of objective functions. Post-Pareto analysis is necessary to select a single solution for implementation.

On the other hand, the most distinguishing feature of single objective evolutionary algorithms (EAs) compared to other heuristics is that EAs work with a population of solutions, and thus are able to search for a set of solutions in a single run. Due to this feature, single objective evolutionary algorithms are easily extended to multiobjective optimization problems. Multiobjective evolutionary algorithms (MOEAs) have become one of the most active research areas in evolutionary computation.

To describe the Pareto optimality of MOP solutions in which we are interested, it needs the following key Pareto concepts: Pareto dominance, Pareto optimality, the Pareto-optimal set, and the Pareto front.

For the convenience of narration, denote $\mathbf{x} = [x_1, \ldots, x_n]^T \in F \subseteq S$ and $\mathbf{x}' = [x'_1, \ldots, x'_n]^T \in F \subseteq S$, where F is the feasible region, in which the constraints are satisfied.

Definition 9.4 (Vector Function Relation [150]) For any two objective vectors $\mathbf{f}(\mathbf{x})$ and $\mathbf{f}(\mathbf{x}')$, their relations are written as

$$\mathbf{f}(\mathbf{x}) = \mathbf{f}(\mathbf{x}') \Leftrightarrow f_i(\mathbf{x}) = f_i(\mathbf{x}'), \ \forall i = 1, \ldots, m; \tag{9.3.2}$$

$$\mathbf{f}(\mathbf{x}) \neq \mathbf{f}(\mathbf{x}') \Leftrightarrow f_i(\mathbf{x}) \neq f_i(\mathbf{x}'), \ \text{for at least one } i \in \{1, \ldots, m\}; \tag{9.3.3}$$

and

$$\mathbf{f}(\mathbf{x}) \leq \mathbf{f}(\mathbf{x}') \Leftrightarrow f_i(\mathbf{x}) \leq f_i(\mathbf{x}'), \ \forall i = 1, \ldots, m; \tag{9.3.4}$$

$$\mathbf{f}(\mathbf{x}) < \mathbf{f}(\mathbf{x}') \Leftrightarrow \mathbf{f}(\mathbf{x}) \leq \mathbf{f}(\mathbf{x}') \wedge \mathbf{f}(\mathbf{x}) \neq \mathbf{f}(\mathbf{x}'); \tag{9.3.5}$$

$$\mathbf{f}(\mathbf{x}) \geq \mathbf{f}(\mathbf{x}') \Leftrightarrow f_i(\mathbf{x}) \geq f_i(\mathbf{x}'), \ \forall i = 1, \ldots, m; \tag{9.3.6}$$

$$\mathbf{f}(\mathbf{x}) > \mathbf{f}(\mathbf{x}') \Leftrightarrow \mathbf{f}(\mathbf{x}) \geq \mathbf{f}(\mathbf{x}') \wedge \mathbf{f}(\mathbf{x}) \neq \mathbf{f}(\mathbf{x}'). \tag{9.3.7}$$

Here the symbol $\wedge$ may be interpreted as logical conjunction operator ("AND").

It goes without saying that the above relations are available for $\mathbf{f}(\mathbf{x}) = \mathbf{x}$ and $\mathbf{f}(\mathbf{x}') = \mathbf{x}'$ as well. However, for two decision vectors $\mathbf{x}$ and $\mathbf{x}'$, these relations are meaningless. In multiobjective optimization, the relation between decision vectors $\mathbf{x}$ and $\mathbf{x}'$ are called the Pareto dominance which is based on the relations between vector objective functions.

Definition 9.5 (Pareto Dominance [150, 171]) For any two decision vectors $\mathbf{x}$ and $\mathbf{x}'$ in minimization problems, $\mathbf{x}$ is said to dominate (or outperform) $\mathbf{x}'$, denoted by $\mathbf{x} \succ \mathbf{x}'$, if and only if their objectives or *outcomes* satisfy jointly the following two conditions:

1. $\mathbf{x}$ is no worse than $\mathbf{x}'$ in all objective functions, i.e., $f_i(\mathbf{x}) \leq f_i(\mathbf{x}')$ for all $i \in \{1, \ldots, m\}$,
2. $\mathbf{x}$ is strictly better than $\mathbf{x}'$ for at least one objective function, i.e., $f_j(\mathbf{x}) < f_j(\mathbf{x}')$ for at least one $j \in \{1, \ldots, m\}$;

9.3 Pareto Optimization Theory

or is equivalently represented as

$$\mathbf{x} \succ \mathbf{x}' \Leftrightarrow \forall i \in \{1, \ldots, m\} : f_i(\mathbf{x}) \leq f_i(\mathbf{x}') \land$$
$$\exists j \in \{1, \ldots, m\} : f_j(\mathbf{x}) < f_j(\mathbf{x}'), \quad (9.3.8)$$

or simply written as

$$\mathbf{x} \succ \mathbf{x}' \Leftrightarrow \mathbf{f}(\mathbf{x}) < \mathbf{f}(\mathbf{x}'). \quad (9.3.9)$$

Similarly, for maximization problems, then

$$\mathbf{x} \succ \mathbf{x}' \Leftrightarrow \forall i \in \{1, \ldots, m\} : f_i(\mathbf{x}) \geq f_i(\mathbf{x}') \land$$
$$\exists j \in \{1, \ldots, m\} : f_j(\mathbf{x}) > f_j(\mathbf{x}'), \quad (9.3.10)$$

or simply written as

$$\mathbf{x} \succ \mathbf{x}' \Leftrightarrow \mathbf{f}(\mathbf{x}) > \mathbf{f}(\mathbf{x}'). \quad (9.3.11)$$

The Pareto dominance can be either weak or strong.

Definition 9.6 (Weak Pareto Dominance [150, 171]) A decision vector $\mathbf{x}$ is said to weakly dominate another decision vector $\mathbf{x}'$, denoted by $\mathbf{x} \succeq \mathbf{x}'$, if and only if

$$\mathbf{x} \succeq \mathbf{x}' \Leftrightarrow \mathbf{f}(\mathbf{x}) \leq \mathbf{f}(\mathbf{x}') \quad \text{(for minimization)}, \quad (9.3.12)$$
$$\mathbf{x} \succeq \mathbf{x}' \Leftrightarrow \mathbf{f}(\mathbf{x}) \geq \mathbf{f}(\mathbf{x}') \quad \text{(for maximization)}. \quad (9.3.13)$$

Definition 9.7 (Strict Pareto Dominance [171]) A decision vector $\mathbf{x}$ is said to strictly dominate $\mathbf{x}'$, denoted by $\mathbf{x} \succ\succ \mathbf{x}'$, if and only if $f_i(\mathbf{x})$ is better than $f_i(\mathbf{x}')$ in all objectives, i.e.,

$$\mathbf{x} \succ\succ \mathbf{x}' \Leftrightarrow \forall i \in \{1, \ldots, m\} : f_i(\mathbf{x}) < f_i(\mathbf{x}') \quad \text{(for minimization)}, \quad (9.3.14)$$
$$\mathbf{x} \succ\succ \mathbf{x}' \Leftrightarrow \forall i \in \{1, \ldots, m\} : f_i(\mathbf{x}) > f_i(\mathbf{x}') \quad \text{(for maximization)}. \quad (9.3.15)$$

Definition 9.8 (Incomparable [150, 171]) Two decision vectors $\mathbf{x}$ and $\mathbf{x}'$ are said to be incomparable, denoted by $\mathbf{x} \parallel \mathbf{x}'$, if neither $\mathbf{x}$ weakly dominates $\mathbf{x}'$ nor $\mathbf{x}'$ weakly dominates $\mathbf{x}$, i.e.,

$$\mathbf{x} \parallel \mathbf{x}' \Leftrightarrow \mathbf{x} \not\succeq \mathbf{x}' \land \mathbf{x}' \not\succeq \mathbf{x}. \quad (9.3.16)$$

Two incomparable solutions $\mathbf{x}$ and $\mathbf{x}'$ are also known as *mutually nondominating*.

Definition 9.9 (Nondominance [144, 171]) A decision vector $\mathbf{x} \in X_f$ is said to be nondominated regarding a set $A \subseteq X_f$ if and only if there is no vector $\mathbf{a}$ in A which dominates **x**; formally

$$\nexists \mathbf{a} \in A : \mathbf{a} \succ \mathbf{x}. \tag{9.3.17}$$

In other words, a solution **x** is said to be a *nondominated solution* if there is no **a** in the subset A that dominates **x**.

A *Pareto improvement* is a change to a different allocation that makes at least one objective better off without making any other objective worse off. An trial solution is known as "*Pareto efficient*" or "*Pareto-optimal*" or "*globally nondominated*" when no further Pareto improvements can be made, in which case we are assumed to have reached Pareto optimality.

Definition 9.10 (Pareto-Optimality [144, 171]) A decision vector $\mathbf{x} \in X_f$ is said to be *Pareto-optimal* if and only if **x** is nondominated regarding the feature region X_f, namely

$$\nexists \mathbf{a} \in X_f : \mathbf{a} \succ \mathbf{x}. \tag{9.3.18}$$

In other words, all decision vectors which are not dominated by any other decision vector are known as "*nondominated*" or Pareto optimal. The phrase "Pareto optimal" means the solution is optimal with respect to the entire decision variable space F unless otherwise specified.

Figure 9.2, by inferencing Economics Wiki on "Pareto Efficient and Pareto Optimal Tutorial," shows the relation of the Pareto efficiency and Pareto improvement for the two-objective minimization problems $\min_\mathbf{x} \mathbf{f}(\mathbf{x}) = [f_1(\mathbf{x}), f_2(\mathbf{x})]^T$.

A nondominated decision vector in $A \subseteq X_f$ is only Pareto-optimal in a local decision space A, while a Pareto-optimal decision vector is Pareto-optimal in entire feature decision space X_f. Therefore, a Pareto-optimal decision vector is definitely nondominated, but a nondominated decision vector is not necessarily Pareto-optimal.

However, to find Pareto-optimal solutions still is not the ultimate aim of multi-objective optimization, this is because:

- There may be a large number of Pareto-optimal solutions. To gain the deeper insight into the multiobjective optimization problem and knowledge about alternate solutions, there is often a special interest in finding or approximating the Pareto-optimal set.
- Due to conflicting objectives it is only possible to obtain a set of trade-off solutions (referred to as Pareto-optimal set in decision space or Pareto-optimal front in objective space), instead of a single optimal solution.

Therefore our aim is to find the decision vectors that are nondominated within the entire search space. These decision vectors constitute the so-called Pareto-optimal set, as defined below.

9.3 Pareto Optimization Theory

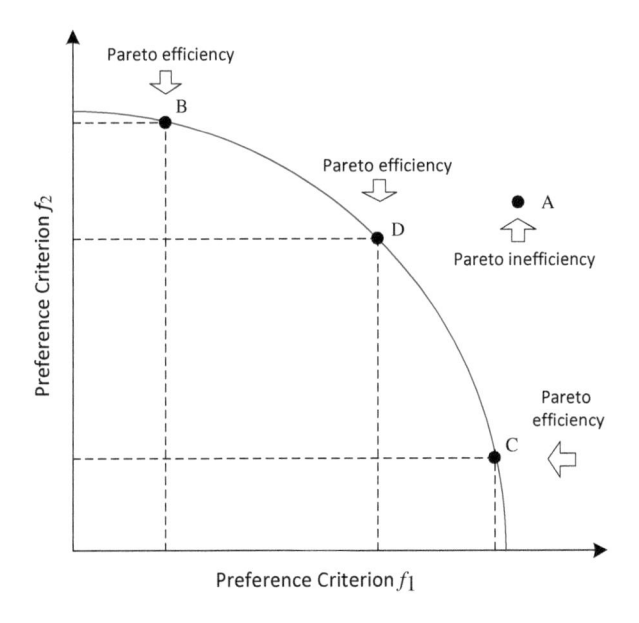

Fig. 9.2 Pareto efficiency and Pareto improvement. Point A is an inefficient allocation between preference criterions f_1 and f_2 because it does not satisfy the constraint curve of f_1 and f_2. Two decisions to move from Point A to Points C and D would be a Pareto improvement, respectively. They improve both f_1 and f_2, without making anyone else worse off. Hence, these two moves would be a Pareto improvement and be Pareto optimal, respectively. A move from Point A to Point B would not be a Pareto improvement because it decreases the cost f_1 by increasing another cost f_2, thus making one side better off by making another worse off. The move from any point that lies under the curve to any point on the curve cannot be a Pareto improvement due to making one of two criterions f_1 and f_2 worse

Definition 9.11 (Nondominated Sets [150]) Let $A \subseteq X_f$ and the function $g(A)$ give the set of nondominated decision vectors in A:

$$g(A) = \{\mathbf{a} \in A | \mathbf{a} \text{ is nondominated regarding } A\}. \tag{9.3.19}$$

The set $g(A)$ is said to be the *nondominated set* (NS) with respect to A. When $A = X_f$, the corresponding set

$$NS_{\min} = \left\{\mathbf{x} \in X_f : \{\mathbf{f}(\mathbf{x}') \in Y : \mathbf{f}(\mathbf{x}') \leq \mathbf{f}(\mathbf{x}) \wedge \mathbf{f}(\mathbf{x}') \neq \mathbf{f}(\mathbf{x})\} = \emptyset\right\}, \tag{9.3.20}$$

$$NS_{\max} = \left\{\mathbf{x} \in X_f : \{\mathbf{f}(\mathbf{x}') \in Y : \mathbf{f}(\mathbf{x}') \geq \mathbf{f}(\mathbf{x}) \wedge \mathbf{f}(\mathbf{x}') \neq \mathbf{f}(\mathbf{x})\} = \emptyset\right\} \tag{9.3.21}$$

is known as the nondominated set in the entire decision space for minimization and maximization, respectively.

Definition 9.12 (Pareto-Optimal Set [150]) The Pareto-optimal set (PS) is the set of all objectives given by nondominated decision vectors **x** in nondominated set NS, namely

$$PS_{\min} = \{\mathbf{f}(\mathbf{x}) \in Y : \{\mathbf{f}(\mathbf{x}') \in Y : \mathbf{f}(\mathbf{x}') \leq \mathbf{f}(\mathbf{x}) \wedge \mathbf{f}(\mathbf{x}') \neq \mathbf{f}(\mathbf{x})\} = \emptyset\}, \quad (9.3.22)$$

$$PS_{\max} = \{\mathbf{f}(\mathbf{x}) \in Y : \{\mathbf{f}(\mathbf{x}') \in Y : \mathbf{f}(\mathbf{x}') \geq \mathbf{f}(\mathbf{x}) \wedge \mathbf{f}(\mathbf{x}') \neq \mathbf{f}(\mathbf{x})\} = \emptyset\}. \quad (9.3.23)$$

Here $PS_{\min}$ and $PS_{\max}$ are the Pareto-optimal sets for minimization and maximization, respectively.

Pareto-optimal decision vectors cannot be improved in any objective without causing a degradation in at least one other objective; they represent globally optimal solutions. Analogous to single-objective optimization problems, Pareto-optimal sets are divided into local and global Pareto-optimal sets.

Definition 9.13 (Local Pareto-Optimal Set [23, 171]) Consider a set of decision vectors $X' \subseteq X_f$. The set X' is said to be a *local Pareto-optimal set* if and only if

$$\forall \mathbf{x}' \in X' | \nexists \mathbf{x} \in X_f : \mathbf{x} \prec \mathbf{x}' \wedge \|\mathbf{x} - \mathbf{x}'\| < \epsilon \wedge$$
$$\forall i \in \{1, \ldots, m\} : \|f_i(\mathbf{x}) - f_i(\mathbf{x}')\| < \delta, \quad (9.3.24)$$

where $\|\cdot\|$ is a corresponding distance metric and $\epsilon > 0, \delta > 0$.

Definition 9.14 (Global Pareto-Optimal Set [23, 171]) The set X' is called a *global Pareto-optimal set* if and only if

$$\forall \mathbf{x}' \in X' : \nexists \mathbf{x} \in X_f : \mathbf{x} \prec \mathbf{x}'. \quad (9.3.25)$$

The set of all objectives given by Pareto-optimal set of decision vectors is known as the *Pareto-optimal front* (or simply called *Pareto front*).

Definition 9.15 (Pareto Front) Given an objective vector $\mathbf{f}(\mathbf{x}) = [f_1(\mathbf{x}), \ldots, f_m(\mathbf{x})]^T$ and a Pareto-optimal set PS, the Pareto front PF is defined as the set of all objective vectors given by decision vectors $\mathbf{x} \in PS$, i.e.,

$$PF \stackrel{\text{def}}{=} \{\mathbf{f}(\mathbf{x}) \in \mathbb{R}^m | \mathbf{x} \in PS\}. \quad (9.3.26)$$

Solutions in the Pareto front represent the possible optimal trade-offs between competing objectives.

For a given system, the *Pareto frontier* is the set of parameterizations (allocations) that are all Pareto efficient or Pareto optimal. Finding Pareto frontiers is particularly useful in engineering. This is because that when yielding all of the potentially optimal solutions, a designer needs only to focus on trade-offs within this constrained set of parameters without considering the full range of parameters.

9.3 Pareto Optimization Theory

It should be noted that a global Pareto-optimal set does not necessarily contain all Pareto-optimal solutions.

Pareto-optimal solutions are those solutions within the genotype search space (i.e., decision space) whose corresponding phenotype objective vector components cannot be all simultaneously improved. These solutions are also called non-inferior, admissible, or efficient solutions [73] in the sense that they are nondominated with respect to all other comparison solution vector and may have no clearly apparent relationship besides their membership in the Pareto-optimal set.

A "current" set of Pareto-optimal solutions is denoted by $P_{current}(t)$, where t represents the generation number. A secondary population storing nondominated solutions found through the generations is denoted by P_{known}, while the "true" Pareto-optimal set, denoted by P_{true}, is not explicitly known for MOP problems of any difficulty. The associated Pareto front for each of $P_{current}(t)$, P_{known}, and P_{true} are termed as $PF_{current}(t)$, PF_{known}, and PF_{true}, respectively.

The decision maker is often selecting solutions via choice of acceptable objective performance, represented by the (known) Pareto front. Choosing an MOP solution that optimizes only one objective may well ignore "better" solutions for other objectives.

We wish to determine the Pareto-optimal set from the set X of all the decision variable vectors that satisfy the inequality constraints (9.2.21) and (9.2.22). But, not all solutions in the Pareto-optimal set are normally desirable or achievable in practice. For example, we may not wish to have different solutions that map to the same values in objective function space.

Lower and upper bounds on objective values of all Pareto-optimal solutions are given by the ideal objective vector $\mathbf{f}^{ideal}$ and the nadir objective vectors $\mathbf{f}^{nad}$, respectively. In other words, the Pareto front of a multiobjective optimization problem is bounded by a nadir objective vector $\mathbf{z}^{nad}$ and an ideal objective vector $\mathbf{z}^{ideal}$, if these are finite.

Definition 9.16 (Nadir Objective Vector, Ideal Objective Vector) The *nadir objective vector* is defined as

$$z_i^{nad} = \sup_{\mathbf{x} \in F \text{ is Pareto optimal}} f_i(\mathbf{x}), \quad \text{for all } i = 1, \ldots, K \quad (9.3.27)$$

and the *ideal objective vector* as

$$z_i^{ideal} = \inf_{\mathbf{x} \in F} f_i(\mathbf{x}), \quad \text{for all } i = 1, \ldots, K. \quad (9.3.28)$$

In other words, the components of a nadir and an ideal objective vector define upper and lower bounds for the objective function values of Pareto-optimal solutions, respectively. In practice, the nadir objective vector can only be approximated as, typically, the whole Pareto-optimal set is unknown. In addition, because of numerical reasons a utopian objective vector $\mathbf{z}^{\text{utopian}}$ with elements

$$z_i^{\text{utopian}} = z_i^{\text{ideal}} - \epsilon, \quad \text{for all } i = 1, \ldots, K, \tag{9.3.29}$$

or in vector form

$$\mathbf{z}^{\text{utopian}} = \mathbf{z}^{\text{ideal}} - \epsilon \mathbf{1}, \tag{9.3.30}$$

is often defined, where $\epsilon > 0$ is a small constant, and $\mathbf{1}$ is an K-dimensional vector of all ones.

Definition 9.17 (Archive) Nondominated set produced by heuristic algorithms is referred to as the *archive* (denoted by A) of the estimated Pareto front.

The archive of estimated Pareto fronts will only be an approximation to the true Pareto front.

When solving multiobjective optimization problems, since there are many possible Pareto-optimal solutions rather than a single optimal solution for possibly contradictory multiobjective functions, we need to organize or partition found solutions for selecting appropriate solution(s). The selection methods for Pareto-optimal solutions are divided to the following four categories:

- *Fitness selection approach*,
- *Nondominated sorting approach*,
- *Crowding distance assignment approach*, and
- *Hierarchical clustering approach*.

9.3.2 Fitness Selection Approach

To fairly evaluate the quality of each solution and make the solution set search in the direction of Pareto-optimal solution, how to implement fitness assignment is important.

There are benchmark metrics or indicators that play an important role in evaluating the Pareto fronts. These metrics or indicators are: hypervolume, spacing, maximum spread, and coverage [139].

Definition 9.18 (Hypervolume [170]) For a nondominated vector $\mathbf{x}^i \in \mathcal{P}_a^*$, its *hypervolume* (HV) is defined as

$$HV = \left\{ \bigcup_i a_i | \mathbf{x}^i \in \mathcal{P}_a^* \right\}, \tag{9.3.31}$$

where $\mathcal{P}_a^*$ is the area in the objective search space covered by the obtained Pareto front, and a_i is the hypervolume determined by the components of $\mathbf{x}^i$ and the origin.

9.3 Pareto Optimization Theory

Definition 9.19 (Spacing) *Spacing* is a measure for the spread (distribution) of nondominated solutions throughout the Pareto front. Actually, it measures the variance of the distance between adjacent nondominated solutions and can be evaluated by Santana et al. [139]

$$S = \sqrt{\frac{1}{m-1} \sum_{i=1}^{m} (\bar{d} - d_i)^2}, \qquad (9.3.32)$$

where $\bar{d}$ is the mean distance between all the adjacent solutions and m is the number of nondominated solutions in the Pareto front, and

$$d_i = \min_j \{|f_1^i(\mathbf{x}) - f_1^j(\mathbf{x})| + |f_2^i(\mathbf{x}) - f_2^j(\mathbf{x})|, \quad i, j = 1, \ldots, m. \qquad (9.3.33)$$

A value of spacing equal to zero means that all the solutions are equidistantly spaced in the Pareto front.

Definition 9.20 (Maximum Spread) This metric, proposed by Zitzler et al. [171], evaluates the maximum extension covered by the nondominated solutions in the Pareto front, and can be calculated by

$$MS = \sqrt{\sum_{i=1}^{K} \left(\max \{f_i^1, \ldots, f_i^m\} - \min \{f_i^1, \ldots, f_i^m\} \right)}, \qquad (9.3.34)$$

where m is the number of solutions in the Pareto front and K is the number of objectives in a given problem. It should be noted that higher values MS indicate better performance.

Definition 9.21 (Coverage) The coverage metric of two sets A and B, proposed by Zitzler and Thiele [170], is denoted by $C(A, B)$, and maps the ordered pair (A, B) to the interval $[0, 1]$ using

$$C(A, B) = \frac{|\{\mathbf{b} \in B; \exists \mathbf{a} \in A : \mathbf{a} \succeq \mathbf{b}\}|}{|B|}, \qquad (9.3.35)$$

where $|B|$ is the number of elements of the set B.

The value $C(A, B) = 1$ means that all solutions in B are weakly dominated by ones in A. On the other hand, $C(A, B) = 0$ means that none of the solutions in B is weakly dominated by A. It should be noted that both $C(A, B)$ and $C(B, A)$ have to be evaluated, respectively, since $C(A, B)$ is not necessarily equal to $1 - C(B, A)$.

If $0 < C(A, B) < 1$ and $0 < C(B, A) < 1$, then neither A weakly dominates B nor B weakly dominates A. Thus, the sets A and B are incomparable, which means that A is not worse than B and vice versa.

Two popular fitness selections are the binary tournament selection and the roulette wheel selection [140].

1. *Binary tournament selection:* k individuals are drawn from the entire population, allowing them to compete (tournament) and extract the best individual among them. The number k is the *tournament size*, and often takes 2. In this special case, we can call it *binary tournament selection. Tournament selection* is just the broader term where k can be any number ≥ 2.
2. *Roulette wheel selection:* It is also known as fitness proportionate selection which is a genetic operator for selecting potentially useful solutions for recombination. In roulette wheel selection, as in all selection methods, the fitness function assigns a fitness to possible solutions or chromosomes. This fitness level is used to associate a probability of selection with each individual chromosome. If F_i is the fitness of individual i in the population, its probability of being selected is

$$p_i = \frac{F_i}{\sum_{j=1}^{N} F_j}, \qquad (9.3.36)$$

where N is the number of individuals in the population.

Algorithm 9.1 gives a roulette wheel fitness selection algorithm.

Algorithm 9.1 Roulette wheel fitness selection algorithm [140]
---
1. **input:** The population size k.
2. **initialization:** Generate randomly the initial population of candidate solutions.
3. Evaluate the fitness F_i of each individual in the population.
4. Compute the probability (slot size) p of selecting each member of the population: $p_i = F_i / \sum_{j=1}^{k} F_j$, where k is the population size.
5. Calculate the cumulative probability q_i for each individual: $q_i = \sum_{j=1}^{i} p_j$.
6. Generate a uniform random number $r \in (0, 1]$.
7. If $r < q_1$, then select the first chromosome $\mathbf{x}_1$, else select the individual $\mathbf{x}_i$ such that $q_{i-1} < r < q_i$.
8. Repeat steps 4–7 k times to create k candidates in the mating pool.
9. **output:** k candidate solutions in the mating pool.

9.3.3 Nondominated Sorting Approach

The nondominated selection is based on nondominated rank: dominated individuals are penalized by subtracting a small fixed penalty term from their expected number of copies during selection. But this algorithm failed when the population had very few nondominated individuals, which may result in a large fitness value for those few nondominated points, eventually leading to a high selection pressure.

The idea behind the nondominated sorting procedure is twofold.

9.3 Pareto Optimization Theory

- Use a ranking selection method to emphasize good points.
- Use a niche method to maintain stable subpopulations of good points.

If an individual (or solution) is not dominated by all other individuals, then it is assigned as the first nondominated level (or rank). An individual is said to have the second nondominated level or rank, if it is dominated only by the individual(s) in the first nondominated level. Similarly, one can define individuals in the third or higher nondominated level or rank. It is noted that more individuals maybe have the same nondominated level or rank. Let i_{rank} represent the nondominated rank of the ith individual. For two solutions with different nondomination ranks, if $i_{\text{rank}} < j_{\text{rank}}$ then the ith solution with the lower (better) rank is preferred.

Algorithm 9.2 shows a fast nondominated sorting approach.

Algorithm 9.2 Fast-nondominated-sort(P) [25]

1. **for** each $p \in P$
2. $S_p = \emptyset$
3. $n_p = 0$
4.   **for** each $q \in P$
5.   **if** ($\mathbf{x}_p \prec \mathbf{x}_q$) **then**   % if p dominates q
6. $S_p = S_p \cup \{q\}$
7.   **else if** ($\mathbf{x}_q \prec \mathbf{x}_p$) **then**
8. $n_p = n_p + 1$
9.   **end if**
10.   **if** $n_p = 0$ **then**
11. $p_{\text{rank}} = 1$
12. $\mathcal{F}_1 = \mathcal{F}_1 \cup \{p\}$
13.   **end if**
14. **end for**
15. $i = 1$
16. **while** $\mathcal{F}_i \neq \emptyset$
17. $Q = \emptyset$
18.   **for** each $p \in \mathcal{F}_i$
19.   **for** each $q \in S_p$
20. $n_q = n_q - 1$
21.   **if** $n_q = 0$ **then**
22. $q_{\text{rank}} = i + 1$
23. $Q = Q \cup \{q\}$
24.   **end if**
25.   **end for**
26.   **end for**
27. $i = i + 1$
28. $\mathcal{F}_i = Q$
29. **end while**
30. **end for**
31. **output:** $\mathcal{F} = (\mathcal{F}_1, \mathcal{F}_2, \ldots) = $ fast-nondominated-sort (P), nondominated fronts of P.

In order to identify solutions of the first nondominated front in a population of size N_P, each solution can be compared with every other solution in the population to find if it is dominated. At this stage, all individuals in the first nondominated level in the population are found. In order to find the individuals in the second nondominated level, the solutions of the first level are discounted temporarily and the above procedure is repeated for finding third and higher levels or ranks of nondomination.

9.3.4 Crowding Distance Assignment Approach

If the two solutions have the same nondominated rank (say $i_{\text{rank}} = j_{\text{rank}}$), which one should we choose? In this case, we need another quality for selecting the better solution. A natural choice is to estimate the density of solutions surrounding a particular solution in the population. For this end, the average distance of two points on either side of this point along each of the objectives needs to be calculated.

Definition 9.22 (Crowding Distance [132]) Let f_1 and f_2 be two objective functions in a multiobjective optimization problem, and let $\mathbf{x}$, $\mathbf{x}_i$ and $\mathbf{x}_j$ be the members of a nondominated list of solutions. Furthermore, $\mathbf{x}_i$ and $\mathbf{x}_j$ are the nearest neighbors of $\mathbf{x}$ in the objective spaces. The *crowding distance* (CD) of a trial solution $\mathbf{x}$ in a nondominated set depicts the perimeter of a hypercube formed by its nearest neighbors (i.e., $\mathbf{x}_i$ and $\mathbf{x}_j$) at the vertices in the fitness landscapes.

The crowding distance computation requires sorting the population according to each objective function value in ascending order of magnitude. Let the population consist of $|I|$ individuals which are ranked according to their fitness values from small to large, and let $f_k^i = f_k^i(\mathbf{x})$ denote the kth objective function of the ith individual, where $i = 1, \ldots, |I|$ and $k = 1, \ldots, K$. Except for boundary individuals 1 and $|I|$, the ith individual in all other intermediate population set $\{2, \ldots, |I|-1\}$ is assigned a finite crowded distance (CD) value given by Luo et al. [98]:

$$CD_i = \frac{1}{K} \sum_{k=1}^{K} \left| f_k^{i+1} - f_k^{i-1} \right|, \quad i = 2, \ldots, |I|-1, \qquad (9.3.37)$$

or [25, 163]

$$CD_i = \sum_{k=1}^{K} \frac{\left| f_k^{i+1} - f_k^{i-1} \right|}{f_k^{\max} - f_k^{\min}}, \quad i = 2, \ldots, |I|-1, \qquad (9.3.38)$$

while the distance values of boundary individuals are assigned as $CD_1 = CD_{|I|} = \infty$. Here, $f_k^{\max} = \max\{f_k^1, \ldots, f_k^{|I|}\}$ and $f_k^{\min} = \min\{f_k^1, \ldots, f_k^{|I|}\}$ are the maximum and minimum values of the kth objective of all individuals, respectively.

The crowded-comparison operator guides the selection process at the various stages of the algorithm toward a uniformly spread-out Pareto-optimal front. Every population has two attributes: nondomination rank and crowding distance. By combining the nondominated rank and the crowded distance, the *crowded-comparison operator* is defined by *partial order* ($\prec_n$) as [25]:

$$i \prec_n j, \quad \text{if } (i_{\text{rank}} < j_{\text{rank}}) \text{ or } (i_{\text{rank}} = j_{\text{rank}}) \text{ and } CD_i > CD_j. \tag{9.3.39}$$

That is, between two solutions with differing nondomination ranks, the solution with the lower (better) rank is preferred. Otherwise, if both solutions belong to the same front, then the solution that is located in a lesser crowded region is preferred. Therefore, the crowded-comparison operator ($\prec_n$) guides the selection process at the various stages of the algorithm toward a uniformly spread-out Pareto-optimal front.

Algorithm 9.3 shows the crowding-distance-assignment (I).

Algorithm 9.3 Crowding-distance-assignment (I) [25]

$l = |I|$
for each i, set $CD_i = 0$
  for each objective j
    $I = \text{sour}(I, j)$
    $CD_1 = CD_l = \infty$
    for $i = 2$ to $(l-1)$
      $CD_i = CD_i + (f_j^{i+1} - f_j^{i-1})/(f_j^{\max} - f_j^{\min})$
    end for
  end for
end for

Between two populations with differing nondomination ranks, the population with the lower (better) rank is preferred. If both populations belong to the same front, then the population with larger crowding distance is preferred.

9.3.5 Hierarchical Clustering Approach

Cluster analysis can be applied to the results of a multiobjective optimization algorithm to organize or partition solutions based on their objective function values.

The steps of hierarchical clustering methodology are given below [108]:

1. Define decision variables, feasible set, and objective functions.
2. Choose and apply a Pareto optimization algorithm.
3. Clustering analysis:

   - *Clustering tendency:* By visual inspection or data projections verify that a hierarchical cluster structure is a reasonable model for the data.
   - *Data scaling:* Remove implicit variable weightings due to relative scales using range scaling.

- *Proximity:* Select and apply an appropriate similarity measure for the data, here, Euclidean distance.
- *Choice of algorithm(s):* Consider the assumptions and characteristics of clustering algorithms and select the most suitable algorithm for the application, here, group average linkage.
- *Application of algorithm:* Apply the selected algorithm to obtain a dendrogram.
- *Validation:* Examine the results based on application subject matter knowledge, assess the fit to the input data and stability of the cluster structure, and compare the results of multiple algorithms, if used.

4. Represent and use the clusters and structure: If the clustering is reasonable and valid, examine the divisions in the hierarchy for trade-offs and other information to aid decision making.

9.3.6 Benchmark Functions for Multiobjective Optimization

When designing an artificial intelligence algorithm or system, we need select the benchmark or test functions to examine its effectiveness in converging and maintaining a diverse set of nondominated solutions. A very simple multiobjective test function is the well-known one-variable two-objective function used by Schaffer [141]. It is defined as follows:

$$\min \mathbf{f}_2(x) = [g(x), h(x)] \quad \text{with} \quad g(x) = x^2, \ h(x) = (x-2)^2, \qquad (9.3.40)$$

where $x \in [0, 2]$.

The following are the several commonly used multiobjective benchmark functions.

1. Benchmark functions for $\min \mathbf{f}(\mathbf{x}) = [g(\mathbf{x}), h(\mathbf{x})]$

    - ZDT1 functions [171]

    $$\left.\begin{aligned} f_1(x_1) &= x_1, \\ g(\mathbf{x}) &= g(x_2, \ldots, x_m) = 1 + 9 \sum_{i=1}^{m} x_i/(m-1), \\ h(\mathbf{x}) &= h(f_1(x_1), g(x_2, \ldots, x_m)) = \sqrt{1 - f_1/g}, \end{aligned}\right\} \qquad (9.3.41)$$

    where $m = 30$, $x_i \in [0, 1]$.

9.3 Pareto Optimization Theory

- ZDT4 functions [171]

$$\left.\begin{array}{l} f_1(x_1) = x_1, \\ g(\mathbf{x}) = g(x_2, \ldots, x_m) = 1 + 10(m-1) + \sum_{i=2}^{m}\left(x_i^2 - 10\cos(4\pi x_i)\right), \\ h(\mathbf{x}) = h(f_1(x_1), g(x_2, \ldots, x_m)) = \sqrt{1 - f_1/g}, \end{array}\right\} \quad (9.3.42)$$

where $m = 10$ and $x_i \in [0, 1]$.

- ZDT6 functions [171]:

$$\left.\begin{array}{l} f_1(x_1) = 1 - \exp(-4x_1)\sin^6(6\pi x_1), \\ g(\mathbf{x}) = g(x_2, \ldots, x_m) = 1 + 9\left(\sum_{i=2}^{m} x_i/(m-1)\right)^{0.25}, \\ h(\mathbf{x}) = 1 - (f_1/g)^2, \end{array}\right\} \quad (9.3.43)$$

where $m = 10$ and $x_i \in [0, 1]$.

2. Benchmark functions for min $\mathbf{f}(\mathbf{x}) = [f_1(\mathbf{x}), f_2(\mathbf{x})]$

- QV functions [126]

$$\left.\begin{array}{l} f_1(\mathbf{x}) = \left(\dfrac{1}{n}\sum_{i=1}^{m}\left(x_i^2 - 10\cos(2\pi x_i) + 10\right)\right)^{1/4}, \\ f_2(\mathbf{x}) = \left(\dfrac{1}{n}\sum_{i=1}^{m}\left((x_i - 1.5)^2 - 10\cos\left(2\pi(x_i - 1.5)\right) + 10\right)\right)^{1/4}, \end{array}\right\} \quad (9.3.44)$$

where $x_i \in [-5, 5]$.

- KUR functions [90]

$$\left.\begin{array}{l} f_1(\mathbf{x}) = \sum_{i=1}^{m-1}\left(-10\exp\left(-0.2\sqrt{x_i^2 + x_{i+1}^2}\right)\right), \\ f_2(\mathbf{x}) = \sum_{i=1}^{m}\left(|x_i|^{0.8} + \sin^3(x_i)\right), \end{array}\right\} \quad (9.3.45)$$

where $x_i \in [-10^3, 10^3]$.

3. *Constrained benchmark functions*
   - Partially separable benchmark functions [23, 171]

   $$\left.\begin{array}{l} \min \ \mathbf{f}(\mathbf{x}) = (f_1(x_1), f_2(\mathbf{x})), \\ \text{subject to } f_2(\mathbf{x}) = g(x_2, \ldots, x_n) h(f_1(x_1) g(x_1, \ldots, x_n)), \end{array}\right\} \quad (9.3.46)$$

   where $\mathbf{x} = [x_1, \ldots, x_n]^T$, the function f_1 is a function of the first decision variable x_1 only, g is a function of the remaining $n - 1$ variables $x_2, \ldots, x_n$, and the parameters of h are the function values of f_1 and g. For example,

   $$\left.\begin{array}{l} f_1(x_1) = x_1, \quad g(x_2, \ldots, x_n) = 1 + \dfrac{9}{n-1} \sum_{i=2}^{n} x_i, \\ h(f_1, g) = 1 - \sqrt{f_1/g} \quad \text{or} \quad h(f_1, g) = 1 - (f_1/g)^2, \end{array}\right\} \quad (9.3.47)$$

   where $n = 30$ and $x_i \in [0, 1]$. The Pareto-optimal front is formed with $g(\mathbf{x}) = 1$.
   - Separable benchmark functions [170]:

   $$\left.\begin{array}{l} \max \ \mathbf{f}(\mathbf{x}) = [f_1(\mathbf{x}), f_2(\mathbf{x})], \quad f_i(\mathbf{x}) = \sum_{j=1}^{500} a_{ij} x_j, \ i = 1, 2 \\ \text{subject to } \sum_{j=1}^{500} b_{ij} x_j \leq c_i, \ i = 1, 2; \quad x_j = 0 \text{ or } 1, \ j = 1, 2, \ldots, 500, \end{array}\right\}$$
   $$(9.3.48)$$

   In this test, $\mathbf{x}$ is a 500-dimensional binary vector, a_{ij} is the profit of item j according to knapsack i, b_{ij} is the weight of item j according to knapsack i, and c_i is the capacity of knapsack i ($i = 1, 2$ and $j = 1, 2, \ldots, 500$).
4. *Multiobjective benchmark functions for* $\min \mathbf{f}(\mathbf{x}) = [f_1(\mathbf{x}), \ldots, f_m(\mathbf{x})]$: SPH-$m$ functions [92, 141]

   $$f_j(\mathbf{x}) = \sum_{i=1, i \neq j}^{n} (x_i)^2 + (x_j - 1)^2, \ j = 1, \ldots, m; m = 2, 3. \quad (9.3.49)$$

   where $x_i \in [-10^3, 10^3]$.
5. *Multiobjective benchmark functions for* $\min \mathbf{f}(\mathbf{x}) = [f_1(\mathbf{x}), f_2(\mathbf{x}), g_3(\mathbf{x}), \ldots, g_m(\mathbf{x})]$

- ISH1 functions [79]

$$\left.\begin{array}{l}g_i(\mathbf{x}) = f_i(\mathbf{x}), \ i = 1, 2, \\ g_i(\mathbf{x}) = \alpha f_1(\mathbf{x}) + (1-\alpha) f_i(\mathbf{x}), \ i = 3, 5, 7, 9, \\ g_i(\mathbf{x}) = \alpha f_2(\mathbf{x}) + (1-\alpha) f_i(\mathbf{x}), \ i = 4, 6, 8, 10,\end{array}\right\} \quad (9.3.50)$$

where the value of $\alpha \in [0, 1]$ can be viewed as the correlation strength. $\alpha = 0$ corresponds to the minimum correlation strength 0, where $g_i(\mathbf{x})$ is the same as the randomly generated objective $f_i(\mathbf{x})$. $\alpha = 1$ corresponds to the maximum correlation strength 1, where $g_i(\mathbf{x})$ is the same as $f_1(\mathbf{x})$ or $f_2(\mathbf{x})$.

- ISH2 functions [79]

$$\left.\begin{array}{l}g_i(\mathbf{x}) = f_i(\mathbf{x}), \ i = 1, 2, \\ g_i(\mathbf{x}) = \alpha_{ik} f_1(\mathbf{x}) + (1-\alpha_{ik}) f_2(\mathbf{x}), \ i = 3, 4, \ldots, k,\end{array}\right\} \quad (9.3.51)$$

where α_{ik} is specified as follows:

$$\alpha_{ik} = (i-2)/k - 1, \quad i = 3, 4, \ldots, k; \ k = 4, 6, 8, 10. \quad (9.3.52)$$

For example, the four-objective problem is specified as

$$\left.\begin{array}{l}g_1(\mathbf{x}) = f_1(\mathbf{x}), \quad g_2(\mathbf{x}) = f_2(\mathbf{x}), \\ g_3(\mathbf{x}) = 1/3 \cdot f_1(\mathbf{x}) + 2/3 \cdot f_2(\mathbf{x}), \\ g_4(\mathbf{x}) = 2/3 \cdot f_1(\mathbf{x}) + 1/3 \cdot f_2(\mathbf{x}).\end{array}\right\} \quad (9.3.53)$$

More benchmark test functions can be found in [149, 165, 172]. Yao et al. [165] summarized 23 benchmark test functions.

9.4 Noisy Multiobjective Optimization

Optimization problems in various real-world applications are often characterized by multiple conflicting objectives and a wide range of uncertainty. Optimization problems containing uncertainty and multiobjectives are termed as uncertain multiobjective optimization problems. In evolutionary optimization community, uncertainty in the objective functions is generally stochastic noise, and the corresponding multiobjective optimization problems are termed as *noisy (or imprecise) multiobjective optimization problems*.

Noisy multiobjective optimization problems are also known as *interval multiobjective optimization problems*, since objectives $f_i(\mathbf{x}, \mathbf{c}_i)$ contaminated by noise vector $\mathbf{c}_i$ can be reformulated, in interval-value form, as $f_i(\mathbf{x}, \mathbf{c}_i) = [\underline{f}_i(\mathbf{x}, \mathbf{c}_i), \overline{f}_i(\mathbf{x}, \mathbf{c}_i)]$.

9.4.1 Pareto Concepts for Noisy Multiobjective Optimization

There are at least two different types of uncertainty that are important for real-world modeling [96].

1. *Noise*, also referred to as *aleatory uncertainty*. Noise is an inherent property of the system modeled (or is introduced into the model to simulate this behavior) and therefore cannot be reduced. By Oberkampf et al. [117], aleatory uncertainty is defined as the "inherent variation associated with the physical system or the environment under consideration."
2. *Imprecision*, also known as *epistemic uncertainty*, describes not uncertainty due to system variance but the uncertainty of the outcome due to "any lack of knowledge or information in any phase or activity of the modeling process" [117].

Uncertainty in objectives can be divided into two kinds [58].

- Decision vector $\mathbf{x} \in \mathbb{R}^d$ corresponding to different objectives is no longer a fixed point, but is characterized by $(\mathbf{x}, \mathbf{c})$ with $\mathbf{c} = [c_1, \ldots, c_d]^T$ that is a local neighborhood of $\mathbf{x}$, where $c_i = [\underline{c}_i, \overline{c}_i]$ is an interval with lower limits $\underline{c}_i$ and upper limit $\overline{c}_i$ for $i = 1, \ldots, d$.
- Objectives f_i is no longer a fixed value $f_i(\mathbf{x})$, but is an interval of objective, denoted by $f_i(\mathbf{x}, \mathbf{c}) = [\underline{f}_i(\mathbf{x}, \mathbf{c}), \overline{f}_i(\mathbf{x}, \mathbf{c})]$.

Therefore, noisy multiobjective optimization problems can be described as the following noisy multiobjective optimization problems [58]:

$$\max/\min \left\{ \mathbf{f} = [f_1(\mathbf{x}, \mathbf{c}_1), \ldots, f_m(\mathbf{x}, \mathbf{c}_m)]^T \right\}, \tag{9.4.1}$$

$$\text{subject to } \mathbf{c}_i = [c_{i1}, \ldots, c_{il}]^T, \ c_{ij} = [\underline{c}_{ij}, \overline{c}_{ij}], \quad \begin{cases} i = 1, \ldots, m, \\ j = 1, \ldots, l, \end{cases} \tag{9.4.2}$$

where $\mathbf{x}$ is a d-dimensional decision variable; X is the decision space of $\mathbf{x}$; $f_i(\mathbf{x}, \mathbf{c}_i) = [\underline{f}_i(\mathbf{x}, \mathbf{c}_i), \overline{f}_i(\mathbf{x}, \mathbf{c}_i)]$ is the ith objective function with interval $[\underline{f}_i, \overline{f}_i]$, $i = 1, \ldots, m$ (m is the number of objectives, and $m \geq 3$); $\mathbf{c}_i$ is a fixed interval vector independent of the decision vector $\mathbf{x}$ (namely, $\mathbf{c}_i$ remains unchanged along with $\mathbf{x}$), while $c_{ij} = [\underline{c}_{ij}, \overline{c}_{ij}]$ is the jth interval parameter component of $\mathbf{c}_i$.

For any two solutions $\mathbf{x}_1$ and $\mathbf{x}_2 \in X$ of problem (9.4.1), the corresponding ith objectives are $f_i(\mathbf{x}_1, \mathbf{c}_i)$ and $f_i(\mathbf{x}_2, \mathbf{c}_i)$, $i = 1, \ldots, m$, respectively.

Definition 9.23 (Interval Order Relation) Consider two objective intervals $\mathbf{f}(\mathbf{x}_1) = [\underline{\mathbf{f}}(\mathbf{x}_1), \overline{\mathbf{f}}(\mathbf{x}_1)]$ and $\mathbf{f}(\mathbf{x}_2) = [\underline{\mathbf{f}}(\mathbf{x}_2), \overline{\mathbf{f}}(\mathbf{x}_2)]$. Their *interval order relations*

9.4 Noisy Multiobjective Optimization

are defined as

$$\mathbf{f}(\mathbf{x}_1) \leq_{IN} \mathbf{f}(\mathbf{x}_2) \Leftrightarrow \underline{\mathbf{f}}(\mathbf{x}_1) \leq \underline{\mathbf{f}}(\mathbf{x}_2) \wedge \bar{\mathbf{f}}(\mathbf{x}_1) \leq \bar{\mathbf{f}}(\mathbf{x}_2), \tag{9.4.3}$$

$$\mathbf{f}(\mathbf{x}_1) \geq_{IN} \mathbf{f}(\mathbf{x}_2) \Leftrightarrow \underline{\mathbf{f}}(\mathbf{x}_1) \geq \underline{\mathbf{f}}(\mathbf{x}_2) \wedge \bar{\mathbf{f}}(\mathbf{x}_1) \geq \bar{\mathbf{f}}(\mathbf{x}_2), \tag{9.4.4}$$

$$\mathbf{f}(\mathbf{x}_1) <_{IN} \mathbf{f}(\mathbf{x}_2) \Leftrightarrow \underline{\mathbf{f}}(\mathbf{x}_1) \leq \underline{\mathbf{f}}(\mathbf{x}_2) \wedge \bar{\mathbf{f}}(\mathbf{x}_1) \leq \bar{\mathbf{f}}(\mathbf{x}_2) \wedge \mathbf{f}(\mathbf{x}_1) \neq \mathbf{f}(\mathbf{x}_2), \tag{9.4.5}$$

$$\mathbf{f}(\mathbf{x}_1) >_{IN} \mathbf{f}(\mathbf{x}_2) \Leftrightarrow \underline{\mathbf{f}}(\mathbf{x}_1) \geq \underline{\mathbf{f}}(\mathbf{x}_2) \wedge \bar{\mathbf{f}}(\mathbf{x}_1) \geq \bar{\mathbf{f}}(\mathbf{x}_2) \wedge \mathbf{f}(\mathbf{x}_1) \neq \mathbf{f}(\mathbf{x}_2), \tag{9.4.6}$$

$$\mathbf{f}(\mathbf{x}_1) \| \mathbf{f}(\mathbf{x}_2) \Leftrightarrow \text{neither } \mathbf{f}(\mathbf{x}_1) \leq_{IN} \mathbf{f}(\mathbf{x}_2) \text{ nor } \mathbf{f}(\mathbf{x}_2) \leq_{IN} \mathbf{f}(\mathbf{x}_1). \tag{9.4.7}$$

In the absence of other factors (e.g., preference for certain objectives, or for a particular region of the trade-off surface), the task of an evolutionary multiobjective optimization (EMO) algorithm is to provide as good an approximation as the true Pareto front. To compare two evolutionary multiobjective optimization algorithms, we need to compare the nondominated sets they produce.

Since an interval is also a set consisting of the components larger than its lower limit and smaller than its upper limit, $\mathbf{x}_1 = (\mathbf{x}, \mathbf{c}_1)$ and $\mathbf{x}_2 = (\mathbf{x}, \mathbf{c}_2)$ can be regarded as two components in sets A and B, respectively. Hence, the decision solutions $\mathbf{x}_1$ and $\mathbf{x}_2$ are not two fixed points but two approximation sets, denoted as A and B, respectively.

Different from Pareto concepts defined by decision vectors in (precise) multiobjective optimization, the Pareto concepts in noisy (or imprecise) multiobjective optimization are called the *Pareto concepts for approximation sets* due to concept definitions for approximation sets.

To evaluate approximations to the true Pareto front, Hansen and Jaszkiewicz [64] define a number of outperformance relations that express the relationship between two sets of internally nondominated objective vectors, A and B. The outperformance is habitually called the dominance.

Definition 9.24 (Dominance for Sets) An approximation set A is said to dominate another approximation set B, denoted by $A \succ B$, if every $\mathbf{x}_2 \in B$ is dominated by at least one $\mathbf{x}_1 \in A$ in all objectives [173], i.e.,

$$A \succ B \Leftrightarrow \exists \mathbf{x}_1 \in A, \mathbf{f}(\mathbf{x}_1) <_{IN} \mathbf{f}(\mathbf{x}_2), \forall \mathbf{x}_2 \in B \text{ (for minimization)}, \tag{9.4.8}$$

$$A \succ B \Leftrightarrow \exists \mathbf{x}_1 \in A, \mathbf{f}(\mathbf{x}_1) >_{IN} \mathbf{f}(\mathbf{x}_2), \forall \mathbf{x}_2 \in B \text{ (for maximization)}, \tag{9.4.9}$$

or written as [64, 89]:

$$A \succ B \Leftrightarrow ND(A \cup B) = A \text{ and } B \setminus ND(A \cup B) \neq \emptyset, \tag{9.4.10}$$

where $ND(S)$ denotes the set of nondominated points in S.

Definition 9.25 (Weak Dominance for Sets) An approximation set A is said to weakly dominate (or weakly outperform) another approximation set B, denoted by

$A \succeq B$, if every $\mathbf{x}_2 \in B$ is weakly dominated by at least one $\mathbf{x}_1 \in A$ in all objectives [173]:

$$A \succeq B \Leftrightarrow \exists \mathbf{x}_1 \in A, \mathbf{f}(\mathbf{x}_1) \leq_{IN} \mathbf{f}(\mathbf{x}_2), \forall \mathbf{x}_2 \in B \text{ (for minimization)}, \quad (9.4.11)$$

$$A \succeq B \Leftrightarrow \exists \mathbf{x}_1 \in A, \mathbf{f}(\mathbf{x}_1) \geq_{IN} \mathbf{f}(\mathbf{x}_2), \forall \mathbf{x}_2 \in B \text{ (for maximization)}, \quad (9.4.12)$$

or written as [64, 89]:

$$A \succeq B \Leftrightarrow ND(A \cup B) = A \quad \text{and} \quad A \neq B. \quad (9.4.13)$$

That is to say, A weakly dominates B if all points in B are "covered" by those in A (here "covered" means is equal to or dominates) and there is at least one point in A that is not contained in B.

Definition 9.26 (Strict Dominance for Sets) An approximation set A is said to strictly dominate (or completely outperform) another approximation set B, denoted by $A \succ\succ B$, if every $\mathbf{x}_2 \in B$ is strictly dominated by at least one $\mathbf{x}_1 \in A$ in all objectives, namely [173]:

$$A \succ\succ B \Leftrightarrow \exists \mathbf{x}_1 \in A, \; f_i(\mathbf{x}_1) <_{IN} f_i(\mathbf{x}_2), \forall \mathbf{x}_2 \in B \text{ (for minimization)}, \quad (9.4.14)$$

$$A \succ\succ B \Leftrightarrow \exists \mathbf{x}_1 \in A, \; f_i(\mathbf{x}_1) >_{IN} f_i(\mathbf{x}_2), \forall \mathbf{x}_2 \in B \text{ (for maximization)}, \quad (9.4.15)$$

for all $i \in \{1, \ldots, m\}$, or written as [64, 89]:

$$A \succ\succ B \Leftrightarrow ND(A \cup B) = A \quad \text{and} \quad B \cap ND(A \cup B) = \emptyset. \quad (9.4.16)$$

Definition 9.27 (Better) An approximation set A is said to be better than another approximation set B, denoted as $A \triangleright B$, and is defined as [173]:

$$A \triangleright B \Leftrightarrow \exists \mathbf{x}_1 \in A, \mathbf{f}(\mathbf{x}_1) \leq_{IN} \mathbf{f}(\mathbf{x}_2), \forall \mathbf{x}_2 \in B \wedge A \neq B \text{ (for minimization)}, \quad (9.4.17)$$

$$A \triangleright B \Leftrightarrow \exists \mathbf{x}_1 \in A, \mathbf{f}(\mathbf{x}_1) \geq_{IN} \mathbf{f}(\mathbf{x}_2), \forall \mathbf{x}_2 \in B \wedge A \neq B \text{ (for maximization)}. \quad (9.4.18)$$

From the above definition of the relation $\triangleright$, one can conclude that $A \succeq B \Rightarrow A \triangleright B \vee A = B$. In other words, if A weakly dominates B, then either A is better than B or they are equal.

Notice that

$$A \succ\succ B \Rightarrow A \succ B \Rightarrow A \triangleright B \Rightarrow A \succeq B. \quad (9.4.19)$$

9.4 Noisy Multiobjective Optimization

Table 9.1 Relation comparison between objective vectors and approximation sets

Relation	Objective vectors		Approximation sets	
Weakly dominates	$\mathbf{x} \succeq \mathbf{x}'$	$\mathbf{x}$ is not worse than $\mathbf{x}'$	$A \succeq B$	Every $\mathbf{f}(\mathbf{x}_2) \in B$ is weakly dominated by at least one $\mathbf{f}(\mathbf{x}_1)$ in A
Dominates	$\mathbf{x} \succ \mathbf{x}'$	$\mathbf{x}$ is not worse than $\mathbf{x}'$ in all objectives and better in at least one objective	$A \succ B$	Every $\mathbf{f}(\mathbf{x}_2) \in B$ is dominated by at least one $\mathbf{f}(\mathbf{x}_1)$ in A
Strictly dominates	$\mathbf{x} \succ\succ \mathbf{x}'$	$\mathbf{x}$ is better than $\mathbf{x}'$ in all objectives	$A \succ\succ B$	Every $\mathbf{f}(\mathbf{x}_2) \in B$ is strictly dominated by at least one $\mathbf{f}(\mathbf{x}_1)$
Better			$A \rhd B$	Every $\mathbf{f}(\mathbf{x}_2) \in B$ is weakly dominated by at least one $\mathbf{f}(\mathbf{x}_1)$ and $A \neq B$
Incomparable	$\mathbf{x} \| \mathbf{x}'$	$\mathbf{x} \not\succeq \mathbf{x}' \wedge \mathbf{x}' \not\succeq \mathbf{x}$	$A \| B$	$A \not\succeq B \wedge B \not\succeq A$

In other words, strict dominance (complete outperformance) is the strongest and weak dominance (weak outperformance) is the weakest among the set dominance relations.

Definition 9.28 (Incomparable for Sets [173]) An approximation set A is said to be incomparable to another approximation set B, denoted by $A \parallel B$, if neither A weakly dominates B nor B weakly dominates A, i.e.,

$$A \parallel B \Leftrightarrow A \not\succeq B \wedge B \not\succeq A. \tag{9.4.20}$$

Table 9.1 compares relation between objective vectors and approximation sets.

For dominance and nondominance, we have the following probability representations.

- A dominates B: the corresponding probabilities are $P(A \succ B) = 1$, $P(A \prec B) = 0$, and $P(A \equiv B) = 0$.
- A is dominated by B: the corresponding probabilities are $P(A \prec B) = 1$, $P(A \succ B) = 0$, and $P(A \equiv B) = 0$.
- A and B are nondominated each other: the corresponding probabilities are $P(A \succ B) = 0$, $P(A \prec B) = 0$, and $P(A \equiv B) = 1$.

The *fitness* of an individual in one population refers to as the direct competition (capability) with some individual(s) from another population. Hence, the fitness plays a crucial role in evaluating individuals in evolutionary process. When considering two fitness values A and B with multiple objectives, under the case without noise, there are three possible outcomes from comparing the two fitness values.

In noisy multiobjective optimization, in addition to the convergence, diversity, and spread of a Pareto front, there are two indicators: hypervolume and imprecision [96]. For convenience, we focus on the multiobjective maximization hereafter.

To determine whether a nondominated set is a Pareto-optimal set, Zitzler et al. [171] suggest three goals that can be identified and measured (see also [89]):

1. The distance of the obtained nondominated front to the Pareto-optimal front should be minimized.
2. A good (in most cases uniform) distribution of the solutions found—in objective space—is desirable.
3. The extent of the obtained nondominated front should be maximized, i.e., for each objective, a wide range of values should be present.

The performance metrics or indicators play an important role in evaluating noisy multiobjective optimization algorithms [53, 89, 173].

Definition 9.29 (Hypervolume for Set [96]) For an approximate Pareto-optimal solution set X of problem (9.4.1), the *hypervolume for set* is defined as

$$H(X) = \left[\underline{H(X)}, \overline{H(X)}\right] = \wedge\left(\bigcup_{\mathbf{x}\in X}\{\mathbf{y} \in \mathbb{R}^n | \mathbf{x} \succ_{IN} \mathbf{y} \succ_{IN} \mathbf{x}_{\text{ref}}\}\right), \quad (9.4.21)$$

where $\mathbf{x}_{\text{ref}}$ is a reference point; $\wedge$ is Lebesgue measure; $\underline{H(X)}$ and $\overline{H(X)}$ are the worst-case and the best-case hypervolume, respectively.

Definition 9.30 (Imprecision for Set) For an approximate Pareto-optimal solution set X of problem (9.4.1), the imprecision of the front corresponding to X is defined as follows [58]:

$$I(X) = \sum_{\mathbf{x}\in X}\sum_{i=1}^{k}\left(\overline{f}_i(\mathbf{x}, \mathbf{c}_i) - \underline{f}_i(\mathbf{x}, \mathbf{c}_i)\right), \quad (9.4.22)$$

where $\mathbf{x}$ is a solution in X, and $f_i(\mathbf{x}, \mathbf{c}_i) = [\underline{f}_i(\mathbf{x}, \mathbf{c}_i), \overline{f}_i(\mathbf{x}, \mathbf{c}_i)]$ is the ith objective function with interval parameters $\mathbf{c}_i = [\underline{\mathbf{c}}_i, \overline{\mathbf{c}}_i], i = 1, \ldots, m$.

The smaller the value of the imprecision, the smaller the uncertainty of the front.

9.4.2 Performance Metrics for Approximation Sets

The performance metrics or indicators play an important role in noisy multiobjective optimization.

9.4 Noisy Multiobjective Optimization

It is usually assumed (e.g., see [2, 14, 53, 75, 76, 89]) that noise has a disruptive influence on the value of each individual in the objective space, namely

$$\tilde{f}_i(\mathbf{x}) = f_i(\mathbf{x}) + N(0, \sigma^2), \qquad (9.4.23)$$

where $N(0, \sigma^2)$ denotes the normal distribution function with zero mean and variance σ^2 representing the level of noise present; and $\tilde{f}_i$ and f_i denotes the ith objective functions with and without the additive noise, respectively. σ^2 is represented as a percentage of $f_i^{\max}$, where $f_i^{\max}$ is the maximum of the ith objective in true Pareto front.

The following are the performance metrics or indicators pertinent to the optimization goals: proximity, diversity, and distribution [53].

1. *Proximity indicator:* The metric of *generational distance* (GD) gives a good indication of the gap between the evolved Pareto front PF_{known} and the true Pareto front PF_{true}, and is defined as

$$\text{GD} = \left(\frac{1}{n_{PF}} \sum_{i=1}^{n_{PF}} d_i^2\right)^{1/2}, \qquad (9.4.24)$$

where n_{PF} is the number of members in PF_{known}, d_i is the Euclidean distance (in objective space) between the member i of PF_{known} and its nearest member of PF_{true}. Intuitively, a low value of GD is desirable because it reflects a small deviation between the evolved and the true Pareto front. However, the metric of GD gives no information about the diversity of the algorithm under evaluation.

2. *Diversity indicator:* To evaluate the diversity of an algorithm, the following modified maximum spread is used as the diversity indicator:

$$\text{MS} = \sqrt{\frac{1}{m} \sum_{i=1}^{m} = \left[\left(\min\{f_i^{\max}, F_i^{\max}\} - \min\{f_i^{\min}, F_i^{\min}\}\right)/(F_i^{\max} - F_i^{\min})\right]^2}, \qquad (9.4.25)$$

where m is the number of objectives, $f_i^{\max}$ and $f_i^{\min}$ are, respectively, the maximum and minimum of the ith objective in PF_{known}; and $F_i^{\max}$ and $F_i^{\min}$ are the maximum and minimum of the ith objective in PF_{true}, respectively. This modified metric takes into account the proximity to FP_{true}.

3. *Distribution indicator:* To evaluate how evenly the nondominated solutions are distributed along the discovered Pareto front, the modified metric of spacing is defined as

$$S = \frac{1}{n_{PF}} \left(\frac{1}{n_{PF}} \sum_{i=1}^{n_{PF}} (d_i - \bar{d})^2\right)^{1/2}, \qquad (9.4.26)$$

where $\bar{d} = \frac{1}{n_{PF}} \sum_{i=1}^{n_{PF}} d_i$ is the average Euclidean distance between all members of PF_{known} and their nearest members of PF_{true}.

9.5 Multiobjective Simulated Annealing

Metropolis algorithm, proposed by Metropolis et al. in 1953 [100], is a simple algorithm that can be used to provide an efficient simulation of a collection of atoms in equilibrium at a given temperature. This algorithm was extended by Hastings in 1970 [65] to the more general case, and hence is also called *Metropolis-Hastings algorithm*. The Metropolis algorithm is the earliest simulated annealing algorithm that is widely used for solving multiobjective combinatorial optimization problems.

9.5.1 Principle of Simulated Annealing

The idea of simulated annealing originates from thermodynamics and metallurgy [153]: when molten iron is cooled slowly enough it tends to solidify in a structure of minimal energy. This annealing process is mimicked by a local search strategy; at the start, almost any move is accepted, which allows one to explore the solution space. Then, the temperature is gradually decreased such that one becomes more and more selective in accepting new solutions. By the end, only improving moves are accepted in practice.

Tools used for generation of efficient solutions in multiobjective combinatorial optimization (MOCO), like single-objective optimization methods, may be classified into one of the following categories [22]:

1. *Exact procedures*;
2. *Specialized heuristic procedures*;
3. *Metaheuristic procedures*.

The main disadvantage of exact algorithms is their high computational complexity and inflexibility.

A *metaheuristic* can be defined as an iterative generation process which guides a subordinate heuristic by combining intelligently different concepts for exploring and exploiting the search space [118]. Metaheuristic is divided into two categories: *single-solution metaheuristic* and *population metaheuristic* [51].

Single-solution metaheuristic considers a single solution (and search trajectory) at a time. Its typical examples include simulated annealing (SA), tabu search (TS), etc. Population metaheuristic evolves concurrently a population of solutions rather than a single solution. Most evolutionary algorithms are based on population metaheuristic.

9.5 Multiobjective Simulated Annealing

There are two basic strategies for heuristics: "divide-and-conquer" and iterative improvement [88].

- The *divide-and-conquer strategy* divides the problem into subproblems of manageable size, then solves the subproblems. The solutions to the subproblems must be patched back together, and the subproblems must be naturally disjoint, while the division made must be appropriate, so that errors made in patching do not offset the gains obtained in applying more powerful methods to the subproblems.
- *Iterative improvement* starts with the system in a known configuration, then a standard rearrangement operation is applied to all parts of the system in turn, until discovering a rearranged configuration that improves the cost function. The rearranged configuration then becomes a new configuration of the system, and the process is continued until no further improvements can be found. Iterative improvement makes a search in this coordinate space for rearrangement steps which lead to downhill. Since this search usually gets stuck in a local but not a global optimum, it is customary to carry out the process several times, starting from different randomly generated configurations, and save the best result.

By metaheuristic procedures, it means that they define only a "skeleton" of the optimization procedure that has to be customized for particular applications. The earliest metaheuristic method is simulated annealing [100]. Other metaheuristic methods include tabu search [52], genetic algorithms [55], and so on.

The goal of multiple-objective metaheuristic procedures is to find a sample of feasible solutions that is a good approximation to the efficient solutions set.

In physically appealing, particles are free to move around at high temperatures, while as the temperature is lowered they are increasingly confined due to the high energy cost of movement. From the point of view of optimization, the energy $E(\mathbf{x})$ of the state $\mathbf{x}$ in physically appealing is regarded as the function to be minimized, and by introducing a parameter T, the computational temperature is lowered throughout the simulation according to an annealing schedule.

Sampling from the equilibrium distribution is usually achieved by Metropolis sampling, which involves making new solutions $\mathbf{x}'$ that are accepted with probability [145]

$$p = \min\left\{1, \exp\left(-\delta E(\mathbf{x}', \mathbf{x}_i)/T\right)\right\}, \qquad (9.5.1)$$

where $\delta E(\mathbf{x}', \mathbf{x}_i)$ is the cost criterion for transition from the current state $\mathbf{x}_i$ to a new state $\mathbf{x}'$, and T is the annealing temperature.

For a single-objective framework, depending on the values of $\delta E(\mathbf{x}', \mathbf{x}_i)$, we have the following different acceptances (or transitions) from the current state (or solution) $\mathbf{x}_i$ to the new state $\mathbf{x}'$ [145]:

- Any move leading to a negative value $\delta E(\mathbf{x}', \mathbf{x}_i) < 0$ is an improving one and always be accepted ($p = 1$).

- If $\delta E(\mathbf{x}', \mathbf{x}_i) > 0$ is an positive value, then the probability to accept $\mathbf{x}'$ as a new current solution $\mathbf{x}_{i+1}$ is given by $p = \exp\left(-\delta E(\mathbf{x}', \mathbf{x}_i)/T\right)$. Clearly, the higher the difference δE, the lower the probability to accept $\mathbf{x}'$ instead of $\mathbf{x}_i$.
- $\delta E(\mathbf{x}', \mathbf{x}_i) = 0$ implies that the new state $\mathbf{x}'$ is the same level of value as the current state $\mathbf{x}$, there may exist two schemes—move to the new state $\mathbf{x}'$ or stay in the current state $\mathbf{x}_i$. The analysis of this problem shows that the move scheme is better than the stay one. In the stay scheme, search will end on both edges of the Pareto frontier not entering the middle of the frontier. However, if the move scheme is used, then search will be continue into the middle part of the frontier, move freely between nondominated states like a random walk when the temperature is low and eventually will be distributed uniformly over the Pareto frontier as time goes to infinity [114]. This result is in accordance with $p = 1$ given by (9.5.1).

The Boltzmann probability factor is defined as $P(E(\mathbf{x})) = \exp(-E(\mathbf{x})/k_B T)$, where $E(\mathbf{x})$ is the energy of the solution $\mathbf{x}$, and k_B is Boltzmann's constant.

At each T the simulated appealing algorithm aims at drawing samples from the equilibrium distribution $P(E(\mathbf{x})) = \exp(-E(\mathbf{x})/T)$. As $T \to 0$, any sample from $P(E(\mathbf{x}))$ will almost surely lie at the minimum of E.

The simulated annealing consists of two processes: first "melting" the system being optimized at a high effective temperature, then lowering the temperature by slow stages until the system "freezes" and no further changes occur. At each temperature, the simulation must proceed long enough so that the system reaches a steady state. The sequence of temperatures and the number of rearrangements of $\{\mathbf{x}_i\}$ attempted to reach equilibrium at each temperature can be thought of as an annealing schedule.

Intuitively, in addition to perturbations which decrease the energy, when T is high, perturbations from $\mathbf{x}$ to $\mathbf{x}'$ which increase the energy are likely to be accepted and the samples can explore the state space. However, as T is reduced, only perturbations leading to small increases in E are accepted, so that only limited exploration is possible as the system settles on (hopefully) the global minimum.

Let $E(\mathbf{x}_i)$ and $E(\mathbf{x}')$ be the cost criterions of the current solution (or state) $\mathbf{x}_i$ and the new solution $\mathbf{x}'$, respectively. The implementation of simulated annealing requires three different kinds of options [153].

1. *Decision Rule:* In this rule, when moving from a current solution (or state) $\mathbf{x}_i$ to the new solution $\mathbf{x}'$, the cost criterion is defined as

$$\delta E(\mathbf{x}', \mathbf{x}_i) = E(\mathbf{x}') - E(\mathbf{x}_i). \tag{9.5.2}$$

2. *Neighborhood* $V(\mathbf{x})$ is defined as a set of feasible solution close to $\mathbf{x}$ such that any solution satisfying the constraints of MOCO problems, $D = \{\mathbf{x} : \mathbf{x} \in LD, \mathbf{x} \in B^n\}$ and $LD = \{\mathbf{x} : \mathbf{Ax} = \mathbf{b}\}$, can be obtained after a finite number of moves.

3. *Typical parameters:* Some simulated annealing-typical parameters must be fixed as follows:

9.5 Multiobjective Simulated Annealing

- The initial temperature parameter T_0 or alternatively an initial acceptance probability p_0;
- The cooling factor α ($\alpha < 1$) and the temperature length step N_{step} in the cooling schedule.
- The stopping rule(s): final temperature T_{stop} and/or the maximum number of iterations without improvement N_{stop}.

Algorithm 9.4 shows a single-objective simulated annealing algorithm.

Algorithm 9.4 Single-objective simulated annealing [153]
---
1. **input:** initial temperature parameter T_0, cooling factor α, temperature length step N_{step}, final temperature T_{stop} and the maximum number of iterations N_{stop}.
2. **initialization:** draw at random an initial solution $\mathbf{x}_0$.
   2.1 Evaluate $z(\mathbf{x}_0) = f(\mathbf{x}_0)$.
   2.2 $X = \{\mathbf{x}_0\}$.
   2.3 $N_{count} = n = 0$.
3. **while** $n = 1, \ldots, N_{stop}$
4.   Draw at random a solution $\mathbf{y}$ in the neighborhood $V(\mathbf{x}_n)$ of $\mathbf{x}_n$.
5.   Evaluate $z(\mathbf{x}_n) = f(\mathbf{x}_n)$.
6.   Compute $\delta z = z(\mathbf{x}') - z(\mathbf{x}_n)$.
7.   The acceptance probability $p = \exp\left(-\frac{\delta z}{T_n}\right)$.
8.   If $\delta z \leq 0$ then we accept the new solution: $\mathbf{x}_{n+1} \leftarrow \mathbf{x}'$.
9.   If $\delta z > 0$ then we accept the new solution with a certain probability
$$\mathbf{x}_{n+1} \begin{cases} \xleftarrow{p} \mathbf{x}', \\ \xleftarrow{1-p} \mathbf{x}_n. \end{cases}$$
10. $n \leftarrow n + 1$.
11.   Update the temperature T_n:
$$T_n = \begin{cases} \alpha T_{n-1}, & \text{if } n \ (\text{mod } N_{stop}) = 0, \\ T_{n-1}, & \text{otherwise.} \end{cases}$$
12.   If $n = N_{stop}$ or $T < T_{stop}$ then exist.
13.   If $n < N_{stop}$ or $T > T_{stop}$ then goto step 4.
14. **end while**
15. **output:** the optimal solution $\mathbf{x}_n$.

Simulated annealing has the following two important features [88]:

1. Annealing differs from iterative improvement in that the procedure need not get stuck since transitions out of a local optimum are always possible at nonzero temperature.
2. Annealing is an adaptive form of the divide-and-conquer approach. Gross features of the eventual state of the system appear at higher temperatures, while fine details develop at lower temperatures.

9.5.2 Multiobjective Simulated Annealing Algorithm

Common ideas behind simulated annealing are [88, 91]:

- the concept of neighborhood;
- acceptance of new solutions with some probability;
- dependence of the probability on a parameter called the temperature; and
- the scheme of the temperature changes.

As comparison with single-objective simulated annealing, population-based simulated annealing (PSA) uses also the following ideas:

1. Apply the concept of a sample (population) of interacting solutions from genetic algorithms [55] at each iteration in simulated annealing. The solutions are called generate solutions.
2. In order to assure dispersion of the generated solutions over the whole set of efficient solutions, one must control the objective weights used in the multiple objective rules for acceptance probability in order to increase or decrease the probability of improving values of the particular objectives.

Different from the single-objective framework, in the multiple-objective framework, when comparing a new solution $\mathbf{x}'$ with the current solution $\mathbf{x}_i$ according to K criteria $z_k(\mathbf{x}', \mathbf{x}_i), k = 1, \ldots, K$, three cases can obviously occur [153].

- If $z_k(\mathbf{x}') \leq z_k(\mathbf{x}_n), \forall k = 1, \ldots, K$ then the move from $\mathbf{x}_i$ to $\mathbf{x}'$ is an improvement with respect to all the objective $\delta z_k(\mathbf{x}', \mathbf{x}_n) = z_k(\mathbf{x}') - z_k(\mathbf{x}_n) \leq 0, \forall k = 1, \ldots, K$. Therefore, $\mathbf{x}'$ is always accepted ($p = 1$).
- An improvement and a deterioration can be simultaneously observed on different cost criteria $\delta z_k < 0$ and $\delta z_{k'} > 0, \exists k \neq k' \in \{1, \ldots, K\}$. The first crucial point is how to define the acceptance probability p.
- When all cost criteria are deteriorated: $\forall k, \delta z_k \geq 0$ at least one strict inequality, a probability p to accept $\mathbf{x}'$ instead of $\mathbf{x}_i$ must be calculated. Let

$$\mathbf{z}(\mathbf{x}') = [z_1(\mathbf{x}'), \ldots, z_K(\mathbf{x}')]^T \quad \text{and} \quad \mathbf{z}(\mathbf{x}_i) = [z_1(\mathbf{x}_i), \ldots, z_K(\mathbf{x}_i)]^T. \tag{9.5.3}$$

For the computation of the probability p, the second crucial point is how to compute the "distance" between $\mathbf{z}(\mathbf{x}')$ and $\mathbf{z}(\mathbf{x}_i)$.

To overcome the above two difficulties, Ulungu et al. [153] proposed a *criterion scalarizing approach* in 1999.

Define the probability from $\mathbf{x}_i$ to $\mathbf{x}'$ with respect to the kth criterion as follows:

$$\pi_k = \begin{cases} \exp\left(-\frac{\delta z_k}{T_i}\right), & \delta z_k > 0, \\ 1, & \delta z_k \leq 0. \end{cases} \tag{9.5.4}$$

9.5 Multiobjective Simulated Annealing

Then, the global acceptance probability p can be determined by

$$p = t(\Pi, \lambda) = \{\pi_1, \ldots, \pi_K\}, \quad \Pi = (\pi_1, \ldots, \pi_K), \tag{9.5.5}$$

where $\Pi = (\pi_1, \ldots, \pi_K)$ is an aggregation of K criteria π_k.

The most intuitive criterion scalarizing functions $t(\Pi, \lambda)$ are the product function

$$t(\Pi, \lambda) = \prod_{k=1}^{K} (\pi_k)^{\lambda_k} \tag{9.5.6}$$

and the minimum function

$$t_{\min}(\Pi, \lambda) = \min_{k=1,\ldots,K} (\pi_k)^{\lambda_k}. \tag{9.5.7}$$

In order to compute the distance between two solutions for each individual and then to evaluate the acceptance probability of the move from $\mathbf{x}_i$ to $\mathbf{x}'$ with respect to this particular criterion, an efficient criterion scalarizing approach is to choose the acceptance probability given by

$$p = \begin{cases} 1, & \text{if } \delta s \leq 0; \\ \exp\left(-\frac{\delta s}{T_i}\right), & \text{if } \delta s > 0. \end{cases} \tag{9.5.8}$$

Here $\delta s = s(\mathbf{z}(\mathbf{x}', \lambda)) - s(\mathbf{z}(\mathbf{x}_i, \lambda))$ may take one of the following two forms:

- *Weighed sum:*

$$s(\mathbf{z}(\mathbf{x}, \lambda)) = \sum_{k=1}^{K} \lambda_k z_k(\mathbf{x}), \tag{9.5.9}$$

where

$$\sum_{k=1}^{K} \lambda_k = 1, \quad \lambda_k > 0, \forall k = 1, \ldots, K. \tag{9.5.10}$$

An equivalent alternative is given by Engrand [39]

$$s(\mathbf{z}(\mathbf{x}, \lambda)) = \sum_{k=1}^{K} \log z_k(\mathbf{x}). \tag{9.5.11}$$

- *Weighed Chebyshev norm L_∞:*

$$s(\mathbf{z}(\mathbf{x}, \lambda)) = \max_{1 \leq k \leq K} \{\lambda_k |\tilde{z}_k - z_k|\}, \quad \lambda_k > 0, \forall k = 1, \ldots, K, \tag{9.5.12}$$

where $\tilde{z}_k$ are the cost (criterion) values of the ideal point $\tilde{z}_k = \min_{\mathbf{x} \in D} z_k(\mathbf{x})$.

It is easily shown [153] that the global acceptance probability p in (9.5.8) is just a most intuitive scalarizing function, i.e., $p = t(\Pi, \boldsymbol{\lambda})$.

Due to conflicting multiple objectives, a good algorithm for solving multiobjective optimization problems needs to have the following four important properties [114].

- *Searching precision:* Because of problem complexity of multiobjective optimization, the algorithm finds hardly the Pareto-optimal solutions, and hence it must find the possible near solutions to the optimal solutions set.
- *Searching-time:* The algorithm must find the optimal set efficiently during searching-time.
- *Uniform probability distribution over the optimal set:* The found solutions must be widely spread, or uniformly distributed over the real Pareto-optimal set rather than converging to one point because every solution is important in multiobjective optimization.
- *Information about Pareto frontier:* The algorithm must give as much information as possible about the Pareto frontier.

A multiobjective simulated annealing algorithm is shown in Algorithm 9.5.

It is recognized (see, e.g., [114]) that simulated annealing for multiobjective optimization has the following properties.

1. By using the concepts of Pareto optimality and domination, high searching precision can be achieved by simulated annealing.
2. The main drawback of simulated annealing is searching-time, as it is generally known that simulated annealing takes long time to find the optimum.
3. An interesting advantage of simulated annealing is its uniform probability distribution property as it is mathematically proved [50, 104] that it can find each of the global optima with the same probability in a scalar finite-state problem.
4. For multiobjective optimization, as all the Pareto solutions have different cost vectors that have a trade-off relationship, a decision maker must select a proper solution from the found Pareto solution set or sometimes by interpolating the found solutions.

Therefore, in order to apply simulated annealing in multiobjective optimization, one must reduce searching-time and effectively search Pareto optimal solutions. It should be noted that any solution in the set PE of Pareto-optimal solutions should not be dominated by other(s), otherwise any dominated solution should be removed from PE. In this sense, Pareto solution set should be the best nondominated set.

9.5 Multiobjective Simulated Annealing

Algorithm 9.5 Multiobjective simulated annealing [153]

1. **input:** initial temperature parameter T_0, cooling factor α, temperature length step N_{step}, final temperature T_{stop} and the maximum number of iterations N_{stop}.
2. **initialization:** draw at random an initial solution $\mathbf{x}_0$.
   2.1 Evaluate $z_k(\mathbf{x}_0) = f_k(\mathbf{x}_0), \forall k = 1, \ldots, K$.
   2.2 The set of potentially optimal solutions PE $= \{\mathbf{x}_0\}$.
   2.3 $N_{\text{count}} = n = 0$.
3. **while** $n = 1, \ldots, N_{\text{stop}}$
4. Draw at random a solution $\mathbf{y}$ in the neighborhood $V(\mathbf{x}_n)$ of $\mathbf{x}_n$.
5. Evaluate $z_k(\mathbf{x}_n) = f_k(\mathbf{x}_n), \forall k = 1, \ldots, K$.
6. Compute $\delta s = \sum_{k=1}^{K} \lambda_k \left(z_k(\mathbf{x}') - z_k(\mathbf{x}_n) \right)$.
7. The acceptance probability $p = \exp\left(-\frac{\delta s}{T_n}\right)$.
8. If $\delta s \leq 0$ then we accept the new solution: $\mathbf{x}_{n+1} \leftarrow \mathbf{x}'$.
9. If $\delta s > 0$ then we accept the new solution with a certain probability
$$\mathbf{x}_{n+1} \begin{cases} \xleftarrow{p} \mathbf{x}', \\ \xleftarrow{1-p} \mathbf{x}_n. \end{cases}$$
10. Remove all the solutions dominated by $\mathbf{x}_{n+1}$ from PE. Add $\mathbf{x}_{n+1}$ to PE if no solutions in PE dominate $\mathbf{x}_{n+1}$.
11. $n \leftarrow n + 1$.
12. Update the temperature T_n:
$$T_n = \begin{cases} \alpha T_{n-1}, & \text{if } n \ (\text{mod } N_{\text{stop}}) = 0, \\ T_{n-1}, & \text{otherwise.} \end{cases}$$
13. If $n = N_{\text{stop}}$ or $T < T_{\text{stop}}$ then exist.
14. If $n < N_{\text{stop}}$ or $T > T_{\text{stop}}$ then goto step 4.
15. **end while**
16. **output:** the Pareto-optimal solutions PE $= \{\mathbf{x}_i \in D\}$.

9.5.3 Archived Multiobjective Simulated Annealing

In multiobjective simulated annealing (MOSA) algorithm developed by Smith et al. [145], the acceptance of a new solution $\mathbf{x}$ is determined by its energy function. If the true Pareto front is available, then the energy of a particular solution $\mathbf{x}$ is calculated as the total energy of all solutions that dominates $\mathbf{x}$. In practical applications, as the true Pareto front is not available all the time, one must estimate first the Pareto front F' which is the set of mutually nondominating solutions found thus far in the process. Then, the energy of the current solution $\mathbf{x}$ is the total energy of nondominating solutions. These nondominating solutions are called the *archival nondominating solutions* or nondominating solutions in Archive.

In order to estimate the energy of the Pareto front F', the number of archival nondominated solutions in the Pareto front should be taken into consideration in MOSA. But, this is not done in MOSA.

By incorporating the nondominating solutions in the Archive, one can determine the acceptance of a new solution. Such an MOSA is known as *archived multiobjective simulated annealing (AMOSA)*, which was proposed by Bandyopadhyay et al. in 2008 [7], see Algorithm 9.6.

Algorithm 9.6 Archived multiobjective simulated annealing (AMOSA) algorithm [7]

1. **input:** $T_{\max}, T_{\min}, HL, SL, iter, \alpha,$ temp $= T_{\max}$.
2. **initialization:**
   2.1 Draw at random an initial Archive.
   2.2 $curren\ pt = \text{random}(Archive)$.
3. **while** $temp > T_{\min}$
4.   **for** $(i = 0; i < \text{iter}, i++)$
5. $new\text{-}pt = pertub(current\text{-}pt)$.
6.   Check the domination status of *new-pt* and *current-pt*.
7.   **if** (*current-pt* dominates *new-pt*) **then**
8. $\Delta dom_{\text{avg}} = \frac{1}{k+1}\left(\left(\sum_{i=1}^{k}\Delta dom_{i,new-pt}\right) + \Delta dom_{current,new-pt}\right)$.

   % $k \geq 0$ is total number of points in the Archive which dominates new-pt.
9. $prob = \frac{1}{1+\exp(\Delta dom_{\text{avg}}\cdot temp)}$.
10.   **if** (*current-pt* and *new-pt* are nondominated each other) **then**
11.   Check the domination status of *new-pt* and *current-pt*.
12.   **if** (*new-pt* is dominated by k ($k \geq 1$) points in the *Archive*) **then**
13. $prob = \frac{1}{1+\exp(\Delta dom_{\text{avg}}\cdot temp)}$.
14. $\Delta dom_{\text{avg}} = \frac{1}{k}\left(\sum_{i=1}^{k}\Delta dom_{i,new-pt}\right)$.
15.   Set *new-pt* as *current-pt* with probability $= prob$.
16.   **if** (*new-pt* is nondominated w.r.t all the points in the *Archive*) **then**
17.   Set *new-pt* as *current-pt* and add it to the *Archive*.
18.   **if** $Archive\ size > SL$ **then**
19.   Cluster *Archive* into HL number of clusters.
20.   **if** (*new-pt* dominates k ($k \geq 1$) points in the *Archive*) **then**
21.   Set *new-pt* as *current-pt* and add it to the *Archive*.
22.   Remove all the k dominated points from the *Archive*.
23.   **if** (*new-pt* dominates *current-pt*) **then**
24.   Check the domination status of *new-pt* and points in the *Archive*.
25.   **if** (*new-pt* is dominated by k ($k \geq 1$) points in the *Archive*) **then**
26. $\Delta dom_{\min}$ = minimum of the difference of domination amount between the *new-pt* and the k points.
27. $prob = \frac{1}{1+\exp(-\Delta_{\min})}$.
28.   Set point in the *Archive* which corresponds to $\Delta dom_{\min}$ as *current-pt* with *prob*,
29.   **else** set *new-pt* as *current-pt*.
30.   **if** (*new-pt* is nondominated w.r.t all the points in the *Archive*) **then**
31.   Set *new-pt* as *current-pt* and add it to the *Archive*.
32.   **if** *current-pt* is in the *Archive*, remove it from the *Archive*.
33.   **else if** $Archive\ size > SL$, then cluster *Archive* into HL number of clusters.
34.   **if** (*new-pt* dominates k other points in the *Archive*) **then**
35.   Set *new-pt* as *current-pt* and add it to the *Archive*.
36.   Remove all the k dominated points from the *Archive*.
37.   **end for**
38. $temp = \alpha \cdot temp$.
39. **end while**
40. **if** $Archive\text{-}size > SL$ **then**
41.   Cluster *Archive* into HL number of clusters.
42. **end if**

The following parameters need to be set a priori in Algorithm 9.6.

- HL: The maximum size of the Archive on termination. This set is equal to the maximum number of nondominated solutions required by the user;
- SL: The maximum size to which the Archive may be filled before clustering is used to reduce its size to HL;
- $T_{\max}$: Maximum (initial) temperature, $T_{\min}$: Minimal (final) temperature;
- iter: Number of iterations at each temperature;
- α: The cooling rate in SA.

Let $\bar{\mathbf{x}}^* = [\bar{x}_1^*, \ldots, \bar{x}_n^*]^T$ denote the decision variable vector that simultaneously maximizes the objective values $\{f_1(\bar{\mathbf{x}}), \ldots, f_K(\bar{\mathbf{x}})\}$, while satisfying the constraints, if any. In maximization problems, a solution $\bar{\mathbf{x}}_i$ is said to dominate $\bar{\mathbf{x}}_j$ if and only if

$$f_k(\bar{\mathbf{x}}_i) \geq f_k(\bar{\mathbf{x}}_j), \quad \forall k = 1, \ldots, m, \tag{9.5.13}$$

$$f_k(\bar{\mathbf{x}}_i) = f_k(\bar{\mathbf{x}}_j), \quad \text{for some } k \in \{1, \ldots, m\}. \tag{9.5.14}$$

Among a set of solutions P, the nondominated set P' of solutions are those that are not dominated by any member of the set P. The nondominated set of the entire search space S is the globally Pareto-optimal set.

At a given temperature T, a new state s is selected with a probability

$$P_{qs} = \frac{1}{1 + \exp\left(\frac{-(E(q,T)-E(s,T))}{T}\right)}, \tag{9.5.15}$$

where q is the current state, whereas $E(q, T)$ and $E(s, T)$ are the corresponding energy values of q and s, respectively.

One of the points, called *current-pt*, is randomly selected from Archive as the initial solution at temperature temp $= T_{\max}$. The current-pt is perturbed to generate a new solution called *new-pt*. The domination status of new-pt is checked with respect to the current-pt and solutions in Archive.

Depending on the domination status between *current-pt* and *new-pt*, the *new-pt* is selected as the *current-pt* with different probabilities. This process constitutes the core of Algorithm 9.6, see Step 4 to Step 37.

9.6 Genetic Algorithm

A genetic algorithm (GA) is a metaheuristic algorithm inspired by the process of natural selection. Genetic algorithms are commonly used to generate high-quality solutions to optimization and search problems by relying on bio-inspired operators such as mutation, crossover, and selection.

In a genetic algorithm, a population of candidate solutions (called individuals, creatures, or phenotypes) to an optimization problem is evolved toward better solutions. Each candidate solution has a set of properties (its chromosomes or genotype) which can be mutated and altered; traditionally, solutions are represented in binary as strings of 0s and 1s, but other encodings are also possible.

The evolution is an iterative process: (a) It usually starts from a population of randomly generated individuals. The population in each iteration is known as a generation. (b) In each generation, the fitness of every individual in the population is evaluated. The fitness is usually the value of the objective function in the optimization problem being solved. (c) The fitter individuals are stochastically selected from the current population, and each individual's genome is modified (recombined and possibly randomly mutated) to form a new generation. (d) This new generation of candidate solutions is then used in the next iteration of the algorithm. (e) The algorithm terminates when either a maximum number of generations has been produced, or a satisfactory fitness level has been reached for the population.

9.6.1 Basic Genetic Algorithm Operations

The genetic algorithm consists of encoded chromosome, fitness function, reproduction, crossover, and mutation operations generally.

A typical genetic algorithm requires three important concepts:

- a genetic representation of the solution domain,
- a fitness function to evaluate the solution domain, and
- a notion of population. Unlike traditional search methods, genetic algorithms rely on a population of candidate solutions.

A genetic algorithm (GA) is a search technique used for finding true or approximate solutions to optimization and search problems. GAs are a particular class of evolutionary algorithms (EAs) that use inheritance, mutation, selection, and crossover (also called recombination) inspired by evolutionary biology.

GAs encode the solutions (or decision variables) of a search problem into finite-length strings of alphabets of certain cardinality. Candidate solutions are called *individuals*, creatures, or phenotypes. An abstract representation of individuals is called *chromosomes*, the genotype or the genome, the alphabets are known as *genes* and the values of genes are referred to as *alleles*. In contrast to traditional optimization techniques, GAs work with coding of parameters, rather than the parameters themselves.

One can think of a population of individuals as one "searcher" sent into the optimization phase space. Each searcher is defined by his genes, namely his position inside the phase space is coded in his genes. Every searcher has the duty to find a value of the quality of his position in the phase space.

To evolve good solutions and to implement natural selection, a measure is necessary for distinguishing good solutions from bad solutions. This measure is called fitness.

9.6 Genetic Algorithm

Once the genetic representation and the fitness function are defined, at the beginning of a run of a genetic algorithm a large population of random chromosomes is created. Each decoded chromosome will represent a different solution to the problem at hand.

When two organisms mate they share their genes. The resultant offspring may end up having half the genes from one parent and half from the other. This process is called *recombination*. Very occasionally a gene may be mutated.

Genetic algorithms are a way of solving problems by mimicking the same processes mother nature uses. They use the same combination of selection, recombination (i.e., crossover), and mutation to evolve a solution to a problem.

Chromosome Code

Each chromosome is made up of a sequence of genes from certain alphabet which can consist of binary digits (0 and 1), floating-point numbers, integers, symbols (i.e., A, B, C, D), etc.

In binary gene representation, a possible solution needs to be encoded as a string of bits (i.e., a chromosome), and these bits need to represent all the different characters available to the solution. For example, four bits are required to represent the range of characters used:

```
0 :    0000
1 :    0001
2 :    0010
3 :    0011
4 :    0100
5 :    0101
6 :    0110
7 :    0111
8 :    1000
9 :    1001
+ :    1010
− :    1011
* :    1100
/ :    1101
```

This shows all the different genes required to encode the problem as described. The possible genes 1110, 1111 will remain unused and will be ignored by the algorithm if encountered.

A string of four bits "7+6*4/2+1" would be represented by nine genes as follows:

```
0111 1010 0110 1100 0100 1101 0010 1010 0001
 7    +    6    *    4    /    2    +    1
```

These genes are all strung together to form the chromosome:

0111 1010 0110 1100 0100 1101 0010 1010 0001

Because the algorithm deals with random arrangements of bits it is often going to come across a string of bits like this:

0010 0010 1010 1110 1011 0111 0010

Decoded, these bits represent:

0010 0010 1010 1110 1011 0111 0010
2   2   +   n/a   −   7   2

Fitness Selection
To evolve good solutions and to implement natural selection, we need a measure for distinguishing good solutions from bad solutions. In essence, the fitness measure must determine a candidate solution's relative fitness, which will subsequently be used by the GA to guide the evolution of good solutions. This is one of the characteristics of the GA: it only uses the fitness of individuals to get the relevant information for the next search step.

To assign fitness, Hajela and Lin [61] proposed a *weighed-sum method*: Each objective is assigned a weight $w_i \in (0, 1)$ such that $\sum_i w_i = 1$, and the scalar fitness value is then calculated by summing up the weighted objective values $w_i f_i(\mathbf{x})$. To search multiple solutions in parallel, the weights are not fixed but coded in the genotype so that the diversity of the weight combinations is promoted by phenotypic fitness sharing [61, 169]. As a consequence, the *fitness assignment* evolves the solutions and weight combinations simultaneously.

Reproduction [164]
Individuals are selected through a fitness-based process, where fitter individuals with lower inferior value are typically more likely to be selected. On the one hand, excellent individuals must be reserved to avoid too random search and low efficiency. On the other hand, these high-fitness individuals are not expected to over breed, avoiding the premature convergence of the algorithm. For simplicity and efficiency, an elitist selection strategy is employed. First, the individuals in the population $p(t)$ are sorted from small to large according to the inferior value. Then the former 1/10 of the sorted individuals are directly selected into the crossover pool. The remaining 9/10 of individuals are gained by random competitions of all individuals in the population.

Crossover and Mutation
Crossover is a unique feature of the originality of GA in evolutionary algorithms. Genetic crossover is a process of genetic recombination that mimics sexual reproduction in nature. Its role is to inherit the original good genes to the next generation

9.6 Genetic Algorithm

of individuals and to generate new individuals with more complex genetic structures and higher fitness value.

The improved *crossover probability* P_{ci} of the ith individual is tuned adaptively as

$$P_{ci} = \begin{cases} P_c \times r_i/r_{\max}, & r_i < r_{\text{avg}}, \\ P_c, & r_i \geq r_{\text{avg}}, \end{cases} \quad (9.6.1)$$

where P_{ci} is the crossover probability of the GA, r_i is the inferior value of the individual, $r_{\max}$ is the maximum inferior value, and r_{avg} is the average inferior value of the population.

By the *crossover rate*, it is simply the chance that two chromosomes will swap their bits. A good value for this is around 0.7. Crossover is performed by selecting a random gene along the length of the chromosomes and swapping all the genes after that point. For example, given two chromosomes

1000 1001 1100 0010 1010

0101 0001 1010 0001 0111

If choosing a random bit along the length, say at position 9, and swap all the bits after that point, then we have

1000 1001 1010 0001 0111

0101 0001 1100 0010 1010

It is seen from Eq. (9.6.1) for crossover probability that when the individual fitness value of the participating crossover is lower than the average fitness value of the population, i.e., the individual is a poor performing individual, a large crossover probability is adopted for it. When the individual fitness value of the crossover is higher than the average fitness value, i.e., the individual has excellent performance, then a small crossover probability is used to reduce the damage of the crossover operator to the better performing individual.

When the fitness values of the offspring generated by the crossover operation are no longer better than their parents, but the global optimal solution is not reached, the GA algorithm will occur premature convergence. At this time, the introduction of the mutation operator in the GA tends to produce good results.

Mutation in the GA simulates the mutation of a certain gene on the chromosome in the evolution of natural organisms in order to change the structure and physical properties of the chromosome.

On one hand, the mutation operator can restore the lost genetic information during population evolution to maintain individual differences in the population and prevent premature convergence. On the other hand, when the population size is large, for one to introduce moderate mutation after crossover operation, one can also improve the local search efficiency of the GA algorithm, thereby increasing the diversity of the population to reach the global domain of the search.

The improved *mutation probability* P_{mi} of the ith individual is tuned adaptively as

$$P_{mi} = \begin{cases} P_m \times r_i/r_{\max}, & r_i < r_{\text{avg}}; \\ P_m, & r_i \geq r_{\text{avg}}. \end{cases} \qquad (9.6.2)$$

The mutation probability P_m of the population is adaptively changed according to the diversity of the current population as

$$P_m = P_{m0}[1 - 2(r_{\max} - r_{\min})/3(\text{PopSize} - 1)], \qquad (9.6.3)$$

where $r_{\max}$ and $r_{\min}$ are, respectively, the maximum and minimum inferior value of the individual, PopSize is the number of individuals, and P_{m0} is the mutation probability of the program.

Mutation rate is the chance that one or more genes of a selected chromosome will be flipped (0 becomes 1, 1 becomes 0). Mutation probability is usually achieved using an adaptive selection:

$$p_m = \begin{cases} k_2 \cdot \frac{f_{\max} - f}{f_{\max} - \bar{f}}, & \text{if } f > \bar{f}; \\ k_4, & \text{if } f \leq \bar{f}. \end{cases} \qquad (9.6.4)$$

Here k_2 and k_4 are equal to 0.5, and f is the fitness of the chromosome under mutation.

It can be seen from the expressions of P_{ci} in (9.6.1) and P_{mi} in (9.6.2) that

- When the individual fitness value is lower than the average fitness value of the population, namely the individual has a poor performance, one adopts a large crossover probability and mutation probability.
- If the individual fitness value is higher than the average fitness value, i.e., the individual's performance is excellent, then the corresponding crossover probability and mutation probability are adaptively taken according to the fitness value.
- When the individual fitness value is closer to the maximum fitness value, the crossover probability and the mutation probability are smaller.
- If the individual fitness is equal to the maximum fitness value, then the crossover probability and the mutation probability are zero.

Therefore, individuals with optimal fitness values will remain intact in the next generation. However, this will allow individuals with the greatest fitness to rapidly multiply in the population, leading to premature convergence of the algorithm. Hence, this adjustment method is applicable to the population in the late stage of evolution, but it is not conducive to the early stage of evolution, because the dominant individuals in the early stage of evolution are almost in a state of no change, and the dominant individuals at this time are not necessarily optimized. The global optimal solution to the problem increases the likelihood of evolution to a locally optimal solution. To overcome this problem, we can use the default crossover probability $p_c = 0.8$ and the default mutation probability $p_m = 0.001$ for each individual, so that the crossover probability and mutation probability of individuals with good performance in the population are no longer zero.

9.6.2 Genetic Algorithm with Gene Rearrangement Clustering

Genetic clustering algorithms have a certain degree of *degeneracy* in the evolution process. The degeneracy problem is mainly caused by the non-one-to-one correspondence between the solution space of the clustering problem and the genetic individual coding in the evolution process [127].

The degeneracy of the evolutionary process usually causes the algorithm to search the local solution space repeatedly. Therefore, the search efficiency of the genetic clustering algorithm can be improved by avoiding the degeneracy of the evolution process.

To illustrate how the degeneracy occurs, consider a clustering problem in which a set of patterns is grouped into K clusters so that patterns in the same cluster are similar and differentiate from those of other clusters in some sense.

Denote two chromosomes as

$$\mathbf{x} = [x_{11}, \ldots, x_{1N}, \ldots, x_{K1}, \ldots, x_{KN}] = [\mathbf{x}_1, \ldots, \mathbf{x}_K], \quad (9.6.5)$$

$$\mathbf{y} = [y_{11}, \ldots, y_{1N}, \ldots, y_{K1}, \ldots, y_{KN}] = [\mathbf{y}_1, \ldots, \mathbf{y}_K], \quad (9.6.6)$$

with $\mathbf{x}_i = [x_{i1}, \ldots, x_{iN}] \in \mathbb{R}^{1 \times N}$, $\mathbf{y}_i = [y_{i1}, \ldots, y_{iN}] \in \mathbb{R}^{1 \times N}$, and $\mathbf{x}$ is called a referenced vector. Let $P_1 = \{1, \ldots, K\}$ be a set of indexes of $\{\mathbf{x}_1, \ldots, \mathbf{x}_K\}$ and $P_2 = \emptyset$. Consider the rearrangement of P_1 to P_2:

    for $i = 1$ to K **do**
        $k = \arg\min_{j \in P_1, j \notin P_2} \|\mathbf{y}_i - \mathbf{x}_j\|^2,$
        $P_1 = P_1 \backslash k$ and $P_2(i) = k$.
    end for

Then, the elements of P_2 will be a permutation of that of P_1, and $\mathbf{y}$ will be rearranged according to P_2, i.e.,

$$\mathbf{y}'_k = \mathbf{y}_{P_2(k)}. \quad (9.6.7)$$

Definition 9.31 (Gene Rearrangement [16]) If $\mathbf{x}$ and $\mathbf{y}$ are two chromosomes in gene representation, and the elements of $\mathbf{y}$ are rearranged according to the indexes in P_2 to get a new vector whose elements are given by

$$\tilde{y}_i = y_{P_2(i)}, \quad i = 1, \ldots, N, \tag{9.6.8}$$

then the new chromosome $\tilde{\mathbf{y}} = [\tilde{y}_i]_{i=1}^N = [y_{P_2(i)}]_{i=1}^N$ is called the gene rearrangement of the original chromosome $\mathbf{y}$.

The *genetic algorithm with gene rearrangement* (GAGR) clustering algorithm is given in Algorithm 9.7.

Algorithm 9.7 Genetic algorithm with gene rearrangement (GAGR) clustering algorithm [16]

1. **initialization:**
   1.1 Generate a group of cluster centers with size NP.
   1.2 Consider only valid chromosomes (that have at least one data point in each cluster).
   1.3 Each data point of the set is assigned to the cluster with closest cluster center using the Euclidean distance.
2. Evaluate each chromosome and copy the best chromosome pbest of the initial population in a separate location.
3. If the termination condition is not reached, go to Step 4; else select the best individual from the population as the best cluster result.
4. Select individuals from the population for crossover and mutation.
5. Apply crossover operator to the selected individuals based on the crossover probability.
6. Apply mutation operator to the selected individuals based on the mutation probability.
7. Evaluate the newly generated candidates.
8. Compare the worst chromosome in the new population with pbest in terms of their fitness values. If the former is worse than the later, then replace it by pbest.
9. Find the best chromosome in the new population and replace pbest.
10. For the new population, select the best chromosome as a reference, which other chromosomes might fall into the gene rearrangement if needed.
11. Go back to Step 3.

The following are the illustration of the (GAGR) steps [16].

1. *Chromosome representation:* Extensive experiments comparing real-valued and binary GAs indicate [102] that the real-valued GA is more efficient in terms of CPU time. In real-valued gene representation, each chromosome constitutes a sequence of cluster centers, i.e., each chromosome is described by $M = N \cdot K$ real-valued numbers as follows:

$$\mathbf{m} = [m_{11}, \ldots, m_{1N}, \ldots, m_{K1}, \ldots, m_{KN}] = [\mathbf{m}_1, \ldots, \mathbf{m}_K], \tag{9.6.9}$$

where N is the dimension of the feature space and K is the number of clusters; and the first N elements denote the first cluster center, the following N elements denote the second cluster center, and so forth.

2. *Population initialization:* In GAGR clustering algorithm, an initial population of size N can be randomly generated and K data points randomly chosen from

the data set, but on the condition that there are no identical points to form a chromosome, presenting the K cluster centers. This process is repeated until NP chromosomes are generated. Only valid strings (i.e., those that have at least one data point in each cluster) are considered to be included in the initial population.

After the population initialization, each data point is assigned to the cluster with closest cluster center using the following equation:

$$\mathbf{x}_i \in C_j \leftrightarrow \|\mathbf{x}_i - \mathbf{m}_j\| = \min_{k=1,\ldots,K} \|\mathbf{x}_i - \mathbf{m}_k\|, \qquad (9.6.10)$$

where $\mathbf{m}_k$ is the center of the kth cluster.

3. *Fitness function:* The fitness function is used to define a fitness value to each candidate solution. A common clustering criterion or quality indicator is the sum of squared error (SSE) measure, defined as

$$SSE = \sum_{C_i} \sum_{\mathbf{x} \in C_i} (\mathbf{x} - \mathbf{m}_i)^T (\mathbf{x} - \mathbf{m}_i) = \sum_{C_i} \sum_{\mathbf{x} \in C_i} \|\mathbf{x} - \mathbf{m}_i\|^2, \qquad (9.6.11)$$

where $\mathbf{x} \in C_i$ is a data point assigned to that cluster. This measure computes the cumulative distance of each pattern from its cluster center of each cluster individually, and then sums those measures over all clusters. If this measure is small, then the distances from patterns to cluster centers are all small and the clustering would be regarded favorably. It is interesting to note that SSE has a theoretical minimum of zero, which corresponds to all clusters containing only a single data point. Then the fitness function of the chromosome is defined as the inverse of SSE, i.e.,

$$f = \frac{1}{SSE}. \qquad (9.6.12)$$

This fitness function will be maximized during the evolutionary process and lead to minimization of the SSE.

4. *Evolutionary operators:*

   - *Crossover:* Let $\mathbf{x}$ and $\mathbf{y}$ be two chromosomes to be crossed, then the heuristic crossover is given by

   $$\mathbf{x}' = \mathbf{x} + r(\mathbf{x} - \mathbf{y}), \qquad (9.6.13)$$

   $$\mathbf{y}' = \mathbf{x}, \qquad (9.6.14)$$

   where $r = U(0, 1)$ with $U(0, 1)$ being a uniform distribution on interval $[0, 1]$ and $\mathbf{x}$ is assumed to be better than $\mathbf{y}$ in terms of the fitness.

   - *Mutation:* Let $f_{\min}$ and $f_{\max}$ be the minimum and maximum fitness values in the current population, respectively. For an individual with fitness value f, a number $\delta \in [-R, +R]$ is generated with uniform distribution, where R is given by Bandyopdhyay and Maulik [5]:

$$R = \begin{cases} \frac{f - f_{min}}{f_{max} - f_{min}}, & \text{if } f_{max} > f, \\ 1, & \text{if } f_{max} = f. \end{cases} \quad (9.6.15)$$

If the minimum and maximum values of the data set along the ith dimension ($i = 1, \ldots, N$) are m^i_{min} and m^i_{max}, respectively, then after mutation the ith element of the individual is given by Bandyopdhyay and Maulik [5]:

$$m^i = \begin{cases} m^i + \delta \cdot (m^i_{max} - m^i), & \text{if } \delta \geq 0, \\ m^i + \delta \cdot (m^i - m^i_{min}), & \text{otherwise.} \end{cases} \quad (9.6.16)$$

One key feature of gene rearrangement is to avoid the degeneracy resulted by the representations used in many clustering problems. The degeneracy mainly arises from a non-one-to-one correspondence between the representation and the clustering result. To illustrate this degeneracy, let $\mathbf{m}_1 = [1.1, 1.0, 2.2, 2.0, 3.4, 1.2]$ and $\mathbf{m}_2 = [3.2, 1.4, 1.8, 2.2, 0.5, 0.7]$ represent, respectively, two clustering results of a two-dimensional data set with three cluster centers in one generation. If $\mathbf{m}_1$ is selected as the referenced vector, then $\mathbf{m}_2$ becomes $\mathbf{m}'_2 = [0.5, 0.7, 1.8, 2.2, 3.2, 1.4]$ after the gene rearrangement described by Definition 9.31.

Figure 9.3 provides a crossover results of $\mathbf{m}_1$ and $\mathbf{m}_2(\mathbf{m}'_2)$.

From Fig. 9.3 we can see that the performance of the chromosome marked by triangles in the figure directly obtained from crossing $\mathbf{m}_1$ and $\mathbf{m}_2$ is poor as two of

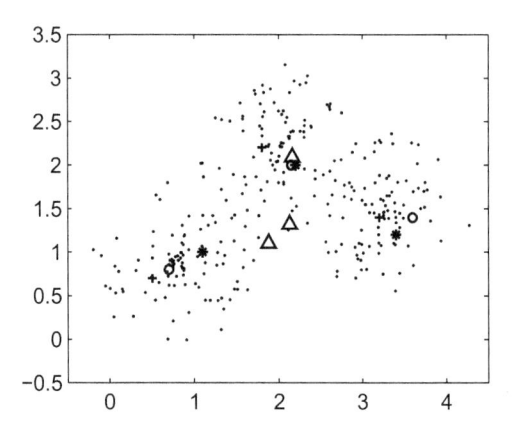

Fig. 9.3 Illustration of degeneracy due to crossover (from [16]). Here, asterisk denotes three cluster centers (1.1, 1.0), (2.2, 2.0), (3.4, 1.2) for chromosome $\mathbf{m}_1$, + denotes three cluster centers (3.2, 1.4), (1.8, 2.2), (0.5, 0.7) for the chromosome $\mathbf{m}_2$, open triangle denotes the chromosome obtained by crossing $\mathbf{m}_1$ and $\mathbf{m}_2$, and open circle denotes the chromosome obtained by crossing $\mathbf{m}_1$ and $\mathbf{m}'_2$ with three cluster centers (0.5, 0.7), (1.8, 2.2), (3.2, 1.4)

the three clusters are very close, producing some confusion, while the chromosome marked by circles in the figure obtained from $\mathbf{m}_1$ and $\mathbf{m}'_2$ with gene rearrangement is more efficient because of no confusion between any pair of clusters.

Since there is a one-to-one correspondence between the representation and the clustering result through the gene rearrangement, there is no degeneracy, which makes the GAGR algorithm more efficient than the classical GA.

9.7 Nondominated Multiobjective Genetic Algorithms

In generation-based evolutionary algorithms (EAs) or genetic algorithms (GAs), the selected individuals are recombined (e.g., crossover) and mutated to constitute the new population. A designer/user prefers the more incremental, steady-state population update, which selects (and possibly deletes) only one or two individuals from the current population and adds the newly recombined and mutated individuals to it. Therefore, the selection of one or two individuals is a key step in generation-based EAs or GAs.

9.7.1 Fitness Functions

In the fields of genetic programming and genetic algorithms, each design solution is commonly represented as a string of numbers (called a chromosome). After each round of testing or simulation, the designer's goal is to delete the "n" worst design solutions, and to breed "n" new ones from the best design solutions. To this end, each design solution needs to be evaluated by a figure of merit in order to determine how close it was to meeting the overall specification. This merit usually uses the fitness function to the test or simulation.

A fitness function, as a single figure of merit, is a particular type of objective function, and is used in genetic programming and genetic algorithms to guide simulations towards optimal design solutions.

According to "survival of the fittest" mechanism, a GA evaluates directly a solution by the fitness value of the solution. The GA requires only the fitness function to satisfy the nonnegativeness. This feature makes the genetic algorithm has wide applicability. Since a feasible solution $\mathbf{x}_1 \in X$ is said to (Pareto) dominate another solution $\mathbf{x}_2 \in X$, if $f_i(\mathbf{x}_1) \leq f_i(\mathbf{x}_2)$ for all indices $i \in \{1, \ldots, k\}$ and $f_j(\mathbf{x}_1) < f_j(\mathbf{x}_2)$ for at least one index $j \in \{1, \ldots, k\}$, the objective function $f_i(\mathbf{x})$ can naturally be taken as the fitness function of a candidate solution $\mathbf{x}$. However, in practical problems, the fitness function is not completely consistent with the objective function of the problem.

In GAs, fitness is the key indicator to describe the individual's performance. Due to selecting the fittest based on fitness, it becomes the driving force of the genetic algorithm. From a biological point of view, the fitness is equivalent to the survival competition. The "survival" biological viability is of great significance in the genetic process. Mapping the objective function of an optimization problem to the individual's fitness can be achieved by optimizing the objective function of the optimization problem in the process of group evolution. The function is also known as the evaluation function, which is a criterion for distinguishing good individuals from bad ones in a group according to the objective function. The fitness is always nonnegative. In any case, it is expected that the larger the fitness value, the better.

In the selection operation, there are two genetic algorithm deceptive problems:

- In the early stage of GA, some supernormal individuals are usually generated. These supernormal individuals will control the selection process due to their outstanding competitiveness, which affects the global optimization performance of the algorithm.
- At the later stage of GA, if the algorithm tends to converge, then the potential for continued optimization is reduced and a certain local optimum solution may be obtained due to the small difference in individual fitness in the population. Therefore, if the fitness function is not properly selected, it will result in more than the problem of deception. The choice of fitness function can be seen as significant for genetic algorithms.

9.7.2 Fitness Selection

EAs/GAs are capable of solving complicated optimization tasks in which an objective function $f : I \to \mathbb{R}$ will be maximized, where $i \in I$ is an individual from the set of feasible solutions. A population is a multi-set of individuals which is maintained and updated as follows: one or more individuals are first selected according to some selection strategy. In generation-based EAs, the selected individuals are then recombined (e.g., via crossover) and mutated in order to constitute the new population. We prefer to select (and possibly delete) only one or two individuals from the current population and adds the newly recombined and mutated individuals to it. We are interested in finding a single individual of maximal objective value for difficult multimodal and deceptive problems.

The following are the two common selection schemes [77]:

1. *Standard Selection Scheme:* This scheme selects all favor individuals of higher fitness, and has the following variants:

   - *Linear Proportionate Selection:* In this method, the probability of selecting an individual depends linearly on its fitness [69].
   - *Truncation Selection:* In this selection the fittest individuals are selected, usually with multiplicity to keep the population size fixed [109].
   - *Ranking Selection:* This selection orders the individuals according to their fitness. The selection probability is, then, a (linear) function of the rank [157].

- *Tournament Selection:* This method selects the best out of individuals. It has primarily developed for steady-state EAs, but can be adapted to generation based EAs [4].

2. *Fitness Uniform Selection Scheme (FUSS):* FUSS is based on the insight that one is not primarily interested in a population converging to maximal fitness but only in a single individual of maximal fitness. The scheme automatically creates a suitable selection pressure and preserves genetic fitness values, and should help to preserve population diversity.

All above four standard selection schemes have the property (and goal) to increase the average fitness of a population, i.e., to evolve the population toward higher fitness [77].

Define the difference or distance between two individuals $f(i)$ and $f(j)$ as [77]:

$$d(i, j) = |f(i) - f(j)|. \tag{9.7.1}$$

The distance is based solely on the fitness function, and is independent of the coding/representation and other problem details, and of the optimization algorithm (e.g., the genetic mutation and recombination operators), and can trivially be computed from the fitness values. If making the natural assumption that functionally similar individuals have similar fitness, then they are also similar with respect to the distance d. On the other hand, individuals with very different coding and even functionally dissimilar individuals may be d-similar.

Two individuals i and j are said to be ϵ-similar if $d(i, j) = |f(i) - f(j)| \leq \epsilon$.

Let the lowest and highest fitness values in the current population be $f_{\min}$ and $f_{\max}$, respectively.

The fitness uniform selection scheme is a two-stage uniform selection process [77]:

- Randomly select a fitness value f uniformly from the fitness values $F = [f_{\min}, f_{\max}]$.
- Select an individual $i \in P$ with fitness nearest to f, and add a copy to the population P, possibly after mutation and recombination.

The probability of selecting a specific individual is proportional to the distance to its nearest fitness neighbor. In a population with a high density of unfitting and low density of fitting individuals, the more fitting ones are effectively favored.

To distinguish good solutions (individuals) from bad solutions, we need to make the fitness selection of individuals.

The fitness selection is a mathematical model or a computer simulation, or even is a subjective function for choosing better solutions over worse ones. In order to guide the evolution of good solutions in GAs, the fitness of every individual in the population is evaluated, multiple individuals are stochastically selected from the current population (based on their fitness), and modified (recombined and possibly mutated) to form a new population. The new population is then used in the next iteration of the GA. This iteration process is repeated until either a maximum number of generations has been produced, or a satisfactory fitness level has been reached for the population. However, if the algorithm has terminated due to a maximum number of generations, a satisfactory solution may or may not have been reached.

In contrast, we may also preserve diversity through deletion rather than selection, because we delete those individuals with "commonly occurring" fitness values. This method is called fitness uniform deletion scheme (FUDS) whose role is to govern actively different parts of the solution space rather than to move the population as a whole toward higher fitness [77]. Thus, FUDS is at least a partial solution to the problem of having to set correctly a selection intensity parameter.

A common fitness selection method is the Goldberg's non-inferior sorting method for selecting the superior individuals [54]. Let $\mathbf{x}_i(t)$ be an arbitrary individual in the population $p(t)$ of tth generation, $r_i(t)$ indicate the number of individuals non-inferior to $\mathbf{x}_i(t)$ in this population. Then, $r_i(t)$ is defined as the inferior value of the individual $\mathbf{x}_i(t)$ in the population. Obviously, individuals with small inferior value are superior, and are close to the Pareto solutions. In the process of evolution, the whole population keeps approaching to the Pareto boundary.

9.7.3 Nondominated Sorting Genetic Algorithms

Different from single-objective optimization problems, the performance improvement of one objective is often inevitable at the cost of descending at least another objective in the multiobjective optimization problems. In other words, it is hard to identify a unique optimal solution but rather a family of optimal solutions called the Pareto-optimal front or nondominated set.

Multiobjective evolutionary algorithms (MOEAs) are a most popular approach to solving multiobjective optimization problems because MOEAs are able to find multiple Pareto-optimal solutions in one single run. Since it is not possible for a multiobjective optimization problem to have a single solution which simultaneously optimizes all objectives, an algorithm is of great practical value if it gives a large number of alternative solutions lying on or near the Pareto-optimal front.

Earlier practical GA, called Vector Evaluated Genetic Algorithm (VEGA), was developed by Schaffer in 1985 [141]. One of the problems with VEGA is its bias toward some Pareto-optimal solutions. A decision maker may want to find as many nondominated points as possible in order to avoid any bias toward such middling individuals.

The number of dominating points $d(\mathbf{s}, P(t))$ is the number of points $\mathbf{y}$ from set $P(t)$ that dominate point $\mathbf{s}$, namely

$$d(\mathbf{s}, P(t)) = |\{\mathbf{y} \in P(t) | \mathbf{y} \prec \mathbf{s}\}|. \tag{9.7.2}$$

The $d(\mathbf{s}, P(t))$ measure favors solutions located in those areas where the better fronts are sparsely populated. When the population may consist of nondominated solutions only, the $d(\mathbf{s}, P(t))$ selection scheme is never available.

Definition 9.32 (Niched Pareto Genetic Algorithm [74]) A *niched Pareto genetic algorithm* combines tournament selection and the concept of Pareto dominance. For two competing individuals and a comparison set consisting of other individuals picked at random from the population, if one of the competing individuals is dominated by any member of the set, and the other is not, then the latter is chosen as winner of the tournament. If both individuals are dominated or not dominated, then the result of the tournament is decided by sharing: the individual which has the least individuals in its niche is selected for reproduction.

The niched Pareto genetic algorithm was proposed by Horn et al. in 1994 [74].

Nondominated sorting genetic algorithm (NSGA) differs from a simple genetic algorithm only in how the selection operator works. The crossover and mutation operators remain as usual: the population is ranked on the basis of an individual's nondomination before the selection is performed. The nondominated individuals present in the population are first identified from the current population, then all these individuals are assumed to constitute the first nondominated front in the population and assigned a large dummy fitness value. If some nondominated individuals are assigned to the same fitness value, then all these nondominated individuals have an equal reproductive potential. To maintain diversity in the population, these classified individuals are then shared with their dummy fitness values.

Let the parameter $d(\mathbf{x}_i, \mathbf{x}_j)$ be the phenotypic distance between two individuals $\mathbf{x}_i$ and $\mathbf{x}_j$ in the current front, and σ_{share} be the maximum phenotypic distance allowed between any two individuals to become members of a niche.

Definition 9.33 (Sharing Function Value [146]) For two individuals $\mathbf{x}_i$ and $\mathbf{x}_j$ in the same front, their *sharing function value*, denoted by $\text{sh}(\mathbf{x}_i, \mathbf{x}_j)$, is defined as

$$\text{sh}(\mathbf{x}_i, \mathbf{x}_j) = \begin{cases} 1 - \left(\frac{d(\mathbf{x}_i, \mathbf{x}_j)}{\sigma_{\text{share}}}\right)^2, & d(\mathbf{x}_i, \mathbf{x}_j) < \sigma_{\text{share}}; \\ 0, & \text{otherwise.} \end{cases} \qquad (9.7.3)$$

For example, $d(\mathbf{x}_1, \mathbf{x}_2) = \sqrt{\sum_{i=1}^{n}(x_{1i} - x_{2i})^2}$ is taken.

Definition 9.34 (Niche Count [56]) Given $X = \{\mathbf{x}_1, \ldots, \mathbf{x}_L\}$ (the union of the parent and offspring populations) and σ (the *niche radius*). Then, the *niche count* of $\mathbf{x} \in X$ is defined by

$$\text{nc}(\mathbf{x}|X, \sigma) = \sum_{i=1, \mathbf{x}_i \neq \mathbf{x}}^{L} \text{sh}(\mathbf{x}, \mathbf{x}_i). \qquad (9.7.4)$$

Sharing is achieved by performing selection operation via degraded fitness values obtained by dividing the original fitness value of an individual by a quantity proportional to the number of individuals around it. This causes multiple optimal points to co-exist in the population. These shared nondominated individuals are ignored temporarily, and are then assigned a new dummy fitness value for keeping smaller than the minimum shared dummy fitness of the previous front. This process is continued until the entire population is classified into several fronts [146].

In multiobjective optimization genetic algorithm, the whole population is checked and all nondominated individuals are assigned rank 1. Other individuals are ranked according to the nondominance of them with respect to the rest of the population as follows. For an individual point, the number of points that strictly dominate the point in the population is first found. Then, the rank of that individual is assigned to be one more than that number. At the end of this ranking procedure, there could be a number of points having the same rank. The selection procedure then uses these ranks to select or delete blocks of points to form the mating pool.

Algorithm 9.8 shows the nondominated sorting genetic algorithm (NSGA) [146].

Algorithm 9.8 Nondominated sorting genetic algorithm (NSGA) [146]

1. **input:** maxgen (maximal generation number).
2. **initialization:** population gen = 0.
3. **while** gen < maxgen
4.   front = 1;
5.   justify population classification;
6.   **if** population is classified **do**
7.   flag_classified = 1;
8.   **else**
9.   flag_classified = 0;
10.   **end if**
11.   **while** flag_classified = 0
12.   identify nondominated individuals;
13.   assign dummy fitness;
14.   sharing in current front;
15.   front = front + 1;
16.   return step 2;
17.   **end while**
18.   **while** flag_classified = 1
19.   reproduction according to dummy fitness;
20.   crossover;
21.   mutation;
22.   **if** gen < maxgen **do**
23.   gen = gen + 1;
24.   return step 1;
25.   **else**
26.   **end if**
27.   **end while**
28. **end while**
29. **output:** nondominated sorting population.

9.7.4 Elitist Nondominated Sorting Genetic Algorithm

The nondominated sorting genetic algorithm (NSGA) proposed by Srinivas and Deb [146] uses nondominated sorting and sharing, and its main criticism are as follows [25]:

- *High computational complexity of nondominated sorting:* Computational complexity of ordinary nondominated sorting algorithm is $O(kN^3)$, where k is the number of objectives and N is the population size.
- *Lack of elitism:* Elitism can speed up the performance of the GA significantly, and also can help to prevent the loss of good solutions once they have been found.
- *Need for specifying the sharing parameter* σ_{share}: A parameterless diversity preservation mechanism is desirable.

To overcome the three shortcomings mentioned above, a fast and elitist multiobjective genetic algorithm (NSGA-II) was proposed by Deb et al. in 2002 [25]. Evolutionary multiobjective algorithms apply biologically inspired evolutionary processes as heuristics to generate nondominated sets of solutions. However, it should be noted that the solutions returned by evolutionary multiobjective algorithms may not be Pareto optimal (i.e., globally nondominated), but the algorithms are designed to evolve solutions that approach the Pareto front and spread out to capture the diversity existing on the Pareto front for obtaining a good approximation of the Pareto front [108].

The NSGA-II features nondominated sorting and a $(\mu + \mu)$ (μ is the number of solution vectors) selection scheme. The secondary ranking criterion for solutions on the same front is called crowding distance. For a solution, the crowding distance is defined as the sum of the side lengths of the cuboid through the neighboring solutions on the front. For extremal solutions it is defined as infinity. The crowding distance value of a solution depends on its neighbors and not directly on the position of the point itself.

At the end of one generation of NSGA-II, the size of the population is twice bigger than the original size. This bigger population is pruned based on the nondominated sorting [80].

The basic NSGA-II steps are as follows [108]:

1. Randomly generate the first generation.
2. Partition the parents and offspring of the current generation into k fronts $F_1, \ldots, F_k$, so that the members of each front are dominated by all members of better fronts and by no members of worse fronts.
3. Calculate the crowding distance for each potential solution: for each front, sort the members of the front according to each objective function from the lowest value to the highest value. Compute the difference between the solution and the closest lesser solution and the closest greater solution, ignoring the other objective functions. Repeat for each objective function. The crowding distance for that solution is the average difference over all the objective func-

tions. Note that the extreme solutions for each objective function are always included.
4. Select the next generation based on nondomination and diversity:

- Add the best fronts $F_1, \ldots, F_j$ to the next generation until it reaches the target population size.
- If part of a front, F_{j+1}, must be added to the generation to reach the target population size, sort the members from the least crowded to most crowded. Then add as many members as are necessary, starting with the least crowded.

5. Generate offspring:

- Apply binary tournament selection by randomly choosing two solutions and including the higher-ranked solution with a fixed probability 0.5–1.
- Apply single-point crossover and the sitewise mutation to generate offspring.

6. Repeat steps 2 through 5 for the desired number of generations.

9.8 Evolutionary Algorithms (EAs)

Optimization plays a very important role in engineering, operational research, information science, and related areas. Optimization techniques can be divided into two categories: derivative-based methods and derivative-free methods. As an important branch of derivative-free methods, *evolutionary algorithms* (EAs) have shown considerable success in solving optimization problems and attracted more and more attention.

Single objective evolutional algorithms are general purpose optimization heuristics inspired by Darwin's principle of natural evolution. Starting with a set of candidate solutions (*population*), in each iteration (*generation*), the better solutions are selected (*parents*) according to their *fitness* (i.e., function values under the objective function) and used to generate new solutions (*offspring*) through recombining the information of two parents in a new way (*crossover*) or randomly modifying a solution (*mutation*). These offspring are then inserted into the population, replacing some of the weaker solutions (*individuals*) [13].

By iteratively selecting the better solutions and using them to create new candidates, the population "evolves," and the obtained solutions become better and better adapted to the optimization problem at hand, just like in the nature, where the individuals become better and better adapted to their environment through evolution.

EAs have successfully employed on a wide variety of complex optimization problems because they can deal with almost arbitrarily complex objective functions and constraints, resulting into very few assumptions and not even requiring a mathematical description of the problem.

9.8.1 (1 + 1) *Evolutionary Algorithm*

The most simple single objective evolutionary algorithm is called (1 + 1) *evolutionary algorithm* or (1 + 1) EA. The basic idea and steps of (1 + 1) EA are as follows [35]:

1. The size of the population is restricted to just one individual and does not use crossover.
2. The current individual is represented as a bit string.
3. Use bitwise mutation operator that flips each bit independently of the others with some probability p_m.
4. The current bit string is replaced by the new one if the fitness of the current bit string is not superior to the fitness of the new string. This replacement strategy selects the best individual of one parent and one child as the new generation.

(1 + 1) EA is shown in Algorithm 9.9.

Algorithm 9.9 (1 + 1) evolutionary algorithm [35]

1. **input:** $p_m = 1/n$.
2. **initialization:** Choose randomly an initial bit string $\mathbf{x} \in \{0, 1\}^n$.
3. **while** $\mathbf{x}$ is not optimal **do**
4.   Compute $\mathbf{x}'$ by flipping independently each bit x_i with probability p_m.
5.   Replace $\mathbf{x}$ by $\mathbf{x}'$ if and only if $f(\mathbf{x}') \geq f(\mathbf{x})$.
6. **end while**
7. **output:** $\mathbf{x}$.

The different choices of selection, crossover, mutation, replacement, and concrete representation of the individuals offer a great variety of different EAs.

The combinatorial optimization problem can be described as follows: given a finite state space S and a function $f(\mathbf{x}), \mathbf{x} \in S$, find the solution

$$\max_{\mathbf{x} \in S} f(\mathbf{x}). \tag{9.8.1}$$

Assume $\mathbf{x}^*$ is one state with the maximum function value, and $f_{\max} = f(\mathbf{x}^*)$. The EA for solving the combinatorial optimization problem can be described as follows:

1. *Initialization:* generate, either randomly or heuristically, an initial population of $2N$ individuals, denoted by $\xi_0 = (x_1, \ldots, x_{2N})$ (where $N > 0$ is an integer), and let $k \leftarrow 0$. For any population ξ_k, define $f(\xi_k) = \max\{f(x_i) : x_i \in \xi_k\}$.
2. *Generation:* generate a new (intermediate) population by crossover and mutation (or any other operators for generating offspring), and denote it as $\xi_{k+1/2}$.
3. *Selection:* select and reproduce $2N$ individuals from populations $\xi_{k+1/2}$ and ξ_k, and obtain another (new intermediate) population ξ_{k+S}.

4. If $f(\xi_{k+S}) = f_{\max}$, then stop; otherwise let $\xi_{k+1} = \xi_{k+S}$ and $k \leftarrow k+1$, and go to step 2 to continue.

9.8.2 Theoretical Analysis on Evolutionary Algorithms

Several different methods are widely used in the theoretical analysis of EAs.

Fitness Partition

Fitness partition method is the very basic method, in which the objective fitness domain is partitioned into several levels.

Theorem 9.1 ([94]) *Consider a maximization problem with fitness function $f : D \to \mathbb{R}$. Suppose D can be partitioned into L subsets and for any solution $\mathbf{x}_i$ in subset i and $\mathbf{x}_j$ in subset j, we have $f(\mathbf{x}_i) < f(\mathbf{x}_j)$ if $i < j$. In addition, subset L only contains optimal solution(s) of f. If an algorithm can jump from subset k to subset $k+1$ or higher with probability at least p_k, then the total expected time for the algorithm to find the optimum satisfies*

$$E(T) \leq \sum_{k=1}^{L-1} \frac{1}{p_k}. \tag{9.8.2}$$

Drift Analysis

By considering the expected progress achieved in a single step at each search point, drift analysis [60, 66] can be used to analyze problems which are hard to be analyzed by fitness partition.

Assume $\mathbf{x}^*$ is an optimal point, and let $d(\mathbf{x}, \mathbf{x}^*)$ be the distance between two points $\mathbf{x}$ and $\mathbf{x}^*$. If there are more than one optimal point (that is, a set S^*), then $d(\mathbf{x}, S^*) = \min\{d(\mathbf{x}, \mathbf{x}^*) : \mathbf{x}^* \in S^*\}$ is used as the distance between individual $\mathbf{x}$ and the optimal set S^*. Denote the distance by $d(\mathbf{x})$. Usually $d(\mathbf{x})$ satisfies $d(\mathbf{x}^*) = 0$ and $d(\mathbf{x}) > 0$ for any $\mathbf{x} \notin S^*$.

Given a population $X = \{\mathbf{x}_1, \ldots, \mathbf{x}_{2N}\}$, let

$$d(X) = \min\{d(\mathbf{x} : \mathbf{x} \in X)\}, \tag{9.8.3}$$

which is used to measure the distance of the population to the optimal solution.

The drift of the random sequence $\{d(\xi_k); k = 0, 1, \ldots\}$ at time k is defined by

$$\Delta(d(\xi_k)) = d(\xi_{k+1}) - d(\xi_k). \tag{9.8.4}$$

Define the stopping time of an EA as $\tau = \min\{k : d(\xi_k) = 0\}$, which is the first hitting time on the optimal solution. The *drift analysis* focuses on the following question [66]: under what conditions of the drift $\Delta(d(\xi_k))$ can we estimate the expect first hitting time $E[\tau]$?

A fitness function $f : \{0, 1\}^n \to \mathbb{R}$ can be written as a polynomial [35]:

$$f(s_1, \ldots, s_n) = \sum_{I \subseteq \{1,\ldots,n\}} c_f(I) \prod_{i \in I} s_i, \qquad (9.8.5)$$

where coefficient $c_f(I) \in \mathbb{R}$.

Let the class of fitness functions $f(s_1, \ldots, s_n)$ satisfy the condition

$$f(s_1, \ldots, s_{k-1}, 0, s_{k+1}, \ldots, s_n) < f(s_1, \ldots, s_{k-1}, 1, s_{k+1}, \ldots, s_n) \qquad (9.8.6)$$

for any $k = 1, \ldots, n$ and fixed $s_1, \ldots, s_{k-1}, s_{k+1}, \ldots, s_n$. This condition shows that if one "0" bit at any position flips into "1," the fitness will increase, and thus $(1, \ldots, 1)$ is the unique maximum point.

Theorem 9.2 ([67]) *For any fitness function (9.8.5) satisfying (9.8.6), the EA with mutation probability $p_m = 1/(2n)$ needs average $O(n \log n)$ steps to reach the optimal solution.*

9.9 Multiobjective Evolutionary Algorithms

Multiobjective evolutionary algorithms (MOEAs) have been shown to be well-suited for solving multiobjective optimization problems (MOPs) with conflicting objectives as they can approximate the Pareto-optimal front with a population in a single run.

9.9.1 Classical Methods for Solving Multiobjective Optimization Problems

All classical methods for solving MOPs are to scalarize the objective vector into one scalar objective. The following are the three commonly used classical methods [146]:

1. *Method of objective weighting:* multiple objective functions are combined into one overall objective function:

$$Z(\mathbf{x}) = \sum_{i=1}^{m} w_i f_i(\mathbf{x}), \quad \mathbf{x} \in \Omega, \qquad (9.9.1)$$

where w_i are fractional numbers ($0 \leq w_i \leq 1$), and all weights are summed up to 1, or $\sum_{i=1}^{k} w_i = 1$. In this method, the optimal solution is controlled by the weigh vector $\mathbf{w} = [w_1, \ldots, w_k]^T$: the preference of each objective $f_i(\mathbf{x})$ can be changed by modifying its corresponding weight w_i.

2. *Method of distance functions:* the scalarization of objective vector $\mathbf{f}(\mathbf{x}) = [f_1(\mathbf{x}), \ldots, f_k(\mathbf{x})]^T$ is achieved by using a demand-level vector $\bar{\mathbf{y}}$ which has to be specified by the decision maker as follows:

$$Z(\mathbf{x}) = \left(\sum_{i=1}^{m} \| f_i(\mathbf{x}) - \bar{\mathbf{y}} \|^r \right)^{1/r}, \quad 1 \leq r < \infty, \tag{9.9.2}$$

where a Euclidean metric $r = 2$ is usually chosen, and the demand-level vector $\bar{\mathbf{y}}$ is individual optima of multiple objectives. The solution given by this method depends on the chosen demand-level vector. Arbitrary selection of a demand-level may be highly undesirable; a wrong demand-level will lead to a non-Pareto-optimal solution.

3. *Min-max formulation:* this method attempts to minimize the relative deviations of the single objective functions from the individual optimum, namely it tries to minimize the objective conflict $\mathcal{F}(\mathbf{x})$:

$$\min \mathcal{F}(\mathbf{x}) = \max Z_j(\mathbf{x}), \quad j = 1, \ldots, m, \tag{9.9.3}$$

where $Z_j(\mathbf{x})$ is calculated for nonnegative target optimal value $\bar{f} > 0$ as follows:

$$Z_j(\mathbf{x}) = \frac{f_j - \bar{f}_j}{\bar{f}_j}, \quad j = 1, \ldots, m. \tag{9.9.4}$$

Drawbacks of the above classical methods are: the optimization of the single objective may guarantee a Pareto-optimal solution, but results in a single-point solution. In real-world situations decision makers often need different alternatives in decision making. Moreover, if some of the objectives are noisy or have discontinuous variable space, then these methods may not work effectively.

Traditional search and optimization methods (such as gradient-based methods) are not available for extending to the multiobjective optimization because their basic design precludes the consideration of multiple solutions. In contrast, population-based methods (such as single objective evolutionary algorithms) are well-suited for handling such situations due to their ability to approximate the entire Pareto front in a single run [20, 24, 110].

The basic difference between multiobjective EAs from single objective EAs is how they rank and select individuals in the conflicting performance measures or objectives, many real-life multiobjective problems must be optimized simultaneously to achieve a trade-off.

In case of single objective, individuals are naturally ranked according to this objective, and it is clear which individuals are best and should be selected as parents. However, if there are multiple objectives, ranking the individuals is no longer obvious. Most people probably agree that a good approximation to the Pareto front is characterized by the following metrics [13]:

- a small distance of the solutions to the true Pareto frontier (such as dominated sorting solutions),
- a wide range of solutions, i.e., an approximation of the extreme values (such as extreme solutions),
- a good distribution of solutions, i.e., an even spread along the Pareto frontier (such as solutions with a larger distance to other solutions).

MOEAs then rank individuals according to how much they contribute to the above goals.

In MOEAs, the ability to balance between convergence and diversity depends on the selection strategy that can be broadly classified as: Pareto dominance-based MOEAs (PDMOEAs), indicator-based MOEAs, and decomposition-based MOEAs [119].

Difficulties in handling multiobjective problems can be roughly classified into the following five categories [87].

1. Difficulties in searching for Pareto-optimal solutions.
2. Difficulties in approximating the entire Pareto front.
3. Difficulties in the presentation of obtained solutions.
4. Difficulties in selecting a single final solution.
5. Difficulties in the evaluation of search algorithms.

Two major problems must be addressed when an evolutionary algorithm is applied to multiobjective optimization [171]:

- How to accomplish fitness assignment and selection, respectively, in order to guide the search towards the Pareto-optimal set.
- How to maintain a diverse population in order to prevent premature convergence and achieve a well distributed trade-off front.

9.9.2 MOEA Based on Decomposition (MOEA/D)

Consider the multiobjective optimization problem (MOP)

$$\min \mathbf{f}(\mathbf{x}) = [f_1(\mathbf{x}), \ldots, f_m(\mathbf{x})]^T, \quad \text{subject to } \mathbf{x} \in \Omega, \tag{9.9.5}$$

where $\mathbf{x} = [x_1, \ldots, x_n]^T$ is the decision (variable) vector, Ω is the decision (variable) space, $\mathbb{R}^m$ is the objective space, and $\mathbf{f} : \Omega \to \mathbb{R}^m$ consists of m real-valued objective functions. If Ω is a closed and connected region in $\mathbb{R}^m$ and all the objectives are continuous of $\mathbf{x}$, then the above problem is known as a continuous MOP.

A point $\mathbf{x}^* \in \Omega$ is called (globally) Pareto optimal if there is no $\mathbf{x} \in \Omega$ such that $\mathbf{f}(\mathbf{x})$ dominates $\mathbf{f}(\mathbf{x}^*)$, i.e., $f_i(\mathbf{x})$ dominates $f_i(\mathbf{x}^*)$ for each $i = 1, \ldots, m$. The set of all the Pareto-optimal points is called the Pareto set. The set of all the Pareto objective vectors, $PF = \{\mathbf{f}(\mathbf{x}) \in \mathbb{R}^m | \mathbf{x} \in \Omega\}$, is called the Pareto front.

Decomposition is a basic strategy in traditional multiobjective optimization. But, general MOEAs treat a MOP as a whole, and do not associate each individual solution with any particular scalar optimization problem. A multiobjective evolutionary algorithm based on decomposition (called MOEA/D) [167] decomposes an MOP by scalarizing functions into a number of subproblems, each of which is associated with a search direction (or weight vector) and assigned a candidate solution.

The followings are the three approaches for converting the problem of approximation of the Pareto front of problem (9.9.5) into a number of scalar optimization problems [167].

1. *Weighted sum approach:* Let $\boldsymbol{\lambda} = [\lambda_1, \ldots, \lambda_m]^T$ be a weight vector with $\lambda_i \geq 0$, $\forall i = 1, \ldots, m$ and $\sum_{i=1}^{m} \lambda_i = 1$. Then, the optimal solution to the scalar optimization problem

$$\max \left\{ g^{\text{ws}}(\mathbf{x}|\boldsymbol{\lambda}) = \sum_{i=1}^{m} \lambda_i f_i(\mathbf{x}) \right\}, \quad \text{subject to } \mathbf{x} \in \Omega \quad (9.9.6)$$

is a Pareto-optimal point to the MOP (9.9.5).

2. *Tchebycheff approach:* This approach [103] can convert the problem of a Pareto-optimal solution of (9.9.5) into the scalar optimization problem:

$$\min \left\{ g^{\text{tc}}(\mathbf{x}|\boldsymbol{\lambda}, \mathbf{z}^*) = \max_{1 \leq i \leq m} \{\lambda_i | f_i(\mathbf{x}) - z_i^* |\} \right\}, \quad \text{subject to } \mathbf{x} \in \Omega, \quad (9.9.7)$$

where $\mathbf{z} = [z_1, \ldots, z_m]^T$ is the reference point vector, i.e., $z_i^* = \max\{f_i(\mathbf{x}) | \mathbf{x} \in \Omega\}$ for each $i = 1, \ldots, m$, and each optimal solution of (9.9.7) is a Pareto-optimal solution of (9.9.5), and thus one is able to obtain different Pareto-optimal solutions by altering the weight vector $\boldsymbol{\lambda}$. Therefore, the Tchebycheff scalar optimization subproblem can be defined as [167]:

$$\min \max_{1 \leq i \leq m} \{\lambda_i | f_i(\mathbf{x}) - z_i^* |\}, \quad \text{subject to } \mathbf{x} \in \Omega, \quad (9.9.8)$$

where λ_i is a weight parameter for the ith objective, and z_i^* is set to the current best fitness of the ith objective. Tchebycheff approach used in MOEA/D can decompose the original MOP into several scalar optimization subproblems in form of (9.9.8), which can be solved with any optimization method for single objective optimization.

9.9 Multiobjective Evolutionary Algorithms

3. *Boundary intersection (BI) approach:* This approach solves the following scalar optimization subproblem:

$$\min \ g^{bi}(\mathbf{x}|\boldsymbol{\lambda}, \mathbf{z}^*) = d, \quad \text{subject to} \ \mathbf{z}^* - \mathbf{f}(\mathbf{x}) = d\boldsymbol{\lambda}, \ \mathbf{x} \in \Omega. \quad (9.9.9)$$

The goal of the constraint $\mathbf{z}^* - \mathbf{f}(\mathbf{x}) = d\boldsymbol{\lambda}$ is to push $\mathbf{f}(\mathbf{x})$ as high as possible so that it reaches the boundary of the attainable objective set. In order to avoid the equality constraint in (9.9.9), one can use a penalty method to deal with the constraint [167]:

$$\min \ g^{bip}(\mathbf{x}|\boldsymbol{\lambda}, \mathbf{z}^*) = d_1 + \theta d_2, \quad \text{subject to} \ \mathbf{x} \in \Omega, \quad (9.9.10)$$

where $\theta > 0$ is a preset penalty parameter, and

$$d_1 = \frac{\|(\mathbf{z}^* - \mathbf{f}(\mathbf{x}))^T \boldsymbol{\lambda}\|}{\|\boldsymbol{\lambda}\|}, \quad (9.9.11)$$

$$d_2 = \|\mathbf{f}(\mathbf{x}) - (\mathbf{z}^* - d_1 \boldsymbol{\lambda})\|. \quad (9.9.12)$$

Multiobjective evolutionary algorithm based on decomposition (MOEA/D), proposed by Zhang and Li [167], uses the Tchebycheff approach to convert the problem of a Pareto-optimal solution of (9.9.5) into the scalar optimization problem (9.9.8).

Let $\boldsymbol{\lambda}^j = [\lambda_1^j, \ldots, \lambda_m^j]^T$ be a set of weight vectors and $\mathbf{z}^*$ be a reference point. The problem of approximation of the PF of (9.9.5) can be decomposed into scalar optimization subproblems by using the Tchebycheff approach and the objective function of the ith subproblem is given by

$$\min \ g^{tc}(\mathbf{x}|\boldsymbol{\lambda}^j, \mathbf{z}^*) = \max_{1 \le i \le m} \left\{ \lambda_i^j |f_i(\mathbf{x}) - z_i^*| \right\}, \quad \text{subject to} \ \mathbf{x} \in \Omega, \quad (9.9.13)$$

where $\boldsymbol{\lambda}^j = [\lambda_1^j, \ldots, \lambda_m^j]^T$.

As g^{tc} is continuous of $\boldsymbol{\lambda}$, the optimal solution of $g^{tc}(\mathbf{x}|\boldsymbol{\lambda}^i, \mathbf{z}^*)$ should be close to that of $g^{tc}(\mathbf{x}|\boldsymbol{\lambda}^j, \mathbf{z}^*)$ if $\boldsymbol{\lambda}^i$ and $\boldsymbol{\lambda}^j$ are close to each other. This makes any information about these g^{tc}'s with weight vectors close to $\boldsymbol{\lambda}^i$ should be helpful for optimizing $g^{tc}(\mathbf{x}|\boldsymbol{\lambda}^i, \mathbf{z}^*)$. This is a basic idea behind MOEA/D.

Let N be the number of the subproblems considered in MOEA/D, i.e., consider a population of N points $\mathbf{x}^1, \ldots, \mathbf{x}^N \in \Omega$, where $\mathbf{x}^i$ is the current solution to the ith subproblem.

In MOEA/D, a neighborhood of weight vector $\boldsymbol{\lambda}^i$ is defined as a set of its closest weight vectors in $\{\boldsymbol{\lambda}^1, \ldots, \boldsymbol{\lambda}^N\}$. The neighborhood of the ith subproblem consists of all the subproblems with the weight vectors from the neighborhood of $\boldsymbol{\lambda}^i$. The population is composed of the best solution found so far for each subproblem. Only the current solutions to its neighboring subproblems are exploited for optimizing a subproblem in MOEA/D.

MOEA/D algorithm is shown in Algorithm 9.10.

Algorithm 9.10 Multiobjective evolutionary algorithm based on decomposition (MOEA/D) [167]

1. **input:**
   1.1 N: the number of the subproblems considered;
   1.2 $\lambda^1, \ldots, \lambda^N$: a set of N weight vectors;
   1.3 T: the number of the weight vectors in the neighborhood of each weight vector;
2. **initialization:**
   2.1 Let external Pareto-optimal set $P = \emptyset$.
   2.2 Compute $d(i, j) = \|\lambda^i - \lambda^j\|_2$ for $i, j \in \{1, \ldots, N\}$ but $i \neq j$. and then work out the T closest weight vectors to each weight vector. For each $i \in \{1, \ldots, k\}$, set $B(i) = \{i_1, \ldots, i_T\}$, where $\lambda^{i_1}, \ldots, \lambda^{i_T}$ are the T closest weight vectors to λ^i.
   2.3 Generate randomly an initial population $\mathbf{x}^1, \ldots, \mathbf{x}^N$ or by a problem-specific method. Let $\mathbf{fv}^i = \mathbf{f}(\mathbf{x}^i)$ and $P = \{\mathbf{x}^1, \ldots, \mathbf{x}^N\}$.
   2.4 Initialize $\mathbf{z} = [z_1, \ldots, z_k]^T$ by a problem-specific method.
3. **for** $i = 1, \ldots, N$ **do**
4.   *Reproduction:* Randomly select two indexes j and l from $B(i)$, and then generate a new solution $\mathbf{y}$ from $\mathbf{x}^j$ and $\mathbf{x}^l$ by using genetic operators.
5.   *Improvement:* Apply a problem-specific repair/improvement heuristic on $\mathbf{y}$ to produce $\mathbf{y}'$.
6.   *Update* $\mathbf{z}$: For each $j = 1, \ldots, k$, if $z_j < f_j(\mathbf{y}')$, then set $z_j = f_j(\mathbf{y}')$.
7.   *Generation:* For each index $j \in B(i)$, if $g^{tc}(\mathbf{y}'|\lambda^j, \mathbf{z}) \leq g^{tc}(\mathbf{x}^j|\lambda^j, \mathbf{z})$, then let $\mathbf{x}^j = \mathbf{y}'$ and update neighboring solutions $\mathbf{fv}^j = \mathbf{f}(\mathbf{y}')$.
8.   *Selection:* Remove all the vectors dominated by $\mathbf{f}(\mathbf{y}')$ from P. Add $\mathbf{f}(\mathbf{y}')$ to P if no vectors in P dominate $\mathbf{f}(\mathbf{y}')$.
9.   *Stopping criteria:* If stopping criteria is satisfied, then stop and output P. Otherwise, go to step 1.
10. **end for**
11. **output:** external Pareto-optimal set P.

However, MOEA/D with simulated binary crossover has the following two shortcomings [93].

- The population in MOEA/D with simulated binary crossover may lose diversity, which is needed for exploring the search space effectively, particularly at the early stage of the search when applied to MOPs with complicated Pareto sets.
- The simulated binary crossover operator often generates inferior solutions in MOEAs.

To overcome these shortcomings, Li and Zhang [93] developed a version of MOEA/D based on differential evolution (DE), called MOEA/D-DE, which uses a differential evolution [124] operator and a polynomial mutation operator [24] to produce a new solution $\mathbf{y} = [y_1, \ldots, y_n]^T$.

1. *Differential evolution operator:* Generate a solution $\bar{\mathbf{y}} = [\bar{y}_1, \ldots, \bar{y}_n]^T$ from three solutions $\mathbf{x}^{r_1}, \mathbf{x}^{r_2}$, and $\mathbf{x}^{r_3}$ selected randomly from P by a DE operator. The elements of $\bar{\mathbf{y}} = [\bar{y}_1, \ldots, \bar{y}_n]^T$ are given by

$$\bar{y}_j = \begin{cases} x_j^{r_1} + F \cdot (x_j^{r_2} - x_j^{r_3}) & \text{with probability } CR, \\ x_j^{r_1} & \text{with probability } 1 - CR, \end{cases} \qquad (9.9.14)$$

for $j = 1, \ldots, n$, where CR and F are two control parameters.
2. *Polynomial mutation operator:* Generate $\mathbf{y} = [y_1, \ldots, y_n]^T$ from $\bar{\mathbf{y}}$ in the following way:

$$y_j = \begin{cases} \bar{y}_j + \sigma_j(b_j - a_j) & \text{with probability } p_m, \\ \bar{y}_j & \text{with probability } 1 - p_m \end{cases} \quad (9.9.15)$$

with

$$\sigma_j = \begin{cases} (2 \cdot rand)^{1/(\eta+1)} - 1, & \text{if } rand < 0.5, \\ 1 - (2 - 2 \cdot rand)^{1/(\eta+1)}, & \text{otherwise,} \end{cases} \quad (9.9.16)$$

for $j = 1, \ldots, n$, where $rand$ is a uniform random number from $[0, 1]$, the distribution index η and the mutation rate p_m are two control parameters, and a_j and b_j are the lower and upper bounds of the kth decision variable, respectively.

At each generation, MOEA/D-DE maintains the following:

- a population of points $\mathbf{x}^1, \ldots, \mathbf{x}^N \in \Omega$, where $\mathbf{x}^i$ is the current solution to the ith subproblem;
- $\mathbf{fv}^1, \ldots, \mathbf{fv}^N$, where $\mathbf{fv}^i$ is the **f**-value of $\mathbf{x}^i$, i.e., $\mathbf{fv}^i = [f_1(\mathbf{x}^i), \ldots, f_m(\mathbf{x}^i)]^T$ for each $i = 1, \ldots, N$;
- $\mathbf{z} = [z_1, \ldots, z_m]^T$, where z_i is the best value found so far for objective f_i.

MOEA/D-DE algorithm in [93] is shown in Algorithm 9.11.

9.9.3 Strength Pareto Evolutionary Algorithm

It is widely recognized (e.g., see [45, 155]) that multiobjective evolutionary algorithms (MOEAs) can stochastically solve multiobjective optimization problems in an acceptable timeframe.

As their name implies, MOEAs are evolutionary computation based multiobjective optimization techniques.

General MOPs are mathematically defined as follows.

Definition 9.35 (General MOP [155]) In general, a multiobjective optimization (MOP) minimizes $\mathbf{f}(\mathbf{x}) = [f_1(\mathbf{x}), \ldots, f_k(\mathbf{x})]$ subject to $g_i(\mathbf{x}) \leq 0$, $i = 1, \ldots, m$, $\mathbf{x} \in \Omega$. An MOP solution minimizes the components of a vector $\mathbf{f}(\mathbf{x})$, where $\mathbf{x}$ is an n-dimensional decision variable vector $\mathbf{x} = [x_1, \ldots, x_n]^T$ from some universe Ω.

Solution to MOP usually consists of both search (i.e., optimization) process and decision process. There are three decision making preferences [78]:

Algorithm 9.11 MOEA/D-DE algorithm for Solving (9.9.5) [93]

1. **input:**
    1.1 a stopping criterion;
    1.2 N: the number of the subproblems considered;
    1.3 $\lambda^1, \ldots, \lambda^N$: a set of N weight vectors;
    1.4 T: the number of the weight vectors in the neighborhood of each weight vector;
    1.5 δ: the probability that parent solutions are selected from the neighborhood;
    1.6 n_r: the maximal number of solutions replaced by each child solution.
2. **initialization:**
    2.1 Compute the Euclidean distances between any two weight vectors and then work out the T closest weight vectors to each weight vector. For each $i = 1, \ldots, N$, set $B(i) = \{i_1, \ldots, i_T\}$ where $\lambda^{i_1}, \ldots, \lambda^{i_T}$ are the T closest weight vectors to λ^i.
    2.2 Generate an initial population $\mathbf{x}^1, \ldots, \mathbf{x}^N$ by uniformly randomly sampling from Ω. Set $\mathbf{fv}_i = [f_1(\mathbf{x}^i), \ldots, f_m(\mathbf{x}^i)]^T$.
    2.3 Initialize $\mathbf{z} = [z_1, \ldots, z_m]^T$ by setting $z_j = \min_{1 \le i \le N} f_j(\mathbf{x}^*)$.
3. **for** $i = 1, \ldots, N$ **do**
4.   *Selection of Mating/Update Range:* Uniformly randomly generate a number *rand* from $[0, 1]$. Then set
$$P = \begin{cases} B(i), & \text{if } rand < \delta, \\ \{1, \ldots, N\}, & \text{otherwise.} \end{cases}$$
5.   *Reproduction:* Set $r_1 = i$ and randomly select two indexes r_2 and r_3 from P, and then generate a solution $\bar{\mathbf{y}}$ from $\mathbf{x}^{r_1}, \mathbf{x}^{r_2}$ and $\mathbf{x}^{r_3}$ by a differential evolution operator (9.9.14), and then perform a mutation operator (9.9.15) on $\bar{\mathbf{y}}$ with probability p_m to produce a new solution $\mathbf{y}$.
6.   *Repair:* If an element of $\mathbf{y}$ is out of the boundary of Ω, its value is reset to be a randomly selected value inside the boundary.
7.   *Update of* $\mathbf{z}$: For each $j = 1, \ldots, m$, if $z_j > f_j(\mathbf{y})$ then set $z_j = f_j(\mathbf{y})$.
8.   *Update of Solutions:* Set $c = 0$ and then do the following:
9.   If $c = n_r$ or P is empty, go to Step 3. Otherwise, randomly pick an index j from P.
10. If $g(\mathbf{y}, \lambda^j, \mathbf{z}) \le g(\mathbf{x}^j, \lambda^j, \mathbf{z})$, then set $\mathbf{x}^j = \mathbf{y}$, $\mathbf{fv}_j = [f_1(\mathbf{y}), \ldots, f_m(\mathbf{y})]^T$ and $c = c+1$.
11. Remove j from P and go to step 9.
12. *Stopping Criterion:* If the stopping criterion is satisfied, then stop and output $\{\mathbf{x}^1, \ldots, \mathbf{x}^N\}$ and $\{(f_1(\mathbf{x}^1), \ldots, f_m(\mathbf{x}^1)), \ldots, (f_1(\mathbf{x}^N), \ldots, f_m(\mathbf{x}^N))\}$. Otherwise go to Step 3.
13. **end for**
14. **Output:**
    Approximation to the Pareto set: $\{\mathbf{x}^1, \ldots, \mathbf{x}^N\}$;
    Approximation to the Pareto front: $\{(f_1(\mathbf{x}^1), \ldots, f_m(\mathbf{x}^1)), \ldots, (f_1(\mathbf{x}^N), \ldots, f_m(\mathbf{x}^N))\}$.

- *A priori preference articulation* (Decide → Search): Decision making combines the differing objectives into a scalar cost function. This effectively makes the MOP a single-objective prior to optimization.
- *Progressive preference articulation* (Search ↔ Decide): Decision making and optimization are intertwined. Partial preference information is provided upon which optimization occurs, providing an "updated" set of solutions for the decision making to consider.
- *A posteriori preference articulation* (Search → Decide): A decision making is presented with a set of Pareto optimal candidate solutions and chooses from that set. However, note that most MOEA researchers search for and present a set of nondominated vectors (PF_{known}) to the decision making.

9.9 Multiobjective Evolutionary Algorithms

As pointed out in [73] and [155], any practical MOEA implementation must include a secondary population composed of all Pareto-optimal solutions $P_{\text{known}}(t)$ found so far during search. This is because the MOEA is a stochastic optimization, which does not guarantee that desirable solutions, once found, remain in the generational population until MOEA termination. For this end, one needs to make *fitness assignment*.

Let P denote the population and P' be an external nondominated set. The fitness assignment procedure is a two-stage process [170].

1. Each solution $\mathbf{x}_i \in P'$ is assigned a real value $s_i \in [0, 1)$, called strength; s_i is proportional to the number of population members for which $\mathbf{x}_i \succeq \mathbf{x}_j$. Let n denote the number of individuals in P that are covered by $\mathbf{x}_i$ and assume N is the size of P. Then s_i is defined as $s_i = \frac{n}{N+1}$. The fitness F_i of $\mathbf{x}_i$ is equal to its strength: $F_i = s_i$.
2. The fitness of an individual $j \in P$ is calculated by summing the strengths of all external nondominated solutions $\mathbf{x}_i \in P'$ that cover $\mathbf{x}_j$. The total fitness is added by 1 in order to guarantee that members of P' have better fitness than members of P (note that fitness is to be minimized, i.e., small fitness values correspond to high reproduction probabilities):

$$F_j = 1 + \sum_{i, \mathbf{x}_i \succeq \mathbf{x}_j} s_i, \quad \text{where } F_j \in [1, N). \quad (9.9.17)$$

These are the following three main techniques used in MOEAs [170].

- Store the nondominated solutions in P' found so far externally.
- Use the concept of Pareto dominance in order to assign scalar fitness values to individuals in P.
- Perform clustering to reduce the number of nondominated solutions stored in P without destroying the characteristics of the trade-off front.

Strength Pareto EA (SPEA) uses the following techniques for finding multiple Pareto-optimal solutions in parallel [170].

(a) It combines the above three techniques in a single algorithm.
(b) The fitness of an individual in P is determined only from the solutions stored in the external nondominated set P'; whether members of the population dominate each other is irrelevant.
(c) All solutions in the external nondominated set P' participate in the selection.
(d) A niching method is provided in order to preserve diversity in the population P; this method is Pareto-based and does not require any distance parameter (like the niche radius for sharing).

Since SPEA uses a mixture of the three basic techniques in MOEAs in (a) and other three techniques (b)–(d), this MOEA method is a strength Pareto variant of MOEAs, and hence is named as SPEA.

Algorithm 9.12 shows the flow of the SPEA algorithm.

Algorithm 9.12 SPEA algorithm [170]

1. **input:** The population size N.
2. **initialization:** Generate an initial population P and create the empty external nondominated set P'.
3. Copy nondominated members of P to P'.
4. Remove solutions within P' which are covered by any other member of P'.
5. If the number of externally stored nondominated solutions exceeds a given maximum N', prune by means of clustering.
6. Calculate the fitness of each individual in P as well as in P'.
7. Select individuals from $P \cup P'$ (e.g., using binary tournament selection), until the mating pool is filled.
8. Apply problem-specific crossover and mutation operators as usual.
9. If the maximum number of generations is reached, then stop, else go to Step 4.
10. **output:** nondominated population P.

The average linkage method [107] can be used for the pruning by clustering in Step 5 in Algorithm 9.12. The average linkage method contains the following steps.

(1) Initialize cluster set C; each external nondominated point $\mathbf{x}_i \in P'$ constitutes a distinct cluster: $C = \bigcup_i \{\mathbf{x}_i\}$.
(2) If $|C| \leq N'$, go to Step (5), else go to Step (3).
(3) Calculate the distance of all possible pairs of clusters. The distance $d(c_1, c_2)$ of two clusters $c_1, c_2 \in C$ is given as the average distance between pairs of individuals across the two clusters:

$$d(c_1, c_2) = \frac{1}{|c_1| \cdot |c_2|} \sum_{\mathbf{x}_{i_1} \in c_1, \mathbf{x}_{i_2} \in c_2} \|\mathbf{x}_{i_1} - \mathbf{x}_{i_2}\|_2^2, \qquad (9.9.18)$$

where $|c_i|$ denotes the number of individuals in class $c_i, i = 1, 2$, the metric $\|\cdot\|_2^2$ reflects the Euclidean distance between two individuals $\mathbf{x}_{i_1}$ and $\mathbf{x}_{i_2}$, and is called as the *Euclidean metric* on the objective space.

(4) Determine two clusters c_1 and c_2 with minimal distance $d(c_1, c_2)$; the chosen clusters amalgamate into a larger cluster $C = C \setminus \{c_1, c_2\} \cup \{c_1 \cup c_2\}$. Go to Step (2).

(5) Compute the reduced nondominated set by selecting a representative individual per cluster. Use the centroid (the point with minimal average distance to all other points in the cluster) as representative solution.

Consider an improvement of SPEA, called SPEA2.

In multiobjective evolutionary optimization, the approximation of the Pareto-optimal set involves itself two (possibly conflicting) objectives: the distance to the optimal front is to be minimized and the diversity of the generated solutions is to be maximized (in terms of objective or parameter values). In this context, there are three fundamental issues when designing a multiobjective evolutionary algorithm [172]: fitness assignment, environmental selection, and mating selection.

Fitness Assignment

To avoid the situation that individuals dominated by the same archive members have identical fitness values, for each individual both dominating and dominated solutions are taken into account in SPEA2. Each individual i in the archive $\bar{P}_t$ and the population P_t is assigned a *strength value* $S(i)$, representing the number of solutions it dominates:

$$S(i) = |\{j \mid j \in P_t \cup \bar{P}_t \wedge i > j\}|, \qquad (9.9.19)$$

where $|\cdot|$ denotes the cardinality of a set, and the symbol $>$ corresponds to the Pareto dominance relation. Use the $S(i)$ values to calculate the *raw fitness value* $R(i)$ of an individual i as follows:

$$R(i) = \sum_{j \in P_t \cup \bar{P}_t,\ j > i} S(j). \qquad (9.9.20)$$

Remark 1 The raw fitness is determined by the strengths of its dominators in both archive and population in SPEA2, while SPEA considers only archive members in this context.

Remark 2 Fitness is to be minimized in SPEA2, i.e., $R(i) = 0$ corresponds to a nondominated individual, while a high $R(i)$ value means that i is dominated by many individuals (which in turn dominate many individuals).

The raw fitness assignment provides a sort of niching mechanism based on the concept of Pareto dominance, but it may fail when most individuals do not dominate each other. To avoid this drawback, additional density information is incorporated to discriminate between individuals having identical raw fitness values. The *density* $D(i)$ corresponding to i is defined by

$$D(i) = \frac{1}{\sigma_i^k + 2}, \qquad (9.9.21)$$

where σ_i^k denotes the distance of i to its kth nearest neighbor in $\bar{P}_{t+1}$, and $k = \sqrt{N + \bar{N}}$.

Finally, adding $D(i)$ to the raw fitness value $R(i)$ of an individual i yields its fitness $F(i)$:

$$F(i) = R(i) + D(i). \qquad (9.9.22)$$

Environmental Selection

This selection involves which individuals in the population to keep during the evolution process. For this end, beside the population, an archive is necessary to contain a representation of the nondominated Pareto front among all solutions considered so far. During environmental selection, the first step is to copy all nondominated individuals having a fitness lower than one from archive and population to the archive of the next generation:

$$\bar{P}_{t+1} = \{i | i \in P_t \cup \bar{P}_t \wedge F(i) < 1\}. \tag{9.9.23}$$

If the nondominated front fits exactly into the archive ($|\bar{P}_{t+1}| = \bar{N}$) the environmental selection step is completed. Otherwise, if the archive is too small ($|\bar{P}_{t+1}| < \bar{N}$), then the best $\bar{N} - |\bar{P}_{t+1}|$ dominated individuals in the previous archive $\bar{P}_t$ and population are copied to the new archive $\bar{P}_{t+1}$. This can be implemented by sorting the multi-set $P_t + \bar{P}_t$ according to the fitness values and copy the first $\bar{N} - |\bar{P}_{t+1}|$ individuals i with $F(i) \geq 1$ from the resulting ordered list to $\bar{P}_{t+1}$. Conversely, if the archive is too large ($|\bar{P}_{t+1}| > \bar{N}$) then an *archive truncation procedure* is invoked which iteratively removes individuals from $\bar{P}_{t+1}$ until $|\bar{P}_{t+1}| = \bar{N}$. In other words, the individual which has the minimum distance to another individual is chosen at each stage; if there are several individuals with minimum distance, the tie is broken by considering the second smallest distances and so forth.

Mating Selection
The pool of individuals at each generation is evaluated in a two stage process. First all individuals are compared on the basis of the Pareto dominance relation, which defines a partial order on this multi-set. Basically, the information which decides how each individual dominates, is dominated by or is indifferent to, is used to define a ranking on the generation pool. Afterwards, this ranking is refined by the incorporation of density information. Various density estimation techniques are used to measure the size of the niche in which a specific individual is located.

Algorithm 9.13 summarizes the improved strength Pareto evolutionary algorithm (SPEA2).

The main differences of SPEA2 in comparison to SPEA are as follows [172].

- An improved fitness assignment scheme is used, which takes for each individual into account how many individuals it dominates and it is dominated by.
- A nearest neighbor density estimation technique is incorporated which allows a more precise guidance of the search process.
- A new archive truncation method guarantees the preservation of boundary solutions.

9.9.4 Achievement Scalarizing Functions

Achievement scalarizing functions (ASFs), introduced by Wierzbicki [158], are denoted by $s_R(\mathbf{f}(\mathbf{x})) : \mathbb{R}^k \to \mathbb{R}$ which map (or scalarize) k objective functions to a scalar. The achievement scalarizing problem is given by

$$\min_{\mathbf{x} \in X} s_R(\mathbf{f}(\mathbf{x})). \tag{9.9.24}$$

9.9 Multiobjective Evolutionary Algorithms

Algorithm 9.13 Improved strength Pareto evolutionary algorithm (SPEA2) [172]

1. **input:** N: population size, $\bar{N}$: archive size, T: maximum number of generations.
2. **initialization:**
   2.1 Generate an initial population P_0 and create the empty archive (external set) $\bar{P}_0 = \emptyset$;
   2.2 Set $t = 0$.
3. **while** $t = 0, 1, \ldots$
   % **Fitness assignment:** Calculate fitness values of individuals in P_t and $\bar{P}_t$
4. Calculate the strength value $S(i) = |\{j | j \in P_t \cup \bar{P}_t \wedge i > j\}|$ for $i = 1, \ldots, N$;
5. Calculate the raw fitness $R(i) = \sum_{j \in P_t \cup \bar{P}_t, j > i} S(j)$ for $i = 1, \ldots, N$;
6. Compute the density $D(i) = \frac{1}{\sigma_i^k + 2}$ for $i = 1, \ldots, N$;
7. Compute the fitness value $F(i) = R(i) + D(i)$ for $i = 1, \ldots, N$.
   % **Environmental selection:**
8. Copy all nondominated individuals in P_t and $\bar{P}_t$ to $\bar{P}_{t+1}$: $\bar{P}_{t+1} = \{i | i \in P_t \cup \bar{P}_t \wedge F(i) < 1\}$;
9. **If** $|\bar{P}_{t+1}| = \bar{N}$ **then** the environmental selection step is completed;
10. **If** $|\bar{P}_{t+1}| < \bar{N}$ **then** sort the multi-set $P_t + \bar{P}_t$ according to the fitness values and copy the first $\bar{N} - |\bar{P}_{t+1}|$ individuals i with $F(i) \geq 1$ from the resulting ordered list to $\bar{P}_{t+1}$.
11. **If** $|\bar{P}_{t+1}| > \bar{N}$ **then** an archive truncation procedure is invoked which iteratively removes individuals from $\bar{P}_{t+1}$ until $|\bar{P}_{t+1}| = \bar{N}$.
    % **Termination:**
12. **If** $t \geq T$ or another stopping criterion is satisfied **then** set A to the set of decision vectors represented by the nondominated individuals in $\bar{P}_{t+1}$. Stop and output.
    % **Mating selection:**
13. Perform binary tournament selection with replacement on P_{t+1} in order to fill the mating pool.
14. **end while**
15. **output:** nondominated set A.

Certain properties of ASFs guarantee that scalarizing problem (9.9.24) yields Pareto-optimal solutions.

Definition 9.36 (Increasing Function [158]) An achievement scalarizing function $s_R(\mathbf{f}(\mathbf{x})) : \mathbb{R}^k \to \mathbb{R}$ is said to be

1. *Increasing:* if for any $\mathbf{y}^1, \mathbf{y}^2 \in \mathbb{R}^k$, $y_i^1 \leq y_i^2$ for all $i \in \{1, \ldots, k\}$, then $s_R(\mathbf{y}^1) \leq s_R(\mathbf{y}^2)$.
2. *Strictly increasing:* if for any $\mathbf{y}^1, \mathbf{y}^2 \in \mathbb{R}^k$, $y_i^1 < y_i^2$ for all $i \in \{1, \ldots, k\}$, then $s_R(\mathbf{y}^1) < s_R(\mathbf{y}^2)$.
3. *Strongly increasing:* if for any $\mathbf{y}^1, \mathbf{y}^2 \in \mathbb{R}^k$, $y_i^1 \leq y_i^2$ for all $i \in \{1, \ldots, k\}$ and $\mathbf{y}^1 \neq \mathbf{y}^2$, then $s_R(\mathbf{y}^1) < s_R(\mathbf{y}^2)$.

Clearly, any strongly increasing ASF is also strictly increasing, and any strictly increasing ASF is also increasing. The following theorems define necessary and sufficient conditions for an optimal solution of (9.9.24) to be (weakly) Pareto optimal.

Theorem 9.3 ([159, 160]) *The necessary and sufficient conditions for an optimal solution being Pareto optimal are as follows.*

1. Let s_R be strongly (strictly) increasing. If $\mathbf{x}^* \in X$ is an optimal solution of problem (9.9.24), then $\mathbf{x}^*$ is (weakly) Pareto optimal.
2. If s_R is increasing and the solution $\mathbf{x}^* \in X$ of (9.9.24) is unique, then $\mathbf{x}^*$ is Pareto optimal.

Theorem 9.4 ([103]) *If s_R is strictly increasing and $\mathbf{x}^* \in X$ is weakly Pareto optimal, then it is a solution of (9.9.24) with the reference point $\mathbf{f}(\mathbf{x}^*)$ and the optimal value of s_R is zero.*

One example of the achievement scalarizing problems can be formulated as

$$\min \max_{i=1,\ldots,k} \left\{ \frac{f_i(\mathbf{x}) - \bar{z}_i}{z_i^{\text{nad}} - z_i^{\text{utopia}}} \right\} + \rho \sum_{i=1}^{k} \frac{f_i(\mathbf{x})}{z_i^{\text{nad}} - z_i^{\text{utopian}}} \qquad (9.9.25)$$

$$\text{subject to } \mathbf{x} \in S, \qquad (9.9.26)$$

where the term $\rho \sum_{i=1}^{k} \frac{f_i(\mathbf{x})}{z_i^{\text{nad}} - z_i^{\text{utopia}}}$ is called the augmentation term, in which $\rho > 0$ is a small constant, and z^{nad} and z^{utopian} are the nadir vector and utopian vectors, respectively. In the above problem, the parameter is the so-called reference point $\bar{z}_i$ which represents objective function values preferred by the decision maker.

The following are the several most well-known ASFs [116].

1. The strictly increasing ASF is of Chebyshev type:

$$s_R^{\infty}(\mathbf{f}(\mathbf{x}), \boldsymbol{\lambda}) = \max_{i \in \{1,\ldots,k\}} \lambda_i (f_i(\mathbf{x}) - f_i(\mathbf{x}^*)), \qquad (9.9.27)$$

where $\mathbf{x}^*$ is a reference point, and $\boldsymbol{\lambda}$ is k-vector of nonnegative coefficients used for scaling purposes, that is, for normalizing objective functions of different magnitudes.

2. The strongly increasing ASF is of augmented Chebyshev type:

$$s_R^{\infty+1}(\mathbf{f}(\mathbf{x}), \boldsymbol{\lambda}) = \rho \sum_{i \in \{1,\ldots,k\}} \lambda_i (f_i(\mathbf{x}) - f_i(\mathbf{x}^*)) + \max_{i \in \{1,\ldots,k\}} \lambda_i (f_i(\mathbf{x}) - f_i(\mathbf{x}^*)), \qquad (9.9.28)$$

where $\rho > 0$ is a small parameter.

3. The additive ASF based on the L_1 metric [137]:

$$s_R^1(\mathbf{f}(\mathbf{x}), \boldsymbol{\lambda}) = \max_{i \in \{1,\ldots,k\}} \left\{ \lambda_i (f_i(\mathbf{x}) - f_i(\mathbf{x}^*)), 0 \right\}. \qquad (9.9.29)$$

9.9 Multiobjective Evolutionary Algorithms

4. The parameterized ASF: Let I_q be a subset of $N_k = \{1, \ldots, k\}$ of cardinality q. The parameterized ASF is defined as follows [116]:

$$\tilde{s}_R^q(\mathbf{f}(\mathbf{x}), \boldsymbol{\lambda}) = \max_{I_q \in N_k : |I_q| = q} \left\{ \sum_{i \in I_q} \max[\lambda_i (f_i(\mathbf{x}) - f_i(\mathbf{x}^*)), 0] \right\}, \qquad (9.9.30)$$

where $q \in N_k$ and $\boldsymbol{\lambda} = [\lambda_1, \ldots, \lambda_k]^T$, $\lambda_i > 0$, $i \in N_k$. Notice that

- for $q \in N_k$: $\tilde{s}_R^q(\mathbf{f}(\mathbf{x}), \boldsymbol{\lambda}) \geq 0$;
- for $q = 1$: $\tilde{s}_R^1(\mathbf{f}(\mathbf{x}), \boldsymbol{\lambda}) = \max_{i \in N_k} \max[\lambda_i (f_i(\mathbf{x}) - f_i(\mathbf{x}^*)), 0] \cong s_R^\infty(\mathbf{f}(\mathbf{x}), \boldsymbol{\lambda})$;
- for $q = k$: $\tilde{s}_R^k(\mathbf{f}(\mathbf{x}), \boldsymbol{\lambda}) = \sum_{i \in N_k} \max[\lambda_i (f_i(\mathbf{x}) - f_i(\mathbf{x}^*)), 0] = s_R^1(\mathbf{f}(\mathbf{x}), \boldsymbol{\lambda})$.

Here, "$\cong$" means equality in the case where there exist no feasible solutions $\mathbf{x} \in X$ which strictly dominate the reference point, i.e., such that $f_i(\mathbf{x}) < f_i(\mathbf{x}^*)$ for all $i \in N_k$.

The performance of the additive ASF can be described by the following two theorems.

Theorem 9.5 ([137]) *Given problem (9.9.24) with ASF defined by (9.9.29), let $\mathbf{f}(\mathbf{x}^*)$ be a reference point such that $\mathbf{f}(\mathbf{x}^*)$ is not dominated by an objective vector of any feasible solution of problem (9.9.24). Also assume $\lambda_i > 0$ for all $i \in \{1, \ldots, k\}$. Then any optimal solution of problem (9.9.24) is a weakly Pareto-optimal solution.*

Theorem 9.6 ([137]) *Given problem (9.9.24) with ASF defined by (9.9.29) and any reference point $\mathbf{f}(\mathbf{x}^*)$, and assume $\lambda_i > 0$ for all $i \in \{1, \ldots, k\}$. Then among the optimal solutions of problem (9.9.24) there exists at least one Pareto-optimal solution. If the optimal solution of problem (9.9.24) is unique, then it is Pareto optimal.*

When adopting the parameterized ASF, the corresponding parameterized achievement scalarizing problem is given by Nikulin et al. [116]:

$$\min_{\mathbf{x} \in X} \tilde{s}_R^q(\mathbf{f}(\mathbf{x}), \boldsymbol{\lambda}). \qquad (9.9.31)$$

For any $\mathbf{x} \in X$, denoting $I_x = \{i \in N_k : f_i(\mathbf{x}^*) \leq f_i(\mathbf{x})\}$, then the following two results are true:

Theorem 9.7 ([116]) *Given problem (9.9.31), let $\mathbf{f}(\mathbf{x}^*)$ be a reference point vector such that there exists no feasible solution whose image strictly dominates $\mathbf{f}(\mathbf{x}^*)$. Also assume $\lambda_i > 0$ for all $i \in N_k$. Then, any optimal solution of problem (9.9.31) is a weakly Pareto-optimal solution.*

Theorem 9.8 ([116]) *Given problem (9.9.31), let $\mathbf{f}(\mathbf{x}^*)$ be any reference point. Also assume $\lambda_i > 0$ for all $i \in N_k$. Then, among the optimal solutions of problem (9.9.31) there exists at least one Pareto-optimal solution.*

Theorem 9.8 implies that the uniqueness of the optimal solution guarantees its Pareto optimality.

9.10 Evolutionary Programming

It is widely recognized [44] that evolutionary programming (EP) was first proposed as an approach to artificial intelligence. The EP has been applied with success to many numerical and combinatorial optimization problems.

9.10.1 Classical Evolutionary Programming

Consider a global minimization problem min $f(\mathbf{x})$ formalized as a pair (S, f), where $\mathbf{x}_{\min} \in S$ with S being a bounded set on $\mathbb{R}^n$ and f is an n-dimensional real-valued function. Our aim is to find a point $\mathbf{x}_{\min} \in S$ such that $f(\mathbf{x}_{\min})$ is a global minimum on S, namely

$$f(\mathbf{x}_{\min}) \leq f(\mathbf{x}), \quad \forall \mathbf{x} \in S, \tag{9.10.1}$$

where f is a bounded function which does not need to be continuous.

Optimization by EP can be summarized into two major steps:

- mutate the solutions in the current population;
- select the next generation from the mutated solutions and the current solutions.

There are the classical evolutionary programming (CEP) with self-adaptive mutation and the CEP without self-adaptive mutation. It is well-known (see, e.g., [3, 42, 165]) that the former usually performs better than the latter.

By Bäck and Schwefel [3], the CEP is implemented as follows (see also [165]).

1. *Initialization:* Generate the initial population of μ individuals, and set $k = 1$. Each individual is taken as a pair of real-valued vectors $(\mathbf{x}_i, \boldsymbol{\eta}_i)$ ($i = 1, \ldots, \mu$), where $\mathbf{x}_i$ are objective variables and $\boldsymbol{\eta}_i$ are standard deviations for Gaussian mutations (also called strategy parameters in self-adaptive evolutionary algorithms).
2. *Evaluation of fitness:* Use the objective function $f(\mathbf{x}_i)$ to evaluate the fitness score for each individual $(\mathbf{x}_i, \boldsymbol{\eta}_i)$, $\forall i = 1, \ldots, \mu$.
3. *Self-adaptive mutation:* Each parent $(\mathbf{x}_i(t-1), \boldsymbol{\eta}_i(t-1))$, $i = 1, \ldots, \mu$ at the tth generation creates a single offspring $(\mathbf{x}_i(t), \boldsymbol{\eta}_i(t))$ at the tth generation:

$$x_i^j(t) = x_i^j(t-1) + \eta_i^j(t-1) N_j(0, 1), \tag{9.10.2}$$

$$\eta_i^j(t) = \eta_i^j(t-1) \exp(\tau' N(0, 1) + \tau N_j(0, 1)), \tag{9.10.3}$$

for $i = 1, \ldots, \mu$; $j = 1, \ldots, n$. Here $x_i^j(t-1)$ and $\eta_i^j(t-1)$ denote the jth components of the parent vectors $\mathbf{x}_i(t-1)$ and $\boldsymbol{\eta}_i(t-1)$ at the $(t-1)$th generations, respectively; and $x_i^j(t-1)$ and $\eta_i^j(t-1)$ are, respectively, the jth components of the offspring $\mathbf{x}_i(t)$ and $\boldsymbol{\eta}_i(t)$ at the tth generation; while $N(0, 1)$ is a standard Gaussian distribution used for all $j = 1, \ldots, n$, and $N_j(0, 1)$ represents the standard Gaussian distribution used just for some given j. The factors are commonly set to $\tau = \left(\sqrt{2\sqrt{n}}\right)^{-1}$ and $\tau' = (\sqrt{2n})^{-1}$.

4. *Fitness for each offspring:* Calculate the fitness of each offspring $(\mathbf{x}_i(t), \boldsymbol{\eta}_i(t))$, $i = 1, \ldots, \mu$.
5. *Pairwise comparison:* Conduct pairwise comparison over the union of parents $(\mathbf{x}_i(t-1), \boldsymbol{\eta}_i(t-1))$ and offspring $(\mathbf{x}_i(t), \boldsymbol{\eta}(t))$ for all $i = 1, \ldots, \mu$. For each individual, opponents are chosen uniformly at random from all the parents and offspring. For each comparison, if the individual's fitness is no smaller than the opponent's, it receives a "win."
6. *Selection:* Select the μ individuals out of $(\mathbf{x}_i(t-1), \boldsymbol{\eta}_i(t-1))$ and $(\mathbf{x}_i(t), \boldsymbol{\eta}_i(t))$, that have the most wins to be parents of the next generation.
7. *Stopping criteria:* Stop if the halting criterion is satisfied; otherwise, $k = k + 1$ and go to Step 3, and repeat the above steps.

The feature of CEP is to use the standard Gaussian mutation operator $N_j(0, 1)$ in self-adaptive mutation (9.10.2).

9.10.2 Fast Evolutionary Programming

To speed up the convergence of the CEP, a fast evolutionary programming (FEP) was suggested by Yao et al. [165]. The one-dimensional Cauchy density function centered at the origin is defined by

$$f_t(x) = \frac{1}{\pi} \frac{t}{t^2 + x^2}, \quad -\infty < x < +\infty, \tag{9.10.4}$$

where $t > 0$ is a scale parameter. The corresponding distribution function is

$$F_t(x) = \frac{1}{2} + \frac{1}{\pi} \arctan\left(\frac{x}{t}\right). \tag{9.10.5}$$

The variance of the Cauchy distribution is infinite.

The FEP is exactly the same as the CEP except for Gaussian mutation operator $N_j(0, 1)$ in (9.10.2) which is replaced by Cauchy mutation operator as follows:

$$x_i'(j) = x_i(j) + \eta_i(j) C_j(0, 1), \tag{9.10.6}$$

where $C_j(0, 1)$ is a standard Cauchy random variable with the scale parameter $t = 1$ and is generated anew for each value of j.

The main difference between the FEP and CEP is: the CEP mutation equation (9.10.2) is a self-adaption controlled by the Gaussian distribution $N_j(0, 1)$; while the FEP mutation equation (9.10.6) is a self-adaption controlled by the Cauchy random variable δ_i.

It is known [165] that Cauchy mutation is more likely to generate an offspring further away from its parent than Gaussian mutation due to its long flat tails. It is expected to have a higher probability of escaping from a local optimum or moving away from a plateau, especially when the "basin of attraction" of the local optimum or the plateau is large relative to the mean step size. Therefore, the FEP is expected to be faster than the CEP from viewpoint of convergence.

The basic steps of the *FEP algorithm* are described as follows [18].

1. Generate an initial population of μ solutions $(\mathbf{x}_i, \boldsymbol{\eta}_i), i \in \{1, \ldots, \mu\}$, and evaluate their fitness $f(\mathbf{x}_i)$.
2. For every parent solution $(\mathbf{x}_i, \boldsymbol{\eta}_i)$ in the population, create an offspring $(\mathbf{x}'_i, \boldsymbol{\eta}'_i)$ using

$$\left.\begin{array}{l} x'_i(j) = x_i(j) + \eta_i(j) C_j(0, 1) \\ \eta'_i(j) = \eta_i(j) \exp(\tau' N(0, 1) + \tau N_j(0, 1)) \end{array}\right\}, \quad i = 1, \ldots, \mu; \; j = 1, \ldots, n,$$

where $x_i(j), \eta_i(j)$ denote the jth components of the parent vectors $\mathbf{x}_i, \boldsymbol{\eta}_i$, and $x'_i(j), \eta'_i(j)$ are the jth components of the offspring $\mathbf{x}'_i, \boldsymbol{\eta}'_i$, respectively; while $N(0, 1)$ is sampled from a standard Gaussian distribution for all $j = 1, \ldots, n$, and $N_j(0, 1)$ is sampled from an another standard Gaussian distribution just for some given j, while $C_j(0, 1)$ is sampled from a standard Cauchy distribution for each i and j. The factors τ and τ' are commonly set to $(\sqrt{2\sqrt{n}})^{-1}$ and $\tau' = (\sqrt{2n})^{-1}$.

3. For every parent solution $(\mathbf{x}_i, \boldsymbol{\eta}_i)$ in the population, create another offspring $(\mathbf{x}''_i, \boldsymbol{\eta}''_i)$ using

$$\left.\begin{array}{l} x''_i(j) = x_i(j) + \eta_i(j) C_j(0, 1) \\ \eta''_i(j) = \eta_i(j) \exp(\tau' N(0, 1) + \tau N_j(0, 1)) \end{array}\right\}, \quad i = 1, \ldots, \mu; \; j = 1, \ldots, n.$$

4. Evaluate the fitness of offspring solutions $(\mathbf{x}'_i, \boldsymbol{\eta}'_i)$ and $(\mathbf{x}''_i, \boldsymbol{\eta}''_i)$.
5. Perform pairwise tournaments over the union of parents $(\mathbf{x}_i, \boldsymbol{\eta}_i)$ and offspring (i.e., $(\mathbf{x}'_i, \boldsymbol{\eta}'_i)$ and $(\mathbf{x}''_i, \boldsymbol{\eta}''_i)$). For each solution in the union, q opponents are chosen uniformly at random from the union. For every tournament, if the fitness of the solution is no smaller than the fitness of the opponent, it is awarded a "win." The total number of wins awarded to a solution $(\mathbf{x}_i, \boldsymbol{\eta}_i)$ is denoted by $\text{win}(\mathbf{x}_i, \boldsymbol{\eta}_i)$.
6. Choose μ individual solutions from the union of parents and offspring at step 5. Chosen solutions will form the population of the next generation.

9.10 Evolutionary Programming

7. Stop if the number of generations exceeds a predetermined number.

By performing on a number of benchmark problems, the experimental results show [165] that FEP with the Cauchy mutation operator $C_j(0, 1)$ performs much better than CEP with the Gaussian mutation operator for multimodal functions with many local minima while being comparable to CEP in performance for unimodal and multimodal functions with only a few local minima.

A comparative study of applications of Cauchy and Gaussian mutation operators in electromagnetics reported in [70] and [72] was also observed that Cauchy mutation operator performs better than its Gaussian counterpart for a number of unconstrained or weakly constrained antenna optimization problems, but it performs relatively poor for highly constrained problems.

9.10.3 Hybrid Evolutionary Programming

Consider evolutionary programming for constrained optimization problems

$$P: \quad \min f(\mathbf{x}) \text{ subject to } g_i(\mathbf{x}) \leq 0, i = 1, \ldots, r, \quad h_j(\mathbf{x}) = 0, j = 1, \ldots, m, \tag{9.10.7}$$

where f and $g_1, \ldots, g_r$ are functions on $\mathbb{R}^n$, and $h_1, \ldots, h_m$ are functions on $\mathbb{R}^m$ for $m \leq n$ and $\mathbf{x} = [x_1, \ldots, x_n]^T \in \mathbb{R}^n$ and $\mathbf{x} \in F \subseteq S$. A vector $\mathbf{x}$ is called a feasible solution to the optimization problem (P) if and only if $\mathbf{x}$ satisfies the r inequality constraints $g_i(\mathbf{x}) \leq 0, i = 1, \ldots, r$ and m equality constraints $h_j(\mathbf{x}) = 0, j = 1, \ldots, m$. When the collection of feasible solutions is empty, $\mathbf{x}$ is said to be infeasible. The set $S \subseteq \mathbb{R}^n$ defines the search space for $\mathbf{x}$ and the set $F \subseteq S$ defines a feasible part of the search space S.

Use the cost function to evaluate a feasible solution, i.e.,

$$\Phi_f(\mathbf{x}) = f(\mathbf{x}), \quad \mathbf{x} \in F, \tag{9.10.8}$$

and define the constraint violation measure for the constraints as follows:

$$\Phi_u(\mathbf{x}) = \sum_{i=1}^{r} [g_i(\mathbf{x})]^+ \sum_{j=1}^{m} |h_j(\mathbf{x})| \tag{9.10.9}$$

or

$$\Phi_u(\mathbf{x}) = \frac{1}{2} \left(\sum_{i=1}^{r} (g_i(\mathbf{x}))^2 + \sum_{j=1}^{m} (h_j(\mathbf{x}))^2 \right), \tag{9.10.10}$$

where $|\cdot|$ denotes an absolute value of the argument, and

$$g_i^+(\mathbf{x}) = \max\{0, g_i(\mathbf{x})\} \tag{9.10.11}$$

is the magnitude of the violation of the ith constraint in (P), where $1 \leq i \leq r$. Then,

$$\Phi(\mathbf{x}) = \Phi_f(\mathbf{x}) + s\Phi_u(\mathbf{x}) \tag{9.10.12}$$

defines the total evaluation of an individual $\mathbf{x}$, where s is a penalty parameter of a positive for minimization or a negative constant for maximization. Clearly, $\Phi(\mathbf{x})$ can be interpreted as the error (for a minimization problem) or fitness (for a maximization problem) of an individual $\mathbf{x}$ to the optimization problem (P).

Therefore, the original constrained optimization problem (P) becomes the following unconstrained optimization problem:

$$\min \{f(\mathbf{x}) + s\Phi_u(\mathbf{x})\}, \tag{9.10.13}$$

where $\Phi_u(\mathbf{x})$ is defined by (9.10.9) or (9.10.10).

Different from CEP and FEP whose each individual is a two-tuples of real-valued vectors $(\mathbf{x}_i, \boldsymbol{\eta}_i)$, Myung et al. [112, 113] proposed an EP to update a triplet of real-valued vectors $(\bar{\mathbf{x}}_i, \bar{\boldsymbol{\sigma}}_i, \bar{\boldsymbol{\eta}}_i), \forall i \in \{1, \ldots, \mu\}$, and named it as the *hybrid evolutionary programming* (HEP).

In HEP, $\bar{\mathbf{x}}_i = [x_i(1), \ldots, x_i(n)]^T$, while $\bar{\boldsymbol{\sigma}}_i$ and $\bar{\boldsymbol{\eta}}_i$ are the n-dimensional solution vector and its two corresponding strategy parameter vectors, respectively. Here $\bar{\boldsymbol{\sigma}}_i$ and $\bar{\boldsymbol{\eta}}_i$ are initialized based on the specified search domains, $\bar{\mathbf{x}}_i \in \{x_{\min}, x_{\max}\}^n$, which may be imposed at the initialized stage.

HEP is named because of applying hybrid (Cauchy + Gaussian) mutation operators as follows [71]:

$$x_i'(j) = x_i(j) + \sigma_i'(j)[N_j(0, 1) + \beta_i'(j)C_j(0, 1)], \tag{9.10.14}$$

$$\beta_i'(j) = \frac{\eta_j'(j)}{\sigma_i'(j)}, \tag{9.10.15}$$

$$\sigma_i'(j) = \sigma_i(j) \exp(\tau' N(0, 1) + \tau N_j(0, 1)), \tag{9.10.16}$$

$$\eta_i'(j) = \eta_i(j) \exp(\tau' N(0, 1) + \tau N_j(0, 1)), \tag{9.10.17}$$

for $i = 1, \ldots, n$, where $x_i(j), \sigma_i(j)$, and $\eta_i(j)$ are the jth components of the vectors $\mathbf{x}_i, \boldsymbol{\sigma}_i$, and $\boldsymbol{\eta}_i$, respectively. $C_j(0, 1)$ in (9.10.14) and $N_j(0, 1)$ in (9.10.14), (9.10.16), and (9.10.17) indicate that the random variables are generated anew for each value of j in each equation.

9.11 Differential Evolution

Problems involving global optimization over continuous spaces are ubiquitous throughout the scientific community [147]. When the cost function is nonlinear and non-differentiable, one usually uses direct search approaches. Most standard direct search methods use the greedy criterion under which a new parameter vector is accepted if and only if it reduces the value of the cost function. Although the greedy decision process converges fairly fast, it is easily trapped in a local minimum.

Users generally demand that a practical minimization technique should fulfill the following requirements [147].

1. Ability to handle non-differentiable, nonlinear, and multimodal cost functions.
2. Parallelizability to cope with intensive computation cost functions.
3. Ease of use, i.e., few control variables to steer the minimization. These variables should also be robust and easy to choose.
4. Good convergence properties, i.e., consistent convergence to the global minimum in consecutive independent trials.

A minimization method called "differential evolution" (DE) was designed by Storn and Price in 1997 [147] to fulfill all of the above requirements.

9.11.1 Classical Differential Evolution

Differential evolution (DE) of Storn and Price [124, 147] is a simple yet effective algorithm for global optimization. DE is a parallel direct search method which utilizes D-dimensional parameter vectors

$$\mathbf{p}_i^G = [P_{1i}^G, \ldots, P_{Di}^G]^T, \quad \forall i \in \{1, \ldots, N_P\} \quad (9.11.1)$$

as a population for each generation G, where N_P is the population size which does not change during the minimization process. The initial vector population is assumed to be a uniform probability distribution for all random decisions.

Solutions in DE are represented by D-dimensional vectors $\mathbf{p}_i, \forall i \in \{1, \ldots, N_P\}$. For each solution $\mathbf{p}_i$, select three parents $\mathbf{p}_{i_1}, \mathbf{p}_{i_2}, \mathbf{p}_{i_3}$, where $i \neq i_1 \neq i_2 \neq i_3$.

Algorithm 9.14 gives the details of the classical DE algorithm, in which BFV: best fitness value so far, NFC: number of function calls, VTR: value to search, and MAX$_{\text{NFC}}$: maximum number of function calls.

The main operations of the classical DE algorithm can be summarized as follows [147].

1. *Mutation:* For each target vector $\mathbf{p}_i^G$, $i = 1, 2, \ldots, N_P$, create the mutant individual $\mathbf{v}_i^{G+1}$ by adding a weighted difference vector between two individuals

Algorithm 9.14 Differential evolution (DE) [128]
1. **input:** Population size N_p, problem dimension D, mutation constant F, crossover rate C_r.
2. **initialization:** Generate uniformly distributed random population $\mathbf{p}_0 = [P_{01}, \ldots, P_{iD}]^T$.
3. **while** (BFV > VTR) and (NFC < MAX$_{\text{NFC}}$) **do**
4.   **for** $i = 0$ to N_p **do**
5.   select three parents $\mathbf{p}_{i_1}, \mathbf{p}_{i_2}, \mathbf{p}_{i_3}$ randomly from current population where $i \neq i_1 \neq i_2 \neq i_3$.
    // Mutation
6. $\mathbf{v}_i \leftarrow \mathbf{p}_{i_1} + F \times (\mathbf{p}_{i_3} - \mathbf{p}_{i_2})$ and denote $\mathbf{v}_i = [V_{1i}, \ldots, V_{Di}]^T$.
    // Crossover
7.   **for** $j = 1$ to D **do**
8.   **if** rand$(0, 1) < C_r$ **then**
9. $U_{ji} \leftarrow V_{ji}$,
10.   **else**
11. $U_{ji} \leftarrow P_{ji}$.
12.   **end if**
13.   **end for**
    // Selection
14.   Denote $\mathbf{u}_i = [U_{1i}, \ldots, U_{Di}]^T$.
15.   **if** $f(\mathbf{u}_i) \leq f(\mathbf{p}_i)$ **then**
16. $\mathbf{p}'_i \leftarrow \mathbf{u}_i$,
17.   **else**
18. $\mathbf{p}'_i \leftarrow \mathbf{p}_i$.
19.   **end if**
20.   **end for**
21. $\mathbf{p}_{i+1} \leftarrow \mathbf{p}'_i$.
22. **end while**
23. **output:** solution vector $\mathbf{p}$.

$\mathbf{p}_{i_2}^G$ and $\mathbf{x}_{i_3}^G$ according to

$$\mathbf{v}_i^{G+1} = \mathbf{p}_{i_1}^G + F \cdot (\mathbf{p}_{i_3}^G - \mathbf{p}_{i_2}^G), \tag{9.11.2}$$

where $F > 0$ is a constant coefficient used to control the differential variation $\mathbf{d}_i^G = \mathbf{p}_{i_3}^G - \mathbf{p}_{i_2}^G$. The evolution optimization based on difference vector between two populations or solutions is known as the *differential evolution*.

2. *Crossover:* Denote $\mathbf{u}_i^G = [U_{1i}^G, \ldots, U_{Di}^G]^T$. In order to increase the diversity of the perturbed parameter vectors, introduce crossover

$$U_{ji}^{G+1} = \begin{cases} V_{ji}^{G+1}, & \text{rand}_j(0, 1) \leq CR \text{ or } j = j_{\text{rand}}, \\ P_{ji}^G, & \text{otherwise}, \end{cases} \tag{9.11.3}$$

for $j = 1, 2, \ldots, D$, where rand$_j(0, 1)$ is an uniformly distributed random number between 0 and 1, and j_{rand} is a randomly chosen index to ensure that the trial vector $\mathbf{u}_i^{G+1}$ does not duplicate $\mathbf{p}_i^G$. $CR \in (0, 1)$ is the crossover rate.

3. *Selection:* To decide whether or not it should become a member of the next generation $G + 1$, the trial vector $\mathbf{u}_i^{G+1}$ is compared to the target vector $\mathbf{p}_i^G$ according to the fitness values of the parent individual $\mathbf{p}_i^G$ and the trial individual

9.11 Differential Evolution

$\mathbf{u}_i^{G+1}$, i.e.,

$$\mathbf{p}_i^{G+1} = \begin{cases} \mathbf{u}_i^{G+1}, & f(\mathbf{u}_i^{G+1}) < f(\mathbf{p}_i^{G}); \\ \mathbf{p}_i^{G}, & \text{otherwise.} \end{cases} \quad (9.11.4)$$

Here $\mathbf{p}_i^{G+1}$ is the offspring of $\mathbf{p}_i^{G}$ for the next generation.

9.11.2 Differential Evolution Variants

There are several schemes of DE based on different mutation strategies [124]:

$$\mathbf{v}_i^G = \mathbf{p}_{r_1}^G + F \cdot (\mathbf{p}_{r_2}^G - \mathbf{p}_{r_3}^G), \quad (9.11.5)$$

$$\mathbf{v}_i^G = \mathbf{p}_{best}^G + F \cdot (\mathbf{p}_{r_1}^G - \mathbf{p}_{r_2}^G), \quad (9.11.6)$$

$$\mathbf{v}_i^G = \mathbf{p}_i^G + F \cdot (\mathbf{p}_{best}^G - \mathbf{p}_i^G) + F \cdot (\mathbf{p}_{r_1}^G - \mathbf{p}_{r_2}^G), \quad (9.11.7)$$

$$\mathbf{v}_i^G = \mathbf{p}_{best}^G + F \cdot (\mathbf{p}_{r_1}^G - \mathbf{p}_{r_2}^G) + F \cdot (\mathbf{p}_{r_3}^G - \mathbf{p}_{r_4}^G), \quad (9.11.8)$$

$$\mathbf{v}_i^G = \mathbf{p}_{r_1}^G + F \cdot (\mathbf{p}_{r_2}^G - \mathbf{p}_{r_3}^G) + F \cdot (\mathbf{p}_{r_4}^G - \mathbf{p}_{r_5}^G), \quad (9.11.9)$$

where $\mathbf{p}_{best}^G$ is the best vector in the population. $\mathbf{p}_{r_1}^G, \mathbf{p}_{r_2}^G, \mathbf{p}_{r_3}^G$, and $\mathbf{p}_{r_4}^G$ are four different randomly selected vectors (where $i \neq r_1 \neq r_2 \neq r_3 \neq r_4$) from the current population.

These schemes can be unified by the notation $DE/a/b/c$ [101]. The meaning of a, b, c in this notation is as follows [101, 128].

- "a" specifies the vector which should be mutated; it can be the best vector ($a =$ "$best$") of the current population or a randomly selected one ($a =$ "$rand$").
- "b" denotes the number of difference vectors which participate in the mutation ($b = 1$ or $b = 2$).
- "c" denotes the applied crossover scheme, binary ($c =$ "bin") or exponential ($c =$ "exp").

The classical DE (9.11.5) with notation $DE/rand/1/bin$ and the scheme (9.11.7) with notation $DE/best/2/exp$ and exponential crossover are the most often used in practice due to their good performance [124, 166].

There are the following DE variants.

1. *Neighborhood search differential evolution (NSDE):* it has been shown to play a crucial role in improving evolutionary programming (EP) algorithm's performance [165]. The neighborhood search differential evolution (NSDE) developed in [162] is the same as the above classical DE except for mutation (9.11.2)

replaced by

$$\mathbf{v}_i^{G+1} = \mathbf{p}_{r_1}^G + \begin{cases} \mathbf{d}_i^G \cdot N(0.5, 0.5), & \text{if rand}(0, 1) < 0.5; \\ \delta \cdot \mathbf{d}_i^G, & \text{otherwise.} \end{cases} \quad (9.11.10)$$

Here $\mathbf{d}_i^G = \mathbf{p}_{r_2}^G - \mathbf{p}_{r_3}^G$, $N(0.5, 0.5)$ denotes a Gaussian random number with mean 0.5 and standard deviation 0.5, and δ denotes a Cauchy random variable with scale parameter 1.

2. *Self-adaptive differential evolution (SaDE):* As compared with the classical DE of Storn and Price [147], the SaDE of Qin and Suganthan [125] has the following main differences: (a) SaDE adopts two different mutation strategies in single DE variant and introduces a probability p to control which mutation strategy to use, and p is gradually self-adapted according to the learning experience. (b) SaDE utilizes two methods to adapt and self-adapt DE's parameters F and CR. The detailed operations of SaDE are summarized as follows:

   - *Mutation strategies self-adaptation:* SaDE selects mutation strategies (9.11.5) and (9.11.7) as candidates to create the mutant individual:

$$\mathbf{v}_i^{G+1} = \begin{cases} \text{Equation (9.11.5)}, & \text{if rand}_i(0, 1) < p; \\ \text{Equation (9.11.7)}, & \text{otherwise.} \end{cases} \quad (9.11.11)$$

   Here, the probability p is updated as

$$p = \frac{ns1 \cdot (ns2 + nf2)}{ns2 \cdot (ns1 + nf1) + ns1 \cdot (ns2 + nf2)}, \quad (9.11.12)$$

   where $ns1$ and $ns2$ are, respectively, the number of offspring successfully entering the next generation while generated by Eqs. (9.11.5) and (9.11.7) after evaluation of all offspring; and both $nf1$ and $nf2$ are the numbers of offspring discarded while generated by Eqs. (9.11.5) and (9.11.7), respectively.

   - *Scale factor F setting:*

$$F_i = N_i(0.5, 0.3), \quad (9.11.13)$$

   where $N_i(0.5, 0.3)$ denotes a Gaussian random number with mean 0.5 and standard deviation 0.3.

   - *Crossover rate CR self-adaptation:* SaDE allocates a CR_i for each individuals:

$$CR_i = N_i(CRm, 0.1) \quad (9.11.14)$$

with the updating formula

$$CRm = \frac{1}{|CRrec|} \sum_{k=1}^{|CRrec|} CRrec(k), \tag{9.11.15}$$

where CRm is set to 0.5 initially, and $CRrec$ is the CR values associated with offspring successfully entering the next generation. $CRrec$ will be reset once CRm is updated. This self-adaptation scheme for CR is denoted as SaCR.

3. *Self-adaptive NSDE (SaNSDE):* This is an algorithm incorporating self-adaptive mechanisms into NSDE, see [166] for details.

On opposition-based differential evolution, an important extension to DE, we will present it in Sect. 9.15.

9.12 Ant Colony Optimization

In addition to evolutionary algorithms, another group of nature inspired metaheuristic algorithms is based on swarm intelligence in biology society.

Swarm intelligence refers to a kind of problem-solving ability through the interactions of simple information-processing units. Swarm intelligence includes two important concepts [85]:

1. The concept of a swarm means multiplicity, stochasticity, randomness, and messiness.
2. The concept of intelligence suggests that the problem-solving method is somehow successful.

The information-processing units that compose a swarm can be animate (such as ants, bees, fishes, etc.), mechanical, computational, or mathematical. Three famous swarm intelligence algorithms are ant colony algorithms, artificial bee colony algorithms, and swarm intelligence algorithms. In this section, we focus on ant colony algorithms.

The original inspired optimization idea behind ant algorithms came from observations of the foraging behavior of ants in the wild, and moreover, the phenomena called *stigmergy* [111]. By stigmergy, it means the indirect communication among a self-organizing emergent system via individuals modifying their local environment [59]. The stigmergic nature of ant colonies was explored by Deneubourg et al. in 1990 [27]: ants communicate indirectly by laying down pheromone trails, which ants then tend to follow.

Depending on the criteria being considered, modern heuristic algorithms for solving combinatorial and numeric optimization problems can be classified into the following four main different groups [82]:

- *Population-based algorithm:* An algorithm working with a set of solutions and trying to improve them is called population based algorithm.
- *Iterative based algorithm:* An algorithm using multiple iterations to approach the solution sought is named as iterative based algorithm.
- *Stochastic algorithm:* An algorithm employing a probabilistic rule for improving a solution is called probabilistic or stochastic algorithm.
- *Deterministic algorithm:* An algorithm employing a deterministic rule for improving a solution is known as deterministic algorithm.

The most popular population-based algorithm is evolutionary algorithms, and the most typical stochastic algorithms are swarm intelligence based algorithms. Both evolutionary algorithms and swarm intelligence based algorithms depend on the nature of phenomenon simulated by algorithms. As the most popular evolutionary algorithm, genetic algorithm attempts to simulate the phenomenon of natural evolution.

The two important swarm optimization algorithms are the ant colony optimization (ACO) algorithms and the artificial bee colony (ABC) algorithms. An ant colony can be thought of as a swarm whose individual agents are ants. Similarly, an artificial bee colony can be thought of as a swarm whose individual agents are bees.

Recently, ABC algorithm has been proposed to solve many difficult optimization problems in different fields such as constraint optimization problem, machine learning, and bioinformatics.

This section discusses ACO algorithms, and the next section discusses ABC algorithms.

9.12.1 Real Ants and Artificial Ants

In nature, the behavioral rules of ants are very simple, and their communication rules via "pheromones" are also very simple. However, a swarm of ants can solve very complex problems such as finding the shortest path between their nest and the food source. If finding the shortest path is looked upon as an optimization problem, then each path between the starting point (their nest) and the terminal (the food source) can be viewed as a feasible solution.

Each real ant in the ant system has the following characteristics [111]:

- By using a transition rule that is a function of the distance to the source, an ant decides which food source to go and the amount of pheromone present along the connecting path.
- Transitions to already visited food sources are added to a tabu list and are not allowed.
- Once a tour is complete, the ant lays a pheromone trail along each path visited in the source.

By Beckers, Deneubourg, and Goss [9], it was demonstrated experimentally that ants are able to find the shortest path.

Artificial ants of the colony have the following properties [21, 31]:

- An ant searches for minimum cost feasible solutions $\hat{J}_\psi = \min_\psi J(L,t)$.
- An ant k has an internal memory $\mathcal{M}^k$ which is used to store the path information followed by the ant k so far (i.e., the previously visited states). Memory can be used to build feasible solutions, to evaluate the solution found, and to retrace the path backward.
- An ant k in state $S_{r,i}$ can move from node i to any node j in its *feasible neighborhood* $\mathcal{N}_i^k$.
- An ant k can be assigned a *start state* S_s^k and one or more *termination conditions* e^k. Usually, the start state is expressed as a unit length sequence, that is, a single component.
- Starting from an initial state S_{initial}, each ant tries to build a feasible solution to the given problem in an incremental way, moving to feasible neighbor states in an iterative fashion through its search space/environment. The construction procedure stops when for at least one ant k, at least one of the termination conditions e^k is satisfied.
- The guidance factors involved in an ant's movement take the form of a transition rule which is applied before every move from state S_i to state S_j. The transition rule may also include additional problem specific constraints and may utilize the ants internal memory.
- The ants' probabilistic decision rule is a function of (a) the values stored in a node local data structure $\mathcal{A}_i = [a_{ij}]$ called *ant-routing table*, obtained by a functional composition of node locally available pheromone trails and heuristic values, (b) the ant's private memory storing its past history, and (c) the problem constraints.
- When moving from node i to neighbor node j the ant can update the pheromone trail τ_{ij} on the arc (i, j). This is called *online step-by-step pheromone update*.
- Once built a solution, the ants can retrace their same path backward and update the pheromone trails on the traversed arcs. This is called *online delayed pheromone update*.
- The amount of pheromone each ant deposits is governed by a problem specific pheromone update rule.
- Ants may deposit pheromone associated with states, or alternatively, with state transitions.
- Once it has built a solution, and, if the case, after it has retraced the path back to the source node, the ant dies, freeing all the allocated resources.

Artificial ants have several characteristics similar to real ants, namely [122]:

(a) Artificial ants have a probabilistic preference for paths with a larger amount of pheromone.
(b) Shorter paths tend to have larger rates of growth in their amount of pheromone.
(c) Artificial ants use an indirect communication system based on the amount of pheromone deposited on each path.

By mimicking bionic swarm intelligence that simulates the behavior of ants and their communication ways in searching for food, ACO algorithms, as a heuristic positive feedback search algorithm, are especially available for solving the optimization problem including two parts: the feasible solution space and the objective function.

The key component of an ACO algorithm is a parameterized probabilistic model, which is called the pheromone model [30].

For a constrained combinatorial optimization problem, the central component of an ACO algorithm is also the pheromone model which is used to probabilistically sample the search space S.

Definition 9.37 (Pheromone Model [30]) A *pheromone model* $P = (S, \Omega, f)$ of a combinatorial optimization problem consists of two parts:

(1) a search (or solution) space S defined over a finite set of discrete decision variables and a set Ω of constraints among the variables and
(2) an objective function $f : S \to \mathbb{R}_+$ to be minimized, which assigns a positive cost function to each solution $s \in S$.

The problem representation of a combinatorial optimization problem which is exploited by the ants can be characterized as follows [33].

- A finite set $C = \{c_1, \ldots, c_{N_c}\}$ of components is given.
- The states of the problem are defined in terms of sequences $x = \{c_i, c_j, \ldots, c_k, \ldots\}$ over the elements of C. The set of all possible sequences is denoted by X. The length of a sequence x, that is, the number of components in the sequence, is expressed by $|x|$. The maximum length of a sequence is bounded by a positive constant $n < +\infty$.
- The set of (candidate) solutions S is a subset of X (i.e., $S \subseteq X$).
- The finite set of constraints Ω defines the set of feasible states $\bar{X}$ with $\Omega \subseteq \bar{X}$.
- A non-empty set S^* of feasible solutions is given, with $\Omega \subseteq \bar{X}$ and $S^* \subseteq S$.
- A cost $f(s, t)$ is associated with each candidate solution $s \in S$.
- In some cases a cost, or the estimate of a cost, $J(x_i, t)$, can be associated with states other than solutions. If x_i can be obtained by adding solution components to a state x_j, then $J(x_i, t) \leq J(x_j, t)$. Note that $J(s, t) \equiv J(t, s)$.

In combinatorial optimization a system may occur in many different configurations. Any configuration has a cost function for that particular configuration. Similar to the simulated annealing of solids, one can statistically model the evolution of the system to be optimized into a state that corresponds with the minimum cost function value.

In order to probabilistically construct solutions, ACO algorithms for combinatorial optimization problems use a pheromone model that consists of a set of *pheromone values*, i.e., a function of the search experience of the algorithm. The pheromone model is used to bias the solution construction toward regions of the search space containing high-quality solutions.

Definition 9.38 (Mixed-Variable Optimization Problem [95]) A pheromone model $R = (S, \Omega, f)$ of a *mixed-variable optimization problem* (MVOP) consists of the following parts.

- A search space S defined over a finite set of both discrete and continuous decision variables and a set Ω of constraints among the variables;
- An objective function $f : S \to \mathbb{R}_0^+$ to be minimized. The search space S is defined by a set of $n = d + r$ variables $x_i, i = 1, \ldots, n$, of which d is the number of discrete decision variables and r is the number of continuous decision variables. A solution $s \in S$ is a complete value assignment, that is, each decision variable is assigned a value. A feasible solution is a solution that satisfies all constraints in the set Ω. A global optimum $s^* \in S$ is a feasible solution satisfying $f(s^*) \leq f(s), \forall s \in S$. The set of all globally optimal solutions is denoted by $s^* \in S^* \subseteq S$. Solving an MVOP requires finding at least one $s^* \in S^*$.

9.12.2 Typical Ant Colony Optimization Problems

Ant colony optimization is so called because of its original inspiration: some ant species forage between their nests and food sources by collectively exploiting the pheromone they deposit on the ground while walking [29]. Similar to real ants, artificial ants in ant colony optimization deposit artificial pheromone on the graph of the problem they are solving.

In ant colony optimization, each individual of the population is an artificial ant that builds incrementally and stochastically a solution to the considered problem. At each step the moves of ants define which solution components are added to the solution under construction.

A probabilistic model is associated with the graph $G(C, L)$ and is used to bias the agents' choices. The probabilistic model is updated online by the agents in order to increase the probability that future agents will build good solutions.

In order to build good solutions, there are three main construction functions (*ACO construction functions* for short) [111]:

- *Ant solution construct:* In the solution construction process, artificial ants move through adjacent states of a problem according to a transition rule, iteratively building solutions.
- *Pheromone update:* It performs pheromone trail updates and may involve updating the pheromone trails once complete solutions have been built, or updating after each iteration.
- *Deamon actions:* It is an optional step in the ant colony optimization and involves applying additional updates from a global perspective (there exists no natural counterpart). An example could be applying additional pheromone reinforcement to the best solution generated (known as offline pheromone trail update).

The following are the three typical optimization problems whose solutions are available for adopting ant colony optimization.

1. *Job Scheduling Problems* (JSP) [105] have a vital role in recent years due to the growing consumer demand for variety, reduced product life cycles, changing markets with global competition, and rapid development of new technologies. The job shop scheduling problem (JSSP) is one of the most popular scheduling models existing in practice, which is among the hardest combinatorial optimization problems. The job scheduling problem consists of the following components:

   - a number of independent (user/application) jobs to be the elements in C,
   - a number of heterogeneous machine candidates to participate in the planning,
   - the workload of each job (in millions of instructions),
   - the computing capacity of each machine,
   - ready time indicates when machine will have finished the previously assigned jobs, and
   - the expected time to compute (ETC) matrix ("n_b" jobs × "n_b" numbers of components machines) in which ETC[i][j] is the expected execution time of job "i" in machine "j."

2. *Data mining* [122]: In an ant colony optimization algorithm, each ant incrementally constructs/modifies a solution for the target problem.

   - In the context of classification task of data mining, the target problem is the discovery of classification rules that is often expressed in the form of IF-THEN rules as follows:

   $$\text{IF } < \text{conditions} > \text{ THEN } < \text{class} > .$$

   The rule antecedent (IF part) contains a set of conditions, usually connected by a logical conjunction operator (AND). The rule consequent (THEN part) specifies the class predicted for cases whose predictor attributes satisfy all the terms specified in the rule antecedent. As a promising research area, ACO algorithms involve simple agents (ants) that cooperate with others to achieve an emergent unified behavior for the system as a whole, producing a robust system capable of finding high-quality solutions for problems with a large search space.
   - In the context of rule discovery, an ACO algorithm is able to perform a flexible robust search for a good combination of terms (logical conditions) involving values of the predictor attributes.

3. *Network coding resource minimization (NCRM) problem* [156]: As a resource optimization problem emerging from the field of network coding, although all nodes have the potential to perform coding would perform coding by default, only a subset of coding-possible nodes suffices to realize network coding-based multicast (NCM) with an expected data rate. Hence, it is worthwhile

to study the problem of minimizing coding operations within NCM. EAs are the mainstream solutions for NCRM in the field of computational intelligence, but EAs are not good for integrating local information of the search space or domain-knowledge of the problem, which could seriously deteriorate their optimization performance. Different from EAs, ACO algorithms are the classes of reactive search optimization methods adopting the principle of "learning while optimizing." ACOs may be a good candidate for solving the NCRM problem.

9.12.3 Ant System and Ant Colony System

As the first ant algorithm, *ant system* was developed by Dorigo et al. in 1996 [34] for applying the well-known benchmark traveling salesman problem.

Let τ_{ij} be the pheromone trail between components (nodes) i and j that determines the path taken by the ants, and $\eta_{ij} = 1/d_{ij}$ be the heuristic values, where τ_{ij} constitute the pheromone matrix $\mathbf{T} = [\tau_{ij}]$. In other words, the shorter the distance d_{ij} between two cities i and j, the higher the heuristic value η_{ij} that constitutes the heuristic matrix.

The values from multiple pheromone (or heuristic) matrices need to be aggregated into a single pheromone (or heuristic) value. The following are the three alternatives of aggregation [97]:

1. *Weighted sum:* for example

$$\tau_{ij} = (1-\lambda)\tau_{ij}^1 + \lambda \tau_{ij}^2 \quad \text{and} \quad \eta_{ij} = (1-\lambda)\eta_{ij}^1 + \lambda \tau_{ij}^2. \tag{9.12.1}$$

2. *Weighted product:* for example

$$\tau_{ij} = (\tau_{ij}^1)^{(1-\lambda)} \cdot (\tau_{ij}^2)^{\lambda} \quad \text{and} \quad \eta_{ij} = (\eta_{ij}^1)^{(1-\lambda)} \cdot (\eta_{ij}^2)^{\lambda}. \tag{9.12.2}$$

3. *Random:* at each construction step, given a uniform random number $U(0, 1)$, an ant selects the first of the two matrices if $U(0, 1) < 1 - \lambda$; otherwise, it selects the other matrix.

The ants exists in an environment represented mathematically as a construction graph $G(C, L)$.

1. $C = \{c_1, \ldots, c_{N_c}\}$ denotes a set of components, where c_i is the ith city and N_c is the number of all cities in tour.
2. L is a set of connections fully connecting.

By Dorigo et al. [34] and Neto and Filho [115], ant model has the following characteristics:

- The ant always occupies a node in a graph which represents a search space. This node is called nf.

- It has an initial state.
- Although it cannot sense the whole graph, it can collect two kinds of information about the neighborhood: (a) the weight of each trail linked to nf and (b) the characteristics of each pheromone deposited on this trail by other ants of the same colony.
- It moves toward a trail C_{ij} that connects nodes i and j of the graph.
- Also, it can alter the pheromones of the trail C_{ij}, in an operation called as "deposit of pheromone levels."
- It can sense the pheromone levels of all C_{ij} trails that connect a node i.
- It can determine a set of "prohibited" trails.
- It presents a pseudo-random behavior enabling the choice among the various possible trails.
- This choice can be (and usually is) influenced by the level of pheromone.
- It can move from node i to node j.

An ant system utilizes the graph representation $G(C, L)$: for each cost measure $\delta(r, s)$ (distance from city r moving to city s), each edge (r, s) has also a desirability measure (called pheromone) $\tau_{rs} = \tau(r, s)$ which is updated at run time by ants.

An ant system works according to the following rules [32]:

1. *State transition rule:* Each ant generates a complete tour by choosing the cities according to a probabilistic state transition rule; ants prefer to move to cities which are connected by short edges with a high amount of pheromone.
2. *Global (pheromone) updating rule:* Once all ants have completed their tours, a global updating rule is used; a fraction of the pheromone evaporates on all edges, and then each ant deposits an amount of pheromone on edges which belong to its tour in proportion to how short its tour was. The process is then iterated.
3. *Random-proportional rule:* It gives the probability with which the kth ant in city r chooses to move to the city s:

$$p_k(r, s) = \begin{cases} \dfrac{[\tau(r, s)] \cdot [\eta(r, s)]^\beta}{\sum_{u \in J_k(r)} [\tau(r, u)] \cdot [\eta(r, u)]^\beta}, & \text{if } s \in J_k(r); \\ 0, & \text{otherwise.} \end{cases} \quad (9.12.3)$$

Here τ is the pheromone, $\eta = 1/\delta$ is the inverse of the distance $\delta(r, s)$, $J_k(r, s)$ is the set of cities that remain to be visited by ant k positioned on city r (to make the solution feasible), and $\beta > 0$ is a parameter for determining the relative importance of pheromone versus distance.

In the way given by (9.12.3), the ants favor the choice of edges which are shorter and have a greater amount of pheromone.

9.12 Ant Colony Optimization

In an ant system, once all ants have built their tours, pheromone is updated on all edges according to the following global updating rule:

$$\tau(r, s) \leftarrow (1 - \alpha) \cdot \tau(r, s) + \sum_{k=1}^{m} \Delta \tau_k(r, s), \tag{9.12.4}$$

where

$$\Delta \tau_k(r, s) = \begin{cases} \dfrac{1}{L_k}, & \text{if } (r, s) \in \text{ tour done by ant } k; \\ 0, & \text{otherwise.} \end{cases} \tag{9.12.5}$$

Here $0 < \alpha < 1$ is a pheromone decay parameter, L_k is the length of the tour performed by ant k, and m is the number of ants.

The *ant colony system (ACS)* differs from the ant system in three main aspects [32]:

- State transition rule provides a direct way to balance between exploration of new edges and exploitation of a priori and accumulated knowledge about the problem.
- Global updating rule is applied only to edges which belong to the best ant tour.
- A *local (pheromone) updating rule* is applied while ants construct a solution.

Different from ant system, ant colony system works according to the following rules [32]:

1. *ACS state transition rule:* In ACS, an ant positioned on node r chooses the city s to move to by applying the ACS state transition rule given by

$$s = \begin{cases} \arg\max_{u \in J_k(r)} \left\{ [\tau(r, u)] \cdot [\eta(r, u)]^\beta \right\}, & \text{if } q \leq q_0 \text{ (exploitation)}; \\ S, & \text{otherwise.} \end{cases} \tag{9.12.6}$$

   Here q is a random number uniformly distributed in $[0, 1]$, $0 \leq q_0 \leq 1$ is a parameter, and S is a random variable selected according to the probability distribution given in (9.12.3). The state transition rule given by (9.12.6) and (9.12.3) is known as *pseudo-random-proportional rule*. This rule favors transitions toward nodes connected by short edges and with a large amount of pheromone.

2. *ACS global updating rule:* In ACS only the globally best ant, constructed the shortest tour from the beginning of the trial, is allowed to deposit pheromone. After all ants have completed their tours, the pheromone level is updated by the following ACS global updating rule:

$$\tau(r, s) \leftarrow (1 - \alpha) \cdot \tau(r, s) + \alpha \cdot \Delta \tau(r, s), \tag{9.12.7}$$

where

$$\Delta \tau(r, s) = \begin{cases} (L_{gb})^{-1}, & \text{if } (r, s) \in \text{global-best-tour}; \\ 0, & \text{otherwise}. \end{cases} \quad (9.12.8)$$

Here $0 < \alpha < 1$ is the pheromone decay parameter, and L_{gb} is the length of the globally best tour from the beginning of the trial.

3. *ACS local updating rule:* While building a solution (i.e., a tour) of the traveling salesman problem (TSP), ants visit edges and change their pheromone level by applying the following local updating rule:

$$\tau(r, s) \leftarrow (1 - \rho) \cdot \tau(r, s) + \rho \cdot \Delta \tau(r, s), \quad (9.12.9)$$

where $0 < \rho < 1$ is a parameter.

9.13 Multiobjective Artificial Bee Colony Algorithms

Artificial bee colony (ABC) algorithm, proposed by Karaboga in 2005 [81], is inspired by the intelligent foraging behavior of the honey bee swarm. Numerical comparisons show that the performance of ABC is competitive to that of other population-based algorithms with an advantage of employing fewer control parameters [82–84]. The main advantages of ABC are its simplicity and ease of implementation, but ABC also faces up to the poor convergence commonly existing in evolutionary algorithms.

9.13.1 Artificial Bee Colony Algorithms

The design of an ant colony optimization (ACO) algorithm implies the specification of the following aspects [11, 99, 122]:

- An appropriate representation of the problem, which allows the ants to incrementally construct/modify solutions through the use of a probabilistic transition rule, based on the amount of pheromone in the trail and on a local problem-dependent heuristic.
- A method to enforce the construction of valid solutions, i.e., solutions that are legal in the real-world situation corresponding to the problem definition.
- A problem dependent heuristic function η that provides a quality measurement of items that can be added to the current partial solution.
- A rule for pheromone updating, which specifies how to modify the pheromone trail τ.

9.13 Multiobjective Artificial Bee Colony Algorithms

- A probabilistic transition rule based on the value of the heuristic function η and on the contents of the pheromone trail τ that is used to iteratively construct a solution.
- A clear specification of when the algorithm converges to a solution.

Unlike ant swarms, the collective intelligence of honey bee swarms consists of three essential components: food sources, employed bees, unemployed (onlooker) bees.

In the ABC algorithm, the position of a food source represents a possible solution of the optimization problem and the nectar amount of a food source corresponds to the quality (fitness) of the associated solution. The *artificial bee colony* is divided into three groups of bees: *employed bees*, *onlooker bees*, and *scout bees*. Half of the colony consists of the employed bees, and another half consists of the onlookers.

Each cycle of the search consists of three steps.

- Employed bees are responsible for exploring the nectar sources and sharing their food information with onlookers within the hive.
- The onlooker bees select good food sources from those foods found by the employed bees to further search the foods.
- If the quality of some food source is not improved through a predetermined number of cycles, then the food source is abandoned by its employed bee, and the employed bee becomes a scout and starts to search for a new food source randomly in the vicinity of the hive.

Each food source $\mathbf{x}_i = [x_{i,1}, \ldots, x_{i,D}]^T$ ($i = 1, \ldots, SN$) is a D-dimensional vector, and stands for a potential solution of the problem to be optimized. Let $P = \{\mathbf{x}_1, \ldots, \mathbf{x}_{SN}\}$ denotes the population of SN food solutions (individuals). The ABC algorithm is an iterative algorithm. At the beginning, ABC generates the population with randomly generated food solutions. The newly generated population P contains SN individuals $\mathbf{x}_i \in \mathbb{R}^D$, where the population size SN is identical to the number of those different kinds of bees. ABC generates a randomly distributed initial population P of SN solutions (food source positions) $\mathbf{x}_1, \ldots, \mathbf{x}_{SN}$ whose elements are as follows:

$$x_{i,j} = x_{\min,j} + U(0,1)(x_{\max,j} - x_{\min,j}), \tag{9.13.1}$$

where $i = 1, \ldots, SN$; $j = 1, \ldots, D$, and D is the number of optimization parameters; $x_{\min,j}$ and $x_{\max,j}$ are the lower and upper bounds for the dimension j, respectively; and $U(0,1)$ is a random number with uniform distribution in $(0,1)$.

After the initialization, the population of solutions (i.e., food sources) undergoes repeated cycles of the search processes of the employed bees, onlooker bees, and scout bees. Each employed bee always remembers its previous best position and produces a new position within its neighborhood in its memory.

After all employed bees finish their search process, they will share the information about nectar amounts and positions of food sources with onlookers. Each onlooker chooses a food source depending on the probability value p_i associated with the food source through

$$p_i = \frac{\text{fit}_i}{\sum_{j=1}^{SN} \text{fit}_j}, \quad (9.13.2)$$

where fit_i represents the fitness value of the ith solution $\mathbf{f}(\mathbf{x}_i)$.

If the new food source has the equal or better quality than the old source, then the old source is replaced by the new source. Otherwise, the old source is retained. ABC uses

$$v_{i,j} = x_{i,j} + \phi_{i,j}(x_{i,j} - x_{l,j}), \quad l \neq i, \quad (9.13.3)$$

to produce a candidate food position $\mathbf{v}_i = [v_{i1}, \ldots, v_{iD}]^T$ from the old solution $\mathbf{x}_i$ in memory. Here, $l \in \{1, \ldots, SN\}$ and $j \in \{1, \ldots, D\}$ are randomly chosen indices, and ϕ_{ij} is a random number in the range $[-1, 1]$.

Then, ABC searches the area within its neighborhood to generate a new candidate solution. If a position cannot be improved further through a predetermined number of cycles, then that food source is said to be abandoned. The value of the predetermined number of cycles is an important control parameter of ABC, and is called the limit for abandonment. When the nectar of a food source is abandoned, the corresponding employed bee becomes a scout. The scout will randomly generate a new food source to replace the abandoned position.

The above framework of artificial bee colony (ABC) algorithm is summarized as Algorithm 9.15.

9.13.2 Variants of ABC Algorithms

The solution search equation of ABC, for generating new candidate solutions based on the information of previous solutions, is good at exploration, but it is poor at exploitation. To achieve good performance on problem optimizations, exploration and exploitation should be well balanced.

To improve the performance of ABC, several variants of ABC were proposed.

1. *GABC:* This a *gbest-guided artificial bee colony* algorithm [168]. By incorporating the information of the gbest solution into the solution search equation to improve the exploitation of ABC. The solution search equation in GABC is given by

$$v_{ij} = x_{ij} + \phi_{ij}(x_{ij} - x_{kj}) + \psi_{ij}(g_j - x_{ij}), \quad (9.13.4)$$

9.13 Multiobjective Artificial Bee Colony Algorithms

Algorithm 9.15 Framework of artificial bee colony (ABC) algorithm [49]

1. **initialization:**
   1.1 Randomly generate SN points in the search space to form an initial population;
   1.2 Evaluate the objective value of the population;
   1.3 $FES = SN$.
   // The employed bee phase:
2. **for** $i = 1, \ldots, SN$ **do**
3.   Generate a candidate solution V_i using (9.13.3).
4.   Evaluate $\mathbf{f}(V_i)$ and set $FES = FES + 1$.
5.   If $f(V_i) < f(X_i)$ then set $X_i = V_i$ and trail$_i = 1$. Otherwise trial$_i = $ trail$_i + 1$.
6.   Calculate the probability p_i using (9.13.2), and set $t = 0$ and $i = 1$.
   // The onlooker bee phase:
7.   **while** $t \leq SN$ **do**
8.   **if** rand$(0, 1) < p_i$ **do**
9.   Generate a candidate solution V_i using (9.13.3).
10.   Evaluate the objective value of the population;
11.   If $f(V_i) < f(X_i)$ then set $X_i = V_i$ and trail$_i = 1$. Otherwise trial$_i = $ trail$_i + 1$.
12.   Set $t = t + 1$.
13.   **end if**
14.   **end while**
15.   Set $i = i + 1$ and if $i = SN$ then set $i = 1$.
   // The scout bee phase:
16.   If max{trial$_i$} > limit then replace X_i with a new randomly generated solution by (9.13.1).
17.   If $FES \geq FES_{\max}$ then stop and output the best solution achieved so far. Otherwise, go to step 1.
18. **end for**
19. **output:** SN solutions (food source positions) $\mathbf{x}_1, \ldots, \mathbf{x}_S N$.

where the third term in the right-hand side is a new added term called the gbest term, g_j is the jth element of the gbest solution vector $\mathbf{g}_{best}$, ϕ_{ij} is a random number in the range $[-1, 1]$, and ψ_{ij} is a uniform random number in $[0, 1.5]$. As demonstrated by experimental results, GABC can outperform ABC in most experiments.

2. *IABC:* This is an *improved artificial bee colony* (IABC) algorithm [48]. In this algorithm, two improved solution search equations are given by

$$v_{ij} = g_{best,j} + \phi_{ij}(x_{ij} - x_{r1,j}), \tag{9.13.5}$$

$$v_{ij} = x_{r1,j} + \phi_{ij}(x_{ij} - x_{r2,j}), \tag{9.13.6}$$

where the indices $r1$ and $r2$ are mutually exclusive integers randomly chosen from $\{1, 2, \ldots, SN\}$, and both of them are different from the base index i; $g_{best,j}$ is the jth element of the best individual vector with the best fitness in the current population, and $j \in \{1, \ldots, D\}$ is a randomly chosen index. Since (9.13.5) generates a candidate solution around $\mathbf{g}_{best}$, it focuses on the exploitation. On the other hand, as (9.13.6) is based on three vectors $\mathbf{x}_i$, $\mathbf{x}_{r1}$, and $\mathbf{x}_{r2}$, and the two latter ones are two different individuals randomly selected from the population, it emphasizes the exploration.

3. *CABC:* The solution search equation is given by Gao et al. [49]

$$v_{ij} = x_{r1,j} + \phi_{ij}(x_{r1,j} - x_{r2,j}), \qquad (9.13.7)$$

where the indices $r1$ and $r2$ are distinct integers uniformly chosen from the range $[1, SN]$ and are also different from i, and ϕ_{ij} is a random number in the range $[-1, 1]$. Since (9.13.7) looks like the crossover operator of GA, the ABC with this search equation is named as *crossover-like artificial bee colony* (CABC) [49].

9.14 Particle Swarm Optimization

Particle swarm optimization (PSO) is a global optimization technique originally developed by Eberhart and Kennedy [36, 86] in 1995. PSO is essentially an optimization technique based on swarm intelligence, and adopts a population-based stochastic algorithm for finding optimal regions of complex search spaces through the interaction of individuals in a population of particles. Different from evolutionary algorithms, the particle swarm does not use selection operation; while interactions of all population members result in iterative improvement of the quality of problem solutions from the beginning of a trial until the end.

PSO is rooted in artificial life and social psychology, as well as in engineering and computer science. It is widely recognized (see, e.g., [19, 37, 85]) that *swarm intelligence (SI)* is an innovative distributed intelligent paradigm for solving optimization problems, and PSO was originally inspired by the swarming behavior observed in flocks of birds, swarms of bees, or schools of fish, and even human social behavior, from which the idea is emerged.

Due to computational intelligence, PSO is not largely affected by the size and nonlinearity of the optimization problem, and can converge to the optimal solution in many problems where most analytical methods fail to converge.

In PSO, the term "particles" refers to population members with an arbitrarily small mass or volume and is subject to velocities and accelerations towards a better mode of behavior.

9.14.1 Basic Concepts

PSO is based on two fundamental disciplines: social science and computer science, and is closely based on the swarm intelligence concepts and principles [26].

1. *Social concepts:* It is known [37] that "*human intelligence* results from social interaction." Evaluation, comparison, and imitation of others, as well as learning from experience allow humans to adapt to the environment and determine optimal patterns of behavior, attitudes, and so on.

2. *Swarm intelligence principles [37, 85, 86]: Swarm Intelligence* can be described by considering the following five fundamental principles.

   - *Proximity principle:* the population should be able to carry out simple space and time computations.
   - *Quality principle:* the population should be able to respond to quality factors in the environment.
   - *Diverse response principle:* the population should not commit its activity along excessively narrow channels.
   - *Stability principle:* the population should not change its mode of behavior whenever the environment changes.
   - *Adaptability principle:* the population should be able to change its behavior mode when it is worth the computational price.

3. *Computational characteristics [37]:* As a useful paradigm for implementing adaptive systems, the swarm intelligence computation is an extension of evolutionary computation and includes the softening parameterization of logical operators like AND, OR, and NOT. In particular, PSO is an extension, and a potentially important incarnation of cellular automata (CA). The particle swarm can be conceptualized as cells in CA, whose states change in many dimensions simultaneously. Both PSO and CA share the following computational attributes.

   - Individual particles (cells) are updated in parallel.
   - Each new value depends only on the previous value of the particle (cell) and its neighbors.
   - All updates are performed according to the same rules.

9.14.2 The Canonical Particle Swarm

Given an unleveled data set $Z = \{\mathbf{z}_1, \ldots, \mathbf{z}_N\}$ representing N patterns $\mathbf{z}_i \in \mathbb{R}^D$, $i = 1, \ldots, N$, each having D features. Then partitional clustering approach aims at clustering the data set into K groups ($K \leq N$) such that

$$C_k \neq \emptyset, \ \forall k = 1, \ldots, K; \tag{9.14.1}$$

$$C_k \cap C_l = \emptyset, \ \forall k, l = 1, \ldots, K; \ k \neq l; \ \bigcup_{k=1}^{K} C_k = Z. \tag{9.14.2}$$

The clustering operation is dependent on the similarity between elements present in the data set. If $\mathbf{f}$ denotes the fitness function, then the clustering task can be viewed as an optimization problem: optimize$_{C_k}$ $\mathbf{f}(Z_k, C_k)$, $\forall k = 1, \ldots, K$. That is to say, the optimization based clustering task is carried out by single objective nature inspired metaheuristic algorithms.

PSO is inspired by social interaction (e.g., foraging process of bird flocking). The original concept of PSO is to simulate the behavior of flying birds and their means of information exchange to solve problems. In PSO, each single solution is like a "bird." Imagine the following scenario: a group of birds are randomly searching food in an area in which only one piece of food is assumed to exist, and the birds do not know where the food is. An efficient strategy for finding the food is to follow the bird nearest to the food.

In a particle swarm optimizer, instead of using genetic operators, each single solution is regarded as a "bird" (particle or individual) in the search space. These individuals are "evolved" by cooperation and competition among the individuals themselves through generations. Each particle adjusts its flying according to its own flying experience and its companions' flying experience. Each individual or particle represents a potential solution to an optimization problem, and is treated as a point in a D-dimensional space.

Define the notations and parameters as follows.

- N_p: the swarm (or population) size;
- $\mathbf{x}_i(t)$: the position vector of the ith particle at time t in the search space;
- $\mathbf{v}_i(t)$: the velocity vector of the ith particle at time t in the search space;
- $\mathbf{p}_{i,best}(t) = [p_{i1}(t), \ldots, p_{iD}(t)]^T$: the *previous best position* (the position giving the best fitness value, simply named pbest) of the ith particle (up to time t) in the search space;
- $\mathbf{g}_{best} = [g_1, \ldots, g_D]^T$: the best particle found so far among all the particles in the population, and is named as the *global best position* (gbest) vector of the swarm;
- $U(0, \beta)$: a uniform random number generator;
- α: *inertia weight* or *constriction coefficient*;
- β: *acceleration constant*.

Position and velocity of each particle are adjusted, and the fitness function with the new position is evaluated at each time step. A population of particles at the time t is initialized with random position of the ith particle at the time t, $\mathbf{x}_i(t) = [x_{i1}(t), \ldots, x_{iD}(t)]^T \in \mathbb{R}^D$, and the velocity of the ith particle at the time t, $\mathbf{v}_i(t) = [v_{i1}(t), \ldots, v_{iD}(t)]^T \in \mathbb{R}^D$.

The most common type of implementation defines the particles' behaviors in the following two formulas [19, 85]. First adjust the step size of the particle:

$$\mathbf{v}_i(t+1) \leftarrow \alpha \mathbf{v}_i(t) + U(0, \beta) \left(\mathbf{p}_{i,best}(t) - \mathbf{x}_i(t)\right) + U(0, \beta) \left(\mathbf{g}_{best} - \mathbf{x}_i(t)\right), \tag{9.14.3}$$

and then move the particle by adding the velocity to its previous position:

$$\mathbf{x}_i(t+1) \leftarrow \mathbf{x}_i(t) + \mathbf{v}_i(t+1), \; i = 1, \ldots, N_p. \tag{9.14.4}$$

9.14 Particle Swarm Optimization

By Clerc and Kennedy [19], though there is variety in the implementations of the particle swarm, the most standard version uses $\alpha = 0.7298$ and $\beta = \psi/2$, where $\psi = 2.9922$.

The personal best position of each particle is updated using

$$\mathbf{p}_{i,best}(t+1) = \begin{cases} \mathbf{p}_{i,best}(t), & \text{if } f(\mathbf{x}_i(t+1)) \geq f(\mathbf{p}_{i,best}(t)), \\ \mathbf{x}_i(t+1), & \text{if } f(\mathbf{x}_i(t+1)) < f(\mathbf{p}_{i,best}(t)), \end{cases} \quad (9.14.5)$$

for $i = 1, \ldots, N_p$, and the global best position $\mathbf{g}_{best}$ found by any particle during all previous steps is defined as

$$\mathbf{g}_{best} = \arg\min_{\mathbf{p}_{i,best}} f(\mathbf{p}_{i,best}(t+1)), \quad 1 \leq i \leq N_p. \quad (9.14.6)$$

The basic particle swarm optimization algorithm is shown in Algorithm 9.16.

Algorithm 9.16 Particle swarm optimization (PSO) algorithm

1. **input:** cost functions f, number of particles N_p, maximum cycles N_c, the dimension D of position and velocity vectors, the inertia weight α, acceleration constant β.
2. **initialization:**
   2.1 Create particle positions $\mathbf{x}_1(1), \ldots, \mathbf{x}_{N_p}(1)$ and particle velocities $\mathbf{v}_1(1), \ldots, \mathbf{v}_{N_p}(1)$.
   2.2 $\mathbf{p}_{i,best}(1) = \mathbf{x}_i(1), i = 1, \ldots, N_p$.
   2.3 $\mathbf{g}_{best} = \mathbf{x}_i(1)$, where $i = \arg\min\{f(\mathbf{x}_1(1)), \ldots, f(\mathbf{x}_{N_p}(1))\}$.
3. **repeat**
4.   **for** $t = 1, \ldots, N_c$ **do**
5.   **for** $i = 1, \ldots, N_p$ **do**
6. $\mathbf{v}_i(t+1) \leftarrow \alpha \mathbf{v}_i(t) + U(0, \beta)\left(\mathbf{p}_{i,best}(t) - \mathbf{x}_i(t)\right) + U(0, \beta)(\mathbf{g}_{best} - \mathbf{x}_i(t));$
7. $\mathbf{x}_i(t+1) = \mathbf{x}_i(t) + \mathbf{v}_i(t+1)$.
8. $\mathbf{p}_{i,best}(t+1) = \begin{cases} \mathbf{p}_{i,best}(t), & \text{if } f(\mathbf{x}_i(t+1)) \geq f(\mathbf{p}_{i,best}(t)), \\ \mathbf{x}_i(t+1), & \text{if } f(\mathbf{x}_i(t+1)) < f(\mathbf{p}_{i,best}(t)). \end{cases}$
9. $\mathbf{g}_{best} = \arg\min_{\mathbf{p}_{i,best}}\{f(\mathbf{p}_{j,best}(t+1))\}, \quad j = 1, \ldots, i.$
10.   **end for**
11.   **end for**
12. **until** $\mathbf{g}_{best}$ is sufficiently good or the maximum cycle is archived.
13. **output:** global best position $\mathbf{g}_{best}$.

9.14.3 Genetic Learning Particle Swarm Optimization

Advantages of PSO over other similar optimization techniques such as GA can be summarized as follows [26]:

- PSO is easier to implement and there are fewer parameters to adjust.

- In PSO, every particle remembers its own previous best value as well as the neighborhood best; therefore, it has a more effective memory capability than the GA.
- PSO is more efficient in maintaining the diversity of the swarm [38] (more similar to the ideal social interaction in a community), since all the particles use the information related to the most successful particle in order to improve themselves, whereas in GA, the worse solutions are discarded and only the good ones are saved; therefore, in GA the population evolves around a subset of the best individuals.

The following are the two main drawbacks of PSO [123].

1. The swarm may prematurely converge.

   - For the global best PSO, particles converge to a single point, which is on the line between the global best and the personal best positions. This point is not guaranteed for a local optimum [154].
   - The fast rate of information flow between particles, resulting in the creation of similar particles with a loss in diversity that increases the possibility of being trapped in local optima.

2. Stochastic approaches have problem-dependent performance. This dependency usually results from the parameter settings in each algorithm. Increasing the inertia weight will increase the speed of the particles resulting in more exploration (global search) and less exploitation (local search) will decrease the speed of the particles resulting in more exploitation and less exploration.

The problem-dependent performance can be addressed via hybrid mechanism combining different approaches in order to be benefited from the advantages of each approach. The following are the comparison between PSO and GA.

- In general PSO, particles are guided by their previous best positions (pbests) and the global best position (gbest) found by the swarm, so the search is more directional than that of GA. Hence, it is expected that GA possesses a better exploration ability than PSO, whereas the latter facilitates faster convergence.
- By applying crossover operation in GA, information can be swapped between two particles to have the ability to fly to the new search area. If applying mutation used in GA to PSO, then it is expected to increase the diversity of the population and the ability to have the PSO to avoid the local maxima.

This hybrid of genetic algorithm and particle swarm optimization is simply called GA-PSO, which combines the advantages of swarm intelligence and a natural selection mechanism in GA, and thus increases the number of highly evaluated agents, while decreases the number of lowly evaluated agents at each iteration step.

An alternative of GA-PSO is the *genetic learning particle swarm optimization* (GL-PSO) in [57], as shown in Algorithm 9.17.

9.14 Particle Swarm Optimization

Algorithm 9.17 Genetic learning PSO (GL-PSO) algorithm [57]

1. **initialize:**
2.   **for** $i = 1$ to M **do**
3.   Randomly initialize $\mathbf{v}_i$ and $\mathbf{x}_i$;
4.   Evaluate $f(\mathbf{x}_i)$;
5. $\mathbf{p}_{i,best} = \mathbf{x}_i$.
6.   **end for**
7. Set $\mathbf{g}_{best}$ to the current best position of particles;
8. **repeat**
9.   **for** $i = 1$ to M **do**
10.   **for** $d = 1$ to D **do**
11.   Randomly select a particle $k \in \{1, \ldots, M\}$;
12.   **if** $f(\mathbf{p}_{i,best}) < f(\mathbf{p}_{k,best})$ **then**
13. $o_{i,d} = r_d \cdot p_{i,d} + (1 - r_d) \cdot g_d$;
14.   **else**
15. $o_{i,d} = p_{k,d}$.
16.   **end if**
17.   **end for**
18.   **for** $d = 1$ to D **do**
19.   **if** rand$(0, 1) < pm$ **then**
20. $o_{i,d} = $ rand(lb_d, ub_d).
21.   **end if**
22.   **end for**
23.   Evaluate $f(O_i)$.
24.   **if** $f(\mathbf{o}_i) < f(\mathbf{e}_i)$ **then**
25. $\mathbf{e}_i = \mathbf{O}_i$.
26.   **end if**
27.   **if** $f(\mathbf{e}_i)$ ceases improving for sg generations **then**
28.   Select $\mathbf{e}_j$ by $20\%M$ tournament;
29. $\mathbf{e}_i = \mathbf{e}_j$.
30.   **end if**
31.   **for** $d = 1$ to D **do**
32. $v_{i,d} = \omega \cdot v_{i,d} + c \cdot r_d \cdot (e_{i,d} - x_{i,d})$;
33. $x_{i,d} = x_{i,d} + v_{i,d}$.
34.   **end for**
35.   Evaluate $f(X_i)$;
36.   Update P_i and G.
37.   **end for**
38. **until** Terminal Condition.

The GL-PSO is an optimization algorithm hybridized by genetic learning and particle swarm optimization hybridized in a highly cohesive way.

9.14.4 Particle Swarm Optimization for Feature Selection

To find the optimal feature subset one needs to enumerate and evaluate all the possible subsets of features in the entire search space, which implies that the search space size is 2^n where n is the number of the original features. Therefore, evaluating the entire feature subset is computationally expensive and also impractical even for a moderate-sized feature set.

Although many feature selection algorithms involve heuristic or random search strategies to find the optimal or near optimal subset of features in order to reduce the computational time, metaheuristic algorithms may clearly be more efficient for feature subset selection.

Consider feature selection in supervised classification. Given the data set $S = \{(\mathbf{x}_1, y_1), \ldots, (\mathbf{x}_n, y_n)\}$, where $\mathbf{x}_i = [x_{i1}, \ldots, x_{id}]^T$ is a multi-dimensional vector sample, d denotes the number of features, n is the number of samples, and $y_i \in I = \{1, \ldots, d\}$ denotes the label of the sample $\mathbf{x}_i$. Let $X = \{\mathbf{x}_1, \ldots, \mathbf{x}_n\}$. The main goal of the supervised learning is to approximate a function $f : X \to I$ to predict the class label for a new input vector $\mathbf{x}$.

In the learning problems, choice of the optimization method is very important to reduce the dimensionality of the feature space and to improve the convergence speed of the learning model. The PSO method would be preferred because it can handle a large number of features.

Potential advantages of PSO for feature selection can be summarized as follows [138]:

- PSO has a powerful exploration ability until the optimal solution is found because different particles can explore different parts of the solution space.
- PSO is particularly attractive for feature selection since the particle swarm has memory, and knowledge of the solution is retained by all particles as they fly within the problem space.
- The attractiveness of PSO is also due to its computationally inexpensive implementation that still gives decent performance.
- PSO works with the population of potential solutions rather than with a single solution.
- PSO can address binary and discrete data.
- PSO has better performance compared to other feature selection techniques in terms of memory and runtime and does not need complex mathematical operators.
- PSO is easy to implement with few parameters, and is easy to realize and gives promising results.
- The performance of PSO is almost unaffected by the dimension of the problem.

Although PSO provides a global search strategy for finding a better solution in the feature selection task, it is infected with two shortcomings: premature convergence and weakness in fine-tuning near local optimum points. To overcome these weaknesses, a hybrid feature selection method based on PSO with local search, called HPSO-LS, was developed in [106].

Define the relevant symbols are as follows. n_f denotes the number of the original features in a given data set, $n_s = |X_s|$ is the number of the selected features, c_{ij} is the Pearson correlation coefficient between two features $\mathbf{x}_i$ and $\mathbf{x}_j$, cor_i is the correlation value for feature i, $\mathbf{x}_i^{\text{best}}$ is the best previously visited position (up to

9.14 Particle Swarm Optimization

time t) of the ith particle, $\mathbf{x}_g^{best}$ is the global best position of the swarm, $x_{id}(t)$ is the position of the dth dimension of the ith particle, $\mathbf{v}_i(t)$ is the velocity of the ith particle, $\mathbf{x}_i(k)$ and $\mathbf{x}_j(k)$ denote the values of the feature vectors i and j for the kth sample.

The pseudo code of the HPSO-LS algorithm is shown in Algorithm 9.18.

Algorithm 9.18 Hybrid particle swarm optimization with local search (HPSO-LS) [106]

1. **input:** $Feature = \{f_1, \ldots, f_D\}$, NC_{max}: the maximum cycles that algorithm repeated, $K \ll D$,
   NP: number of particles.
2. **begin initialize:**
3.   **for** $i = 1$ to NP **do**
4. X_{ij} = create random particle $\in \{0, 1\} \Rightarrow \mathbf{x}_i = [X_{i1}, \ldots, X_{iD}]^T$
5. V_{ij} = create random velocity $\sim U[0, 1] \Rightarrow \mathbf{v}_i = [V_{i1}, \ldots, V_{iD}]^T$
6.   **end for**
7.   **for** all feature i **do**
8. $c_{ij} = \frac{\sum_{k=1}^{m}(\mathbf{x}_i(k)-\bar{\mathbf{x}}_i)^T(\mathbf{x}_j(k)-\bar{\mathbf{x}}_j)}{\sqrt{\sum_{k=1}^{m}\|\mathbf{x}_i(k)-\bar{\mathbf{x}}_i\|^2}\sqrt{\sum_{k=1}^{m}\|\mathbf{x}_j(k)-\bar{\mathbf{x}}_j\|^2}}$
9. $cor_i = \frac{1}{n_f - 1}\sum_{j=1, j\neq i}^{n_f}|c_{ij}|, \quad i \neq j$
10. **end for**
11. Similar feature set $\leftarrow f_i$ if $cor_i \geq cor_{mid}$
12. Dissimilar feature set $\leftarrow f_i$ if $cor_i < cor_{mid}$
13. **end initialize**
14. **for** $i = 1$ to NC_{max} **do**
15. $\hat{\mathbf{v}}_i(t+1) = \mathbf{v}_i(t) + c_1 \cdot r_1 \cdot \text{rand}\left(\mathbf{x}_i^{best} - \mathbf{x}_i(t)\right)\mathbf{v}_i(t) + c_2 \cdot r_2 \cdot \text{rand}\left(\mathbf{x}_g^{best} - \mathbf{x}_i(t)\right)$
16.   **for** $d = 1$ to D **do**
17. $v_{id}(t+1) = \text{sign}\left(\hat{v}_{id}(t+1)\right)\min\{|\hat{v}_{id}(t+1)|, v_{i,max}\}$
18. $s(v_{id}(t+1)) = \frac{1}{1+\exp(-v_{id}(t+1))}$
19. $x_{id}(t+1) = \begin{cases} 1č & \text{if } ranm < s(v_{id}(t+1)); \\ 0, & \text{otherwise.} \end{cases}$
20.   **end for**
21. X_s = Similar feature set
22. X_d = Dissimilar feature set
23. $\mathbf{x}'_i = \mathbf{x}_i$
24.   **Remove** all feature in X_d that is 0 in particle $\mathbf{x}_i$
25.   **Remove** all feature in X_s that is 0 in particle $\mathbf{x}_i$
26. $n_s = |X_s|$ and $n_d = |X_d|$
27.   **Perform** "particle movement" on each position of particle and replace it
28. $\bar{f}_i = fitness(\mathbf{x}'_i)$
29. $f'_i = fitness(\mathbf{x}'_i) = \begin{cases} fetness(\mathbf{x}_i^{best}), & \text{if } \bar{f}_i > fitness(\mathbf{x}_i^{best}) \\ fetness(\mathbf{x}_g^{best}), & \text{if } \bar{f}_i > fitness(\mathbf{x}_g^{best}) \\ \bar{f}_i, & \text{otherwise.} \end{cases}$
30. **end for**
31. **output:** feature$' = \{f'_1, \ldots, f'_k\}$

9.15 Opposition-Based Evolutionary Computation

It is widely recognized that hybridization with different algorithms is another direction for improving machine intelligence algorithms.

Learning, search, and optimization are fundamental tasks in the machine intelligence: Artificial intelligence algorithms learn from past data or instructions, optimize estimated solutions, and search for an existing solution in large spaces. In many cases, the machine learning starts at a random point, such as weights of a neural network, initial population of soft computing algorithms, and action policy of reinforcement agents. If the starting point is close to the optimal solution, then its convergence is faster. On the other hand, if it is very far from the optimal solution, such as opposite location in the worst case, the convergence will take much more time or even the solution can be intractable. In these bad cases, looking simultaneously for a better candidate solution in both current and opposite directions may help us to solve the present problem quickly and efficiently, which could be beneficial.

Opposition-based evolutionary computation is inspired in part by the observation that opposites permeate everything around us, in some form or another.

9.15.1 Opposition-Based Learning

There are several definitions on opposition of an existing solution **x**. Focused on metaheuristics and optimization, opposition is presented as a relationship between a pair of candidate solutions. In combinatorial and continuous optimization, each candidate solution has its own particularly defined opposite candidate.

Definition 9.39 (Opposite Number [151]) Let $x \in \mathbb{R}$ be a real number defined on a certain interval: $x \in [a, b]$. The *opposite number* $\check{x}$ is defined as

$$\check{x} = (a + b) - x. \tag{9.15.1}$$

The opposite number in a multidimensional case can be analogously defined. Figure 9.4 shows opposite number of a real number x.

Definition 9.40 (Opposite Point [131, 151]) Let $\mathbf{x} = [x_1, \ldots, x_D]^T$ be a point (vector) in D-dimensional space, where $x_1, \ldots, x_D \in \mathbb{R}$ and $x_i \in [a_i, b_i]$, $i = 1, \ldots, D$. The *opposite point* $\check{\mathbf{x}} = [\check{x}_1, \ldots, \check{x}_D]^T$ is completely defined by its

Fig. 9.4 Geometric representation of opposite number of a real number x

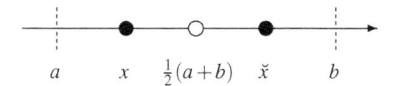

9.15 Opposition-Based Evolutionary Computation

components

$$\check{x}_i = \check{a}_i + \check{b}_i - x_i, \quad i = 1, \ldots, D. \tag{9.15.2}$$

Theorem 9.9 (Uniqueness [130]) *Every point* $\mathbf{x} = [x_1, \ldots, x_D]^T \in \mathbb{R}^D$ *of real numbers with* $x_i \in [a_i, b_i]$ *has a unique opposite point* $\check{\mathbf{x}} = [\check{x}_1, \ldots, \check{x}_D]^T$ *defined by* $\check{x}_i = a_i + b_i - x_i, i = 1, \ldots, D$.

Definition 9.41 (Opposite Solution) Given a candidate solution $\mathbf{x} = [x_1, \ldots, x_D]^T$ in a d-dimensional vector space, its opposition solution $\bar{\mathbf{x}} = [\bar{x}_1, \ldots, \bar{x}_D]^T$ is defined as element form:

$$\bar{x}_i = (a_i + b_i) - x_i, \quad i = 1, \ldots, D; \tag{9.15.3}$$

and *generalized opposition solution* $\tilde{\mathbf{x}}_g = [\tilde{x}_1, \ldots, \tilde{x}_D]^T$ is defined as

$$\tilde{\mathbf{x}}_g = [\tilde{x}_i]_{i=1}^d = [k \cdot (a+b) - x_i]_{i=1}^D, \tag{9.15.4}$$

where k a random number in $[0, 1]$; while a_i and b_i are the minimum and maximum values for ith element of $\mathbf{x}$.

Center-based sampling (CBS) [129] is a variant of opposition solution similar to generalized opposition solution.

Definition 9.42 (Center-Based Sampling [129, 135]) Let $\mathbf{x} = x_1, \ldots, x_D]^T$ be a solution in a D-dimension vector space. An opposite candidate solution $\hat{\mathbf{x}}_c = [\hat{x}_1, \ldots, \hat{x}_D]^T$ is defined by a random point between $\mathbf{x}$ and its opposite point $\bar{\mathbf{x}}$ as follows:

$$\hat{x}_i = \text{rand}_i \cdot (a_i + b_i - 2x_i) + x_i, \quad i = 1, \ldots, D, \tag{9.15.5}$$

where rand_i is a uniformly distributed random number in $[0, 1]$; and a_i and b_i are the minimum and maximum values of the ith component of $\mathbf{x}$.

The objective of center-based sampling is to obtain opposite candidate solutions closer to the center of the domain of each variable.

Definition 9.43 (Opposition-Based Learning [151]) Let $\mathbf{x} = [x_1, \ldots, x_D]^T$ and $\bar{\mathbf{x}} = [\bar{x}_1, \ldots, \bar{x}_D]^T$ be a solution point and its opposition solution point in D-dimensional vector space, respectively. If the opposition solution has the better fitness, i.e., $f(\bar{\mathbf{x}}) < f(\mathbf{x})$, then the original solution $\mathbf{x}$ should be replaced with the new candidate solution $\bar{\mathbf{x}}$; otherwise, $\mathbf{x}$ is continue to be used as a candidate solution. This machine learning based on opposition is called *opposition-based learning* (OBL). The optimization using the candidate solution and its opposition solution simultaneously for searching the better one is referred to as the *opposition-based optimization* (OBO).

The basic concept of OBL was originally introduced by Tizhoosh in 2005 [151]. The OBL has been applied in evolutionary algorithms, differential evolution, particle swarm optimization, harmony search, among others (see, e.g., Survey Papers [135, 161]).

9.15.2 Opposition-Based Differential Evolution

For evolutionary optimization methods, they start with some initial solutions (initial population) and try to improve them toward some optimal solution(s). In the absence of a priori information about the solution, some initial solutions are needed to be randomly guessed. Then, evolutionary optimization is related to the distance of these initial guesses from the optimal solution. An important fact is [131] that evolutionary optimization can be improved, if it starts with a closer (fitter) solution by simultaneously checking the opposite solution.

Let $\mathbf{p} = [p_1, \ldots, p_D]^T$ be a candidate solution in D-dimensional space, and $\check{\mathbf{p}} = [\check{p}_1, \ldots, \check{p}_D]^T$ be the *opposite solution* in D-dimensional space. Assume that $f(\mathbf{p})$ is a fitness function for point $\mathbf{p}$, which is used to measure the candidate's fitness. If $f(\check{\mathbf{p}}) \geq f(\mathbf{p})$, then point $\mathbf{p}$ can be replaced with $\check{\mathbf{p}}$; otherwise, continue with $\mathbf{p}$. Hence, the point $\mathbf{p}$ and its opposite point $\check{\mathbf{p}}$ are evaluated simultaneously in order to continue with the fitter one.

By applying the OBL to the classical DE, Rahnamayan et al. [131] developed an opposition-based differential evolution (ODE).

The ODE consists of the following two OBL parts [131].

1. *Opposition-based population initialization:*

   - Initialize the population $\boldsymbol{P} = (\mathbf{p}_1, \ldots, \mathbf{p}_{N_p})$ randomly, here $\mathbf{p}_i = [P_{i1}, \ldots, P_{iD}]^T$.
   - Calculate opposite population by $OP_{ij} = a_j + b_j - P_{ij}$ for $i = 1, \ldots, N_p$; $j = 1, \ldots, D$. Here, P_{ij} and OP_{ij} denote jth variable of the ith vector of the population and the opposite-population, respectively.
   - Select the N_p fittest individuals from $\{P \cup OP\}$ as initial population.

2. *Opposition-based generation jumping:* This part forces the evolutionary process to jump to a new solution candidate which ideally is fitter than the current population. Unlike opposition-based initialization that selects $OP_{ij} = a_j + b_j - P_{ij}$, opposition-based generation jumping selects the new population from the current population as $OP_{ij} = \text{MIN}_j^p + \text{MAX}_j^p - P_{ij}$ for $i = 1, \ldots, N_p$; $j = 1, \ldots, D$. Here, MIN_j^p and MAX_j^p are minimum and maximum values of the jth variable in the current population.

In essence, opposition-based differential evolution is an evolution optimization based on both difference vector between two populations or solutions for mutation and opposition-based learning for initialization and generation jumping.

9.15 Opposition-Based Evolutionary Computation

Opposition-based differential evolution is shown in Algorithm 9.19. As compared with the classical differential evolution Algorithm 9.14, the opposition-based differential evolution Algorithm 9.19 adds one pre-processing (*Opposition-Based Population Initialization*) and one post-processing (*Opposition-Based Generation Jumping*).

Algorithm 9.19 Opposition-based differential evolution (ODE) [130]

1. **input:** Population size N_p, problem dimension D, mutation constant F, crossover rate C_r.
2. Generate uniformly distributed random population $\mathbf{p}_0$.
   // Begin of Opposition-Based Population Initialization
3. **for** $i = 0$ to N_p **do**
4.   **for** $j = 0$ to D **do**
5. $OP_{0i,j} \leftarrow a_j + b_j - P_{0i,j}$.
6.   **end for**
7. **end for**
8. Select N_p fittest individuals from the set $\{P_0, OP_0\}$ as initial population P_0.
   // End of Opposition-Based Population Initialization
   // Begin of DE's Evolution Steps
9. Call Algorithm 9.14 Differential evolution (DE)
   // End of DE's Evolution Steps
   // Begin of Opposition-Based Generation Jumping
10. **if** rand$(0, 1) < J_r$ **then**
11.   **for** $i = 0$ to N_p **do**
12.   **for** $j = 0$ to D **do**
13. $OP_{i,j} \leftarrow MIN_j^p + MAX_j^p - P_{i,j}$.
14.   **end for**
15.   **end for**
16.   Select N_p fittest individuals from the set $\{P, OP\}$ as current population P.
17. **end if**
   // End of Opposition-Based Generation Jumping
18. **end while**

9.15.3 Two Variants of Opposition-Based Learning

In opposition-based learning, an opposite point is deterministically calculated with the reference point at the center in the fixed search range. There are two variants of opposite point selection.

The first variant is called *quasi opposition-based learning* (quasi-OBL), proposed in [130].

Definition 9.44 (Quasi-Opposite Number [130]) Let x be any real number between $[a, b]$ and $x^o = \check{x}$ be its opposite number. The *quasi-opposite number* of x, denoted by x^q, is defined as

$$x^q = \text{rand}(m, x^o), \qquad (9.15.6)$$

where $m = (a+b)/2$ is the middle of the interval $[a, b]$ and rand(m, x^o) is a random number uniformly distributed between m and the opposite point x^o.

Definition 9.45 (Quasi-Opposite Point [130]) Let $\mathbf{x}_i = [x_{i1}, \ldots, x_{iD}]^T \in \mathbb{R}^D$ be any vector (point) in D-dimensional real space, and $x_{ij} \in [a_j, b_j]$ for $i = 1, \ldots, N_p$; $j = 1, \ldots, D$. The *quasi-opposite point* of $\mathbf{x}_i$, denoted by $\mathbf{x}_i^q = [x_{i1}^q, \ldots, x_{iD}^q]^T$, is defined in its element form as

$$x_{ij}^q = \text{rand}(M_{ij}, x_{ij}^o), \tag{9.15.7}$$

where $i = 1, \ldots, N_p$; $j = 1, \ldots D$, $M_{ij} = (a_j + b_j)/2$ is the middle of the interval $[a_j, b_j]$, and rand(M_{ij}, OP_{ij}) is a random number uniformly distributed between the middle M_{ij} and opposite number $x_{ij}^o = \text{OP}_{ij}$.

The calculation of the quasi-opposite point in (9.15.7) depends on the comparison of the values between the opposite-point and the original point as follows:

$$x_{ij}^q = \begin{cases} M_{ij} + \text{rand}(0, 1) \cdot (x_{ij}^o - M_{ij}), & \text{if } x_{ij} < M_{ij}; \\ x_{ij}^o + \text{rand}(0, 1) \cdot (M_{ij} - x_{ij}^o), & \text{otherwise}. \end{cases} \tag{9.15.8}$$

Figure 9.5 shows the opposite and quasi-opposite points of a solution (point) $\mathbf{x}$.

Theorem 9.10 ([130]) *Given a guess point $\mathbf{x} \in \mathbb{R}^D$, its opposite point $\mathbf{x}^o \in \mathbb{R}^D$ and quasi opposite point $\mathbf{x}^q \in \mathbb{R}^D$, and given the distance from the solution $d(\cdot)$ and probability function $P_r(\cdot)$, then*

$$P_r[d(\mathbf{x}^q) < d(\mathbf{x}^o)] > 1/2. \tag{9.15.9}$$

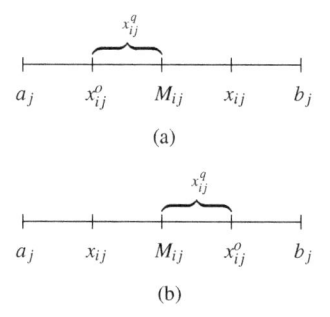

Fig. 9.5 The opposite point and the quasi-opposite point. Given a solution $\mathbf{x}_i = [x_{i1}, \ldots, x_{iD}]^T$ with $x_{ij} \in [a_j, b_j]$ and $M_{ij} = (a_j + b_j)/2$, then x_{ij}^o and x_{ij}^q are the jth elements of the opposite point $\mathbf{x}_i^o$ and the quasi-opposite point $\mathbf{x}_i^q$ of $\mathbf{x}_i$, respectively. (**a**) When $x_{ij} > M_{ij}$. (**b**) When $x_{ij} < M_{ij}$

9.15 Opposition-Based Evolutionary Computation

The distance in the above theorem can take the Euclidean distance

$$d(\mathbf{x}^o) = \|\mathbf{x}^o - \mathbf{x}\|_2 \quad \text{and} \quad d(\mathbf{x}^q) = \|\mathbf{x}^q - \mathbf{x}\|_2 \quad (9.15.10)$$

or other distances.

Theorem 9.10 implies that for a black-box optimization problem (which means solution can appear anywhere over the search space), the quasi opposite point $\mathbf{x}^q$ has a higher chance than the opposite point $\mathbf{x}^o$ to be closer to the solution.

The quasi opposition-based learning consists of Quasi-Oppositional Population Initialization and Quasi-Oppositional Generation Jumping [130].

1. *Quasi-oppositional population initialization:* Generate uniformly distributed random populations $\mathbf{p}_i^0 = [\mathrm{P}_{i1}^0, \ldots, \mathrm{P}_{iD}^0]^T$ for $i = 1, \ldots, N_p$, where $\mathrm{P}_{ij}^0 \in [a_j, b_j]$ for $j = 1, \ldots, D$.

    - Compute the opposite populations $\mathbf{p}_i^o = [\mathrm{OP}_{i1}^0, \ldots, \mathrm{OP}_{iD}^0]^T$ associated with $\mathbf{p}_i^0$ in element form:

    $$\mathrm{OP}_{ij}^0 = a_j + b_j - \mathrm{P}_{ij}^0, \quad i = 1, \ldots, N_p; j = 1, \ldots, D. \quad (9.15.11)$$

    - Denote the middle point between a_j and b_j as

    $$M_{ij} = (a_j + b_j)/2, \quad i = 1, \ldots, N_p; j = 1, \ldots, D. \quad (9.15.12)$$

    - If letting the quasi-opposition of $\mathbf{p}_i$ be $\mathbf{p}_i^q = [\mathrm{QOP}_{i1}^0, \ldots, \mathrm{QOP}_{iD}^0]^T$, then its elements are given by

    $$\mathrm{QOP}_{ij}^0 = \begin{cases} M_{ij} + \mathrm{rand}(0,1) \cdot (\mathrm{OP}_{ij}^0 - M_{ij}), & \text{if } \mathrm{P}_{ij}^0 < M_{ij}; \\ \mathrm{OP}_{ij}^0 + \mathrm{rand}(0,1) \cdot (M_{ij} - \mathrm{OP}_{ij}^0), & \text{otherwise.} \end{cases} \quad (9.15.13)$$

    Here, $i = 1, \ldots, N_p; j = 1, \ldots, D$.
    - Select N_p fittest individuals from the set $\{\mathbf{p}_i^o, \mathbf{p}_i^q\}$ as initial population $\mathbf{p}^0$.

2. *Quasi-oppositional generation jumping:* Let J_r be jumping rate, MIN_j^p be minimum value of the jth individual in the current population, and MAX_j^p denote maximum value of the jth individual in the current population.

    - If $\mathrm{rand}(0,1) < J_r$ then calculate

    $$\mathrm{OP}_{ij} = \mathrm{MIN}_j^p + \mathrm{MAX}_j^p - \mathrm{P}_{ij}, \quad (9.15.14)$$

    $$M_{ij} = (\mathrm{MIN}_j^p + \mathrm{MAX}_j^p)/2, \quad (9.15.15)$$

    $$\mathrm{QOP}_{ij} = \begin{cases} M_{ij} + \mathrm{rand}(0,1) \cdot (\mathrm{OP}_{ij} - M_{ij}), & \text{if } \mathrm{P}_{ij} < M_{ij}; \\ \mathrm{OP}_{ij} + \mathrm{rand}(0,1) \cdot (M_{ij} - \mathrm{OP}_{ij}), & \text{otherwise.} \end{cases} \quad (9.15.16)$$

Here, $i = 0, \ldots N_p - 1$; $j = 0, \ldots, D - 1$.
- Select N_p fittest individuals from set the $\{P_{ij}, QOP_{ij}\}$ as current population **p**.

If the oppositional population initialization and the oppositional generation jumping in Algorithm 9.19 are, respectively, replaced by the above quasi-oppositional population initialization and quasi-oppositional generation jumping, then the opposition-based differential evolution (ODE) described in Algorithm 9.19 becomes the *quasi-oppositional differential evolution* (QODE) in [130].

The second variant of the OBL is quasi reflection point-based OBL, which was proposed in [40].

Definition 9.46 (Quasi-Reflected Point [40]) Let $\mathbf{x}_i = [x_{i1}, \ldots, x_{iD}]^T$ be any point in D-dimensional real space, and $x_{ij} \in [a_j, b_j]$. Then the *quasi-reflected point* of $\mathbf{x}_i$, denoted by $\mathbf{x}_i^{qr} = [x_{i1}^{qr}, \ldots, x_{iD}^{qr}]^T$, is defined in its element form as

$$x_{ij}^{qr} = \text{rand}(M_{ij}, x_{ij}), \tag{9.15.17}$$

where $M_{ij} = (a_j + b_j)/2$, and $\text{rand}(M_{ij}, x_{ij})$ is a random number uniformly distributed between M_{ij} and x_{ij}.

Figure 9.6 shows the opposite, quasi-opposite, and quasi-reflected points of a solution (point) $\mathbf{x}_i$.

There are the following relations between the quasi-opposite point and quasi-reflected opposite point.

- The quasi-opposite point is generated by $x_{ij}^q = \text{rand}(M_{ij}, x_{ij}^o)$, whereas the quasi-reflected opposite point is defined by $x_{ij}^{qr} = \text{rand}(M_{ij}, x_{ij})$.
- The quasi-reflected opposite point is the mirror image of the quasi-opposite point with regard to the middle point M_{ij}, and hence is named as the quasi-reflected opposite point.

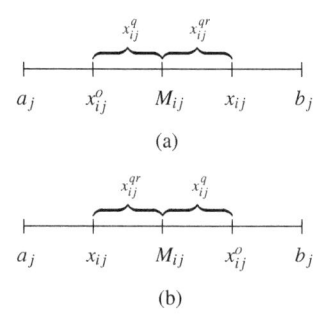

Fig. 9.6 The opposite, quasi-opposite, and quasi-reflected points of a solution (point) $\mathbf{x}_i$. Given a solution $\mathbf{x}_i = [x_{i1}, \ldots, x_{iD}]^T$ with $x_{ij} \in [a_j, b_j]$ and $M_{ij} = (a_j + b_j)/2$, then x_{ij}^o, x_{ij}^q, and x_{ij}^{qr} are the jth elements of the opposite point $\mathbf{x}_i^o$, the quasi-opposite point $\mathbf{x}_i^q$, and the quasi-reflected opposite point $\mathbf{x}_i^{qr}$ of $\mathbf{x}_i$, respectively. (**a**) When $x_{ij} > M_{ij}$. (**b**) When $x_{ij} < M_{ij}$

After calculating the fitness of opposite population generated by quasi-reflected opposition, the fittest N_p individuals in P and OP constitute the updated population. The application of the quasi-opposite points or the quasi-reflected opposite points will result into various variants and extensions of the basic OBL algorithm. For example, a stochastic opposition-based learning using a Beta distribution in differential evolution can be found in [121].

Brief Summary of This Chapter

- This chapter presents evolutionary computation tree.
- Evolutionary computation is a subset of artificial intelligence, which is essentially a computational intelligence for solving combinatorial optimization problems.
- Evolutionary computation, starting from a group of randomly generated individuals, imitates the biological genetic mode to update the next generation of individuals by replication, crossover, and mutation. Then, according to the size of fitness, the individual survival of the fittest improves the quality of the new generation of groups, after repeated iterations, gradually approaching the optimal solution. From the mathematical point of view, evolutionary computation is essentially a search optimization method.
- This chapter focuses on the Pareto optimization theory in multiobjective optimization problems existing in coevolutionary computation.
- Evolutionary computation has been widely used in artificial intelligence, pattern recognition, image processing, biology, electrical engineering, communication, economic management, mechanical engineering, and many other fields.

References

1. Agrawal, R., Imielinski, T., Swami, A.N.: Mining association rules between sets of items in large databases. In: Proceedings of ACM SIGMOD International Conference on the Management of Data, pp. 207–216 (1992)
2. Bäck, T., Hammel, U.: Evolution strategies applied to perturbed objective functions. In: Proceedings of 1st IEEE Conference Evolutionary Computation, vol. 1, pp. 40–45 (1994)
3. Bäck, T., Schwefel H.-P.: An overview of evolutionary algorithms for parameter optimization. Evol. Comput. **1**(1), 1–23 (1993)
4. Baker, J.E.: Adaptive selection methods for genetic algorithms. In: Proceedings of the First International Conference on Genetic Algorithms and their Applications, Pittsburgh, pp. 101–111 (1985)
5. Bandyopdhyay, S., Maulik, U.: An evolutionary technique based on K-means algorithm for optimal clustering in $\mathbb{R}^N$. Inform. Sci. **146**(1–4), 221–237 (2002)
6. Bandyopadhyay, S., Maulik, U., Holder, L.B., Cook, D.J.: Advanced Methods for Knowledge Discovery From Complex Data (Advanced Information and Knowledge Processing). Springer, London (2005)

7. Bandyopadhyay, S., Saha, S., Maulik, U., Deb, K.: A simulated annealing-based multiobjective optimization algorithm: AMOSA. IEEE Trans. Evol. Comput. **12**(3), 269–283 (2008)
8. Basu, M.: Dynamic economic emission dispatch using non-dominated sorting genetic algorithm-II. Electr. Power Energy Syst. **30**, 140–149 (2008)
9. Beckers, R., Deneubourg, J.L., Goss, S.: Trails and U-turns in the selection of the shortest path by the ant Lasius Niger. J. Theor. Biol. **159**, 397–415 (1992)
10. Benson, H.P., Sayin, S.: Towards finding global representations of the efficient set in multiple objective mathematical programming. Naval Res. Logist. **44**, 47–67 (1997)
11. Bonabeau, E., Dorigo, M., Theraulaz, G.: Swarm Intelligence: From Natural to Artificial Systems. Oxford University Press, New York (1999)
12. Box, G.E.P.: Evolutionary operation: a method for increasing industrial productivity. Appl. Stat. **VI**(2), 81–101 (1957)
13. Branke, J.: Multi-objective evolutionary algorithms and MCDA. European Working Group "Multiple Criteria Decision Aiding", ser. 3, vol. 25, pp. 1–3 (2012)
14. Branke, J., Schmidt, C., Schmeck, H.: Efficient fitness estimation in noisy environments. In: Proceedings of the Genetic and Evolutionary Computation, pp. 243–250 (2001)
15. Bremermann, H.J.: Optimization through evolution and recombination. In: Yovits M.C., et al. (eds.) Self-Organizing Systems. Spartan, Washington (1962)
16. Chang, D.X., Zhang, X.D., Zheng, C.W.: A genetic algorithm with gene rearrangement for K-means clustering. Pattern Recognit. **42**, 1210–1222 (2009)
17. Charnes, A., Cooper, W., Niehaus, R., Stredry, A.: Static and dynamic model with multiple objectives and some remarks on organisational design. Manag. Sci. **15B**, 365–375 (1969)
18. Chen, G., Low, C.P., Yang, Z.: Preserving and exploiting genetic diversity in evolutionary programming algorithms. IEEE Trans. Evol. Comput. **13**(3), 661–673 (2009)
19. Clerc, M., Kennedy, J.: The particle swarm - explosion, stability, and convergence in a multidimensional complex space. IEEE Trans. Evol. Comput. **6**(1), 58–73 (2002)
20. Coello Coello, C.A., Lamont, G.B., van Veldhuizen, D.A.: Evolutionary Algorithms for Solving Multi-Objective Problems (Genetic and Evolutionary Computation), 2nd edn. Springer, Berlin (2007)
21. Cordon, O., Herrera, F., Stutzle, T.: A review on the ant colony optimization metaheuristic: basis, models and new trends. Mathware Soft Comput. **9**(2–3), 141–175 (2002)
22. Czyzak, P., Jaszkiewicz, A.: Pareto simulated annealing—a metaheuristic technique for multiple-objective combinatorial optimization. J. Multi-Crit. Decis. Anal. **7**, 34–47 (1998)
23. Deb, K.: Multi-objective genetic algorithms: problem difficulties and construction of test problems. Evol. Comput. **7**(3), 205–230 (1999)
24. Deb, K.: Multi-Objective Optimization Using Evolutionary Algorithms. Wiley, London (2001)
25. Deb, K., Agarwal, S., Pratap, A., Meyarivan, T.: A fast and elitist multiobjective genetic algorithm: NSGA-II. IEEE Trans. Evol. Comput. **6**(2), 182–197 (2002)
26. del Valle, Y., Venayagamoorthy, G.K., Mohagheghi, S., Hernandez, J.-C., Harley, R.G.: Particle swarm optimization: basic concepts, variants and applications in power systems. IEEE Trans. Evol. Comput. **12**(2), 171–195 (2008)
27. Deneubourg, J.L., Aron, S., Goss, S., Pasteels, J.M.: The self-organizing exploratory pattern of the argentine ant. J. Insect Behav. **3**, 159 (1990)
28. Dorigo, M.: Optimization, learning and natural algorithms, Ph.D.Thesis, Politecnico diMilano (1992)
29. Dorigo, M., Birattari, M.: Ant colony optimization. In: Sammut, C., Webb, G.I. (eds.) Encyclopedia of Machine Learning, pp. 37–40. Springer, Berlin (2011)
30. Dorigo, M., Blumb, C.: Ant colony optimization theory: a survey. Theor. Comput. Sci. **344**, 243–278 (2005)
31. Dorigo, M., Caro, G.D.: The ant colony optimization meta-heuristic. In: Corne, D., Dorigo, M, Glover, F. (eds.) New Ideas in Optimization, chap. 2. McGraw-Hill, New York (1999)
32. Dorigo, M., Gambardella, L.M.: Ant colony system: a cooperative learning approach to the traveling salesman problem. IEEE Trans. Evol. Comput. **1**(1), 53–66 (1997)

33. Dorigo, M., Stützle, T.: The ant colony optimization metaheuristic: algorithms, applications, and advances. In: Glover, F., Kochenberger, G.A. (eds.) Handbook of Metaheuristics, chap. 9. Kluwer Academic, New York (2003)
34. Dorigo, M., Maniezzo, V., Colorni, A.: The ant system: optimization by a colony of cooperating agents. IEEE Trans. Syst. Man Cybern. B **26**(2), 29–41 (1996)
35. Droste, S., Jansen, T., Wegener, I.: On the analysis of the $(1+1)$ evolutionary algorithms. Theor. Comput. Sci. **276**, 51–81 (2002)
36. Eberhart, R., Kennedy, J.: A new optimizer using particle swarm theory. In: Proceedings of the 6th International Symposium on Micro Machine and Human Science (MHS), pp. 39–43 (1995)
37. Eberhart, R., Shi, Y., Kennedy, J.: Swarm Intelligence. Morgan Kaufmann, San Francisco (2001)
38. Engelbrecht, A.P.: Particle swarm optimization: where does it belong? In: Proceedings of the 2003 IEEE Swarm Intelligence Symposium, pp. 48–54 (2006)
39. Engrand, P.: A multi-objective approach based on simulated annealing and its application to nuclear fuel management. In: 5th International Conference on Nuclear Engineering, Nice, pp. 416–423 (1997)
40. Ergezer, M., Simon, D., Du, D.: Oppositional biogeography-based optimization. In: Proceedings of IEEE International Conference on Systems, Man and Cybernetics (SMC), San Antonio, pp. 1009–1014 (2009)
41. Fogel, L.J.: Autonomous automata. Ind. Res. **4**, 14–19 (1962)
42. Fogel, D.B.: An introduction to simulated evolutionary optimization. IEEE Trans. Neural Netw. **5**, 3–14 (1994)
43. Fogel, D.B.: Evolutionary Computation: Toward a New Philosophy of Machine Intelligence. IEEE Press, Piscataway (1995)
44. Fogel, L.J., Owens, A.J., Walsh, M.J.: Artificial Intelligence Through Simulated Evolution. Wiley, New York (1966)
45. Fonseca, C.M., Fleming, P.J.: An overview of evolutionary algorithms in multiobjective optimization. Evol. Comput. **3**(1), 1–16 (1995)
46. Friedberg, R.M.: A learning machine: part I. IBM J. **2**(1), 2–13 (1958)
47. Friedberg, R.M., Dunham, B., North, J.H.: A learning machine: part II. IBM J. **3**(7), 282–287 (1959)
48. Gao, W.F., Liu, S.Y.: Improved artificial bee colony algorithm for global optimization. Inf. Process. Lett. **111**(17), 871–882 (2011)
49. Gao, W.F., Liu, S.Y., Huang, L.L.: A novel artificial bee colony algorithm based on modified search equation and orthogonal learning. IEEE Trans. Cybern. **43**(3), 1011–1024 (2013)
50. Geman, A., Geman, D.: Stochastic relaxation, Gibbs distributions, and the Bayesian restoration of images. IEEE Trans. Pattern Anal. Mach. Intell. **6**(6), 721–741 (1984)
51. Gendreau, M., Potvin, J.-Y.: Metaheuristics in combinatorial optimization. Ann. Oper. Res. **140**(1), 189–213 (2005)
52. Glover, F.: Tabu search — Part I. ORSA J. Comput. **1**, 190–206 (1989)
53. Goh, C.K., Tan, K.C.: An investigation on noisy environments in evolutionary multiobjective optimization. IEEE Trans. Evol. Comput. **11**(3), 354–381 (2007)
54. Goldberg, D.E.: Genetic Algorithms in Search, Optimization, and Machine Learning. Addison-Wesley, Reading (1989)
55. Goldberg, D.E., Holland, J.H.: Genetic algorithms and machine learning. Mach. Learn. **3**(2), 95–99 (1988)
56. Goldberg, D.E., Richardson, J.: Genetic algorithms with sharing for multimodal function optimization. In: Proceedings of the Second International Conference on Genetic Algorithms on Genetic Algorithms and Their Application, Cambridge, pp. 41–49 (1987)
57. Gong, Y.-J., Li, J.-J., Zhou, Y., Li, Y.,, Chung, H.S., Shi, Y.-H., Zhang, J.: Genetic learning particle swarm optimization. IEEE Trans. Cybern. **46**(10), 2277–2290 (2016)
58. Gong, D., Sun, J., Miao, Z.: A set-based genetic algorithm for interval many-objective optimization problems. IEEE Trans. Evol. Comput. **22**(1), 47–60 (2018)

59. Grasse, P.P.: La reconstruction du nid et les coordinations interindividuelles chez bellicositermes natalensis et cubitermes sp. la theorie de la stigmergie: Essai dinterpretation du comportement des termites constructeurs. Insectes Sociaux **6**, 41–81 (1959)
60. Hajek, B.: Hitting-time and occupation-time bounds implied by drift analysis with applications. Adv. Appl. Probab. **14**(3), 502–525 (1982)
61. Hajela, P., Lin, C.-Y.: Genetic search strategies in multicriterion optimal design. Struct. Optim. **4**, 99–107 (1992)
62. Han, J., Kamber, M.: Data Mining: Concepts and Techniques. Morgan Kaufmann, San Francisco (2000)
63. Hans, A.E.: Multicriteria optimization for highly accurate systems. In: Stadler, W. (ed.) Multicriteria Optimization in Engineering and Sciences, Mathematical concepts and methods in science and engineering, vol. 19, pp. 309–352. Plenum Press, New York (1988)
64. Hansen, M.P., Jaszkiewicz, A.: Evaluating the quality of approximations to the non-dominated set. Technical Report IMM-REP-1998-7, Technical University of Denmark (1998)
65. Hastings, W.K.: Monte Carlo sampling methods using Markov chains and their applications. Biometrika **57**(1), 97–109 (1970)
66. He, J., Yao, X.: Drift analysis and average time complexity of evolutionary algorithms. Artif. Intell. **127**(1), 57–85 (2001)
67. He, J., Yao, X.: Erratum to: drift analysis and average time complexity of evolutionary algorithms. Artif. Intell. **140**, 245–248 (2002)
68. Holland, J.H.: Outline for a logical theory of adaptive systems. J. Assoc. Comput. Mach. **3**, 297–314 (1962)
69. Holland, J.H.: Adaption in Natural and Artificial Systems. University of Michigan Press, Ann Arbor (1975)
70. Hoorfar, A.: Mutation-based evolutionary algorithms and their applications to optimization of antennas in layered media. In: Proceedings of IEEE Antennas and Propagation Society International Symposium, Orlando, pp. 2876–2879 (1999)
71. Hoorfar, A.: Evolutionary programming in electromagnetic optimization: a review. IEEE Trans. Antennas Propag. **55**(3), 523–537 (2007)
72. Hoorfar, A., Liu, Y.: A study of Cauchy and Gaussian mutation operators in evolutionary programming optimization of antenna structures. In: Proceedings of 16th Annual Applied Computational Electromagnetics Conference, Monterey, pp. 63–69 (2000)
73. Horn, J.: Multicriterion decision making. In: Bäck, T., Fogel, D., Michalewicz, Z. (eds.) Handbook of Evolutionary Computation, vol. 1, pp. F1.9:1–F1.9:15. Oxford University Press, Oxford (1997)
74. Horn, J., Nafpliotis, N., Goldberg, D.E.: A niched Pareto genetic algorithm for multiobjective optimization. In: Proceedings of the First IEEE Conference on Evolutionary Computation, IEEE World Congress on Computational Intelligence, vol. 1, pp. 82–87. IEEE Press, Piscataway (1994)
75. Hughes, E.J.: Evolutionary multi-objective ranking with uncertainty and noise. In: Proceedings of first International Conference on Evolutionary Multi-Criterion Optimization, Zürich, pp. 329–343 (2001)
76. Hughes, E.J.: Constraint handling with uncertain and noisy multi-objective evolution. In: Proceedings of 2001 Congress on Evolutionary Computation, vol. 2, pp. 963–970 (2001)
77. Hutter, M., Legg, S.: Fitness uniform optimization. IEEE Trans. Evol. Comput. **10**(5), 568–589 (2006)
78. Hwang, C.-L., Masud, A.S.M.: Multiple Objective Decision Making-Methods and Applications. Springer, Berlin (1979)
79. Ishibuchi, H., Akedo, N., Nojima, Y.: Behavior of multiobjective evolutionary algorithms on many-objective knapsack problems. IEEE Trans. Evol. Comput. **19**(2), 264–283 (2015)
80. Jensen, M.T.: Reducing the run-time complexity of multiobjective EAs: the NSGA-II and other algorithms. IEEE Trans. Evol. Comput. **7**(5), 503–515 (2003)

81. Karaboga, D.: An idea based on honey bee swarm for numerical optimization. Erciyes University, Kayseri, Tech. Rep.-TR06 (2005)
82. Karaboga, D., Basturk, B.: A powerful and efficient algorithm for numerical function optimization: artificial bee colony (ABC) algorithm. J. Global Optim. **39**, 459–471 (2007)
83. Karaboga, D., Basturk, B.: On the performance of artificial bee colony (ABC) algorithm. Appl. Soft Comput. **8**(1), 687–697 (2008)
84. Karaboga, D., Basturk, B.: A comparative study of artificial bee colony algorithm. Appl. Math. Comput. **214**(1), 108–132 (2009)
85. Kennedy, J.: Swarm intelligence. In: Zomaya, A.Y. (ed.) Handbook of Nature-Inspired and Innovative Computing, pp. 187–219. Springer, New York (2006)
86. Kennedy, J., Eberhart, R.: Particle swarm optimization. In: Proceedings of IEEE International Conference on Neural Networks (ICNN), vol. IV, pp. 1942–1948 (1995)
87. Kim, J.-H., Han, J.-H., Kim, Y.-H., Choi, S.-H., Kim, E.-S.: Preference-based solution selection algorithm for evolutionary multiobjective optimization. IEEE Trans. Evol. Comput. **16**(1), 20–34 (2012)
88. Kirkpatrick, S., Gelatt, C.D., Vecchi, M.P.: Optimization by simulated annealing. Science **220**, 671–680 (1983)
89. Knowles, J.D., Corne, D.W.: On metrics for comparing nondominated sets. In: Proceedings of the Congress on Evolutionary Computation, vol. 1, pp. 711–716 (2002)
90. Kursawe, F.: A variant of evolution strategies for vector optimization. In: Schwefel, H.-P., Manner, R. Parallel Problem Solving from Nature, 193–197. Springer, Berlin (1991)
91. Laarhoven, P.J.M., Aarts, E.H.L.: Simulated Annealing: Theory and Applications. Reidel, Dordrecht (1987)
92. Laumanns, M., Rudolph, G., Schwefel, H.-P.: Mutation control and convergence in evolutionary multi-objective optimization. In: Proceedings of the 7th International Mendel Conference on Soft Computing (MENDEL 2001), Brno (2001)
93. Li, H., Zhang, Q.: Multiobjective optimization problems with complicated Pareto sets, MOEA/D and NSGA-II. IEEE Trans. Evol. Comput. **13**(2), 284–302 (2009)
94. Li, Y.-L., Zhou, Y.-R., Zhan, Z.-H., Zhang, J.: A primary theoretical study on decomposition-based multiobjective evolutionary algorithms. IEEE Trans. Evol. Comput. **20**(4), 563–576 (2016)
95. Liao, T., Socha, K., Montes, M.A., Stützle, T., Dorigo, M.: Ant colony optimization for mixed-variable optimization problems. **18**(4), 503–518 (2014)
96. Limbourg, P., Aponte, D.E.S.: An optimization algorithm for imprecise multi-objective problem function. In: Proceedings of IEEE Congress on Evolutionary Computation, Edinburgh, pp. 459–466 (2005)
97. López-Ioáñez, M., Stützle, T.: The automatic design of multiobjective ant colony optimization algorithms. IEEE Trans. Evol. Comput. **16**(6), 861–875 (2012)
98. Luo, B., Zheng, J., Xie, J., Wu, J.: Dynamic crowding distance - a new diversity maintenance strategy for MOEAs. In: Fourth International Conference on Natural Computation, pp. 580–585 (2008)
99. Martens, D., Backer, M.D., Haesen, R., Vanthienen, J., Snoeck, M., Baesens, B.: Classification with ant colony optimization. IEEE Trans. Evol. Comput. **11**(5), 651–665 (2007)
100. Metropolis, N., Rosenbluth, A.W., Rosenbluth, M.N., Teller, A.H., Teller, E.: Equations of state calculations by fast computing machines. J. Chem. Phys. **21**(6), 1087–1092 (1953)
101. Mezura-Montes, E., Velázquez-Reyes, J., Coello, C.A.C.: A comparative study of differential evolution variants for global optimization. In: Proceedings of the 2006 Conference on Genetic and Evolutionary Computation (GECCO-2006), Seattle, pp. 485–492 (2006)
102. Michalewicz, Z.: Genetic Algorithms+Data Structures=Evolution Programs. AI Series. Springer, New York (1994)
103. Miettinen, K.: Nonlinear Multiobjective Optimization. Kluwer, Norwell (1999)
104. Mitra, D., Romeo, F., Sangiovanni-Vincentelli, A.: Convergence and finite-time behavior of simulated annealing. Adv. Appl. Probab. **18**, 747–771 (1986)

105. Mohan, B.C., Baskaran, R.: A survey: ant colony optimization based recent research and implementation on several engineering domain. Exp. Syst. Appl. **39**, 4618–4627 (2012)
106. Moradi, P., Gholampour, M.: A hybrid particle swarm optimization for feature subset selection by integrating a novel local search strategy. Appl. Soft Comput. **43**, 117–130 (2016)
107. Morse, J.N.: Reducing the size of the nondominated set: pruning by clustering. Comput. Oper. Res. **7**(1–2), 55–66 (1980)
108. Moulton, C.M., Roberts, S.A., Calatn, P.H.: Hierarchical clustering of multiobjective optimization results to inform land-use decision making. URISA J. **21**(2), 25–38 (2009)
109. Mühlenbein, H., Schlierkamp-Voosen, D.: The science of breeding and its application to the breeder genetic algorithm (BGA). Evol. Comput. **1**(4), 335–360 (1994)
110. Mukhopadhyay, A., Maulik, U., Bandyopadhyay, S., Coello Coello, C.A.: A survey of multiobjective evolutionary algorithms for data mining: part I. IEEE Trans. Evol. Comput. **18**(1), 4–19 (2014)
111. Mullen, R.J., Monekosso, D., Barman, S., Remagnino, P.: A review of ant algorithms. Exp. Syst. Appl. **36**, 9608–9617 (2009)
112. Myung, H., Kim, J.-H.: Hybrid evolutionary programming for heavily constrained problems. BioSystems **38**, 29–43 (1996)
113. Myung, H., Kim, J.-H., Fogel, D.B.: Preliminary investigations into a two-stage method of evolutionary optimization on constrained problems. In: McDonnell, J.R., Reynolds, R.G., Fogel, D.B. (eds.) Proceedings of the Fourth Annual Conference Evolutionary Programming, pp. 449–463. MIT Press, Cambridge (1995)
114. Nam, D.K., Park, C.H.: Multiobjective simulated annealing: a comparative study to evolutionary algorithms. Inf. J. Fuzzy Syst. **2**(2), 87–97 (2000)
115. Neto, R.F.T., Filho, M.G.: A software model to prototype ant colony optimization algorithms. Exp. Syst. Appl. **38**, 249–259 (2011)
116. Nikulin, Y., Miettinen, K., Mäkelä, M.M.: A new achievement scalarizing function based on parameterization in multiobjective optimization. OR Spectr. **34**, 69–87 (2012)
117. Oberkampf, W.L., Helton, J.C., Joslyn, C.A., Wojtkiewicz, S.F., Ferson, S.: Challenge problems: uncertainty in system response given uncertain parameters. Reliab. Eng. Syst. Saf. **85**, 11–19 (2004)
118. Osman, I.H., Laporte, G.: Metaheuristics: a bibliography. Ann. Oper. Res. **63**(5), 511–623 (1996)
119. Palakonda, V., Mallipeddi, R.: Pareto dominance-based algorithms with ranking methods for many-objective optimization. IEEE Access **5**, 11043–11053 (2017)
120. Papadimitriou, C.H., Steiglitz, K.: Combinatorial Optimization: Algorithms and Complexity. Dover, New York (1982)
121. Park, S.-Y., Lee, J.-J.: Stochastic opposition-based learning using a Beta distribution in differential evolution. IEEE Trans. Cybern. **46**(10), 2184–2194 (2016)
122. Parpinelli, R.S., Lopes, H.S., Freitas, A.A.: Data mining with an ant colony optimization algorithm. IEEE Trans. Evol. Comput. **6**(4), 321–332 (2002)
123. Premalatha, K., Natarajan, A.M.: Hybrid PSO and GA for global maximization. Int. J. Open Problems Compt. Math. **2**(4), 597–608 (2009)
124. Price, K., Storn, R., Lampinen, J.: Differential Evolution: A Practical Approach to Global Optimization. Springer, Berlin (2005)
125. Qin, A.K., Suganthan, P.N.: Self-adaptive differential evolution algorithm for numerical optimization. In: Proceedings of the 2005 IEEE Congress on Evolutionary Computation, vol. 2, pp. 1785–1791 (2005)
126. Quagliarella, D., Vicini, A.: Coupling genetic algorithms and gradient based optimization techniques. In: Quagliarella, D., Periaux, J., Poloni, C., Winter, G. (eds.) Genetic Algorithms and Evolution Strategy in Engineering and Computer Science – Recent Advances and Industrial Applications. Wiley, Chichester (1997)
127. Radcliffe, N., Surry, P.: Fitness variance of formae and performance prediction. In: Foundations of Genetic Algorithms 3, pp. 51–72. Morgan Kaufmann, San Mateo (1995)

128. Rahnamayan, S.: Opposition-based differential evolution. Thesis for Doctor of Philosophy, University of Waterloo (2007)
129. Rahnamayan, S., Wang, G.G.: Center-based sampling for population-based algorithms. In: 2009 IEEE Congress on Evolutionary Computation, pp. 933–938 (2009)
130. Rahnamayan, S., Tizhoosh, H.R., Salama, M.: Quasi-oppositional differential evolution. In: Proceedings of the IEEE Congress on Evolutionary Computation (CEC), Singapore, pp. 2229–2236 (2007)
131. Rahnamayan, S., Tizhoosh, H.R., Salama, N.M.M.: Opposition-based differential evolution. IEEE Trans. Evol. Comput. **12**(1), 64–79 (2008)
132. Rakshit, P., Konar, A.: Differential evolution for noisy multiobjective optimization. Artif. Intell. **227**, 165–189 (2015)
133. Rechenberg, I.: Cybernetic solution path of an experimental problem. Royal Aircraft Establishment, Library translation No. 1122, Farnborough, Hants (1965)
134. Revelle, C., Cohon, J.L., Shobys, D.: Multiple objectives in facility location: a review. In: Beckmann, M., Kunzi, A.P. (eds.) Lecture Notes in Economics and Mathematical Systems, vol. 190, pp. 321–337. Springer, Berlin (1981)
135. Rojas-Morales, N., Riff Rojas, M.-C., Ureta, E.M.: A survey and classification of opposition-based metaheuristics. Comput. Ind. Eng. **110**, 424–435 (2017)
136. Rosenthal, R.E.: Principles of multiobjective optimization. Decis. Sci. **16**, 133–152 (1985)
137. Ruiz, F., Luque, M., Miguel, F., del Mar Muñoz, M.: An additive achievement scalarizing function for multiobjective programming problems. Eur. J. Oper. Res. **188**(3), 683–694 (2008)
138. Sakri, S., Rashid, N.A., Zain, Z.M.: Particle swarm optimization feature selection for breast cancer recurrence prediction. IEEE Access **6**, 29637–29647 (2018)
139. Santana, R.A., Pontes, M.R., Bastos-Filho, C.J.A.: A multiple objective particle Swarm optimization approach using crowding distance and roulette wheel. In: Ninth International Conference on Intelligent Systems Design and Applications, pp. 237–242 (2009)
140. Sastry, K., Goldberg, D., Kendall, G.: Genetic algorithms. In: Burke, E.K., Kendall, G. (eds.) Search Methodologies: Introductory Tutorials in Optimization and Decision Support Techniques. Springer, New York (2005)
141. Schaffer, J.D.: Multiple objective optimization with vector evaluated genetic algorithms. In: Proceedings of the First International Conference on Genetic Algorithms (ICGA'85), pp. 93–100 (1985)
142. Schapire, R.E.: The strength of weak learnability. Mach. Learn. **5**, 197–227 (1990)
143. Schwefel, H.P.: Numerical Optimization of Computer Models. Wiley, Hoboken (1981)
144. Slater, M.: Lagrange multipliers (revisited). Cowles Commission Discussion Paper: Mathematics 403 (1950)
145. Smith, K., Everson, R., Fieldsend, J.: Dominance measures for multi-objective simulated annealing. In: Proceedings of the 2004 IEEE Congress on Evolutionary Computation, pp. 23–30 (2004)
146. Srinivas, N., Deb, K.: Multiobjective optimization using nondominated sorting in genetic algorithms. Evol. Comput. **2**(3), 221–248 (1995)
147. Storn, R., Price, K.: Differential evolution - a simple and efficient heuristic for global optimization over continuous spaces. J. Global Optim. **11**, 341–359 (1997)
148. Tan, P.-N., Kumar, V., Srivastava, J.: Selecting the right interestingness measure for association patterns. In: Proceedings of the 8th ACM SIGKDD International Conference on KDD, pp. 32–41 (2002)
149. Tang, K., Li, X., Suganthan, P.N., Yang, Z., Weise, T.: Benchmark functions for the CEC'2010 special session and competition on large-scale global optimization. Tech. Rep. (2009)
150. Teich, J.: Pareto-front exploration with uncertain objectives. In: Zitzler, E. et al. (eds.) Evolutionary Multi-Criterion Optimization (EMO) 2001. Lecture Notes in Computer Science, vol. 1993, pp. 314–328 (2001)

151. Tizhoosh, H.R.: Opposition-based learning: a new scheme for machine intelligence. In: Proceedings of the International Conference on Computational Intelligence for Modelling, Control and Automation, and International Conference on Intelligent Agents, Web Technologies and Internet Commerce, 28–30 November, Vienna, vol. 1, pp. 695–701 (2005)
152. Ulungu, E.L., Teghem, J.: Multi-objective combinatorial optimization problems: a survey. J. MultiCrit. Decis. Anal. **3**, 83–101 (1994)
153. Ulungu, E.L., Teghem, J., Fortemps, P., Tuyttens, D.: MOSA method: a tool for solving multiobjective combinatorial optimization problems. J. MultiCrit. Decis. Anal. **8**, 221–236 (1999)
154. van den Bergh, F., Engelbrecht, A.P.: A cooperative approach to particle swarm optimization. IEEE Trans. Evol. Comput. **8**(3), 225–239 (2004)
155. Van Veldhuizen, D.A., Lamont, G.B.: Multiobjective evolutionary algorithms: analyzing the state-of-the-art. Evol. Comput. **8**(2), 125–147 (2000)
156. Wang, R., Zhang, Q., Zhang, T.: Decomposition-based algorithms using Pareto adaptive scalarizing methods. IEEE Trans. Evol. Comput. **20**(6), 821–837 (2016)
157. Whitley, D.: The GENITOR algorithm and selection pressure: why rank-based allocation of reproductive trials is best. In: Proceedings of the Third International Conference on Genetic Algorithms, San Mateo, pp. 116–123 (1989)
158. Wierzbicki, A.P.: The use of reference objectives in multiobjective optimization. In: Fandel, G., Gal, T. (eds.) Multiple Criteria Decision Making Theory and Applications. MCDM Theory and Applications Proceedings. Lecture Notes in Economics and Mathematical Systems, vol. 177. Springer, Berlin, pp. 468–486 (1980)
159. Wierzbicki, A.P.: A methodological approach to comparing parametric characterizations of efficient solutions. In: Fandel, G. et al. (eds.) Large-Scale Modeling and Interactive Decision Analysis. Lecture Notes in Economics and Mathematical Systems, vol. 273, pp. 27–45. Springer, Berlin (1986)
160. Wierzbicki, A.P.: On the completeness and constructiveness of parametric characterizations to vector optimization problems. OR Spectr. **8**, 73–87 (1986)
161. Xu, Q., Wang, L., Wang, N., Hei, X., Zhao, L.: A review of opposition-based learning from 2005 to 2012. Eng. Appl. Artif. Intell. **29**, 1–12 (2014)
162. Yang, Z., He, J., Yao, X.: Making a difference to differential evolution. In: Michalewicz, Z., Siarry, P. (eds.) Advances in Metaheuristics for Hard Optimization, pp. 397–414. Springer, Berlin (2008)
163. Yang, L., Guan, Y., Sheng, W.: A novel dynamic crowding distance based diversity maintenance strategy for MOEAs. In: Proceedings of the 2017 International Conference on Machine Learning and Cybernetics, Ningbo, pp. 211–216 (2017)
164. Yang, D., Liu, Z., Shu, T., Yang, L., Ouyang, J., Shen, Z.: An improved genetic algorithm for multiobjective optimization of helical coil electromagnetic launchers. IEEE Trans. Plasma Sci. **46**(1), 127–133 (2018)
165. Yao, X., Liu, Y., Lin, G.: Evolutionary programming made faster. IEEE Trans. Evol. Comput. **3**(2), 82–102 (1999)
166. Yao, X., Liu, Y., Lin, G.: Self-adaptive differential evolution with neighborhood search. In: Proceedings of the 2008 Congress on Evolutionary Computation (CEC2008), pp. 1110–1116 (2008)
167. Zhang, Q., Li, H.: MOEA/D: a multiobjective evolutionary algorithm based on decomposition. IEEE Trans. Evol. Comput. **11**(6), 712–731 (2007)
168. Zhu, G.P., Kwong, S.: Gbest-guided artificial bee colony algorithm for numerical function optimization. Appl. Math. Comput. **217**(7), 3166–3173 (2010)
169. Zitzler, E., Thiele, L.: Multiobjective optimization using evolutionary algorithms - a comparative case study. In: Eiben, V.A.E. et al. (eds.) Parallel Problem Solving From Nature. Springer, Berlin, 292–301 (1998)
170. Zitzler, E., Thiele, L.: Multiobjective evolutionary algorithms: a comparative case study and the strength Pareto approach. IEEE Trans. Evol. Comput. **3**(4), 257–271 (1999)

171. Zitzler, E., Deb, K., Thiele, L.: Comparison of multiobjective evolutionary algorithms: empirical results. Evol. Comput. **8**(2), 173–195 (2000)
172. Zitzler, E., Laumanns, M., Thiele, L.: SPEA2: improving the strength Pareto evolutionary algorithm for multiobjective optimization. In: Proceedings of the Evolutionary Methods for Design, Optimization and Control with Applications to Industrial Problems (EUROGEN), pp. 95–100 (2002)
173. Zitzler, E., Thiele, L., Laumanns, M., Fonseca, C.M., Fonseca, V.G.: Performance assessment of multiobjective optimizers: an analysis and review. IEEE Trans. Evol. Comput. **7**(2), 117–132 (2003)

Index

Symbols
K-sparse approximation, 194
L-Lipschitz continuous function, 113
L-Lipschitz continuously differentiable, 114
ℓ_0-norm, 15
ℓ_0-norm minimization, 193
ℓ_1-norm minimization, 194
ℓ_1-penalty minimization, 195
ℓ_p pooling, 505
ℓ_p-metric distance, 555
ϵ-neighborhood, 16, 559
k-nearest neighbor (kNN), 298
k-nearest neighborhood, 559
n-dimensional hypermatrix, 302
p-Dirichlet norm, 365

A
Acceleration constant, 782
Achievement scalarizing functions (ASFs), 754
   additive ASF, 756
   parameterized ASF, 757
   strictly increasing ASF, 756
   strongly increasing ASF, 756
Achievement scalarizing problem, 754
ACO construction functions, 771
   ant solution construct, 771
   deamon actions, 771
   pheromone update, 771
A-conjugacy, 163
Action space, 391
Action-value function, 393
Activation function, 452
   exponential linear unit (ELU), 509
   Kullback-Leibler (KL) divergence, 452
   logistic function, 452
   maxout, 509
   nonlinear activation function, 543
   probout, 509
   rectified linear unit (ReLU), 457, 507
      leaky rectified linear unit (Leaky ReLU), 458, 508
      noisy rectified linear unit (NReLU), 508
      parametric rectified linear unit (PReLU), 458, 509
      randomized rectified linear unit (RReLU), 509
   sigmoid function, 452
   softmax function, 455
   Softplus function, 457
   Softsign function, 456
   Soft-step function, 452
   tangent (tanh) function, 456
Active constraint, 141
Active learning (AL), 378, 380
   co-active learning, 382
   membership query learning, 380
   pool-based active learning, 381
   selective sampling, 381
   sequential learning, 381
   stream-based active learning, 381
Active learning-extreme learning machine (AL-ELM) algorithm, 386
Active set, 141
Adaptive boosting (AdaBoost), 252
Adaptive convexity parameter, 117
Additive logistic regression model, 249
Adversarial discriminator, 599
Affine function, 103

Aggregate functions, 581
  LSTM aggregate function, 582
  mean aggregate function, 582
  pooling aggregate function, 582
Agree, 248
Aleatory uncertainty, 708
Alleles, 724
All-sharing, 541
Alternating direction multiplier method (ADMM), 144
Alternating least squares (ALS) method, 182
Alternative weak distance metrics, 333
Ant colony system (ACS), 775
  global updating rule, 775
  local updating rule, 775, 776
  state transition rule, 775
Anti-Tikhonov regularization method, 180
Ant system, 773
  global updating rule, 774
  random-proportional rule, 774
  state transition rule, 774
APPROX coordinate descent method, 236
Approximation, 90
Approximation error, 618
Archival nondominated solutions, 721
Archived multiobjective simulated annealing (AMOSA), 721
Archive truncation procedure, 754
Artificial ants
  ant-routing table, 769
  feasible neighborhood, 769
  online delayed pheromone update, 769
  start state, 769
  step-by-step pheromone update, 769
  termination conditions, 769
Artificial bee colony (ABC), 777
  abandoned position, 778
  ABC algorithms, 778
  crossover-like artificial bee colony (CABC), 780
  employed bees, 777
  food sources, 777
  gbest-guided artificial bee colony (GABC), 778
  improved artificial bee colony (IABC), 779
  nectar amounts, 778
  onlooker bees, 777
  scout bees, 777
Augmented Lagrange multiplier method, 135
Augmented matrix, 158
Autocorrelation matrix, 21
Autocovariance matrix, 21
Autoencoder, 525
  code prediction energy, 528
  contractive autoencoder (CAE), 532
  convolutional autoencoder (CAE), 538
  decoder network, 527
  denoising autoencoder (DAE), 535
  encoder network, 526
  nonnegativity constrained autoencoder (NCAE), 541
  reconstruction energy, 528
  sparse autoencoder (SAE), 533
  stacked autoencoder, 532
  stacked convolutional denoising autoencoder (SCDAE), 539
  stacked sparse autoencoder (SSAE), 534
Auxiliary set, 339
Average pooling, 506

B

Backpropagation through time (BPTT), 465
Backtracking line search, 295
Backward iteration, 129
Barrier function, 134
  exponential barrier function, 134
  Fiacco-McCormick logarithmic barrier function, 134
  inverse barrier function, 134
  logarithmic barrier function, 134
  power-function barrier function, 134
Barrier method, 134
Basis pursuit (BP), 195
Basis pursuit denoising (BPDN), 195
Batch gradient, *see* Barch method
Batch method, 228
Batch normalization (BatchNorm), 459, 589
  mini-batch mean, 458
  mini-batch normalization, 458
  mini-batch variance, 458
  parameters, 459
Batch normalizing transform, 589
Bayesian classification rule, 492
Bayesian classification theory, 491
Bayes' rule, 490
Bellman equation, 393
Benchmark function, 236
Bernoulli distribution, 516
Bernoulli vector, 516
Between-class scatter matrix, 310
Between-class variance, 262
Between-sets covariance matrix, 348
Bidirectional recurrent neural network (BRNN), 468
Binary mask vector, 521
Boltzmann machine, 477, 478
Boltzmann probability factor, 716

Index   807

Boltzmann's constant, 716
Boosting, 247
Boosting algorithm, 251
Boundary intersection (BI) approach, 747
Broyden-Fletcher-Goldfarb-Shanno (BFGS)
    method, 148

C
Cannot-Link, 333
Canonical correlation analysis (CCA), 343
    canonical correlation, 344
    canonical variants, 344
    canonical weight vectors, 344
    CCA algorithm, 347
    diagonal penalized CCA, 352
    Kernel canonical correlation analysis, 347,
        349
    penalized canonical correlation analysis,
        352
    penalized (sparse) CCA algorithm, 353
    score variants, 344
    sparse canonical variants, 352
    sparse CCA, 352
Canonical decomposition (CANDECOM), 306
Cartesian product, 12
Cauchy graph embedding, 566
Cauchy mutation operator, 759
Cauchy-Riemann condition, 72
Cauchy-Riemann equations, 72
Center-based sampling (CBS), 789
Centered kernel function, 266
Centered kernel matrix, 266
Chain rule, 60
Characteristic equation, 205
Characteristic function, 25
Characteristic matrix, 205
Characteristic polynomial, 205
Chromosomes, 724
City-block distance, 555
Class discrimination space, 219
Classification, 257, 259
    supervised classification, 257
    unsupervised classification, 257
Classification problem, 445
Classification vertices, 375
Closed ball, 103
Closed neighborhood, 91
Cluster analysis, 314
Cluster indicator vector, 333
Cofactor, 28
Cogradient matrix, 59
Cogradient operator, 77
Cogradient vector, 59, 77

Coherent, 23
Column space, 33
Community membership, 569
Community membership vector, 570
Community preserving network embedding,
    571
Community representation matrix, 572
Commutation matrix, 49
Complex analytic function, 72
Complex conjugate partial derivative, 74
Complex conjugate transpose, 7
Complex differentiable, 72
Complex Gaussian random vector, 25
Complex gradient, 71
Composite mirror descent method, 448
Composite optimization, 228
Concept, 248
Conditional risk function, 492
Condition number, 169
    ℓ_1 condition number, 169
    ℓ_2 condition number, 170
    ℓ_∞ condition number, 170
    Frobenius-norm condition number, 170
Confidence, 254, 299
Confidence parameter, 251
Conic hull, 102
Conjugate cogradient operator, 77
Conjugate cogradient vector, 77
Conjugate gradient, 71
Conjugate gradient algorithm, 163
Connectionist temporal classification (CTC),
    474
Consistent, 248
Consistent equation, 33
Constant Error Carousel (CEC), 477
Constrained clustering, 339
Constrained convex optimization problem, 132
Constrained minimization problem, 132
Constrained spectral clustering, 334
Constraint matrix, 333
    normalized constraint matrix, 334
Constriction coefficient, *see* Inertia weight
Contrastive divergence (CD), 484, 485
Contribution ratio, 271
Convergence rate, 110
    local convergence rate, 111
        linear convergence rate, 112
        quadratic convergence rate, 112
        sublinear convergence rate, 112
    logarithmic convergence rate, 111
    quotient-convergence rate, 110, 111
        cubic convergence rate, 111
        linear convergence rate, 111
        quadraticr convergence rate, 111

sublinear convergence rate, 111
superlinear linear convergence rate, 111
Conves hull, 102
Convex cone, 103
Convex function, 103
Convexity parameter, 104
Convex programming problem, 446
Convex relaxation, 194
Convolutional neural networks (CNNs), 497
  convolutional layers, 497
  downsampling layers, 497
  full-connected layers, 497
  input layers, 497
  loss layers, 497
  pooling layers, 497
  rectified linear unit (ReLU), 497
Convolutions, 498, 501
  multi-input multi-output (MIMO) convolution, 502
  multi-input single-output (MISO) convolution, 503
  single-input multi-output (SIMO) convolution, 502
  single-input single-output (SISO) convolution, 502
  2-D convolution, 500
Convolution theorem in frequency domain, 584
Convolution theorem in time domain, 584
Co-occurring data, 341
Coordinate descent methods (CDMs), 232
  accelerated coordinate descent method (ACDM), 235
  cyclic coordinate descent, 233
  randomized coordinate descent method (RCDM), 235
  stochastic coordinate descent, 233
Coordinate-wise descent algorithms, 291
Core tensor, 305
Correlation coefficient, 23, 264
Correlation method, 262
Cost, 447
Cost function, 89
Co-training, 342
Covariant operator, 59
Coverage, 699
Criterion scalarizing approach, 718
Criterion scalarizing functions, 719
Cross-correlation matrix, 22
Cross-covariance matrix, 22
Cross-entropy, 453
Crossover, 727
Crossover probability, 727
Crossover rate, 727
Crowded-comparison operator, 703

Crowding distance assignment approach, 702
Crowding distance (CD), 702
Cumulative probability distribution, 253
Cut, 371

D

Darwin's principle of natural evolution, 740
Data bag, 342
Data centering, 258
Data scaling, 258
Data zero-meaning, *see* Data centering
Davidon-Fletcher-Powell (DFP) method, 148
Decision boundary, 299
Decision functions, 297, 650
Decision making approach, 691
Definition domain, 101
Degeneracy, 729
Democratic co-learning, 343
Density, 753
Descent step, 106
Determinant, 27
Differential evolution, 764
Differential evolution (DE) operator, 748
Dimensionality reduction, 258, 269
Direct acyclic graph (DAG), 655
  directed direct acyclic graphs (DDAGs), 655
  rooted binary DAGs, 655
Directed acyclic graph SVM (DAGSVM) method, 655
Direct sum, 43
Disagree, 248
Discount factor, 392
Discrepancy, *see* Loss
Discriminative model, 598
Discriminator, 600
Distributed nonnegative encoding, 540
Distributed optimization problems, 144
Distribution indicator, 713
Divergence, 324
Diversity indicator, 713
Divide-and-conquer strategy, 715
Domain, 248, 403
Domain adaptation, 410
  cross-domain transform method, 421
  feature augmentation method, 418
  transfer component analysis method, 423
Double Q-learning, 397
Downsampling factor, 504
Drift analysis, 742
DropConnect, 520
  binary mask matrix, 522
  feature extractor, 522

Index 809

layer, 522
  softmax classification layer, 522
DropConnect Network, 522
Dropout, 514
Dropout neural network model, 516
Dropout spherical K-means, 519
Duality gap, 139
Dual residual, 146
Dyadic decomposition, 171

E

Echelon matrix, 159
Eckart-Young theorem, 173
Edge derivative, 361
Effective rank, 178
EigenCluster, 417
Eigenpair, 204
Eigensystem, 210
EigenTransfer, 416
Eigenvalue, 203
  algebraic multiplicity, 206
  multiple eigenvalue, 206
  single eigenvalue, 206
Eigenvalue decomposition (EVD), 204
Eigenvalue-eigenvector equation, 204
Eigenvector, 203
  left eigenvector, 208
  right eigenvector, 208
Elastic net, 279
Elementary row operations, 158
  Type I elementary row operation, 158
  Type II elementary row operation, 158
  Type III elementary row operation, 158
Elemenwise division, 44
Elemenwise product, 44
Empirical risk, 226, 253
Empirical risk minimization (ERM), 254, 619
Environmental selection, 753, 754
Episode, 388
Epistemic uncertainty, 708
Equality constraints, 101
Equivalent subspaces, 273
Equivalent systems, 158
Error constrained ℓ_1-norm minimization, 195
Error for hidden nodes, 466
Error for output nodes, 465
Error function, 464
Error parameter, 251
Estimation error, 618
Euclidean distance, 16
Euclidean leangth, 16
Euclidean measures, 354
Euclidean metric, 752

Euclidean metric distance, 555
Euclidean structure, 354
Evalustion functional, 624
Evolutionary algorithms (EAs), 740
  crossover, 740
  1 + 1 evolutionary algorithm, 741
  fitness, 740
  generation, 740
  individual, 740
  mutation, 740, 741
  offspring, 740
  parents, 740
  population, 740
  replacement, 741
Example vertices, 375
Exhaustive enumeration method, 262
Expected risk, 227, 253
Explicit constraints, 101
Expression ratio criterion, 263
Expression ratio method, 263
Extrema/extreme, 92
  global maximum, 92
  global minimum, 90, 92
  global minimum point, 90
  local maximum, 92
  local minimum, 91
  strict global maximum, 92
  strict global minimum, 92
  strict global minimum point, 90
  strict local maximum, 92
  strict local minimum, 91
  strict local minimum point, 91
Extreme learning machine (ELM), 542
  algorithm, 546
  binary classification, 547
  multiclass classification, 551
  regression, 547
Extreme point, 92
  global maximum point, 92
  global minimum point, 92
  isolated local extreme point, 92
  local maximum point, 92
  local minimum point, 91
  strict global maximum point, 92
  strict local maximum point, 92

F

Fast evolutionary programming (FEP), 759
Feasible point, 101
Feasible set, 101, 136
Feature augmentation, 418
Feature extraction, 269
Feature learning, 260

Feature ranking, 262
Feature selection, 258, 260, 261
Feature vector, 258
FEP algorithm, 760
Finiteness, 620
First-order differential, 64
First-order necessary condition, 94
Fisher discriminant analysis (FDA), 310, 324
Fisher measure, 217, 324
Fisher's criterion, 263
Fitness, 711
Fitness assignment, 726, 751, 753
Fitness selection, 734
    fitness uniform selection scheme, 735
    standard selection scheme, 734
        linear proportionate selection, 734
        ranking selection, 734
        tournament selection, 735
        truncation selection, 734
Fitness selection approach, 698
Flipped block-structured matrix, 501
Flipped matrix, 501
Flipped vector, 501
Formal partial derivatives, 73
Forward iteration, 129
Forward problem, 14
Frobenius norm ratio, 179
Fully-nonseparable function, 237
Function approximation, 390
Function vector, 4

G

Gain shape vector quantization, 329
Gated graph neural networks (GG-NN), 577
Gauss elimination method, 160
Gauss elimination method for matrix inversion, 161
Gaussian approximation, 666
Gaussian mutation operator, 759
Gaussian process classification, 666
Gaussian process regression, 663
Gaussian process regression algorithm, 666
Gaussian random vector, 24
Gauss-Markov theorem, 177
Gauss-Seidel method, 182
    circle phenomenon, 183
    swamp, 183
Generalized characteristic equation, 211
Generalized characteristic polynomial, 211
Generalized eigenpair, 211
Generalized eigenvalue, 210
Generalized eigenvalue decomposition (GEVD), 210
Generalized eigenvector, 210
Generalized opposition solution, 789
Generalized total least squares (GTLS), 191
Generational distance, 713
Generative Adversarial Networks (GANs), 598
Generative model, 598
Generators, 327, 599
Gene rearrangement, 729
Genes, 724
Genetic algorithm (GA), 723
Genetic algorithm with gene rearrangement (GAGR), 730
Global acceptance probability, 719
Global best position (gbest), 782
Global output function, 578
Global transition function, 578
Gradient aggregation, 230
Gradient computation, 59
Gradient descent algorithm, 100
Gradient flow direction, 59
Gradient matrix, 58
Gradient matrix operator, 58
Gradient projection, 295
Gradient projection for sparse reconstruction (GPSR), 293
Gradient-projection method, 108
Gradient vector operators, 57
Gradient vectors, 58
Gram matrix, 625
Graph, 355
    adjacency matrix, 355, 356
    affinity matrix, 356
    r-ball neighborhood graph, 373
    complete weighted graph, 373
    degree, 356
    degree matrix, 356
    directed graph, 355
    edge set, 355
    edge weighting functions, 356
    edge weights, 356
        dot-product weighting, 357
        heat kernel weighting, 357
        thresholded Gaussian kernel weighting, 357
        0-1 weighting, 357
    first-order proximity, 552
    fully connected graph, 358
    ℓ_1 graph, 378
    higher-order proximity, 553
    higher-order proximity matrix, 553
    mutual K-nearest neighbor graph, 358
    k-nearest neighbor graph, 358, 373
    ϵ-neighborhood graph, 358
    k-order proximity, 553

Index 811

proximity matrix, 553
second-order proximity, 553
undirected graphs, 355
vertex set, 355
weighted adjacency matrix, 356
Graph convolution, 364
Graph convolution theorem, 585
Graph convolutional networks (GCNs), 577, 583
Graph embedding, 554
Graph filtering, 363
Graph Fourier basis, 360
Graph Fourier transform, 362, 585
Graph inverse Fourier transform, 363
Graph kerners, 366
Graph K-means clustering, 361
Graph Laplacian matrices, 358
  combinatorial graph Laplacian, 359
  normalized graph Laplacian matrices, 359
  random walk normalized graph Laplacian, 359
  symmetric normalized graph Laplacian, 359
Graph mincut learning algorithm, 375
Graph minor component analysis (GMCA), 361
Graph principal component analysis (GPCA), 361
GraphSAGE, 581
  aggregator architectur, 581
  embedding generation, 581
  parameter learning, 581
Graph signal processing, 363
Graph signals, 354
Graph spectrum, 360
Graph structure, 354
Graph-structured data, 354
Graph Tikhonov regularization, 366
Grassmann manifold, 273
Greedy method, 262
Greedy projection algorithm, 447
Grouped Lasso, 292
Group normalization (GN), 597
Group-sharing, 541

H

Hadamard inequality, 31
Hadamard product, 44
Half-space, 109
Hankel matrix, 498
  extended Hankel matrix, 503
  wrap-around Hankel matrix, 499
Hankel structured matrix, 498
Harmonic function, 73
Heat kernel, 374, 565
Heavy ball method (HBM), 115
Hermitian conjugate, *see* Complex conjugate transpose
Heterogeneous domain adaptation (HDA), 418
Heterogeneous feature augmentation (HFA), 419
Heterogeneous transfer learning, 407
Heuristic algorithms, 767
  deterministic algorithm, 768
  iterative based algorithm, 768
  population-based algorithm, 768
  stochastic algorithm, 768
Heuristic procedures, 714
Hidden layer output matrix, 545, 546
Hidden nodes, 543
Hierarchical clustering, 320
  array of similarity measures, 320
  distance matrix, 322
  hierarchical clustering scheme (HCS), 320
  hierarchical system of clustering representations, 320
  maximum method, 324
  minimum method, 323
  ultrametric inequality, 322
Hierarchical clustering approach, 703
Higher-order matrix algebra, 301
Higher-order proximity, 574
Higher-order proximity matrix, 574
  Adamic-Adar (AA), 575
  common neighbors, 575
  Katz index, 574
  Rooted PageRank, 575
Higher-order singular value decomposition, 306
Higher rank tensor ridge regression (hrTRR), 311
Holomorphic complex matrix functions, 75
Holomorphic function, 72
Homogeneity constraint, 272
Homogeneous transfer learning, 407
Hopfield network, 477
Horizontal unfolding, 303
  Kiers horizontal unfolding method, 303
  Kolda horizontal unfolding method, 304
  LMV horizontal unfolding method, 303
Human intelligence, 780
Hybrid evolutionary programming (HEP), 762
Hybrid mutation operators, 762
Hyperplane, 109, 618
Hypervolume for set, 712

Hypervolume (HV), 698
Hypothesis, 619
Hypothesis space, 411, 619

I
Ideal objective vector, 697
Idempotent matrix, 207
Ill-conditioned problem, 167
Immediate reward, 392
Imprecision, 708
Imprecision for set, 712
Inactive constraint, 141
Inconsistent, 248
Inconsistent equation, 33
Increasing, 755
Indefinite matrix, 27
Independence assumption, 60
Independent and identically distributed (i.i.d.), 296
Independent identically distributed (iid), 24
Individuals, 724
Inductive learning, 338
Inductive transfer learning, 408
Inequality constraints, 101
Inertia weight, 782
Inexact line search, 149
Infeasible point, 101
Infimum, 139
Infinite kernel learning (IKL), 420
Information network, 355, 568
Information theoretic metric learning (ITML), 423
Inner relation, 285
Instance, 296
Instance distribution, 248
Instance normalization, 596
Instance space, 248
Intercept, 297
Interlacing theorem for singular values, 174
Interpolation, *see* Regression
Interval multiobjective optimization problems, 708
Interval order relation, 709
Interval vector, 708
Invariant feature learning, 525
Invariant features, 525
Inverse affine image, 102
Inverse Euclidean distance, 374
Inverse graph Fourier transform, 585
Inverse matrix, 9
Inverse problem, 14
Iterate averages, 229
Iteration matrix, 163

Iterative improvement, 715
Iterative soft thresholding method, 131

J
Jacobian matrix, 56
Jacobian operator, 56
Jensen inequality, 103
Jensen-Shannon (JS) divergence, 514
Job scheduling problems (JSP), 772
Joint probability distribution, 618

K
Karush-Kuhn-Tucker (KKT) conditions, 140
Kernel alignment, 267
Kernel functions, 628, 663
    additive kernel function, 629
    B-splines kernel function, 628
    exponential radial basis function, 628
    Gaussian radial basis function, 628
    linear kernel function, 628
    multi-layer perceptron kernel function, 628
    polynomial kernel function, 628
Kernel K-means clustering, 627
Kernel partial least squares regression, 632
Kernel PCA, 627
Kernel reproducing property, 626
K-means algorithm, 328
    random K-means algorithm, 328
Krylov subspace, 163
Krylov subspace method, 163
Kullback-Leibler (KL) divergence, 513

L
Label, 297
Labeled set, 223
Lagrange dual method, 139
Lagrange multiplier vector, 135
Laplace equations, 72
Laplace's method, 666
Laplacian embedding, 563
Laplacian regularized least squares (LapRLS) solution, 631
Laplacian support vector machines, 633
LAR algorithm, 199
Lasso
    distributed Lasso, 197
    generalized Lasso, 197
    group Lasso, 197
    MRM-Lasso, 197
Layer normalization (LN), 596
Leading-1 entry, 158

Leading entry, 158
Learning machine, 253, 254
Learning step, 100
Least absolute shrinkage and selection operator (Lasso), 279, 290
Least angle regressions (LARS), 198
Least square regression error, 264
Leave-one-out cross-validation (LOOCV), 375
Left generalized eigenvector, 212
Left inverse, 39
Left Kronecker product, 46
Left pseudo-inverse matrix, 39
Linear discriminant analysis, 556
Linear inverse transform, 9
Linear regression, 278
Linear rule, 59
Linear transform matrix, 9
Lipschitz constant, 113
Lipschitz continuous, 113
Lipschitz continuous function, 113
Locality preserving projections (LPP), 565
Local output function, 577
Local topology preserving, 555
Local transition function, 577
Logistic function, 249, 452
Logistic regression, 537
LogitBoost, 250
Longitudinal unfolding, 303
Long short-term memory (LSTM), 471
    backward pass, 473
    forget gate, 472
    forward pass, 472
    input gate, 472
    memory blocks, 472
    output gate, 472
    peephole LSTM, 477
Loss, 618
Loss function, 227, 464, 510
    contrastive loss, 511
    coupled clusters loss, 512
    cross-entropy loss function, 452, 527
    double-contrastive loss, 512
    hinge loss, 510
        ℓ_1-loss, 510
        ℓ_2-loss, 511
        square hinge loss, 511
    large-margin softmax (L-Softmax) loss function, 513
    logistic loss function, 452, 523
    softmax loss, 511
Lower unbounded, 101, 139
Low-rank approximation, 178

M
Machine learning (ML), 223
    expected performance, 256
    accuracy, 256
    complexity, 256
    convergence reliability, 256
    convergence time, 256
    response time, 256
    scalability, 256
    training data, 256
    training time, 256
Mahalanobis metric learning, 423
Majority voting, 248
Majorization-minimization (MM), 241
    majorization step, 242
    majorizer, 242
    minimization step, 242
Majorization-minimization (MM) algorithm, 244
    monotonicity decreasing, 244
    stationary point, 245
Manhattan metric, 555
Manifold learning, 559
    isometric map (Isomap), 559
        ϵ-Isomap, 559
        k-Isomap, 559
        conformal Isomap, 560
    Laplacian eigenmap method, 564
    Laplacian eigenmaps, 563
    locally linear embedding (LLE), 560
Mapping, 12
    codomain, 12
    domain, 12
    image, 12
    injective mapping, 13
    inverse mapping, 13
    linear mapping, 13
    one-to-one mapping, 13
    range, 12
    surjective mapping, 13
Margin, 618, 621
Markov decision process (MDP, 391
Mating selection, 754
Matricization, 50
    matricization of column vector, 50, 51
Matrix, 3
    block matrix, 6
    broad matrix, 5
    diagonal matrix, 5
    full column rank matrix, 33
    full rank matrix, 33
    full row rank matrix, 33

Hermitian matrix, 7
identity matrix, 5
nonsingular matrix, 28
rank-deficient matrix, 33
square matrix, 5
submatrix, 6
symmetric matrix, 7
tall matrix, 5
zero matrix, 5
Matrix diffferential, 62
Matrix equation, 3
Matrix inversion lemma, 36
Matrix norm
   entrywise norm
      p-norm, 19
   Frobenius norm, 19
   Mahalanobis norm, 20
   max norm, 19
   ℓ_1-norm, 19
Matrix pair, 210
Matrix pencil, 210
Matrix transpose, 7
Max-flow method, 376
Maximal margin hyperplane, 652
Maximin problem, 138
Maximum ascent rate, 59
Maximum descent rate, 59
Maximum margin, 622
Maximum mean discrepancy embedding (MMDE), 424
Maximum spread, 699
Max pooling, 506
Max Wins strategy, 655
Mean tensor, 310
Measure, 316
   nonnegativity, 316
   symmetry, 316
   triangle inequality, 316
Mercer kernel, 627
   exponential radial basis function kernel, 627
   Gaussian radial basis function (GRBF) kernel, 627
Mercer's theorem, 626
Mesoscopic structure, 571
Message passing neural networks (MPNNs), 577
Metaheuristic, 714
   metaheuristic procedures, 714
   population metaheuristic, 714
   single-solution metaheuristic, 714
Method of distance functions, 744
Method of objective weighting, 744
Metropolis algorithm, 714

Metropolis-Hastings algorithm, *see* Metropolis algorithm
Microscopic structure, 571
Minimax problem, 138
Minimum cut (min-cut), 371
Minkowski p-metric distance, 555
Min-max formulation, 744
Minor component analysis (MCA), 271
Minor components, 270
Minor subspace, 271
Minor subspace analysis (MSA), 271
Misclassification rate, 492
Misclassification tolerance, 637
Mixed pooling, 506
Mixed-variable optimization problem (MVOP), 771
Model mismatch, 618
Modified connectionist Q-learning, 399
Modularity, 570
Modularity matrix, 570
Momentum, 116
Moore-Aronszajn Theorem, 626
Moore-Penrose conditions, 41
Moore-Penrose inverse, 41
Moreau decomposition, 124
Multiclass classifier, 299
Multidimensional scaling (MDS), 558
Multilinear data analysis, 541
Multiobjective combinatorial optimization (MOCO), 684
   multiobjective assignment problems, 684
   multiobjective min sum location problem, 685
   multiobjective network flow, 685
   multiobjective transportation problem, 684
   unconstrained combinatorial optimization problem, 685
Multiobjective evolutionary algorithm based on decomposition (MOEA/D), 746
Multiobjective optimization problem (MOP), 686
   decision space, 686
   decision vector, 686
   feasible set, 687
   ideal solution, 688
   ideal vector, 688
   objective space, 686
   objective vector, 686
Multi-output nodes, 549
Multiple kernel learning (MKL), 421
Must-Link, 333
Mutation, 727
Mutation probability, 728
Mutation rate, 728

Index

N
Nadir objective vector, 697
Naive Bayesian classification, 490
Nash equilibrium, 600
Natural basis vector, 233
Nearest neighbor, 317
Nearest neighbor classification, 317
Negative definite matrix, 27
Negative semi-definite matrix, 27
Negative transfer, 406
Neighborhood, 93
Neighborhood preserving embedding (NPE), 562
Network ambedding, 568
Network coding resource minimization (NCRM), 773
Network embedding
    large-scale information network embedding, 568
    network inference, 568
    network reconstruction, 568
Neural network tree, 444
Newton method, 107
    modified Newton method, 148
    truncated Newton method, 147
Niche count, 737
Niched Pareto Algorithm, 737
Niched Pareto genetic algorithm, 737
Niche radius, 737
Node embedding, 577
Node vector, 577
Noise projection matrix, 272
Noise subspace, 271
Noisy multiobjective optimization problems, 707
Nondominated sorting approach, 700
Nondomination level, 703
Nondomination rank, 703
Non-Euclidean structure, 354, 355
Nonlinear iterative partial least squares (NIPALS), 632
Nonlinear iterative partial least squares (NIPALS) algorithm, 286
Nonnegative constraints, 540
Nonnegative matrix factorization (NMF), 540
Nonnegative orthant, 103
Non-pivot features, 414
Nonseparable case, 623
Nonseparable function, 237
Nonsmooth convex optimization, 128
Nonstationary iterative method, 163
Normalized singular values, 178
Numerically stable, 168

Numerical stability, 168
Nyström r-rank approximation, 335

O
Objective function, 89
One-against-all (OAA) method, 653
One-against-one (OAO) method, 654
One-versus-one (OVO) method, *see* One-against-one (OAO) method
One-versus-rest (OVR) method, *see* One-against-all (OAA) method
Online convex optimization, 446
Online convex programming problem, 446
Open ball, 103
Open neighborhood, 91
Opposite number, 788
Opposite pair, 383
Opposite point, 789
Opposition-Based Generation Jumping, 791
Opposition-based learning (OBL), 789
Opposition-based optimization (OBO), 789
Opposition-Based Population Initialization, 791
Opposition number, 788
Opposition solution, 789
Optimal behavior models, 390
    average-reward model, 391
    finite-horizon model, 390
    infinite-horizon discounted model, 390
Optimal class discrimination matrix, 325
Optimal experimental design, *see* Active learning
Optimal primal value, 138
Optimal rank tensor ridge regression (orTRR), 312
Optimal solution, 101
Optimal unbiased estimator, 177
Optimization vector, 89
Ordered n-ples, 12
Original residual, 146
Orthogonality constraint, 272
Outer relations, 285
Overfitting, 251, 617

P
Pairwise constraints, 333
Parallel factors decomposition (PARFAC), 306
Parameterized achievement scalarizing problem, 757
Pareto concepts, 691, 692
    archive, 698

global Pareto-optimal set, 696
globally nondominated, 694
incomparable, 693
local Pareto-optimal set, 696
mutually nondominating, 693
nondominated sets, 695
nondominated solution, 694
outcomes, 692
Pareto dominance, 692
Pareto efficient, 694
Pareto front, 696
Pareto frontier, 696
Pareto improvement, 694
Pareto-optimal, 694
Pareto-optimal front, 696
Pareto-optimality, 694
Pareto-optimal solutions, 691
strict Pareto dominance, 693
weak Pareto dominance, 693
Pareto concepts for approximation sets, 709
incomparable dominance for approximation sets, 711
strict dominance for approximation sets, 710
Pareto concepts for sets
dominance for approximation sets, 709
weak dominance for approximation sets, 710
Pareto optimization approach, 691
Pareto optimization theory, 691
Partial labeling, 333
Partial least squares (PLS), 285
Partial least squares (PLS) regression, 287
Partial order, 703
Particle swarm
canonical particle swarm, 781
Particle swarm optimization (PSO), 780
genetic learning particle swarm optimization (GL-PSO), 784
Particle swarm optimization (PSO) algorithm [19], 783
Partition function, 479, 481
Partitional clustering, 781
Parts combination, 541
Passive learning, 378
Pattern, 257
Pattern recognition, 257
Pattern vectors, 258
Pearson correlation coefficients, 262
Penalized regression, 290
Penalty function
exterior penalty function, 133
interior penalty function, 133
Penalty function method, 133, 134

Penalty parameter vector, 135
Pheromone model, 770
Pheromone values, 770
Piecewise continuous, 543
Pivot column, 159
Pivot features, 414
Pivot position, 159
Policy, 393
Polynomial mutation operator, 749
Polynomially evaluatble, 249
Pool-based active learning (PAL), 382
Pooling stride, 504
Positive definite matrix, 27
Positive definition kernel, 625
Positive semi-definite cones, 110
Positive semi-definite matrix, 27
Preconditioned conjugate gradient (PCG), 166
Prediction, *see* Regression
Prediction error, 227
Prediction function, 226, 227
Predictors, 259, 278
Primal cost function, 137
Primal-dual subgradient method, 448
Principal component analysis (PCA), 270
Principal component pursuit (PCP), 276
Principal subspace, 271
Principal subspace analysis (PSA), 271
Probably approximately correct (PAC) learnable, 251
efficiently PAC learnable, 251
Probably approximately correct (PAC) learning, 251
Probably approximately correct (PAC) model, 251
Product rule, 59
Projected gradient update, 446
Projection approximation subspace tracking (PAST), 274
Projection operator, 108
Proximal operator, 123
Proximity indicator, 713
Pythagorean theorem, 18

Q

Q-function, 393, 394
Q-learning, 394
QR factorization, 170
Quadratically constrained linear program (QCLP), 293
Quadratic form, 27
Quadratic program (QP), 293
bound-constrained quadratic program (BCQP), 294

Quadratic programming (QP) problem, 195
Quasi-convex, 104
Quasi-opposite number, 791
Quasi-opposite point, 792
Quasi-oppositional differential evolution (QODE), 794
Quasi opposition-based learning (quasi-OBL), 791
Query learning, *see* Active learning
Query strategies, 381
　balance exploration and exploitation, 381
　expected model change, 381
　exponentiated gradient exploration, 381
　least certainty, 381
　middle certainty, 381
　query by committee, 381
　uncertainty sampling, 381
　variance reduction, 381
Quotient rule, 60

R
Random vector, 4
Random walks, 580
Range, 33
Rank, 33
Rank-one decomposition, 307
Raw fitness value, 753
Rayleigh quotient, 215
Rayleigh-Ritz theorem, 215
Recombination, 725
Rectangular set, 109
Recurrent neural networks (RNNs), 460
　backward mechanism, 467
　bidirectional mechanism, 468
　forward mechanism, 467
　recurrent computation, 467
Recursive feature elimination (RFE), 652
Reduced row-echelon form (RREF) matrix, 159
Reference point vector, 746
Reformulation approach, 691
Regression, 259, 282
Regression intercept, 260
Regression problem, 445
Regression residual, 260
Regret, 447
Regularization parameter, 180, 637
Regularization path, 180
Regularized Gauss-Seidel method, 184
Regularized least squares cost function, 180
Reinforcement learning, 387
　action, 388
　agent, 388
　　autonomy, 388
　　pro-activeness, 388
　　reactivity, 388
　　social ability, 388
　control policy, 389
　　deterministic policy, 389
　　probabilistic policy, 389
　improvement, 392
　payoff, 389
　return, 389
　reward function, 388
　state, 388
　state-action space, 388
Relative entropy, *see* Kullback-Leibler (KL) divergence
Relative feasible set, 142
Relative interior points, 142
Relaxation, 90
Relevance feedback, 380
Relevance vector machine (RVM), 667
Representation choice, 260
Representation learning, 260
Representer theorem for semi-supervised learning, 630
Representer theorem for supervised learning, 629
Reproducing kernel, 624, 625
Reproducing kernel Hilbert space (RKHS), 624
Reproducing property, 624, 626
Residual variance, 265
Restricted Boltzmann machine (RBM), 481
Return, 393
Reward function, 392
Rewards, 389
Richardson iteration, 163
Ridge regression, 290
Right generalized eigenvector, 212
Right Kronecker product, 46
Right pseudo-inverse matrix, 39
Risk function, 618
Robust principal component analysis, 276
Roulette wheel selection, 700
　roulette wheel fitness selection algorithm, 700
Row equivalent, 158
Row partial derivative operator, 55
Row partial derivative vector, 55

S
Second-order necessary condition, 94
Second-order sufficient condition, 94
Self-taught transfer learning, 409

Semi-orthogonal matrix, 272
Semi-supervised classification, 339
Semi-supervised clustering, 341
    propagating 1-nearest neighbor clustering, 341
Semi-supervised learning, 337
    bootstrapping, 340
    co-training, 342
    self-teaching, 340
    self-training, 340
    semi-supervised inductive learning, 338
    semi-supervised transductive learning, 338
Separable case, 621
Separable function, 237
Set, 10
    difference set, 11
    intersection set, 11
    null set, 11
    proper subset, 11
    singleton, 10
    subset, 11
    sum set, 11
    superset, 11
    union set, 11
Sharing function value, 737
Shrinkage operator, 125
Signal projection matrix, 271
Signal subspace, 271
Similarity, 316
    combined similarity, 320
    dissimilarity, 316
    distance matrix, 319
    Euclidean distance, 316
    Mahalanobis distance, 317
    normalized Euclidean distance, 317
    regularized Euclidean distance, 316
    semantic similarity, 320
    similarity matrix, 319
    similarity strength, 319
    syntax similarity, 320
    Tanimoto measure, 318
    word similarity, 320
Similarity score, 299
SinC function, 641
Single-hidden layer feedforward networks (SLFNs), 542
Single-output node, 546
Singular value decomposition (SVD)
    full singular value decomposition, 172
    left-singular vector matrix, 171
    left-singular vectors, 171
    right-singular vector matrix, 171
    right-singular vectors, 171
    singular values, 171

    truncated singular value decomposition, 172
Singular value thresholding (SVT), 130, 174
Slackness parameters, 637
Slack variable vector, 143
Slater condition, 142
Smooth function, 113
Soft thresholding, 174
Soft thresholding operation, 129, 175
Soft thresholding operator, 125
Soft threshold value, 125
Source domain, 403
Source-domain data, 403
Source embedding vectors, 574
Source learning task, 405
Spacing, 699
Sparse Bayesian classification, 672
Sparse Bayesian learning, 667, 669
Sparse constraints, 541
Sparse principal component analysis (SPCA) algorithm, 280
Sparse restruction, 293
Sparsest solution, 194
Sparsity, 194, 668
Spatial pyramid pooling, 507
Spectral convolution, 583
Spectral graph theory, 354
Spectral pooling, 507
Spherical K-means, 329, 518
    code vector, 518
    codebook matrix, 518
    dictionary, 518
Stagewise regression, 198
Staionary iterative method, 163
State embedding, 577
State transition function, 391
State value function, 393
State vector, 577
Stationary point, 90
Statistically uncorrelated, 23
Statistical uncorrelation, 23
Steepest descent direction, 107
Steepest descent method, 107
Step direction, 106
Stepwise regression, 198
Stochastic average gradient aggregation (SAGA) method, 232
Stochastic average gradient (SAG), 230
Stochastic gradient (SG) method, 228
Stochastic pooling, 506
Stochastic variance reduced gradient (SVRG) method, 231
Strength value, 753
Stress, 557

Strictly convex function, 103
Strictly increasing, 755
Strictly quasi-convex, 104
Stride of convolution, 500
Strong duality, 140
Strong learning algorithm, 247
Strong PAC-learning algorithm, 251
Strongly convex, 104
Strongly increasing, 755
Strongly quasi-convex, 104
Structural correspondence learning (SCL), 414
Structural learning, 412
Subdifferentiable, 121
Subdifferential, 120
Subgradient vector, 120
Subsec-mil, 35
Subspace analysis methods, 273
Superdiagonal line, 302
Supersymmetric tensor, 302
Supervised feature selection, 258
Supervised learning, 224
Supervised learning classification, 297
Supervised tensor learning, 307
 alternating projection algorithm, 309
Support, 254
Support vector expansion, 638
Support vector machine binary classification, 641
Support vector machine binary classifier, 642
 least squares support vector machine (LS-SVM) classifier, 647
 proximal support vector machine binary classifier, 649
 v-support vector binary classifier, 645
Support vector machine multiclass classification, 653
 least-squares support machine multiclass classifier, 656
 proximal support machine multiclass classifier, 659
Support vector machine regression, 635
 ϵ-support vector machine regression, 636
 ϵ-insensitive tube, 637
 Wolfe dual ϵ-support vector machine regression, 638
 v-support vector machine regression, 639
 Wolfe dual v-support vector machine regression, 641
Support vector machines (SVMs), 617
 B-splines SVM, 628
 least-squares support vector machine, 647
 linear SVM, 628
 multi-layer SVM, 628
 polynomial SVM, 628
 proximal support vector machine, 649
 radial basis function (RBF) SVM, 628
 summing SVM, 629
Supremum, 138
Surrogate function, 242
Swarm intelligence, 781
System of linear equations, 3

T

Target distribution, 248
Target domain, 403
Target-domain data, 403
Target embedding vectors, 574
Target learning task, 405
Task, 404
Tchebycheff approach, 746
Tchebycheff scalar optimization subproblem, 746
Tensor, 301
 Frobenius norm, 304
Tensor algebra, 301
Tensor classifier, 307
Tensor distance (TD), 314
Tensor dot product, 304
Tensor fibers, 302
 column fibers, 302
 row fibers, 302
 tube fiber, 302
Tensor Fisher discriminant analysis (TFDA), 310
Tensor inner product, 304
Tensor K-means clustering, 314
Tensor matrixing, 303
Tensor outer product, 304
Tensor predictor, 307
Tensor regression, 311
Tensor unfolding, 303
Tensor vectorization, 303
Tessellation, 327
Testing phase, 259
Test set, 223
Thinned dictionary, 520
Third-order tensor, 302
Tied weight matrix, 536
Tied weights, 527
Tikhonov regularization method, 180
Tikhonov regularization solution, 180
Tournament selection, 700
 binary tournament selection, 700
 tournament size, 700
TrAdaBoost, 410
Trained machine, 253
Training output matrix, 546

Training phase, 259
Training sample, 296
Training set, 223
Transductive transfer learning, 408
Transfer AdaBoost learning framework, 411
Transfer component analysis (TCA), 426
Transfer learning, 405
Transfer of knowledge, 333
Transformation metric, 419
Transitivity, 573
Travelling salesman problem (TSP), 689, 773
  asymmetric TSP, 689
  dynamic TSP (DTSP), 689
  Euclidean TSP, 689
  symmetric TSP, 689
Triplet loss, 512
Tri-training, 343
Trivial solution, 35
Tucker decomposition, 305
Tucker mode-1 product, 305
Tucker mode-2 product, 305
Tucker mode-3 product, 305
Tucker operator, 305
Two-view data, 341
Type II maximum likelihood, 671

U

Unconstrained optimization problem, 89
Under-regression, 198
Unique solution, 35
Unit ball, 102
Unit coordinate vector, 233
Unit tensor, 302
Unlabeled set, 223
Unsupervised feature selection, 258
Unsupervised learning, 224
Unsupervised transfer learning, 408

V

Value function, 89
Variable ranking, 261
Variable selection, 260
Variance reduction, 231
VC-dimension, 619, 620
VC-theory, 617
Vector, 3
  algebraic vector, 4
    constant vector, 4
  basic vector, 5
  dot product, 6
  geometric vector, 4
  inner product, 6
  normalized vector, 16
  outer product, 7
  physical vector, 4
  vector addition, 6
Vector cross product
  outer product, 7
Vectorization, 48
  column vectorization, 48
  row vectorization, 48
Vector multiplication, 6
Vector norm, 14
  Euclidean norm, 15
  ℓ_1-norm, 15
  ℓ_2-norm, 15
  ℓ_∞-norm, 15
  ℓ_p-norm, 15
  unitary invariant norm, 16
Vector quantization (VQ), 540
Violated constraint, 141
Visible binary rating matrix, 487
Voronoi set, 327
Voronoi tessellation, 327
Voting strategy, 655

W

Weak duality, 140
Weak learning algorithm, 247
Weak PAC learning algorithm, 251
Web-page, 341
Weighed Chebyshev norm, 719
Weighed sum, 719
Weighed-sum method, 726
Weight-delay (wd), 531
Weighted boundary volume, 374
Weighted sum approach, 746
Weight mask matrix, 521
Well-conditioned problem, 167
Winner-take-all, 541
Wirtinger partial derivatives, 73
Within-class scatter matrix, 218, 310
Within-class variance, 262
Within-sets covariance matrix, 348
Wolfe dual optimization problem, 622
Woodbury formula, 36

Z

Zero-phase component analysis (ZCA), 536
Zero vector, 5